MICROCHIP

MANUFACTURING

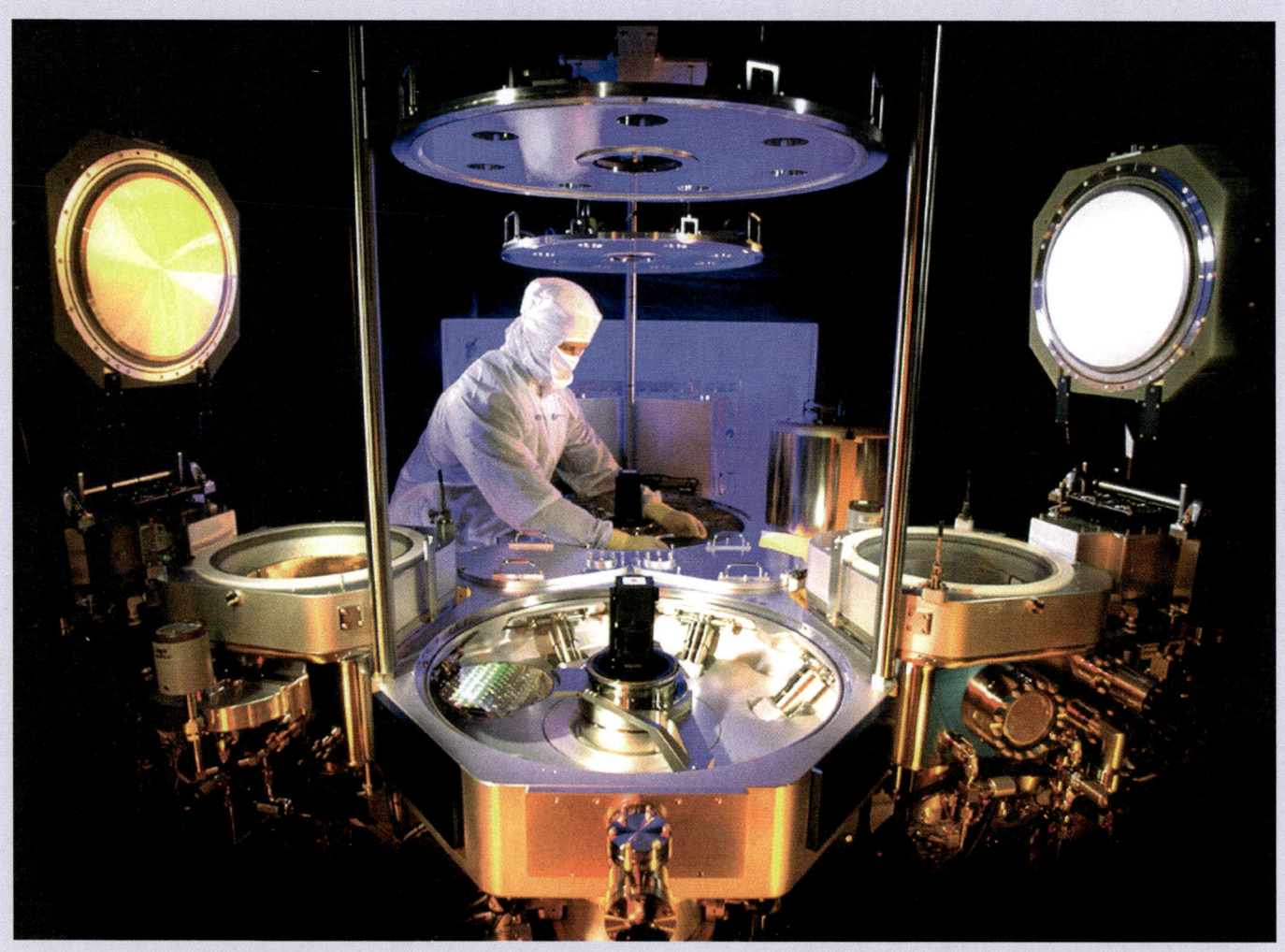
Technician servicing a sputtering tool. Photograph courtesy of Applied Materials, Inc.

MICROCHIP

MANUFACTURING

Stanley Wolf Ph.D.

LATTICE PRESS
Sunset Beach, California

DISCLAIMER

This publication is based on sources and information believed to be reliable, but the authors and LATTICE PRESS disclaim any warranty or liability based on or relating to the contents of this publication.

Published by:

LATTICE PRESS
Post Office Box 340
Sunset Beach, California 90742, U.S.A.
http://www.latticepress.com

Cover design by: LATTICE PRESS Graphic Arts Department
Cover image: Lam's 2300 Excelan® dielectric etch system, shown against a SEM taken following an in-situ photoresist strip and dual-damascene etch. Courtesy of Lam Research Corporation.

Copyright © 2004 by LATTICE PRESS.
All rights reserved. No part of this book may be reproduced or transmitted in any form or by any means, electronic or mechanical, including photocopying, recording or by any information storage and retrieval system without written permission from the publisher, except for the inclusion of brief quotations in a review. Requests for permission or further information should be addressed to the Permissions Department, LATTICE PRESS.

Library of Congress Cataloging in Publication Data
Wolf, Stanley, 1943 -

MICROCHIP MANUFACTURING

Includes Bibliographical references and Index
1. Integrated circuits - Very large scale
  integration - Design and construction. 2. Silicon.
3. Metal-oxide semiconductors. 4. Silicon technology. I. Title

ISBN 0-9616721-8-8

Printed in the United States of America

Printing (last digit): 9 8 7 6 5 4 3 2 1

CONCISE TABLE OF CONTENTS

Chap. 1 - The Semiconductor Industry	1
Chap. 2 - Semiconductors and Basic Materials Science	15
Chap. 3 - Semiconductor Devices and Integrated-Circuit Types	33
Chap. 4 - Overview of Semiconductor Manufacturing	51
Chap. 5 - Gases, Liquid Chemicals, and Ultrapure Water for ULSI	73
Chap. 6 - Vacuum Technology for ULSI Applications	87
Chap. 7 - The Basics of Thin-Films	111
Chap. 8 - Contamination-Control and Cleaning-Technology for ULSI	121
Chap. 9 - Silicon: Single-Crystal Growth	145
Chap. 10 - Silicon Wafer Production: From Ingot to Finished Wafer	163
Chap. 11 - Diffusion	175
Chap. 12 - Ion Implantation	189
Chap. 13 - Oxidation	213
Chap. 14 - Plasmas Used in Microchip Manufacturing	237
Chap. 15 - Aluminum Thin-Films and Sputter-Deposition in ULSI	247
Chap. 16 - CVD of Amorphous and Polycrystalline Thin Films	269
Chap. 17 - Silicon Epitaxy and Silicon-On-Insulator	305
Chap. 18 - Lithography I: Photoresist Materials & Process	321
Chap. 19 - Lithography II: Image-Formation and Optical-Hardware	341
Chap. 20 - Lithography III: Photomasks and Next-Generation Lithography (*NGL*)	359
Chap. 21 - Terminology of Etching and Wet-Chemical Etching	381
Chap. 22 - Dry-Etching for ULSI	391
Chap. 23 - Chemical-Mechanical Polishing	417
Chap. 24 - Multi-Level Interconnects: Copper, Low-*k*, and Dual-Damascene	437
Chap. 25 - Materials Characterization Techniques	455
Chap. 26 - Wafer and Chip Testing	475
Chap. 27 - Assembly and Packaging for ULSI	483
Chap. 28 - Wafer-Fab Operations and Yield	503
Chap. 29 - Environmental, Health, and Safety Issues	517
APPENDICES: A - Modeling Thermal Oxidation of Silicon	529
B - Mathematical Models of Diffusion in Silicon	535
C - Modeling of Impurity-Profiles of Implanted Ions	543
D - Mathematical Modeling of CVD Processes	547
E - Arrhenius Behavior	550
F - Error-Function (erf) Tables	551
INDEX	553

DETAILED TABLE OF CONTENTS

PREFACE

Chap. 1 - THE SEMICONDUCTOR INDUSTRY 1

1.1 21st-CENTURY ELECTRONICS INDUSTRY 1
   1.1.1 Electronics Manufacturing Facilities
   1.1.2 IC-Manufacturing Overview
   1.1.3 The Semiconductor Industry

1.2 THE HISTORY OF ELECTRONICS 4
   The Science of Electricity
   The History of Electricity
   Vacuum Tubes: Dawn of the Age of Electronics

1.3 THE TRANSISTOR:
THE AGE OF SOLID-STATE ELECTRONICS 8

1.4 THE AGE OF INTEGRATED-CIRCUITS 9

1.5 THE GROWTH OF THE
SEMICONDUCTOR INDUSTRY 11

REFERENCES 13
PROBLEMS 13

Chap. 2 - SEMICONDUCTORS AND BASIC MATERIALS SCIENCE 15

2.1 INTRODUCING SEMICONDUCTORS 15

2.2 THE NATURE OF MATTER:
ATOMIC STRUCTURE 15
   2.2.1 The Structure of the Atom
   2.2.2 The Periodic Table
      Ions

2.3 MOLECULES & COMPOUNDS 19

2.4 SOLIDS, LIQUIDS, AND GASES 19
   Units of Energy

2.5 CLASSIFYING MATERIALS BY THEIR
ELECTRICAL PROPERTIES 20
   2.5.1 Electrical-Conductors (and Resistivity)
   2.5.2 Electrical-Insulators (Dielectrics)
   2.5.3 Semiconductors
   2.5.4 Chemical-Bonding Model for Explaining
      the Electrical Properties of Solids
      Ionic Bonding
      Metallic Bonding
      Covalent Bonding

   2.5.5 Energy-Band Model for Explaining
      the Electrical Properties of Solids

2.6 SEMICONDUCTOR SILICON 25
   2.6.1 Intrinsic-Silicon (Pure-Silicon)
   2.6.2 Extrinsic-Silicon (Doped-Silicon)
   2.6.3 Dopants in Silicon:
      (Donors and Acceptors)
   2.6.4 n-Type Silicon
   2.6.5 p-Type Silicon (Holes)
   2.6.6 Dependence of Resistivity on
      Doping-Concentration in Silicon
   2.6.7 Carrier Mobility (and Its Impact on
      Transistor Performance)

2.7 OTHER SEMICONDUCTORS 30
REFERENCES 31
PROBLEMS 31

Chap. 3 - SEMICONDUCTOR DEVICES AND INTEGRATED-CIRCUIT TYPES 33

3.1 INTEGRATED-CIRCUIT RESISTORS 34
   3.1.1 Diffused Resistors
   3.1.2 Thin-Film Resistors

3.2 INTEGRATED-CIRCUIT CAPACITORS 35
   3.2.1 Metal-Oxide-Semiconductor
      Capacitors (MOS-Cs)
   3.2.2 Memory-Cell Capacitors for DRAMs

3.3 pn-JUNCTION DIODES 36
   3.3.1 pn-Junction with No Bias ($V_A = 0$)
   3.3.2 Forward-Biasing a pn-Junction ($V_A > 0$)
   3.3.3 Reverse-Biasing a pn-Junction ($V_A < 0$)
   3.3.4 pn-Junction Photodiodes

3.4 BIPOLAR JUNCTION TRANSISTOR (BJT) 39

3.5 METAL-OXIDE-SEMICONDUCTOR 42
FIELD-EFFECT TRANSISTORS (MOSFETs)
   3.5.1 MOSFET Operating Principles

3.6 INTEGRATED-CIRCUIT TYPES 44
   3.6.1 Analog Integrated Circuits
   3.6.2 Microprocessors
   3.6.3 ASICS

3.6.4 Memory Circuits
    DRAMs
    SRAMs
    EPROM, EEPROM, and Flash Memory

REFERENCES   47

PROBLEMS   48

Chap. 4 - OVERVIEW OF SEMICONDUCTOR MANUFACTURING   51

4.1 DESIGN-PHASE OF IC MANUFACTURE   52

4.2 CMOS PROCESS-FLOW   55
    4.2.1 Starting Material for CMOS ICs
    4.2.2 Formation of Active Regions, Channel Stops, & LOCOS-Isolation Structures
    4.2.3 Well-Formation
    4.2.4 Threshold-Adjust Implantation Step
    4.2.5 Gate-Oxide Growth
    4.2.6 Polysilicon Deposition and Patterning
    4.2.7 Formation of Source/Drain Regions
    4.2.8 Formation of $TiSi_2$ Salicide
    4.2.9 Premetal-Oxide Deposition Planarization, and Contact-Formation
    4.2.10 Metal-1 Deposition and Patterning
    4.2.11 Intermetal-Dielectric Deposition and CMP, Via-Patterning, and Metal-2 Deposition
    4.2.12 Passivation-Layer and Pad-Mask

4.3 SHALLOW-TRENCH ISOLATION   68

REFERENCES   69

PROBLEMS   70

Chap. 5 - GASES, LIQUID CHEMICALS, AND ULTRAPURE WATER FOR ULSI   73

5.1 BASIC PROPERTIES OF MATERIALS   73
    5.1.1 Temperature
    5.1.2 Thermal Expansion of Materials
    5.1.3 Density
    5.1.4 Specifying the Purity of Chemicals Used in IC Fabrication

5.2 LIQUID CHEMICALS USED IN IC FABRICATION   75
    5.2.1 Acids Used in IC Fabrication
    5.2.2 Bases Used in IC Fabrication
    5.2.3 Solvents Used in IC Fabrication
    5.2.4 pH-Scale
    5.2.5 Liquid Chemical Distribution

5.3 GASES USED IN IC FABRICATION   79
    5.3.1 Bulk-Gas Chemicals Used in IC Processing
    5.3.2 Distribution of Bulk Gases
    5.3.3 Specialty-Gases Used in IC Fabrication
    5.3.4 Gas Sources and Delivery Systems for Specialty-Gases
    5.3.5 Gas Stick, Gas Panel, and Gas Manifold
    5.3.6 Procedures for Changing Gas Cylinders

5.4 ULTRAPURE WATER (UPW)   83

REFERENCES   85

PROBLEMS   85

Chap. 6 - VACUUM TECHNOLOGY FOR ULSI APPLICATIONS   87

6.1 BASIC CONCEPTS OF GASES AND VACUUMS   87
    6.1.1 Gas Pressure and Vacuum

6.2 PRESSURE UNITS   89

6.3 VACUUM-PRESSURE RANGES   89
    6.3.1 Mean-Free-Path
    6.3.2 Closed and Open Gas-Systems and Gas-Flow Regimes

6.4 THE LANGUAGE OF GAS/SOLID INTERACTIONS   92
    6.4.1 Load Locks

6.5 TERMINOLOGY OF VACUUM PUMPS   93
    6.5.1 Vacuum Pump Types

6.6 ROUGH PUMPS   96
    6.6.1 Oil-Sealed Rotary Mechanical Pumps
    6.6.2 Roots-Blower/Booster Pumps
    6.6.3 Dry Mechanical Rough Pumps

6.7 HIGH-VACUUM PUMPS I: CRYOGENIC PUMPS   99
    6.7.1 Cryopump Operation
    6.7.2 Cryopump Regeneration

CONTENTS

6.8 HIGH-VACUUM PUMPS II: TURBOMOLECULAR PUMPS — 101

6.9 VACUUM GAUGES: (TOTAL-PRESSURE MEASUREMENT) — 103
   6.9.1 Capacitance Manometers
   6.9.2 Thermocouple Gauges
   6.9.3 Pirani Gauges
   6.9.4 Ionization (High-Vacuum) Gauges

6.10 MEASUREMENT OF PARTIAL-PRESSURE: RESIDUAL-GAS ANALYZERS (RGAs) — 106
   6.10.1 Operation of RGAs
   6.10.2 Interpretation of RGA Spectra

6.11 MASS-FLOW CONTROLLERS — 108

REFERENCES — 108

PROBLEMS — 109

Chap. 7 - THE BASICS OF THIN FILMS — 111

7.1 STAGES OF GROWTH OF AMORPHOUS AND POLYCRYSTALLINE THIN FILMS — 113
   7.1.1 Various Stages of Thin-Film Formation
   7.1.2 The Structure of Thin Films
   7.1.3 Factors that Influence Epitaxial Deposition of Thin Films

7.2 MECHANICAL-PROPERTIES OF THIN FILMS — 115
   7.2.1 Adhesion
   7.2.2 Stress in Thin Films

7.3 MEASUREMENT OF ELECTRICAL PROPERTIES OF THIN-FILMS — 117

REFERENCES — 118

PROBLEMS — 118

Chap. 8 - CONTAMINATION CONTROL AND CLEANING TECHNOLOGY FOR ULSI — 121

8.1 CONTAMINATION TYPES IN IC FABRICATION — 122

8.2 CONTAMINATION SOURCES IN IC PROCESSING — 122

8.3 EFFECTS OF CONTAMINATION ON ULSI DEVICES — 123

8.4 CONTAMINATION PREVENTION — 125
   8.4.1 Cleanroom Design and SMIF
   8.4.2 Gowning Procedures
   8.4.3 Cleanroom Protocols
   8.4.4 Ultra-Pure Chemicals
   8.4.5 Machine-Design & Wafer-Handling
   8.4.6 Process Modifications

8.5 WAFER CLEANING TECHNIQUES — 133
   8.5.1 Wet-Chemical Removal of Film Contaminants
   8.5.2 FEOL Wet-Chemical Cleaning: RCA Clean
   8.5.3 Spray Processing
   8.5.4 Photoresist Removal (Resist Stripping)

8.6 PARTICLE REMOVAL — 135
   8.6.1 Vibrational Scrubbing: (Ultrasonic and Megasonic)
   8.6.2 Particle Removal by Brush Scrubbing

8.7 RINSING AND DRYING WAFERS — 137
   8.7.1 Rinsing
   8.7.2 Wafer Drying after Rinse
   8.7.3 Spin-Dryer
   8.7.4 Isopropyl-Alcohol (IPA) Vapor Dryers
   8.7.5 Wet Cleaning Systems

8.8 PARTICLE DETECTION ON WAFER-SURFACES — 140
   8.8.1 Particle-Per-Wafer-Per-Pass
   8.8.2 Microscopy Tools for Particle Detection
   8.8.3 Automatic Laser Particle Counters for Detecting Particles on Wafers

REFERENCES — 141

PROBLEMS — 142

Chap. 9 - SILICON: SINGLE-CRYSTAL GROWTH — 145

9.1 CRYSTAL TERMINOLOGY — 145

9.2 MANUFACTURING SINGLE-CRYSTAL SILICON — 147
   9.2.1 From Raw Material to Electronic Grade Polysilicon (EGS)

9.3 CZOCHRALSKI CRYSTAL GROWTH — 149
   9.3.1 Czochralski Crystal-Growth Sequence
   9.3.2 Incorporation of Impurities in the Crystal

Solid-Solubility
9.3.3 Czochralski-Silicon Growing Equipment
    Furnace
    Crystal Pulling Mechanism
    Ambient Control
    Control System
9.4 CRYSTAL-DEFECTS IN SILICON   154
9.5 POINT-DEFECTS IN SILICON   155
9.6 ONE-DIMENSIONAL DEFECTS (DISLOCATIONS)   155
9.7 AREA-DEFECTS IN SILICON-CRYSTALS   157
9.8 BULK-DEFECTS & PRECIPITATION   158
9.9 OXYGEN IN SILICON   159
9.10 GETTERING   159
    9.10.1 Basic Gettering Principles
REFERENCES   161
PROBLEMS   161

Chap. 10 - WAFER PRODUCTION: FROM INGOT TO FINISHED WAFER   163

10.1 INGOT-EVALUATION   163
10.2 INGOT-GRINDING   163
10.3 FLAT (OR NOTCH) GRINDING   163
10.4 INGOT-SAWING (WAFERING)   164
10.5 LASER-MARKING OF WAFERS   166
10.6 WAFER-LAPPING AND GRINDING   166
10.7 EDGE-ROUNDING AND POLISHING OF WAFERS   166
10.8 REMOVAL OF SURFACE MECHANICAL DAMAGE BY CHEMICAL ETCHING   167
10.9 CHEMICAL-MECHANICAL POLISHING OF THE WAFER-SURFACE   167
10.10 CLEANING THE WAFERS   167
10.11 DEPOSITING EPITAXIAL-SILICON LAYERS ON THE WAFERS   168
10.12 SPECIFICATIONS OF SILICON WAFERS FOR ULSI   168
    10.12.1 Electrical Specifications
    10.12.2 Mechanical/Dimensional Specifications
    10.12.3 Chemical/Structural Specifications
    10.12.4 Surface/Near Surface Specifications
10.13 ECONOMICS OF SILICON WAFERS   171
REFERENCES   172
PROBLEMS   172

Chap. 11 - DIFFUSION   175

11.1 CONCEPT OF DIFFUSION   176
11.2 ATOMIC-SCALE MODELS OF DIFFUSION IN SOLID SILICON   178
11.3 CONCENTRATION VERSUS DEPTH GRAPHS: DOPING-PROFILES AND JUNCTION-DEPTHS   179
11.4 PRINCIPLES OF DIFFUSION-PROCESSES IN SILICON-DEVICE FABRICATION   180
    11.4.1 Diffusivity of Impurities in Silicon
    11.4.2 Pre-Dep & Drive-In Diffusion Steps
        Pre-Dep
        Drive-In
11.5 DIFFUSION IN SILICON-DIOXIDE   182
    11.5.1 Lateral Diffusion Under Oxide Windows
11.6 DIFFUSION SYSTEMS AND DIFFUSION SOURCES   182
    11.6.1 Gaseous Dopant-Sources
    11.6.2 Liquid Dopant-Sources
    11.6.3 Solid Dopant-Sources
11.7 MEASUREMENT TECHNIQUES FOR DIFFUSED LAYERS   184
    11.7.1 Sheet-Resistance Measurements
    11.7.2 Spreading-Resistance for Measuring Doping-Profiles
11.8 MEASURING JUNCTION-DEPTHS   185
    11.8.1 Angle-Lap (or Groove) and Stain for Junction-Depth Measurements
REFERENCES   186
PROBLEMS   186

Chap. 12 - ION IMPLANTATION   189

12.1 IMPURITY-PROFILES OF IMPLANTED IONS   191
    12.1.1 Ion Range
    12.1.2 Ion-Stopping Mechanisms

12.1.3 Implantation Into Single Crystals: Channeling

12.2 ION-IMPLANTATION DAMAGE-ACCUMULATION & ANNEALING IN SILICON — 195
12.2.1 Implantation Damage in Silicon
12.2.2 Primary Crystalline Defect Damage
12.2.3 Amorphous Layer Damage
12.2.4 Electrical Activation & Dopant Diffusion
    Electrical Activation of Implanted Impurities
    Diffusion of Implanted Impurities

12.3 ION-IMPLANTATION EQUIPMENT — 199
12.3.1 Ion-Implanter Types
12.3.2 Medium-Current Implanters
12.3.3 High-Current Implanters
12.3.4 Low-Energy Implanters
12.3.5 High-Energy Implanters
12.3.6 High-Angle Implanters
12.3.7 Ion Implantation Equipment System Limitations
    Elemental & Particulate Contamination
    Dose Monitoring Inaccuracies due to Beam Charge-State Changes
    Wafer Charging During Implantation

12.4 CHARACTERIZING IMPLANTS — 207
12.4.1 Measurement of Implanted Dose and Dose Uniformity
    Implantation Dose Measurements
    Implantation Dose Uniformity and Diagnosis of Implanter Performance
12.4.2 Measuring Implantation Depth Profiles
12.4.3 Measuring Implant Damage

12.5 ION-IMPLANTATION PROCESS APPLICATIONS — 208
12.5.1 Selecting Masking Layer Materials and Thickness
12.5.2 Threshold-Voltage Control in MOSFETs
12.5.3 Shallow Junction Formation by Ion Implantation
12.5.4 High-Energy Implantation

REFERENCES — 210
PROBLEMS — 210

Chap. 13 - OXIDATION — 213

13.1 APPLICATIONS OF SiO_2 — 214
13.2 PHYSICAL PROPERTIES OF SiO_2 — 214
13.3 THE THEORY OF SILICON-DIOXIDE GROWTH — 216
13.3.1 Deal-Grove (Linear-Parabolic) Model
    Linear and Parabolic Growth-Rate Regimes
    Factors that Impact B and B/A

13.4 SECONDARY-FACTORS THAT AFFECT ONE-DIMENSIONAL OXIDATION — 220
13.4.1 Dopant Effects on Oxidation Growth Rates
13.4.2 Dependence on Pressure
13.4.3 Dependence on Chlorine

13.5 THE GROWTH OF THIN-OXIDES — 221
13.6 THE Si/SiO_2 INTERFACE — 222
13.7 DOPANT-REDISTRIBUTION DURING OXIDATION — 223
13.8 OXIDATION OF POLYSILICON — 224
13.9 OXIDATION SYSTEMS — 225
13.9.1 Oxidation Furnaces
13.9.2 Horizontal-Furnaces
    Wet-Oxidation Gas Sources
13.9.3 Vertical-Furnaces
13.9.4 The Construction and Operation of Vertical-Furnaces
    Fast-Ramp, Small-Batch Furnaces
13.9.5 Rapid Thermal Processing
13.9.6 RTP Systems

13.10 OXIDATION PROCESS-RECIPE — 229
13.11 OXIDE-FILM THICKNESS MEASUREMENTS — 230
13.11.1 Color Chart
13.11.2 Optical-Interference for Measuring Oxide-Film Thickness
13.11.3 Ellipsometry
13.11.4 Transmission-Electron-Microscopy for Measuring Oxide-Film-Thickness

REFERENCES — 233
PROBLEMS — 233

Chap. 14 - PLASMAS USED IN MICROCHIP-MANUFACTURING — 237

14.1 INTRODUCTION TO GLOW-DISCHARGES — 238

CONTENTS

14.1.1 The Creation of DC-Glow-Discharges
  Summary of Electron/Gas-Atom and Electron/Gas-Molecule Inelastic-Collision Events
14.1.2 The Structure of Self-Sustaining Discharges and Their Dark-Spaces
14.2 RADIO-FREQUENCY (RF) GLOW-DISCHARGES — 243
14.3 HIGH-DENSITY PLASMAS — 244
REFERENCES — 245
PROBLEMS — 245

Chap. 15 - ALUMINUM THIN-FILMS AND SPUTTER-DEPOSITION FOR ULSI — 247

15.1 ALUMINUM THIN-FILMS IN ULSI — 248
15.2 SPUTTER-DEPOSITION FOR ULSI — 249
15.3 STEPS OF THE SPUTTER-DEPOSITION PROCESS — 249
15.4 THE PHYSICS OF SPUTTERING — 249
  15.4.1 The Billiard-Ball Model of Sputtering
  15.4.2 Sputter-Yield
  15.4.3 Selection Criteria for Process Conditions and Sputter-Gas Type
  15.4.4 Secondary-Electron Production for Sustaining the Glow-Discharge
  15.4.5 Sputter-Deposited Film Growth
  15.4.6 Species That Strike the Wafer During Film Deposition
15.5 MAGNETRON SPUTTERING — 253
  15.5.1 Magnetron Sputter-Sources for ULSI
    Evolution of Planar-Circular Sputtering-Sources
    Deposition-Rate and Thickness Uniformity with Circular Planar-Magnetrons
15.6 SPUTTER-DEPOSITION EQUIPMENT — 256
  15.6.1 The Components of Sputtering Systems
    Vacuum-Pumps for Sputtering Systems
    Power-Supplies for Sputtering Systems
  15.6.2 Commercial Sputtering Systems for 150-mm Wafers
  15.6.3 Sputtering Systems for 200-mm and 300-mm Wafers
15.7 SPUTTER-PROCESS CONSIDERATIONS — 258
  15.7.1 Sputter-Deposition of Alloy Films
  15.7.2 Effects on the Sputter-Process of the Transport of Vaporized Atoms between the Target and Substrate
  15.7.3 Faceting
  15.7.4 Particle-Generation in Sputtering
  15.7.5 Reactive-Sputtering
15.8 STEP-COVERAGE AND VIA/CONTACT-HOLE FILLING BY SPUTTERING — 261
  15.8.1 Sputter-Deposition of Barrier-Layer-Films into Contact-Holes and Vias
    Sputter-Deposition with Collimators
    Long-Throw Collimated Sputtering
    Ionized-Sputter Deposition
15.9 METAL FILM-THICKNESS MEASUREMENTS — 265
REFERENCES — 266
PROBLEMS — 266

Chap. 16 - CHEMICAL VAPOR DEPOSITION OF AMORPHOUS & POLYCRYSTALLINE THIN-FILMS — 269

16.1 BASIC ASPECTS OF CHEMICAL VAPOR DEPOSITION — 270
  16.1.1 Simplified CVD Film-Growth Model
16.2 CVD SYSTEMS — 273
  16.2.1 Components of CVD Systems
  16.2.2 Gas-Sources & Delivery-Systems for CVD
  16.2.3 Heating-Sources for CVD Reaction-Chambers
  16.2.4 Terminology of CVD-Reactor Design
  16.2.5 Atmospheric-Pressure-CVD Reactors
  16.2.6 Low-Pressure-CVD (LPCVD) Reactors
    Horizontal LPCVD Batch Reactors (Hot-Wall)
  16.2.7 Plasma-Enhanced CVD: Physics, Chemistry, & Reactor Designs
    Parallel-Plate, Cold-Wall Batch PECVD Reactors
    Mini-Batch, Radial Cold-Wall PECVD Reactors
    Single-Wafer, Cold-Wall PECVD Reactors
16.3 POLYCRYSTALLINE SILICON: PROPERTIES AND CVD-METHODS — 286
  16.3.1 Properties of Polysilicon Thin Films
    Physical & Mechanical Properties of Poly-Si
    Electrical Properties of Polysilicon

Deposition Parameters
- **16.3.2** Chemical Vapor Deposition of PolySi
- **16.3.3** Doping Techniques for Polysilicon
  - Diffusion Doping of Polysilicon
  - Ion Implantation Doping of Polysilicon
  - In Situ Doping of Polysilicon

16.4 PROPERTIES AND DEPOSITION OF CVD SiO_2 — 290
- **16.4.1** Chemical-Reactions for CVD SiO_2
- **16.4.2** Low-Temperature Silane-Based CVD-SiO_2
- **16.4.3** Medium-Temperature LPCVD TEOS SiO_2
  - Low-Temperature PECVD TEOS
  - Ozone TEOS
- **16.4.4** Step Coverage of CVD-SiO_2 Films
- **16.4.5** Applications of Undoped and Doped CVD-SiO_2 Films
  - Undoped CVD SiO_2
  - Phosphosilicate Glass
  - Borophosphosilicate Glass

16.5 PROPERTIES & CHEMICAL VAPOR DEPOSITION OF SILICON-NITRIDE — 296

16.6 SILICON-OXYNITRIDES DEPOSITED BY CVD — 297

16.7 CVD OF METALS, SILICIDES, AND NITRIDES — 298
- **16.7.1** CVD of Tungsten (W)
- **16.7.2** CVD Tungsten Chemistry
- **16.7.3** Blanket CVD W and Etchback
- **16.7.4** CVD of Titanium Nitride (TiN)

REFERENCES — 301
PROBLEMS — 302

Chap. 17 - SILICON EPITAXY AND SILICON-ON-INSULATOR — 305

17.1 EPITAXY DEVICE APPLICATIONS — 306
17.2 GROWTH OF EPITAXIAL LAYERS — 307
17.3 CHEM-REACTIONS OF Si-EPITAXY — 308
17.4 PROCESS CONSIDERATIONS FOR EPITAXIAL DEPOSITION — 308
- **17.4.1** Intentional Doping of Epi-Films
- **17.4.2** Unintentional Doping of Epi Layers (Autodoping and Solid-State Diffusion)
- **17.4.3** Wafer Cleaning Prior to Epi Deposition
- **17.4.4** Batch Epitaxy Process Sequence

17.5 EPITAXIAL PROCESS-EQUIPMENT — 311
- **17.5.1** Batch Horizontal-Tube Epi-Reactors
- **17.5.2** Vertical Pancake Epi-Reactors
- **17.5.3** Cylindrical Epi-Reactors
- **17.5.4** Single-Wafer Epitaxial-Systems

17.6 CHARACTERIZING EPI-LAYERS — 314
- **17.6.1** Optical Inspection of Epi Film Surfaces
- **17.6.2** Electrical Characterization
- **17.6.3** Epi-Film-Thickness Measurements
- **17.6.4** Infrared-Reflectance Techniques

17.7 SILICON-ON-INSULATOR (SOI) — 315
- **17.7.1** Separation by Implanting Oxygen (SIMOX)
- **17.7.2** Wafer-Bonding (Smart-Cut)

REFERENCES — 317
PROBLEMS — 318

Chap. 18 - LITHOGRAPHY I: PHOTORESIST MATERIALS AND PROCESSING — 321

18.1 PHOTORESISTS — 321
- **18.1.1** Basic Photoresist Terminology

18.2 RESIST MATERIAL PARAMETERS — 323
18.3 OPTICAL RESIST TYPES — 323
- **18.3.1** Positive Optical Photoresists
- **18.3.2** Negative Optical Photoresists
- **18.3.3** Chemically-Amplified Resists

18.4 PHOTORESIST PROCESSING — 327
- **18.4.1** Resist Processing: Dehydration Baking and Priming
- **18.4.2** Resist Processing: Spin Coating
- **18.4.3** Resist Processing: Soft-Bake
- **18.4.4** Resist Processing: Exposure
  - Standing Waves
  - Anti-Reflective Coatings (ARCs)
- **18.4.6** Resist Processing: Development
- **18.4.7** Resist Processing: After-Develop Inspection
  - Linewidth Variation and Control
  - Linewidth Measurements
- **18.4.8** Resist Processing: Post-Development-Bake

18.5 RESIST PROCESSING SYSTEMS 337
REFERENCES 339
PROBLEMS 339

Chap. 19 LITHOGRAPHY II: IMAGE FORMATION AND OPTICAL HARDWARE 341

19.1 PRELIMINARIES: WAVE-MOTION AND THE BEHAVIOR OF LIGHT 342
   19.1.1 Refraction and Diffraction of Light
19.2 RESOLUTION IN MICROLITHOGRAPHY APPLICATIONS 346
   19.2.1 Depth of Focus
   19.2.2 Definition of Lithographic Resolution
19.3 LIGHT-SOURCES FOR LITHOGRAPHY 348
   19.3.1 Mercury-Arc Lamps
      Arc-Lamp Illumination Systems
   19.3.2 Excimer-Laser DUV Light-Sources
19.4 LITHOGRAPHIC EXPOSURE TOOLS 350
   19.4.1 Contact-Printing
   19.4.2 Proximity-Printing
19.5 PROJECTION-PRINTERS 351
   19.5.1 Scanning Projection-Printing
   19.5.2 Step-and-Repeat Projection-Printing
   19.5.3 Step-and-Scan Projection-Printing
19.6 OVERLAY AND WAFER-STAGES
19.7 OFF-AXIS ILLUMINATION 355
REFERENCES 357
PROBLEMS 357

Chap. 20 - LITHOGRAPHY III: PHOTOMASKS AND NEXT-GENERATION LITHO (NGL) 359

20.1 MASK (or RETICLE) FABRICATION 359
   20.1.1 Terminology and History of Photomasks
   20.1.2 Fabrication of Photomasks
     Glass Quality and Preparation
     Glass-Coating (Chrome)
     Mask Imaging (Resist Application and Processing)
     Pattern-Generation
   20.1.3 Mask and Reticle Defects and Their Detection and Repair
     Repairing Defects in Masks & Reticles
   20.1.4 Pellicles
   20.1.5 Critical-Dimension and Registration Inspection of Masks & Reticles
   20.1.6 Storage, Transport, Loading of Reticles
20.2 RESOLUTION-ENHANCEMENT TECHNIQUES (RET) 369
   20.2.1 Optical Proximity Correction (OPC)
   20.2.2 Phase-Shift Masks (PSM)
20.3 MICROLITHOGRAPHY TRENDS 374
   20.3.1 The Limits of Optical-Lithography
20.4 NEXT-GENERATION LITHOGRAPHIC TECHNOLOGIES (NGL) 376
   20.4.1 Extreme Ultra-Violet Projection-Lithography (EUV)
   20.4.2 Electron-Beam Projection-Lithography
REFERENCES 378
PROBLEMS 379

Chap. 21 - TERMINOLOGY OF ETCHING AND WET-CHEMICAL ETCHING 381

21.1 THE TERMINOLOGY OF ETCHING 382
21.2 ETCH PARAMETERS 382
   21.2.1 Etch-Rate
   21.2.2 Etch-Rate Uniformity
   21.2.3 Etch Profile
   21.2.4 Selectivity
   21.2.5 Etch Bias
21.3 WET-ETCHING TECHNOLOGY 385
   21.3.1 Wet-Etching Silicon
   21.3.2 Wet-Etching Silicon Dioxide
   21.3.3 Wet-Etching Silicon Nitride
   21.3.4 Wet-Etching Aluminum
REFERENCES 389
PROBLEMS 389

Chap. 22 - DRY ETCHING FOR ULSI 391

22.1 TYPES OF DRY-ETCHING PROCESSES 392
22.2 THE PHYSICS AND CHEMISTRY

OF PLASMA-ETCHING　393
   22.2.1 The Reactive-Gas Glow Discharge
   22.2.2 Electrical Aspects of Glow Discharges
   22.2.3 Heterogeneous Reaction Considerations
   22.2.4 Parameter Control in Plasma Processes
22.3 ETCHING SILICON AND SILICON DIOXIDE IN FLUOROCARBON PLASMAS　396
   The Etching of Silicon by Molecular Fluorine
   Dry-Etching Silicon with O_2 Added to CF_4 Plasmas
   Dry-Etching Silicon with H_2 Added to CF_4 Plasmas
   Dry-Etching SiO_2 with CF_4 Plasmas
22.4 ANISOTROPIC ETCH MECHANISMS　399
22.5 DRY-ETCHING OF VARIOUS MATERIALS IN ULSI PROCESSING　401
   22.5.1 Dry-Etching of Silicon Dioxide (SiO_2)
      Shaping the Sidewalls of Contact-Holes and Vias by Dry-Etching
      Via-Veil-Removal After Via-Etching
   22.5.2 Dry-Etching of Silicon Nitride
   22.5.3 Dry-Etching of Polysilicon
   22.5.4 Dry-Etching Aluminum and Aluminum Alloys
      Dry-Etching of Al:Cu Alloys
      Post-Dry-Etch Aluminum Corrosion
   22.5.5 Dry-Etching of Organic Films
22.6 PROCESS MONITORING: ENDPOINT DETECTION　408
   22.6.1 Laser Interferometry & Laser Reflectance
   22.6.2 Optical-Emission Spectroscopy
22.7 BATCH DRY-ETCH EQUIPMENT CONFIGURATIONS　409
   22.7.1 Barrel Etchers
   22.7.2 Parallel Electrode (Planar) Reactors
   22.7.3 Cylindrical Batch Etch Reactors: (Hexode Etchers)
22.8 SINGLE-WAFER ETCHERS　412
   22.8.1 Single-Wafer Parallel-Plate Reactors
   22.8.2 Magnetic-Enhanced Reactive Ion Etchers (MERIE)
22.9 HIGH-DENSITY PLASMA SOURCES　414

   22.9.1 HDP Sources in ULSI Fabrication
   22.9.2 Electrostatic Chucks
22.10 DAMAGE FROM DRY-ETCHING　414
   22.10.1 Oxide Damage During Polysilicon or Metal-Etch Processes
REFERENCES　415
PROBLEMS　415

Chap. 23 - CHEMICAL-MECHANICAL POLISHING (CMP)　417

23.1 THE HISTORY OF CMP　418
23.2 THE MECHANISMS OF CMP　419
   23.2.1 Metal CMP Mechanisms
   23.2.2 Silicon Dioxide CMP Mechanisms
   23.2.3 CMP of Low-k Dielectrics
23.3 CMP EQUIPMENT　422
23.4 CMP POLISHING TOOLS　423
23.5 CMP POLISHING PADS　425
23.6 CMP CONSUMABLES (SLURRIES)　428
   23.6.1 Slurry-Distribution Systems
23.7 CMP ENDPOINT DETECTION　430
23.8 CLEANING ISSUES IN CMP　431
23.9 CMP METROLOGY　432
23.10 DISHING PROBLEMS IN CMP　433
23.11 THICKNESS NON-UNIFORMITY WITHIN A WAFER AFTER CMP　434
23.12 ECONOMIC CONSIDERATIONS: THROUGHPUT AND COST OF OWNERSHIP　435
REFERENCES　435
PROBLEMS　435

Chap. 24 - MULTILEVEL-INTERCONNECTS: COPPER, LOW-k DIELECTRIC & DAMASCENE　437

24.1 THE NEED FOR MULTILEVEL-INTERCONNECT TECHNOLOGY　437
   24.1.1 Interconnect Limitations of ULSI
   24.1.2 Terminology of Multilevel-Interconnect Structures
24.2 COPPER FOR ULSI INTERCONNECTS　440
   24.2.1 Process Integration Issues of Copper
   24.2.2 Electroplating of Copper

CONTENTS

24.3 LOW-k DIELECTRICS   **444**
    24.3.1 Process-Integration of Low-k Dielectrics
    24.3.2 First Gen Low-k Dielectrics ($2.8 < k < 3.5$)
    24.3.3 Second-Gen Low-k Dielectrics:
        ($2.5 < k < 2.8$)
        2nd-Gen Spin-On Dielectrics with $2.5 < k < 2.8$
        2nd Gen-CVD Dielectrics with $2.5 < k < 2.8$
    24.3.4 Ultra-Low-k Dielectrics ($k < 2.0$)
24.4 DAMASCENE & DUAL-DAMASCENE   **449**
INTERCONNECT STRUCTURES

REFERENCES   **452**

PROBLEMS   **452**

Chap. 25 - MATERIALS CHARACTERIZATION TECHNIQUES   455

25.1 WHAT IS BEING DETECTED,
AND HOW IS IT DONE?   **456**
    25.1.1 Energy Regimes and Energy Levels in
        Materials Characterization
    25.1.2 Definitions of Material
        Characterization Terminology
    25.1.3 Vacuum Requirements
        of Compositional Analysis
25.2 MICROSCOPY FOR ANALYZING IC
FEATURES   **458**
25.3 OPTICAL MICROSCOPES   **458**
    25.3.1 Resolution, Magnification, and Numerical
        Aperture
    25.3.2 Brightfield & Darkfield Illumination
    25.3.3 Television System Interface Capability
25.4 SCANNING ELECTRON
MICROSCOPY (SEM)   **460**
    25.4.1 Production & Failure Analysis SEMs
25.5 TRANSMISSION ELECTRON   **463**
MICROSCOPY
    25.5.1 Sample Preparation
25.6 ATOMIC-FORCE MICROSCOPY (AFM)   **464**
25.7 ELECTRON/X-RAY COMPOSITIONAL
ANALYSIS TECHNIQUES   **466**
25.8 AUGER EMISSION
SPECTROSCOPY (AES)   **466**
25.9 X-RAY EMISSION
SPECTROSCOPY (XES)   **468**
25.10 X-RAY PHOTOELECTRON
SPECTROSCOPY (XPS, ESCA)   **469**
25.11 X-RAY FLUORESCENCE (XRF)   **469**
25.12 ION-BEAM-EXCITED COMPOSITIONAL
ANALYSIS   **470**
25.13 SECONDARY-ION
MASS-SPECTROSCOPY (SIMS)   **470**
    25.13.1 Laser Ionization Mass Spectroscopy
        (LIMS) and Time of Flight SIMS (TOF)
25.14 FOCUSED-ION-BEAM ANALYSIS   **472**

REFERENCES   **473**

PROBLEMS   **473**

Chap. 26 - WAFER AND CHIP TESTING   475

26.1 PURPOSE AND PHILOSOPHY OF
TESTING INTEGRATED CIRCUITS   **475**
26.2 IN-LINE PARAMETRIC TESTING   **476**
26.3 WAFER SORT   **478**
    26.3.1 Wafer-Sort Test Procedure
26.4 BURN-IN   **479**
26.5 FINAL FUNCTIONAL TESTING   **480**
26.6 BINNING   **480**
26.7 MARK, PACK, AND SHIP   **481**

REFERENCES   **481**

PROBLEMS   **481**

Chap. 27 - ASSEMBLY & PACKAGING   483

27.1 DIE SEPARATION   **484**
27.2 DIE-ATTACH   **485**
    27.2.1 Die-Attach-Related Reliability Issues
    27.2.2 Die-Attach Materials
    27.2.3 Die Attach Equipment
27.3 BOND-PAD/PACKAGE CONNECTIONS   **487**
    27.3.1 Wire Bonding
    27.3.2 Wire Bond Spacing
    27.3.3 Tape Automated Bonding (TAB)
    27.3.4 Flip-Chip Bonding
27.4 INTRODUCTION TO CHIP PACKAGES   **492**

**27.5 PACKAGING TECHNOLOGY I:
HERMETIC PACKAGES (CERAMIC)** — 493
**27.6 PACKAGING TECHNOLOGY II:
PLASTIC PACKAGES** — 494
   27.6.1 Plastic Package Reliability Problems
27.7 ULSI PACKAGE TYPES — 497
   27.7.1 Through-Hole (TH) IC Packages
   27.7.2 Surface Mount (SM) IC Packages
27.8 TRENDS IN ULSI IC PACKAGES — 499
   27.8.1 Packageless Technologies

REFERENCES — 500
PROBLEMS — 500

Chap. 28 - WAFER-FAB OPERATIONS AND YIELD — 503

28.1 WAFER-FAB OPERATING COSTS — 504
   28.1.1 Overhead-Costs
   28.1.2 Materials Costs
   28.1.3 Equipment Costs
   28.1.4 Labor Costs
**28.2 WAFER-FAB LAYOUT AND
WAFER-TRANSPORT** — 505
**28.3 STRATEGIES FOR IMPROVING THE
EFFICIENCY OF IC-MANUFACTURING** — 507
   28.3.1 The Economic Impact of Wafer-Size
   28.3.2 IC Fabrication-Yield
      Definition of Yield
      Definition of a Yield Model
   28.3.3 Basic Mathematical Yield-Models
   28.3.4 Yield-Models for Large-Die-Size ICs
   28.3.5 Defect-Detection and Analysis
**28.4 YIELD-MANAGEMENT AND
THE "LEARNING-CURVE"** — 511
   28.4.1 Yield Ramps
      Yield Ramps - Stage-1
      Yield Ramps - Stage-2
      Yield Ramps - Stage-3
      Yield Ramps - Stage-4
      Yield Ramps - Stage-5
      The Time-Compression of Yield-Ramps

   28.4.2 Statistical Process-Control
**28.5 ORGANIZATIONAL STAFFING
OF WAFER FABS** — 514

REFERENCES — 515
PROBLEMS — 515

Chap. 29 - ENVIRONMENTAL, HEALTH, AND SAFETY ISSUES — 517

**29.1 SAFETY RECORD OF THE
SEMICONDUCTOR INDUSTRY** — 517
29.2 SAFETY-HAZARDS IN WAFER FABS — 518
   Hazardous Materials & Physical Conditions
   Hazardous Work Operations
29.3 GENERAL SAFETY PROCEDURES — 518
   Warning Labels
   Hazard Information Labels
   Material Safety Data Sheets
29.4 HAZARDOUS PROCESS GASES — 520
   29.4.1 Reducing Toxic-Gas Hazards
29.5 HAZARDOUS PROCESS CHEMICALS — 522
   Handling Corrosive Chemicals
   Handling Solvents
   Hydrofluoric Acid
**29.6 ELECTRIC SHOCK HAZARDS
AND ION-IMPLANTER SAFETY** — 523
29.7 WAFER-FAB RADIATION HAZARDS — 524
   DUV Laser-Light Sources
   X-Ray Radiation
29.8 FIRES IN WAFER FABS — 525
29.9 ENVIRONMENTAL SAFETY ISSUES
   29.9.1 Scrubbed Exhaust Systems
      Combustion Reactors
      Scrubbers
   29.9.2 Chemical Recycling
   29.9.3 Perfluorocarbon Compounds
      (Global Warming Gases)

REFERENCES — 528
PROBLEMS — 528

APPENDICES

APPENDIX A - MODELING THERMAL OXIDATION OF SILICON   529

APPENDIX B - MATHEMATICAL MODELS OF DIFFUSION IN SILICON   535

APPENDIX C - MODELING OF IMPURITY-PROFILES OF IMPLANTED IONS   543

APPENDIX D - MATHEMATICAL MODELING OF CVD-PROCESSES   547

APPENDIX E - ARRHENIUS BEHAVIOR   550

APPENDIX F - ERROR-FUNCTION (*erf*) TABLES   551

INDEX   553

PREFACE

Microchip Manufacturing is written for a wider audience than our previous books. The four-volume series *Silicon Processing for the VLSI Era* has long been popular among process engineers working in the field, and with students training for carrers in microelectronics. However, for readers studying the topic for the first time (or those not needing such in-depth information), these books may have been too daunting.

Thus, we set out to write a book that provides a more introductory and accessible treatment of this vital and fascinating technology. The chief innovation that makes the material easier to grasp is the use of *color illustrations* throughout. In fact, to our knowledge, it is the *first full-color technician/engineering-level textbook* ever produced on IC-fabrication technology.

The book also conveniently serves semester-long classes on IC fabrication in two academic curricula:

1. Introductory courses in undergraduate, four-year engineering programs of electrical, chemical, and mechanical engineering, materials science, and physics. The mathematical underpinning for such courses is provided in the Appendices. At least half the problems at the end of each chapter have been selected with such courses in mind.

2. As a basic book for two-year technician-training programs. Without referring to the Appendices, the remainder of the material contains only the level of mathematics that can be handled by students enrolled in a two-year Semiconductor Manufacturing Technology program.

Figures from the text (on CD), and a Solution Manual for the problems, are available to all adopting faculty.

The text should also appeal to many semiconductor-industry professionals, especially those peripherally involved with wafer-fabs. If you sell equipment, materials, and services to semiconductor manufacturers, it provides an understanding of the overall picture, and how your products fit in. IC designers, product engineers, test and field-service engineers, IP-lawyers, quality-control and reliability personnel, and sales and marketing workers will all benefit from its contents.

Many people helped make the publication of this book possible, and we thank and acknowledge their assistance here. First, several faculty members of the Digital Arts Department at Golden West College taught me the graphic-design programs needed to produce this book. Foremost among them was Sean Glumace, who was also an invaluable consultant when production-problems arose. Various companies (and people within) provided many of the beautiful photographs. They included: Chris Moran, Fred Helmrich, and Patricia Quon at Applied Materials; Shawn Lynch at Lam Research; Bob Klimo and Page Jasienski at Novellus Systems; Maureen Hart and Dan Ferrin at Axcelis Technologies; Chuck Gwynn at Intel Corp; and T.C. Smith (formerly of Motorola). Other individuals read parts of the manuscript and gave insightful feedback, including David Hata and Joseph McGuire. Special thanks goes to Jerry Healey, who offered significant technical help with computer issues, read several chapters, and afforded access to his technical archives. Helga Janssen sustained the backup-services that kept Lattice Press operating smoothly during the production-phase.

My family also delivered both direct and indirect assistance. My son, Ross, served as a hardware and software troubleshooter many times in a pinch. My daughter Jennifer gave marketing advice and an enthusiastic ear when spirits were low. My wife, Carrol Ann, not only helped prepare many of the graphic illustrations, proof-read many chapters, and typed the Index, but also took on the burden of handling the household administration, as I toiled over the manuscript. Without all of the above contributions, the book could not have been completed.

Stanley Wolf Ph.D.

P.S. Additional copies of this book (as well as copies of any of the volumes of the series - *SILICON PROCESSING FOR THE VLSI ERA*), can be ordered directly from the LATTICE PRESS website:

www.latticepress.com

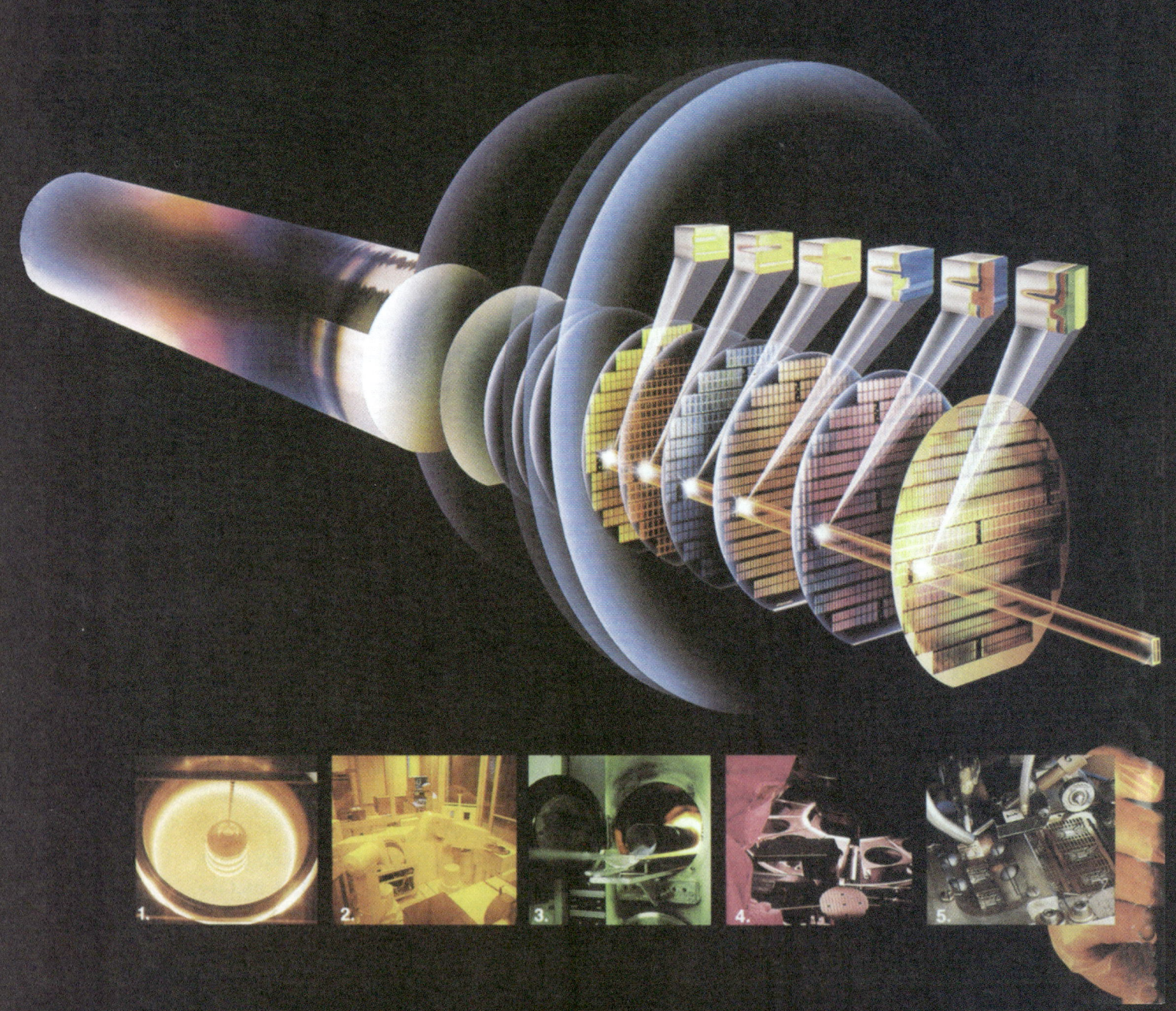

CHAPTER 1

THE SEMICONDUCTOR INDUSTRY

CHAPTER CONTENTS

1.1 THE WORLDWIDE ELECTRONICS INDUSTRY AT THE START OF THE 21st CENTURY

1.2 THE HISTORY OF ELECTRONICS

1.3 THE TRANSISTOR: THE AGE OF SOLID-STATE ELECTRONICS

1.4 THE AGE OF INTEGRATED-CIRCUITS

1.5 THE GROWTH OF THE SEMICONDUCTOR INDUSTRY

"The best way to predict the future is to invent it"

Alan Kay

Illustration compliments of Texas Instruments Incorporated.

The *semiconductor-industry* was born with the invention of the *transistor* in 1947 at ATT Bell Labs. The debut of the *integrated-circuit* (*IC*) at the beginning of the 1960s was another critical event in the growth of this industry. These breakthroughs spawned a technology that by 2002 could integrate a *billion components* onto a single-piece of silicon having an area of one-square centimeter (the 1-Gb DRAM)! From a beginning in which ICs were used in only a limited number of specialized applications, has grown a technology that is pervasive in today's world (Fig. 1-1). It has also spurred related technologies, such as software and the Internet.

At the beginning of the 2000-millennium, the electronics-industry exceeded $1 trillion in sales per year, and semiconductors made up between $150-$200-billion of that number. The industry has grown so large in part because the progress in technology has allowed a con-

Fig. 1-1 The heart of the Electronics Industry (whose sales exceeded one trillion/ dollars per year at the start of the 21st century) is *microchip manufacturing*. Photo courtesy of UMC.

1

2 MICROCHIP MANUFACTURING

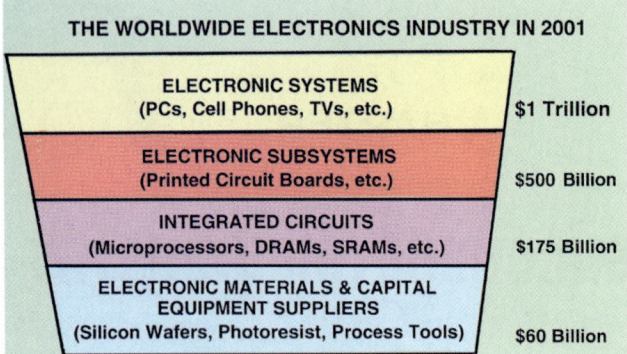

Fig. 1-2 Structure of the worldwide Electronics Industry at the start of the 2000 decade.

tinual increase in semiconductor-performance while decreasing price. In this chapter we introduce the semiconductor-industry as a prelude to the rest of the book, which covers IC-manufacturing-technology.

1.1 THE WORLDWIDE ELECTRONICS-INDUSTRY AT THE START OF THE 21ST-CENTURY

Figure 1-2 illustrates the makeup of the worldwide electronics-industry in 2001. At the top of the chart are the *end-product electronic-systems* sold to the final customers. Their annual-sales in 2002 exceeded $1-trillion. Examples of such electronic-systems are: computers and peripherals (Fig. 1-3a); telephones and telecommunications-equipment (Fig. 1-3b); consumer-electronics-products (Fig. 1-3c); defense and space systems; and industrial-electronics (Fig. 1-3d). Hardware is coupled with software to integrate and control the IC-functions in electronic-products.

On the rung below the system-level are the electronic-subsystems, mainly *printed-circuit-boards* (*PCBs*) that are found in virtually every electronic product (Fig. 1-4). These PCBs are plastic-boards onto which are mounted packaged-ICs and other components interconnected by wiring-paths fabricated on the PCB. The value of all the PCBs produced in 2001 was about $500-billion.

The ICs mounted on the PCBs are at the next level below the PCBs (Fig. 1-5). Their sales in 2001 were about $175-billion. It is the manufacture of these ICs that is the main subject of this book.

The last-level on the chart in Fig. 1-2 represents companies that supply the equipment and materials used to manufacture ICs. The sales of these suppliers in 2001 was about $60-billion.

1.1.1 Electronics Manufacturing Facilities

The products manufactured in each of the levels shown in Fig. 1-2 are done in different manufact-

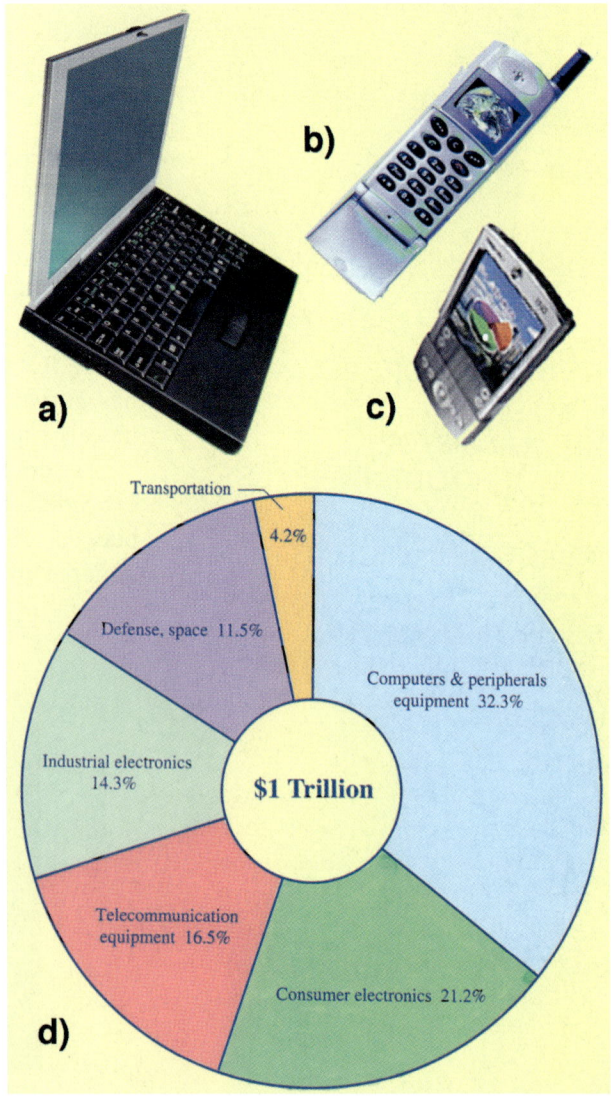

Fig. 1-3 Examples of modern electronic products: (a) laptops; (b) cell-phones; and (c) personal-digital assistants. (d) Market share of the six major electronics applications.

CHAPTER 1   THE SEMICONDUCTOR INDUSTRY

Fig. 1-4 Photograph of a printed-circuit board (PCB).

uring-facilities, as shown in Fig. 1-6. The end-product electronic-systems are assembled in a final-product assembly-plant (for example, a *PC assembly-plant*) from subsystems and components (typically purchased from subsystem-manufacturers). The packaged-chips are mounted and connected onto the circuit-boards in *PCB-factories*. The ICs are manufactured in a *wafer-fabrication-facility* (or *wafer-fab*), and the finished-wafers are sent to another plant, called an *assembly/packaging-plant* where the wafers are cut apart and mounted into chip-packages. The packaged-chips are sent to the PCB-factory for mounting onto the PCBs.

1.1.2 Integrated-Circuit Manufacturing

The *integrated-circuits* used in the electronic-systems are fabricated on silicon-wafers in a *wafer-fab*. A number of the main-processes used to fabricate them are shown in Fig. 1-7. The manufacturing-flow for making ICs is basically a sequential-layering-process, much like that used to make a submarine-sandwich (Fig. 1-8a). That is, such a sandwich is built layer-by-layer on a base of bread (the *substrate*). Then layers of cheese, meat, lettuce, tomatoes, mustard, pickles, and spices (salt and pepper) are added to make the end-product – the *submarine-sandwich*.

In an IC, a silicon-wafer is the substrate, and layers of silicon-dioxide, polysilicon, metal-layers, and dop-

Fig. 1-5 (a) Microchips (ICs) on a finished wafer. (b) Individual chip before being packaged. (c) ICs mounted in their protective packages. (d) Fully-packaged ICs.

ants (the "spices") are added, and are covered with an overcoat to make the end-product - the *integrated circuit* (Fig. 1-8b). In making ICs, a lithography-process is also needed to pattern the layers and to make certain they are properly aligned to one another.

1.1.3 The Semiconductor Industry

As shown in Fig. 1-9, the semiconductor industry consists of many groups of companies and institutions, all of which contribute to its vitality. At the center are the chip-manufacturers, but they are supported by a large number of outside organizations, including the following: manufacturers of chip-processing and metrology-tools; suppliers of materials and chemicals; analytical-laboratories; industry-associations that provide manufacturing-standards and organize co-operative research efforts; and colleges and universities that provide technically-trained workers.

4 MICROCHIP MANUFACTURING

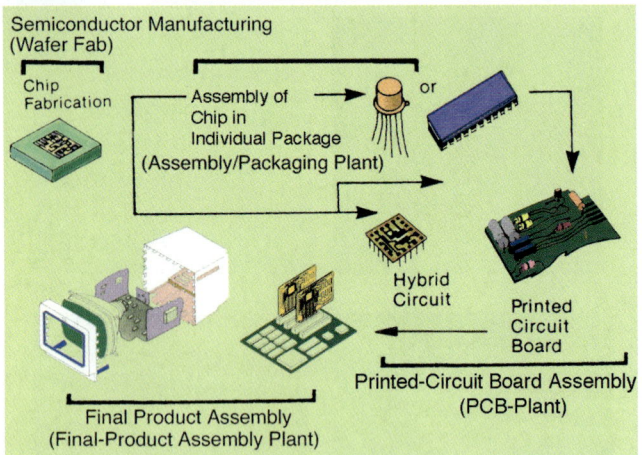

Fig. 1-6 The various stages in the building of an electronic-system. The figure also indicates the manufacturing facility in which each stage is carried out.

The chip manufacturers are also suppliers to customers in many groups, including automotive-electronics, computers, communications-electronics, medical-instrumentation, and industrial-electronics. These are fed through the intermediate levels of the PCB-manufacturers and the final-system-assemblers (as discussed in Sect. 1.1.1).

1.2 THE HISTORY OF ELECTRONICS

Electronics is a branch of science and engineering that deals with devices which control the flow of electrons (*electronic-devices*). While *electronics* is closely related to the science of *electricity*, electronics does jobs that electricity alone cannot do. That is, *electricity-based (electrical) devices* (such as motors, electric-lights, and electric-heaters), primarily deal with electricity as a way to deliver energy. *Electronic devices* manipulate electric-currents mainly in the form of *signals*. Thus, electronics lends itself to *communications-applications*, making possible such modern wonders as computers, television, radio, radar, and wireless-phones. While this section traces the history of electronics, it also begins with a brief history of electricity.

The Science of Electricity: Electricity is a form of energy that arises from the electrical-nature of matter. As discussed in Chap. 2, atoms contain electrically-charged subatomic-particles (*protons* and *electrons*). An electron has one-unit of *negative-charge*, and a proton one-unit of *positive-charge*. If an atom gains some electrons, it becomes negatively-charged. If it loses some, it becomes positively-charged. Atoms with a *net electric-charge* are called *ions*.

Every charged-particle also creates an *electric-field*, and because of this field, a force is exerted on other charged particles - even when not in physical contact. Charged-objects with *unlike-charges*, attract one another (Fig. 1-10b), but those with *like-charges* (Fig. 1-10c) repel one another (for instance, electrons repel other electrons). *Static-electricity* consists of the forces among electrons and ions that *do not move*. *Current-electricity* is made up of the *organized motion of electrons or ions*. Thus, a substance that conducts electricity must contain charged-particles that are free to move. The electric-current flowing in a metal-wire

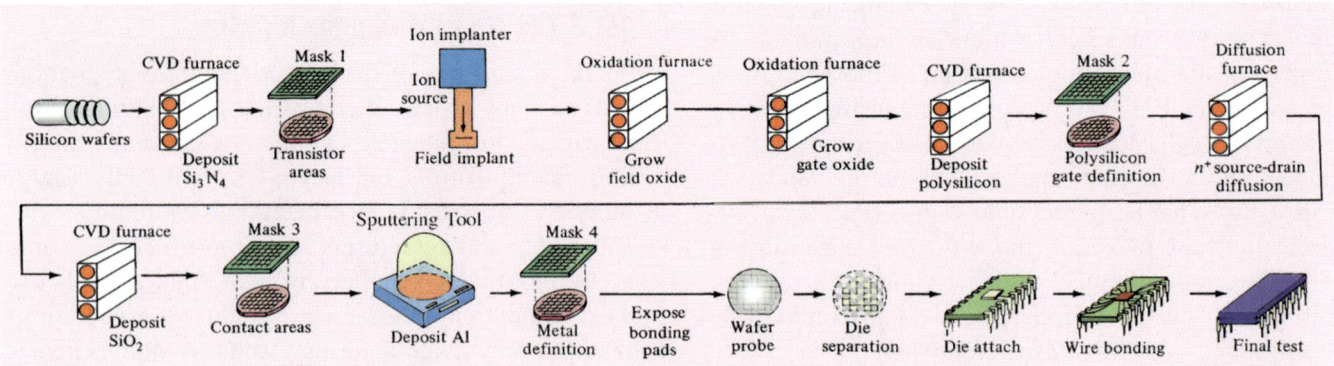

Fig. 1-7 Overview of the sequence of process-steps involved in manufacturing microchips.

CHAPTER 1   THE SEMICONDUCTOR INDUSTRY

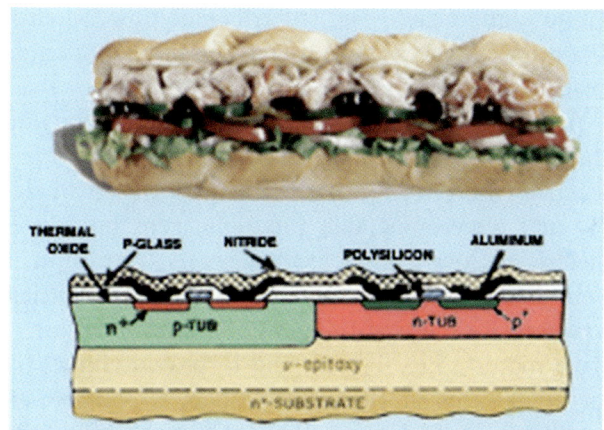

Fig. 1-8 *Making a submarine sandwich,* and *manufacturing a microchip* both use some of the same approaches.

consists of moving electrons. Almost all the electricity we use is in the form of electric-current.

Many effects of electricity can be seen in nature. For example, lightning is a huge flash of light caused by electricity (Fig. 1-11a), and certain types of eels can produce lethal electric-shocks (Fig. 1-11b).

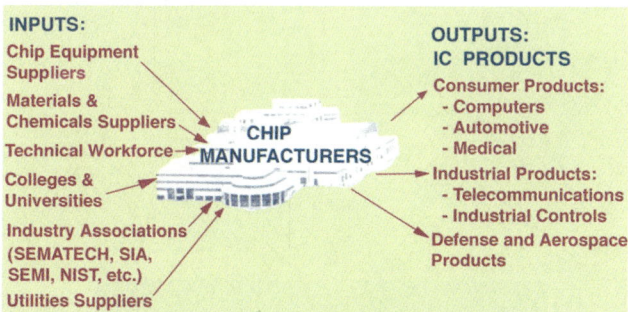

Fig. 1-9 The infrastructure of the Semiconductor Industry.

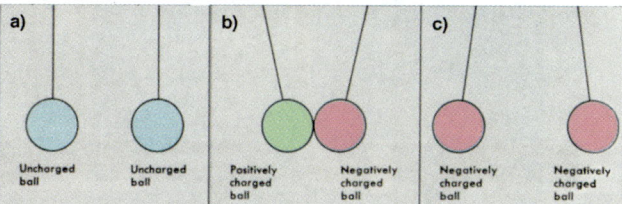

Fig. 1-10 (a) Electrically-uncharged objects exert no forces (i.e., they neither attract nor repel one another). (b) Objects with *unlike* electrical-charge attract one another. (c) Objects with *like* electrical-charge repel one another.

The History of Electricity: Electricity was first mentioned in history by the Greek-philosopher Thales, who in 600 BCE observed the electrical-effect of amber attracting small-bits of straw after being rubbed with cloth. In 1600 CE, Gilbert (a physician to Queen Elizabeth I of England), discovered that other materials (including glass and diamond) also behaved in this manner - that is - like amber). He called these

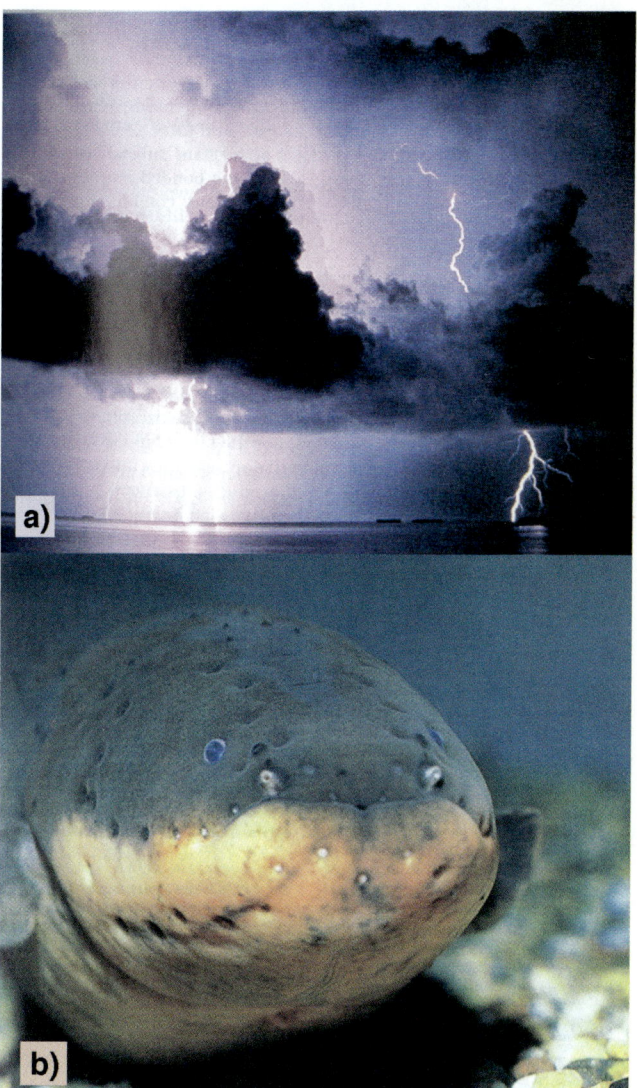

Fig. 1-11 Electrical-phenomena observed in nature: (a) Lightning; and (b) Current from an electric-eel (which can produce lethal electric-shocks).

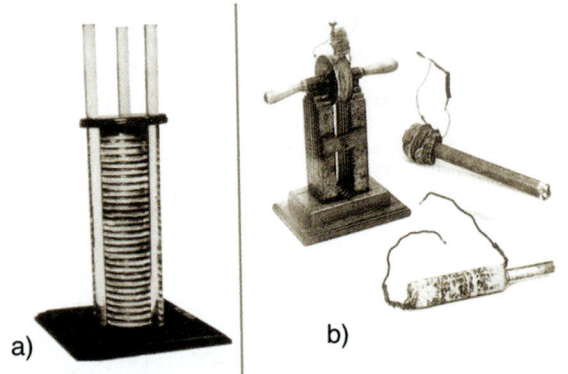

Fig. 1-12 (a) The *voltaic pile*, invented in the late 1790's was the first battery. (b) Faraday's experiments with crude inductors led to his discovery of the Law of Induction (and an understanding of electromagnetism). Unable to purchase electrical materials, Faraday is said to have insulated wires by wrapping them with strips of his wife's petticoat.

materials *electrics* (based on *electrum*, the Latin word for amber). In 1646, the English physician Browne devised the word *electricity*.

Electrical-conductivity in materials was discovered in 1729, and the attractive and repulsive nature of electric-charge was found at about the same time. In 1746 Benjamin Franklin proposed that electricity consisted of a fluid (electrons were then still unknown). In 1785 the French physicist, Coulomb, formulated the laws of attraction and repulsion between electrically-charged bodies.

In 1786, Galvani performed one of the first experiments with electric-current. He hung a freshly-killed-frog by a copper-hook. The legs twitched when they touched an iron-railing. Volta correctly explained this effect in 1790. He proposed that the chemical-action of moisture and the two metals (copper and iron) produced the electricity that caused the twitching. Using this concept, he built the first battery from stacked-pairs of metal disks (called a *voltaic-pile*, Fig. 1-12a). Such voltaic-piles (or *batteries*) were the first-sources of steady electric-current. Later experiments with magnetism (Fig. 1-12b) gave Faraday an understanding of *electromagnetism*. His work led to the development of electric-circuits. *Electric-circuits* are closed loops of conductors (typically metal-wires) and other circuit-components. The latter include power-sources (batteries or generators) and passive-components (resistors, capacitors, and inductors).

The formulation of the principles of electricity and magnetism by the mid-1800's ushered in the *Age of Electricity*. From about 1850 onward a variety of electric and electromagnetic products were introduced. Electric-motors using electromagnets were built, as were electric-generators and transformers. Electricity was used for the first time to do mechanical work (to drive motors, Fig. 1-13a), and to provide light (first for lighthouses in 1856, powered by *arc-lamps*, and later using *incandescent-filament lamps*, perfected by Edison in 1879, Fig. 1-13b). In the mid-1880s, the California Electric Light Company began operating the world's first central-power-plant that sold electricity to private-customers. By 1895, hydroelectric generators at Niagara Falls provided local industries with electric power. (Fig. 1-13c shows a modern motor.)

Vacuum-Tubes and the Dawning of the Age of Electronics:
In the late 1890's scientists experimented

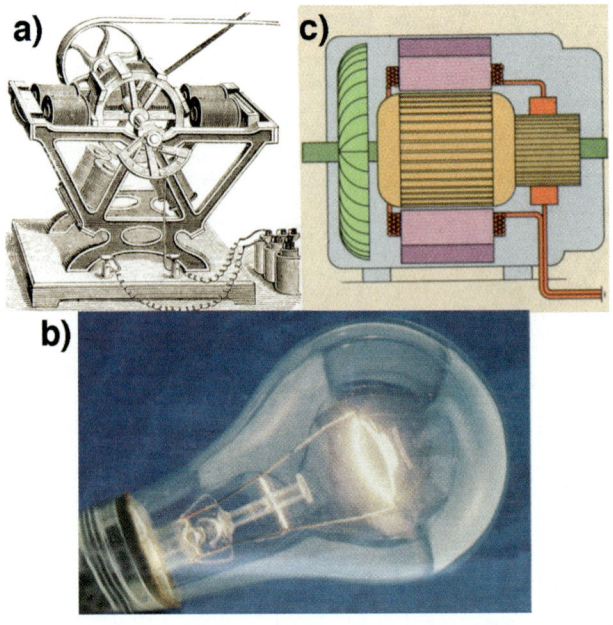

Fig. 1-13 (a) An early electric motor, developed in 1845, which included electromagnets. It was used to operate telegraphs. (b) An incandescent light-bulb. (c) Drawing of a modern electric motor.

CHAPTER 1   THE SEMICONDUCTOR INDUSTRY

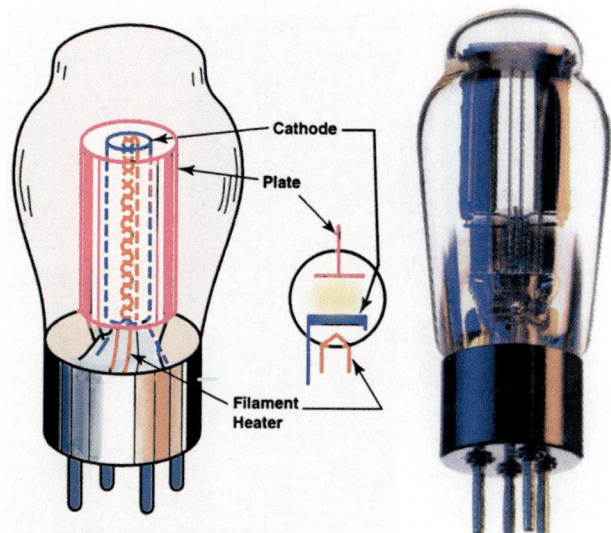

Fig. 1-14 Vacuum tubes.

with the first electronic devices, *vacuum-tubes* (Fig. 1-14). These experiments showed that electron-flow could be controlled in a *vacuum*, specifically in these devices. When vacuum tubes were added to *electric-circuits*, the *electronic-circuit* was born. Further experiments showed that such electronic-circuits could enhance the sending and receiving of radio signals (for instance, they could *amplify* weak radio-signals). This enabled wireless-communication to be developed, such as radios (Fig. 1-15).

This was the dawn of the *Age of Electronics*. By the 1920's factories began to mass-produce vacuum-tubes for radios, and later for television sets. In 1947 the world's first *electronic-computer* was built using vacuum tubes (the ENIAC, Fig. 1-16). Vacuum-tubes ruled the electronics-world until the 1950's.

Vacuum-tubes, however, have several severe-limitations. They are fragile, they burn out after several thousand hours of use, are bulky, and they give off considerable heat. For example, the ENIAC computer (built with 19,000 vacuum-tubes), occupied 3000 ft^2, weighed 50-tons, generated large quantities of heat, and cost $400,000 (in 1940 dollars). Despite its mammoth size and component count, its computing-power was less than that of one of today's programmable engineering-calculators. Thus, smaller, more reliable, and more rugged electronic-devices were sought. This effort led to the invention of the *transistor*.

1.3 THE TRANSISTOR: THE AGE OF SOLID-STATE ELECTRONICS

The transistor was invented in 1947 by Bardeen, Brattain, and Shockley at ATT Bell Labs (Fig.1-17b). This was the beginning of the *Age of Solid-State*

Fig. 1-15 (a) Vacuum-tube radio. (b) Tube-radio chassis.

Fig. 1-16 The first electronic-computer (the ENIAC) was built using vacuum-tubes.

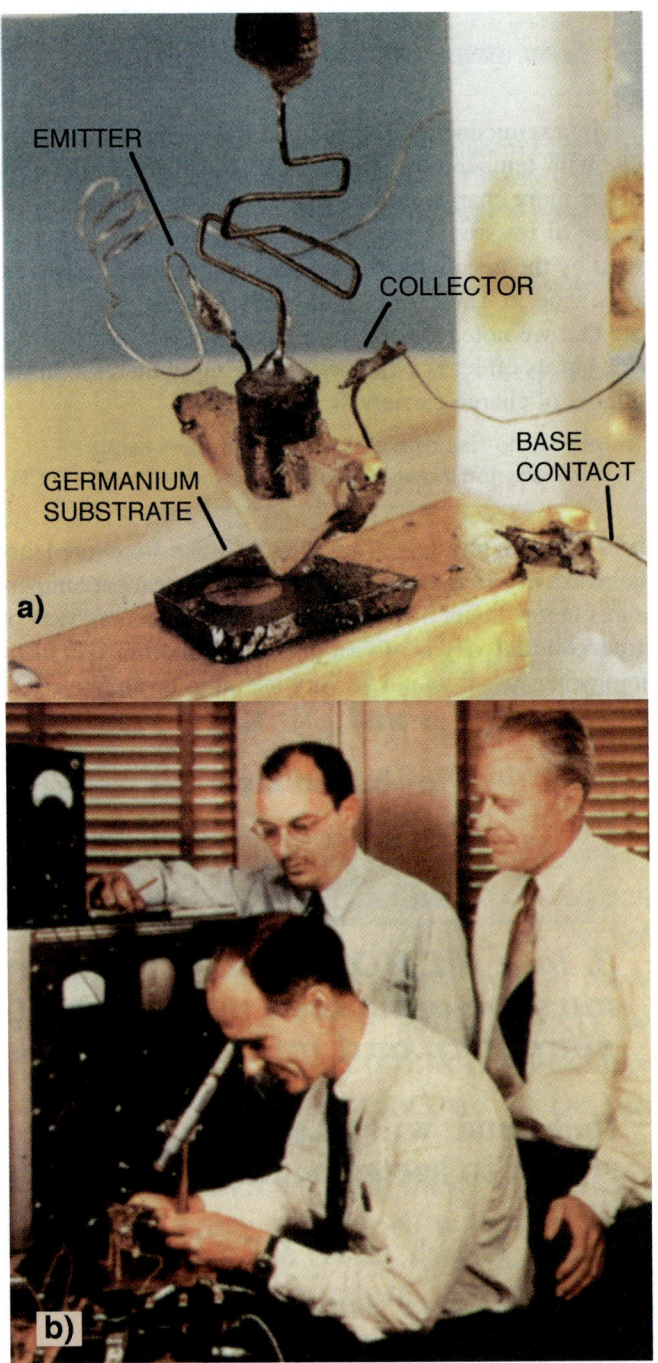

Fig. 1-17 (a) The point-contact transistor invented at Bell Telephone Laboratories in 1947, by (b) Bardeen, Brattain, and Shockley. Reprinted with permission of Lucent Technologies, Bell Labs Innovations.

Electronics, and also the start of the semiconductor-industry. Figure 1-17a is a photo of the first transistor (built using *germanium*). The three scientists shared the Nobel Prize in physics for this invention in 1956.

The *transistor* is a device in which the flow of electrons is controlled in a *solid-material* (a semiconductor), not in *vacuum* (as occurs in vacuum-tubes). Such solid-state electronic-devices perform the same electrical-functions as do vacuum-tubes, but they can be made much-smaller, more-reliable, and rugged. Instead of electrons jumping between elements of the device through a vacuum, electrons in transistors cross electrical-junctions formed in the solid-material. Besides transistors, solid-state technology is also used to fabricate diodes, capacitors, and resistors (as discussed in *Chap. 3*).

Fueled by military and civilian demand for electronic-devices, the semiconductor industry-developed rapidly in the 1950's. Silicon-transistors soon replaced their germanium-counterparts, and silicon remains the most important material for making solid-state devices (including integrated-circuits) to this day.

The solid-state-devices of the 1950's (diodes and transistors) were all *discrete-devices*. (Discrete-devices are electronic-devices that contain only one device per package.) That is, each transistor is cut from a wafer containing many transistors, and put into a package (Figs. 1-18a and 18b). Then this packaged

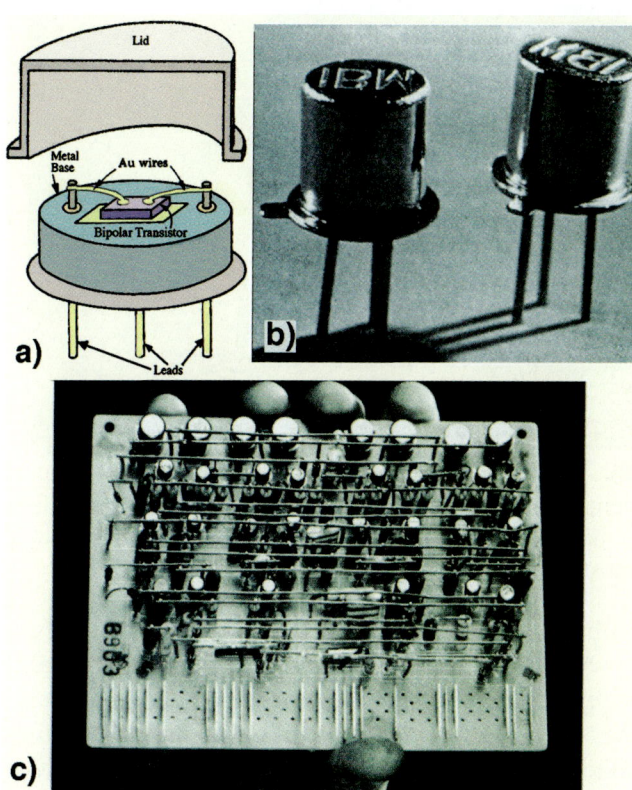

Fig. 1-18 (a) A bipolar transistor mounted on a TO-5 header. (b) Photograph of transistors in TO-5 packages. (c) Circuit board with transistors replacing vacuum-tubes.

transistor is connected into a circuit (Fig. 1-18c). By replacing vacuum-tubes in electronics-products with such discrete-transistors, it became possible to design products that could not be built with vacuum-tubes. These products could be much smaller, sturdier, and could run on portable-power (batteries). One such product was the SONY portable pocket-radio (Fig. 1-19). In many such products, transistors serve as amplifying devices, which make weak-electrical input-signals much stronger (Fig. 1-20).

The structure of the transistor evolved with continuing-progress in semiconductor process-technology. Eventually, the planar-transistor structure used in ICs emerged (Fig. 1-21). It was invented by Hoerni at Fairchild Semiconductor in 1957, and was built using *planar-processing* (see Fig. 3-1 in Ch. 3).

In the *planar-process*, dopants that form the various regions of a transistor are diffused into the silicon-wafer from its top-surface. Thermal-oxide on the silicon acts both as a diffusion-mask (that allows the dopants to be selectively introduced into the silicon-surface) and as a passivating-layer. Aluminum-thin-films are used as the metal-interconnect layer on top of the insulating silicon-dioxide. *Planar-processing* and the *planar-transistor* were important steps toward the invention of the integrated circuit.

1.4 THE AGE OF INTEGRATED-CIRCUITS

By the late 1950's it became apparent that cutting up transistors from a wafer on which many existed, and reconnecting them to make a circuit was not an efficient way to fabricate electronic-circuits (Fig. 1-22). Instead, why not leave them on the wafer and interconnect them there to form the circuit? Pursuit of this idea lead to the invention of the *integrated (monolithic) circuit*. (The term monolithic means *single-stone*,

Fig. 1-19 SONY pioneered the manufacture of portable radios, built using transistor circuits.

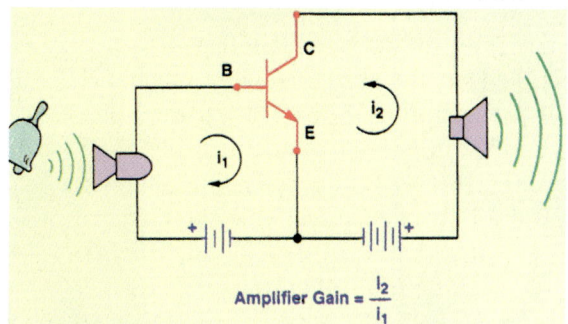

Fig. 1-20 Early electronic circuits were chiefly used to amplify weak electromagnetic signals (such as those generated by radio and TV transmitters).

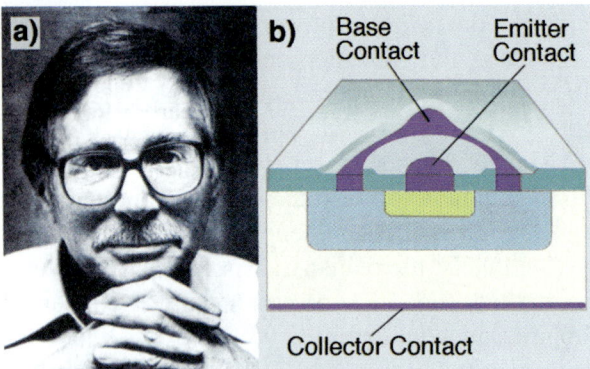

Fig. 1-21 (a) Jean Hoerni (Fairchild) invented the *planar process*, and (b) used it to build the first *planar transistor*.

and refers to the fact that an entire circuit can exist on a single piece of silicon - the stone).

The patent for the invention of the integrated-circuit was given to two engineers who did their work separately in 1958 at two different companies - Robert Noyce (then at Fairchild Semiconductor, and later a founder of Intel) - and Jack S. Kilby of Texas Instruments. Figures 1-23a-c show Noyce's integrated circuit, and Figs. 1-23d-g shows Kilby's. The IC of

Fig. 1-22 Cutting transistors from wafers and assembling them into circuits is inefficient. It was soon discovered how to make *integrated circuits,* each containing a multitude of transistors on a single-piece of silicon.

Kilby's embodiment interconnected several transistors (made on a single-piece of germanium) with gold wires. Noyce's version, however, used planar-processing, with interconnects (made of thin aluminum-lines adherent to the underlying silicon-dioxide) to wire

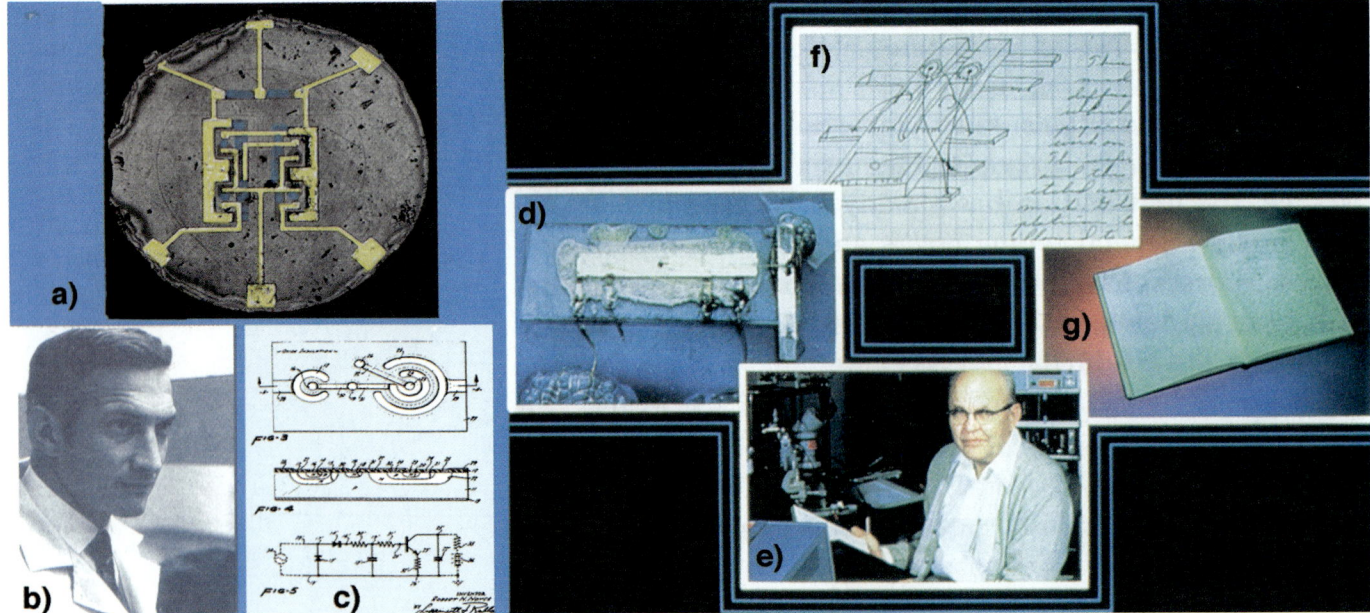

Fig. 1-23 (a) Photomicrograph of the IC invented in 1961 by (b) Robert Noyce, while at Fairchild Semiconductor. (c) Noyce's patent-application. The IC was co-invented by Jack Kilby at Texas Instruments. (d) His invention. (e) Photograph of Jack Kilby. (f) Kilby's lab notes. (g) Kilby's patent-application. Courtesy of Texas Instruments.

CHAPTER 1   THE SEMICONDUCTOR INDUSTRY

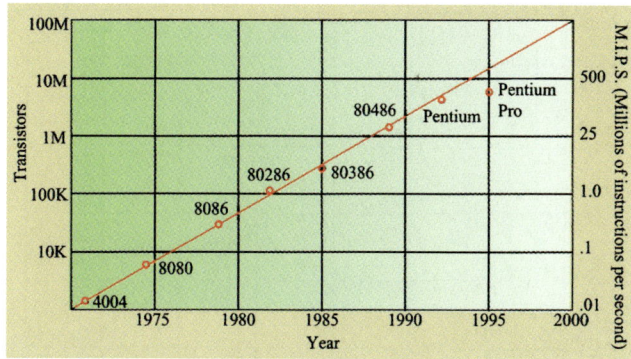

Fig. 1-24 Moore's Law[4] (© IEEE 1998).

TABLE 1-1 INTEGRATION-LEVELS IN ICs		
Integration-Level	Time-Period	Number of Devices per Chip
Small-Scale Integration (SSI)	1960-65	2 - 100
Medium-Scale Integration (MSI)	1965-75	100 – 10,000
Large-Scale Integration (LSI)	1975-85	10,000 – 500,000
Very-Large-Scale Integration (VLSI)	1985-95	500,000 – 5,000,000
Ultra-Large-Scale Integration (ULSI)	1995–Present	> 5,000,000

together several transistors fabricated in a silicon-substrate. Noyce's version is much more like the ICs still being made today. The smallest-dimensions on the earliest ICs were about 20-µm. As time has passed, ICs have become ever-more-complex electronic-components. It became possible to make monolithic-ICs that were equivalent to those that formerly required a multitude of discrete-devices (Fig. 1-22).

1.5 THE GROWTH OF THE SEMICONDUCTOR-INDUSTRY

The first ICs had only a few interconnected-transistors. Soon, it was learned how to make the devices smaller (also referred to as *scaling* the transistors), thus enabling ICs with more integrated-components to be built. The levels of integration were classified roughly as follows (see also Table 1-1). The first ICs (early 1960s) were grouped under the term *small-scale-integration* (*SSI*), and it meant that 2 to 100 components were on such ICs. The next level of integration (1965-1975) was *medium-scale-integration* (*MSI*), which had 100 to 10,000 devices. Then came *large-scale- integration* (*LSI*), which contained 10,000 to 500,000 transistors (1975-1985). Eventually *very-large-scale- integration* (VLSI) was achieved (1985-1995), which had 500,000 to 5,000,000 components. Finally, in about 1995 *ultra-large-scale-integration* (*ULSI*) was reached, in which more than 5,000,000 devices are on a single-chip. By 2002, DRAMs with 1-billion memory-cells were being manufactured!

This growth in the number of components on a chip was foreseen by Gordon Moore (one of the founders of Intel). In 1964 he predicted that the number of transistors on a chip would roughly double every 18-months, and this prediction is known as *Moore's Law*. Figure 1-24 plots the growth in the number of transistors on a microprocessor-chip as a function of time. It shows that Moore's Law has been surprisingly accurate for more than 30 years!

The ability to increase the component count on a chip relies on being able to reduce the size of the minimum-feature-dimension with time. (The semiconductor-industry expresses the minimum-feature-size in *microns*, which are equal to 10^{-6} of a meter, and are written as 1-micron = 1-µm. Note that a human-hair has a diameter of 50-100-µm). Such feature-size-reduction has also relentlessly continued (Fig. 1-25),

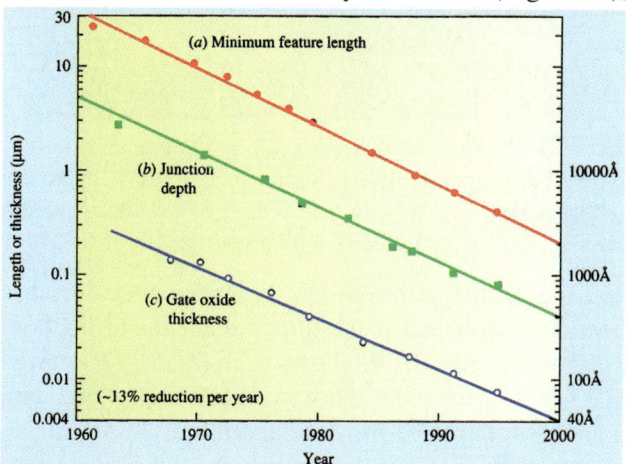

Fig. 1-25 Exponential decrease of: (a) minimum-feature dimension; (b) junction-depth; and (c) gate-oxide-thickness of MOSFETs in integrated circuits.

12 MICROCHIP MANUFACTURING

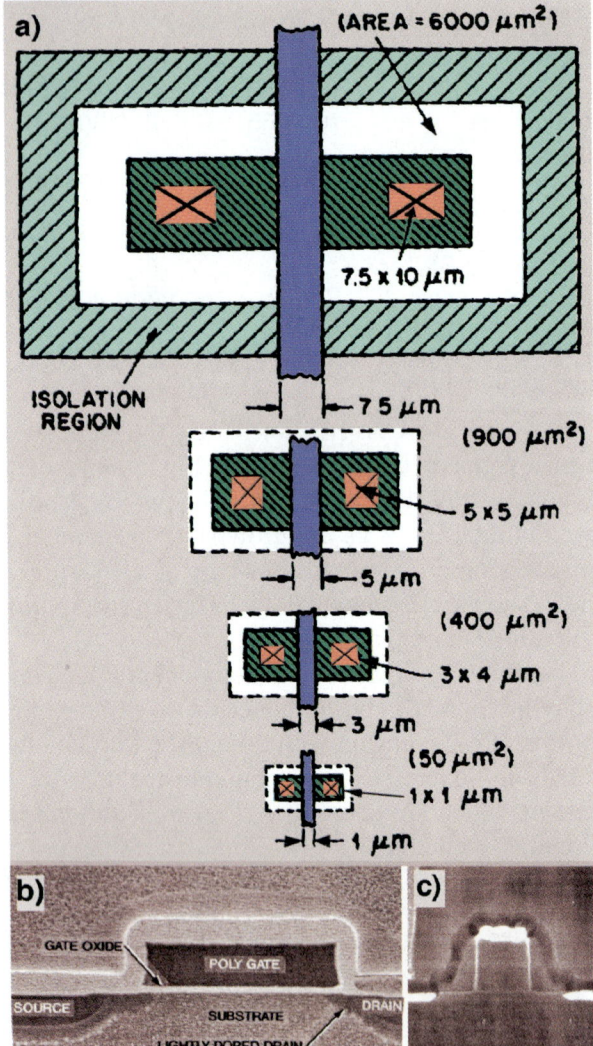

Fig. 1-26 (a) Drawing showing the reduction in area of a MOSFET as gate-length is reduced. (b) SEM cross-section of a MOSFET with a gate-length of 2.0-μm. (c) SEM cross-section of a MOSFET with a gate-length of 0.35-μm.

and the details of how such dimensional-scaling has been accomplished is the thread that runs throughout our book. Figure 1-26a shows such MOSFET scaling, and Figs. 1-26b and c show two MOSFETs - one with a gate-length of 2.0-μm, and another of 0.35-μm.

At the time ICs were invented (early 1960s) the minimum feature-sizes on chips were in the range of 10-20-μm. By the 2002-time-frame, the minimum-feature-size on the most advanced ICs had shrunk to

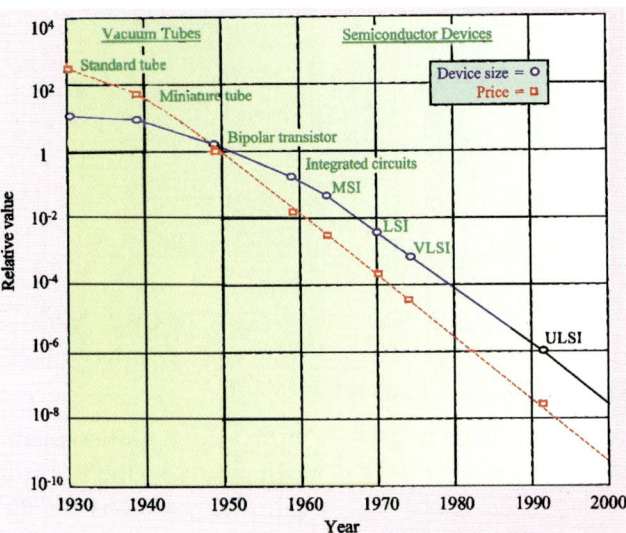

Fig. 1-27 Price decrease of semiconductor-components as a function of time. The drop in price accelerated with the invention of the IC.[6] (© 1991 IEEE)

below 0.1-μm (100-nm), and the end to this scaling is not yet in sight! (Note that Fig. 1-25 also illustrates the reduction in the junction-depth and the shrinking thickness of gate-oxides in MOSFETs.)

However, perhaps the most important reason that the IC-industry has grown so rapidly is that the progress in semiconductor-manufacturing-technology has continuously decreased the cost of chips, relative to their performance and functionality. This factor - increased-performance at lower-cost - has caused many older products to be designed using solid-state

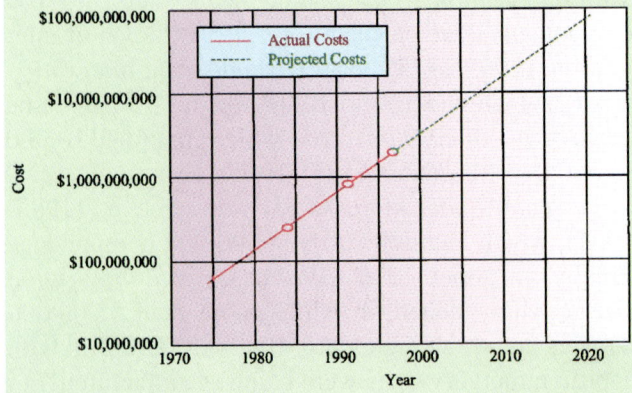

Fig. 1-28 Start-up cost of wafer fabs as a function of time. (© IEEE 1998)

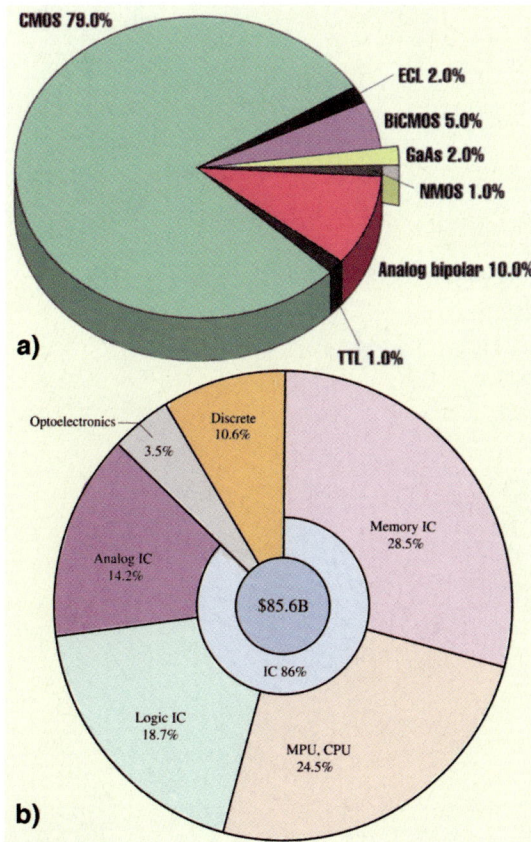

Fig. 1-29 (a) Estimated market share of IC technologies in 2000. (b) Pie-chart showing the dominant types of ICs manufactured in the late 1990's.

electronics, and the invention of many new products.

As shown in Fig. 1-27, the price of microchips has steadily declined. *From 1950 to 2000, it decreased by more than a factor of 100 million!* In 1958 a single-transistor cost $10. Today, $10 would buy two 256-Mb DRAM-chips. Besides the decreasing feature-size, this price-reduction has been driven by economies of scale of manufacturing. The introduction of manufacturing-improvements to the equipment and processes used to fabricate ICs also reduced costs.

One of the major challenges being faced by the semiconductor-industry is that the cost of the wafer-fab has also been steadily rising (as seen in Fig. 1-28). A new 300-mm fab in the 2002-timeframe could cost in excess of $3-billion. To overcome this high cost-of-entry into advanced-technology, many U.S. compa-

nies have tried new approaches, such as joint-venture agreements with companies from other countries, or using *silicon-foundries* (instead of building their own fabs). This subject is discussed further in Chap. 28.

The breakdown of the types of ICs sold in 2000 (as a percentage of total sales) is shown in Fig. 1-29. It can be seen that CMOS is by far the most dominant IC-device-technology in current use.

REFERENCES

1. M. Riordan and L. Hoddeson, *Crystal Fire,* W.W. Norton & Company, New York, 1997.
2. M.S. Malone, *The Microprocessor: A Biography,* TELOS/Springer-Verlag, 1995, Santa Clara, CA.
3. G. Moore, "The Role of Fairchild in Silicon Technology in the Early Days of Silicon Valley," *Proc. IEEE,* **86-1**, January 1998, p. 59.
4. G. Moore, "Cramming More Components onto Integrated Circuits," *Proc. IEEE,* **86-1**, January 1998, p. 84. Reprinted from *Electronics,* April 19, 1965.
5. Texas Instruments, *The Chip That Jack Built Changed the World,* Texas Instruments, Sept. 1997.
6. K. Shoda, "Home Electronics in the 1990's" *Proc. Intl. Smp. on VLSI, Syst. & Applications* (1991).

PROBLEMS

1. When, where, and by whom was the *transistor* invented?
2. What is a *monolithic integrated circuit*? Who were it's inventors, and when was the IC invented.
3. Give a brief description of the history of the electronics era from 1945 to the present.
4. By what factor has the price per bit of DRAM memory chips decreased since they were introduced in about 1965?
5. What is a wafer-fab expected to cost to build and equip in: (a) 10 years; (b) 20 years?
6. In 1985, 64-kb DRAM devices were being introduced. According to Moore's Law, how many DRAM bits would have been available on a chip by 2000.
7. In 2002, the size of DRAMs had grown to 256-Mb. According to Moore's Law, how many DRAM bits will be on a chip by 2020?
8. How many transistors can fit on the head of a pin?

CHAPTER 2

SEMICONDUCTORS AND BASIC MATERIALS SCIENCE

CHAPTER CONTENTS

2.1 INTRODUCING SEMICONDUCTORS

2.2 THE NATURE OF MATTER: ATOMIC STRUCTURE

2.3 MOLECULES & COMPOUNDS

2.4 SOLIDS, LIQUIDS, AND GASES

2.5 CLASSIFYING MATERIALS BY THEIR ELECTRICAL PROPERTIES

2.6 SEMICONDUCTOR SILICON

2.7 OTHER SEMICONDUCTORS

"According to the theory of aerodynamics, and as may be readily demonstrated through experiments, the Bumble Bee is unable to fly.

This is because the size, the weight, and the shape of his body in relation to his wingspread make flying impossible."

BUT, the Bumble Bee, being ignorant of these scientific truths goes ahead and flys anyway - and makes a little honey every day!"

Author Unknown

A wafer-transport carrier provides wafer protection, enabling greater fab productivity. Photo courtesy of Entegris Inc.

Silicon is the most important semiconductor-material in use today. Electronic-devices in integrated-circuits are built on silicon wafers because silicon has a unique set of properties that make it best suited for making such devices.

In this chapter the basics of atomic-structure are first presented. This becomes the foundation for explaining the properties of silicon (and how it becomes possible to use it to build semiconductor-devices). This background information enables readers to better understand the details of the different steps used throughout wafer-fabrication.

2.1 INTRODUCING SEMICONDUCTORS

The integrated-circuits of so many of today's electronic-products are based on such semiconductor materials as silicon, gallium-arsenide, gallium-phosphide, etc. The unique physical and electrical properties of semiconductor-materials permits fabrication of such ICs. Thus, it is useful to introduce the fundamentals of semi conductors here. However, this subject rests on an understanding of the structure and bonding of atoms. Thus, both subjects make up the contents of this chapter.

2.2 THE NATURE OF MATTER: ATOMIC-STRUCTURE

The properties of semiconductor materials are based on the *structure of the atom.* All matter in the universe is made from approximately 100 materials known as *elements*. Each element has its own atomic-structure, and this uniqueness gives rise to the different properties of the elements.

An *atom* is the smallest object of matter that still retains the unique characteristics of each element. As an example, consider the element *gold*, whose properties include having a bright yellow-color and high electrical-conductivity; and being a soft, heavy metal. If a piece of gold is broken into smaller and smaller fragments, eventually an object is reached – a *gold-atom* - that still has the properties of gold. The gold-

MICROCHIP MANUFACTURING

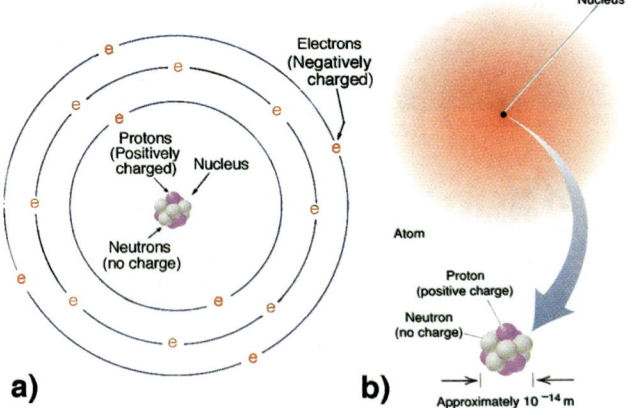

Fig. 2-1 (a) Schematic drawing of Bohr's model of the structure of the atom. (b) Illustration of the Quantum Mechanical model of the structure of the atom.

atoms themselves (as well as all other atoms) consist of *subatomic-particles* – protons, electrons, and neutrons. Each of these subatomic-particles also have their own properties. But, it is only when they are assembled together in a particular combination and configuration (an atom) that they form gold (or some other element).

2.2.1 The Structure of the Atom

The structure of atoms has only been understood since the beginning of the 20th century. The model of the atom proposed by Bohr in about 1920 was the first one that correctly explained the most important aspects of atomic-structure (Fig. 2-1a). While the model of the atom based on Quantum Mechanics (introduced in the 1930's) is even more correct (and its depiction of how electrons surround the nucleus is shown in Fig. 2-1b), Bohr's model of the atom is suitable for our level of discussion, and it is the one we will use for our discussions of atoms and the bonding that occurs among atoms.

Bohr's model says that protons and neutrons (which are 2000 times more massive than electrons) exist at the center of the atom, in the *nucleus*. The electrons, being much lighter, exist in stable orbits around the nucleus, somewhat like the planets that orbit the sun. Electrons and protons are also electrically-charged-particles. Protons have a positive-charge and electrons a negative-charge, but of equal magnitude. Since positive and negative charges attract one another, this force of electrical-attraction maintains the stability of the atomic-structure. Thus, in a neutral-atom, there are equal numbers of electrons and protons. For example, the simplest atom (*hydrogen*) has one of each.

Every element has a specific and unique number of protons. For instance, all helium (He) atoms have exactly two protons, while all oxygen (O) atoms have exactly eight. The number of protons in an atom is called its *atomic-number*.

The electron-orbits of an atom are called *shells*, and each shell can only contain up to a specific maximum number of electrons. For example, the first *atomic-shell* can contain up to only 2 electrons, while the second only up to 8. Once a shell is filled, addi-

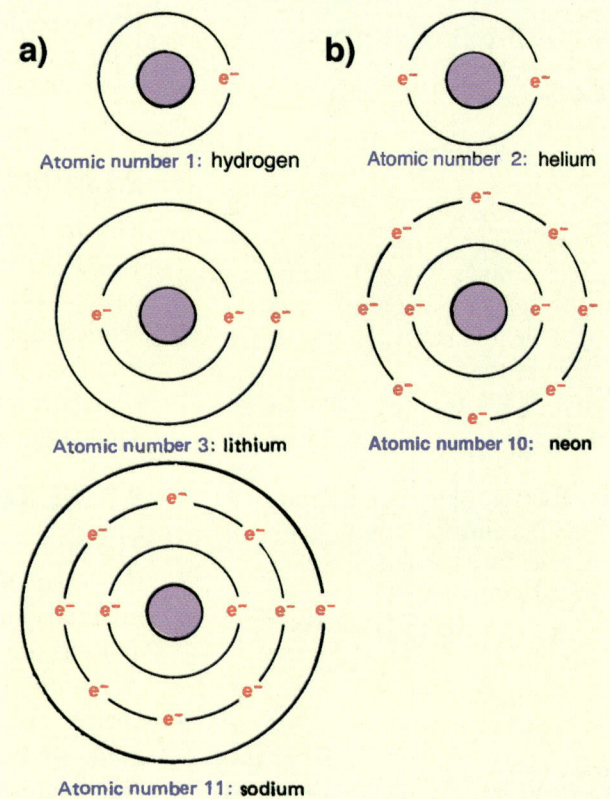

Fig. 2-2 The electronic structure of several atoms: (a) Left column – hydrogen, lithium, and sodium. (b) Right column – helium and neon.

tional electrons must go into the next one. This means that *hydrogen* (atomic-number = 1) has one electron in the first-shell, and *helium* (atomic-number = 2) has two. But, in *lithium* (atomic-number = 3), the third electron must occupy a place in the second shell (see Fig. 2-2). Likewise, in *neon* (with an atomic-number = 10), the second-shell has eight electrons, but in sodium (whose atomic-number = 11), the 11th electron must take a place in the third-shell (Fig. 2-2). The *outermost shell* is called the *valence-shell*.

Another useful fact to note is that hydrogen, lithium, and sodium have a common property - namely that each atom has only one-electron in its valence-shell. This illustrates another aspect of atoms, namely that elements with the same number of valence-shell-electrons have similar properties. This also implies that the electrons in the valence-shell generally have the most influence on the physical, chemical, and electrical-properties of an element. Thus, it is no accident that the three best electrical-conductors (copper, silver, and gold) have the same number of valence-shell-electrons.

2.2.2 The Periodic Table

The *Periodic Table* is a catalog of all of the elements, from the simplest to the most complex (Fig. 2-3). Each *element-box* of the Periodic Table provides information about the element. First, it identifies the element by an abbreviation of one or two letters, known as the *atomic-symbol*. (For instance, the atomic-symbol for hydrogen is H, and that for lithium is Li.) In the upper-left corner is the atomic-number (the number of protons), and at the bottom is the mass (see Fig. 2-4), given in *atomic-mass-units* (or *amu*). For example, oxygen has an atomic-number of 8, and an atomic-mass of ~16 (amu = 15.99), indicating that oxygen has 8 neutrons (as well as 8 protons) in its nucleus.

We are most interested in that part of the Table

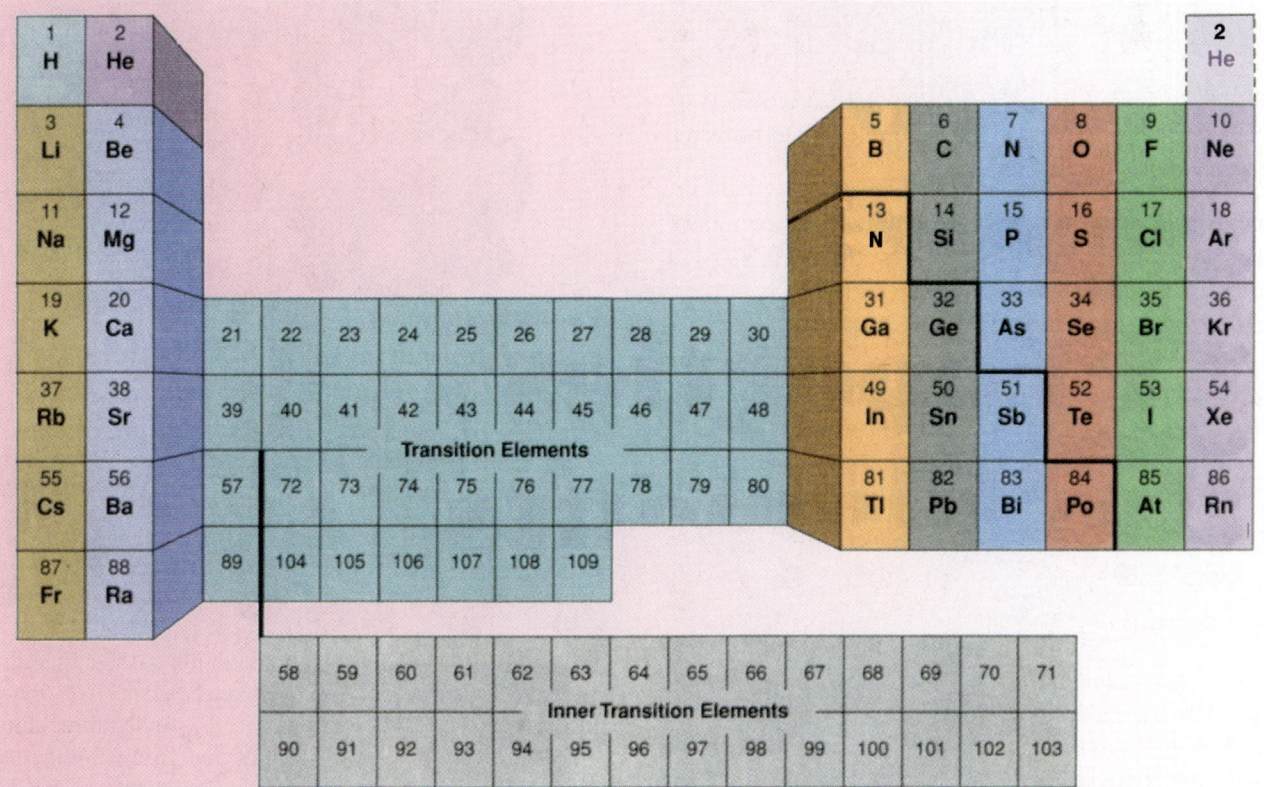

Fig. 2-3 The complete Periodic Table.

18 MICROCHIP MANUFACTURING

III^A	IV^A	V^A	VI^A
5 **B** 10.81	6 **C** 12.01	7 **N** 14.01	8 **O** 16.00
13 **Al** 26.98	14 **Si** 28.09	15 **P** 30.97	16 **S** 32.06
31 **Ga** 69.72	32 **Ge** 72.59	33 **As** 74.92	34 **Se** 78.96
49 **In** 114.8	50 **Sn** 118.7	51 **Sb** 121.8	52 **Te** 127.6
81 **Tl** 204.4	82 **Pb** 207.2	83 **Bi** 209.0	84 **Po** 209

(Additional columns on left: II^B with 30 Zn 65.39, 48 Cd 112.4, 80 Hg 200.6)

Fig. 2-4 Portion of the Periodic Table relevant to semiconductor materials and doping.

Ions: An ion is formed when a (neutral) atom gains or loses one or more electrons. A neutral-atom becomes positively-charged when it loses an electron (forming a *positive-ion*), and negatively-charged when it gains one (forming a *negative-ion*). Ions with opposite charge attract one another, and this can lead to the formation of a chemical-bond, which results in an ionic-compound (see Sect. 2.4.4). As an example, sodium-atoms (Na) have a tendency to form positive-ions, since their lone valence-electron is easily lost (Fig. 2-5a). On the other hand, chlorine-atoms (Cl) are more likely to become negative-ions, as they tend to accept an extra electron to complete the filling of their valence-shell (Fig. 2-5b). Such Na$^+$ and Cl$^-$ ions in close proximity will attract one another and arrange themselves in an ionic-compound NaCl (the common substance, *table-salt*, Fig. 2-6).

shown in Fig. 2-4, since this contains those elements most widely found in the devices of silicon ICs.

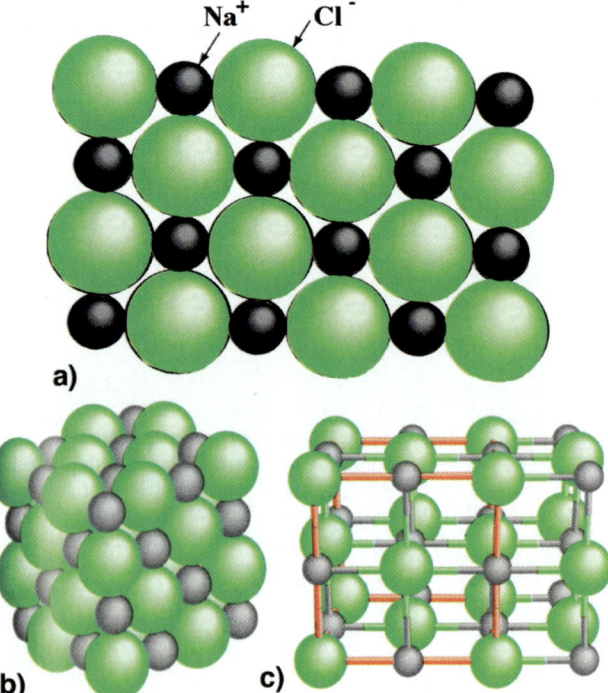

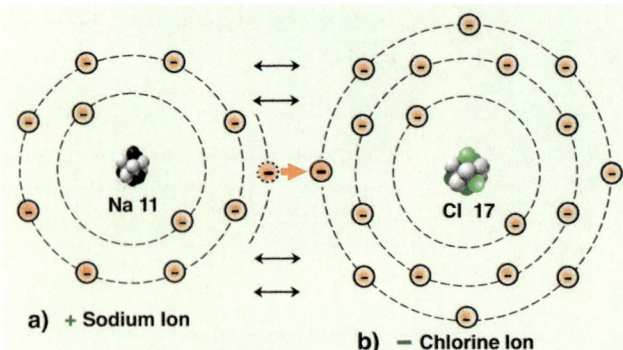

Fig. 2-5 (a) Sodium (Na) ion. A *positive* or "+" ion is formed when a neutral-atom loses one (or more) electrons. (b) Chlorine (Cl) ion. A *negative* or "-" ion is formed when a neutral-atom gains one (or more) electrons. Being of unlike-charge, the Na$^+$ and Cl$^-$ ions attract one another.

Fig. 2-6 (a) Two-dimensional illustration of the lattice structure of NaCl. (b) Three-dimensional perspective drawing of the lattice structure of NaCl. (c) Expanded three-dimensional drawing of the structure of NaCl showing that each Na-ion has six surrounding Cl-ions and that each Cl-ion likewise has six surrounding Na-ions.

CHAPTER 2   SEMCONDUCTORS AND BASIC MATERIALS SCIENCE   19

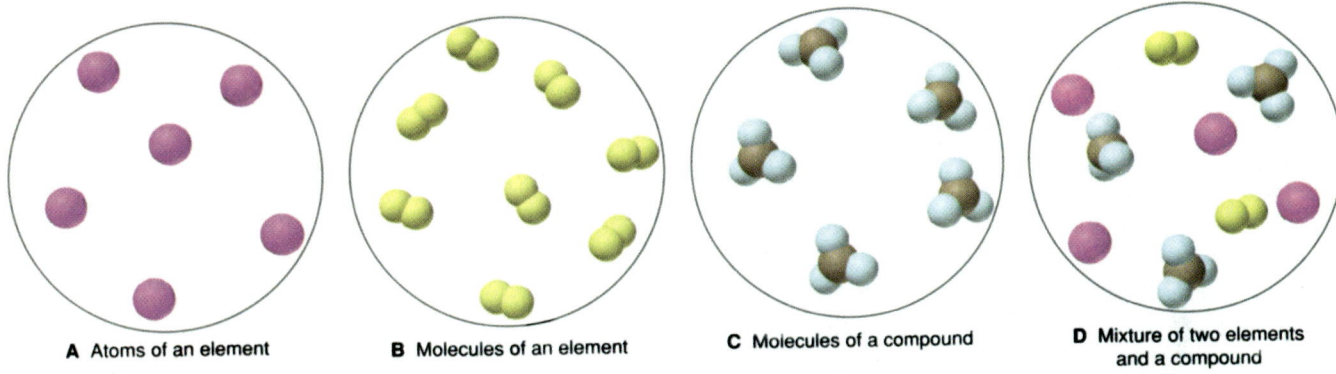

A Atoms of an element **B** Molecules of an element **C** Molecules of a compound **D** Mixture of two elements and a compound

Fig. 2-7 (a) An *element* consists of identical atoms, but only some elements occur as individual atoms. (b) Some elements occur instead as molecules. (c) A *molecule* of a compound consists of a characteristic-number of atoms of two or more elements, chemically-bound together. (d) A *mixture* contains the individual units of two or more elements and/or compounds that are physically intermixed.

2.3 MOLECULES AND COMPOUNDS

Substances formed from a single one of the 100 element contain only one type of atom. These are the simplest materials that have unique physical and chemical properties (Fig. 2-7a). However, there are many more than 100 substances in existence, with the others being non-elemental materials. Such non-elemental substances are called *compounds*, and they consist of combinations of two or more elements.

The basic-unit of a *compound-material* is the *molecule*. Molecules are objects that consist of two or more atoms, chemically-bonded in a structure that behaves as an independent-unit (Figs. 2-7b and c). For example, consider the non-elemental-compound, *water*. The basic-unit of matter that retains the properties of water is a molecule consisting of one oxygen-atom and two hydrogen-atoms (Fig. 2-8b). The multiplicity of materials comes from the ability of atoms to bond together to form molecules.

Although pictorial diagrams are used to depict particular molecules (Fig. 2-8), the more common way they are represented is with a *molecular-formula*. For instance, the molecular-formula of *water* is the familiar H_2O. Molecular-formulas indicate exactly the elements and their number in a compound material.

Note that some elements combine into *diatomic-molecules* (that is, composed of two atoms of the same element, Fig. 2-7b). These diatomic-molecules are examples of substances having a molecular-structure but are not compounds. The familiar process-gases hydrogen, nitrogen, and oxygen, in their natural state, are composed of diatomic-molecules, with molecular-formulas being H_2, N_2, and O_2, respectively.

2.4 SOLIDS, LIQUIDS, AND GASES

Matter can exist in three states: *solid*, *liquid*, or *gas* (Fig. 2-9). A clear distinction can be made between these various states on the basis of how substances behave when placed in an empty container. A *solid* is *rigid*, having a definite shape and volume. Its shape does not conform to that of its container. A *liquid* substance has a definite volume that assumes the shape of its container, but it will only fill the container to the extent of its volume. A *gas* occupies the entire space available to it, and its shape *and* volume are determined by its container. Most substances can

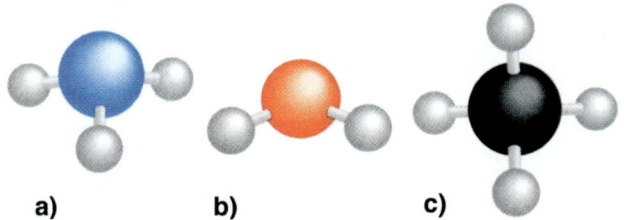

a)   b)   c)

Fig. 2-8 Molecules of three familiar substances, drawn in ball-and-stick fashion. (a) Ammonia, NH_3. (b) Water, H_2O. (c) Methane, CH_4.

exist in any of these three states, depending on the temperature, T, and pressure, P. For instance, water at atmospheric-pressure undergoes a change of state (from solid to liquid) at 0°C, and then turns from liquid to gas at 100°C. Most substances behave in a similar manner, although they change states at other specific values of T and P.

This book primarily describes the fabrication of electronic-devices built in *solid* materials (i.e., those in which electron-flow is controlled in semiconductors - the so-called *solid-state electronic-devices*). Examples of such devices include *pn*-diodes, bipolar-transistors, MOSFETs, ICs made on silicon-substrates (chips), and light-emitting diodes (LEDs). However, IC-fabrication also uses *liquid* chemicals and *gases*. Thus, knowledge about the materials science of all three states of matter is needed.

Units of Energy: Energy is transferred among substances during the various IC-fabrication-processes. It is important to understand the units of energy used when describing the events occurring in each process-step. The SI-unit of energy is the *joule (J)*, which is defined (in an electrical-context), as the amount of energy required to move *one-coulomb (C)* of *electric-charge* through *one-volt (V)* of *potential-difference*.

However, *one-coulomb* of electric-charge corresponds to the charge on 6.2×10^{18} electrons. Thus, when events on the atomic-scale are being described (i.e., one-atom or one-electron at a time), it would be more convenient if they were expressed in units of energy more appropriate to the scale of such events (i.e., in units that are much smaller than a *joule*).

Such a smaller unit of energy is the *electron-volt (eV)*. It corresponds to the energy needed to move a single-electron (which has only *one-electronic-charge q*) through a potential-difference of one-volt. Since *one-electronic-charge* $= 1.6 \times 10^{-19}$ C, then:

$$1 \text{ eV} = 1.6 \times 10^{-19} \text{ C} \cdot 1\text{V} = 1.6 \times 10^{-19} \text{ J} \quad (2.1)$$

The electron-volt is used to describe the energy levels of electrons in atoms and the energies involved in various processes (e.g., the bombardment-energy of ions in etching-processes, the energy of ion-beams in ion-implantation-processes, and the energy of electron-beams in scanning-electron-microscopes).

2.5 CLASSIFYING MATERIALS BY THEIR ELECTRICAL-PROPERTIES

Electricity was introduced in Chap. 1. Here we add some important supplementary information with respect to the electrical-properties of solids. This is required for the topic of semiconductor-manufacturing, because the devices that are fabricated rely on all aspects of the electrical-conductivity of matter.

Solids can be categorized with respect to their ability to conduct electricity (i.e., to support an electrical current-flow), into three groups: **1)** conductors; **2)** insulators; and **3)** semiconductors.

2.5.1 Electrical-Conductors and Resistivity

Materials are classified as being *conductors* if large electrical-currents can easily be established in them by the application of an electric-field. A material is a good conductor only if a large number of electrons easily flow in it as electric-current. This circumstance occurs if the protons of the nucleus maintain a weak hold on the valence-electrons of an element. This condition exists in most metals (although other materials can also exhibit good electrical-conductivity).

The property that quantifies the degree to which a material is a good conductor is called its *conductivity* σ. The reciprocal of the conductivity is its *resistivity* $\rho = 1/\sigma$. The lower the resistivity of a material, the better is its conducting ability. The units of ρ are ohm-centimeters (Ω-cm). For example, aluminum (a very-good conducting metal) has a very-low value of ρ ($\rho_{Al} = 2.7 \times 10^{-6}$ Ω-cm, or 2.7 $\mu\Omega$-cm). Note that ρ is strictly a property that depends on the material and not the geometry of the electrical-component made from it. Figure 2-10 shows the conductivity of a variety of materials.

The property of a component made from a specific material that includes the geometrical-form of the material (and its resistivity) is called its *resistance (R)*, and the units of resistance are *ohms* (Ω). The value of R of a resistor made of a specific material and

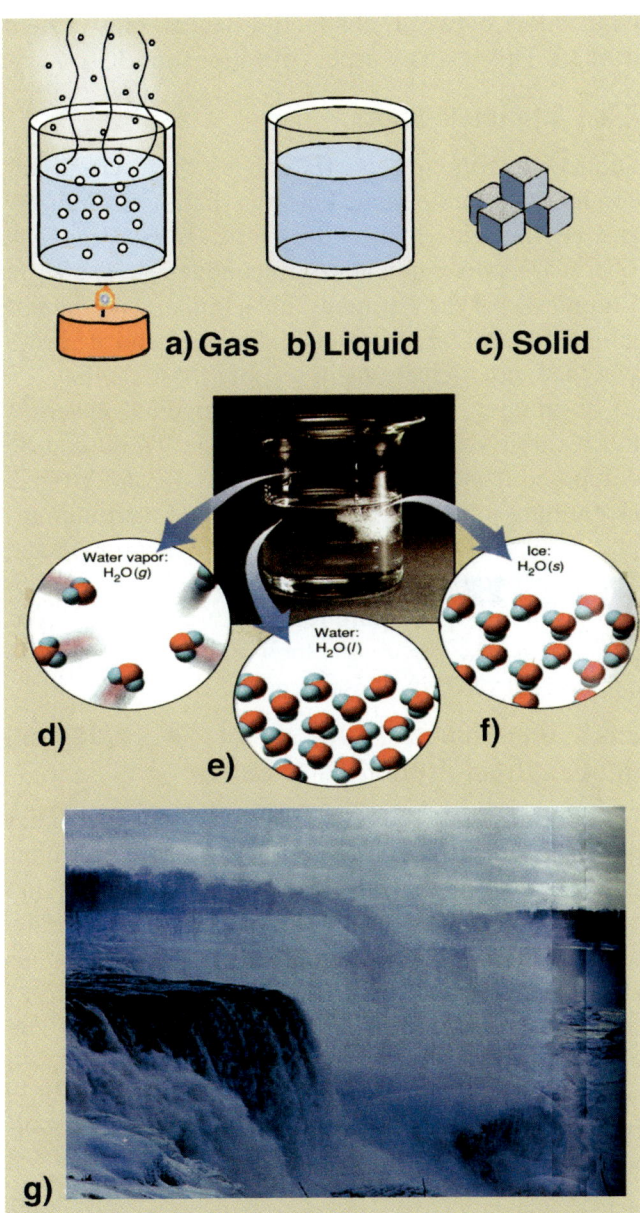

Fig. 2-9 Matter can exist in three states: (a) gas; (b) liquid; and (c) solid. For example, water can take three physical states: (d) water-vapor; (e) liquid water; or (f) ice. Despite the physical differences of water on the macroscopic level, all three states consist of the same individual H_2O molecules. (g) Niagara Falls in winter. Humid-wind, churning-water and rigid-ice display the power and beauty of the three states of water, co-existing at once.

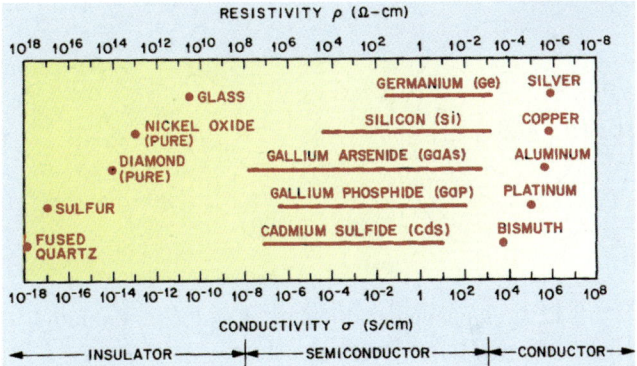

Fig. 2-10 Typical range of conductivities for insulators, semiconductors, and conductors.

having a particular geometrical-form is given by:

$$R (\Omega) = \rho \, l/A \qquad (2.2)$$

where l is the length of the resistor and A is the cross-sectional area (i.e., its width x thickness, see Fig. 2-11). The technology of fabricating the resistors used as circuit-components in integrated-circuits is described in Chap. 3. However, it should be noted here that the interconnect-wires in ICs are also made from thin-films of materials that have a low, but still finite, value of resistivity. As the feature-sizes of ICs get smaller, the resistance of these interconnect-wires (or *lines*) becomes a limiting-factor in the speed-performance of advanced ICs. The lower value of ρ of Cu (ρ_{Cu} = 1.7 $\mu\Omega$-cm) is the main reason why Cu-interconnects are replacing Al-interconnects in modern ICs (see Chap. 24).

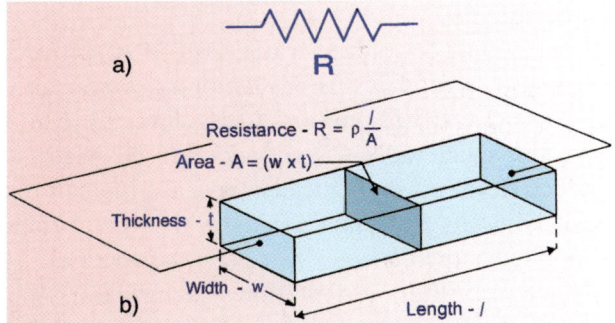

Fig. 2-11 (a) Circuit symbol of a *resistor*. (b) Relationship used to calculate the resistance of a circuit element from: (**1**) its physical dimensions (l, w, and t); and (**2**) the resistivity (ρ) of the material of which it is made.

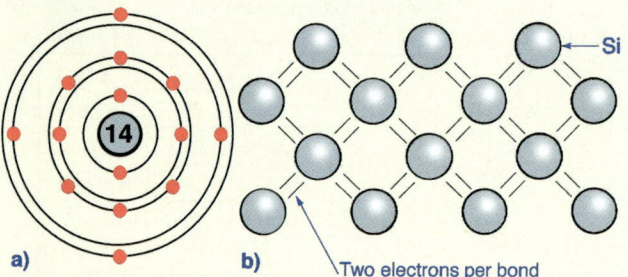

Fig. 2-12 (a) Conceptual drawing of a silicon-atom and its electronic structure. The four outermost electrons are the valence-electrons that participate in covalent-bonds. (b) A two-dimensional drawing of the covalent-bonding arrangement in a Si-lattice [i.e., viewed along a (100) direction].

2.5.2 Electrical-Insulators (Dielectrics)

Insulators (also called *dielectrics*) are materials at the other extreme of the conductivity-scale from metals and other conductors. In an insulator there are extremely-few mobile charge-carriers available to move in response to an applied electric-field. This is because the atoms of dielectric-materials hold their valence-electrons tightly, resulting in an absence of charge-transport (and thus, very-little current-flow).

Insulators therefore exhibit very-low conductivities (and very-high resistivities). For example, the resistivity of silicon-dioxide is $> 10^{18}$ Ω-cm - almost 24 orders-of-magnitude higher than the resistivity of aluminum! Other examples of insulators in everyday life are glass, rubber, plastics, and ceramics. Insulators used in IC-fabrication include silicon-dioxide and silicon-nitride.

Purified and de-ionized-water used in IC-processing is also a good insulator, exhibiting a resistivity $\rho_{DI\text{-water}} = 18 \times 10^6$-$\Omega$-cm (or 18 megohm-cm). But, if impurities (such as NaCl) are added to DI-water, its conductivity rises dramatically (because the salt dissociates into its Na^+ and Cl^- ions, which then behave as mobile charge-carrying species in the water).

Since very-little current flows in insulators, this property is exploited in ICs to isolate the various layers of conducting-interconnects. That is, thin-films of SiO_2 are deposited on the Si-substrate and between interconnect-layers. Openings are created to allow connection between conducting-layers only where desired. This is a key aspect of fabricating ICs.

2.5.3 Semiconductors

Semiconductors are materials that exhibit electrical-conductivity in the range between that of a conductor and an insulator (hence the name, *semiconductor*). There are two elemental-semiconductors, found in Column-IV of the Periodic-Table, *silicon* and *germanium*. Figure 2-12a depicts a Si-atom, and it shows that there are 4-electrons in its outermost-shell.

In addition, there are some compound-materials that exhibit semiconducting-behavior. These can be composed from elements of Columns-III and V (such as gallium-arsenide, gallium-phosphide, and indium-phosphide), or from elements from Columns-II and VI. Semiconductors display one set of electrical-properties when they are pure (*intrinsic*), but if they contain impurities (*dopants*), their electrical properties can be controlled by these dopants.

2.5.4 Chemical-Bonding-Model for Explaining the Electrical Properties of Solids

Up to this point we have only identified the three-classes of materials according to their ability to conduct current. In the next two sections, the models that account for such electrical-behavior in solid-materials are covered. There are two such models: **1)** the *chemical-bonding-model* (discussed in this section); and **2)** the *energy-band-model* (covered in Sect. 2.5.5). The chemical-bonding-model explains conductivity based on three-types of bonding found in solids: **a)** ionic-bonding; **b)** metallic-bonding; and **c)** covalent-bonding. Each type is described next.

Ionic-Bonding: The interaction of electrons in neighboring-atoms of a solid serves the important function of holding the solid together. As noted, at the end of Sect. 2.1, one type of such bonding is ionic-bonding, and the compound NaCl is held together in this way. As shown in Figs. 2-6b and c (and in Fig. 2-13a), each Na-ion is surrounded by 6 nearest-neighbor Cl^- ions, and vice-versa. That is, the positive Na^+-ions attract the negative Cl^--ions, and these coulombic-forces pull

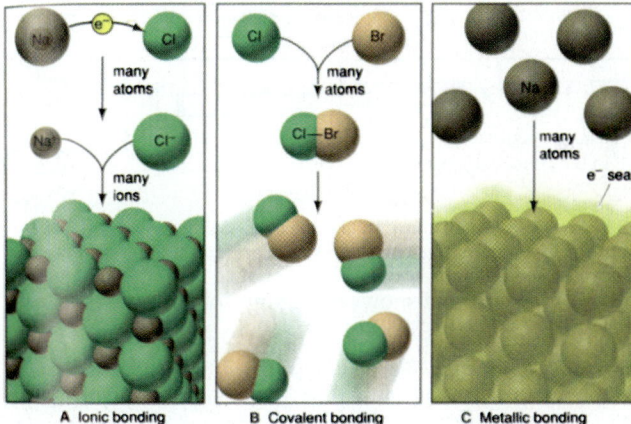

Fig. 2-13 Drawing showing the three types of bonding in solids (from the *chemical-bonding model* view): (a) Ionic-bonding. (b) Covalent-bonding. (c) Metallic-bonding.

the lattice together until the ions arrange themselves in a simple-cubic-lattice structure. An important consequence of this ionic-bonding in the NaCl-solid is that all electrons are tightly bound to atoms. There are no loosely-bound electrons available to participate in current-flow, making NaCl a good insulator.

Metallic-Bonding: In metals, the valence-shell is only partially filled, often with only one or two electrons (such as Na, which has only one valence-electron). These electrons in a metal can be easily removed from the valence-shell. In fact, in a pure metallic-substance, each metal-atom contributes a valence-electron to the crystal as a whole. Thus, the metal is a solid consisting of fixed (i.e., non-movable) ions with closed, completed outer-shells, immersed in a cloud of mobile-electrons (Fig. 2-13c). The forces that hold a metal-lattice together arise from an interaction between the *ionic-cores* and the *mobile-electrons*. This is termed *metallic-bonding*. In any case, the result is that in metals there are many mobile-electrons. These are free to move about the crystal when an electric-field is applied, giving rise to electric-current.

Covalent-Bonding: The third type of chemical-bonding is called *covalent-bonding*, and it occurs in elemental-solids formed of materials from Column-IV of the Periodic Table (C, Si, and Ge). That is, an iso-lated-atom of an element from Column-IV has four valence-electrons. Thus, each such atom is capable of *contributing* one valence-electron to form a bond with a neighboring-atom of the same type (and also *accepting* one electron for each such bond). Thus, a lattice-structure as shown in Fig. 2-13b (and in two-dimensional-form in Fig. 2-12b) can arise. The interaction between the electrons of neighboring-atoms forms so-called *covalent-bonds*. In this structure each atom (e.g., Si) has four nearest-neighbors, with whom a covalent-bond exists (and the structure is called a *diamond-lattice*, see also Chap. 9). Each electron-pair constitutes a covalent-bond. (Note that covalent-bonding also occurs in certain molecules, such as H_2.)

As occurs in the case of ionic-bonding, no mobile-electrons are available to the lattice in the covalent diamond-structure. From this argument, Si and Ge should be perfect-insulators. However, the situation shown in Fig. 2-12b strictly applies only to a Si (or Ge) lattice when it is at 0 K (i.e., at absolute-zero). But, if the lattice is at higher-temperature than 0 K (for instance, at room-temperature, 300 K), some of the covalent-bonds are broken by the thermal-energy of the lattice. This produces some mobile-electrons, which become available to participate in conduction. For the case of a pure-Si lattice at 300 K, there are only 10^{10} mobile-electrons/cm^3 created by such thermal bond-breaking (meaning that just *one in 10 trillion* electrons/cm^3 become available for conduction in pure-Si at room-temperature). Hence, the resistivity of pure-Si at 300 K is high ($\rho_{Si} = 2.2 \times 10^5$ Ω-cm).

2.5.5 Energy-Band-Model for Explaining the Electrical-Properties of Solids

While the *chemical-bonding-model* offers a simple way to introduce the concepts involving electrical-conductivity in solids, it does not provide insight about the energies involved in creating mobile-electrons (i.e., breaking covalent-bonds, etc.). On the other hand, the *energy-band-model* is capable of giving insight into the energies of such events, while still providing the same basis for conductivity-phenomena in solids as does the chemical-bonding-model. Thus, it is the more useful and more widely-employed model for describing conduction in solids.

The energy-band-model is based on the premise that when atoms bond-together to form a crystalline-solid, the energy-levels associated with their outer-orbits may (or may not) overlap as shown in Fig. 2-14. (This phenomenon arises from the Quantum Mechanical aspects of atomic-structure.) Since there are so many atoms in close-proximity in a solid, these energy-levels are so close together that they actually appear as *bands of energies* (as shown in Fig. 2-14 and 15). Diagrams like those of Figs. 2-14 and 15 are thus called *energy-band-diagrams*. The upper energy-bands are called *conduction-bands* and the lower energy-bands are called *valence-bands*. Electrons in a conduction-band (see Fig. 2-14c) can move freely in the solid (i.e., they are mobile-electrons), and can hence conduct current when an electric-field is applied. Electrons in the valence-band remain bonded to their respective atoms and are not capable of participating in electrical-conduction. Since the valence-band has a lower potential-energy, electrons tend to stay in it. Figure 2-16 shows the correlation between the *chemical-bonding* and *energy-band* models.

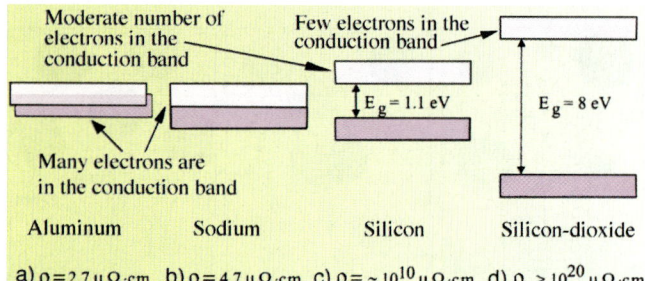

a) $\rho = 2.7\ \mu\Omega\cdot cm$ b) $\rho = 4.7\ \mu\Omega\cdot cm$ c) $\rho = \sim 10^{10}\ \mu\Omega\cdot cm$ d) $\rho > 10^{20}\ \mu\Omega\cdot cm$

Fig. 2-15 This figure shows the energy-bands, the band-gap, and the resistivities (ρ) of: (a) Al; (b) Na; (c) Si; and (d) SiO_2. Note that the *band-gap* E_g of insulators (like SiO_2, or diamond) is wide, and thus there are few-electrons in the conduction-band (and ρ is very-high). In metals there is either no band-gap, or it is very small. Thus, many electrons are always in the conduction-band (and ρ is very-small). In semiconductors (like Si, Ge, or GaAs), the band-gap is smaller than in insulators. Thus, thermal-excitation of electrons into the conduction band is moderately easy, and this makes ρ in-between the values of a metal and insulator.

Note that there can be a separation (or *energy-gap*, E_g) between the two bands. In fact, the upper-edge of the valence-band is denoted in the energy-band diagram as E_V, and the lower edge of the conduction-band as E_C. Their difference is the energy-gap, E_g ($E_g = E_c - E_V$).

However, in *metals* there is either no energy-gap between these bands, or the valence-band and conduction-band may even overlap (Figs. 2-15a and 2-15b). In either case, the thermal-energy of the lattice can easily cause electrons to move from the valence-band up into the conduction-band. Thus, at 300 K the conduction-band of metals has many mobile-electrons, and a large current can flow, even when only a very-small electric-field is applied.

In the case of *insulators* and *semiconductors*, a finite energy-gap exists between the valence-band and conduction-bands (also referred to as the *forbidden-gap*). In these materials, an electron in the valence-band must receive enough energy from some source (e.g., thermal or optical) to allow it to jump-up into the conduction-band. The larger the energy-gap, the fewer the number of electrons that will be able to be

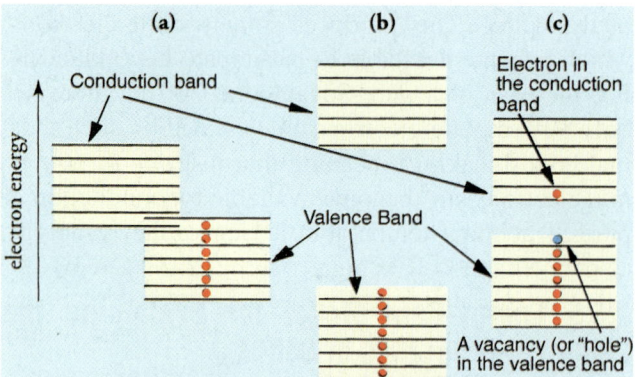

Fig. 2-14 Energy-levels in solids form bands of energies. These bands are depicted using *energy-band diagrams* as shown in this figure: (a) The valence-band and conduction-band overlap one another; (b) The valence-band and conduction-band are separated by a wide *energy-gap*, which prevents electrons from easily getting into the conduction-band. In this drawing, the conduction-band is empty; (c) The valence-band and conduction-band are separated by a smaller *energy-gap*, allowing electrons to more easily get excited into the conduction-band.

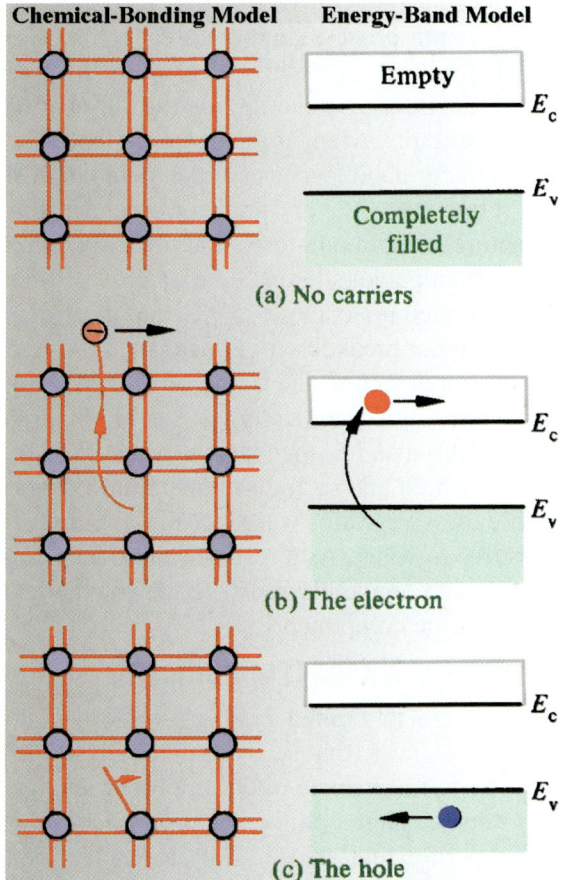

Fig. 2-16 Correlation between the *chemical-bonding model* (CB) and the *energy-band model* (EB) with respect to electrical-conduction. (a) *All bonds are intact* (CB), or *there are no electrons in the conduction-band* (EB). In either case, there are no carriers, so no current-flow is possible. (b) Electrons become available to carry current either by *breaking a bond to become "free"* (CB) or by *getting excited into the conduction-band to become "free."* (EB). (c) Holes can carry current by *making a vacancy among the bonds* (CB) - to allow other electrons to rearrange themselves - or by *the creation of a vacancy in the energy-levels of the valence-band* (EB).

excited into the conduction-band of that material.

For *insulators*, such as glass or plastic, the energy-gap is so wide (E_{gSiO_2} = 8-eV, see Fig. 2-15d), that very-few electrons end-up being excited into its conduction-band. Thus, very-little current can flow in insulators. Note that photons of visible-light have energies in the range of 1.8-2.8-eV. Thus, when they pass through a diamond (which is an insulator whose E_g = 5.0-eV), their energy is too-small to excite electrons into the conduction-band of the diamond. Thus, they pass through diamond without being absorbed. This is why diamonds are transparent.

For semiconductors, the energy-gap is smaller than in an insulator. For example, the energy-gaps are 1.1-eV in silicon, 0.66-eV in germanium, and 1.40-eV in gallium-arsenide (Fig. 2-15c). Since the energy-gap of silicon is smaller than the energy of visible-light photons, such photons have enough energy to excite electrons into the conduction-band of silicon. Thus, visible-light photons are absorbed as they pass through silicon, making silicon an opaque-material.

2.6 SEMICONDUCTOR SILICON

Silicon (Si) is presently the most important semiconductor-material for the electronics-industry, with VLSI and ULSI technology being based almost entirely on this substance. Silicon is the second-most-abundant element (after oxygen), making up about 25% of the earth's crust. Its atomic-number is 14, and it is found in Column-IV of the Periodic Table. Silicon has a high melting-point (1420°C). It is opaque (with a silvery-gray color), and is a hard and brittle material that fractures easily.

A key-fact to remember is that there are about 5×10^{22} atoms/cm³ in single-crystal-silicon (Fig. 2-17). This number is important because it gives a benchmark for comparison when dealing with the concentration of impurity-dopants in silicon. The parameter-values of some of the other most useful properties of silicon are given on the endsheet.

Solid-state-electronics was launched with the invention of the bipolar-transistor in 1947. Germanium (Ge) was the first semiconductor-material used to make transistors, but it was largely replaced by silicon for the following reasons: The narrow-bandgap (0.66-eV) limits the operation of Ge-devices to temperatures below 100°C. (Ge also melts at the relatively-low-temperature of 937°C.) In addition, IC planar-processing needs a layer on

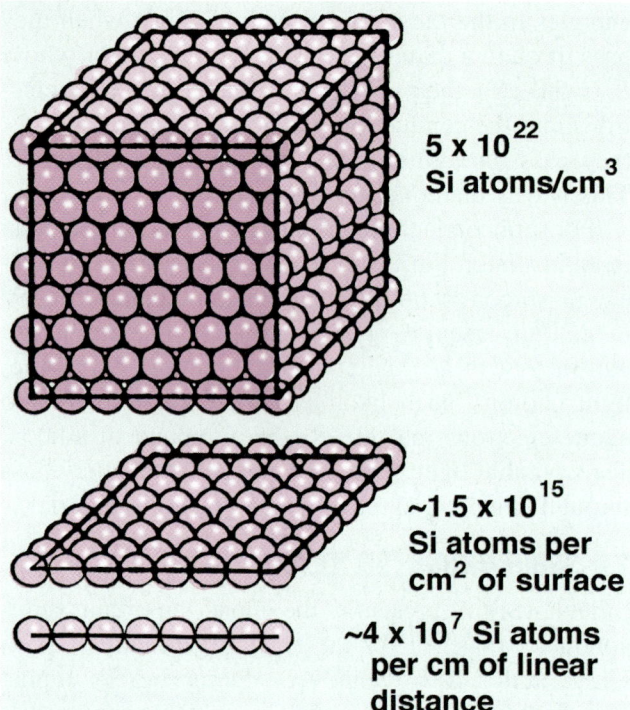

Fig. 2-17 Drawing showing the number of atoms contained in one cubic-centimeter of solid-silicon.

5×10^{22} Si atoms/cm^3

$\sim 1.5 \times 10^{15}$ Si atoms per cm^2 of surface

$\sim 4 \times 10^7$ Si atoms per cm of linear distance

unstable at high-process-temperatures (> 800°C), but worse, it dissolves in water. All of these limitations make Ge an inferior material for the fabrication of integrated-circuits, when compared to silicon.

On the other hand, because of the abundance of Si, cost to refine it to the very-pure-form needed for IC-manufacture is about one-tenth that of Ge. In addition, the larger-bandgap of silicon (1.1 eV) allows Si-devices to be operated up to about 150°C, and these devices also have higher breakdown-voltages than those made from Ge. Furthermore, the oxide of silicon, SiO$_2$, is easy to form and is chemically very-stable (up to over 1500°C). SiO$_2$ also has mechanical-properties similar to silicon, which allows high-temperature processing without excessive wafer-warpage. Finally, SiO$_2$ does not dissolve in water. As a result of these advantages, Si has almost completely replaced Ge for fabricating microelectronic components.

2.6.1 Intrinsic-Silicon (Pure-Silicon)

Pure-silicon (also termed *intrinsic-silicon*) contains no other substances (that is, no contaminants or impurities). The term *intrinsic* refers to properties that are natural to a substance. Since intrinsic-silicon is pure, it has properties that are due only to itself. Some of the properties exhibited by *impure-silicon* (also called *extrinsic* or *doped* silicon) can be due to the impurities it contains.

The semiconductor properties of intrinsic crystal-

the semiconductor-surface that can serve as a good chemical-barrier (to protect the wafer-material from external-contaminants) and as a passivation-layer. While the oxide of germanium (GeO$_2$) could serve as such a layer, it has some severe limitations. First, it is

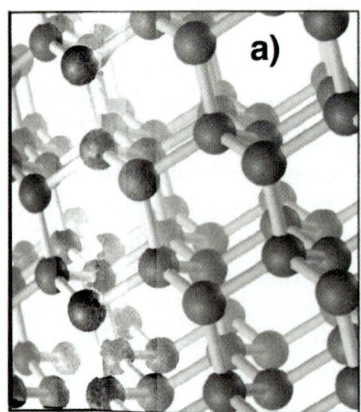
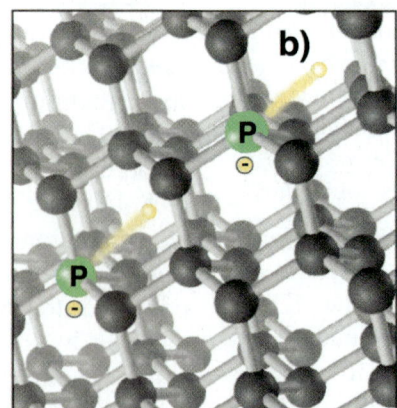
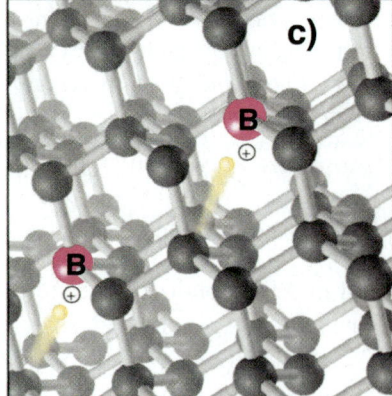

Fig. 2-18 (a) Illustration of a pure (intrinsic) Si-lattice, showing the covalent-bonds. (b) A Si-lattice doped with *phosphorus* (P) atoms, forming *n*-type Si. (c) A Si-lattice doped with boron (B) atoms, forming *p*-type Si.

line-Si arise from the fact that its atoms have four electrons in their valence-shell. It shares each of them with one neighboring-Si-atom to form four covalent-bonds (as shown in Figs. 2-12b and 2-18a). The atoms in crystalline-silicon thus bond together through covalent-bonds to share electrons. This bonding also completes the valence-shell in all of the Si-atoms.

Many of silicon's properties arise from such covalent-bonding. These bonds are strong, making silicon a stable-solid at room-temperature. (They also hold Si together up to high-temperatures, giving it the high melting-point of 1420°C). It is also hard to remove electrons from these strong-bonds. In fact, at room-temperature nearly all of the Si-valence-shells are fully occupied. (As noted earlier, at room-temperature, there are only 1×10^{10} mobile-electrons per cm^3 out of the 2×10^{23} valence-electrons per cm^3 in Si.) Thus, intrinsic-silicon at room-temperature is a poor-conductor (it's resistivity is high, with a value of 220,000 Ω-cm.). In fact, in its intrinsic-form, Si is not of much practical use in semiconductor-technology.

It is also important to point out that when an electron escapes from its covalent-bond (and becomes a mobile-electron), it leaves behind a vacant spot in that covalent-bond (called a *hole*, Fig. 2-19). Thus, intrinsic-silicon contains an *equal number* of mobile-electrons and holes (that is, neither mobile-electrons nor holes are in the majority).

2.6.2 Extrinsic-Silicon (Doped-Silicon)

The high-resistivity of intrinsic-silicon can be decreased by adding small quantities of certain other elements (*impurities*) through a process called *doping*. This process allows such *doped-silicon* to become suitable for fabricating semiconductor-devices. Even the presence of very-small quantities of such impurities can dramatically decrease the resistivity of intrinsic-silicon. For example, by adding only one dopant-atom (e.g., boron or phosphorus) for each million silicon-atoms, the resistivity will drop from 220,000 Ω-cm to 0.2 Ω-cm. This represents a decrease in the resistivity by a factor of over one-million! The more impurity-atoms that are added, the better the conduc-

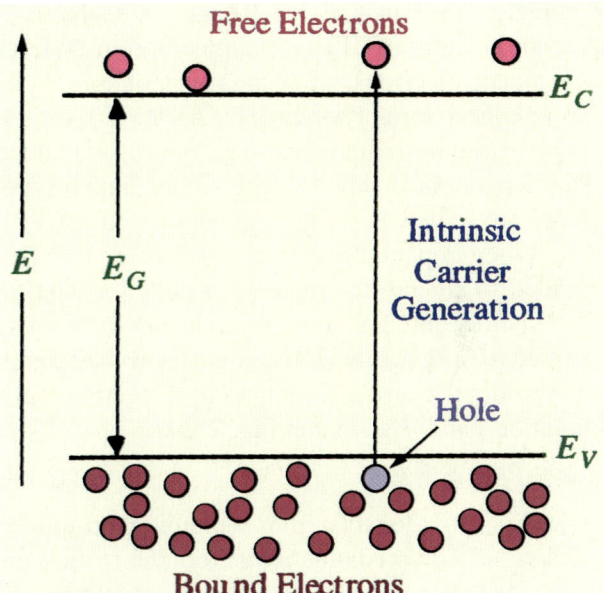

Fig. 2-19 Simple band-diagram representation of intrinsic (pure) silicon. The process of *intrinsic-carrier-generation* is also illustrated.

tivity of the Si. As noted earlier, *doped-silicon* is also called *extrinsic* or *impure silicon*.

Doping of silicon (and other semiconductors) has another important benefit, namely that current can take place through the flow of either negative-charge (electron-movement), or positive-charge (i.e., through the flow of *holes* – a subject described in more detail in Sect. 2.6.5). The latter phenomenon is called *hole-flow*, and is unique to semiconductors. Thus, not only can silicon be doped into a much wider, useful resistivity-range (from 0.001-Ω-cm to 1000-Ω-cm), but by choosing the right *type* of dopant, it can be made either electron-rich (*n*-type), or hole-rich (*p*-type). The ability to create both *p*-type and *n*-type regions allows such devices as *pn*-diodes, bipolar-transistors, NMOSFETs and PMOSFETs to be built in silicon.

2.6.3 Dopants in Silicon (Donors and Acceptors)

As noted previously, Si is found in Column-IV of the Periodic Table, and thus has four valence-electrons. Elements from the two adjacent-Columns in the Periodic Table are commonly used for doping

Si, namely those in Column-III and in Column-V. Elements in Column-III (for example boron, B) have three valence-electrons, while those in Column-V (for example, phosphorus P, or arsenic, As) have five.

When Column-III-dopants (e.g., boron) are added to Si, the material is called *p*-type-silicon, and *p*-type-dopants are called *acceptors* (as discussed in Sect. 2.6.5). When dopants from Column-V (e.g., P or As) are added to silicon, the material is called *n*-type-Si, and *n*-type-dopants are called *donors* (for reasons also described in Sect. 2.6.4). The energy-band-diagram of an *n*-type-dopant such as P, donating an electron to the conduction-band is shown in Fig. 2-20a.

2.6.4 *n*-Type Silicon

The addition of dopants from Column-V to silicon increases silicon's-conductivity for the following reason: Phosphorus (or arsenic) atoms in silicon can replace a Si-atom at a given lattice-site. However, such dopants have an extra-electron in their valence-shell, and these extra-electrons cannot participate in covalent-bonding. In addition, such extra-electrons are easily broken away from the P (or As) atom (see Fig. 2-18b and 2-20a) to become a *mobile-electron* (or *carrier*) - which then can contribute to current-flow. Such an electron is said to be *donated* to the silicon-material for the purpose of electrical-conduction. If $10^{15}/cm^3$ (or more) donors are added to the silicon, and each donates one mobile-electron, the number of donated-carriers will far exceed the number present in intrinsic-silicon at room-temperature ($10^{15}/cm^3 \gg 10^{10}/cm^3$). In this way, the addition of even a small-percentage of *n*-type dopants dramatically increases the concentration of mobile-electrons (10^{15} dopant atoms/cm^3 represents about 1-dopant-atom per 10-million-silicon-atoms, or 0.0001%), and yet this vastly increases the conductivity of silicon.

Electrons will be attracted to the positive-terminal of a battery that is connected to the silicon (Fig. 2-21a). Any *mobile-electrons* move in the direction toward the positive-terminal, and this organized-movement of charge-carriers creates an electron-current in the semiconductor. Note that since there are many more mobile-electrons created by the donor dopant-atoms in *n*-type-silicon than there are holes, the mobile-electrons are called the *majority-carriers* in *n*-type silicon. On the other hand, holes present in *n*-type-Si are called *minority-carriers*.

2.6.5 *p*-Type-Silicon (Holes)

The introduction of dopants from Column-III (*p*-type dopants) contributes to the conductivity of silicon in a somewhat more complex-manner than does the presence of *n*-type-dopants in Si. That is, boron-atoms can also replace Si-atoms on lattice-sites. However, B only has three electrons in its valence-shell. Thus, it can *accept* an electron from a neighboring silicon-covalent-bond (Fig. 2-18c and 2-20b). But this creates a vacancy in that bond (which, as we noted earlier, is called a *hole*). If a voltage is applied to *p*-type silicon (for example using a battery), an electron near the hole will be attracted toward the positive-terminal of the battery and will jump to fill the hole (Fig. 2-21b). However, this will leave the hole in the spot from which this electron jumped. It is seen that the hole moves continuously in the opposite-direction to that of electron-movement. When all is said and done, the hole will have moved to the edge of the semiconductor near the negative-terminal. For someone measuring current-flow in this circuit with just an ammeter, however, it appears that there is a current due to the movement of positive-charge carriers (the holes),

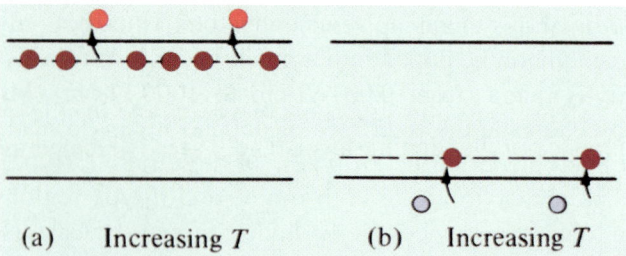

Fig. 2-20 Energy-band diagram of silicon that has donors and acceptors on lattice-sites. The energy-levels that these impurities produce are also shown in the diagram. The mechanisms of how mobile-electrons arise from the presence of donor-atoms and holes from the presence of acceptor-atoms in the lattice is also shown from the perspective of the energy-band model.

while it is actually due to the sum of the electrons moving in the opposite-direction. Nevertheless, the apparent-current due to hole-movement can be treated just like another-component of the total-electric-current present in a semiconductor. An energy-band diagram of a *p*-type dopant (acceptor) accepting an electron from the valence-band, is shown in Fig. 2-20b.

The notion of treating the movement of vacancies in the valence-band of silicon (i.e., holes) as one form of electric-current may seem odd. However, there are other everyday-phenomena we routinely accept in the same spirit. For example, if a nearly-full bottle of shampoo is turned upside down, we say that a bubble inside the bottle rises to the top. In fact, what is occurring is that the shampoo is flowing down to the bottom. But it is easier to see this event as a rising bubble. Hole-movement in the silicon is much like the bubbles that rise in upside-down shampoo-bottles.

In any case, each B-atom causes the creation of a hole. If there are a substantial-number of B-atoms in a Si-lattice, this will produce a situation where the number of holes exceeds the number of intrinsic mobile-electrons. Thus, the current in *p*-type-semiconductors will be predominantly due to *hole-flow*. Furthermore, since there are many more valence-band-holes in *p*-type silicon than there are mobile-electrons, holes are the *majority-carriers* in *p*-type-silicon (and the mobile-electrons are the *minority-carriers*).

2.6.6 Dependence of Resistivity on Doping-Concentration in Silicon

Metals have a resistivity (ρ) range limited to 10^{-4} to 10^{-6} Ω-cm (only a factor of 100). Given a specific metal, this means there is only one practical way to change the resistance of a device made from this metal to a greater-degree than allowed by the resistivity-range. For example, if a resistor was being designed, the physical-dimensions of the resistor would have be changed (according to Eq. 2-2) to allow a larger-resistance-value to be obtained.

On the other hand, semiconductor-materials allow a change in the value of the resistivity (through the addition of dopant) over a much wider-range (i.e.,

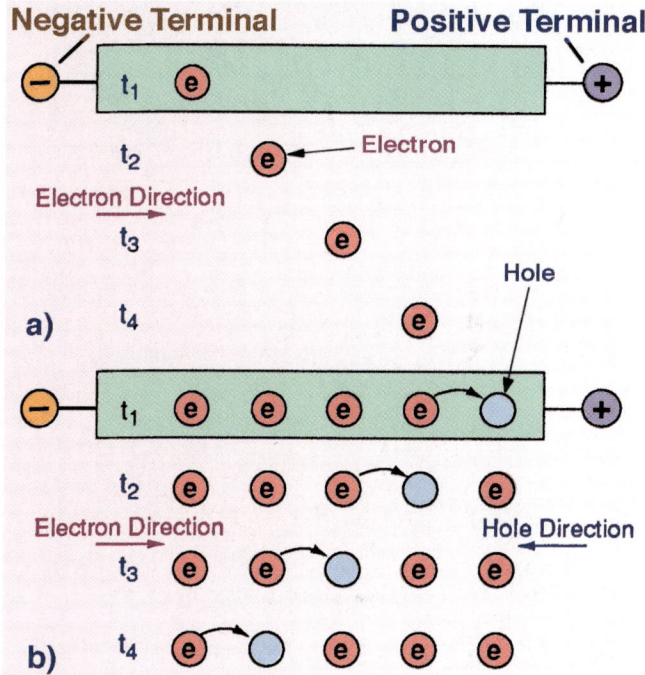

Fig. 2-21 (a) Conceptual drawing of current due to *electron-flow* in a semiconductor. Note that the electrons move toward the positive-terminal of the power-supply. (b) Conceptual drawing of current due to *hole-flow* in semiconductors. Holes move toward the negative-terminal, as electrons move toward the positive-terminal.

from 0.001-Ω-cm to 1000-Ω-cm – a factor of over 1,000,000). Thus, this allows much-greater flexibility in the design of resistors built in silicon than in metal. (It can also be seen that the higher the doping-concentration in Si, the lower the resistivity-value.)

Figure 2-22 illustrates how the resistivity in silicon depends on the doping-level. The *y*-axis is labeled as the *dopant-concentration* (in units of #/cm^3). The *x*-axis gives the resistivity (in units of Ω-cm). Note that there are two curves in this figure: one for *n*-type-doping and the other for *p*-type. They show that for the same value of doping-concentration, *n*-type-doped-silicon has a lower resistivity than *p*-type. As discussed in the next section, this occurs because electrons (the dominant-carriers in *n*-type-silicon) move more rapidly than do holes (the dominant-carriers in *p*-type-silicon).

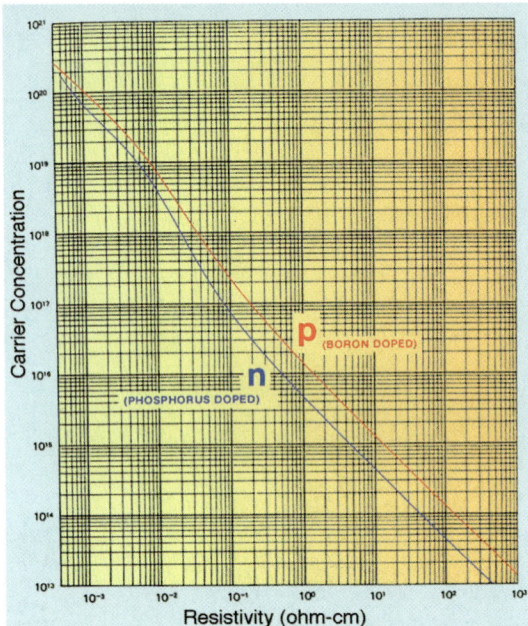

Fig. 2-22 Resistivity vs. doping for *n*- and *p*-type silicon.

The data in this figure also indicate that it takes the presence of only 0.000001% to 0.1% of a dopant to cause Si to have a useful-range of resistivity. Information in this figure is used to select the correct doping-concentration in Si to give the desired resistivity in a particular device-region. This amount of dopant is then carefully introduced into the silicon during the semiconductor-manufacturing process to achieve this precise-resistivity. The process-technologies used to introduce dopants into Si are *chemical-diffusion* (see Chap. 11), and *ion-implantation* (see Chap. 12).

2.6.7 Carrier-Mobility (and Its Impact on Transistor Performance)

We have shown that there are two-types of charge-carriers in a semiconductor: *electrons* and *holes*. Both holes and mobile-electrons are caused to move in silicon when acted on by an electric-field. For the same value of electric-field-strength, electrons travel more easily (and therefore more rapidly) through silicon than do holes. The speed of movement of a charge-carrier (either an electron or hole) for a given electric-field-strength is called the *carrier-mobility*, with holes having a lower-mobility than electrons in Si. This means that if the same number of electrons and holes are flowing in an identical electric-field, the electron-flow will produce the larger-component of electric-current. This effect is important in the selection and design of transistors. That is, it is the basis of why *npn*-bipolar-transistors are generally preferred over their *pnp*-counterparts, and why digital-logic-circuits built with NMOSFETs operate at higher clock-speeds than those built with PMOSFETs.

2.7 OTHER SEMICONDUCTOR MATERIALS

Silicon (Si) and germanium (Ge) are the two *elemental*-semiconductor-materials from Column-IV of the Periodic Table. It was noted that Si is the most widely-used material for making semiconductor-devices today, but it should also be mentioned that the first transistors were fabricated in 1947 using Ge. Germanium is still used in some specialty-semiconductor-devices, such as high-power *pn*-diodes.

The other types of semiconductor-materials are *compound-semiconductors*. They are materials consisting of two (or more) elements. The applications of devices using these alternative-semiconductor-materials are many, including optical-devices (LEDs, infrared-detectors, solar-cells, and semiconductor-lasers), high-frequency communications-chips (rf- and microwave-circuits for wireless-applications), and special high-power-devices. Although these devices are important, the total-sales each year for such alternative-semiconductor devices-still makes it a niche-market. That is they consist of only about 2% of the total-sales of semiconductor-devices (with the rest going to silicon-based devices).

The most-widely used alternative-semiconductors consist of materials from Columns III and V of the Periodic Table (the so-called III-V compounds, such as GaAs, GaP, and InP). GaAs is the most widely used alternative-semiconductor-material. It is primarily employed in high-frequency-devices in communications-systems. The high-mobility of electrons in GaAs allows such devices to operate at speeds about 3-times faster than comparable silicon-devices. This makes them suitable for such high-frequency operation. GaP (and GaAsP) is used to make light-emitting diodes (LEDs). GaN is used to make blue and white LEDs.

Other compound-semiconductors are II-VI -materials, such as CdSe, and ZnSe. These make up a small-market ($179 million in 2001). CdTe is used to make infrared-detectors, and ZnSe to make blue-LEDs.

REFERENCES

1. M. Silberberg, *Chemistry: The Molecular Nature of Matter and Change*, Mosby, St. Louis, 1996.
2. D. A. Neamen, *Semiconductor Physics and Devices*, 3rd Ed. McGraw-Hill, New York, 2003.
3. W.E. Beadle et al., *Quick Reference Manual for Silicon Integrated Circuit Technology*, Wiley, 1985.

PROBLEMS

1. Explain the electrical difference between a conductor, an insulator, and a semiconductor.
2. What is a *semiconductor*? List three of the most commonly used semiconductor materials.
3. Why is it necessary to dope silicon in order to build solid-state devices? How does the resistivity of a semiconductor change with dopant concentration?
4. Which substances have higher resistivities, metals or intrinsic semiconductors?
5. What is the majority carrier in *p*-type semiconductors? Name 2 elements that are *p*-type dopants in silicon.
6. Does a *p*-type semiconductor exhibit negative or positive current? Explain your answer.
7. Define the term *minority-carrier*. What carriers are the minority-carriers in *n*-type silicon?
8. At the same doping concentration-value, which has the higher electrical-conductivity, *n*-doped or *p*-doped silicon?
9. What is the distinction between the concentration N_D of an *n*-type impurity in a wafer, and the concentration n of majority carriers in the same wafer?
10. A 1-mm thick silicon wafer having a diameter of 20-cm contains 6.77 mg of boron ($_{11}B^5$), uniformly distributed in substitutional sites. Find: (a) the boron concentration, in atoms/cm^3, and (b) the average distance between boron-atoms.
11. Aluminum has an atomic density of 6.02×10^{22} atoms/cm^3. What is the mass-density (g/cm^3)? If the Al atoms are stacked uniformly in a cubic-array, how many atoms/cm^2 are in the top atomic-layer. What is the center-to-center spacing of the Al atoms?
12. Find the concentration of electrons and holes in silicon doped as follows: (a) undoped (intrinsic) silicon; (b) arsenic doped with 10^{16} atoms/cm^3; (c) phosphorus doped with 10^{20} atoms/cm^3. For this problem, use the expression $pn = n_i^2$, where n_i is the intrinsic carrier-concentration at 300 K.
13. Find the electron and hole concentrations in silicon at 300 K (a) for 1×10^{15} boron atoms/cm^3, and (b) for 1×10^{16} atoms/cm^3. Note, that in solving part (b), that each boron dopant atom creates one hole, and each arsenic atom creates one electron. Their presence cancels each other out, and the net majority carrier-concentration that results is $N_A - N_D$.
14. The intrinsic carrier-concentration n_i in silicon (cm^{-3}) is given by:

$$n_i = 3.1 \times 10^{16} \, T^{3/2} \exp(-6.03 \text{ eV}/kT)$$

where T is in kelvin, and k is Boltzmann's constant. Use this equation to find n_i at: (a) room temperature; and (b) 800°C, and (c) 1000°C.
15. Plot the value of n_i versus T from 600°C to 1200°C, making the calculation every 100°C.
16. Find the resistivity of pure-silicon (in Ω-cm) at: (a) 300 K; (b) 77 K; and (c) 1000 K.
17. If a piece of intrinsic silicon is 100-μm in length, and has a cross-sectional area of 1-μm^2, how much current would flow in such a "resistor" at room-temperature in response to an applied voltage of 1-volt?

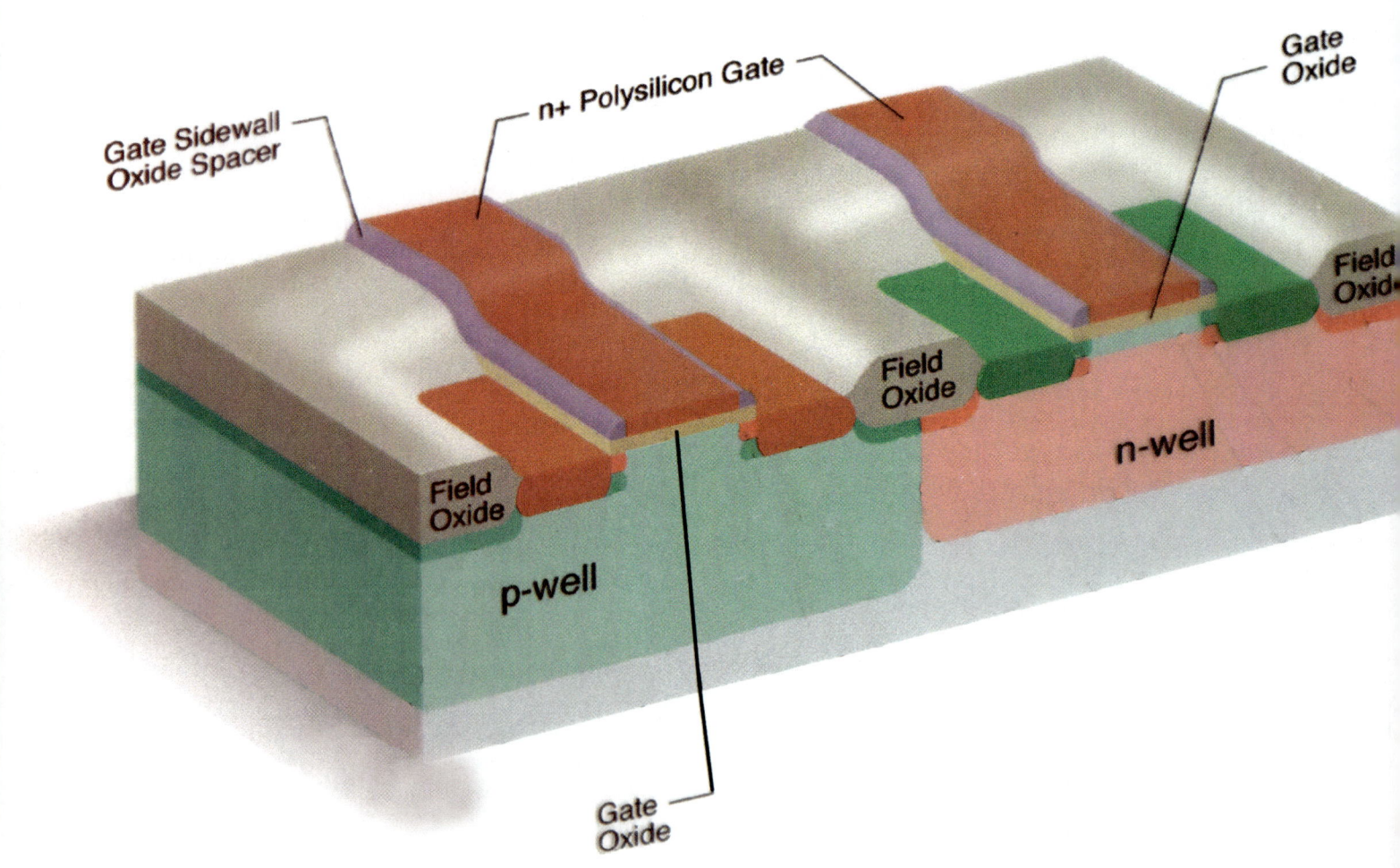

CHAPTER 3

SEMICONDUCTOR DEVICES AND INTEGRATED-CIRCUIT TYPES

CHAPTER CONTENTS

3.1 INTEGRATED-CIRCUIT RESISTORS
3.2 INTEGRATED-CIRCUIT CAPACITORS
3.3 *pn*-JUNCTION DIODES
3.4 BIPOLAR JUNCTION TRANSISTORS (BJTs)
3.5 METAL-OXIDE-SEMICONDUCTOR FIELD-EFFECT TRANSISTORS (MOSFETs)
3.6 INTEGRATED-CIRCUIT TYPES

"In our description of nature, the purpose is not to disclose the real essence of the phenomena but only to track down, as far as it is possible, relations between the manifold aspects of our experience."

Niels Bohr

A perspective drawing of a CMOS device structure. Courtesy of Axcelis Technologies Inc.

Integrated-circuits are built from individual devices on a single-piece of semiconductor material (usually silicon), and these devices are then interconnected with conducting-lines fabricated on the chip. Figure 3-1 shows the basic *planar-processing steps* used to build all IC devices.

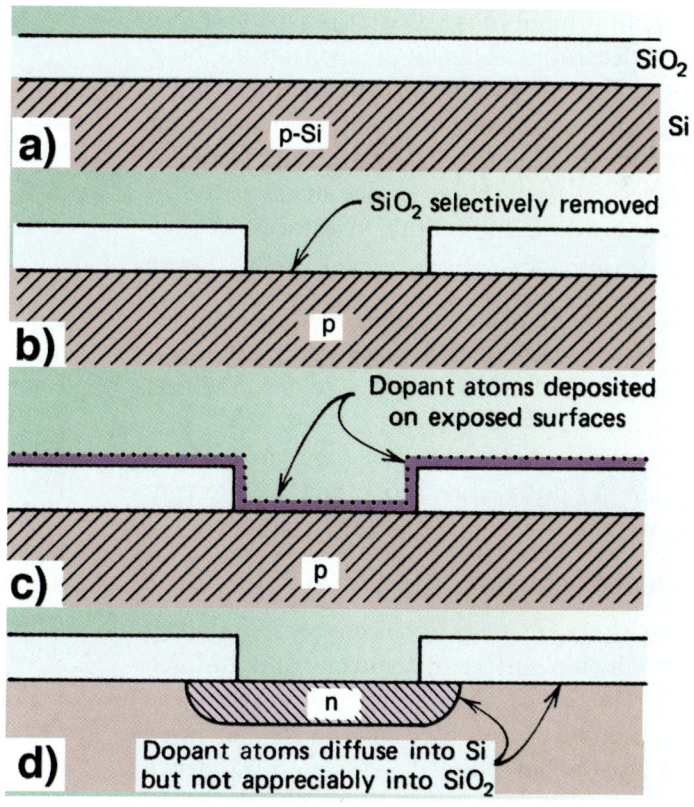

Fig. 3-1 The basic fabrication-steps of the *planar-process* used to fabricate monolithic IC-components in silicon-wafers: (a) A thermal-oxide is grown on a Si wafer; (b) This oxide is removed in selected areas of the top-surface of the wafer - to expose the silicon surface; (c) Dopant-atoms are deposited into the exposed-silicon; (d) These dopant-atoms are diffused to the desired depths in the silicon.

In this chapter the various device-types that are the building blocks of ICs are described, including:

1. IC Resistors;
2. IC Capacitors;
3. *pn* Diodes;
4. Bipolar Junction Transistors (BJTs);
5. MOS-Field-Effect Transistors (MOSFETs).

Also discussed are the most common types of integrated circuits being manufactured today.

3.1 INTEGRATED-CIRCUIT RESISTORS

Resistors are circuit-elements that restrict the flow of electric current. They obey Ohm's Law, R (Ω) = V/I, and the circuit symbol of the resistor is shown in Fig. 2-8 (Ch. 2). Resistors as circuit-components are built from materials with specific resistivities, and these are then shaped into structures that yield the desired resistance-value. As noted in Chap. 2, the equation that predicts *resistance* based on the materials resistivity ρ, and the dimensions of the resistor is:

$$R (\Omega) = \rho\, l\, / A = \rho\, l\, / w\, t \qquad (3.1)$$

where *l* is the length of the resistive region, and A is its cross-section (see Fig. 2-8). The area, A, is $w \times t$, where *w* is the width of the resistive region, and *t* its thickness.

The two most commonly used *integrated-circuit resistors* are: **1)** diffused resistors; and **2)** thin-film resistors. We describe how both of these are made.

3.1.1 Diffused Resistors

Diffused resistors are made by selectively introducing dopant into the top-surface of a silicon-wafer through an opening in the oxide-layer on the surface. The sequence of process-steps used to produce a diffused resistor is also shown in Fig. 3-1. A pattern is opened in the surface-oxide, and the square ends serve as the contact regions (as shown in Fig. 3-2a). The long, thin region shown in Fig. 3-2a between them serves as the resistive region. Following diffusion of the dopant and subsequent-reoxidation of the surface, contact holes are opened to the square end-regions to allow

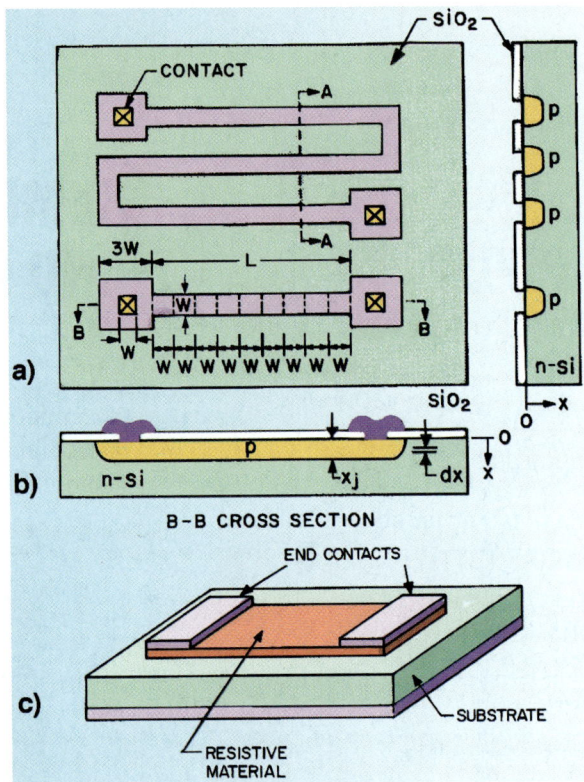

Fig. 3-2 (a) Top view of a zig-zag and a straight *diffused resistor*. (b) Side view of the straight resistor. (c) *Thin-film resistor* (typically, it's resistor-material is polysilicon).

the resistor to be connected to the IC-circuit. The dopant can be introduced either by chemical-diffusion or by ion-implantation, but both resultant-structures are still called *diffused resistors*. The depth of the *pn*-junction formed by introduction of dopant becomes the thickness *t* used in Eq. 3-1 to calculate R, and the resistivity ρ is the average-resistivity that is due to the dopants that have been put into the silicon. Normally, these two factors are measured by finding the sheet-resistance of the diffused region (as described in Chap. 11). The shape of the resistor can be a straight path, or it can have a serpentine shape. The latter allows longer resistive-regions to be placed on a small area of silicon. (Both kinds are shown in Fig. 3-2a.)

3.1.2 Thin-Film Resistors

Integrated-circuit resistors can also be fabricated from thin-films deposited on the wafer-surface. Figure 3-2c

CHAPTER 3 SEMICONDUCTORS DEVICES AND INTEGRATED CIRCUIT TYPES

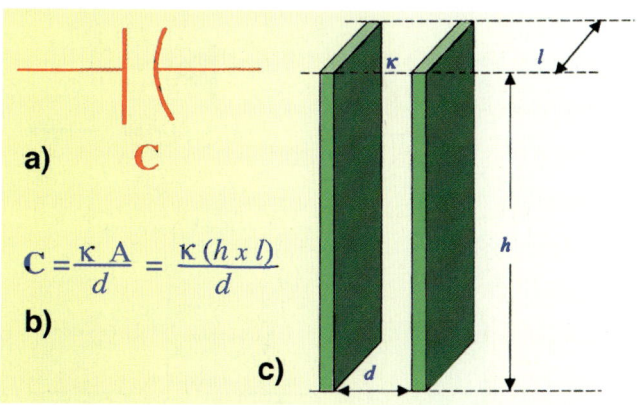

Fig. 3-3 (a) Circuit-symbol for a capacitor. (b) Formula for calculating the capacitance of a parallel-plate capacitor in terms of the area of its plates, A, the dielectric constant of the material separating the plates, κ, and the distance between the plates, d. (c) Parallel-plate capacitor.

shows an example of such a *thin-film resistor*. The most common material used to make such resistors is polysilicon. The polysilicon is appropriately doped to give the film the specific-resistivity needed for the resistor being built. Thin-film resistors allow the value of the resistance to be controlled more tightly than in diffused resistors, and are preferred for ICs.

3.2 INTEGRATED-CIRCUIT CAPACITORS

Capacitors are circuit elements used to store charge. *Capacitance* (C) is defined as the amount of charge, Q, that can be stored per volt, V (C = Q/V), and the unit of capacitance is the *farad* (F). The circuit-symbol of a capacitor is shown in Fig. 3-3a. Structures are built to allow specific amounts of capacitance to exist in specific locations in a circuit, and these components are called *capacitors*. The parallel-plate structure shown in Fig. 3-3c is the type most-commonly used to build capacitors. Two conducting parallel-plates are separated by a distance d, with a dielectric of some kind (with dielectric constant $\kappa = k\varepsilon_o$) between them. The capacitance in such a structure is given by:

$$C\ (F) = \kappa A / d = k \varepsilon_o A / d \qquad (3.2)$$

where A is the area of the plates, d is the distance between them, ε_o is the permittivity of free space (ε_o = 8.85x10^{-12} F/m), and k is the relative dielectric constant of the particular dielectric being used in the capacitor. Note that k of SiO$_2$ (the most common dielectric used in IC capacitors) is 3.6.

There are two types of integrated-circuit capacitors in common use: **1)** *metal-oxide-semiconductor capacitors* (MOS-C); and **2)** *memory-cell capacitors*, used in dynamic-random-access memories (DRAMs). We describe both of these here.

3.2.1 Metal-Oxide-Semiconductor Capacitors

The *metal-oxide-semiconductor capacitor* (*MOS-C*) structure uses the silicon-wafer as the lower plate, and a metal-film (or heavily-doped polysilicon-film) as the top-plate of the planar, parallel-plate capacitor-structure in integrated circuits. The dielectric material between the plates is thermally-grown silicon-dioxide, hence the name "MOS-C." A process-sequence for forming the sandwich structure of the MOS-C is shown in Fig. 3-4a - d.

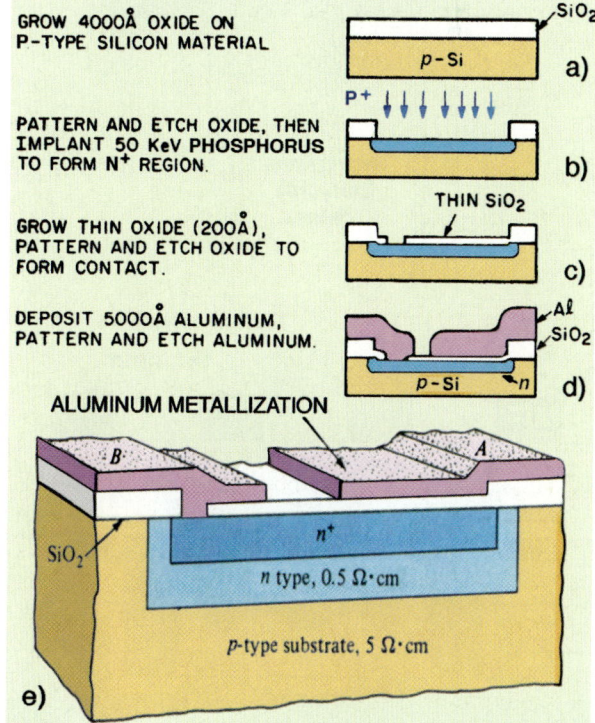

Fig. 3-4 (a) - (d) Sequence of steps used to fabricate a basic planar MOS-capacitor structure. (e) Perspective drawing of a planar MOS-capacitor structure.

3.2.2 Memory-Cell Capacitors for DRAMs

The *memory-cell capacitors* used in DRAMs store the charge that represents the data present in the DRAM memory-cell. The same general materials used in the MOS-C are used in DRAM memory-cell capacitors, but the plates may have a different geometrical configuration. That is, the quantity of charge stored in a single memory-cell must remain constant, even as the area of silicon allotted per cell shrinks with each new-generation of memory device. To allow the area of the cell-plates to remain the same, they are either fabricated in a trench (*trench-capacitor cell*, Fig. 3-5a) or in a stacked configuration above the wafer-surface (*stacked-capacitor cell*, Fig. 3-5b). The trenches are etched anisotropically in a silicon-wafer, and the trench-sidewalls (which serve as the first plate of the

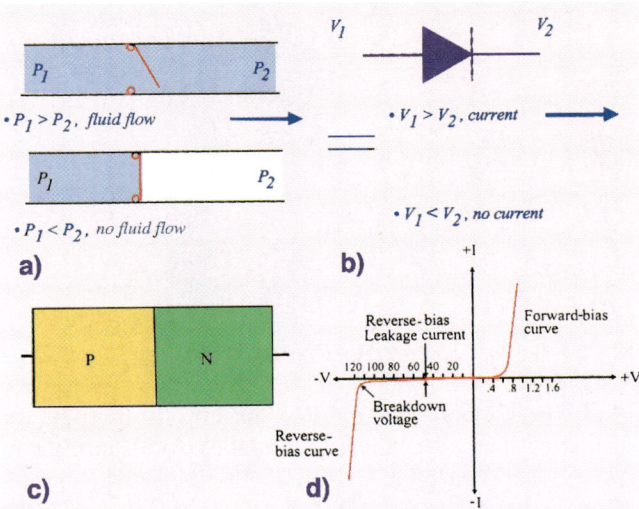

Fig. 3-6 (a) Pressure-valve analogy to illustrate the operation of a rectifying-diode. (b) Circuit symbol of a rectifying-diode. (c) Schematic drawing of a *pn*-junction rectifying-diode. (d) I-V curve of a *pn*-junction diode.

capacitor) are oxidized. The center of the trench is filled with polysilicon, which becomes the second capacitor-plate. By making the trench sufficiently deep, the area of the plates can be made large enough to store sufficient charge. In *stacked-capacitor cells*, the plates are made from polysilicon structures having cylindrical-, crown-, or fin-shapes.

3.3 *pn*-JUNCTION DIODES

A *rectifying-diode* in its ideal-mode of operation is a two-terminal circuit-element that allows current to flow in one direction in a circuit branch, but completely blocks its flow in the other (Fig. 3-6a). The circuit symbol of an ideal rectifying-diode is shown in Fig. 3-6b. Devices that behave like rectifying-diodes can be made from vacuum-tubes and *pn*-junctions. Here we only describe *pn*-junction diodes because they are the type used in integrated circuits (Fig. 3-6c). In the planar-process used to fabricate ICs, a *pn*-junction is formed if dopant of the opposite-type of that present in a silicon-wafer is introduced into the top-surface of the wafer (see Fig. 3-1). The I-V characteristic of a *pn*-junction diode is shown in Fig. 3-6d.

The I-V curve shown in Fig. 3-6d shows that cur-

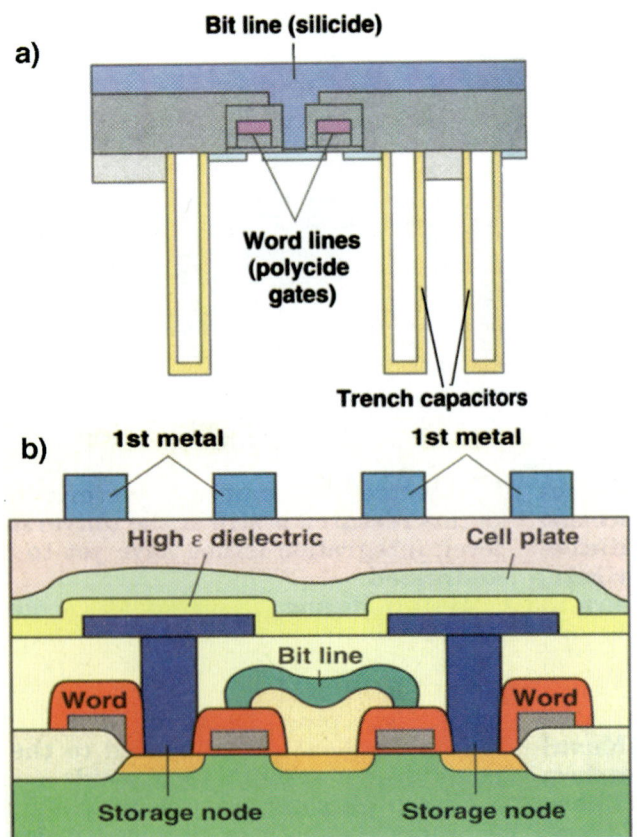

Fig. 3-5 (a) Trench-capacitor DRAM-cell. (b) Stacked-capacitor DRAM-cell.

CHAPTER 3 SEMICONDUCTORS DEVICES AND INTEGRATED CIRCUIT TYPES

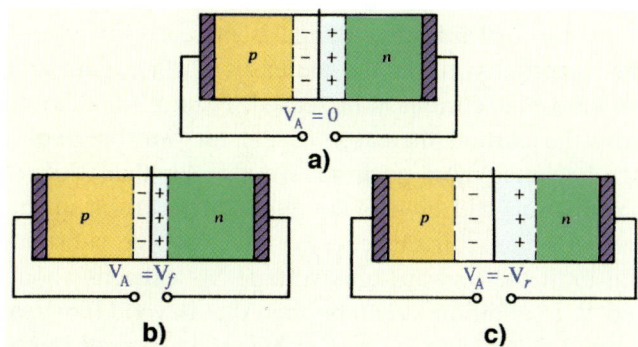

Fig. 3-7 *pn*-junction diode with: (a) no externally-applied voltage; (b) forward-bias; (c) reverse-bias.

rent in a *pn*-junction diode flows if a voltage with the correct polarity is applied to this device. As shown in Fig. 3-7b, this is called *forward-biasing the pn-junction*. On the other hand, current in a *pn*-junction diode is blocked if a voltage with the opposite polarity is applied. As shown in Fig. 3-7c, this is called *reverse-biasing the pn-junction*. Note that if no voltage is applied, $V_A = 0$, and current does not flow in the branch with the diode (Fig. 3-7a).

Diodes are used in many ways in ICs. First, in some circuits they are utilized to steer current. By proper choice of the voltage-polarities applied to *pn*-junctions, current is allowed to pass in some branches and is blocked out of others. Second, the operation of bipolar-junction-transistors (BJTs) relies on *pn*-junctions. That is, the BJT is a device whose transistor-action occurs because there are two *pn*-junctions very close to one another. This will be discussed further in Sect. 3.4. Finally, a *pn*-junction exists at each edge of a MOSFET channel, and thus they are also an integral part of MOSFET device-structures.

3.3.1 *pn*-Junction Operation: No Bias ($V_A = 0$)

The basis of the operation of *pn*-junction diodes is now briefly explained: A piece of silicon-doped *p*-type is rich in *holes,* while one doped *n*-type is rich in *mobile-electrons* (Fig. 3-8a). If two such separated pieces of silicon are pushed into contact with one another, but the joined-structure is not connected into a closed-circuit, there is no voltage applied across the junction ($V_A = 0$, Fig. 3-8b).

However, several important phenomena arise when such *n*-type and *p*-type pieces of silicon are joined. First, where the two pieces meet, a *pn*-junction is formed. Second, since the holes and mobile-electrons are both free to move, each species diffuses into the region where there are fewer numbers of such carriers. That is, mobile-electrons diffuse across the junction into the *p*-side, and (mobile) holes move into the *n*-side (see Chap. 11 for a detailed discussion of the phenomenon of diffusion). Since such movement of charge constitutes electric-current - and there is an open-circuit - this movement can only go on for a very-short time (i.e., it must quickly cease). One may ask: What causes the movement of these charges to stop, and current flow to cease, making I = 0?

The answer is that an electric-field automatically arises to stop further charge-movement. That is, when the electrons leave the *n*-region, they recombine with holes on the *p*-side and leave behind positive (e.g., phosphorus) ions. The holes that have moved to the *n*-side recombine with electrons there and they leave behind negative (e.g., boron) ions in regions near the junction (Fig. 3-9a-1). These resulting ionic charges set up an electric-field that points in a direction that repels electrons and holes, keeping additional majority carriers from crossing the junction. In this manner

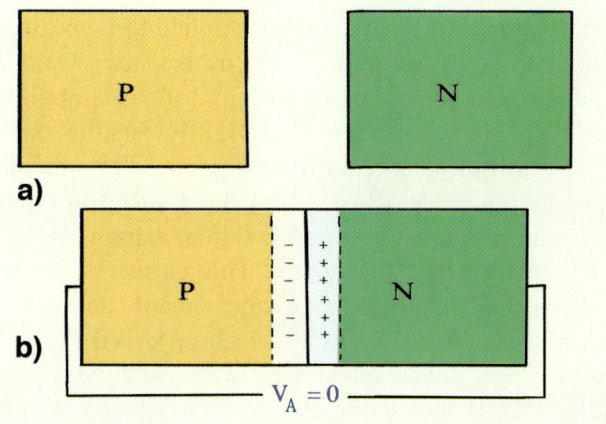

Fig. 3-8 (a) A piece of *p*-doped Si is rich in holes, while a piece that is *n*-doped is rich in electrons. (b) If the *p*- and *n*-doped pieces are joined together, but no voltage is applied, current does not flow in this circuit.

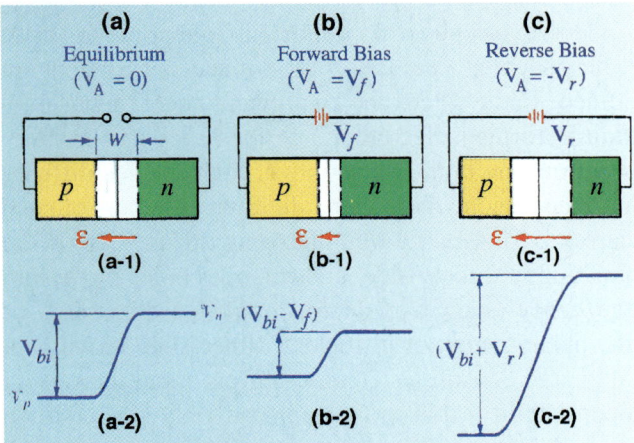

Fig. 3-9 (a-1) *pn*-junction with no applied-voltage ($V_A = 0$). (a-2) The built-in junction-voltage V_{bi} is shown. (b-1) Forward-biased *pn*-junction ($V_A = V_f$). (b-2) The junction-voltage is reduced to $V_j = (V_{bi} - V_f)$. (c-1) Reverse-biased *pn*-junction ($V_A = -V_r$). (c-2) The junction-voltage is increased to $V_j = V_{bi} + V_r$.[2] Reprinted with permission.

further current-flow is halted. The region(s) where the ions exist are also depleted of mobile-carriers. Hence, this region is called the *depletion-region* (or *space-charge region*). The voltage drop across the depletion-region, V_j, when the externally-applied voltage $V_A = 0$, is called the *built-in voltage*, as shown in Fig. 3-9a-2. Note that in this case $V_j = V_{bi}$.

3.3.2 Forward-Biasing a *pn*-Junction ($V_A > 0$)

When a forward-bias voltage is applied to a *pn*-junction, current flows. This happens because when a positive voltage (as shown in Fig. 3-9b-1) is applied to the *pn*-junction (that is $V_f > 0$), it lowers the value of the built-in voltage, or $V_j = V_{bi} - V_f$. This smaller V_j-value allows electrons from the *n*-side and holes from the *p*-side to cross the junction, setting up current-flow across the junction. This current flows as long as the forward-bias voltage remains applied to the junction. Figure 3-9b-2 shows how V_j is decreased by this externally-applied forward-biasing. Note that the electrons that cross the junction into the *p*-side leave the Si through the "+" terminal, and then travel around the closed circuit and re-enter the Si at the "-" terminal. This movement of carriers keeps a continuous-current flowing throughout the circuit.

If the forward-bias voltage is increased in value, the junction-voltage V_j decreases further, and the forward-bias current is increased. Figure 3-6d shows how the current increases as the forward-bias voltage becomes more positive. We also see that while a *pn*-junction will always conduct some current under forward-bias, this current does not become substantial until the forward-bias-voltage V_f exceeds about 0.6-V. In addition, it can be seen that beyond 0.6-V, a very-small change in forward-bias-voltage produces a large change in the forward-bias diode-current.

3.3.3 Reverse-Biasing a *pn*-Junction ($V_A < 0$)

When a reverse-bias voltage is applied to a *pn*-junction, almost no current flows. (Only a very-small reverse-bias leakage-current flows.) This occurs because when a negative voltage (as shown in Fig. 3-9c-1) is applied to the *pn*-junction (that is $V_r < 0$), it raises the value of V_j to a value higher than that of the built-in voltage, or $V_j = V_{bi} + V_r$ (Fig. 3-9c-2). Since carriers cannot cross the junction when $V_j = V_{bi}$, they certainly cannot cross the junction if V_j gets even larger. Figure 3-6d shows that when V_r is negative, the current through the *pn*-junction is a very-small negative current (the *reverse-bias leakage-current*).

Note, however, that if the *reverse-bias voltage* gets too large, the junction *breaks down*. That is, it starts conducting a large current even under reverse-bias. This breakdown-effect is also shown in Fig. 3-6d.

3.3.4 *pn*-Junction Photodiodes

A *pn*-junction photodiode is merely a *pn*-junction upon which light is shined while it is operated in reverse-bias (Fig. 3-9a). Light shining on a reverse-biased *pn*-junction causes the reverse-bias leakage-current to increase. The value of the leakage-current also increases as the intensity of the light is raised. It is useful to understand why this effect occurs, because it helps explain the basis of transistor-action in BJTs.

We start our description by recalling that the leakage-current of a reverse-biased *pn*-junction is not zero, but is instead has a very-small negative-value. This current flows because there are minority carriers

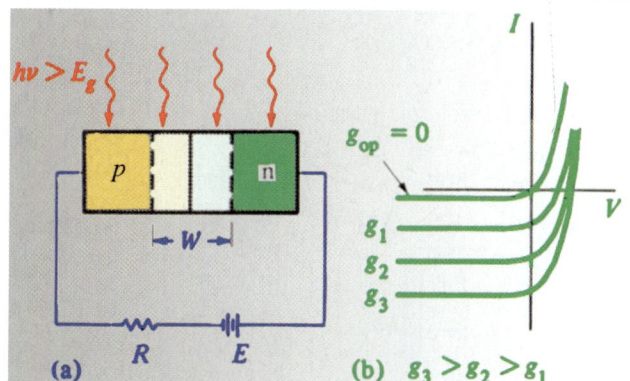

Fig. 3-10 (a) Light shining on a photodiode. (b) I-V curves of an illuminated photodiode (g is the light-intensity).

in doped-semiconductors, as well as the majority carriers. (That is, there are also a small-number of holes on the *n*-side of the junction, and a small-number of electrons on the *p*-side.) While the electrons on the *n*-side are prevented from crossing the junction under reverse-bias by the large value of $V_j = V_{bi} + V_r$, the electrons on the *p*-side (minority carriers) are not. In fact, they are attracted by this field, and flow across the junction. The same situation applies to the minority holes on the *n*-side. Thus, these carriers cross the junction and are the cause of the reverse-bias leakage-current. The reason that the leakage-current is so small is that its value is limited by the small-number of minority-carriers available to cross the junction.

If the quantity of minority-carriers could be increased, the reverse-bias leakage-current would rise. Such an increase in minority-carriers can be caused by shining light on the junction. The energy from the light-photons produces electron-hole pairs, thereby increasing both the majority and minority carrier concentrations. Only the extra minority-carriers cross the junction, leading to an increase in the *reverse-bias* leakage current. Figure 3-10b shows this increase in the photodiode-current. Note that as the intensity of the light is raised, the leakage-current value also rises. This is because more photons are entering the silicon, producing a larger number of electron-hole pairs.

3.4 BIPOLAR-JUNCTION-TRANSISTORS (BJT)

The *transistor* is an *active* electronic device. This means it can perform such functions as *amplifying electrical-signals*, or *switching electrical-signals at high speeds*. With the exception of the vacuum tube (which suffers the drawbacks of fragility and unreliable operating-behavior) these feats cannot be performed by any other electrical device. Thus, the invention of the transistor was a revolutionary event, and it ushered-in the modern electronic and computer age we live-in today. The two main-types of transistors used to make integrated circuits are the *bipolar-junction-transistor* (*BJT*) and the *MOS-Field-Effect Transistor* (*MOSFET*). We describe BJTs in this section, and MOSFETs in Sect. 3.5.

The *BJT* was invented in 1947. It is a three-terminal device (with the terminals called the *emitter*, *base*, and *collector*). The BJT also has two *pn*-junctions very-close to one another. In fact, it is this structure that is the basis of *transistor-action* in the BJT – namely, the ability to control the current flowing in one *pn*-junction by varying the bias-voltage on the neighboring *pn*-junction.

As described in the section on *pn*-junction photodiodes, the reverse-bias current of a *pn*-junction can be controlled by varying the intensity of the light shining on it. In a BJT, such control of the reverse-bias current can be done with an electrical-signal (instead of by light). This capability makes the BJT useful in electrical circuits, because it does not rely on light to perform the same controlling role.

Specifically, as shown in Sect. 3.3.3, if a *pn*-junction is under reverse-bias, the reverse-bias current can be increased if the supply of minority-carriers available to cross that junction is increased. In the BJT, this supply comes from a nearby forward-biased *pn*-junction. The carriers crossing from this nearby junction into the region between the two junctions (called the *base* of the BJT), become minority-carriers in this base. (For example, as shown in Fig. 3-11, in an *npn*-BJT the electrons entering the base from the *n*-region of the forward-biased neighboring-junction [called the *emitter* of this transistor], become minority-carriers in the *p*-type base region of the BJT.) If the separation between these two junctions is small (i.e., the *base-region of the BJT is narrow*), most of these *injected*

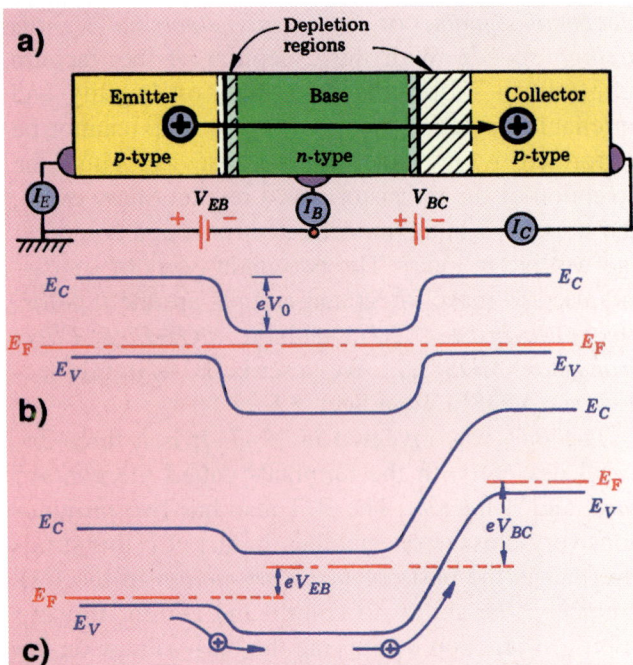

Fig. 3-11 A *pnp*-bipolar-transistor with forward-biased emitter-to-base voltage V_{EB} and reverse-biased base-to-collector voltage V_{BC}: (a) Schematic with shaded depletion-regions; (b) Energy-levels at equilibrium with junction-potential $V_j = V_{bi}$; (c) Energy-levels under operating voltage showing holes injected into (and collected from) the base.

electrons will manage to travel into the vicinity of the reverse-biased junction (before they are recombined in the base with holes). At the reverse-biased junction they will be swept across to become current flowing out of the third terminal of the BJT (i.e., its *collector*), and become the *collector-current* I_C.

IN SUMMARY: If there is forward-biased *pn*-junction close to a reverse-biased *pn*-junction, the output (*collector*) current of such a device (the reverse-bias current of the second (*collector-base*) *pn*-junction can be controlled by changing the forward-bias voltage on the first (*emitter-base*) junction. That is, if the emitter-base junction has a forward-bias voltage greater than about 0.6-V, a small change in this voltage (which is the *input-signal* to the BJT) will cause a much larger change in the collector-current (the output-signal of the BJT). *This is the basis of the amplifying transis-*

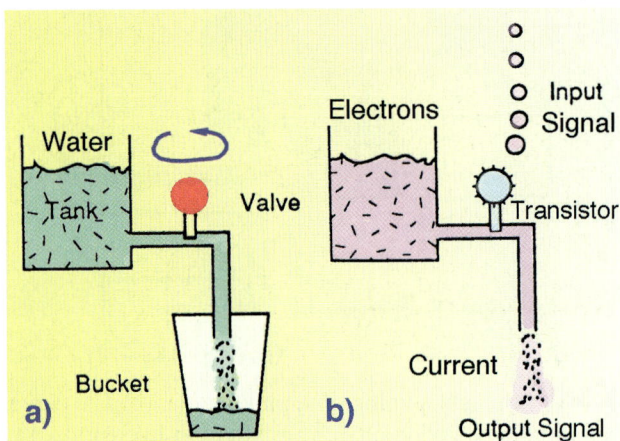

Fig. 3-12 (a) Water-flow analogy of the operation of a bipolar transistor. Only a small amount of energy is needed to open and close a valve, which in turn can control the flow of a large amount of water. (b) In a *bipolar-transistor*, a small change in the electrical-current input-signal controls a larger output-signal (having the form of an electric-current).

tor action of BJTs. Note that as shown in Fig. 3-12, this effect is analogous to applying a small force to open a valve to control a much-larger flow of water. The transistor allows large-variations in an electrical signal to be controlled by a much smaller one.

The output characteristics for a *pnp*-BJT (i.e., its I_C vs. V_{CE} curves) are shown in Fig. 3-13. Note that if the characteristics of the photodiode were flipped from the 3rd quadrant of the *x-y* axes into the 1st quadrant, they would have the same form as those of the BJT shown here (except that the parameter that controls the value of each curve would be the light intensity, instead of I_B).

The amplification property of BJTs is more commonly expressed in terms of a quantity called its *current-gain*, or *beta* (β). This is the ratio of the output-current (*collector-current*) to the input-current (*base-current*). The β of modern BJTs ranges from 30-300.

Although BJTs with a *pnp*-structure can also be made, most BJTs are built using an *npn*-structure. The *npn*-structure has better performance-characteristics because electrons in the *p*-type base have a higher-mobility than holes in the *n*-type base of *pnp*-transistors. The circuit symbols for *npn*- and *pnp*-BJTs are

CHAPTER 3 SEMICONDUCTORS DEVICES AND INTEGRATED CIRCUIT TYPES

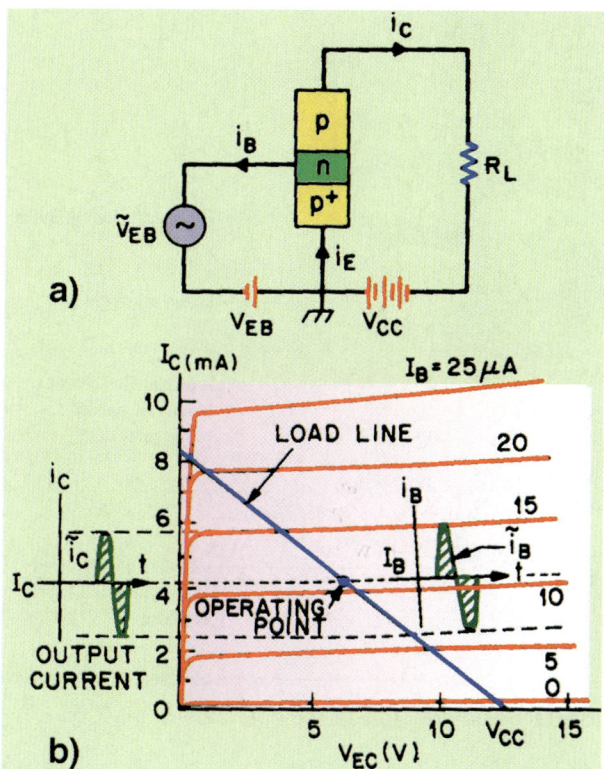

Fig. 3-13 (a) Bipolar transistor connected in the common-emitter (CE) configuration. (b) Output-characteristics of the bipolar transistor operating in this CE-configuration.[5]

shown in Fig. 3-14a. Most BJTs are made using the *standard-buried collector* (*SBC*) structure. Figure. 3-15 shows the masking steps and process-flow used to fabricate such SBC-BJT structures.

Bipolar transistors were used to make the first ICs, and they dominated the semiconductor industry from 1950 to the late 1970s. Today they are used mostly in linear IC-applications (amplifying circuits in radios, operational amplifiers, automotive electronics, and

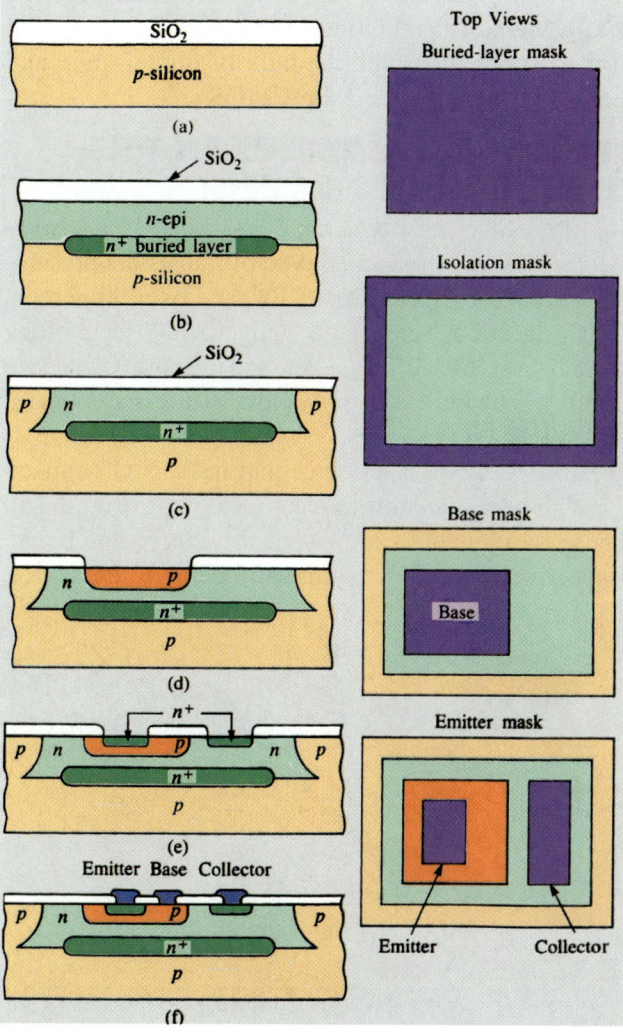

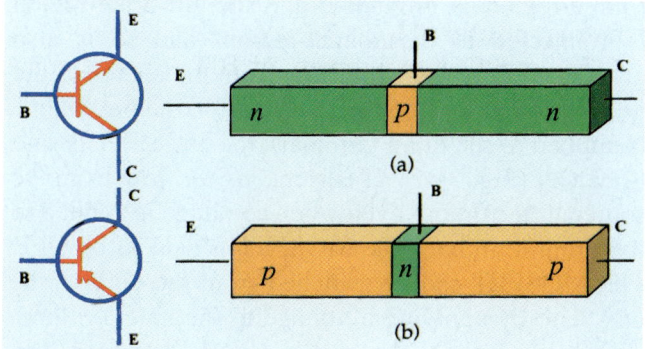

Fig. 3-14 Circuit symbols for: (a) *npn*; and (b) *pnp* bipolar transistors.

Fig. 3-15 Cross-sectional view of the major-steps in a basic SBC-bipolar-transistor process: (a) Wafer with silicon-dioxide layer; (b) After the buried-layer diffusion (using first mask), and subsequent epitaxial-layer growth and oxidation; (c) After deep-isolation diffusion (using the second mask); (d) After boron-base diffusion (using the third mask). (e) The fourth mask defines the emitter and collector contact-regions; (f) The final structure after contact and metal-mask steps.[6] (Reprinted with permission.)

biomedical instrumentation) and in some high-speed digital circuits. They are also used in BiCMOS ICs and in advanced BiCMOS-processes that incorporate Si:Ge BJTs. However, the power-dissipation of bipolar switching-devices is much higher than those made with CMOS. Thus, bipolar-based digital-circuits find limited application, and the bulk of digital-logic and memory-ICs are made using CMOS.

3.5 METAL-OXIDE-SEMICONDUCTOR, FIELD-EFFECT TRANSISTORS (MOSFETS)

Metal-Oxide-Semiconductor, Field-Effect Transistors (MOSFETs) are the second type of transistors presently used in integrated-circuits. In fact, over 90 percent of ICs in 2000 were made using CMOS-technology (which uses MOSFETs). We will discuss CMOS in Chap 4, but here we will explain the operation of MOSFETs.

The MOSFET is a device that has a MOS-capacitor with a *pn*-junction on each side of it (Fig. 3-16). These two junctions are called the *source* and *drain*, respectively, and they are integral parts of the device.

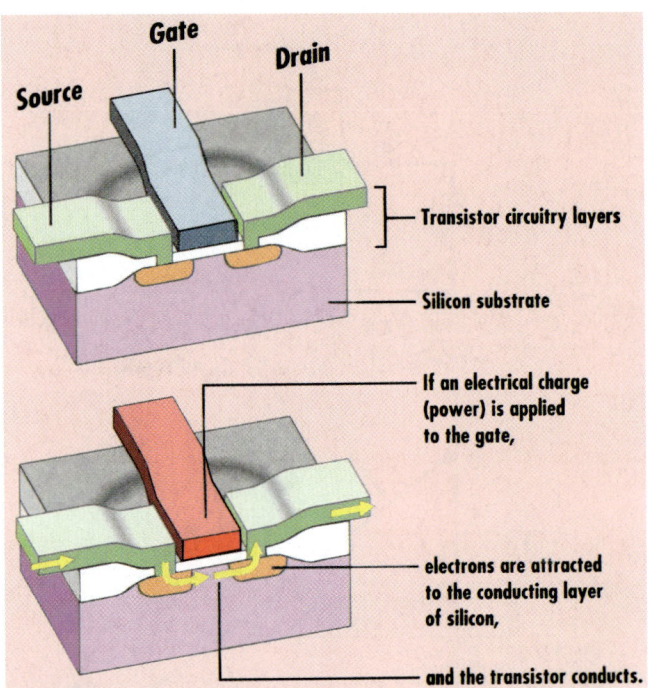

Fig. 3-17 Basic principle of the operation of a MOSFET.

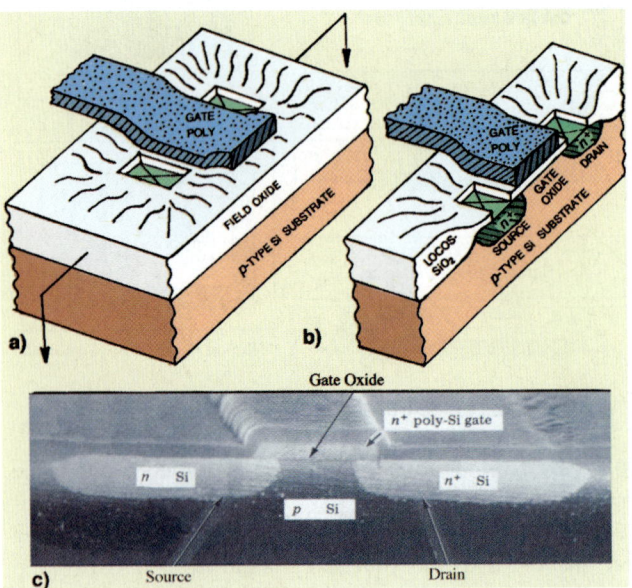

Fig. 3-16 (a) MOSFET from a 3-D perspective; (b) MOSFET from a 3-D perspective, but sliced down the middle of the channel so that a cross-section of the device is also visible; (c) SEM of a cross-section of a MOSFET.

The region in the silicon substrate under the MOS-C oxide (between the edges of the source and drain junctions), is called the *channel-region* of the MOSFET. The metal-film (or in modern MOSFETs, a polysilicon-film) above the oxide of the MOS-C part of the device, is called the *gate* of the MOSFET.

External-connections are made to the source, drain, gate, and substrate of the MOSFET (which makes this structure a *4-terminal device*). Normally, the source-region is grounded, and the substrate-region is connected to the source-region (and so is also grounded). If current *can* flow from the source to the drain through the MOSFET-channel (when a voltage is applied to the drain terminal), the MOSFET is said to be ON (Fig. 3-17). If current *cannot* flow from the source to the drain - even when a voltage is applied to the drain-terminal - the MOSFET is said to be OFF. The MOSFET can be switched from the OFF to the ON state by applying a voltage to the gate-terminal. This voltage establishes an electric-field that turns an OFF-MOSFET to the ON-state. Since such switching in this device is controlled by an electric-field, it is

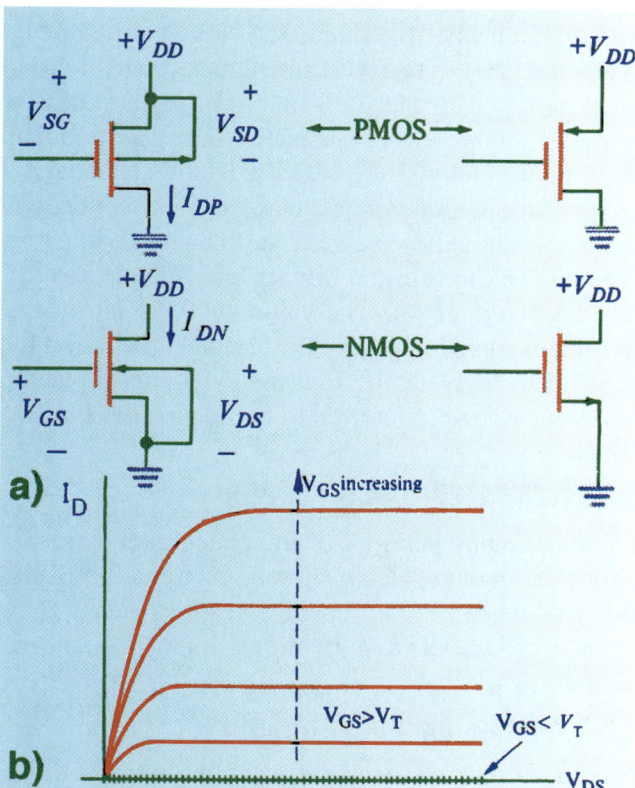

Fig. 3-18 (a) MOSFET circuit symbols. (b) Output-characteristics of an NMOSFET.

given the name the name MOS *field-effect transistor* (or *MOSFET*, for short).

Integrated circuits can be built using MOSFETs with a *p*–type channel (PMOS technology), with an *n*-type channel (NMOS), or with both types (CMOS). The circuit symbols of PMOSFETs and NMOSFETs are shown in Fig. 3-18a, and the output-characteristics of an NMOSFET are shown in Fig. 3-18b. For a PMOSFET, the source and drain regions are heavily *p*-doped, and the silicon-substrate is lightly *n*-doped. In NMOSFETs, the opposite doping is used. The first microprocessors were built using PMOS. However, NMOS offers better device-performance. Thus, when the technical difficulties of manufacturing NMOS were solved, it displaced PMOS. Finally, CMOS replaced NMOS in the mid-1985s (for reasons discussed in Chap. 4). CMOS remains the dominant IC technology today. The process-flow for producing NMOS ICs is shown in Fig. 3-19. Figure 3-20 shows a perspective-drawing of an *inverter* logic-circuit-

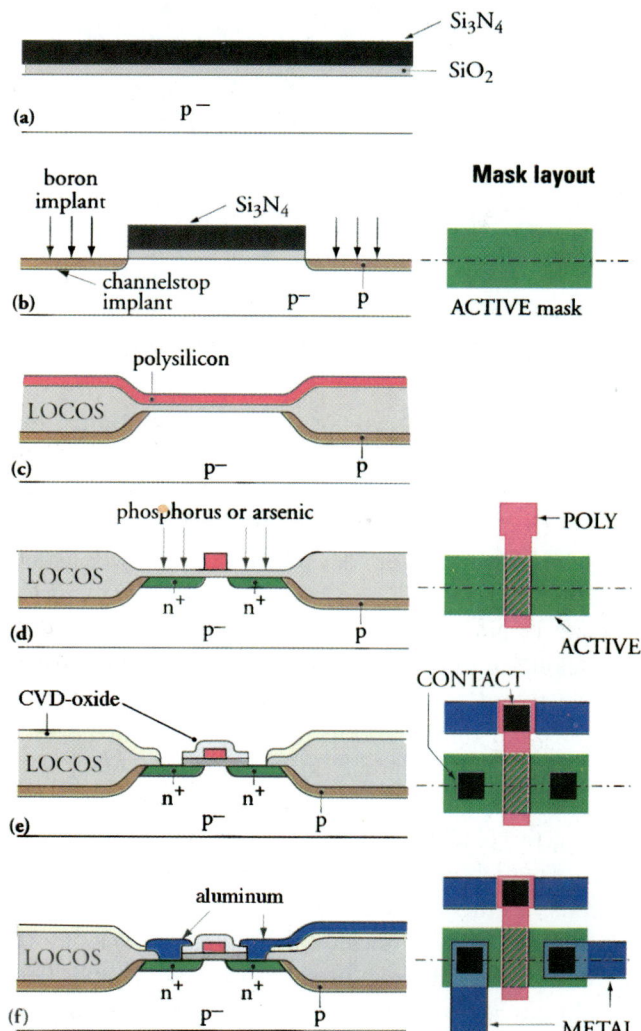

Fig. 3-19 The basic silicon-gate *n*MOS process-flow: (a) A silicon-wafer is deposited with a silicon-nitride film over a thin pad-layer of silicon-dioxide; (b) Etched-wafer after first mask-step. A boron-implant is used to set the field-region threshold-voltage (*channel-stop implant*); (c) After the nitride is removed, a layer of polysilicon is deposited by CVD; (d) Wafer after second mask-step, etching of the polySi, and S/D implants; (e) A third mask is used to open contact-windows after deposition of a CVD-oxide; (f) The final structure following metal-deposition and patterning with a fourth mask.

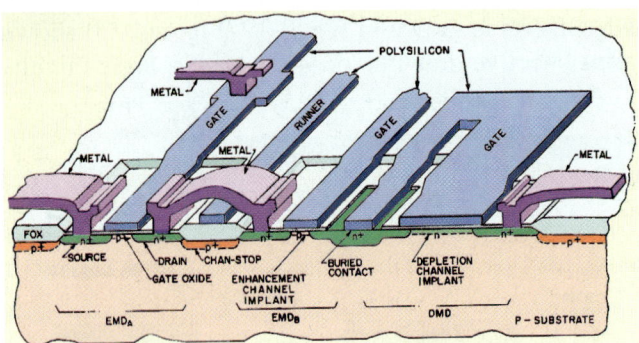

Fig. 3-20 Three-dimensional view of an NMOS logic-circuit containing two enhancement-mode MOSFETs in series with a depletion-mode MOSFET. Courtesy of Lucent Technologies - Bell Laboratories Inovations.

implemented using NMOS technology. For clarity, the intermediate dielectric is not shown.

3.5.1 MOSFET Operating Principles

The basic principles of MOSFET operation are described next - using an NMOSFET as the example device. In NMOSFETs, the source and drain regions are heavily n-doped, making them electron-rich. They are separated from one another by an electron-poor region - the lightly p-doped substrate (which forms the NMOSFET channel). In the normally-OFF state (i.e., when there is no voltage applied to the gate-terminal), the channel behaves like an open switch, the ends of which are the electron-rich source and drain-regions. In this condition, no current can flow from the source to drain, because the switch is open.

The gate is used to close and open the switch. That is, although the gate is near the MOSFET channel, it is kept electrically-isolated from it by the gate-oxide. However, when a positive-voltage is applied to the gate-terminal, it produces a positive-charge on the gate (Fig. 3-17a). This positive-charge attracts electrons in the substrate, and they move as close to the gate as they can get (i.e., to the surface of the Si substrate, right under the gate oxide - which also happens to be the location of the MOSFET-channel). Their presence in the channel creates a conductive-path between the source and drain. In effect, this *closes the switch*, allowing current to flow through the MOSFET channel (i.e., the MOSFET is turned ON). When the gate-voltage is returned to zero, the transistor is returned to the OFF-state. For PMOSFETs, a *negative voltage* must be applied to the gate-terminal to turn a normally-OFF PMOSFET to it's ON-state.

The amount of voltage that must be applied to the gate-terminal to create a sufficiently-conductive channel (i.e., to turn the MOSFET ON) is called the *threshold voltage* V_T. The value of V_T is set by the amount of p-type dopant in the channel under the gate oxide. This level can be adjusted by an ion-implantation step, a process referred to as the *threshold-adjust implant*.

3.6 INTEGRATED-CIRCUIT TYPES

There are many varieties of integrated-circuits incorporated in today's electrical and electronic products. We will describe their main categories here. They include: **a)** analog-ICs; **b)** digital-logic ICs, and **c)** memory-ICs.

3.6.1 Analog Integrated-Circuits

Analog circuits were the first types of ICs to be offered commercially. The most widely used application of analog-circuits is *amplification*. Amplifier circuits are designed in a variety of configurations (and for many different applications). However, all have the same basic goal – to amplify incoming signals. As an example, audio-systems require amplification of a weak-signal from a magnetic-tape or CD in order to produce a signal-level strong enough to drive a speaker. Most amplifier-circuits used in ICs are of the *differential-amplifier* type, and these are embedded in

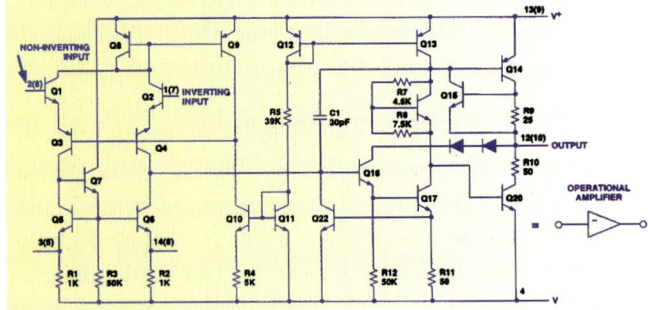

Fig. 3-21 Circuit-diagram of an IC operational-amplifier.

Fig. 3-22 Photograph of an Intel®Pentium® 4 microprocessor die on 0.13 micron. Courtesy of Intel Corp.

the overall design of *operational-amplifiers* (Fig. 3-21). Bipolar devices are best for such analog ICs.

Other applications of analog-circuits include radio communications, industrial controls, instrumentation, aerospace, and automotive electronics. These may need other analog circuits such as: **a**) *voltage regulators* (used in computers, computer peripherals, and various kinds of instrumentation); and **b**) *stepper-motor drivers* (used in laser-printers, photocopiers, and scanners).

As mentioned in Chap. 1, analog-data must also often be converted into digital-form. Circuits for doing this are called *analog-to-digital converters* (A/D). When the digital-data is converted back into analog-form, circuits that perform the reverse process are termed *digital-to-analog circuits* (D/A).

3.6.2 Microprocessors

Microprocessors (μP) are complex, general-purpose processing chips (or *central processing units [CPU]*). They use digital logic-gates as their building blocks. The most common application for microprocessors is to serve as the central-processor in personal computers (Fig. 3-22). These types of processors are capable of executing-instructions programmed into a separate (or internal) read-only memory (ROM). The μP can perform both logic and arithmetic functions.

3.6.3 ASICS

The term ASIC is the acronym for *application-specific integrated circuit*. These are ICs designed and manufactured for a single end-user and are dedicated to a particular application. ASICs therefore implement customer-specified functions and various possibilities are available for the associated customization. That is, this can be an integral part of an IC's-design or production-process, or it can be accomplished by programming special-devices. Many chips belong to this category, including digital-signal-processing (DSP) chips, power devices, and chips for the Internet, automobile, and telecommunications.

3.6.4 Memory Integrated-Circuits

Memory chips are used to store (or *memorize*) digital-data by storing and discharging electrical charges. They are widely used in computers and other electronic products for data-storage. There are a number of different types of memory-chips (Fig. 3-23), the most important of which are described here. They all

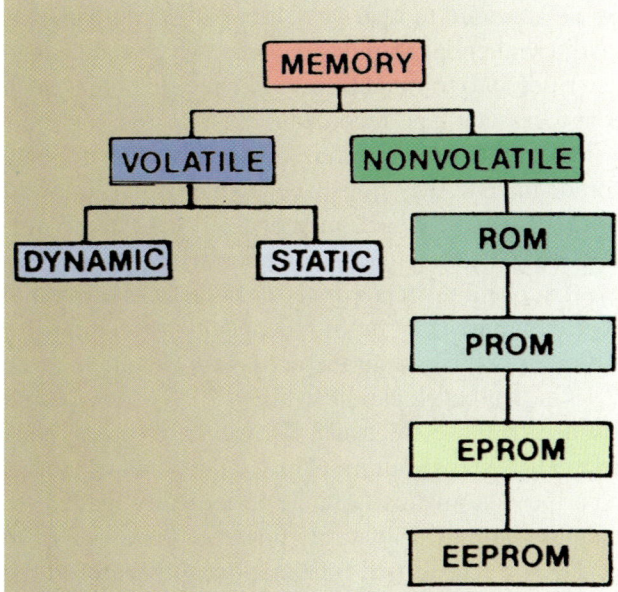

Fig. 3-23 Types of memory integrated-circuits.

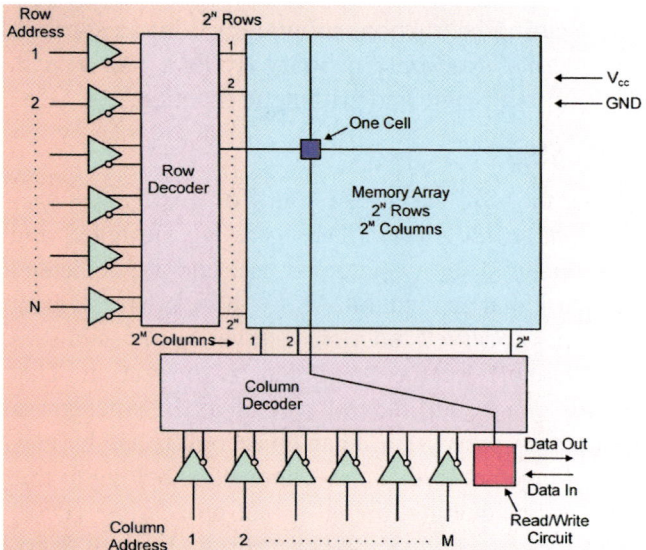

Fig. 3-24 Organization of random-access-memory (RAM).

while editing a file on a computer, all of the input-materials (including words, graphics, and symbols) are stored in the DRAM of the computer before the "Save" command writes them permanently into the hard-disk. It is therefore important to Save files being edited-frequently when working on lengthy documents. Otherwise, in the event of a power-outage, significant work may be lost. Because of their relative simplicity, they are the least-expensive, and most widely-used memory chips. By 2002, 256-Mb DRAMs were routinely commercially available.

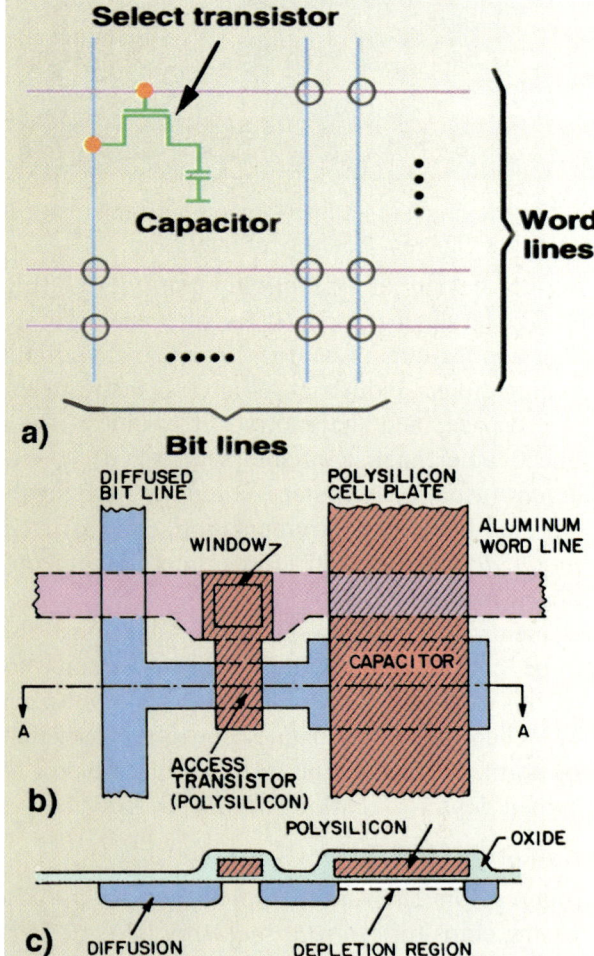

Fig. 3-25 Single-transistor DRAM-cell with storage capacitor: (a) Circuit schematic; (b) Cell layout; (c) Cross-section through A-A.

have a large-array of memory-cells arranged in rows and columns. Individual cells can be selected to *write a value to* (or *read a value from*) by selecting a specific-row and column.

DRAMS: DRAM stands for *dynamic random-access memory*. As described above, the term *random-access* (Fig. 3-24), implies that each memory-cell in the chip can be accessed to read or write in any order (in contrast to sequential memory-devices, where data has to be written into [or read from] in a specific order - such as in a cassette tape-recorder).

The term *dynamic* refers to the fact that the data stored in DRAMs must be dynamically-renewed (*refreshed*) within each memory-cell every few milliseconds. That is, each cell consists of an *access-transistor* and a *MOS-capacitor* (Fig. 3-25). Enough electrons must flow onto the capacitor-structure of a particular cell (through the access-transistor) to store a "1" in that cell-location (while a "0" is stored as an absence of these electrons). However, electrons leak from DRAM capacitors. Thus, additional electrons must be automatically and periodically added to maintain stored "1"-data. If power is removed from the DRAM, the stored-information will therefore be lost, as refreshing ceases to take place. For example,

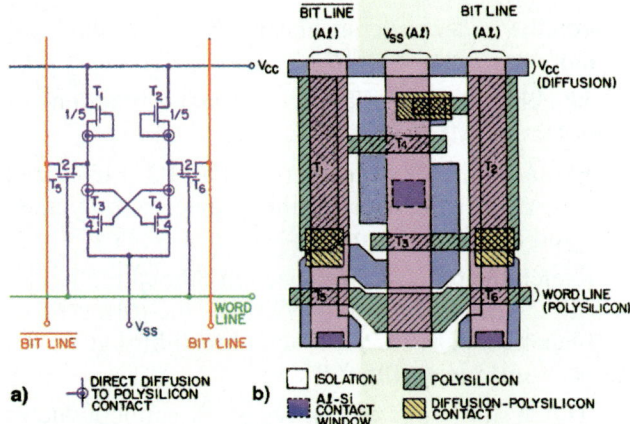

Fig. 3-26 Example of an SRAM cell: (a) Circuit diagram; (b) Layout.

SRAMs: SRAM means *static random-access memory*. In SRAMs, a cell contains six components (four transistors and two resistors, or six transistors) instead of just two in the DRAM-cell (Fig. 3-26). However, in SRAMs there is no need to periodically-recharge a capacitor (although data will still be lost if power is turned off). SRAMs have much faster access-times than do DRAMs, and thus are used in computers as cache memory (i.e., to store the most frequently used instructions). However, since an SRAM-cell has six components, instead of the two in a DRAM-cell, fewer SRAM-cells can be built on the same-size piece of silicon as a DRAM. Hence, SRAM-memory having the same number of cells is more expensive than DRAM. Thus, DRAM is used for the main memory.

EPROM, EEPROM, and FLASH MEMORY: DRAMs and SRAMs both need to have power continuously applied to retain their stored-data. Thus, both types are included in the category of *volatile-memory*. There is another category of memory called *non-volatile memory* and two kinds of such chips are *erasable programmable read-only memory* (*EPROM*) an *electrically-erasable programmable read-only memory* (*EEPROM*). They are mainly used for permanent storage of data and instructions without a power-supply.

In the *EPROM*, data can be written into it electrically (this is called *programming the memory*). Electrons are electrically injected onto a floating poly gate to program a cell. But in EPROMs, the stored charge on the floating-gate can only be erased (discharged) using ultra-violet light. The entire memory-data is erased during such a UV-erase step.

In EEPROMs, data on a floating-gate can be electrically-erased (UV-erase is not needed). Individual-cells can thus be erased, instead of the entire-chip at once. *Flash-memory* is similar to the EEPROM. It is used in memory-cards for digital cameras and hand-held-computers. Embedded flash-memory can help achieve the goal of a *system-on-a-chip*.

REFERENCES

1. W. Shockley, *Electrons and Holes in Semiconductors*, Van Nostrand, Princeton, 1950.

2. B.G. Streetman *Solid-State Electronic Devices*, Prentice-Hall, Upper Saddle River, NJ, 1995.

3. S. Wolf, *Silicon Processing for the VLSI Era: Vol. 3 Submicron MOSFET*, Lattice Press, Sunset Beach CA 1995

4. S.M. Sze, *Semiconductor Devices: Physics & Technology*, John Wiley & Sons, New York, 1985.

5. R.C. Jaeger, *Intro. to Microelectronic Fabrication*, 2nd Ed., Prentice-Hall, Englewood Cliffs, NJ, 2001.

PROBLEMS

1. What types of materials are used to make IC resistors? What factors determine the resistance of such resistors. ?

2. What is the minimum number of mask-steps needed to fabricate ICs based on PMOSFET-technology? Repeat the question for ICs made using bipolar transistors. Based on your answers, with which of these technologies is it cheaper to build ICs?

3. Describe what is meant by the term *self-aligned gate*. Why is this approach preferred over a non-self-aligned gate structure. Why was it not possible to implement the self-aligned gate process until polysilicon was adopted as the gate material?

4. Describe how NMOSFETs are turned ON & OFF.

5. Calculate the capacitance for a parallel-plate capacitor, the area of whose plates is 100-μm^2. Assume the dielectric between the plates is silicon-dioxide with a thickness of 1000Å.

6. By reducing the distance between the plates, the capacitor plate area can be decreased, while still maintaining the same capacitance value. For example, if the plate area is reduced to 1-μm^2, what would the oxide thickness have to be to produce a structure with the same capacitance-value as in Prob. 5?

7. Another way to alter the capacitance is to change the dielectric constant. What would have to be the dielectric constant in a capacitor with a plate area of 1-μm^2 to produce a structure with the same capacitance-value as in Prob. 5?

8. Calculate the capacitance of an MOS-C that has a cross-section of 2-μm by 2-μm if the gate oxide is 400Å thick (assuming there are no interface or oxide charges, see Chap. 13).

9. An aluminum line on a silicon wafer is 500-μm long, and 10-μm wide, and it is running over a field oxide that is 5000Å thick. What is the capacitance of this line to the substrate (assuming you can use the parallel-plate capacitance formula)?

10. In order to increase the capacitance of MOS capacitors in ULSI DRAMs, first an oxide/nitride/oxide (ONO) layer was used to replace SiO_2 as the DRAM-cell capacitor dielectric material. Later, Ta_2O_5 replaced the ONO layer. The dielectric constants of SiO_2, nitride, and Ta_2O_5 are about 3.9, 7.6, and 25, respectively. What is the capacitance ratio for the capacitors with Ta_2O_5 and ONO dielectrics for the same dielectric thickness, provided that the ONO has a thickness ratio of 1:1 for the oxide to the nitride?

11. Assume you were asked to design a MOS-Capacitor to act as a test-structure for monitoring the threshold-voltage of a silicon-gate NMOSFET. Formulate a detailed process flow of the fabrication steps needed to manufacture such an *n*-channel MOS-Capacitor.

12. The material on one-side of an abrupt silicon *pn* junction is 1-ohm-cm *n-type*, and on the other side it is 1-ohm-cm *p-type*. What is the impurity concentration in atoms/cm^3 on each side?

13. Explain what is meant by the *built-in voltage* that exists at a *pn*-junction.

14. Explain the origin of the *reverse-bias leakage current* in *pn*-diodes, and explain why it remains essentially under reverse biasing of the diode until breakdown.

15. Sketch and name the three parts and functions of a bipolar and MOS transistor.

16. Explain the difference between a MOSFET made with a metal-gate, a polysilicon-gate, a polycided gate, and a salicided gate.

17. Sketch the process steps used in fabricating devices with the *planar process*. Why was the invention of the planar process such a crucial breakthrough in the quest to build monolithic ICs?

18. Name the three main types of semiconductor memory circuits, and explain how they differ from one another in their operation.

19. Define the term RAM used in describing DRAMs and SRAMs. (Do not just give the meaning of the acronym, but explain how these memory chips operate, with respect to this term.)

CHAPTER 3 SEMICONDUCTORS DEVICES AND INTEGRATED CIRCUIT TYPES

Drawings depicting the dopant concentrations in a MOSFET device. Lower figure is a contour plot representation, and the Upper figure is a perspective representation. From the cover of *Silicon Processing for the VLSI Era - Vol. 3 - The Submicron MOSFET,* by S. Wolf. Published by LATTICE PRESS in 1995.

CHAPTER 4

OVERVIEW OF CMOS PROCESS-INTEGRATION

CHAPTER CONTENTS

4.1 DESIGN-PHASE OF IC MANUFACTURE

4.2 CMOS PROCESS-FLOW

4.3 SHALLOW-TRENCH ISOLATION

"Big fleas have little fleas,
Upon their backs to bite 'em
Little fleas have lesser fleas,
And so, *ad infinitum*"

     Author Unknown

Illustration of a cross-section of a CMOS structure with six-levels of metal interconnect. Taken from the cover of *Silicon Processing for the VLSI Era:* Vol. 4, Lattice Press, 2002. Graphic design by S. Ross Wolf.

The design, fabrication, test, and packaging of integrated-circuits is arguably the most complex manufacturing-sequence being carried out on the planet today. Figure 4-1 illustrates the entire sequence of steps that occurs in the course of manufacturing an integrated-circuit. These steps can be grouped into two main phases: **1)** the *design-phase*; and **2)** the *fabrication-phase* (Fig. 4-2). This book is concerned primarily with the fabrica-

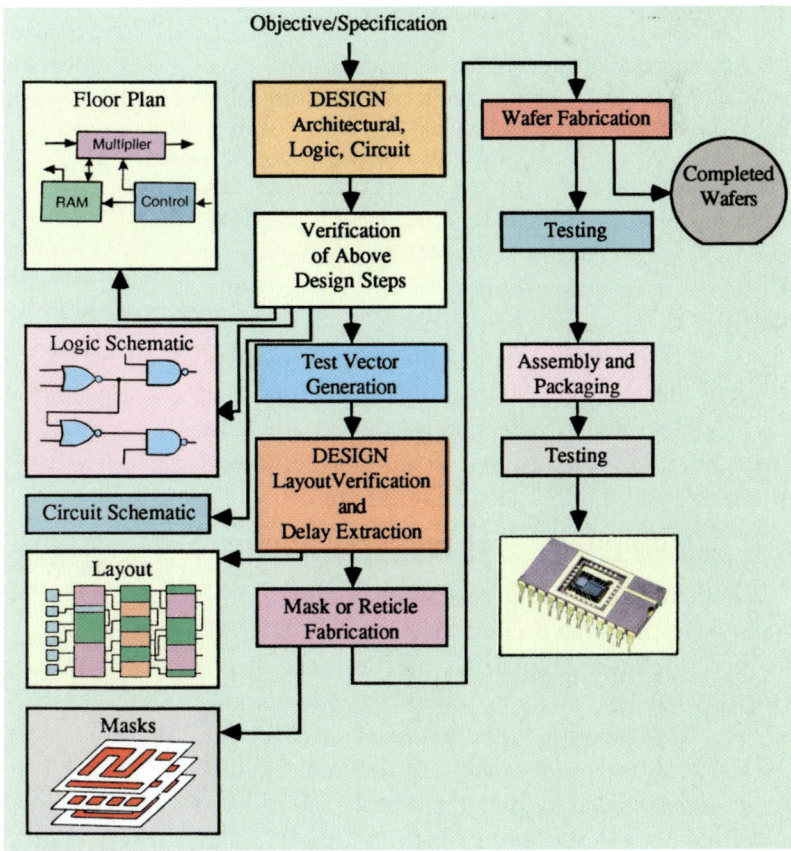

Fig. 4-1 Overview of the steps required for the manufacture of integrated-circuits, expressed in a flow-diagram format.

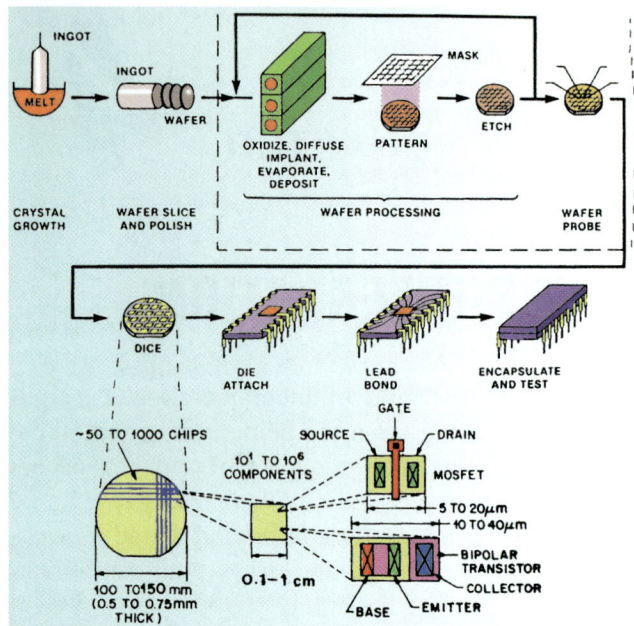

Fig. 4-2 An expanded diagram of the fabrication steps of the IC manufacture-sequence shown in Fig. 4-1 (starting from the *Wafer-Fabrication* block and concluding with *Testing*).

tion-phase, and the main subject of this chapter is an example of a process-flow used in fabricating CMOS ICs (which is presented in Sects. 4.2 and 4.3).

However, it is also useful to briefly outline the steps of the design-phase here. This is done in Sect. 4.1, and it provides the context that allows readers to perceive the role of silicon-processing within the totality of integrated-circuit manufacturing. Readers wishing to explore steps of the design-phase in more detail are referred to other technical literature.[3]

4.1 THE DESIGN-PHASE OF IC MANUFACTURING

The desired functions and necessary operating-specifications of an integrated-circuit are initially defined in the *design-phase*. Digital-ICs are designed from the "top down." That is, the required large *functional-blocks* are first identified (the *Floor-Plan* block in Fig. 4-1). Next, their *sub-blocks* are defined. Finally the *logic-gates* needed to implement these sub-blocks are specified (*Logic-Schematic* block in Fig. 4-1, and Fig. 4-3). Each logic-gate is designed by appropriately connecting devices that are slated for fabrication on

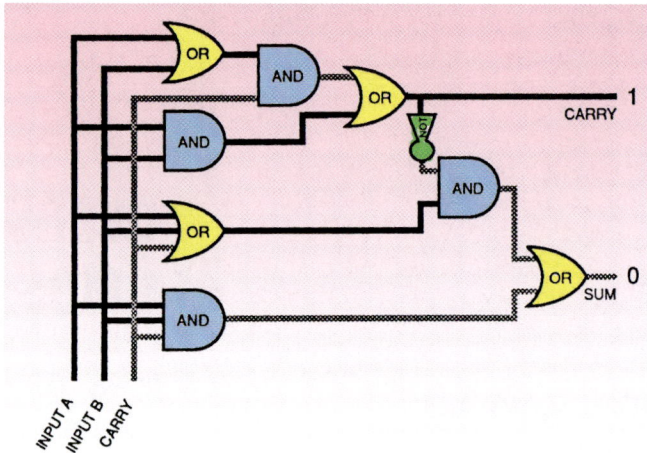

Fig. 4-3 The *logic-design* of a digital-IC is expressed in *logic-schematic* form. As shown here, the components in such schematics are switches, inverters, and logic-gates.

the Si wafer (the *Circuit Schematic* block in Fig. 4-1). Figure 4-4a shows a segment of a bipolar-transistor IC expressed in circuit-schematic form, and Fig. 4-7a depicts a CMOS-inverter in such form.

On completion of these various levels of design, each level is rechecked for correct functionality. When all aspects of the circuit-design are correct to the designer's satisfaction, *test-vectors* (that will be used to test the manufactured-circuits), are generated

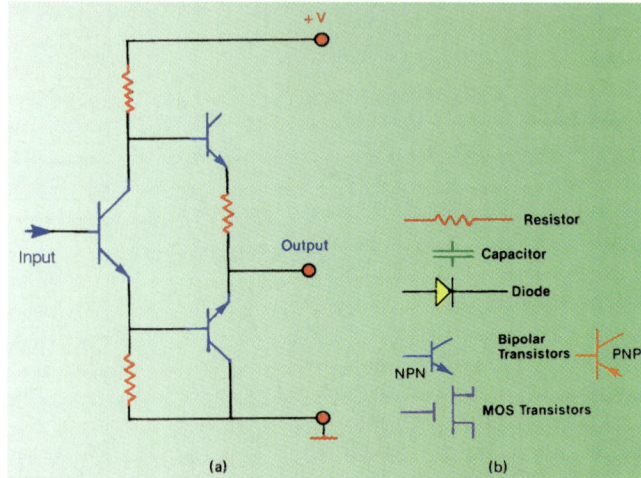

Fig. 4-4 (a) Example of a circuit schematic of a segment of an IC based on bipolar-transistor technology. (b) Circuit-symbols of components used in IC circuit-schematics.

CHAPTER 4 OVERVIEW OF CMOS PROCESS INTEGRATION

from the logic schematic.

The circuit is then layed out. The *layout* consists of sets of patterns that will be transferred to the silicon wafer (the *Layout-block* of Fig. 4-1, and Fig. 4-5). These patterns correspond to device-regions or interconnect-structures, and such patterns are sequentially transferred to the wafers as part of the wafer-fabrication sequence (through the use of photolithographic processes and a set of masks, as shown in Fig. 4-6). The result of each pattern-transfer-step is a set of features created on the wafer surface. These features are generally either in the form of: **a**) an *etched-opening* in a film (or region of the substrate); or **b**) a *patterned-feature* of a film present on the surface (e.g., an interconnect-line or a pad). After openings (or *windows*) are created by the pattern-transfer-step, either controlled quantities of dopant are added to the Si-substrate through the openings, or another layer is

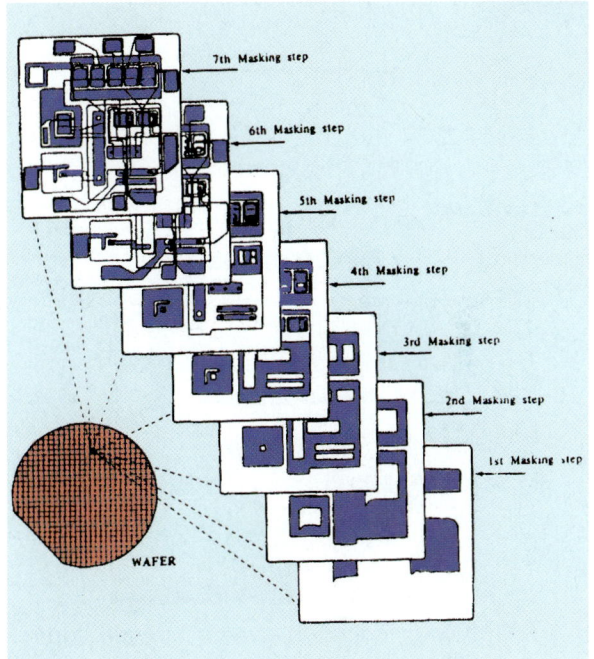

Fig. 4-6 Example of the patterns transferred to a wafer during a seven-mask process-sequence.

deposited that makes contact to the underlying-layer through the opening. In either case, device-regions (and interconnect-structures), are produced by the patterning-processes and associated fabrication-steps.

While the circuit is *designed* from the "top down," *creation of the layout* proceeds from the "bottom up." A variety of typical devices (e.g., transistors and resistors) are first layed-out. Then, a set of cells representing the required primitive logic-gates are created by interconnecting appropriate devices (see Figs. 4-4 and 4-7a). Next, sub-blocks are generated by connecting these logic gates (Fig. 4-3), and finally the functional blocks are layed out by connecting the sub-blocks.

Additional items required by the circuit-design are also incorporated during the layout-process (e.g., power-busses, clock-lines, and input-output-pads). The completed layout is then subjected to a set of *design-rule-checks* and *propagation-delay-simulations* to verify that correct implementation of the circuit has been achieved in layout-form. Upon completion of this checking procedure, the layout informa-

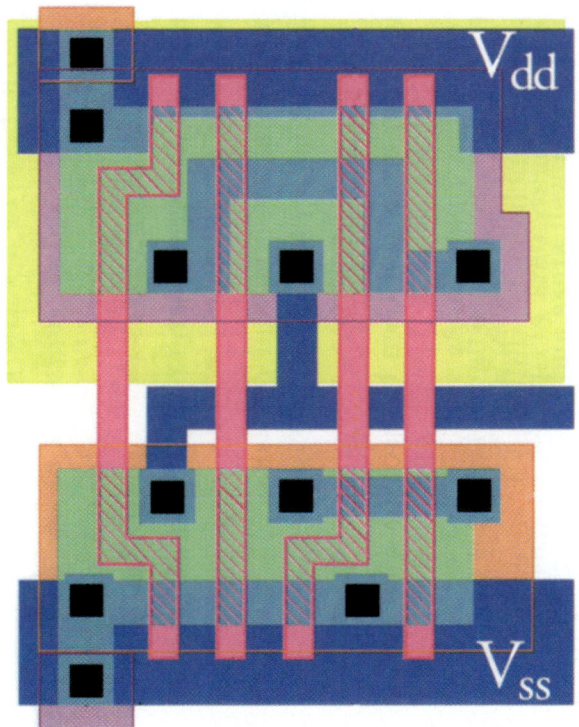

Fig. 4-5 Example of a *layout* of a circuit-schematic of a logic sub-block.

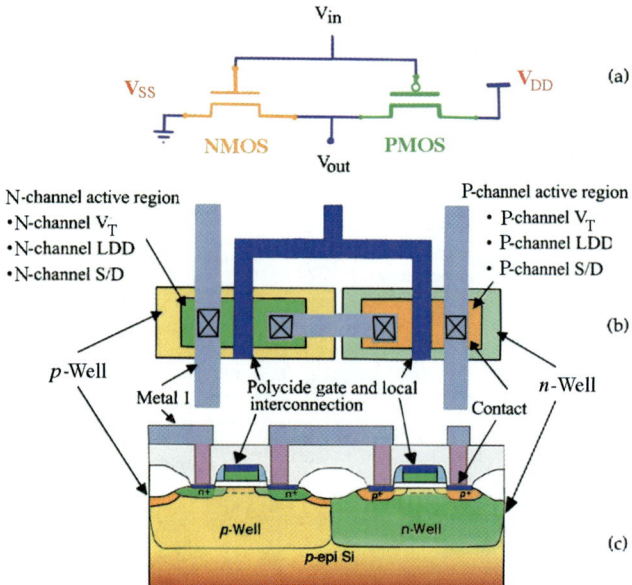

Fig. 4-7 CMOS inverter: (a) *Circuit-diagram* form; (b) *Layout* form; and (c) *Physical* form (in cross-section).

tion is ready to be used to generate a set of masks that will serve as tools for specifying the circuit-patterns on silicon-wafers. This layout information is stored on a computer. A photograph of a completed (unpackaged) IC is shown in Fig. 4-9.

In this book the details of the integrated-circuit manufacturing steps summarized in Fig. 4-2 are discussed. These steps start when the *layout-information* has been finalized. At that point procedures are utilized to convert the layout-information stored on the computer, into a set of *masks* or *reticles*. This procedure is described in Chap. 20. The individual fabrication process-modules associated with creating-patterns, introducing-dopants, and depositing-films on silicon-substrates (to form the integrated-circuit features) are also subjects of this volume.

In this chapter, information is presented about how such individual process-modules are combined to create a complete CMOS chip (a subject referred to as *process-integration*).[1,2] Note that examples of process-flows for fabricating bipolar and NMOS ICs are given in Chap. 3 (see Figs. 3-15 and 3-19, respectively). The CMOS-inverter shown in Fig. 4-7 can be used as an example of the type of IC-structure created by such a CMOS process-flow. In this figure, the CMOS-inverter is represented in several ways. First, it is shown in its *circuit-schematic* form (Fig. 4-7a). Next the layout of the completed CMOS-inverter is

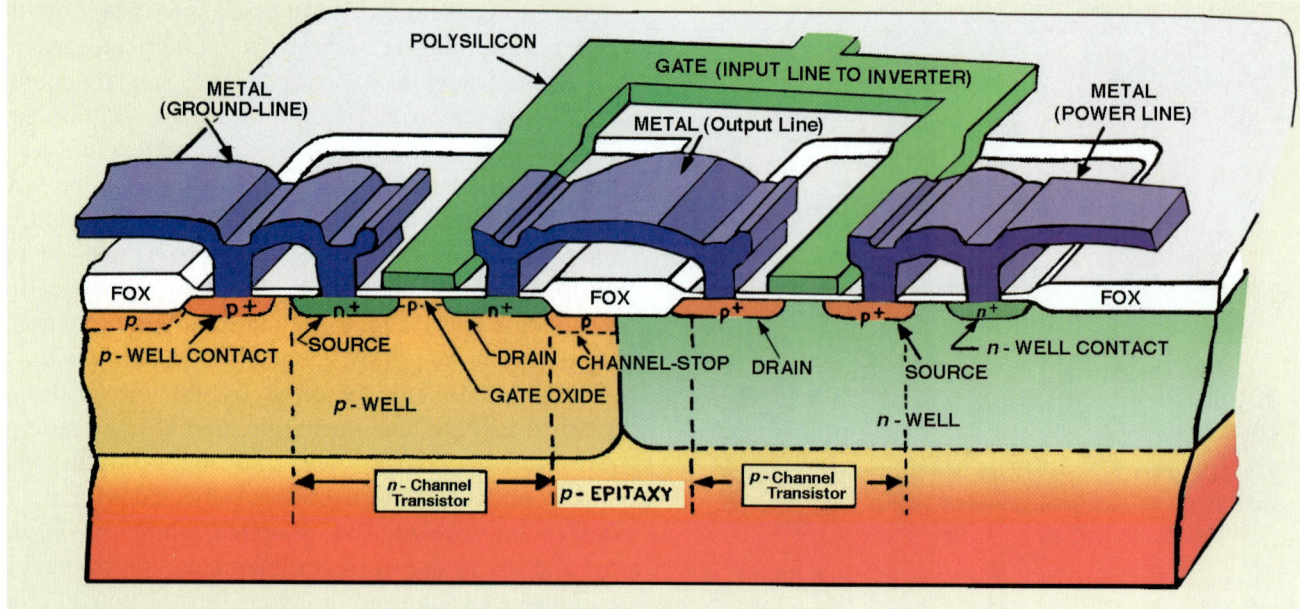

Fig. 4-8 Perspective-view of an inverter logic-gate fabricated using a twin-well CMOS technology.

CHAPTER 4   OVERVIEW OF CMOS PROCESS INTEGRATION

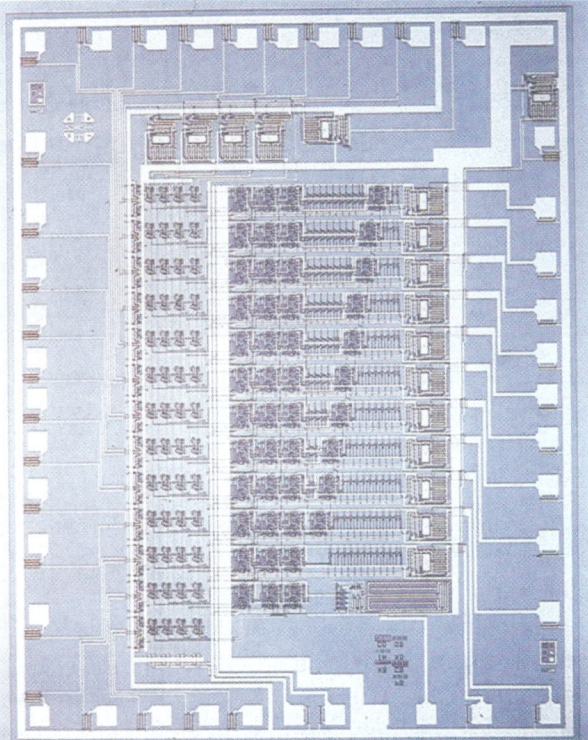

Fig. 4-9 Photograph of a completed IC-chip (unpackaged).

given (Fig. 4-7b). Then a *cross-sectional view* of the inverter structure on an IC-chip is depicted (Fig. 4-7c). It is this cross-sectional form that will be used to step through the *CMOS process-flow*. Finally, Fig. 4-8 depicts a *perspective-view* of such a CMOS-inverter IC-structure. This view includes the isolation-structures, power-lines, signal-lines, and contacts to the silicon-substrate needed in CMOS.

4.2 CMOS PROCESS-FLOW

A modern CMOS-IC process-flow involves 350 (or more) process steps and may take six-to-eight weeks to complete. The process-flow is a sequence of the chemical and physical operations that are performed on the silicon-wafer.

Here we present a 15-mask, twin-well, 2-level-metal CMOS process-flow. (Details about each of the individual processes are given in later chapters.) This flow is representative of process sequences used to fabricate CMOS-ICs from 1.2-μm down to about 0.35-μm. It provides an overview of the series of steps employed in fabricating such CMOS-ICs. A cross-section (at completion of Metal-1 processing) using this process-flow is shown in Fig. 4-7c.

Twin-well CMOS-technology is used for CMOS generations below 1-μm because two separate-wells can be formed in the lightly-doped substrate region. Well-profiles can be tailored independently so that neither device suffers from excessive-doping effects.

Note that in this process-flow, LOCOS-isolation is used (defined in Sect. 4.2.2). In more advanced ICs (i.e., for CMOS-technologies of 0.25-μm and smaller), an alternative isolation approach, namely *shallow-trench-isolation* (*STI*), is implemented. The details of forming such STI-structures is described in Sect. 4.3.

4.2.1 Starting Material for CMOS-ICs

All MOS technologies use silicon-wafers with a <100>-orientation. Typically, in twin-well CMOS the starting material is also either a lightly p-doped wafer (*p-bulk-wafer*), or a heavily p-doped wafer on which a thin, lightly p–doped epi-layer is grown (Figs. 4-10a and 10b, respectively). The concentration range of the p-doping in both the bulk-wafer and the surface p–epi

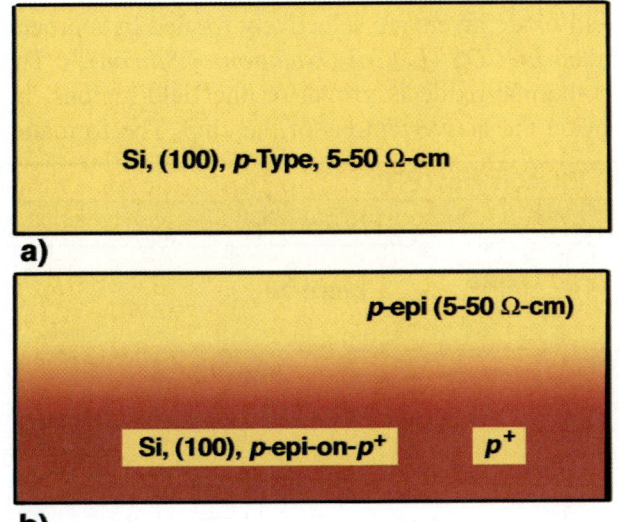

Fig. 4-10 Silicon starting substrates for twin-well CMOS: (a) p-bulk wafer; (b) p-epi-on-p^+ wafer.

layer is 8×10^{14}-1.2×10^{15} cm^{-3}, while the p^+ substrate beneath the epi-layer is B-doped to ~10^{20} cm^{-3}. The p-epi-on-p^+ wafers (see Chap. 17) provide several advantages including: **1**) improved latchup-protection (see Ref. 2); **2**) gate-oxides with better dielectric-reliability are grown on epi-layers than on bulk-Si surfaces; and **3**) improved gettering capability is obtained (see Chap. 9).

The chief disadvantage of epi-wafers is their higher cost. For example, in 2000, the price of a 200-mm bulk-starting-wafer was ~$80 (US), while that of a 200-mm epi-wafer was ~$150. Nevertheless, for our process-flow, p-epi-on-p^+ wafers will be used. The exact doping-concentration of the epi-layer is chosen to provide the best overall set of the following device-characteristics: low source/drain-to-substrate capacitance; high source/drain-to-substrate break-down voltage; high carrier-mobility; and low-sensitivity to source-substrate bias-effects.

4.2.2 Formation of Active-Regions, Channel-Stops, and LOCOS-Isolation Structures

The first features to be fabricated with this process-flow are the *isolation structures* of the CMOS-IC. These consist of a *thick field-oxide*, together with *channel-stop dopants* beneath this field-oxide. The thick field-oxide layers are selectively formed by a process called *LOCOS* (*LOCal Oxidation of Silicon*).[2,4] That is, thermal-oxide is grown on the field-regions, but not on the active-regions of the chip. The formation

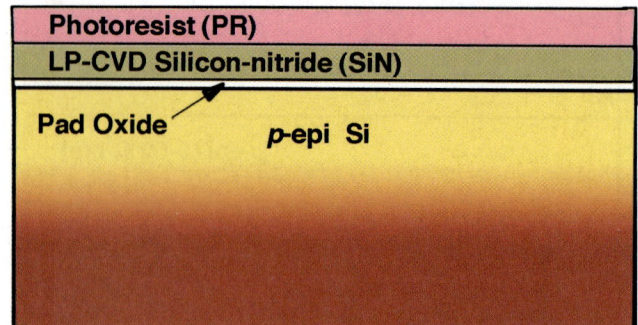

Fig. 4-11 Forming active and field regions using LOCOS isolation, Step **1**: Grow a pad-oxide film, deposit a silicon-nitride (SiN) film by LPCVD, and apply photoresist.

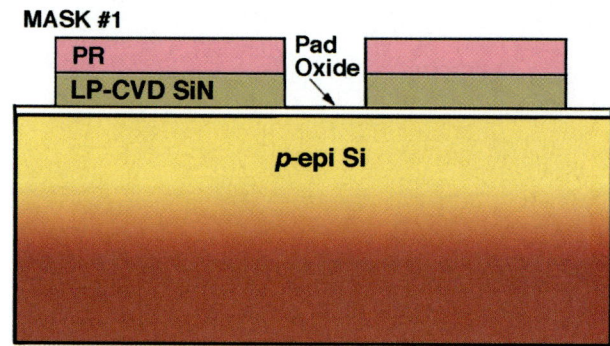

Fig. 4-12 Forming active and field regions using LOCOS isolation, Step **2**: Use *Mask #1* to define the active regions (i.e., leave them covered with a resist-mask). Then etch-away the SiN-film that lies over the field-regions.

of the *isolation* (or *field*) regions thus simultaneously defines the *field* and *active regions* of the circuit (the latter being where the transistors will reside).

The combination of a thick field-oxide - and channel-stop dopant beneath it - forms a *parasitic-MOSFET* in the field-regions with a threshold-voltage made deliberately too-high to be turned-on by the normal voltages used to run the IC. This electrically isolates *active-devices* on the chip from one another.

The LOCOS-process for forming isolation-structures consists of four steps, as shown in Figs. 4-11 to 4-14. The first step begins by cleaning the starting-wafers and growing a thin thermal-oxide on them to a thickness of 40–50-nm (called a *pad-oxide*). The pad-oxide serves as stress-relief layer between the silicon-nitride film (deposited next) and the silicon-substrate. The silicon-nitride is deposited on the pad-oxide (to a thickness of about 100-nm) with an LPCVD process, and a layer of photoresist is applied (see Fig. 4-11).

In Step-**2**, the resist layer is exposed using *Mask #1*. This mask is used to create the patterns that define the active and field regions of the circuit. After exposure and development, the nitride is dry-etched. Figure 4-12 shows the wafer after this nitride-etching step.

In Step-**3**, a *channel-stop dopant-layer* is formed in the field-regions of the p-wells. This is done by implanting a layer of boron-atoms (close to the Si surface) in the field-regions (Fig. 4-13). This layer of

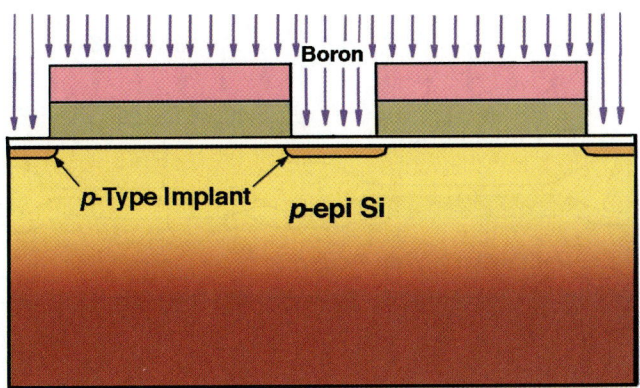

Fig. 4-13 Forming active and field regions using LOCOS isolation, Step **3**: Implant B into the field-regions (this process is known as a *channel-stop implant*).

under the edges of the nitride film. That is, a tapered field-oxide layer extends for some distance under the nitride edge. This means that the field-oxide is actu-

implanted B-atoms increases the threshold-voltage of the parasitic *p*-channel field-region MOSFETs. Boron channel-stop dopant is ion-implanted into the surface with a dose of 10^{12}-10^{13}-atoms/cm^2 at 40-80-keV, using a medium-current implanter.

Finally, a thermal-oxidation step is carried out to grow the field-oxide layer, using wet-oxidation. The required oxide thickness (350–400-nm) can be grown at 1000°C in about 90-min in H$_2$O. Since silicon-nitride acts as a diffusion-barrier, oxide grows only over the regions not covered by nitride (Fig. 4-14). The nitride is next stripped using hot phosphoric-acid (an etchant that does not attack the thick thermal-oxide grown over the field regions).

Note, however, that some oxidation does occur

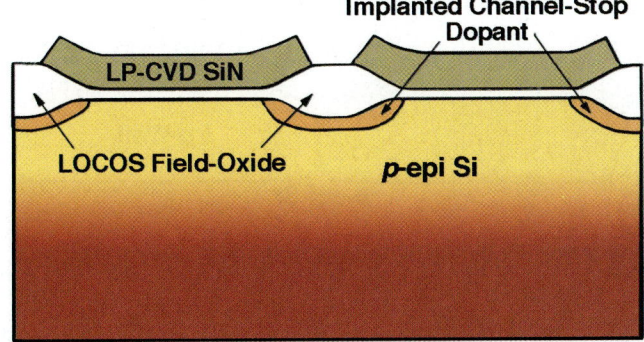

Fig. 4-14 Forming active and field regions using LOCOS isolation, Step **4**: Selectively grow a thick field-oxide with a wet-oxidation process (3000-5000-Å thick).

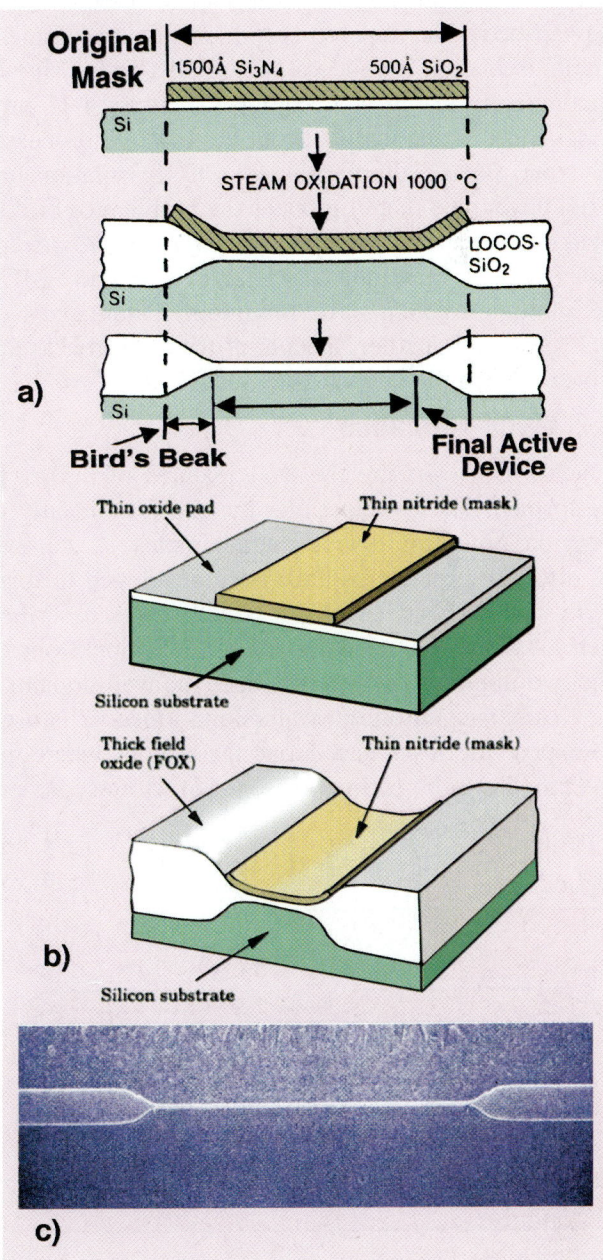

Fig. 4-15 (a) Schematic of *Bird's Beak* encroachment arising from the LOCOS process. (b) Perspective drawing of LOCOS. (c) SEM of a wafer with semi-recessed LOCOS field-oxide.

ally formed over a larger wafer-surface-region than the mask-pattern used to define the nitride-film. This becomes a major concern when small features must be defined. The tapered-shape of this two-dimensional oxidation is called a *bird's-beak*, and is shown qualitatively in Fig. 4-14 and 4-15. Figure 4-15 provides more visual details about the LOCOS process.

Note that an alternative to LOCOS isolation has been developed that eliminates the problem of bird's-beak-encroachment. It is called *shallow-trench-isolation* (STI). As mentioned earlier, STI is used in CMOS technologies which have minimum feature-sizes of 0.25-μm, and smaller. Details of the STI-process are given later in the chapter (see Sect. 4-3).

4.2.3 Well-Formation

The *CMOS wells* are the next features of the IC to be formed. These wells tailor the substrate locally to provide optimum device-characteristics. A number of different procedures have been developed to form twin-wells. The most obvious method (as described here) is to use two masking-steps. Each blocks one of the two implants used to introduce the well-dopants.

Resist is spun on the wafers and *Mask #2* is used to expose the resist and define the regions where the *p*-wells are to be formed (Fig. 4-16). Boron ions are

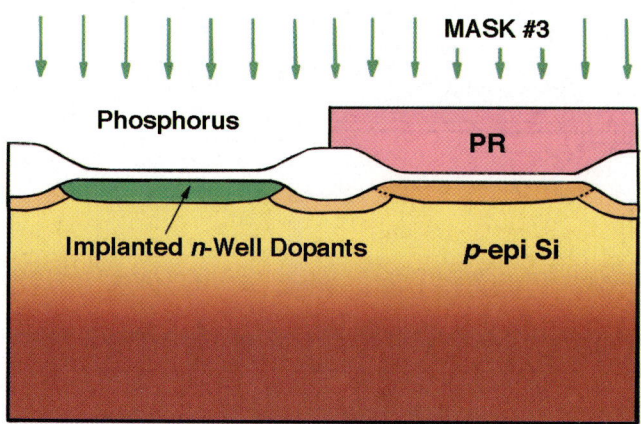

Fig. 4-17 Photoresist is applied, and *Mask #3* is used to cover regions where NMOS-devices will be built with resist. A phosphorus-implant provides dopants for the *n*-wells of the twin-well CMOS structure. After this implant is completed, the resist is stripped.

implanted with an energy of about 50-keV and a dose of about 1×10^{13}-B-atoms/cm². When this implant is completed, the resist is stripped and another layer of resist is applied. *Mask #3* is then used to expose the resist and define the regions where the *n*-wells are to be formed (Fig. 4-17). The required *n*-dopants are introduced into these regions by implanting phosphorus. The implant is performed at 80-keV and a dose of $\sim 1 \times 10^{13}$ cm⁻², using a medium-current implanter.

Next the well-dopants are diffused into the substrate using a long, high-temperature drive-in step (e.g., 1100–1200°C for 3–20 hours). At the conclusion of the drive-in step (Fig. 4-18), the dopant concentra-

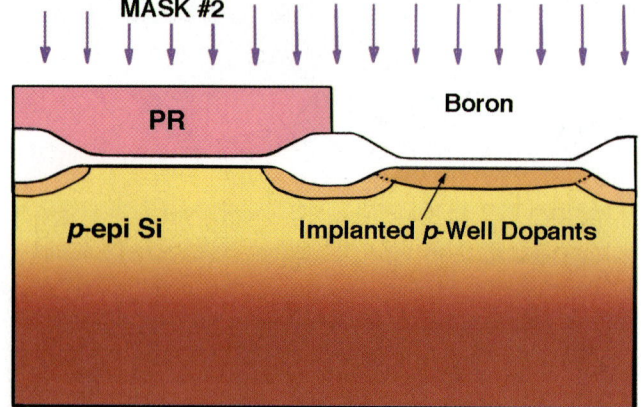

Fig. 4-16 Photoresist is applied, and *Mask #2* is used to cover regions where PMOS-devices will be built. A boron-implant provides doping for the *p*-wells of the twin-well CMOS structure. After the implant is completed, the resist is stripped.

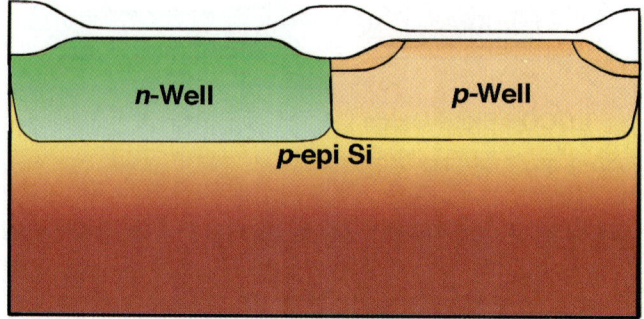

Fig. 4-18 A high-temperature drive-in step completes the formation of the *n*-wells and *p*-wells.

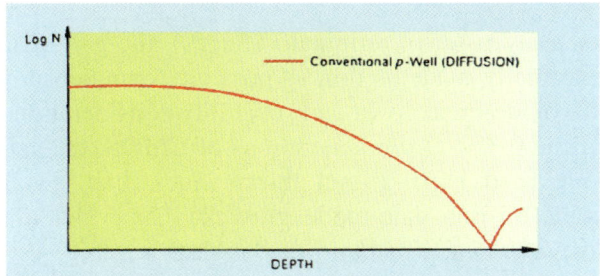

Fig. 4-19 Doping profile in a conventional-well formed by a shallow implant, then a long, high-temperature drive-in step.

tion at the Si-surface in the wells is ~1×10^{16}/cm^3 for the *p*-well, and ~3×10^{16}/cm^3 for the *n*-well. With the isolation-regions and well-regions in place, the wafer is now ready for *active-device* fabrication.

Note that when wells are formed in this manner, the well-doping profile is at a maximum at the surface and then monotonically decreases with depth (Fig. 4-19). Well surface-doping-concentrations should be at least an order of magnitude higher than the epi-layer doping (so the expected variations in epi-doping do not affect the well-concentration). The junction-depth of the wells is about ~2-μm deep. Finally, we observe that the thin pad-oxide is retained over the active-regions in preparation for the next process-step.

4.2.4 Threshold-Adjust Implantation Step

The V_T-adjust implant steps set the threshold-voltage of both the NMOSFETs and PMOSFETs, typically

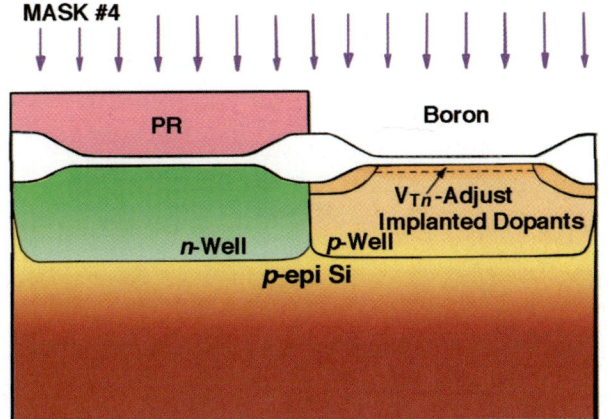

Fig. 4-20 Photoresist is applied, and *Mask #4* defines the NMOS transistors. A B-implant adjusts the *n*-channel-V_T (V_{Tn}). After this implant, the resist is stripped.

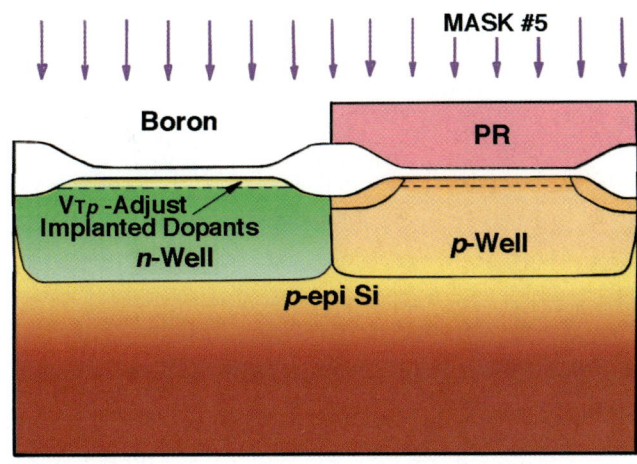

Fig. 4-21 Photoresist is applied, and *Mask #5* is used to define the PMOS transistors. A boron-implant adjusts the *p*-channel-V_T (V_{Tp}). After the implant is completed, the resist is stripped.

to values of 0.6-V and –0.6-V, respectively. The V_T-value for each type of MOSFET is set separately, and thus two masking-steps are needed. Figure 4-20 shows the step used to adjust the NMOS V_T. *Mask #4* is used to expose the resist and define the regions where the NMOS devices are located. After developing, boron ions are implanted with an energy of 50-75-keV and a dose of about 10^{12} B-atoms cm^{-2} to adjust V_{Tn}. The energy is chosen to be high-enough to get the implant-dose through the thin surface-oxide, but low-enough to keep the boron near the surface.

When this V_{Tn}-implant is completed, the resist is stripped and another layer of resist is applied in preparation for doing the PMOS V_T-implant. *Mask #5* is used to expose the resist and define the regions where the PMOSFETs are located (Fig. 4-21). Usually a B-implant is again used, with a different dose and energy than was used to adjust V_{Tn}. (The reasons why a B-implant is used for these CMOS generations is discussed further in Refs. 1 and 2.) After completing this V_{Tp}-implant, the resist is stripped, and the pad-oxide on the active-regions is etched away.

A separate *punchthrough-prevention implant* for NMOS-devices is also usually needed, especially when the minimum feature size decreases below about 1-μm (see Ref. 3). The punchthrough-implant can be

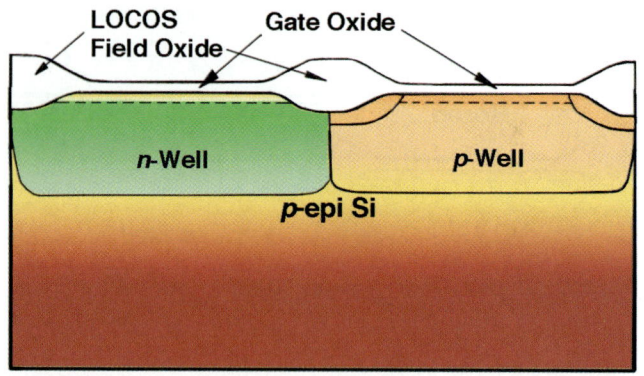

Fig. 4-22 The silicon wafer surface after: (a) etching back the pad-oxide to bare silicon; and (b) growing the *gate-oxide*.

performed sequentially using the same implanter, along with the NMOSFET V_T-adjust implant.

4.2.5 Gate-Oxide Growth

In the next step, the *gate-oxide* is formed. The growth of the gate-oxide is critical. A defect-free, very-thin (6-20-nm), high-quality oxide without contamination is essential for proper device operation. The gate-oxide is grown only on the exposed *active-regions* (following a careful cleaning of the wafer surface just prior to an oxide growth-process in dry-O_2).

Since the drain-current in an MOS transistor is inversely proportional to the gate-oxide thickness (for a given set of terminal voltages), the gate-oxide is normally made as thin as possible (commensurate with oxide-breakdown and reliability considerations). For the 0.35-1.0-μm generations of CMOS, the gate-oxide used is 10-20-nm thick. (See Chap. 13 for details on thin gate-oxide growth.) The wafer after this *gate-oxide growth process* is shown in Fig. 4-22.

4.2.6 Polysilicon-Deposition and Patterning

Once the gate-oxidation step is completed, a heavily *n*–doped polysilicon-gate structure is fabricated. Polysilicon is the preferred gate material for several reasons. First, it can withstand the high-temperature steps required to form the source/drain junctions. Second, the poly/SiO_2 interface is well understood and electrically stable. The gate-formation procedure begins with the deposition of a 0.4–0.5-μm-thick, undoped polysilicon-film by LPCVD (Fig. 4-23, see also Chap. 16). This layer is then doped with phosphorus by ion-implantation or chemical-doping, producing a film with a sheet resistance of 20–30-Ω/sq.

The gate-structure (and polysilicon-interconnect-structures) are then patterned using *Mask #6*. Following exposure and development of the resist, the polysilicon-film is dry-etched (Fig. 4-24). This is also a critical etch-step for several reasons. First, due to the self-aligned nature of silicon-gate technology,

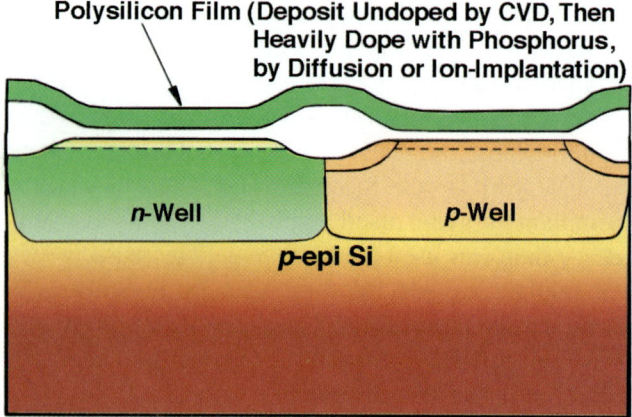

Fig. 4-23 A layer of undoped-polysilicon is next blanket-deposited using LPCVD. Diffusion or ion-implantation is used to heavily-dope the polySi layer with phosphorus.

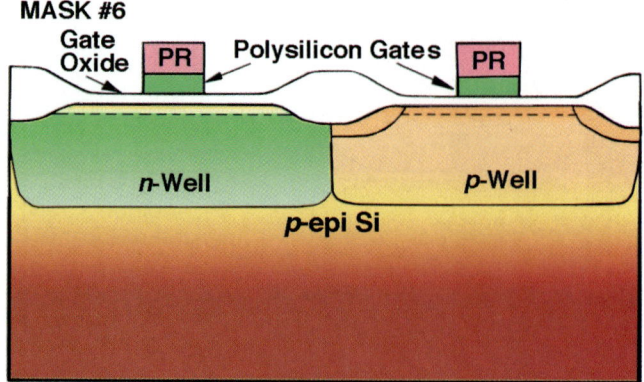

Fig. 4-24 Photoresist is applied, and *Mask #6* is used to define the gates of the MOSFETs in the polysilicon-film. An anistoropic polySi dry-etch-step defines their gatelength. Following this etch, the resist is stripped.

the channel-length of the device depends on the width of the polysilicon-line. Hence, the gate-length dimension must be precisely maintained across the entire wafer (and from wafer-to-wafer). If the gate-length is too long, the drain-current of the MOSFETs will decrease, slowing-down the performance of the IC. If the gate-length is too short, the source and drain may *punch-through*. In addition, the profile of the etched poly-gate structure should be vertical. This prevents variation of channel-lengths (due to penetration of the ions of the thinner regions of the gate sidewalls during formation of the source/drain regions by ion-implantation). Finally, to achieve the above goals, an anisotropic polysilicon-etch-process must be employed. This process, however, requires overetching to remove the locally thicker regions of polysilicon that exist wherever it crosses steps on the wafer surface (see Fig. 22-13). During this *overetch-time*, areas of the thin gate-oxide are exposed to the etchants. Thus, it is necessary to use a polysilicon etch-process that is highly-selective with respect to SiO_2 (see Chap. 22).

4.2.7 Formation of Source/Drain Regions

The next step in the process-flow is the formation of source and drain (S/D) regions of the MOSFETs. Such regions are paths of current-flow in the silicon

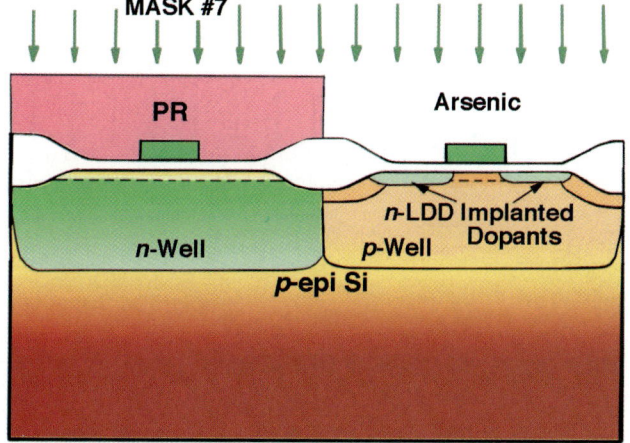

Fig. 4-25 Photoresist is applied, and *Mask #7* is used to cover the regions where PMOSFETs exist. A shallow arsenic-implant provides the doping for the *lightly-doped-drain* (*LDD*) regions of the NMOSFETs.

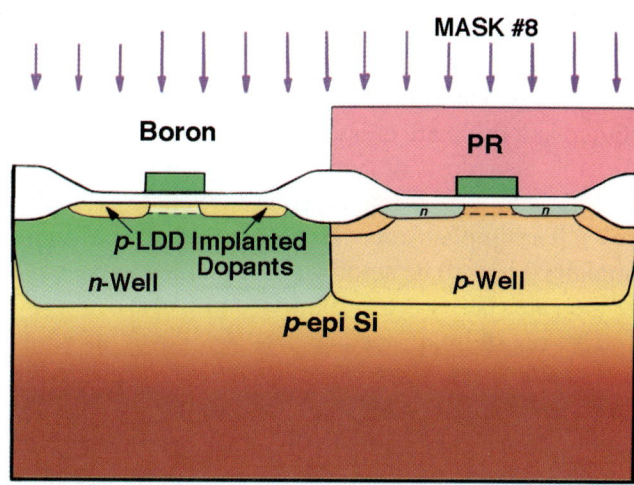

Fig. 4-26 Photoresist is applied, and *Mask #8* is used to cover the regions where NMOSFETs exist. A shallow boron-implant provides the doping for the *lightly-doped-drain* (*LDD*) regions of the PMOSFETs.

between the metal interconnect-lines and the channel of the transistor. As such, it is important that they have the lowest possible resistance. In addition, submicron-MOSFETs require S/D junctions to be as shallow as possible to suppress such short-channel effects as punchthrough (see Ref. 2). To obtain low resistivity, the S/D regions are doped as heavily as possible (typically using ion-implantation, with a dose on the order of ~10^{15} cm^{-3}). The NMOS S/D regions are doped with arsenic because it has a high-solubility, low-diffusivity, and a shallow projected-range at the low ion-implantation energies that are used. Boron or BF_2^+ are used to dope the PMOSFET S/D regions. However, boron has a higher-diffusivity in Si than does arsenic. Hence, shallow S/D-junctions in PMOSFETs are harder to achieve than in NMOSFETs.

In submicron-CMOS processes, gate-lengths become so small that *lightly-doped-drain* (*LDD*) structures must be used to minimize hot-electron effects, especially in NMOS devices. Thus, procedures are integrated into the CMOS process-flow to fabricate such LDD-structures. If LDD-structures are needed for both PMOS and NMOS devices, two more masking-layers are needed.

The LDD-structures are formed in the following way. Resist is spun on, and *Mask #7* is used to protect all the devices except the NMOS-transistors (Fig. 4-25). The lightly-doped regions of the NMOS source and drain are created with an ion-implant step. Arsenic at a dose of approximately 3×10^{13}–3×10^{14} cm^{-3} dopant-ions is implanted at low-energy (30–50-keV). The implant-process causes the edge of these implanted ions to be automatically aligned to the edge of the gate (i.e., it is a *self-aligned process*).

The resist is stripped and a new layer of resist is spun-on, and *Mask #8* is used to protect all the devices except the NMOS transistors (Fig. 4-26). The lightly-doped regions of the PMOS source and drain are created with an ion-implant step. Boron at a dose of between 3×10^{13}-3×10^{14}/cm^2 ions is implanted at low-energy (30–50-keV), and the resist is stripped.

A conformal layer of dielectric material (usually SiO$_2$ or silicon-nitride) is then deposited over the entire wafer (Fig. 4-27). An anisotropic-etch process is used to clear the oxide in the flat areas while leaving *spacers* on the sidewalls of the poly gates (Fig. 4-28). These spacers cover and protect the regions beneath them from the subsequent *high-dose implants* that form the rest of the S/D regions.

A photoresist layer and *Mask #9* is used to define the areas where the heavily-doped regions of the NMOS source and drain will be located. (Figure 4-29). These are formed with a heavy-dose arsenic implant-step (dose = 2×10^{15}-4×10^{15} As-ions/cm^2 at 40–80-keV). A photoresist layer and *Mask #10* is then

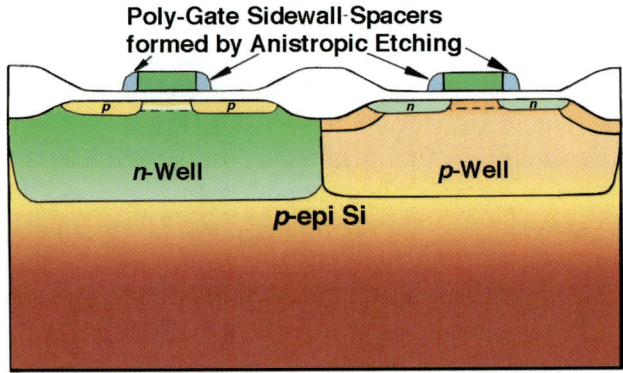

Fig. 4-28 The deposited dielectric-layer is etched-back anisotroppically, leaving sidewall-spacers along the edges of the polysilicon gate-structures.

used to define the areas where heavily-doped PMOS source/drain regions will be located (Figure 4-30). These are formed with a heavy-dose boron implant-step (dose = 1×10^{15}-3×10^{15} B-ions/cm^2 at 50-keV).

In the final step of the active-device formation process, a furnace anneal (typically at ~900°C for 30 min, or a rapid-thermal-anneal [RTA] for ~1 min at 1000-1050°C) is carried out. This thermal-step activates all the implants, anneals the implant-damage, and drives the junctions to their final depths.

4.2.8 Formation of TiSi$_2$ Salicide

After stripping the thin-oxide on the active-Si regions and cleaning the wafer surface (to ensure that no

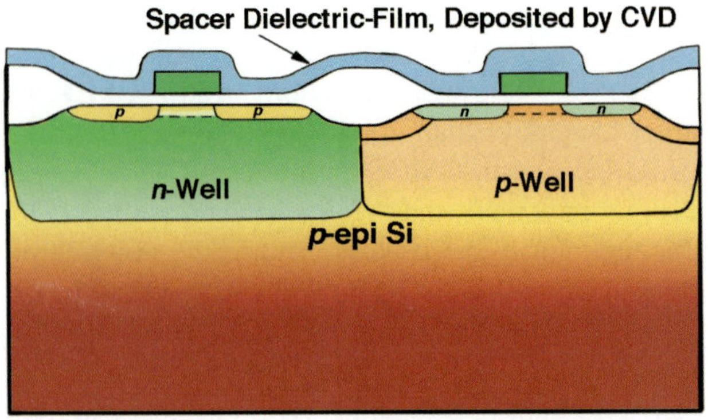

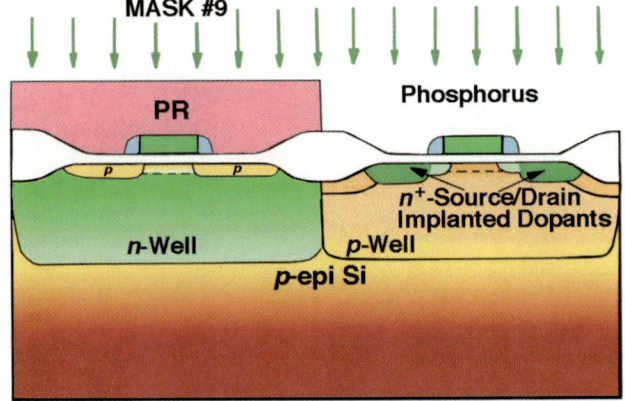

Fig. 4-27 A conformal layer of SiO$_2$ or SiN is deposited by CVD in preparation for the sidewall-spacer formation.

Fig. 4-29 After applying resist, *Mask #9* is used to cover the regions with PMOSFETs. A phosphorus-implant is used to form the n^+-source/drain regions of the NMOSFETs.

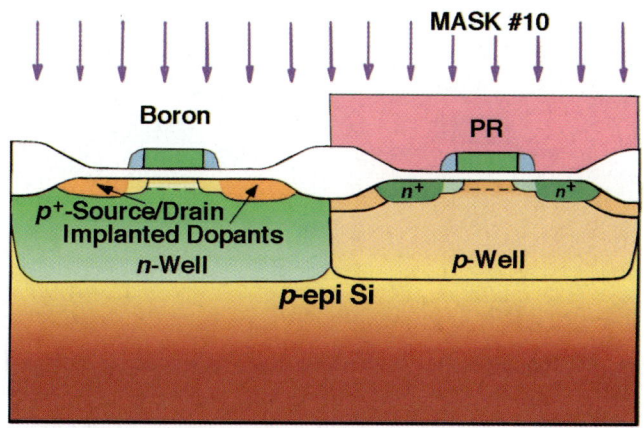

Fig. 4-30 After applying resist, *Mask #10* is used to cover the regions with NMOSFETs. A boron-implant is then used to form the p^+-source/drain regions of the PMOSFETs.

native-oxide exists on the exposed silicon of the active regions and on the polysilicon-gates, Fig. 4-31), a thin-layer of titanium (Ti) is deposited by sputtering (50-100-nm thick, Fig. 4-32). The next step makes use of two chemical reactions. The wafers are first heated to 600°C in an N_2-ambient for a short time (about 1-min). At this temperature, the Ti reacts with Si where they are in contact to form $TiSi_2$. $TiSi_2$ forms a low-resistance-contact to silicon, and it is also an excellent conductor. The wafer is then immersed in a solution of $NH_4OH:H_2O_2:H_2O$ (1:1:5) to selectively remove the unreacted Ti (i.e., on regions of the wafer covered with SiO_2), but the $TiSi_2$ remains in regions over Si where it had reacted (Fig. 4-33). After this

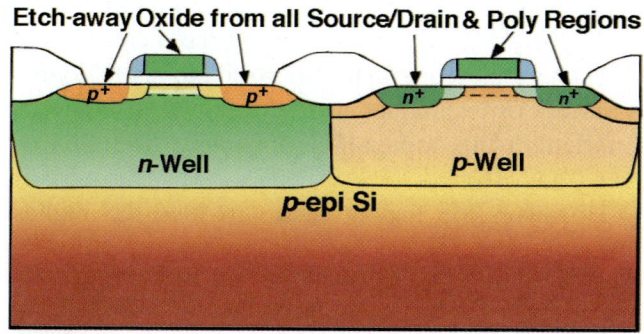

Fig. 4-31 An etch-step is used to remove the thin-oxide layer that covers the source/drain and gate-poly regions of all MOSFETs, in preparation for salicide formation.

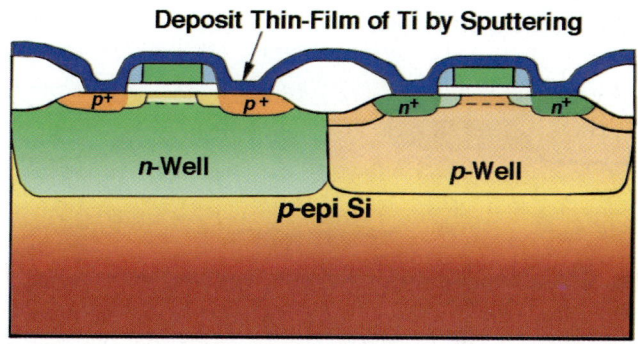

Fig. 4-32 A thin-layer of titanium (Ti) is blanket-deposited over the wafer-surface by sputtering.

Ti-etch-step, the wafer is heated to 800°C for about 1-minute in N_2 (to reduce the resistivity of the $TiSi_2$-layer to its final value of ~1 Ω/sq). Figure 4-34 shows the details of the *Ti-salicide* formation process.[5,6]

4.2.9 Premetal-Oxide Deposition and Planarization, and Contact-Formation

Following formation of the source and drain regions and the Ti-salicide structure, a doped dielectric-film is deposited by CVD. This layer is known as an *interlevel dielectric* (*ILD*). Contact-windows are opened in this dielectric-layer to allow electrical connections to be made between Metal-1 and the following structures:

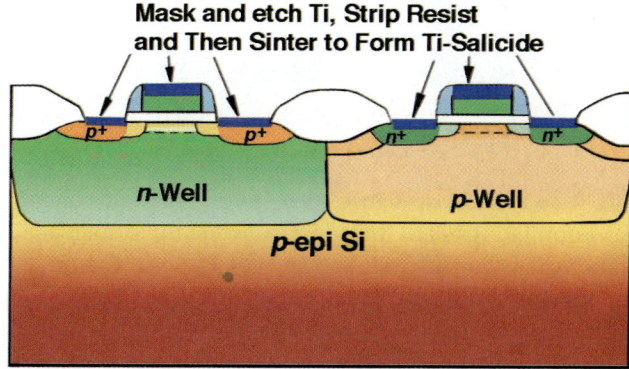

Fig. 4-33 A lower-temperature anneal-step (600°C, 1-min RTP in N_2) converts the Ti to $TiSi_2$ on regions where Ti is contact with Si. Elsewhere, Ti does not react, and can therefore be selectively removed with a wet-etch step. A second, higher-temperature anneal-step (800°C, 1-min RTP in N_2) converts the $TiSi_2$ to its final, lower-resistance form.

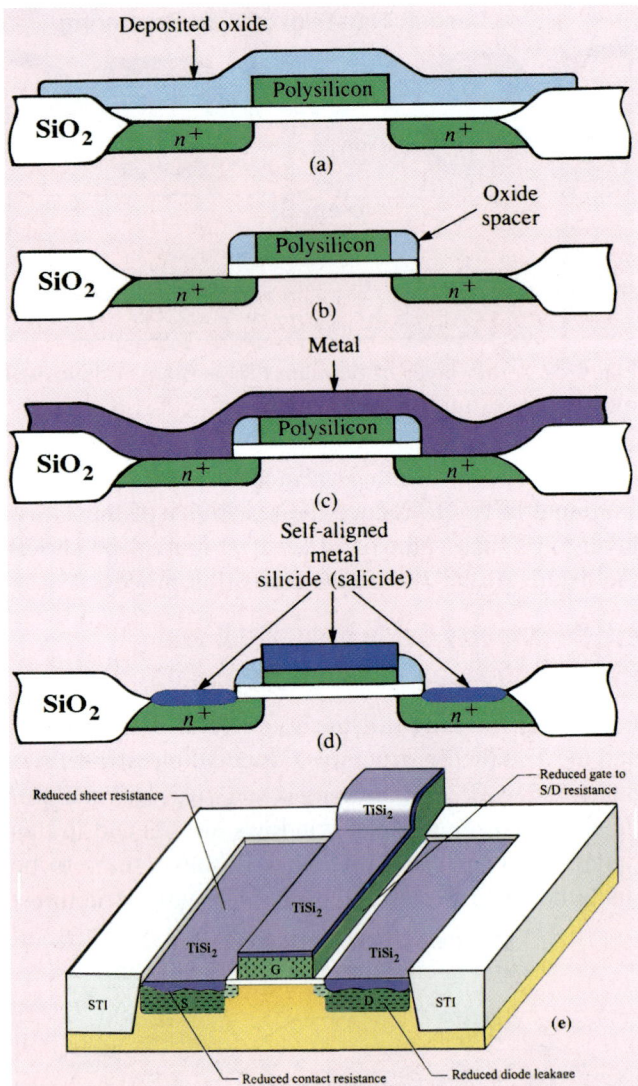

Fig. 4-34 Process sequence used to form titanium-salicide; (a) Form the standard MOSFET structure up to source and drain; (b) Form sidewall-spacers; (c) Deposit Ti-film and react to form TiSi$_2$ in regions where Ti is in contact with Si; (d) Selectively remove unreacted Ti film; (e) a perspective drawing of the final salicide-structure.

1) source/drain contact regions: **2**) gate contacts; **3**) substrate-contact regions; and **4**) well-contact regions. A CVD-process is used to deposit this doped SiO$_2$-film (about 1-μm thick), onto the wafers (see Chap. 16 for details of this process). The dopant in the SiO$_2$

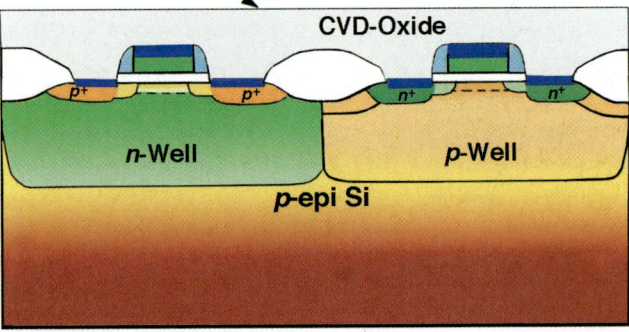

Fig. 4-35 A conformal interlevel-dielectric layer (typically CVD-SiO$_2$) is deposited. Then chemical-mechanical polishing (CMP) is used to planarize the steps of the ILD.

is either phosphorus (P - in which case the material is referred to as *phosphosilicate glass* or *PSG*), or both P and B (*borophosphosilicate glass* or *BPSG*).

The doped CVD SiO$_2$-layer plays several roles in the fabrication and operating aspects of the circuit. First, it acts as an *insulating-layer* between polySi and Metal-1. Second, it reduces the *parasitic-capacitance* between Metal-1 and the substrate. Third, adding P to the glass makes the layer an *excellent getter of Na ions* (contamination by Na can destabilize the V$_T$ of a MOSFET). The PSG (or BPSG) binds otherwise mobile Na-atoms in the doped-glass layer, preventing them from reaching the gate-oxide and altering V$_T$.

Note that the surface of the wafer after the ILD deposition is highly nonplanar. For the sake of reducing potential problems with metal discontinuities, it would be preferable not to deposit the metal directly on such rough topography. To avoid having to do this, a variety of techniques were developed to *planarize* (or flatten) this topography, including *BPSG-reflow* (see Chap. 16).

However, the method that provides the highest level of planarity is *chemical-mechanical polishing* (*CMP*), and is described here (see also Chap. 23). The ILD-layer is deposited thicker than the largest steps present on the wafer surface (that is, thicker than ~1-μm). The wafer is then placed face-down in a CMP-tool, and its upper-surface is polished-flat using a high-pH silica-slurry. This CMP-process results in a structure that is shown in Fig. 4-35.

Contact-openings are next created by a lithography-and-etch step (Fig. 4-36). After applying resist, *Mask #11* is used to define contact-patterns in a photoresist-film. A dry-etch process is then used to open the contact-windows through the ILD to the underlying polySi and the source/drain regions in the silicon.

This contact-opening step can be critical, as the contact size and its alignment to underlying patterns limit the minimum-size of the device. The source/drain regions must also be large enough for the contact to fit, with an allowance for alignment tolerance. If the contact opening exposes a part of the substrate, the drain or source will be shorted to it. Likewise, if the contact-opening overlaps the both the source/drain and the gate, a short will be created between them.

4.2.10 Metal-1 Deposition and Patterning

After the contacts have been opened, the metallization layer is deposited. Because the metal-layer is highly conductive, it is used whenever possible to interconnect circuit-elements and to carry large amounts of supply-current. Metal interconnect-lines must have sufficient thickness, width, and step-coverage to keep the current-density in each line below the level that

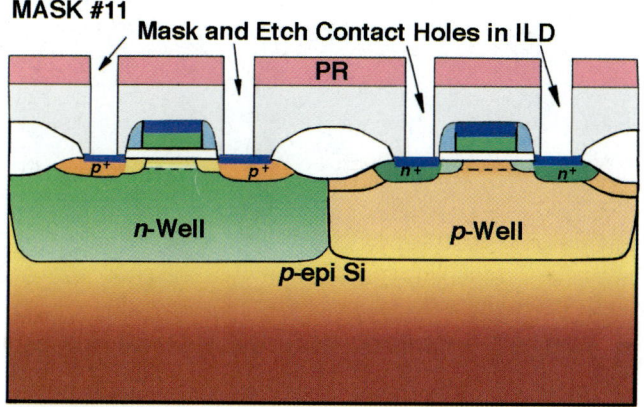

Fig. 4-36 After applying resist, *Mask #11* is used to pattern the contact-opening regions. A dry-etch step is used to anisotropically etch the ILD layer to allow connections to be made to the silicon-substrate (and polysilicon layers).

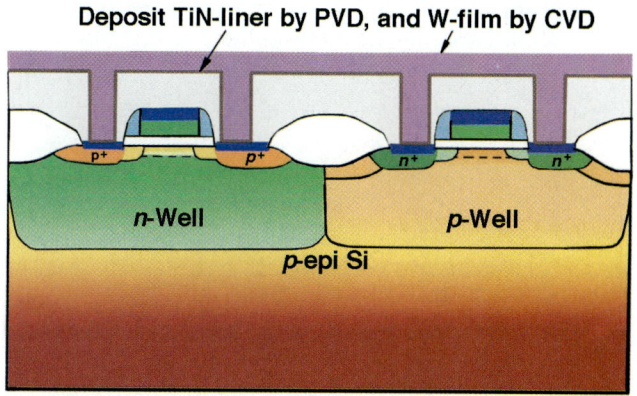

Fig. 4-37 To form the tungsten-contact holes, an adhesion-layer of titanium-nitride (TiN) is first blanket-sputter-deposited onto the wafer. Then a CVD-process is used to blanket-deposit a film of tungsten. This W-film completely fills the contact-holes.

could produce electromigration failure (see Ref. 2). In addition, the space between adjacent metal-lines must be large enough so that the lines never touch, even under worst-case process variations.

The metallization-structure is formed using two separate processes: **1)** *W-plug formation*; and **2)** *main interconnect-line formation*, using an Al:Cu film. In the W-plug formation-process a *thin barrier/glue layer* of Ti/TiN is first blanket-deposited by sputtering (a few tens-of-nm thick). It provides good adhesion to the SiO_2 and other underlying materials, as well as serving as an effective barrier-layer between the upper and lower metal-layers. The next step is deposition of a blanket-W layer by CVD, as shown in Fig. 4-37. CMP is then used to planarize the wafer surface, and to remove the W and Ti/TiN everywhere but in the contact-holes. Thus, W-plugs are formed, as shown in Fig. 4-38.

The main-interconnect lines are formed from Al:Cu-alloy films that are deposited next by sputtering. A resist-layer and *Mask #12* are used to define these main interconnect-line patterns (Fig. 4-39), which are then formed using a dry-etching process (Fig. 4-40). Note that Cu is replacing Al:Cu in some of the most advanced IC-technologies, because of its higher-conductivity and better electromigration resis-

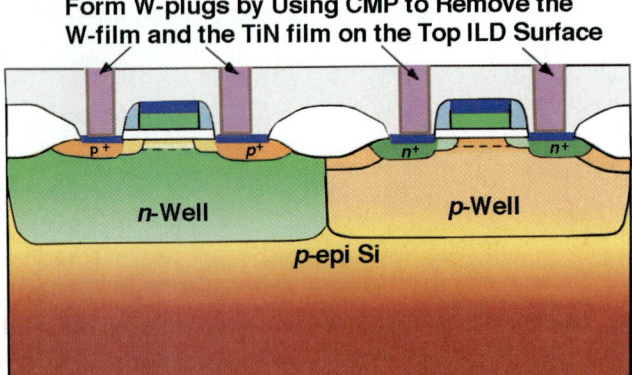

Fig. 4-38 The W and TiN that is deposited on the top surface of the ILD is removed by a CMP process, leaving just W-plugs.

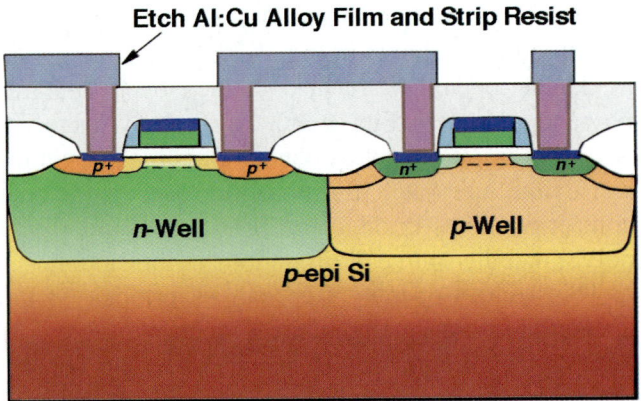

Fig. 4-40 The Al:Cu Metal-1 lines are created by using an anisotropic Al dry-etch process. The resist then stripped.

tance (see Chap. 24). Figure 4-41 shows a SEM of Al:Cu-interconnect-lines (Metal-1) and W-plugs.

If a single-level of metal is used in the CMOS process, a *sintering-step* occurs after the metal has been patterned. This step brings the metal and the n^+ and p^+ regions in the silicon into intimate contact. Such intimate contact between the metal and the heavily-doped Si regions establishes low-resistance ohmic contacts. The annealing-process also exposes the wafer to a 375–500°C temperature in an H_2, or N_2 + H_2 (5%) ambient for about 30 minutes. This step also serves as the annealing-process for reducing the interface-trap-density in the gate-oxide that was introduced by earlier processing steps (see Chap. 13).

4.2.11 Intermetal-Dielectric Deposition, Via-Patterning, and Metal-2 Deposition and Etch

Most modern ULSI technologies use more than one level of wiring on the wafer surface. This is because in complex circuits it is usually very difficult to completely interconnect all the devices in the circuit without using multiple interconnect-levels. The processes used to deposit and define each level are similar to those we described for Metal-1. Here we show a two-

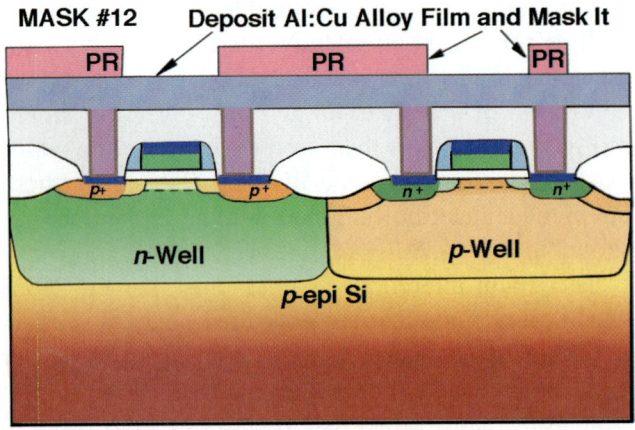

Fig. 4-39 An Al:Cu alloy film is sputter deposited onto the wafer. Photoresist is applied and *Mask #12* is used to define the Metal-1 lines that be formed from the Al:Cu film.

Fig. 4-41 A SEM of a Metal-1 interconnect-structure formed with W-plugs and Al:Cu interconnect-lines. Note that the ILD has been removed to make it possible to see the W-plugs. Photograph courtesy of ChipWorks.

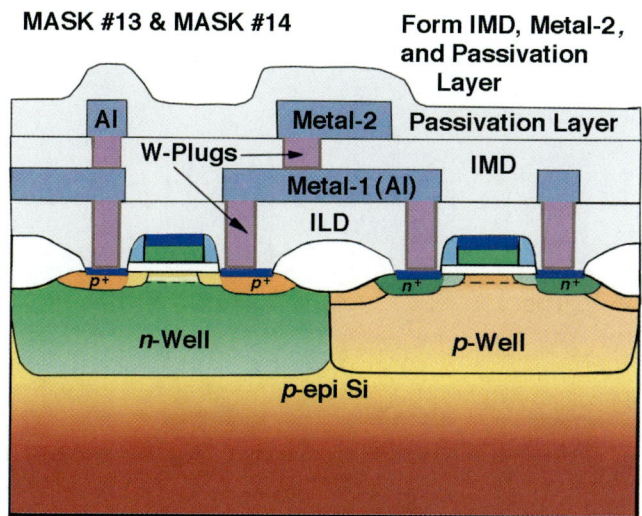

Fig. 4-42 Metal-2 is formed with the same steps as those shown in Figs. 4-35 to 4-40. *Mask #13* is used to define the via-holes between Metal-1 and Metal-2. *Mask #14* is used to define Metal-2. A final passivation-layer (typically CVD-oxide, or CVD-nitride - or both) covers the chip.

level-metal interconnect structure, but the faceplate of the chapter shows a cross-sectional drawing of an IC with 6-levels of metal.

To create the second interconnect-level, an *intermetal dielectric* (*IMD*) must first be deposited. It electrically isolates the Metal-1 layer from the Metal-2 layer. Next, vias must be opened in this IMD-layer so that electrical connections can be established between Metal-2 and Metal-1 at desired locations (*Mask #13*).

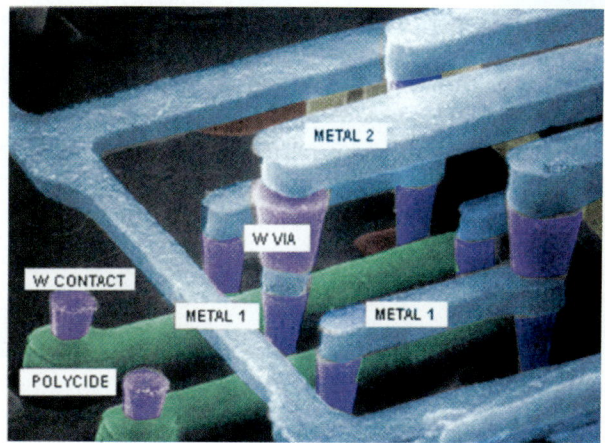

Fig. 4-43 SEM of an IBM circuit showing a two-level-metal interconnect structure formed with W-plugs and Al-lines. Photograph courtesy of Chip Works.

W-plugs and an Al:Cu-layer and a lithographic-step using *Mask #14* are used to form the Metal-2 interconnect-structures (Fig. 4-42). Figure 4-43 shows a SEM of a two-level-metal structure formed using Al-lines and W-plugs.

As noted above, more advanced ICs now use more than 2-levels of metal interconnects. Figure 4-44 shows a SEM of a four-level interconnect-structure. More details about fabricating such multilevel interconnect-structures are found in Chap. 24 and Ref. 3.

4.2.12 Passivation Layer and Pad Mask

Finally, a *passivation* (or *overcoat*) layer, such as CVD-PSG or plasma-enhanced-CVD silicon-nitride

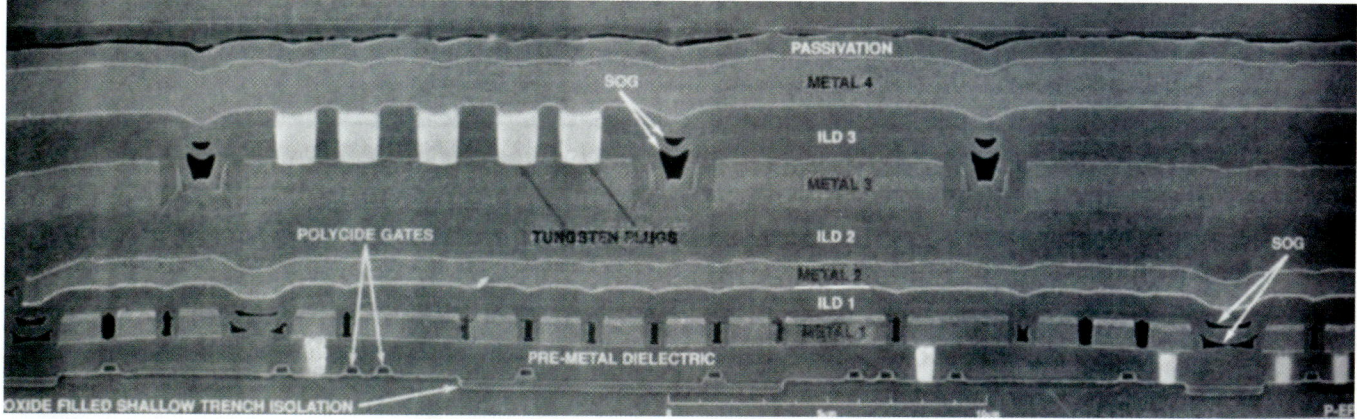

Fig. 4-44 SEM of 4-level-metal interconnect structure. Courtesy of ChipWorks.

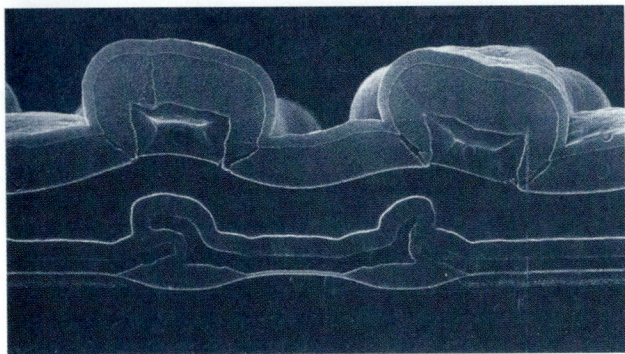

Fig. 4-45 Passivation layer SEM. Courtesy of ChipWorks.

(or both), is put down onto the wafer surface (Fig. 4-45). This layer seals the device structures on the wafer, protecting them from contaminants and moisture. It also serves as a scratch-protection layer.

Openings are etched into this layer so that a set of special metallization-patterns under the passivation layer is exposed. These metal-patterns are normally located in the periphery of the circuit and are called *bonding-pads* (Fig. 4-46). Bonding-pads are typically about 100x100-μm in size and are separated by a space of 50 to 100-μm. Wires are connected (bonded) to the metal of the bonding-pads and are then bonded to the chip-package. In this way connections are established from the chip to the package-leads (see Chap. 27 for details on chip-bonding).

The bond-pad openings are created by patterning the passivation-layer with *Mask #15*. If a PSG-layer is used, the phosphorus (2-6-wt%) in the glass not only causes it to act as a getter for Na, but also prevents the film from cracking. Care must be taken to ensure that not more than 6%-phosphorus is incorporated into the PSG, as this can cause corrosion of the underlying metal if moisture enters the circuit-package (see Chap. 22). When silicon-nitride is used, care must be taken to ensure that the deposited nitride-film exhibits low-stress (either tensile or compressive), so it will not crack. Cracking would compromise the sealing capability of the final passivation-film.

4.3 SHALLOW-TRENCH ISOLATION (STI)

Despite advances made to decrease bird's beak penetration in LOCOS isolation, such techniques

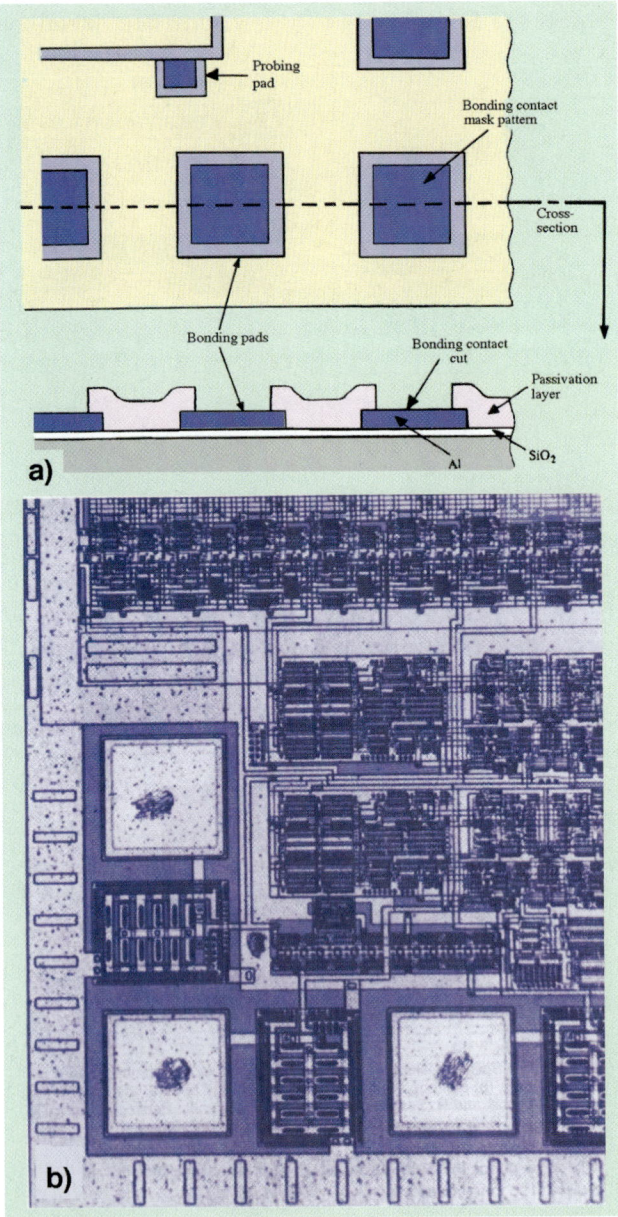

Fig. 4-46 (a) Drawing, and (b) SEM of *bonding-pads*.

eventually became inadequate for deep-submicron IC-technologies. The successor to LOCOS in CMOS is *shallow-trench isolation* (*STI*).[7,8] Note that STI is not only is a different type of isolation-structure but it is also formed prior to the well-structures. In addition, STI provides a planar surface for further processing.

The sequence of steps for forming STI-structures

begins with the growth of a pad oxide, followed by an LPCVD nitride layer. Then resist is applied and a Mask is used to pattern the STI trench-openings. The nitride and pad-oxide are etched first. Then the trench is anisotropically dry-etched to a depth of 400–500-nm (Fig. 4-47a). The trench-etching process should yield smooth trench-sidewalls with angles of 70–85°, rounded bottom-corners (to minimize stress), and a residue-free silicon-surface after etch. After the trenches are etched, the resist is stripped and a thin thermal-oxide is grown on the trench walls (Fig. 4-47b). Next, a CVD dielectric-film is used to fill the trench. This layer also covers the areas of the wafer where the nitride remains (Fig. 4-47c). A CMP step is used to polish back the dielectric-layer until the nitride is reached (the nitride acts as a CMP-stop layer, Fig. 4-47d). The dielectric material is densified at about 900°C. Then the nitride is stripped, leaving the STI structure in place (Fig. 4-47e). Figure 4-47f shows a cross-section of a CMOS-inverter built using

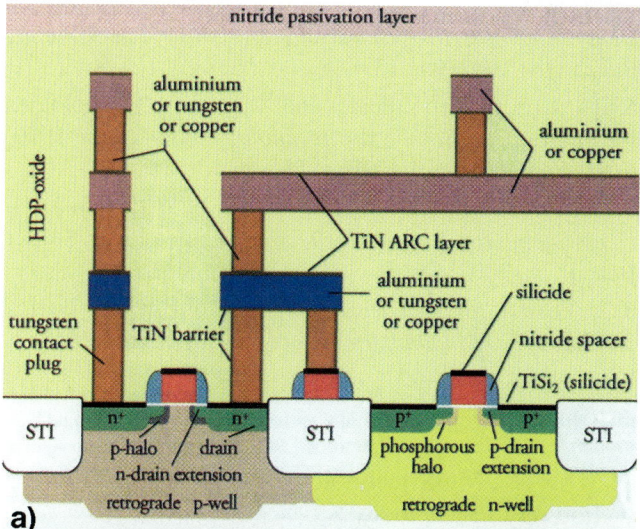

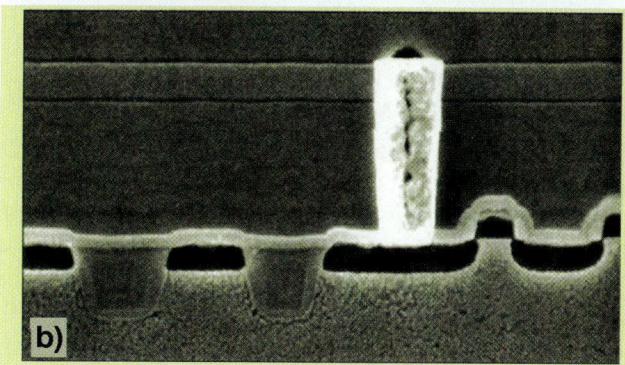

Fig. 4-48 (a) Schematic of STI integrated into a complete CMOS IC technology. (b) SEM of an STI structure.

STI. In Fig. 4-48a, we see how an STI-structure is integrated in a completed CMOS-technology, and Fig. 4-48b shows a SEM of an actual STI-structure.

REFERENCES

1. S. Wolf, *Silicon Processing for the VLSI Era - Vol. 2: Process Integration,* Lattice Press Sunset Beach CA 1990.

2. S. Wolf and R.N. Tauber, *Silicon Processing for the VLSI Era, Vol. 1 - Process Technology,* 2nd Ed., Lattice Press, Sunset Beach CA 2000, Ch 16.

3. R.J. Baker, H. Li, and D. Boyce, *CMOS Circuit Design, Layout, and Simulation,* IEEE Press, New York, 1998.

4. E. Kooi, *The Invention of LOCOS,* IEEE Press, New York, 1991.

5. S. Wolf, *Silicon Processing for the VLSI Era, Vol. 4 -*

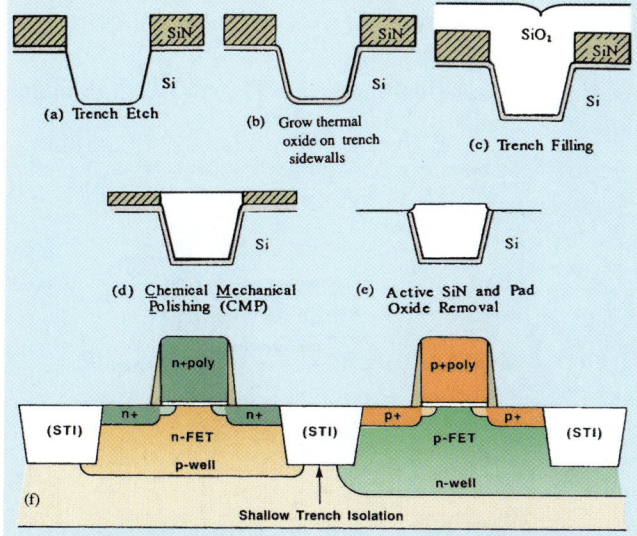

Fig. 4-47 Process-sequence used to form *shallow-trench isolation* structures (*STI*): (a) Etch trench in silicon substrate; (b) Grow thermal-oxide trench-liner to improve trench Si/SiO$_2$ interface (and for trench corner-rounding); (c) Fill trench with CVD oxide; (d) Use CMP to etch-back the oxide to the nitride layer; (e) Strip nitride to leave the STI structure; (f) Schematic of a CMOS inverter fabricated with STI.

Deep-Submicron Process Technology, Lattice Press, Sunset Beach CA, 2002. Ch. 13.

6. J.A. Kittl et al., "Salicides and Alternative Technologies for Future ICs," *Solid State Technology,* Part-I, June 1999, p. 81; Part-II, August 1999, p. 55.

7. S. Wolf, *Silicon Processing for the VLSI Era, Vol. 4 - Deep-Submicron Process Technology,* Lattice Press, Sunset Beach CA, 2002. Ch. 9.

8. L. Peters, "Choices and Challenges for Shallow Trench Isoolation," *Semicondutor International,* April 1999, p. 69.

PROBLEMS

1. Why is CMOS technology the most dominant IC technology in use today?

2. When ICs were first introduced, the main technology used to build them was based on bipolar transistors. Today MOS transistors rule. Why did the transition from bipolar to MOS technology occur?

3. Draw the layout of an NMOS and CMOS NAND gate. (Use the layout of the CMOS-inverter gate shown in Fig. P4-1 as a guide.) Use the same scale for both gates. Label your layouts. Estimate roughly how much more area is required for the CMOS NAND gate than for the NMOS-gate.

4. Why is the <100>-orientation preferred for the starting wafers used to fabricate silicon ICs today?

5. What problems can arise if too-thin a field-oxide is used when building MOS-based ICs?

6. Why is the formation of the gate-oxide in MOS technology such an important step, and why is control of this oxide thickness so critical.

7. What is the difference between the forming the source/drain regions of a MOSFET with an LDD structure and a single-drain structure? Why did LDDs get incorporated in MOSFET device structures?

8. Explain the self-aligned silicide (*salicide*) process. Why is cobalt-silicide used instead of titanium-silicide when IC feature-sizes get smaller than 0.25-μm? (You may need to consult some external references to answer the latter question.)

9. What materials are used for the final passivation-layer on ICs?

10. Silicon-nitride is a good diffusion barrier material and it is used in such applications as a final passivation-layer, as sidewall-spacer material, and as a dielectric-layer in DRAM capacitors. Yet, it is never used as a dielectric-layer between metal-layers in interconnect-structures. Why not?

11. Assume you wanted to fabricate a simple diffused-resistor consisting of a long, rectangular *p*-type region in an *n*-type substrate, with contacts formed on either end (see Fig. 3-2): (a) What are the minimum number of masking steps needed to fabricate such a structure?; (b) What would the dimensions (length *l*, and width *w*) of a 500-Ω resistor have to be if the sheet-resistance of the *p*-region is 50-Ω/square, and the minimum feature-size is 1-μm? (Ignore the resistance at the contacts.)

12. Sketch a process-flow that would result in the single-well CMOS-structure shown in Fig. P4-1. The drawings in this chapter can serve as a guide. You only need to describe the flow up through the stage at which the active device formation starts, since from that point on, the process is similar to that given.

13. Describe the main differences between the formation of LOCOS and STI.

14. Explain the advantages of STI over LOCOS isolation structures.

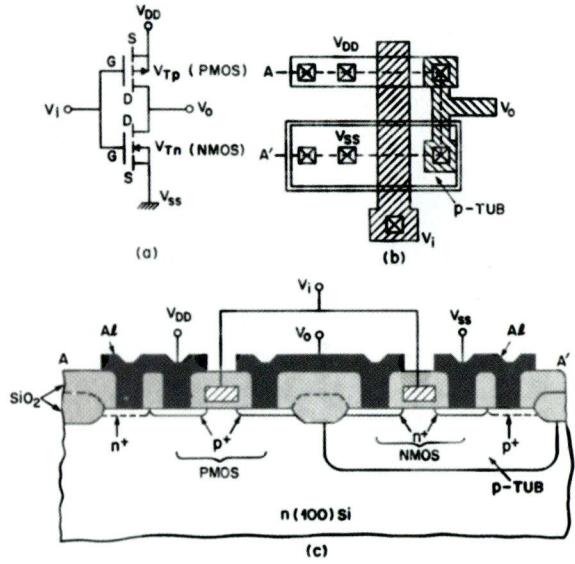

Fig. P4-1 (a) Circuit schematic. (b) Layout. and (c) Cross-section of a single-well (*p*-well) CMOS inverter.

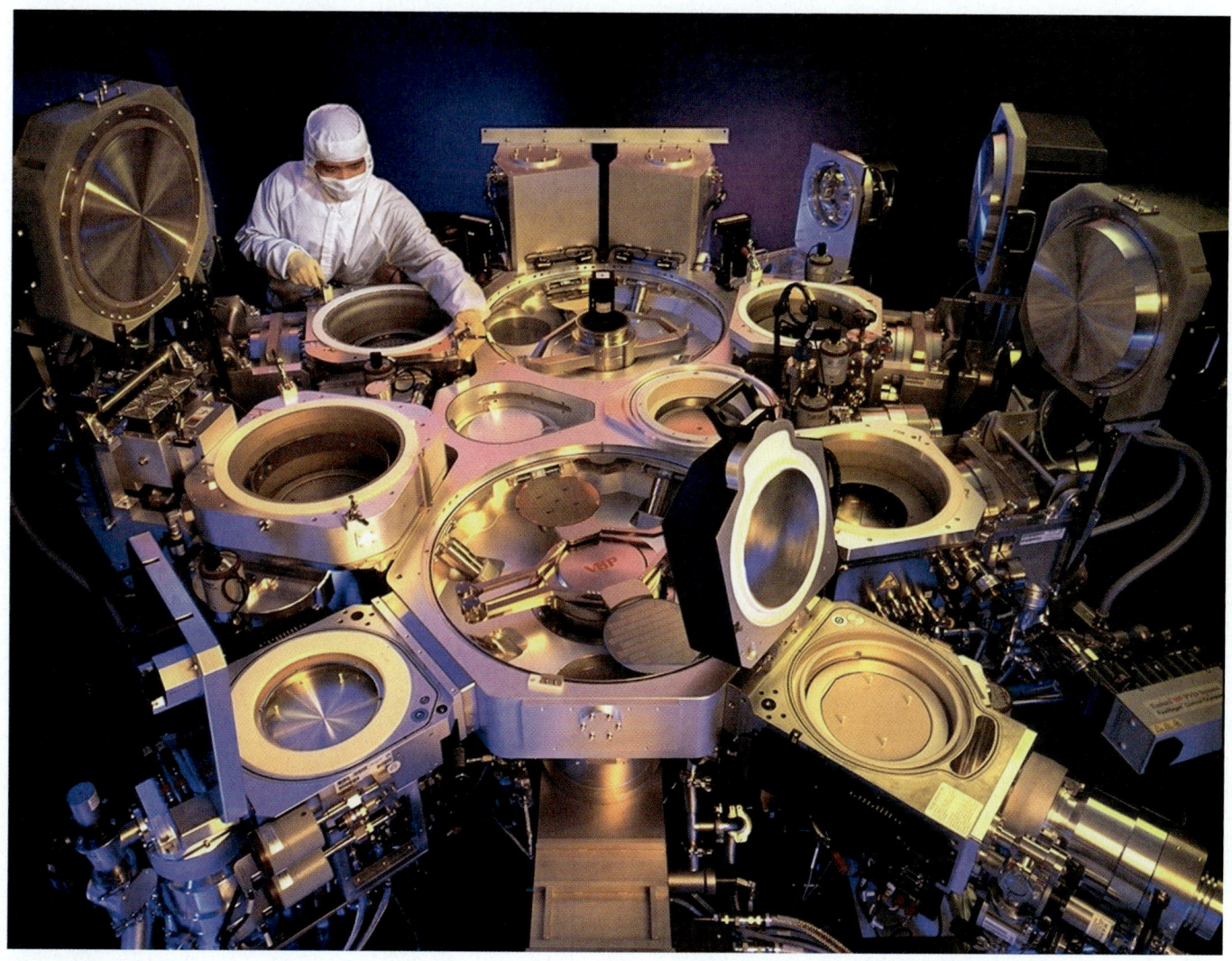

Cluster tool for integrated metallization process. Photograph courtesy of Applied Materials, Inc.

CHAPTER 5

GASES, LIQUID-CHEMICALS, AND ULTRAPURE-WATER FOR ULSI

CHAPTER CONTENTS

5.1 BASIC PROPERTIES OF MATERIALS

5.2 LIQUID-CHEMCIALS USED IN IC FABRICATION

5.3 GASES USED IN IC FABRICATION

5.4 ULTRAPURE WATER (UPW)

"If you're not busy bein' born,
   You're busy dyin."
                    Bob Dylan

Ultra-pure chemicals are needed to manufacture ULSI and VLSI microchips. Photograph courtesy of Musashino Chemical Laboratory Ltd., Tokyo, Japan.

The manufacture of integrated-circuits involves many chemical-processes. A variety of chemicals of ultrahigh-purity (called the *process-chemicals*) are introduced to the surface of silicon-wafers during these processes, and chemical-reactions occur to form the IC-features. Chemicals are also used to clean the wafers. Here the liquid and gaseous chemicals used in IC-fabrication are introduced. Also discussed is the important process-material *ultrapure-water* (*UPW*).

5.1 BASIC PROPERTIES OF MATERIALS

There are a number of basic-characteristics of matter that are useful to define prior to discussing the gases and liquids used in IC-fabrication. These are covered in this section.

5.1.1 Temperature

Temperature is a property of matter associated with its "hotness" or "coldness," as measured on some defined scale. It is also a measure of the thermal-energy (atomic or molecular) of a substance. The transfer of such energy between material-objects represents *heat-flow*. Temperature can be perceived by touch, but is measured more precisely by instruments called *thermometers* (Fig. 5-1).

Wafer-fabrication uses many processes that are carried out at high-temperatures (in the range of 400-1200°C). Some examples of why such high-temperatures are needed include: **1)** they allow the rate of chemical-reactions that drive processes to be increased; and **2)** they permit atoms in solid-silicon to move about and rearrange their crystal-structure (perhaps after being damaged by an ion-implantation process).

There are three temperature-scales: the *Celcius-scale* in °C (also often called *centigrade*), the *Fahrenheit-scale* in °F, and the *absolute-temperature* (or *Kelvin*) scale in K. Fig. 5-2 shows how these scales are related.

The *Celcius-scale* is often used in scientific work, and is based on the changes of state of water. It sets 0°C as the freezing-point of water,

MICROCHIP MANUFACTURING

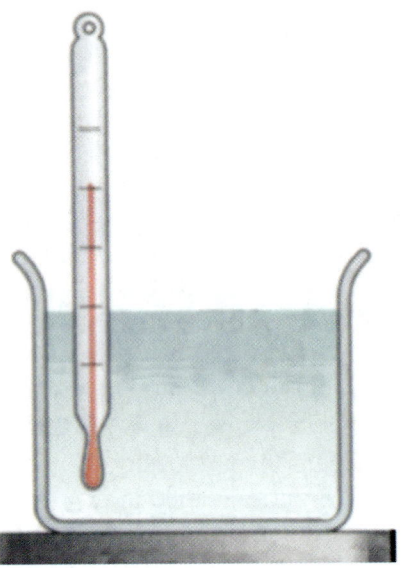

Fig. 5-1 Drawing of a mercury thermometer, used to measure temperature.

5.1.2 The Thermal-Expansion of Materials

When a material is heated it expands, due to the increase in the vibrational-energy of its atoms. Thus, the size of a body increases as it gets hotter. For example, a tight metal-jar-lid can often be loosened by holding it under a stream of hot-water. The metal-lid expands slightly more than the glass-jar as its temperature rises. Likewise, when a substance is cooled, it contracts. An object cooled to its initial-temperature returns to the size it was prior to heating.

However, not all materials expand to the same extent when heated. That is, some expand more than others. The extent to which a material expands is called its *coefficient-of-thermal-expansion,* or *CTE*, expressed for example, in terms of units of $(°C)^{-1}$. Figure 5-3 shows the CTEs of some of the materials used in semiconductor-devices.

The phenomenon of thermal-expansion is important in IC-fabrication, especially when two different materials are in contact. One example is when an aluminum-film is deposited on a silicon-wafer. When the wafer is heated, the aluminum (which has a higher CTE than Si) would like to expand more than is allowed (because it is held tightly to the silicon-wafer, Fig. 5-4). Thus, the Al-film comes under stress in this

and 100°C as its boiling-point (when measured at the pressure of one-atmosphere at sea-level). There are exactly 100 Celcius-degrees between the two points.

The *Fahrenheit-scale* is based on the behavior of a specific water-and-salt solution. In the particular salt-water solution chosen by the developer of this scale (Gabriel Fahrenheit), the freezing-point of water corresponds to 32°C, and its boiling-point to 212°F, with 180-degrees between the two points.

To convert between Celcius and Fahrenheit:

$$°C = 5/9 \ (°F - 32) \quad (5.1)$$

$$°F = 9/5°C + 32 \quad (5.2)$$

The *Kelvin-scale* is the third, and it is the base-unit of temperature for the metric-system. The temperature of 0 K is *absolute-zero,* and it corresponds to −273.15°C. The Kelvin-scale uses the same scale-factor as the Celcius-scale. Thus, in the Kelvin-scale, water freezes at 273 K and boils at 373 K. *Room-temperature* is about 300 K. To convert between Kelvin and Celcius;

$$K = °C + 273 \quad (5.3)$$

$$°C = K - 273 \quad (5.4)$$

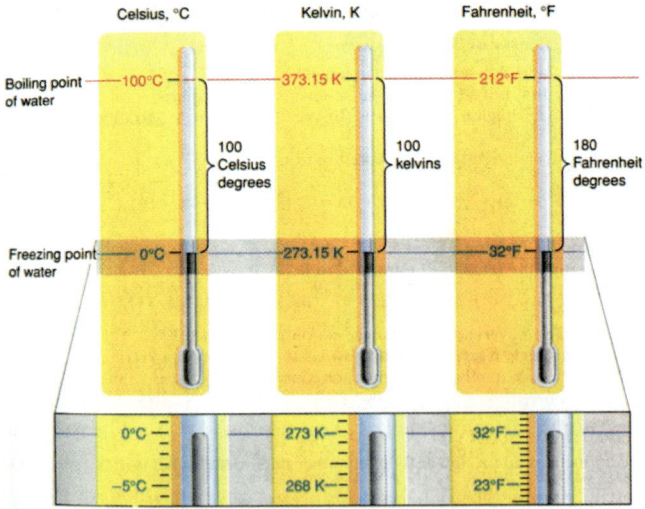

Fig. 5-2 The freezing- and boiling-points of H_2O in Celcius, Kelvin (absolute), and Fahrenheit temperature-scales.

CHAPTER 5   GASES, LIQUID CHEMICALS, AND ULTRAPURE WATER FOR ULSI

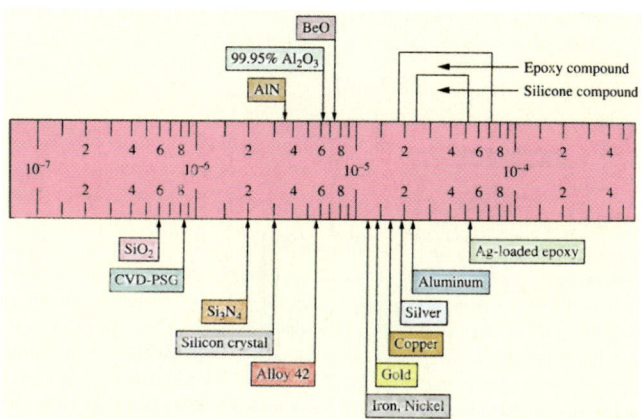

Fig. 5-3 The temperature-coefficients-of-expansion (TCE) of materials used in semiconductor devices (in °C^{-1}).

situation. Excessive-stress will cause a thin-film to undergo deformation, which is an undesirable result.

5.1.3 Density

The *density* of a material is defined as its mass divided by its volume:

$$\text{density} = \text{mass/volume} \quad (5.5)$$

An example of units commonly used to express the density is g/cm^3.

A dense-material is considered as being *heavy*. If two objects have the same volume, the denser one will be heavier. For example, a steel ball-bearing is denser than a cork. Water is the standard for density, having a density of 1-g/cm^3. Silicon has a density of 2.3-g/cm^3, making it denser than water.

5.1.4 Specifying the Purity of Chemicals Used in IC-Fabrication

All chemicals used in ULSI manufacture must be extremely-free of such contaminants as particulates, metallic-ions, or unwanted-chemicals. The purity of chemicals is a relative-property of the chemical, depending on the application. However, for semiconductor-manufacture, a unit that is commonly used to describe small-quantities of an impurity in a liquid- or gaseous-chemical is *parts-per-million* (*ppm*), which is measured either in terms of volume or mass. In fact, the chemicals used in wafer-fabrication are the pur-

est form of such chemicals sold. They have their own designation: *electronic-grade* or *semiconductor-grade chemicals*.

As processes have entered the deep-submicron-regime, chemical-purity-specifications have become even more stringent. The chemicals used for these advanced-processes are termed *ultra-high-purity chemicals*, in which the levels of contaminants are controlled to less than *parts-per-billion* or even *parts-per-trillion*.

5.2 LIQUID-CHEMCIALS USED IN IC-FABRICATION

Liquids can be *pure-substances* (such as pure-water), or they can be *mixtures* (such as *wine* - a mixture of water and alcohol and other grape-substances). If a chemical-mixture is so thoroughly blended that the molecules or atoms of the components are the same throughout (i.e., it is *homogeneous*), this called a *chemical-solution*. Gasoline (a chemical-mixture of hydrocarbons and additives) is a chemical-solution. Another example is the first-aid antiseptic hydrogen-peroxide. In fact, this product only contains about 5% hydrogen-peroxide (H_2O_2), with the remainder being water. The major-component of the solution (the one present in greatest quantity) is called the *solvent*, while the minor-components (i.e., the dissolved-substances) are called the *solutes*. Solutions in water (such as hydrogen-peroxide [H_2O_2] in H_2O) are referred to as *aqueous-solutions*.

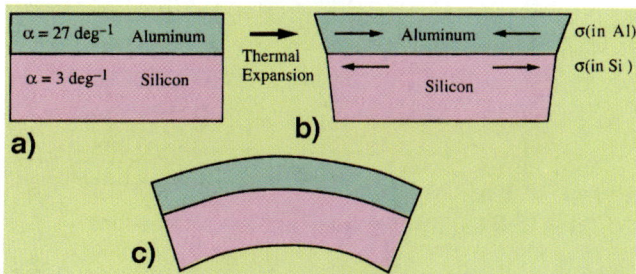

Fig. 5-4 Thermal-expansion differences between Al and Si, shown in (a), lead to stresses in the Al thin-film and Si substrate, shown in (b). In this case, tensile-stress develops on top of the Si-substrate and compressive-stress develops in the Al thin-film. (c) This drawing shows the bending of the structures (exaggerated) that occurs due to these forces.

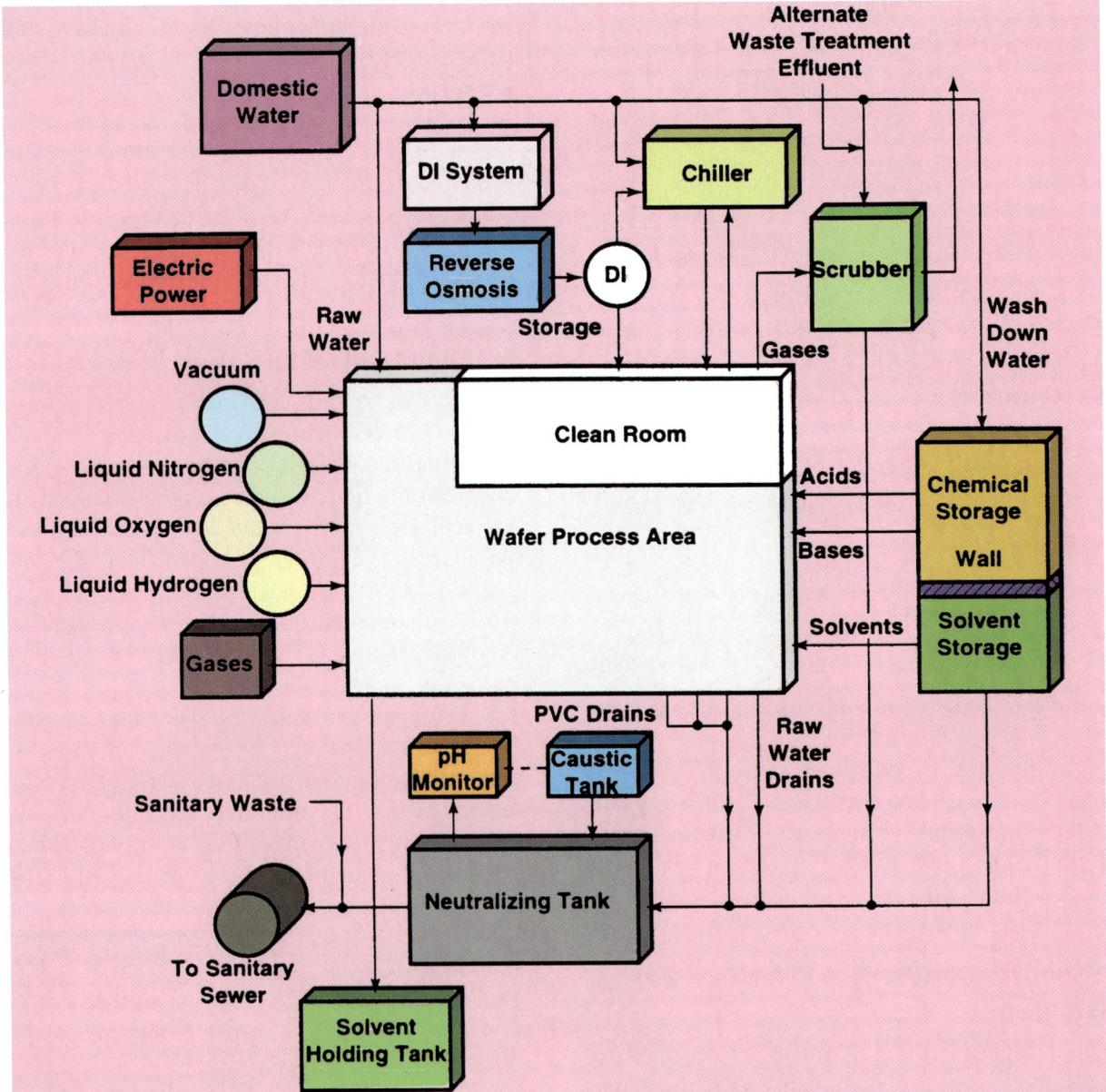

Fig. 5-5 Utility flow-diagram for a wafer-fab.

Liquid-chemicals used in IC-fabrication can be classified into three groups:

1. Acids
2. Bases
3. Solvents

Note that both acids and bases are reactive with skin and other chemicals, and should thus be stored and handled with all the prescribed-safety-precautions. Figure 5-5 shows a schematic of how the cleanroom of a wafer-fab is supplied with the liquid-chemicals and gases needed to manufacture ICs.

5.2.1 Acids Used in IC-Fabrication

In pure-water (H_2O) there are small (but equal) concentrations of hydrogen-ions (H^+) and hydroxide (or hydroxyl) ions (OH⁻). These ions arise from the partial-ionization of water. An *acid* is defined as a substance that, when dissolved in water, increases the concentration of hydrogen-ions over that which exists in pure-water. For example, HCl dissolved in water breaks up (dissociates) into:

$$HCl \rightarrow H^+ + Cl^- \qquad (5.6)$$

The fact that an acid contributes additional hydrogen-ions to an aqueous-solution is often acknowledged in it's chemical formula. That is, acids usually (but not always) have an H as the first elemental-symbol (i.e., nitric-acid [HNO_3], or hydrofluoric-acid [HF]).

A *base* is defined as a substance, when dissolved in water, increases the concentration of hydroxyl-ions that would exist in pure-water. For example, sodium-hydroxide (NaOH) dissolves in water according to the equation:

$$NaOH \rightarrow Na^+ + OH^- \qquad (5.7)$$

and is a strong-base. The fact that a *base* contributes additional hydroxyl-ions to an aqueous-solution is acknowledged in the chemical-formula of many bases (i.e., they have an OH in them, such as NaOH).

Acids and bases are commonly found in the home, with lemon-juice and vinegar being acids, while ammonia and baking-soda in solutions with water are bases. There are also many types of acid used in semiconductor-manufacturing. Table 5.1 lists some of the most common ones, as well as an example application. Note that there are *inorganic* and *organic* acids. Organic-acids contain hydrocarbons, while inorganic-acids do not. Acetic-acid (CH_3COOH) is an example of an organic-acid used in IC-manufacturing.

5.2.2 Bases Used in IC-Fabrication

Bases are compounds that have an OH-group in their chemical-formula. Examples include KOH (potassium-hydroxide) and NH_4OH (ammonium-hydroxide). Their molecules dissociate in water to produce the OH⁻ ion (hydroxyl). Thus, a base is an aqueous-solution in which the OH⁻-concentration is higher than in pure-water. A base is also often referred to as an *alkaline-substance* (or an *alkali*). Table 5-2 lists some of the common-bases used in IC-manufacture, together with a sample application.

5.2.3 Solvents Used in IC-Fabrication

As noted earlier, a *solvent* is a material that is capable of dissolving another substance to form a mixture called a *solution*. A material that is referred to as a "good solvent" is one that will dissolve a broad-range of other substances.

Water is an excellent solvent of many substances, especially ionic-materials (such as table-salt, NaCl). Water-molecules dissolve ionic-compounds present in the water by separating the ions, then surrounding and dispersing them into the liquid. Water does this by overcoming the electrostatic force-of-attraction between the ions. Other solvents used in IC-manufacture, such as alcohol and acetone, are volatile and flammable. Common-solvents used in the wafer fab are listed in Table 5-3.

Table 5.1 Acids Used in Chip-Making

Hydrofluoric-acid	HF	Used to etch SiO_2-films and clean furnace-tubes
Buffered-Oxide-Etch	BOE	Used to etch SiO_2-films
Nitric-acid	HNO_3	Used in mixtures of HF & HNO_3 to etch silicon
Sulfuric-acid	H_2SO_4	Used in a mixture of H_2SO_4 and H_2O_2 to clean wafers
Hydrochloric-acid	HCl	Used to clean wafers
Phosphoric-acid	H_3PO_4	Used to etch SiN & Al
Acetic Acid	CH_3COOH	Used in Al-etch-baths

Table 5.2 Bases Used in Chip-Making

Ammonium-hydroxide	NH_4OH	Used in wafer-cleaning-processes
Potassium-hydroxide	KOH	Positive-photoresist developer
Tetramethyl-ammonium-hydroxide	TMAH	Positive-photoresist developer

Table 5.3 Solvents Used in Chip-Making

Ultrapure-Water (H_2O)	UPW	Widely used to rinse-wafers and dilute-aqueous chemical solutions
Isopropyl-Alcohol (C_3H_8O)	IPA	General-purpose cleaning-solvent
Trichloroethylene	(TCE)	Cleaning-solvent
Acetone		General-purpose-cleaner
Xylene		PR edge-bead-removal

5.2.4 The pH-Scale

Acids and bases may be strong or weak. Their relative-strength is indicated by the pH-scale. That is, the pH of a solution indicates whether it is acidic or basic, and how strong is this characteristic. The pH-scale ranges from 0 to 14, with 7 being the neutral-point. Water is neutral, being neither an acid nor a base. Acids have pH-values from 0 to below 7, and bases have values above 7 to 14. Strong-acids have pH values from 0 to 3 (for example, sulfuric-acid, H_2SO_4), and strong bases have pH-values approaching 14 (for example, NH_4OH). Fig. 5-6 shows some common-chemicals and their place on the pH-scale.

5.2.5 Liquid-Chemical Distribution

As we see, there are a large number of chemicals used in the manufacture of ICs (Fig. 5-8a). Many of them are toxic or otherwise hazardous. It must be possible to bring these chemicals in the desired-quantity from the storage-vessel to the work-area of the fab in a safe manner that preserves their purity. In advanced-wafer-fabs this is typically done using a *bulk-chemical-distribution* (*BCD*) system. (An introduction to safety-issues is found in Chap. 29.)

Such a BCD-system has storage vessels, chemical-delivery subsystems (consisting of filters, pumps, and mixing-units), and piping for bringing chemicals to the individual process-stations (Fig. 5-8b). Modern BCDs are integrated into a fully-computerized and networked system for real-time chemical-monitoring and control. The BCD-storage-vessels and chemical-delivery subsystems are housed in the space beneath the main production-floor of the fab.

Some chemicals may not be suitable for BCD. That is, they may have a limited shelf-life, or may be used in such small quantities that it is impractical to store them in a BCD-system. Such chemicals now utilize special packaging-systems designed for *point-of-use* (*POU*) delivery (which means they are stored and used at the process-station). The *photoresist* employed in photolithography is an example of such a chemical.

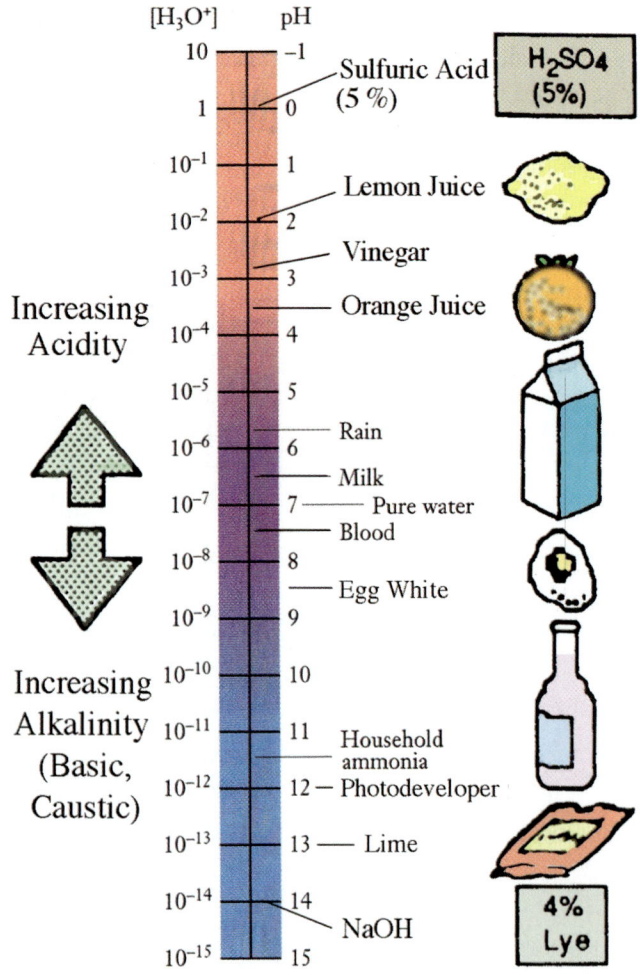

Fig. 5-6 Many everyday-substances are acidic or basic aqueous-solutions with a wide-range of pH values.

5.3 GASES USED IN IC-FABRICATION

There are over 50 types of gases used in the manufacture of modern-ICs. The types and quantities employed are always changing because new materials are continually being introduced into the IC-process-sequence.[2,3] Gases are classified into two groups: **1)** *bulk-gases*; and **2)** *specialty-gases*. Bulk-gases are used in larger quantities. The specialty-gases (also called *process-gases*) are used in smaller quantities, and there are many more kinds than the bulk-gases (see Table 5.5).

There are a number of important criteria that must be met by the gases used in semiconductor manufacturing, including high-purity and low-particulate levels. Bulk-gas purity is controlled to seven-nines-purity (99.99999% pure), and specialty-gases must be controlled to a purity better than four-nines (99.99%).[4] Contaminants that must be excluded from process-gases to extremely low-levels include oxygen, water, and metals. Particulates in process-gases must also not be allowed to exceed 0.1-μm in size. Finally, many of the process-gases are hazardous, and workers must be kept from being exposed to them. For all of the above reasons, these gases must be contained in a gas-delivery system that can distribute them in a safe, clean, and precise manner to the appropriate work-locations in the cleanroom (Fig. 5-7a).

5.3.1 Bulk-Gases Used in IC Fabrication

As noted above, the five most-common bulk-gases used in a wafer fab are N_2, O_2, H_2, Ar, and He. These can be grouped into three categories: **1)** inert-gas; **2)** oxidizing-gas; and **3)** reducing-gas. These groupings are given in Table 5-4, which also lists some applications of each type of bulk-gas.

5.3.2 Distribution of Bulk-Gases

Bulk gases are stored in large storage-tanks (or large 1000-lb tube-trailers) outside of the wafer fab (Fig. 5-7b). As mentioned earlier, the bulk-gases are supplied to the work-stations through a *bulk-gas distribution* (*BGD*) *system*. The advantages of using such a centralized distribution-system include lower-cost, less human labor, reduced likelihood of particulate-contamination, and better safety and reliability of the bulk-gas delivery.

Table 5.4 Bulk-Gases Used in Chip-Making

Inert Gases		Example Applications
Nitrogen	N_2	Purge gas-lines of water-vapor and residual-gas.
Argon	Ar	The glow-discharge gas used in PVD is argon-based.
Helium	He	Vacuum-leak checking. Also used as a process-gas in dry-etching.
Oxidizing Gas		
Oxygen	O_2	Process-chamber gas.
Reducing Gas		
Hydrogen	H_2	Carrier-gas in epitaxial-Si process. Also used in combination with O_2 to form H_2O in wet-oxidation.

5.3.3 Specialty-Gases Used in IC-Fabrication

There are many specialty-gases used in silicon IC-fabrication. Some of those most commonly used are listed in Table 5.5, with an example of the their use.

Table 5.5 Specialty-Gases of Chip-Making

Hydrides

Silane	SiH_4	Pyrophoric, flammable & explosive	Source of Si in deposition processes
Arsine	ArH_3	Extremely toxic	Dopant-gas for ion implanting arsenic
Phosphine	PH_3	Extremely toxic	Dopant-gas for ion-implanting phosphorus
Diborane	B_2H_6	Extremely toxic	Dopant-gas for ion-implanting boron

Silicon-Halogen Compounds

Silicon-tetrachloride	$SiCl_4$	Source-gas used in Si-deposition processes
Trichlorosilane	$SiHCl_3$	Source-gas used in Si-deposition processes

Dichlorosilane SiH$_2$Cl$_2$	Source-gas used in Si-deposition processes

Fluorinated Compounds

Nitrogen-fluoride NF$_3$	Etch-gas that produces fluorine-ions
Tungsten-hexafluoride WF$_6$	Tungsten-source-gas used in W-CVD processes
Carbon-tetrafluoride CF$_4$	Etch-gas that produces fluorine-ions
Tetrafluoromethane C$_2$F$_4$	Etch-gas that produces fluorine-ions
Silicon-tetrafluoride SiF$_4$	Source-gas for Si & F-ions used in etch & implant steps
Chlorinetrifluoride ClF$_3$	Etch-gas that produces fluorine and chlorine-ions
Sulfur-hexafluoride SF$_6$	Etch-gas that produces fluorine-ions

Acid Gases

Boron-trifluoride BF$_3$	Dopant-gas of boron-ions used when implanting-B
Boron-trichloride BCl$_3$	Etch-gas used in dry-etch-processes of Al-films
Chlorine Cl$_2$	Etch-gas used in dry-etch-processes of Al-films
Hydrogen-chloride HCl	Furnace-tube cleaning-gas
Hydrogen-bromide HBr	Etch-gas used in dry-etch processes of polySi-films
Boron-tribromide BBr$_3$	Etch-gas used in dry-etch-processes of polySi-films

Miscellaneous Specialty-Gases

Tetraethooxysilane TEOS Si(OC$_2$H$_5$)	Source-gas used in CVD of SiO$_2$-films
Phosphorus-oxychloride POCl$_3$	Source-gas of phosphorus for doping Si-wafers
Ozone O$_3$	Used together with TEOS for CVD of oxide-films
Ammonia NH$_3$	Used with dichlorosilane for CVD of silicon-nitride
Nitrous-oxide N$_2$O	Source-gas for growing nitrided-oxide-films
Nitric-oxide NO	Source-gas for growing nitrided-oxide-films
Carbon-monoxide CO	Used in etch-processes
Titanium-tetrachloride TiCl$_4$	Titanium-source-gas used in Ti-CVD processes
Hexamethyldisilazane (HMDS)	Used as an adhesion-promoter of PR on oxides
TDMAT Ti(N[CH$_3$]$_2$)$_4$	Used with ammonia in CVD of titanium-nitride (TiN)
Hfac(Cu)tmvs	Copper-source-gas used Cu-CVD processes

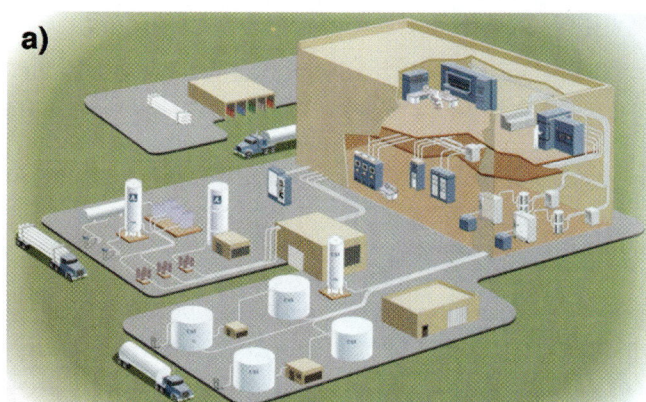

Fig. 5-7 (a) Schematic of a gas-management system for a wafer fab. (b) Bulk-gas tanks typically hold thousands of cubic-feet of such gases as oxygen, nitrogen, hydrogen, argon, and helium. Courtesy of Air-Liquide.

CHAPTER 5 GASES, LIQUID CHEMICALS, AND ULTRAPURE WATER FOR ULSI

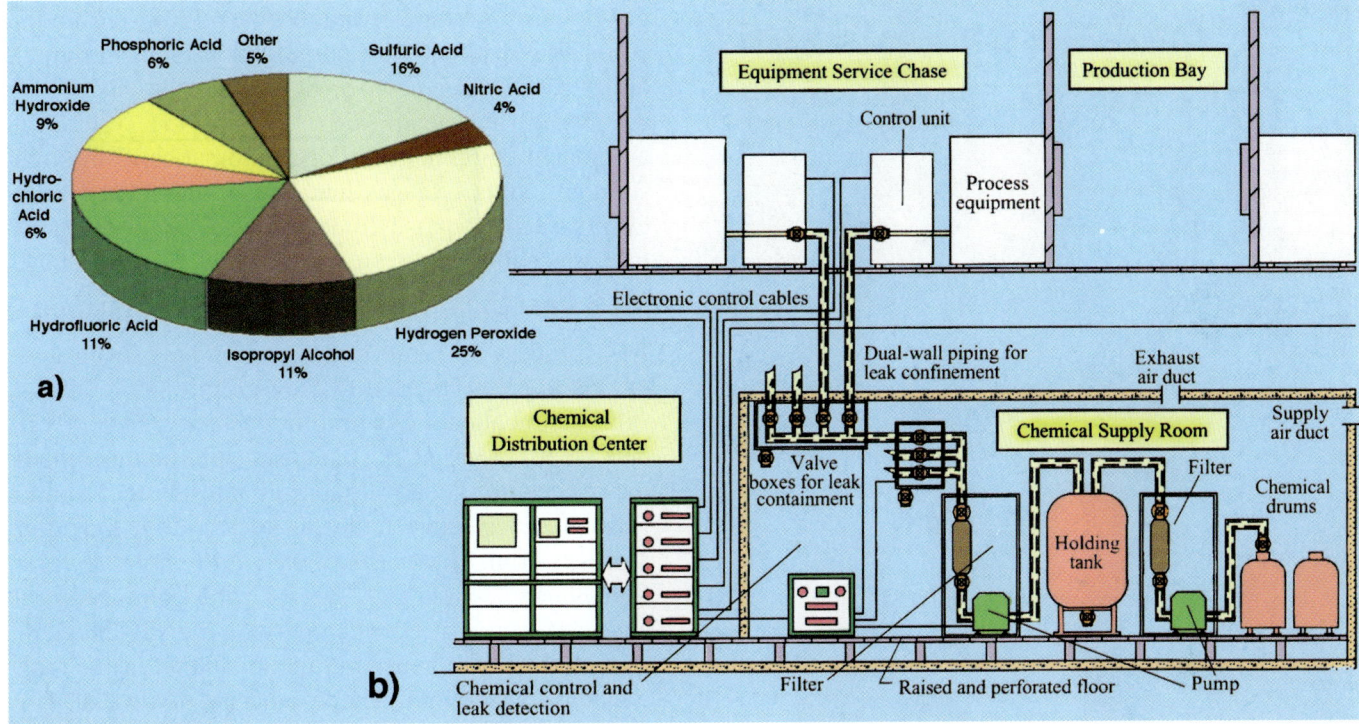

Fig. 5-8 (a) Percentage-breakdown of the most widely used chemicals in microchip-manufacturing. (b) Bulk-chemical distribution. Courtesy of Semiconductor International.

5.3.4 Gas-Sources and Delivery-Systems for Specialty-Gases

Specialty-gases (which are typically more hazardous than bulk-gases) are supplied and delivered to process-chambers with special gas-handling systems. Such gas-delivery systems must satisfy three important criteria: **1)** they must protect personnel and the environment from the hazardous properties of the reactant-gases; **2)** they must provide a steady and known flow of each gas; and **3)** they must minimize contamination in the process-gas stream.

To protect workers from toxic gases, gas-delivery systems must be able to detect such hazardous conditions as excess-pressure or leaks before they become catastrophic. In addition, automatic-shutdown of the system must take place if dangerous conditions are detected. Specialty-gas distribution systems should have multiple safety-features for this purpose, including excess-flow switches that detect any unusually high gas-flow and alert the system-controller, which may close the valve closest to the cylinder. Often, a flow-restricting orifice is installed at the cylinder to limit maximum-flow in case of valve failure.

To prevent corrosion by the reactive-gases they carry, these delivery-systems utilize electropolished 316L stainless-steel piping and valving (316L is a special type of stainless steel). They are welded in construction and all valves are pneumatically-operated for positive action to prevent system contamination due to leakage from outside the system. This piping can be hundreds of feet long with many welds, fittings, and bends, and is connected to the tool through a *gas-stick* (see Sect. 5.3.5) at the tool "drop."

The gases are transported to the fab and stored in 100-lb metal-containers called *gas-cylinders* (see Fig. 5-9) usually in *gas-cabinets* (Fig. 5-10) connected to an exhaust-duct to minimize personnel-exposure in the event of a leak. (The cylinders are usually strapped to the cabinet, to prevent them from falling over.) The

Fig. 5-9 (a) Specialty compressed-gas cylinders. (b) Cutaway view of a gas-cylinder valve. (c) Gas-flow regulator connected to a gas-cylinder.

cabinet has a control-panel with *regulators* to control pressure and flow, *shut-off valves*, and a *purge-panel* that enables the purging-cycles to be controlled when performing gas-cylinder changes. The cabinet also contains safety-equipment, such as fire-sensors and leak-detection sensors. The gas supply-lines downstream from the gas-cabinets are often double-walled, allowing any leaking gases to be directed into exhausted enclosures (Fig. 5-11).

Gases must also be properly purged from a gas system in certain situations. A *gas-purge* is a technique that flushes hazardous residual-gases, atmospheric-gases, or water-vapor from a process-chamber and the gas-delivery system. It involves replacing the undesirable-gas with an inert purge-gas (such as nitrogen) either by displacement, or by pulling the gas out of the system by vacuum-flow. Purging the system on automated-equipment is done automatically through software-control of gas-line valves before and after events such as gas-cylinder changes or opening of process chambers.

5.3.5 Gas-Stick, Gas-Panel, & Supply-Manifold

At a process-tool, the incoming gas-lines from gas-cylinders are connected to the tool through a special subsystem called a *gas-drop* or *gas-stick* (Figs. 5-12a and 5-12b). A gas-stick contains the plumbing components needed to control gas-flow into the tool and to flush the lines between the gas-stick and the tool after a process run. It consists of an *on/off valve*, a *mass-flow-controller* (*MFC*, Chap. 6), *filters* and *isolation* and *vent-valves* (used when purging the gas-lines).

Gas-sticks are housed in the *gas-panel* of the tool, which is located immediately adjacent to the tool. The gas-panel normally contains a number of gas-sticks, each one controlling one-type of incoming-gas. The number of incoming gas-lines to a process-tool depends on the process. Multi-chamber process-tools may have in excess of 30 incoming lines.

Premixing of the gases must be done upon (or just

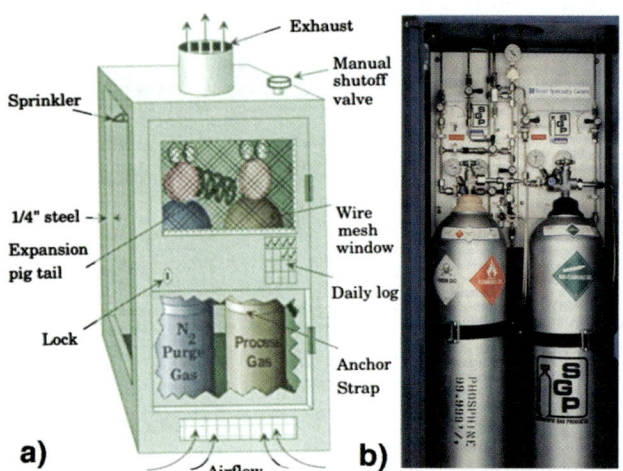

Fig. 5-10 a) Drawing of a gas storage-cabinet. b) Photograph of a gas-cabinet. Courtesy of Scott Specialty Gases.

before) entry into the reaction-chamber. This is done in the process tool itself, in the *supply-manifold* (see Fig. 16-6a). The supply-manifold is a chamber with multiple-openings through which incoming-gases are fed. After being well-mixed in the manifold, the gases are fed into the process chamber. Use of supply-manifolds is necessary because laminar flow-conditions are generally used in the chamber, and good gas-mixing does not occur under laminar-flow conditions.

5.3.6 Procedures for Changing Gas-Cylinders

When a specialty gas-cylinder is empty, it must be replaced with a full-one (often as frequently as several times a week). This job is called *cylinder change-out*, and is performed by technicians (Fig. 5-13).

Because specialty-gases can be toxic or otherwise hazardous, care must taken to follow correct safety-procedures during such change-outs. One safety-concern involves incorrect-purging of the lines before a cylinder-change, which could allow residual-gas to escape, leading to a flame or toxic vapor-cloud. Another safety-issue is an improperly-supported cylinder that may fall over, causing a leak. Due to such safety hazards there has been a trend to convert some specialty-gases to bulk-distribution.

5.4 ULTRAPURE-WATER (UPW)

Large quantities of water are needed during microchip-fabrication, mostly for rinsing wafers after wet

Fig. 5-11 Specialty-gas and chemical distribution-system. Courtesy of BOC Edwards.

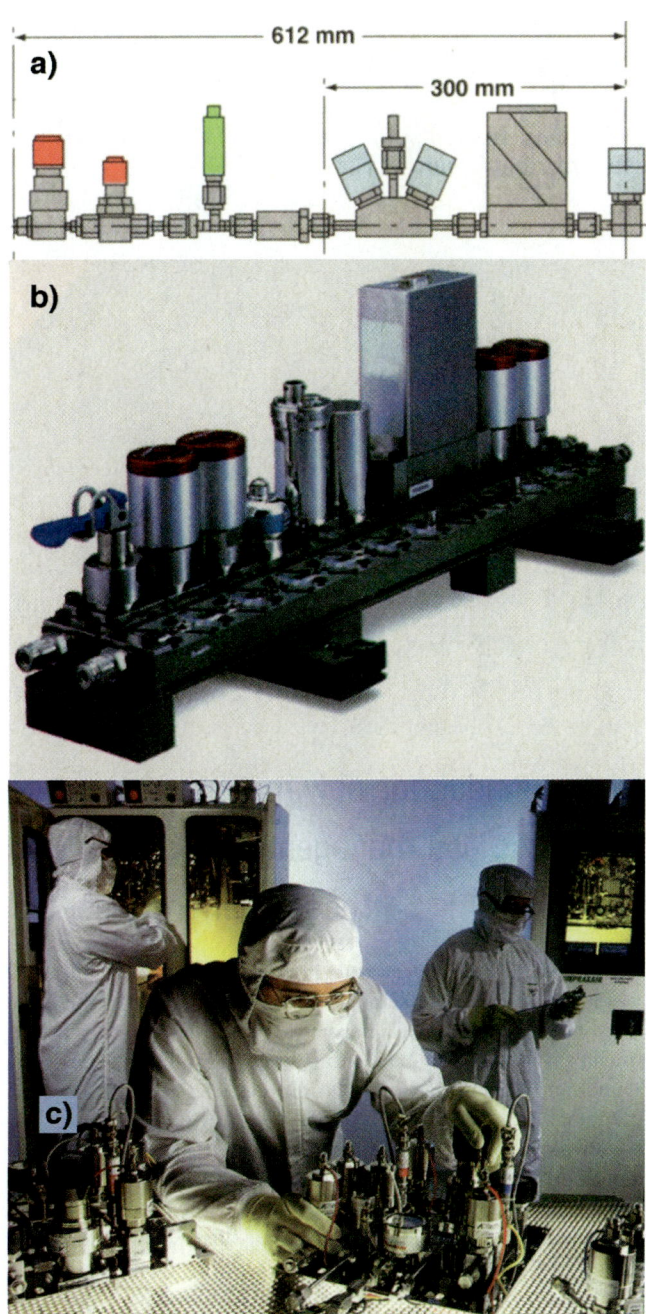

Fig. 5-12 (a) Drawing of *gas-stick*. (b) Photograph of a gas-stick. (c) Technician calibrating a gas-panel. Courtesy of Praxair.

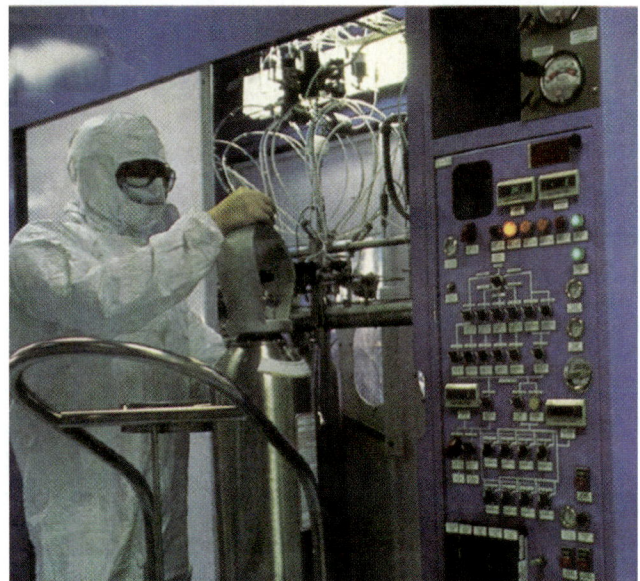

Fig. 5-13 Technician performing a cylinder changeout. Courtesy of Praxair.

cleaning and etching. About 2000-gallons of water are consumed in the course of passing a single 200-mm-wafer through its 450 process-steps. A 200-mm IC fab with 5000-wafer-starts/week is estimated to require at least 3-million gallons of UPW-water per day (about as much water as used by a town of 60,000-people).

However, by about the year 2000, semiconductor manufacturers in many locations were being denied requests for the additional water needed to support plans for expanding production (or even to build a new fab at a site where it would become a dominant water user). This means that in the future, less water will have to be used per wafer. Efforts to reduce water-consumption depend on decreasing its usage, recycling, and reclamation. In modern fabs, over 70% of the water is reclaimed after use.[5]

Pure-water is a substance that is neutral, having a pH of 7 (being neither an acid nor a base). This property of chemical-neutrality is important for the applications in which water is used in a fab. Thus, fab-water must be a chemically-pure liquid, free of particulates. Raw (untreated) water from a city-system contains many unacceptable contaminants including: organic-material, inorganic-material, ionic-impurities, dissolved-gases, and bacteria (as well as particles). These contaminants must be removed to make the water suitable for IC processing. Such cleaned water is termed *de-ionized* (*DI*) *water* or *ultrapure* (*UPW*).

But, the task of purifying and supplying UPW-water is expensive and challenging. A special water-treatment facility must be installed at each wafer fab to do this job.[6] It consists of de-gasifiers, filters, and reverse-osmosis units (R/O) and ion-exchange columns (Fig. 5-14).[7] Water passes sequentially through these stages to emerge as suitably-pure water (which is then stored in an appropriate tank). When UPW-water is delivered to wafers, it is subjected to a final point-of-use (POU) filtering-step (with a preceeding UV-irradiation-process to destroy bacteria).

The purity-specifications of UPW-water include: **a)** *high-resistivity* (a resistivity of 18-MΩ-cm is desired, as this is the value exhibited by water with no impurities present at 25°C); **b)** *low particle-concentration*; **c)** *low bacteria-concentration*; **d)** *small-concentrations of the following contaminants* (in the parts-per-billion [ppb] range) – oxygen, dissolved-silica, Na, Cl, metal-ions, and total organic-concentration (TOC). The exact specifications depend on the feature-size being fabricated, but these specifications continue to be tightened as critical dimensions are reduced.

Reverse-osmosis-units (R/Os) typically remove more than 90-percent of the contaminants in the water

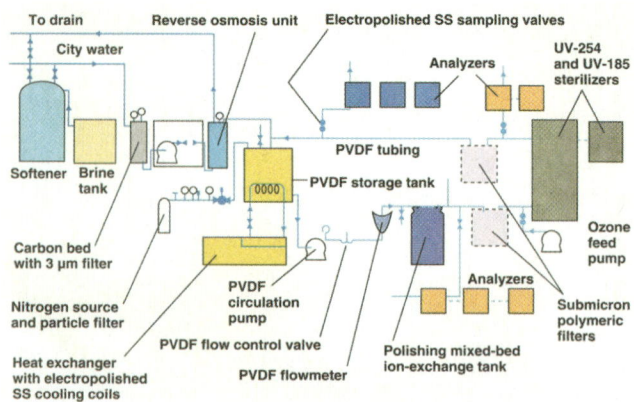

Fig. 5-14 Diagram of an ultra-pure-water (UPW) treatment-system.[7] Courtesy of Semiconductor International.

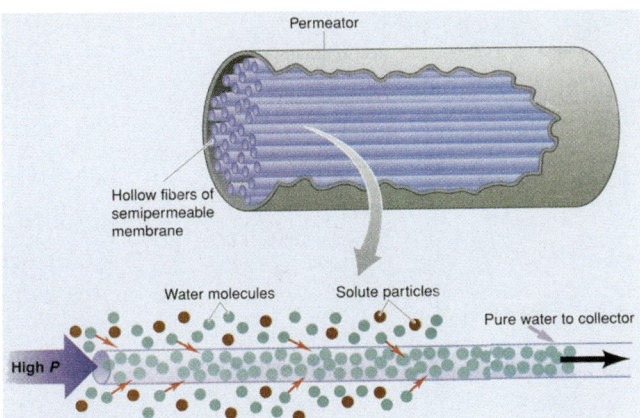

Fig. 5-15 The principle of *reverse-osmosis*.

and this enables the downstream ion-exchange columns to bring the water delivered from an R/O to the quality needed by the wafer fab. *Osmosis* is the transport of a liquid across a semipermeable-membrane, so that the concentration of a solute (contaminant) on the "dirty-side" of the membrane is diluted. That is, the solute-concentration approaches the concentration-level that exists on the purer-side. In the *reverse-osmosis-process* used in UPW-systems, an external-pressure is applied to the contaminated-water entering on the "dirty-side" of the membrane (Fig. 5-15). This externally-applied-pressure increases the component of pure-water-flow from the contaminated-side to the purer-side, well above that required for balancing the osmotic-component flowing in the opposite-direction. Thus, the quality of the pure-water on the "clean-side" increases, while the quality decreases on the "dirty" (or "contaminated") side (i.e., in the *reject-water*).

REFERENCES

1. M. Silberberg, *Chemistry: The Molecular Nature of Matter and Change,* Mosby, St. Louis, 1996.
2. P. Singer, "Trends in Gas Management and Use," *Semiconductor International,* April 1998, p. 112.
3. A.E Braun, "Gas Management Confronts Materials, Cost & Service Issues," *Semiconductor Intl.,* Dec. 1998, p. 60.
4. R. Iscoff, "Process Gas Purity: The 'Nines' Game," *Semiconductor International.,* March 1995, p. 65.
5. L Peters, "Ultrapure Water: The Rewards for Recycling," *Semiconductor International,* February 1998, p. 71.
6. D. Tolliver, Ed. *Handbook of Contamination Control in Microelectronics,"* Noyes, Norwich NY, 1998, Ch. 6 (Ultra High Purity Water), Ch. 7 (DI-Water Filtration Technol).
7. R.A. Governal, "Ultrapure Water: A Battle Every Step of the Way," *Semiconductor International*, July 1994, p. 176.

PROBLEMS

1. If the temperature is 32°F, what would it be if expressed using: (a) the *celcius* scale (°C); and (b) the *absolute-scale* (*kelvin*, K). Repeat for 100°F. Convert 500°C and 1000°C to temperature in *kelvin*.
2. What is a *chemical-solution*, and what are it's components? What is an *aqueous-solution*?
3. What do the abbreviations *ppm* and *ppb* represent?
4. Define an *acid*. What are two acids commonly used in IC fabrication?
5. Define a *base*. Specifically explain how a base differs from an acid.
6. Explain the *pH-scale*, and how it expresses acidity and alkalinity. Acids have a pH between ___ and ___. Bases have a pH between ___ and ___. What is the pH of water? What are the pH values of solvents?
7. What is a *solvent*? List three common solvents used in microchip manufacturing.
8. Name the two groups of gases used in wafer fabs?
9. Describe a *gas stick*.
10. The temperature of a 200-mm wafer is heated 400°C above room-temperature. By how much will this temperature-rise cause the diameter of the wafer to increase? If an adherent-aluminum film is present on the wafer, it will only be able to laterally expand by this same amount. If an aluminum-film of the same dimensions was not attached to the silicon-wafer, and was raised to the same temperature, by how much would its diameter expand?
11. Explain the principle of *reverse-osmosis* and how it is exploited to purify water for use in wafer-fabs.
12. Explain why bacteria are a particularly vexing contaminant in ultra-pure water (UPW) supplies of wafer-fabs. How do bacteria survive (and multiply) in the apparently sterile-UPW environment?
13. What is the *resistivity-specification* of UPW?

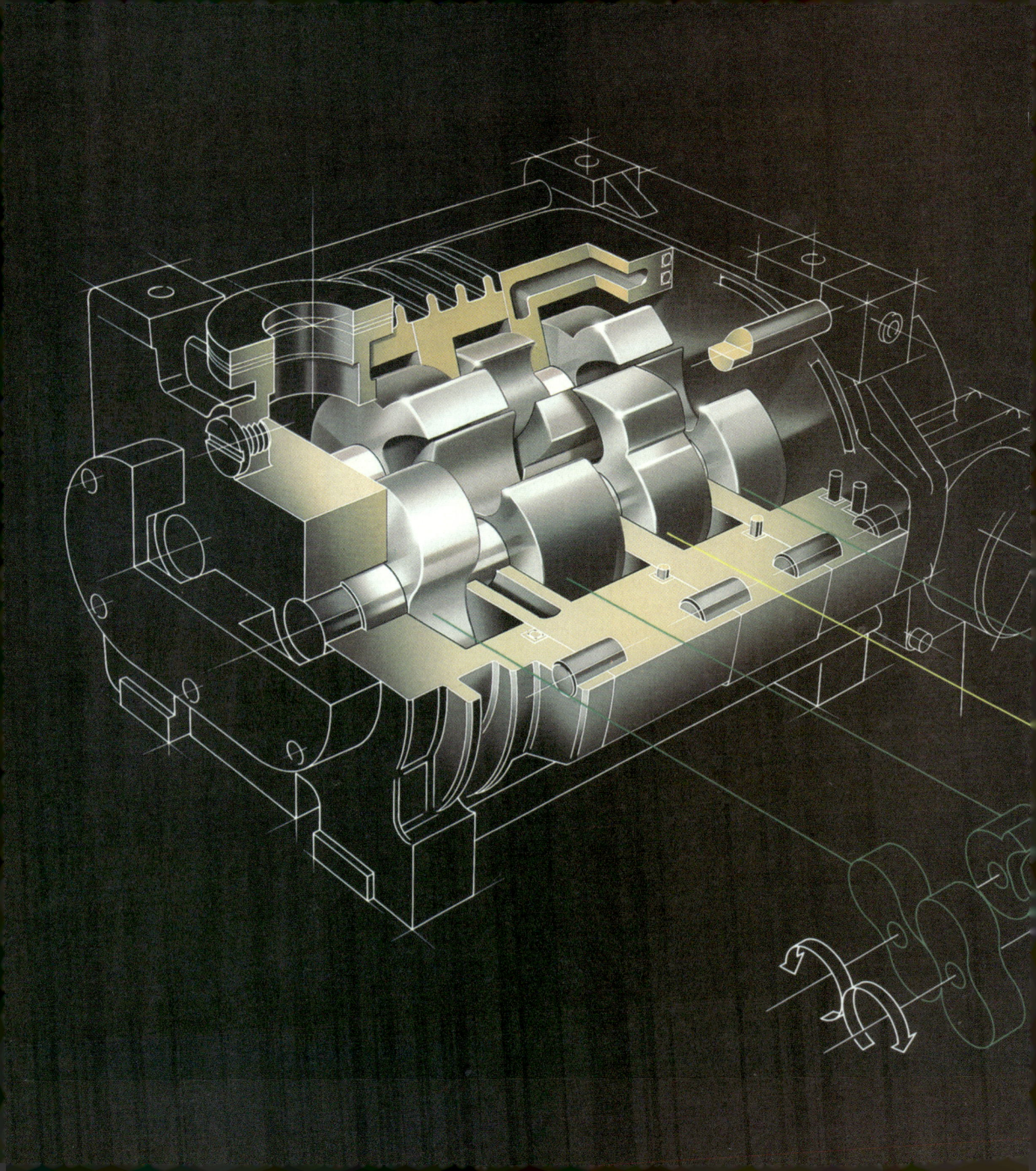

CHAPTER 6

VACUUM TECHNOLOGY FOR ULSI APPLICATIONS

CHAPTER CONTENTS

6.1 BASIC CONCEPTS OF GASES AND VACUUMS

6.2 PRESSURE UNITS

6.3 VACUUM-PRESSURE RANGES

6.4 THE LANGUAGE OF GAS/SOLID INTERACTIONS

6.5 TERMINOLOGY OF VACUUM PUMPS

6.6 ROUGH PUMPS

6.7 HIGH-VACUUM PUMPS I: CRYOGENIC PUMPS

6.8 HIGH-VACUUM PUMPS II: TURBOMOLECULAR PUMPS

6.9 VACUUM GAUGES: (TOTAL-PRESSURE MEASUREMENT)

6.10 MEASUREMENT OF PARTIAL-PRESSURE: RESIDUAL-GAS ANALYZERS (RGAs)

6.11 MASS-FLOW CONTROLLERS

"Much Ado About Nothing."

            William Shakespeare

Illustration of the inner mechanisms of a dry mechanical-pump. Courtesy of BOC Edwards.

Vacuum-technology is an important aspect of many semiconductor manufacturing-processes. A variety of fabrication steps are carried out under reduced pressure (i.e., *vacuum conditions*), including: sputtering, plasma-etching, ion-implantation, low-pressure CVD, and plasma-enhanced CVD. In addition, semiconductor process-equipment incorporates more and more vacuum-technology with each generation.

Cassette-to-cassette wafer-handling, load-locks, and the transport of wafers through evacuated central-chambers in cluster-tools all require vacuum pumps and controls. Furthermore, the properties of the materials formed using vacuum-based processes are critically dependent on the conditions of the vacuum. For example, vacuum environments from which water and oxygen can be excluded promote the deposition of purer metal-layers.

Since vacuum-systems play such a significant role in IC-processing, the job tasks of many technicians include maintenance and repair of vacuum systems. For these reasons semiconductor workers should understand vacuum-technology and its relationship to IC fabrication processes.

In this chapter we describe the most important aspects of vacuum-systems in microelectronic applications, including: basic definitions; pump technology; vacuum measurement; and residual-gas analysis.

6.1 BASIC CONCEPTS OF GASES AND VACUUMS

As noted in Chap. 2, matter exists in solid, liquid, or gaseous forms. In the gaseous state, molecules are most independent of one another. The properties of real gases (including their behavior in vacuums) is simply and accurately predicted by assuming that gas-molecules act like an *ideal gas*. The *ideal-gas model* assumes that: **1**) Gas-molecules are minute spheres which do not interact with each other, unless they collide; **2**) Between collisions, gas-molecules travel along straight paths in a completely random fashion (Fig. 6-1); **3**) They make perfectly elastic collisions.

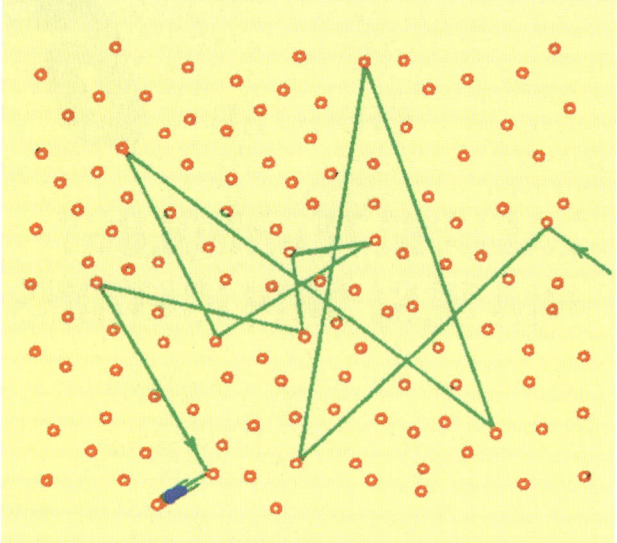

Fig. 6-1 Molecule traveling in a gas, colliding with other molecules in its path. All other molecules move in a similar fashion.

Many useful properties of gases can be derived using these assumptions, including the fact that the gas-molecules in a container have an *average speed* **c** that depends on the gas temperature according to the following equation:

$$c = \left[\frac{8kT}{\pi m}\right]^{1/2} \quad (6.1)$$

where k is Boltzmann's constant, T is the absolute temperature (in kelvins, K), and m is the mass of a molecule. Equation 6-1 indicates that the average speed **c** of the molecules in a gas gets larger as T is increased (and that it gets smaller if T is lowered).

A *gram-molecule* is the quantity of material corresponding to the molecular weight in grams, and *one gram-molecule* of material contains 6.023×10^{23} molecules (Avogadro's number). A gram-molecule of any gas at *standard temperature* and *pressure, STP* (T = 0°C = 273 K, and P = 1 atmosphere, at sea level) contains 6.023×10^{23} molecules, and occupies a fixed volume - which is 22.4 liters. From this information it can be calculated that *one cubic-centimeter* of a volume of a gas (at STP) contains 2.7×10^{19} molecules.

6.1.1 Gas-Pressure and Vacuum

Gas fills the entire volume of a container and exerts uniform pressure on its walls. This occurs because gas molecules each have some mass and are continuously moving. Thus, not only will gas-atoms collide with each other, they will strike and collide with any surface in contact with the gas (Fig. 6-2). The net effect on such surfaces is that a force is exerted upon them by the sum of these collisions. If the solid-surface struck by the gas-atoms can move (such as a piston in an engine), to keep the piston stationary in the face of the gas pressure, an external, equal force must be exerted in the opposite direction.

The force exerted by a contained gas-per unit-area of the container walls is called the *gas-pressure, P*. Gas-pressure depends on two factors: **1)** the number of collisions that take place per unit time; and **2)** the *average-momentum* possessed by the molecules striking the walls. For a given type of gas, the momentum depends on the average speed of the gas molecules - which in turn is controlled by the *gas-temperature*.

Assume that only *factor-***1)** is important (that is, the gas is being held at a constant-temperature). Then the pressure inside a container can be altered only by changing the number of molecules inside the container (by removing or adding some of them), because this changes the number of collisions per unit-area of the walls. For example, if gas in a container is at *atmospheric-pressure*, and more molecules are added to it, the pressure will be increased above that

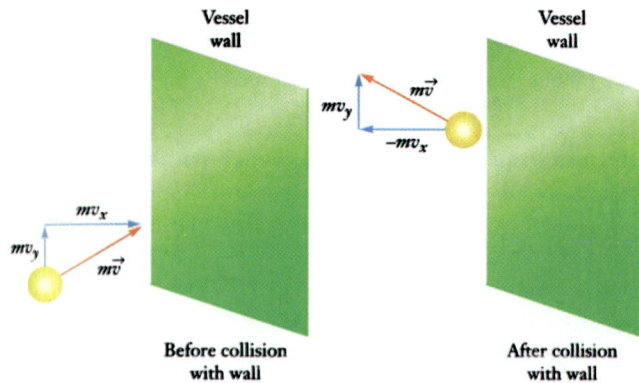

Fig. 6-2 An elastic-collision of a molecule with a wall.

CHAPTER 6 VACUUM TECHNOLOGY FOR ULSI APPLICATIONS

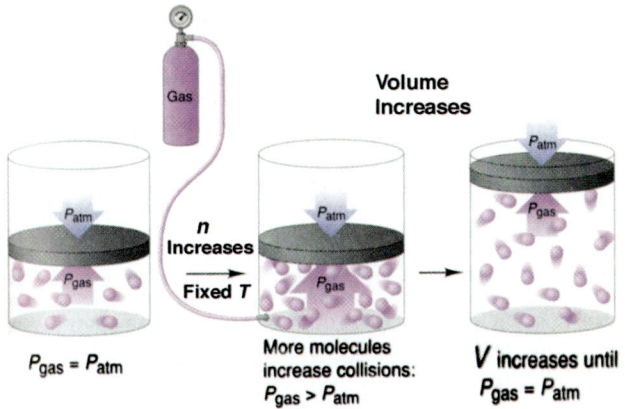

Fig. 6-3 Assume that at a fixed temperature (T), *n*-molecules of gas give rise to a pressure P_{gas}. If more gas is added, *n* increases. Thus, collisions with the walls are more frequent and P_{gas} increases.

of atmosphere (Fig. 6-3). Conversely, if molecules are removed from within the container, the pressure will be *reduced* below that of atmospheric-pressure. A *vacuum* is generally defined as being *any pressure lower than that of atmospheric-pressure*.

6.2 PRESSURE UNITS

Vacuum-pressure is expressed according to a number of different systems of units. These various units evolved for certain historical reasons. The most "correct" pressure-unit today is based on System Internationale (SI). This means it is derived from *meter-kilogram-second (MKS) units*. This pressure-unit is the *pascal* (Pa). One-Pa corresponds to 1-newton/m². The pressure at 1-*atmosphere* expressed in Pa is 1.013×10^5-Pa. The *bar* is also an accepted SI unit and 1-*bar* corresponds to 10^5-N/m² = 10^5-Pa.

However, another set of pressure-units that was not derived from MKS units (but instead from measurements of the height of a column of liquid, mercury) were the earliest pressure-units used. In spite of having been officially declared obsolete, they still find wide use, and thus also need to be defined here.

That is, it was observed that the pressure of *one-atmosphere* (at sea level, and at 0°C) was able to balance a column of mercury (Hg) 760-mm in height (Fig. 6-4). The unit of 1-mm (or 1-*torr*) of pressure is understood as that gas-pressure which is able to balance a column of Hg which is 1-mm high. The unit of the *millitorr of Hg* (10^{-3}-torr) is also frequently encountered. Another familiar unit of pressure is given in *pounds per square inch* (*psi*). The pressure of *one-atmosphere* (at sea level and at 0°C) in psi units = 14.7-psi. In other words, 14.7-psi = 29.92 inches Hg = 760-mm Hg = 760-torr. The following relationships are used to convert pressure values from Pa to torr units, and back (see Fig. 6-5):

1-pascal (Pa) = 7.5×10^{-3}-torr = 7.5-mtorr Hg (6.2a)
1-torr = 133.3-Pa (6.2b)
1-bar = 1×10^5-Pa = 750-torr (6.2c)
1-atm = 1.013×10^5-Pa = 760-torr (6.2d)

6.3 VACUUM-PRESSURE RANGES

As noted earlier, *vacuum* is defined as a region of space in which the pressure is below that of atmospheric-pressure. It has been customary to divide the pressure-scale below atmospheric-pressure into several ranges that denote the *degree* of vacuum. These ranges are listed in Table 6-1.

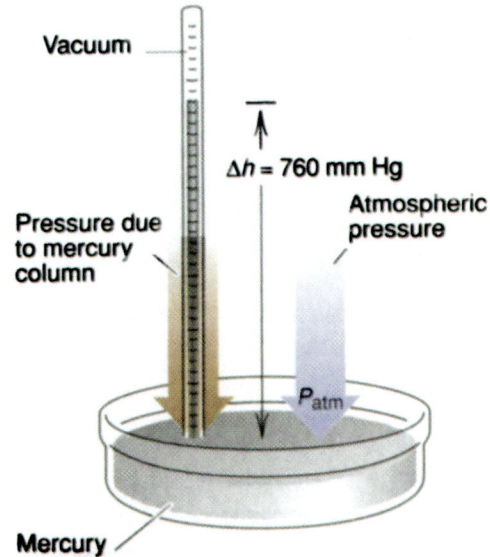

Fig. 6-4 A *mercury-barometer*. The pressure of the mercury-column is balanced by the pressure of the atmosphere. At sea-level and 0°C, the average atmospheric-pressure can support a column of mercury that is 760-mm high.

Table 6-1 VACUUM RANGES

One-Atmosphere = 1.013×10^5 Pa (760 torr)
Low-Vacuum = 10^5 Pa (750 torr) to 3×10^3 Pa (25 torr)
Medium-Vacuum = 3×10^3 Pa (25 torr) - 10^{-1} Pa (7×10^{-4} torr)
High-Vacuum = 10^{-1} Pa (7×10^{-4} torr) - 10^{-4} Pa (7×10^{-7} torr)
Very-High-Vacuum = 10^{-4} Pa (7.5×10^{-7} torr)
 to 10^{-7} Pa (7.5×10^{-10} torr)
Ultra-High Vacuum = 10^{-7} Pa (7.5×10^{-10} torr)
 to 10^{-10} Pa (7.5×10^{-13} torr)

In semiconductor processes, the most common vacuum-ranges encountered are the *medium-vacuum* and *high-vacuum* ranges (see Fig. 6-6).

6.3.1 Mean-Free-Path, λ

Each molecule in a gas is randomly located in the volume and is moving at a different speed and direction than the others. This means that each molecule will move a different straight-line distance before colliding with another one (Fig. 6-1). The distance between collisions is called the *free-path*, and it will vary because of the randomness of the particle locations and velocities. However, an *average*, or *mean* of the free paths can be determined (and it is called the *mean-free-path, λ*). For air at room-temperature (26°C = 300 K) a simple and convenient relation for

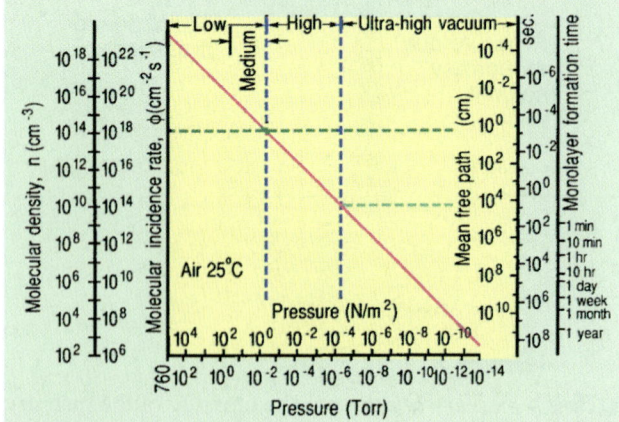

Fig. 6-5 Histogram that plots: *molecular-density, incidence-rate, mean-free-path,* and *monolayer-formation-time* as a function of *gas-pressure*.

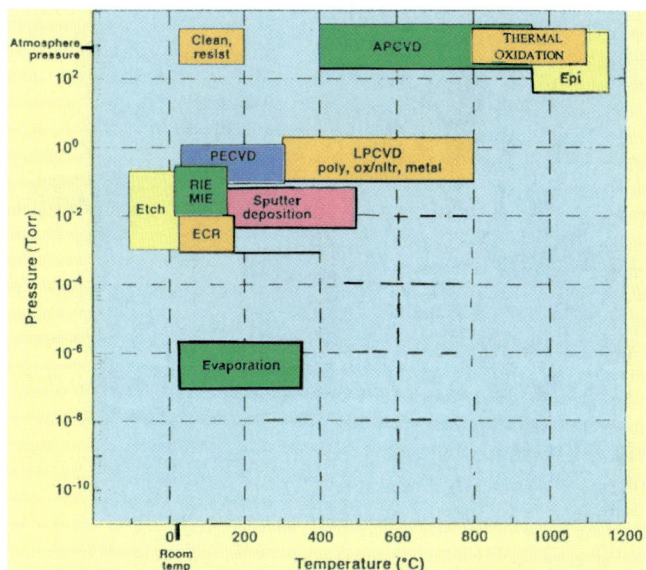

Fig. 6-6 Semiconductor-processes as a function of pressure and temperature.

λ (in millimeters) is given by

$$\lambda(\text{mm}) = \frac{6.6}{P(\text{Pa})} = \frac{0.05}{P(\text{torr})} \qquad (6.3)$$

EXAMPLE 6-1: Given a vacuum-chamber that contains argon-gas (atomic-weight = 39.94-amu), at 300 K, and pressures of: **a)** 200-Pa (1.5-torr); and **b)** 0.2-Pa (1.5-mtorr). Find: 1) the average-velocity of the gas-molecules; 2) their mean-free-paths.

SOLUTION: 1) The average-velocity **c** is found from Eq. 6-1, where: $k = 1.38 \times 10^{-23}$ J/K, and m_{Ar} = 39.94 x 1.66×10^{-27} kg.

$c = (8 \, k \, T / \pi \, m)^{1/2}$

$= (8 \cdot 300 \cdot 1.38 \times 10^{23} / \pi \cdot 39.94 \cdot 1.66 \times 10^{-27})^{1/2}$

$\cong 4 \times 10^2$ m/sec

2) The mean-free-path λ is found from Eq. 6-2:

$\lambda(\text{mm}) = 6.6/P(\text{Pa})$

$\lambda_{200 \, Pa} = 3.3 \times 10^{-2}$ mm, $\lambda_{0.2 \, Pa} = 333$ mm.

6.3.2 Closed and Open Gas-Systems and Gas-Flow Regimes

In some processes (e.g., the deposition of metal-films by evaporation) gases may exist in a *closed-system* (e.g., a vacuum bell-jar). However, in many other semiconductor manufacturing-processes there is a steady number of atoms being continuously fed by gas-flow (and simultaneously being pumped-out of the chamber) by a *vacuum-system*. These are termed *open-flow systems*. Thus, a brief mention of the behavior of flowing-gases should be made.

An important aspect of gas-flow is that its nature can vary considerably, depending upon the pressure and the dimensions of the vessel in which it is moving. At high-pressures, the mean-free-path λ is short, and therefore the behavior of the particles is completely governed by collisions with other gas-particles. In gas-flows of this type, the particles move along as a stream, and such motion is called *viscous-flow* (Fig. 6-7a and b). Such flow can be either *laminar* (in which the gas streams are orderly, and move parallel to one another, Fig. 6-7b), or *turbulent* (which is more disorganized flow, see Fig. 6-7a and Fig. 6-8).

However, at sufficiently low-pressures the mean-

Fig. 6-8 At a certain point, the rising flow of smoke and heated-gas changes from *laminar* to *turbulent*.

free-path can be quite long, and molecules will encounter collisions with the walls of the container much more frequently than with each other. The particles arrive at a wall, stick, and then rebound in a direction independent of their arrival-velocity. This form of gas-flow is called *molecular-flow*. Because gas-molecules in molecular-flow scarcely ever collide with each other, two gases can move in opposite directions in the system, with neither being affected by the presence of the other (Fig. 6-7c).

In general, when the characteristic dimension, d, of a vacuum-chamber or pipe in which the gas is flowing is much larger than the mean-free-path of a gas ($\lambda \ll d$), *viscous-flow* occurs. When the mean-free-path is much larger than d ($\lambda \gg d$), the flow regime is *molecular* (see Fig. 6-9). In the region where $0.01d < \lambda < d$, the flow is governed by both viscous and molecular phenomena, and it is referred to as *intermediate*, *transition*, or *Knudsen* flow.

6.4 THE LANGUAGE OF GAS/SOLID INTERACTIONS

When two-states of matter, a *condensed-phase* (that is, a solid or liquid) and its *vapor-phase*, co-exist at the same temperature, and remain in contact with each

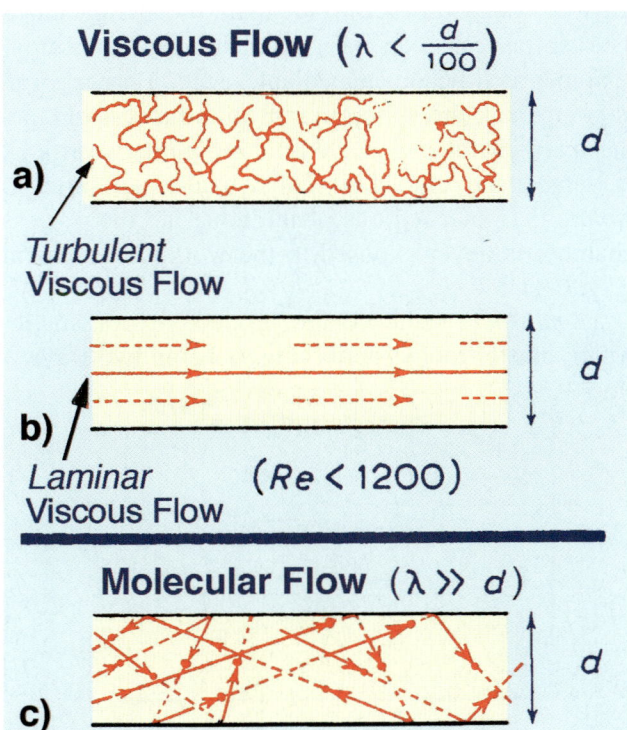

Fig. 6-7 Types of gas-flow through cylinder tubes.

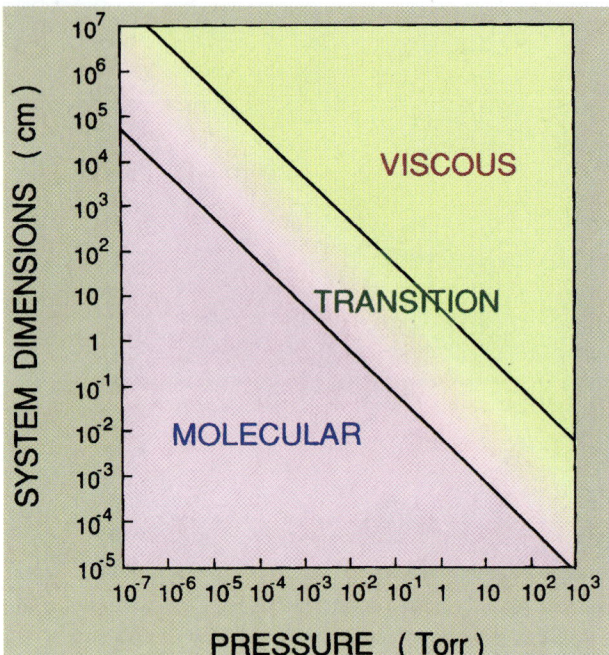

Fig. 6-9 Dominant gas-flow regimes as a function of system dimensions and pressure.

atoms, ions, or electrons. *Vaporization* is the process in which a condensed-vapor is thermally transformed into a vapor. *Evaporation* is the conversion of a substance from the liquid-state to a gas, while *sublimation* is the process of transition directly from a solid to a vapor-phase (without passing through the intermediate liquid-phase).

The release of gas (and/or vapor) by a solid is important in several aspects of vacuum-processing, including the phenomenon of *outgassing*. This effect is observed as the slow evolution of gases and vapors from interior-surfaces of a vacuum-container and its contents (see Fig. 6-10).

6.4.1 Load-Locks

A load-lock is a small loading-chamber that is attached to, but is separately pumped-from, the main process-chamber. Wafers enter the tool by being transported from the cassette into the load-lock, which is then sealed and pumped-down. When fully-evacuated, the load-lock is opened to the main process-chamber. The wafers can then be moved into the deposition-chamber as it becomes available. After a process-run is completed, the wafers return to the load-lock, and the main-chamber is sealed-off before the load-lock is vented and unloaded. Thus, load-locks serve to *isolate* the inner-regions of the tool, and the process chamber is never exposed to the workplace ambient (Fig. 6-11).

Load-locks enhance the productivity of single-wafer, cluster-tool systems (Fig. 6-12) in two ways:

other without undergoing net changes, they are said to be in *thermodynamic equilibrium*. Under such conditions the same amount of material is evaporated from the surface of the condensed-phase as is condensing upon it. The pressure of the vapor in the container is called its *vapor-pressure*. The vapor-pressure of a substance increases as the temperature is increased.

When a vapor (or gas) and a condensed-phase interact, gas or vapor molecules which strike the condensed-phase may become bound to it. If the binding occurs only at the *surface*, the process is called *adsorption*. A familiar example of adsorption is the condensation of water on an automobile's glass-windshield at night. The water does not penetrate the glass, but remains *adsorbed* on its surface. On the other hand, if the molecules penetrate the surface and become bound or captured in the bulk of a solid (or liquid), the process is known as *absorption*. When adsorbed molecules are released from a surface, the process is called *desorption*. Desorption can be speeded up by supplying energy to the surface from such sources as thermal energy, photo energy, or impact by

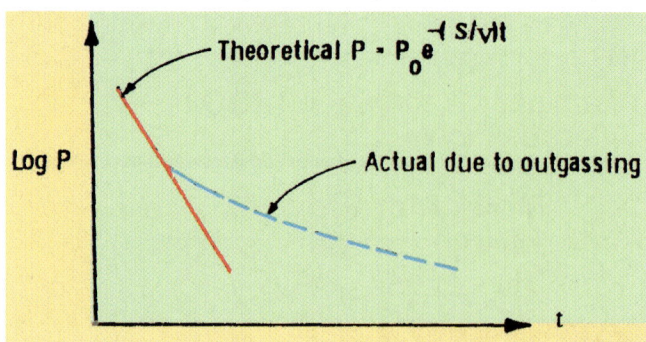

Fig. 6-10 Pressure in a vacuum-chamber as a function of pumping time, showing the effects of *outgassing*.

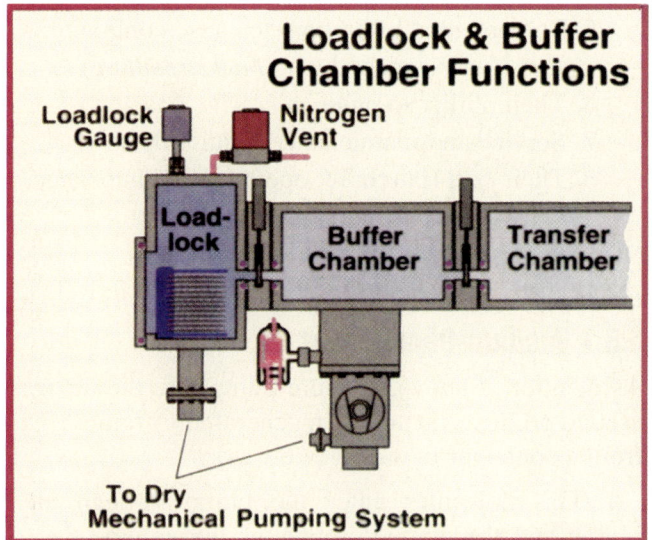

Fig. 6-11 By using a controlled-pumpdown-sequence for load-lock-chambers and buffer-chambers, the transfer of contaminant-gases into the process-chamber is minimized.

1. An entire-cassette of wafers can be placed into a load-lock and then pumped down to the process-pressure simultaneously (even if the process-chamber only processes one wafer at a time). This reduces the average time that each wafer is exposed to the pumpdown cycle. The load-lock chamber should thus have as small a volume as possible (to minimize *pump-down time*). In addition, heating of the wafer-holder (and wafers) in the load-lock can reduce the water-vapor outgassing-rate in the deposition-chamber.

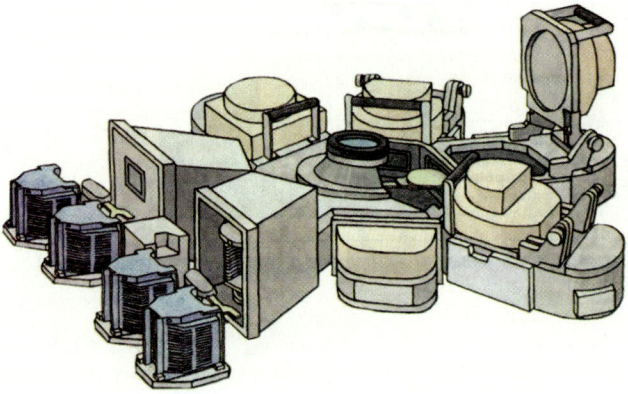

Fig. 6-12 Drawing of a *cluster-tool system*.

2. Because the process-chamber is kept under vacuum at all times, and is never exposed to the fab-ambient, its inner-walls do not adsorb water-vapor. Thus, the outgassing-problem (which significantly increases the time to reach the process-pressure in chambers that are vented to atmosphere) is eliminated. This shortens wafer-turnaround time.

6.5 TERMINOLOGY OF VACUUM-PUMPS

As discussed earlier, if a vessel or chamber contains air at atmospheric-pressure, some of the gas molecules within it must be removed to create a vacuum there. *Vacuum-pumps* are used for this purpose.

If no gas is allowed to flow into the chamber after it is pumped, this is called a *closed vacuum-system* (e.g., a vacuum-evaporation bell-jar, as shown in Fig. 6-13). Creation of *high-vacuum* within a closed vacuum-system would be relatively simple if no gases are adsorbed on the surfaces of its interior chamber-walls. In that case, gases could be pumped out, and the vacuum would be maintained indefinitely (assuming the system has no leaks). However, if gases are adsorbed on the inner-walls, after high-vacuum has been attained within the closed chamber, such gases can desorb from these surfaces and evolve into the vacuum atmosphere (a phenomenon called *outgassing*). Since gas-molecules are commonly adsorbed (and then desorb into the vacuum-environment) by solid-surfaces, continuous pumping is needed to keep a high-vacuum, even in closed vacuum-systems.

In many IC fabrication-processes, however, gases are deliberately flowed into a process-chamber though an inlet (and simultaneously removed from the outlet) while a specified-value of reduced-pressure (vacuum) is continuously maintained in the process-chamber. Such process-environments are referred to as *open-flow vacuum-systems* (Fig. 6-14). The situation of *continuous-pumping* described for the closed-vacuum systems is likewise encountered in open-flow systems, although the gas-flow within the system is now deliberate and usually involves much larger gas-flows than that due to outgassing. The issues of pumping

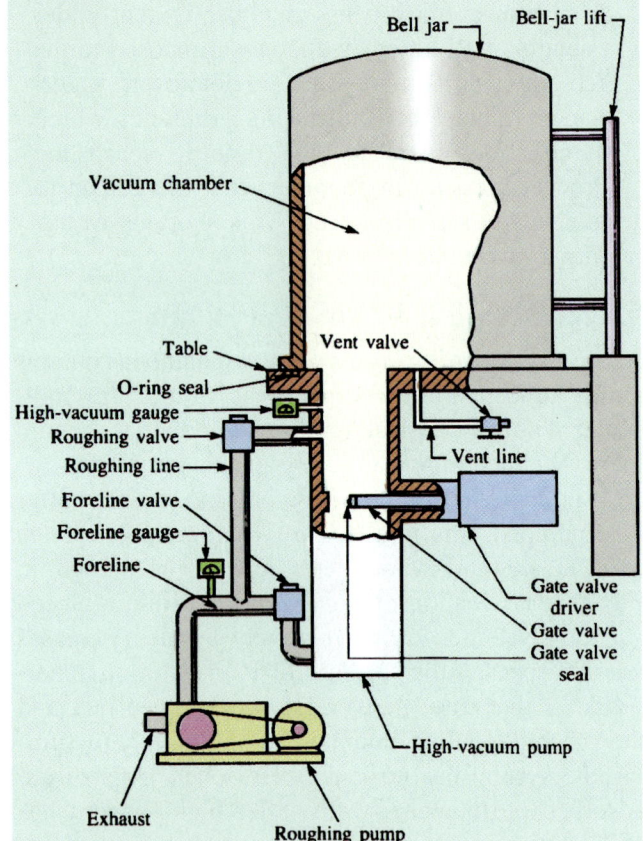

Fig. 6-13 Schematic of a typical *closed vacuum-system*.

open-flow vacuum-systems in IC-fabrication will be discussed in later sections of this chapter.

Vacuum-pumps are the machines used to create (and maintain) vacuum environments in process-chambers. There are many types of vacuum-pumps that have been developed, although only a few kinds are used in most IC-processes. We will restrict our discussion to these sorts of vacuum-pumps.

Vacuum-pumps are selected and used based on a number of criteria, including:

1. The *vacuum-range* required
2. The *pumping-speed* needed
3. The *maximum continuous-throughput*
4. Their ability to handle *impulsive gas-loads* during startup (and periodic outgassing events)
5. The potential of a pump to cause contamination (for instance, due to *backstreaming*)
6. Their ability to *pump corrosive-gases*
7. Service and maintenance requirements
8. Their *cost* (purchase, operating, maintenance, and repair costs)

Note that more details on such topics as *pumping speed, throughput,* and *flow-rate* are found in Ref. 1.

6.5.1 Vacuum-Pump Types

As mentioned above, vacuum-pumps are the devices used to produce vacuums. In general, they remove gas from a container in one of two ways:

1. Gas-molecules enter the pump through an *inlet* and the pump compresses the volume they occupy. Such compression increases the pressure exerted by these molecules. When sufficient compression has occurred so that the pressure exceeds that which exists at its *outlet*, the gas exits. Pumps operating on this principle are called *compression, discharge,* or *throughput* pumps.

2. Gas-molecules enter the pump, where they condense (or are immobilized in some other way) on a solid-surface within the pump. These are known as *entrapment* or *capture* pumps.

The total pressure on the outlet side of a *compression pump* is called the *backing-, exhaust-, discharge-,* or *forepressure. Entrapment pumps*, on the other hand, have an *inlet*, but no outlet.

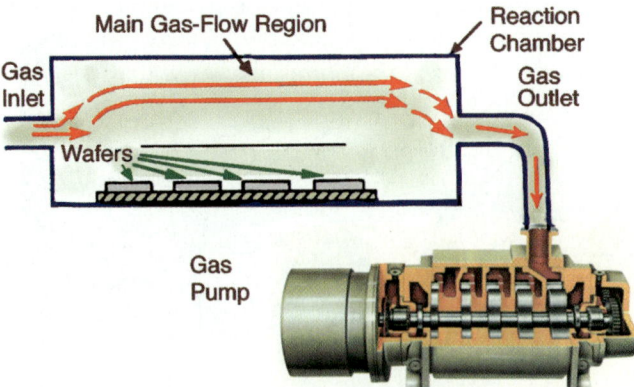

Fig. 6-14 Schematic of an *open-flow vacuum-system*.

CHAPTER 6   VACUUM TECHNOLOGY FOR ULSI APPLICATIONS

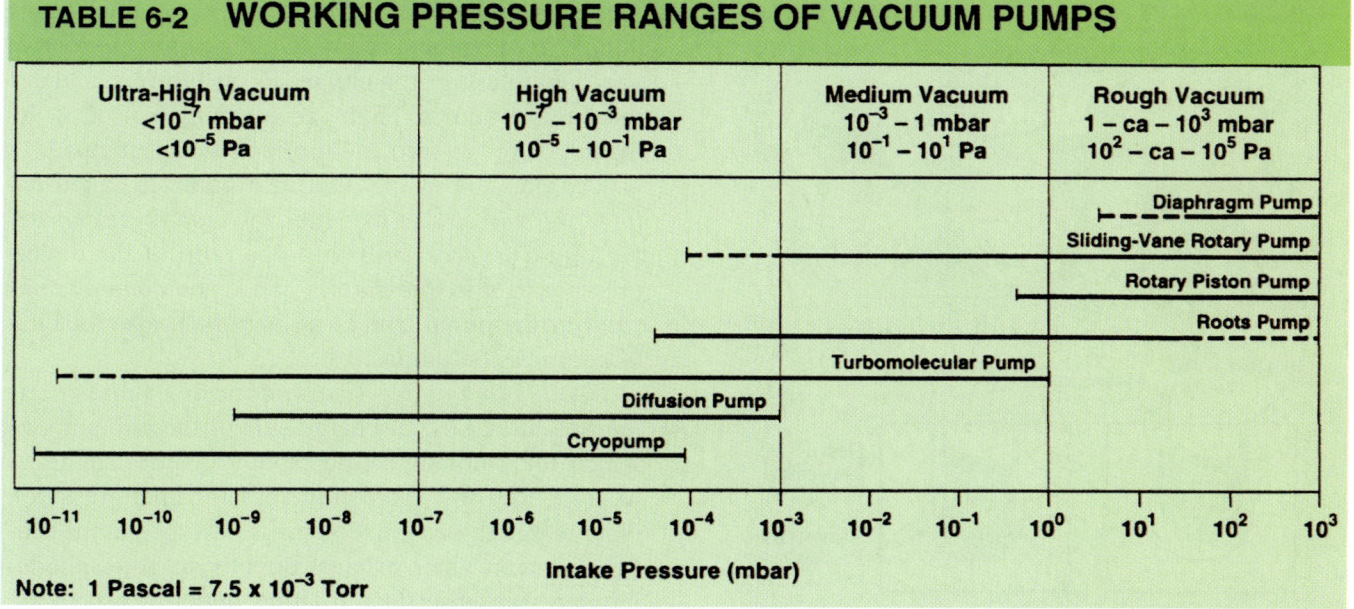

TABLE 6-2   WORKING PRESSURE RANGES OF VACUUM PUMPS

Note: 1 Pascal = 7.5 × 10⁻³ Torr

The *vacuum-environment* is divided into several ranges (as defined earlier), and no single pump exists which is capable of pumping from atmospheric pressure to the high (or ultra-high) vacuum-range. As a result, many pumps have been invented, each of which can operate effectively between specific pressure-levels. Table 6-2 shows the operating pressure-ranges for a variety of pumps used in IC production.

For ULSI applications, it is useful to identify two broad vacuum-ranges: **a)** the *low-to-medium vacuum range* (10^1–10^{-3} torr); and **b)** the *high-vacuum range* (~10^{-3}–10^{-6} torr). Most microelectronic processes are operated in one or the other of these ranges. Pumps that produce vacuums in the low-to-medium vacuum range are called *rough-pumps*, and those that maintain high-vacuum are called *high-vacuum pumps*.

Recall that when the characteristic dimension, d, of a vacuum-chamber or pipe in which the gas is flowing is much larger than the mean-free-path λ of a gas ($d \gg \lambda$), *viscous-flow* occurs (Figs. 6-7a and 6-7b). When the mean-free-path is much larger than d ($\lambda \gg d$), *molecular-flow* conditions exist. In viscous-flow, the short mean-free-path makes collisions among gas molecules frequent, and large pressure-differences exist between the chamber and pump-inlet. In *viscous-flow regimes,* movement of gas-molecules in the chamber toward the (rough) pump-inlet follows the direction of reducing pressure.

In *molecular-flow regimes*, pressure-differences are much smaller, and there are few collisions among molecules (Fig. 6-7c). Thus, molecules only enter inlets of high-vacuum pumps as a result of their random thermal-motion. Also, when molecules rebound from a pump-surface in molecular-flow, there is no preference for them to move in the down-stream direction. They have an equal chance of moving toward or away from the pump-inlet. Substances originating within a pump can likewise travel back to the chamber if pressure in the pump allows molecular-flow. This gives rise to *backstreaming*. In pumps which rely on oil to establish their pumping-capability (such as oil-sealed rough-pumps, and high-vacuum diffusion pumps), backstreaming of pump-oil vapors represents a possible contamination-hazard, and limits their use in ULSI fabrication. *Very broadly speaking, rough-pumps are associated with viscous-flow, and high-vacuum pumps with molecular-flow.*

Since no one pump-type can pump from atmospheric-pressure to high-vacuum, a rough-pump must first evacuate chambers to the medium-vacuum range

MICROCHIP MANUFACTURING

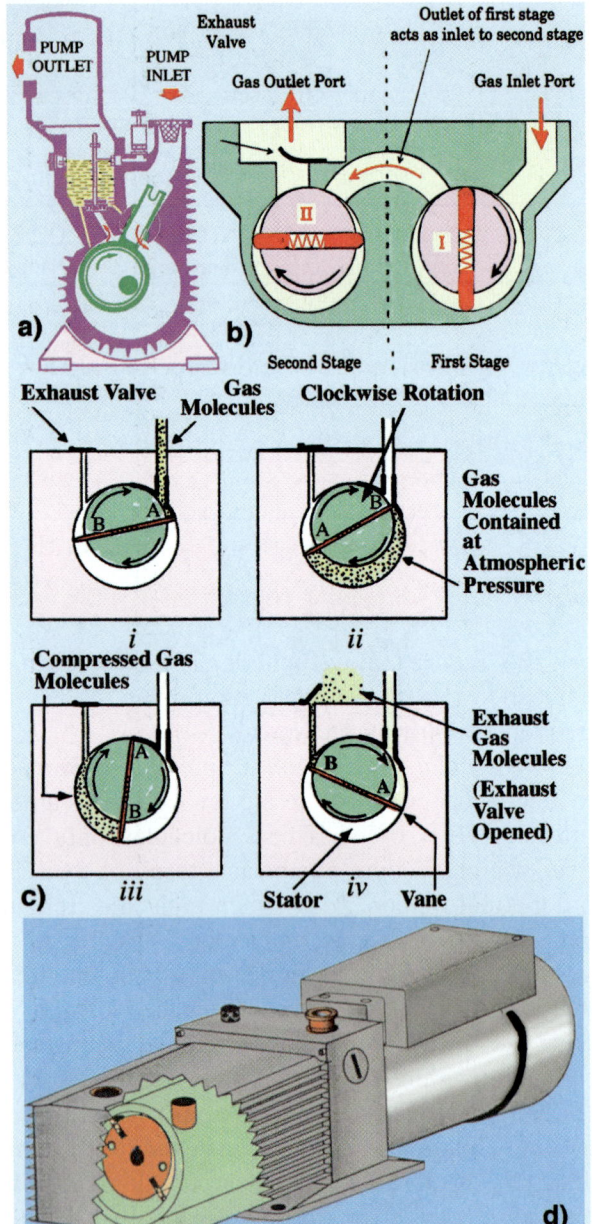

Fig. 6-15 (a) Cross-section of a single-stage rotary-piston pump (b) Cross-section of a two-stage sliding-vane rotary pump. Courtesy of Leybold-Heraeus Vacuum Products, Inc. (c) Principle of the pumping-action of a rotary-vane pump. (d) Cutaway-view of a rotary-vane pump.

(see Table 6-2). Then a high-vacuum pump takes over. The pressure at which high-vacuum pumps can be connected to the vacuum-system following initial roughing is called the *crossover pressure*. The rough-pumps most widely used in IC-fabrication are rotary-vane and rotary-piston pumps, Roots blowers, and dry mechanical pumps. High-vacuum pumps used in IC manufacture are cryo and turbomolecular pumps.

The pressure at the inlet of a pump is called the *inlet pressure,* and at its outlet, the *outlet- (discharge-backing-,* or *fore-) pressure*. The ratio of the outlet-pressure to inlet-pressure is called the *compression-ratio* of the pump, and is an important specification of compression-pumps.

The *limiting outlet- (backing-* or *fore-) pressure* is the pressure at the discharge-side of the pump above which the pumping-action rapidly deteriorates (e.g., as evidenced by a sudden increase of inlet-pressure). Note that high-vacuum compression-type pumps cannot discharge their exhaust directly to atmospheric-pressure (because their limiting outlet-pressures are much lower than 1-atm). Thus, they must discharge their pumped-gases into a volume whose pressure is also maintained at less than 1-atm. It is thus the function of the *fore-pump* (or *backing-pump*) to produce a forepressure value lower than the limiting outlet-value. We see that rough-pumps thus play a *dual-role* in high-vacuum systems (in which a high-vacuum pump is a compression-type pump): **1)** During evacuation of the chamber from atmosphere, it serves as a rough-pump; but **2)** When the high-vacuum compression-pump is engaged, it becomes a backing-pump.

6.6 ROUGH-PUMPS
6.6.1 Oil-Sealed Rotary Mechanical Pumps
Oil-sealed rotary mechanical-pumps are compression pumps used almost exclusively to perform two important functions in semiconductor vacuum-systems: **1)** to pump vacuum-chambers from atmosphere to medium-vacuum (0.1-10^{-3} torr); and **2)** to serve as forepumps for turbomolecular high-vacuum pumps.

There are two types of oil-sealed pumps: *rotary-piston* (Fig. 6-15a), and *rotary-vane* pumps (see Figs. 6-15b, c and d). Note that (b) shows a two-stage version. In both types, the gas enters the suction cavity, whose volume is compressed by the rotor and piston (or vane) as they rotate. At the same time a reduced-pressure region in the cavity is created behind the gas-volume being compressed. The compressed-gas is expelled to atmosphere through the outlet valve, and another valve opens to the reduced-pressure region of the cavity (allowing gas from the chamber to enter). As the pump rotates, more and more gas is removed from the process-chamber, causing its pressure to drop. The outlet-valves open automatically when the pressure in the pump exceeds atmospheric-pressure. Thus, the lowest pressure these pumps produce occurs when the inlet-pressure is so low that compression in the pump does not rise above atmospheric-pressure. This occurs at inlet-pressures of a few-millitorr. However, continuous-pumping at such low pressures leads to oil-backstreaming. Thus, oil-sealed pumps are rarely used to pump below ~0.1 torr.

In *rotary-vane pump*s the vanes are spring-loaded and press against the inner-surfaces of the stator. An oil film serves to lubricate all the parts of the pump and also seals the space between the vanes and the housing. The whole rotor-stator assembly in rotary-vane pumps is also submerged in pump oil. Because of the tight fit between the vane and stator wall, rotary-vane pumps can reach lower ultimate pressures than rotary-piston pumps, and are most widely used for non-corrosive gas-pumping applications.

In *rotary-piston pumps* the piston is rotated on an eccentric rotor and slides along the wall of the stator, although it is constrained from contacting the stator. (Oil is used to seal the space between the fixed and moving components.) The clearances of piston pumps are greater than those of vane-pumps making them more tolerant of particle contamination. Also, their simplicity helps increase ruggedness. Since there are no vanes to stick or rub against the stator housing, they pump as long as the rotor can be turned. In addition, pistons are more corrosion-resistant than vanes. Thus, piston pumps are better for applications that require pumping of reactive gases.

6.6.2 Roots-Blower/Booster Pumps

Roots-pumps (also often referred to as *Roots-blowers*, or *boosters*) are single-stage mechanical pumps typically used in series with rotary pumps. In processes in which the pressure is >150-mtorr (or the pumping-speed is less than ~60-L/sec), rotary-pumps can be used alone. However, if pressures lower than 150-mtorr are needed, it is necessary to use a combination of a Roots-pump and a rotary-pump. Base pressures of 10–20-mtorr can be obtained in this manner. In addition, if higher throughputs at a constant pressure are needed, these can likewise be achieved by using a Roots-pump and an oil-sealed backing-pump, because together they increase the system pumping-speed. These pumps are also well-proven, sturdy, and reliable. Roots-pumps contain two counter-rotating lobes mounted on parallel shafts that rotate synchronously in opposite directions at speeds of 3000–3500 rpm (Fig. 6-16a and Fig. 6-17). The lobes do not touch each other or the housing, with clearances being about

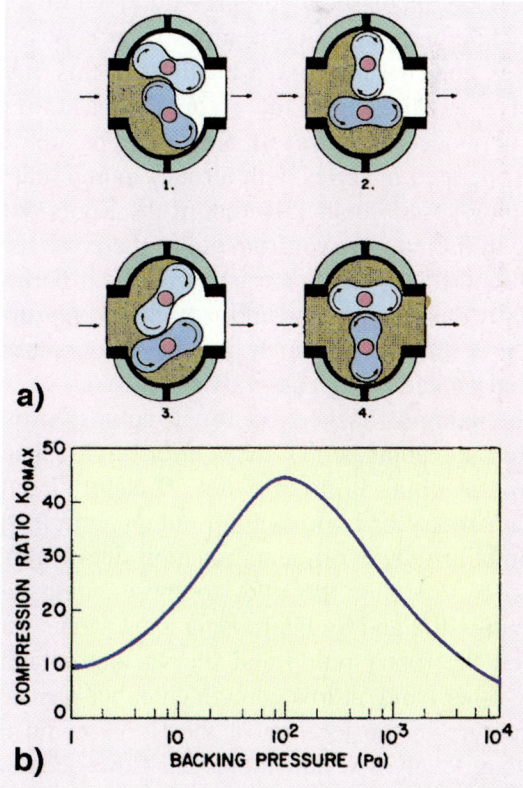

Fig. 6-16 (a) Operating-principle of the Roots-pump. (b) Dependence of the *compression-ratio* of a Roots pump on the *outlet*- (or *backing*-) pressure.

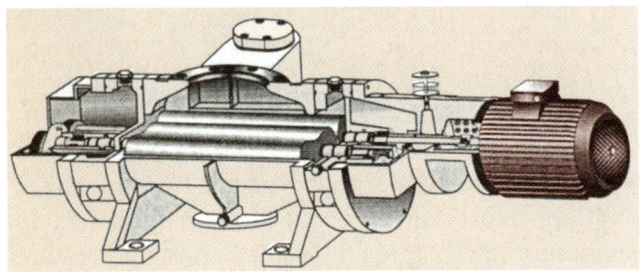

Fig. 6-17 Cutaway view of a Roots-pump.

0.1-mm. Since no oil is used to seal the gaps, such pumps are termed *dry* pumps. However, oil *is* used to lubricate pump-bearings.

The compression-ratio of the Roots-pump is *pressure-dependent*, generally having a maximum (between 10 and 100) near 1-torr. At higher pressures, the compression-ratio is lower because the gap-conductance increases with pressure (Fig. 6-16b). Thus, while pumping against atmosphere is possible, it causes considerable pump-heating. If this causes excessive expansion of the rotors, damage to the pump may occur. As a result, Roots-pumps are commonly operated in series with a rotary-pump that has a pumping-speed about 1/10 that of the Roots-pump. During initial-pumpdown from atmosphere, a bypass-line around the Roots-pump is opened. All pumping is done by the rotary-pump until a backing-pressure of ~15-torr is reached, at which time the Roots-pump is switched-on and the bypass-valve is closed.

Roots-pumps (backed by rough-pumps) provide two key advantages: **1)** they offer high-capacity, low-cost pumping in a range not efficiently handled by either rotary or high-vacuum pumps; and **2)** they substantially reduce oil-contamination due to rotary-pumps, by keeping the inlet to these pumps at a higher-pressure and by interposing a dry Roots-pump between the rotary-pump and the vacuum-chamber. On the other hand, at low enough chamber-pressures, the pressure in the foreline of the Roots-pump may drop to a level at which molecular-flow occurs. In such cases, oil can backstream to the Roots-pump, creep around its interior-surfaces, and finally enter the process-chamber. This limits the use of Roots-plus oil-sealed rotary-pump combinations to conditions at which the Roots-forepressure is kept above 0.1-torr. Replacing the oil-sealed rotary backing-pump with a dry-pump eases this restriction.

6.6.3 Dry Mechanical Rough-Pumps

Oil-sealed mechanical-pumps have some serious drawbacks, and these led vacuum-pump manufacturers to design rough-pumps that do not rely on oil-sealing. These are called *compression-type "dry" pumps* (Fig. 6-18). Such pumps can operate effectively from atmosphere down to about 10^{-2}-torr and have replaced oil-sealed rotary-pumps in many applications. Such replacements have been made mainly in processes that present harsh pumping-environments for oil-sealed pumps - such as the dry-etching of Al-films and the deposition of silicon-nitride by CVD. In these applications, expensive pump-oils have to be frequently replaced when they become saturated with toxic and corrosive gases (and their by-products). In addition, solid-particle by-products of such processes become trapped in the pump-oils and grind away on the pumps themselves. Damage can be severe enough so the pumps may have to be replaced.

Toxic and corrosive materials do not build-up in dry-pumps (Fig. 6-19), since there is no oil in them.

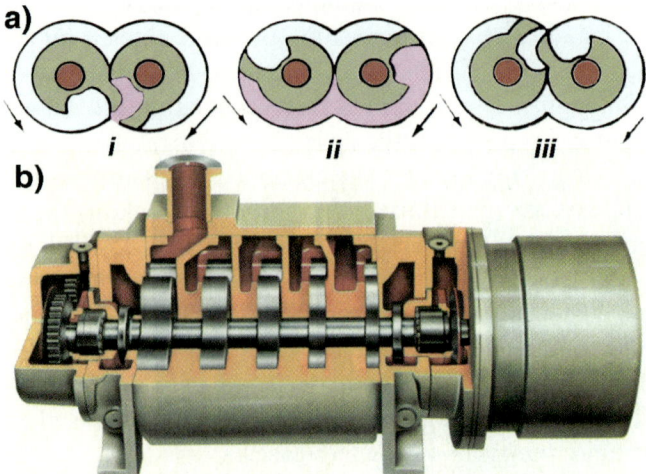

Fig. 6-18 (a) Principle of operation of a *hook-and-claw* dry-pump vacuum stage. (b) Cutaway-view of a mechanical dry-pump. Courtesy of Ebara Technologies.

CHAPTER 6 VACUUM TECHNOLOGY FOR ULSI APPLICATIONS

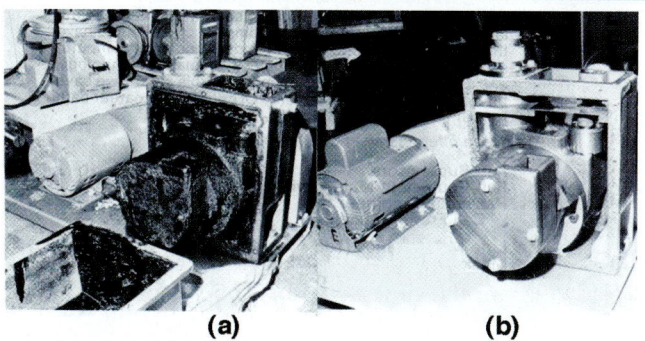

Fig. 6-19 (a) Sludge generated on the inside of a rotary-pump from hydrocarbon-pump-fluid. (b) Same pump-insides when operated with perfluorinated-polyether fluid.

Instead, these harsh chemicals simply pass through the pump to a scrubber. Also, the cost of initially filling and periodically replenishing the pump with expensive oil is eliminated. Finally, risk of process contamination from pump-oil vapors is significantly reduced (though not entirely eliminated). That is, oil is still used in the gear-boxes of these pumps, but it is separated from the vacuum-side by a shaft-seal.

The main disadvantage of such dry pumps is their initial expense. (Dry-pumps cost three or four times as much as the oil-sealed rotary-pumps they replace.) However, if the replacement involves a process where the cost of the pump-oil is as high as $2000 per month (and maintenance costs are on top of that), dry-pumps will eventually pay for themselves (frequently within one to two years). Nevertheless, not all applications can justify replacing an oil-sealed pump with a dry-pump for economic reasons alone, especially if the application requires the pump to handle only non-corrosive environments.

6.7 HIGH-VACUUM-PUMPS I: CRYOGENIC PUMPS

Cryogenic-pumps use the fact that gas-molecules "freeze-out" on cold surfaces. (The frost that collects on the inside of a refrigerator is a cryogenic phenomenon.) High-vacuum cryogenic-pumps remove gases from vacuum-chambers by capturing these gases on cold surfaces. They efficiently pump all gases in the pressure-range of 5×10^{-3} to 10^{-10}-torr. Since cryo-pumps ("cryos") are *entrapment-pumps* they do not need backing-pumps. They are also highly reliable, being able to run for two to five years without maintenance (except regeneration). In the semiconductor-industry they are most widely used in sputtering-tools and on the end-stations and beamlines of ion-implanters. One of their main advantages over turbopumps is that they pump water-vapor (and H_2) much faster. Water-vapor is often the most prevalent residual-gas in vacuum environments.

6.7.1 Cryopump Operation

The housing of cryopumps is typically a stainless-steel or aluminum vacuum-vessel with a high-vacuum flange for mounting the pump directly onto the high-vacuum valve of a process-chamber (Fig. 6-20). The refrigerators have two stages so that they can cool two different surfaces, with each being held at a specific temperature (i.e., one at 65 K and the other at about 12 K). They operate much like household-refrigerators, in that the refrigerant (helium in this case) is compressed, cooled, and then allowed to expand. The closed-loop refrigeration-system is usually split into two parts: **1)** a two-stage, *mechanical reciprocating-expander* (or *cold-head*) mounted within the vacuum-vessel itself; and **2)** a remotely-located compressor (Fig. 6-20). The role of the compressor is to take helium-gas returning from the expander at 100-psi and boost this pressure to 300-psi before sending it back to the pump. The compressor is connected to the

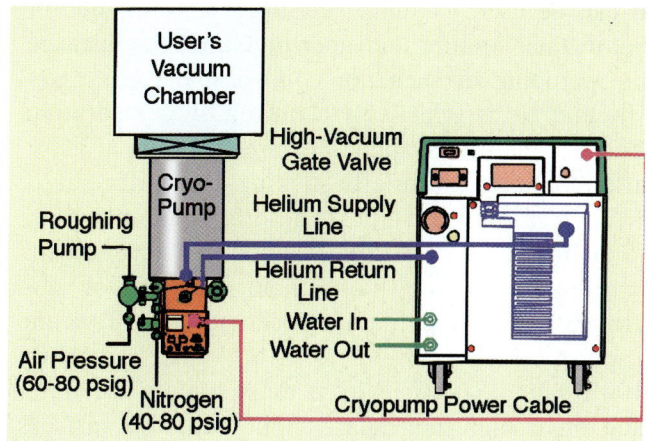

Fig. 6-20 Cryopump and refrigerator-unit.

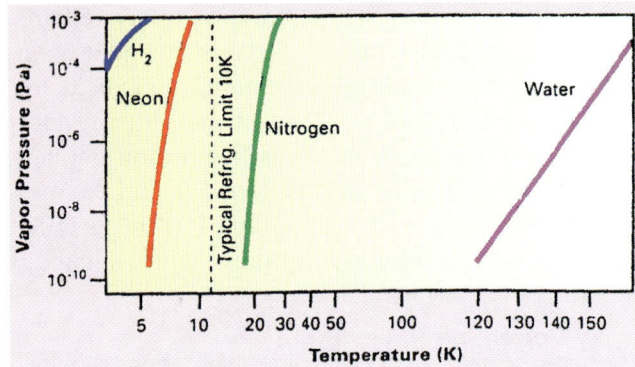

Fig. 6-21 Vapor-pressures of various gases vs temperature.

expander by helium lines typically 10–60 feet long. Cooling occurs by *adiabatic-expansion*.

The process of gas capture by cryopumps involves both cryocondensation and cryosorption. *Cryocondensation* refers to the condensation of a vapor on a cold surface so that its vapor-pressure becomes low enough that it is effectively removed from the process-chamber. Cryocondensation is thus utilized to pump *condensable-gases*.

Cryosorption involves the *adsorption without condensation* of a gas on a cold-surface. A solid-surface always exhibits a weak attraction for at least a few monolayers of gas or vapor. After a few monolayers have been deposited, the force decreases and the surface ceases to act as an effective pumping-mechanism. Cryosorption is an important phenomenon, as it can be used to pump substances to pressures far below their equilibrium vapor-pressures. For instance, the pumping of such non-condensable gases as H, He, and Ne is effectively achieved by cryosorption. Because their vapor-pressures are still high at 20 K, after colliding and briefly sticking to a surface at 20 K, such gases would quickly desorb. Thus, attempting to pump them by cryocondensation will not succeed.

Figure 6-21 shows the vapor pressure of several gases as a function of temperature. At 130 K the vapor-pressure of water is $\sim 10^{-10}$-torr. Essentially 100% pumping-efficiency is thus obtained for water and other high-molecular-weight gases by surfaces cooled to temperatures ≤ 130 K. However, temperatures below 20 K are necessary for condensing such gases as O_2, N_2, and Ar. Note that at such temperatures these gases become dense, ice-like solids.

Cryopumps use three pumping zones to entrap gases (Fig. 6-22). The *first stage* of the refrigerator (which is a surface held at 65 K), is connected to the inlet array of the pump (or *baffle*) and to the thermal radiation shield. These two fixtures are thus kept at 65 K, and are the *first pumping-zone*. The main function of this zone is to condense water and other Type-I gases (i.e., those that condense above 80 K).

Within the pump-volume enclosed by these two 65 K components are the other two pumping-zones, held at 12 K by the *second-stage* of the refrigerator (a 12 K surface). The role of the first 12 K-zone (the *cryocondensing-zone*) is to entrap N_2, O_2, Ar, and other Type-II gases (i.e., those that condense only when T is below 80 K). The function of the third

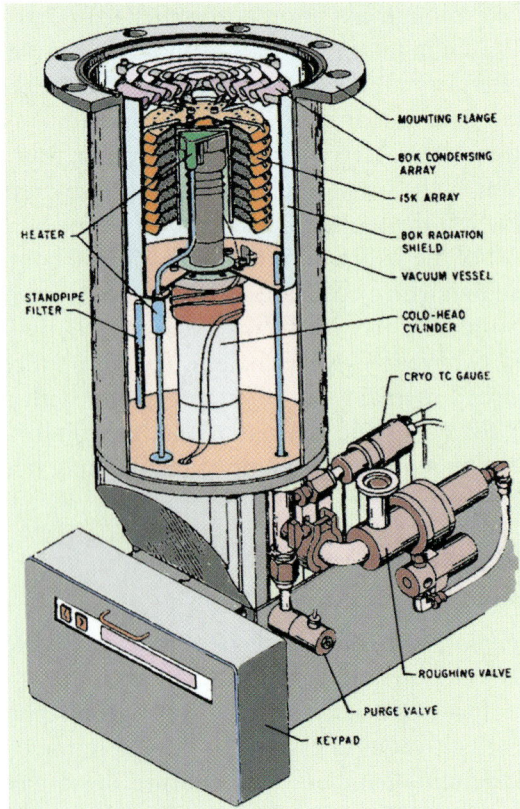

Fig. 6-22 Cutaway view of a cryopump. (On-Board 8" cryopump, courtesy of CTI-Cryogenics.)

zone (the *cryosorbing-zone*) is to immobilize H, He, and Ne (i.e., the Type-III, or *non-condensable* gases). The cryosorption-zone is shielded as much as possible from the inlet to lengthen and randomize the path that a molecule entering the pump must traverse before striking the cryosorption-surface. This increases the probability that Type-II gases will cryo-condense on the first 12 K zone (allowing the limited 12 K adsorbent-surface to remain available for pumping Type-III gases).

Activated-charcoal is the most common cryosorption material used, because it has a very-large surface-to-volume ratio (and it can be regenerated at ambient temperature). It consists of porous-granules of charcoal arranged as an interconnected network of minute-channels with diameters ranging from 10-30-Å.

6.7.2 Cryopump Regeneration

Even under normal-operation the surfaces of the cryopump will eventually become less-effective at capturing gases. When the pump becomes saturated (full of either solidified or crysorbed gas-species) the base-pressure in the chamber increases. At that point, the accumulated gases must be removed from the pump, a process referred to as *regeneration*.

Regeneration, is carried out in three steps: **1)** warmup; **2)** pumpout; and **3)** chilldown. Fully-automated, single-push-button *regeneration-controllers* are available, designed to minimize regeneration-time and reduce labor costs. Regeneration is usually done on a regular basis (such as once every week or month), scheduled to coincide with other routine system-maintenance. Generally, cryopumps are not used for processes that use large quantities of toxic or corrosive gases, because they could be accidentally released if there was a loss of cooling-power.

6.8 HIGH-VACUUM PUMPS II: TURBOMOLECULAR PUMPS

The *turbomolecular-pump* (*TMP*) is a compression-type pump that functions by imparting momentum from high-speed rotating blades to the gas-molecules. It is a very popular high-vacuum pump as it provides clean-vacuum with high reliability and ease of opera-

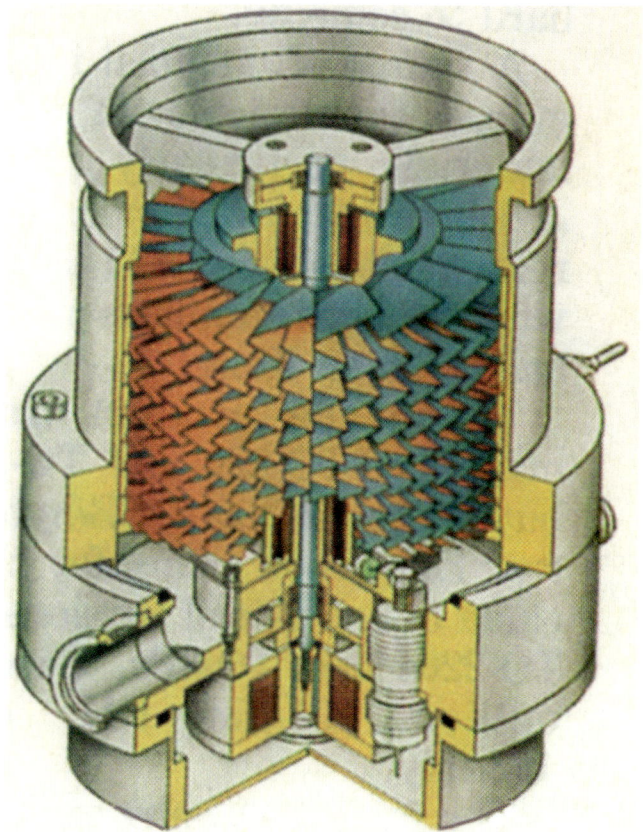

Fig. 6-23 Cut-away view of a turbopump.

tion down to pressures of 10^{-10} torr. TMPs with pumping-speeds up to about 3000-L/sec are available.

A series of blades (much like those used in turbines) are assembled on a single shaft supported by two-sets of bearings and driven by a high-rpm motor (Fig. 6-23). This assembly is designed to rotate at speeds ranging from 24,000–90,000-rpm, making the rotor-speeds comparable to the speed of the molecules of the pumped-gas. The spaces along the axis of the shaft are alternated between rotating-blades (*rotors*) and stationary-blades (*stators*), with clearances between them being less than a millimeter (Fig. 6-24). The rotor-impulse transmitted to the molecules changes their motion from being randomly-directed to one with the same direction as the blade (Fig. 6-25a). The deflected-molecules are thus driven toward the interior of the pump, which is the basis of the pump-

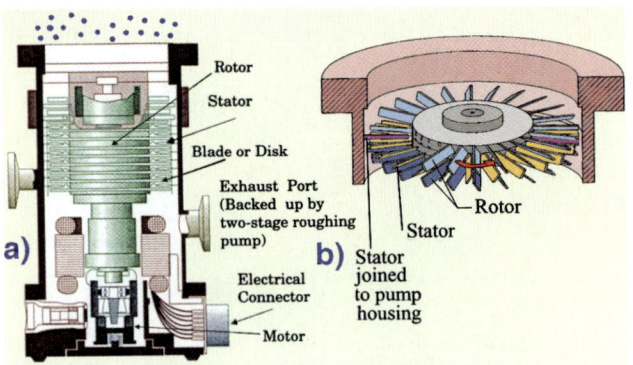

Fig. 6-24 (a) Cross-sectional view of a turbopump. (b) Single rotor-and-stator stage of a turbopump.

ing action. When the mean-free-path of the molecules is larger than the spacing between the rotor and stator blades, molecular-flow exists ($\leq 10^{-3}$-torr). In this flow-regime, gas-molecules collide mostly with the pump-blades (and not with each other), resulting in an efficient pumping-process. As the pumped-gas moves from the inlet further into the pump, it is continually-compressed by impact from the rotor-blades. Since each pair of rotor and stator blades represents a single pumping-stage with a relatively small compression-ratio, a large number of stages (9 to 13) is cascaded to achieve adequate pumping-capability. For a series of stages, the total pump-compression-ratio is approximately the product of the compression-ratios of each stage. Because the throughput of each stage is constant (and is also given by the product of the pressure and the pumping-speed), the blades nearest the inlet of the pump are designed to have a high pumping-speed and low-compression. Thus, the blades near the pump-inlet are strongly inclined (to make the pump inlet more "open" to incoming molecules, see Fig. 6-23b). Conversely, those blades nearest the outlet of the pump are designed for high-compression and low-pumping speed (and hence are low-angle-inclined blades). The compression-ratio is higher for heavy molecules, and hence the overall-compression of TMPs of light-gases (H and He) is smaller than that of heavier gases such as nitrogen (see Fig. 6-25b).

If the pressure throughout the entire pump is kept within the range of molecular-flow, efficient transfer of momentum takes place from the pump to the gas. On the other hand, if viscous-flow prevails (i.e., due to higher-pressures developing near the outlet of the pump), intermolecular-collisions will dominate over collisions with the blades. In this case, momentum is not efficiently transferred to the pumped gas, and the pump operates poorly. As a result, TMPs require a backing-pump to keep the forepressure low-enough to sustain molecular-flow everywhere in the pump. If the forepressure is allowed to rise to the point of viscous-flow ($\sim 10^{-2}$ to 10^{-1}-torr), the pump experiences a sudden decrease in rotor-speed (and thereby also pumping-speed). TMPs will operate at constant speed over their normal operating-ranges. Their ultimate-pressure ($\sim 10^{-10}$-torr) is limited by their compression-ratio for light-gases and the effects of outgassing.

Most commercial-TMPs are equipped with either mechanical rotor-bearings or magnetically-suspended bearings. *Ceramic-balls* are widely used today in the mechanical-bearings, as they are harder and their surfaces smoother than those of metal-balls, making them more reliable (even under lubrication-starved conditions). In TMPs with magnetically-suspended bearings the rotor is levitated by electro-magnets

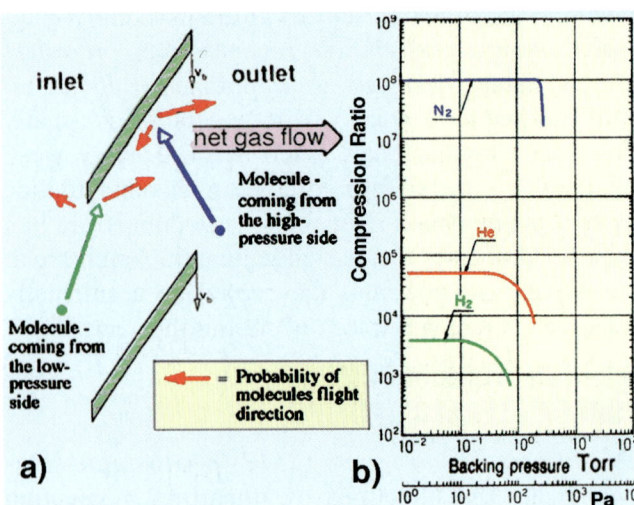

Fig. 6-25 (a) Physical basis for the pumping-mechanism in turbomolecular pumps. (b) Pumping-speed curves of a turbomolecular pump.

(Mag-lev), so there is no mechanical friction. Hence, no wear occurs (although back-up or touch-down bearings are typically used to protect the pump in case of an inadvertent power-failure). These back-up bearings have a finite-lifetime of 3 or 4 touchdowns (since they are not lubricated). Mag-lev bearings also eliminate the risk of oil-backstreaming from the lubricant materials of mechanical bearings. In non-corrosive pumping-applications such Mag-lev TMPs can operate almost indefinitely with little or no maintenance. Mag-lev TMP pumps are somewhat more expensive than those with mechanical-bearings.

Turbomolecular-pumps offer several advantages over cryopumps including: **a**) continuous pumping with low-maintenance (e.g., no regeneration is required); and **b**) the ability to pump toxic, corrosive, or explosive gases (because there is no material buildup within the pump). Although they were originally designed with high-vacuum applications in mind, TMPs are now also finding use in newer CVD and etch-processes involving high-density-plasma sources. These require process-pressures in the lower-end of the medium-vacuum range (about 40×10^{-3}-torr), together with high pumping-speeds and high-throughputs. TMPs (often with a molecular-drag stage) are employed as upstream-boosters for Roots-blowers (which by themselves would not have sufficient throughput at such pressure-ranges). To limit particle formation, some TMPs can be heated. This helps prevent reactant and by-product gases from condensing into particles in the pump. However, since three pumps are used for a single vacuum-application (i.e., a turbopump, Roots-blower, and mechanical backing-pump) this is quite an expensive solution.

Potential limitations of TMPs include: **a**) poor compression-ratios for pumping light-gases; **b**) relatively high initial-cost (i.e., they are more costly than comparably-sized cryopumps); **c**) special design of the rotors and bearing-assembly to resist corrosion if reactive-gases must be pumped; **d**) potential-damage or destruction of the rotor by particles falling into the pump (this is minimized by the placement of fine screens over the pump inlet, but pumping speed is thus decreased); **e**) bearing maintenance; **f**) the possibility of backstreaming of mechanical-pump oil during roughing or turbopump-shutdown (these can be prevented by proper venting-procedures initiated in the event of pump-shutdown or power-outage); and **g**) the possibility of pump-failure during a sudden in-rush of gas (caused by rotor-blade upward-flexing during a large gas-impulse, resulting in contact of the high-speed rotor-blades with the stator).

With respect to pumping corrosive-gases, early TMPs had some problems. That is, their blades were made of Al-alloys and they were originally designed for handling noncorrosive-gases. Thus, such pumps had lifetimes of only about 50-h when they were used to pump chlorinated-gases. By redesigning them for such applications, these difficulties were overcome. The chief improvements involved utilizing anodized-blades and the use of a continuous-flow of inert-gas (to protect the ball-bearings of the rotor against back-streaming of pumped-gases into the chamber containing the bearings).

The maintenance of turbopumps involves the periodic-lubrication and replacement of any mechanical bearings. The turbo design and lubrication-system varies widely from brand to brand and among models, as do the lubrication-intervals and difficulty of bearing-change. Pump maintenance must be performed by skilled workers (preferably trained by the pump manufacturer), under cleanroom conditions. Typical bearing-lifetimes are in the two-to-five year ranges (15,000–50,000-hours), depending mainly on the process-application.

6.9 VACUUM GAUGES: (TOTAL-PRESSURE MEASUREMENT)

The total gas-pressure is the quantity most often measured to characterize the degree of vacuum existing in a system. The pressures measured in vacuum-technology cover the range from atmospheric-pressure to 10^{-9}-torr - over 11 orders of magnitude! Instruments used to measure total-pressure are known as *vacuum gauges*. Since it is not possible to build a vacuum gauge which can give quantitative-measurements

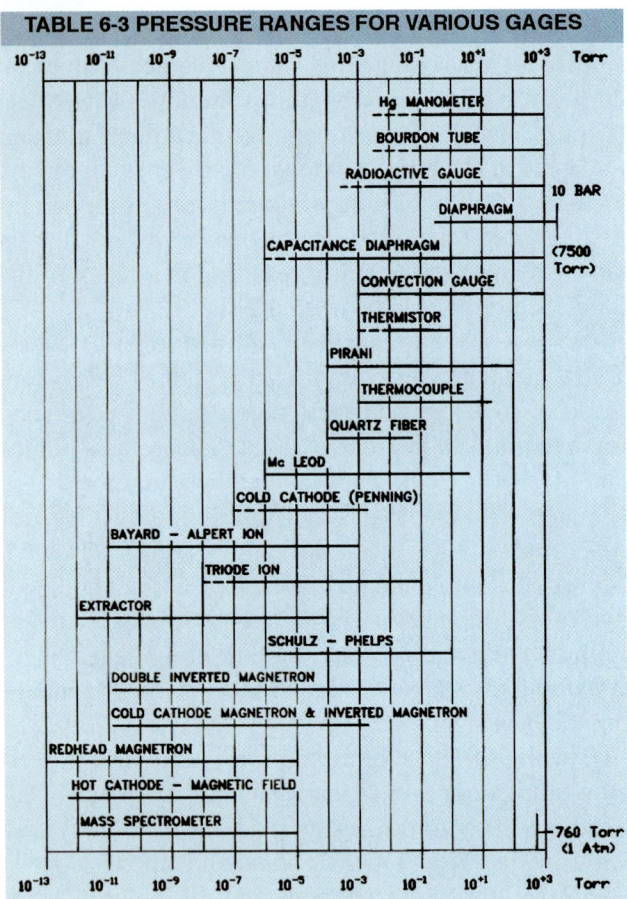

TABLE 6-3 PRESSURE RANGES FOR VARIOUS GAGES

6.9.1 Capacitance-Manometers

Capacitance-manometers (Fig. 6-26a) provide an accurate measure of pressure (e.g., to within ±0.1%) from atmosphere down to about 10^{-2}-torr - a range of about four or five orders of magnitude (Fig. 6-26b). For processes carried out at these pressures, capacitance-manometers are the pressure-gauges of choice. Two of their major advantages are that they detect pressures independent of gas-species and over their useful pressure-range they are the only type of gauge that provides direct and accurate absolute-pressure measurement. All other pressure-gauge measurements are indirect and are accurate only to ±10–15%. Capacitance-manometers are also quite reliable.

Changes in pressure produce a change in the position of a flexible-diaphragm relative to that of a fixed-plate. The capacitance between the diaphragm and the fixed-plate depends on the distance between them. Thus the diaphragm-movement produces a capacitance-change that is converted to a pressure-reading. However, even small changes in ambient-temperature can cause errors in its readings. Temperature-controlled heads or correction-tables built-into the electronics have been used to minimize this problem.

6.9.2 Thermocouple Gauges

Thermocouple (and Pirani) gauges are widely used to measure total pressure in the range from 760-torr to 10^{-3}-torr ($\sim 10^5$–1-Pa). For example, they are employed to measure: **a**) the backing-pressures of turbo-pumps; **b**) the pressure inside a cryopump during chilldown; and **c**) chamber-pressures as they are being rough-pumped during pumpdown.

Within the gauge a filament is heated and an attached thermocouple is used to measure the filament-temperature (Fig. 6-27b). The temperature increases with decreasing-pressure over the pressure-range of the gauge (for constant input-power to the filament). Eventually, at low-enough pressures (e.g., 10^{-3}-torr), the radiative heat-loss from the filament becomes dominant. Since this mechanism is pressure-independent, the gauge ceases to be effective.

It should be noted that high-accuracy from ther-

over the entire ranges of vacuum, a great variety of vacuum-gauges have been developed. Each type has a characteristic measuring-range, which mostly extends over a few orders of magnitude (see Table 6-3).

Since the factors involved in selecting and using vacuum gauges are not as critical as other components of vacuum systems, our discussion will be brief and limited to the following gauges (which are most commonly encountered when measuring total-pressure in microelectronic vacuum-environments):

Low-vacuum gauges (1-torr-1-atm) - Capacitance manometers
Medium-vacuum gauges (10^{-2}-1-torr) - Thermocouple/Pirani gauges
High-vacuum gauges (10^{-6}-10^{-2}-torr) - Ionization gauges

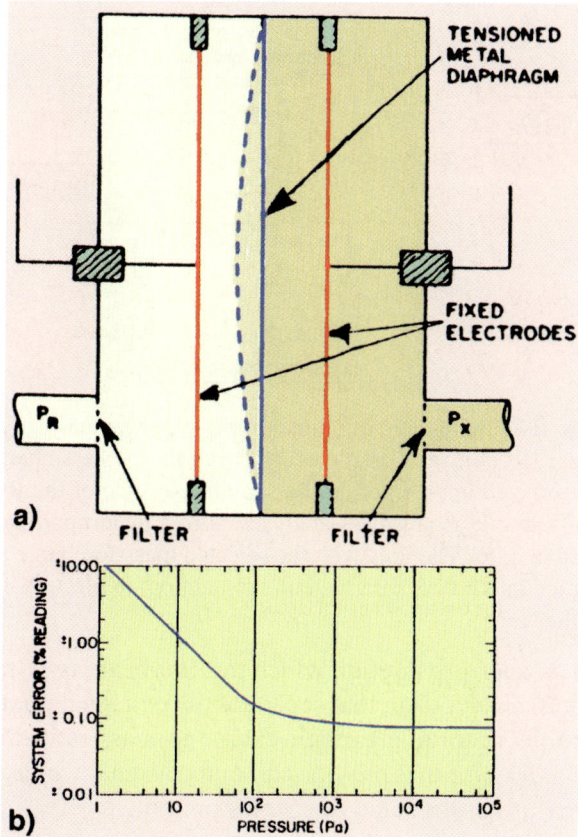

Fig. 6-26 (a) Capacitance-manometer. (b) Performance curve for a capacitance-manometer.

mocouple and Pirani gauges is rarely needed, since it is not called-for in the above applications. Although thermocouple and Pirani gauges are not as accurate over the range where a capacitance-manometer can operate, they are nevertheless much-less expensive, and are thus used where the added accuracy does not justify the extra cost (e.g., while monitoring chamber-pressures during rough-pumping from atmosphere).

6.9.3 Pirani-Gauges

Pirani-gauges measure pressure over the same general range as thermocouple-gauges, although their upper pressure-limit is lower (1×10^{-2}-torr). They utilize a resistance-wire (enclosed in a glass-envelope) which is exposed to the pressure in the vacuum-system while being part of a Wheatstone-bridge circuit (Fig. 6-27a). A fixed-current is passed through the wire-resistor, causing it to be heated. The temperature of the *sensing-wire* decreases as the thermal-conductivity of the gas rises, which decreases the resistance of the wire. Hence, the current in the unbalanced-bridge circuit is an indirect-measure of the pressure.

6.9.4 Ionization (High-Vacuum) Gauges

In ionization-gauges a hot-filament provides electrons that are attracted by a positively-charged grid (Fig. 6-28). These electrons strike and ionize residual gas-atoms in the vacuum. The ions flow to a negatively-charged cathode, and the resulting current is related to the ion-density (or to the pressure in the chamber).

The two most popular ionization-gauges are: **1)** the Bayard-Alpert gauge; and **2)** the Schultz-Phelps gauge. The *Bayard-Alpert gauge* is used for vacuum-levels from 10^{-8}–10^{-3}-torr. Such gauges have a tube containing either a double tungsten-filament (one is a spare), or a single thoriated-tungsten filament. The double-tungsten-filament design is more accurate, but is subject to burn-out if exposed to air at high-pressure. The thoriated-tungsten filament can tolerate many exposures to air.

The *Schultz-Phelps* type gauges are used for reading higher pressures (e.g., in the ranges in which sputtering is carried out, up to ~500 mtorr) because they can tolerate these higher pressures and still operate at pressures as low as 10^{-5} torr.

6.10 MEASUREMENT OF PARTIAL PRESSURE: RESIDUAL-GAS-ANALYZERS

In many vacuum processes it is as important to know the composition of the residual-gas mixture in a pro-

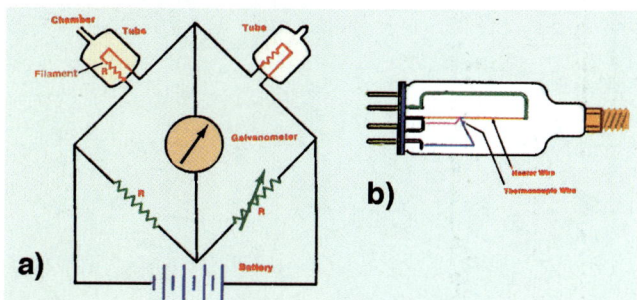

Fig. 6-27 (a) Pirani gauge. (b) Thermocouple gauge.

cess-chamber (in terms of the *partial-pressures*), as it is to know the *total-pressure*. For example, knowledge of partial-pressures can provide information about gas-purity and identify sources of outgassing in process-chambers. Such data can relate film properties to the presence of specific-gases, and can help evaluate the effects of such procedures as baking or glow-discharge cleaning.

Measurement of partial-pressures in high-vacuum systems is typically performed by *residual-gas-analyzers* (*RGA*). RGAs are also used for leak-detection and troubleshooting in vacuum-chambers. Relatively-inexpensive RGAs are now available to measure the presence of contamination in gases over a wide range of pressures. Such RGAs find widespread use as diagnostic and process-control tools in sputtering, CVD, and ion-implantation processes, sometimes even being permanently-mounted onto process-equipment.

6.10.1 Operation of RGAs

RGAs determine the gases present in vacuum-environments by separating, identifying, and measuring the quantity of all gas-molecules in the vacuum-system (Fig. 6-29). There are four parts to an RGA: **1)** an *ionizer* that is responsible for producing a beam of ions from samples of the gas in the process-chamber;

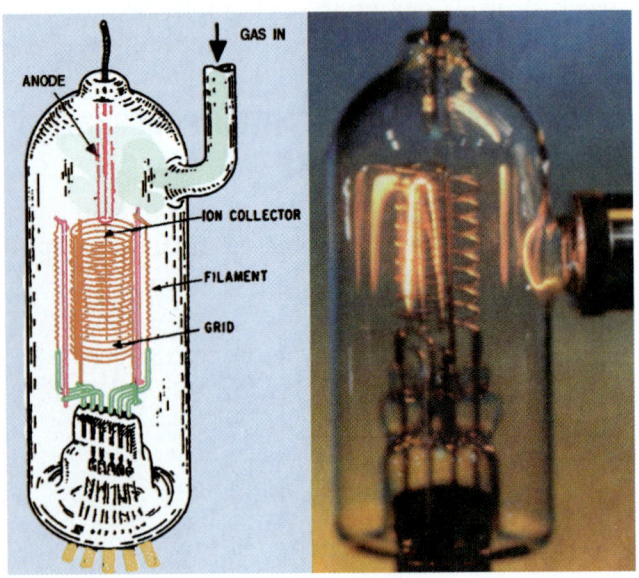

Fig. 6-28 Ionization gauge.

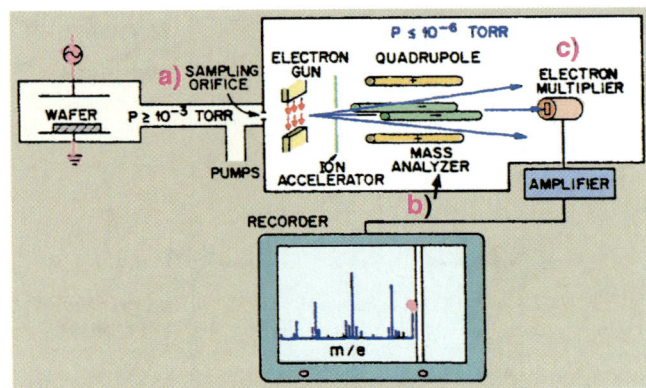

Fig. 6-29 Schematic of a quadrupole-type residual-gas-analyzer (RGA): (a) Gas-molecules from the process-chamber are ionized upon entering the sampling-orifice of the RGA; (b) These ions are mass-analyzed with a quadrupole-mass-analyzer; (c) The ions with the selected mass-to-charge ratio are counted by a Faraday-cup, containing an electron-multiplier-tube.

2) an *aperture* through which this ion-beam is extracted; **3)** an analyzer, that separates the extracted mixture of ions according to their charge-to-mass ratios; and **4)** a detector that provides an output-signal which is a measure of the relative-species present.

6.10.2 Interpretation of RGA-Spectra

The output-data from an RGA is called a *mass-scan*, or *mass-spectrum*. It is represented on a chart with the mass-to-charge ratio on the horizontal-axis, and the relative-intensity on the vertical-axis (Fig. 6-30). Signals from the chamber-gas being analyzed (which correspond to specific mass-to-charge ratios, and their relative intensities) are plotted on the chart. The resulting mass-spectra need to be interpreted so that the recorded peaks can be correlated to the gas-species from which they originated. Ambiguity about the origin of peaks can arise when different molecules have the same mass. This problem can be resolved by considering the dissociation and double-ionization behavior of molecules, processes which are specific for different species. That is, when molecules of a gas are struck by electrons whose energy can cause ionization, fragments of several mass-to-charge ratios are created. This pattern of fragments, called a *crack-*

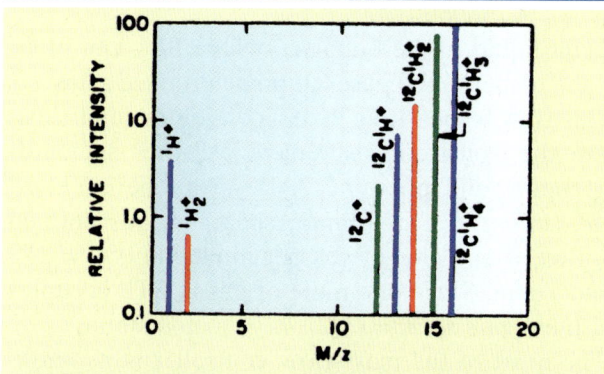

Cracking pattern of methane illustrates the distinctive localization peaks.

Fig. 6-30 Resultant mass-spectrum of the detected gas molecules.

ing-pattern, forms a fingerprint that may be used for absolute identification of a gas or vapor.

Cracking-patterns of hydrocarbon-fragment ions characteristic of common pump-oils have been compiled to allow identification of backstreaming sources and to discriminate against other sources of system contamination. In fact, one method for determining the nature of organics in systems is to become adept at recognizing the cracking-patterns of commonly used solvents, pump-fluids, and elastomers.

Qualitative-analysis of the types of gas and vapor in a vacuum-system are generally of most use to those employing an RGA. In many instances, the detection of a particular species points the way to fixing a leak or correcting a problematic process-step. An inexpensive-RGA tuned to the mass of the offending-vapor is used to indicate when that vapor has exceeded a predetermined partial-pressure. Many gases, cleaning-solvent residues, and traces of pumping-fluids can be readily identified by their spectra.

Simple RGAs have a mass-range of 1-50-amu and a resolving-power of 1–3-amu. Such instruments are capable of performing routine-monitoring of background-gases (fixed gases up to mass 44 and hydrocarbons at mass numbers 39, 41, and 43). More elaborate instruments are needed for detailed RGA applications. For example a mass-range of 1–200-amu permits identification of most pump-oils and many heavy-solvents.

6.11 MASS-FLOW CONTROLLERS (MFC)

CVD reactors, dry-etchers and diffusion-furnaces operate as open-flow systems. Thus, they require that process-gases be fed into the reaction-chamber in a controlled manner. This metering of gases is done most commonly using an instrument called a *mass-flow controller* (*MFC*). Mass-flow controllers consist of a mass-flowmeter, a controller, and a valve. They are located between the gas-source and chamber (or before the bubbler if they meter a carrier-gas into the

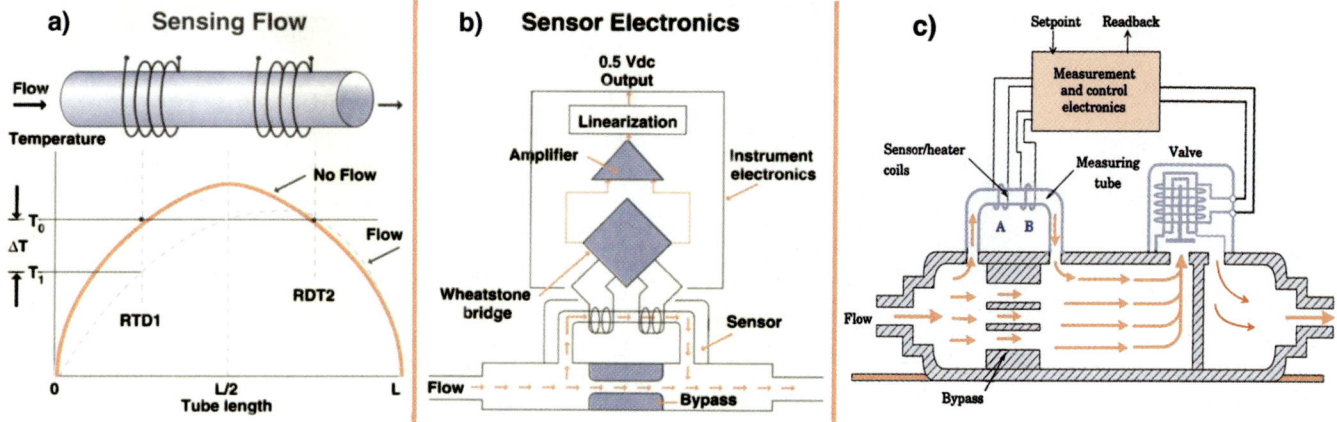

Fig. 6-31 Operating principle of a mass-flow controller (MFC): (a) The difference in temperature ΔT between the two resistance-temperarature-detectors (RTDs) corresponds to the gas flow-rate. (b) MFC sensor-electronics. (c) Schematic-diagram showing gas flowing through a MFC.

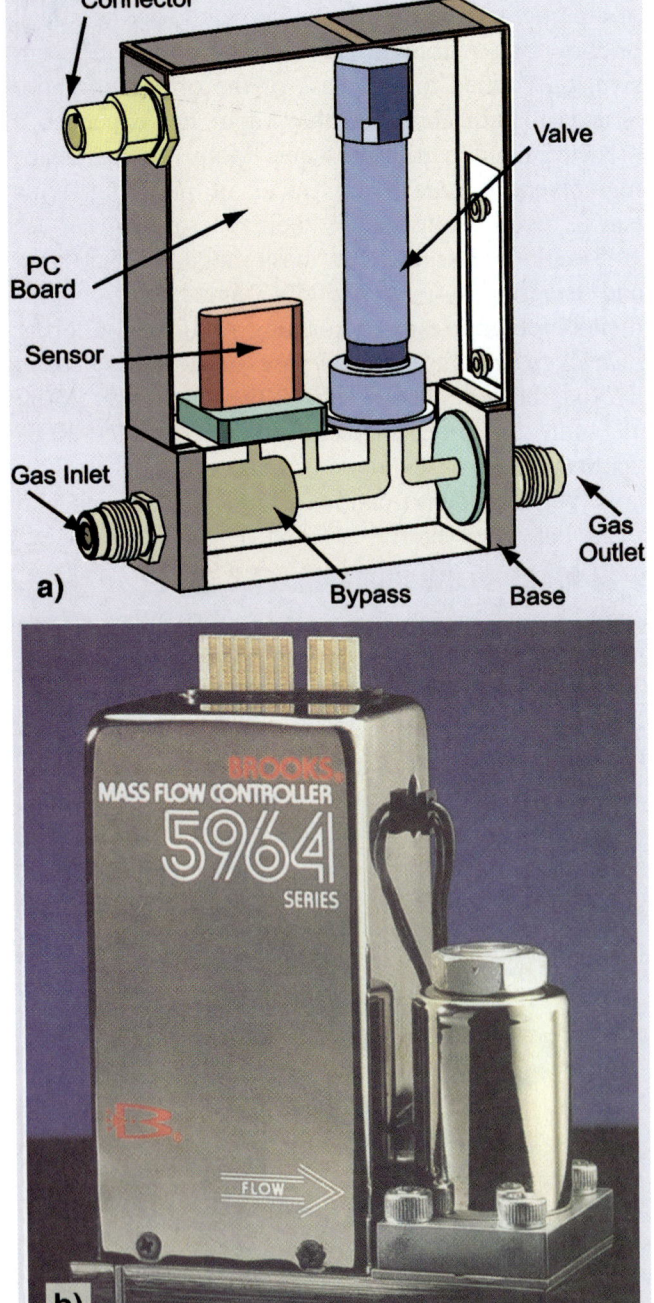

Fig. 6-32 (a) Cross-section; and (b) Photograph of a *mass-flow-controller*. Courtesy of Brooks Instrument.

reactant-liquid in the bubbler), where they can monitor and dispense the gases at predetermined rates.

Gas-flow is measured in units of volume/unit-time, where the volume-measurement assumes standard-temperature and pressure. A flow of one *standard cubic-centimeter per minute* (sccm) is thus defined as a flux of one cm^3 of gas per minute at 0°C = 273 K and 1-atm. Since one-mole of gas at STP occupies 22.4 liters, *one sccm* = $(1/22.4 \times 10^3)$ moles/min.

The heart of the mass-flow controller is its *mass-flowmeter*, of which there are two types: **1)** the *differential-pressure type*; and **2)** the *thermal-type*. In semiconductor applications, the thermal-type is more widely used, as a result of its relatively-rapid response time (e.g., 2–3 sec), and its lower cost.

The operation of thermal-MFCs relies on the ability of a flowing-gas to transfer heat. A schematic of thermal mass-flow-controller operation is shown in Fig. 6-31a. The MFC contains a heated gas-measuring tube with two temperature-sensors. When no gas is flowing, these sensors are at the same temperature. With the introduction of a gas-flow, the downstream sensor reads higher. The temperature difference between the two sensors is related to the amount of gas that has moved downstream. The MFC has a feedback-mechanism and flow-control valve to keep a steady amount of material flowing through it (Fig. 6-31c and Fig. 6-32).

REFERENCES

1. S. Wolf and R.N. Tauber, *Silicon Processing for the VLSI Era, Vol. 1 - Process Technology,* 2nd Ed., Lattice Press, Sunset Beach CA 2000, Ch. 3 (Vacuum Technology).

2. *Handbbok of Vacuum Science and Technology,* Eds. D,M Hoffman et al., Academic Press, San Diego CA, 1998.

3. J. F. O'Hanlon, *A Users Guide to Vacuum Technology,* John Wiley & Sons, New York, 2nd Ed. 1989.

4. M.H. Hablanian, *High-Vacuum Techniques: A Practical Guide,* M. Dekker, New York 1990.

5 A. Roth, *Vacuum Sealing Techniques,* American Instiute of Physics, New Yory, 1994.

6. L Peters, "Residual Gas Analysis: Technology at a Crossroads," *Semiconductor International,* Oct. 1997, p. 95.

CHAPTER 6 VACUUM TECHNOLOGY FOR ULSI APPLICATIONS

PROBLEMS

1. Define the *average-speed* and the *mean-free-path* of a molecule in a gas.

2. How many molecules of O_2 are in a room temperature 10-liter chamber at a pressure of 1-mtorr? Repeat the calculation for 1-cm^3 and 1-μm^3.

3. An ultra-high vacuum system operates at a pressure of 10^{-8} Pa. What is the concentration of residual air molecules in the chamber at 300 K?

4. The *mean-free-path* of a molecule is given by $\lambda = kT/(\sqrt{2}\pi d^2 P)$, where P is the gas pressure, d is the diameter of the molecule, T is the temperature (in K), and k is Boltzmann's constant. Estimate the mean-free-path for: (a) a sputtering system with an Ar-pressure of 4-mtorr (d = 0.36-nm); (b) a CVD-W deposition-system with a hydrogen-pressure of 30-mtorr (d_{WF_6} = 0.56-nm, d_{H_2} = 0.28-nm) and (c) an O_3: TEOS system with an N_2-pressure of 600-torr (d_{N_2} = 0.37-nm, and d_{TEOS} = 2-nm).

5. The distance between the source and the wafer in a deposition chamber is 20-cm. Estimate the pressure at which this distance becomes 1% of the mean-free-path of the source molecules.

6. Define: (a) viscous flow; (b) molecular flow; (c) laminar flow; and (d) turbulent flow.

7. If the transition between viscous flow and molecular flow occurs when the mean-free-path is 1-cm, at what pressure will this occur? (Assume the molecule diameter is 0.3-nm).

8. Assume the residual-pressure of oxygen in a vacuum system is 1-Pa. How long does it take to deposit one atomic-layer of oxygen on the surface of the wafer at 300 K? The impingement rate z is given by:
$$z = 2.63 \times 10^{20} \ (P/\sqrt{MT})$$
where P is the gas-pressure, T is temperature in K, and M is the molecular weight (e.g., the molecular weight of O_2 = 32). The radius of an O_2 molecule is 0.36-nm.

9. A silicon wafer sits on a bench in a laboratory at a temperature of 300 K, and a pressure of 1-atm. Assume the air consists of 100% oxygen. How long does it take to deposit one atomic-layer of oxygen on the wafer surface, assuming 100% adhesion?

10. Explain why thermal-conductivity-type pressure gauges will not work in an ultra-high vacuum.

11. Why are entrapment pumps generally not used to pump toxic-gases?

12. Find the average molecular-speed of air molecules at 300 K. The molecular-weight of air is 29.

13. For a gas of 0.3-nm-diameter molecules at 1-bar pressure at 25°C: (a) What fraction of the total volume is occupied by the molecules? and (b) What is the ratio of the mean-free-path to the molecular-diameter?

14. Explain why cryocondensation is ineffective for pumping the gases of He and H_2 at 15-22 K.

15. Cite at least 2 specific applications in which each of the two high-vacuum pumps described earlier would be best suited for the required task. Cite two disadvantages of each of the high-vacuum pump types.

16. Convert the following pressure-values from torr-units to Pa-units: (a) 1-torr; (b) 3-mtorr; (c) 5×10^{-6}-torr. Convert the following pressure values from Pa to torr: (d) 150-Pa; and (e) 2×10^{-2}-Pa.

Single Atom Arrives

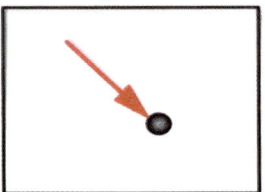

Surface Migration & Evaporation

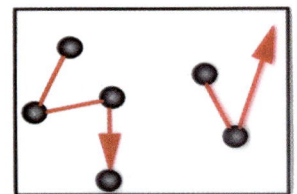

Coating Flux

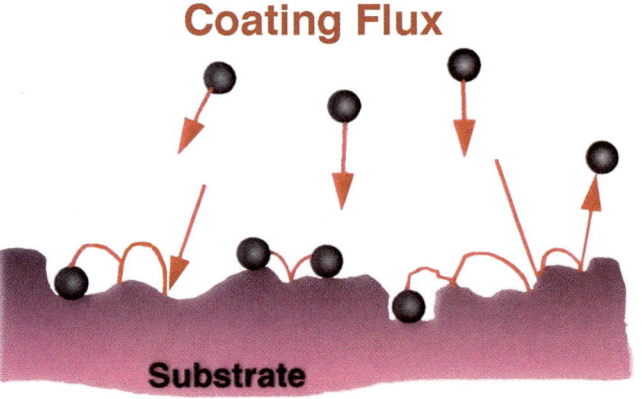

Collision and Combination of Single Atoms

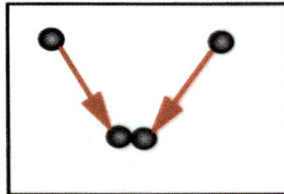  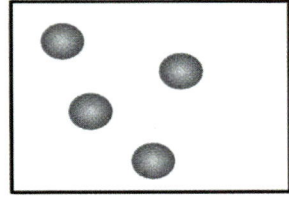

Nucleation of Islands

Mass Transport

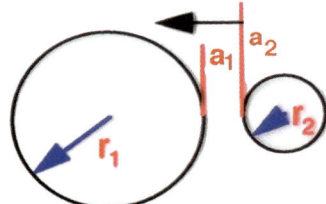

Coalescence of Islands is Due to Ostwald Ripening

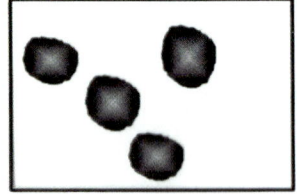 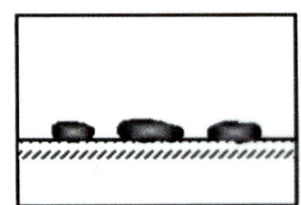 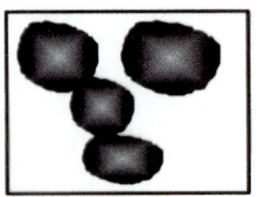 

Top View   **Side View**   **Coalescence**   **Continuity**

Columnar Film

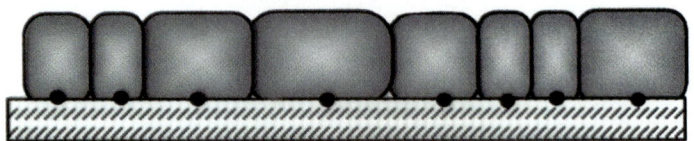

Substrate

Rougher Film

Substrate

CHAPTER 7

BASICS OF THIN-FILMS

CHAPTER CONTENTS

7.1 STAGES OF GROWTH OF AMORPHOUS AND POLY-CRYSTALLINE THIN-FILMS

7.2 MECHANICAL-PROPERTIES OF THIN FILMS

7.3 MEASUREMENT OF ELECTRICAL PROPERTIES OF THIN-FILMS

Various *thin-films* are used in the fabrication of ULSI devices.[1,2,3,5] Examples of some are illustrated in Fig. 7-1, which depicts the cross-section of a MOSFET and interconnect-layers on an IC. It can be seen that thin-films used in integrated-circuits include metals, semiconductors, and insulators. Their structure can be *crystalline* (e.g., epitaxial-silicon layers), *polycrystalline* (polysilicon, aluminum, tungsten, and copper films), or *amorphous* (silicon-dioxide or silicon-nitride films). They can also be deposited using a variety of different techniques (the subjects of later chapters of the book).

Since thin-films are so widely used in VLSI and ULSI devices, it is important to have an understanding of their chemical and physical properties, and the mechanisms by which they are formed. The purpose of this chapter is to provide such background information on thin-films.

The thin-films used in ULSI fabrication must satisfy many chemical, structural, and electrical requirements. For example, film-composition and thickness must be strictly controlled to allow effective-etching of IC structures. Very-low densities of both particulate-defects and film-imperfections (such as pin-holes), become critical for the small-linewidths,

"To render a problem tractable, we must concentrate on its most important features."

Andrew. S. Grove, *Physics and Technology of Semiconductor Devices,* Wiley, 1967.

Illustration of the various stages in the growth of polycrystalline and amorphous thin-films.

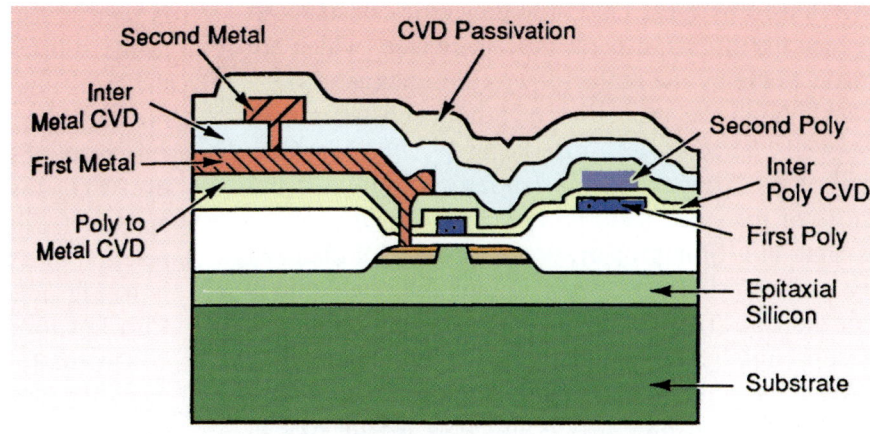

Fig. 7-1 Cross-section of an IC device-structure showing examples of the many kinds of thin-films used to build them.

111

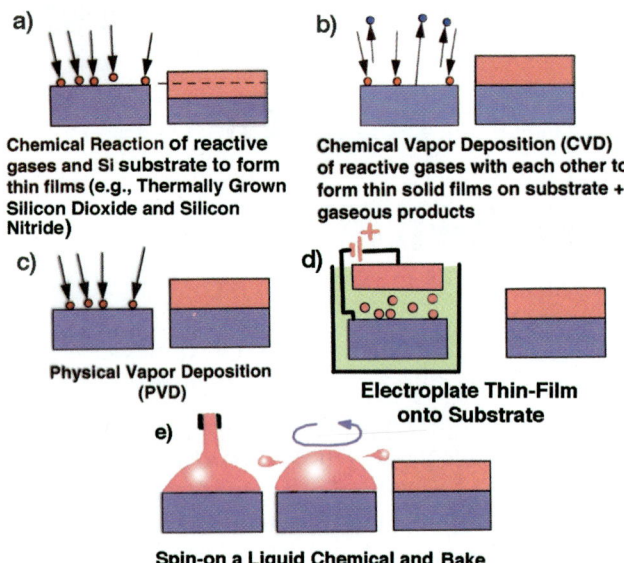

Fig. 7-2 Various methods used to form thin-films in ICs include: (a) Thermal-oxidation; (b) CVD; (c) PVD (d) Electroplating; and (e) Spin-On Liquid-Chemical & Bake.

high-densities, and large areas necessary for ULSI. These small geometries also create highly rugged topography for overlying films to cover. Therefore, excellent-adhesion, low-stress, and conformal-step coverage are required of a ULSI thin-film. If possible, its own surface contours should reduce or even planarize the underlying-steps. Finally, non-conducting thin-films must have low dielectric-constants to reduce the effects of parasitic-capacitance (which are made worse by the scaled-down film-thicknesses).

Although the properties of a bulk-substance may be well characterized, material in its *thin-film form* may have properties substantially different from those of the same material in *bulk-form*. One reason is that thin-film properties are strongly influenced by surface-properties, while in bulk-materials this is not the case. (A thin-film, by its very definition, has a substantially higher surface-to-volume ratio than does a bulk-material.) The structure of thin-films and their method of preparation also play a vital role in determining the film-properties.

The formation of such films is accomplished by a large variety of techniques that can be divided into three groups (see Fig. 7-2):

Group-1. *Film-growth* by interaction of a vapor-deposited species with the substrate (Fig. 7-2a);

Group-2. *Film-formation* by deposition without causing changes to the substrate material (Figs. 7-2b, 7-2c, and 7-2d);

Group-3. Coating of the substrate with a liquid, which is then dried to form the solid thin-film (Fig. 7-2e).

Group-1 includes thermal-oxidation and nitridation of single-crystal (and polycrystalline) silicon (Chap. 13), and the formation of silicides by direct-reaction of a deposited-metal and the substrate.

We see that **Group-2** includes another three sub-classes of deposition. The first of these is *chemical vapor deposition* (CVD, Fig. 7-2b), in which solid-films are formed on a substrate by the *chemical-reaction* of vapor-phase species that contain the required constituents. (When a CVD-process is used to form *single-crystal* thin-films, the process is termed *epitaxy* and this is the subject of Chap. 17. The formation of *amorphous* and *polycrystalline* thin-films by CVD is discussed in Chap. 16.) The second of these is *physical vapor deposition* (PVD), in which the species of the thin-film are *physically* dislodged from a source to form a vapor (Fig. 7-2c). This vaporized-material is then transported across a reduced-pressure region to the substrate (wafer), where it condenses to form the thin-film. PVD includes the processes of sputtering and evaporation (Chap. 15). Note that no chemical- reactions occur during formation of thin-films by PVD. The last subclass is *electroplating*. In this approach, films are formed by an electrochemical-reaction of a liquid-electrolyte and a wafer immersed in it (Fig. 7-2d). Electroplating is used to deposit thin Cu-films for advanced interconnects (see Chap. 24).

Group-3 involves *deposition of a liquid on a substrate to form a thin-film* (most commonly by spin-coating, Fig. 7-2e). This is discussed in Ch. 18 as part of the *photoresist-processing* topic, and in Chap. 24, Sect. 24.3.3, which covers *spin-on low-k dielectrics*.

The purpose of this chapter is to cover the terminology and properties of thin-films, in general. That is, basic background-information on the growth-mechanisms, structure, mechanical properties, and electrical properties of thin-films is provided for readers not well acquainted with materials-science concepts. By defining these properties early-on, it is not necessary to repeat such information each time a new thin-film deposition method is introduced. The specific formation and associated processing of particular films is described in detail in later chapters.

7.1 STAGES OF GROWTH OF AMORPHOUS AND POLYCRYSTALLINE THIN-FILMS

There are many types of *amorphous* and *polycrystalline* thin-films used in IC processing. Deposition of such thin-films on a substrate different from the film itself (for example, Al on SiO_2), involves the condensation of atoms in vapor-form (in this case Al) onto a solid-substrate. Such solid-film formation processes take place in three stages (see the chapter faceplate):

1. Nucleation
2. Coalescence
3. Continuous film-growth.

Since important properties of such polycrystalline films depend on how these stages occur, it useful to discuss each of them in more detail.

Condensation of vapor-atoms on a surface is initiated by formation of small *clusters* (or *nuclei*) through the random gathering of several adsorbed-atoms. The enlargement of these nuclei to form a continuous thin-film is termed *growth*. Often nucleation and growth can occur simultaneously during film-formation.

7.1.1 The Various Stages of Thin-Film Formation

In order for atoms in a vapor-phase to turn into a solid-film, they must condense on the solid-surface. Such condensation requires that there is an attraction between the vapor-atoms and the surface when the surface is struck by them. If sufficient attraction exists to allow an atom to become attached (*adsorbed*) onto a surface, it is then called an *adatom*. Such adatoms may continue to move about the surface by hopping from one location to another (*surface-migration*). The energy for surface-migration can be supplied by a heated substrate-surface. However, in the course of such hopping, some adatoms may also re-evaporate from the surface back into the vapor-phase.

Such migrating-adatoms may encounter other adatoms on the surface. If they are sufficiently attracted to one another, small clusters of adatoms will start to form, called *nuclei*, and their formation is known as *nucleation*. Nuclei are much less mobile than individual-adatoms, and they tend to remain at the locations where they are formed. The onset of *condensation* is thus marked by the initial formation of such nuclei.

After the formation of sufficiently large-sized nuclei, the *film-growth stage* begins. Film-growth has been studied with the aid of electron-microscopy, and several-substages of growth have been observed (see Faceplate). The *island-stage* occurs when nuclei grow in three-dimensions. This stage is followed by *coalescence*, where nuclei contact each other and form new larger and rounded shapes. Eventually during coalescence, larger-islands take on a crystallographic appearance, in the form of hexagonal-shapes. The large islands continue to grow, leaving behind channels or holes of exposed substrate (*channel-stage*). Finally, a *continuous-film* is formed.

7.1.2 The Structure of Thin-Films

Thin-films generally have smaller grain-sizes than do bulk polycrystalline materials. *Grain-size* is a function of the deposition-conditions and annealing-temperatures (Fig. 7-3). Larger grains are observed in thicker-films. This effect is enhanced by raising the substrate-temperature during deposition. Larger grains are expected for increased substrate and annealing temperatures as a result of the increased surface-mobility of adatoms. Annealing a film at a temperature equal to that at which it was deposited does not produce as large grains.

Surface-roughness of films (see Fig. 7-4) occurs as a result of the randomness of the deposition-process.

Real films almost always show surface-roughness, even though this represents a higher energy-state than that of a perfectly-flat film. Depositions at high-temperatures tend to show less surface-roughness. This is because increased surface-mobility from the higher substrate temperatures can lead to leveling of the peaks and valleys. On the other hand, higher-temperatures can lead to the development of crystal-facets, which may continue to grow in favored-directions, leading to *increased* surface-roughness. At low-temperatures, surface-roughness (as measured by surface area) tends to increase with increased film-thickness.

The *density* of a thin-film provides insight about its physical structure. Density is usually found by weighing a film and measuring its volume (area x thickness). If a film is porous after being deposited, it usually has a lower-density than does bulk-material.

The *crystallographic-structure* of thin-films depends on the adatom-mobility, and can vary from a highly disordered (or amorphous-like) state, to a well-ordered state (e.g., *epitaxial-growth* on a single-crystal substrate). *Amorphous-structures* are frequently observed for deposited dielectrics such as SiO_2 and Si_3N_4, while most metal thin-films have a *polycrystal-*

Fig. 7-4 CVD tungsten-film, deposited under conditions that produces severe surface-roughness.

line-structure. Silicon-films can be either amorphous, polycrystalline, or single-crystal, depending on the deposition-parameters and the substrate material.

7.1.3 Factors that Influence Epitaxial-Deposition of Thin-Films

Epitaxial-growth is another way that thin-films can be formed. However, these growth-mechanisms are quite different from the polycrystalline and amorphous film-growth described above. That is, *epitaxial-growth* is the phenomenon that can occur on a crystal-surface such that the deposited-film has an ordered structure. Typically, it is formed with a single-crystal structure of its own. If the film and the substrate are composed of the same material, this growth is termed *homoepitaxy* (e.g., Si-on-Si, and GaAs-on-GaAs), while if the film and substrate are different materials, such growth is termed *heteroepitaxy* (e.g., single-crystal Si on crystalline Al_2O_3 [sapphire], also called *silicon-on-sapphire* [SOS]).

A *single-crystal substrate* is the most important factor needed for epitaxial-growth to occur. However, some *crystal-symmetry* must also exist between the deposited-film and the substrate, and the *lattice parameters* of each material must also be close to the same value. That is, to achieve epitaxy, the lattice-misfit (which is the percentage-difference between the lattice-constant of the deposit and that of the substrate-plane in contact), must also be small. The *deposition-temperature* also plays a role in epitaxy. That is, the likelihood that epitaxial-growth will occur

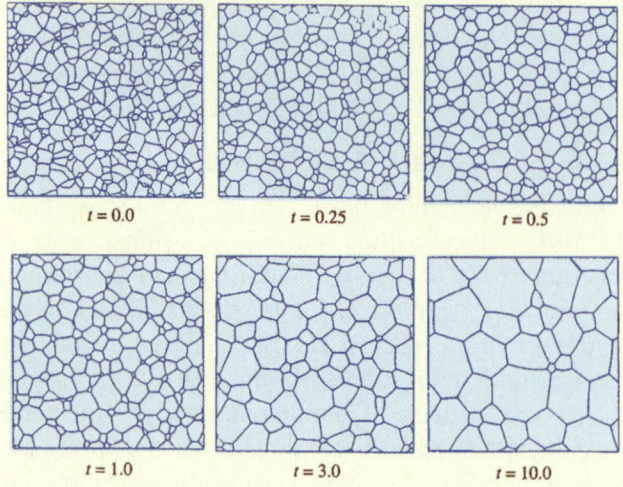

Fig. 7-3 Simulation of normal grain-growth in a polycrystalline thin-film at elevated temperature with time.[7] Reprinted with permission of TMS (The Minerals, Metals, and Materials Society).

is increased by elevated temperatures, clean substrate-surfaces, and increased adatom-energy.

The *deposition-rate* also plays a role in epitaxy. Lower deposition-rates favor epitaxy, while higher deposition-rates lead to polycrystalline or amorphous films. Three growth-regimes are observed for the deposition of Si on a clean Si-substrate: **1**) at low-temperature and high deposition-rates the deposits are amorphous; **2**) at high substrate-temperatures and low deposition-rates, they tend to be single-crystal; and **3**) at intermediate-conditions, polycrystalline-films tend to form. Chapter 17 discusses the growth of epitaxial silicon-films in more detail.

7.2 MECHANICAL PROPERTIES OF THIN-FILMS
7.2.1 Adhesion of Thin-Films

The adhesion of grown and deposited films used in ULSI processing must be excellent (both *as-deposited*, and after subsequent processing). Film-delamination from the substrate would result in device-failure. Thus, excellent-adhesion is critical to maintaining the electrical and mechanical integrity of the structure, particularly in multilevel-metal interconnects.

A simple *qualitative* technique for measuring the adhesion of thin-films is the *tape-test*. It consists of pressing a piece of adhesive-tape to the film (Fig. 7-5b). When the tape is pulled off, the film is either removed (in whole or in part), or remains on the substrate. This method is qualitative, and if the film remains on the substrate, this test does not provide any data about the magnitude of the adhesion-forces. If the film is removed, a *quantitative-value* of the adhesion can be obtained by measuring the force to remove it. Nevertheless, for most films used in VLSI processing, failure of the tape-test implies that the film is probably unsuitable for device-fabrication.

A relatively simple *quantitative* pull-test for measuring adhesion involves attaching a small metal-pin to the film-surface (after surrounding film-material has been removed, and with an epoxy that does not react with the film material, Fig. 7-5a). An increasing force is applied normal to the substrate-surface, until either the *film-substrate* or *film-epoxy* bond fails.

Adhesion is strongly effected by the cleanliness of the substrate. Contamination generally results in reduced-adhesion, as does an adsorbed gas-layer. Effective-cleaning of the substrate prior to deposition is therefore important to insure good film-adhesion.

In some applications it is necessary to include a layer of a strong oxide-forming element between an oxide-substrate and such metal thin-films as tungsten, copper, and gold (which all adhere-poorly to oxide surfaces). Various layers have been used to serve as such intermediate "adhesion" (or "glue") layers, including titanium and chromium.

7.2.2 Stress in Thin-Films

Nearly all thin-films are found to be in a state of internal stress, regardless of the means by which they have been produced. The stress may be *compressive* or *tensile*. Such stresses arise because films are usually deposited at elevated-temperatures. Generally, the film and the substrate have different *coefficients-of-thermal-expansion* (CTE). Thus, when they cool down to room-temperature, the film (which must tightly-adhere to the substrate) does not have the same length as it would have if could have cooled without being attached to the substrate.

By analogy, an Al-thin-film and a Si-wafer can be considered to be two springs made of Al and Si, respectively (and each having the same initial length, as shown in Fig. 7-6a). For this example, we assume

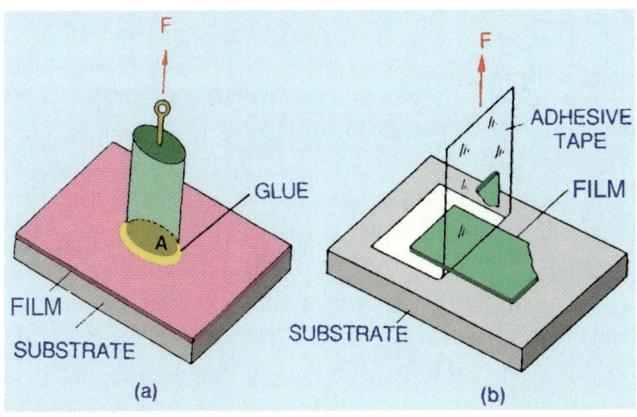

Fig. 7-5 Measuring adhesion of a thin-film to a substrate: (a) pull-off test; (b) tape-test.

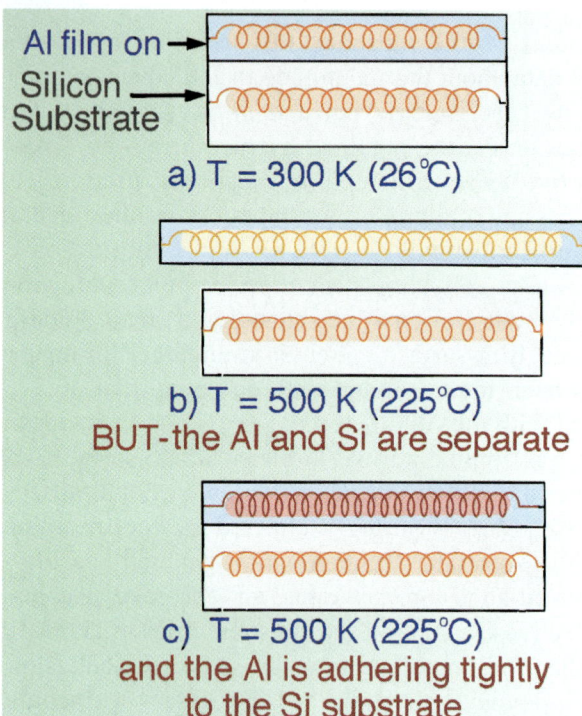

Fig. 7-6 An example of how *compressive-stress* arises in a thin-film adhering to a thicker substrate, using coil-springs as an analogy: (a) An Al-thin film (1-μm thick) is deposited on a silicon wafer that is much thicker (500-μm thick). Assume that at room-temperature there is no stress in the Al film, and both the Al-film and the Si wafer are like coil-springs at their natural-length at that temperature; (b) If the temperature is raised, the materials will expand. Since the CTE of Al (23×10^{-6} °C^{-1}) is much larger than that of Si (2×10^{-6} °C^{-1}), if the Al-film and the Si were not attached to one another, the Al would expand more than the Si-wafer in the lateral direction; (c) But since the Al-film adheres tightly to the Si wafer, it cannot expand as much it could, and thus ends up being like a compressed coil-spring.

that at *room-temperature* the springs are not stressed (i.e., neither spring is compressed or extended). Let us consider what happens if we heat the springs to the same elevated-temperature while they are unattached. In this case, each is free to expand to the degree dictated by their respective CTE. (This implies that the Al-spring - with a larger CTE - becomes longer than the Si-spring, see Fig. 7-6b). Nevertheless, such

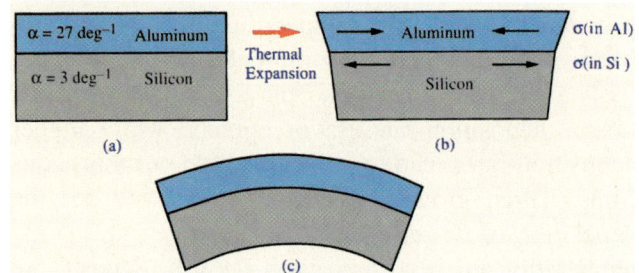

Fig. 7-7 Thermal-expansion-differences between Al and Si, shown in (a), lead to stresses in the Al thin-film and Si substrate, shown in (b). In this case, tensile-stress develops on top of the Si substrate and compressive-stress develops in the Al thin-film. (c) Illustrates the bending of the structures (exaggerated) that occurs due to these forces.

unconstrained, heated-springs are not yet stressed by heating. (In other words, "free" springs do not become compressed or extended when heated.) However, if two such *tightly-adherent* springs are heated, the Al-spring cannot expand to the length it could when not-attached to the Si-spring. Instead, its tendency to expand is constrained by the thicker Si-wafer, putting it into a compressed state (Fig. 7-6c). Note that even if the ends of such compressed-springs remain fixed, their stress can be released if the spring pops outward somewhere along its length. Such stress-relief events occur in thin-films on Si-wafers, as discussed next.

From the above analogy, we see that if thin-films formed on a Si-wafer become compressively stressed, they would like to *expand* parallel to the substrate surface (Fig. 7-7). On the other hand, such films in *tensile-stress* would like to *contract* parallel to the substrate-surface. If the *compressive-stress* becomes too great, it will cause the film deform in one of several ways. First, the film may buckle-up from the substrate in a wave-like manner, as shown in Fig. 7-8a. Or, protrusions (small hills, or *hillocks*) may appear on the surface of the film (Fig. 7-8b). On the other hand, if the *tensile-stress* in a thin-film exceeds the elastic limit of the material, the film may crack. In general, the stresses in thin-films range from 10^8–5×10^{10}-dynes/cm^2.

Highly-stressed films of either kind thus are generally undesirable for VLSI applications. Some of the

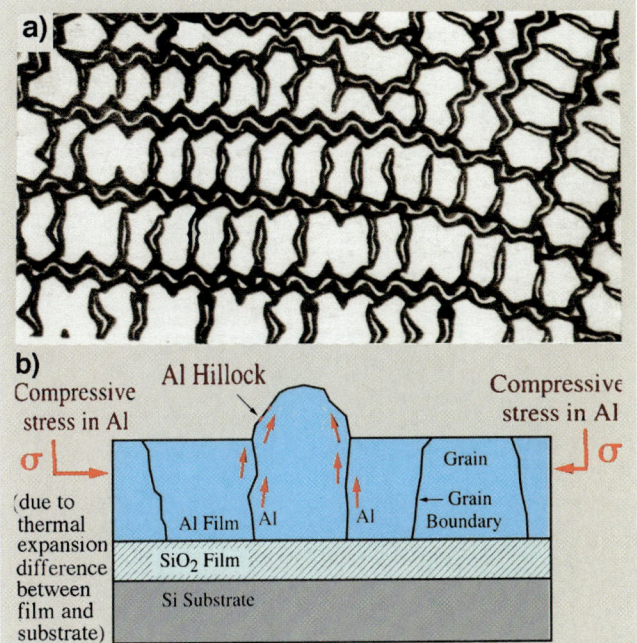

Fig. 7-8 (a) Tungsten thin-film on mica at 300°C, showing buckle waves. (b) Schematic-illustration of *hillock-formation* due to compressive-stress in an Al-film on a silicon-wafer (with an SiO$_2$ layer grown on it's surface). Aluminum-diffusion along grain-boundaries is indicated.[4] Reprinted with permission.

problems they cause are: **1)** they are more likely to exhibit poor-adhesion; **2)** they are more susceptible to corrosion; **3)** brittle films, such as inorganic-dielectrics, may undergo cracking in tensile-stress; and **4)** the resistivity of stressed metallic-films is higher than that of their annealed-counterparts.

Measuring stress in thin-films is a relatively straightforward procedure, based upon beam-bending. It is performed on commercial laser-based wafer-flatness measuring-equipment. This technique allows stress to be directly determined on the substrate that the film is deposited upon. The procedure is based upon the phenomenon that a stressed-film on a thin-substrate will cause bending of the substrate.

Compressively-stressed films will bend the substrate so that it becomes *convex* (Fig. 7-9a), while tensile-stressed films will make it *concave* (Fig. 7-9b). With Si-wafers, the film is deposited on one side, and stress is determined by measuring the deflection of the wafer-center. Use of laser flatness-equipment allows direct-measurement of the wafer-deflection.

7.3 MEASUREMENT OF ELECTRICAL PROPERTIES OF THIN-FILMS

The *sheet-resistance* R_S of a conducting thin-film is one of its most important electrical-characteristics. This is because R_S is frequently used to monitor the deposition of conducting-film processes and deposition-chamber performance. The concept behind the sheet-resistance parameter can be explained as follows. The resistance of a rectangular-shaped section of film of length = l, width = b, and thickness = t (measured in the direction parallel to the film), is given by:

$$R = \frac{\rho l}{t b} \quad (7.1)$$

where R is the resistance of the film and ρ is its' resistivity. If the film-length equals its width and (i.e., l = b), Eq. 7-1 then becomes:

$$R = \rho/t = R_S \quad (7.2)$$

where R_S is defined as the sheet-resistance in units of Ω/*square* of the film. Note that R_S is independent of the size of the square, and depends only on the film-resistivity and thickness of the film. As shown in Fig. 7-10, for a given thin-film, the sheet-resistance is the same for each of the squares. If the thickness and sheet-resistance are known, the film-resistivity can be calculated from:

$$\rho = R_S t \quad (7.3)$$

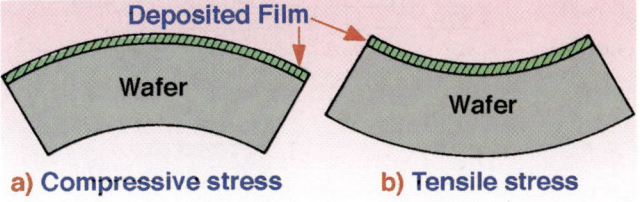

Fig. 7-9 (a) Compressive-stress causes *convex* bending of the substrate. (b) Tensile-stress causes *concave* bending.

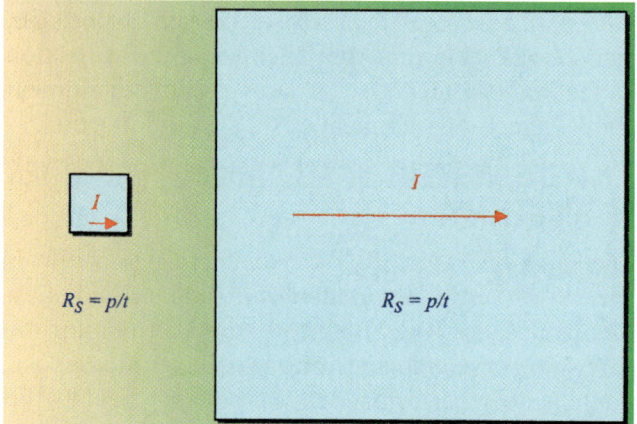

Fig. 7-10 The *sheet-resistance* R_S is the same for these two squares.

The sheet-resistance, R_S, can be measured by several techniques. In the direct-measurement method, a conductor strip is fabricated (Fig. 7-11). The voltage-drop across the strip-length is measured while a current is driven through the conductor. The resistance of the strip is given by $R = V/I$. The number of squares are counted between the voltage terminals, and $R_S = R/N$ (where N is the number of squares, i.e., $N = l/b$).

A second method which does not require the use of a conducting strip is the *4-point-probe method* (Fig. 7-12).[6] It is the more commonly-used method

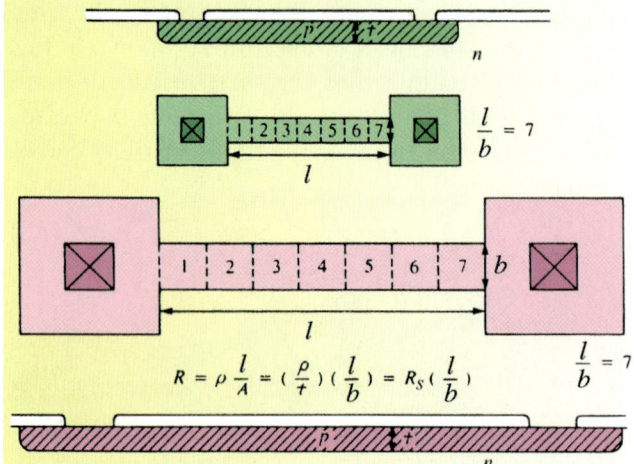

Fig. 7-11 Top- and side-views of two diffused-resistors of different physical size, but with equal values of resistance. Each resistor has a ratio of l/b equal to 7 squares.

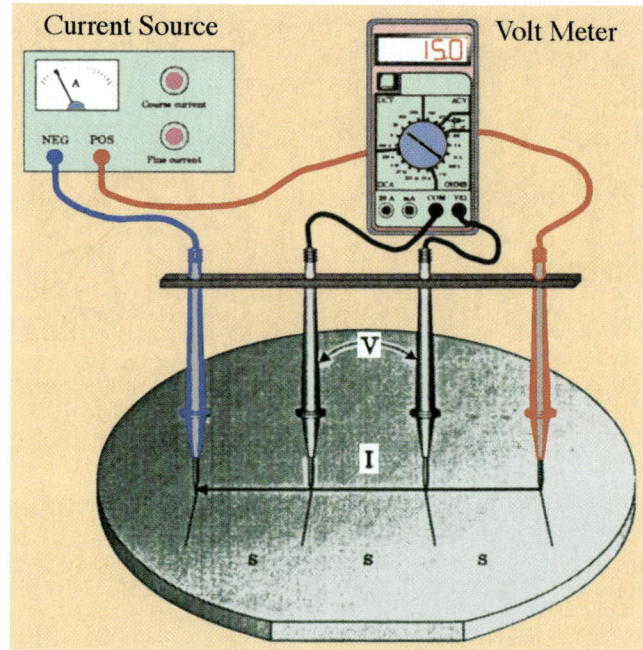

Fig. 7-12 Schematic of a *4-point-probe* measurement setup.

of measuring R_S. A fixed-current is passed between the two outer-probes, and the voltage across the inner two probes is measured. If the sample is semi-infinite (with regards to the probe spacings), and the probe-spacings, s, are equal, then the resistivity ρ is given by:

$$\rho = \frac{t V 2\pi}{I \ln(2)} = 4.532 \frac{t V}{I} \text{ (ohm-cm)} \quad (7.4)$$

If the material on which the probe is placed is an infinitely-thin slice on an insulating substrate, then Eq. 7-4 reduces to an equation for R_S;

$$R_S = 4.53 \frac{V}{I} \text{ (ohms/square)} \quad (7.5)$$

Commercial equipment for measuring R_S in thin metal-films and semiconductor-layers is available.

REFERENCES

1. *Thin-Film Processes II*, Eds. J.L. Vossen and W. Kern, Academic Press, San Diego CA, 1991.
2. M. Ohring, *The Materials Science of Thin-films*, Academic Press, San Diego CA, 1992.

3. D.L. Smith *Thin-Film Deposition: Principles and Practice,* McGraw-Hill, New York, 1995.

4. J.D. Plummer, M.D. Deal, and P.B. Griffin, *Silicon VLSI Technology,* Prentice-Hall, 2000, p. 698.

5. L.I. Maissel and R. Glang, *Handbook of Thin Films,* McGraw-Hill, New York, 1970.

6. L.B. Valdes, *Proc. IRE,* **42**, 420, (1954).

7. H.J. Frost and C.V. Thompson, "Computer Simulation of Microstructural Evolution in Thin-Films," *J. of Electronic Materials,* **17**, 447 (1988).

PROBLEMS

1. If a thin-film has a sheet-resistance of 1-kΩ/square, find the maximum resistance that can be fabricated on a 2.5 x 2.5-mm chip, for 2-μm lines and 4-μm *pitch* (i.e., the distance between the centers of the lines).

2. What is the total-number of squares in the straight diffused resistor depicted in Fig. 3-2?

3. What is the sheet resistance of a 1-μm-thick Al:Cu line with a resistivity of 3.2-$\mu\Omega$-cm?

4. Repeat Problem **3** for a polysilicon-line with a resistivity of 500-$\mu\Omega$/square.

5. The *coefficient of thermal expansion* (CTE) of Si at 100°C is 2.6x10^{-6}/°C. How much would a 10-mm chip increase in length if heated by 300°C (assuming that the CTE is constant)?

6. A thin-film of Al (ρ = 2.5x10^{-6} Ω-cm) is deposited, with a length of 0.1-cm, a thickness of 1-μm, and a width of 2.5-μm: (a) What is the resistance R (in Ω) of the strip; (b) For a current of 10-mA flowing along the strip, what voltage would you measure along the length, and how many watts are dissipated? (c) If the thickness of the Al strip is doubled, what is the resistivity? (d) If the ρ of half the length of the strip is doubled, what is the resistance of the strip?

7. Metal lines of varying widths are to be fabricated. Before being patterned, the sheet resistance of the metal film is measured and found to be 2.5-Ω/sq. Find the total resistance of 0.5, 1.0, 2.0, and 5.0 μm wide and 1-cm long lines. What other information is still needed before it can be determined what type of metal makes up these lines?

8. Explain why *compressive-stresses* in thin-films can cause *convex-bending,* and *tensile-stresses* can cause *concave-bending* of the substrates on which they are deposited.

9. The linear thermal-coefficient-of-expansion of Al is of the order of 25x10^{-6} cm/cm/°C, and that for Si is 2.6x10^{-6} cm/cm/°C. Using this information, offer an explanation as to why thin-films of Al on Si substrates exhibit hillocks, after being subjected to elevated temperatures (e.g., > 250°C).

CHAPTER 8

CONTAMINATION CONTROL AND CLEANING TECHNOLOGY FOR ULSI

CHAPTER CONTENTS

8.1 CONTAMINATION TYPES IN IC FABRICATION

8.2 CONTAMINATION SOURCES IN IC PROCESSING

8.3 EFFECTS OF CONTAMINATION ON ULSI DEVICES

8.4 CONTAMINATION PREVENTION

8.5 WAFER CLEANING TECHNIQUES

8.6 PARTICLE REMOVAL

8.7 RINSING AND DRYING WAFERS

8.8 PARTICLE DETECTION ON WAFER-SURFACES

"It has long been an axiom of mine that the little things are infinitely the most important."

Sir Arthur Conan Doyle

Operations taking place in the cleanroom of a 300-mm wafer-fab. Shown is the SEMICONDUCTOR300 (SC300) fab in Dresden, Germany. Photograph courtesy of Infineon Technologies Corporation.

The need for extremely clean wafers has been recognized since the dawn of semiconductor-manufacturing technology. Clean substrate-surfaces are critical for obtaining maximum device-performance, long-term reliability, and high yields. Cleaning techniques are used to remove particulate and chemical impurities so contamination-free surfaces can be obtained. However, such cleaning-methods must also be able to do this without damaging the wafer surface. Cleaning procedures should also be safe, simple, economical, and produce minimum hazardous waste-products.[1]

There are many aspects involved in obtaining clean wafers. Numerous kinds of contamination exist, and each of the hundreds of processing steps in ULSI fabrication can add one type or another. There are also

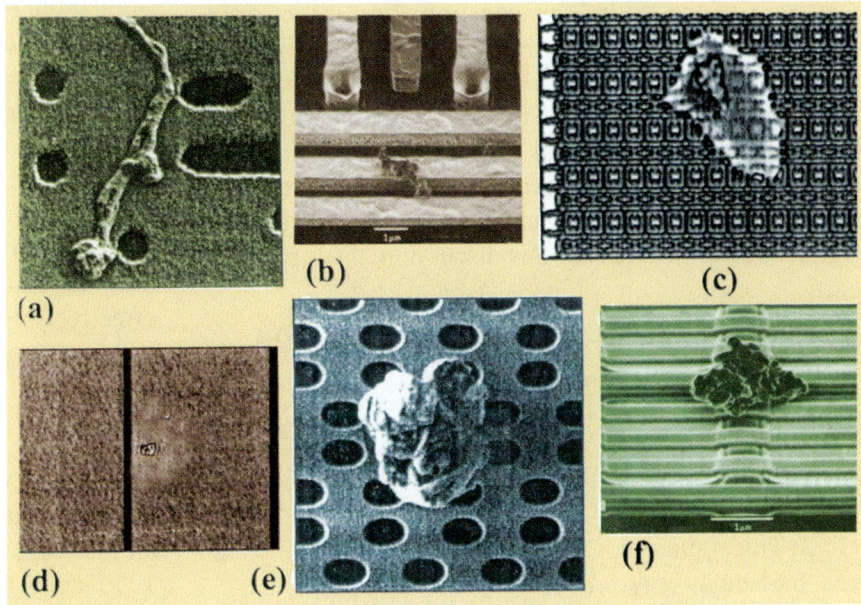

Fig. 8-1 SEM-photographs of examples of particle-types found on IC wafers: (a) extra material; (b) embedded particle; (c) missing pattern; (d) poly-flake; (e) surface particle; (f) missing pattern.[13] Reprinted with permission of M. Dekker.

many techniques used to prevent contaminants from invading the fabrication environment.

A variety of cleaning-processes have also been developed to remove contamination when it arises, and many such cleaning steps are incorporated in an IC process-flow. For instance, a 64-Mb, 0.25-μm DRAM-process-flow has 60 to 70 cleaning-steps. Advanced CMOS-technology employs about 80 cleaning-steps out of the total 400 process-steps. Stricter requirements are also being imposed on cleaning-technologies, including:

1. The concentration of contaminants allowed to remain after cleaning is being reduced.

2. The tolerable particle-size and their quantity per wafer are being decreased.

3. Constraints are being increased on the permitted consumption of cleaning-chemicals and ultra-pure (de-ionized) water.

In this chapter the various types of contamination encountered in microelectronic-fabrication (see also Fig. 8-1) are first listed. Next, the damaging effects that each of these kinds of contamination inflict on integrated-circuits is described. The most important sources of such contamination are then identified. (If the source responsible for the contamination is known, steps can be taken to minimize such intrusion into the fabrication cycle.) Finally, the various techniques used to remove contamination from wafers are presented. As a part of this discussion, the topic of particle-detection on wafers is also covered.

8.1 CONTAMINATION-TYPES IN IC FABRICATION

There are three general categories of *wafer-contamination*:

1. Particles;
2. Films;
3. Trace-quantities of contaminants in molecular or atomic form.

Particles are any bits of material present on a wafer surface that have readily-definable boundaries. As feature-sizes shrink, the sizes of particles that can cause defects also decrease. Sources of particles include silicon-dust, quartz-dust, and atmospheric-dust. Particles also originate from cleanroom-personnel and processing-equipment, lint (from street clothing that escapes protective cleanroom-garments), photoresist "chunks" (which flake-off during wafer-handling and clamping), and bacteria (which can grow in ultra-pure water [UPW] supply-systems).

Layers of foreign-material on wafer surfaces are sources of *film-contamination*. Portions of films may break loose and become particles, as often happens with photoresist-residues. Examples of films that contaminate wafers include: *residues from solvents* (such as isopropyl- or methyl-alcohol, and xylene); *residues from the resist-development step* (arising from dissolved photoresist in the developer, or from inadequate post-development rinsing); and *oil-films* (introduced through improperly-filtered air or gas lines, or from the use of oil-sealed vacuum-pumps). *Metallic* and *ionic contaminants* are often deposited during immersion of wafers in etchants or resist-strippers, both of which may contain metal-ions and traces of metal in solution.

The chemical-cleaning and photoresist-stripping operations used to remove film-contamination have also been identified as significant sources of particle-contamination. Thus, one stage of a cleaning-process may, in fact, reduce the effectiveness of other cleaning-procedures. To prevent particle-contamination during chemical-cleaning, the chemical-processes must be carefully monitored and particle-deposition effectively controlled during their use.

8.2 CONTAMINATION-SOURCES IN IC-PROCESSES

Contaminants that arise in IC-fabrication can come from many sources, including process-equipment, process-chemicals, ultrapure-water, chemical storage-containers (and piping), cleanroom-personnel, and the air (see Fig. 8-2). *Process-equipment* leads the list. In 1990 it was believed to be the source of about 75% of particle-contamination in IC fabrication.[2]

Liquid and gaseous chemicals have tended to be more responsible for forming contaminant-films.

Chemical contamination can arrive at the wafer-surface from *ultrapure-water* in the form of ions, bacteria, organic-particles, and low-level inorganic-residues (such as silica). For example, bacteria are inevitably found in ultrapure-water. They have a unique capability to reproduce in otherwise contaminant-free water, and must therefore be removed. This is done using UV-light sterilizers to kill and break them up into fragments, and filters to remove them.

Polymeric photoresist liquids can leave behind organic film-residues. They can also potentially contain significant levels of metal contaminants. The lithographic developer-liquids originally contained ionic species, but now *metal-ion-free* developers are mandated for advanced processes. Organic-films can also be residues of such solvents as acetone and isopropyl alcohol.

The *containers of chemicals* (and the *cassettes* that hold wafers) have historically been a significant source of particles. Particles are generated as the wafers are transported, and also from mechanical-vibration as these cassettes sit in process-tools.

Human-beings give off between 100,000 and 1,000,000 particles per minute, even when sitting still after just having showered. This number increases dramatically when a person is moving. Particles come from such sources as hair, skin-flakes, cosmetics, and hair-spray. Street-clothing adds more millions of particles. Human breath contains high-levels of contamination. Every exhalation puts many water droplets (that also contain sodium) and particles into the air.

Ordinary air is so full of particles that it must be treated before being allowed to enter a cleanroom. That is, normal air contains a large number of small dust-particles that "float" and remain in the air for long periods of time. The air in a typical city may contain up to 5-million particles per cubic-foot.

8.3 EFFECTS OF CONTAMINATION ON ULSI DEVICES

Contaminants are a major concern in IC-manufacture because they reduce the fraction of the properly functioning chips at the conclusion of the fabrication sequence.[2] It is estimated that over fifty-percent of the yield-losses in semiconductor-manufacture are due to micro-contamination. The effects of contaminants on semiconductor-devices depend on the nature and quantity of the specific-type of contaminant.

Contaminant-films on a wafer-surface can impact ICs in a number of harmful ways. In general, they reduce the effectiveness of cleaning and rinsing processes, they prevent good-adhesion of deposited-films to the wafer-surface, and they may decompose into harmful by-products.

Contaminants adsorbed onto the wafer-surface in trace quantities (i.e., not in the form of a particle or continuous thin-film) also cause serious problems. *Metallic-contaminants* (such as Fe and Cu) on the surface of a wafer rapidly distribute themselves throughout the silicon-bulk during high-temperature processing steps. Their presence then degrades device-performance and dramatically increases *pn-junction reverse-bias leakage-currents*. The presence of trace-quantities of metals near the wafer-surface also decreases the reliability of MOSFET gate-oxide films grown on the wafer.

Ionic-contaminants (such as sodium, Na) cause another set of problems with respect to device-operation. That is, Na spreads rapidly throughout surface SiO_2 films and along the Si/SiO_2 interface. Small

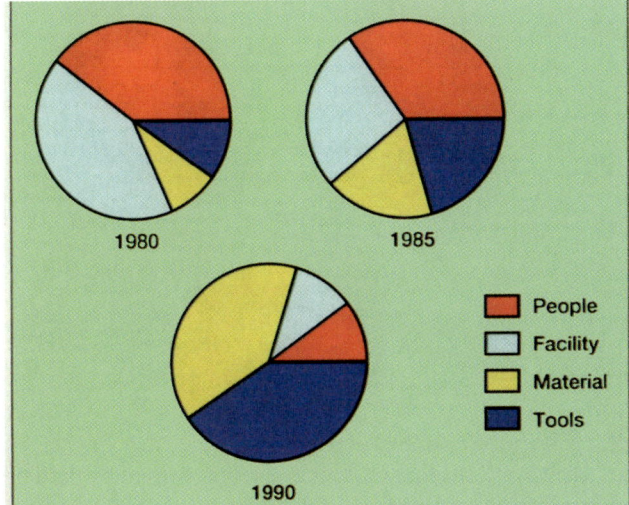

Fig. 8-2 Sources of contamination in IC-fabrication (from 1980 to 1990).

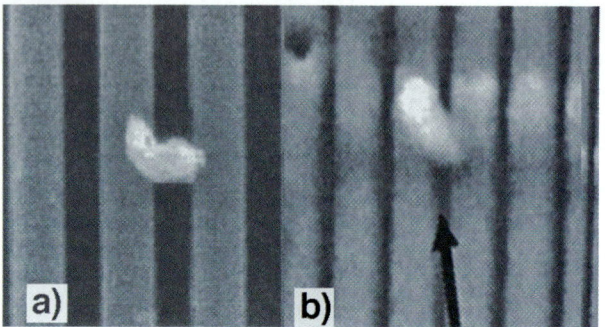

Fig. 8-3 (a) SEM of a particle on a wafer (uncovered). (b) SEM of the same particle covered by an overlying-film.

quantities of Na ions in the oxide give rise to threshold-voltage instabilities in MOSFETs and cause a reduction in the breakdown-voltage of gate-oxides grown on surfaces containing Na.

Particles on the wafer-surface can result in the blocking or masking of various processing-operations. For example, a particle may block an implant or locally disrupt pattern-development during a lithography step (Fig. 8-3). Particles may also stubbornly cling to the wafer-surface and become embedded during film-deposition. Particles present during deposition can lead to pinholes, microcracks, or thinning in gate-oxides and intermetal-dielectric layers (as well as producing the masking-defects described above). In later stages of the process-flow (when interconnects are formed), particles can cause shorts between adjacent conductor-lines or opens in a conductor line.

Since the size of the features used in ICs is so small, the presence of even tiny particles on a wafer during fabrication can cause IC-failure. Figure 8-4 shows a

Figure 8-4 SEM-photograph of a human hair-on an IC surface, with metal-lines having a minimum feature-size of about 5-μm. Reprinted with permission of Oxford Press.

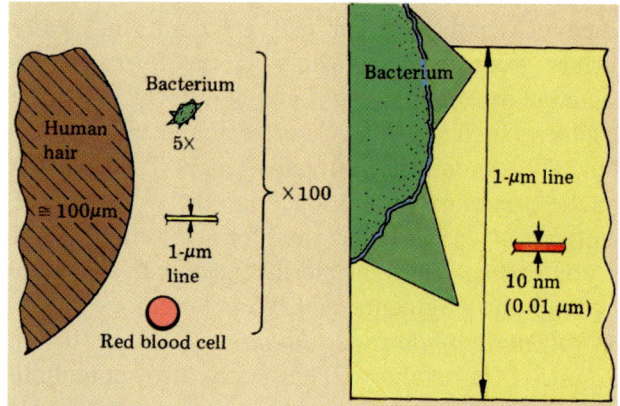

Fig. 8-5 Relative sizes of some microscopic objects as compared to IC dimensions.

human hair in comparison to IC patterns fabricated in the mid-1980s, and Fig. 8-5 compares the size of a human hair and a bacterium to the dimensions of IC features. The minimum particle size that can induce a "killer" defect (i.e., one that causes a chip to fail)

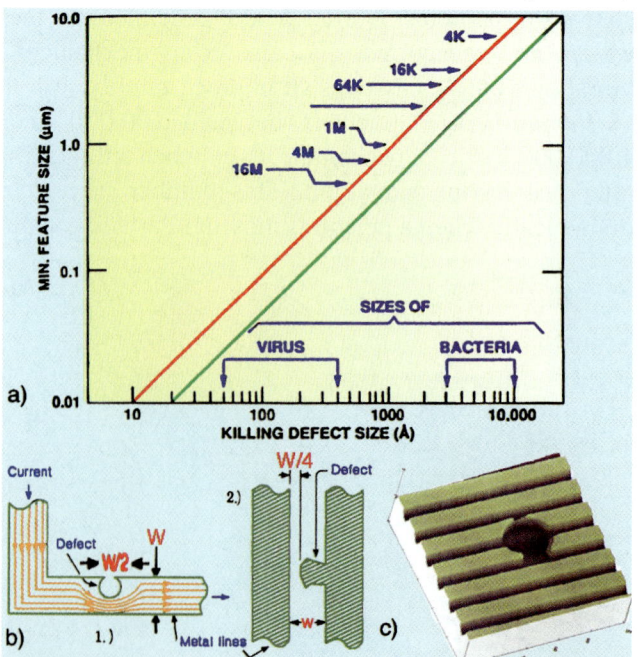

Fig. 8-6 (a) Shrinking design-rules and the impact of defects. (b) Illustration of "killer-defects" caused by particle contamination: (b1) Missing-material in a conductor-line; (b2) Extra-material deposited between lines. (c) SEM-photo of missing-material in a conductor-line.

CHAPTER 8   CONTAMINATION CONTROL AND CLEANING TECHNOLOGY FOR ULSI

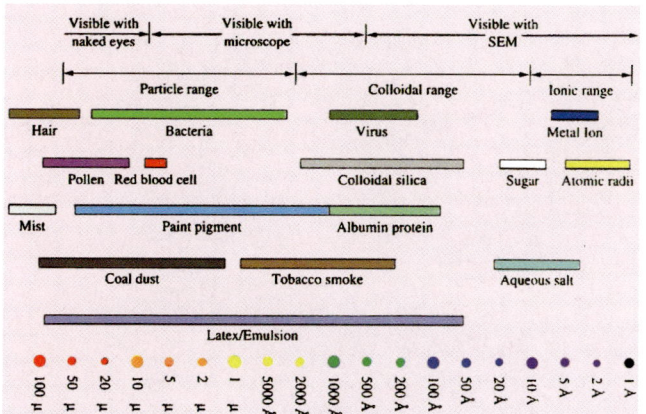

Fig. 8-7 Sizes of various types of particles.

depends on the minimum feature-size and on the region where the particle falls. As a rule-of-thumb, a particle that exceeds 20–50% of the minimum feature-size has the potential for causing a fatal defect. Figure 8-6a shows the decrease in killer-defect size as a function of the minimum feature-size of an IC. For 0.25-μm CMOS, the size of a killer-defect is thus only 0.06-μm. Figures 8-6b and 8-6c illustrate how particles can distort the patterns of IC-features. Note that bacteria are much larger than the killer-defect size for submicron IC-technologies. Figure 8-7 shows the sizes of many types of particles.

Note that particles between 10-nm and 10-μm are those of most concern. In room-air, particles smaller than 10-nm tend to clump together and form particles of larger sizes, and those larger than 10-μm tend to be heavy enough to fall to the floor quickly. But particles between 10-nm and 10-μm can remain suspended in air for very-long periods of time. Such particles can thus end up depositing on such surfaces as a wafer.

8.4 CONTAMINATION-PREVENTION MEASURES

"An ounce of prevention is worth a pound of cure," also applies to cleaning strategies for IC fabrication. It is more effective to eliminate a source of contamination, than it is to try to remove the contaminant after it has adhered to the wafer. While cleaning procedures can remove an immediate problem, they represent additional processing-steps that may create more contamination.

8.4.1 Cleanroom Design, Minienvironments, and SMIF

The first-line of defense against contamination (especially that due to the generation and deposition of particles onto wafers) involves carrying-out IC fabrication in a *cleanroom* (Fig. 8-8). Cleanrooms are special facilities designed to reduce the presence of particles and other contaminants to very-low levels.[3] Performing wafer-fabrication processes in such cleanrooms also allows the environment of the wafer to be tightly specified. That is, many important environmental conditions can be well-controlled within a cleanroom, including: temperature, relative-humidity, the number of airborne particles and gases, chemical

Fig. 8-8 Worker transporting wafers in an IC-cleanroom. Courtesy of Shuttleworth Corp.

MICROCHIP MANUFACTURING

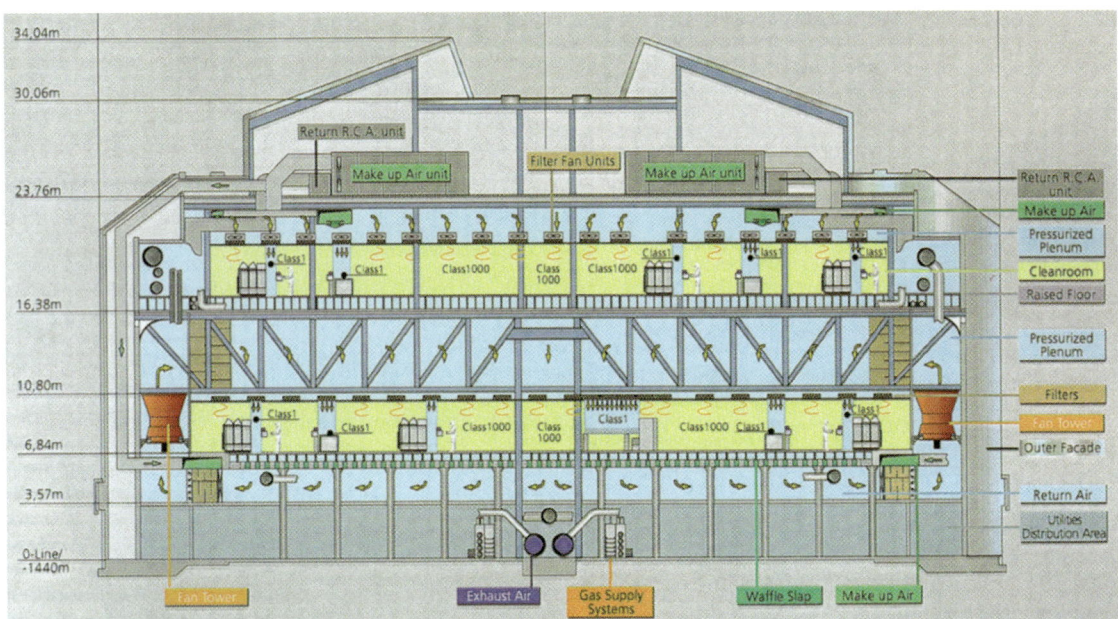

Fig. 8-9 Cross-section of an IC fabrication-facility. Courtesy of Messner & Wurst.

contamination, and vibration. The overall-principle is to build a sealed-room that is supplied with clean air, and is built with materials that are non-contaminating. The building that houses the cleanroom should include systems to prevent accidental-contamination from outside, and from personnel working within it.

In the 1970's (when the IC manufacturing-process began to require strict-control of the wafer-environment), the "ballroom-type cleanroom" concept was created. In this approach, all manufacturing-steps are carried out in one large room. The cleanroom itself has a raised-floor with grid-panels. The air flows from the ceiling (Fig. 8-9) through *high-efficiency particulate attenuation* (*HEPA*) filters (Fig. 8-10) through the floor, to the space beneath the floor. To achieve bet-

ter than Class-100 cleanliness, linear-airflow (called *laminar-flow*) must be maintained, and air-turbulence avoided (see Sect. 6.3.2). With laminar-flow, airborne particles are carried down and away. The cleanroom is always kept at higher-pressure than non-clean areas. This creates outward air-flow against airborne particles when a door is opened.

The *ballroom-concept* evolved into the *cleanroom* and *chase* (*service-area*) configuration (see Chap. 28). In this approach a corridor separates the process-area from the service-area (whose cleanliness-level is not as high as that of the cleanroom). The wafers are handled outside of process-tools only within the cleanroom-area, whereas the majority of the tool-maintenance functions are performed behind the

CHAPTER 8 CONTAMINATION CONTROL AND CLEANING TECHNOLOGY FOR ULSI

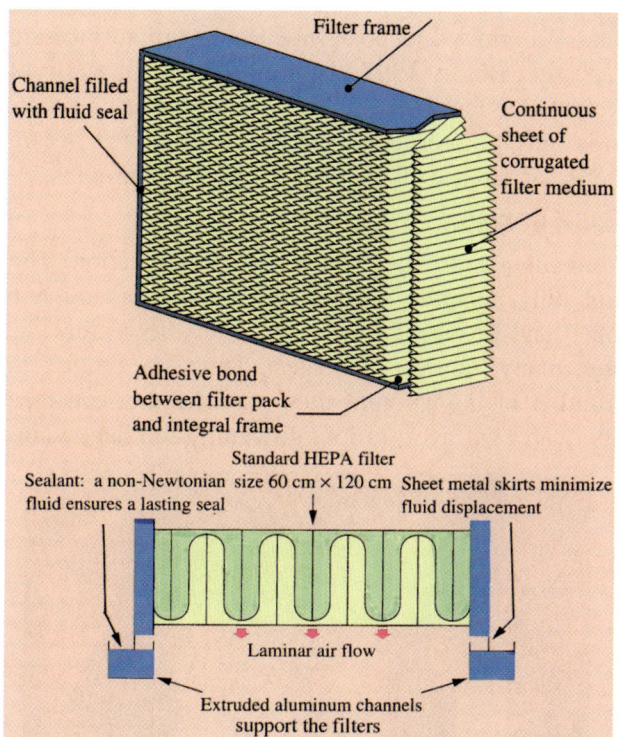

Fig. 8-10 Drawing of a high-efficiency-particulate-attenuation (HEPA) filter.

bulkhead (in the chase-areas, Fig. 8-11a). This allows the costly, highly-clean areas to be reduced in size.

As the minimum feature-size on wafers has decreased below 0.25-μm, the environmental-cleanliness conditions have become so stringent that the *cleanroom/chase configuration* may need to be replaced. That is, each process-tool may need to be completely housed in its own enclosure (Fig. 8-11c). This permits an even-smaller volume of cleanroom space to be separately and more tightly controlled (i.e., a set of *minienvironments* is created). Within such enclosures a processing-environment of extremely-high cleanliness can be established and thereafter maintained.[8] Thus, the volume within the rest of the building that contains these minienvironments does not have to be as clean. The transport of wafers outside the minienvironments (i.e., between tools) is done by means of SMIF-pods (*Standard Mechanical InterFace*, Fig. 8-12). This technique was pioneered by Hewlett Packard, and SMIF-pods are now supplied by Asyst Technologies. New designs for transporting 300-mm wafers, such as *Front-Opening Unified Pods* (FOUPs) are evolving (Fig. 8-13).

The degree of cleanliness is defined by a U.S.-government-standard (known as *Federal Standard 209*). It specifies *airborne particulate-cleanliness* in terms of how many particles are allowed to exist per unit volume of cleanroom-air. In the past three-decades this standard has undergone several revisions, and the latest one, Fed. Std. 209E was released in 1992.

It is common to characterize the cleanliness of the air (in English units) by the designation "Class-10" or "Class-100," etc. "Class-X" simply means that in each cubic-foot of air, there are less than X total-particles greater than 0.5-μm in size.

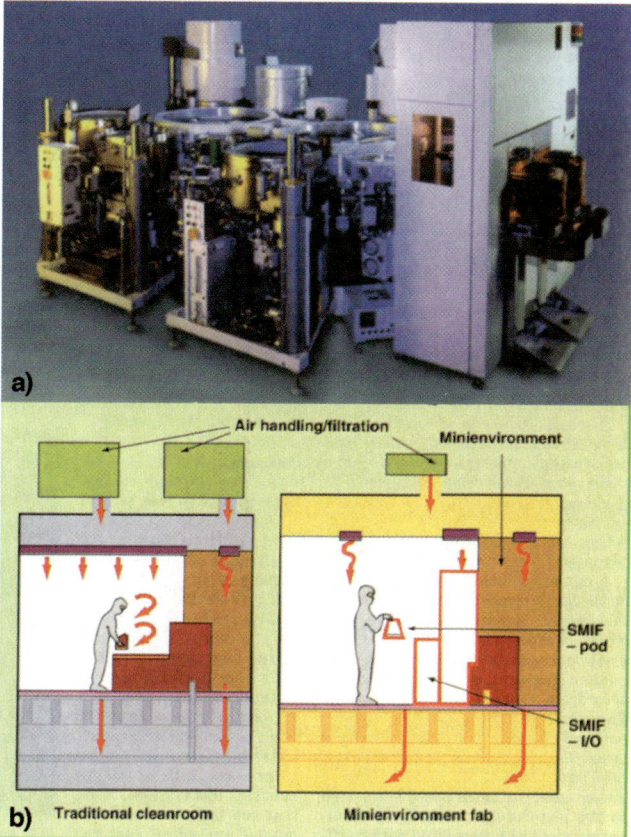

Fig. 8-11 (a) Example of *bulkhead-installation* of a wafer-fabrication tool. Courtesy of Novellus Systems. (b) Comparison of a conventional *laminar-flow-hood* workspace-approach versus a *minienvironment* approach.

than 0.1-μm will deposit on each cm^2 of surface-area per hour. But by 1997, the total-number of allowable defects having a size of 0.1-μm on a starting wafer was only 0.6-per-cm^2. This explains why it became necessary to use Class-10 or Class-1 cleanrooms.

8.4.2 Gowning Procedures

Humans are a major source of contamination. They not only generate a large number of contaminants, but they are also in close-proximity to wafers during many fabrication-stages. Thus, to reduce the number of human-generated particles that can reach the wafer surfaces, proper gowning (and degowning)

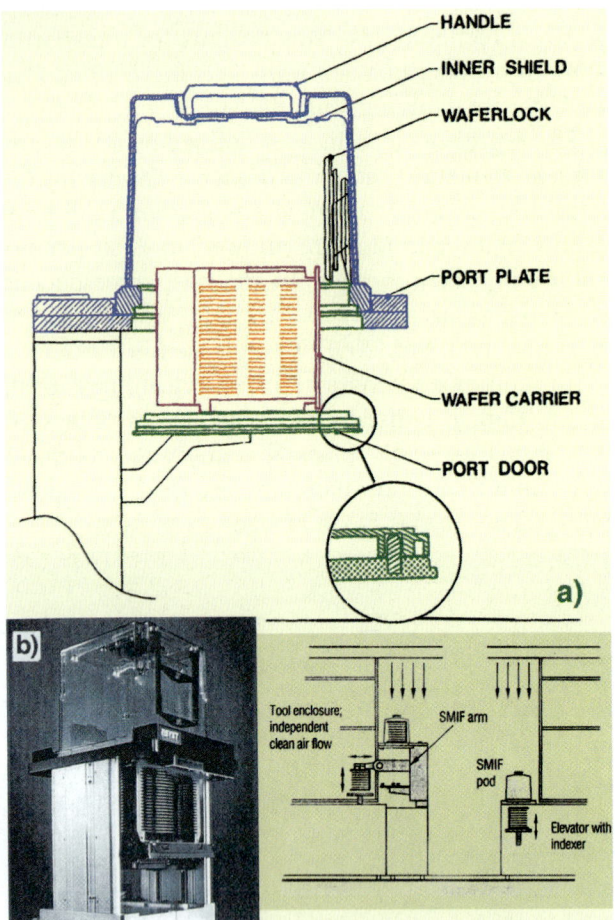

Fig. 8-12 a) Creating an ultraclean minienvironment for a process-tool by using the *standard-mechanical interface* (*SMIF*). b) Photo of a SMIF box. Courtesy of Asyst Technologies, Inc.

For example, as shown in Fig. 8-14a, a Class-1 cleanroom is permitted to have only *one*-particle greater than 0.5-μm in size per cubic-foot of air (and only 35-particles *per cubic-foot* greater than 0.1-μm in size). A Class-100 cleanroom is allowed to have up to 100-particles per cubic-foot greater than 0.5-μm. The numerical-designation of the Class in SI-units (e.g., Class-M2) is taken from the logarithm (base 10) of the maximum allowable-number of particles, 0.5-μm and larger, per *cubic-meter*. Class-1 is equivalent to Class-M 1.5 (see Fig. 8-14a).

In a Class-100 cleanroom about 5 particles larger

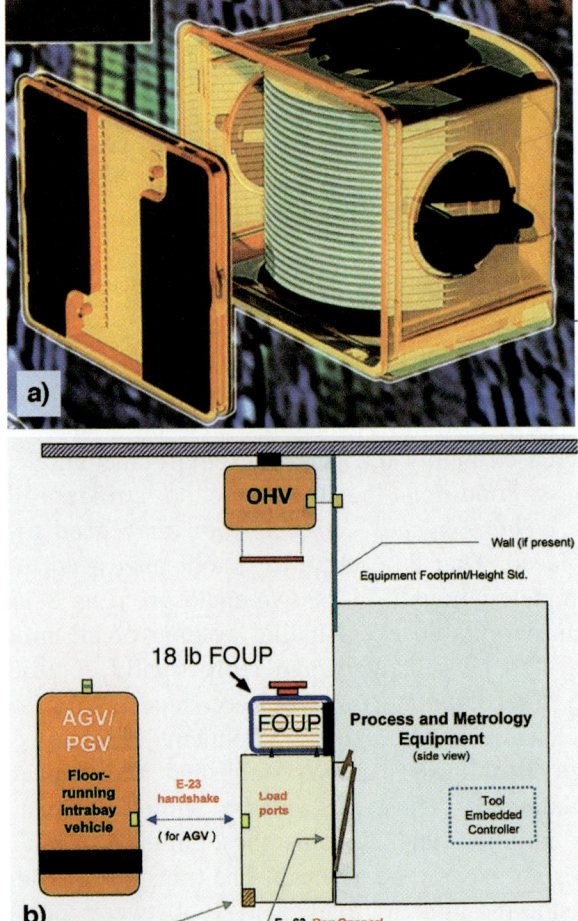

Fig. 8-13 (a) *Front-opening unified-pod* (*FOUP*). (b) Loading wafers into tool minienvironment from a FOUP.

CHAPTER 8 CONTAMINATION CONTROL AND CLEANING TECHNOLOGY FOR ULSI   129

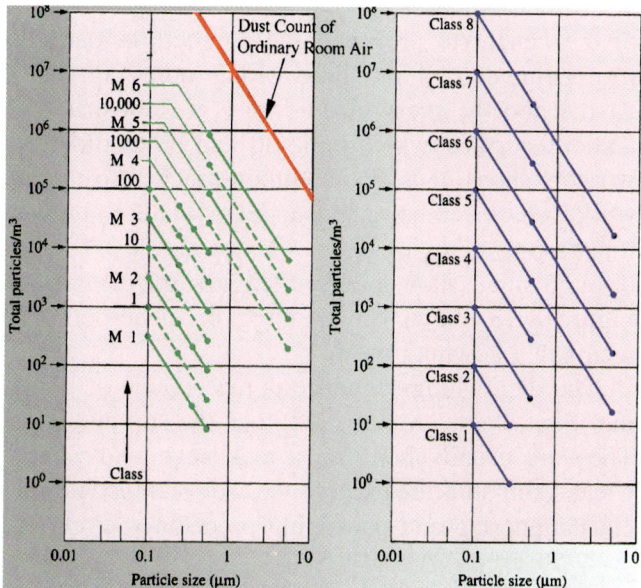

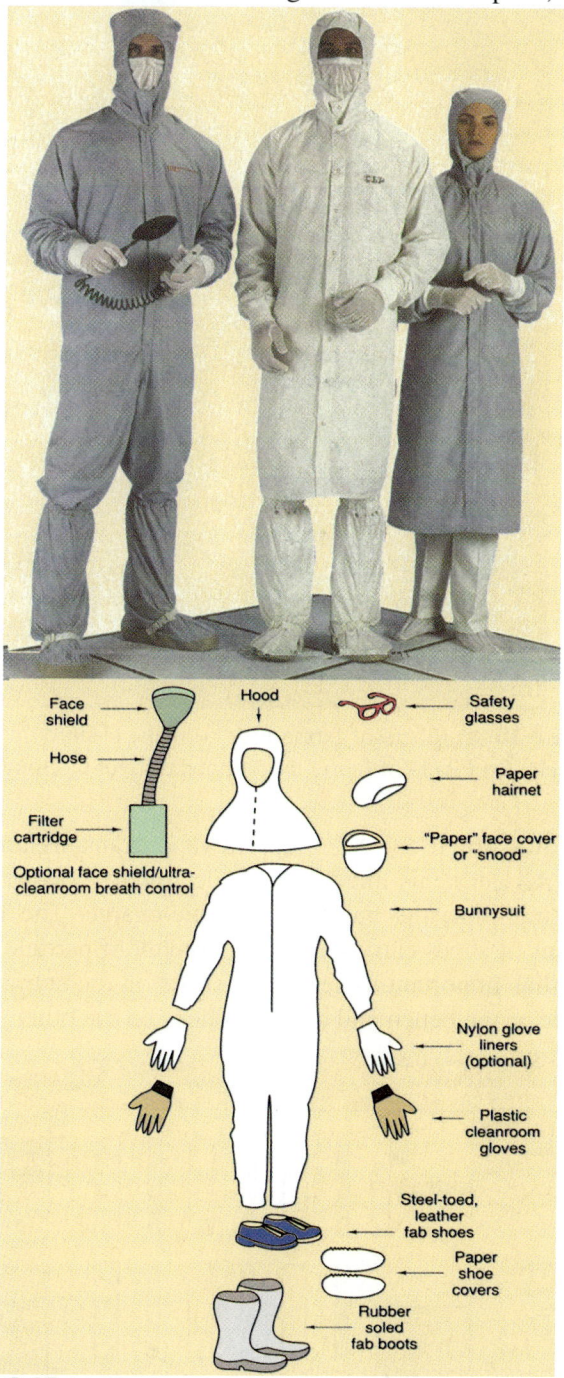

Fig. 8-14 (a) Cleanroom Class-limits according to Fed. Std. 209E. These limits are expressed both in maximum-number of particles per cubic-foot (Ex: Class-10,000 in diagram), and per cubic-meter (Ex: class-M 7 on diagram) equal to, or greater than, the particle-size shown. (b) Cleanroom Class-limits according to Japanese Std. B9920 rev.

procedures are imposed on all persons entering the cleanroom. This suppresses the number of particles in the cleanroom-ambient that are either shed by humans or emanate from their street-clothing.

All personnel that enter a cleanroom must therefore wear company-provided cleanroom-apparel.[10] This includes a body-suit, gloves, hair-net and inner-cap, hood, and booties, to achieve as complete body coverage as possible (Fig. 8-15a). A face-mask with a forced-air breathing-unit covers the nose, mouth, and even the eyes. Figure 8-15b shows the various-articles of cleanroom-apparel. Some fabs require people to don linen-gloves before entering the gowning-room. These prevent sodium and particles on hands from contaminating the cleanroom-garments.

The garment-materials are made of woven fabrics of long synthetic-fibers (Fig. 8-16a), covered with PTFE (polytetrafluoroethylene), as opposed to conventional-fabrics, which would shed particles (Fig. 8-16b). Such fabric also prevents particles generated within the suit from entering the cleanroom space, but

Fig. 8-15 (a) Workers in clean-room garments. b) Articles of cleanroom attire.

130 MICROCHIP MANUFACTURING

Fig. 8-16 (a) Fibers of fabric used to make cleanroom garments. (b) Fibers of fabric used to make street-clothing.

allows a high-degree of air and vapor transmission

All garments must also be electrically-conductive to minimize the build-up of static-charge. The bottoms of shoes carry the largest amount of particles. It is thus important to keep bare-shoes on the entrance-side of the bench and covered-shoes on the other side

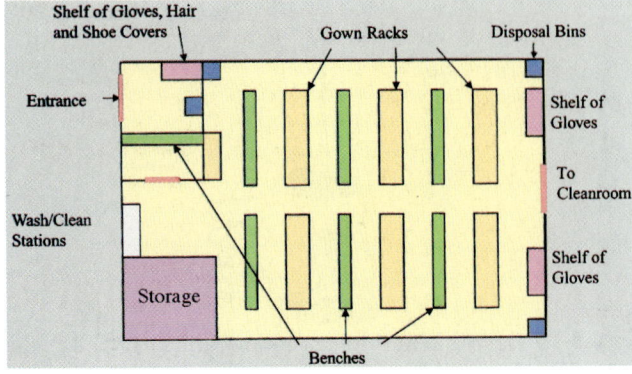

Fig. 8-17 Layout of a *gowning-room*.

of the bench (to prevent large numbers of particles from being carried into the gowning room, Fig. 8-17). At the door to every cleanroom is a floor-mat with a sticky surface, which pulls off and holds dirt from workers shoes. It is also common practice to require workers to wash their hands in UPW-water to prevent transferring particles to the gloves, and also to pass through an air-shower before entering the cleanroom. Figure 8-18 shows the donning of cleanroom-garments in a gowning-room.

The degowning sequence is the reverse of gowning. First, boots are removed, then the suit and hood. They are usually hung on a rack next to the entry. Cleanroom-suits are usually washed weekly, with special laundering and packaging procedures employed

Fig. 8-18 Donning cleanroom garments.

CHAPTER 8 CONTAMINATION CONTROL AND CLEANING TECHNOLOGY FOR ULSI

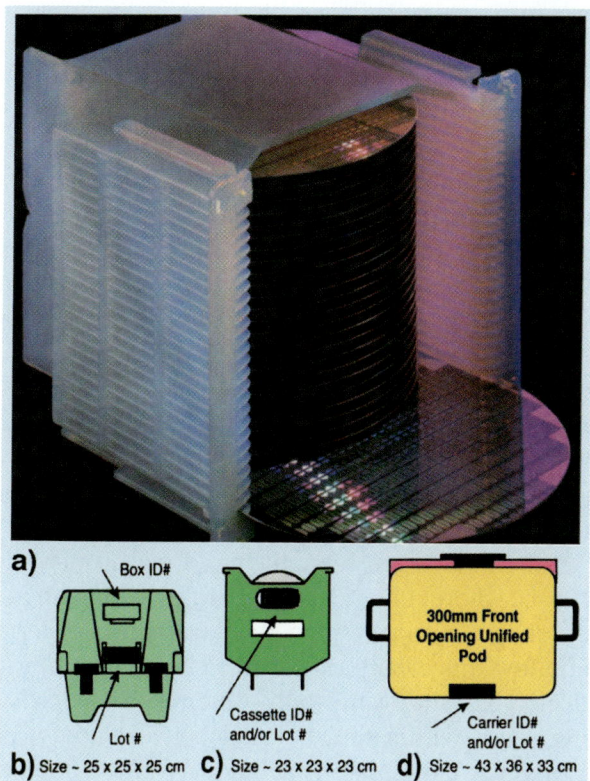

Fig. 8-19 (a) Wafers in a cassette. Comparison of: (b) 200-mm wafer-box; (c) cassette and; (d) 300-mm POD.

cleanrooms, all records are kept electronically, using handheld and desktop computers.

Wafers are normally moved as a group in containers called *cassettes* (Fig. 8-19). However, if wafers must be individually moved, this must only be done using a vacuum-wand. They are picked up by the wand only from the backside (Fig. 8-20).

Fab workers are not allowed to wear makeup, cologne, or after-shave products, because these substances emit particles. Workers cannot wear contact-lens

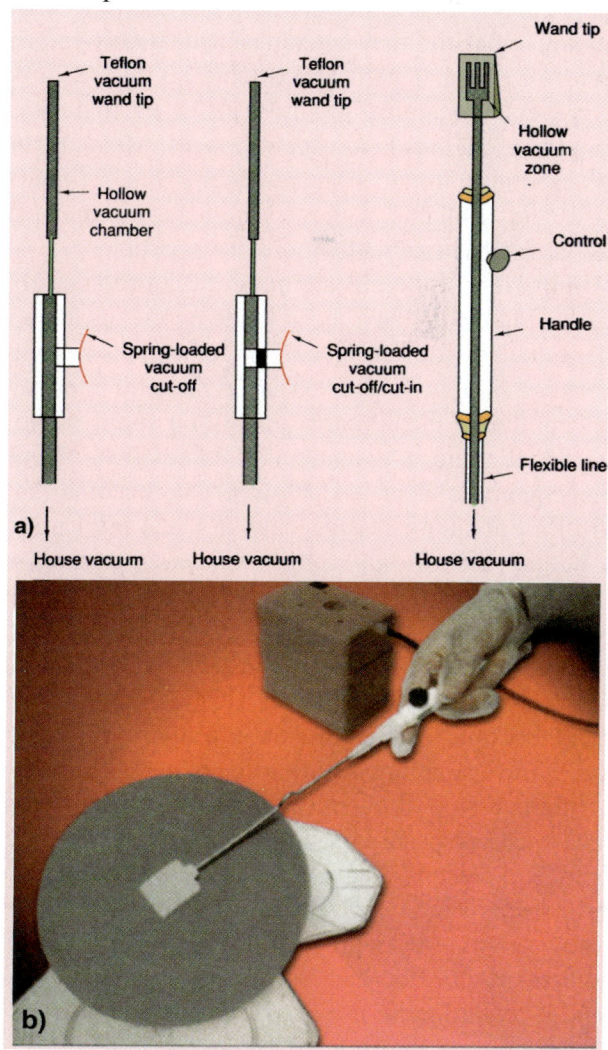

8-20 (a) Vacuum-wand schematic drawings. (b) Holding a wafer from its backside with a vacuum-wand. Courtesy of Fluoro-Mechanic Co. Inc., Tokyo.

to limit contamination of the apparel while it is being washed, transported, and stored between uses. Some fabs use disposable-garments that are thrown away after use. Once outside the cleanroom, workers take off and dispose of hair-covers and latex-gloves. Shoe-covers are usually the last items to be removed.

8.4.3 Cleanroom-Protocols

Cleanroom-protocols have been developed to keep particles from becoming airborne, and to prevent contaminants from reaching wafers.[9] Once inside a cleanroom, workers must walk steadily. Running or jumping could disturb particles on the surfaces of the floor, walls, or ceiling. There are few chairs in a cleanroom, as particles on the chair-surfaces could be dislodged when a person sits or stands up. For the same reason, sitting on tables or leaning on walls is not permitted. Since ordinary-paper contains small fiber-fragments, only special cleanroom-paper is allowed. In Class-1

es because trace-amounts of chlorine in the fab may react with the lenses and cause eye-injury. Smoking is forbidden in the cleanroom and adjacent buildings. Smokers who work in the cleanroom are strongly encouraged to quit, since smokers continue to emit particles they have inhaled even after they have finished a smoke. Eating and drinking are also strictly forbidden in cleanrooms.

8.4.4 Ultra-Pure Chemicals

Many chemicals are used in the fabrication of ULSI, and any impurities they contain can negatively-impact device-quality and yield. Strict-control is thus required to prevent contamination by transfer of impurities from these chemicals to wafer surfaces. Most chemicals used in IC-production are either gases or liquids. Gases are available with very-high purity, and can also be filtered to effectively remove particles.

Impurities in liquids are more difficult to control. Substantial efforts have been made by chemical-suppliers to provide the semiconductor-industry with *ultrapure chemicals* (defined as having total impurity-concentrations in the parts per billion [ppb] range). Use of ultrapure, low-particulate liquids (i.e., chemicals, organic solvents, and DI water) has become standard for critical processing steps in ULSI fabrication.

Future use of *point-of-use chemical-generation* (POUCG) is expected to become more widespread. In this technology high-purity, anhydrous-gases are piped into pure DI-water to form such wafer-cleaning chemicals as NH_4OH, HF, and HCl, as needed. This offers benefits of precise-concentrations, recirculation, filtration, and point-of-use guarantee of purity.

Reprocessing of chemicals will also be mandated, since waste-disposal will no longer be acceptable for ecological reasons. Recycling will also become more cost-effective as a high-degree of ultra-purification is now possible by reprocessing. Since gases can be more-easily purified and less quantity is needed, the trend is toward more-use of gaseous reactants.

8.4.5 Machine-Design and Wafer-Handling

Processing-equipment is a major source of particulate-contamination, and it is predicted that the percentage of particles generated by the equipment and the process itself will rise as advances continue in controlling externally-introduced particle-contamination. While mechanical operations, such as wafer-handling and transport, pumping and venting, gas-flow, and valve operation can all produce particles, various measures have been implemented to control this problem. Scheduled-maintenance helps reduce particle-generation, as does prevention of electrostatic charge-build-up within the machines (because electrostatic-attraction is one of the most common-ways particles are transported to the wafer-surface). Minimizing friction of moving equipment-parts and reducing the number of moving-parts within vacuum-chambers are also effective particle-control techniques.

As new tools are brought into a cleanroom, they must be thoroughly-cleaned before shipment, and they should also be shipped in wrapping materials that allow this cleanliness-level to be retained during transport to the wafer-fab. Automated wafer-handling (both within the tool and between tools) can also reduce particulate-generation. Wafer-movement within a tool is done by robots. The newer robots are fast, reliable, and clean.

8.4.6 Process Modifications

Particle-generation within a process-tool can also be effectively reduced by optimizing the process conditions for film-deposition, plasma-etching, ion implantation, and thermal-treatments. The following are some of many examples of process-conditions that give rise to particles. In CVD-processes, solid particles can form by reactions in the gas-phase. These can then fall down onto the wafer-surfaces. Thus, process-conditions should be developed that avoid such particle-formation. In PVD and CVD processes, films can also build-up on the inner-walls of the process-chamber. If these films become too thick, they can delaminate (and thus become particles within the chamber). Scheduled-shutdown of the tool to clean the walls, or periodic self-cleaning steps can minimize such film-buildup. In ion-implantation processes, material sputtered from the beam-line (or from the wafer-holder, and the wafer itself) may redeposit and form highly-stressed, brittle films that can flake-off and become particles. Metals sputtered from beam-line fixtures can also contaminate wafers.

8.5 WAFER-CLEANING TECHNIQUES FOR ULSI[3]

Since several classes of contamination exist, there are separate cleaning-procedures required to remove them. Some of them are effective in removing more than one group of contaminants. However, both chemical cleaning-procedures and particulate removal-techniques may need to be employed to obtain a completely-clean surface. When one-technique follows another, the latter-steps must not re-contaminate the surface, nor degrade the effectiveness of the former cleaning-procedures.

8.5.1 Wet-Chemical Removal of Film Contaminants

Wet-chemical cleaning has been the standard technique used to remove chemically-bonded films and contaminant molecules and atoms from wafer-surfaces.[6,7,11] Some wet-cleaning procedures are also effective in removing particles. The goals of chemical-cleaning are defined quite clearly in the International Semiconductor Technology Roadmap (ITRS), which separately defines wafer-surface preparation require-

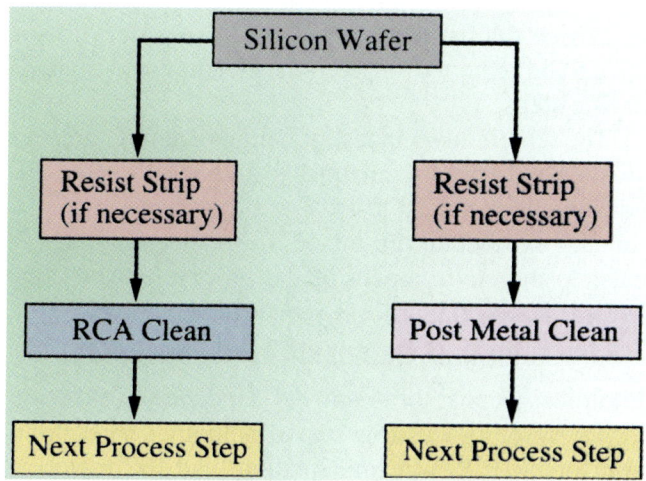

Fig. 8-21 Cleaning strategies in modern microchip manufacturing processes. Left. FEOL; Right, BEOL.[12]

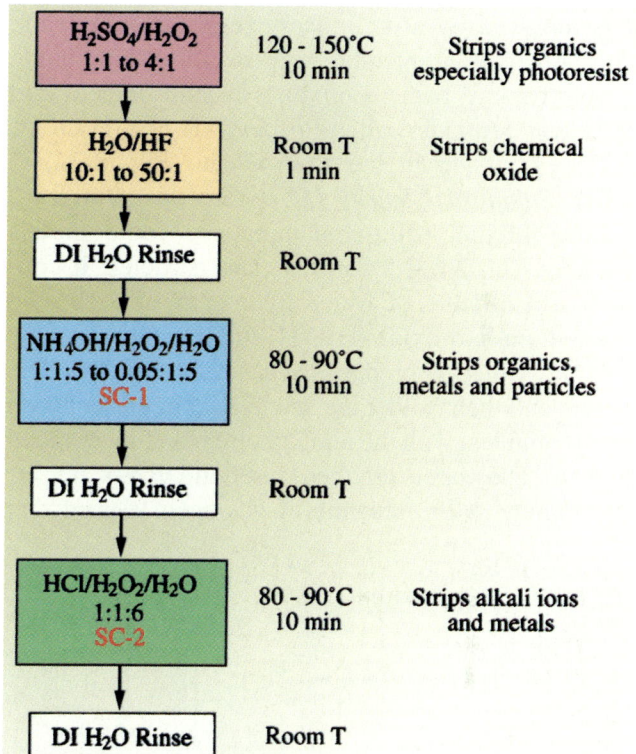

Fig. 8-22 Cleaning-step sequence used in the RCA-clean.[12] Reprinted with permission.

ments for *front-end-of-line* (*FEOL*) and *back-end-of-line* (*BEOL*) processes (Fig. 8-21). FEOL is defined in this context as the steps that begin with a starting wafer, up to the first-metal contact-cut. BEOL is defined as all process steps from that point on.

8.5.2 FEOL Wet-Chemical Cleaning – RCA Clean

Wafers processed up to the end of FEOL generally contain only silicon, silicon dioxide, or silicon nitride on their surfaces prior to high-temperature operations (such as oxidation, diffusion, epitaxy, or ion-implant anneal). Consequently, wafer-cleaning operations can utilize highly-reactive chemicals that do not attack these corrosion-resistant materials.

When bare-silicon (or a silicon-wafer with only thermally-grown oxide), is chemically-cleaned prior to a furnace-process, a two-step cleaning-procedure (known as the *RCA-Clean*) is widely used (Fig. 8-22).[4] Note that if photoresist is present on wafers, it is removed by a preliminary resist-stripping step. This involves immersing wafers in an inorganic resist-stripping bath (a mixture of sulfuric-acid and hydrogen-peroxide, called *Piranha* - H_2SO_4-H_2O_2).

The first-step in the RCA-Clean (Fig. 8-22) is called *Standard Clean-1* (*SC-1*). Its function is to remove organic film-contamination, some metals (Au, Ag, Cu, Ni, Cd, and Cr), and particles. It consists of a mixture of water, hydrogen-peroxide, and ammonium-hydroxide (H_2O:H_2O_2:NH_4OH - 5:1:1 by volume) which is prepared and heated to 70°C. The wafers and their holder are submerged in this solution for 10 minutes, with the temperature being maintained at 70°C. The wafers are then rinsed in UPW-water for one minute. More recently, it has been learned that lower-concentrations of NH_4OH have been found to be as-effective in cleaning the wafers and also produce less surface roughening. Furthermore, since this results in less chemical-consumption, many fabs are incorporating this *diluted SC-1 cleaning recipe*.

The second step is called *Standard Clean-2* (*SC-2*). Its function is to remove inorganic-ions, alkali-ions, and heavy-metals. It consists of a mixture of H_2O:HCl:H_2O_2 (6:1:1 by volume), which is prepared and heated to 70°C. The still-wet rinsed-wafers from Step-1 are submerged into the solution for 10 minutes. The wafers are then rinsed in UPW-water and dried.

8.5.3 Spray-Processing

The use of centrifugal spray-cleaning (instead of immersion in cleaning-solutions), is also possible. In the batch-spray method, wafers are enclosed in a chamber purged with N_2. A sequence of fine sprays of cleaning-solutions and high-purity water from a stationary center-post continuously sprays freshly-blended chemicals across the wafers. The wafers reside in a cassette that is loaded onto a turntable that rotates them past the post (Fig. 8-23). Spent chemicals are drained continuously through the bottom of the bowl, so that fresh-chemicals always contact the wafer. This eliminates the problems of contamination and/or degradation of the cleaning-solution that occur in immersion-cleaning. Rinsing and drying can also be carried-out in the same processor, so wafers do not have to be removed between the cleaning and rinsing/drying steps.

The chief advantages of spray-cleaning are: **1)** smaller volumes of chemicals and UPW-water are consumed (about 2/3 less); **2)** wafer surfaces are continually exposed to pre-mixed, fresh, cleaning-chemicals; **3)** the environment of the process is well controlled; and **4)** the process-sequence is automated.

8.5.4 Photoresist Removal (Resist Stripping)

Photoresist must be removed (*stripped*) following a variety of processing steps, including: **1)** etching (wet or dry); **2)** ion implantation; and **3)** "rework." The main objective in resist-stripping is to insure that all the photoresist is removed as quickly as possible

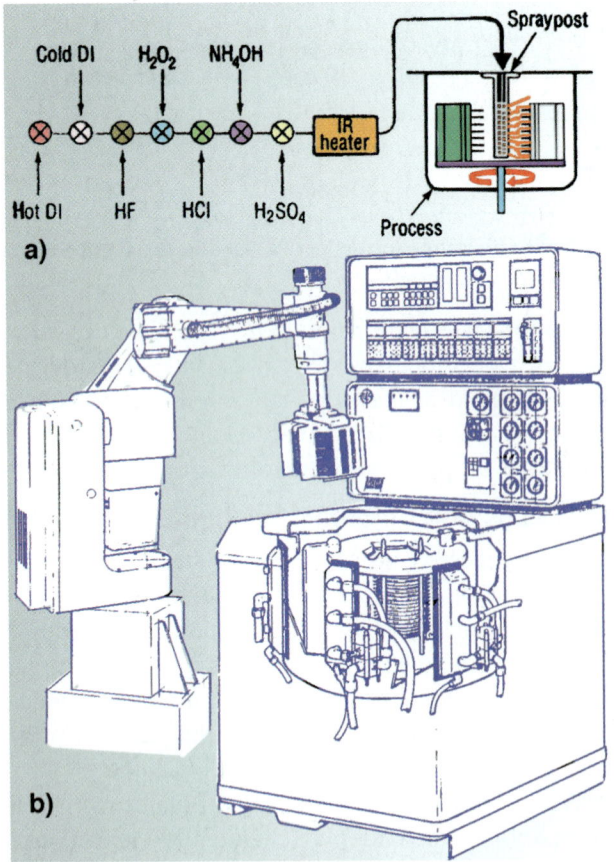

Fig. 8-23 (a) Schematic diagram of a spray-tool. (b) with robot. Courtesy of FSI International.

without attacking any underlying surface-materials. In fact, no single resist-stripping chemical or technique is suitable for all applications. Resist-stripping processes are thus divided into three classes: **1)** wet inorganic-strippers; **2)** organic strippers; and **3)** dry-type stripping techniques.

Wet inorganic-strippers are used when the underlaying material is SiO_2, but they cannot be used if metal-films are present on the wafer surface (since the metal will also be stripped by these mixtures). The most widely-used wet inorganic-stripper is a mixture of 7-parts sulfuric-acid (98% H_2SO_4) to 3-parts 30% H_2O_2 heated to ~125°C. It has earned the name "Pirhana clean" because it attacks organic-materials so aggressively. It performs its function by oxidizing the carbon in the resist to form CO_2, which exits from the bath as a gas. Wafers are immersed in Pirhana at 100–130°C for about 10-min. On removal, they are rinsed in 18–23°C UPW water.

Organic-strippers are used for removing resist from wafers having metal-films on their surface. They remove resist by breaking down the structure of the resist-layer. Early organic-strippers were phenol-based (such as J-100) and they enjoyed wide popularity. However, their use became limited due to their relatively short operating-life and the problem of phenol disposal. A class of *low-phenol* and *phenol-free* organic-strippers have been formulated to overcome these drawbacks. Organic strippers based on n-methyl-pyrrolidone (and other solvents that are biodegradable, disposable by dilution, or recyclable), are also available. The latter are used to clean wafers in a wet-bench after the bulk of the resist is removed by an oxygen-plasma.

Dry-etching of resist is done in plasma resist-stripping tools, with oxygen being the process-gas. This offers several advantages over wet resist-stripping, including safer operating-conditions, no metal-ion contamination, reduced pollution problems, and no attack of most underlying substrate-materials. Dry-etching of organic-films, including resist, is discussed in further detail in Chap. 22.

8.6 PARTICLE REMOVAL

Particles are deposited on wafers as aerosols from the air, or as particles present in the liquids in which wafers are immersed. The basis of their removal rests on an understanding of how such deposited-particles adhere to wafer-surfaces.

If a solid-particle wanders close enough to the vicinity of a solid surface, the *van-der-Waals force* can attract and capture it. The particle is then *physisorbed* on the surface. The van-der-Waals force is universal, and is effective in trapping all types of solid particles (and gas-molecules, as well). Other forces that physisorb particles include *static-charge on the particle*, *electrostatic double-layer repulsion* (EDR), and *capillary-adhesion* (the force exerted by a liquid-film between the particle and the surface), but the van-der-Waals (VDW) force is the strongest.

There are three methods used to remove physisorbed particles: **1)** *chemically-assisted removal* (particle dissolution, particle oxidation [and then dissolution, and lift-off by slight-etching of the surface]); **2)** *electrical-repulsion of the particle and surface*; and **3)** *mechanical-dislodgement* (sliding and rotating a particle by applying a shear-force).

Chemically-assisted particle-removal techniques are implemented in some wet-cleaning procedures (Fig. 8-24). For example, in the SC-1-step of the RCA-Clean, several such mechanisms are believed to work in tandem. That is, the SC-1 solution removes some particles by dissolution (i.e., organic particles). It removes other kinds by slightly simultaneously oxi-

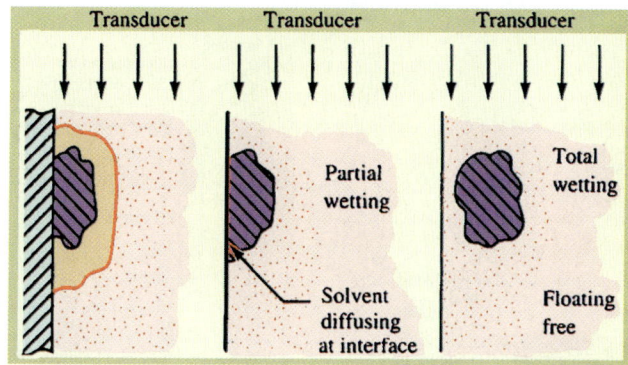

Fig. 8-24 Models for the mechanisms of particle removal.

dizing and etching the Si-surface. That is, it undercuts the oxide under a particle, which reduces the contact-area between the particle and surface. This makes the particle more prone to being lifted off. The SC-2 process only dissolves certain types of metal-particles. However, since it does not attack the oxide, it does not remove particles as well as SC-1.

The *mechanical-removal of particles* generally proceeds by momentum transfer. Mechanical forces impact the particle and dislodge it, either by sliding it off the surface, or by rotating it until it detaches. As particle-sizes decrease, they become more difficult to dislodge by mechanical-force. Mechanical particle-removal techniques have traditionally involved either vibrational-scrubbing (*ultrasonic* or *megasonic*) or a technique that combines high-pressure liquid-spraying and brush-scrubbing.

8.6.1 Vibrational Scrubbing (Ultrasonic and Megasonic)

Vibrational scrubbing of wafers is a non-contact and brushless technique for removing particles. In *ultrasonic scrubbing*, wafers are immersed in a suitable liquid-medium to which sonic-energy at 10–100-kHz is applied. High-intensity sound-waves generate pressure-fluctuations that produce microscopic bubbles in the liquid. These bubbles rapidly form and collapse (a mechanism called *cavitation*), producing shock-waves which strike the wafer-surfaces. These shock waves displace or loosen particulates. UPW water is an effective ultrasonic cleaning-liquid for removing polymeric-particles, while ethanol-acetone (1:1) is better for removing inorganic-particles. However, ultrasonic scrubbing can damage some structures on the surface of wafers.

Cleaning systems using higher-frequency sonic waves (700–1000-kHz) are also available. Because such megasonic-frequencies are much higher than ultrasonic frequencies, large cavitation-bubbles do not have time to form and surface damage is reduced. The energy (350–800-W) is produced by an array of piezoelectric-transducers mounted on the side or bottom of the tank. The sonic pressure-waves travel through the liquid, and cleaning is accomplished by the impact of these pressure-waves on the particles. Megasonic vibrations can be generated in tanks containing SC-1 and SC-2 solutions, and it has been reported that this approach is more effective than performing such cleans without megasonic-scrubbing. Figure 8-25 shows a schematic drawing of a megasonic cleaning-tank. In such systems, chemical cleaning and contaminant desorption can be accomplished, while simultaneously removing particulates.

8.6.2 Particle-Removal by Brush-Scrubbing

Particulate-removal by a combination of high-pressure spraying and brush-scrubbing has also been carried out *after* a variety of process steps (sawing, lapping, polishing, and CMP), and *before* others (metallization, CVD, and epitaxy [double-sided scrubbing]).

The scrubbing-process operates by rotating a polyvinyl-alcohol (PVA) brush across the surface of a wafer (Fig. 8-26) while directing a high-pressure water-jet at the surface. The brush material should not come into contact with the wafer-surface. Instead, the layer of fluid between the brush and wafer-surface dislodges the particles hydrodynamically as the liquid is both compressed and pushed along by the brush. Careful mechanical-adjustment must be employed to ensure that the brush hydroplanes over, but does not actually contact, the wafer-surface. Early brush-cleaning systems did not always offer such control, and they were notorious for damaging wafer-surfaces.

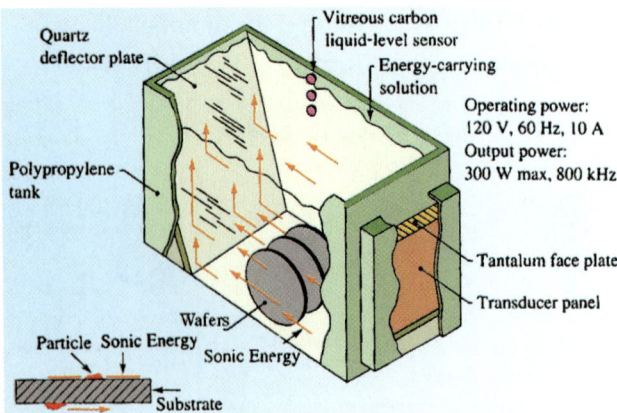

Fig. 8-25 Schematic of megasonic cleaning-tank.

Improved system-design and better brush-materials have overcome this problem.

The PVA brush-material is a soft, sponge-like material that is highly compressible. This material allows the mechanical component of cleaning to occur without damaging the surface of wafers. The widespread acceptance of chemical-mechanical polishing has propelled brush-scrubbing into the mainstream of wafer cleaning. It is considered to be the most effective method of removing slurry from the wafer-surface in the CMP cleaning-process. Simultaneous double-sided scrubbing is used in CMP-cleaning. Reportedly, particles as small as 0.12-μm have been effectively removed by brush-scrubbing. Particles are dislodged and they then stay suspended in the liquid until they are pushed over the edge of the wafer into the drain (as liquid is continuously being delivered to the wafer surface). Dislodged-particles, however, must be prevented from re-adhering to the wafer-surface or the brush-material. This is done by controlling the electrostatic forces between the particles and wafer or brush surfaces. Keeping the pH-values of the cleaning liquid greater than 7 seems to work best for this.

8.7 RINSING AND DRYING WAFERS

After the cleaning-chemicals perform their function, they must be removed. This is typically done by rinsing the wafers with DI-water and then drying them. Rinsing and drying represent critical steps in the cleaning-sequence since they are performed so frequently. Rinse-tanks and dryers may become major sources of particulate-contamination unless they are monitored and properly maintained. In addition, complete removal of the cleaning-chemicals cannot be done just by dunking the wafers in a tank of water. Thorough-rinsing requires a continuous-supply of clean-water to the wafer surface. The resistivity of the effluent rinse-water is normally monitored, and rinsing is continued until it exhibits the 18-MΩ-cm resistivity of pure DI-water.

8.7.1 Rinsing

Wafer-rinsing in DI-water is performed in one of several types of rinsers. These include: *centrifugal spin

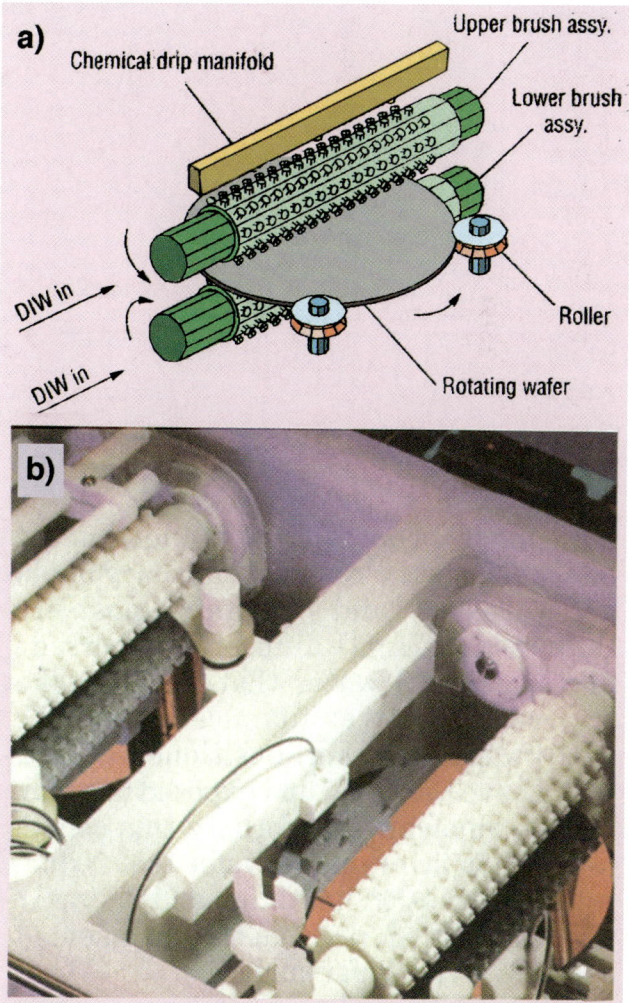

Fig. 8-26 (a) Schematic; and (b) Photograph of a double-sided brush-scrubber. Courtesy of Lam Research.

rinser/dryers; *overflow rinsers*; *dump rinsers*; and *cascade rinsers*. In the centrifugal *spin-rinse/dryer* (SRD), water is sprayed on wafers as they rotate in a chamber. Holders mounted along the inner-periphery of a drum hold the wafers. A perforated-tube located along the axis of the drum is connected to a source of UPW-water and heated N_2-gas. Water is first sprayed on the wafers as the drum holding the wafers is rotated. After the rinse-cycle is finished, the water is turned off and the drying-cycle starts. The drum is slowly accelerated to higher-speed as hot N_2-flows into the drum. The high-speed rotation throws water

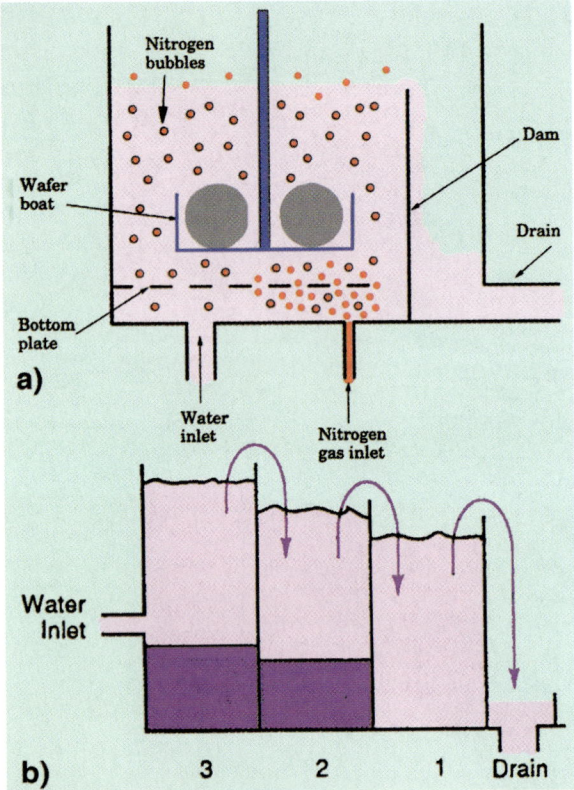

Fig. 8-27 (a) Schematic of an overflow-rinser. (b) Schematic of a cascade rinser.

off the wafer surfaces and drying is assisted by the hot-N_2. The process completely dries wafers in a short time.

An *overflow-rinser* is a tank recessed into the cleaning-station deck. The wafers in a boat are loaded into the tank. UPW-water enters the bottom of the tank and flows through and around the wafers, exiting over a dam into a drain system (Fig. 8-27a). The rinsing action that removes chemicals from the wafer surface is enhanced by a stream of nitrogen bubbles introduced into the rinser from the bottom. Rinsing is carried out for about 5-min with high flow-rate of UPW-water. For instance, if the tank has a volume of 3-L, the UPW flow-rate should be at least 15-L/min. Megasonic-rinsing improves the overflow and cascade rinser (see below) characteristics.

The *cascade-rinse system* consists of two or three overflow-rinsers connected to each other (Fig. 8-27b). Clean UPW-water enters only the rinser on the left of this figure, but it then cascades through the downstream rinsers. Wafers start the rinse process in the downstream-rinser and are moved sequentially to the rinser with the direct water-supply. When several boats are being rinsed simultaneously this system is more efficient in its use of water. The cascade overflow system adds the least number of particles to the wafer surface, as there are no sprays.

The *dump-rinser* is like an overflow-rinser with a spray-capability. Wafers are placed into the empty tank and DI-water is sprayed down from nozzles onto them. This quickly removes chemicals on the wafer surface. While they are being sprayed, water is also rapidly injected from an inlet near the bottom to fill the tank. The tank has a false-bottom, and as the water overflows the top, a trap-door swings open and the spent water is dumped instantly into the drain (Fig. 8-28). This fill-and-dump action is performed a number of times until the wafers are thoroughly rinsed. However, spray-nozzles tend to generate particles and grow bacteria. Also, quick-dump rinsers produce turbulent convection-currents in the water, which increase particle mobility to the wafer-surface.

8.7.2 Wafer Drying after Rinse

The drying-process is also a key step in achieving an effective cleaning-sequence. Most of the water should be *physically* removed from the surface, leaving only a thin-layer to evaporate. Thus, residue from evaporation is minimized. One widespread residue-species is the *water-spot*. Water-spots (1-10-μm in size) come from dissolved material in UPW water (often silica) left behind as water droplets evaporate. Water-spotting can cause such problems as poorly-adhering films, high contact-resistance, and gate-oxide defects.

8.7.3 Spin-Dryer

The spin-rinser/dryer (SRD) described in the previous section has been the workhorse of wafer-drying in most fabs (Fig. 8-29). SRDs provide the safest, most cost-effective means of wafer-drying. In the past, a persistent problem of spin dryers was the electrical

CHAPTER 8 CONTAMINATION CONTROL AND CLEANING TECHNOLOGY FOR ULSI

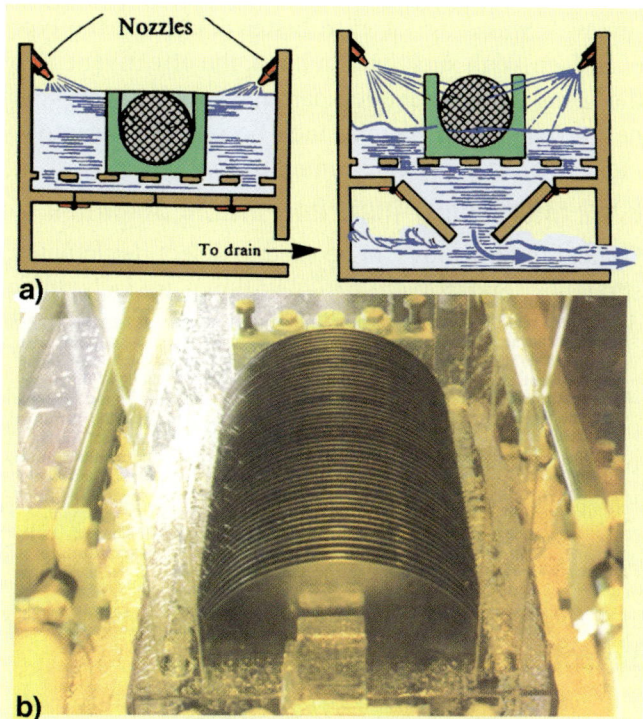

Fig. 8-28 (a) Schematic of a quick-dump rinser. (b) Photograph of a quick-dump rinser.

charging of cassettes and wafers, causing them to attract particles upon exposure to the cleanroom. This charging is now suppressed by flowing ionized-nitrogen into the drying-chamber. The newer models are also designed to produce less vibration, which in turn helps reduce particulates.

8.7.4 Isopropyl-Alcohol (IPA) Vapor-Dryers

Isopropyl-alcohol vapor-dryers were introduced in the mid-1980s. In vapor-dryers, wafers still wet with DI-water are suspended in an enclosed chamber, the ambient of which is a cloud of solvent-vapor (most commonly isopropyl-alcohol [IPA], heated to about 80°C). The IPA displaces the water on the wafer, leaving a "IPA-vapor-coated" surface. When the displacement is completed, the chamber doors are opened and the wafers are raised out of the IPA-vapor. They dry quickly as they cool because of the high-volatility of the IPA. A series of cooling-coils condenses IPA that attempts to escape as the wafers are being raised, and

drains it away (Fig. 8-30). In a final step, a nitrogen-purge removes any IPA from the volume in which the wafers are dried. The IPA-drying approach is attractive because it results in lower particulate-densities than spin-drying. IPA-dryers are becoming more common in newer fabs.

8.7.5 Wet-Cleaning Systems

The wet-cleaning process is carried out on systems called *wet-benches*. Figure 8-31 shows a schematic drawing of an immersion wet-bench.

8.8 PARTICLE-DETECTION ON WAFER SURFACES

Various techniques are to detect the presence of particles on wafer surfaces. These include: optical microscopes; scanning-electron microscopes (SEMs), and laser particle-counters.

8.8.1 Particles-Per-Wafer-Per-Pass

The concentration-levels of particles needs to be quantifiably-measured after each process-step. A *par-*

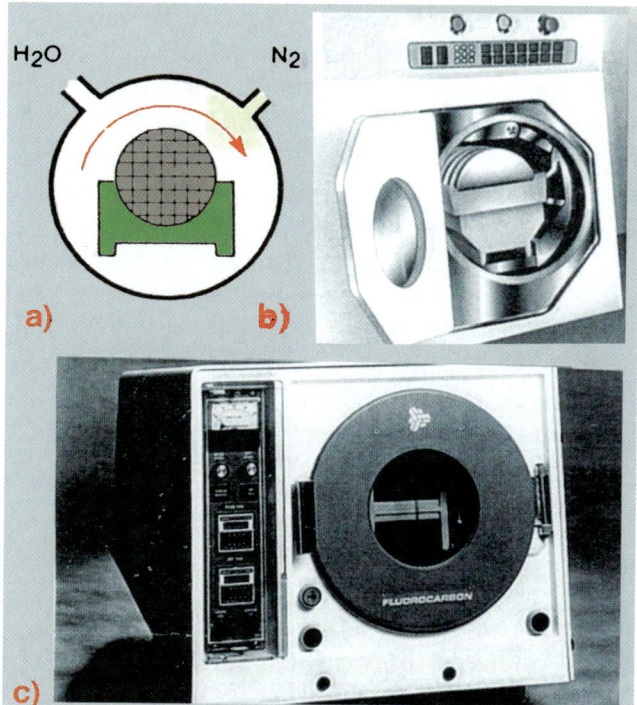

Fig. 8-29 (a) Schematic of a spin-rinser/dryer. (b) Single-boat axial rinser/dryer, door-open, and (c) door-closed.

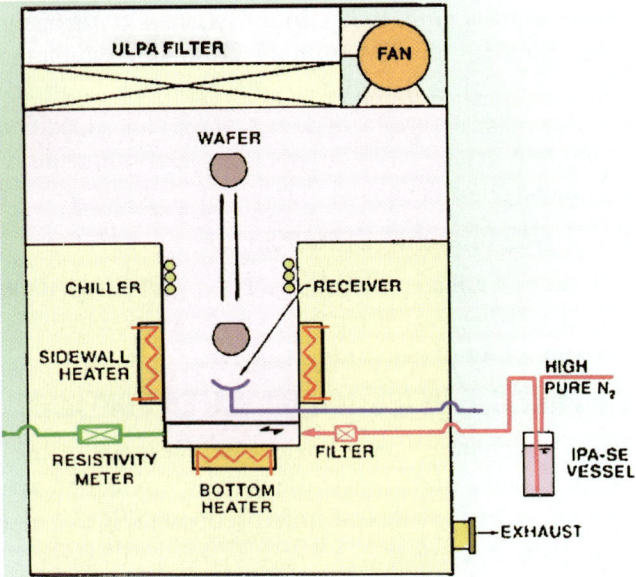

Fig. 8-30 Cross-section of an *isopropyl alcohol* (*IPA*) *dryer*. (© IEEE 1989)

review stations are used to better determine the type of defect. With such data in hand, the effectiveness of the steps taken to reduce particulate-counts can also be evaluated, and it may then be possible to implement improved wafer-cleaning procedures.

8.8.2 Microscopy Tools for Particle Detection

Optical microscopy can be used to detect particulates, scratches, and solvent-residues down to 1–2-μm in diameter. SEMs are used to analyze contaminants whose dimensions are smaller than 1.0-μm. However, in many cases, one is interested in the number of particles per unit area. In such cases, the particles must be counted. Since microscopic counting is extremely laborious and dependent on trained operators, *particle-counting* is now routinely performed using automatic particle-counters.

8.8.3 Automatic Laser Particle-Counters for Detecting Particles on Wafers

Automatic laser particle-counters (also termed *surface scanners*) count particles on wafer-surfaces, and such measurements are most often carried out on bare-silicon (i.e., *unpatterned*) wafers. Such inspection tools rely on the optical-scattering of light-beams by particles. Figure 8-33a shows their principle of operation. A beam of laser-light (originally 550-nm xenon-lasers, but now 488-nm Ar-laser sources) illuminates the wafer at an angle to the surface. This beam is scanned across the wafer surface, and an integrating light-collector is used to collect any scattered-light. On a clean, smooth-surface, the incident

particles-per-wafer-per-pass (*PWP*) method is used to make these measurements. PWP is the number of particles added to wafers in one pass through one or more pieces of process-equipment. Relatively-clean wafers are initially scanned to obtain a baseline particle-count. These are then passed through the process-equipment, and the resulting counts measured. The difference between the *pre* and *post* counts is the number of particles added (or removed from) the wafer by the process-equipment (Fig. 8-32). The PWP test is used as part of day-to-day monitoring operations as a "go/no-go" test to ensure that any catastrophic, "yield-busting" particle problems do not occur. If there is a problem, more advanced defect-

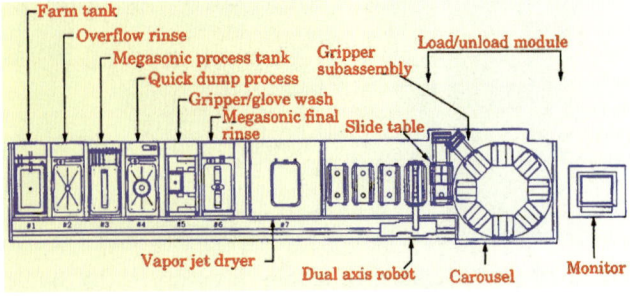

Fig. 8-31 Schematic drawing of a modern wet-bench.

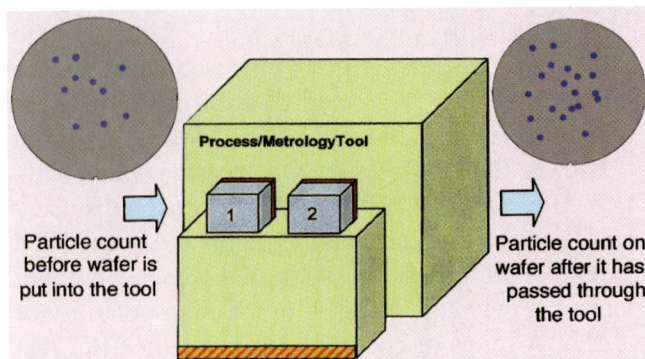

Fig. 8-32 Concept of *particles-per-wafer-per-pass* (*PWP*).

laser-light will be reflected at the same angle, and will thus escape between the light-collectors without being detected. However, light that hits a particle is scattered in a broad direction. Some of this scattered-light is collected by the high-efficiency light-collection system. This signal is amplified by a photo-multiplier tube. After the wafer is scanned, a graphic display (such as a wafer-map, Fig. 8-33c) is produced showing the number and type of defects. By the late 1990's such scanners were able to detect particles (and pits, epi spikes, protrusions, cracks, and scratches) on well-polished unpatterned Si-wafers smaller than 0.1-μm in diameter. Total scanning-time for unpatterned surfaces is tens of seconds for 200-mm wafers.

The problem of counting-particles on unpatterned-surface-layers other than bare, smooth silicon, is more difficult. Such surfaces are generally rougher, and this creates a condition known as *haze,* which effectively obscures the detection of particles below a minimum-dimension. Figure 8-34 shows the mini-

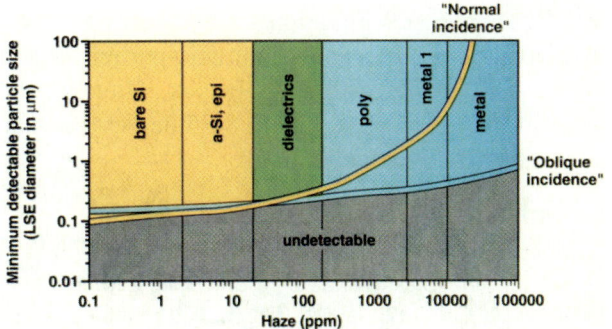

Fig. 8-34 Minimum detectable particle-size on different surfaces. Courtesy of Semiconductor International.

mum-detectable particle-size for various surfaces. For some types of film-surfaces (i.e., polysilicon and metals), the wafer-surface is illuminated with an oblique laser-light-beam (i.e., instead of with a beam that is

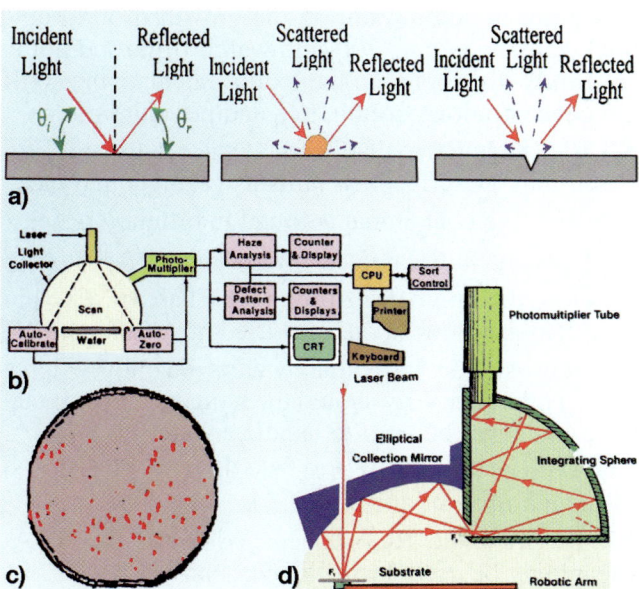

Fig. 8-33 (a) Light-reflection and scattering from the wafer-surface and a particle. (b) Functional block-diagram of automatic defect-detection-system, utilizing a laser-scanning detector. (c) Typical surface-pattern from an automatic laser-scanning system. (d) the optical-unit of the system. Courtesy of KLA-Tencor.

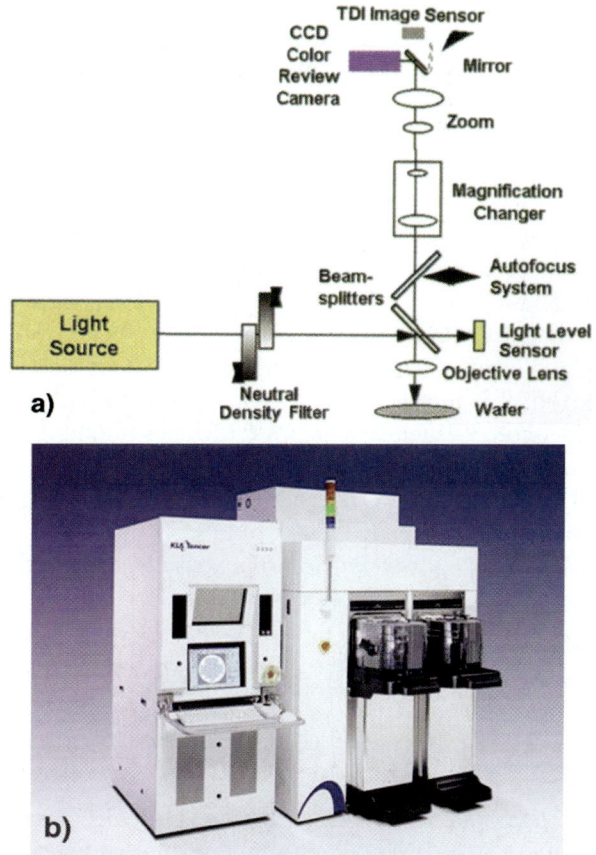

Fig. 8-35 Automatic particle-detection tool. Courtesy of KLA-Tencor

normally-incident). This allows particles of smaller-dimensions to be detected. Automatic-particle-detectors are sold by KLA-Tencor, Inspex, Hitachi, and Applied Materials/Orbot. Figure 8-35 depicts such an automatic particle-detection tool.

REFERENCES

1. S. Wolf and R.N. Tauber, *Silicon Processing for the VLSI Era, Vol. 1 - Process Technology*, 2nd Ed., Lattice Press, Sunset Beach CA 2000, Ch. 5 (Cleaning Technology).
2. *Handbook of Contamination Control in Microelectronics"* D. Tolliver, Ed., Noyes, Norwich NY, 2nd Ed. 1998.
3. H.P. Tseng and R. Jansen, "Cleanroom Technology," *ULSI Technology*, Ed. C. Chang and S.M. sze, McGraw-Hill, New York, 1996, p. 1.
4. *Handbook of Semiconductor Wafer Cleaning Technology*, Ed. W. Kern, Noyes Publications, Norwich NY, 1993.
5. W. Kern, "The Evolution of Silicon Wafer Cleaning Technology," *J. Electrochem. Soc.* **137**:1887 (1990).
6. C. Chang and T. Chao, "Wafer-Cleaning Technology," *ULSI Technology*, Ed. C. Chang and S.M. sze, McGraw-Hill, New York, 1996, p. 67.
7. T. Hatori, *Ultraclean Surface Processing of Silicon Wafers*, Springer-Verlag, New York, 1998.
8. A. Braun, "Cleanroom Technologies Keep Contamination at Bay," *Semiconductor International*, April 1998, p. 112.
9. R. Kraft, "Proper Cleanroom Protocol," *Semiconductor International*, March 1998, p. 73.
10. R. Iscoff, "Cleanroom Apparel: A Question of Tradeoffs," *Semiconductor Internatl*, March 1994, p. 66.
11. R. Iscoff, "Wafer Cleaning: Wet Methods Still Lead the Pack," *Semiconductor International*, July, 1993, p. 58.
12. J.D. Plummer, M. Deal, and P.R. Griffin, *Silicon VLSI Technology*, Prentice-Hall, 2000, p. 159.
13. *Handbook of Silicon Semiconductor Metrology*, Ed. A.C. Diebold, Marcel Dekker, New York, 2001.

PROBLEMS

1. For a Class-100 cleanroom, find the number of dust particles per cubic-meter with particle sizes: (a) between 0.5 and 1.0 μm; (b) between 1 and 2 μm; and (c) above 2-μm.
2. Semiconductor process tools are typically assembled in a Class-1000 cleanroom. Calculate how many particles larger than 0.5-μm per cubic-ft are the air.
3. Explain why the processing equipment itself is kept in the chase areas of the cleanroom (with a Class-1000), instead of in the Class-1 processing areas.
4. The Class-1 definition according to Japanese Std. B9920 rev. describes air-cleanliness with no more than 10 particles/m^3 of a size 0.1-μm or larger. Determine the equivalent air cleanliness according to U.S. Fed. Std. 209E.
5. The cleanliness of outside ambient air is around M7 according to Fed. Std. 209E. Estimate the cleanliness of air leaving the make-up air unit having a HEPA filter of 99.97% filtration efficiency. What is the minimum filtration efficiency of a ULPA filter required for the process-area to reach a Class-M1 (at 0.1-μm)?
6. Sodium is a mobile-ion in silicon dioxide, and its presence in trace quantities in the wafer-fab environment can destroy the reliable operation of microelectronic devices. Name four potential sources of sodium in the cleanroom environment. List five measures employed to keep to concentrations below those that will impair device operation.
7. Assume a wafer is contaminated with one billionth of a gram of sodium during the growth of a 10-nm-thick gate oxide on a 200-mm wafer. If the sodium is uniformly distributed in the oxide, calculate the sodium concentration per unit area and per unit volume.
8. List 5 techniques that keep contamination out of a cleanroom. Describe how humans generate particles.
9. Name five contaminants found in ordinary water.
10. Define a "killer defect." What is PWP?
11. Describe laminar air-flow (see Chap. 6). Explain why laminar-flow hoods are used in work-stations.
12. How should individual wafers be moved by a human operator. Why should the human hand or metal tweezers never be used to handle a wafer?
13. Define the terms: (a) SMIF; (b) FOUP; (c) Minienvironment; and (d) HEPA.
14. Describe the RCA-Clean. Why can the RCA-Clean only be used during FEOL, and not BEOL?
15. Describe modifications to the RCA-Clean.
16. A filter with a thickness of 1-cm is available. Its efficiency for removal of 0.5-μm particles is 0.993. What efficiency would be expected of the same filter material with a thickness of 2-cm?

CHAPTER 8 CONTAMINATION CONTROL AND CLEANING TECHNOLOGY FOR ULSI

143

A 300-mm wafer processing tool with FOUPs mounted on its load-ports.
Photograph courtesy of Tokyo Electron-America.

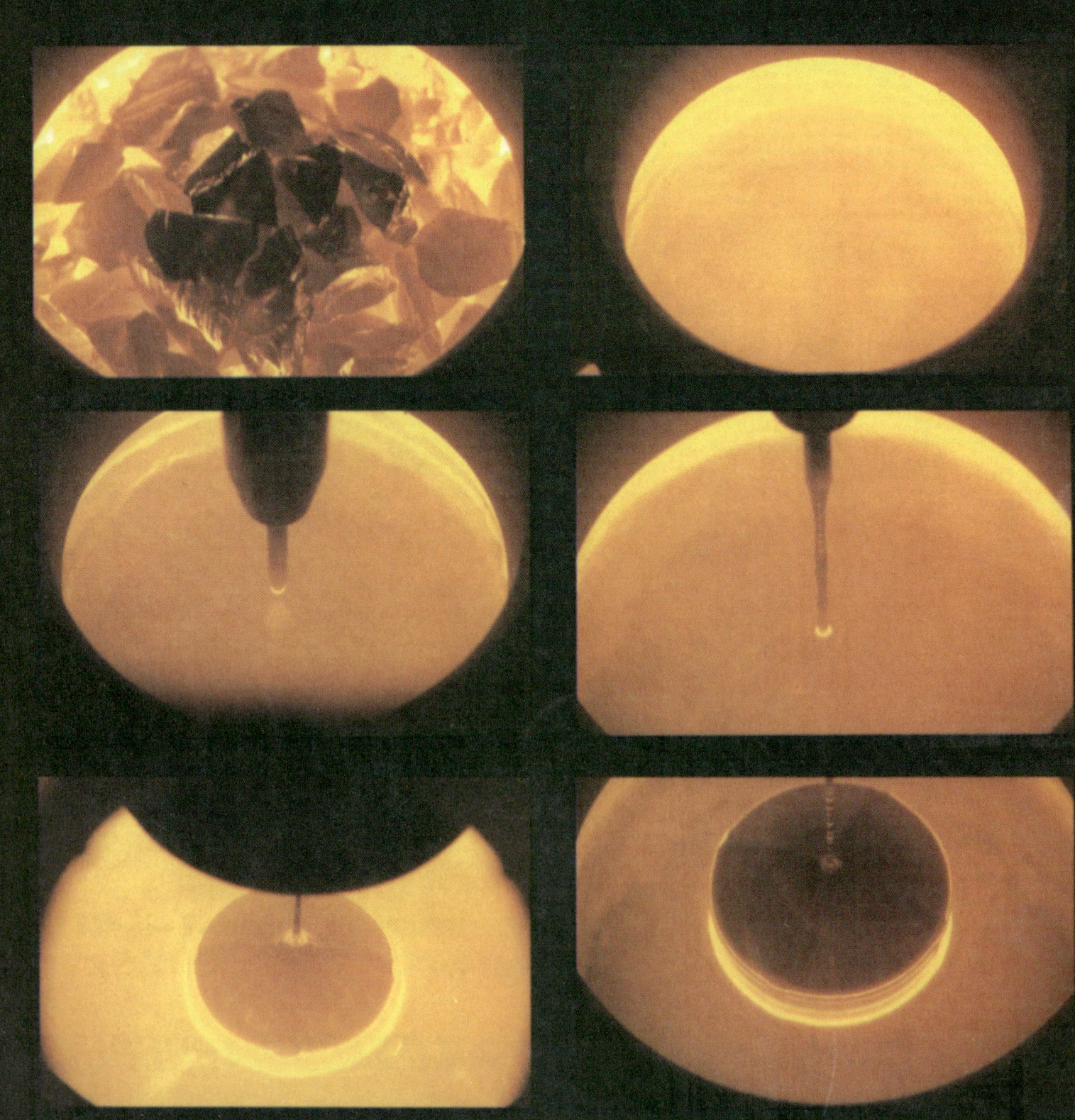

CHAPTER 9

SILICON: SINGLE-CRYSTAL GROWTH

CHAPTER CONTENTS

9.1 CRYSTAL TERMINOLOGY

9.2 MANUFACTURING SINGLE-CRYSTAL SILICON

9.3 CZOCHRALSKI CRYSTAL GROWTH

9.4 CRYSTAL-DEFECTS IN SILICON

9.5 POINT-DEFECTS IN SILICON CRYSTALS

9.6 ONE-DIMENSIONAL DEFECTS (DISLOCATIONS)

9.7 AREA-DEFECTS IN SILICON CRYSTALS

9.8 BULK-DEFECTS AND PRECIPITATON

9.9 OXYGEN IN SILICON

9.10 GETTERING

"To see a world in a grain of sand,
and Heaven in a wild flower,
Hold infinity in the palm of your hand,
and Eternity in an Hour."

Robert Blake

Illustration of several process steps during Czochralski (CZ) crystal growth.

In the next two chapters we show how *single-crystal silicon starting-material* is produced.[1-6] In Chap. 9, we first explain how raw-material is processed to obtain *electronic-grade polysilicon* (*EGS*). Then the Czochralski-method of growing single-crystal silicon from such EGS is described. The crystalline-defects that occur in silicon (and techniques for suppressing their formation) are also covered in this chapter.

The technology for forming *silicon wafers* from single-crystal ingots is covered in Chap. 10. This includes a discussion of the material, electrical, and structural characteristics that such wafers must possess in order to be suitable for VLSI (and ULSI) fabrication.

9.1 TERMINOLOGY OF CRYSTAL-STRUCTURE

Solid-matter exists in crystalline and amorphous forms (Fig. 9-1). In crystalline-solids, the atoms which make up the solid are spatially arranged in a periodic-fashion. If this periodic-arrangement exists throughout the entire solid, the substance is defined as being formed of a *single-crystal*. If the solid is composed of many small single-crystal regions, the solid is referred to as *polycrystalline-material*. Nevertheless, within any single-crystal region, the structure appears exactly the same at one point as it

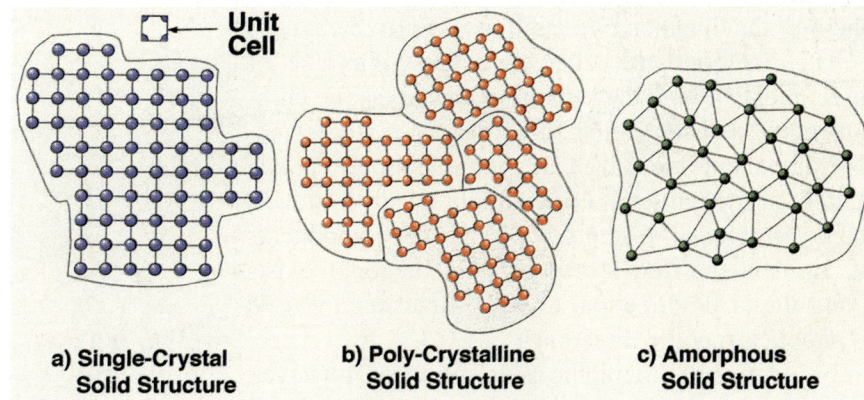

Fig. 9-1 The three different forms of solid-material: (a) Single-Crystal. (b) Poly-Crystalline. (c) Amorphous.

145

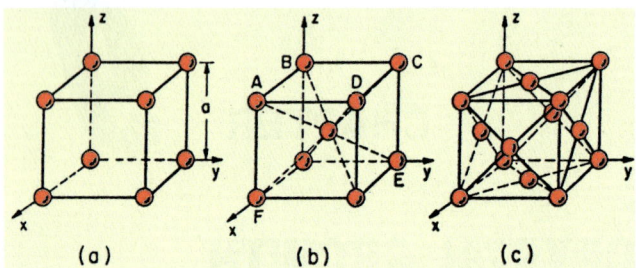

Fig. 9-2 Cubic-crystal unit-cells: (a) Simple Cubic SC; (b) Body-Centered Cubic BCC; (c) Face-Centered Cubic FCC.

does at a series of equivalent points. An *amorphous-material* has no such long-range periodic-structure. The *starting Si-material* (*wafers*) on which VLSI and ULSI circuits are built, have single-crystal form.

The periodic-arrangement of atoms in a crystal is called the *lattice*. The crystal-lattice always contains a volume representative of the entire lattice (referred to as a *unit-cell*), which is regularly repeated throughout the crystal (Fig. 9-1a). The importance of unit-cells is that crystals as a whole can be studied by analyzing such representative volumes. For example, distances of nearest-neighbor atoms in a lattice can be found.

The simplest crystal-lattices are the three *cubic lattices* seen in Fig. 9-2: (a) *Simple Cubic* [SC]; (b) *Body-Centered Cubic* [BCC], and (c) *Face-Centered Cubic* [FCC]. The dimension **a** for a cubic-cell is called the *lattice-constant*, but the distance of nearest-neighbor atoms may be smaller than **a**. For example, in an FCC-lattice the nearest-neighbor distance is one-half the diagonal of a cube face, or $(a\sqrt{2})/2$.

The directions in a lattice are expressed as a set of three integers, with the same relationship as the components of a vector in that direction. These three vector-components are given in multiples of the *basis-vectors* of the unit-cell. For example in cubic lattices, the *body-diagonal* (see Fig. 9-3c) has the components of 1a, 1b, and 1c. Therefore, this diagonal exists along the [111]-direction. (The [ ]-brackets are used to denote a specific direction.)

From a crystallographic point of view, however, many directions in a crystal are equivalent, depending only on the arbitrary choice of orientation for the axes. Such equivalent directions are expressed with angular brackets < >. For example, the crystal-directions [100], [010], and [001] in a cubic-lattice are all crystallographically equivalent, and are collectively referred-to as <100> directions.

It is also useful to describe *planes* in a crystal (see also Fig. 9-3). Note that ( ) parentheses are used to denote a specific plane. For *cubic-lattices*, direction [h k l] is perpendicular to a plane with the identical three integers (h k l). This fact makes it convenient to analyze cubic-lattices. That is, if either a *direction or a plane* in a cubic-lattice is known, its perpendicular-counterpart can be quickly determined without further calculation. For example, the (100)-plane is perpendicular to the [100]-direction.

Silicon has a *diamond cubic-lattice*, a structure which can be represented as two interpenetrating face-centered-cubic lattices (Fig. 9-4a). Thus, the simplicity of analyzing and visualizing cubic-lattices can be extended to characterization of silicon-crystals. The three-dimensional representation of a diamond-lattice unit-cell is shown in Fig. 9-4b and 9-4c.

In studying the diamond-lattice unit-cell, it is evident that each atom has four nearest-neighbors (see Fig. 9-4c). The importance of this fact becomes evident when the electronic-structure of silicon is considered. That is, since Si is a Column-IV element, and has four valence-electrons, each of these electrons in the Si crystal-lattice is shared with one of its four nearest-neighbor Si-atoms. (This nearest-neighbor atom likewise contributes one of its electrons to the

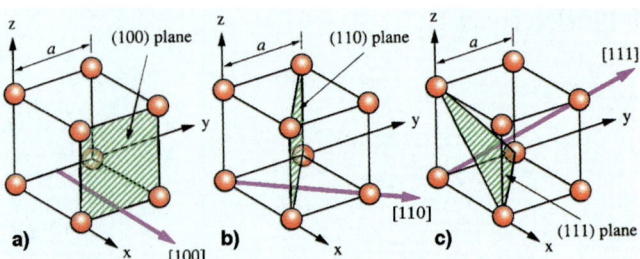

Fig. 9-3 Crystal-planes and major directions for a cubic-lattice. The heavy-arrows in each figure correspond to crystal-directions designated by the brackets [hkl]. The shaded-areas are the corresponding crystal-planes, designated by the brackets (hkl).

bond.) As discussed in Chap. 2, the two valence-electrons shared between nearest-neighbor atoms form a *covalent-bond* and such interactions of electrons in neighboring-atoms of a solid serve the function of holding the crystal together. Since all valence-electrons in a perfect Si-crystal are involved in such shared bonding-pairs, none are available as free-electrons to conduct electrical-current (as long as they remain localized in the bonding region). However, covalently-bonded electrons can be excited, becoming free to participate in conduction. (Note in Si, **a** = 5.43-Å, and the nearest-neighbor distance is 2.43-Å.)

In several processing technologies, reference is made to various planes and directions in silicon-crystals - especially to the (100), (110), and (111) planes (and directions). This is because various structural-properties of Si depend on crystal-orientation, as do many ULSI fabrication-steps. For example {111}-planes have the highest-density of atoms. Therefore {111}-planes oxidize more rapidly than do {100}-planes (since more atoms per unit surface-area are available for the oxidation-reaction). Also, all MOS-devices are fabricated on {100}-wafers, since the smallest surface-state densities are observed on such orientations. Finally, Fig. 9-4d illustrates the open structure of the silicon-lattice. Unobstructed channels can be seen to exist when crossing the Si-crystal along <110>-directions (i.e., perpendicular to the {110}-plane shown in Figs. 9-3b and 12-9d).

Actual solid-materials in the form of single-crystals differ to some extent from the ideal crystal-structure discussed in this section. To begin with, even ideal-crystals possess *surfaces* at their boundaries, and at a surface, the atoms are incompletely-bonded. In addition, real crystals have a variety of imperfections (called *crystalline-defects*) which significantly alter the properties of the crystal. (More information about defects in Si and their impact on crystal and device properties is given later in the chapter.)

It is useful to note the units used to specify dopant and contaminant-concentrations in Si. Usually they are given in *atoms/cm³*, but sometimes in *parts per million atoms* (*ppma*). Since there are about 5×10^{22} Si

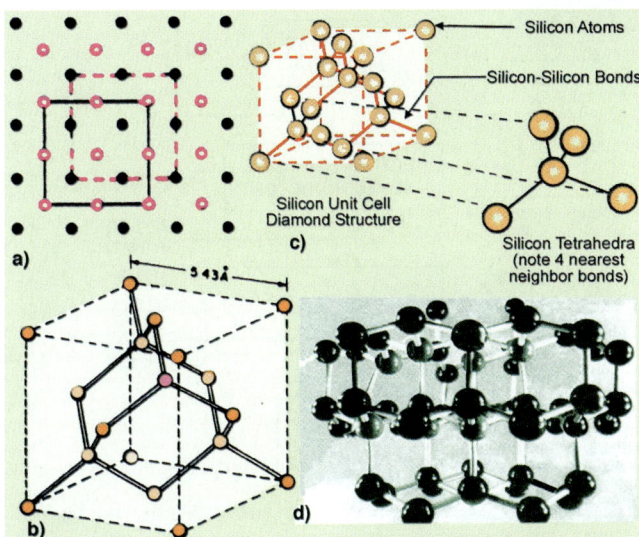

Fig. 9-4 (a) Top-view (along any <100>-direction) of an extended diamond-lattice unit-cell. The open-circles indicate one FCC sub-lattice and the solid-circles indicate the interpenetrating FCC-lattice; (b) Schematic-drawing of the diamond-lattice unit-cell; (c) Schematic drawing of the diamond-lattice unit-cell that also shows the four nearest-neighbors of each silicon-atom. (d) Photograph of the diamond-lattice showing an example of the open channels that occur along some crystal-directions.

atoms/cm³ in single-crystal-Si, an impurity concentration of 5×10^{16}/cm³ is equivalent to 1-ppma. Other concentration levels can likewise be converted.

9.2 MANUFACTURING SINGLE-CRYSTAL SILICON

The fabrication of VLSI and ULSI takes place on silicon-substrates (wafers) possessing very-high crystalline perfection. Polycrystalline-material cannot be used, since it would exhibit inadequately-short minority-carrier-lifetimes, due to the defects occurring at the grain-boundaries of the polycrystalline-grains. The method of obtaining such highly-pure single-crystal Si for ICs, involves several steps (Fig. 9-5):

1. Raw-material (*sand*), is refined by a complex, multi-stage process which produces *electronic-grade polysilicon*, EGS.

2. This polysilicon is used to grow single-crystal silicon by *Czochralski (CZ)* growth or *float-zone (FZ)* growth. Single-crystal Si is commercially

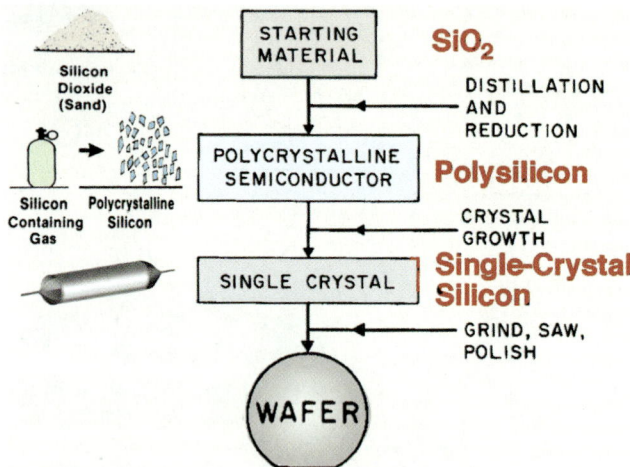

Fig. 9-5 Process sequence from starting-material to polished-wafer.

available in either {100}- or {111}-orientations.

In CZ-growth, single-crystal ingots are pulled from molten-silicon contained in a crucible. Czochralski-silicon is preferred for ULSI applications since it can withstand thermal-stresses better than FZ-material, and it is able to offer an internal-gettering mechanism that can remove unwanted impurities from the device-structures on wafer-surfaces. Float-zone crystals, because they are grown without making contact to any container or crucible, can attain higher-purity (and thereby higher-resistivity) than CZ-silicon. Devices and circuits calling for ultra-high-purity starting material (e.g., high-voltage or high-power devices) are sometimes fabricated from FZ-silicon.

9.2.1 From Raw Material to Electronic-Grade Polysilicon (EGS)

Single-crystal silicon is grown from melts of *electronic-grade polycrystalline silicon* (*EGS*). Since the CZ-process (which is used to grow most single-crystal silicon) adds impurities to the resultant ingot, the EGS must be extraordinarily pure. In fact, in order to achieve controlled-doping during subsequent single-crystal growth, EGS must have impurity-levels in the parts-per-billion atoms (*ppba*) range (or $10^{13}/cm^3$), making it the *purest-material routinely available on earth*. The raw-sand from which EGS is refined, contains high-levels of impurities (e.g., aluminum levels of $\sim 3 \times 10^{20}/cm^3$). Thus, the refining-process must reduce these impurities by approximately *eight orders of magnitude!* Such a radical refining-procedure involves four major stages (Fig. 9-6):

1. Reduction of *silica* (*sand*) to *metallurgical-grade silicon* (*MGS*) with a purity of ~98%;
2. Converting *MGS* to *trichlorosilane* (*SiHCl₃*);
3. Purification of this *SiHCl₃* by distillation;
4. Chemical vapor deposition (CVD) of Si from the purified *SiHCl₃* as *EGS*.

Metallurgical-grade silicon is produced by heating silica with carbon in a reducing ambient (Fig. 9-6a):

$$SiO_2(solid) + 2C(solid) \rightarrow Si\,(solid) + 2CO(gas) \quad (9.1)$$

Metallurgical-grade silicon is mainly used in the manufacture of aluminum or for producing silicone-polymers. That is, most silicon manufactured every year is consumed by such applications, and only a

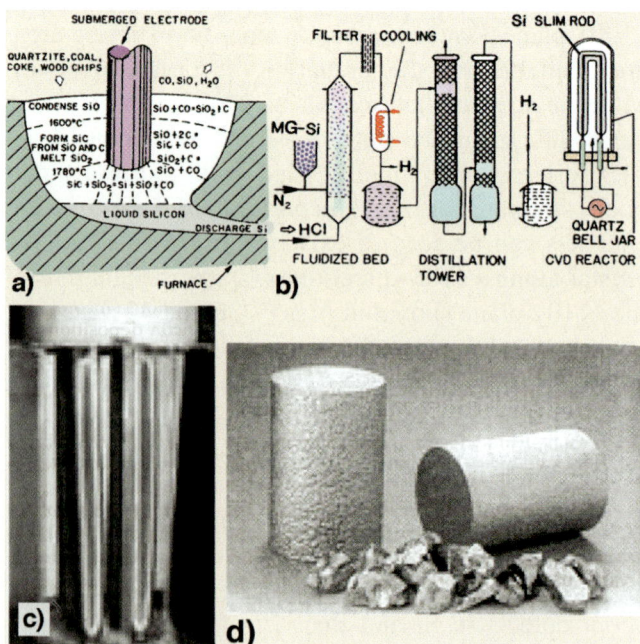

Fig. 9-6 (a) Schematic of submerged-electrode arc-furnace for production of MGS. Reprinted with permission of the publisher, the Electrochemical Society. (b) Schematic of fluidized-bed, distillation-tower, and CVD-reactor developed by Siemens. (c) and (d) EGS in polysilicon-form.

small-fraction is further refined into EGS for silicon-wafers. The main impurities in MGS are Al and Fe.

These are removed in *Stage* **2** (Fig. 9-6b) in which trichlorosilane (SiHCl$_2$) is formed by the reaction of HCl and Si (MGS):

$$Si(sol) + 3HCl\ (gas) \rightarrow SiHCl_3\ (liq) + H_2\ (gas) \quad (9.2)$$

SiHCl$_3$ is a liquid at room-temperature (boiling point 31.8°C), and can be purified by distillation (*Stage* **3**), i.e., in a distillation-tower, as shown in Fig. 9-6b).

In *Stage* **4**, the highly-purified SiHCl$_3$ is converted back into polycrystalline-Si (*EGS*) by CVD in the presence of hydrogen:

$$2SiHCl_3\ (gas) + 2H_2\ (gas) \rightarrow 2Si\ (sol) + 6HCl \quad (9.3)$$

The process (first proposed by Siemens GmbH in the late 1950s) takes place in a reactor (rightmost chamber in Fig. 9-6b). The starting-surface is a thin silicon-rod which serves as a nucleation-surface for the depositing silicon. For large rods (200-mm in diameter and several meters long), the deposition process takes several-hundred hours. Figure 9-6c shows some examples of such EGS polysilicon-rods. In 1995 the worldwide-consumption of EGS was ~12-million kilograms (1.2x10^{10} gm).

9.3 CZOCHRALSKI (CZ) CRYSTAL-GROWTH

Czochralski (CZ) growth, named for its inventor, involves the crystalline-solidification of atoms from a liquid-phase at an interface. The basic production-process for producing such CZ-Si has undergone remarkably little change since it was pioneered in the early 1950s.

The simplicity of the process and the relatively high-degree of crystal-purity helped establish the CZ-process as the dominant Si-crystal-growing technology in the early-years of the semiconductor industry. The development of dislocation-free ingot growth and automatic diameter-control in the late 1960s lead to a rapid growth in the sizes of ingot-diameters and crystal-melts. Further increases in charge-size and diameters were driven by the economic-advantages of even larger ingots. As shown in Fig. 9-7, by 1998 300-mm diameter crystals had become available, grown with

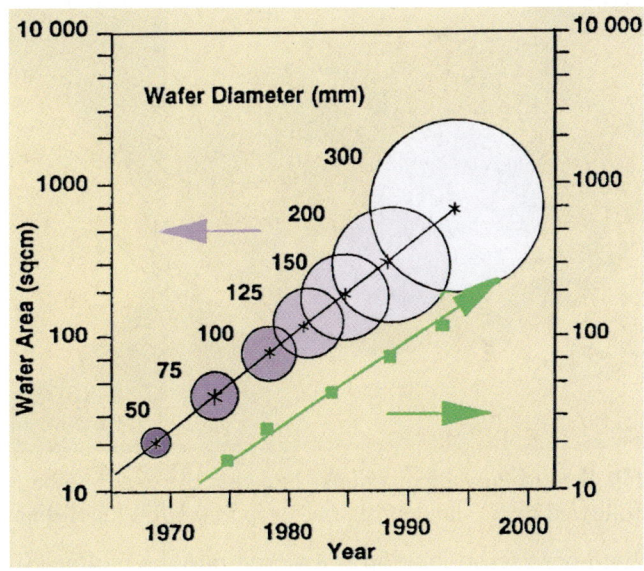

Fig. 9-7 The increase with time of CZ-silicon diameter.

150-kg charges. Typical commercial-ingots are grown 0.5–2.5-m in length, and a 100-kg charge can grow an ingot with a diameter of 200-mm, roughly 1-m long.

In this section we introduce the most important aspects of CZ-growth, focusing on the details that impact the material-properties of the silicon-wafers for ULSI. These include: **a**) the sequence of steps in the growth of the ingot; **b**) how impurities are incorporated into the ingot; **c**) the components of CZ crystal-growing equipment; and **d**) the evaluation of CZ silicon-crystals in ingot-form.

9.3.1 Czochralski Crystal-Growth Sequence

The steps used in growing a CZ Si-crystal are shown in the photographs on the chapter faceplate (p. 144). A schematic diagram of a CZ crystal-grower in which this process is performed is given in Fig. 9-14.

A fused-silica crucible is first loaded with a charge of undoped-EGS, together with a precise-amount of diluted silicon-alloy (Fig. 9-8a). The growth-chamber is then pumped out, backfilled with an inert-gas (to limit incorporation of atmospheric-gases into the melt during growth), and the charge is melted (the melting-point of silicon is 1421°C). Next, a slim seed-crystal of silicon (≈5-mm in diameter and 100–300-mm long)

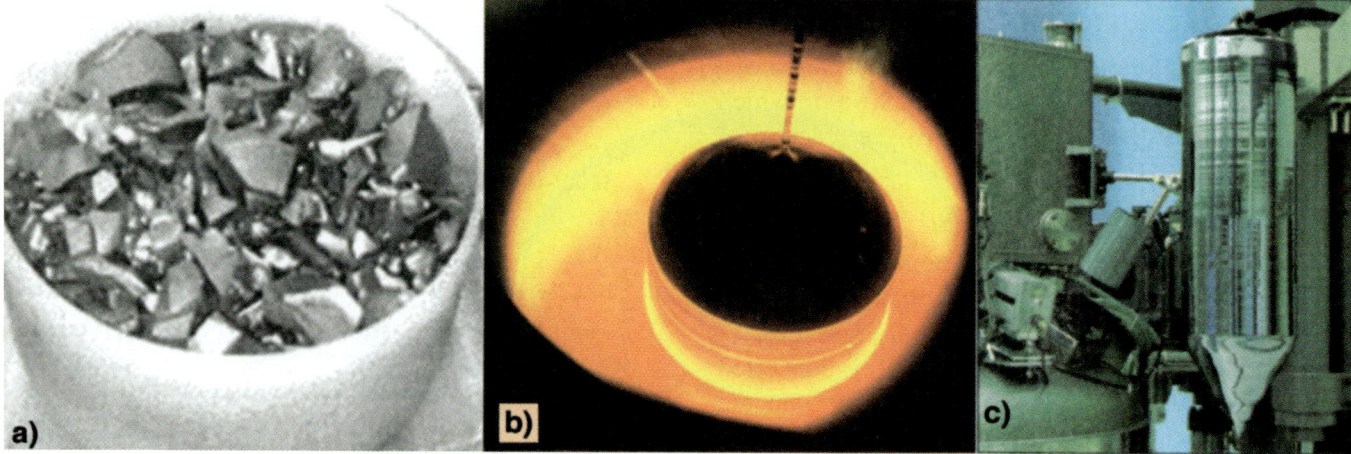

Fig. 9-8 (a) Polysilicon chunks in a crucible, ready to be loaded into CZ-crystal-puller. (b) A thin neck-region is first grown, followed by a "shouldering-out" step to the full ingot-diameter. (c) 300-mm single-crystal CZ Si-ingot.

with precise orientation is lowered into the molten-Si. The seed-crystal is withdrawn at a controlled rate. Both the seed-crystal and crucible are rotated during the pulling-process, but in opposite directions.

The initial pull-rate is relatively rapid so that a thin neck (~3-mm in diameter) is produced (Fig. 9-8b). As was first discovered by Dash, by forming this thin neck, the growing-ingot can achieve a macroscopically dislocation-free crystalline state. At that point, the melt-temperature is reduced and stabilized so that the desired ingot-diameter can be formed (*shouldering-out step*). This diameter is subsequently maintained by precise-monitoring of the pull-rate. The pulling continues until the charge is nearly exhausted, at which point the ingot is withdrawn to form the *tang* (*tail-off*). A shutdown-procedure of the furnace completes the process. Figure 9-8c shows a 300-mm ingot after it has been removed from the crystal-puller.

When CZ Si-crystals were first grown, the process required skilled operators to control all the important growth-parameters, including melt-temperature, ingot-diameter, pull-rate, and rotation. Today, such parameters are automatically-monitored, resulting in more reproducible-control of these aspects of the pulling-process. A variety of growth-process sequences can be carried out by modern crystal-growth systems, with "one-button" operations that automatically take the system from *meltdown* to *shutdown*.

9.3.2 Incorporation of Impurities into the Crystal

Impurities are incorporated into the CZ-crystal as it solidifies, together with the silicon. The desired impurities are included in the melt by adding precise amounts of silicon-alloy (heavily doped with the impurity of choice) into the crucible with the EGS. As the silicon grows, however, only a fraction of the impurity-concentration that was present in the molten volume that solidified (as specified by the value of the *segregation-coefficient* k_o of the impurity), gets incorporated into the crystal. The remaining-fraction remains in the melt (Fig. 9-9). As the ingot grows, the melt-volume is consumed, and therefore decreases. However, a fraction of the impurity arriving at the solidifying-interface between the melt and solid is continuously-rejected by the crystal and stays in the melt. If k_o is less than 1 (as is the case for most impurities in silicon, as shown in Table 9-1), the melt becomes progressively more concentrated with the impurity. As a result, as the ingot grows, the molten-material arriving at the melt-crystal interface contains an ever-larger proportion of impurity-atoms. Since the same *fraction* of total-impurities present in the arriving materials gets incorporated into the crystal, the ingot also becomes more heavily doped along its length.

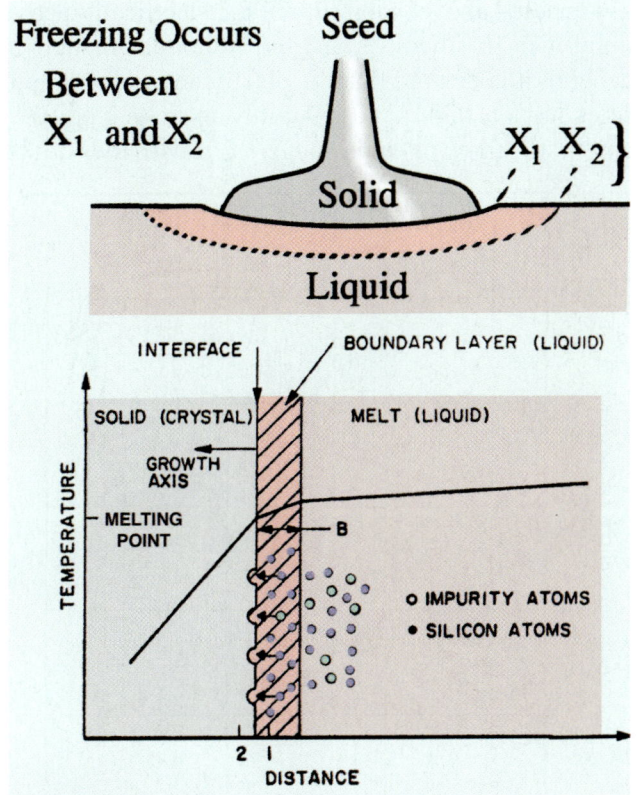

Fig. 9-9 *Impurity-segregation phenomena* near the melt-solid interface during CZ Si-growth.

Figure 9-10 shows how the predicted impurity-concentration in an ingot will vary for various k_o-values (including those of B and P). The information in Fig. 9-10 indicates that when k_o is close to 1 (e.g., for B, $k_o = 0.8$), it is less-difficult to achieve relatively-level resistivity values along the length of the ingot. But when $k_o \ll 1$ (e.g., for P, $k_o = 0.3$, or for Sb, $k_o = 0.023$), the impurity-concentration along the ingot length will vary much more.

Starting-wafers are available with specific impurity-concentrations from 10^{14}-10^{19}-atoms/cm^3 to meet the needs of various device-applications. The amount of an impurity that can be added into the melt is constrained by its *solubility-limit*. As shown in Fig. 9-12b, two *n*-type dopants (arsenic and phosphorus) have solubility-limits in Si that exceed 10^{20}/cm^3 at the Si melting-temperature, but boron is the only *p*-type dopant whose solubility-limit in Si that is greater than 1×10^{20} cm^3. Note that solubility is an issue for the electrically-active impurities in silicon for another reason. In some device-regions, high-concentrations of *p* or *n* dopants are required. Only those impurities with sufficiently-high solubilities can be used for such applications. That is why boron is used almost exclusively as the *p*-type dopant in silicon-devices.

Figure 9-11 shows how impurity-segregation during ingot-growth impacts the fraction of the ingot-length that can meet a given resistivity-specification. In this example, wafers must meet a 8–12 Ω-cm resistivity-range specification (~0.8–1.2x10^{15}/cm^3 impurity-concentration). About 80% of the ingot-length can provide such wafers if it is B-doped, but only 50% can be used if the ingot is P or As doped. If we also include the loss of ingot-material due to removal of grinding damage, the crystal-material-yields drop to 66% and 40% respectively.

Solid-Solubility: As defined in Chap. 5, a *solvent* is a substance capable of dissolving another substance

Table 9-1 SEGREGATION COEFFICIENTS FOR COMMON IMPURITIES IN SILICON

IMPURITY	Al	As	B	O	P	Sb
k_o	0.002	0.3	0.8	1.25	0.35	0.023

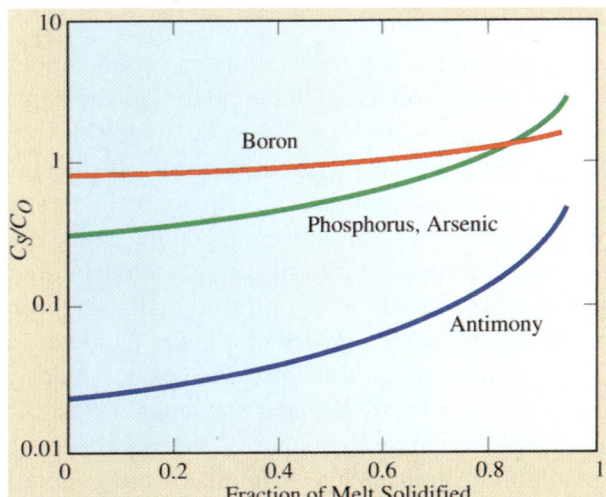

Fig. 9-10 Curves of growth from the melt, showing the doping-concentration in a solid as a function of the fraction solidified, with examples of some impurities in Si.

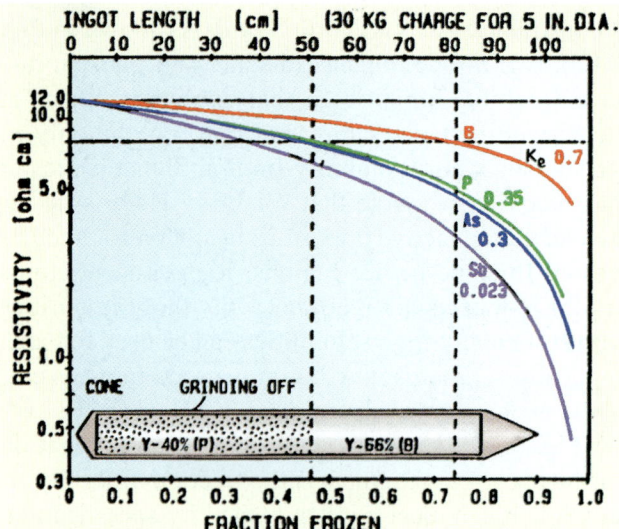

Fig. 9-11 Examples of resistivity distributions for several dopants along the ingot-length.

to form a *solution*. In Chap. 5, the solvents discussed were liquids, but solutions can also exist in solids. In single-crystal silicon (e.g., Si-wafers), the silicon is the *solvent*, and the impurities are the *solutes*. One important aspect of solutes in a solid-solvent that needs to be addressed here is *solid-solubility*. This is the maximum concentration an impurity can be dissolved under equilibrium conditions. (A familiar analogy is the maximum *liquid-solubility* of sugar in coffee. The coffee can only dissolve a certain amount of sugar before collecting in the bottom of the cup as a solid.) Figure 9-12a shows the solid-solubilities versus temperature for a variety of elements in silicon.[8]

For almost all impurities in silicon, solid-solubility decreases with decreasing temperature. Thus, if an impurity is introduced into silicon at a temperature at the maximum-concentration allowed by its solubility, and the crystal is then cooled to a lower-temperature, a *supersaturation-condition* is said to exist. The crystal ceases to be supersaturated and achieves equilibrium by precipitating the impurity-atoms in excess of the solubility-limit. While precipitates are generally undesirable, in some cases their presence can be a benefit. This is the case for oxygen-precipitates in silicon wafers, as covered in Sects. 9.9 and 9.10.

As noted above, solubility of the electrically-active impurities in silicon is also an important matter. In general, it is desired that the electrically-active impurities have a high solid-solubility in Si (e.g., $>10^{20}/cm^{-3}$), to allow a wide-range of resistivities to be

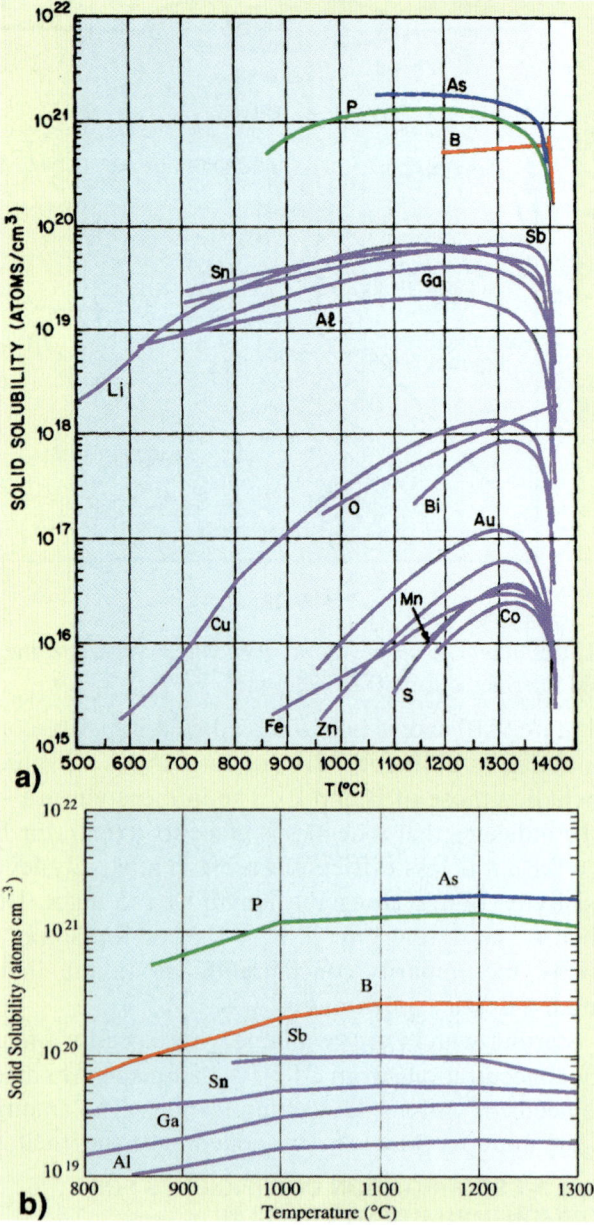

Fig. 9-12 (a) *Solid-solubility* of impurity elements in Si. (b) Details of the curves shown in (a) for the impurity dopants of Si.[8] Reprinted permission from Lucent Technologies.

formed in the various regions of devices. However, as discussed earlier (and as shown in Fig. 9-12b), there are only a few electrically-active elements that have high solid-solubilities in silicon.

9.3.3 Czochralski-Silicon Growing Equipment

The systems used to grow Czochralski Si-crystals (also known as *pullers* (Figs 9-13 and 9-14), consist of four subsystems as shown in Fig. 9-13:

1. *Furnace*: crucible, susceptor, rotation mechanism, heater and power supply, & growth chamber.
2. *Crystal-pulling mechanism*: seed-cable or chain, rotation-mechanism, seed-chuck, and crystal-handling device.
3. *Ambient control*: chamber gas-source, flow controller, and vacuum/exhaust system.
4. *Control system*: computer-based controller, and sensors.

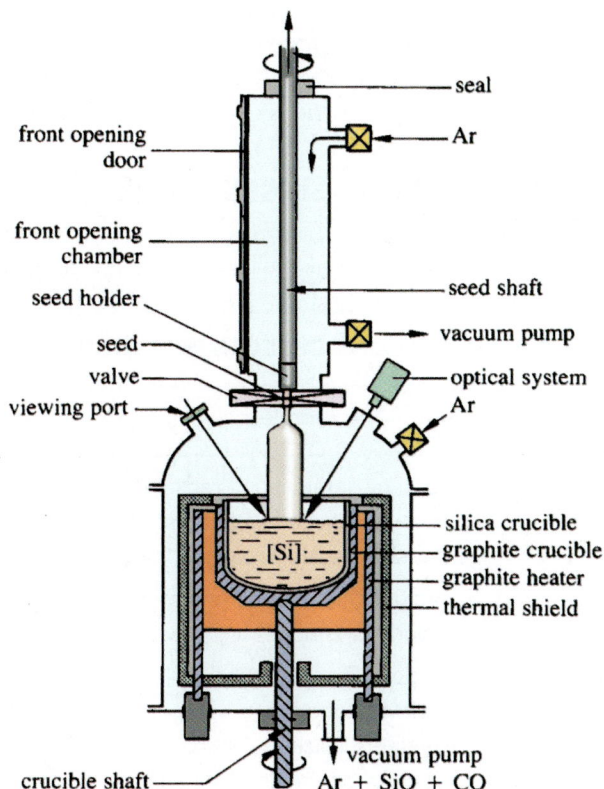

Fig. 9-13 Schematic drawing of CZ-ingot-puller.[4]

Fig. 9-14 (a) Modern CZ-pullers for 200-mm and 300-mm wafers. Courtesy of Ferrofluidics/Kayex.

Furnace: The material chosen for CZ-silicon crucibles is fused-silica. It has a high melting-point, good thermal-stability, and is relatively non-reactive with molten-silicon. The molten-silicon to some extent does corrode the fused-silica crucible, especially during the higher temperature melt-in period. Thus, substantial quantities of oxygen become dissolved into the melt. Some of this oxygen becomes incorporated into the crystal, initially taking the form of an interstitial-impurity. The presence of this oxygen can have potentially good and bad effects on the crystal properties. The purity of the crucible-silica is also important, since any impurities can limit the upper values of resistivity of ingots grown in the crucible (the levels of B and P in Si are typically 8–10 ppma). In addition, silica has no distinct melting-point, but gradually softens with increasing temperature.

Near the melting-point of silicon, the silica-material of the crucible becomes so soft that it requires the support of a heat-resistant and rigid outer-crucible (*susceptor*). Such susceptors are fabricated from high-purity graphite, which exhibits adequate temperature and cleanliness properties. The other components in the chamber that are subjected to high-temperatures are also made from graphite, including the *pedestal* (on which the susceptor sits, and which can rotate as well as raise or lower the susceptor [and thereby also the crucible and melt]), *heater elements*, *insulation*, and a *heat shield*. The silicon-charge is melted by graphite resistance-heaters connected to a power-supply. Large pullers require tens of kilowatts of power for operation.

Crystal-Pulling Mechanism: The mechanism which pulls the crystal from the melt (and simultaneously rotates it), must do this with minimal vibration and sufficient precision. Most systems use a cable or chain to pull the crystal. The hot-crystals have become too large to be manually handled. Thus, automated unloading-systems are used to transport ingots from the furnace to the cropping and evaluation areas.

Ambient Control: The environment in the chamber must be kept free of reactive-gases (O, CO, etc.). This is accomplished by filling the chamber with an inert-gas (e.g., Ar). During the process, this gas is swept out of the chamber, together with any CO and SiO that may be evolving in the chamber, and replaced with fresh inert-gas. A typical consumption-rate of inert-gas for this purpose is ~1500 liters/kg of silicon grown. Ingot-growth at a reduced-pressure in the chamber can also be used to alter evaporation from the melt.

Control System: The process-parameters, such as pull-rate, crucible-rotation, seed-rotation, melt-temperature, gas-flow, and crystal-diameter, are controlled in modern pullers with digital-computer systems. As mentioned earlier, completely automated cycles (one-button operation) require manual-intervention only for loading the crucible with EGS. Recipes that result in various crystal-properties are stored in the computer memory, allowing the system to produce nearly identical crystals from run-to-run. Other control-system features may include: central-computer interfacing capability, data-logging and printing, CRT terminals, and tape or magnetic-disk program storage.

Ingot-diameters during growth are controlled by focusing an infrared temperature-sensor on the melt-crystal interface, and monitoring changes of the meniscus-temperature. The pull-rate mechanism and chamber heater is slaved to the sensor-output, and the diameter is adjusted by changes in the pull-rate. The level of the melt is detected by laser-beam reflection.

9.4 CRYSTALLINE-DEFECTS IN SILICON

Real crystals differ from the ideal in that they possess *imperfections* or *defects*.[7] Some, due to impurity dopant-atoms, are absolutely necessary for creating devices in the crystal. Other defects may be helpful if present in moderate density. Most however, are undesirable, regardless of the density in which they are found in a crystal. Table 9-2 lists crystalline defects according to their geometry. There are *zero-dimensional* (or *point-defects*), *one-dimensional* (or *line-defects*), *two-dimensional* (or *area-defects*), and *three-dimensional* (or *volume-defects*), as shown in Fig. 9-15.

9.5 POINT-DEFECTS IN SILICON CRYSTALS

The various forms of zero-dimensional or point-defects are shown in Fig. 9-16. In crystal regions where no point-defects exist, each site on the lattice

Table 9-2 CRYSTALLINE-DEFECTS IN SILICON

Type	Dimension	Examples
Point	0	Vacancy, Interstitial, Frenkel Defects
		(Intrinsic - silicon self-interstitials)
		(Extrinsic - dopants, oxygen, carbon, metals)
Line	1	Straight Dislocations (edge or screw)
		Dislocation Loops
Area	2	Stacking-Faults
		Grain-Boundaries
		Crystal-Surfaces
Volume	3	Precipitates, Voids (Oxygen Precipitates,
		Metal Precipitates)

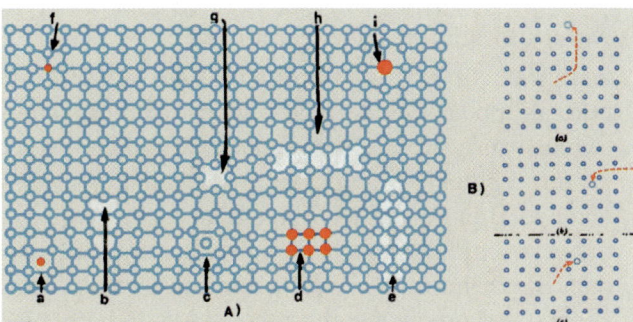

Fig. 9-15 (A) Various crystal defects in a simple-cubic lattice: (a) interstitial impurity-atom; (b) edge-dislocation; (c) self-interstitial; (d) coherent-precipitate of substitutional atoms; (e) dislocation-loop formed by agglomeration of self-interstitials; (f) substitutional-atom widening the lattice; (g) vacancy; (h) dislocation-loop formed by agglomeration of vacancies; (i) substitutional-impurity-atom compressing the lattice. (B) Point-defects: (a) Schottky-defect; (b) interstitial arriving from surface; (c) Frenkel-defect.

is occupied by an atom. If an atom is missing from one of these sites a *vacancy* exists in the lattice. In the event that an atom leaves a lattice site and *moves to the external surface* of the crystal, a *Schottky-defect* is created (Fig. 9-15Ba). That is, in a Schottky-defect a vacancy is created in the lattice, but the displaced atom does not remain in the bulk of the crystal. It moves instead to a location on its surface. If an atom is found at some non-lattice site in the crystal, it is said to lie at an *interstitial-site*. Silicon-atoms at interstitial-sites in a Si-crystal are called *self-interstitials*, and self-interstitial atoms may originate from a lattice-site in the crystal, or from the crystal-surface. Vacancies and self-interstitials are classified as *intrinsic point-defects*.

In any solid, lattice-atoms are in continuous vibration about their equilibrium positions. The amplitude of these vibrations increases as the temperature is raised and may become large enough for an occasional atom to leave its equilibrium-position entirely and move to a location at the surface (Schottky-defect). This requires energy to be supplied to the solid since the bonds holding the atom to its neighbors must be broken, and additional energy is also needed to create

new surface. At room-temperature (300 K) only a tiny vacancy-concentration exists, whereas at wafer-processing temperatures, much-higher concentrations of vacancies exist. As discussed in more detail in Ref. 1, such higher concentrations of point-defects play critical roles in the diffusion of impurities in silicon.

9.6 ONE-DIMENSIONAL DEFECTS (DISLOCATIONS)

One-dimensional (or *line*) defects in crystals take the form of *dislocations*. They may be *edge*-dislocations, *screw*-dislocations, or *mixed*-dislocations (which contain both screw and edge components). Dislocations are an important class of defects in silicon crystals, and a full-treatment of their properties, formation, growth, and movement is a complex subject. Readers are directed to Ref. 1 for more details.

An *edge-dislocation* is shown in Fig. 9-17a. Note that the upper-half of the crystal contains an *extra half-plane* of atoms squeezed into the same space normally occupied by the atoms of an ideal crystal. The *dislocation* occurs at the termination of the sheet of extra-atoms. An edge-dislocation could be theoretically created by slicing the lattice and inserting an extra half-plane of atoms into the slice as shown in Fig. 9-17b. In the example shown in Fig. 9-17b stresses exist in the crystal as a result of the dislocation. The lattice planes in the upper-half of the lattice near the dislocation are in *compressive* stress, while those in the lower half are in *tensile* stress. The *dislocation line* in an edge-dislocation is the line connecting all

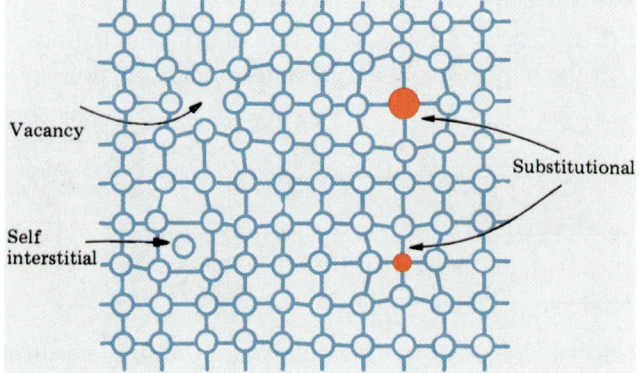

Fig. 9-16 Various *point-defects* in a simple-cubic lattice.

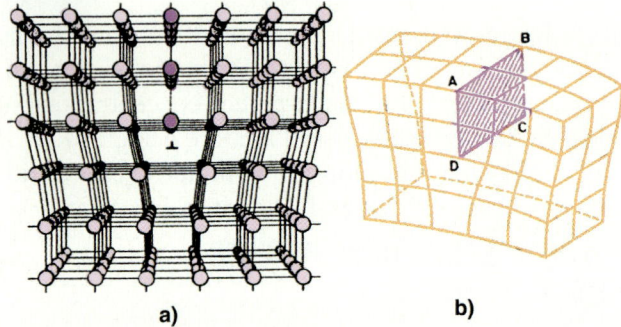

Fig. 9-17 (a) Model of an edge-dislocation. (b) Positive edge-dislocation formed by inserting an extra half-plane of atoms in ABCD.

the end-atoms on the extra half-plane.

The *dislocation-line* must either terminate on an external surface of the crystal (see Fig. 9-18 for example), or form a closed-curve if the dislocation lies entirely within the crystal (a *dislocation-loop*). That is, in a dislocation-loop an extra-plane of atoms (or a missing-plane of atoms) that exists entirely within the interior of the crystal must have a dislocation along the outer-edge of the plane (Fig. 9-18).

Dislocations can form during ingot-growth or can be introduced into Si wafers by subjecting them to various other non-equilibrium process-conditions. That is, dislocations can arise during thermal-processing by the growth and multiplication of the microscopic dislocation-loops formed during ingot cooling. They can also be generated from dislocations originating at the surface in response to stresses created in the wafers during later process-steps. Such stresses can arise in a number of ways including:

1. Differences in the expansion of different regions of the lattice due to temperature-variations among various locations on the wafer itself.

2. Introduction of high-concentrations of substitutional-impurities in local-regions of a crystal (which cause stresses between doped and undoped crystal-regions). Impurity-atoms smaller than Si (such as B or P) cause shrinkage in regions where they exist, producing tensile-stresses. Larger atoms than Si put into the crystal cause these regions to expand and produce compressive-stress. If present in high-enough concentrations, stresses caused by substitutional-impurities produce dislocations, and these are known as *misfit-dislocations*;

3. Compressive-stresses from volume mismatches that arise from the formation of SiO_2 precipitates in CZ-silicon.

4. Stresses caused by differences in the coefficient of thermal-expansion between a layer present on the surface and that of the crystal. An example is the stress caused during the formation of LOCOS isolation-structures. LOCOS-isolation involves the growth of thermal-oxide on local-regions of the wafer surface, and suppression of the oxidation on other areas (by the presence of nitride layers). Dislocation-causing stresses result from tensile-stress intrinsic to SiN, and from the volume-expansion of oxide formed in the field-regions.

Dislocations in *wafers* can be induced by thermal-stress during furnace-operations. Upon removing heated-wafers from a furnace, the edges of the wafers cool faster than their centers (especially when wafers are held vertically and spaced closely in wafer-boats). Wafer-edges radiate heat to the cooler surroundings while their centers are surrounded by the heated adjacent-wafers. The resultant temperature-difference from edge-to-center causes the edges to undergo relatively more contraction than the centers. If the induced-stress exceeds the *yield-strength* of the silicon, dislocations (seen as *slip-defects*) are formed.

The thermally-induced dislocations as a result of wafer-cooling upon being withdrawn from a furnace are generated at the wafer-edge, where surface irregularities and lattice-damage act preferentially as

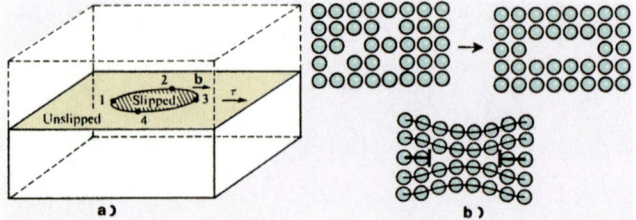

Fig. 9-18 (a) A *dislocation-line* entirely within a lattice, forming a closed loop. (b) Formation of a *dislocation-loop* by agglomeration and collapse of *vacancies*.

the dislocation-sources. The dislocations propagate from the edge toward the center of the wafer, relieving some of the mechanical-stress and producing slip. Figure 9-19a shows an *x*-ray topograph of slip-patterns in a <111> wafer, and Fig. 9-19b the slippage-distribution of 9-19a. To prevent such slip-defects (and resultant wafer-warpage), wafers are pushed very-slowly into and out of heated furnaces.

The distortion of the lattice at dislocations causes them to become sites that are energetically-favored to be occupied by impurity-atoms. That is, diffusing impurity-atoms are more likely to be *captured* at dislocation-sites because they distort the lattice less at such sites than elsewhere in a perfect-lattice. This results in a lower free-energy of the crystal. Impurity-atoms larger than Si occupy regions nearby the dislocation that are in compression, while those smaller than silicon locate in the tensile-stressed portions. In addition, there is also a chemically-active dangling-bond sticking out from each atom on the edge of a dislocation, which also helps trap impurities.

These phenomena are very important in silicon processing, as dislocations can trap fast-diffusing metal-impurities in silicon. The effect can either be harmful to device operation - or if correctly harnessed - can be used to advantage (that is, through the intentional application of gettering techniques, see Sect 9.10). When impurities have segregated themselves at dislocation-sites, the dislocation is said to be *decorated* with them, while *pure* dislocations are free of any impurities.

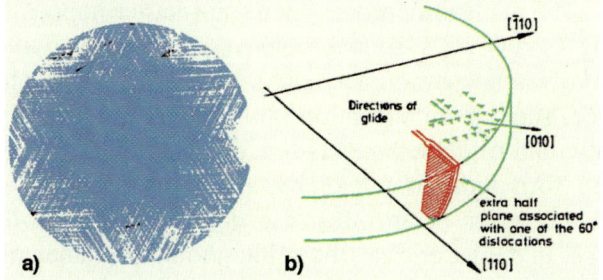

Fig. 9-19 (a) A (111)-oriented Si wafer in which the severe slip has been delineated by etching to show the emergence of the dislocations. (b) Schematic of the slippage-distribution of (a).

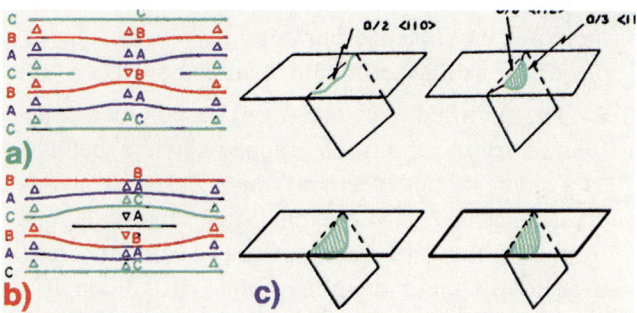

Fig. 9-20 (a) Intrinsic and (b) extrinsic-stacking-faults in a FCC lattice. (c) Generation of an extrinsic-stacking-fault from a dislocation at a surface.

9.7 AREA-DEFECTS IN SILICON CRYSTALS

Area-defects in single-crystals include *stacking-faults*, *crystal-surfaces themselves*, and *grain-boundaries*.

Stacking-faults are the most important area-defects that degrade the device-performance of integrated-circuit components. As shown in Fig. 9-20a, the removal of an extra-plane of Si-atoms creates an *intrinsic* stacking-fault, while the insertion of an extra-plane creates an *extrinsic* stacking-fault (Fig. 9-20b). Dislocations formed during oxidation are known as *oxidation-induced stacking-faults* (OISF). Stacking-faults cause problems in silicon-devices because they are bounded by dislocations, and fast-diffusing impurities from the surrounding lattice can be absorbed at these dislocations. Various device-degrading effects are caused when such stacking-faults become electrically-active as the result of absorbing metallic-impurities. For example, excess reverse-bias currents in *pn*-junctions, and storage-time degradation in DRAMs, can be caused by electrically-active stacking-faults.

The *surface* of a crystal is also an area-defect, as the periodicity of the lattice abruptly terminates at a surface. Thus, a surface-atom no longer has all four of the nearest-neighbors that it would have in the bulk. There are two important consequences of such disruptions in lattice-periodicity.

1. Energy-states exist in the band-gap of silicon at the surface that do not exist in the bulk. Such energy-states in the band-gap provide generation-recombination (g-r) centers for carriers, which

increase the leakage-currents in reverse-biased *pn*-junctions that terminate at silicon-surfaces.

2. The unfilled (or *dangling*) bonds can trap free- carriers, and such trapped-carriers behave as a layer of net electrical-charge present at the crystal-surface. To prevent such charge-layers from forming on silicon surfaces, techniques are used to fill these dangling-bonds. In silicon IC-processing the most important of these techniques involves the growth of thermal-SiO_2 on the wafer-surface (which is effective in satisfying the vast majority of such bonds), together with hydrogen-annealing to tie up as many as possible of those bonds remaining after SiO_2 growth.

Grain-boundaries are another important area-defect (Fig. 9-21). Si-atoms with only three nearest neighbors are more chemically-reactive than those having all-four bonds completed. They are also more prone to trap free-carriers or to attach an impurity-atom to complete their last unfilled-bond. This makes silicon with many grain-boundaries (i.e., polySi films) full of sites that can: **1**) trap mobile-carriers originating in the bulk-regions of the crystal-grains; or **2**) segregate electrically-active impurities from the bulk to the grain-boundaries. The electrical-conductivity of poly-Si is significantly impacted by the extent to which grain-boundaries exist in such films (see Chap. 16).

It is also easier for substitutional-impurities to move along grain-boundary diffusion-paths than along diffusion-paths in the bulk for two reasons. The grain-boundary region is a more open structure that enhances the diffusivity of impurities. Also, fewer bonds must be broken by an impurity-atom to enable it to diffuse if located on a grain-boundary site than if it occupies a site in the bulk-crystal. Grain-boundary diffusion thus allows uniform distribution of dopants in polySi-films to be achieved at lower-temperatures than in single-crystal layers. Grain-boundary diffusion is also an important phenomenon in other polycrystalline-films encountered in IC fabrication, including Al, Ti, WSi_x, and TiN films (see Chap. 15).

9.8 BULK-DEFECTS AND PRECIPITATION

Bulk (or volume) defects in crystals include *voids* and *local amorphous-regions*, but the most important volume defects are *precipitates* of extrinsic or intrinsic point-defects. The formation and growth of precipitates is a complex subject, but basically precipitates first form (or *nucleate*) as tiny-particles called *nuclei*. The nature of the structure of such nuclei is difficult to determine because they are so small.

The nucleation process in a solid-lattice happens in one of two ways: **a**) *heterogeneous-nucleation*, which involves nucleation at a *crystalline-defect* of the lattice, such as a dislocation, grain-boundary, or impurity; and **b**) *homogeneous-nucleation*, which involves the formation of nuclei by a random assembly of solute-atoms in an otherwise *defect-free crystal-lattice*. In the latter case, the solute-atoms coincidentally cluster at some location, thereby forming a *nucleus*. Homogeneous-nucleation is more likely when a large supersaturation of solute-atoms exists in the crystal. In addition, the more-rapidly solute-atoms can dif-

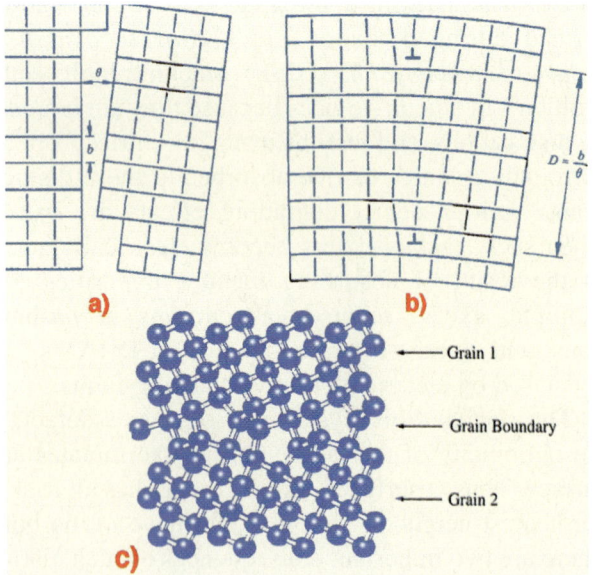

Fig. 9-21 Dislocation model of grain-boundary. Two grains with an angle of misorientation θ are shown in (a). As the two grains form a wedge-shaped intersection, certain rows of atoms cannot extend their full lengths and the dislocation (b) results. (c) 3-D perspective drawing of grain-boundary.

fuse, the more-quickly will precipitation take place. In dislocation-free Si-ingots and starting-wafers, homogeneous-nucleation is the dominant nucleation-mechanism of bulk (i.e., three-dimensional) defects.

9.9 OXYGEN IN SILICON

Other than in heavily-doped crystals, the most abundant impurity in Czochralski (CZ) Si crystals is oxygen, with concentrations ranging from 4.5×10^{17}-1×10^{18} atoms/cm^3 (9–20 ppma). Oxygen is incorporated into CZ-Si as a result of crucible-erosion during CZ growth. The majority of the oxygen (~95%) in the as-grown crystal is atomically-dissolved and occupies interstitial-sites. The diffusivity of oxygen is moderately-high, whereas its solubility decreases strongly at lower-temperatures (see Fig. 9-12a). These two properties make oxygen the most-important precipitate-forming element in CZ-silicon. Precipitation-behavior impacts all of the important effects influenced by the presence of oxygen in silicon: including: **a)** *surface oxygen-precipitates*; and **b)** *bulk oxygen-precipitates* (that are the basis of intrinsic-getting).

The presence of SiO$_2$-precipitates near the wafer-surface degrades the quality of thin gate-oxides and causes oxygen-induced stacking-faults (which can induce device failures by excessive leakage-currents). Such SiO$_2$-precipitates must be eliminated from regions near the wafer-surface. The procedure employed to prevent oxygen-precipitates from forming near the wafer-surface involves the creation of a *denuded-zone*, or a zone that contains less interstitial-oxygen than required to form oxygen-precipitates (Fig. 9-22). Special thermal-steps at the beginning of a wafer-fabrication run are used to form such denuded-zones.

Bulk oxygen-precipitates can act as sites that trap fast-diffusing metallic-impurities *(gettering-sites)*, and thus their presence can be used as a helpful effect. Such oxygen-precipitates can be created in bulk-silicon containing appropriate oxygen-levels, through the use of a sequence of thermal-steps. Details about this process are found in Ref. 1.

9.10 GETTERING

Unwanted crystalline-defects and impurities can be introduced during silicon crystal-growth or subsequent wafer-fabrication processes. The presence of crystalline-defects and impurities is undesirable because both can degrade device-characteristics and overall-yield. The degradation of device-properties by *extended crystalline-defects* (dislocations, precipitates, and stacking-faults) was described earlier. The problems caused by *impurities* are discussed here.

The impurities that create problems are primarily metals, and include Cu, Ni, Au, and Fe. These metallic-impurities are highly-mobile and diffuse long distances in Si-crystals at moderate process-temperatures. Thus, there is high-probability they will migrate to extended defect-sites, and be captured there by them. Extended-defects in device-regions of a wafer

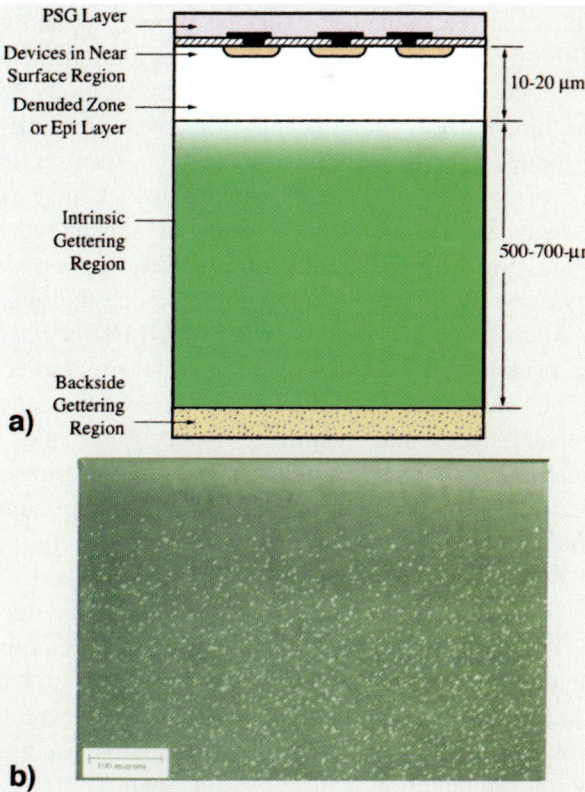

Fig. 9-22 a) Wafer-gettering strategies showing a *surface-denuded-zone, intrinsic-gettering-region,* and *backside-gettering*.[9] (b) SEM photograph of a cross-section of a Si wafer with a denuded-zone at the surface and oxygen-precipitates in the bulk (Courtesy of MEMC).

which capture metal-impurities give rise to larger leakage-currents and lower breakdown voltages.

Eliminating the effects of impurities and defects involve four approaches: **1)** keep new extended-defects from forming; **2)** prevent extended-defects from occurring in the device-regions; **3)** keep metallic impurities from getting into the wafer; and **4)** stop any metallic-impurities that have gotten in from ending-up in device-regions (Fig. 9-23). Technique-4 is referred to as *gettering,* a term originally used to describe the process in which an appropriate material (such as cesium), was used to getter (remove) the last traces of gas in vacuum-tubes, so that the required high-vacuums within the glass tube could be attained.

By combining the four approaches, product-yields can be enhanced. That is, proper application of the first-two approaches minimizes the number of extended-defects that exist in the active-regions, and the third reduces the concentrations of the impurities introduced into the Si. *Gettering* is then used as a final line-of-defense to remove any impurities that still managed to be incorporated during crystal-growth or processing. This reduces their probability of poisoning any extended-defects formed in the device-regions.

Methods used to suppress extended-defect nucleation include: **a)** control of the thermal process-conditions during wafer fabrication that lead to excess thermal-stress and resultant dislocations; and **b)** creation of oxygen denuded-zones at the wafer-surface regions in which active-devices are fabricated. The latter procedure eliminates oxygen-precipitates in these device-regions. *Elimination of pre-existing or process-induced extended-defects* such as stacking-faults, is achieved by using various techniques, including elevated temperatures (>1200°C) and the addition of HCl to the growth-ambient during the growth of SiO_2. Healing crystalline-damage caused by ion implantation is discussed in Chap. 12.

The use of Technique-**3**, namely *the control of unwanted-impurities* (especially metals), involves identification and removal of the sources of contamination. Some examples of contaminant-sources that have been identified include: **a)** stainless-steel fix-

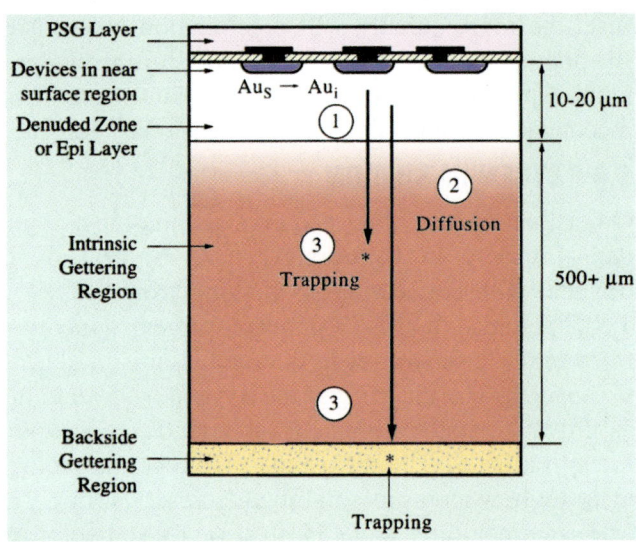

Fig. 9-23 Three-step gettering process: (1) release; (2) diffusion; and (3) trapping. (Gold [Au] is used as an example of the impurity being gettered.)[9] Used with permission.

tures in ion-implanters that ended up being sputtered during the ion-implant process; **b)** impurities originating from the sputtering of chamber-walls and fixtures of reactive-ion-etching systems; **c)** the diffusion of heavy-metals from the heater-coils through the fused-silica walls of diffusion-furnaces; and **d)** release of transition-metals (e.g., Fe and Ti) by the SiC-coated susceptors of epitaxial-reactors.

9.10.1 Basic Gettering Principles

Gettering removes transition-metal impurities from device-regions by 3 steps (Fig. 9-23):

1. The metal-impurities to be gettered are released into solid-solution from any stable-precipitate form they have assumed.

2. They then diffuse through the silicon.

3. They are then "captured" at a position away from the device-regions by either: **a)** the formation of chemical-pairs (e.g., Fe-B pairs) with other impurities, such as boron present in the crystal, or **b)** deliberately created extended-defects (dislocations or precipitates). Metal-impurities are prevented from being released into the wafer during subsequent heat-treatment.

Gettering processes are divided into two categories: extrinsic and intrinsic. *Extrinsic-gettering* exploits damage or stress on the surface of the silicon-wafer. This creates extended-defects or chemically-reactive sites at which the mobile-impurities are captured. Normally such sites are generated on the backside of bare starting-wafers using ion-implant damage or the deposition of a polysilicon-layer. *Intrinsic-gettering* involves localizing-impurities at extended-defects or regions of high-concentration of electrical-impurities (such as boron) which exist within the bulk-material of the Si-wafer, and whose origin is due to an "intrinsic" property of the starting-wafer, such as its oxygen or boron content acquired during CZ crystal-growth. More details on gettering are found in Refs. 1 and 2..

REFERENCES

1. S. Wolf and R.N. Tauber, *Silicon Processing for the VLSI Era, Vol. 1 - Process Technology,* 2nd Ed., Lattice Press, Sunset Beach CA 2000, Chaps. 1 and 2.

2. S. Wolf, *Silicon Processing for the VLSI Era, Vol. 4 - Deep-Submicron Process Technology,* Lattice Press, Sunset Beach CA 2000, Ch. 2 (300-mm Wafers).

3. *Handbook of Semiconductor Silicon Technology,* W.C. O'Mara, R.B. Herring, and L.P. Hunt, Eds., Noyes, 1990.

4. W. Zuhlehner and D. Huber, "Czochralski Grown Silicon," in *Crystals 8,* Springer-Verlag, Berlin, 1982.

5. T. Abe, "Crystal Fabrication," in N.G. Einspruch And H. Huff, Eds, VLSI Microstructure Science, Vol. 12, Academic Press, Orlando, FL., 1985, Chap. 1, p. 3.

6. W. Lin and H. Huff, "Silicon Materials," in *Handbook of Semiconducor Manufacturing Technology,"* Eds. Y. Nishi & R. Doering, M. Dekker, New York, 2000, Ch. 3, p. 35.

7. K.V. Ravi, *Imperfections and Impurities in Semiconductor Silicon,* Wiley-Interscience, New York, 1981.

8. F.A. Trumbore, "Solid Solubilities of Impurity Elements in Ge & Silicon," *Bell Sys. Tech. J,,* **39**, 205 (1960).

9. J.D. Plummer, M. Deal, and P.B. Griffin, *Silicon VLSI Technology,* Prentice-Hall, 2000, p. 186.

PROBLEMS

1. Why is it necessary to use single-crystal wafers on which to build microchips?

2. List 5 advantages that silicon has over other semiconductor materials.

3. Define a *unit-cell*. Draw a cubic unit-cell, and the <100> and <111> planes of the cubic lattice. Show atoms present on a <110>-plane of an FCC unit cell.

4. Describe the operation of a crystal puller. Why do wafers made with CZ process contain large concentrations of oxygen, while those made with FZ do not?

5. A CZ grown crystal is doped with boron. Why is the boron-concentration higher at the tail-end of the ingot than at the seed-end?

6. The equation that gives the impurity concentration along the length of a CZ ingot C_s is given as:

$$C_s = k_o C_o (1 - X)^{k_o - 1}$$

where k_o is the segregation-coefficient (see Table 9-1), X is the melt-fraction that has solidified, and C_o is the initial impurity-concentration in the melt. Using this equation, plot the doping distribution of: (a) arsenic; and (b) phosphorus, at distances of 20, 40, 60, 80, and 90 cm from the seed-end of a 100-cm long ingot that has been pulled from a melt with an initial doping concentration of 10^{16} cm^{-3}.

7. Describe the differences between *extrinsic* and *intrinsic* gettering.

8. A silicon crystal doped with boron is to have a resistivity of 4.0 Ω-cm when one half of the ingot is grown. If the Si charge weighs 60 kg, how many grams of boron-doped Si with a resistivity of 0.02-Ω-cm should be introduced into the melt to achieve the desired ingot resistivity?

9. (a) The *lattice-constant* of Si is 5.43-Å. Find the nearest-neighbor distance in a diamond-lattice having this lattice-constant. (b) Calculate the density of Si from the lattice constant, the atomic weight, and Avogadro's number. Compare the results with those given on the Table found on the back book-endsheet.

10. Si has an atomic-density of 5×10^{22} atoms/cm^3. For this problem, assume they are packed in a simple cubic-lattice structure (and not in a diamond lattice) with 5×10^{19} As atoms/cm^3 uniformly distributed. (a) What is the average-spacing between Si atoms? (b) How many atoms would lie on the outermost layer (a monolayer), (c) What would be the average spacing between As atoms?

11. Explain why a dislocation must terminate on itself or on a surface (or on a volume-defect).

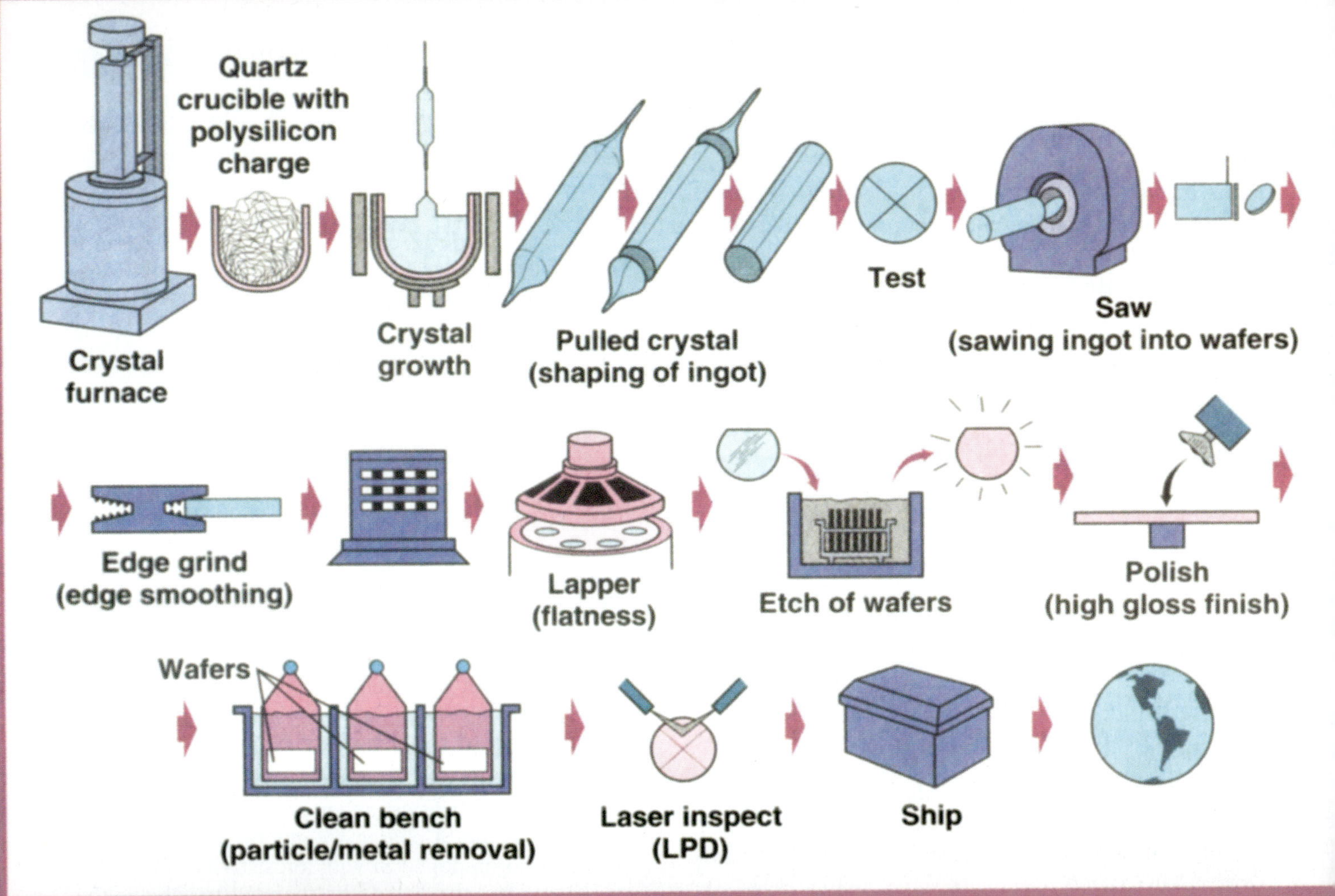

Wafer manufacturing process

- **Crystal furnace** → **Quartz crucible with polysilicon charge** → **Crystal growth** → **Pulled crystal (shaping of ingot)** → **Test** → **Saw (sawing ingot into wafers)** →
- **Edge grind (edge smoothing)** → **Lapper (flatness)** → **Etch of wafers** → **Polish (high gloss finish)** →
- **Clean bench (particle/metal removal)** (Wafers) → **Laser inspect (LPD)** → **Ship**

CHAPTER CONTENTS

10.1 INGOT-EVALUATION
10.2 INGOT-GRINDING
10.3 FLAT (OR NOTCH) GRINDING
10.4 INGOT-SAWING (WAFERING)
10.5 LASER-MARKING OF WAFERS
10.6 WAFER-LAPPING AND GRINDING
10.7 EDGE-ROUNDING AND POLISHING OF WAFERS
10.8 REMOVAL OF SURFACE MECHANICAL DAMAGE BY CHEMICAL ETCHING
10.9 CHEMICAL-MECHANICAL POLISHING OF THE WAFER-SURFACE
10.10 CLEANING THE WAFERS
10.11 DEPOSITING EPITAXIAL-SILICON LAYERS ON THE WAFERS
10.12 SPECIFICATIONS OF SILICON WAFERS FOR ULSI
10.13 ECONOMICS OF SILICON WAFERS

"Just as the Industrial Revolution enabled man to apply and control greater physical power than his own muscle, so electronics has extended its intellectual power."

Dr. Robert Noyce,
Co-Founder,
Fairchild Semiconductor
Intel

eps in the wafering process
and
rowth in the size of wafers used
microchip manufacturing.

CHAPTER 10

SILICON WAFER PRODUCTION: FROM INGOT TO FINISHED WAFER

After a single-crystal ingot has been grown, a complex sequence of shaping and polishing steps are performed to produce the starting material suitable for fabricating ULSI-devices (*silicon wafers*). Figure 10-1 shows a flow-chart of the wafer-shaping steps. The chapter faceplate also illustrates this *wafering procedure*, which includes the steps covered next.

10.1 INGOT EVALUATION

After it is grown, a single-crystal ingot undergoes evaluation of resistivity, impurity-content, crystal-perfection, size, and weight. Ingot sections that are crystallographically-defective, irregularly-shaped, or undersized, are cut-off and discarded (as are the *seed* and *butt* [or *tang*] ends of the ingot, Fig. 10-2). Loss of ingot-material may approach 30% at this point.

10.2 INGOT-GRINDING

Since ingots do not grow perfectly-round, nor with sufficiently-uniform diameters, they must shaped to the desired form and dimension. Thus, they are deliberately grown larger than the final desired wafer-diameter. A grinding-operation then removes the excess-material to reduce the as-grown crystals to a cylindrical-shape of precise-diameter. (Exact-diameter dimensions are required to make wafers compatible with automated wafer-handling equipment.) Since silicon is a hard, brittle material, grinding machines with diamond-wheels are used to perform this shaping process (Fig. 10-3). The grinding-step is followed by an etch-step to remove the grinding work-damage. The ingot diameter is reduced 0.25-1.0-cm by these grinding and etching steps.

10.3 FLAT (OR NOTCH) GRINDING

For wafers up to 150-mm, *flats* are used for aligning and crystal orientation. For larger wafers (≥ 200-mm), *notches* are mostly used (see Fig. 10-4). For 300-mm wafers, notches are used exclusively. Such flats or notches are ground along the length of the ingot. The largest flat, called the *primary-flat*, is usually positioned relative to a specific crystal-direction. For example, the flat or notch is placed perpendicular to a <110>-direc-

163

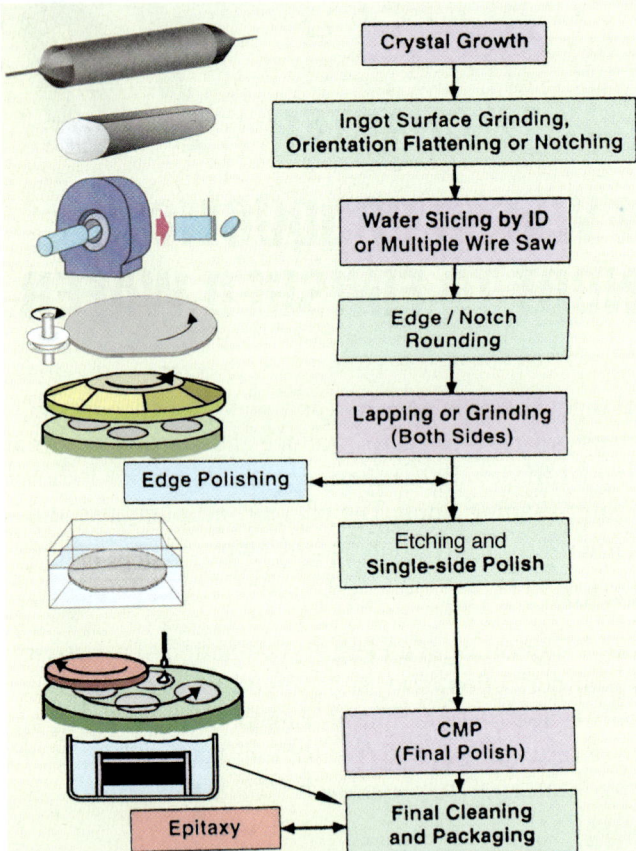

Fig. 10-1 Flow-chart depicting the steps involved in wafer preparation, starting with an intact Si-ingot.

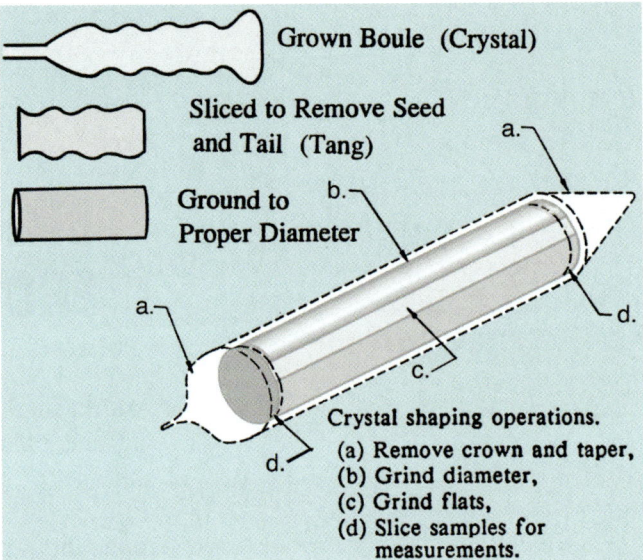

Fig. 10-2 Ingot-shaping operations.

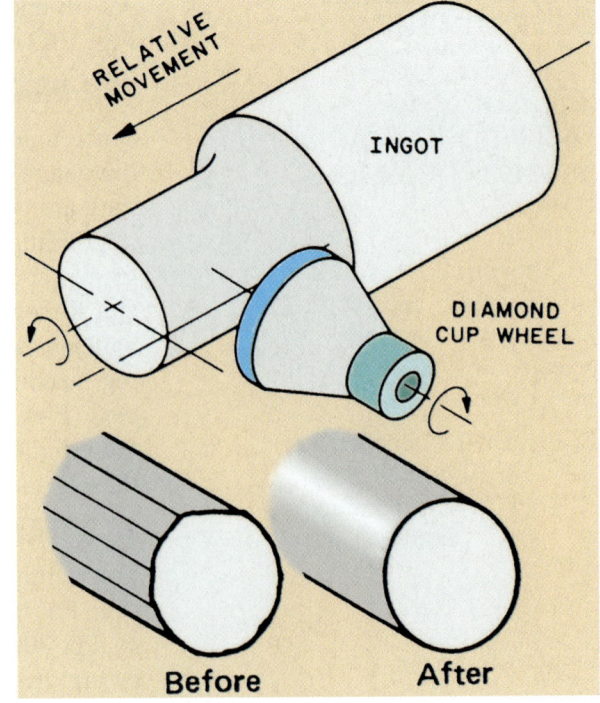

Fig. 10-3 Schematic of the Si-ingot grinding-process.

is rigidly-mounted to maintain exact crystallographic-orientation during the sawing-process. Wafers of <100>-orientation are normally cut "on-orientation" tion on a (100)-wafer. (The primary-flat orientation is found by using x-ray-analysis.) The flat (or notch) is used for several purposes. Automated wafer-handling equipment utilizes the primary-flat to obtain correct alignment, and devices on the wafer can be oriented to specific crystal-directions with this flat as a reference. Smaller flats are called *secondary-flats*, and they are utilized to identify the orientation and conductivity-type of the wafer. Since automated-equipment relies on the flats for correct operation, the flat-dimensions must be precisely machined.

10.4 INGOT-SAWING (WAFERING)

The sawing operation that produces *wafers* (or *slices*) from the shaped-ingot also defines the surface-orientation, thickness, taper, and bow of the slice. The ingot

CHAPTER 10 SILICON WAFER PRODUCTION: FROM INGOT TO FINISHED WAFER

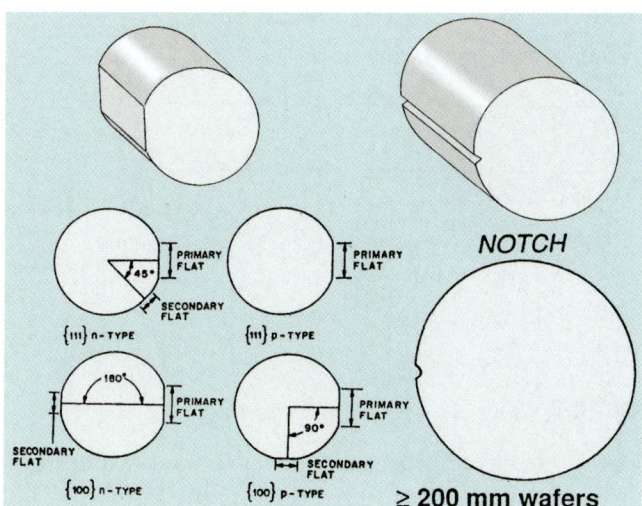

Fig. 10-4 (a) Identifying-flats on a silicon-wafer. (b) Notched wafers.

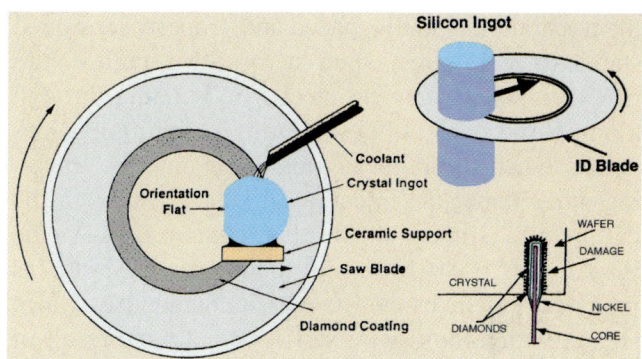

Fig. 10-5 Inner-diameter (ID) saw-geometry.

(e.g., within ± 0.5° of the <100>-orientation).

Wafer-thickness is primarily set by the sawing operation, although some material is also removed by subsequent operations (e.g., for a 200-mm-wafer of 725-μm final-thickness, a total of ~1250-μm of ingot-silicon is needed: 725-μm final-wafer-thickness; 400-μm for kerf-loss during sawing; 50-μm lapping-loss; 50-μm etching-loss; 25-μm polishing-loss. Thus, about 8, 200-mm wafers per linear-cm of ingot can be produced. As wafers get larger they must be thicker (Table 10-2) to allow them to withstand thermal-processes (e.g., epitaxy and oxidation) and handling during fabrication. Thus, fewer wafers per unit-length of ingot are possible than for smaller-wafers.

Until recently, the most common method of slicing has been *inner-diameter* (ID) *slicing* using stainless-steel blades with diamond-particles bonded to their edges (Fig. 10-5). Continuous-monitoring with blade-deflection sensors is required to assure that slices are sawn adequately-free of bow, taper, work-damage, and saw-marks. Cutting-speeds are in the neighborhood of 0.05-cm/sec. Since one-slice is cut at a time by the saw, this is a relatively-slow process.

Wafers with 300-mm-diameters, however, are sawn with slurry-coated wires instead of with inner-diameter saws. More slices-per-inch of ingot are possible due to smaller kerf-loss. In addition, wire-sawing enables the silicon-slicing to be carried out at higher-throughput and superior mechanical-properties (such as reduced bow and warp). Figure 10-6 shows a schematic of a multiple-wire saw. In this arrange-

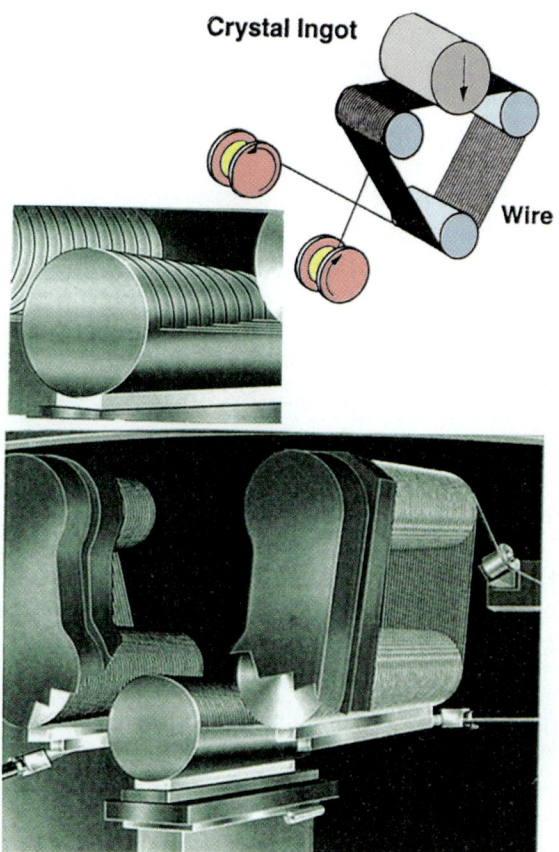

Fig. 10-6 Wire-sawing an ingot to produce wafers.

ment, parallel, equally-spaced and properly-tensioned stainless-steel wires, spun across two pulleys are part of a single stainless-steel-wire winding through a complex set of pulleys. Cutting of multiple-slices results when the ingot is pressed against the traveling wires under the injection of slurry. Although the cutting-rate is much slower than with an ID-saw (the ID-saw has a 80-100-times-higher sawing-rate), as many as 300-slices can be cut simultaneously. Since slicing with a multi-wire-saw is actually the result of low-speed grinding-action by the slurry, flatter wafers are obtained than with ID-sawing.

10.5 LASER-MARKING OF WAFERS

A technique using a *laser* is employed next to create an *alphanumeric* (or *bar-code*) identification-mark on the front of the wafer near the primary-flat or notch (Fig. 10-7). *Laser-marking* specifies an 18-character field to identify the wafer-manufacturer, conductivity-type, resistivity, flatness, wafer-number, and device-type. This allows each wafer to be identified. Wafer-traceability, correlation of the device-characteristics produced on a wafer with the wafer-properties, and the ability to prevent in-process-rejected wafers from being re-inserted into the fab-line are all benefits of utilizing wafer-marking. Automatic-code readers at

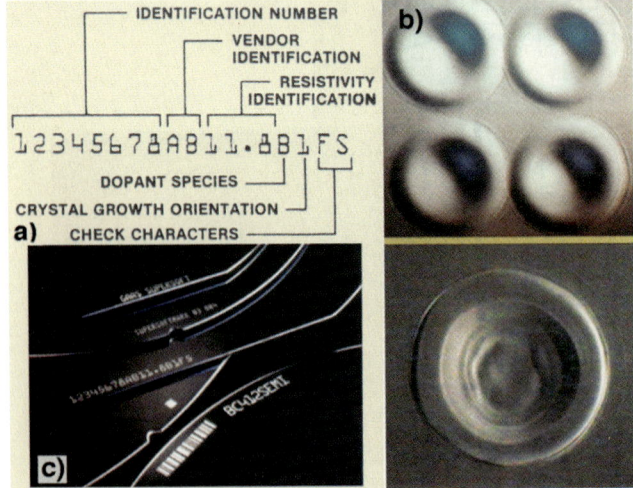

Fig. 10-7 (a) SEMI-specifications for 18-character vendor code. (SEMI Standard, used with permission.) (b) Laser-marks. (c) Photo of laser-marks on a wafer.

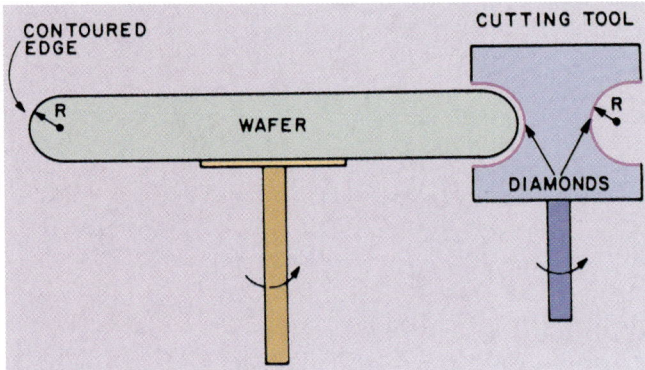

Fig. 10-8 Wafer edge-rounding-process.

various stages of processing can also track wafers and provide data on wafer-movement in the fab.

10.6 WAFER-LAPPING AND GRINDING

The thickness of the as-sawn-slices is sufficiently variable that a *lapping-*and-*grinding-*step is next employed to bring all slices to within the specified thickness-tolerance. The lapping takes place between two counter-rotating cast-iron plates in the presence of an abrasive-slurry, usually a mixture of micron-sized alumina or silicon-carbide particles suspended in a solution. This step also serves to reduce bow and warp, and to increase slice-flatness. Both sides of the slice are lapped using a mixture of Al_2O_3 and glycerine, and a succession of increasingly-finer polishing-grits is used in multiple-lapping steps. This process can achieve flatness-uniformities of ± 2-μm.

10.7 EDGE-ROUNDING & POLISHING OF WAFERS

Particle-contamination reduces yield. In advanced, submicron IC-processing some of the major sources of particles are the process equipment and the wafer itself. Defects found on the wafers during the device-fabrication sequence are often caused by tiny-chips of silicon from the wafer, or by particles removed from the process-equipment. This abrasive-effect is caused by contact of silicon-wafers with the process-tools, and can be influenced by the smoothness of the contact-area, primarily the wafers-edge.

The *wafer-edges* of as-sawn wafers are sharp and prone to chipping (Fig 10-9a). As a result, after

CHAPTER 10 SILICON WAFER PRODUCTION: FROM INGOT TO FINISHED WAFER

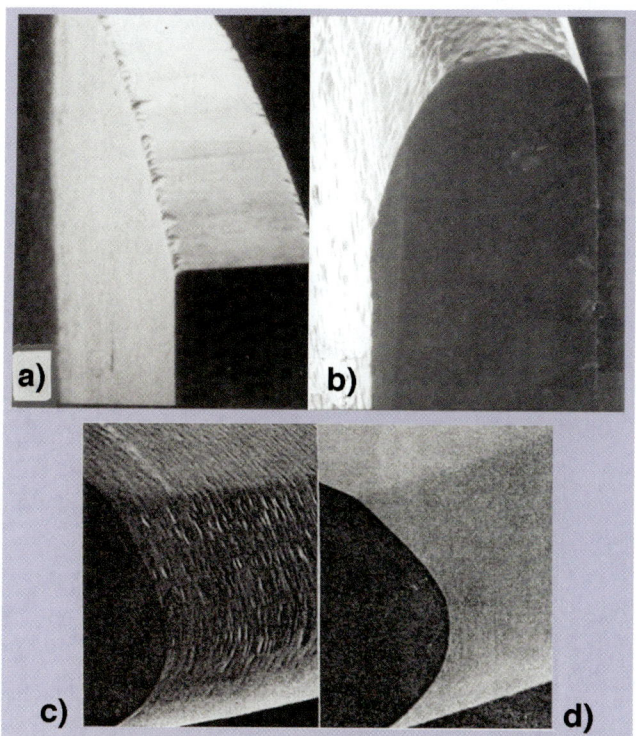

Fig. 10-9 (a) Non-edge-rounded wafer; (b) Edge-rounded wafer; (c) Edge-rounded wafer before edge-polishing; (d) Edge-rounded wafer after edge-polishing.

sawing, wafers are generally subjected to a grinding-process that rounds the wafer edge (Fig. 10-8). Such *edge-rounding* substantially reduces the incidence of chipping during normal wafer-handling and helps minimize film-buildup on the edge of the wafer during photoresist and epitaxial processing. (Edge-chip defect-sites are known to cause *slip-defects* on wafers [see Chap. 9] during subsequent thermal-processing steps.) The shape of the rounded edge-profile usually follows an industry standard (SEMI) in which the profile fits within the boundary of a standard template.

While this rounds wafer-edges, they are not yet perfectly smooth (as shown in Fig. 10-9c). Such relatively-rough edges can still produce enough particles to reduce the yield of ICs (compared to yields obtained when these edges are polished by etching to remove the grinding-damage). That is, after *edge-polishing* the wafer-edge is perfectly smooth (see Fig. 10-9d). Reports indicate that the defect-density on edge-polished wafers is reduced. For future generations of ICs that will be produced on 300-mm-wafers, the benefit of edge-polishing should be even more pronounced. That is, the area of the wafer-edge increases with increasing wafer-radius and thickness. In addition, it seems that other particle-sources on the wafer-topography (such as the notch and deep laser-scribe-marks), must also be polished (or in some other way smoothed) to further-reduce defect-density.

10.8 REMOVAL OF SURFACE MECHANICAL-DAMAGE BY CHEMICAL-ETCHING

The work-damage and contamination caused by the previous shaping-steps is next removed. This is done by *chemically-etching-away* the resultant damaged- and contaminated-layer. The wet-etch procedure for this step typically utilizes an etchant-solution of hydrofluoric (HF) and nitric (HNO_3) acid (with modifiers such as acetic-acid), to attack the silicon. A relatively non-porous and clean wafer-backside is also produced by this step. Note that etching of the wafer may be repeated after the subsequent operation of edge-rounding.

10.9 CHEMICAL-MECHANICAL POLISHING OF THE WAFER-SURFACE

A *chemical-mechanical polishing-step* is used to produce the highly-reflective, scratch and damage-free surface on one or both sides of the wafer. This is accomplished by mounting unpolished-slices onto a carrier, and then putting them on a polishing-machine. There, a powered-platen drives an appropriate polishing-pad material across the wafer-surface. A colloidal-silica slurry of sodium-hydroxide and fine (~10-nm) SiO_2-particles is dripped onto the table. The frictional-heat of the sliding mounted-wafers causes the sodium-hydroxide to oxidize the Si (i.e., the chemical-part of the process). The oxide is then abraded away by the silica-particles (i.e., the mechanical-part). Figure 10-10 shows the wafer-thickness and surface-roughness changes as it undergoes the various grinding, lapping, and polishing processes.

For 300-mm-wafers, double-sided polishing will be standard (instead of the single-sided polishing used up to the 200-mm wafer-sizes). Double-sided polishing (CMP on both the front- and back-surfaces simultaneously) has been found to result in superior flatness than single-sided polishing (Fig. 10-11).

10.10 CLEANING THE WAFERS

Following the polishing-cycle (in which a layer of silicon about 25-μm-thick is removed) wafers are subjected to a series of chemical-dips and rinses to remove the polish-slurry. A cleaning-process concludes this sequence. This is a key step as the Si-wafer must be free of contamination before it is shipped to the wafer-fab. Chemical-cleaning is effective in removing contaminants from the wafer surface. The "RCA-Clean" (see Ch. 8) is widely used to clean wafers before shipping. Inspection (and packing in a particle-free package) finish the wafering sequence.

10.11 DEPOSITING EPITAXIAL-SILICON-LAYERS ON THE WAFERS

Epitaxial-layers are required for some ULSI process-technologies. Wafer-manufacturers now supply silicon-wafers with a variety of epitaxial-layers for CMOS ULSI-technologies (see Chap. 17 for details of epitaxial-deposition). Another option being offered is a layer of polysilicon or mechanical-damage on the wafer-backside for *extrinsic-gettering* purposes.

10.12 ULSI SILICON-WAFER SPECIFICATIONS

There are many properties that a wafer must exhibit in order to be an appropriate starting-substrate for the fabrication of VLSI and ULSI (see Table 10-1). In this section the terms used to specify these properties are defined. Readers are also directed to recommended test-procedures that have been developed to verify that received silicon-wafers conform to purchase-specifications. Many of these tests have been standardized by the American Society for Testing and Materials (ASTM) and the Semiconductor Equipment and Materials Institute (SEMI). The standard test-procedures are available (updated when appropriate) in

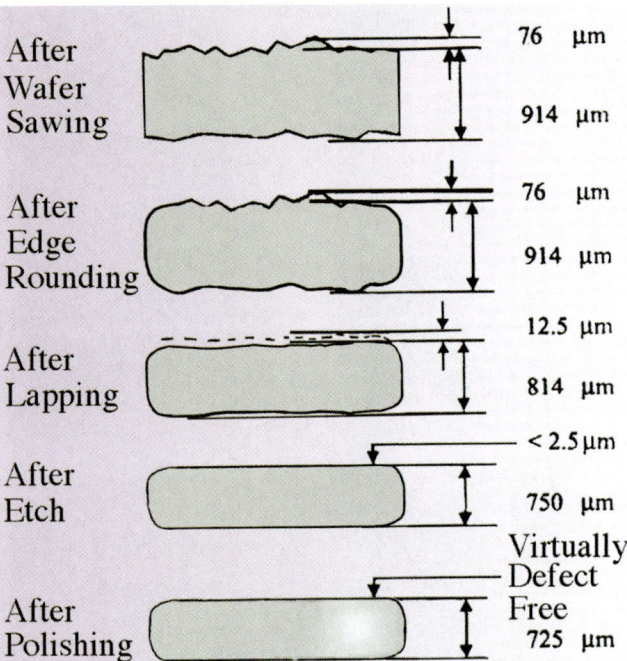

Fig. 10-10 Wafer-thickness and surface-roughness changes during grinding and polishing.

publications annually-issued by these organizations. Table 10-2 provides some typical specification-ranges for 125-, 150-, and 200-mm-diameter CZ-wafers. Note that specifications have become tighter with increasing wafer-size, due to automatic wafer-handling and photolithographic requirements. ADE Corporation is a major supplier of tools used to measure many of these specifications.

10.12.1 Electrical-Specifications

Conductivity-Type - Specifies whether wafers are *n*- or *p*-type. Information about which element was used to dope the wafer should also be provided.

Resistivity or Resistivity-Ranges (Ω-cm) - specifies the average (or range of) resistivities of the wafers. The resistivity is related to the doping-density, as is discussed in Chap. 2.

10.12.2 Mechanical/Dimensional Specifications

Diameter - specifies the linear-dimension across the surface of the wafer (passing through the center of the wafer), excluding flats. This specification has

Table 10-1 COMPARISON OF MATERIAL PROPERTIES AND REQUIREMENTS FOR VLSI AND ULSI

Property	Czochralski	Float Zone	Requirements for ULSI
Resistivity (phosphorus) n-type (ohm-cm)	1–50	1–300 and up	5–50 and up
Resistivity (antimony) n-type (ohm-cm)	0.005–10	–	0.001–0.02
Resistivity (boron) p-type (ohm-cm)	0.005–50	1–300	5–50 and up
Resistivity gradient (four point probe) (%)	5–10	20	< 1
Minority carrier lifetime (μs)	30–300	50–500	300–1000
Oxygen (ppma)	5–25	Not detected	Uniform and controlled
Carbon (ppma)	1–5	0.1–1	< 0.1
Dislocation (before processing) (per cm^2)	≤ 500	≤ 500	≤ 1
Diameter (mm)	up to 300	up to 100	up to 300
Wafer bow (μm)	≤ 25	≤ 25	< 5
Wafer taper (μm)	≤ 15	≤ 15	< 5
Surface flatness (μm)	≤ 5	≤ 5	< 1
Heavy-metal impurities (ppba)	≤ 1	≤ 0.01	< 0.001

been continuously tightened to improve pre-alignment placement of wafers onto steppers and to facilitate automated wafer-handling. For 200-mm-wafers the diameter-specification is 200 ±0.2 mm.

Thickness - must be controlled. If wafers are too thin, they may break or warp during normal-processing. If too thick, they may not work in all process-equipment and fixtures. Thickness is measured at the wafer-center. 200- and 300-mm-wafers are 725 ±20-μm and 775 μm ±25-μm-thick, respectively.

Total-Thickness-Variation (TTV) - Excessive thickness-variations may cause problems in mechanical-handling and lithographic processes, especially if stepper-processing employs back-surface-referencing. TTV is the difference between the maximum and minimum values of thickness on a wafer (Fig. 10-12), and is mostly a function of the sawing-process.

Bow is the concavity due to sawing, or the deformation from thermal-processing of the wafer-centerline (Fig. 10-12). Bow is a *bulk*-property of the wafer, not of the wafer-surface, and units of bow are μm. Parameters such as bow and warp are measured on free wafers (rather than on wafers flattened by a vacuum-chuck).

Warp - is the deviation (difference between maximum and minimum distance) exhibited by the centerline of a wafer from a planar-condition, when such a deviation includes both concave- and convex-regions (Fig. 10-12). Warp is also a *bulk*-property (expressed in μm) and should not be confused with flatness. Warp can occur during wafer processing, as discussed in Chap. 9.

Flat-dimensions - must be machined to a precise-length and orientation (Fig. 10-3a). Edge-rounding of the wafer can degrade the ends of the flatted-area, and can thus change the effective flat-length.

Fig. 10-11 Photograph of a double-sided wafer-polisher.

Table 10-2 SPECIFICATIONS FOR SINGLE-CRYSTAL SILICON WAFERS

Parameter	150 mm	200 mm	300 mm
Diameter (mm)	150 ± 0.2	200 ± 0.2	300 ± 0.1
Thickness (μm)	675 ± 25	725 ± 20	775 ± 25
Primary flat length (mm)	55–60	notch	notch
Oxygen content (ASTM '79) (ppma)		≥ 23 ± 2	≥ 22 ± 1.5
Flatness (μm) (GTIR)	1.5	1.5	1.5
Site Flatness (μm) (SFQD)		± 0.17 (1998)	± 0.12 (2001)
Surface orientation	(100)	(100)	(100)

Notches - Notches are features machined on the edge of large (≥ 200-mm) wafers for alignment and crystal-orientation purposes (Fig. 10-3b). Notches are 1-mm-deep, and the notch-axis (a line through the pin-center to the center of the wafer) is a <110> direction. A pin is used to find the notch, and thus align the wafer in a fixture.

Edge-contour - the wafer edges are contoured (rounded) to minimize occurrence of chipping and to help reduce film-buildup at the wafer edge during photoresist- and epitaxial-deposition. The contour quality is measured by comparison to a template.

Laser-Marking - as described earlier, a laser can be used to create a dot-matrix, alphanumeric-code on the wafer surface.

10.12.3 Chemical/Structural Specifications

Surface-Orientation - specifies the orientation of the surface of the semiconductor-wafer and the allowed-degree of misorientation. Normally <100> material is cut *on-orientation* ±0.5°.

Oxygen-Content - is important because it affects wafer-strength and impurity-gettering-processes. It is measured using FTIR. The interstitial oxygen-content will be between 13 and 19 ppma, and should be tailored to the customer's-specification. A tolerance of 2-ppma may be acceptable.

10.12.4 Surface/Near Surface Specifications

Flatness - is the maximum peak-to-valley deviation of a wafer surface as measured from a reference-plane (Fig. 10-12). Flatness is a *surface* property, expressed in μm. Flatness-parameters, wafer-thickness, and TTV are usually measured with the wafer pulled-flat on a vacuum-chuck. Global-flatness (GTIR - global-total-indicator-reading) of approximately 1.5-μm will be required for small feature-size (≤ 0.35-μm) applications. A local-site-flatness (SFQD) of ±0.12-μm over a 26x32-mm field-of-view will ensure that each step-and-repeat field is sufficiently flat, and thus

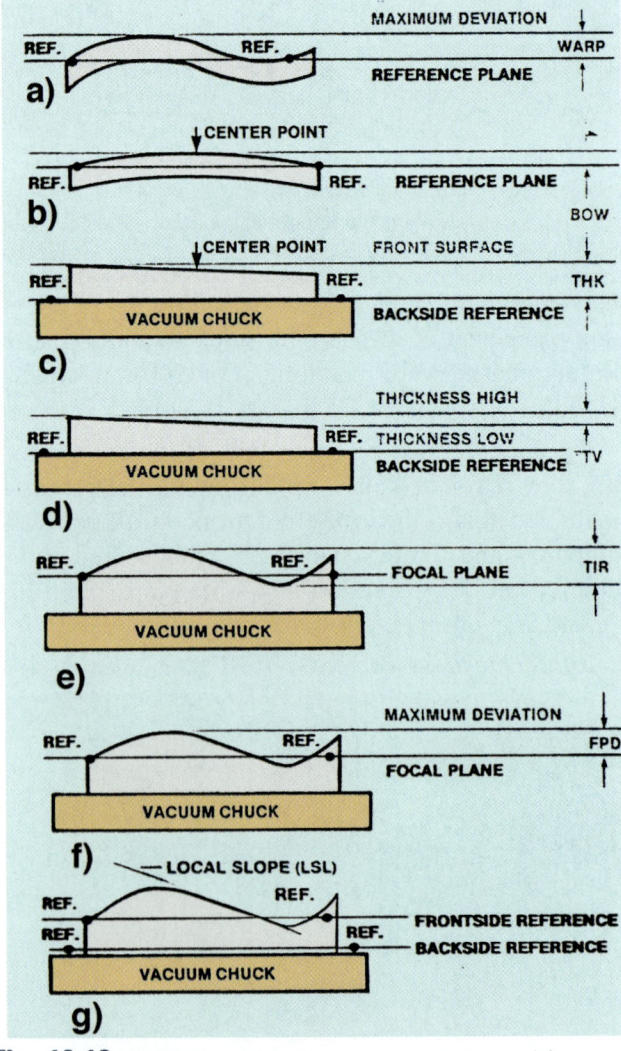

Fig. 10-12 Typical wafer-flatness parameters: (a) warp; (b) bow; (c) thickness; (d) total-thickness-variation, TTV; (e) total-indicator-reading, TIR; (f) focal-plane-deviation, FPD; (g) local-slope uniformity (LSU).

CHAPTER 10 SILICON WAFER PRODUCTION: FROM INGOT TO FINISHED WAFER

Diameter	Area (Sq. Inches)
2 Inch	3.1
3 Inch	7.1
100mm	12.6
125mm	19.6
150mm	28.3
200mm	50.2
300mm	109.5

Diameter Increase	Percentage Area Increase
2" → 3"	125%
3" → 100 mm	78%
100 mm → 125 mm	56%
125 mm → 150 mm	44%
150 mm → 200 mm	78%
200 mm → 300 mm	125%

Fig. 10-13 Percentage increase in wafer-area with increase in wafer-diameter.

remains within the depth-of-focus of lenses designed to resolve 0.18-μm dimensions.[6]

10.13 THE ECONOMICS OF SILICON-WAFERS

The size of silicon-wafers has been continuously increasing. The motivation for using larger wafers is economic. The benefits of larger wafers are gained from the additional silicon-area they provide (Fig. 10-13). If the increase in the manufacturing-cost per wafer is smaller than the percentage-increase in Si-area, manufacturing-costs per chip are then lowered.

Let us look at the issue of silicon-wafer costs. From Fig. 10-14 it can be seen that the cost per square-inch of silicon on a given wafer-size is highest when the new wafer-size is first introduced, but then drops with time. For 125-mm, prime device-wafers, the minimum silicon-cost reached ~$1/in^2. For each succeeding wafer-size the minimum cost/in^2 is expected to be 1.4x the cost of the previous generation. Thus, for 200-mm-wafers the silicon-cost is $2/in^2, resulting in the minimum cost of prime bulk-CZ wafers of this size being $70–$80. The cost of 300-mm-wafers is expected to eventually decline to somewhere above

$3/in^2, with a single bulk-CZ wafer thus eventually costing $250–$300. Note that adding an epitaxial layer to the wafer approximately doubles the wafer cost, and the cost of 200-mm epi-wafers in 2000 was $120–$150.

The cost/in^2 increases with wafer-size for a number reasons. First, larger-wafers are thicker, meaning fewer-wafers per cm of ingot-length are produced. Second, a smaller-percentage of the electronic-grade-polysilicon (EGS) ends up in prime-wafers. (For 150-mm-wafers this percentage was 30%, for 200-mm-wafers it dropped to 17%, and for 300-mm-wafers it is expected to be only ~10%.) Third, wafer-specs are tighter for each generation, reducing the "sweet-spot" in the middle of the ingot that yields prime-wafers. Finally, epi may be required for DRAM-production on 300-mm wafers, which will create a steep-rise in average-price of prime-wafers.

Note that the consumption of wafers of a given size also varies with time (see Fig. 10-15), and each new wafer-size remains in production for about 24 years. For example, production of 150-mm wafers began in the late 1970s, and their consumption-rate peaked in 1997. Their use in production is expected to be phased-out by about 2010. Next, 200-mm wafers were first introduced in 1988, and they are expected to reach their peak consumption-rate sometime between 2005 and 2010 (depending on the rate of ramp-up of the new 300-mm wafer-generation). By 2002, however, full-scale production IC-factories using 300-mm wafers were not yet running. As a result, well-

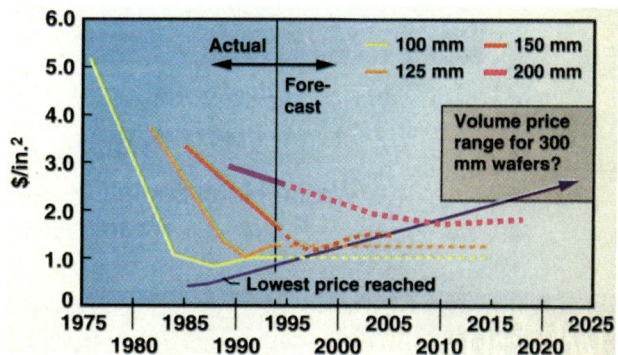

Fig. 10-14 Actual wafer-pricing-cycle by wafer-diameter.[4]

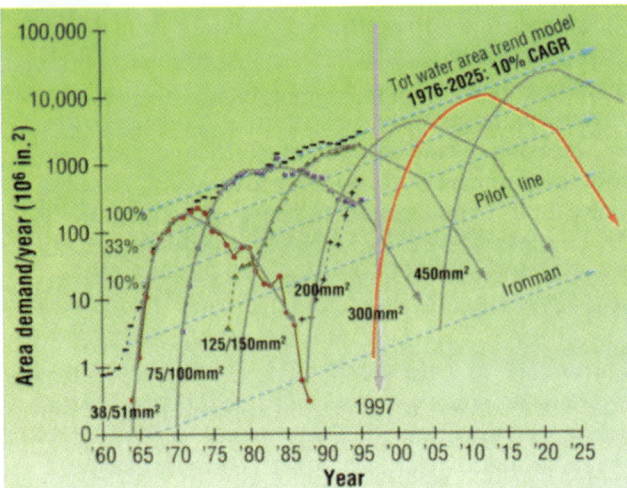

Fig. 10-15 The 24-year single-diameter-cycle of silicon wafers.[5]

designed 200-mm-equipment is predicted to have a long working-life before it is made obsolete by future wafer-fabs running larger wafer-sizes.

The worldwide-consumption of single-crystal silicon has been growing at 15–20% per year. In 1992, about 1.9-billion in² of silicon were sold. By 1994 this number was 2.7-billion in², and by 1999 it had increased to ~4.5 billion in². In 1994 this translated to a total-production of about 100-million wafers. These sales generated an income of $5.9-billion US in 1999. There are a relatively large-number of silicon-wafer suppliers. In the late 1990's the largest were: Shin-Etsu Handotai (SEH); MEMC; Wacker-Siltronic; Sumitomo-Sitix; Mitsubishi Materials Silicon; and Komatsu Electric.

REFERENCES

1. S. Wolf and R.N. Tauber, *Silicon Processing for the VLSI Era, Vol. 1- Process Technology*, Lattice Press, Sunset Beach CA 2000, Ch. 1.

2. *Handbook of Semiconductor Silicon Technology*, W.C. O'Mara, R.B. Herring, and L.P. Hunt, Eds., Noyes, 1990.

3. Semiconductor Equipment and Materials Institiue (SEMI), San Jose CA.

4. W.H. Reed, "Large Diameter Wafer Update," *Semiconductor International*, Nov. 1994, p. 140.

5. D. Anderson, "Stoking the Productivity Engine with New Materials and Larger Wafers," *Solid State Technology*, March 1997, p. 57.

6. W. Lin and H. Huff, "Silicon Materials," in *Handbook of Semiconducor Manufacturing Technology*, Eds. Y. Nishi & R. Doering, Marcel Dekker, New York, 2000, Ch. 3, p. 35.

PROBLEMS

1. Present two reasons why semiconductor wafers must have an extremely-flat surface.

2. List three advantages that wire sawing of silicon ingots has over inner-diameter-blade sawing.

3. Why are wafers: (a) *edge-rounded* and (b) in 300-mm wafers, also *edge-polished*?

4. 300-mm wafers are 775-μm thick. Explain why only 17 wafers/linear-inch of ingot length are obtained.

5. Using the value for density of silicon, and value of the thickness of 300-mm wafers, find the weight of 26, 300-mm silicon wafers (the standard large-batch size of wafers used in 300-mm manufacturing).

6. List some of the reasons for the continued increase in wafer-diameter.

7. What is the purpose of grinding a CZ silicon ingot before sawing it into wafers?

8. Why have *notches* replaced *flats* on larger diameter wafers (200-mm and larger)?

9. Define the terms: (a) lapping; (b) chemical etching; and (c) chemical-mechanical polishing with reference to the wafering process-flow described in this chapter.

10. What is *wafer-flatness*, and how it measured?

11. Using the specification for wafer-flatness for 300-mm wafers (within ± 1.5-μm across the wafer), to within what linear-dimension would it be necessary to control the flatness of a 5000-foot-long airport runway to make it as flat as a 300-mm wafer.

12. Explain the difference between *bow* and *TLV* in silicon wafers.

CHAPTER 10 SILICON WAFER PRODUCTION: FROM INGOT TO FINISHED WAFER

Photograph of a 300-mm plasma resist-stripping tool, which features a multi-station sequential processing architecture for high throughput. Courtesy of Novellus Systems.

CHAPTER 11

DIFFUSION

CHAPTER CONTENTS

11.1 CONCEPT OF DIFFUSION
11.2 ATOMIC-SCALE MODELS OF DIFFUSION IN SOLID SILICON
11.3 CONCENTRATION VERSUS DEPTH GRAPHS: DOPING-PROFILES & JUNCTION-DEPTHS
11.4 PRINCIPLES OF DIFFUSION-PROCESSES IN SILICON-DEVICE FABRICATION
11.5 DIFFUSION IN SILICON-DIOXIDE
11.6 DIFFUSION SYSTEMS AND DIFFUSION SOURCES
11.7 MEASUREMENT TECHNIQUES FOR DIFFUSED LAYERS
11.8 MEASURING JUNCTION-DEPTHS

"The Devil is in the Details."

Inscription over the entrance of the Physics Building at the University of Tennessee

Photograph of a process tool used to carry out RTP-diffusion processes. Shown is the AMAT Radiance® system. Courtesy of Applied Materials, Inc.

The devices built in integrated-circuits on silicon-wafers use an *isoplanar-processing* approach. That is, dopants are selectively introduced into the wafer from the top surface through windows in an oxide masking-layer. The processes of *chemical-diffusion* and *ion-implantation* are employed for introducing these dopants. Figure 3-1 (Chap. 3) shows how a single diffusion-step (for example, introducing a heavier-concentration of *n*-type dopant into a more lightly-doped *p*-type Si-wafer), creates a *pn*-junction near the surface of the wafer. Figure 11-1 illustrates how *two* diffusions (e.g., a *p*-type-diffusion, followed by an *n*-type-diffusion,

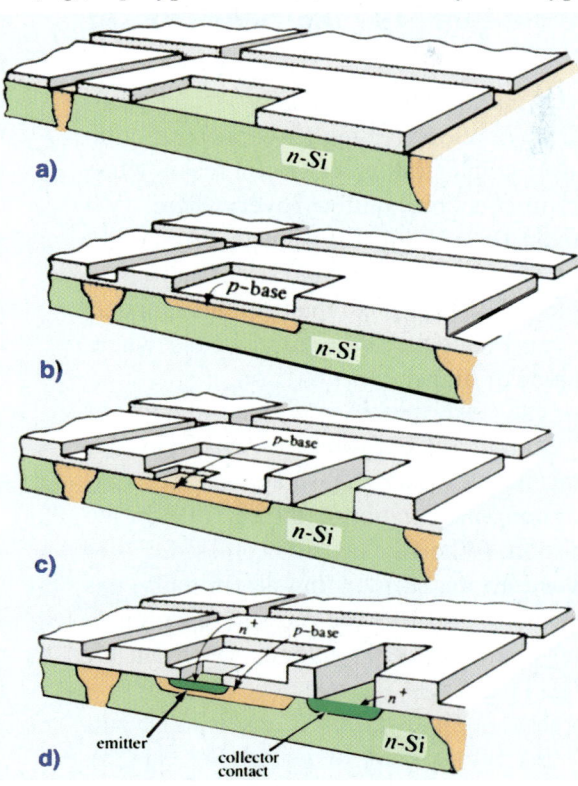

Fig. 11-1 A basic planar-bipolar-transistor structure can be fabricated with two successive, selective diffusions into a Si-substrate from the top-surface. (a) Openings in the surface-oxide. (b) First-diffusion is *p*-type through the oxide-openings to form a *p*-type *base-region*. (c) Openings in the surface-oxide for the second-diffusion. (d) The second-diffusion (*n*-type), forms the *emitter* (and collector-contact) regions.

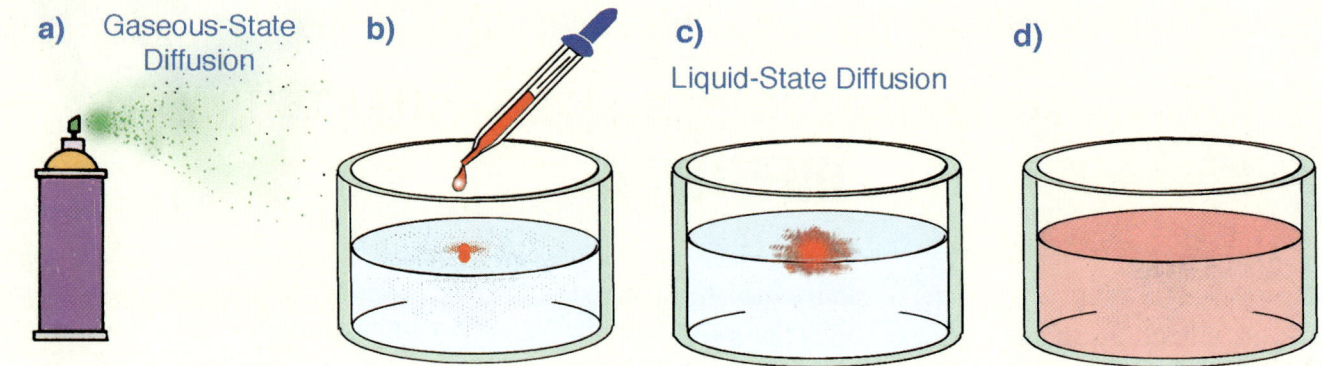

Fig. 11-2 (a) Example of diffusion involving gases. (b) - (d) Example of diffusion involving liquids.

- through windows in the top-surface oxide-layer) are used to form an isoplanar *bipolar-junction-transistor* (*BJT*) structure.

11.1 INTRODUCTION TO THE CONCEPT OF DIFFUSION

Diffusion is a natural-phenomenon by which substances move from regions of higher-concentration to regions of lower-concentration. Diffusion ceases when the concentration becomes uniform everywhere. Two conditions must be met for diffusion to take place: **1)** a material must exist in higher-concentrations at some locations within a volume than at others; and **2)** this material must have the ability to move elsewhere from the regions of higher-concentration.

Diffusion can take place in a gas, liquid, or solid. As an example of *diffusion in a gas*, consider deodorant-gas in a pressurized spray-can. The gas exists at a higher-concentration in the can than in the surrounding-room. When the nozzle is pressed, gas in the can moves into the surrounding-air. Initially, gas is driven to leave the can by the pressure-difference, but thereafter it disperses throughout the room by diffusion (Fig. 11-2a).

An example of *liquid-diffusion* is observed by putting drop of ink into a glass of water (Fig. 11-2b). The ink is initially more concentrated in the drop, but spreads out by diffusion. (Fig. 11-2c) until its' concentration eventually becomes uniform throughout the water - Fig. 11-2d - at which point, diffusion stops. If an ink-drop is put into hot-water, diffusion occurs more rapidly, as molecules move faster than in colder-water.

The concept of *diffusion in a solid* is illustrated by the column of balls in Fig. 11-3 (which represent impurity-atoms highly localized in a solid). If these impurity-atoms are able to move, their localized-distribution will spread out with time, until the concentration becomes uniform everywhere. The graphs in this figure also show how the slope of the concentration-level changes as time goes by. At first, slopes are steep (*high-concentration-difference*), but eventually, when the concentration becomes uniform (*no-concentration-difference*), the lines of the graph no longer have any *slope* (or *gradient*). As shown in Fig. 11-4, if the number of diffusing-objects is too great to allow them to be drawn individually on a drawing, the concept of diffusion can instead be illustrated by using continuous-curves.

Impurities are introduced into silicon-wafers in isoplanar-processing from the *top-surface* of the wafer. If a chemical-diffusion step is used, a high-concentration of the impurity is created at the surface of the wafer, and this impurity diffuses downwards into the silicon (Fig. 11-5).

The controlled-diffusion of specific-impurities (or dopants) in solid-Si is the basis of device-fabrication in semiconductor-processing. Fortunately, the impurities used to make the *n*- and *p*-regions of solid-state devices in silicon (B, P, and As) diffuse slowly, even at high-process-temperatures (900-1100°C). This allows precise-control of where such impurities end-up after

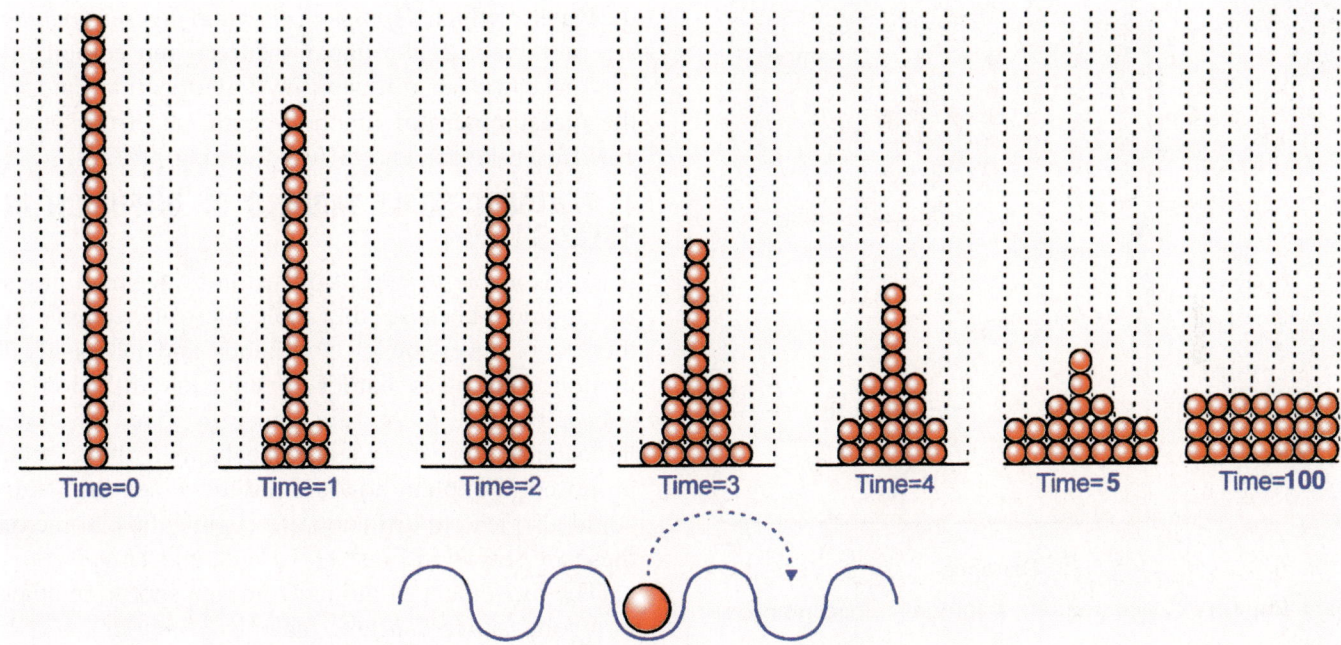

Fig. 11-3 Concept of diffusion in solids: Diffusion is taking place in both $+x$ and $-x$ directions from a high-concentration region at $x = 0$. Diffusion ceases when the concentration is uniform (*concentration-gradient* = 0). Courtesy of *www.icknowledge.com*.

a diffusion-step. When the wafers are returned to 300 K, the impurities no longer diffuse, because their movement (i.e., their *diffusivity*) is negligibly-small at 300 K (room-temperature).

In the early days of transistor- and IC-processing, the B, P, and As dopants were supplied to the silicon by *chemical-sources* (available in gas, liquid, or solid form). These dopants were then diffused to the desired depths by subjecting the wafers to elevated-temperatures (900-1100°C, Fig. 11-6). More recently, ion-implantation has become the primary-method of introducing impurities into silicon (see Chap. 12). While ion-implantation is able to place the impurities near the silicon-surface, a diffusion-step is still required to drive-them to the appropriate depths. In addition, ion-implanted-devices must also be subjected to elevated-temperatures to activate the implanted-impurities, and to heal the defects of implant-damage by annealing (see Chap. 12). When these high-temperature-steps are performed, diffusion of the impurities occurs at the same time. Thus, the process of impurity-diffusion in silicon remains a topic that must be understood.

In the past, junction-depths in devices were often in the range of 1-3-μm. In ULSI-devices, however, these depths have decreased to 0.05-0.5-μm. For example, in advanced-CMOS-devices it is common

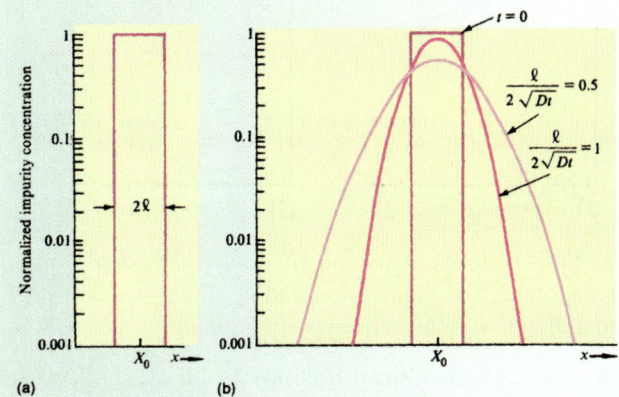

Fig. 11-4 If the number of balls shown in Fig. 11-3 gets to be too great to be able to draw them individually on a diagram, the concept of diffusion in two directions from a region of high-concentration can instead be illustrated using continuous-curves, as is shown in this figure.

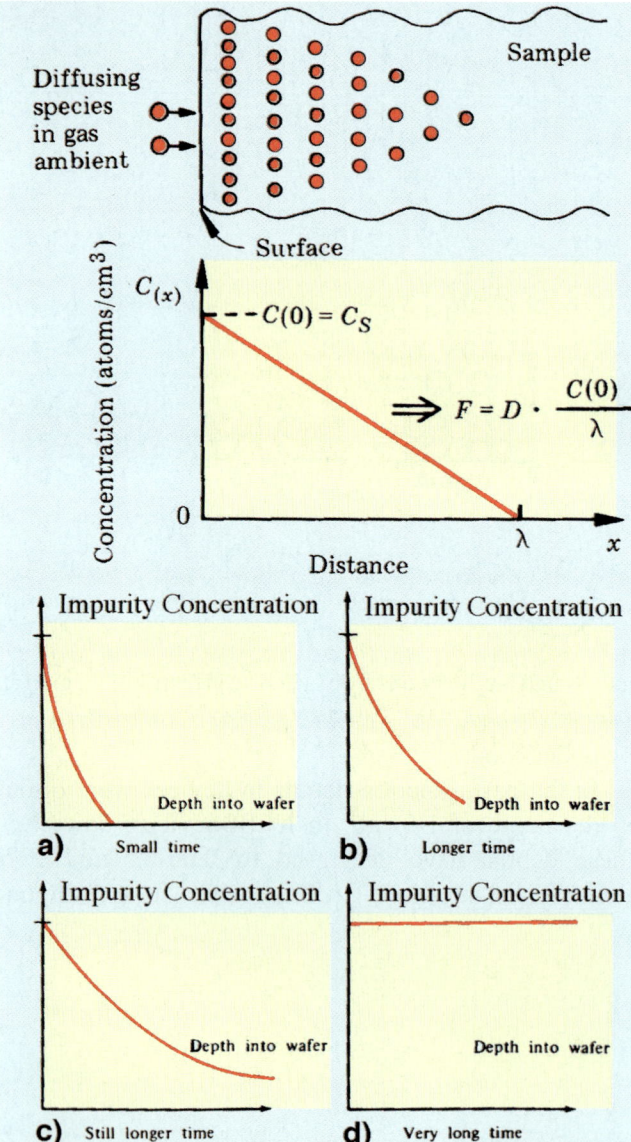

Fig. 11-5 Pictorial depiction of *dopant-diffusion* from a region of high dopant-concentration at the wafer-surface into a region of lighter (or opposite) doping.

for source- and drain-diffusions to be as shallow as 0.05-μm. Control of such shallow-junctions is more difficult to achieve, and their measurement requires advanced techniques.

In this chapter we describe: **1)** the mechanisms that allow diffusion of dopants to occur in solid-silicon; **2)** the ways in which *doping-profiles* and *junction-depths* are portrayed; **3)** the dopant-sources and equipment used to carry out diffusion in IC-fabrication; and **4)** the measurement of junction-depth, sheet-resistance of diffused-layers, and diffused doping-profiles.

11.2 ATOMIC-SCALE MODELS OF DIFFUSION IN SOLID-SILICON

It is reasonable to expect diffusion to occur in gases and liquids because the molecules and atoms of these substances are in continuous and independent motion. But, it is harder to envision diffusion in solids, because atoms are generally fixed on their lattice-sites. However, some mechanisms have been proposed to explain how atomic-movement in solids could also lead to diffusion-effects in solids. Some of these are shown in Figs. 11-7, 11-8, and 11-9.[1-3]

The movement of atoms from one spot to another in a solid-lattice can take place by *direct* or *indirect* mechanisms. Impurity-atoms that do not have a strong bonding-interaction with Si atoms, and are located instead in the spaces between lattice-sites (i.e., in the *interstices* of the lattice), can jump directly between the interstices (*interstitial-impurities*). As discussed previously, species such as H and O, and metal-atoms such as Cu, Fe, and Au are believed to diffuse in this manner (Fig. 11-7). The considerable space that exists between atoms in the silicon-lattice permits such *interstitial-diffusion* to occur much more rapidly than does *substitutional-diffusion* (discussed in the next paragraph). As a result, metallic-atoms such as Cu, Fe, Au, and Na rapidly distribute themselves throughout a silicon-lattice at relatively low-temperatures.

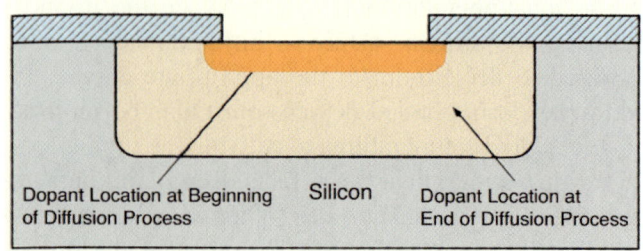

Fig. 11-6 Schematic drawing showing dopant-locations after *pre-dep* and *drive-in* diffusions.

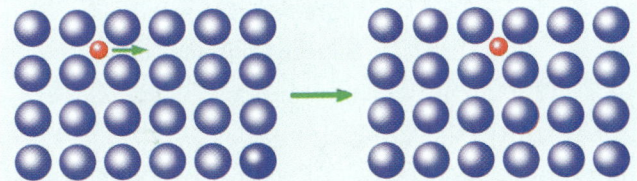

Fig. 11-7 Drawing showing the *interstitial-impurity diffusion mechanism*.

The diffusion of impurities that replace silicon-atoms on lattice-sites (*substitutional-impurities*, such as B, P, and As) requires one of a more-complex set of events, including: **1)** *direct-exchange*; **2)** *vacancy-assisted-movement*; and/or **3)** *interstitial-assisted-movement*.

The *direct-exchange-mechanism* involves an impurity-atom exchanging places with a silicon-atom, as shown in Fig. 11-8a. While such an event is possible, it is not likely to occur, because at least six covalent-bonds must be broken for the host-atom and impurity-atom to exchange positions. On the other hand, this exchange is considerably easier if the adjacent lattice-site is empty (i.e., a *vacancy* exists). Then, only three bonds must be broken. Vacancy-exchange is thus one of the most likely diffusion-mechanisms for substitutional-impurities. Figure 11-8b depicts the movement of atoms via the *vacancy-mechanism*. Note, however,

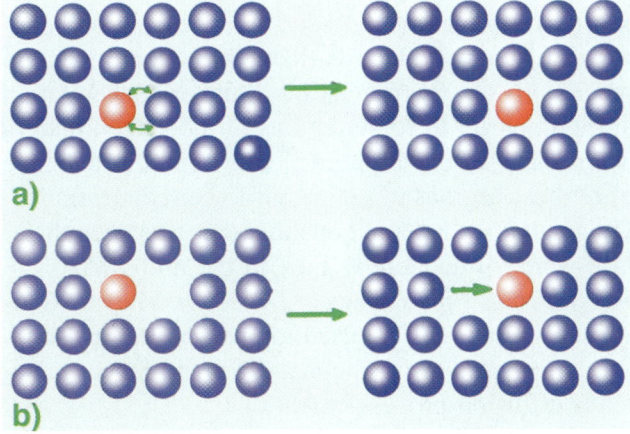

Fig. 11-8 Drawing showing two proposed mechanisms that could give rise to diffusion in solids: (a) *Direct exchange* of dopant and Si atoms; (b) *Vacancy-aided* mechanism.

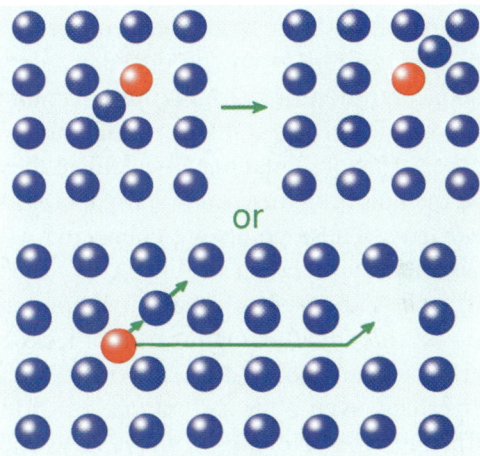

Fig. 11-9 Drawing showing the proposed *"kick-out"* mechanisms that could give rise to diffusion in solids.

that vacancy-assisted diffusion is much slower than interstitial-diffusion (as shown in Fig. 11-7), due to the limited-supply of vacancies.

Other studies indicate that *silicon-interstitials* also play a role in substitutional-diffusion of P, B, and As. Figure 11-9 shows a silicon-interstitial atom "kicking-out" an impurity-atom, making the impurity-atom an interstitial. The impurity-atom then moves rapidly through the lattice to another lattice-site, where it "kicks-out" a silicon-atom, taking its place in the lattice, and creating another silicon-interstitial.

11.3 CONCENTRATION VERSUS DEPTH GRAPHS: DOPING-PROFILES AND JUNCTION-DEPTHS

A common way to represent the concentration of dopants in a silicon-wafer is with a *concentration-versus-depth graph*, which is also called a *doping-profile*. This type of graph could plot both the *concentration* (vertical-axis) and *depth into the silicon-wafer* (horizontal-axis) with *linear-scales* (Fig. 11-10). But, since the concentration varies over such a large range, it is more useful to plot the doping-concentration on a *logarithmic* (or *log*) *scale* (Fig. 11-11), and the depth into the silicon-wafer on a *linear-scale*. The example doping-profile shown in Fig. 11-11 is that of the concentration of boron vs depth, after a chemical-pre-deposition-process (as described in Sect. 11.4).

Such *doping-profiles* are produced by determining the concentration of boron-atoms at a number of points below the surface (either experimentally or by solving the diffusion-equations to calculate these values), and then plotting them on graph-paper with such *log-linear axes*. These points are connected, yielding curves like the ones shown in Fig. 11-11. Such doping-profile plots are used to show the *doping-concentration-versus-depth* after a number of processes have been performed (including diffusion or ion-implantation, or perhaps even after a combination of implant plus diffusion steps). *Doping-profile plots* as shown in Fig. 11-11 show the variation of doping in one-dimension (along the depth of the wafer at one point on the wafer-surface). Two-dimensional doping-profiles can also be plotted, as shown in Fig. 11-12. They depict the doping-concentration in two-dimensions, with the lines representing the locations of a constant-level of doping-concentration.

Often the doping-concentration-level of an impurity present in the starting-wafer (called the *background* or *[substrate] doping-concentration*) is also plotted on such a curve (Fig. 11-11). If a process is performed in which the dopant introduced into the Si is of the opposite-type to the background-dopant (and exceeds the background-doping-levels at some locations), the doping-profiles will look something like

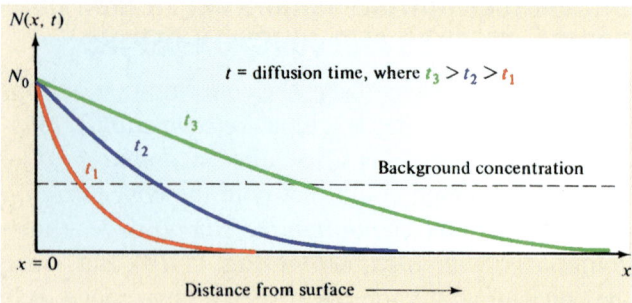

Fig. 11-10 *Complimentary error-function (erfc) doping-profile (in one-dimension) vs. time. In this figure, a linear-scale is used to plot the doping-concentration on the vertical-axis versus depth in the silicon.*

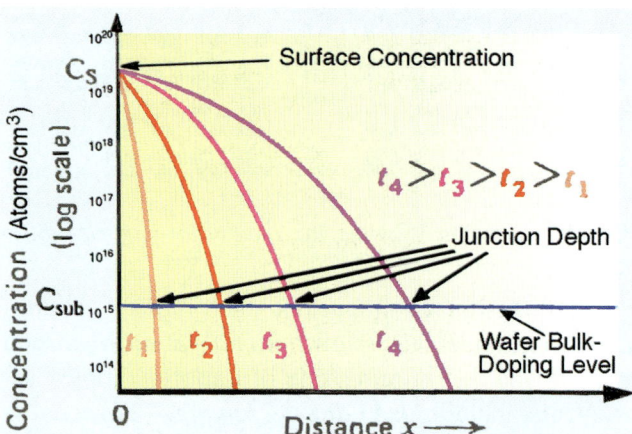

Fig. 11-11 *Complimentary error-function (erfc) doping-profile (in one dimension) vs. time. In this figure, a logarithmic-scale is used to plot the doping-concentration on the vertical-axis versus depth in the silicon (linear scale).*

the curves shown in Fig. 11-11. That is, at some depth the concentration of the two dopants will be equal. The *junction-depth* x_j is defined as the point where the *diffused* dopant-concentration and the *background* dopant-concentration are equal. Figure 11-11 also shows the junction-depth.

11.4 PRINCIPLES OF DIFFUSION-PROCESSES IN SILICON DEVICE-FABRICATION

As noted in Sect. 11.1, two general-conditions must be met for diffusion to occur. For the case of impurity-diffusion in silicon-lattices, these conditions are specifically: **1)** there must be a higher-concentration of the impurity in some regions of the silicon than in others; and **2)** the impurities must be able to move from those regions to regions of lower-concentration. The mathematical-equations that describe such diffusion were formulated by Fick in the mid-1800s, and are known as *Fick's Laws of Diffusion*. The solutions to Fick's-equations are used to provide the final-dopant-concentrations in silicon as a result of a particular set of diffusion-process-conditions.

The first Ficks-Law mathematically restates the above principles, saying that the tendency for an impurity to diffuse is given by the *product* of the two conditions. (More mathematical-details on Fick's

CHAPTER 11   DIFFUSION

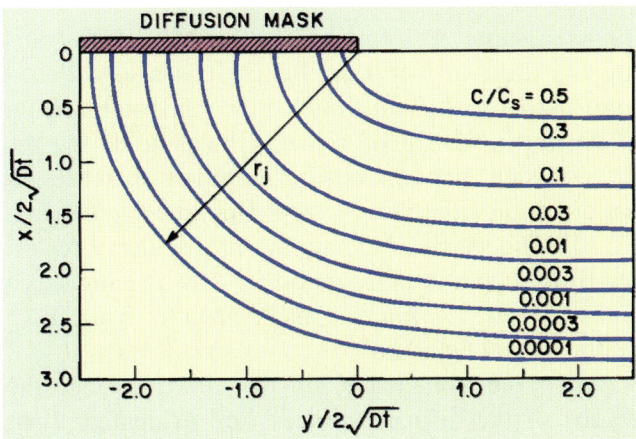

Fig. 11-12 Two-dimensional doping-profile used to show the extent of lateral-diffusion under an oxide-mask.[4]

Laws and how they apply to impurity-diffusion in IC-fabrication are given in Appendix-B.) Here the discussion of impurity-diffusion in silicon remains *qualitative* (i.e., non-mathematical).

11.4.1 The Diffusivity of Impurities in Silicon

The first-term in Fick's First-Law indicates the ease with which impurities can move in the lattice. This characteristic is called the *diffusivity* (or *diffusion-coefficient*) D of the impurity, and is usually expressed in units of cm^2/sec. The higher the value of D, the faster the impurity is able to move. Note that the electrically-active-impurities in silicon (both the *n*-type [P and As] and *p*-type [B] dopants) have such small-diffusivities at room-temperature that they are considered immobile (no diffusion occurs at 300 K, even when large-concentration-differences exist).

However, diffusivity increases exponentially with temperature ($D \propto \exp[-E_A/kT]$ where E_A is the *activation-energy*, k is *Boltzmann's-constant*, and T is the temperature). For sufficiently high-temperatures (800-1100°C), the electrically-active dopants in Si (i.e., B, P, and As) can be made to diffuse over the distances needed to form the regions of silicon-devices in a reasonable time. The conditions of a diffusion-step (temperature and time) are chosen for each particular-impurity such that it will diffuse the necessary distance to produce the required-device-region. Process-simulation-programs, such as SUPREM III,

are used to determine the extent of diffusion when process-flows for device-structures are developed.

11.4.2 Pre-Dep and Drive-In Diffusion-Steps

Most diffusion-processes in IC-fabrication are done in two-stages. The first-step is a short, *constant-source-diffusion* called *pre-deposition* (or *pre-dep*). It is followed by a longer, *limited-diffusion-step*, called *drive-in* (Fig. 11-13). Use of a two-step process gives more flexibility in setting the doping-concentrations and the final-junction-depths of the impurities than if just a single-step diffusion-process was used.

Pre-Dep: In the predep-step, a dopant-oxide is formed on the silicon-surface by flowing the dopant-containing gas (e.g., B_2O_6 or $POCl_3$), oxygen, and a carrier gas, into a heated furnace-tube containing the wafers. The dopant-gas and oxygen react to form an oxide-layer on the silicon-surface containing the dopant to be diffused (e.g., B_2O_3 or P_2O_5). Such dopant-rich oxide-films become high-concentration-sources of dopant on the top-surface of the silicon. During the short, high-temperature pre-dep-step (e.g., 950°C for 30-min), dopant from this surface-oxide-layer enters the silicon at the solid-solubility-concentration of the process-temperature (see Sect. 9.3), and diffuses a very-short distance into the silicon.

Note that the value of the *dopant-surface-concen-*

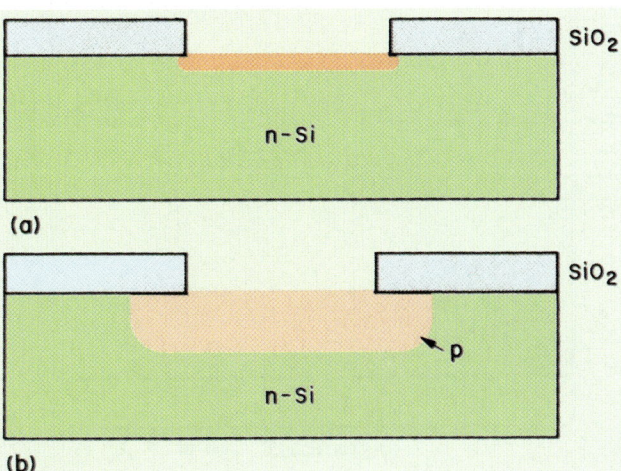

Fig. 11-13 Location of dopants after: (a) a pre-dep step; and (b) after a drive-in step.

tration remains constant during the pre-dep-process (i.e., at its *solid-solubility-value*), and hence this step is also called a *constant-source diffusion-step*. But the maximum solid-solubility limit also increases with temperature. Thus, the amount of dopant that enters the silicon is a function of the temperature of the pre-dep step (and its duration). As a result, the longer the pre-dep process proceeds, the further the diffusion-front progresses into the wafer (Fig. 11-11). The doping-profile that results from such constant-source diffusion has a specific-shape, known in mathematics as the *complementary-error-function* (*erfc*). This is described in more detail in Appendix B.

Drive-In: Despite the fact that the diffusion-front moves deeper into the silicon as time progresses, in practice the pre-dep-step is kept short. This is intended to limit the distance that the dopant moves in the silicon during pre-dep. A second longer, and higher-temperature step (called the *drive-in*, or *redistribution-diffusion*), is used to move the dopants to their desired final-locations (Fig. 11-12).

Prior to the drive-in step, the dopant-oxide-layer is stripped from the silicon surface. Thus, there is no longer a supply of dopants entering the silicon during the drive-in step (and hence it is also called a *limited-source diffusion-step*). Since the total-amount of dopant in the silicon remains the same during the drive-in-step, the surface-concentration decreases as the diffusion-front moves deeper into the silicon (Fig. 11-14). The shape of the doping-profile curve resulting from such limited-source-diffusions is known as a *Gaussian-distribution* (see also Appendix B for more mathematical details).

After the conclusion of the drive-in-step, the results of the diffusion-process are evaluated. Test-wafers are measured for *surface-concentration* with a four-point-probe (after pre-dep), and for determining the *junction-depth* after the drive-in-step.

11.5 DIFFUSION IN SILICON-DIOXIDE

Silicon-dioxide serves as a diffusion-masking layer in chemical-diffusion steps. Its role is to prevent dopants from reaching locations on the wafer where they are not wanted. SiO_2 is effective for this task because the diffusivities of the common dopants are much lower in SiO_2 than in silicon.

The Group III and V elements are known to form glassy-networks with SiO_2, and as a result their diffusivity strongly depends on their concentration.

11.5.1 Lateral-Diffusion Under Oxide-Windows

Since most diffusions are performed through openings in SiO_2 surface-films, the extent of lateral-diffusion under the oxide-edge is an important consideration. Such lateral incursion is ~75-85% of the vertical diffusion-depth (see Fig. 11-11).

11.6 DIFFUSION SYSTEMS AND DIFFUSION-SOURCES

Diffusion processes are conducted in systems known as *diffusion-furnaces*, which provide controlled high-temperature and gas-flow conditions. A typical *diffusion-system* consists of a heating-element, diffusion-tube, boat, and dopant-delivery system. The components of diffusion-furnaces are essentially identical to those used to grow thermal SiO_2-films, and their

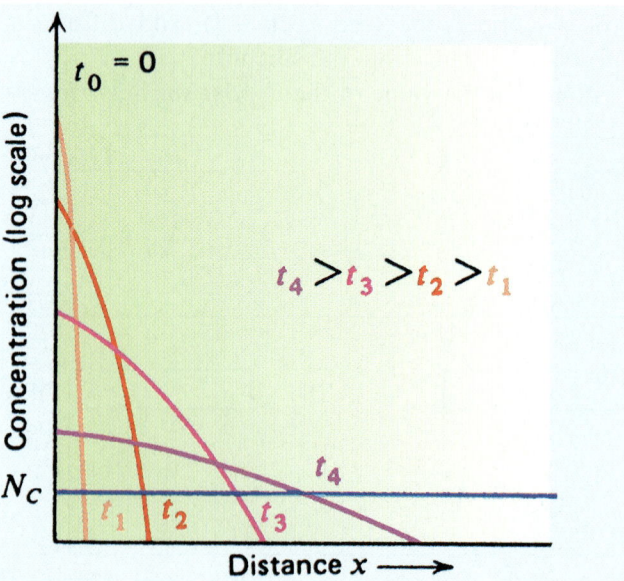

Fig. 11-14 Doping-profiles that result from drive-in steps (Gaussian curves).

details are described in Chap. 13. In the past, most impurities were introduced into silicon-wafers with chemical-sources, and the processes for such steps are briefly described in this chapter.

Since about the mid-1980's, ion-implantation has taken over as the preferred means of introducing dopants into silicon (Chap. 12). Nevertheless, in some current processes, drive-in diffusion-steps are still used for redistributing impurities (e.g., the well-drive steps in CMOS), or for achieving electrical-activation after ion implantation. Ambient-gases in furnace-tubes during such drive-ins can be *oxidizing* (O_2, H_2O), *reducing* (H_2), or *neutral* (Ar, He, N_2).

The major classes of chemical-sources for chemical diffusions include: **a**) *gaseous*; **b**) *liquid* (bubbler and spin-on); and **c**) *solid* (tablet, powder, or disc) sources. The various types of sources available for each of the major dopant-atoms used in silicon-processing (As, P, B, and Sb) are described next.

11.6.1 Gaseous Dopant-Sources

Gaseous dopant-sources have been the most widely used types in silicon-fabrication. This is especially true in ion-implantation applications, where 90% of the sources used are gaseous. Gaseous-sources have the advantages of convenience, direct-supply to the furnace-tube in a gaseous-state from a pressurized-cylinder, and the ability to accurately measure the gas flow-rate with a *mass-flow controller* (see Chap. 6). Gases also retain their purity for longer time periods than do corrosive liquid-sources.

The gas-species most widely used for sources of boron are *boron-trifluoride* (BF_3), *boron-trichloride* (BCl_3), and *diborane* (B_2H_6). Gas types used as phosphorus sources include *phosphine* (PH_3), and *phosphorous-pentafluoride* (PF_5). Arsenic sources include *arsine* (AsH_3), and *arsenic-trifluoride* (AsF_3).

When these materials are used for chemical-diffusion processes, they are normally mixed with a *carrier-gas* (that also serves as *diluent*) such as N_2, and perhaps a reactive-gas such as O_2. The source-gas reacts with the oxygen at the wafer-surface (Fig. 11-15) to form a dopant-oxide (e.g., P_2O_5 or B_2O_3).

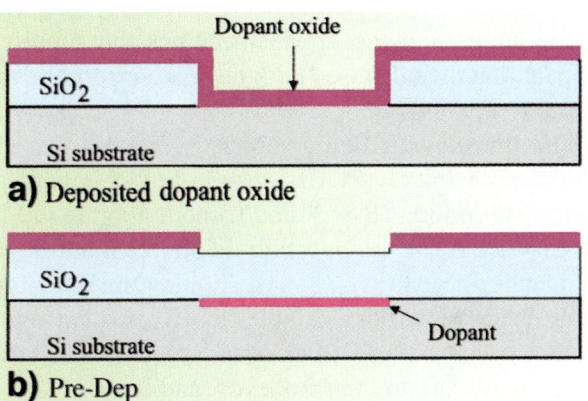

Fig. 11-15 Diffusion doping-process sequence.

The dopant then diffuses from this oxide-film into the top-surface of the silicon wafer, resulting in a uniform dopant-concentration across the surface.

Since these gases are all highly-toxic (or otherwise hazardous), additional input-purging and trapping systems are needed to ensure that all of the source-gas is removed from the system before wafer-loading or removal. The unused dopant-gases must also be processed by burning, or by chemical or water scrubbing (to destroy their toxicity) before being exhausted into the atmosphere (see Chap. 29).

11.6.2 Liquid Dopant-Sources

Liquid-sources are normally used with *bubblers*. Bubblers are quartz-flasks that hold the liquid-sources and can be heated. They are used to convert the liquid-source to a vapor by flowing an inert carrier-gas (such as N_2) through the heated liquid-dopant (e.g., $POCl_3$, or BBr_3), held at a constant-temperature. The bubbles of the carrier-gas pick-up the dopant-vapor, and carry it into the diffusion-furnace. The bubbler-temperature determines the vapor-pressure of the liquid-source, and thus the concentration of dopant reaching the wafer. The process is easily controlled by starting and stopping the gas-flow to the bubbler. More information about bubblers is found in Chap. 16.

In a manner similar to gas-source doping, vapors from liquid-sources react with oxygen also flowing in the furnace tube to form dopant-oxides on the wafer surface (e.g., $4POCl_3 + 3O_2 \rightarrow 2P_2O_5 + 6Cl_2$). The oxide containing the dopant becomes the supply of dopant that then diffuses into the silicon from the top-surface (Fig. 11-15).

The *phosphorus* liquid-source is phosphorus-oxychloride, or *POCL* ($POCl_3$). *Boron* liquid-sources are boron-tribromide (BBr_3), and trimethylborate (TMB) ($[CH_3O]_3B$). The *antimony* liquid-source is antimony-pentachloride (Sb_3Cl_5). These liquids are also typically hazardous. For example, POCL is a corrosive-liquid, causing burns to skin and eyes. POCL-vapor is also irritating to the skin, eyes, and lungs.

11.6.3 Solid Dopant-Sources

The most widely used *solid-sources* are *discs* that contain the desired dopant. Disc-type sources are used for boron, phosphorus, and arsenic diffusion. Boron-nitride (BN) discs are the most widely-used variety. Such discs are slightly larger than the Si-wafers, and they are stacked in a wafer-boat together with the wafers. Such sources provide good-uniformity for large-area wafers because the disc-area is comparable to that of the wafer. The BN is oxidized at 750-1100°C to form a thin-skin of B_2O_3. This oxide serves as the diffusion-source. A small-flow of N_2 is used to prevent backstreaming of airborne-contaminants into the diffusion-tube. The discs are safer to use since they produce no toxic-vapors at room-temperature.

11.7 MEASUREMENT TECHNIQUES FOR DIFFUSED (AND ION-IMPLANTED) LAYERS

The diffusion process on a production-line must be continuously monitored to maintain process-control. This is accomplished by measuring the junction-depth and sheet-resistance after each of the diffusion steps.[5,6] In this section, measurement-techniques for finding these quantities are given, as well as those for determining the doping-profiles of junctions formed by diffusion and ion implantation.

11.7.1 Sheet-Resistance Measurements

The *sheet-resistance* R_S of a diffused-layer is the resistance exhibited in a square (i.e., a region of equal-length and width) of diffused-material which has a thickness x_j (junction-depth). The value of R_S is expressed in units of (Ω/sq), and is related to the resistivity of a diffused-layer by:

$$R_S (\Omega/sq) = \rho (\Omega\text{–cm})/x_j (cm) \qquad (11.1)$$

where ρ is the volume-resistivity of the diffused-layer. The value of R_S is obtained by measuring the resistance of the diffused-layer using a *4-point-probe* technique. This measurement-approach is described in more detail in Chap. 7. Automated 4-point-probe-equipment is currently available from several vendors. Such equipment is capable of measuring the sheet-resistance at many places on a wafer, thereby providing a *contour-map* of the sheet-resistance.

11.7.2 Spreading-Resistance for Measuring Doping-Profiles

Using a *spreading-resistance-probe* (*SRP*) to measure doping-profiles was pioneered by Mazur and Dickey,[6] who applied it to determining the thickness of dif-

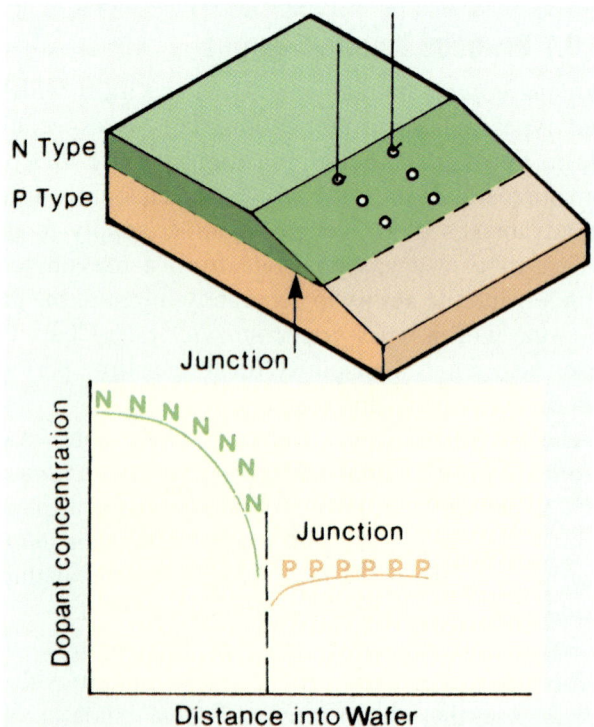

Fig. 11-16 *Spreading-resistance-profile measurement* showing the capability of measuring very-shallow doping-concentration profiles.

CHAPTER 11   DIFFUSION

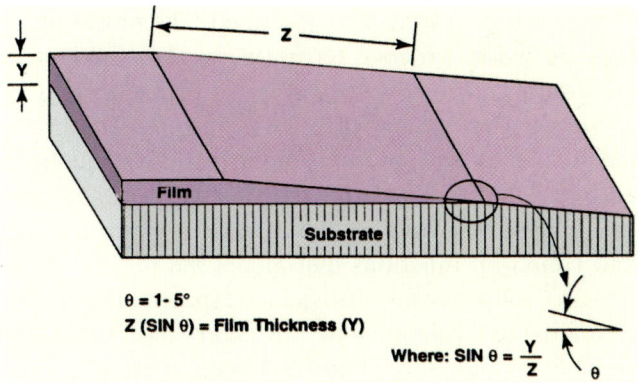

Fig. 11-17 Schematic of the *angle-lap-and-stain* junction-profiling technique.

fused and epitaxial-layers in silicon. They found that it could also reveal the impurity-profile and resistivity of very-shallow *pn*-junctions. Note, however, that SRP only measures the concentrations of *electrically-active dopants*. Other methods (such as SIMS), must be used to determine the *total-doping-concentration profile* (active + inactive dopants).

To make SRP-measurements, a wafer with the junction to be measured is carefully angle-lapped to less than 1°. Then SRP-measurements are taken

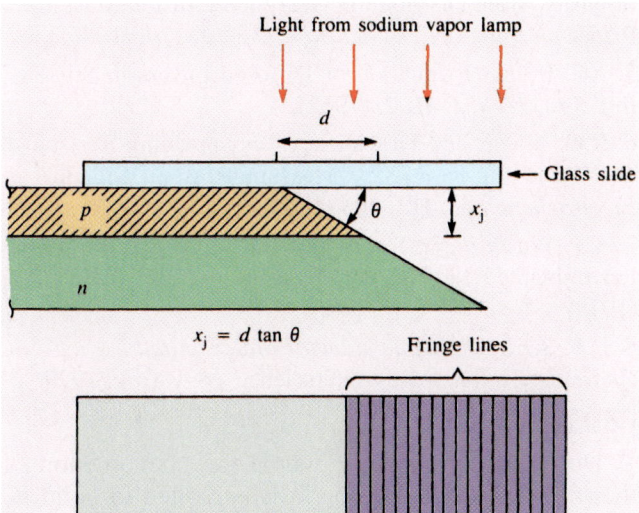

Fig. 11-18 After angle-lapping-and-staining, the interference fringe-lines are used to measure the distance *d* (which is related to the junction-depth).

along the length of the lapped-surface. That is, two carefully-aligned probes are moved in steps along the beveled-surface. A known-current is applied between the two probes, and the voltage-drop is measured across them to obtain a spreading-resistance (R_{SR} = V/I, Fig. 11-16). The resistivity ρ of the silicon at each particular-depth is related to the value of R_{SR} by:

$$R_{SR} = \rho / 4r \qquad (11.2)$$

where r is the distance between the probes.

The junction-depth can be determined from these measurements if the angle of the bevel is known. Elaborate correction-factors have also been derived which accurately allow the values of R_{SR} to be converted into carrier-concentration as a function of depth. The values of these correction-factors depend on the specific-nature of the layers. The SRP-technique, however, has several limitations: **1)** it requires a skilled and experienced operator to obtain accurate-results; **2)** it is time-consuming; and **3)** it is a destructive-measurement.

11.8 JUNCTION-DEPTH MEASUREMENTS

ULSI junction-depths, x_j, are often shallow (< 0.5-μm). In order to be useful, measurement-techniques used to determine x_j should be accurate to ±20-nm. Several methods are used to measure junction-depth, including: angle-lap-and-stain, groove-and-stain, spreading-resistance, and secondary-ion mass-spectroscopy (SIMS). The first three of these are described here, and SIMS is discussed in Ch. 25.

11.8.1 Angle-Lap-and-Stain (or Groove-and-Stain) for Junction-Depth Measurements

The *angle-lap-and-stain method* is an effective technique for measuring the junction-depth on relatively deep, highly-doped junctions, but it is accurate to only ± 0.5-μm.

The angle-lapping approach requires the grinding of a small-angular-bevel on the silicon-wafer. Small-angles (≤ 2°) of good quality must be accurately produced to obtain accurate-results (Fig. 11-17). The junction does not become visible unless it is chemi-

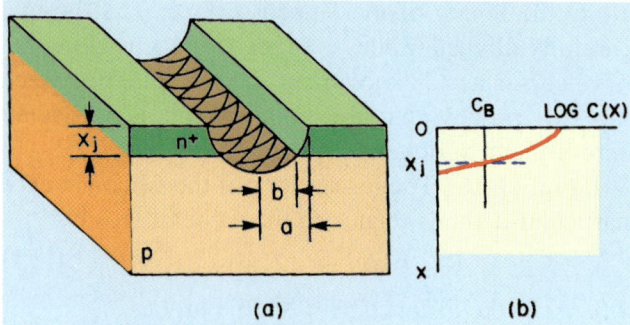

Fig. 11-19 Schematic of the groove-and-stain junction-depth measurement technique.

cally-stained. Thus, after lapping has been performed, a staining-solution is used to delineate either the n- or p-type area. Copper-sulfate ($CuSO_4$) solutions are used to stain n-type-regions, and HF and water are used to stain p-type-regions, making them appear darker. Exposing the junction to a bright light-source enhances the chemical staining-action. After the staining has occurred, the junction-depth is measured by counting interference-fringes with the aid of an optical-microscope (Fig. 11-18).

A related technique for measuring x_j is that of *groove-and-stain*. This method rapidly exposes the junction by rotating a 1-inch-diameter diamond-grit-impregnated-wheel against the wafer. This cuts a groove in the wafer-surface (Fig. 11-19). By making the groove sufficiently-deep, the junction is exposed. After grooving, the junction is stained using the same methods as in angle-lapping. Note that since both of these staining-methods do not give accurate results for junctions less than about 0.5-μm in depth, they are not as widely used as they once were.

SUMMARY

Diffusion is an important method of introducing dopants into silicon wafers. This chapter considered a number of basic aspects of this process. The concept of *diffusion in solids* was first introduced, together with the diffusion-mechanisms in silicon-crystals that occur at the atomic-scale. The techniques used to express doping-concentrations in silicon-devices (i.e., *doping profiles*) were also described. The most important diffusion process terms were also explained, including: *diffusion coefficient, diffusion-length, pre-dep-diffusion,* and *drive-in diffusion.* (The mathematical details of solutions of the diffusion equations are presented in Appendix B at the back of the book, including a derivation of the *Fick's Laws and* the *erfc* and *Gaussian* functions that model the results of the pre-dep and drive-in diffusions, respectively.) It also covered the evaluation of diffusion steps by sheet-resistance and spreading-resistance measurements, as well-as determination of the junction-depths by angle-lap-and-stain.

REFERENCES

1. R.B. Fair, "Concentration Profiles of Diffused Dopants," in F.Y. Wang, Ed. *Impurity Doping Processes in Silicon*, North-Holland, Amsterdam, 1981.

2. P.B. Griffin & J.D. Plummer, "Advanced Diffusion Models for VLSI," Solid State Technl., *May 1988,* p. 171.

3. P.M. Fahey, P.B. Griffin, and J.D. Plummer, "Defects and Dopant Diffusion in Silicon," *Reviews of Modern Physics,* **61**, 289 (1989).

4. D.P. Kennedy and R.R. O'Brien, "Analysis of the Impurity Atom Distribution Near the Diffusion Mask for a Planar *pn* Junction," *IBM J. Res. Dev.*, **9**, 179, (1965).

5. J.C. Irvin, "Evaluation of Diffused Layers in Silicon," *Bell Syst. Tech. J.* **41**, 2, (1962).

6. R.G. Mazur and P.H. Dickey, "A Spreading Resistance Technique for Resistivity Measurements on Silicon," *J. Electrochem. Soc.* **113**, 255 (1966).

7. C.P. Wu *et al.*, "Techniques for Lapping and Staining Ion-Implanted Layers," *J. Electrochem. Soc.*, **126**, 1982, (1979).

8. D.K Schroder, *Semiconductor Material and Device Characterization,* Wiley-Interscience, New York, 1990.

PROBLEMS

1. Describe in your own words the phenomenon of thermal diffusion, and how it is exploited to build IC device structures.

2. Explain the difference between a *pre-dep* and a *drive-in* diffusion.

3. Describe two evaluation measurements made after

the drive-in diffusion process.

4. Define the term *junction depth*.

5. Express Fick's First and Second Laws in words.

6. Determine the solid-solubility and the diffusion coefficient of: (a) boron at 950°C; (b) phosphorus at 1050°C; and (c) arsenic at 1100°C.

7. During predeposition, what parameter controls the concentration of dopant at the surface of the wafer?

8. Estimate the time needed to form a junction 2-μm deep by performing a constant source, solid-solubility limited boron diffusion into silicon at 900°C, 1000°C, and 1100°C, assuming a background arsenic concentration of 10^{16} cm^{-3}.

9. It is observed that the diffusion coefficient of boron is a factor of 10 greater than that of arsenic at 1150°C, and that the solid-solubility (assumed to be equal to the surface-concentration) of arsenic is a factor of 10 higher than that of boron: $D_B = 10 D_{As}$; $C_o(As) = 10\, C_o(B)$. For the same diffusion time, t:

(a) Which species (boron or arsenic), has the larger diffusion length?

(b) Which has the largest number of atoms/cm^2 diffused into the Si?

(c) What would lead to the greatest change in diffusion length: a 20% change in *time*, *temperature*, or *surface-concentration*?

10. For a boron diffusion into silicon at 1000°C, the surface concentration is maintained at 10^{19} cm^{-3} and the diffusion time is 1-hour. Find $Q(t)$ and the gradient at $x = 0$, and at a location where the dopant-concentration reaches 10^{15} cm^{-3}.

11. A solar-cell is fabricated by diffusing phosphorus from a constant surface-source of 10^{20} atoms/cm^3 into a *p*-type Si wafer with a concentration of 10^{16} B atoms/cm^3. The diffusivity of phosphorus is 10^{-12} cm^2/sec, and the diffusion time is 1-hour. How far from the surface is the junction-depth?

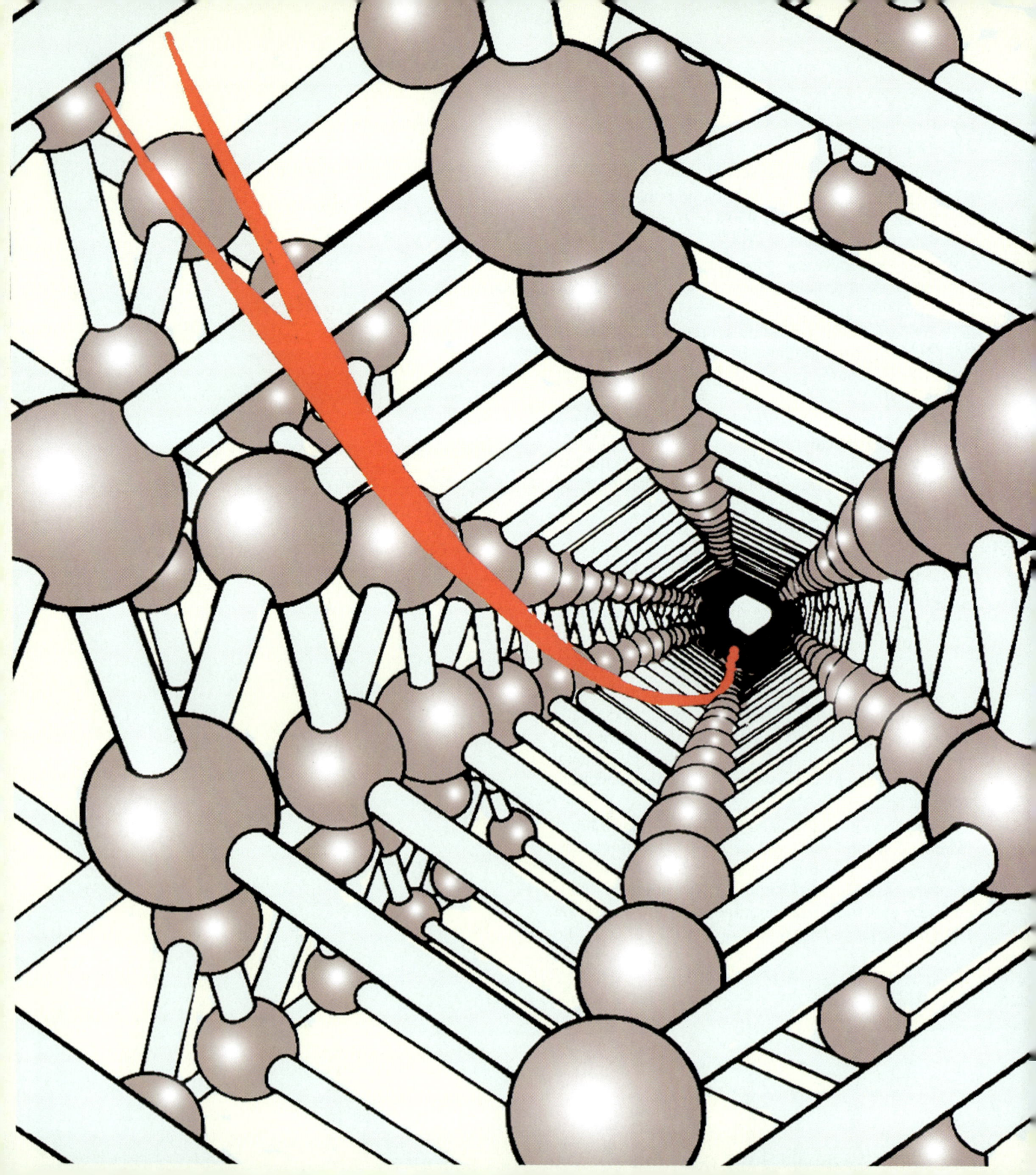

CHAPTER 12

ION IMPLANTATION

CHAPTER CONTENTS

12.1 IMPURITY-PROFILES OF IMPLANTED IONS

12.2 IMPLANTATION DAMAGE-ACCUMULATION & ANNEALING

12.3 ION-IMPLANTATION EQUIPMENT

12.4 CHARACTERIZING IMPLANTS

12.5 ION-IMPLANTATION PROCESS APPLICATIONS
  Masking Layer Materials and Thickness
  Threshold-Voltage Control in MOSFETs
  Shallow Junction Formation by Ion Implantation
  High-Energy Implantation

"The trouble with doing something right the first time, is that nobody appreciates how difficult it was."

Anon.

Model for a diamond lattice, viewed along a <110>-axis, with an ion whistling down one of the channels.

Ion implantation is a process in which energetic, charged-atoms (or molecules) are directly introduced into a substrate (Fig. 12-1b). In IC-fabrication, ion implantation is primarily used to add dopant-ions (most often selectively) into the surface of silicon-wafers.[1,2] Ion-energies for CMOS doping-applications range between 0.2-keV (for *shallow source/drain junctions*) to ≈2-MeV (for *retrograde-well doping*). The ions with such energies attain a high-velocity (on the order of 10^7-cm/sec). As described below, ion-implantation has replaced diffusion for almost all ULSI doping processes.

Ion implantation offers the following advantages:

1. Ionic-species can be implanted with high-accuracy (over many orders-of-magnitude of doping levels), whereas dopant-control by chemical-diffusion is far less exact.

2. The desired depth-profiles of the implanted-species can be obtained by controlling *ion-energy* and *channeling-effects*.

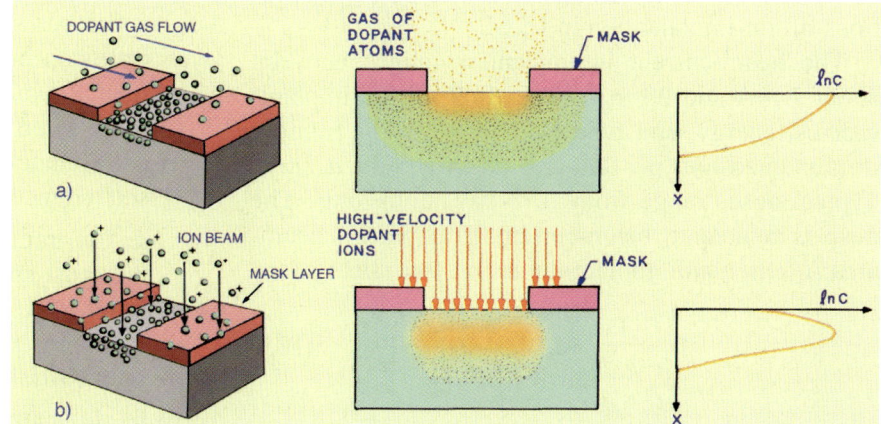

Fig. 12-1 Comparison of: (a) *diffusion*; and (b) *ion-implantation*, as techniques for the selective-introduction of dopants into a semiconductor-substrate.

189

Table 12-1 ION IMPLANTATION APPLICATIONS IN ULSI CMOS

Junction Formation
(Used to fabricate both MOS and bipolar devices)

CMOS Fabrication
- Threshold Voltage Control/Adjustment
- Channel-Stop Implantation
- Source/Drain Formation
- Well Formation
- Punchthrough Stopper Implantation
- Graded Source and Drain Formation

Bipolar Fabrication
- Predeposition
- Base Formation Implantation
- Arsenic-Implanted Poly-Si Emitter
- High Value Resistor Formation Implantation

Formation of Silicon-on-Insulator Materials
- High Dose Oxygen Implantation (SIMOX - see Ch.17)
- High Dose Hydrogen Implantation (Smart-Cut - see Ch.17)

Miscellaneous Process Applications
- Backside Damage Layer Formation for Gettering (see Chap. 9)
- High-Energy Implantation to Form Buried Layers for Gettering
- Ion Beam Mixing to Promote Silicidation Reactions
- Buried Insulator Layer Formation
- Nitrogen Implantation into Poly-Si Gates to Reduce Boron Diffusion
- Polysilicon Resistor Formation
- High Energy Implantation to Form Buried Collectors
- Emitter Formation Implantation

---

3. By using photoresist as the masking-material, dopants can be introduced into selected regions of the wafer-surface at near-room temperatures.

4. Both *p*-type and *n*-type dopants can be implanted into silicon-wafers.

5. The crystalline-structure of implant-damaged silicon can be restored through subsequent thermal-annealing cycles.

Table 10-1 lists some of the most important applications of ion-implantation in ULSI.

The beam-current in implanters ranges between about 1-mA and 30-mA, depending on the implant-species, energy, and type of implanter. The number of implanted-ions per unit area, is termed the *dose*, ϕ. Typical doses range from 10^{11}-10^{16}-atoms/cm^2. The dose is related to beam-current I (amperes), beam-area A (cm^2), and implant-duration t (sec) by:

$$\phi = \frac{I\,t}{q_i\,A} \qquad (12.1)$$

where q_i is the charge per ion (normally equal to one *electronic-charge* = 1.6×10^{-19} coulomb).

Ion implantation also has limitations, including:

1. It disorders the crystalline-structure of Si-wafers, creating such damage as crystalline-defects and amorphous-layers. To restore wafers to their pre-implanted condition, thermal-processing after implantation must be performed. In some cases, not all implantation-damage can be healed.

2. The maximum implant-depth using conventional implanters is relatively shallow (≈ 1.0-μm).

3. While the lateral-distribution of implanted-species is smaller than the lateral-movement in a diffusion-process, it is not zero. This is a fundamental limiting-factor in fabricating some minimum-sized device-structures (such as the electrical channel-length between source and drain in self-aligned MOSFETs).

4. Ion-implanters are complex machines, among the most-sophisticated systems in a wafer-fab. In order to be effectively utilized, they must be conscientiously operated, monitored, and maintained by well-trained personnel.

5. Ion-implantation equipment contains such potential safety-hazards as high-voltage, radiation, and toxic-gases.

12.1 IMPURITY-PROFILES OF IMPLANTED IONS

To select the appropriate implant-doses and energies for particular device-applications, it is necessary to know where the implanted atoms are located after implantation (i.e., it must be possible to predict the *depth-distribution*, or *impurity-profile* of the as-implanted atoms). Here we discuss how this is done.

12.1.1 Ion-Range

As energetic-ions penetrate a solid target-material, they collide with target-atoms (which slows them down), and eventually they come to rest. The total distance an ion travels in the target is termed the *range*, R. As a result of the collisions between the ions and the nuclei of the target, this trajectory is a zig-zag path. But, what is of greater interest than R, is the *projection* of this range on the direction parallel with the incident-beam (since this represents the penetration-depth of the implanted-ions along the implantation direction). This quantity is called the *projected-range*, R_p (Fig. 12-2a). Note that the higher the energy of an ion, the deeper it can penetrate into the substrate. Thus, the greater is the value of R_p. Since the number of collisions and the energy-lost per collision by the penetrating-ion are random-variables, ions having the same initial-energy and mass, will end up spatially-distributed in the target.

That is, some ions undergo fewer scattering events than the average, and come to rest more deeply. Others suffer more collisions and come to rest closer to the surface. When a large numbers of ions are implanted,

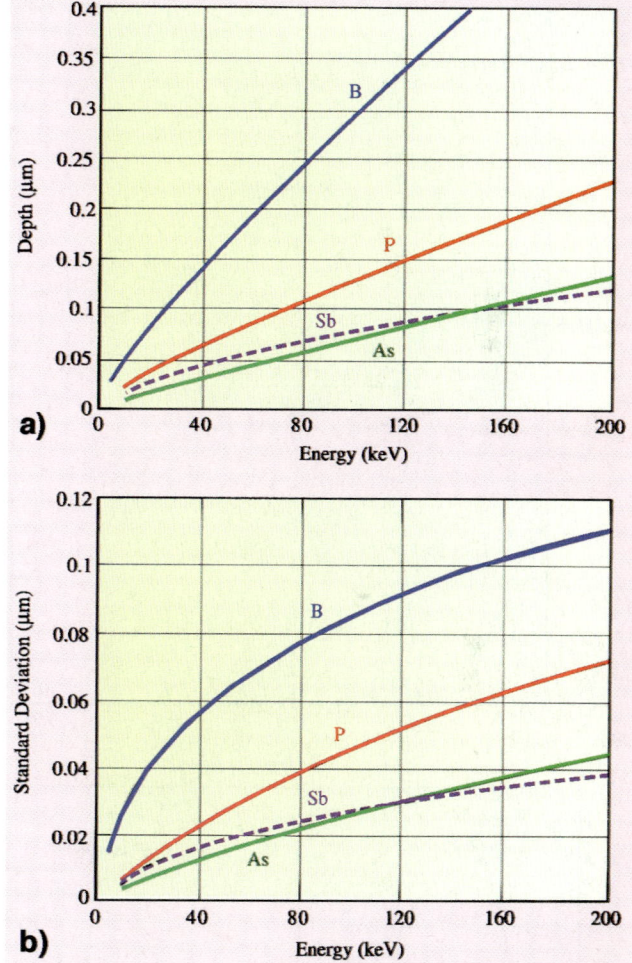

Fig. 12-3 Graphs of: (a) the *projected-range*, R_p; and (b) *straggle*, ΔR_p, for B, P, As and Sb in amorphous-silicon.[3]

R_p corresponds to the depth at which most stop, and it is the distance at which the profile has it's maximum value. The value of R_p is an important-parameter because it dictates the energy that a specific dopant must be given to produce the desired junction-depth.

The values of R_p for various dopants in silicon have been determined as a function of implant-energy. This data is shown in Fig. 12-3a. That is, Fig. 12-3a illustrates the *projected-range* as a function of *implantation-energy* for B, P, As and Sb in silicon. It can be seen that lighter-ions penetrate deepest into the wafer, and the heavier the ion, the smaller the value of R_p.

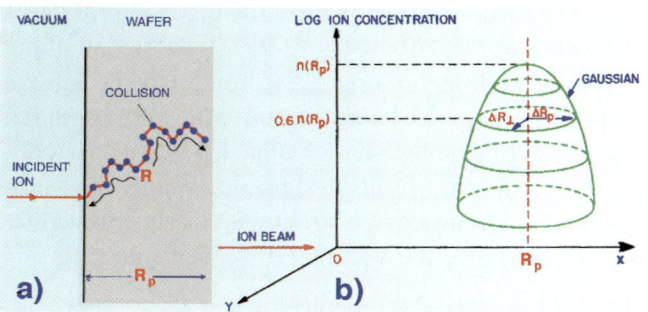

Fig. 12-2 (a) Schematic of the *ion-range*, R, and *projected-range*, R_p. (b) Two-dimensional distribution of the implanted atoms.

MICROCHIP MANUFACTURING

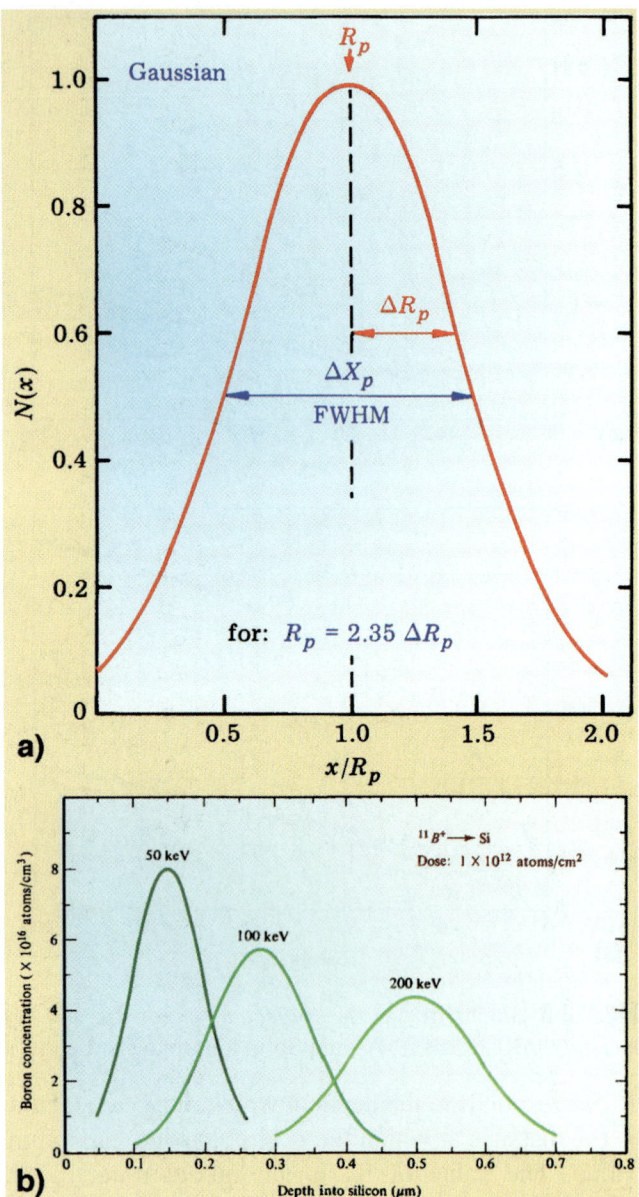

Fig. 12-4 (a) Gaussian-curve. (b) Calculated Gaussian profiles of boron, implanted into silicon using the range parameters of Fig. 12-3 for different energies.

The mathematical shape of the doping-profile of an ion-implantation process is a *Gaussian-curve* (also known as a *normal-distribution*), as shown in Fig. 12-4 (see Appendix-C for mathematical details of the Gaussian curve). The junction between the implanted

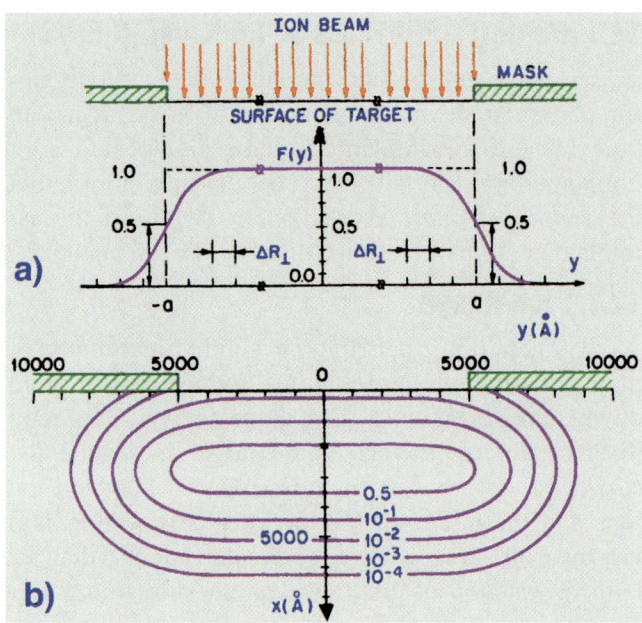

Fig. 12-5 Illustration of lateral implant-profiles. (a) Ion concentration along the lateral direction (y) for a mask of dimension 2**a**, with **a** $\gg \Delta R_\perp$ and infinite extension in the x-direction. (b) Contours of equal-ion concentrations for 70-keV B^+ (R_p = 2710Å, ΔR_p = 824Å, and $\Delta R_\perp$ = 1006Å) incident into silicon through a 1-μm slit.

ions and the wafer-doping occurs where the ion-concentration equals the bulk doping-concentration.

The statistical-fluctuation along the direction of the projected-range is given by the quantity known as the *projected-straggle*, ΔR_p (Fig. 12-3b). The ions are also scattered to some degree along the direction perpendicular to the incident direction, and the statistical-fluctuation along this direction is called the *projected-lateral-straggle*, $\Delta R_\perp$ (Fig. 12-3b). Information about the projected-lateral-straggle is important because it tells the extent of lateral-penetration of ions under the edges of implant-masks (Fig. 12-5). Such lateral-penetration may limit design-dimensions in some integrated-circuit device structures. In general, values of ΔR_p and $\Delta R_\perp$ are within ±20% of one another.

12.1.2 Ion-Stopping Mechanisms

As an ion travels through a solid-target, it loses energy by collisions with the target nuclei (*nuclear collisions*,

Fig. 12-6) and by interaction with the negatively charged electrons in the target material (Fig. 12-17a). As noted earlier, such *energy-losses* gradually slow-down an ion, eventually bringing it to a stop.

This ion stopping-process can be compared to firing cannonballs into a muddy river-bank, with stones about the same mass as the cannonballs also being embedded in the mud (Fig. 12-7b). The mud by itself will gradually decelerate a cannonball until it comes rest after travelling a straight line in the mud. However, if a cannonball hits a stone, it can get deflected, and its path in the mud will not be straight. The stones can also be displaced from their original positions when they are hit by a cannonball. The path that each cannonball takes and its final resting place will be a somewhat random event. To predict the outcome of a large number of such events requires the use of *statistical-methods*. Figure 12-7c shows the result of such statistical-calculations used to determine the locations where ions (phosphorus) come rest after having been implanted into silicon. This explains why the profile of the implanted-ions is spread out with a Gaussian distribution, as described previously.

12.1.3 Ion-Implantation Into Single Crystals: Channeling

In semiconductor manufacturing, ion implantation is carried out by directing an ion-beam into single-crys-

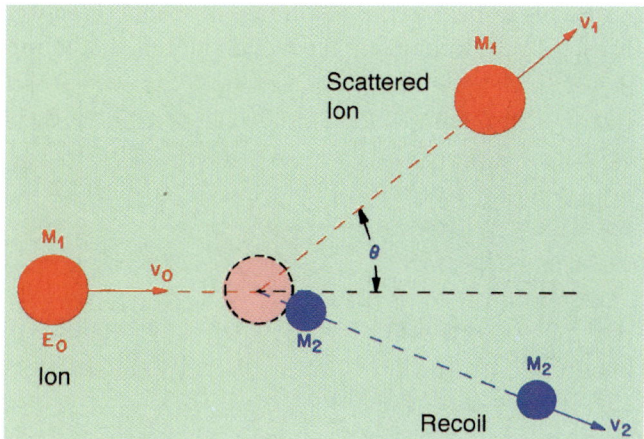

Fig. 12-6 Collision between two particles treated as if they are hard spheres.

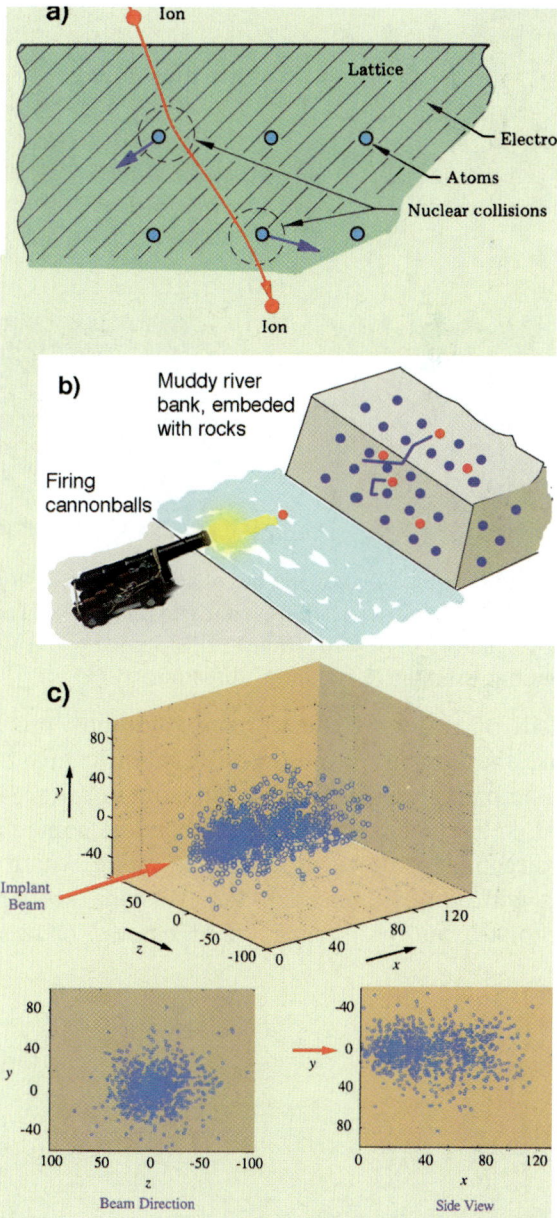

Fig. 12-7 (a) Schematic of nuclear- and electronic-stopping mechanisms in ion-implantation. (b) Ion-implantation stopping mechanisms - using an analogy of firing cannonballs into a muddy river-bank that also has rocks about the size of the cannonballs embedded in it. (c) Monte-Carlo simulation of the 3D-distribution of 1000 phosphorus-ions, implanted at 35-keV. The 2-D projection (side-view) and top-view are also shown. The axes are in units of nm.[10] Reprinted with permission of Prentice-Hall.

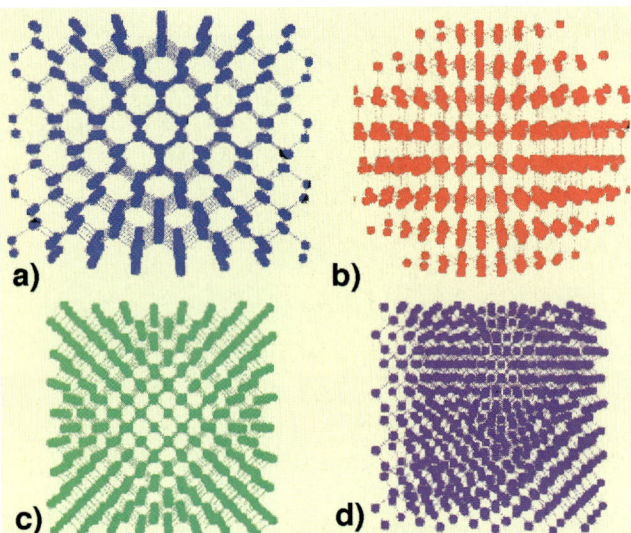

Fig. 12-8 Ball-model showing the relative degree of "openness" of the Si-lattice when traversing in the: (a) <111>; (b) <100>; and (c) <110> directions. (d) Si-lattice with tilt and rotation to simulate a "random" direction.

tal silicon-wafers. Under certain conditions, this can cause the implant-profile to differ significantly from the simple Gaussian-distribution shown in Fig. 12-4. That is, in single-crystal lattices there are some crystal-directions (known as *channels*) along which the ions will not encounter any target-nuclei (see Chap. Faceplate), and will instead be *channeled*, or *steered* along such open-channels of the lattice. Figure 12-8 illustrates the most likely channeling-directions. They are (in order), <110>, <111>, and <100>. These are the directions that exhibit channels of decreasing "openness." As the implanted-atoms travel along channels, the slowing-down is accomplished mainly by electronic-stopping, and the ions can penetrate the lattice much deeper than in amorphous-targets (Fig. 12-9). At first, this might appear to offer the advantages of being able to produce deeper implanted-junctions (and less lattice-damage). It turns out, however, that the large sensitivity to incident-beam-direction, and the unpredictable effects of *de-channeling* (which cause anomalous profile-tails), make channeling-effects difficult to control.

Instead, techniques have been developed to prevent (or at least minimize) channeling-effects. The most widely-adopted procedure is to *tilt* the wafer surface relative to the incident beam direction (most commonly by ~7°), so that the lattice presents a dense orientation to the incident-beam (i.e., the approximate <763> direction). This approach *reduces* (but does not entirely eliminate) channeling because even when atoms enter the lattice in an apparently random, "dense-appearing" direction, they can be diverted into one of the open-channels after entering the lattice (Fig. 12-10). Thus, in addition to tilting the wafers, they must also be oriented with an appropriate *twist* with respect to the incoming ion-beam direction (Figs. 12-8d). Implanters have end-stations that allow

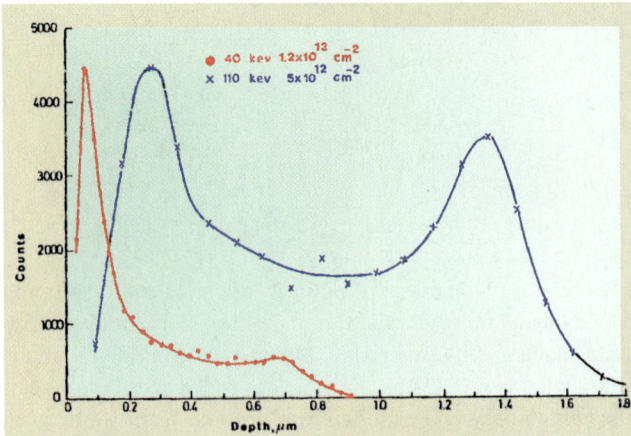

Fig. 12-9 Variation of P^{32}-concentration-profiles with ion energy for implantation along <110> axis, plotted on a linear-scale. By permission of Canadian Jnl. of Physics.

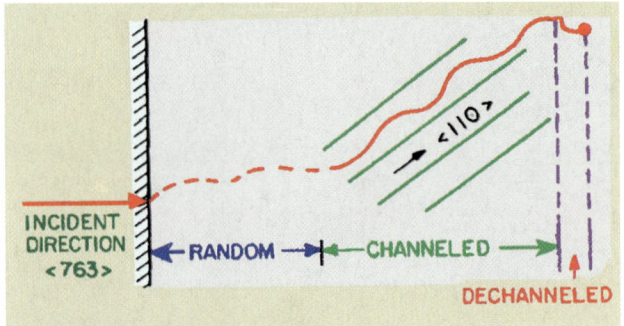

Fig. 12-10 Schematic diagram of an ion-path in a single-crystal for an ion incident in a "dense" <763>-direction. The path shown has non-channeled and channeled behavior.

CHAPTER 12   ION IMPLANTATION   195

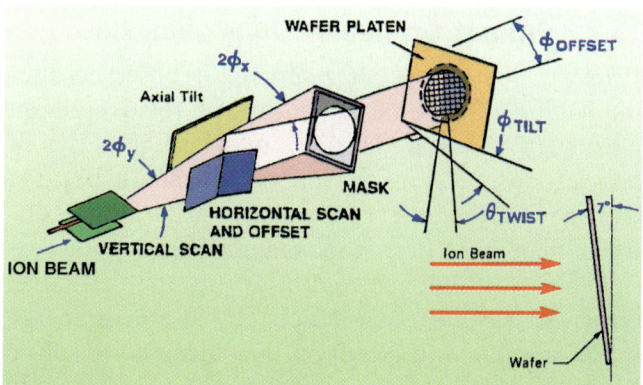

Fig. 12-11 Channeling-avoidance by using *axial-tilt* and *azimuthal-twist*.

control of tilt-and twist-angles (Fig. 12-11).

Other methods to further reduce channeling-effects (see Fig. 12-12) include: **a)** implanting at a non-90° angle to the Si-surface; **b)** pre-amorphizing the lattice by a prior-implantation (e.g., with Si or Ge); and **c)** implanting through a surface-oxide (to randomize the directions of the ions as they enter the lattice). Details of these approaches are found in Ref. 1.

12.2 ION-IMPLANTATION DAMAGE-ACCUMULATION AND ANNEALING IN SILICON

One drawback of ion-implantation is that it damages the substrate material. That is, when high-energy ions collide with substrate-atoms they knock them from their lattice-sites in large numbers. Furthermore, only a small percentage of the as-implanted atoms end-up on electrically-active lattice sites.

In order to successfully fabricate devices, the damaged substrate-regions must be restored to their pre-implanted condition *and* the implanted-species must be electrically-activated. In this section we describe methods for healing such damage, and activating the implanted-dopants.

12.2.1 Implantation-Damage in Silicon

When energetic-ions strike a silicon-substrate they lose their energy in a series of nuclear and electronic collisions, and come to rest some hundreds of atomic-layers below the surface. Only the nuclear-collisions result in *displaced silicon-atoms* (a condition also referred to as *damage* or *disorder*). Different types of damage can occur from these events, including:

Type-1. *Isolated point-defects* or *point-defect clusters* are created in essentially crystalline-silicon. This type of damage results when light-ions are implanted (Fig. 12-13a).

Type-2. *Local-zones of completely-amorphous material* are produced in an otherwise crystalline layer. An *amorphous-region* is defined as a region

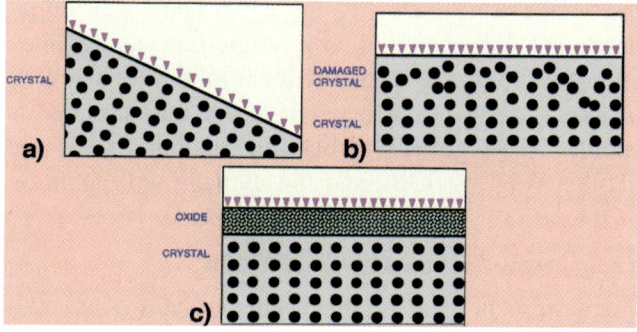

Fig. 12-12 Techniques developed to suppress channeling during ion-implantation: a) implanting at a non-90°-angle to the silicon-surface; b) performing a prior implantation which pre-amorphizes the surface-silicon region; c) implanting through a surface-oxide (to randomize the directions of the ions as they enter the lattice).

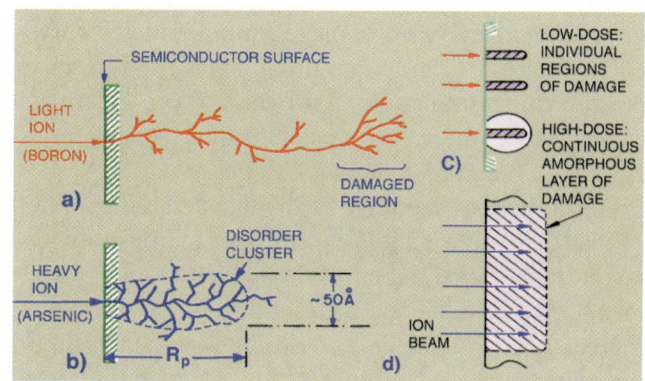

Fig. 12-13 Schematic representation of the disorder produced by ion implantation: (a) Low-dose and light-ions - individual regions with degree of disorder increasing as ions penetrate deeper into substrate; (b) and (c) Low-dose and heavy ions - individual regions of more uniform disorder along the entire ion trajectory. (d) Heavy-doses - formation of an amorphous layer.

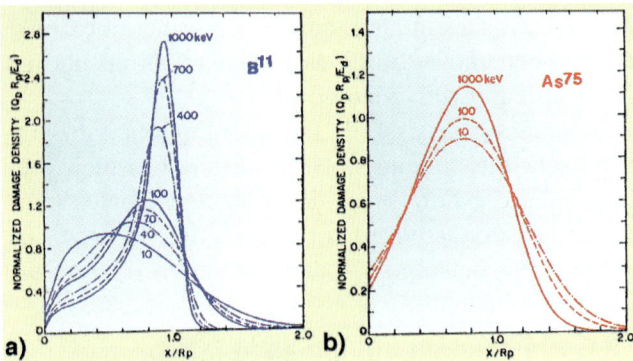

Fig. 12-14 Calculated damage-density profiles of: (a) boron; and (b) arsenic.

in which the displaced-atoms per unit volume approach the atomic-density of the semiconductor. Local-zones of amorphous-damage are associated with low-dose implants of heavy-ions (Figs. 12-13b and 12-13c).

Type-3. *Continuous-amorphous-layers* form as damage from the ions accumulates. As the dose (typically of heavy-ions) increases, the local-amorphous-regions eventually overlap, forming a continuous-amorphous-layer (Fig. 12-13d).

In our discussion, Type-**1** and Type-**2** damage are grouped into the category of *primary-crystalline-defect damage*. Type-**3** damage is referred to as *amorphous-layer damage*. Such grouping is used because the annealing-strategy for Type-**1** and Type-**2** damage is the same, but a different annealing procedure is employed for Type-**3** damage.

Regardless of the form of the damage, the number of displaced-atoms after an implant is almost always larger than the number of implanted-atoms. These displaced-atoms reduce the mobility in the damaged-regions and produce defect-levels in the band-gap of the material (i.e., deep-level-traps, for both electrons and holes) which have a strong tendency to capture free-carriers from the both the conduction-band and valence-band. Due to the relatively large-number of displaced-atoms (and the fact that few *as-implanted* impurities occupy substitutional-sites), non-annealed implanted-regions exhibit high-resistivity.

12.2.2 Primary Crystalline-Defect Damage

Primary crystalline-defect damage is observed after implanting a Si-crystal (28-amu) with relatively-light ions, such as boron (11-amu), or with light-doses of heavier-ions, such as phosphorus (31-amu). But, the damage configurations from light-ions are quite different than those from implantation of heavy-ions.

When fast-moving *light-ions* (in this example, boron) first enter the Si-lattice they lose their energy by interaction with electrons. This does not displace Si-atoms from their sites. However, as B-ions penetrate deeper into the lattice, their energy is steadily reduced, and eventually they are slowed-down enough so that nuclear-stopping predominates (at about an energy of 10-keV for B). As shown in Fig. 12-13a, most lattice-damage occurs in the part of the light-ion trajectory beyond that point. Furthermore, each ion produces a trail of well-separated primary recoiled-Si-atoms in the wake of the implanted-ion. The damage from boron implantations is thus characterized by *primary crystalline-defects*. Damage-density is distributed versus depth as shown in Fig. 12-14a, which shows a sharp buried-peak-concentration and qualitatively fits the description of the damage-creation process.

However, when *heavy-ions* are implanted, the energy-loss is predominantly due to nuclear-collisions over the entire-range of energies experienced by the decelerating-ions (Fig. 12-14b). Thus, substantial damage is expected. Even a single-ion can produce enough damage to cause a local-amorphous region in the silicon-solid. However, if the implant-dose is small, these regions will be well-separated from one another (Fig. 12-13c), and the damage will again be of the *primary-crystalline-defect* kind.

12.2.3 Amorphous-Layer Damage

Bombardment by a larger number of heavy-ions will lead to the formation of *continuous amorphous-layers*. That is, heavy lattice-damage accumulates with ion-dose through an increase in the density of localized amorphous-regions. Eventually these regions overlap, and a continuous amorphous-layer is formed. The evolution of a continuous amorphous-layer from

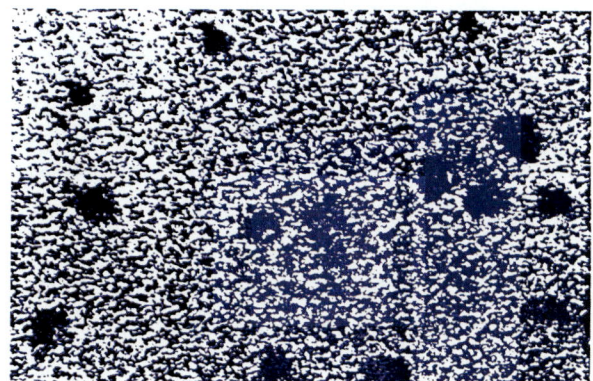

Fig. 12-15 Bright-field electron micrographs of amorphous regions in Si produced by bombardment with doses of 3×10^{11} cm^{-2}, using 10-keV bismuth ions.

the accumulation and overlap of damage formed by individual-atoms has been observed using TEM. In Fig. 12-15 it can be seen that the individual 10-keV Bi$^+$ ions create localized amorphous-zones in Si. As the dose is increased, a continuous amorphous-layer extends to the surface. The amorphization begins at the depth of the maximum nuclear-collision energy-deposition (at slightly less than the projected-range) and spreads towards the surface. In addition, primary crystalline-damage exists in the silicon immediately beneath the amorphous-layer.

There is a minimum (or *threshold*) dose required to convert a crystalline-material to an amorphous-layer. This depends on the type of ion being implanted as well as its energy. For 100-keV As-ions, this threshold dose is 6×10^{13} As-ions/cm^2.

12.2.4 Electrical-Activation and Dopant-Diffusion

In this section the electrical-activation of the implanted impurity atoms is described. Also covered is the diffusion of such impurities during the thermal-activation steps. Additional information about the annealing of implant-damage can be found in Ref. 1.

Electrical-Activation of Implanted Impurities: Since most as-implanted impurities do not occupy substitutional-sites, a subsequent thermal-step is employed to bring about electrical-activation. The degree to which the thermal procedure is effective in electrically-activating impurities, is determined either by *Hall-effect measurements* or from the *sheet-resistance*, R_S.

Electrical-activation of implanted impurities in amorphous-layers proceeds differently than in layers with primary crystalline-damage. Electrical-activation in amorphous-layers occurs as the impurities in the layer are incorporated onto lattice-sites during recrystallization.

Electrical-activation in primary crystalline-damaged regions is more complicated. Figure 12-16 shows the electrical-activation behavior of implanted-boron versus temperature. In this curve, the measured *surface-carrier-concentration* is used to indicate *the degree-of-activation*. That is, full-activation is reached when (p_{Hall}/ϕ) on this curve reaches 1.

We see that in the temperature-range up to 500°C (Region-**1** of Fig. 12-16) there is a steady increase in electrical-activation. This is because at such tem-

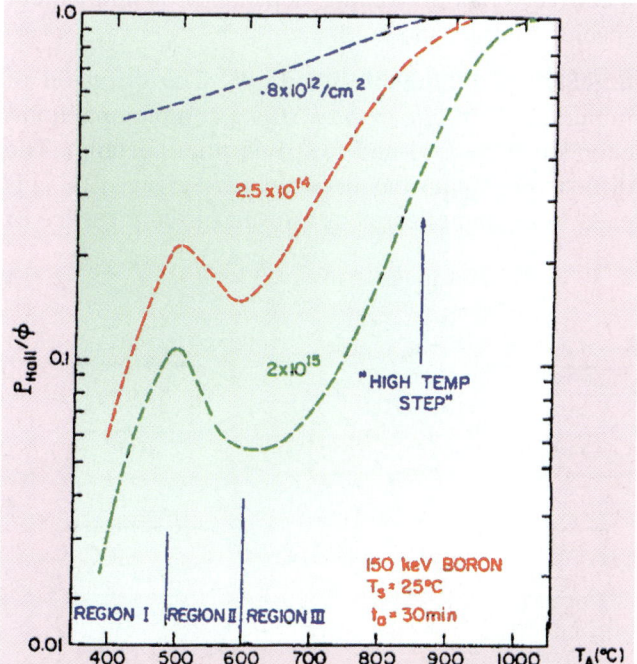

Fig. 12-16 Graph of the fractional-activation p_{Hall}/ϕ of a boron-implant as a function of anneal-temperature, showing a region of reverse-activation. *Isochronal-anneals* (30-min at each temperature) were used in these experiments.

peratures, trapping-defects are annealed out (i.e., they disappear). This causes a large increase in free-carrier concentration (as carriers are released from these traps). In Region-2 (500–600°C) activation is observed to *decrease* because the concentration of substitutional-boron declines. This occurs because dislocation-defects in the silicon are formed at these temperatures. Some boron-atoms that were already on substitutional-sites precipitate on dislocations, becoming *electrically-inactive*. In Region-3 (>600°C) electrical-activation again increases. Full-activation is achieved at temperatures ~800-1000°C. The higher the dose, the more the disorder, and thus the higher the final-temperature needed for full-activation.

Activation of implanted-impurities by *rapid-thermal-processing* (RTP) has been found to be an effective method for electrically-activating impurities. The time-temperature cycle to reach minimum sheet-resistance for As, P, and B is ~5-10-sec at 1000–1200°C, the exact condition being dependent on implanted-species, energy, and dose.

Diffusion of Implanted-Impurities: The diffusion of impurities in *single-crystal Si* is a complex phenomenon (see Chap. 11 and Ref. 1 for more details). The diffusion of impurities in *implanted-Si* (Fig. 12-17) is even more complicated as a result of the presence of

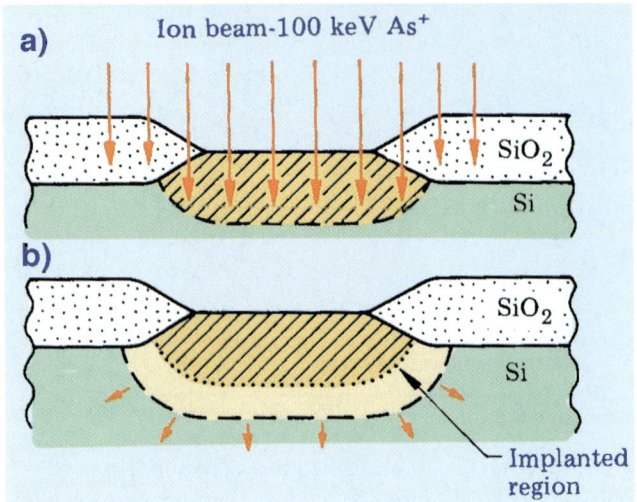

Fig. 12-17 (a) *Predeposition* using ion-implantation. (b) After *drive-in diffusion*.

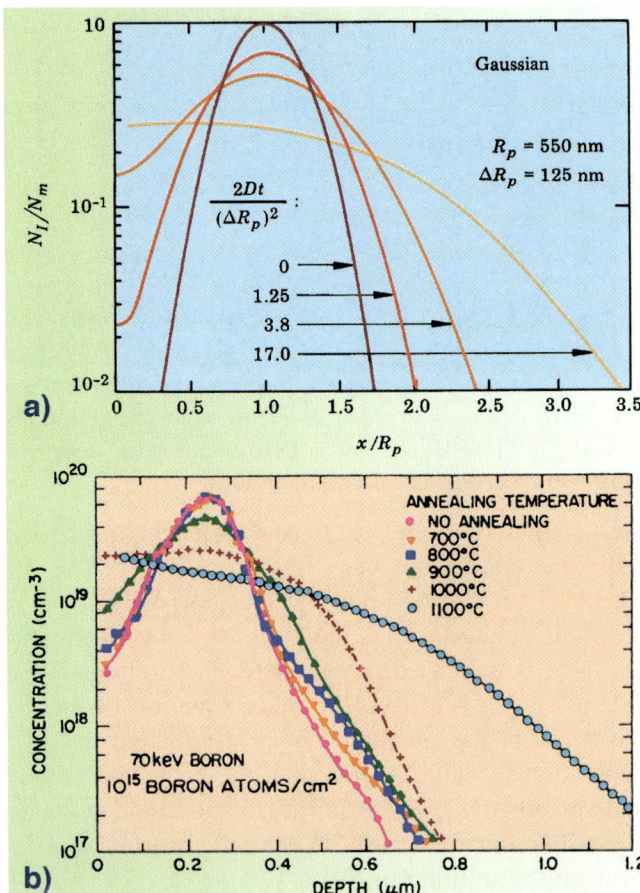

Fig. 12-18 Boron-concentration as a function of annealing at various temperatures for 35 minutes: (a) Calculated; (b) Actual experimental data.

implantation-damage. As an example, studies of the diffusion of boron in implanted single-crystal Si indicate that at high-temperatures (≥1000°C), the results appear to obey ordinary diffusion-theory (see Fig. 12-18, in which part-**a** shows the calculated-diffusion based on a Gaussian implant-distribution, and part-**b** shows actual experimental-data). But, at lower temperatures, the diffusion behavior is not correctly predicted. That is, at 900°C the predicted boron-profile fits the experimental-data only if a diffusion-constant is used that is about three times the value observed under a chemical 900°C-diffusion. In addition, at 700–800°C, the *depth of the profile-peak* remains fixed, but the *concentration in the tail* is much deeper than is predicted from normal boron-diffusion-constants. This increase in the boron-diffusion (by as much as 4x) in the tail of implanted-profiles occurs during the initial-stage of the thermal-anneal, but slows to the expected diffusion-rates for the rest of the anneal. Consequently, this phenomenon is called *transient-enhanced-diffusion* (or *TED*). TED is driven by the *extra-defects* and *Si interstitial-atoms* that are present after the implant, but which are rapidly annihilated during the first-stages of thermal anneals. More details about this complex (but increasingly-serious problem in ULSI fabrication) are found in Refs. 1 and 2.

In ULSI-fabrication, RTP is used to anneal implants with minimal impurity-redistribution. RTP cycles of ~1000°C for 10-sec can activate implanted-layers as effectively as 30-minute furnace-anneals at 1000°C (Fig. 12-19), but with impurity-redistribution-distances of only a few-hundred angstroms (compared to several-thousand angstroms for furnace-anneals). RTP-cycles for shallow-junction annealing often use *spike-anneal* temperature profiles, where a fast temperature-ramp-up is followed immediately by a cool-down after the anneal-temperature is reached.

12.3 ION-IMPLANTATION EQUIPMENT

Ion implanters are large, high-performance tools. In

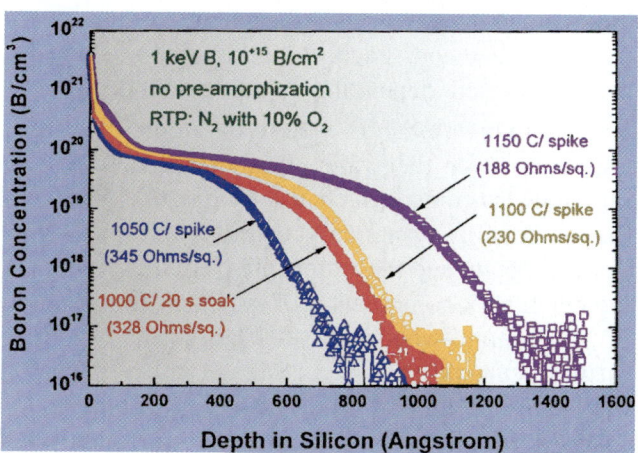

Fig. 12-19 Atomic-profiles for 1-keV B-implants after an anneal at 1000°C for 20-sec, or at *spike-anneals* at 1050°C, 1100°C, and 1150°C, showing transient-enhanced diffusion (TED).

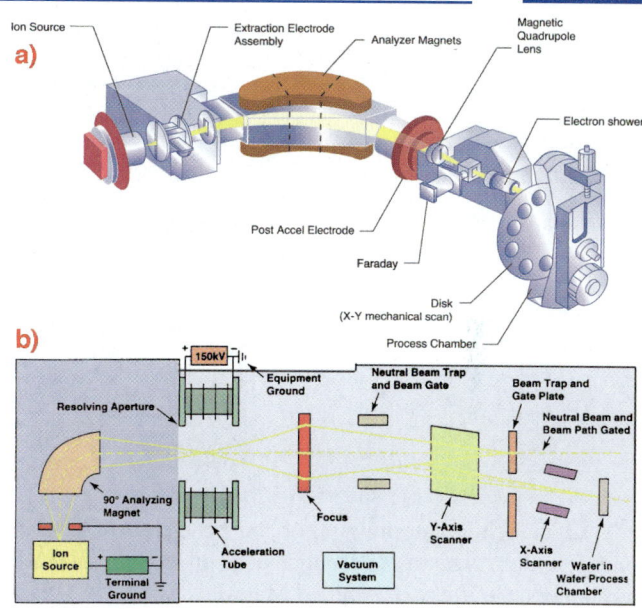

Fig. 12-20 Configuration of a typical medium-current implanter: (a) perspective view (Courtesy of Axcelis Technologies); (b) top-down view.

fact they are among the most complex systems used in ULSI-fabrication. As is shown in Fig. 12-20 they contain the following subsystems:

1. A *feed-source* that contains the species to be implanted. The ions most commonly implanted into silicon are As^+, P^+, and B^+, but the elements As, P, and B are solids at room-temperature. To allow these elements to be fed to the implanter in gaseous-form, they are supplied as part of the gaseous hydrides (AsH_3, PH_3, and BF_3, respectively). These (highly-toxic) gases are stored at high-pressure in small gas-cylinders. The risk associated with inadvertent gas-releases is reduced by limiting the cylinder-size and by diluting these gases with hydrogen. Recent advances in toxic-gas storage-techniques have resulted in the development of a so-called *Safe Delivery System* (SDS, see Chap. 29), which further reduces the hazard of using of these toxic-gases.

2. An *ion-source* is employed to ionize the feed-gases. An arc-discharge plasma (produced at pressures of ~10^{-3} torr in the arc-chamber) performs

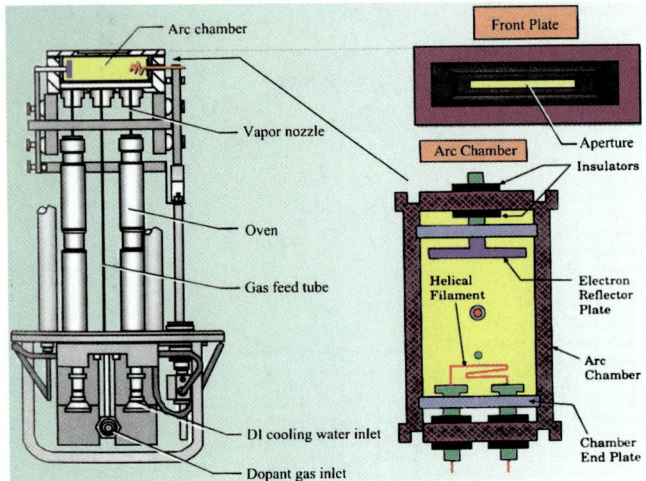

Fig. 12-21 Schematic drawing of the *Bernas arc-chamber*, one common type of ion-source used in commercial ion implanters. Courtesy of Applied Materials

the ionization through the collision of high-energy electrons (present in the arc-discharge) with neutral gas-atoms. The ion-source is provided with its own power-supply and vacuum-pump (typically a turbo-molecular-pump). Figure 12-21 shows one type of ion-source used in commercial ion implanters.

3. An *ion-extraction* and *analyzing-device* (Fig. 12-22) performs the task of selecting only the desired ion-species according to their mass (and rejecting all others). In a typical implanter, ions are

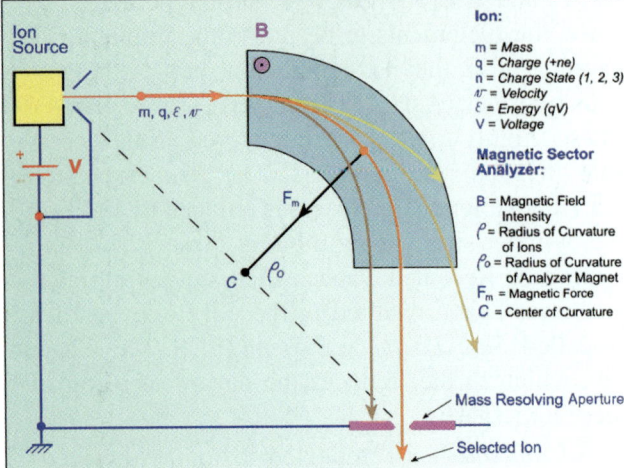

Fig. 12-22 Mass-analyzer of an ion implanter. Courtesy of T.C. Smith.

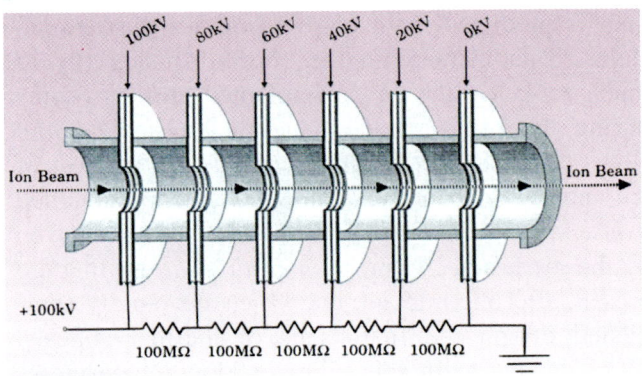

Fig. 12-23 Acceleration-column of an ion implanter.

first *extracted* from the ion-source with voltages of 15-40 kV, and then are *analyzed*. The extracted ion-beam must be spatially-separated into several beams according to their mass and charge-state, because the ion-beam as extracted from the source is a mixture of different types of ionized-molecules and atoms of the source-feed material. By adjusting the magnetic-field-strength, only the ionic-species of interest will be given the radius-of-curvature that allows it to pass through the resolving-slit (aperture) and into the acceleration-tube.

4. An *acceleration-tube* (and its power-supplies) creates the acceleration-field needed to increase the ion-energy to the desired value (Fig. 12-23). The ion-beam is focused in the acceleration-tube to a particular size and shape. Round or ribbon-shaped beams are used, depending on equipment design.

5. A *scanning-system* is used to distribute the ions uniformly over the wafer. In most high-current-implanter systems (which use beams of >5-mA), the ion-beam is a stationary, cylindrical (or ribbon-shaped) beam, and the wafers are passed in front of the beam by a combination of mechanical motions of a spinning-wheel (or disk, Fig. 12-24). In medium-current-implanter designs (which employ ion-beams of <2-mA), the ion-beam is typically scanned in one or two dimensions with either a stationary-wafer (for two-directional beam-scanning), or with a combination of a moving-wafer and a scanning-beam (Fig. 12-25).

Fig. 12-24 Spinning-wheel disk of a *batch* ion-implanter. Courtesy of Axcelis Technologies.

6. A *system end-station* includes a Faraday-cup and current-integrator (which directly measures the implant-dose by collecting the beam-current and integrating it over the implant-time), together with a subsystem that loads, holds, and positions the wafers (Fig. 12-26). For implantation systems where the ion-beam is stationary, the end-station also contains a mechanism for scanning one or more wafers in front of the beam. Medium-current implanters use combinations of scanned-beams and moving-wafers to distribute a uniform-dose of ions across the wafer-surface.

Beam-current measurements (used to control the dose of implanted-ions), are made with *Faraday cups*. They contain electrostatic or magnetic fields at the cup-entrance to prevent background-electrons and ions from entering the cup (and thus interfering with the ion-beam-current measurement).

7. A *high-vacuum pumping-system* serves to evacuate the magnetic-analysis area, the acceleration-column, and the end-station to a background pressure of $<10^{-6}$-torr. Cryopumps are commonly used in modern beamline and end-station regions.

8. A *computer control-system* provides recipe-driven operation of the implanter. Linkage between the implanter control-system and the fab host-computer-network, is used to track wafer-lots, specify implant process-conditions, and forward run-sta-

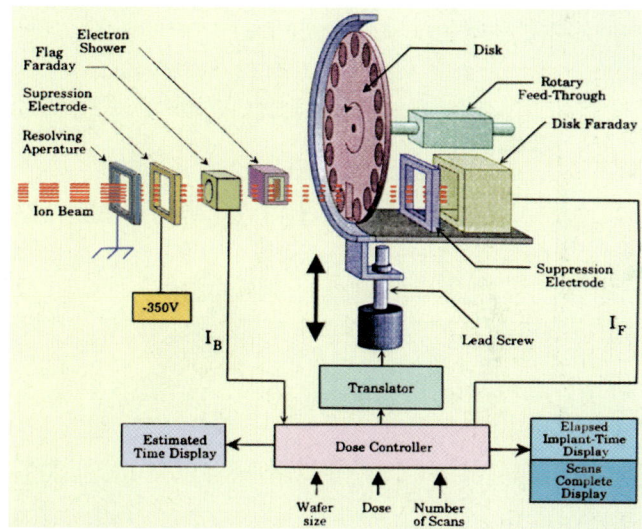

Fig. 12-25 *Implant-wheel* and *dose-control* subsystems of an implanter.

tistics to process-control and tool-performance stations.

12.3.1 Ion-Implanter Types

Ion-implantation systems have evolved into a number of distinct tool-types because of the wide range of ion-implantation process-applications (Table 12-2). When ion implanters were first used in IC-fabrication in the 1980's, two types were commercially offered:

1. *Medium-current implanters*, having beam-currents from a few-μA up to about 1-mA, and which operated over a useful energy range of 20–200-keV. Such machines produced the lower-dose implants used in the majority of doping-steps.

2. *High-current implanters*, which produce beam-currents up to 30-mA and operate at maximum energies from 80-200-keV. High-current implanters are used for doping-steps that require doses above 10^{15}-ions/cm^2. Such high-doses are needed to form source/drain-regions in MOSFETs and collector/emitter-regions of bipolar transistors.

As integrated-circuit fabrication migrated toward the ULSI era (and to deep-sub-micron feature-sizes), designs for implanters divided functionality in a different way. Implanter categories are now split into

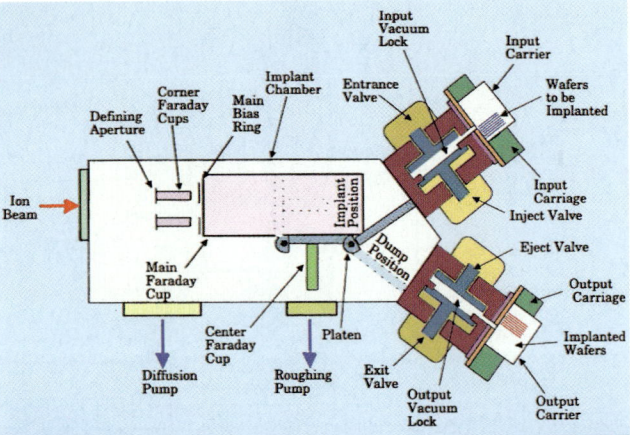

Low-energy-machines are used to form the source/drain and channel-regions of the transistor (with junction-depths of 30-to-100-nm), and high-energy machines are used to produce deeper (1-2-μm) CMOS-wells and isolation-junctions. Each of these four implanter-types will now be described in detail.

12.3.2 Medium-Current Implanters

Medium-current implanters are *single-wafer-processing machines,* having maximum beam-currents of ~2-mA. They scan the ion-beam in a square pattern in the *x* and *y* directions (Fig. 12-27). Modern medium-current machines use electrostatic and mechanical wafer-scanning designs that allow for a parallel-beam path (i.e., constant-beam-incidence-angle on the wafer) as the beam passes over the wafer. Such systems are typically used to perform low-dose and medium-dose implants. In CMOS applications these include threshold-voltage-adjust implants, well-implants, and isolation-implants. In bipolar-technology they are used to perform the resistor, base, and isolation implants.

The *throughput* (in wafers-per-hour) of a single-wafer implanter for a specific-process and wafer-size is calculated by finding the *flux-density of the machine* Φ in [ions/(cm² sec)]:

$$\Phi = \frac{6.24 \times 10^{15} (\text{ions/sec mA})}{I_s A_s} \quad (12.2)$$

where: I_s is the scanned beam current in mA; and A_s is the scanned-area in cm². The *throughput per hour* θ is found from:

$$\theta = \frac{3600}{T_i + T_t} \quad (12.3)$$

where T_i is the *implantation-time* (in sec), given by (T_i = dose/Φ), and T_t is the *wafer-transfer-time* (or the time from the end of one implantation to the beginning of the next, in sec), as specified by the manufacturer. A typical implantation-time is ~10-sec per wafer. Using vacuum-loadlocks with cassette-to-cassette loading, the wafer-handling time is ~2-4 sec. As wafer sizes have increased, the beam-current capabilities of medium-current systems have been increased

Fig. 12-26 Example of an end-station for a *serial* ion-implanter. Courtesy of Axcelis Technologies.

two groups, according to their useful energy-range. These two new implanter-classifications are:

3. *Low-energy implanters,* which are high-current tools capable of operating with maximum beam-currents of 1-to-20-mA over an energy range of 0.2–80-keV.

4. *High-energy implanters,* which can implant ions at energies from 0.5-MeV to 2-3-MeV.

Table 12-2 ION IMPLANTATION SYSTEM TYPES

Years	Era	Machine Type	Beam Energy (keV)	Beam Current (mA)	Key Application	Tool Examples
1970-1985	LSI	Medium Current	20–200	< 1	Threshold Voltage Junction Isolation	Varian DF-4
		"Pre-Dep" High Current	20–80	1–10	S/D Contact	Eaton NV-10-80/160 Varian 120-10
1985-1995	VLSI	Medium Current/ High Tilt (40-60°)	20–200	< 1	Gate-Overlap LDD (GOLDD)	Varian E-220
		High Current	20–200	0.01–25	Source/Drain; Channel; Poly	Applied Pl900/9200 Eaton NV-20A Varian E-1000
1995-2005	ULSI	High-Energy	400–3000	0.02–1	CMOS Wells	Eaton GSD-HE
		Low-Energy	0.2–80	0.01–25	S/D Extensions/ Contacts	Applied xR LEAP Eaton ULE
		High-Current/ High Tilt (40-60°)	0.5–80	0.01–10	FLASH S/D	Varian VIISta-810

to keep the implant time at (or under) 1-sec for doses of less than 10^{15}-ions/cm^2. Note that the throughput of implanters is typically much higher than that of other front-end process tools.

12.3.3 High-Current Implanters

High-current implanters have been developed for applications that require doses greater than 10^{15} ions/cm^2 (the maximum doping-application dose is $\approx 10^{16}$/cm^2). These include: source/drain and polysilicon-doping implantation in MOS devices; and emitter formation or buried-layer-doping in bipolar devices.

The high-current ion-beams of these machines (up to 30-mA) enables them to perform these doping-tasks with adequate throughput by processing the wafers in a batch-mode. The batch-size varies with wafer-size and system-design, ranging from 13 to 25 wafers. An advantage of implanting several wafers at a time is that the beam-power is distributed over the entire load of wafers, which simplifies the engineering challenge of controlling the heating of the wafers by the beam.

In most batch-implant systems, the ion-beam is stationary and the wafers are scanned in front of the beam on a rotating-wheel (see Figs. 12-24, 12-25 and 12-28b). The wheel spins at 800 to 1250-rpm, depending on the machine design. In order to distribute the ion-dose uniformly in the radial-direction, the entire spinning-wheel is itself scanned back and forth across the beam-location by an additional mechanical-assembly. The wafer-motion from the spinning can be as fast as 90 m/s, while the wheel-scanning-motion is much slower, varying from 1-to-10-cm/s.

The wafer-handling time of high-current implanters is ~10-sec/wafer (giving "mechanical" [i.e., low-dose] throughputs of $\approx$250 wafers/hr for 200-mm wafers). The throughput θ in wafers per hour of high-current implanters can be calculated from:

$$\theta = \frac{3600}{T_i + T_t} \quad (12.4)$$

where T_i is the *batch-implant time* for the dose being implanted (in sec), and T_t is the *loading and vacuum-pumping time* (in sec).

The challenge of controlling wafer-heating in high-current implanters still remains. Even though the power is distributed across a batch of 13-to-25-wafers

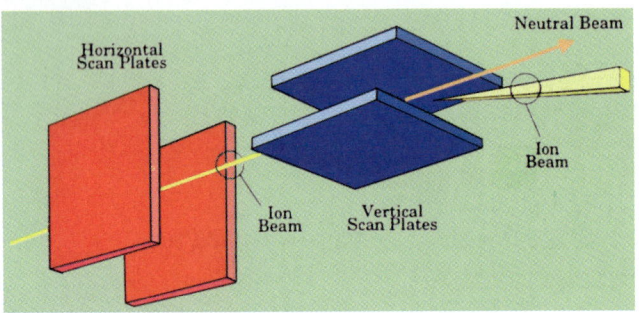

Fig. 12-27 Electrostatic scanning system of a medium-current implanter.

per implant, the beam-currents are considerably higher than in medium-current implanters. Water-cooled heat-sinks are used to maintain wafer-temperatures within acceptable limits, usually in the range of 40–80°C for high-power implants (i.e., 1–3-kW).

12.3.4 Low-Energy Implanters

Low-energy implanters have evolved from the high-current machines, and they have special capabilities for implanting light-ions (e.g., boron) at energies below 1-keV with adequate beam-currents for high-dose, shallow-junction processes. The production expectation is a throughput of ≈50-wafers/hour for source/drain-extensions (with a dose of 5×10^{14} cm^2) and contacts. This corresponds to ion-beam-currents of 1–5-mA at energies of 0.2-to-2-keV. Batch end-station designs are used for some low-energy-implanters because of the difficulties of scanning a low-energy, high-current ion-beam (Fig. 12-29). Others use a single-wafer design (Varian VIISta-810, Fig. 12-30) in which a stationary ribbon-ion-beam is used, and a single-wafer is mechanically-scanned across the beam in a direction perpendicular to the ion-beam. The highest-energy at which low-energy implanters can operate at is limited to about 80-keV.

12.3.5 High-Energy Implanters

High-energy implanters (Fig. 12-31) are machines which operate with ion-beam energies from ≈400-keV up to several-MeV, but with relatively-low beam-currents (0.005-0.5-mA).[11]

High-energy implanters are widely used for the formation of retrograde-wells and buried-layers for control of latchup in CMOS devices. One key-benefit of MeV-implantation is that most of the ion-damage is buried deep within a region near the bottom of the CMOS wells. Thus, near-surface regions (such as the MOSFET-channel), remain relatively undisturbed.

12.3.6 High-Angle Implanters

High-angle implanters are a variation on the standard machine-types described above, in which the beam-incidence-angle on the wafer can be varied from the usual 7° or 0°. For several batch-machines the implant-angle can be varied from 0° up to 7–10°. In the Axcelis GSD (Gyroscopic Super Disk) machines,

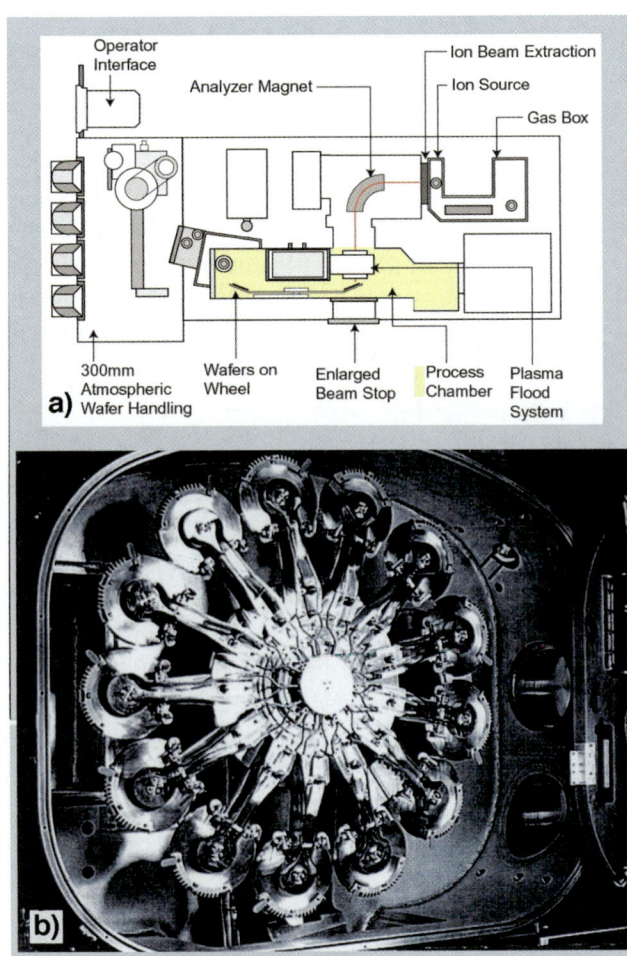

Fig. 12-28 (a) Schematic of a high-current implanter. (b) Photograph of a high-current implanter wheel. Courtesy of Applied Materials, Inc.

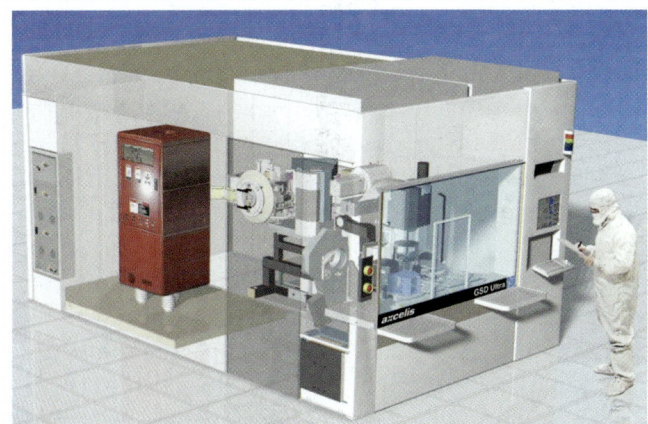

Fig. 12-29 Axcelis HC3® low-energy implanter. Courtesy of Axcelis Technologies Inc.

the beam-angle can be varied by tilting the wheel up to ±10° along both the horizontal and vertical directions, achieving a beam-strike from four separate directions (known as a "quad-implant," Fig. 12-32). Modern single-wafer machines have *wafer-tilt* and *rotation*-fixtures that allow implantation-angles up to 60-degrees from the wafer-normal, and full-rotation of the wafer in the ion-beam. These implanters are used to form regions under the edges of the gate-mask (e.g., source/drain-extensions called *LATID* [*L*arge-*A*ngle-*TI*lt *D*rain], and punchthrough-stoppers called *LATIPS* [*L*arge-*A*ngle-*TI*lt *P*unchthrough *S*topper]).

12.3.7 Limitations of Ion Implanters

Some limitations of ion-implant equipment will be described here, including: **a**) elemental and particulate-contamination; **b**) dose-monitoring inaccuracies due to beam charge-state-change effects; **c**) implantation-mask issues; and **d**) wafer-charging during implantation.[13]

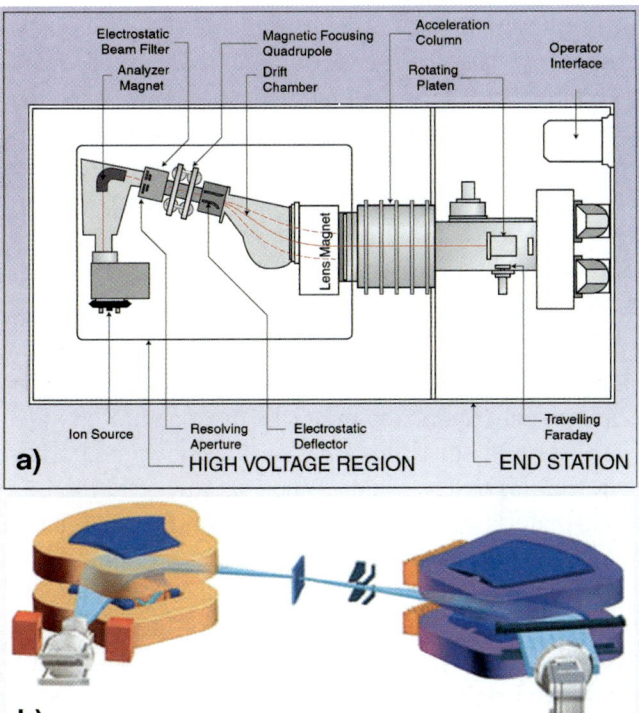

Fig. 12-30 (a) Drawing of a Varian E500® medium-current implanter. (b) Beamline for a Varian VIISta-80, ribbon-beam implanter with a mechanical single-wafer scan. (b) Courtesy of Varian Semiconductor Equipment Associates.

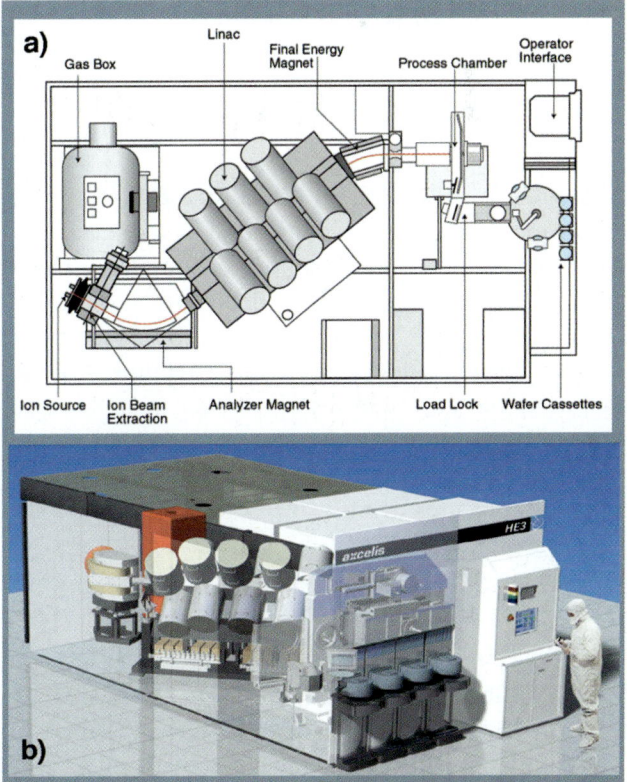

Fig. 12-31 (a) Beamline for an Axcelis GSD-VHE,® a high-energy implanter using a LINAC (*LI*Near *AC*celerator) beamline. (b) Axcelis HE3 high-energy implanter. Courtesy of Axcelis Technologies Inc.

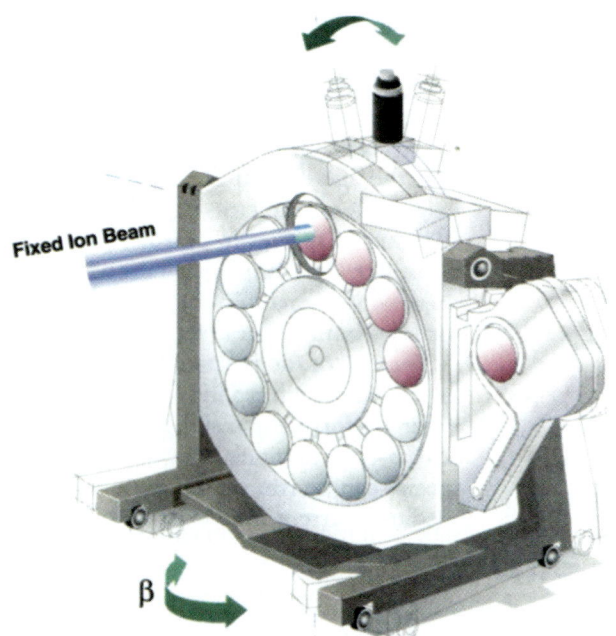

Fig. 12-32 *In-situ* quad-implants can be run on a batch end-station with orthogonal tilt axes (designated α and β). Courtesy of Axcelis Technologies Inc.

Elemental and Particulate Contamination: Although ion implantation is a considerably cleaner process compared to other doping-techniques, several sources of contamination still need to be controlled. These include: **1)** metals sputtered from the apertures or wafer-holders; **2)** cross-contamination by dopant-atoms from previous implant runs; **3)** sodium-contamination from filament-impurities and maintenance-procedures; **4)** high-molecular-weight organic-molecules from vacuum-oils and improper cleaning-procedures; and **5)** particles generated by arc-discharges in the source and beamline, as well as by abrasion of machine-parts and wafer-contact-areas. More details about these issues are given in Ref. 1.

Dose-Monitoring Inaccuracies due to Beam Charge-State Changes: The implanted-dose is measured by integrating the total-charge collected by the electrically-isolated collector that is located within a *Faraday cup*. Modern Faraday-cup designs with magnetic shielding at the entrance of the collection-column provide accurate measures of the net charge-flux. However, Faraday-cups do not measure the charge-state of individual-ions and do not count dopant-atoms which arrive as neutrals at the Faraday-cup (and at the wafer-surface). The charge-state of ions in the beam can change as a result of collisions with residual gas-atoms in the beamline and wafer-chamber. This can cause ions with an altered charge-state to strike the wafer and Faraday-cup collectors. This leads to errors in the measured-dose (since the dose is inferred by the charge collected by the Faraday-cup, assuming that each ion has a specific charge-state [usually +1]).

The sources of residual-atoms in the beam-path include poorly-designed pumping lines and pumps, leaks of process-gases from control-valves, and poor vacuum-to-atmosphere seals. But most residual-atoms in the beamline come from the outgassing of photoresist-masking material present on wafers. Even though modern cryopumps are very-effective in pumping away photoresist outgassing-products (which consist mainly of hydrogen and organic molecules), such cryopumps must be regenerated at regular intervals in order to maintain adequate vacuum-levels, and this impacts the productivity of the implantation process (see Chap. 6). Beam-neutralization effects are controlled in some implant system by use of a feedback loop which "corrects" the observed current-signal at the Faraday-cup in response to changes in the beamline pressure.

Implantation-Mask Problems: The impact of the ion-beam on photoresist material (the most commonly utilized mask against implantation into unwanted areas on a wafer), is significant. When an energetic-ion hits resist-material, polymer bonds are broken and a large number of gas-atoms per incident-ion are released. If not pumped-away by the vacuum system, such outgassing lead to increased neutralization of the ion-beam. Also, the implanted-surface of the resist is converted to a carbon-rich layer, and the resist-layer thickness is decreased. The resulting "crust" of carbon-rich, dense material causes the resist to become increasingly difficult to strip as the ion-dose and energy increase. If the wafer-temperature is not controlled, the polymer resist-material can flow, crack, or blister, leading to loss of dimension-control and breakdown of the masking process.

Wafer-Charging During Implantation: The control of charges on a wafer-surface during implant is a complex process. Positive-charge comes from the ion-beam, and secondary-electrons are also ejected from its surface. During high-current-implantation this can lead to positive charge build-up on the wafer, particularly when implanting into electrically-isolated films (such as poly-Si on SiO_2). Charge build-up on the wafer can alter the charge-balance in the ion-beam and lead to significant dose-variations across the wafer. To overcome this effect, many implanter systems direct a stream of low-energy-electrons at the wafer during implantation from an *electron-flood-gun* (helping improve doping-uniformity, Fig. 12-33).

12.4 CHARACTERIZATION OF ION-IMPLANTATION

Various characterization-tools are used to monitor and control ion-implantation processes. Such characterization-techniques must be employed both during process-development and in production. The parameters determined from such measurements include: **a)** *implanted-dose*; **b)** *uniformity* of the *implanted-dose across the wafer*; **c)** *implant depth-profiles*; **d)** *implantation-damage*; and **e)** the *effectiveness of implantation-damage annealing methods*.

12.4.1 Measurement of Implanted-Dose and Dose-Uniformity

Although monitoring during the implant-process can determine the total-dose received by each wafer, a measurement after implantation should also be done to insure that equipment problems did not cause *dose-errors*. Post-implantation dose-measurement techniques include: **a)** *four-point-probe sheet-resistance*; **b)** *optical-density changes*; **c)** *optical-reflectivity*; and **d)** *secondary-ion mass-spectroscopy*.

In addition to total-dose checking, it must be verified that the process has produced a uniform-deposit across the wafer. Dose-uniformity is monitored by wafer-mapping techniques, based on any of the above measurements.

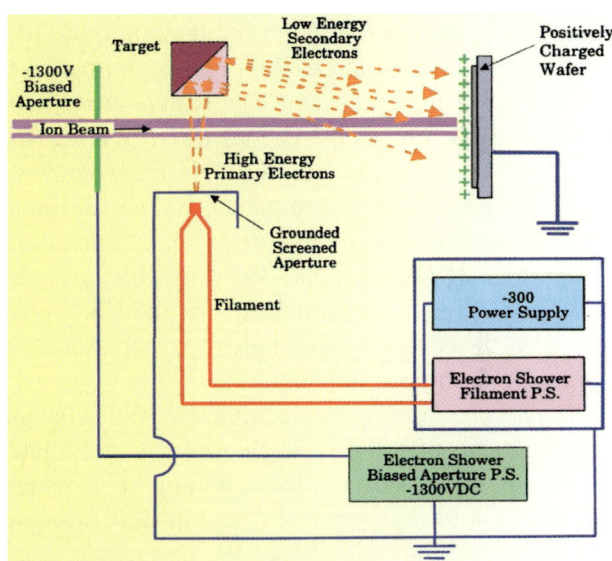

Fig. 12-33 Typical *wafer-charging neutralization scheme.* Courtesy of Axcelis Technologies Inc.

Post-implantation total-dose and dose-uniformity measurements are normally conducted on special monitor or test wafers. *Monitor-wafers* do not have device structures, while *test-wafers* contain arrays of test-structures designed for measuring the parameters of interest.

Implantation-Dose Measurements: Modern ULSI processes employ doses ranging from $\sim 10^{11}$–10^{16} ions/cm^2. Each of the dose-measurement techniques listed above finds best use over specific ranges. The *four-point-probe sheet-resistance* method is the most commonly-utilized technique because of its wide dose and energy-range. Doses from 10^{12}-ions/cm^2 and upwards can be directly measured with routine accuracy of ~0.25%.

The *optical-density technique* is based on change in the optical-transparency of a dye-loaded polymer film. That is, the change in optical-transparency is proportional to the ion-dose. Doses down to the 10^{11} ions/cm^2 range can be monitored with this method. Monitor-wafers are fabricated by spin-coating a dye-loaded polymer-film onto transparent glass-substrates. Prior to implanting, the film is scanned by an *optical-densitometer*, and its optical-absorption-pattern stored in a computer file. After implantation, the film is scanned again, and the background is subtracted to yield optical-density changes. The results are usually displayed as contour-maps. The *sensitivity (change-in-signal per change-in-implant-dose)* is somewhat better than sheet-resistance measurements, and no additional post-implantation processing-steps (e.g., annealing) are required. However, *optical-density-changes* saturate at doses above 10^{13}-ions/cm^2 for the films presently being used, so optical-density is not useful for monitoring high-dose implants.

The *Thermal Wave* (TW) technique detects changes in the optical-reflectivity of Si under a modulated laser-beam which heats a local region. It is widely used for monitoring implants with ion-doses ranging from 10^{12}-10^{14}-ions/cm^2. Since the probe-laser spot-size is so small (≈ 1-μm), this technique can measure local-regions on product wafers (Fig. 12-34).

Implantation-Dose-Uniformity and Diagnosis of Implanter Performance: The *contour-mapping technique* is very useful for routinely monitoring implantation-dose-uniformity. The causes of non-uniformity are related to limitations of implantation-equipment design and to machine-malfunction. Contour-map data can be used to track down their origin. Specifically, such malfunctions as *scan non-uniformity, scan lock-up, beam-neutralization problems, charging*, and *channeling-effects* can be identified (Fig. 12-35).

12.4.2 Measuring Implantation Depth-Profiles

The concentration of implanted-species as a function of depth (*atomic depth-profiles*), and the *depth-profiles of impurities that have been electrically-activated* must both be determined. The former are measured with *secondary-ion-mass-spectroscopy* (SIMS), and the latter are characterized with *spreading-resistance* (*SR*) measurements (see Chap. 11). The junction-depths of implanted-layers can also be measured by lapping-and-staining (see Chap. 11).

12.4.3 Measuring Implant Damage and Annealing

Both *as-implanted damage* and *damage that remains after various annealing-cycles* are determined with the same set of characterization methods. Damage resulting in defects intersecting the surface (e.g., *oxidation-induced stacking faults* and *dislocations* that intersect the surface) can be detected using chemical-etches and inspection with optical-microscopy or SEM.

Damage that produces defects and amorphous-regions in the shallow-layers below the surface can be detected by a variety of techniques, including *transmission electron microscopy* (*TEM*), *x-ray topography* (which also images distortion of lattice-planes), and *Rutherford backscattering spectroscopy* (which yields information on the location of some impurity-atoms in the Si-lattice, as well as the degree of crystalline perfection present after an annealing-step).

12.5 ION-IMPLANT PROCESS APPLICATIONS[14]

12.5.1 Selecting Masking-Layer Materials and Thickness

A masking-layer needs to be present on the wafer surface to keep ions from being implanted into unwanted substrate-regions. A number of materials can be employed, but photoresist and SiO_2 are the most commonly-used materials for masking-layers.

Typically, at least 99.99% of the implanted-ions should be stopped by the masking-layer. The minimum material-thickness needed to stop such a percentage of incident-ions is illustrated in Fig. 12-36 for SiO_2 and photoresist (as a function of ionic-type and energy). A problem with using photoresist as a masking-layer is that it will outgas-atoms when struck by an ion-beam (mostly hydrogen and various hydrocarbons, see Sect 12.3.7).

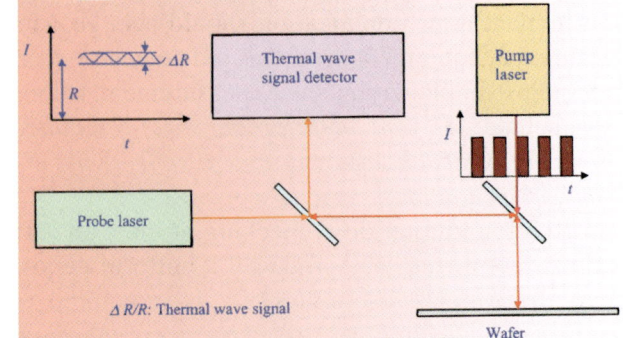

Fig. 12-34 Schematic of a Thermal-wave probing technique.

12.5.2 Threshold-Voltage Control in MOSFETs

One of the most-important applications of ion implantation is the *control of the threshold-voltage V_T of MOS devices*. To achieve adequate control over V_T, the ion-dose and energy need to be controlled to within a few percent at a dose of $\approx 10^{12}$-ions/cm^2. The ability to achieve this level of control was a critical factor in bringing ion-implantation into use in IC-fabrication processes.

Since such V_T-implants are done relatively early in the process-flow (before the gate-oxidation, gate-formation and source/drain doping-steps), it is important to use dopants that do not diffuse far from their as-implanted locations during the later thermal-processes. For this reason, indium (In) for *p*-channels and antimony (Sb) for *n*-channels, are replacing B and As in V_T-implants in advanced CMOS. An example of an In channel-implant profile is shown in Fig. 12-37.

12.5.3 Shallow-Junction Formation by Ion Implantation

As transistor lateral-dimensions are reduced to achieve higher speed-performance and greater functional-density (most notably by shrinking the size of the MOSFET-gate), the source/drain junction-depths must also become shallower (Fig. 12-38). The depths

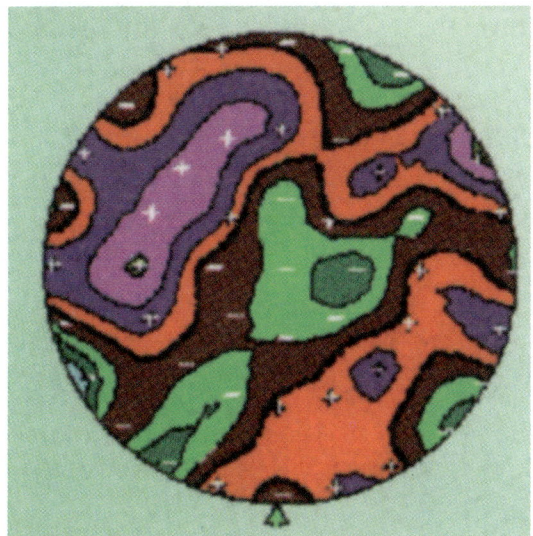

Fig. 12-35 Sheet-resistance contour-map of an ion implanted wafer.

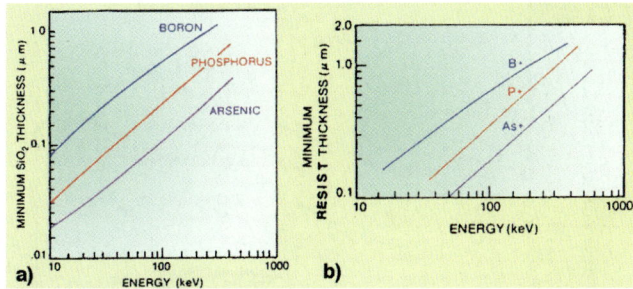

Fig. 12-36 Minimum-thickness to stop 0.9999 of incident-ions as a function of energy for: (a) silicon-dioxide; and (b) photoresist.

of these junctions in 0.1-μm (100-nm) MOSFETs must be in the range of $\approx$30-nm. Low-energy ions must be used to make such shallow-junctions (below 1-keV in the case of boron-implants). This must be combined with a short RTA-step ($\approx$1-sec) at a temperature of $\approx$1050°C to anneal the lattice-damage created by the implant and to electrically-activate the dopant-atoms - without allowing appreciable diffusion of the implanted-species.

12.5.4 High-Energy Implantation

In conventional CMOS-processes, well-dopants are diffused deeply into the substrate from their near-surface implanted locations. This is because the maximum ion-energies in older implanters was about 200-

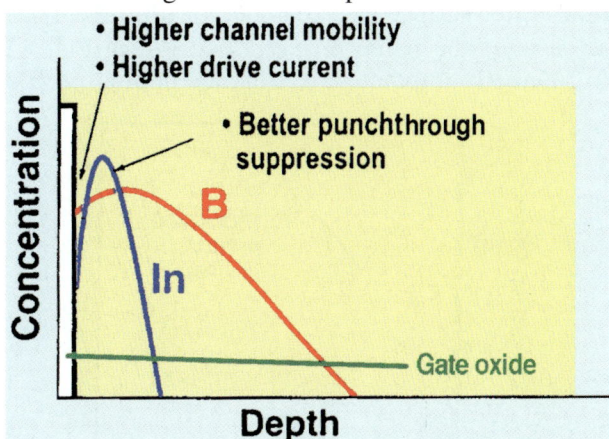

Fig. 12-37 As-implanted doping-concentration profiles of threshold-adjust implants with boron (B) and indium (In) as the implant species. The indium-implant produces a super-steep retrograde (SSR) profile.

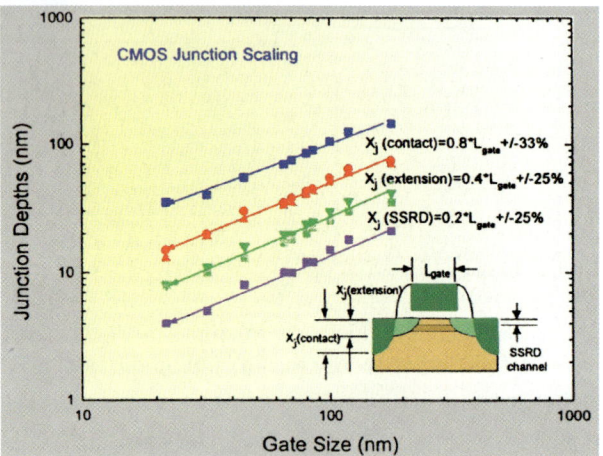

Fig. 12-38 Source/drain and channel-doping junction-depths for planar CMOS on bulk-Si.[12]

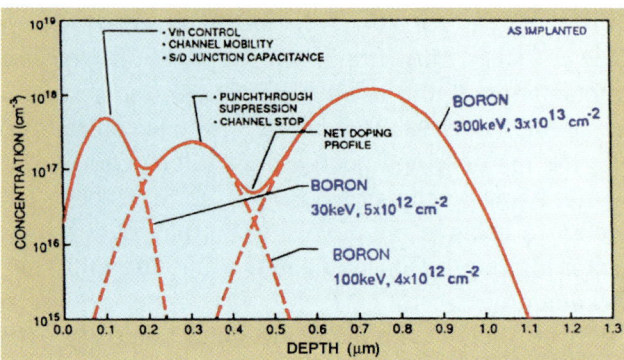

Fig. 12-40 Doping-profile in the *p*-well showing the as-implanted: (a) retrograde *p*-well-implant (boron @ 300-keV); (b) the anti-punchthrough-implant (boron @ 100-keV); and (c) the V_T-implant (boron @ 30-keV).[15]

keV, restricting implants to near-surface depths. The long drive-in diffusions (typically 12-hr @ 1100°C) resulted in CMOS well-doping profiles with a maximum concentration at the wafer-surface and a monotonic decrease in doping-concentration with depth.

On the other hand, advanced CMOS-processes use *high-energy, directly-implanted profiles* to set doping-levels in the wells. The ion-energies for such well-implants range from a few-hundred-keV to several-MeV. The shape of such high-energy implantation-profiles (which rises from a low-concentration at the wafer-surface to a peak deep within the Si), is the basis of the name *retrograde-well* (see Fig. 12-39). Directly-implanted, retrograde-doping-profiles provide much better protection against "latch-up" events in CMOS. The directly-implanted processes also allow for separate-doping of the channel (for threshold-voltage-control) and the region beneath the channel (anti-punchthrough-implant - see Fig. 12-40). In addition to these device benefits, using high-energy implantation eliminates the long-duration well-diffusion step, which significantly reduces process-time.

REFERENCES

1. S. Wolf and R.N. Tauber, *Silicon Processing for the VLSI Era, Vol. 1 - Process Technology,* 2nd Ed., Lattice Press, Sunset Beach CA 2000, Chap. 12 (*Ion Implantation*).

2. *Ion Implantation 2000: Science and Technology,* J.F. Ziegler, Ed., Ion Implantation Technology Co., 2000.

3. B. Smith, *Ion Implantation Range Data for Silicon and Germanium Device Technologies,* Research Studies, Forest Grove Or, 1977.

4. G. Dearnaley, *et al., Canadian J. of Phys.,* **46**, 587 (1968).

5. D.K. Brice, "Recoil Contribution to Ion Implant Energy Deposition Distributions," *J. Appl. Phys.,* **46**, 3385 (1975).

6. T.E. Seidel and A.U. MacRae, in F. Eisen and L. Chadderton, Eds., *First International Conf. on Ion Implantation,* Thousand Oaks, Gordon and Breach, 1971.

7. W.K. Hofker, "Implantation of Boron in Silicon," *Philips Res. Reports Suppl.,* No. 8 (1975).

8. M.I. Current *et al.,* "Ultra-Shallow Junction Technology for 100-nm CMOS: xR Leap Implanter and RTP Centura

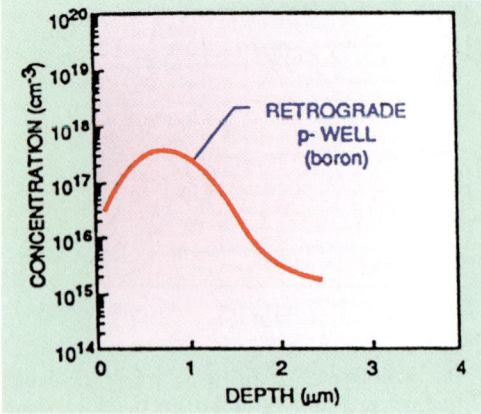

Fig. 12-39 Example of the *doping-profile* of a retrograde *p*-well (boron) implant. (400-keV).

RTP," *Materials Chem. & Phys,* Elsevier, **54**, 33 (1998).
9. E. Rimini, *Ion Implantation: Basics to Device Fabrication,* Klewer Academic, Boston, 1995.
10. J.D. Plummer, M.D. Deal, and P.B. Griffin, *Silicon VLSI Technology,* Prentice-Hall, 2000, p. 455.
11. L. Rubin and W. Morris, "High-Energy Ion Implanter Applications," *Semiconductor Interntl.,* April 1997, p. 77.
12. M. Foad and D. Jennings, "Formation of Ultra-Shallow Junctions by Ion Implantation and RTA," *Solid State Technology,* December 1998, p. 43.
13. D. Duff and L. Rubin, "Ion-Implant Equipment Challenges for 0.18-μm and Beyond," *Solid State Technology,* June 1998, p. 90.
14. R. Simonton and F. Sinclair, "Ion Implantation Applications," *Handbook of Ion Implantation Technology,* 1st Ed., J.F. Ziegler, Elsevier, Amsterdam, 1992.
15. T. Tsukamoto et al., "High-Energy Ion Implantation for ULSI Well-Engineering and Gettering," *Solid State Technology,* June 1992, p. 49.

PROBLEMS

1. Cite the advantages of ion implantation as a doping process compared to a chemical-diffusion process.
2. Name five ion-implantation applications in silicon IC fabrication.
3. Name the two ion-stopping mechanisms.
4. The two most important characteristics of a doping process are the peak dopant concentration and the junction depth. What factors in ion-implantation control these characteristics?
5. If two identical ions have the same energy and incident angle into silicon, will they both necessarily stop at the same depth in the silicon. Explain your answer.
6. Describe the relationship between the *ion projected range*, and the ion energy and ion species.
7. Why are the wafers usually tilted with respect to the ion-beam direction, during an ion-implantation step?
8. Name the subsystems of a generic ion-implanter. In which subsystem are the dopant ions selected?
9. Why does a wafer need to be annealed at high temperature after implantation? What are the benefits of using an RTP anneal for this step?
10. Plot the vertical-implanted ion-profiles about the value of R_p for a dose of 3×10^{14} P atoms/cm^2 and an implant energy of 80-keV. Assume a Gaussian distribution for the implanted ions and a thick mask-layer with a vertical mask-edge. What are the *projected-range* and *straggle* associated with this implant?
11. How do the *range & projected range* differ?
12. Explain why lighter ions penetrate more deeply into a target than do heavier ones. (Hint: lighter ions have fewer electrons.)
13. A *pn*-junction is formed by implanting As at 100-keV through an opening in a layer of thermal SiO$_2$. If a dose of 5×10^{15} ions/cm^2 is used and a background concentration of 1×10^{15}/cm^3 exists in the substrate, calculate the depth of the junction. Assume a Gaussian distribution for the implanted dose. Sketch the junction, and also sketch the junction that would be created if a diffusion process was used instead. In what ways would these junctions differ?
14. A high-current ion implanter has a beam current of 30-mA. The wafer-holder can accommodate 30 100-mm diameter wafers. Assume a 130-keV energy and a total implant time of 5-minutes. What is the dose received by the wafers? (see Appendix C)
15. A threshold-voltage-adjust implant is made by implanting through a 30-nm SiO$_2$ gate oxide. If the field-oxide is 500-nm thick, what percentage of the implanted atoms penetrates this field-oxide. Assume that $\Delta R_p \sim 0.3 R_p$.
16. What are the relative energies of a boron atom accelerated through a 10-keV beamline in the form of: a) a single ion; b) as part of a BF$_2$ ion; and c) as part of a B$_{10}$H$_{14}$ ion? What are the ranges in Si for these ion types for a 10 keV-beam?
17. Find the boron dose and energy required to form a buried *p*-type layer with a peak concentration of 10^{19}/cm^3 located at a depth of 2.0 μm.
18. The allowable depth of source/drain junctions as MOSFETs are scaled is being reduced. For devices with features-sizes below 0.18-μm, these depths will need to be less than 0.1-μm. Is this a significant issue for ion-implantation? Explain your answer for both n^+/p and p^+/n junctions. What are the main problems encountered in trying to form such junctions?
19. For 100-keV implant of As ions into Si at a current of 1-mA over an area of 200-cm^2 for 10-min, what is the number of implanted ions/cm^2? For a beam current of 10-μA, how long does it take to implant a B dose of 10^{15}/cm^2 into a wafer with a 125-mm diameter?

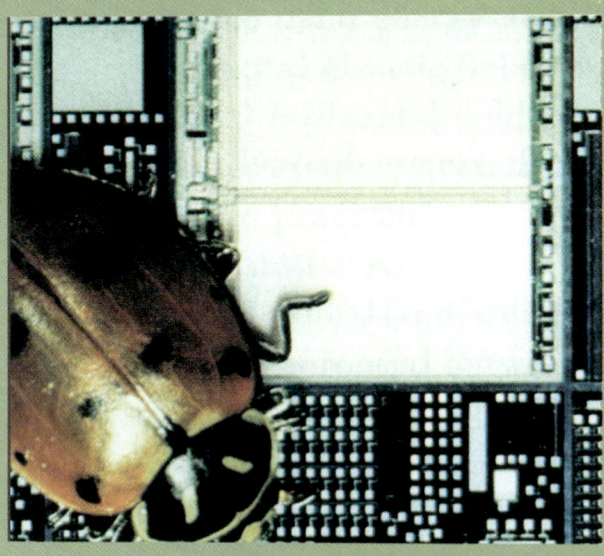

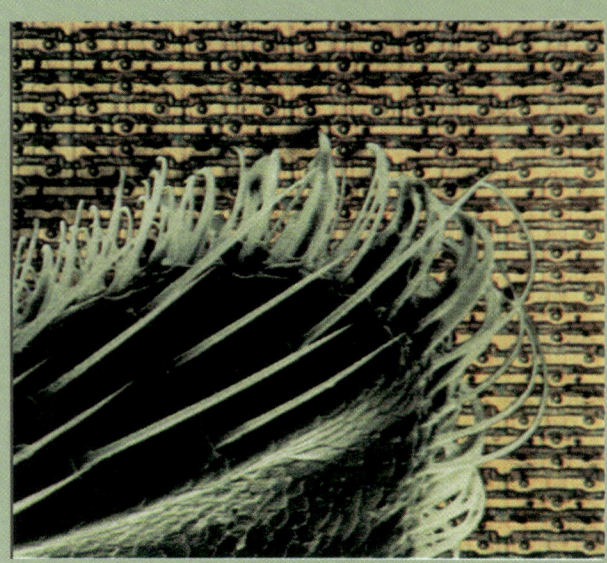
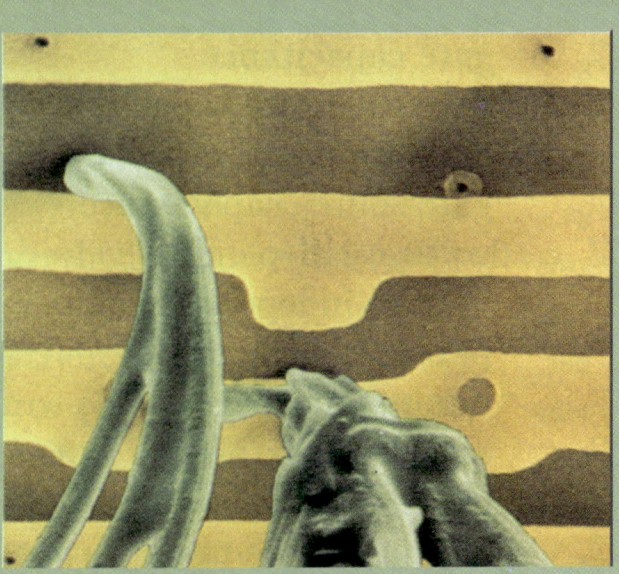

CHAPTER 13

OXIDATION

CHAPTER CONTENTS

13.1 APPLICATIONS OF SiO_2

13.2 PHYSICAL PROPERTIES OF SiO_2

13.3 THE THEORY OF SILICON-DIOXIDE GROWTH

The Deal-Grove (Linear-Parabolic) Model

13.4 SECONDARY-FACTORS WHICH AFFECT ONE-DIMENSIONAL OXIDATION

13.5 THE GROWTH OF THIN-OXIDES

13.6 THE Si/SiO_2 INTERFACE

13.7 DOPANT-REDISTRIBUTION DURING OXIDATION

13.8 OXIDATION OF POLYSILICON

13.9 OXIDATION SYSTEMS

13.10 OXIDATION PROCESS-RECIPE

13.11 OXIDE-FILM THICKNESS MEASUREMENTS

"Silicon-dioxide is God's Gift to the Semiconductor Industry ... and Cleanliness is next to Godliness."

Dr. Bruce Deal
Co-author of the Deal-Grove Oxide-Growth Model

From a *wafer* to *on-chip connections*. The details compared to those of a ladybug. (Photograph Texas Instruments/Koning & Hartman).

If a bare-silicon surface is exposed to oxygen, the silicon and oxygen will react to form a surface-oxide-film (SiO_2). Such silicon-dioxide films are stable (both chemically and thermally), and adhere tightly to the silicon. They are high-quality insulators that can also be used as barriers during impurity-diffusion. In addition, the Si/SiO_2 interface exhibits high electrical-stability, a critical-factor needed for building reliable MOS-devices in silicon. These attributes of SiO_2 are among the main reasons why silicon is the dominant material used to fabricate integrated-circuits.

This chapter discusses the properties of thermal-SiO_2, the theory and technology of growing thin oxide-films, factors affecting their growth-rate, impurity-redistribution during oxidation, process-equipment for thermal-oxidation, and methods for determining the thickness of silicon-dioxide films. More details on thermal oxidation are found in Refs. 1-3.

13.1 INTRODUCTION

Silicon readily reacts with oxygen to form silicon-dioxide. While silicon in *elemental form* is not found in nature, it is routinely encountered in *compound forms* such as *sand* (which is comprised of silicon-dioxide and impurities). If a bare silicon-wafer is exposed to either oxygen or water-vapor at room-temperature, a very-thin (1.0–2.0-nm) oxide-layer (called a *native oxide*) will form on it. However, such native oxides have poor electrical and mechanical properties, and are not thick enough to be used for integrated-circuit applications. To form the SiO_2 films needed in ICs, it is necessary to expose Si-wafers to oxygen-ambients at much-higher temperatures (i.e., 800-1200°C). This process allows high-quality oxide-films of appropriate thickness to be grown on the silicon-surface, and is called *thermal-oxidation*. The resulting SiO_2 films have a very-high melting-point (about 1732°C), even higher than that of silicon itself. In this chapter, the details of such thermal-oxidation processes are described.

Note that while chips are made on silicon-wafers, some people maintain that SiO_2 is the true "magical-material" of integrated-circuits. Indeed,

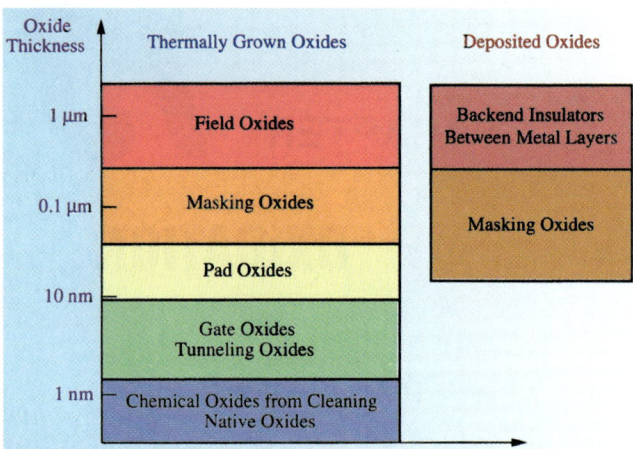

Fig. 13-1 Applications of SiO_2 in silicon-technology.

it offers many wondrous properties for this role. First, it is compatible with Si in many respects. That is, it adheres tightly to Si, does not dissolve in water, and is chemically-stable up to the melting-point of Si. Second, it is an excellent dielectric with a high-electric-field breakdown-strength, and it exhibits a stable-interface with Si (critical for MOSFET-operation). Third, SiO_2 passivates the silicon-surface, preventing surface leakage-paths among the devices on the same piece of silicon. Finally, it serves as an effective diffusion-mask against chemical-diffusion.

13.1.1 APPLICATIONS OF THERMAL SiO$_2$

Thermally-grown SiO_2 is employed in IC-applications in the thickness-range of 3.0-400-nm. Some of the roles that SiO_2-films play in ICs include: **a)** gate-oxides in MOSFETs; **b)** capacitor-dielectrics in DRAMS; **c)** passivation-layers on silicon-devices; **d)** tunnel-oxides in non-volatile-memory devices; and **e)** field-oxides in LOCOS (*LOCal Oxidation of Silicon*) isolation structures (see Fig. 13-1).

Although all thermal-oxides used in IC-devices must be carefully grown, the most critical-process involves forming the *gate-oxide* of MOSFETs. The gate-oxide-thickness used in the most advanced processes has become ever smaller as device feature-sizes have been scaled. This thinning-trend is likely to continue with each technology-node (Fig. 13-2).

As an example, the gate-oxide-thickness used in MOSFETs in 1989 was about 250-Å (25-nm), but by 2002 it had been reduced to 15-Å (1.5-nm). *Note that this is less than 10-atomic-layers.* During the shrinkage from 3.0-μm to 0.6-μm CMOS-technologies, the power-supply voltage remained at 5-V. However, at gate-lengths below 0.5-μm this voltage had to be reduced below 5-V. For example at 0.35-μm, it was 3.3-V, and at 0.13-μm only 2.2-V. This decrease in power-supply voltage is necessary because the gate-oxide in these devices is so thin that they cannot be operated reliably if higher voltages are used.

One of the most significant difficulties in scaling MOSFETs is the ability to grow high-quality gate-oxides that are only a few atomic-layers-thick. However, significant-improvements in wafer-cleaning and oxidation-equipment have allowed continuously-thinner oxides to be manufactured.

Table 13-1 is extracted from the International Technology Roadmap for Semiconductors (ITRS) published in 2001 by the International SEMATECH.[4] It shows the equivalent gate-oxide thickness that will be required for each new technology-generation from 2001 to 2012. In general, it also indicates that thinner gate-oxide films will needed in microprocessors than in such memory devices as DRAMs.

13.2 THE PHYSICAL PROPERTIES OF SiO$_2$

Silicon-dioxide (SiO_2) can exist in either *crystalline* (e.g., quartz), or *non-crystalline* (amorphous) forms.

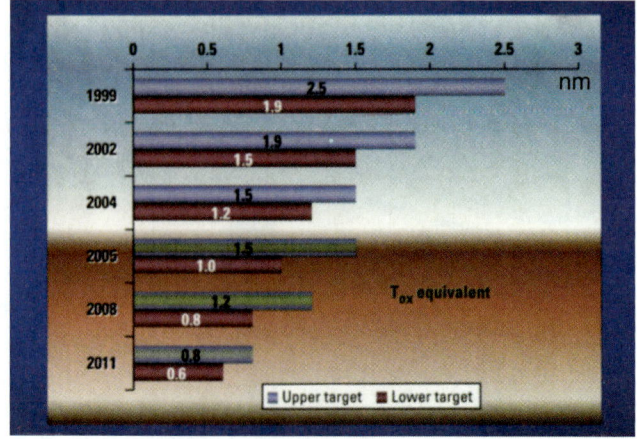

Fig. 13-2 Gate-oxide thickness-reduction-progression according to ITRS.[4]

Table 13-1 Gate Oxide Thickness for Deep-Submicron CMOS						
	1999 (180-nm)	2001 (150-nm)	2003 (130-nm)	2006 (100-nm)	2009 (70-nm)	2012 (50-nm)
Oxide thickness (nm)	3–4	2–3	2–3	1.5–2	<1.5	<1.0

The thermal-oxide on silicon wafers grows as an amorphous, glassy-film because there are no crystalline-forms of SiO_2 whose lattice-size closely matches that of the silicon substrate.

Nevertheless, in all forms of SiO_2 (crystalline or amorphous) the silicon and oxygen are regularly arranged within a short-range structure. That is, a silicon atom is at the center of 4 oxygen-atoms, arranged in a tetrahedron, as shown in Fig. 13-3a. This structure satisfies the valence-shell of the silicon, and such SiO_2 tetrahedra are the basic-units from which SiO_2 is configured. These tetrahedra also bond together by sharing oxygen-atoms. In *crystalline-SiO_2* (quartz) each oxygen-atom belongs to two tetrahedra, and is thus also bonded to 2 silicon-atoms. Oxygen-atoms that link two tetrahedra are called *bridging-oxygen atoms,* and quartz contains only bridging-oxygen atoms. As a result quartz exhibits long-range (crystalline) order as shown in Fig. 13-3b.

On the other hand, as shown in Fig. 13-3c, thermally-grown SiO_2 (also called *fused-silica*) has an amorphous-structure. The melting point of fused-silica is very high (about 1732°C - even higher than that of silicon). Long-range order is absent because the tetrahedra are not arranged in a regular, three-dimensional array (as exists in crystalline-quartz). As also shown in this figure, some of the oxygen-atoms are not shared between two tetrahedra, and these are called *non-bridging-oxygen-atoms*.

If an oxygen-atom in SiO_2 is not shared, its tetrahedron remains disconnected from neighboring-tetrahedra. Since the tetrahedra in fused-silica are not all connected, they also end up being more randomly arranged in a *glassy-network*. This more-disorganized-structure has some important properties. First, the glassy-state is less-dense than the crystalline-state (i.e., the density of *fused-silica* is 2.15–2.25-g/cm³, while that of *crystalline-quartz* is 2.65-g/cm³). Second, the glassy-state has a more open-structure, which allows a variety of impurity-atoms (including O_2, H_2O, and Na) to easily enter and diffuse through the oxide-films. Third, the ratio of *non-bridging-to-bridging-oxygen-atoms* is important. That is, the larger the fraction of non-bridging-oxygen-atoms compared to bridging-oxygen atoms, the poorer is the quality of the SiO_2-film.

Various impurities can also exist in thermal-oxides (Fig. 13-4). The presence of some of these can significantly affect the oxide-properties. For example, if boron is introduced into SiO_2, some of the boron-atoms will replace the Si-atom within an SiO_2 tetrahedron. Since the boron-atom has one less valence-electron, this tetrahedron will have one oxygen-atom that cannot bond with a neighbor-tetrahedron. Thus, a non-bridging-oxygen-atom is created. As a result, introducing boron into silicon-dioxide layers will weaken the glassy-network. This effect can be exploited to reduce the temperature at which oxide-films flow. That is, boron can be intentionally added

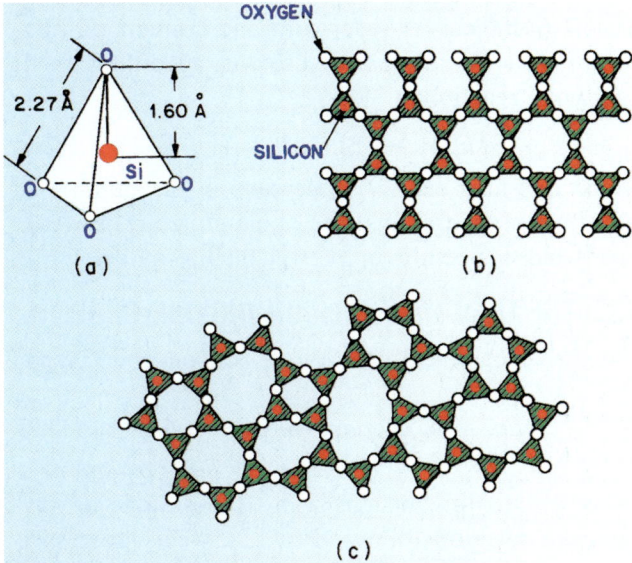

Fig. 13-3 Schematic representation of: (a) the $(SiO_4)^{4-}$ tetrahedron; (b) $(SiO_4)^{4-}$ tetrahedra showing long-range order and periodicity (*crystalline quartz*); c) $(SiO_4)^{4-}$ tetrahedra arranged in a random-pattern (amorphous - as in the *fused-silica* form of (SiO_2).

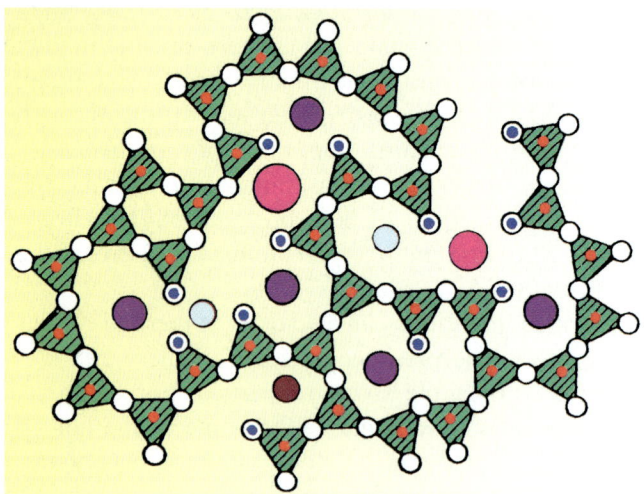

Fig. 13-4 Impurity-incorporation model in *fused-silica*.[5]

to deposited-oxide-films that are used for *reflow-applications* (i.e., to smooth the surface topography of wafers, see Chap. 16, Sect 16.4.5).

Another impurity that affects oxide-quality is the *hydroxyl* (OH). Hydroxyls form in SiO_2 when water (H_2O) molecules are introduced into the network. The H_2O-molecules react-with and convert a bridging-oxygen site into two-hydroxyls, according to the following reaction:

$$H_2O + Si:O:Si \rightarrow Si:O:H + H:O:Si \quad (13.1)$$

Since this also causes a decrease in the number of bridging-oxygen-atoms, the oxide-structure is weakened, and this results in a lower-quality-oxide.

13.3 THE THEORY OF SILICON-DIOXIDE-GROWTH

The basic chemical-reaction of silicon and oxygen is:

$$Si\ (solid) + O_2\ (vapor) \rightarrow SiO_2\ (solid) \quad (13.2)$$
Dry Oxidation

When SiO_2-films are grown by exposing Si to oxygen, the reaction follows Eq. 13-2. This process is called *dry-oxidation*. However, another way to grow SiO_2 is to expose silicon to water-vapor (also at high-temperatures). This approach is called *wet-oxidation* or *steam-oxidation*. The chemical-reaction of wet-oxidation proceeds by the following equation:

$$Si\ (solid) + H_2O\ (vapor) \rightarrow SiO_2\ (solid) + 2H_2\ (gas)$$
Steam- or Wet-Oxidation $\quad (13.3)$

In both dry- and wet-oxidation, the reactions that form SiO_2 occur at the Si/SiO_2-interface. The Si for the reaction is provided by the wafer itself. Thus, as more oxide is grown, Si is consumed and the layer of SiO_2 grows *into* the Si-surface. Figure 13-5 schematically shows these phenomena. Calculations based on the relative-densities and molecular-weights of Si and SiO_2 indicate that the amount of silicon consumed is 44% of the final-oxide-thickness. That is, if 1000-Å of oxide is grown, 440-Å of Si is consumed.

Although Eqs. 13-2 and 13-3 show how silicon and oxygen react, they do not provide any information about the *mechanisms* by which oxide-films grow. To get such insight, a model for oxide-growth is needed. Having such a model is useful because IC-applications call for oxide-films with a variety of thicknesses. It must be possible to specify the process-conditions needed to grow oxide-films for each specific-thickness. A model of the oxide-growth-mechanisms has been developed which accurately allows the process-conditions to grow an oxide of any thickness (except very-thin oxides) to be chosen. It is called the *Deal-Grove* or *linear-parabolic* (L-P) *model*.

13.3.1 Deal-Grove (or Linear-Parabolic) Model

The oxide-growth model was published by Deal and Grove in the early 1960's.[6] It is based on the following assumptions (Fig. 13-6): **a)** the reaction of the silicon and oxygen takes place at the silicon-surface; and **b)** there are two events that impact the oxide growth-

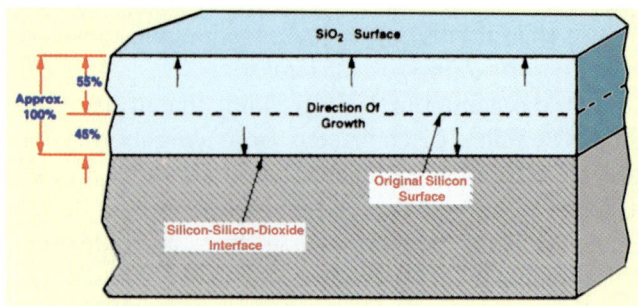

Fig. 13-5 Silicon-dioxide growth by thermal oxidation.

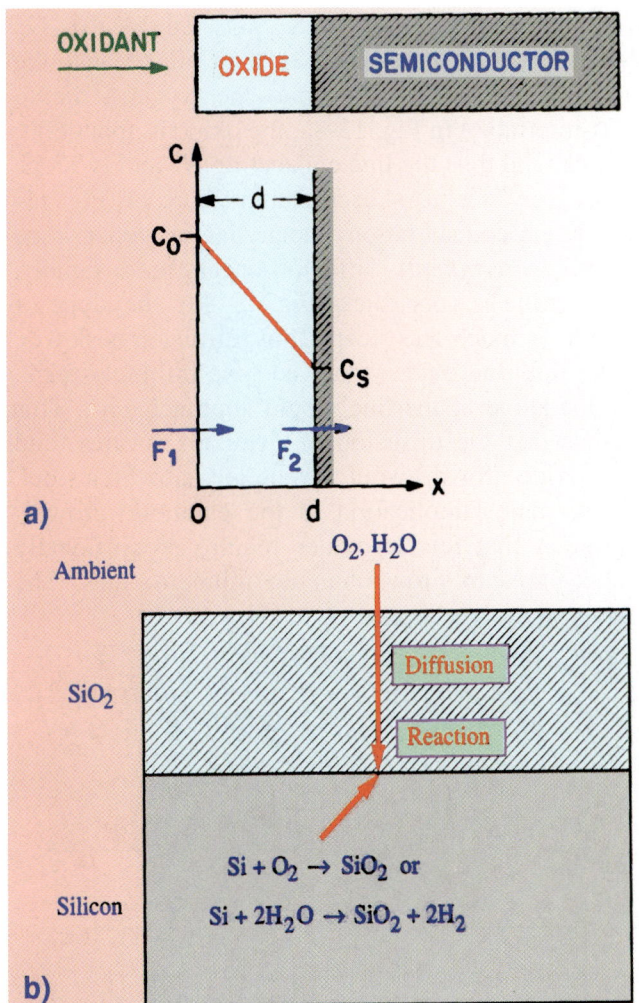

Fig. 13-6 The Deal-Grove model of thermal-oxidation of silicon.[6] Reprinted with permission of the American Physical Society.

rate: **1)** the rate at which O_2 or H_2O molecules can cross the oxide film on the wafer surface (F_1 in Fig. 13-6); and **2)** the rate at which the silicon and oxygen can react at the silicon surface (F_2 in Fig. 13-6).

Assumption **a)** is important for two reasons:

1. Since the reaction takes place at the bottom of the growing oxide, the new oxide being formed is always protected from any contaminants that exist on the oxide-surface (and which can be removed after the oxide process is completed). This helps ensure the integrity and purity of thermally-grown oxide films.

2. It implies that in order to react, the oxidizing species (O_2 or H_2O) must migrate (the formal term is *diffuse*) through the oxide to reach the silicon-surface. As the oxide gets thicker, the time to cross the oxide-film gets longer, which slows down the oxide growth-rate.

The equation of the Deal-Grove model that predicts the oxide thickness $x_{ox}(t)$ for a particular set of oxidation conditions is given by:

$$x_{ox}(t) = (A/2)\left[\left(1 + [t+\tau]/[A^2/4B]\right)^{1/2} - 1\right] \quad (13.4)$$

where t is the time of the oxidation-step, B and B/A are the *parabolic* and *linear* rate-constants, respectively, and τ is a factor used to account for any oxide present at the start of the oxidation. (Note that B and B/A will be defined in more detail in upcoming paragraphs, and reasons for why they are so-named will also be given. Furthermore, Eq. 13-4 is derived in Appendix A, and more details about it are found in Ref. 1.)

Note that Eq. 13-4 can be re-arranged and written in another form that helps give insight about the terms B and B/A:

$$(x^2_{ox}/B) + (x_{ox}/[B/A]) = t + \tau \quad (13.5)$$

Linear and Parabolic Growth-Rate Regimes: There are two limiting-forms of Eq. 13-5. These occur when one of the two terms B or B/A dominates. That is, when B/A is much larger than B, the second term of Eq. 13-5 is smaller. Then, Eq. 13-5 is approximately:

$$x^2_{ox} \cong B(t+\tau) \quad (13.6)$$

On the other hand, when B is much larger than B/A, Eq. 13-5 becomes approximately:

$$x_{ox} \cong (A/B)(t+\tau) \quad (13.7)$$

The reasons for calling B and B/A the *parabolic* and *linear* rate constants are evident when Eq. 13-5 is expressed as Eq. 13-6 or Eq. 13-7. That is, when Eq. 13-7 is obeyed (Fig. 13-7a), the oxide-thickness is *linearly-proportional* to the oxidation-time (see also Fig. 13-7c), and the ratio B/A is thus a *linear* proportionality-constant. When oxide-growth obeys Eq.

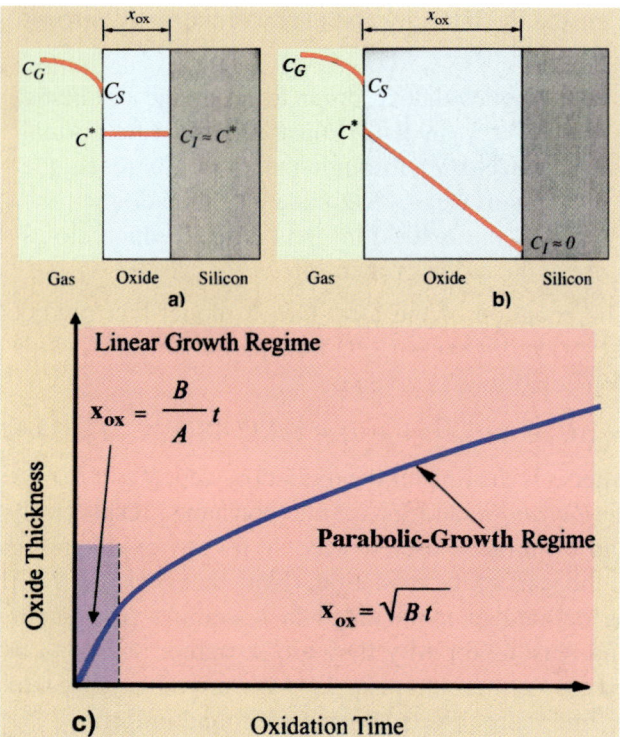

Fig. 13-7 (a) Oxide growth-mechanisms in thin-oxides (*linear-growth-regime*). (b) Oxide growth-mechanisms in thicker oxides (*parabolic-growth regime*). (c) The relationship governing the thermal oxidation of silicon and its two limiting forms.

13-6 (Fig. 13-7b), the oxide-thickness is proportional to the *square-root* of the oxidation time (see also Fig. 13-7c). Thus, B is called the *parabolic* rate-constant. The parabolic-term of Eq. 13-5 will dominate for larger x_{ox} values (thick-oxide growth-processes), and the linear-term will dominate for smaller ones (thin-oxide growth-processes). This means that a thick-oxide grows in two stages: **1)** a linear growth-regime occurs when a bare Si-wafer is first exposed to oxygen; and **2)** the parabolic-regime takes over after about 500-Å of oxide has been grown (Fig. 13-7c).

The slow-down in the oxide growth-rate in the parabolic-growth regime can also be explained with the aid of Fig. 13-8. In Fig. 13-8, the concentration of the oxidizing-species in the gas-phase (and thus also at the top-surface of the oxide) is given by N_o, and its concentration at the Si/SiO$_2$ interface by N_i. (Note that N_i is much smaller than N_o because the SiO$_2$-forming-reaction consumes the oxygen at the Si/SiO$_2$-interface.) In Fig. 13-8a, the oxide is about 500-Å thick, and thus the line connecting N_o and N_i slope is relatively steep. As discussed in Chap. 11 (Fig. 11-3), when a concentration-change has a steeper slope (or *steeper gradient*), diffusion occurs more rapidly. If the oxide is very thick (Fig. 13-8b), the slope of this line is much-less steep. Thus, diffusion is slower. In the limit, an oxide would become infinitely-thick, and the slope of the line would approach zero. This implies that the diffusion of oxidizing-species (and thus, oxide-growth) would eventually slow to a stop.

The major implication of the parabolic growth-regime is that thicker-oxides require proportionally longer-times to grow, than do thinner-oxides. The

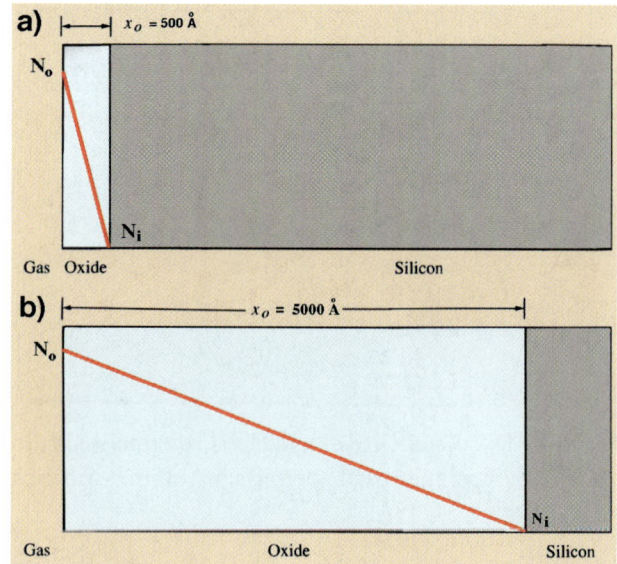

Fig. 13-8 Schematic-drawing that explains why the oxide growth-rate decreases with increasing oxide-thickness in the *parabolic-growth regime*. (a) When the oxide is thinner (e.g., 500 Å-thick), the gradient of the concentration of oxidizing-species is steep, and thus their transport across the oxide-layer is rapid, making F_1 of Fig. 13-6 large. Hence, the growth-rate is also faster. (b) When the oxide is thicker (e.g., 5000 Å-thick), this gradient is not as steep, making the transport of the oxidizing-species across the oxide slower. This makes the oxide growth-rate in this case smaller.

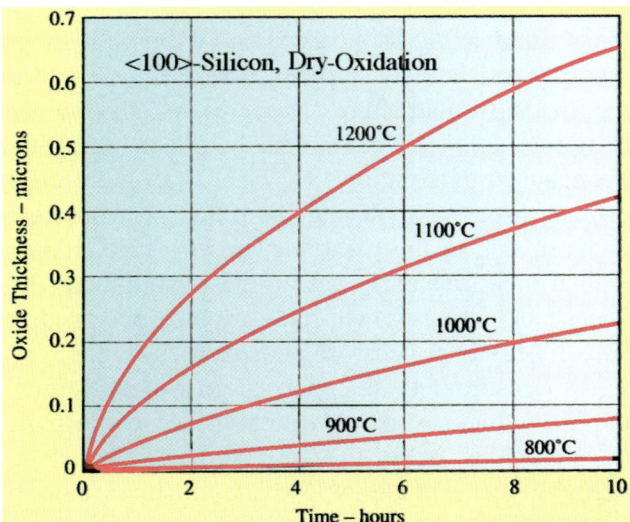

Fig. 13-9 Calculated oxide thicknesses for (100)-silicon in dry-O_2, based on the Deal-Grove model. The initial, fast-oxidation for the first 20-nm is not included ($\tau = 0$).

oxide-thickness versus time calculated using Eq. 13-4 is shown in Figs. 13-9 and 13-10 for dry-oxidation and wet-oxidation, respectively, for different growth-temperatures on (100)-silicon.

Factors that Impact B and B/A: If the values of B and B/A are known, Eq. 13-4 can be used to calculate the oxide-thickness versus oxidation-time. Their values have been experimentally determined, and have been found to depend on many factors, the two most important being the *oxidation-temperature*, and whether a *dry-* or *wet-oxidation* is being performed. (Note that the oxide growth-rate increases with increasing temperature.) However, there are also other secondary-factors that impact the values of B and B/A, including: the doping-concentration at the surface of the silicon-wafer; the oxidant-pressure during the growth process; and the presence of chlorine in the oxidizing ambient. These are discussed further in Sect. 13-4.

Tables 13-2 and 13-3 give the measured values for B, B/A, and τ for wet-oxidation and dry-oxidation at several temperatures. Note that the B/A and B constants are much larger for wet-oxidation than for dry-oxidation. Hence, the oxidation-rate for wet-oxidation is much higher at equivalent-temperatures. In fact, wet-oxidation produces films about 5 to 10 times thicker than those grown using dry-oxidation.

Two factors contribute to this growth-rate enhancement during wet-oxidation. First, the solubility of H_2O in SiO_2 is much higher ($3 \times 10^{19}/cm^3$) than O_2 in SiO_2 ($5.2 \times 10^{16}/cm^3$). Second, the smaller H_2O molecules move more easily through the SiO_2 network than do the larger O_2 molecules. Dry-oxidation is thus generally used for producing oxide-films up to 1000-Å, while thicker films are normally grown using H_2O ambients. The values of B, B/A, and τ for various process-conditions are also stored in such process simulators as SUPREM-III and SUPREM-IV.

Example 3-1: (a) We wish to grow a 2000-Å-thick (0.2-μm) oxide on a (100) silicon-wafer. How long will it take to grow this oxide at 1100°C in dry-oxygen? (b) If the wafer with $x_{ox} = 0.2$-μm is put back into the furnace in wet-oxygen at 1000°C, how long will it take to grow an additional 3000-Å of oxide?

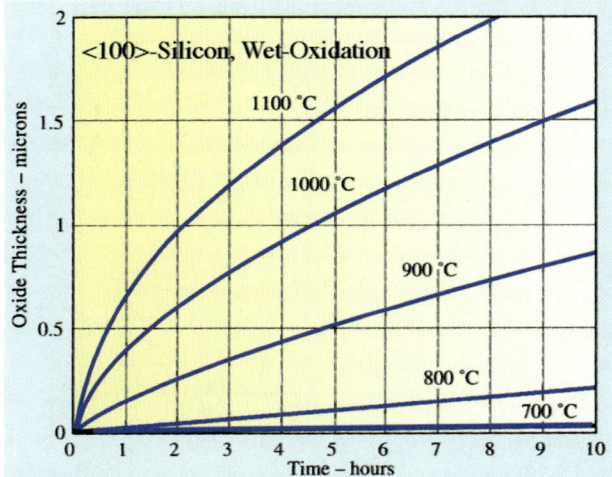

Fig. 13-10 Calculated oxide-thicknesses for (100)-silicon in H_2O, based on the Deal-Grove model.

Table 13-2	RATE CONSTANTS FOR WET OXIDATION OF (100) SILICON		
Oxidation Temperature (°C)	B ($\mu m^2/hr$)	B/A ($\mu m/hr$)	τ (hr)
1200	0.82	8.6	0
1100	0.58	2.76	0
1000	0.316	0.75	0
920	0.223	0.24	0

Table 13-3 RATE CONSTANTS FOR DRY OXIDATION OF (100) SILICON

Oxidation Temperature (°C)	B ($\mu m^2/hr$)	B/A ($\mu m/hr$)	τ (hr)
1200	0.045	0.667	0.027
1100	0.027	0.178	0.067
1000	0.0117	0.042	0.37
920	0.0049	0.013	1.40
800	0.0011	0.0018	9.0
700	—	0.00015	81.0

Solution: (a) Using Fig. 13-9, we see that it would take 2.75-hours to grow the initial 0.2-μm oxide.

(b) From Table 13-2, at 1000°C for wet-oxidation, $B = 0.316$-$\mu m^2/hr$ and $B/A = 0.75$-$\mu m/hr$. Thus, using these values in Eq. 13-5, and a value of $x_{OX} = 0.2$-μm and t = 0, we calculate the value of τ:

$$[(0.2)^2/0.316] + [0.2/0.75] = 0.4 \text{ hr} = \tau$$

Using this value of τ and the final-value of x_{OX} (i.e., 0.5-μm), we calculate the time, t, needed to grow the additional 0.3-μm of oxide using Eq. 13-5:

$$t = [(0.5)^2/0.316] + [0.5/0.75] - 0.4\text{-hr} = 1.05\text{-hr}$$

13.4 SECONDARY-FACTORS WHICH AFFECT THE ONE-DIMENSIONAL OXIDATION-RATE

Several secondary-factors affect the oxidation-rate (in addition to the primary-factors of temperature and type of oxidant). The most important of the former are covered here, including: **a)** dopant-type and dopant-concentration at the silicon-surface; **b)** oxidant-pressure in the growth-ambient; and **c)** the presence of a chlorine-gas in the growth ambient. All of these factors (including the primary ones) are modeled using process simulators, such as SUPREM3 and SUPREM4.

13.4.1 Dopant-Effects on Oxidation Growth-Rates

Both *p*-type (e.g., B) and *n*-type (e.g., P, As) dopants present at high-concentrations near the surface of a silicon-wafer will increase the growth-rate of the oxide. Since SiO_2 can accommodate high-concentrations of boron, a high-level of B at the silicon-surface means that a high-concentration of boron will get incorporated into the growing oxide. This weakens the oxide film-structure by increasing the number of non-bridging oxygen atoms (see Sect.13.2). Such a weakened-oxide allows O_2 or H_2O to more easily diffuse through it, causing the oxide growth-rate to increase during the parabolic growth-stage. Experimental measurements confirm these predictions. Enhanced oxidation-growth-rates are also observed for silicon heavily-doped with phosphorous (Fig. 13-11).

As a result of these enhanced-growth-rate effects, the oxide-thickness on the wafer-surface over heavily-doped regions will be greater than over lightly-doped regions. Such variations in oxide-thickness must be taken into account when specifying an etch-process (to insure that the oxide on the thickest-regions of the device are completely removed).

13.4.2 Dependence of Oxidation-Rates on Pressure

The oxidant-pressure in the growth-ambient also impacts the oxide-growth-rate. That is, the oxide-growth-model predicts that the growth-rate is directly-proportional to the oxidant-pressure. For example, the oxide grows more rapidly at higher oxygen-pressures. This effect was exploited in earlier generations of IC-technology to grow thick-oxide films under high-pressure-oxidation conditions. However, since

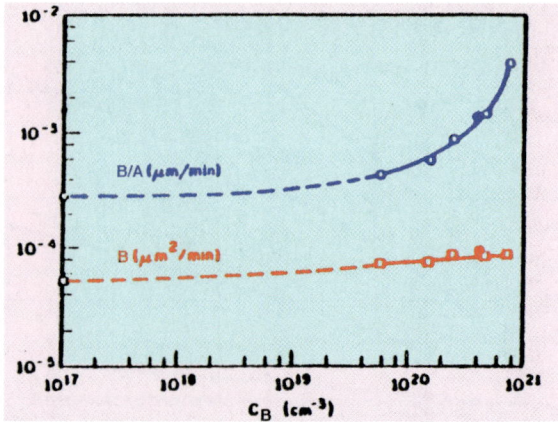

Fig. 13-11 Oxidation rate-constants for dry-oxidation as a function of phosphorus-doping level at 900°C.[7] Reprinted with permission of the Electrochemical Society.

thick-oxides no longer find much application in advanced ICs, further development of this approach has not been pursued.

On the other hand, the oxide-growth-rate slows down under reduced oxygen-pressure. This effect can be used to controllably-grow very-thin-oxides. Two methods can be used to reduce the oxygen-pressure. The first is to operate the process in pure-oxygen, but at reduced-pressure. The second way is to operate the process at atmospheric-pressure, but to dilute the oxygen with an inert-gas (such as Ar or N_2). That is, the *partial-pressure* of the oxygen in the ambient is thus reduced, which has the same effect as operating at a reduced-total-pressure (Fig. 13-12).

13.4.3 Dependence of Oxidation-Rates on Chlorine in the Growth Ambient

A chlorine-containing gas in small concentrations (1 to 3%) is often mixed with oxygen in a dry-oxidation process to improve the characteristics of the oxide film. These improvements include: **1)** a reduction in the concentration of sodium in the oxide; and **2)** a reduction in the number of defects in gate-oxides grown in Cl-containing ambients (which strengthens the oxide-breakdown-properties).

These benefits arise because chlorine reacts with metallic impurities (e.g., Na, Fe, and Ni) present on the wafer-surface during oxide growth. Chlorine with Na forms volatile NaCl, and with Fe and Ni it forms other volatile chlorides - all of which can be pumped-out of the furnace-tube. Removing these metals is critical. Measurable instabilities in the threshold-voltage of MOSFETs are caused by mobile-ion contamination (due to Na), even at low (e.g., ppm) concentration-levels. The presence of Fe and Ni seriously degrades the integrity of gate-oxide films (i.e., their *gate-oxide-integrity,* or *GOI*) is reduced.

Since including chlorine during oxide-growth can be beneficial, most furnace-systems are designed to pipe a chlorine-containing gas into the process tube. Early workers used HCl to introduce Cl, but HCl is such a corrosive-gas that now dichloroethane (DCE) is more widely used. Even when Cl is not used in the oxidation ambient, furnace-tubes are routinely cleaned *in-situ* by flowing Cl and steam through them at elevated temperatures.

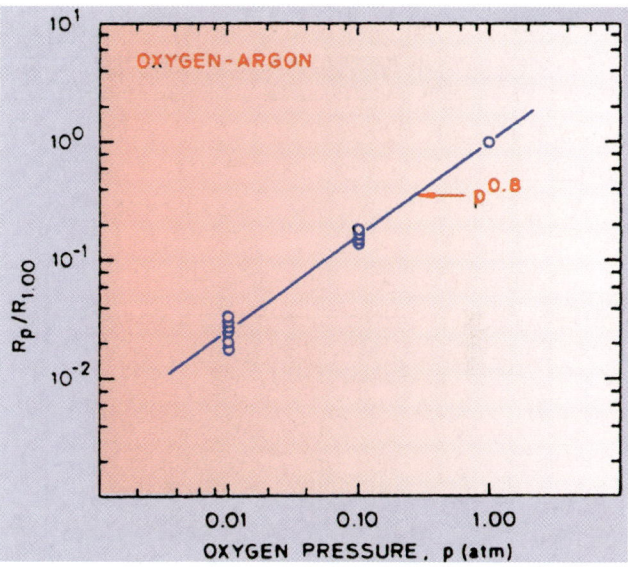

Fig. 13-12 Normalized oxidation-rate as a function of oxygen partial-pressure for (100), (110), and (111), lightly doped silicon in dry O_2-Ar mixtures at 800–1000°C.[8]

Note that the use of Cl in dry-oxidation also causes an increase of about 10-15% in the oxidation-rate. The reason why the growth-rate is enhanced by the presence of chlorine is not yet well-understood.

13.5 THE GROWTH OF THIN-OXIDES

While the Deal-Grove model accurately predicts oxidation-growth for thicknesses greater than 300-Å, it is not accurate for thinner-oxides grown by dry-oxidation. (The model significantly underestimates the growth-rate for *thin dry-oxides.*) Since the majority of gate-oxides grown in ULSI devices are thinner than 100-Å, this is a serious limitation. Although considerable effort has been expended to develop a model to handle such thin-oxide-growth, this has not yet been successfully accomplished. Instead, the enhanced growth-rates of such thin-oxides are accounted for empirically, and correction terms to the Deal-Grove model must be used (see Ref. 1). However, when these are added, good estimates of the experimental-values of all oxide-thicknesses are obtained. This

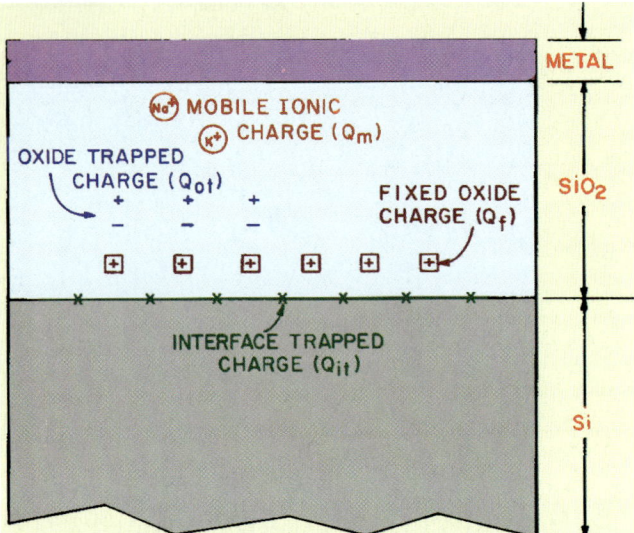

Fig. 13-13 Standardized terminology for oxide-charges associated with thermally-oxidized silicon.[9] (© 1980 IEEE).

makes it possible to reproducibly grow high-quality gate-oxides as thin as 3.0-nm in a manufacturing environment.

13.6 THE Si/SiO₂ INTERFACE

In addition to the characteristics of the bulk-SiO_2 structure, the properties of the Si/SiO_2 interface also play a crucial role when SiO_2 functions as the gate-dielectric layer of a MOSFET. Hence, the nature of this interface must also be understood. The Si/SiO_2-interface and the bulk-SiO_2 contain both charges and traps, and their presence has a profound impact on the characteristics of devices fabricated in the underlying silicon. For instance, the transistor-behavior of submicron-MOSFETs is dominated by the thin gate-oxide. The types of charges and traps present at the Si/SiO_2 interface are listed here. For more details about them, readers should consult Refs. 1, 2, 9, or 10.

There are four-types of charges associated with the oxide and interface as shown schematically in Fig. 13-13. In general, it is desired that they not be present, but unfortunately, this is not possible. Thus, efforts are made to keep the values of these charges as small as practical. The standardized-terminology and symbols used to represent the various individual charge types are as follows:

1. *Fixed oxide-charge* per unit area, Q_f (C/cm²). This charge is thought to be due to ionic Si-atoms left near the interface when the oxidation is halted. Such positive-charge is located within 30-Å of the interface, and the silicon-wafer orientation impacts its value. That is, the lowest Q_f values exist on <100>-Si-surfaces. This is why all MOSFET ICs are built on <100>-Si wafers. Optimized oxidation and annealing steps are also used to suppress Q_f to levels of about 10^{10}/cm².

2. *Mobile ionic-charge* per unit area, Q_m (C/cm²). These charges are due to the presence of sodium or other alkali atoms in the oxide. Such atoms are mobile within the oxide under high-temperature or high-voltage operations. Their movement within the oxide under these conditions gives rise to shifts in the threshold-voltages of MOSFETs. This results in unreliable device-operation (even if the concentration-level of sodium-atoms in the oxide is as small as in the ppm range). Thus, extraordinary measures are taken to exclude sodium from the wafer fab-environment (see Chap. 8).

3. *Interface-trap charge* per unit area, Q_{it} (C/cm²). The *number of interface-trap charges* per unit area *and* energy, is denoted by D_{it} (#/cm²eV). The interface-trap charge arises because at a single-crystal Si-surface there are unsatisfied covalent bonds (*called dangling bonds*, see Fig. 13-14).

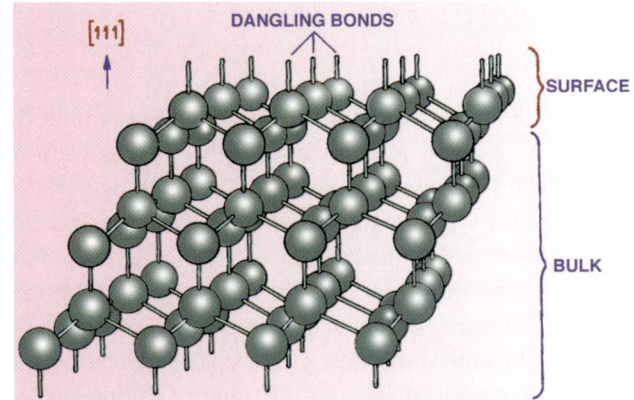

Fig. 13-14 Schematic diagram of the (dangling) bonds at a clean Si-surface.

CHAPTER 13 OXIDATION

trapped at defects in the bulk-regions of the oxide. Normally, the level of such defects is insignificant in freshly-grown SiO_2 layers, but they can be created later if the oxide is exposed to X-ray radiation, high-energy electrons, or gate-current.

13.7 DOPANT-REDISTRIBUTION IN OXIDATION

The thermal-oxidation process impacts the doping-concentration near the silicon-surface.[11] This change in dopant-concentration is called *dopant-redistribution*. It is an important effect since the doping-concentration at the surface of the silicon controls such device parameters as the threshold-voltage (V_T) of MOSFETs. Dopant-redistribution during oxidation is due to two main factors. The first is the *segregation coefficient*, m, defined as:

$$m = \frac{\text{equilibrium concentration of impurity in Si}}{\text{equilibrium concentration in SiO}_2} = \left(\frac{C_{Si}}{C_{SiO_2}}\right) \quad (13.8)$$

That is, as oxide is grown, both silicon and any impurities in the silicon get incorporated in the SiO_2-film. However, SiO_2 is only able to accept an impurity up its solid-solubility (see Chap. 9). If $m > 1$, this solubility is smaller than that of the silicon. Thus, all of the impurity in the silicon cannot be accepted by the oxide. Hence, the impurity-concentration in the oxide will be lower than in the silicon ($C_{SiO_2} < C_{Si}$). If $m < 1$, all of the impurity in the silicon *can* be incorporated into the growing oxide (and $C_{SiO_2} > C_{Si}$). Table 13-4 shows the range of m values for selected dopant-impurities in silicon.

Dopant-redistribution is also impacted by a second factor, namely the *diffusivity of the dopant-impurity in the Si and the oxide*. If the dopant quickly diffuses away from the interface (and then through the oxide), the concentration in the Si can be further reduced. Even if impurities segregate into the oxide, they may escape from it through the SiO_2 surface. Specifically, if impurities have a high-diffusivity in oxide, their concentration in the oxide will be low due to such escape. Gallium has high diffusivity in SiO_2 (as does B if subjected to H_2-ambients at high-temperature). A pictorial representation of the four dopant-redis-

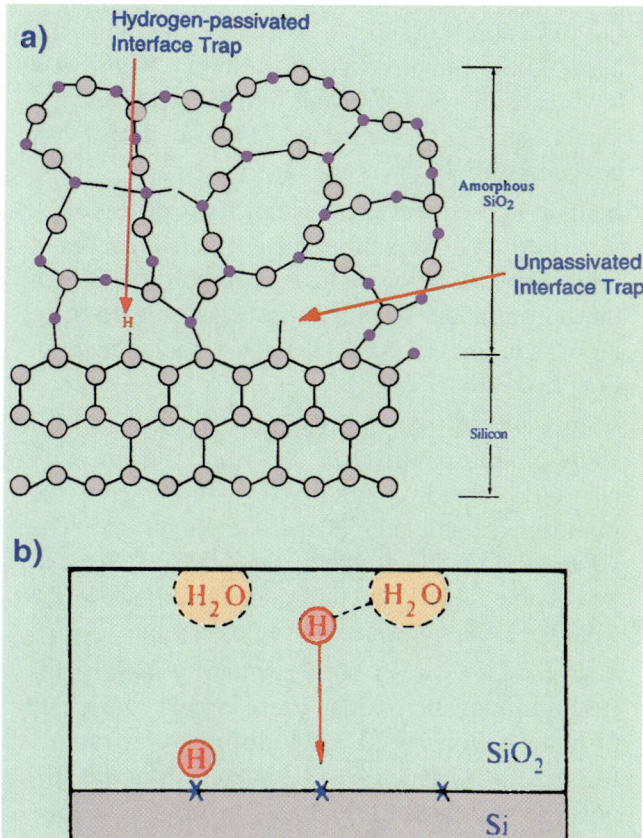

Fig. 13-15 (a) How hydrogen can passivate the dangling-bonds at the silicon-surface that have not been completed by SiO_2. (b) Conceptual-picture of how hydrogen is thought to penetrate the gate-oxide and travel to the oxide-Si interface, and passivate dangling-bonds.

Although oxidation ties up most of them, a few are left (Fig. 13-15a). These remaining dangling-bonds trap electrons, and this trapped-charge gives rise to Q_{it}. An anneal in a hydrogen-ambient at 450°C is normally the final-step prior to assembly and packaging of an IC. This is done to allow hydrogen to penetrate the gate-oxide and to try to tie-up those dangling-bonds not satisfied by oxidation (see Fig. 13-15b). This is able to lower the value of Q_{it} to about $10^{10}/cm^2$.

4. *Oxide trapped-charge* per unit area, Q_{ot} (C/cm^2). These charges are due to electrons or holes

Table 13-4 VALUES OF THE SEGREGATION COEFFICIENT OF IMPURITIES AT THE Si/SiO$_2$ INTERFACE

Impurity	Thermodynamic Estimates	Redistribution Experiments	Oxide Masking Experiments
Ga	> 1000		20.0
B	0.001–1000	0.3, 0.1	0.01
In	> 1000		
P, Sb, As	>1000		≈10

tribution cases is shown in Fig. 13-16. Note that two general circumstances exist: **1)** $m < 1$ (which means that the oxide takes up all the impurity present at the Si/SiO$_2$ interface); and **2)** $m > 1$ (which means that the SiO$_2$ rejects some of the impurity). The resulting dopant profiles in both the Si and SiO$_2$ then depend largely on how rapidly the impurity diffuses through the oxide, as discussed next (see also Fig. 13-16):

a. In *Case #1* (shown in Fig. 13-16a), $m < 1$ and the diffusion through the oxide is slow (as is the case of boron in an oxidizing-ambient, where $m = 0.3$). Thus, boron segregates into the oxide and the silicon-surface gets depleted of boron. The boron concentration in the oxide also increases.

b. If in subsequent processing, the oxide is subjected to a hydrogen-ambient (which causes a rapid-increase in the diffusivity of boron in SiO$_2$), the boron at the silicon-surface is depleted even more than in *Case #1* (i.e., see *Case #2*, shown in Fig. 13-16b).

c. In *Case #3* (see Fig. 13-16c) the oxide rejects some of the impurity ($m > 1$) *and* the impurity also diffuses slowly in the oxide. In this case, the impurity accumulates at the silicon-surface. The value of m for the *n*-type dopants in Si (P, As, and Sb) is about 10. Thus, these dopants pileup in the Si at the silicon-surface during oxidation.

d. In *Case #4*, $m > 1$ but the impurity rapidly diffuses through the oxide. For example, the value of m for gallium is 20, but it diffuses very rapidly through SiO$_2$. Thus, Ga is depleted from the silicon-surface, but little remains in the oxide (Fig. 13-16d).

The effects of *B-depletion*, and *P-* and *As-pileup*, are particularly important in both bipolar and MOS processing, Process-design must take both effects into account, and it may be necessary to add or change process steps to overcome the effects of these phenomena.

13.8 THE OXIDATION OF POLYSILICON

Oxide-layers will also grow on polysilicon if it is exposed to oxygen at high-temperatures. Poly-silicon-oxides are important device-layers in non-volatile memory-structures. The oxidation-rate of polySi is impacted by some factors that do not apply to oxides grown on single-crystal wafers. These include: poly grain-size; the type of dopant in the poly; and its concentration. Heavily-doped-polysilicon typically oxidizes more rapidly than lightly-doped-polysilicon.

The quality and reliability of poly oxides is cru-

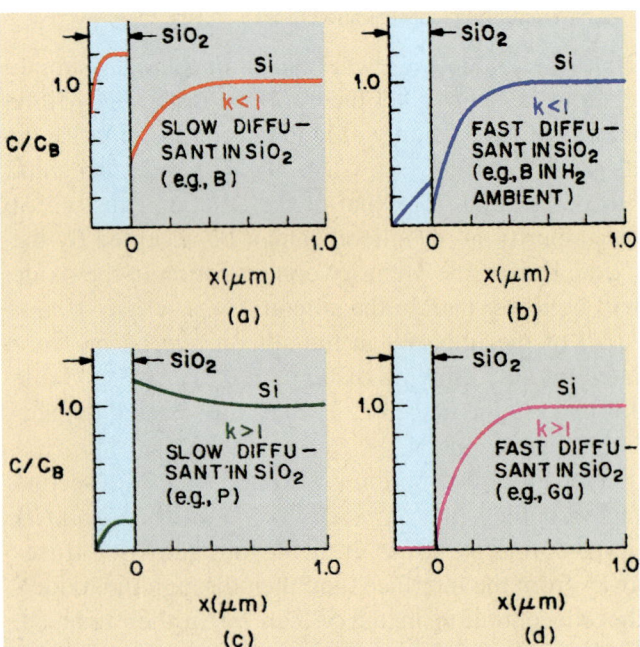

Fig. 13-16 The effect of thermal-oxidation of silicon on the *impurity-segregation* at the Si/SiO$_2$ interface: (a) slow-diffusion in oxide, $m < 1$ (boron); (b) fast-diffusion in oxide, $m < 1$ (boron in H$_2$-ambient); (c) slow-diffusion in oxide, $m > 1$ (phosphorus); and (d) fast-diffusion in oxide, $m > 1$ (gallium).[11] Reprinted with permission of the American Physical Society.

13.9 OXIDATION SYSTEMS

In this section we discuss the role of *thermal processing tools* (also termed *furnaces*) in the manufacture of ICs.[12] Conventional horizontal- and vertical-furnaces, and rapid-thermal-processors (RTP) are described.

13.9.1 Oxidation-Furnaces

Oxide film-growth is still mainly a *batch-process*. That is, many wafers (up to 150 in a batch) are loaded into an oxidation-furnace, where they are all oxidized at once. Since the oxide growth-rate is highly-dependent on temperature, this parameter must be tightly controlled over a long-length of the oxidation tube. The constant-temperature-region of the furnace is called its *flat-zone* (Fig. 13-17a).

The hot-wall *horizontal-diffusion-furnace* was the workhorse until the mid-1980s. Figure 13-17b shows wafers loaded on a *quartz-boat*. Figure 13-17c shows the wafers on a boat residing in a *white-elephant*, waiting to be inserted into a horizontal furnace-tube.

In the late 1980s, *vertical*-furnaces began to replace horizontal-furnaces in production. Vertical-furnaces provide technological advantages, while horizontal-furnaces offer high-productivity at lower cost. More recently *rapid-thermal-processors* (RTP) have begun to displace vertical-furnaces in some applications. Single-wafer, RTP-chambers are used when the process-ambient must be tightly-controlled (for instance, when oxygen must be excluded from the processing ambient), and when process-times are very short.

Vertical-furnace technology has also continued to evolve. Fast-ramp, small-batch vertical-furnaces now compete with RTP systems in some processes where thermal-budget and throughput are key factors. There are a large number of conventional furnace suppliers, including: Tokyo Electron (TEL); Kokusai; Silicon Valley Group; ASM International (ASMI); and Semitherm. Suppliers of RTP systems include: Applied Materials; Steag; and Axcelis.

13.9.2 Horizontal-Furnaces

Most horizontal-furnaces are configured as a 4-stack system (i.e., with four high-purity fused-silica furnace tubes stacked above one another (see Figs. 13-18 and

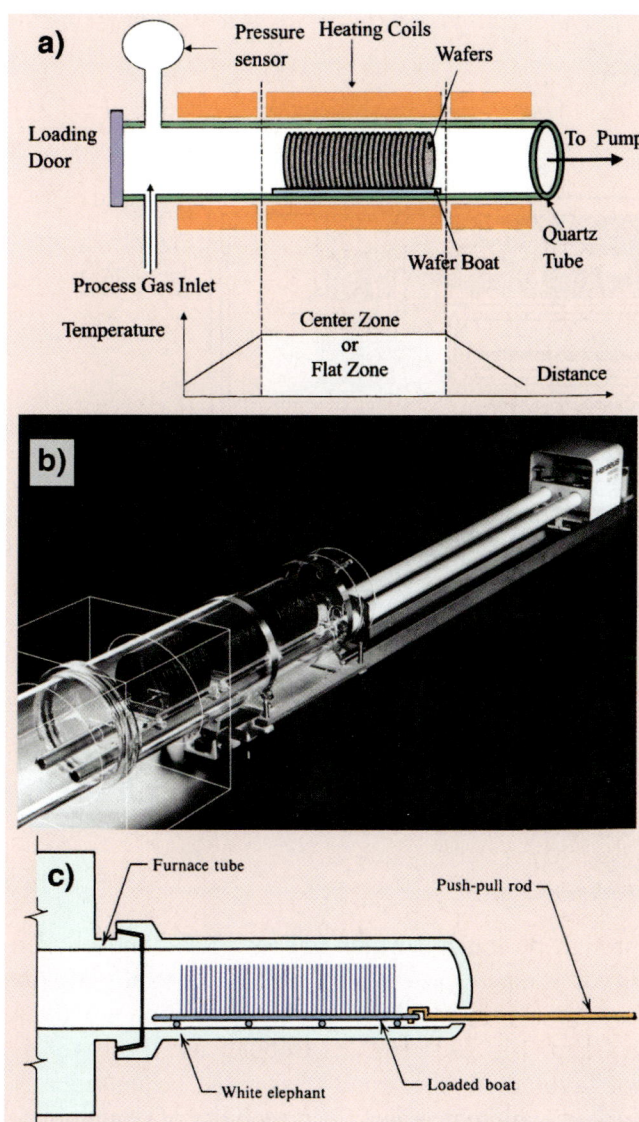

Fig. 13-17 (a) Flat-zone of an oxidation tube-furnace. (b) Wafers on a boat, ready to be loaded into a furnace-tube. (c) Loaded diffusion-boat ready to be inserted into a furnace. The use of push-pull rods and a removable tube-extension (called a *white elephant*), are also depicted in (c).

cial. One important quality, its *breakdown-strength*, is dependent on the smoothness of the polysilicon surface. Typically, the rougher the surface, the lower the breakdown-strength (since such roughness enhances the local electric-field at some spots). This leads to poly-oxide-breakdown at lower voltages.

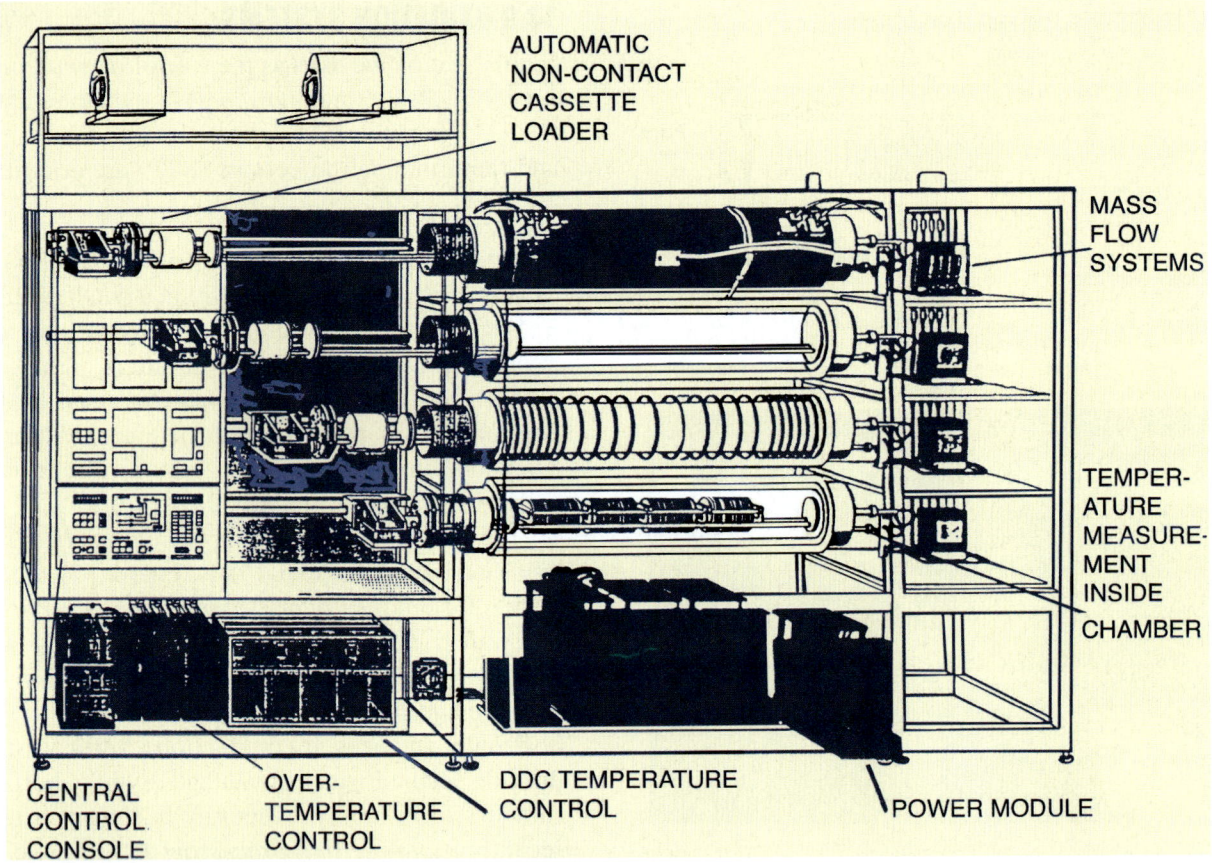

Fig. 13-18 Schematic drawing of a *4-stack, horizontal-tube oxidation-system*.

13-19). A horizontal-furnace has the following sub-systems: **a)** a furnace-cabinet; **b)** heating-elements; **c)** measuring and control thermocouples; **c)** fused-silica process-tubes; **e)** fused-silica paddles and boats; **f)** a temperature-control system; **g)** a load-station; and **h)** a source gas-cabinet and gas-delivery-system (often called a *jungle,* Fig. 13-20). For additional information, readers should consult Ref. 2 and product-literature distributed by horizontal-furnace suppliers.

Wet-Oxidation Gas Sources: Wet-oxidation is typically used to grow thicker oxides, as the oxide growth-rate is faster than if dry-oxidation is used. The H_2O species (*water-vapor*) needed for the wet-oxidation reaction (see Eq. 13-3) can be obtained from either a bubbler, or from a pyrogenic reaction of H_2 and O_2.

If a *bubbler* is used (Fig. 13-21a), oxygen is passed through de-ionized water (DI) at 95°C. The oxygen serves as a *transport-gas* to carry water-vapor into the furnace. (More details on bubbler operation are given in Chap. 16, Sect. 16.2.2.) Bubbler use, however, is not favored, as the water-flask must be continually refilled, which increases the possibility of contaminating the water (and thus, also the oxide being grown).

In the *pyrogenic-steam* method, hydrogen and oxygen gases are separately fed directly into the entrance of the heated furnace-tube (Fig. 13-21b). Pure gases are used, and this also eliminates the handling problems of using bubblers. Hydrogen ignites spontaneously in the presence of oxygen at about 400°C to form steam. A burn-box must be used at the tube-exhaust to burn any residual-hydrogen before releasing the process-gas effluents into the atmosphere (as hydrogen is a flammable and explosive gas).

13.9.3 Vertical-Furnaces

Despite their higher cost, *vertical-furnaces* are used instead of horizontal-furnaces in many applications because they offer the following benefits: **1)** superior process-control; **2)** fewer particles-per-wafer; and **3)** better automation-compatibility. These are three of the major economic drivers in advanced wafer-fabs.

Process control is improved for several reasons: **a)** the temperature within the vertical-furnace is more uniform; **b)** the wafers are held flat on their backs and are well-centered in the *tower*. The tower (and the wafers) can thus be more easily centered in the *furnace-tube*. This allows gas-flow dynamics in the furnace to be optimized; **c)** the wafer-spacing is equal between all wafers in the load; **d)** since the tower is loaded vertically into the tube, it is easier to keep it centered, than in a horizontal furnace (in which a heavy boat must be loaded and then maintained in position by a long cantilever; and **e)** the tower can be rotated (which averages out variations in the temperature and gas-flow).

Fewer particles are generated in vertical-furnaces for the following reasons: **1)** Vertical-furnaces use automated loading and unloading; **2)** The wafers are shielded from falling particles by the wafers above them. (Any articles flaking off the tube-walls tend to fall into the space between the walls and the tower);

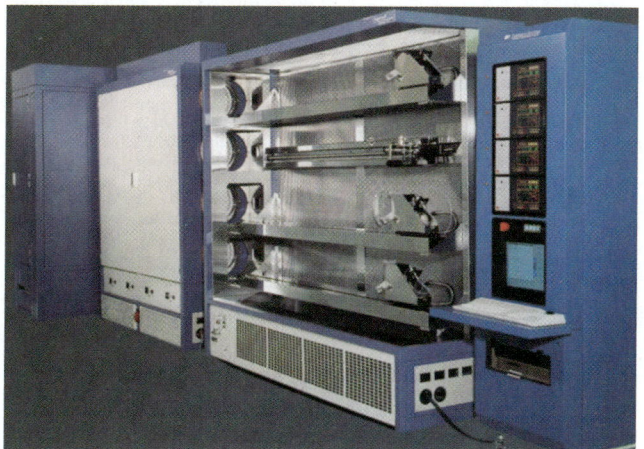

Fig. 13-19 Photograph of a *4-stack, horizontal-tube oxidation-system*. Courtesy of ASM Lithography.

Fig. 13-20 Gas-panels used with a horizontal-tube furnaces.

3) An *ultra-low-particle-filter air-handling system* (*ULPA*) can be integrated into the load-station; and **4)** Many furnaces have an automated wafer-stocking feature, in which multiple-cassettes of 25-wafers can be maintained in an ultra-clean ambient.

13.9.4 The Construction and Operation of Vertical-Furnaces

Figure 13-22a shows a vertical-furnace. The *process tube* is made of fused silica. In some such furnaces a double-walled tube is used. This allows inert or chlorinated-gases to be flowed between the inner and outer-tubes, a technique employed to keep impurities from diffusing through the inner-tube and entering the process-region. A *fused-silica* (or *silicon*) *tower* is used to hold the wafers in a horizontal position (Fig. 13-22b). Up to 150-wafers per batch can be processed in such tools.

Wafers are loaded onto the tower from a cassette by a robot (Fig. 13-22c). The tower is next moved into

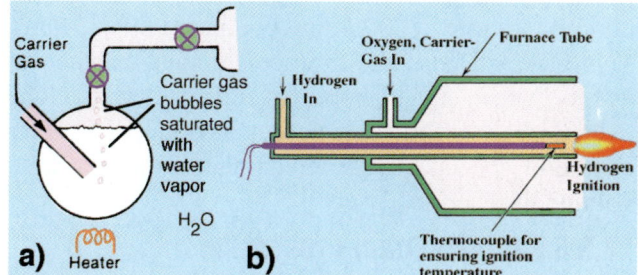

Fig. 13-21 (a) *Bubbler* that provides water-vapor for wet-oxidation. (b) *Hydrogen-injection system* for steam oxidation.

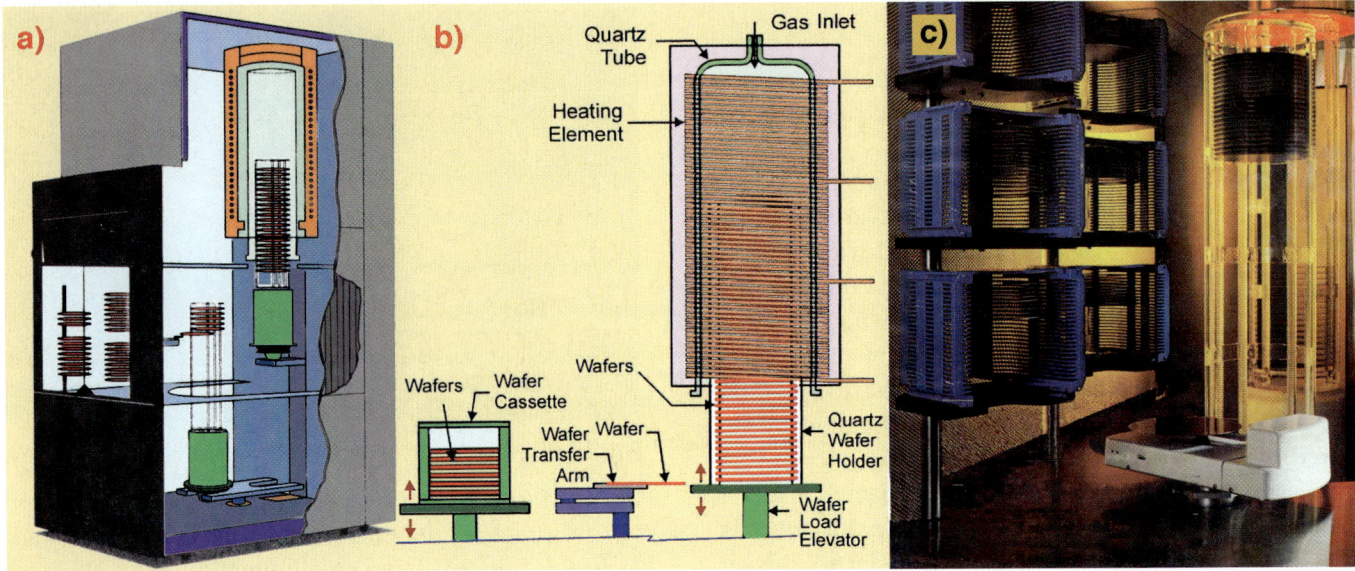

Fig. 13-22 (a) Cutaway drawing of a *vertical-furnace*. (b) Cross-section of a vertical furnace-tube and boat-loading system; (c) Photograph of a vertical furnace-tube and boat.

the hot-zone of the furnace, which is held between process steps at lower-temperature (700–800°C). Then the temperature is slowly raised to the process value. When the process is complete, the wafer temperature is slowly reduced until the wafers are cool enough to be withdrawn from the furnace.

Fast-Ramp, Small-Batch Furnaces: Recently, a new type of vertical-furnace called the *fast-ramp, small-batch furnace* has been introduced. These can rapidly raise the temperature of a batch of 50-wafers to the process temperature (*ramp-up time*), and can also quickly cool the wafers after the process is completed (*ramp-down time*). The ramp-up-rate is increased to 100°C/min from 10-20°C/min in conventional-furnaces, and the ramp-down-rate to 60°C/min (compared to 5°C/min). This makes it possible to process a smaller batch of wafers with a much-shorter overall cycle-time. The overall process-time can be cut in about half. This can be of special benefit to companies that run low-volume production.

13.9.5 Rapid-Thermal-Processing

Rapid-thermal-processing (RTP) is a single-wafer, thermal-processing method in which the process-temperature is ramped up and down very rapidly (e.g., at a rate of 75–200°C per second, compared to less than 1°C per second in a furnace). This means, for example, that RTP can heat a wafer from room-temperature to 1100°C in a few seconds. The use of RTP entered mainstream IC-processing in the mid-1990's, and its use since then has increased rapidly.[13]

RTP offers a number of advantages over furnaces, including: **1)** reduced thermal-budget; **2)** single-wafer processing; **3)** higher-temperature processing; **4)** better control of the process-ambient; **5)** shorter process-time-per-wafer; and **6)** ease of clustering multiple-tools. RTP is the preferred-approach for such applications as: **a)** post-ion-implantation annealing and activation; **b)** titanium- and cobalt-silicide annealing; and **c)** *rapid-thermal-oxidation (RTO)* to form ultrathin gate-oxide layers.

In spite of these benefits, RTP has not yet completely replaced furnaces. Furnaces still perform well enough for some applications and are used in them because they are less-expensive than RTP. However, RTP will be required for even more of the production-processes used to make feature sizes smaller than 0.18-μm on 300-mm wafers.

13.9.6 RTP Systems

RTP systems are available in a number of configurations. The RTP-chamber can be operated at atmospheric- or reduced-pressure (Fig. 13-23). The wafer in the chamber can be heated in a number of ways (including graphite-heaters), but the most common method uses *tungsten-halogen lamps*. These lamps are usually configured in an array, with one array above, and one below the wafer. The lamp-array can be arranged in a *linear-configuration* (Fig. 13-24), or in a *honeycomb-manner* (Fig. 13-25). The honeycomb-arrangement offers the benefit of close-packing and fine spatial-resolution of the lamp-output. The lamps generate intense-heat in the form of infra-red radiation, and efficiently transfer it to the wafers

Fig. 13-24 RTP-chamber heated with a *linear lamp-array*.

without causing much heating of the transparent chamber-walls.

The wafer-temperature is precisely measured by several infrared-pyrometers within the system. Feedback signals are used to control the power of the lamps in different heating-zones to achieve very precise and uniform wafer-heating. In addition, the wafer can be rotated in the chamber to improve heating and gas-flow uniformity. Figure 13-26 is an example of a cross-section of an RTP-system.

13.10 OXIDATION PROCESS-RECIPE

The goal of a thermal-oxidation process is to grow a defect-free, uniform layer of SiO_2 to the specified thickness. In today's processes, thin-oxides are most commonly grown using dry-oxidation. Certain process conditions are followed in a specific sequence, called a *process recipe*. Here, an example of such a growth sequence for producing oxides ~50-100-Å thick in a batch-oxidation-furnace is given.

The wafers are cleaned just prior to loading into the oxidation-furnace (using an RCA-clean, see Chap. 8). While the furnace is in *idle-mode* waiting for the next wafer-batch, nitrogen purge-gas is continuously flowed through it, and it sits at an elevated temperature (~800°C). The wafers are loaded onto the fused-

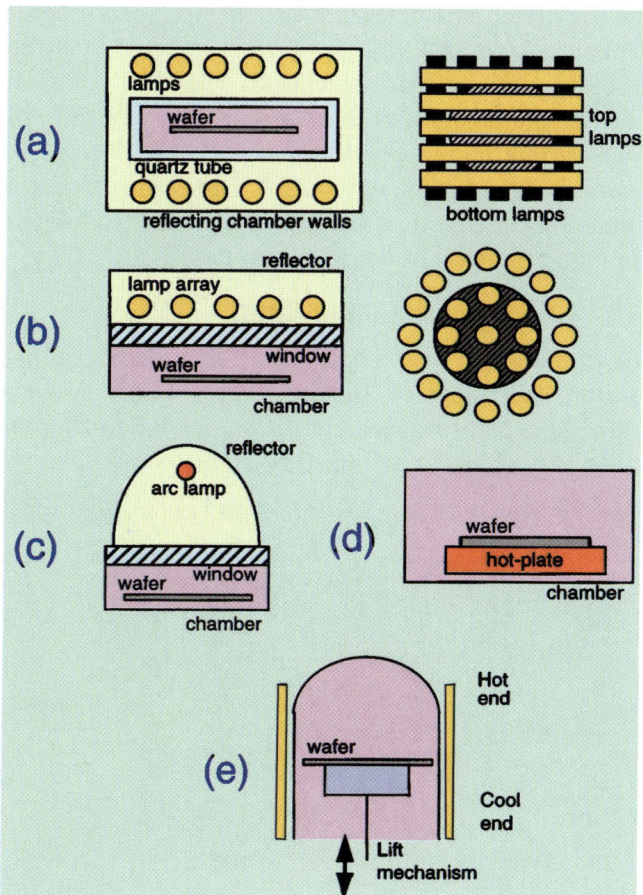

Fig. 13-23 Drawings of various RTP reaction-chambers.

230 MICROCHIP MANUFACTURING

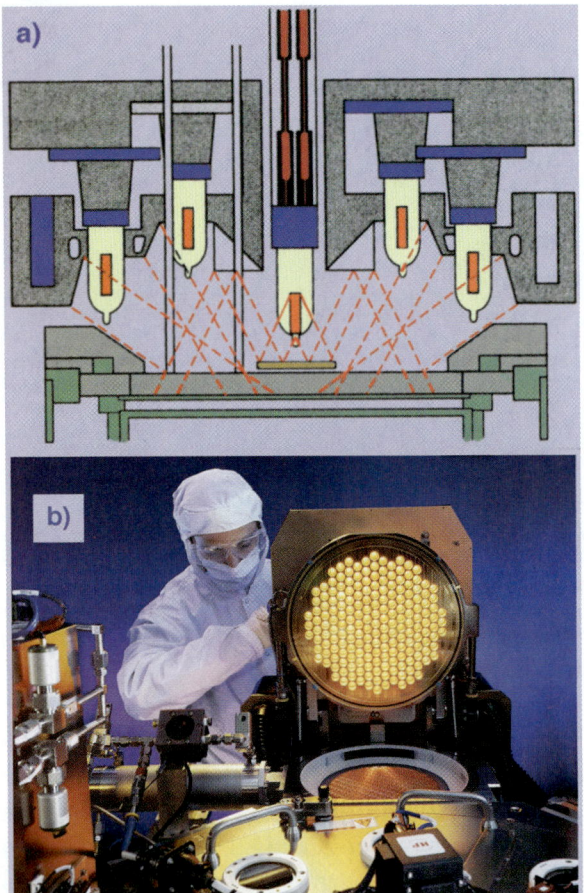

Fig. 13-25 RTP reaction-chamber heated with a honeycomb-array of lamps. Courtesy of Applied Materials.

silica or silicon-carbide *boat* (or *tower*), and the boat with the wafers is slowly pushed into the chamber (with process-N_2-gas flowing). It takes several minutes to push the wafer-holder into the furnace-flat-zone to prevent wafer-warping that would occur due to a more-sudden temperature-change (see Chap. 9).

When the wafers arrive in the flat-zone, the furnace-temperature is ramped-up at a rate of 10-20°C/min. After the process-temperature is reached (e.g., 950°C), the furnace is given a few minutes to stabilize under N_2-flow. Then dry-oxygen (and perhaps a chlorine-containing gas) are turned on, and the nitrogen is turned off. The oxide grows while the oxygen flows. After the oxide has been grown to the specified-thickness, the oxygen and Cl-gas are turned off.

The wafers then undergo a post-oxidation anneal for about another 30-min in nitrogen at the process-temperature (to reduce the oxide fixed-charge, Q_f). Next, the furnace-temperature is slowly ramped back-down (~30 min). When the idle-temperature is reached, the tower is withdrawn slowly from the furnace. The total time of this process (from loading to unloading the wafers) is about 2-hours. To allow adequate throughput (*wafers-per-hour*), a large number of wafers (up to 150 per batch) must be processed at once.

13.11 OXIDE FILM-THICKNESS MEASUREMENTS

The thickness of oxide-films must be accurately measured during IC-fabrication to ensure adequate process-control.[14-16] For example, the gate-oxide thickness must be kept within a tight range, as it is an important parameter that establishes the threshold-voltage V_T of MOSFETs. Various methods can be used to measure oxide-thickness. The most common are: **a)** examination of the *color of the oxide-film*; **b)** *optical-interference*; **c)** *ellipsometry*; and **d)** *transmission electron microscopy* (*TEM*). Optical-interference and ellipsometry are the most widely used techniques in modern production environments.

13.11.1 Oxide-Film Color-Chart

Silicon-dioxide layers exhibit different colors on the wafer, depending on their thickness While silicon-dioxide itself is transparent (glass), an oxide-film on a silicon wafer has a color. The color seen is actually

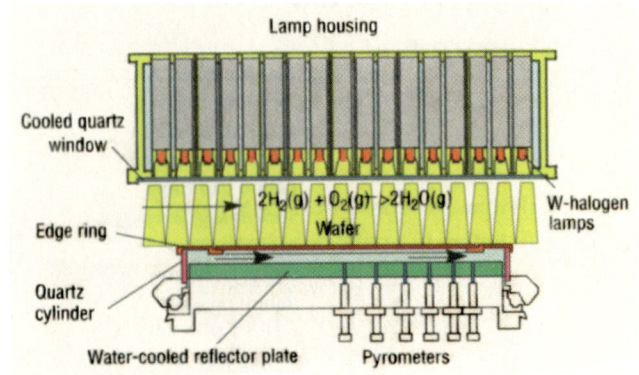

Fig. 13-26 Schematic drawing of a *rapid-thermal processor*. Courtesy of Applied Materials, Inc.

Color	Thickness (Å)			
	1	2	3	4
Grey	100			
Tan	300			
Brown	500			
Blue	800			
Violet	1000	2800	4600	6500
Blue	1500	3000	4900	6800
Green	1800	3300	5200	7200
Yellow	2100	3700	5600	7500
Orange	2200	4000	6000	
Red	2500	4400	6200	

Fig. 13-27 Color-chart for thermally-grown SiO_2 films observed perpendicularly under fluorescent lighting.

the result of an interference phenomenon, the same as the one that creates the colors of a rainbow. The SiO_2-film is actually a thin transparent-film on a reflecting-substrate. Some of the light impinging on the wafer will reflect-off of the oxide-surface, while the rest will pass through the transparent-oxide and reflect-off the mirrored wafer-surface. When the reflected-light rays exit the film, they combine with the surface-reflected ray, giving the appearance of having a color.

The exact color is a function of three factors: **1)** the *index-of-refraction* of the film; **2)** the *viewing-angle*; and **3)** the *film-thickness*. The color of the film becomes an indication of the thickness when the type of viewing-light and viewing-angle are specified (e.g., daylight or fluorescent). A color-versus-thickness chart is then produced to provide a standard from which to determine the thickness (Fig. 13-27).

Note that as the film gets thicker, the colors change in a specific-sequence, and then repeat themselves. Each repetition of the color is called an *order*. To determine the exact film-thickness, a knowledge of the *color-order* is necessary. A principal-use of the color chart is process-control. A typical color-chart is accurate to ±30-nm. However, with the advent of high-speed, microprocessor-controlled techniques, the color-chart method does not find as much use as it once did.

13.11.2 Optical-Interference for Measuring Oxide Film-Thickness

The *optical-interference* technique is a non-destructive, non-contact method widely used for determining the thickness of single-layer, transparent, dielectric-films on silicon. SiO_2-films are the kind most frequently measured with this method. However, it also works well for other transparent-films, as long as their index-of-refraction is known. Optical-interference can be employed to obtain the thickness of oxide-films from 100-Å to 10,000-Å (1.0-μm).

Monochromatic-light of different-wavelengths is directed at nearly-normal incidence to the surface of a wafer with an oxide layer of unknown thickness (Fig. 13-28). The reflected-light intensity is measured as a function of its wavelength.

Some of the light is reflected from the top of the oxide-film, while the rest travels through the film and is reflected from the silicon-surface (which is at the bottom of the oxide-film). If the wavelength of the light is such that the light-beams reflected from the top and bottom of the oxide-film interfere-construc-

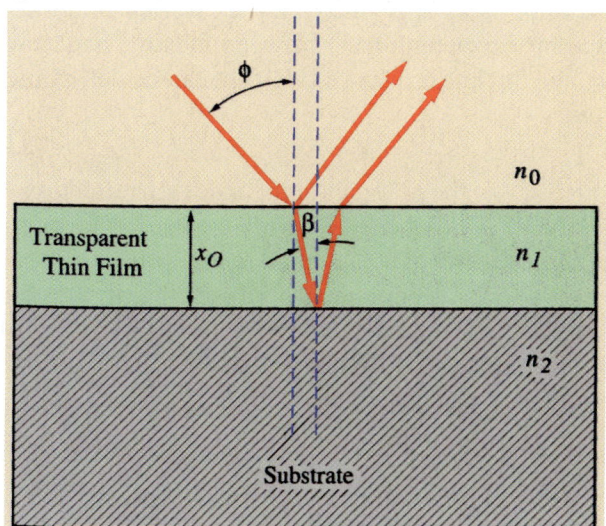

Fig. 13-28 Using *reflected-light* and *interference* to measure oxide film-thickness.

232 MICROCHIP MANUFACTURING

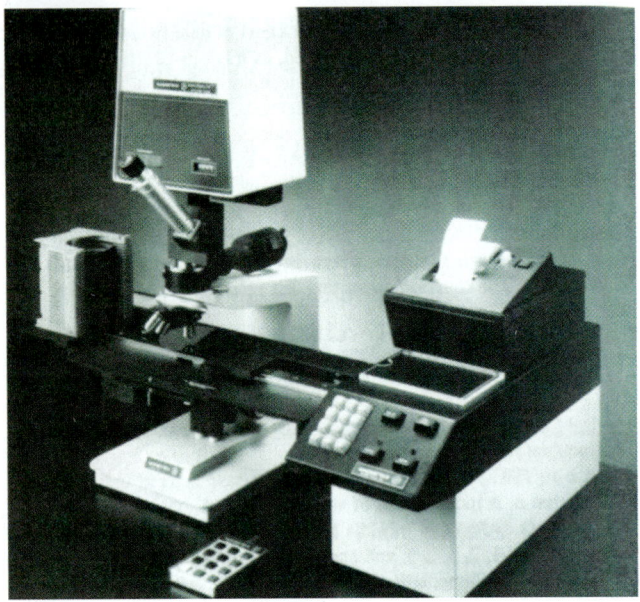

Fig. 13-29 Automated thin-film thickness measuring-system, based on interference. Courtesy of Nanometrics.

tively, an intensity-maximum will occur. If the waves interfere-destructively, an intensity-minimum occurs. For some values of λ the two reflected-waves will interfere-constructively, but for other values of λ, destructive-interference occurs. The result is that the intensity of the reflected-light will go though maxima and minima as λ is varied. By measuring the difference in wavelength (Δλ) between maxima and minima, the thickness of the oxide-film can be determined from:

$$t_{ox} = \Delta\lambda / 2n_{ox} \quad (13.9)$$

where n_{ox} is the real-part of the index-of-refraction of the oxide. A number of instruments that perform this technique are available commercially (Fig. 13-29).

While this technique works reliably if the film thickness is greater than a few-hundred angstroms, for thinner films it is difficult to detect the first minimum unless very-short-wavelength light is used

13.11.3 Ellipsometry

A sophisticated optical-technique for measuring oxide-thickness and index-of-refraction is *ellipsometry*. It is better than optical-interference for measuring very-thin oxide-films. By using ellipsometry the thickness of oxide-films as thin as 20-Å (and in some cases only 10-Å thick) can be reliably determined.

The ellipsometry technique makes use of the fact that when a light-beam is reflected, its polarization changes (as shown in Fig. 13-30a). By monitoring this change, data can be obtained about the refractive-index and the thickness of the dielectric-film on a wafer-surface.

A polarized-beam of light is reflected-off the oxide-surface at some angle, with a He-Ne laser being the common light-source. The reflected-light-intensity is measured as a function of the polarization-angle (Fig. 13-30b). By comparing the incident and reflected light-intensity, and the change in polarization-angle, the film-thickness and the index-of-refraction can be found. In modern ellipsometers, light with

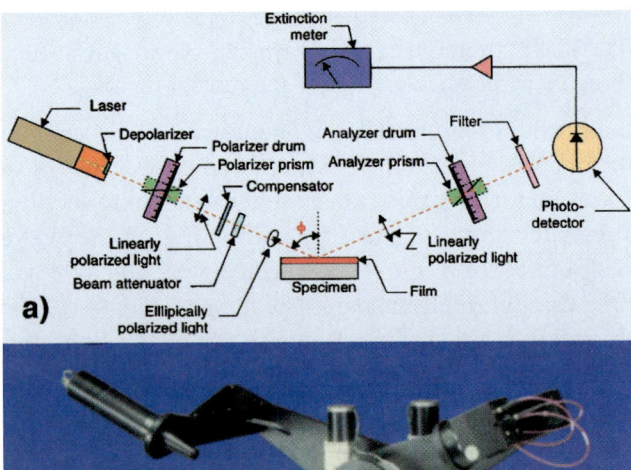

Fig. 13-30 (a) Drawing showing the principles of ellipsometry. (b) Photograph of an ellipsometer. Courtesy of J.A. Woollam Corp.

four different wavelengths is used (*multi-wavelength ellipsometry*). This allows determination of thickness for unknown, multilayer-samples, better sensitivity to film-thickness changes, and improved data-averaging.

In the past, ellipsometry-measurements required skilled and experienced workers to obtain accurate results. However, automated-equipment has since been developed (Fig. 13-31). Usually these instruments are computer-controlled so that the film properties are calculated automatically. Such instruments are capable of measuring oxide-film-thicknesses in the range of 20-Å to 20,000-Å thick.

13.11.4 Transmission-Electron-Microscopy Techniques for Measuring Oxide-Film-Thickness

Oxide film-thickness is most accurately measured from *high-resolution transmission-electron-microscope* (*HRTEM*) images of oxide-films fabricated in a *MOS-capacitor structure*. The lattice spacing of the Si substrate of the MOS-capacitor is clearly visible in such HRTEM-images. Thus, the distance between the silicon lattice-planes provides an accurate and automatic thickness-standard. By using this standard, the

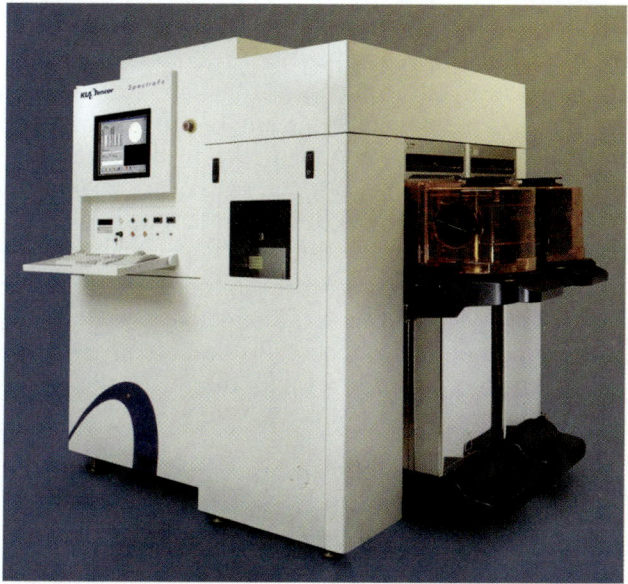

Fig. 13-31 Photograph of an automated thin-film thickness-measuring system. Courtesy of KLA-Tencor.

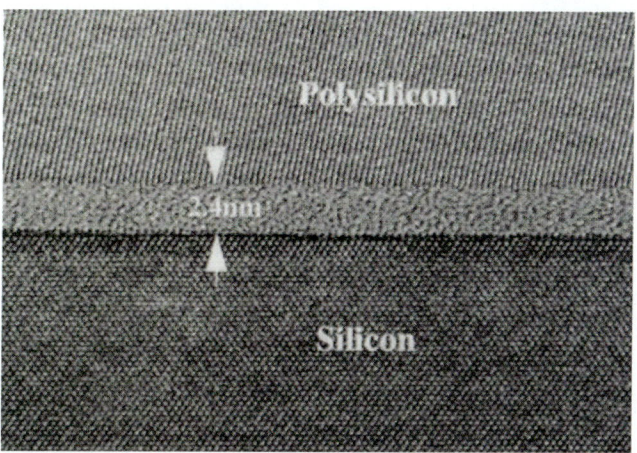

Fig. 13-32 High-resolution-TEM showing a crystalline silicon/thin-oxide/polysilicon structure. The lattice planes in the crystal are defined. Courtesy of Applied Materials.

thickness of the oxide can be accurately determined.

However, preparing samples for HRTEM is time-consuming and expensive. Hence, this technique is mainly used for research and development applications. Figure 13-32 shows an HRTEM cross-section of a thin oxide-film in a MOS-C device-structure.

REFERENCES

1. S. Wolf and R.N. Tauber, *Silicon Processing for the VLSI Era - Vol. 1: Process Technology,* 2nd-Ed., Lattice Press, Sunset Beach CA, 2000, Ch. 8 (*Thermal Oxides*).

2. S. Wolf, *Silicon Processing for the VLSI Era - Vol. 4: Deep-Submicron Process Technology,* Lattice Press, Sunset Beach CA 2002, Ch. 3 (*Thin Gate-Oxides*).

3. S. Wolf, *Silicon Processing for the VLSI Era Vol. 3: The Submicron MOSFET,* Lattice Press, Sunset Beach CA 1995 Ch. 7 (*Thin Gate-Oxides - Growth and Reliability*).

4. *The International Technology Roadmap for Semiconductors (ITRS) 2002.* International SEMATECH.

5. A. Revesz, "Defect-Structure of Grown Silicon Dioxide Films," *IEEE Trans. Electron Dev.*, **ED-12**, 97 (1965).

6. B. Deal and A. Grove, "General Relationship for Thermal Oxidation of Silicon," *J. Appl. Phys.*, **36**, 3770 (1965).

7. C.P. Ho, *et al.*, "Thermal Oxidation of Heavily Phosphorus Doped Silicon," *J. Electrochem. Soc.*, **125**, 665 (1978).

8. H.Z. Massoud, J.D. Plummer, and E.A. Irene, "Thermal

Oxidation of Silicon in Dry Oxidation: Growth-Rate Enhancement in the Thin Regime," I. Experimental Results, II. Physical Mechanisms," *J. Electrochem. Soc.*, **132**, 2685 (1985).

9. B.E. Deal, "Standardized Terminology for Oxide Charges Associated with Thermally Oxidized Silicon," *IEEE Trans. Electron Dev.*, **ED-27,** 606 (1980).

10. E.H. Nicollian and J.R. Brews, *MOS (Metal Oxide Semiconductor) Physics and Technology,* John Wiley & Sons, Inc., New York, 1982, Chap. 12.

11. A.S. Grove, O. Leistiko, and C.T. Sah, "Redistribution of Acceptor and Donor Impurities During Thermal Oxidation of Silicon," *J. Appl. Phys.*, **35**, 2629 (1964).

12. P. Singer, "Furnaces Evolving to Meet Device Thermal Processing Needs," *Semiconductor Intl.,* Mar. 1997, p. 85.

13. P.J. Timans, R. Sharangpani, and R.P.S. Thakur, "Rapid Thermal Processing," in *Handbook of Semiconductor Manufacturing Technology,* Y. Nishi and R. Doering, Eds., Marcel Dekker, New York, 2000, Ch. 9, p. 201-286.

14. D.K. Schroder, *Semiconductor Material and Device Characterization*, 2nd Ed., John Wiley, New York (1998).

15. R. DeJule, "Advances in Thin-Film Measurement," *Semiconductor International*, May 1998, p. 52.

16. T. Hori, *Gate Dielectrics and MOS ULSIs*, Springer-Verlag, Heidelberg (1997), p. 41.

PROBLEMS

1. What are the differences between a thermally-grown and a CVD-deposited oxide layer?
2. Describe the microscopic-structure of *fused silica* (i.e., thermally-grown silicon-dioxide).
3. Why are the gate-oxides of MOSFET devices thermally-grown (i.e., instead of being deposited)?
4. Explain *gate-oxide integrity* (GOI).
5. Explain why steam (wet) oxidation proceeds at a faster-rate than dry-oxidation.
6. List six applications of thermally-grown silicon-dioxide in ICs, and give a purpose for each one.
7. List the four types of *oxide-charges* associated with the Si/SiO$_2$ interface. Are such oxide charges desirable? Why, or why not?
8. Under what conditions is the thermal-oxide growth-rate of silicon-dioxide *linearly-proportional* to time?
9. (a) If a layer of silicon-dioxide of thickness x, is thermally grown, derive the thickness of the silicon consumed (starting with the density of Si and SiO$_2$, and the chemical-reaction equation of Si and O$_2$). (b) If a (111) silicon-wafer is oxidized, how much silicon is consumed while 3000-Å of thermal oxide is grown? How much is consumed if the orientation is (100)>

10. A silicon-wafer has a 0.3-μm-thick layer of thermally-grown silicon-dioxide on it. Calculate the weight of this SiO$_2$ layer.

11. How does doping in the Si affect SiO$_2$ growth?

12. A *p*-type <100>-oriented, silicon wafer of resistivity 10-Ω-cm is placed in a wet-oxidation system to grow a field-oxide of 4500-Å at 1050°C. Determine the time required to grow this oxide film.

13. A <100>-silicon-wafer has a 0.3-μm layer of thermally-grown SiO$_2$ on its surface. Find the time required to increase the SiO$_2$ to a thickness of 0.5-μm by wet-oxidation at 1100°C. Repeat the calculation for dry-oxidation at 1100°C.

14. A two-step process is used to grow a 0.1-μm thick oxide. The first step grows the first 500-Å, and the second step the remaining 500-Å. If each of the oxidations are carried out by dry-oxidation at a temperature of 1000°C, find the time required for each step (and assume <100> silicon substrates are used).

15. Determine the SiO$_2$ thickness which would result from the following oxidation-sequence: 30-min dry O$_2$, followed by 30-min in wet-O$_2$, both at 1050°C. Assume <100>-Si is being used.

16. A cut is made in an SiO$_2$ film that is 0.4-μm in thickness. An oxide of 150-nm is grown over this cut. Sketch a cross-section of the resulting Si/SiO$_2$ structure, showing: (a) the thickness of the oxide in all locations; and (b) the Si/SiO$_2$ interface at all locations. (Assume a <100> silicon-wafer is being used.)

17. If a *sidewall-oxide-isolated* transistor is to be made with an isolating-oxide thickness of 1.2-μm, how deep should the recess in the silicon be if the oxide surface is to be level with the main silicon surface? Draw a sketch to show the steps.

18. Give some reasons why vertical-furnaces have advantages over horizontal furnaces for growing thermal oxides, and also list some possible disadvantages.

19. What are the temperature ramp-rates in RTP systems? How do they compare to those of furnaces? Why can't the temperature in a furnace be ramped as fast as in an RTP system?

Photograph of a CVD tool for depositing high-*k* dielectrics. Courtesy of Applied Materials, Inc.

CHAPTER 14

PLASMAS USED IN SEMICONDUCTOR-MANUFACTURING

CHAPTER CONTENTS
14.1 INTRODUCTION TO GLOW-DISCHARGES
14.2 RADIO-FREQUENCY (RF) GLOW-DISCHARGES
14.3 HIGH-DENSITY PLASMAS

"I see the atoms, free and fine,
 That bubble like a sparkling wine;
 I hear the songs electrons sing,
 Jumping from ring to outer ring:"

Lister, *The Physicist*

Clouds of gas surround very hot stars in these galactic clusters. The red color arises from radiation emitted by hydrogen plasmas. (Copyright Anglo-Australian Observatory/David Malin Images.)

Matter can exist in three states: solid, liquid, or gas. However, if enough energy is added to a gas, it will assume a so-called *fourth-state of matter*, that of a *plasma*. Plasmas are important in IC-fabrication, being an integral aspect of many processes, including: sputtering; dry-etch; and plasma-enhanced-CVD. They are also used to create ions for ion implantation and to produce the radiant-energy emitted by mercury-arc lamps used in lithography. Since plasmas are encountered in so many ULSI processes, it is useful to provide background-information about the basics of plasmas. This will help readers gain a deeper grasp of the processes that rely on their use. More details about them can be found in Refs. 1 - 5.

To begin, it is assumed that all of the atoms or molecules in an (ideal) gas are *electrically-neutral*. On the other hand, charged-particles exist in a plasma together with neutral gas-atoms. That is, a *plasma* is a gaseous state of matter containing *an equal number of positively and negatively charged particles per unit volume* (electrons and ions) – usually in addition to neutral gas-atoms (Fig. 14-1a). The energy that must be added to

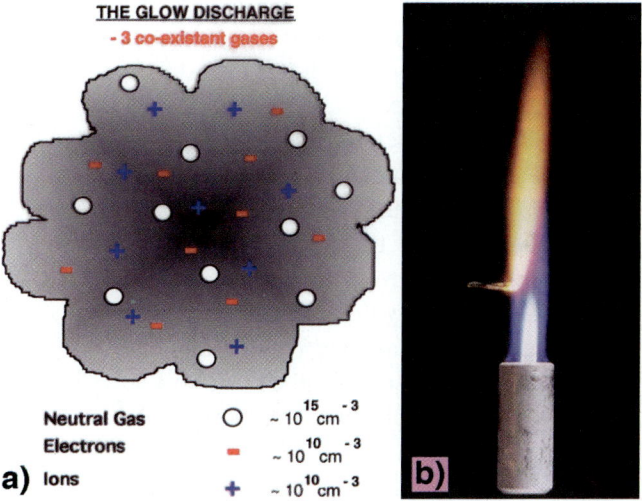

Fig. 14-1 (a) Schematic-diagram of a plasma. (b) A flame is an example of a glowing plasma.

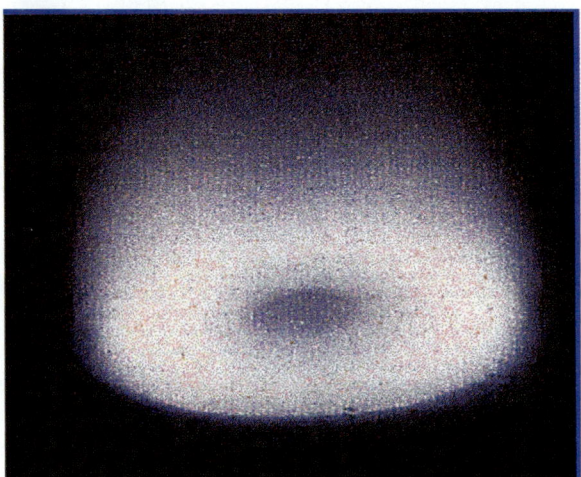

Fig. 14-2 Photograph of a glow-discharge that is excited in a sputtering-tool.

gas to turn it into a plasma can be provided in various ways, but the simplest way is by *heating* the gas.

Plasmas created by adding heat to a gas are found in nature. That is, the matter in stars, sparks, flames, and lightning are all examples of plasmas found in nature (Fig. 14-1b). A gas could be turned into such a *thermal-plasma* if it was heated to a high-enough temperature (dependent on the kind of atoms that constitute the gas). That is, a gas at room-temperature (300 K) contains a negligibly-small concentration of independent charged-particles. To produce a sufficiently-high concentration of charged particles to create a *thermal-plasma*, the temperature of the gas has to be raised to 4,000 K (for such easy-to-ionize gases as cesium) or to 20,000 K (for such hard-to-ionize gases as helium). At these high-temperatures, free-electrons within the gas have high enough kinetic-energies to ionize a sufficient fraction of the gas-atoms to produce a stable plasma-state. However, trying to use such hot gases (plasmas) for wafer-fabrication applications would be impractical, as they would vaporize any thin-films present on the surface of the wafers.

Thus, in order to create plasmas that have gas-temperatures compatible with the thin-films used in IC processing, it must be possible to add energy to a gas in another way to establish a plasma. In arc-welding, an *arc-discharge plasma* is initiated by touching together two electrodes and then separating them while applying a large voltage. This ionizes some of the air molecules (*break down*), causing the air-gas to become conductive. This allows current to flow between the electrodes and to sustain the intensely-heated arc-discharge plasma used for welding.

The plasmas used in semiconductor manufacturing are more *weakly-ionized plasmas* than such highly-ionized arc-discharges, and are produced within process chambers (Fig. 14-2). They are also created by adding energy from electric and magnetic fields, instead of by heating the gas. As will be seen, this allows only the charged-particles in the plasma to gain energy (i.e., the neutral gas-atoms are not directly impacted by the electric and magnetic fields). By gaining sufficient-energy in this way, free-electrons are not only able to ionize the gas-atoms and produce a plasma, but they can also cause other events needed by particular processes (such as the creation of reactive-species in a process-gas used for dry-etching). At the same time, the neutral gas-atoms in such process plasmas do not become heated, but remain very close to the temperature present in the process-chamber. The details of how energy is added to gases using electric (and magnetic) fields is covered in the following sections.

14.1 INTRODUCTION TO GLOW-DISCHARGES

The type of plasma used most commonly in IC-processes is the glow-discharge. A *glow-discharge* is a self-sustaining, weakly-ionized plasma that emits light (i.e., it *glows*). A neon-light is an example of a glow-discharge encountered in daily life (see Fig. 14-3). Similar kinds (excited by dc and rf electric-fields) are created in process-chambers for IC-processing.

Figure 14-4 shows a simple *dc-diode-type* system. We use it to explain how glow-discharges are established. The dc-diode system consists of a glass-tube from which the air has been evacuated. It is then re-filled with a specific-species of gas at low-pressure. Within the tube are two electrodes (a positively-charged *anode* and a negatively-charged *cathode*) and a dc-potential-difference is applied between them.

CHAPTER 14 PLASMAS USED IN SEMICONDUCTOR MANUFACTURING

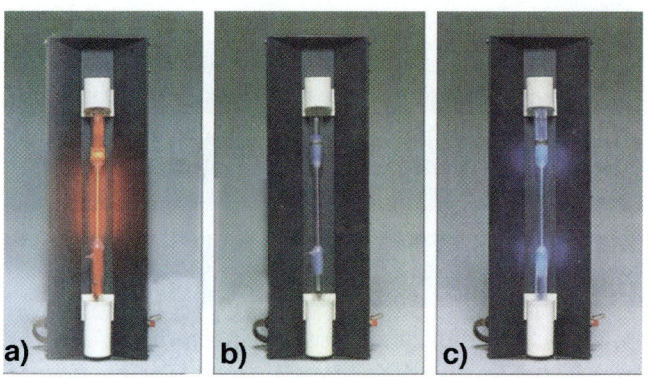

Fig. 14-3 When a gas is excited in a glow-discharge, it glows as it emits light. The colors of the light emitted by three gases – (a) neon, (b) argon, and (c) mercury – are shown. Each emission consist of several-wavelengths of light, and the perceived color depends on which wavelength predominates.

14.1.1 The Creation of DC Glow-Discharges

Assume in Fig. 14-4 that the glass-tube is: **1)** filled with argon-gas at a pressure of 1-torr; **2)** the distance between the electrodes is 15-cm; and **3)** a 1.5-kV dc potential-difference is applied between them. At the outset no current flows in this circuit - because all of the Ar gas-atoms are neutral, and thus there are no charged-particles in the gas. The full 1.5-kV is then dropped entirely between the two electrodes. However, if a free-electron enters the tube (most likely created from ionization of an Ar-atom by a passing cosmic-ray), it will be accelerated by the electric-field E existing between the electrodes (and whose magnitude is: E = V/d = 1.5-kV/15-cm = 100-V/cm).

The average-distance that a free-electron travels at a pressure $P = 1$-torr before colliding with an Ar-atom (i.e., the *mean-free-path* λ) is 0.0122-cm (see Ch. 6).

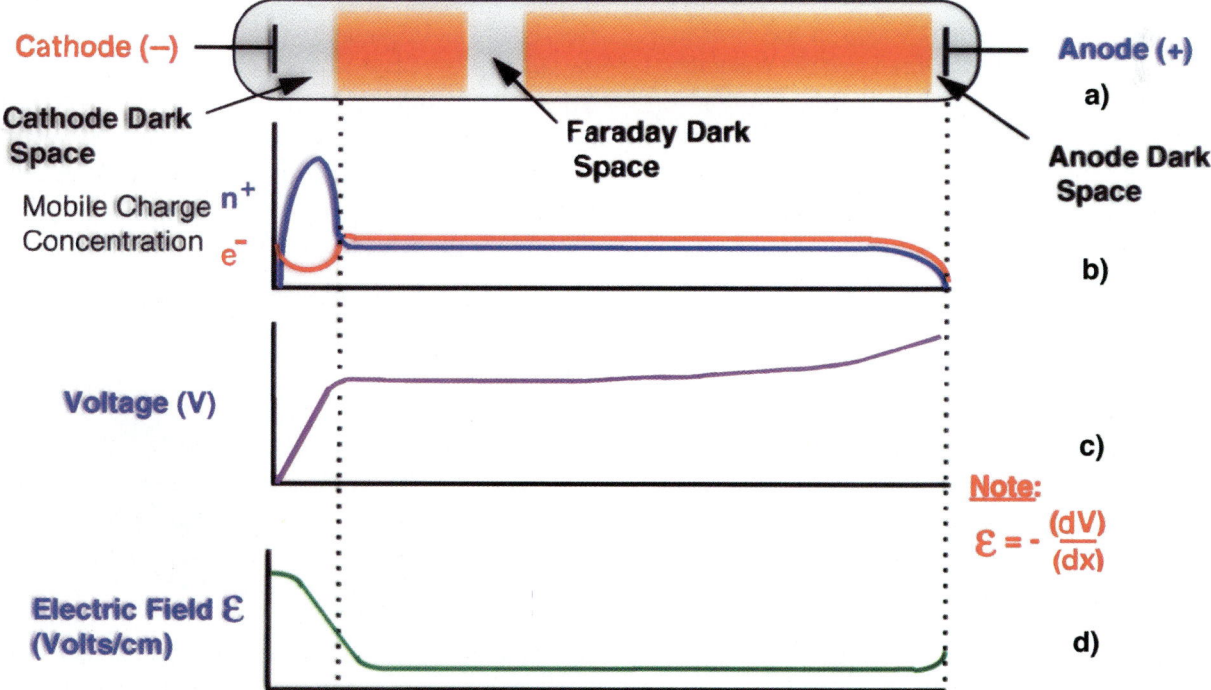

Fig. 14-4 (a) Schematic drawing of a dc glow-discharge established in a glass tube containing a gas under reduced-pressure, and two electrodes. Also shown are the most prominent regions of discharge. (b) Positive and negative charge-densities present in the dc glow-discharge along the length of the tube. (c) The voltage as function of position in a dc glow-discharge along the length of the tube. (d) Electric-field strength in the dc glow-discharge along the length of the tube. Note that the electric field is strongest immediately adjacent to the *cathode*.

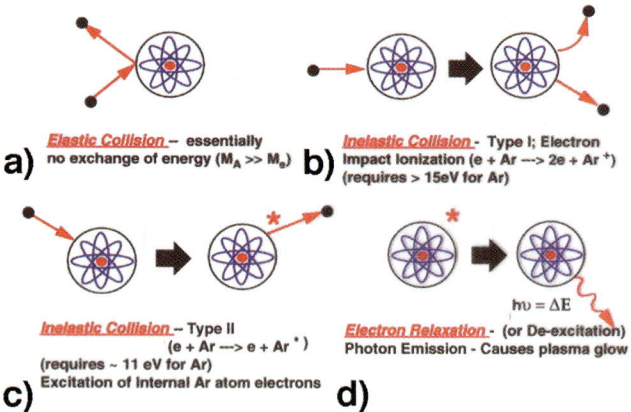

Fig. 14-5 Events associated with electron-gas collisions in a glow-discharge: (a) Elastic collision, almost no energy is lost by the electron as a result of the collision; (b) Inelastic collision – Type-I – Electron *impact-ionization* causes an electron in the atom to escape from the atomic-structure, creating a free-electron and an ion; (c) Inelastic collision – Type-II – This event causes an electron in the atom to be *excited into a higher energy energy-level*; (d) When the excited electron resulting from a Type-II collision returns to its original energy-level (*electron relaxation*), it emits a photon – which is responsible for the glow of the discharge.

Most such electron-atom collisions are *elastic*. This means that virtually no energy is transferred between the electron and gas-atom as a result of the collision (Fig. 14-5a). Such *elastic-collisions* occur because the mass of the electron is much smaller than that of the atom. (By analogy, think of a basketball being thrown against the side of an ocean-liner, as shown in Fig. 14-6. After the hitting the ship, the ball rebounds with almost its initial-energy, while the ocean-liner hardly moves at all.) Thus, the minimum-distance an electron must travel before it can undergo an *inelastic-collision* (in which significant-energy *is* transferred to the atom - either by the excitation of an atomic-electron to a higher-energy-level, or the causing of its escape from the atom) is about ten-times λ, or 0.122-cm. If this is the minimum-distance that must be traveled by electrons between inelastic-collisions, there must be a significant number of *electron-path-lengths* longer than λ (i.e., in the range of 0.5–1.0-cm).

If a free-electron travels 1-cm in the 100-V/cm electric-field, it will have picked up 100-eV of kinetic energy. With this amount of energy, the free-electron *can* transfer enough energy to an electron in an argon gas-atom to cause it to become excited or ionized. If this transferred energy E is less than the *ionization-potential* (e.g., 11.5-eV < E < 15.7-eV for Ar), one of the atom's orbital-electrons will be excited to a higher energy-state for about 10^{-8} sec (Fig. 14-5c). It will then return (*relax*) to the ground-state (emitting a visible-light photon when it returns to a lower energy-state, Fig. 14-5d). Such *excitation-events* are the source of the characteristic-glow of the discharge (from where it also gets its name). Glow-discharges of different gases (e.g., oxygen, argon, fluorine) emit light of different colors (see also Fig. 14-3).

If the energy transferred by the inelastic-collision of an electron and a gas-atom is greater than the ionization-potential (i.e., >15.7-eV for Ar), a second free-electron (together with a positive-ion) will be created (Fig. 14-5b). Subsequently, *both* free-electrons will become accelerated again, and the opportunity for cascading the number of free-electrons (and creating a condition known as *gas-breakdown*) exists. Figure 14-7 shows the breakdown-voltage required to initiate discharge, as a function of the product of the

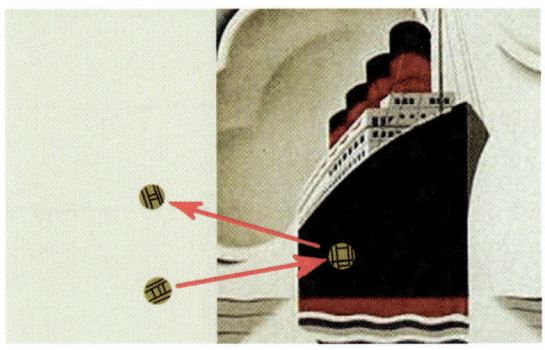

Fig. 14-6 When a low-energy electron collides with an atom, very-little energy is transferred (as the electron has a much-smaller mass than the atom). Thus, the electron rebounds with almost all of its pre-collision energy - and this is called an *elastic-collision*. Such elastic-collisions also occur if a basketball is thrown against the side of an ocean-liner. The baskeball rebounds with most its initial velocity (& kinetic energy), while the ship hardly moves.

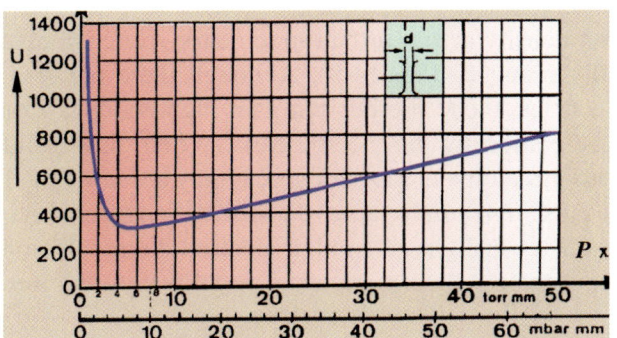

Fig. 14-7 Breakdown-voltage, U, between two parallel plane-electrodes in a homogeneous electric-field as a function of gas-pressure, P, and electrode-distance, d, for air. Such a plot is referred to a *Paschen curve*.

gas-pressure, P, and the electrode-spacing, d. When the condition of gas-breakdown is reached, current flows in the external circuit as the collision-generated free-electrons are collected by the anode. Each new ionization-event, however, will take place closer to the positively-charged anode (as the electrons are accelerated in its direction, see Fig. 14-8). Therefore, unless there is a mechanism available for generating additional free-electrons (to sustain the current flow), the current will increase to a maximum and quickly decay to zero. When a sufficient number of electrons are available to maintain the discharge, it is said to be *self-sustained*. The source of such glow-discharge-sustaining-electrons is considered later.

In addition to excitation and ionization of *gas-atoms,* glow-discharges can rupture the bonds of *gas molecules* to create chemically-reactive molecular-fragments, including *reactive-ions* and *radicals*. This event of breaking a chemical-bond by the collision of an electron and gas-molecule is called *dissociation*.

Summary of Electron/Gas-Atom and Electron/Gas-Molecule Inelastic-Collision Events: There are a number of events that can occur when an energetic-electron undergoes an inelastic-collision with a gas-atom or gas-molecule. These are summarized in Table 14-1. (Note that the superscript "*" refers to species whose energy is much larger than the ground-state.) Dissociated-atoms or molecular-fragments are called *radicals*. Radicals have an incomplete bonding-state

and thus are *extremely-reactive*. In *atomic-plasmas* only gas-atoms are present, but no molecules (for instance, in an Ar-based plasma), and thus there are no radicals.

It should also be mentioned that the collision processes that require the least-amount to energy to trigger, are the ones most likely to occur. Thus, since it generally takes less-energy to rupture molecular-bonds than it does to ionize an atom or molecule, the concentration of radicals in glow-discharges is greater than the concentration of ions. In fact, in simple glow-discharges, radicals constitute about 1% of the total-species present in the plasma, while ions may be present in concentrations less than 0.01%. The remainder of the species in the glow-discharge are the neutral molecules of the feed-gas (see Fig. 14-1).

14.1.2 The Structure of Self-Sustaining DC-Glow-Discharges and Their Dark Spaces

A self-sustaining discharge in a system exhibits certain characteristics. That is, the discharge has a particular structure, as is shown in Fig. 14-4a. The most

TABLE 14-1 Events That Can Occur When Electrons Collide with Gas Atoms and Molecules

Events Resulting from Electron/Gas-Atom Collisions

Type	Example
Atomic Excitation	$e^* + Ar \leftrightarrow Ar^* + e$
Atomic Ionization	$e^* + Ar \rightarrow Ar^+ + e + e$

Events Resulting from Electron/Molecule Collisions

Type	Example
Molecular Excitation	$e^* + O_2 \leftrightarrow O_2^* + e$
	$e^* + CF_4 \leftrightarrow CF_4^* + e$
Molecular Ionization	$e^* + O_2 \leftrightarrow O_2^+ + e + e$
	$e^* + CF_4 \leftrightarrow CF_4^+ + e + e$
Dissociation	$e^* + O_2 \leftrightarrow O + O + e$
	$e^* + CF_4 \leftrightarrow CF_3 + F + e$
Dissociation +Ionization	$e^* + O_2 \leftrightarrow O + O^+ + e + e$
	$e^* + CF_4 \leftrightarrow CF_3^+ + F + e + e$

- Impact ionization
- Photon emission by excited gas atoms
- An avalanche of free electrons and ions *(Breakdown)*

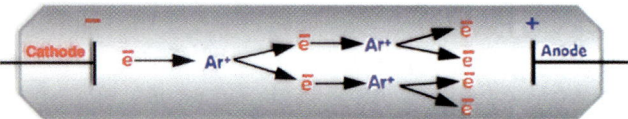

Fig. 14-8 When a colliding-electron creates a second free-electron in a *Type-I-collision*, both can be accelerated again, allowing each one to create another free-electron. Under the right conditions, this kind of "avalanche" effect can be exploited to establish a self-sustaining dc glow-discharge in the tube.

important region of the glow-discharge is the *dark-space* between the negative-glow and the cathode (also called a *sheath*).

Figure 14-9 shows the arrangement of charge near the cathode. It is seen that the positive-ions of the discharge are present in higher-density in front of the negatively-charged cathode, resulting in a localized positive *space-charge* there. Thus, any electrons near the cathode are rapidly accelerated *away* from it, due to their negative-charge and relatively small mass. The much-more-massive ions are accelerated *toward* the cathode, but less quickly. On average, it takes ions longer time to cross the dark-space than it does electrons. Thus, at any instant the concentration of ions in the dark-space is greater than that of electrons.

The presence of this *space-charge* has the effect of greatly increasing the electric-field immediately in front of the cathode. This implies that the electric-field in the rest of the discharge is rather weak (and uniform, see Fig. 14-4d). The largest-fraction of the voltage dropped between the anode and cathode is thus across the dark-space (see Fig. 14-4c). Therefore, charged-particles (ions and electrons) experience their largest acceleration in this region.

Less light is emitted from the dark-space region due to the presence of the strong electric-field. That is, since electrons in the dark-space are strongly accelerated and traverse through the space quickly, the electron density at any instant in the dark-space is drastically reduced (Fig. 14-9). Although there are fewer electrons, those present can gain high-energies from acceleration by the electric field. When they collide with gas atoms in the dark space, they are more likely to cause *ionization*, rather than *excitation* events. As a result fewer light-generating electron-atom collisions occur than in the negative and positive glow-regions.

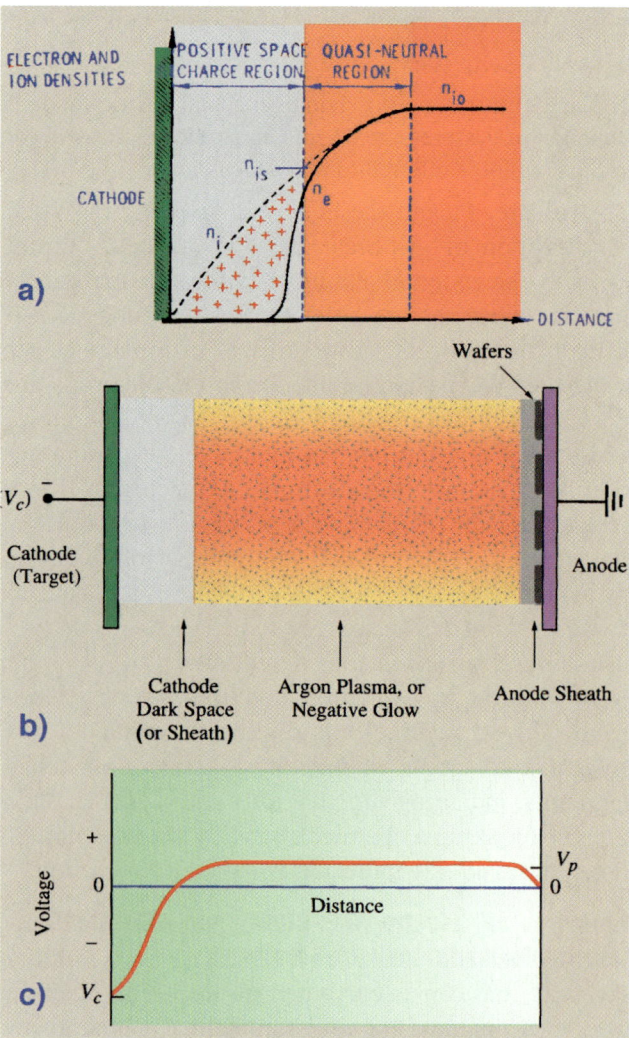

Fig. 14-9 (a) Schematic representation of the positive space-charge-region near a cathode. (b) DC glow-discharge used in a sputtering tool, with the sputtering-target being the cathode, and the wafers placed on the anode. (c) The voltage as a function of position in such a dc glow-discharge.

The source of electrons that sustains dc-diode discharges is the *cathode*, which emits *secondary-electrons* when struck by ions. Thus, upon entering the dark-space (after being ejected from the cathode), these secondary-electrons are accelerated in the dark-space field toward the anode. As they accelerate, they gain energy, and eventually have enough to cause ionizing and light-emitting collisions with the gas-atoms. The edge of the glow-region therefore provides an indication of where such events begin to occur. If the pressure in the tube is reduced, the probability of such collisions is reduced, and the dark-space gets wider. If the pressure gets too-low, the dark-space extends the full-length between the electrodes, and the glow-discharge is extinguished. For dc-diode sputtering-systems, this limits the minimum practical-pressure for sputter-deposition to the range of 10–40-mtorr (1.3–5.2-Pa).

In dc-diode configurations there must be a sufficient number of secondary-electrons emitted from the cathode to keep the discharge self-sustaining. Furthermore, these electrons must undergo an adequate number of ion-producing collisions with the sputter-gas. In turn, enough ions must be created in the dark-space and glow-regions to generate these secondary-electrons in sufficient number to continuously sustain the discharge.

14.2 RADIO-FREQUENCY (RF) GLOW-DISCHARGES

One of the limitations of dc-diode systems is that they cannot sputter insulators (dielectrics). That is, glow-discharges cannot be maintained with a dc-voltage if the electrodes are covered with insulating-layers. This represents a significant drawback, as there are many important applications that call for maintaining a continuous discharge in the face of dielectric-covered electrodes, including:

a) The *in-situ* sputter-removal of thin *native-oxides* from Si and Al-surfaces prior to deposition of overlying films (*in-situ* implies that sputter-etch and sputter-deposition are performed sequentially without breaking vacuum).

b) A glow-discharge must be continuously maintained in dry-etching process-chambers that are equipped with electrodes coated with insulating materials. In these chambers, wafers with insulating materials on their surfaces (e.g., photoresist and oxide layers) are mounted on the electrodes.

That is, when a negatively-biased cathode in a dc-diode system is bombarded by ions, an electron is stripped from the cathode-surface each time an impinging positive-ion is neutralized. If the cathode material is a *conductor*, such electrons can be replaced by electrical conduction, and the cathode surface maintains the negative-potential required to sustain the discharge.

However, if the electrode surface material is an *insulator*, the electrons lost from the cathode-surface are not replaced, since electrical conduction from the insulator interior to the sputtered surface is not possible. Thus, the front-surface accumulates a positive-charge that increases with time of ion-bombardment. This causes the potential-difference between the cathode-surface and that of the anode to decrease (Fig. 14-10). As soon as it drops below the value needed to sustain the discharge, the discharge extinguishes. In practice, the time for the insulator surface to acquire this charge is very short (about 1–10-μs).

A technique was developed to allow replenishment of such lost electrons to the insulator surface. This involves the application of an ac-voltage (rather than a dc-voltage) to the electrodes, and glow-discharges based on applying this method are known as *rf glow-discharges*.[6] Figure 14-11 shows the dc-voltage as a function of position across the chamber. Superimposed on this dc-level is the rf-signal. Since the plasma is conductive, the voltage-drop across the glow-discharge is small. Due to electron-depletion, however, large dc-voltage-drops exist between the plasma and the electrodes.

Most rf-glow-discharges are operated at 13.56-MHz (i.e., for sputtering-processes and plasma/RIE dry-etch-processes). The power applied to the glow-discharge (supplied by an rf-power-supply and matching-network, Fig. 14-12) has a frequency in the radio-

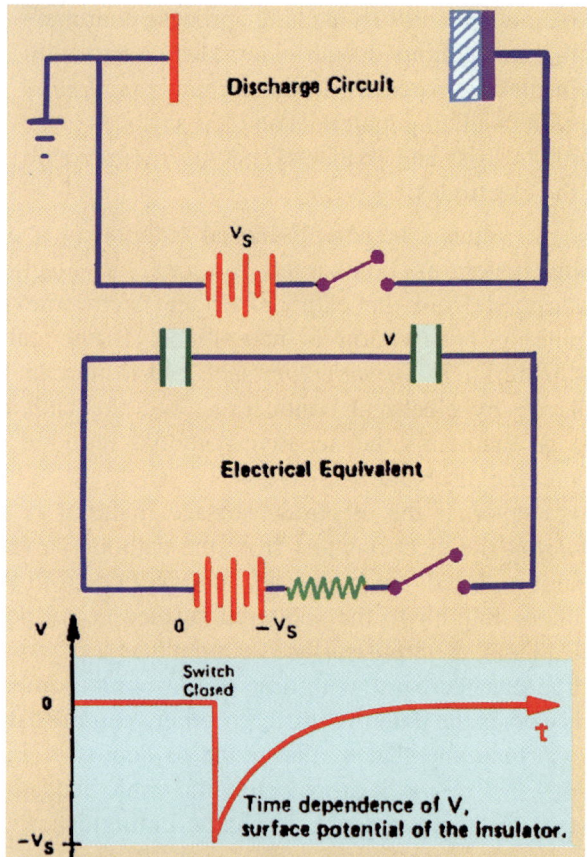

Fig. 14-10 Surface-charging of an insulating cathode. Reprinted with permission of John Wiley & Sons.[2]

frequency range (hence the name *rf glow-discharge*), and the 13.56-MHz frequency is used because it is one of the frequencies set aside by the FCC and other international communications-authorities for industrial applications. That is, a limited amount of electromagnetic-energy at such frequencies is permitted to be radiated because it will not cause interference with other radio-frequency transmitted signals. Rf-glow-discharges are used in *dry-etching* and *plasma-enhanced CVD (PECVD)*. More details about rf glow-discharges can be found in Refs. 1, 2, and 6.

14.3 HIGH-DENSITY PLASMAS

As noted earlier, the concentrations of ions and electrons in dc and rf glow-discharges are a small-fraction

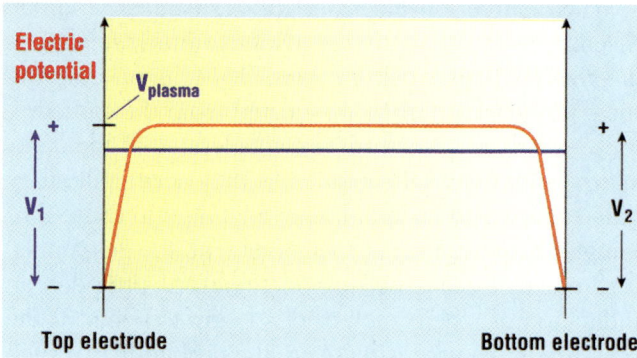

Fig. 14-11 Typical plot of dc-voltages as a function of position in an rf-plasma.

of the total gas atoms present. Since ion-bombardment of surfaces in the chamber can often be exploited to improve the processes, it is desirable to increase the relative concentrations of the ions in the plasma. Various techniques have been developed to produce plasmas with higher ion-concentrations, including *magnetron-plasmas* and *high-density-plasmas (HDP)*. Magnetron-plasmas for sputter-deposition were developed in the late 1970's and became the dominant source used in sputter-deposition applications. They are discussed in Chap. 15. Magnetron-plasmas were adapted for dry-etching tools in the late-1980s and these are called *magnetically-enhanced RIE systems – or MERIE*). They are described in Chap. 22.

More recently, new HDP reactor-configurations were introduced and applied to etching and to chemical-vapor deposition (CVD) processes.[7] The term *high-density plasma* refers to a plasma in which ion-concentrations are increased from 0.01% to about 1-10% ($\sim 10^{13}$-ions/cm^3). All the HDP plasma-sources

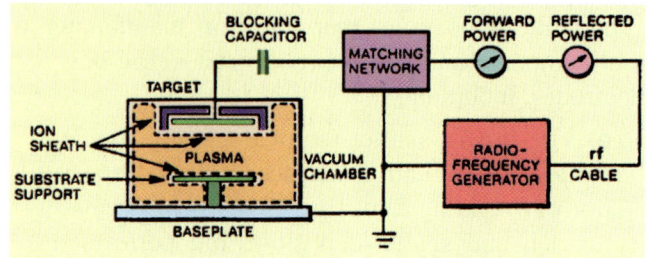

Fig. 14-12 Radio-frequency (rf) sputtering system.

use magnetic and electric fields together to boost radical and ion-densities in the plasma. In addition, such sources are operated at lower-pressures (typically below 10-mtorr). Ions can be extracted from the plasma-region and directed at the wafer at low-energies. This allows large numbers of ions to strike wafer-surfaces at low-energy, improving the process characteristics without increasing damage to the silicon wafers. The drawbacks include higher-cost and non-uniformity of etching/deposition across a wafer.

REFERENCES

1. S. Wolf and R.N. Tauber, *Silicon Processing for the VLSI Era - Vol. 1: Process Technology, 2nd Edition,* Lattice Press, Sunset Beach CA, 2000, Ch. 15.

2. B. Chapman, *Glow-Discharge Processes,* John Wiley & Sons, New York, 1980.

3. J.L Vossen and J.J. Cuomo, "Glow-Discharge Sputter Deposition," in *Thin-Film Processes,* Eds. J.L. Vossen and W. Kern, Academic Press, 1978. Ch. II-1, p. 11-73.

4. S. Rossnagel, "Glow-Discharge Plasmas and Sources for Etching and Deposition," in *Thin-Film Processes,* 2nd Ed., J.L. Vossen & W. Kern, Eds., Academic Press, San Diego, 1991, Ch. II-1, pp. 11-78.

5. F.F. Chen, *Introduction to Plasma Physics and Controlled Fusion,* Vol. 1: Plasma Physics, 2nd Ed., Plenum Press, New York, 1984.

6. H.R. Koenig and L.I. Maissel, "Applications of RF Discharges to Sputtering," *IBM J. Res. Dev.,* **14**, 168 (1970).

7. O.A. Popov, *High Density Plasma Sources: Design, Physics and Performance,* Noyes Publications, Park Ridge, NJ, 1995.

PROBLEMS

1. What is a *plasma?* What is the most common form of plasma used in microchip manufacturing applications? Name the three species present in a plasma. Which ones are not present in a neutral-gas?

2. Explain why using plasmas is beneficial for some IC processes.

3. Explain the difference between the terms *plasma* and *glow-discharge.*

4. List three important electron-atom collision events that may give rise to plasmas.

5. Explain the importance of plasma ion-bombardment for etching, PECVD, and sputtering PVD.

6. What does the term *radical* refer to in the context of a glow-discharge environment? How are such radicals created in a glow-discharge?

7. It is possible to determine the chemical composition of a glow-discharge by measuring the emission spectrum (i.e., the optical intensity versus wavelength of the glowing-light being emitted). Explain why?

8. In some processes a dc power-supply is used to establish and sustain a glow-discharge, while in others an rf-power supply is needed. Explain when one is type is used, and then when the other is required. If one type (i.e., the rf power-supply) could work for all processes, why is it not used exclusively? Why can't a dc glow-discharge be used for plasma processing when a dielectric-surface exists on one or both of the electrodes?

9. Explain why the *potential* of a *glow-discharge* used in dry-etching or sputter-deposition, is positive relative to ground.

10. If a target is made with fused-silica with a target thickness of 2.5-mm, it has a capacitance of ~1-pF/cm^2. If the current-density during rf-sputtering of this target is 1-mA/cm^2, and the applied voltage is 1000-V, calculate the time required to charge this capacitance. This roughly represents the time-interval during which a glow-discharge can remain "ON."

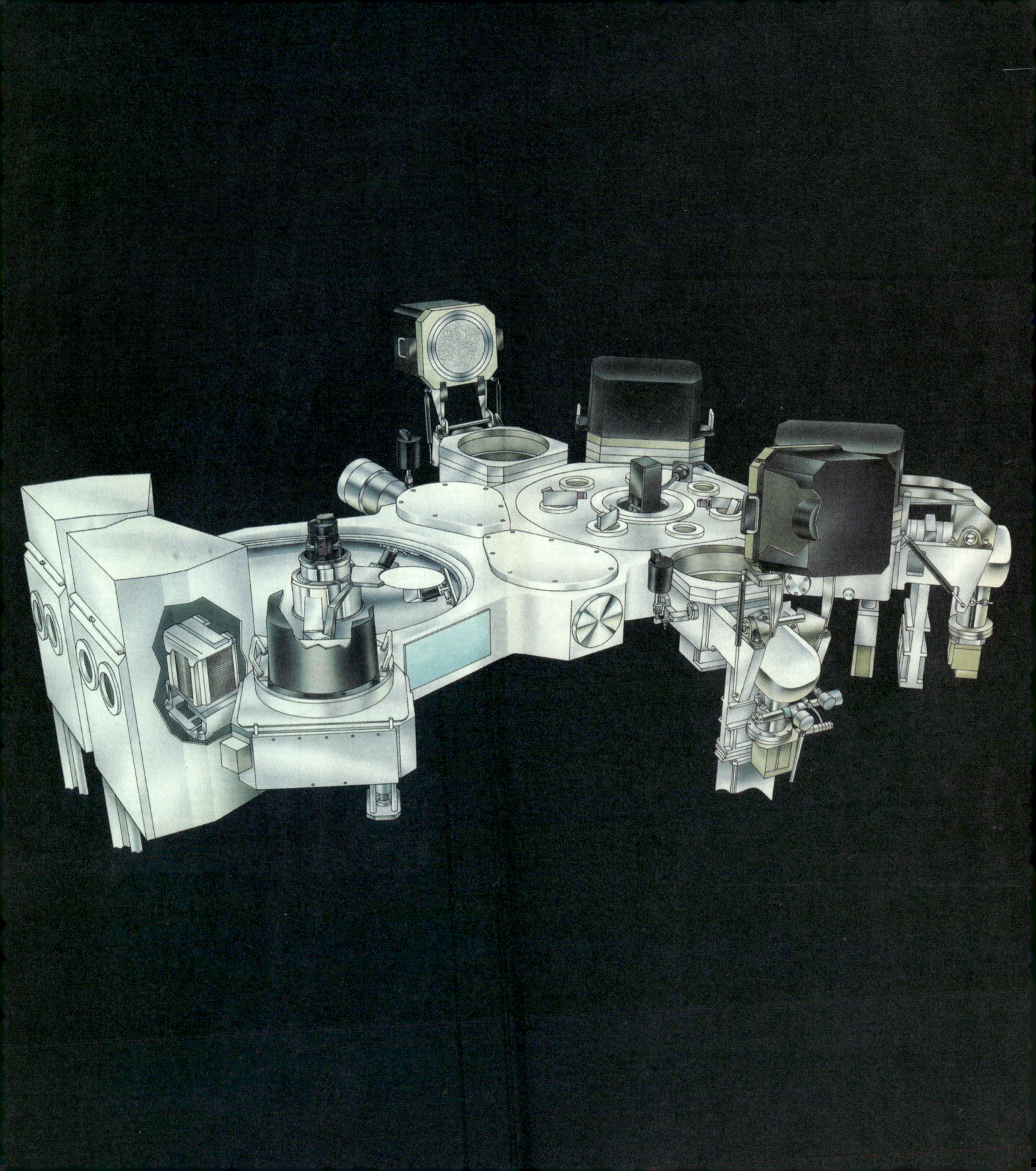

CHAPTER **15**

ALUMINUM THIN-FILMS AND SPUTTER-DEPOSITION IN ULSI

CHAPTER CONTENTS

15.1 ALUMINUM THIN-FILMS IN ULSI

15.2 SPUTTER-DEPOSITION FOR ULSI

15.3 STEPS OF THE SPUTTER-DEPOSITION PROCESS

15.4 THE PHYSICS OF SPUTTERING

15.5 MAGNETRON SPUTTERING

15.6 SPUTTER-DEPOSITION EQUIPMENT

15.7 SPUTTER-PROCESS CONSIDERATIONS

15.8 STEP-COVERAGE AND VIA/CONTACT-HOLE FILLING BY SPUTTERING

15.9 METAL FILM-THICKNESS MEASUREMENTS

"The essence of engineering is knowing what variables you can afford to ignore."

W.C. Randels,
In *Transistors for Integrated Circuit Engineering,* 1983.

Cluster tool configured for deposition. Shown is the AMAT Endura.® Courtesy of Applied Materials.

This chapter covers three major-topics involved in the formation of inter-connect-structures in ULSI:[1]

1. The properties of aluminum (Al) and Al-alloy thin-films in ULSI.[2]
2. Thin-film formation by *physical-vapor-deposition* (*PVD*).
3. Evaluating the thickness of opaque (metallic) thin-films.

The main topic is *physical-vapor-deposition* (*PVD*) - by *sputtering*. Thus, it is useful to define the technique of PVD, and to introduce the common characteristics shared by PVD-processes. That is, all *PVD-processes* proceed according to a sequence of the following three-steps (Fig. 15-1):

1. The material to be deposited (from a *solid* or *liquid* source), is first physically-converted to the vapor-phase.
2. This vapor is transported across a region of *reduced-pressure,* which exists between the *source* and the *substrate*.
3. If the substrate is located close to the surface being vaporized, some of the vapor condenses on it to form a thin solid-film.

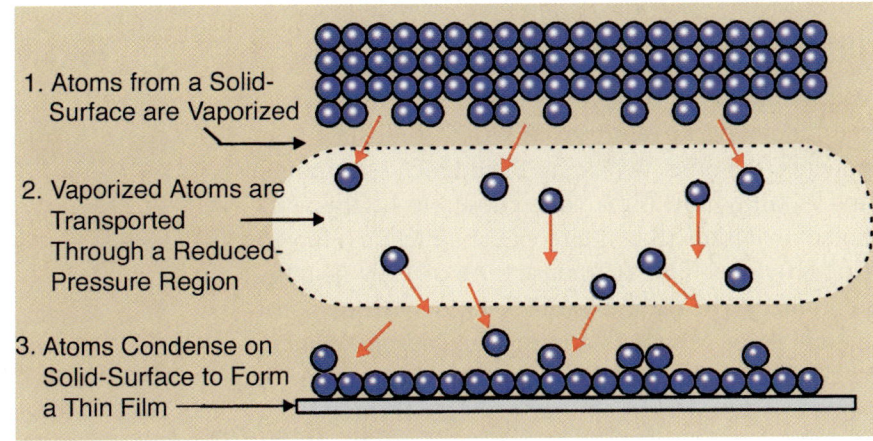

Fig. 15-1 The 3 steps of *physical-vapor deposition* (*PVD*).

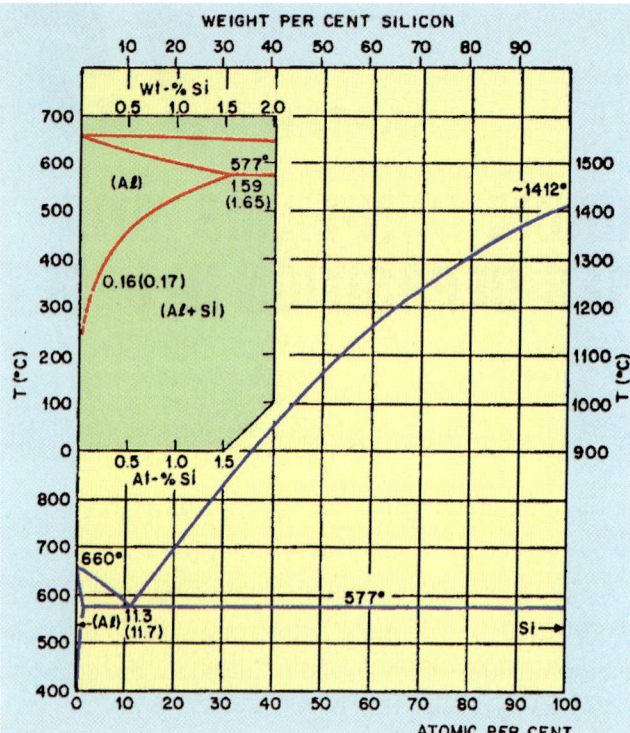

Fig. 15-2 *Phase-diagram* of the aluminum-silicon system.[3]

In sputter-deposition, the conversion to a gaseous-phase (Step-**1**) is done by the physical-dislodgment of surface-atoms by momentum transfer. (Note that many concepts used to describe PVD-processes, including *vacuum-pressure*, *mean-free-path*, *vapor-pressure*, and *gas-flow regimes* are covered in Ch. 6.)

15.1 ALUMINUM THIN-FILMS IN ULSI

Aluminum has historically been the third major-material employed to fabricate Si-based solid-state components (the other two being Si and SiO_2). Aluminum and Al-alloy thin-films were chosen to be the workhorse materials for interconnect-structures for the first 30-years of the IC industry. As of 1999 Al-alloys continued to be the most widely-used materials for IC metallization (although copper-films began to emerge as probable replacements). The dominance of Al for interconnect-applications arose because of its low-resistivity ($\rho_{Al} = 2.7\ \mu\Omega$-cm), and its compatibility with Si and SiO_2. Aluminum thin-films adhere well to SiO_2, and make low-resistance contacts to heavily doped *n*- and *p*-type silicon. Note that the relatively-low melting-point of Al (660°C), and the Al-Si eutectic-temperature (577°C) restrict the value of subsequent processing-temperatures once Al-films have been deposited. Thin-films of sputtered-Al are typically deposited in the thickness range of 500-1500-nm (0.5-1.5-μm), and have a polycrystalline structure.

Aluminum-alloys (rather than pure-Al) are used for IC interconnects. The two most-widely-used Al-alloys are Al:1wt%Si and Al:2wt%Cu. Early microelectronic devices utilized evaporation to deposit pure-Al and Al:Si- or Al:Cu-alloys. However, the stringent alloy-composition requirements of advanced devices (as well as some other limitations of evaporation) gave sputtering an advantage over evaporation for depositing metal-films. The development of *magnetron-sputtering* (which allows Al deposition-rates of up to 1000-nm/min), caused sputtering to displace evaporation for Al-deposition.

The alloy *Al:Si* replaced *pure-Al* because *junction-spiking* occurs at the interface of pure-Al and Si. That is, the solubility of silicon in aluminum (Fig. 15-2) rises as the temperature increases (e.g., to about 0.5% at 400°C).[3] When a Si-substrate is put

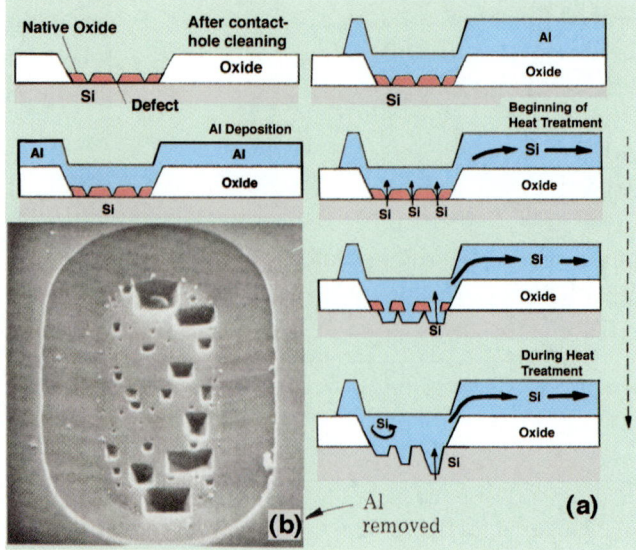

Fig. 15-3 (a) *Junction-spiking* and *silicon-migration* during contact sintering. (b) Pit formation in an Al contact to Si.

into intimate contact with a pure Al-film, the wafer becomes a source of Si that can dissolve in the solid Al-solvent. In addition, the diffusivity of silicon in Al-films at 400°C is high. Thus, a significant quantity of Si moves from the region beneath the Al/Si-interface into the Al-film. Simultaneously, Al from the metal-film moves to fill the voids created by the departing Si (Fig. 15-3). If Al penetrates deeper than the *pn-junction-depth* beneath the contact, the junction will be electrically-shorted (*junction-spiking*). One way to suppress this set of events is to use an Al-alloy with a concentration of Si that exceeds the Si solubility-limit at the maximum process-temperature. In that case, when the (Al:Si-alloy)/Si interface is heated, the Al:Si-film cannot dissolve any more silicon originating from the substrate - and junction-spiking is avoided.

However, such use of Al-Si alloy films to combat junction-spiking in ULSI devices was eventually abandoned in favor of contact-structures employing *barrier-layers*. The most widely used barrier-layer for Al-films in ULSI is titanium-nitride (TiN).

It has been found that the addition of small quantities of other materials, primarily copper (Cu, 0.5-4wt%) to Al-thin-films improves their electromigration-resistance and reduces the formation of *hillocks* (protrusions on the Al-film surface, see Fig. 7-8b, Ch. 7). However, the addition of Cu increases the film-resistivity by 10–30%. In addition, it is more difficult to dry-etch Al:Cu alloy-films than pure Al-films.

15.2 SPUTTER-DEPOSITION FOR ULSI

Sputtering is a term used to describe the mechanism in which atoms are ejected from the surface of a material when it is struck by sufficiently-energetic particles. Sputtering has become the dominant-technique for depositing a variety of metallic-films in VLSI and ULSI fabrication, including: *aluminum-alloys*; *titanium*; *titanium:tungsten*; *titanium-nitride*; *tantalum*, and *cobalt*.[4,5,7,9] Sputtering was chosen over the original PVD technique for depositing metal-films (evaporation) for the following reasons:

1. Sputtering can be accomplished from large-area targets, simplifying the problem of depositing films with uniform-thickness over large wafers.

2. The alloy-composition of sputter-deposited films can be more tightly (and easily) controlled than in evaporated-films.

3. Many important film-properties, such as step-coverage and grain-structure can be improved through sputtering.

4. The surface of the substrates can be sputter-cleaned in vacuum prior to initiating film-deposition (and the surface is not exposed again to ambient after such cleaning).

5. There is sufficient material in most sputter-targets to allow many deposition-runs before target-replacement is necessary.

As is true with other processes, however, sputtering also has its drawbacks. They include:

1. Sputtering-processes involve high capital-equipment costs.

2. Since the process is carried out in low-to-medium vacuum-ranges (compared to the high-vacuum conditions under which evaporation is conducted), there is a greater possibility of incorporating impurities into the deposited-films.

3. Better step-coverage may be achieved by CVD.

15.3 STEPS OF THE SPUTTER-DEPOSITION PROCESS

There are four stages to a *sputter process* (Fig. 15-4):

1. Ions are generated and directed at a target.
2. These ions sputter atoms from the target.
3. Ejected (sputtered) atoms are transported to the substrate.
4. There they condense and form a thin-film.

15.4 THE PHYSICS OF SPUTTERING

When a solid-surface is bombarded by atoms, ions, or molecules at moderate energies (between about 0.1-2-keV) two effects arise: **1**) some fraction of the energy of the impinging-ions is transferred to the solid as *heat* and *lattice-damage*; and **2**) another fraction causes atoms from the surface to be dislodged and ejected into the gas-phase (*sputtering*).

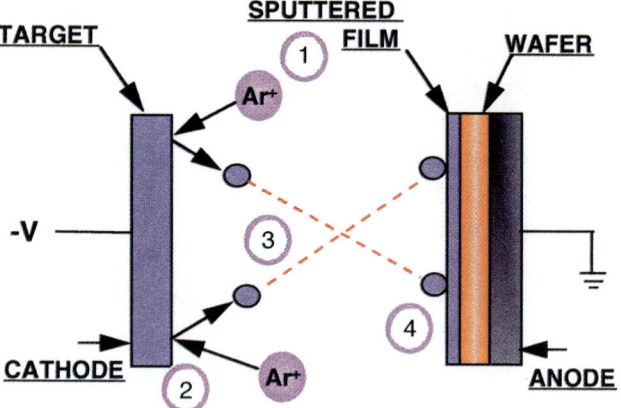

Fig. 15-4 The four steps of a sputtering-process: (a) Ions are generated by a glow-discharge and are directed at the sputtering target; (b) These ions strike the target and eject target-atoms (*sputtering*); (c) The sputtered-atoms are transported across the reduced-pressure space that separates the target from the wafers; and (d) The sputtered target-atoms condense on the wafer-surface to form a thin-film.

15.4.1 The Billiard-Ball Model of Sputtering

The exact mechanisms which lead to the ejection of atoms under ion-bombardment are complex, but some of the details can be described by viewing it as a game of *three-dimensional billiards* - played with atoms (Fig. 15-5a). Using this analogy, it is possible to visualize how atoms may be ejected from a surface as the result of two collisions (Figs. 15-5b) when a surface (called the *target*) is struck by an atom with a velocity perpendicular to the surface (atom-A in Fig. 15-5b). When atom-B is struck by atom-A, atom-B may leave the point of impact at an angle greater than 45°. If atom-C is then struck by atom-B, the angle at which atom-C leaves the secondary impact-point may again be greater than 45°. Thus, it is possible that atom-C can have a velocity-component greater than 90° (and be directed-away from the surface). As a result, atom-C may be ejected from the surface as a result of the surface being struck by atom-A.

Note that the energy-range of sputtered-atoms leaving the target is 3–10-eV, and some of the bombarding-atoms rebound with substantial-energy after hitting the surface. Thus, the target-surface is a source of sputtered-atoms *and* energetic backscattered-atoms (Fig. 15-5c).

For the case when the surface is bombarded by ions at an oblique angle (i.e., 45°–90°), there is a higher probability that the primary-collision between incident-ions and surface-atoms will lead to sputtering events. Furthermore, oblique-incidence confines the action closer to the surface, and thus sputtering is enhanced, and sputtered-atoms are ejected strongly in the forward direction. In addition, the sputter-yield is much greater than that resulting from normal incidence by bombarding ions (Fig. 15-6). This effect also leads to *faceting* (see Sect. 15.7.4).

15.4.2 Sputter-Yield

The *sputter-yield* (defined as *the number of atoms ejected per incident ion*), is an important factor in determining the rate of sputter-deposition. It depends on a number of things besides the direction of incidence of the ions, and these include: **a)** the target material; **b)** the mass of the bombarding ions; and **c)** their energy. There is a minimum energy-threshold for

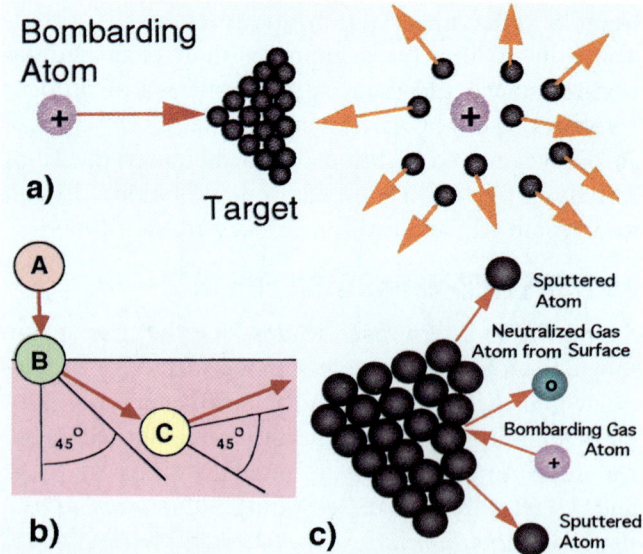

Fig. 15-5 (a) *Billiard-ball model* of sputtering. (b) Binary-collision between atoms A and B, followed by a binary-collision between atoms B and C. (c) Events that take place when energetic-ions strike a sputtering-target surface – target-atoms are ejected, and the ions become neutralized.

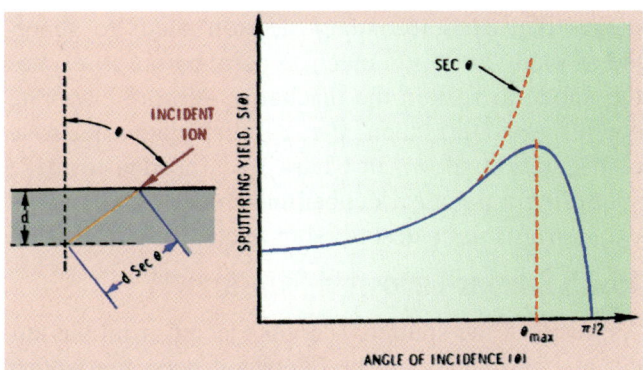

Fig. 15-6 Schematic-diagram showing the variation of *sputtering-yield* with the *angle-of-incidence* of the ion.

Table 15-1	SPUTTER YIELDS FOR METALS IN ARGON (ATOMS/ION)					
Target	At.Wt./Dens.	100 eV	300 eV	600 eV	1000 eV	2000 eV
Al	10.0	0.11	0.65	1.2	1.9	2.0
Au	10.2	0.32	1.65	2.8	3.6	5.6
Cu	7.09	0.5	1.6	2.3	3.2	4.3
Ni	6.6	0.28	0.95	1.5	2.1	
Pt	9.12	0.2	0.75	1.6		
Si	12.05	0.07	0.31	0.5	0.6	0.9
Ta	10.9	0.1	0.4	0.6	0.9	
Ti	10.62	0.08	0.33	0.41	0.7	
W	14.06	0.12	0.41	0.75		

sputtering that is approximately equal to the heat of sublimation (e.g., 13.5-eV for Si). In the energy-range of sputtering (10–5000-eV), the yield increases with ion-energy and mass. Figure 15-7 shows the sputter-yield of Cu as a function of energy for various noble-gas ions.[5] The sputter-yields of various materials in argon-gas at different energies is given in Table 15-1.

One other important matter related to sputter-yield should be noted. Although the sputtering-yields of various materials are different, as a group they are much closer in value to one another than, for example, the vapor-pressure of comparable materials. This makes the deposition of multilayer or multi-component films much more controllable by sputtering than by evaporation.

15.4.3 Selection-Criteria for Sputtering Process-Conditions and Sputter-Gas Type

The information learned from sputter-yields and the physics of sputtering can be applied toward an understanding of how process-conditions and materials are selected for sputtering, including: **a)** the type of sputtering-gas; **b)** the pressure-range of operation; and **c)** the electrical conditions for the glow-discharge.

In purely physical-sputtering (as opposed to reactive-sputtering), it is important that the ions or atoms of the sputtering-gas *not react* with the growing-film. This limits the selection of sputtering-species to the *noble-gases*. Hence, *argon (Ar)* is generally the gas of choice, since it is easily available (making it low in cost). Also, its mass is a good match to those elements most frequently sputtered (Al, Cu, Si, and Ti), giving adequate sputtering-yields for these elements.

The *pressure range of operation* is set by the glow-discharge (lower-limit ~2–3-mtorr for magnetron-sputtering) and the scattering of sputtered-atoms by the sputter-gas (upper-limit ~100-mtorr).

Finally, a desired goal of sputter-deposition is to obtain maximum deposition-rates. As a result, *electrical conditions* are selected to give a maximum sputter-yield per unit-energy. That is, as energy is increased, each energy-increment gives a progres-

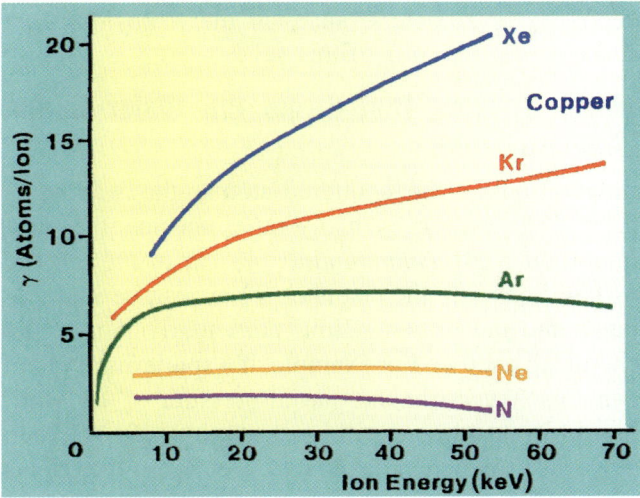

Fig. 15-7 Sputtering-yields of the noble gases on copper, as a function of energy.[6]

sively smaller increase in sputter-yield. This occurs because higher-energy ions implant themselves, and thus end-up dissipating a greater proportion of their energy via *non-sputtering* processes. The most efficient ion-energies for sputtering are typically obtained for electrode-voltages of several-hundred volts.

It should also be noted that sputtering is a *highly-inefficient process*. Over ~70% of the energy consumed during the sputtering-process is dissipated as heat in the target (and another ~25% by emission of secondary-electrons and photons). The heating can raise target-temperatures to levels capable of damaging the target (or the material that bonds a target to the backing-electrode). Thus, the target must be cooled to avoid such problems. This is normally done by bonding targets to water-cooled Cu backing-plates.

15.4.4 Secondary-Electron Production for Sustaining the Glow-Discharge

As discussed in Chap. 14, glow-discharges must be continuously provided with free-electrons to keep them sustained. In most dc sputtering-systems, the source of such electrons is *secondary-electron emission* from the target. An important mechanism that generates such secondary-electrons is *Auger-emission*. The details of the generation of Auger-emission are found in Ref. 1. Nevertheless, it should be mentioned that the *secondary-electron yield* per bombarding-ion is quite small (about 0.05–0.1 for metal-targets). This means that each secondary-electron needs to generate as many as 20 ion-electron pairs before it reaches the anode to sustain the discharge. A more elaborate sputtering-source than the parallel-plate electrode configuration shown in Chap. 14 is needed to make sputtering a practical deposition-process (such as the *magnetron-source* described in Sect. 15.5).

15.4.5 Sputter-Deposited Film-Growth

As many of the sputtered-atoms ejected from the target as possible should be deposited upon the wafers (i.e., and not upon the sputter-chamber walls) to form the thin-film. One way to accomplish this is to keep the target and wafers closely spaced (with target-to-wafer spacings of 5-10-cm being typical).

However, this means that since the *mean-free-path*, λ, of sputtered-atoms at typical sputter-pressures is less than 5–10-cm (e.g., at 5-mtorr, $\lambda \cong 1$-cm), it is likely that sputtered-atoms will still suffer one or more collisions with the sputter-gas atoms before reaching the wafer (Fig. 15-8). The sputtered-atoms may therefore: **a)** arrive at the substrate with reduced energy (~1–2-eV); **b)** be backscattered to the target or the chamber walls; or **c)** lose enough energy so that thereafter they move by diffusion in the same manner as neutral sputter-gas atoms. These events imply that the *sputtering gas-pressure* can impact various film-deposition parameters, such as the *deposition-rate* and *composition* of the film.

15.4.6 Species That Strike the Wafer During Film-Deposition

In addition to the sputtered-atoms, the wafer-surface is also struck by many other species, the most important being *contaminants present in the form of residual-gases* in the chamber. These can be incorporated into the growing-film, particularly if they are chemically active. For example, if a sputtering-system contains oxygen in the partial-pressure of 10^{-6}-torr, the substrate-surface will be struck by $\sim 10^{15}$-oxygen atoms/sec (or 1-monolayer/sec). Details on the effects of residual-gases on sputter-deposited films is given in the section on *Sputter-Processing Issues* (Sect.

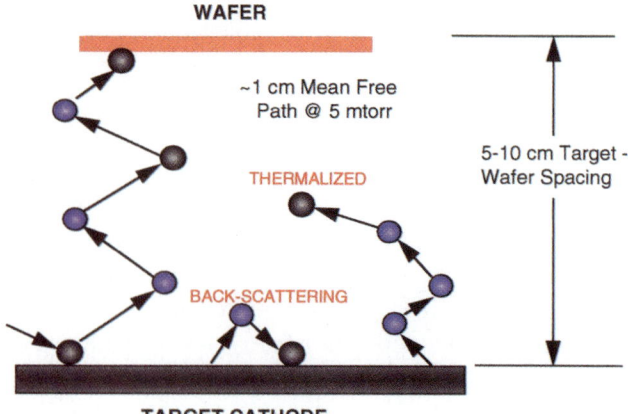

Fig. 15-8 *Gas-scattering events* in sputter processes.

CHAPTER 15 ALUMINUM THIN-FILMS AND SPUTTER-DEPOSITION FOR ULSI

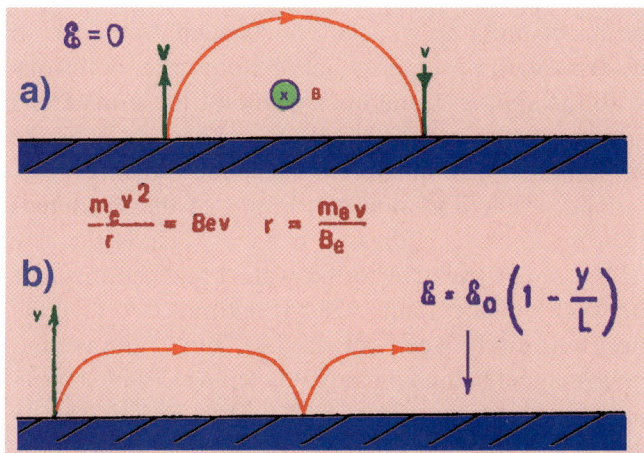

Fig. 15-9 (a) Motion of an electron ejected from a surface with velocity **v** into a region having a magnetic-field **B** parallel to the surface, with no electric-field; and (b) with a linearly-decreasing electric-field.[4]

15.7).

15.5 MAGNETRON-SPUTTERING

As noted in Sect. 15.4.4, secondary-electrons emitted from the target must produce about 20 ion-electron pairs before reaching the anode in order to sustain the glow-discharge. In dc-diode configurations it is usually necessary to operate at fairly-high pressures (20–200-mtorr) to have any reasonable ion-current. Such high gas-pressures result in slow-transport of the sputtered-atoms away from the cathode (see Sect 15.7.2), and consequently result in very-low deposition-rates. As a result, dc-diodes are rarely used in IC production. To make sputtering a viable production-process, an alternative sputtering-source which gives significantly-higher deposition-rates is needed.

The alternative sputter-source that satisfies these requirements relies on a mechanism termed *magnetron-sputtering*. Magnetron-sources increase the percentage of electrons that cause ionizing-collisions by utilizing magnetic fields to help confine the electrons near the target-surface (see Fig. 15-9). As a result, in magnetron-sputtering processes, current-densities at the target are increased to 10–100-mA/cm^2, compared to about only 1-mA/cm^2 for dc-diode configu-

rations. In addition, magnetron-sources can sustain a glow-discharge at much lower-pressures than can dc-diodes (typically pressures as small as only 1-mtorr are needed, instead of 20-mtorr for dc-diodes). Since the mean-free-path of gas-atoms, λ, at P = 1-mtorr is ~6.6-cm (see Eq. 6-5), this distance is comparable to the distance between the target and the wafer (which is 4–10-cm in modern sputter-chambers). Thus, most sputtered-atoms in magnetron-sources arrive at the wafer having traveled in a straight-line path (i.e., without having undergone any collisions with the sputter-gas atoms). Magnetron sputtering-sources were introduced in 1974, and because of their clear superiority, quickly became (and have remained) the dominant source-configuration for sputtering applications. The principles of magnetron-sputtering are detailed in Ref. 1.

15.5.1 Magnetron Sputter-Sources for ULSI

The *circular planar-magnetron sputter-source* configuration is the most widely used magnetron-source for ULSI sputtering (Fig. 15-10). In earlier generations of IC-technology, however, other magnetron source-configurations found much wider use than they do today. These included the *S-Gun* (or *Sputter Gun*) and the *rectangular planar-magnetron*. More information about these two magnetron source-configurations can be found in Refs. 4 and 5. Here our discussion is confined to the circular planar-magnetron sputter-source.

Evolution of Planar-Circular Sputtering-Sources: The circular planar-magnetron sputter-sources originally used a ring of permanent bar-magnets (plus one central-one) located behind the target (see Fig. 15-11). Using the strongest Nd-Fe-B magnets, this magnetic-field approached 1-kilogauss (or 0.1-T in SI-units). Although this source-configuration improves upon the dc-diode configuration (chiefly in terms of much-higher sputter-deposition rates), it also has several drawbacks. Namely, in a magnetron-source with permanent-magnets, the glow-discharge region of maximum-intensity is an annular-ring above the target-surface. Hence, most of the sputtering also occurs from a ring-shaped region of the target immediately beneath

254 MICROCHIP MANUFACTURING

Fig. 15-10 An example of a sputtering-target for a planar circular-magnetron source. Courtesy of Tosoh SMD.

this intense plasma-ring (Fig. 15-12).

This causes the following problems. First, the deposition-rate across the wafer is less-uniform than with a dc-diode source. Second, target-utilization is poor. The target erodes rapidly in the ring-region, but not elsewhere. The target must thus be replaced before most of its material is used, driving up the cost of the sputter-process. Finally, on regions of the target-surface away from the ring-region, there can be a net *build-up* of sputter-deposited film (due to re-deposition of some sputtered-atoms after collisions with the sputter-gas atoms). Some of these materials adhere poorly, and may peel off and become a source of particle-contamination in the sputter-chamber.

The problem of non-uniformity of the sputtered-film thickness is addressed in two ways: **1**) the size of the target (relative to the wafer) and the distance between the target and wafer is optimized to give the best uniformity (see Sect. 15.5.2); and **2**) the permanent-magnet behind the target is replaced with a movable-magnet array (Fig. 15-13). This array is slowly spiraled behind the target during sputter-deposition, and the peak magnetic-field is not kept fixed just above the annular ring. More of the target-surface is thus exposed to strong ion-bombardment, and much-more area of the target-surface ends up being eroded. The circular planar-magnetron source with movable magnets is said to exhibit *full-face erosion*. However, as shown in Fig. 15-14 (where a new and a spent circular-magnetron sputtering target are shown), the entire-surface of the target may indeed be eroded to

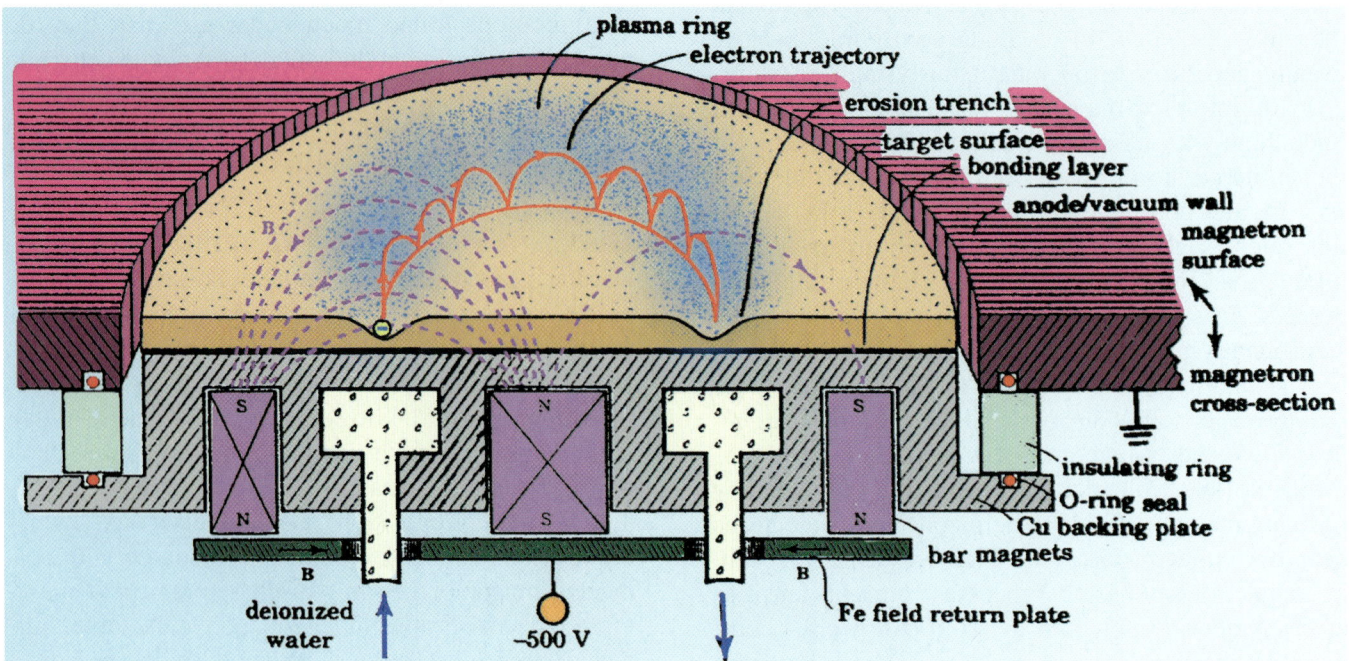

Fig. 15-11 Perspective drawing of a *planar, circular-magnetron sputter-source*.[8]

CHAPTER 15 ALUMINUM THIN-FILMS AND SPUTTER-DEPOSITION FOR ULSI

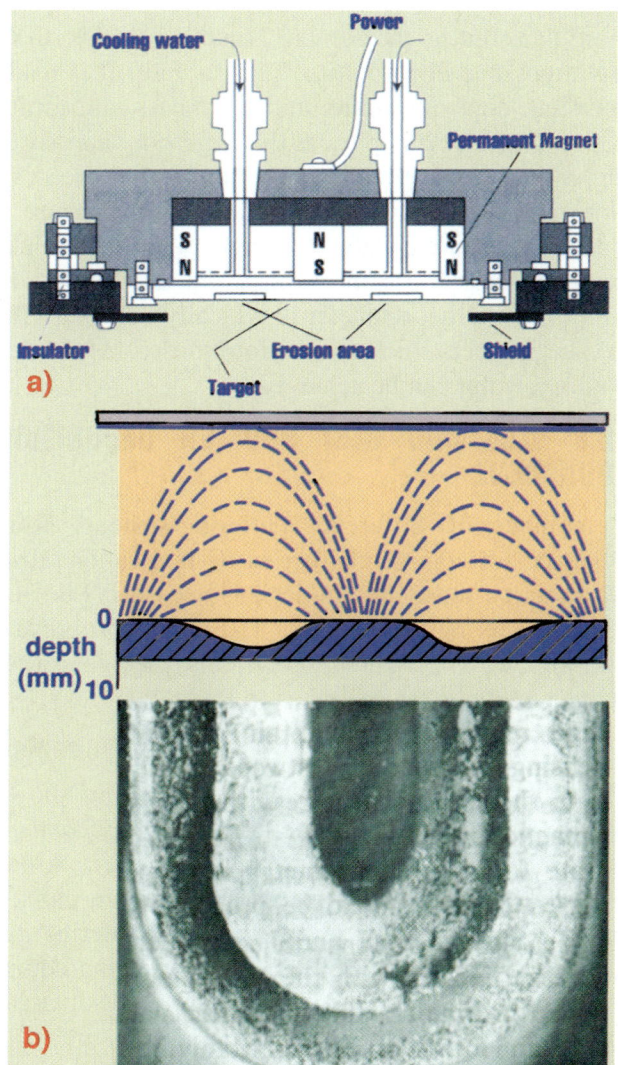

Fig. 15-12 (a) Cross-section of a planar circular-magnetron source with permanent magnets behind the target. (b) Example of the "racetrack" configuration of the erosion pattern from a planar magnetron sputtering-source.

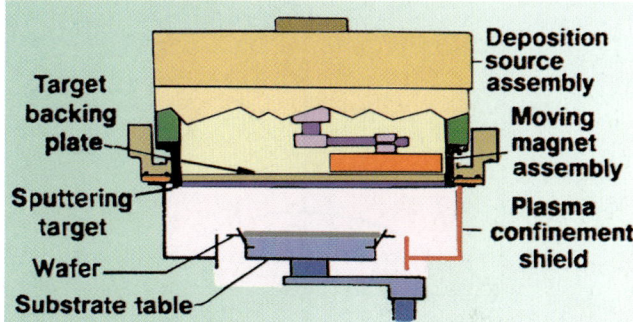

Fig. 15-13 Cross-section of a circular planar-magnetron-source, with rotating movable-magnets behind the target.

Deposition-Rate and Thickness-Uniformity with Circular Planar Magnetrons: Two of the most important characteristics of a deposition process are *film-deposition-rate* and *film-thickness-uniformity across a wafer*. A sufficiently-high deposition-rate is needed to make the process economically-viable. The thickness must be sufficiently-uniform to allow control of other processes such as etching, and to guarantee that electrical-specifications of features formed from deposited-films are met.

Deposition-rates of 1200-nm/min can be achieved in modern sputtering-tools for Al:(0.5%)Cu films on 200-mm wafers. The maximum sputter-deposition-rates are usually limited by the ability to cool the cathode-target to keep it from melting.

Since the annular-shape of the plasma in a circular planar-magnetron results in a non-uniform erosion of the target, this also causes the deposition onto nearby wafers to be non-uniform. In fact, the deposition-profile mirrors the race-track when the wafer is close to the target, but this deposition profile becomes some degree, but erosion is not uniform across the entire area. That is, a spent-target has a central mound of thicker material, with two or three eroded rings surrounding it. Nevertheless, *full-face erosion* consumes more of the target, which improves *target-utilization,* and fewer particles are formed (because re-deposition of material on the target-face is suppressed).

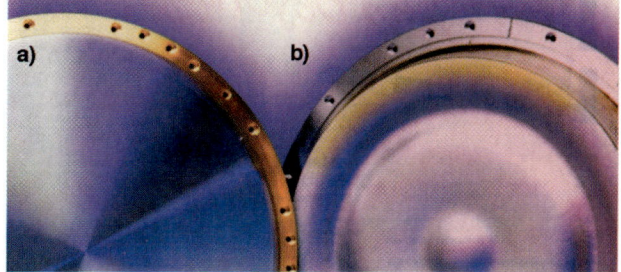

Fig. 15-14 (a) *New*; and (b) *Spent* circular-planar-magnetron targets.

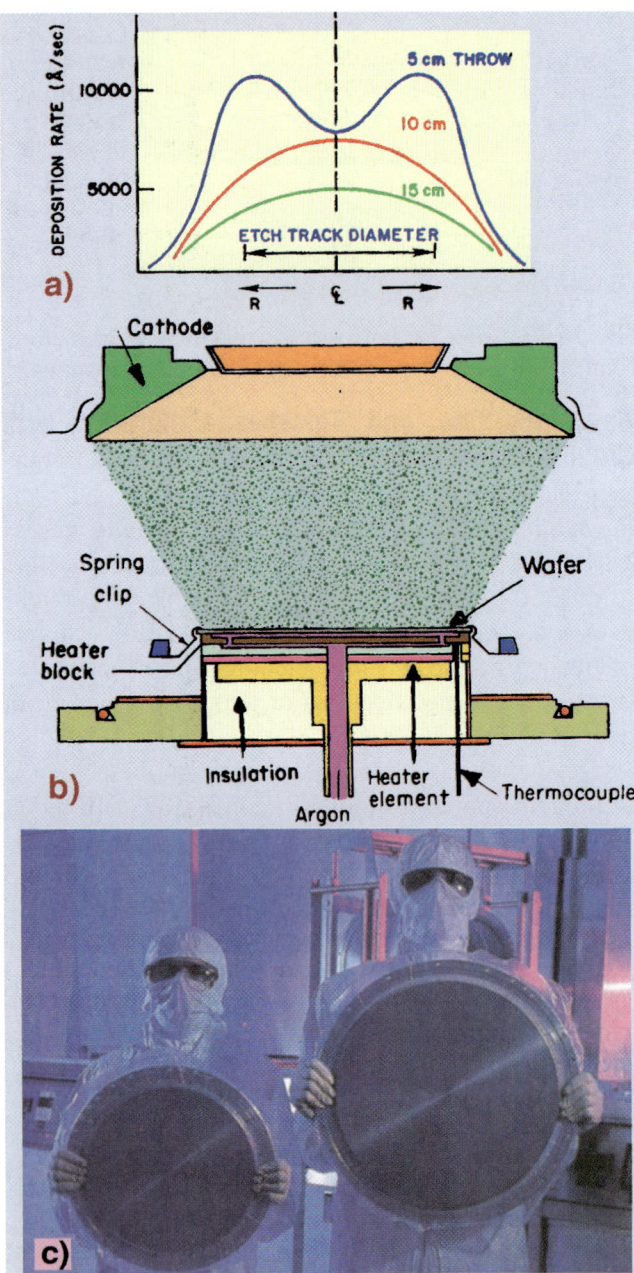

Fig. 15-15 (a) The *deposition-profile* and *deposition-rate* on a wafer in front of a magnetron-cathode as a function of wafer-distance. (b) Sputter-deposition from a magnetron-target whose diameter is larger than that of the wafer. (c) Photograph of 200-mm and 300-mm circular sputtering-targets. (Photo courtesy of Johnson Matthey Electronics.)

more smeared as the sample is moved further from the target (see Fig. 15-15a). Thus, a tradeoff is made between deposition-rate and thickness-uniformity (i.e., uniformity improves as the target-to-wafer spacing is increased, but the deposition-rate decreases). Note from Fig. 15-15b that the target-diameter is also larger than the wafer-diameter (to improve thickness uniformity). As an example, for 200-mm (8-in) wafers, the target-diameter is typically 330-mm (14-in). In practice, thickness uniformities of ±5% (3σ) within a wafer can be achieved.

15.6 VLSI AND ULSI SPUTTER DEPOSITION EQUIPMENT

A number of different sputtering-systems were designed for commercial use in IC fabrication. Here we describe those system-designs used in VLSI and ULSI sputtering-processes for 150-mm, 200-mm, and 300-mm wafers. Sputtering-systems used in earlier IC-technologies are described in Refs 4 and 5.

15.6.1 Generic Sputtering-System Components

The subsystems of a basic sputtering-system are shown in Fig. 15-16. They include: **a**) a *sputter-chamber*, where the wafer-holder and sputtering-source reside (the source also holds the *target*); **b**) vacuum-pumps; **c**) power-supplies (dc and/or rf); **d**) sputtering-gas supply and flow-controllers; **e**) monitoring-equipment (pressure-gauges, voltmeters, and residual-gas analyzers); **f**) wafer-handling mechanisms; and **g**) a computer-controller.

Sputtering-Targets: *Sputtering-targets* consist of the material that is to be sputter-deposited. Targets in circular-magnetrons are *circular-discs*, 3-to-10-mm thick, as shown in Fig. 15-10. These are bonded to a water-cooled *copper backing-plate* for good thermal-contact (see Fig. 15-11). Targets must be adequately cooled to prevent warpage, delamination from the backing-plate, or even target melting. De-ionized water is used as the cooling-liquid.

Targets consisting of metals with relatively-low melting-points (e.g., Al, Cu, and Ti) are fabricated by melting and casting the metals in either vacuum or

CHAPTER 15 ALUMINUM THIN-FILMS AND SPUTTER-DEPOSITION FOR ULSI

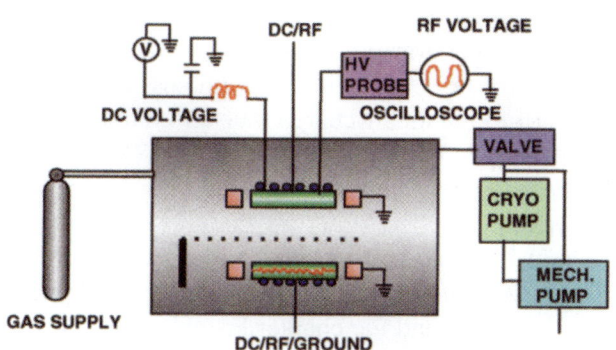

Fig. 15-16 Schematic-drawing showing the components of a generic sputtering-system.

inert ambients. The starting-materials to fabricate the targets must be very pure (e.g., up to 99.9999%-pure for aluminum targets). Targets of high-melting-point metals and compound materials are generally formed from powders by hot-pressing. Some device applications (e.g., memory-circuits) also demand material very-low in alpha-particle-emitting elements (less than 0.01-ppb U and Th content).

Vacuum-Pumps for Sputtering-Systems: Modern sputtering-systems use cryopumps to keep the sputter-chambers under high (or ultra-high) vacuum between times of sputtering. Cryopumps offer higher water-pumping speeds than do turbo-pumps (see Chap. 6) and clean operation. State-of-the-art sputtering chambers typically operate with base-pressures of 1×10^{-8}-torr, or less. Prior to beginning the sputter-process, the chamber is backfilled with Ar-gas. Since an open-flow process-configuration is used, once the sputter-process is initiated, the cryopump is used to continually pump the sputter-gas (Ar) out of the chamber. That is, during sputtering, Ar is metered into the chamber at a controlled flow-rate, and the pump simultaneously evacuates it, while maintaining the pressure in the chamber at the desired process-pressure (e.g., 2-5-mtorr). Such continuous renewal of the Ar-gas is another measure taken to ensure that the partial-pressures of contaminant-gases in the chamber are kept very low.

Power-Supplies for Sputtering-Systems: Sputtering-systems are typically provided with both *dc* and *rf power supplies*. Dc power-supplies can be built to supply up to 20-kW of power, whereas rf power-supplies are limited to ~3-kW. As a result, dc magnetron-sputtering can provide higher deposition-rates than rf magnetron-sputtering. Consequently dc-magnetron-sputtering is the operational-mode of choice for high-deposition-rate processes.

15.6.2 Commercial Sputtering-Systems for 150-mm Wafers

The most widely-used sputtering-tools for 150-mm wafers were the Varian-3180 (for wafers up to 125-mm in diameter) and the Varian-3290 models (for wafer-diameters up to 150-mm). They employed a *static* sputtering mode (the wafer and target remain stationary during the sputtering-process, Fig. 15-17).

The sputter-sources in the 3180 were initially S-Guns® (see Ref. 4 for data on S-Guns), but later, circular planar-magnetron sources were used. The wafers and target were both vertically-positioned and target-to-wafer distance was 5-10-cm. One wafer at a time was sputter-deposited by each target, and there were three sputtering-target stations. Up to 3-wafers could be sputtered simultaneously. If a different target material was placed at each station, multi-layered

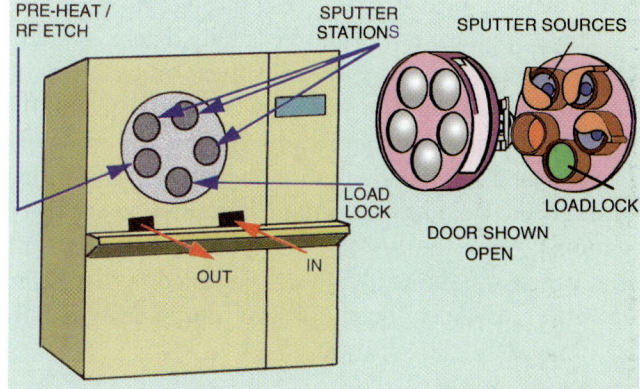

Fig. 15-17 The dominant sputtering-system for 125-mm and 150-mm wafer-processing generations: the Varian-3180® (for wafers up to 125-mm in diameter), and the Varian-3290® (for wafers up to 150-mm in diameter).

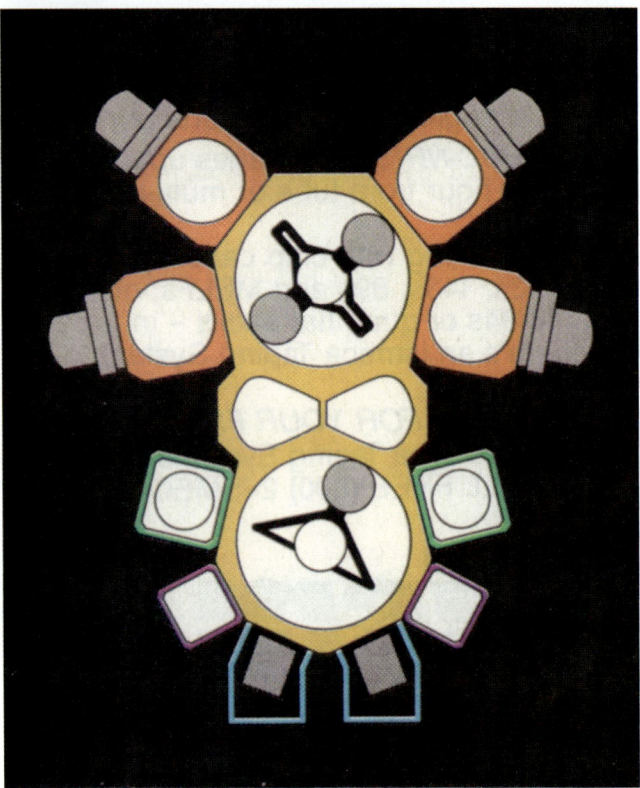

Fig. 15-18 Layout of an Endura 5500® sputtering-system. Courtesy of Applied Materials.

films could be deposited by moving a wafer between stations. All three stations could also be equipped with identical targets to increase throughput.

These sputtering-tools also contained an rf sputter-etch/heating station and were designed for cassette-to-cassette operation. The wafers were fed sequentially through the process-chamber by means of a transfer-plate, which rotated clockwise and stepped the wafer through the heating-station and the three sputtering-stations. After the wafers were stepped through all four process-stations, they were returned to the wafer cassette. However, this tool was not redesigned to accommodate 200-mm wafers.

15.6.3 Sputtering-Systems for 200-mm and 300-mm Wafers

The sputtering-systems for 200-mm (and larger) wafers are, by and large, configured as *cluster-tools*.

In a cluster-tool a number of sequential process-steps can be carried-out without exposing the wafer to the atmosphere between steps. This is accomplished by moving the wafer from one process-chamber to another through an evacuated transfer-chamber, using a robotic transfer-mechanism (Fig. 15-18). The sputtering-system that came to dominate 200-mm technology was the Applied Materials Endura.® It was introduced in 1990, and by 1995 had captured almost 80% of the market-share for 200-mm sputtering-systems (see Fig. 15-19 and the chapter Faceplate, p. 246).[10]

The Endura® won this large market-share for ULSI sputtering-applications for several reasons, but probably the foremost was that it proved to be a highly-reliable, multichamber process-tool.[10] In one reported manufacturing-evaluation, the Endura® demonstrated a long mean-time-between-failure (MTBF) record (i.e., > 120 hrs) and had an uptime that exceeded 85%. Details about the operation of the Endura are found in Ref. 1. Other suppliers of 300-mm, ULSI sputtering-systems include Novellus, Ulvac, Anelva, and TEL.

15.7 PROCESS-CONSIDERATIONS IN SPUTTER-DEPOSITION

A deposited interconnect-film must satisfy a large number of requirements in order to effectively perform its role in ULSI integrated-circuits. The most

Fig. 15-19 Changing a target in an Endura® sputtering-system. Courtesy of Applied Materials.

important ones impacted by the deposition-process itself include: **a**) correct nominal-thickness; **b**) thickness uniformity of at least ± 5% (within a wafer, and wafer-to-wafer); **c**) adequately-low resistivity; **d**) resistivity uniform to at least ± 5%; **e**) good adhesion to underlying and overlying layers; **f**) step-coverage of at least 50%); **g**) high electromigration-resistance (and good-resistance to hillock-formation); **h**) low-resistance contacts to Si and/or other interconnect-layers can be made; **i**) films can be deposited as an alloy with tightly-controlled composition; and **j**) stresses in the deposited-film are low.

15.7.1 Sputter-Deposition of Alloy-Films

Many sputtering-applications call for the deposition of alloys rather than pure-films (e.g., Al:Cu or Ti:W). Multi-component targets are usually used to deposit such alloys. The elements are sputtered from the target in the same ratio as are present in the alloy. This is one of the main reasons why sputtering was chosen over evaporation for VLSI and ULSI PVD-depositions. It is much more difficult to control the alloy-composition of deposited-films on large wafers (and from wafer-to-wafer) with evaporation.

15.7.2 The Effects on the Sputter-Process of the Transport of Vaporized-Atoms Between the Target and Substrate

Events that occur as the sputtered-atoms are transported in the reduced-pressure gas-phase between the target and wafer can impact the deposited-film characteristics. That is, atoms ejected from the target that collide with the sputter-gas atoms change the direction of their trajectories (see Fig. 15-7). When the pressure in the sputter-chamber is high (as is the case for dc-diode sputtering), the mean-free-path is short, and many such collisions can occur. A large-fraction of the sputtered-atoms thus get scattered to the sidewalls of the chamber (or are re-deposited onto the target surface). This contributes to the low deposition-rates seen with dc-diode sputtering. Since lower-pressures can be used with magnetron-sputtering, this issue is less severe. However, another problem arises.

That is, atoms whose masses are smaller than

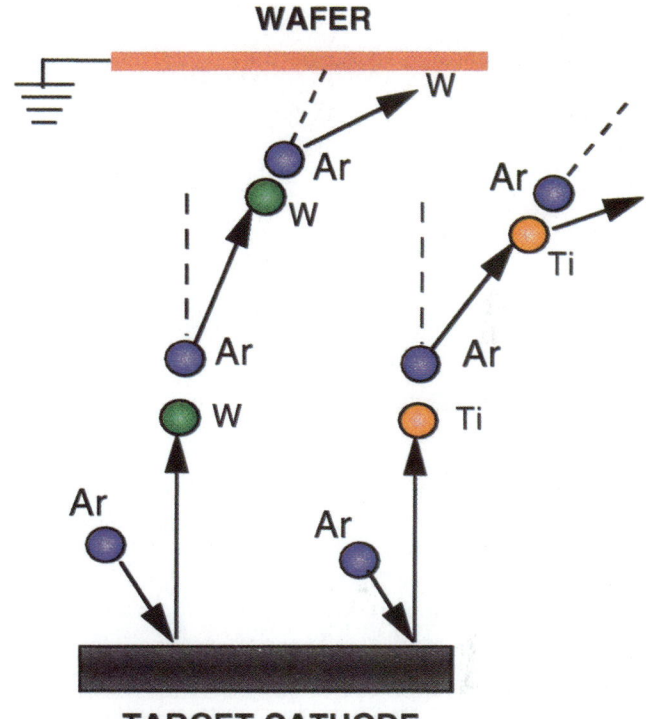

Fig. 15-20 Titanium:tungsten gas-scattering model.

(or are close to) that of Ar, will suffer larger directional-changes than those whose mass exceeds that of Ar. This is important when a binary-alloy material is being sputtered (e.g., Ti:W), and the mass of the atoms of one of the alloy-materials (Ti) is much smaller than the other (W). The smaller-mass atoms will be scattered more during vapor-transport, and thus a smaller concentration of them will end up in the deposited-film than are present in the target. The collisions will cause more of the smaller-mass atoms to be scattered toward and deposited on the chamber-sidewalls.[11]

In the case of Ti:W, the deposited-film contains less Ti than the target (~50% less). The mass of a Ti-atom is much less than that of W (48 vs. 184 amu). Since the mass of Ti is only slightly-larger than that of Ar (40-amu), Ti-atoms are strongly scattered by Ar gas-atoms (Fig. 15-20). But, the trajectories of the heavier W-atoms are less impacted by collisions with Ar. Thus, to deposit films having 10% Ti and 90% W (by

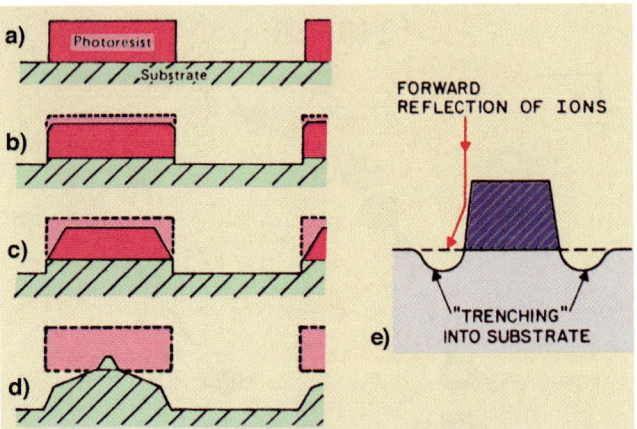

Fig. 15-21 *Faceting* results from the dependence of sputter-etch-rate on the angle-of-incidence of ions striking the surface: (a) prior to etching; (b) initiation of the *facet*; (c) facet intersects the substrate-surface; (d) substrate is exposed and forms its own facet; (e) *Trench-formation* arises from an excess-flux of ions resulting from reflection off of the sidewall.

weight) at a pressure of ~5 mtorr, the target must contain ~19% Ti. This has two important implications:

1. The resistivity of the deposited-film depends on the Ti-concentration since $\rho_{Ti} = 48\ \mu\Omega$-cm, and $\rho_W = 5.5\ \mu\Omega$-cm. In some applications the resistivity of Ti:W films is a critical parameter (e.g., in *fuse-links* of some memory devices).

2. Some back-scattered Ti atoms re-deposit on the target-surface. If these deposits get too thick, they can flake-off, and become particles in the sputter-chamber.

15.7.3 Faceting

Faceting is an effect arising from sputtering that can impact the results of sputter-etch processes, as well as the reactive-ion etching processes discussed in Chap. 22. The effect of *faceting* is illustrated in Fig. 15-21a-d, in which the sidewall of a feature is seen to develop an increasingly larger facet as the sputtering continues. (The term *facet* originated as a description of the *small faces* that are cut from the surface of a jewel.) The faceting-effect arises from the fact that sputtering-yield is greater from surfaces which are inclined at a non-90° angle to the incoming-ions (see Fig. 15-5). The facet is inclined in the direction of the incident-angle corresponding to the angle of maximum sputtering-yield. The faceting-effect can *directly* alter substrate step-profiles if an unprotected-substrate is sputter-etched. On the other hand, if a protective layer is used during a sputter-etch or RIE process, the facet developed in the masking layer can still be *indirectly* transferred into the etched-film - if etching proceeds long enough that the facet intersects the surface.

15.7.4 Particle-Generation in Sputtering

By 1994, for ULSI applications it was required that less-than $0.07/cm^2$ particles (0.3-μm or smaller) be produced by a sputtering-process. Particles arise in sputtering-processes primarily from the buildup of barrier-layer material (e.g., TiN or Ti:W) into films on the shields, and on the target-surface itself (due to re-deposition of sputtered-material onto areas of the target-surface that subsequently get little erosion). That is, the sputtering-shields get coated with highly-stressed, brittle barrier-metal films (i.e., TiN). This buildup is continuous and increases as the number of wafers is processed. Ultimately, such films delaminate, flake-off, and shower the substrate with particles. To reduce particulate-generation, equipment-vendors offer *process-specific shields*, designed to improve the adhesion of films that condense and build-up on them. The shields must nevertheless be replaced on a regular-basis, normally after processing a pre-determined number of wafers. To prevent film-buildup on the target-surface, magnetron-sources that produce *full-target-surface erosion* should be used.

15.7.5 Reactive-Sputtering

The introduction of reactive-gases during the deposition process allows *compound-films* to be deposited, a process termed *reactive-sputtering*. The main applications of reactive-sputtering in ULSI involve the sputter-deposition of TiN and TaN. Here a target of pure Ti or Ta is sputtered in a glow-discharge containing a *gas-mixture* of Ar and N_2. By using metal-only targets, higher-purity can be achieved, and the target fabrication-cost is lowered. The TiN and TaN compounds are formed when sputtered Ti or Ta atoms adsorb on the wafer-surface, and then react with the nitrogen gas-atoms that impinge on the wafer surface. The plasma provides enough energy to dissociate the nitrogen-molecules into atomic-nitrogen. The concentration of N_2 determines the stoichiometry of the deposited-film (Fig. 15-22). A typical N_2:Ar ratio of the sputter-gas is around 1:1.

15.8 STEP-COVERAGE AND VIA/CONTACT-HOLE FILLING BY SPUTTERING

In most applications it is desired that thin-films maintain a uniform-thickness and freedom from cracks or voids. As thin-films go across steps that occur on the surface of the underlying-substrate, they may suffer unwanted deviations from the ideal (such as *thinning* or *cracking*). A measure of how well a film maintains its nominal-thickness is expressed by the ratio of the *minimum-thickness* of a film as it crosses a step, t_s, to the *nominal-thickness* of the film on flat-regions, t_n (Fig. 15-23). This film-property is referred to as it's *step-coverage,* and is expressed as the percentage of the nominal-thickness that occurs at the step:

Step Coverage (%) = (t_s/t_n) x 100%   (15.1)

Step-coverage of 100% is ideal, but each process is normally specified by a lesser minimum-value that is acceptable for a given application. The height of the

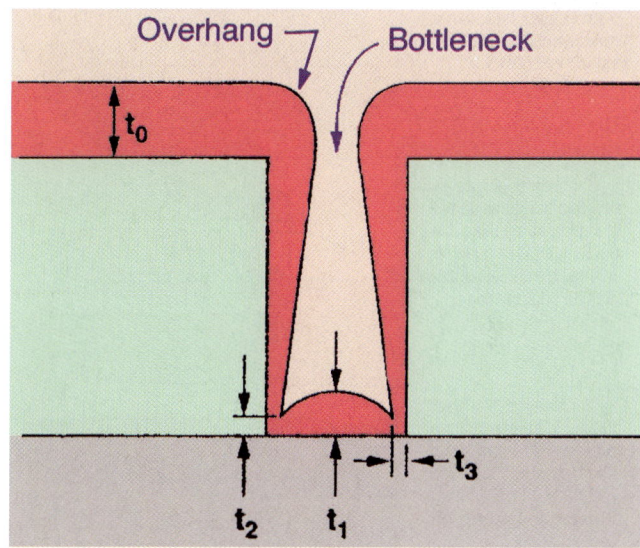

Fig. 15-23 Depiction of *step-coverage*. The figure also shows the development of a bottleneck due to the buildup of material on the top corners.

step and the *aspect-ratio* of the feature being covered (the height-to-spacing ratio of two adjacent steps) also determine the expected step-coverage. The greater the step-height or aspect-ratio, the more difficult it is to cover the step without thinning of the film. In addition, the smoother the contour and the smaller the slope of the step, the better the expected coverage.

Heating the wafers also improves step-coverage during sputter-deposition. This effect occurs because the surface-migration of adsorbed atoms is faster at higher substrate-temperatures. For Al-films, the wafers must be heated to > 250°C before there is a significant improvement in step-coverage.

An important point to note when considering the deposition of a film into a trench or contact-hole is the following: Assume a contact-hole has the shape of a cube with side **a**, and the top-face is open. This represents a contact-hole with an *aspect-ratio of 1:1*.

In such a case there are five inner-surfaces of the cube that must be covered by material which would otherwise deposit on an area $\mathbf{a}^2$ on the top-surface of the cube (see Fig. 15-24). However, if the top of the contact-hole is open, the total-area inside the hole that must be covered now is $5\mathbf{a}^2$. Thus, if the material is

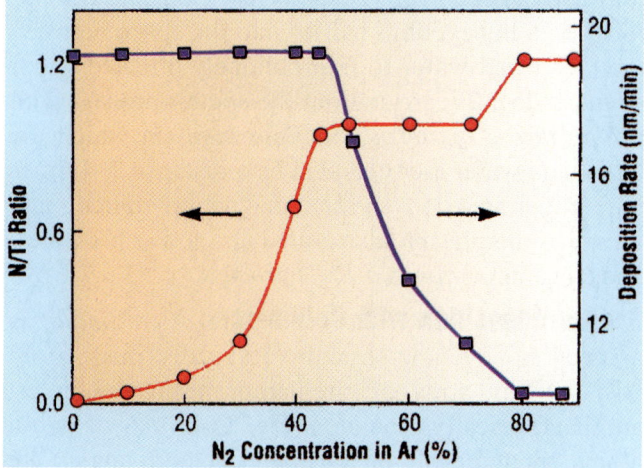

Fig. 15-22 Deposition-rate and N:Ti ratio for Ti sputtered in a reactive-ambient of N_2:Ar ratios.[12]

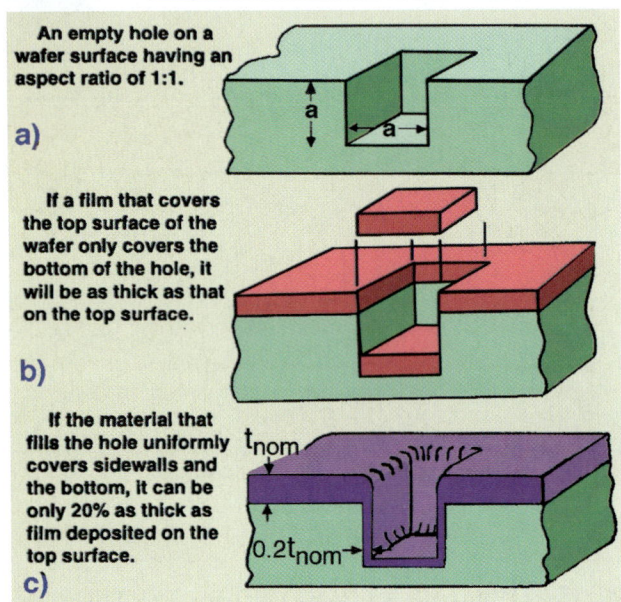

Fig. 15-24 One of the dilemmas faced when depositing films into high-aspect-ratio holes by PVD.

evenly-distributed onto all five inner-sides of the contact-hole, the film-thickness will only be 20% as thick as that on the flat-areas outside of the contact-hole.

Obviously, if the area of the opening remains constant while its depth is increased, the aspect-ratio will increase. For instance, if the square contact-hole-opening-size remained at a^2, but the depth became $2a$, the aspect-ratio will increase to 2:1, and the area that will have be covered by the material of a^2 will now be $9a^2$. Thus, for the case of a physical-vapor-deposition process, as the aspect-ratio of a recessed-feature is increased, the material entering the feature must be spread over a larger-area, making the thickness per unit-area decrease. Furthermore, if the film on the upper flat-regions must be etched or polished-away after the hole has been filled, a much thicker-film now has to be removed (which makes the job of the etching- or CMP-process significantly more time consuming and difficult).

15.8.1 Sputter-Deposition of Barrier-Layer-Films into Contact-Holes and Vias

It becomes progressively harder to fill contact-holes as the aspect-ratio increases because the material that enters the hole must be shared by the increasingly large total-area of the contact-hole sidewalls and bottom. Another related problem makes the task of filling high-aspect-ratio holes even more difficult. Sputtered atoms arrive at the upper-corners of a trench or contact-hole on the wafer from a larger arrival-angle than they do at locations within the hole. Thus, the film deposition-rate at the top-corners is higher than along the hole sidewalls or bottom. As a result, the top opening of the hole begins to close-off (i.e., an *overhang* forms), as shown in Fig. 15-23. This causes the deposition-rate onto the sidewalls and bottom to continually decrease as time goes on. If the overhang gets thick enough, the hole closes-off entirely, leaving a *void* or *keyhole* in the metal that fills a contact-hole.

Even if the deposition does not continue long-enough to completely close-off the hole, the overhang will shadow the *bottom-corners* of the contact-hole, where the arrival-angle for sputtered-atoms is already the smallest (see Fig. 15-23). Thus, the deposited film will be thinnest at the bottom-corners of the hole. Since a barrier-film must have a minimum-adequate-thickness along the entire bottom in order to function effectively, *thin-spots* at the corners can jeopardize the *integrity* of the barrier-film (e.g., TiN).

Three modifications to the sputtering-process have been developed to overcome such thinning at the bottom-corners of contact-holes: **1)** insertion of a *collimator* (i.e., a honeycomb-baffle) into the space between the target and wafer to trap obliquely-directed vapor (Fig. 15-25); **2)** carrying out the sputter-process with a *long-throw-sputtering configuration* (in which the target and wafer are separated by a distance 2–4 times longer than in a conventional magnetron-source - and in which the plasma is operated at a lower pressure); and **3)** using an *ionized-PVD process*.

Sputter-Deposition with Collimators: A *collimator* is a metal honeycomb-structure (typically made from Al), with an array of circular or hexagonal holes, and is electrically grounded (Fig. 15-26).[13,15] Neutral atoms sputtered-off of the target at high-angles are intercepted by the collimator, where they adhere to its sidewalls (vanes). As a result, only atoms ejected

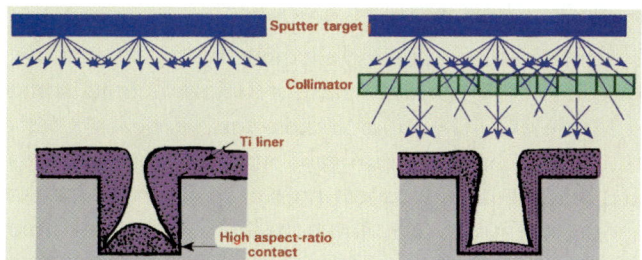

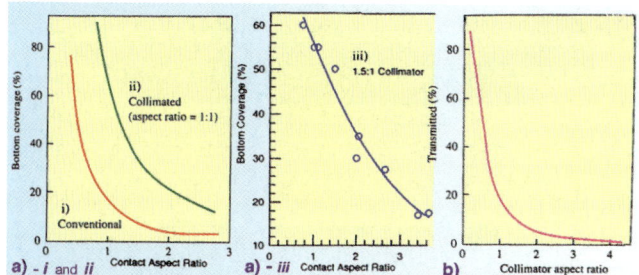

Fig. 15-25 Insertion of a collimator in the space between the target and the wafer, and its effect on the deposition of a sputtered-film into a via.

from the target at small-angles are able to reach the wafer. These atoms have a much-higher chance of being deposited on the bottom of the contact-hole than those ejected at higher-angles.

The aspect-ratio of the collimator-openings can be varied. As the aspect-ratio is increased (i.e., the collimator is made thicker for the same-size opening) collimation increases and the bottom-coverage of the contact-holes is improved. However, the deposition-rate of the film on the wafer also decreases as more of the sputtered-flux is intercepted. Typical

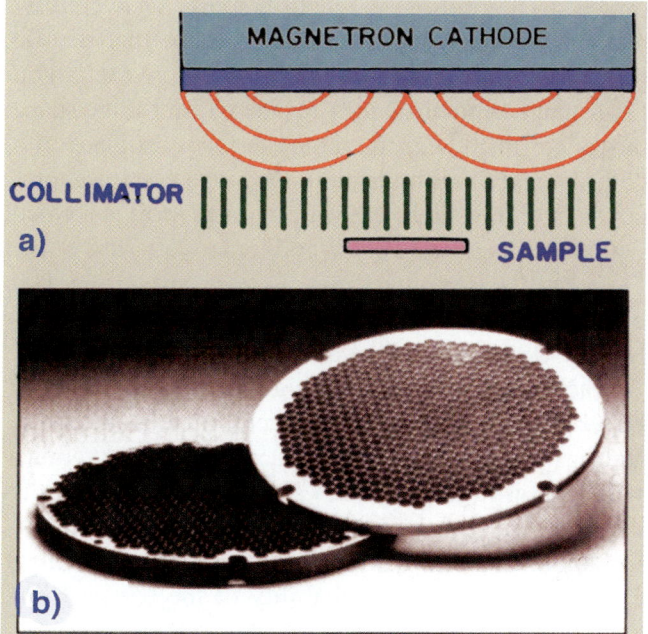

Fig. 15-26 (a) Schematic of a *collimator* in a sputtering tool. (b) Photograph of collimators.

Fig. 15-27 Characteristics of collimated-sputtering: (a) Bottom-coverage of holes versus the hole aspect-ratio, for i) conventional magnetron-sputtering, ii) collimated-sputtering with an aspect-ratio of 1:1, iii) collimated-sputtering with an aspect-ratio of 1.5:1. (b) Deposition rate of Ti and TiN as a function of the aspect-ratio of the collimator.[14] (© IEEE 1996).

collimators employ aspect-ratios ranging between 1:1–3:1. Figure 15-27a shows the bottom-coverage of collimated-sputtering compared to conventional (non-collimated) versus *contact-hole aspect-ratio* for 1:1 *collimator aspect-ratio*.[14] The increase in bottom-coverage is significant, and goes from <10% to >25% for a contact-hole with an aspect-ratio of 2:1 Thus, for contact and barrier-layer films (such as Ti and TiN), where bottom-coverage is critical, collimation provides a simple extension of an existing process that is easily integrated into IC sputter-technology. Collimated-sputtering has been successfully adopted for 0.35-μm and 0.25-μm CMOS generations.

Besides reduced deposition-rates, however, the use of a collimator does have some other significant drawbacks. First, the deposition-rate further decreases as more wafers are sputtered (since the openings of the collimator continuously close-up as the film deposited on the collimator-sidewalls gets thicker). The deposition-rate thus varies with time, and this effect must be controlled in a production-line to maintain a uniform barrier-layer thickness from wafer-to-wafer as the collimator "ages." The high-percentage of metal intercepted also increases the cost of the barrier-film deposition-step because the throughput of the sputtering-tool and the target-utilization are both significantly reduced. In addition, the collimator must eventually be replaced (e.g., once transmission has decreased to

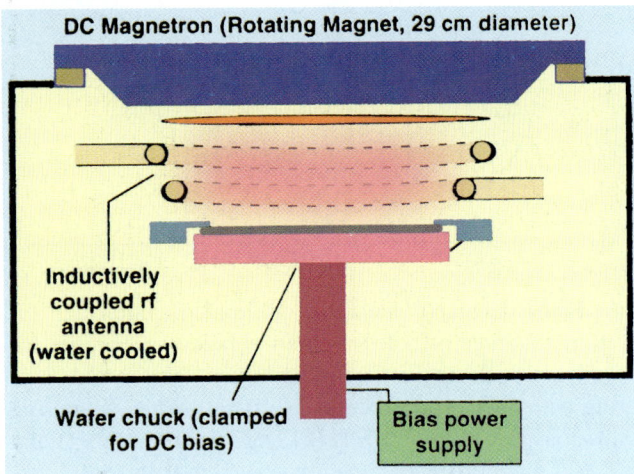

Fig. 15-28 Schematic of a magnetron sputter-chamber fitted with an rf-coil which allows an *ionized-magnetron sputtering-process* to be implemented.[16]

50% of its initial value). The expense of the collimator, and the attendant tool-downtime further raises the cost. The presence of the collimator also increases the potential for particle-generation, as the brittle barrier-layer films deposited on the collimator may eventually begin to flake-off and shower the wafer below.

Long-Throw Collimated-Sputtering: Another collimated-sputtering technique that does not employ a collimator-plate has also been pursued, namely *long-throw collimated-sputtering*. In this approach, a much-longer distance exists between the target and wafer than in a conventional magnetron-sputtering-system (25–30-cm vs. about 5-cm). The plasma is also operated at a lower-pressure (e.g., 0.1-mtorr). The sputtered-atoms thus undergo few (if any) collisions and arrive at the wafer by a straight-line (line-of-sight) trajectory after being ejected from the target. Atoms ejected at large-angles strike the chamber-walls (or shields), and condense upon them. Only those ejected at small-angles arrive at the wafer-surface.

Long-throw sputtering is similar to a collimator with a single opening (i.e., the chamber itself). Particle-generation from the larger wall-surfaces, and asymmetry of the step-coverage in holes away from the center of the wafer, are drawbacks of such tools.

Ionized Sputter-Deposition: While the two techniques described above produce adequate barrier-layer films for contact-holes and vias for 0.35-μm (and some 0.25-μm) CMOS-technologies, they appear to fall-short for other 0.25-μm (and smaller) contact-holes and vias when their aspect-ratio exceeds about 3:1. As shown in Fig. 15-27b, the bottom-coverage in a hole with an aspect-ratio of 3:1 is only 15%, even when a collimator with an aspect-ratio of 1.5:1 is used. An alternative to collimated-sputtering, *ionized sputtering*,[16] or I-PVD (also termed *ion metal-plasma [IMP]* by Applied Materials), has been developed to deposit the next-generation of ULSI barrier-layers with PVD. In this method, a large-fraction of the neutral sputtered-atoms (e.g., Ti) are ionized while in flight between the target and wafer. A bias on the wafer results in an electrical-acceleration of these ion-species to the wafer-surface with a controlled-energy and direction (i.e., perpendicular to the wafer-surface).

Figure 15-28 shows a sputtering-chamber with an rf-coil that produces the secondary-plasma needed to implement an ionized-magnetron sputter-deposition process. The original magnetron-source (powered by a dc-power-supply) remains in place so that neutral metal-atoms are sputtered (as before) from the target. (The magnetron produces a plasma and dark-space that is confined near the target-surface.) In place of a collimator (but with virtually the same system-geometry), an rf-powered (and water-cooled) coil is located within the chamber. This rf-coil ionizes 50-70% of the Ti-atoms. These ionized metal-atoms are accelerated by a third negatively-biased dc or rf-power-supply toward the wafer-surface at controlled-energies.

The bottom-coverage of high-aspect-ratio holes is better than that obtained with collimated-sputtering. Bottom-coverage of 40-80% occurs on 0.18-μm holes with aspect-ratios of up to 8:1. Resputtering of the deposited-film at the bottom of the via by the incoming-ions helps refill the thinner corner-regions (Fig. 15-29), improving the reliability of the barrier-films even more.

The I-PVD process offers several other benefits. First, much better target-utilization is obtained than

CHAPTER 15 ALUMINUM THIN-FILMS AND SPUTTER-DEPOSITION FOR ULSI

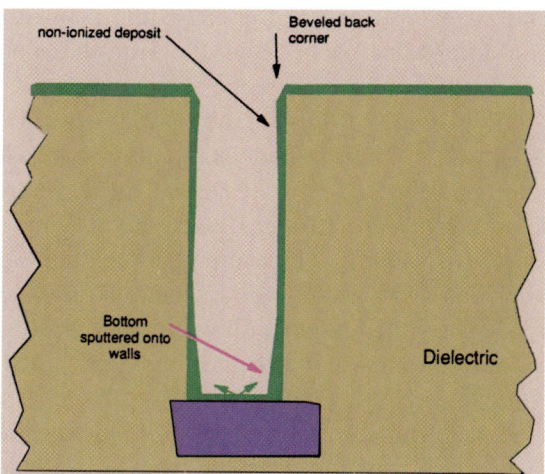

Fig. 15-29 Drawing of an *ionized-PVD deposition* showing the effects of re-sputtering of the deposited-film.

with collimated-sputtering (as almost all of the sputtered-atoms arrive at the wafer-surface, instead of being intercepted by the collimator). Sputtering-tool throughputs are thereby increased. Second, the high flux of metal-atoms directed perpendicular to the wafer-surface also re-sputters atoms that would otherwise form an overhang at the top-corners of the contact-hole. This keeps the top-opening of the contact-hole from closing-off, and maintains the thickness of the film on the hole-sidewalls at a small value. Figure 15-30 shows the coverage of the sidewall and bottom of a high-aspect-ratio hole with I-PVD.

The *hollow-cathode source* used in the Novellus Systems INOVA® sputtering-system (Fig. 15-31) employs a similar technique to ionize sputtered-atoms as they are transported to the substrate. A uniform plasma throughout the volume of the hollow-cathode provides sputtering from a large-fraction of the area of the target-surface. This provides the benefit of high target-utilization. A bias on the wafer-chuck accelerates the ionized sputter-atoms to the wafer in a direction perpendicular to its surface, allowing good coverage of high-aspect-ratio holes.

15.9 METAL FILM-THICKNESS MEASUREMENT

The thickness of sputtered metal-films is measured after the sputtering-process is complete. Metal-film

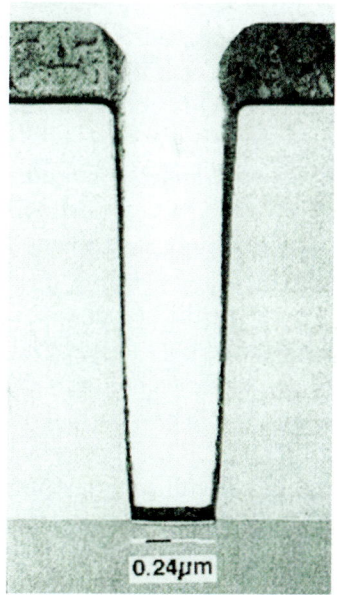

Fig. 15-30 SEM of a hole with an aspect-ratio of 8:1 deposited with a TiN-layer using an ionized-metal-plasma (IMP) process. Courtesy of Applied Materials.

thickness (on flat-areas of a wafer) is measured in one of three ways: **1**) directly, with a surface-profiling device (known as a *stylus-profilometer*); **2**) indirectly, by electrical measurement of the sheet-resistivity of

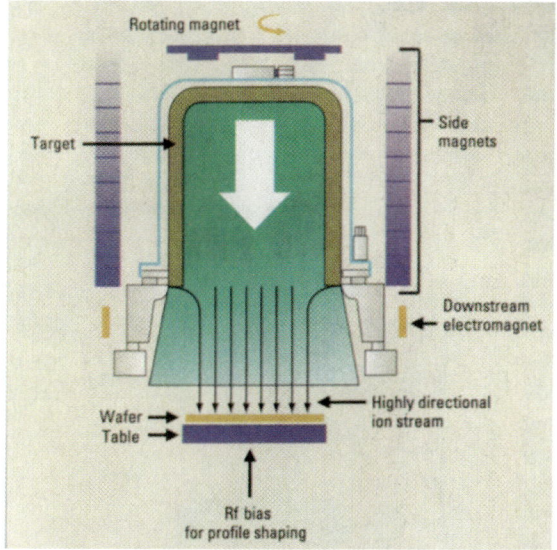

Fig. 15-31 Drawing of the *hollow-cathode-source* used in the Novellus INOVA® sputtering system. Courtesy of Novellus Systems, Inc.

blanket-deposited layers; and **3**) indirectly, by measurement of reflectivity-variations induced by short sound-pulses (picoseconds long) in opaque metal-films (*photoacoustics*).[17]

When the *stylus-profilometer* is used, a step in the deposited-film is created by a post-deposition etch process. Next, the profiling-instrument draws a fine stylus along the surface containing the stepped-film (Fig. 15-32). Whenever the stylus encounters a step, a signal variation (based on a differential-capacitance or differential-inductance technique), yields an indication of the step-height. The information is displayed on a chart-recorder or CRT. Films less than 100-nm thick can be measured with such instruments, but for films much-thinner than 100-nm, sheet-resistivity techniques are used. Some newer surface-profilers also use atomic-force microscopy (AFM).

The sheet-resistance of metal-films is measured with a four-point probe, and the film-thickness is then inferred from a knowledge of the material's bulk resistivity. Details of this method are given in Ch. 7. The thinnest films that can be accurately measured depend on the metal. For Al, Ti, and Cu respectively, these minimums are 30–40, 10–20, and 100-nm.

The *photoacoustic* measurement-technique is a non-contact, non-destructive method that uses a small spot-size (less than 8-μm), allowing it to be used on a test-structure that is < 20-μm in diameter. Such test structures can be included in the scribe-lines of product wafers, eliminating the need for monitor-wafers.

Picosecond sound-pulses are launched vertically into the metal-film. When reflected back, a slight change in the reflectivity occurs at the reflecting interface. A probe-pulse, split off from the excitation pulse, is delayed, and the reflecting light from the probing pulse is measured. The time it takes for the reflecting pulse to come back (echo) to the top-surface of the metal-film is used to deduce its thickness.

REFERENCES

1. S. Wolf and R.N. Tauber, *Silicon Processing for the VLSI Era, Vol. 1 - Process Technology,* 2nd Ed., Lattice Press, Sunset Beach CA, Chap. 11 (*Physical Vapor Deposition*).

2. D. Pramanik and A.N. Saxena, "VLSI Metallization Using Aluminum and Its Alloys, pt. I," *Solid State Technol.*, Jan. 1983, p. 127, and pt. II, March 1983, p. 131, and "Aluminum Metallization for ULSI," *Solid State Technology,* March 1990, p. 73.

3. M. Hansen & A. Anderko, *Constitution of Binary Alloys,* McGraw-Hill, New York, 1958.

4. B. Chapman, *Glow Discharge Processes*, John Wiley & Sons, New York, 1980.

5. J.L. Vossen and J.J. Cuomo, "Glow Discharge Sputter Deposition," in T*hin Film Processes*, Eds. J. Vossen & W. Kern, Academic Press, New York, 1978, Ch. II-1, p. 11.

6. O. Almen and G. Bruce, *Nuclear Instrumentation Methods,* **11**, 257 (1961).

7. D.M. Mattox, *Handbook of Physical Vapor Deposition (PVD) Processing,* Noyes Pubs., Park Ridge, N.J., 1998.

8. D.L. Smith, *Thin-Film Deposition: Principles and Practices,* McGraw-Hill, New York, 1995.

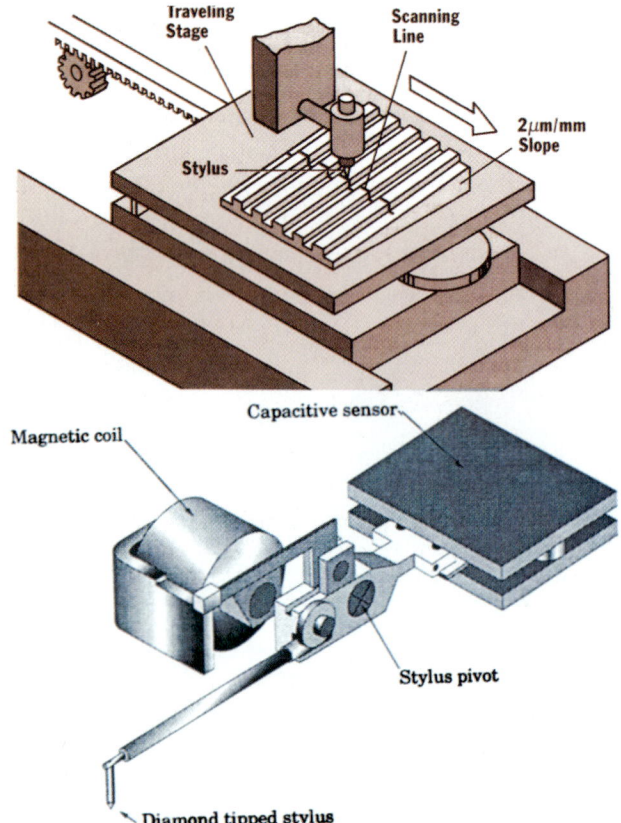

Fig. 15-32 Schematic drawings of *surface-profilometers.*

9. S. Rossnagel, "Glow Discharge Plasmas and Sources for Etching and Deposition," in *Thin Film Processes II*, Eds. J.L. Vossen and W. Kern, Academic Press, San Diego, 1991, Ch. II-1, p. 11.

10. "The Endura 5500's PVD Metallization Process," *Semiconductor International,* 1990, p. 30.

11. S.M. Rossnagel, I. Yang, and J.J. Cuomo, "Compositional Changes During Magnetron Sputtering of Alloys," *Thin Solid Films,* **199**, 59 (1991).

12. D. Pramanik and V. Jain, "Barrier Metals for ULSI: Deposition and Manufacturing," *Solid State Technology*, January 1993, p. 73.

13. S.M. Rossnagel, D. Mikalsen, H. Kinoshita, and J. Cuomo, "Collimated Magnetron Sputter Deposition," *J. Vac. Sci. Technol.* **A-9**, 261 (1991).

14. R.V. Joshi and S. Brodsky, "Collimated Sputtering of Ti/TiN Liners Into Sub-Half-Micron High-Aspect-Ratio Contacts/Lines," *Proc. VMIC*, 1992, p. 253.

15. P. Burggraaf, "Straightening Out Sputter Deposition," *Semiconductor International,* August 1995, p. 69.

16. S.M. Rossnagel, "Ionized Magnetron Sputtering for Lining and Filling Trenches and Vias," *Semiconductor International,* February 1996, p. 99.

17. R. De Jule, "Advances in Thin-Film Measurement," *Semiconductor International*, May 1998, p. 5.

PROBLEMS

1. Explain why aluminum was chosen as the interconnect metal for ICs when they were first invented, and why it has remained the main interconnect metal for this application for over 30 years (until recently).

2. Why have small quantities of each of the following elements been added to the Al used for IC interconnect lines: (a) silicon; and (b) copper. Explain why each one is added.

3. Sketch and name the major subsystems of a basic sputtering system.

4. Describe the problem of *Al junction-spiking*, and two methods that have been developed to control it.

5. What tare he main advantages of sputtering over evaporation for fabricating silicon ICs (such that sputtering is now used almost exclusively to deposit PVD metal-layers).

6. Give a brief description of: (a) how the phenomenon of sputtering occurs; and (b) how sputtering is exploited to deposit metal films for IC applications.

7. Define the concept of *sputter-yield*. Discuss why sputter-yield reaches a maximum at some energy, and then declines beyond that energy.

8. Explain why *argon* is the most-widely used chamber-gas for sputtering applications.

9. Why are pure targets perferred for sputtering over compound targets (e.g., even when TiN is deposited by PVD, a pure Ti target is used. The TiN is formed by reaction of the sputter-deposited Ti and the nitrogen-gas that is present in the process chamber during the process.)

10. Explain why *magnetron-sputtering* is an improvement over *dc-diode* sputtering (including a brief description of the basis of magnetron-sputtering).

11. Discuss the phenomenon of *electromigration*, explaining how it can impact the reliability of metal interconnect lines in ICs. What design and process measures are used to control the problem.

12. Define the term *collimated sputtering*. What advantages are obtained in the sputtering process by inserting a collimator between the target and wafer in a sputter chamber? What are some of the drawbacks involved if collimated sputtering is employed.

13. Despite its drawbacks, collimated sputtering finds use in some sputter-applications. Name them and explain why these applications require the use of collimators.

14. Describe the process of *ionized-PVD*. Explain how ionized-PVD improves the properties of metal films that are sputter-deposited into high-aspect ratio recesses (even more than is possible with the use of collimators)?

15. Calculate the *mean-free-path* of a sputtered atom in an Ar glow-discharge at 10-mtorr pressure. If the cathode-to-substrate spacing is 5-cm, how many gas-phase collisions is a sputtered atom likely to encounter as it traverses this distance?

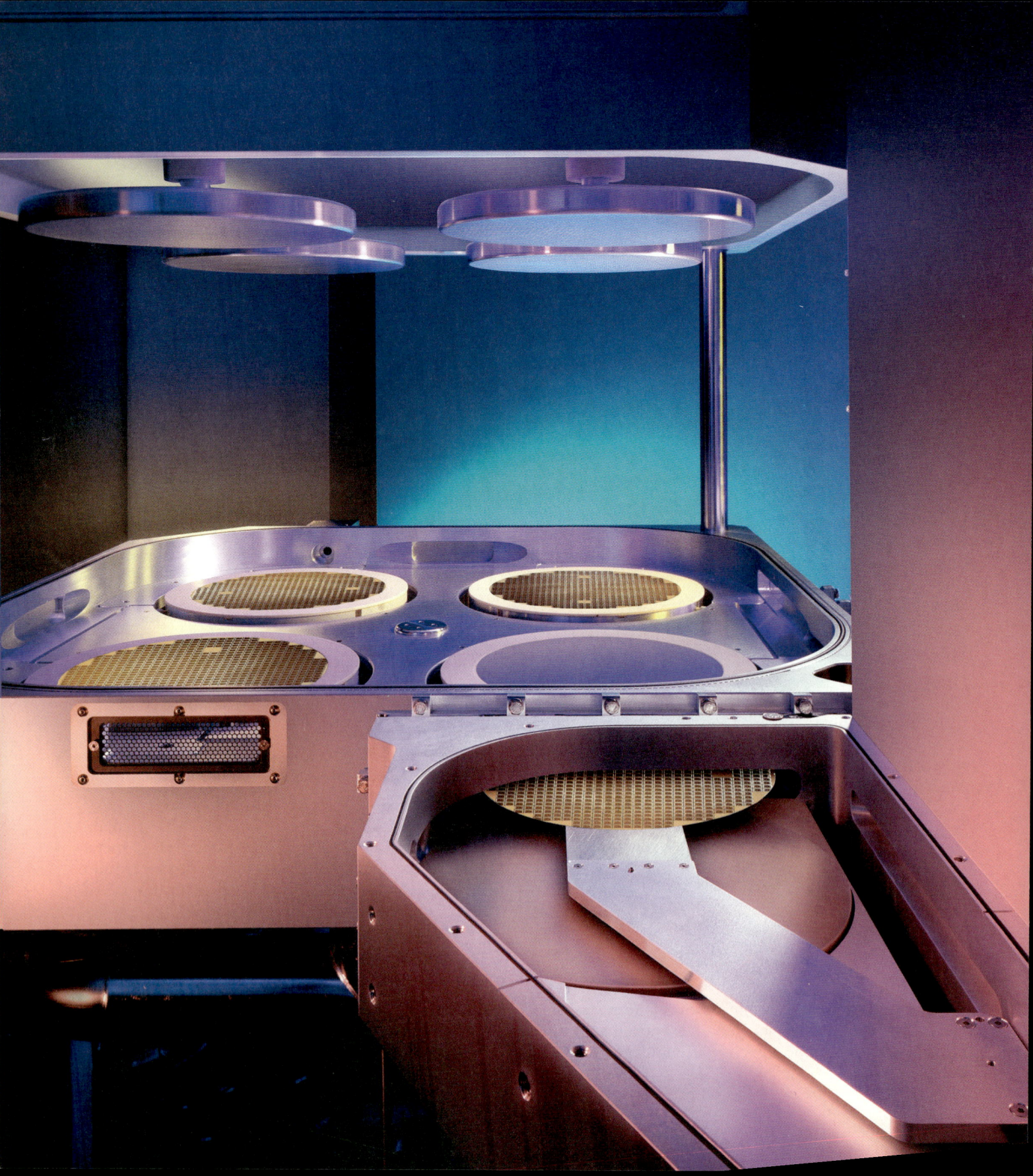

CHAPTER 16

CVD OF AMORPHOUS AND POLYCRYSTALLINE THIN-FILMS

CHAPTER CONTENTS

16.1 BASIC ASPECTS OF CHEMICAL VAPOR DEPOSITION

16.2 CVD SYSTEMS

16.3 POLYCRYSTALLINE SILICON: PROPERTIES AND CVD-METHODS

16.4 PROPERTIES AND DEPOSITION OF CVD SiO_2

16.5 PROPERTIES & CHEMICAL VAPOR DEPOSITION OF SILICON-NITRIDE

16.6 SILICON-OXYNITRIDES DEPOSITED BY CVD

16.7 CVD OF METALS, SILICIDES, AND NITRIDES

"A well-designed apparatus will do its work without effort. It will cost considerably less than one improperly planned and in whose design established principles have been overlooked. It will also perform with greater economy of steam, which is always an important consideration."

A.L. Weber,
Evaporation, The Chemical Catalog Co. New York, 1926.

PECVD sytsem for 300-mm wafers. Shown is the Novellus VECTOR® PECVD tool. Courtsey of Novellus Systems, Inc.

Chemical vapor deposition (*CVD*) is a process by which a solid-film is deposited on a substrate by reacting vapor-phase chemicals that contain the constituents of the film. The reactant-gases are fed into a reaction chamber. There they decompose and/or react at a heated-surface to form the thin-film (Fig. 16-1). Note that in CVD the reactant-gases do not react with (and therefore do not consume) any substrate-surface-material.

A wide-variety of the thin-films used in ULSI fabrication are formed by CVD. The deposition of amorphous and polycrystalline thin-films is the subject of this chapter. The growth of single-crystal silicon films by epitaxial CVD-techniques is described in Chap. 17.

CVD involves many scientific disciplines, including gas-phase reaction-chemistry, thermodynamics, kinetics, heat-transfer, fluid-mechanics, surface-reactions, plasma-reactions, film-growth phenomena, and reactor-engineering. Obviously these interdisciplinary principles can only be introduced here. Interested readers are advised to consult the references

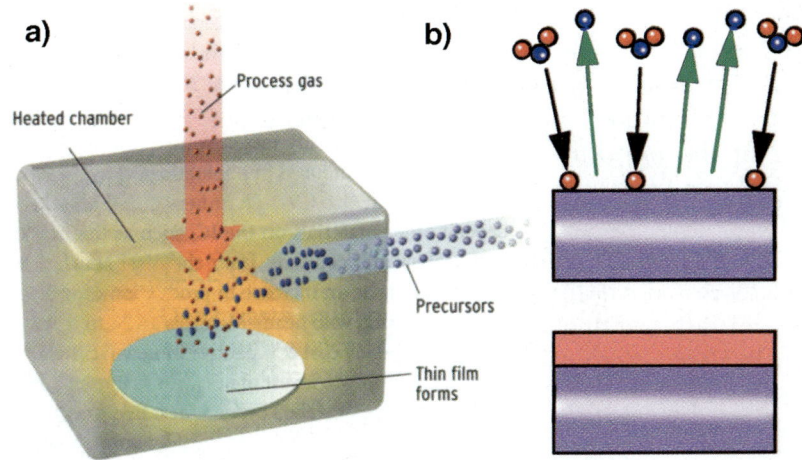

Fig. 16-1 Chemical-vapor-deposition (CVD) produces a solid-film on a substrate by reacting vapor-phase chemicals that contain the constituents of the film. The reactant-gases do not react with (and therefore do not consume) any substrate surface-material: (a) Macroscopic perspective; and (b) Microscopic perspective of CVD.

269

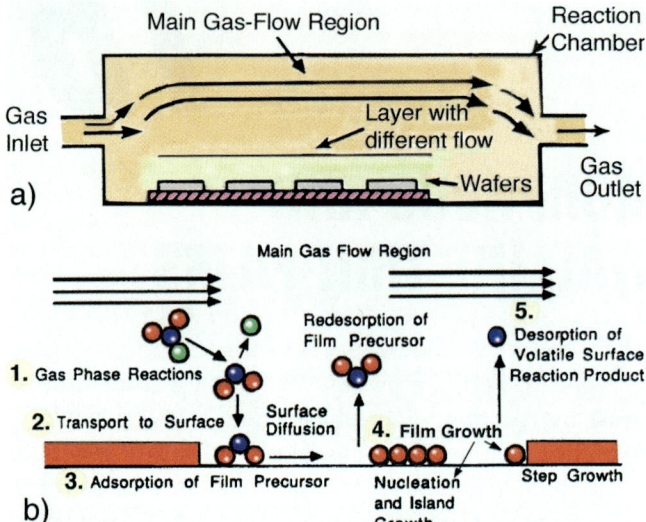

Fig. 16-2 The CVD-process from: (a) a *macroscopic perspective*; and (b) a *microscopic perspective*. In part (a) a CVD reaction-chamber is shown with gases flowing into and out of it. In part (b) the five sequential-steps of a CVD-process are depicted.

for more information on these topics.[1-4]

In this chapter, the mechanisms and a model for thin-film growth in CVD are discussed first. Next, the deposition-technology and equipment used to prepare such films by CVD are covered. Finally, the properties and deposition conditions of some of the most widely-used films deposited by CVD (including polycrystalline-silicon, silicon-dioxide, silicon-nitride, tungsten, and titanium-nitride) are covered. Measurement of many of the properties of CVD-films is virtually identical to the measurement of the same properties in other thin-films. Consult the index to find information on various generic thin-film measurement methods.

As discussed in Chap. 7, thin-films are used in a host of different applications in ULSI fabrication, and can be prepared using a variety of techniques. Regardless of the method by which they are formed, the process must be economical, and the resultant films must exhibit the following characteristics: **a**) good thickness-uniformity; **b**) high-purity and density; **c**) controlled-composition and stoichiometry; **d**) a high-degree of structural-perfection; **e**) good electrical-properties; **f**) excellent adhesion; **g**) good step-coverage; and **h**) low defect-density.

Specific deposition methods have been developed to fabricate such thin-films, based on the required capabilities for satisfying these demanding criteria. CVD-processes are often selected over competing techniques because they offer the following advantages: **a**) high-purity deposits can be achieved; **b**) a great variety of chemical-compositions can be deposited; **c**) some films cannot be deposited with acceptable film properties by any other method; and **d**) good economy and process-control are possible for many CVD films.

16.1 THE BASIC ASPECTS OF CHEMICAL VAPOR DEPOSITION (CVD)

A CVD-process consists of the following five steps:

1. Reactant-gases (frequently diluted by mixing with a *carrier-gas* [or *diluent-gas*]) are introduced into a reaction-chamber. They move from the inlet to the outlet in the *main gas-flow region* (see Fig. 16-2a). While flowing through the chamber, they come into the vicinity of the wafers in it.

2. The gases are transported by *diffusion* to the wafer-surface through a sector called the *boundary-layer*. That is, a *boundary-layer* in the gas exists between the main gas-flow region and the surface of the wafer (see Figs. 16-2a and b).

3. The reactants are *adsorbed* on the substrate-surface (whereupon they are called *adatoms*).

4. These adatoms migrate to growth-sites, where the *film-forming chemical-reactions* take place. These reactions create the *solid-film* and *gaseous by-products*.

5. The *gaseous by-products* of the reaction desorb from the surface, diffuse through the boundary-layer near the surface, into the main- gas-flow region, and are removed from the chamber as the main-gas-flow moves toward the outlet.

Energy to drive the film-forming reactions is supplied by one or more of the following energy-sources: thermal energy; photons; or electrons. *Thermal energy* is by far the most common energy-source used.

Many types of chemical-reactions are employed in CVD to fabricate the various kinds of films needed in ICs, including: *pyrolysis* (thermal-decomposition); *reduction*; and *oxidation* of the reactants.

Several other practical aspects of CVD processes should be mentioned. First, the deposition-time needed to form the film with the desired thickness must be sufficiently-short to permit adequate wafer-throughput. Second, the process-temperature must be low-enough so that it does not adversely impact the stability of previously-deposited layers (e.g., low-melting-point metal layers such as Al may already be on a wafer). Third, the CVD-process should not allow by-products of the reactant-gases to become incorporated into the growing films (but this is sometimes unavoidable, such as when a significant-amount of hydrogen is incorporated in PECVD silicon-nitride films - see Sect. 16.5).

The chemical-reactions leading to the formation of the solid that makes up the deposited-film may, in practice, take place not only on (or very-close-to) the wafer-surface (*heterogeneous-reaction*), but also in the gas-phase (*homogeneous-reaction*). Heterogeneous-reactions are preferred, as they occur selectively only on heated-surfaces, and produce good-quality films. Homogeneous-reactions are undesirable because they form solid-clusters of the depositing material in the gas-phase. These clusters can rain-down on the film being grown on the wafers, causing defects in this film (as well as other problems, including poor-adhesion and low film-density). In addition, homogeneous-reactions also consume reactants, and thus cause the deposition-rate to decrease. As a result, one important characteristic of a chemical-reaction for CVD-application is the degree to which heterogeneous-reactions are favored over gas-phase (homogeneous) reactions.

16.1.1 A Simplified CVD Film-Growth Model

Since the steps of a CVD process are sequential, the one which occurs at the slowest-rate will determine the overall-rate of film deposition. This slowest-step is referred to as the *growth-rate-limiting step*. Since

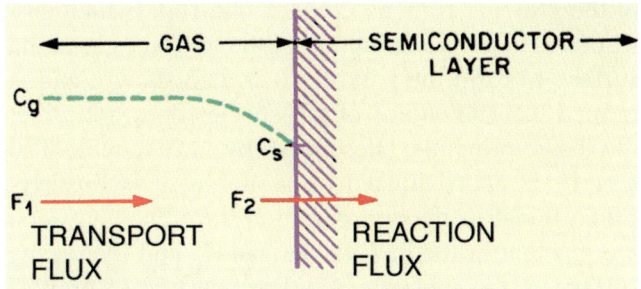

Fig. 16-3 Grove's model of the CVD-process depicting the *transport-flux* F_1 and the *reaction-flux* F_2.[5]

high growth-rates are essential for making a deposition-process economically feasible, knowing which one limits the growth-rate can be valuable. Once identified, process-modifications can be explored to speed-up that step, perhaps allowing the overall deposition-rate to be increased.

A model that permits the growth-rate of CVD-films to be predicted is also a useful tool for developing CVD-processes. However, deriving a simple mathematical-relationship which predicts CVD growth-rates based on all five-steps of the above CVD process-sequence has proven difficult, even up to this time. Instead, less-complex growth-rate models have been created, based on the observation that the 5-steps of the CVD-process can be grouped into two-categories: **1)** those that occur in the gas-phase (*gas-phase processes*); and **2)** those that occur on the substrate surface or chamber-wall surface (*surface processes*). By using this observation, a simple CVD growth-rate model can be established.

This model assumes that *only one of the gas-phase growth steps* (i.e., transport of the reactants across the boundary-layer - Step **2**), or *only one of the surface-processes* (i.e., the surface chemical-reaction - Step **4**) *is the rate-limiting step*.[5] This model explains many phenomena observed in CVD-processes, and predicts the growth-rates of many films quite accurately.

Figure 16-3 is a schematic depicting the essentials of this model. It shows the concentration distribution of the reactant-gas and the *flux* (number of atoms or molecules crossing a unit-area in a unit-time; e.g., atoms/cm² sec) from the bulk of the gas to the surface of the growing-film, as F_1. A second-flux (which corresponds to the consumption of the reactant-gas at the surface to form the solid-film) is also shown, and is termed F_2. The choice of gas-species does not affect the basic-principles described by the model. The flux F_1 is approximated by assuming it is linearly-proportional to the concentration-difference between the reactant in the bulk of the gas C_g and that at the surface of the substrate, C_s. The constant of proportionality is termed the *gas-phase mass-transfer-coefficient*, h_g. The relationship for F_1 is thus:

$$F_1 = h_g (C_g - C_s) \quad (16.1)$$

The flux F_2 is assumed to be linearly-proportional to the surface-concentration of the reactant, and the constant of proportionality is the *chemical surface-reaction-rate-constant*, k_s. The expression for F_2 is:

$$F_2 = k_s C_s \quad (16.2)$$

Under *steady-state-deposition* conditions the two-fluxes must be equal: $F_1 = F_2 = F$. As noted earlier, in many CVD-processes the reactant-gas is diluted in a *carrier* (or *diluent*) gas. In such cases, the concentration of the reactant in the gas-phase can be defined as $C_g = C_T Y$, where Y is the mole-fraction of the reaction-species and C_T is the total-number of molecules (including the carrier-gas) per cm^3 in the chamber. In this case, an equation that calculates the *growth-rate of the film G* can be derived from this model, and is given by:

$$G = \frac{k_s h_g}{k_s + h_g} \frac{C_T}{N_1} Y \quad (16.3)$$

where N_1 is the number of atoms incorporated into a unit-volume of the film. (Recall that the value of N_1 for silicon is 5×10^{22}-atoms/cm^3.) The detailed derivation of this equation is given in Appendix D.

There are two limiting-cases of Eq. 16-3. If $k_s \ll h_g$, this is termed the *surface-reaction-rate-limited* regime. On the other hand, if $h_g \ll k_s$, this is termed the *mass-transfer-rate-limited* regime. In either of these two regimes, *the growth-rate at constant C_g (or Y) is controlled (in the limits) by the smaller-value of k_s and h_g*. This means that in these limiting-cases, the growth-rates are given by:

$$G = (C_T k_s Y)/N_1 \quad k_s \ll h_g$$

(*surface-reaction-rate-limited case*) (16.4)

or by:

$$G = (C_T h_g Y)/N_1 \quad h_g \ll k_s$$

(*mass-transfer-rate-limited case*) (16.5)

Basically, if the CVD-process is being operated in the surface-reaction-rate-limited regime, then the growth-rate is very sensitive to variations in the temperature. This also implies that as the temperature is reduced, the surface-reaction-rate is reduced. At sufficiently-low temperatures, the arrival-rate of reactants eventually exceeds the rate at which they are consumed by the surface-reaction, and the deposition-rate becomes *surface-reaction-rate-limited*.

The surface reaction-rate also increases exponentially with increasing-temperature (see Appendix D). For a given surface-reaction, if the temperature rises high enough, the reaction-rate will exceed the rate at which reactant-species arrive at the surface. In such cases, the reaction cannot proceed any more rapidly than the rate at which reactant-gases are supplied to the substrate by mass-transport, no matter how-high the temperature is increased. Then the growth-rate becomes *mass-transport-limited*.

This CVD-growth-model also gives insight into CVD-processes. For example, in CVD-processes run under *surface-reaction-rate-limited conditions* the *process-temperature* is the key-parameter. In such processes, uniform deposition-rates throughout a reactor require conditions that maintain a constant surface reaction-rate. This, in turn, implies that a constant-temperature must also exist everywhere at all wafer-surfaces. Thus, controlling the temperature becomes the main-issue in the reactor-design. Under such conditions, the rate at which reactant-species arrive at the surface is not as important (since their concentration does not limit the growth-rate). Thus, it is not as critical that a reactor be designed to supply an equal-flux of reactants to all-locations of a wafer-surface. It will be seen that in low-pressure-CVD (LPCVD) reactors, wafers can be stacked at very-close-spacing because such-systems operate in

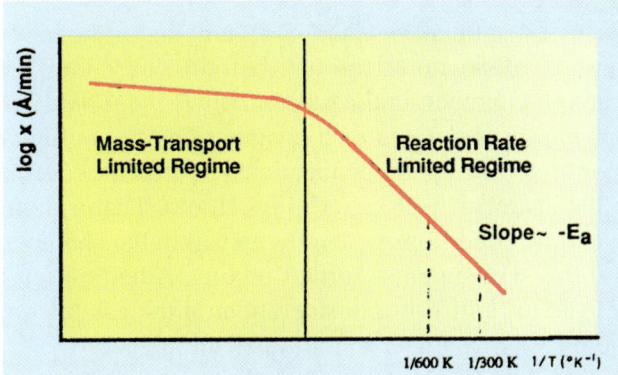

Fig. 16-4 Temperature-dependence of the *growth-rate* for CVD films.

a surface-reaction-rate-limited mode.

In deposition-processes that are *mass-transport-limited*, the temperature-control is not nearly as critical because the mass-transport process that limits the growth-rate is only weakly-dependent on temperature. In these processes, it is very important that the *same concentration of reactants* be present in the main (or bulk) gas-flow-regions adjacent to all locations of a wafer, since the arrival-rate is directly proportional to the concentration in the bulk-gas. Thus, to insure uniform film-thickness across a wafer, reactors which are operated in the mass-transport-limited regime must be designed so that all locations of wafer-surfaces are supplied with an equal-flux of reactant-species. Atmospheric-pressure-CVD (APCVD) reactors that deposit SiO_2 at ~400°C, and epitaxial-reactors operating at ≥1000°C, operate in the mass-transport-limited regime. The most widely-used APCVD-SiO_2 reactor-designs provide a uniform-supply of reactants by horizontally-positioning the wafers and moving them under a stream of reactant gases.

To Summarize: *At high-temperatures*, the deposition is usually *mass-transport-limited*, while at *lower-temperatures* it is usually *surface-reaction-rate-limited* (Fig. 16-4). In actual-processes, the temperature at which the deposition-condition moves from one of these growth-regimes to the other is dependent on the activation-energy of the reaction and the gas-flow-conditions in the reactor.

16.2 CHEMICAL-VAPOR-DEPOSITION SYSTEMS

In this section the equipment used in CVD-processes is described. The general terminology of CVD-tools is first presented, and then the particular CVD-reactor types are discussed.

16.2.1 Components of CVD-Systems

Figure 16-5 shows some of the reactor types that have been used for CVD-processes. CVD-reactors are generally *open-flow* systems, in which gases continuously flow into the reaction-chamber (where the deposition occurs). Reactant-gases are frequently carried by *diluent* (or *carrier*) gases such as H_2, N_2, or Ar, depending on the specific CVD-process. Gaseous by-products are exhausted together with the diluent-gas (and any unused reactant-gases). Gas-flows in the chamber are slow enough so that the pressure can be considered to be uniform. Corrosive and hazardous-gases pumped from the chamber are removed from the exhaust-gas flow by a *scrubber* (see Chap. 29), and the remainder

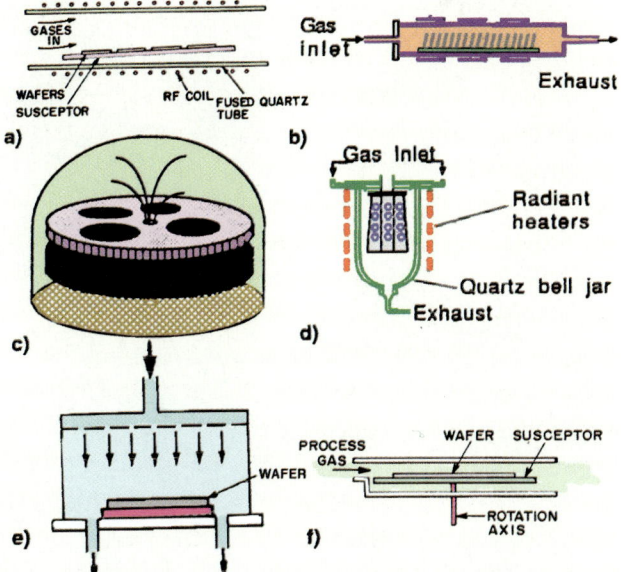

Fig. 16-5 CVD-reactor types: (a) Horizontal-flow reactor with tilted susceptor; (b) Horizontal-flow reactor with wafers stacked vertically; (c) Pancake reactor; (d) Barrel reactor; (e) Single-wafer reactor with showerhead gas-injector; (f) Horizontal-flow single-wafer reactor with rotating susceptor.

are vented to the atmosphere.

CVD-systems usually contain the following subsystems: **a**) gas-sources; **b**) gas feed-lines; **c**) mass-flow controllers (for metering the gases into the system); **d**) a reaction-chamber (or reactor); **e**) a method for heating the wafers onto which the film is to be deposited (and in some types of systems, for adding additional-energy by other means); and **f**) temperature sensors. LPCVD and PECVD systems also contain pumps for establishing the reduced-pressure and for exhausting gases from the chamber. Vacuum considerations and mass-flow-controllers (MFCs) are covered in Chap. 6. In this section gas-sources and heating-techniques in CVD-reactors are covered. CVD reactor-configurations are discussed in later sections.

16.2.2 Gas Sources & Delivery Systems for CVD

Gas-handling systems are used to supply and deliver the reactant-gases to the CVD-reactor, and these are

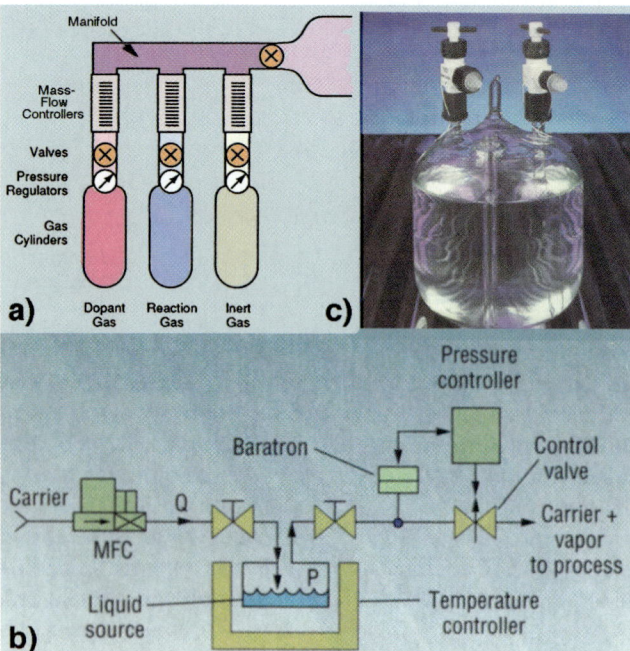

Fig. 16-6 (a) Gas-source manifold. (b) Bubbler-system for vapor-flow control. A mass-flow controller (MFC) delivers a carrier-gas into a temperature-controlled container holding the liquid-source, where the gas becomes saturated with the reactant-vapor.[6] (c) Photograph of a bubbler vapor-source. Courtesy of Schumacher.

discussed in detail in Chap. 5. Premixing of the gases must be done upon (or just before) entry into the reaction-chamber, in the reactor *supply manifold* (i.e., a *manifold* is defined as a chamber having multiple-apertures for making connections) so that they arrive at the chamber well-mixed (Fig. 16-6a). That is, laminar-flow conditions generally prevail in the chamber, and this will suppress further mixing. Attention must be paid to fluid-flow considerations at the gas-inlet of the reaction-chamber to minimize turbulence.

In CVD both gaseous and liquid sources are available. For example, both silane SiH_4 (a gas) and tetraethoxysilane TEOS (a liquid) are reactants used in CVD-SiO_2. The gaseous forms of the CVD reactants have been widely-used since the outset of the IC industry, but they are being replaced by liquid sources. Gaseous-precursors can be metered directly into the process chamber with a mass-flow controller. However, liquid-sources are preferred for a number of reasons. First, many of the gases used in CVD are hazardous-materials, posing one or more of the following dangers: being toxic, pyrophoric, or corrosive (see Chap. 29). Such hazardous compounds are inherently safer if they are liquids at room-temperature than if they are in gaseous form. That is, the vapor-pressure of the liquid is much lower than that of the gas, and hence the danger of inhaling a lethal-dose in case of a leak is smaller for liquids. In addition, liquid-spills are limited to a defined area, and in most cases there is no plume of toxic-gas. (An exception involves chlorinated liquids. These can react with water in the air to form a hazardous plume of HCl.) Besides safety considerations, many liquid-sources produce CVD-films with better characteristics, as will be discussed later.

However, if the source of the reactant-gas at the elevated process temperatures is a liquid at room-temperature, it must be vaporized before being delivered to the reaction-chamber. The lower a material's vapor-pressure, the more difficult it is to deliver it to the reactor. This, perhaps more than any other factor, determines which of the seemingly-endless number of possible organic-compounds are suitable for IC production use. Delivery of the vapors of a liquid-source is usually done by one of several methods, including: **1)** *bubblers*; **2)** heated source-containers; and **3)** direct-liquid-injection systems.

A common technique for delivering material from a liquid source to the CVD-chamber is with a *bubbler*. Here, a carrier-gas (e.g., N_2, H_2, or Ar) is metered through a mass-flow controller, and this gas is bubbled through the liquid to sweep the liquid-source molecules into the reaction-chamber (Fig. 16-6b). The carrier-gas becomes saturated with the reactant-vapor at the temperature of the liquid in the carefully-temperature-controlled bubbler. A control-loop on the downstream-side of the bubbler controls exhaust-pressure and vapor-flow into the reactor.

The main disadvantage of a bubbler is that the flow of the precursor-species is only indirectly controlled via the control of the carrier-gas flow. This can represent a significant control problem if the precursor has a steep vapor-pressure curve. For example, the vapor-pressure of TEOS will fluctuate by 32% if the bubbler-temperature fluctuates from 60 to 62°C. Bubblers also have problems in delivering materials with very-low vapor-pressures. These tend to condense between the source and the chamber. To avoid this problem the lines from the bubbler to the reaction-chamber must be heated to prevent deposition on the walls of the tubing by condensation.

Several alternatives to conventional bubbler-technology are available, including the two listed above. In the first, the source is heated and the vapor is drawn off and controlled directly by a vapor mass-flow controller. It also requires heated gas-supply-lines. Such *direct-vapor-pressure systems* are capable of delivering materials with vapor-pressures from 1-torr up to atmospheric-pressure.

The second alternative-delivery method is *direct-liquid-injection* (Fig. 16-7). In this method, the liquid remains at room-temperature until it is pumped into the vaporizer. There it is vaporized and then injected into the process-chamber. This feature is needed when the precursors are temperature-sensitive or when they decompose after long periods of heating (such as Cu [I][hfac], a liquid-source used in the CVD of Cu).

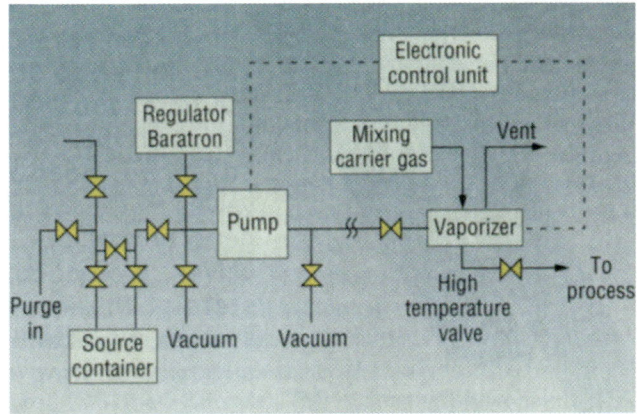

Fig. 16-7 Direct-liquid-injection (DLI) precursor delivery system.[6]

Such alternative-delivery systems offer the promise of being cleaner, cost-effective, and more reliable than bubbler-systems.

16.2.3 Heating-Sources for CVD Reaction-Chambers

In essentially all CVD-processes, films are deposited at temperatures significantly higher than room-temperature. The walls of the reaction-chamber are maintained at T_W. The wafers rest on a holder maintained at temperature T_S. In some cases $T_W = T_S$, and these are called *hot-wall-CVD-reactors*. In others $T_W < T_S$, and these are termed *cold-wall-CVD reactors*. Note that just because the temperatures of the walls of cold-wall reactors are lower than that of the wafers, they may still be significantly hotter than room-temperature. In fact, in some cold-wall systems, significant chamber-wall-heating can still take place. To limit reactions or depositions from occurring there, provisions for cooling the walls must be implemented (e.g., by water-cooling). The temperature of the gas as it enters the reaction-chamber is assumed to be equal to T_W. There are a number of methods used to raise the temperature to that required by the process, as described here.

In the first, coils of resistive-elements are wrapped around the outside of the reactor-tube. Since the hot-coils heat the entire volume they surround, the reactor-walls (as well as the wafers) have the same

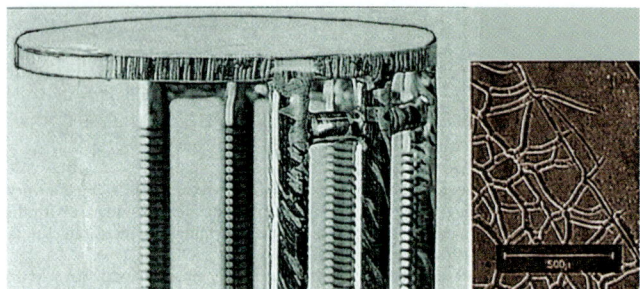

Fig. 16-8 Polysilicon film build-up and cracking on the wafer holder of a vertical LPCVD polysilicon-reactor.

temperature ($T_W = T_S$), making this a *hot-wall system*. Since the CVD-processes that are run in such systems are typically surface-reaction-rate-limited, the temperature must be uniform throughout the chamber and tightly-controlled. Control of better than 5°C at 1000°C can be achieved with such coil-resistive-heating methods.

Resistive-heating can also be used to heat *only the wafer holder* by providing electrical-contacts to elements in the heater-block beneath the wafer-holder to external power-supplies. Since heat is transferred directly to the wafer-holder, it and the wafers are hotter than the reactor-walls, making this a *cold-wall* configuration ($T_W < T_S$).

Heating can also be performed *inductively* or with *high-intensity-radiation lamps*. Both of these methods primarily heat only the wafers and the wafer-susceptor, again resulting in cold-wall reactor configurations. Inductive-heating of a conductive-susceptor (such as graphite), is done by wrapping cooled rf-coils outside and around the chamber. The rf-power coupled from these coils induces eddy-currents in the susceptor, which cause its temperature to increase. The non-conductive (e.g., fused-silica) chamber-walls are not heated by the rf-field, and they remain substantially cooler than the susceptor. For the lamp-heated case, chamber-walls are made of materials that are largely transparent to light. Thus, they are not heated as much as the opaque wafers and susceptor, which absorb more of the incident-light-energy.

Hot-wall and cold-wall reactor-configurations both have advantages and disadvantages. In hot-wall reactors $T_W = T_S$. Since the reaction is thermally driven, CVD film-forming-reactions (and hence, film deposition) can take place on their chamber-walls and wafer-holders (as well as on the wafer-surfaces). While many films grown in hot-wall reactors form dense, adherent deposits, after a number of wafers are processed, these deposits on the walls can get thick enough to begin flaking-off. To prevent particle-contamination from such sources, hot-wall systems require periodic cleaning. Figure 16-8 shows the film-buildup and cracking on a wafer-holder of a vertical hot-wall LPCVD polysilicon-reactor.

In cold-wall reactors, less-deposition occurs on the reactor-walls, since their temperature is lower than that of the susceptor and wafers. However, some material is still deposited on the cooler walls, and it is often more porous and less adherent. Thus, particle detachment can still be a problem and cleaning must still be employed. Some single-wafer reactors use an *in-situ* cleaning-process after each film-deposition step to keep the film-buildup on the walls to a minimum, and to provide the same chamber-surface conditions for each wafer.

16.2.4 The Terminology of CVD-Reactor-Design

The design and operation of CVD-reactors depends on a variety of factors, and hence they can be categorized in several ways. Figure 16-9 illustrates one way of grouping CVD reactor types. The first distinction between reactor types (i.e., *hot-wall* versus *cold-wall* reactors), has already been indicated. The next criterion used to distinguish reactor-types is their pressure-

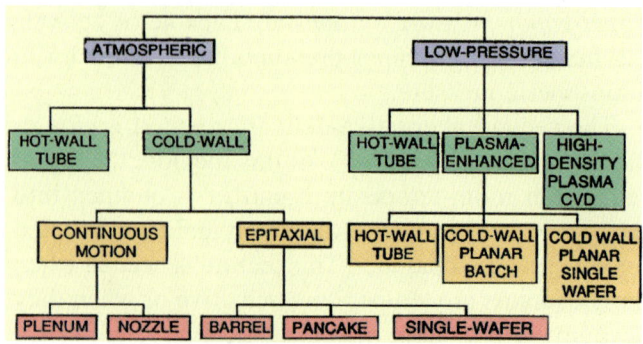

Fig. 16-9 Chart of CVD-reactor types.

Table 16-1. CHARACTERISTICS and APPLICATIONS OF CVD-REACTORS

PROCESS	ADVANTAGES	DISADVANTAGES	APPLICATIONS
APCVD	Simple Reactor, Fast Deposition, Low Temperature	Poor Step Coverage, Particle Contamination	Low Temperature Oxides, both doped and undoped; Epi films
LPCVD	Excellent Purity and Uniformity, Conformal Step Coverage, Large Wafer Capacity	High Temperature Low Deposition Rate Source Depletion	High Temperature Oxides, both doped & undoped, Silicon Nitride, Poly-Si, W, WSi_2
PECVD	Low Temperature, Fast Deposition, Good Step Coverage	Chemical (e.g., H_2) and Particulate Contamination	Low Temperature Insulators over Metals, Passivation (Nitrides)

regime of operation (i.e., *atmospheric-pressure* versus *reduced-pressure* reactors). Finally, the reduced pressure group is split into: (**a**) low-pressure reactors (the so-called *low-pressure-CVD*, or *LPCVD*-reactors), in which the energy-input is entirely thermal; and (**b**) those in which energy is partially supplied by a plasma as well as by thermal-energy. The latter are known as *plasma-enhanced-CVD*, or *PECVD-reactors*. Newer high-density-plasma (HDP) reactors for CVD are also now available (see Chap. 22). Each of the reactor-types in the two pressure-regimes are further divided into sub-groups, defined by reactor-configuration and method of heating (see Fig. 16-9).

Reactor-geometry is constrained by the pressure-regime and energy-source used, and is an important factor in wafer-throughput. Since atmospheric-pressure-reactors by and large operate in the mass-transport-limited regime, they must be designed so that an equal flux of reactants is delivered to each wafer, and to all locations of the wafer. As a result, wafers in APCVD reactors are never stacked at close spacing, but are laid flat on a horizontal-surface. An undesirable consequence of this design is that the growing films on the wafer-surfaces can incorporate particles that fall upon them. LPCVD hot-wall-reactors are not constrained by the mass-transfer-rate limitation, allowing designs that accommodate a larger number of wafers per run. That is, wafers can be stacked side-by-side, only a few-mm apart in a quartz reaction-tube. Quartz wafer-holders (boats) can hold up to 200-wafers. Since LPCVD-reactors operate in the surface-reaction-rate-limited mode, they must, however, be capable of precise temperature-control. Table 16-1 summarizes the characteristics and applications of the various CVD reactor-designs.

16.2.5 Atmospheric-Pressure (APCVD) Reactors

Atmospheric-pressure-CVD reactors (*APCVD*) were the first CVD-reactors to be used by the microelectronics industry. Their early applications included CVD-oxide and epitaxial film-deposition, and they are still used for these purposes. Operation at atmospheric-pressures keeps reactor-design simple and allows high film-deposition-rates. APCVD, however, is susceptible to gas-phase reactions, and silane-chemistry-based SiO_2-films deposited by APCVD exhibit poor step-coverage. Since APCVD-processes are also generally conducted in the mass-transport-limited regime, the reactant-flux to all parts of every substrate in the reactor must be well controlled. This places constraints on reactor-geometry and gas-flow patterns. Although APCVD-processes for deposition

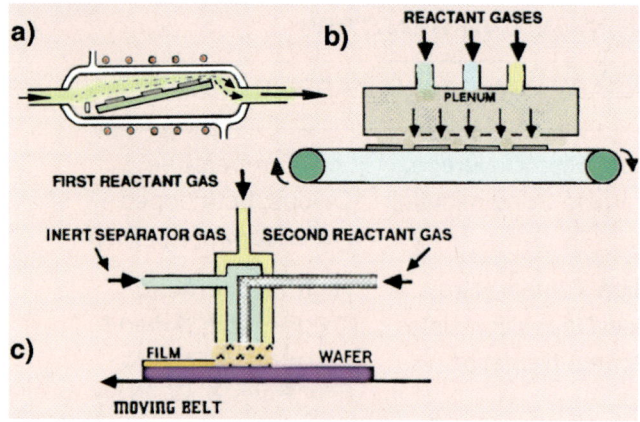

Fig. 16-10 (a) Horizontal-tube APCVD-reactor. (b) Plenum-type continuous-processing APCVD-reactor (c) Gas-injector type continuous-processing APCVD-reactor.

of nitrides and polysilicon were also developed, they were replaced by LPCVD-processes (see below).

Figure 16-10 shows schematics of three types of atmospheric-CVD systems. The first is the *horizontal-tube* (Fig. 16-10a). Such systems have a horizontal quartz-tube, with the wafers lying flat on a fixed horizontal plate, while gas in the main-flow region moves parallel to the wafer-surface. Reactant-gases are metered into one end of the tube, and unused (or by-product) gases are exhausted through the other. Energy is supplied by radiant-heating from resistance-heated coils that surround the tube (hot-wall), or inductive-coils that heat only the susceptor (cold-wall). It is possible to deposit a variety of films in these systems, but since they suffer from low-throughput, poor-uniformity and particulate-contamination, such systems find little use in VLSI and ULSI fabrication today.

The second type is the *continuous-processing APCVD-reactor* (Figs. 16-10b and 16-10c). This configuration was widely-used design for depositing low-temperature CVD-SiO_2-films in production applications in the 1980's. Continuous-processing APCVD-reactors move the wafers at constant-speed through a heated reaction-chamber, either on a moving-plate or a continuous conveyor-belt. The deposition-region is carefully isolated from outside air by curtains of flowing inert-gas. In the continuous-reactor, the reaction-chamber is at steady-state, with reactant-gases and wafers being continuously introduced and removed from the reaction-area. Reactant-gases in earlier systems were mixed in a *plenum* (i.e., from the Latin word *full*), which is a confined-volume filled with flowing process-gases at a positive pressure (slightly greater than atmospheric-pressure). But this approach causes excessive particulate-formation from gas-phase nucleation.

In a third type of APCVD-reactor (sold by Watkins-

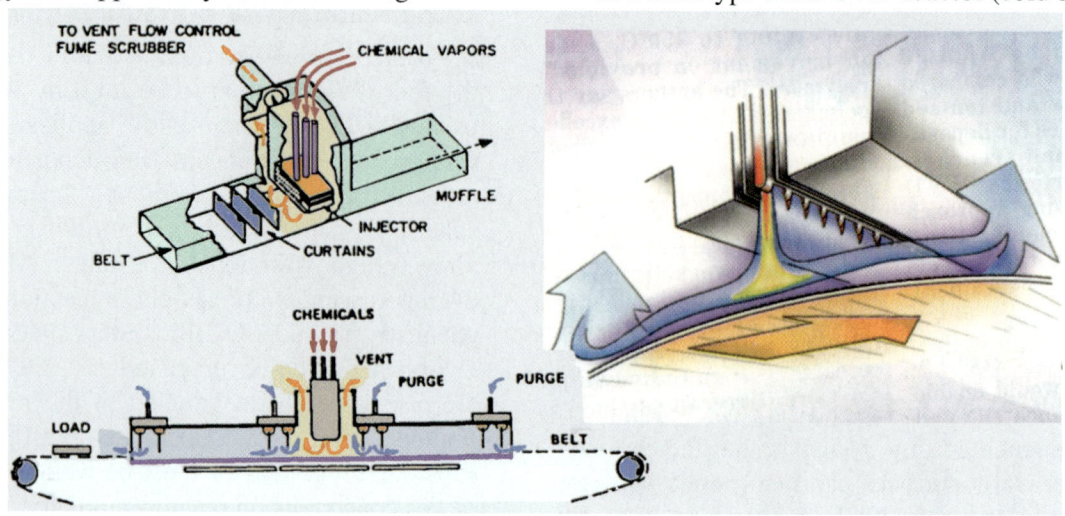

Fig. 16-11 Schematic of a conveyorized APCVD reactor with gas injection.

Johnson), gases are injected from cooled, nitrogen-shrouded nozzles. Mixing occurs millimeters above the wafer-surface, reducing gas-phase reactions.

The latter type of APCVD reactor (Fig. 16-11) was widely-used for depositing CVD-oxide films up until the mid-1990's. The gases pass through narrow-slots of an injector head and are exhausted through vents by high-N_2 flow-rates to avoid re-circulation and gas-phase-nucleation reactions. The wafers are heated by a resistive-heater situated beneath the reaction surface. Despite the fact that the nozzles are cooled, some deposits still build-up on them, and they must be frequently cleaned to keep particles from flaking-off and falling on the wafers moving below them.

Film-deposition also takes place on the plate or belt, as well as on the wafers, so frequent cleaning of the system is necessary. In the Watkins-Johnson systems, an *in-situ* HF-vapor-step cleaned the belt continuously, without disassembly. Such reactors were capable of good thickness-uniformity, low contamination, and high-throughput. However, they have lost ground to single-wafer reactors, which deposit SiO_2 films at low-temperatures with PECVD and O_3:TEOS, and HDP-CVD.

16.2.6 Low-Pressure-CVD (LPCVD) Reactors

Low-pressure chemical vapor deposition (*LPCVD*) of some types of films produces better uniformity and step-coverage (and lower particulate-contamination), than those deposited in APCVD-systems. LPCVD-reactors (Fig. 16-12) operate in the *surface-reaction-rate-limited regime*, under medium-vacuum (30–250-Pa [0.25–2.0-torr]) and moderate-temperatures (550–700°C).

At such reduced-pressures, the diffusivity of the reactant-gas molecules is sufficiently increased so that mass-transfer to the substrate no longer limits the growth-rate. The surface-reaction-rate is very sensitive to temperature, but precise temperature-control is relatively easy to achieve. For example, near the temperature commonly-used to deposit polysilicon by LPCVD (625°C), the deposition-rate changes by 2–2.5% for every degree-change in temperature. The elimination of mass-transfer constraints on reac-

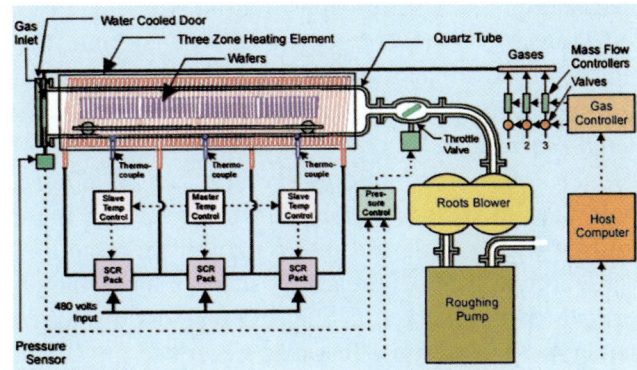

Fig. 16-12 Schematic of an LPCVD process-tool and vacuum pumps. Courtesy of *www.icknowledge.com*.

tor-design allows batch-reactors to be optimized for high wafer-capacity. Low-pressure operation also decreases gas-phase reactions, making LPCVD-films less subject to particle-contamination. LPCVD is used for depositing many types of films, including poly-Si, Si_3N_4, SiO_2, PSG, BPSG, and W.

Under the low-pressure of an LPCVD-reactor (~1 torr), the diffusivity of the gas-species is increased by a factor of 1000 over that at atmospheric-pressure. Thus, there is more than an order-of-magnitude increase in the transport of *reactants to* (and of *by-products away from*) the substrate-surface, and the rate-limiting step thus becomes the rate of the surface-reaction. However, since the surface reaction-rate is also directly-proportional to the surface-concentration (see Eq. 16-2), non-uniform gas-phase concentrations produced by local-depletion of reactants within the reactor can result in deposition non-uniformities. An example of such an effect is the depletion of reactants from a gas by their deposition on wafers located nearer to the inlet of an *end-feed reactor-tube*. Wafers near the outlet-end of such tubes are exposed to lower-concentrations of reactants than those at the inlet-end.

The two main disadvantages of LPCVD-processes are their relatively-low deposition-rates and their relatively-high operating-temperatures. Attempting to increase deposition-rates by increasing the reactant partial-pressures tends to initiate gas-phase reactions. Attempting to operate at lower-temperatures results in unacceptably-slow film-deposition-rates.

Figure 16-12 shows a generic LPCVD-system. The reactors used in such LPCVD-systems are designed in four primary configurations: **1)** *horizontal-tube batch-reactors*; **2)** *vertical-tube batch-reactors*; **3)** *stepped, mini-batch radial-reactors*; and **4)** *single-wafer low-pressure-reactors*. Only the first of these configurations will be discussed in this section. Vertical-tube LPCVD-reactors share many characteristics of the horizontal-LPCVD reactors discussed here (and the vertical-furnaces described in Chap. 13). The stepped, mini-batch-radial LPCVD-reactor has the same configuration as the stepped-mini-batch PECVD-reactor described in Chap. 16, Sect. 16.2.7. Finally, single-wafer LPCVD-reactors can have one of several configurations. Single-wafer tools for depositing polysilicon by LPCVD have a similar configuration to the single-wafer epi-reactors described in Chap. 17. Others, such as reactors used to deposit LPCVD-W have configurations similar to the single-wafer PECVD-reactors described in Chap. 16, Sect. 16.2.7. Both configurations may also share some design features with the RTP-reactors described in Chap. 13 (e.g., lamp-heating methods).

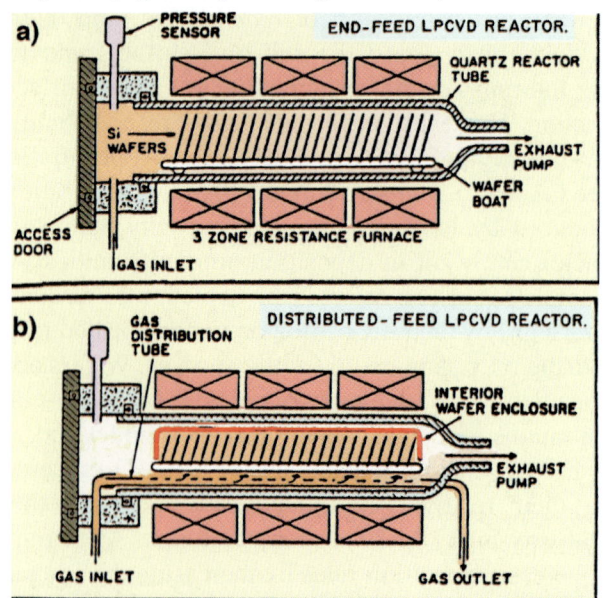

Fig. 16-13 (a) *End-feed* LPCVD reactor. (b) *Distributed-feed* LPCVD reactor.

Horizontal-Tube LPCVD Batch-Reactors (Hot-Wall):

Horizontal-tube (and more recently, vertical-tube), hot-wall batch-reactors are the most widely used LPCVD-reactors in VLSI-processing. They are primarily employed for depositing poly-Si and silicon-nitride films. They find such broad applicability mainly because of their superior economy, throughput, uniformity, and ability to accommodate large-diameter wafers. Their main disadvantages are susceptibility to particulate-contamination and low deposition-rates.

Conventional LPCVD-tube-reactors are very similar to the APCVD-tube-furnaces used to grow thermal SiO_2 and to carry out chemical-diffusion. The wafers are radiantly-heated by resistive-heating coils surrounding the tube, and reactant-gases are metered into one end of a horizontal quartz-tube (using mass-flow-controllers), and reaction by-products, unreacted-gases, and carrier-gases are pumped out the other end. Vacuum-pumps are used to establish the required reduced-pressure in the tube (typically 0.25–2.0-torr). The pumping-system includes a mechanical-pump augmented with a Roots-pump. A pressure-controller, with feedback-control of gas-flow (employing a throttle-valve at the pump inlet) can also be used.

The fact that LPCVD-processes operate in the surface-reaction-rate-limited regime allows wafers to be arranged differently inside the tube than in APCVD-tube-reactors (since equal mass-transport to all areas of every wafer is no longer critical). That is, in horizontal LPCVD-reactors, the wafers stand upright on edge at very-close spacing (e.g., 5–6-mm). They sit in a *quartz-boat*, similar to the way they are stacked in a diffusion-furnace (see Fig. 16-13). In both horizontal and vertical LPCVD-batch-reactors, the wafers are also positioned perpendicular to the direction of the main-gas-flow. The gases flow by forced-convection through the annular-space between the wafers and the walls of the circular cross-sectioned tube. From 100 to 150, 200-mm wafers can be processed per run (with fewer per run for 300-mm wafers). The large number of wafers processed in a single process-cycle allows comparable throughput to APVCD-reactors, despite the lower deposition-rates of LPCVD-processes (10–50-nm/min). In addition, it is relatively easy to scale-up the tube-diameter of LPCVD-reactors to accommodate larger wafer-sizes (which cannot be done as easily for the chambers of the non-continuous APCVD-reactors).

Gas-depletion effects (which reduce gas-phase concentrations as reactants are consumed by reactions on wafer surfaces), still occur in such end-fed reactors. That is, wafers near the inlet are exposed to higher concentrations of reactant-gases. Furthermore, the evolution of by-products dilutes reactant-gas concentrations even more along the length of the tube (since the pressure throughout the reactor is constant). For example, in polysilicon-deposition, as the silane reacts, two-moles of hydrogen are produced for every mole of silane consumed. This further decreases the silane partial-pressure along the length of the tube. Finally, both sides of the wafers are exposed, and get coated with the depositing-film. This makes the gas-depletion-problem even more severe.

Several methods have been adopted to reduce this problem. First, since the reaction-rates increase with increasing temperature, these deposition-rate variations can be minimized by gradually increasing temperature along the flat-zone of the tube (e.g., by 0.6–1.0°C/in). Such temperature-ramping compensates for reactant-gas depletion, and better thickness-uniformity is obtained (Fig. 16-13a). Nonuniform-temperatures along the tube-length, however, will affect film-properties that are dependent on the deposition-temperature. This is a particular problem for poly-Si and low-temperature (~400°C) deposition of doped-SiO_2. Hence, end-feed LP-CVD reactors with *temperature-ramping* are only used for high-temperature deposition (800–850°C) of silicon nitrides and oxynitrides, and for undoped poly-Si films (~600°C), in which the temperature-differential does not significantly impact film-properties.

The second technique which controls gas-depletion effects in hot-wall LPCVD batch-reactors, is the use of a *distributed feed*. In this approach, special injection-systems supply fresh-reactants at a number of

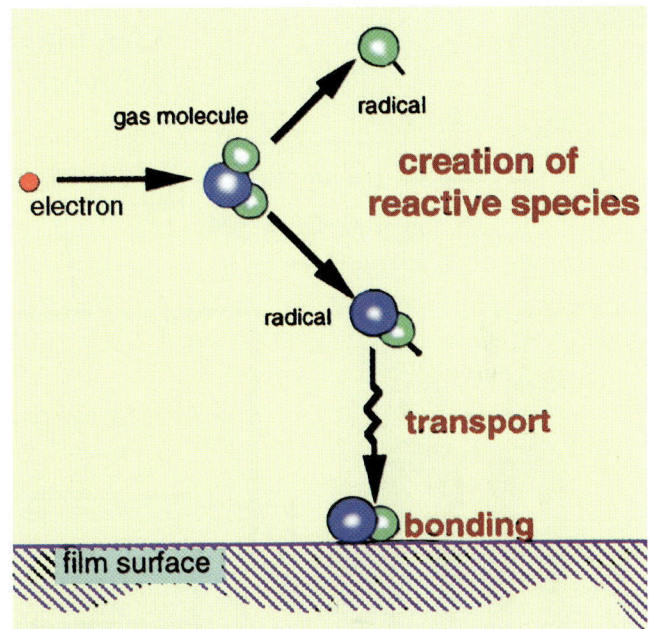

Fig. 16-14 The basic elements of a PECVD-process: (a) creation of reactive-species in the gas-phase by electron impact (e.g., molecular-dissociation); (b) transport of the reactive-species to the film-surface; and (c) bonding to the film-surface.

points along the length of the tube (Fig. 16-13b). But, this technique requires the use of specially-designed quartz containers to confine these injected-gases and produce a so-called *cross-flow* effect. In this distributed-feed method, the temperature along the tube is kept constant. Distributed-feed LPCVD hot-wall reactors are used for low-temperature (<600°C) deposition of SiO_2, PSG, BPSG films, and poly-Si films.

16.2.7 Plasma-Enhanced CVD: Physics, Chemistry, and Reactor Designs

The third (and last) major CVD-method is categorized not only by the pressure-regime, but also by its method of energy-input. Rather than relying solely on thermal-energy to initiate and sustain chemical reactions, *plasma-enhanced-CVD* (or *PECVD*) uses an rf-induced glow-discharge to transfer energy into the reactant-gases, allowing higher deposition-rates at lower-temperatures than either APCVD or LPCVD. Lower substrate-temperature is the major advantage

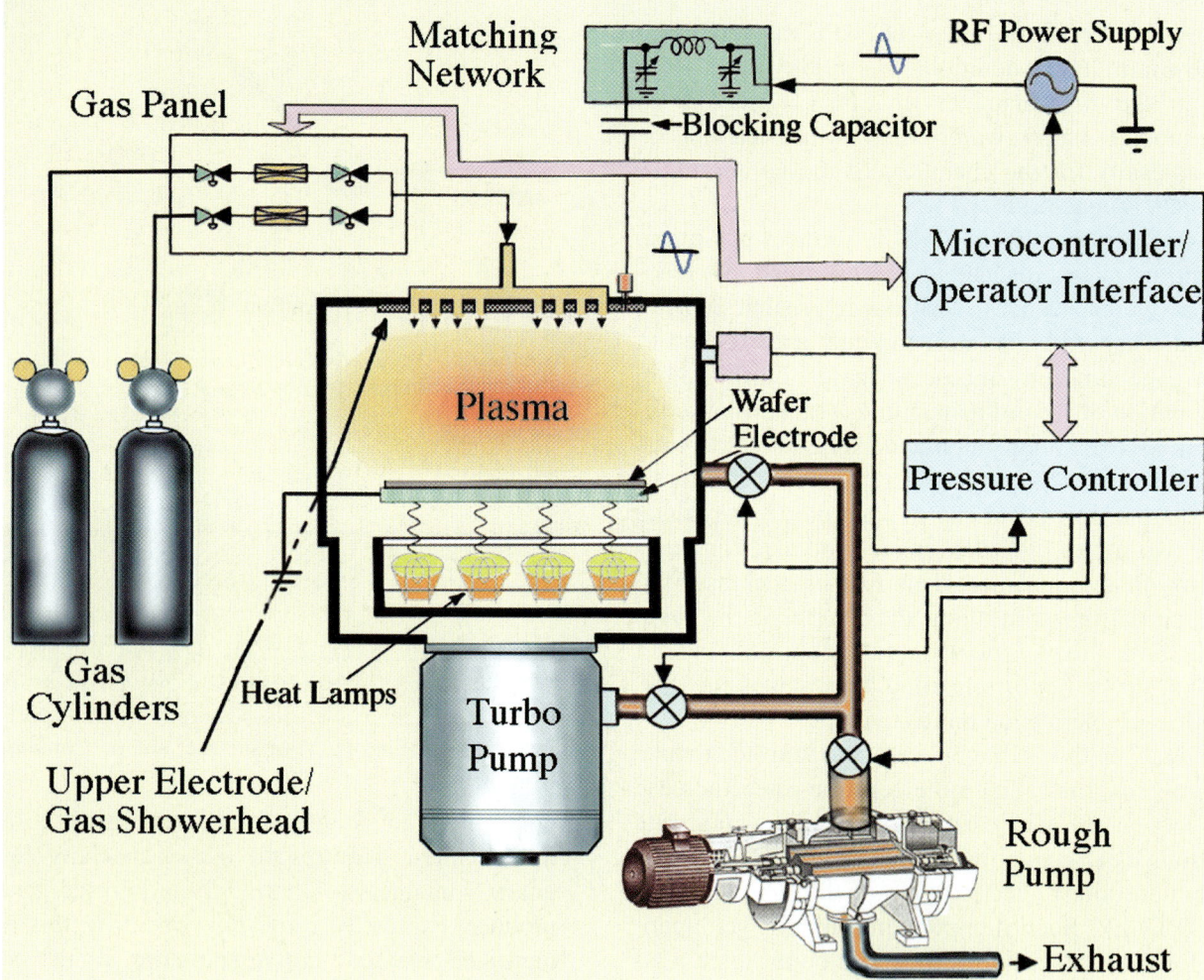

Fig. 16-15 Schematic drawing of a PECVD process system.

of PECVD, and in fact, PECVD provides a method of depositing films on substrates that do not have the thermal-stability to accept coating by other methods (the most important being the formation of SiO_2 and silicon-nitride films over aluminum). In addition, PECVD can increase the deposition-rate when compared to thermal-reactions alone, and produce films of unique compositions and properties. Desirable properties such as excellent-adhesion, low pinhole-density, good step-coverage, adequate electrical-properties, and compatibility with fine-line pattern-transfer processes, have led to the application of these films in VLSI and ULSI.

The plasma (more correctly a *glow-discharge*) is generated by applying an rf-field to a low-pressure gas, thereby creating free-electrons within the discharge-region. (A more detailed-discussion of glow-discharge physics is given in Chap. 14.) The electrons gain sufficient-energy from the electric-field so that when they collide with the reactant-gas molecules (Fig. 16-14), these molecules decompose into a variety of species, including: radicals, ionic-species, atoms, and molecules in excited-states. The net-effect of the interactions among these reactive-molecular fragments is to cause chemical-reactions at much lower-temperatures than in conventional CVD-reactors (i.e., operated without the presence of a plasma).

The energetic-species (predominantly radicals, as described in Chap. 14), are then adsorbed on the film-surface. Note that the radicals tend to have high sticking-coefficients because they are more reactive than the feedstock-gas molecules, and thus tend to form stronger bonds to the surface when they are adsorbed. They also appear to migrate easily along the surface after adsorption. These two factors help produce excellent film-conformality in deposited-films.

Upon being adsorbed on the substrate, the adatoms are subjected to ion and electron bombardment, reactions with other adsorbed-species, and film-formation and growth. The fact that the radicals formed in the glow-discharge are highly-reactive, presents some options as well as some problems to process engineers. PECVD films, in general, are less stoichiometric because the deposition-reactions are so varied and complicated. Moreover, in addition to the desired products, other by-product gases (such as H_2, N_2, and O_2) may get incorporated into the growing films. Excessive incorporation of these contaminants may lead to various problems, including: **1**) outgassing and bubble-formation; **2**) cracking or peeling during later thermal-cycling; **3**) shifts in the threshold-voltage of MOSFETs; and **4**) increased susceptibility to degradation from hot-carrier-effects in MOSFETs.

Figure 16-15 shows a schematic of a PECVD tool, along with its peripheral supporting-equipment. The heart of these systems is the reaction-chamber, and three types of PECVD-reactors are currently available: **1**) *parallel-plate batch*; **2**) *stepped mini-batch-radial*; and **3**) *single-wafer*. It should be noted that the discussions on glow-discharges and rf-diode sputtering in Chap. 14 and rf-plasmas for dry-etching in Chap. 22 apply to the production of rf-generated plasma in PECVD reactors. Readers are referred to these chapters for more information on rf-plasmas.

Parallel-Plate Cold-Wall Batch PECVD-Reactors: The first commercially important PECVD-reactor was the radial parallel-plate type described by Reinberg in 1974. It became available for IC-production in 1976 (see schematic in Fig. 16-16). The reaction-chamber is a short, vertically-oriented cylinder, typically constructed of stainless-steel. The rf-power (which establishes the plasma), is applied to the upper-electrode and the wafers reside on the bottom grounded-electrode (which can be rotated for improved uniformity, and heated up to 400°C). The most common rf-frequency is 13.56 MHz, but as described later, other frequencies are also used. The electrode-spacing was typically 5-10-cm, and such systems operated in the pressure range of 0.1-5-torr. Reactants were introduced either through the center and removed from the periphery, or from the periphery and removed through the center. Despite depletion-effects, uniform deposition can be achieved by correctly-balancing the plasma-density and gas-flows. Parallel-plate systems, however, suffered from low-throughput for larger-diameter wafers. In addition, particulates flaking-off walls or the upper electrode, could fall onto the horizontally-positioned wafers. Thus, use of the Reinberg reactor for batch-processing ceased for wafer-diameters beyond 100-mm. However, the modern PECVD-systems that create a plasma between two parallel-electrodes still essentially use the Reinberg configuration (albeit for depositing the film on one-wafer-at-a-time).

Mini-Batch Radial Cold-Wall PECVD-Reactors: In 1986, a new stepped-mini-batch cold-wall PECVD-system designed for depositing dielectric-films called the

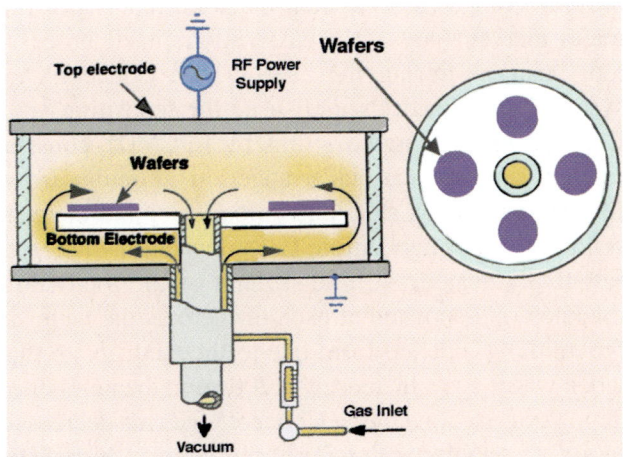

Fig. 16-16 Parallel-plate, radial-flow, batch-type CVD reactor (Reinberg design).[7]

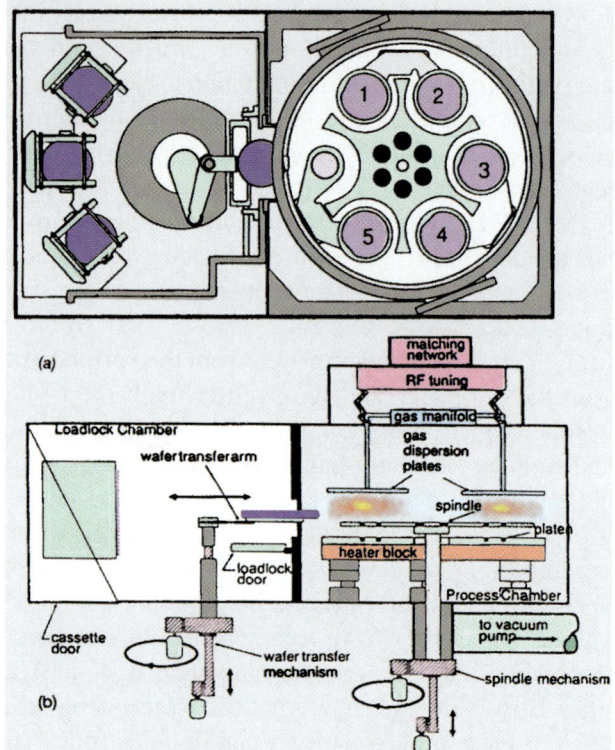

Fig. 16-17 (a) Wafer-transport sequence in the Novellus Concept One (200-mm) CVD-system. (b) Schematic of the Concept One tool. Courtesy of Novellus Systems, Inc.

Concept One® was introduced by Novellus Systems (Fig. 16-17). The Concept One® eventually displaced the hot-wall PECVD-batch-reactors that held the major market share of the passivation nitride process-tool business during the 1980's. The Concept One system (which is also widely used for depositing SiO_2 films by PECVD and W-films by LPCVD), consists of eight separate stations arranged in a circular-array. One station serves as a load/unload site, and the others are used for deposition. Each deposition-station is resistively-heated and one-seventh of the total film-thickness is deposited at each of them as the wafer is sequentially indexed from one to the next. (Note that when wafer-sizes increased to 200-mm, the number of deposition-stations in the Concept One was decreased to six, to enable processing these larger wafers. The footprint of the overall 200-mm wafer-tool was not changed.) With deposition-rates of 400-800-nm/min for SiO_2-films, the deposition-time for a typical CVD-SiO_2 film using this system is less than two minutes.

Wafers are loaded into the system in a cassette that is placed in a vacuum-loadlock-chamber. They are then moved one-by-one from the cassette into and out of the process-chamber by a robotic transfer-arm. The only moving part in the chamber is a rotating-spindle, which consists of eight forks. The spindle is lifted and the wafer coming into the chamber is placed on one of the forks waiting at the loading-site. The spindle is rotated by 45° to the first deposition-station, and then lowered so that the forks recess into the heating block. The deposition is then started. Gases are introduced from the showerhead located above each station.

This showerhead also serves as one of the electrodes that produces the glow-discharge plasma (so that each station represents a single-wafer parallel-plate rf-plasma-reactor). The plasma is thereby selectively created only above each wafer. In this way, deposition of the films is largely limited to the wafer area. Rf-power for creating the plasma is supplied by standard generators. For oxide-deposition, a single 13.56-MHz rf-power-supply is used (with an electrode-spacing of 0.5-inches), but for silicon-nitride and silicon-oxynitride films, two rf-supplies are used (i.e., a 13.56-MHz supply and a 300-400-kHz supply). Using a dual-frequency power-supply allows stress in SiN and oxynitride films to be controlled. Heating is from resistance-coils in the heater-block.

The uniformity of the deposition across each wafer is improved because deposition-anomalies at any individual station are averaged out. Film-thickness uniformities of ~1% (1σ) across the wafer, and wafer-to-wafer, can be achieved. The reaction-chamber is kept at a constant-temperature and pressure (because the wafer cassettes reside in a loadlock). This reduces flaking of deposited-material from chamber-surfaces due to temperature or pressure cycling. In addition, the cassette-exchange-time is used to perform an *in-situ* chamber-clean. Mechanically-generated particles are reduced by keeping the number of moving parts in the reaction-chamber to a minimum. The automatic process-chamber-cleaning during cassette-exchange

CHAPTER 16 CVD OF AMORPHOUS AND POLYCRYSTALLINE THIN-FILMS

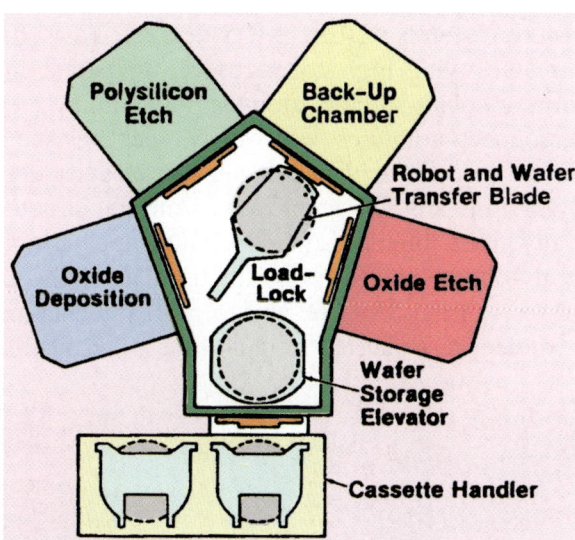

Fig. 16-18 Schematic top-view of the multi-chamber Applied Materials Precision 5000® cluster-tool. Courtesy of Applied Materials, Inc.

also reduces particulate-contamination and downtime. Hillock-growth in aluminum-films during the deposition of intermetal or passivation dielectrics is suppressed because the first deposition creates a cap-layer on the aluminum-film within 10-seconds of the start of the 400°C deposition-process.

Single-Wafer Cold-Wall PECVD-Reactors: Single-wafer PECVD-reactors became commercially available in the late 1980's. A multi-chamber, single-wafer system called the Precision-5000® was introduced by Applied Materials in 1987. It was also the first "cluster-tool" to gain widespread industry acceptance. It can have up to four individual process-chambers, with each chamber being self-contained and bolted onto a central loadlock/wafer-handling subsystem (Fig. 16-18). The automatic wafer-handler transfers wafers under low-pressure into any of the four chambers. By connecting different types of reactor-chambers, this system enables processes previously requiring separate systems to be integrated. For example, both CVD and etch processes can be performed sequentially. For simpler applications, the four chambers can be programmed for parallel-execution of identical processes. Thus, higher throughput can be obtained.

The PECVD-reactor used in the Precision 5000 platform was designed to deposit both doped and undoped SiO_2-films with TEOS (although silane-chemistries could also be used), as well as low-temperature silicon-nitride and oxynitride films. Each CVD-chamber contains a gas-dispersion-head that also serves as the powered-electrode (which is powered by a 13.56-MHz rf-generator). The wafers are loaded onto a susceptor that is automatically-adjusted to provide various electrode-spacings as required by different processes. The heater-module consists of an array of lamps which provides rapid radiant-heating of each wafer (Fig. 16-19). The electrodes also permit an *in-situ* plasma-cleaning step to be automatically performed after each run, so each wafer is exposed to the same processing environment. Wafers ≥ 200-mm can be accommodated, and film-thickness uniformities of 1–3% are possible. The Precision-5000 CVD and etch chambers can also be connected to a cluster-tool platform introduced by Applied Materials in 1992, called the Centura® (see Fig. 16-20a). In 1998 Applied Materials introduced a tool (which they named the Producer®) that performs simultaneous processing in two chambers (see Fig. 16-20b).

16.3 POLYCRYSTALLINE SILICON: PROPERTIES AND CVD-DEPOSITION METHODS

Polycrystalline silicon (also called *polysilicon, poly-Si,* or *poly*) in thin-film form has many important

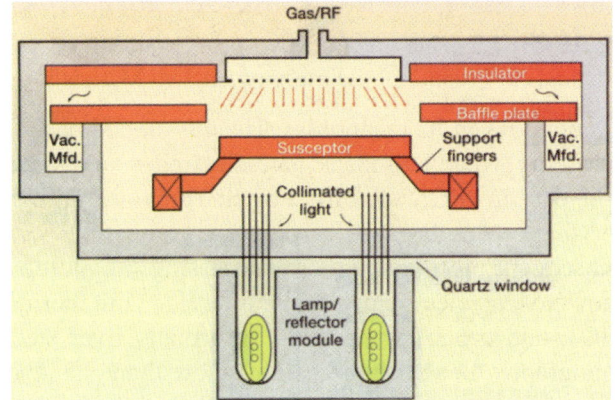

Fig. 16-19 Side view of the Applied Materials Precision 5000® PECVD single-wafer chamber, with lamp-heating. Courtesy of Applied Materials, Inc.

More recently, ULSI MOS-devices have been fabricated with high-conductivity tungsten-, titanium-, or cobalt-silicides atop the poly-film (to create interconnect-structures with lower sheet-resistances than are possible with polySi alone). Heavily-doped polySi-films are also employed as emitter-structures in advanced bipolar and BiCMOS technologies. Lightly-doped polySi-films are used as high-value load-resistors in SRAMs, and to refill trenches used in dielectric-isolation technologies. Polysilicon

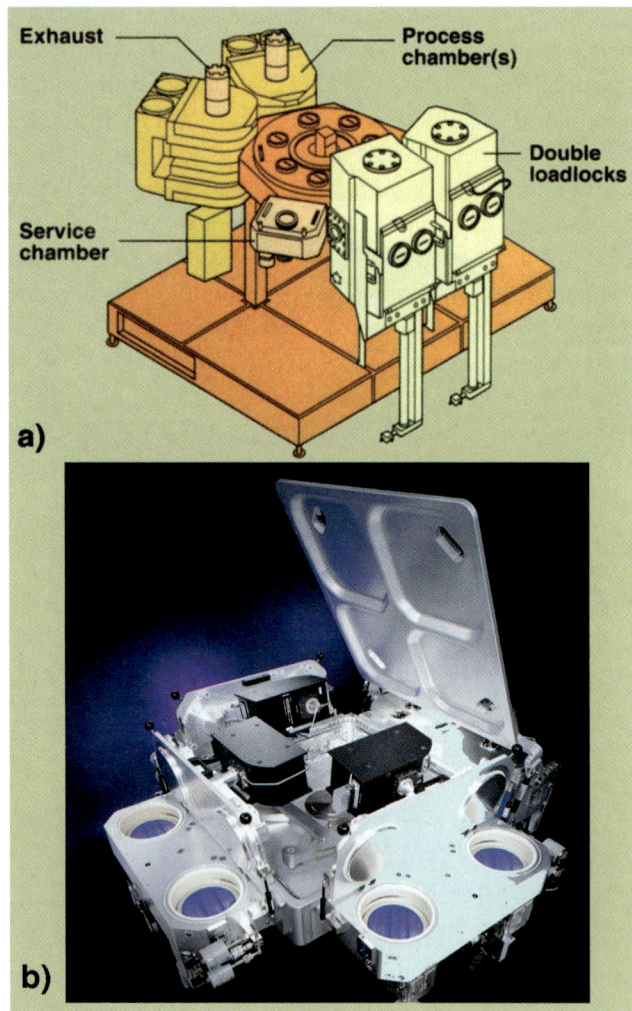

Fig. 16-20 Schematic of the AMAT: (a) Centura® CVD system; and (b) Producer.® Courtesy of Applied Materials.

applications in IC technology. Heavily-doped polySi-films are widely used as gate-electrodes and interconnects in MOS circuits. PolySi films are utilized for these roles because of they are compatible with subsequent high-temperature processing. In addition, they have an electrically-stable interface with thermal SiO_2, and can be deposited conformally over steep topography. In some applications more than one layer of polySi is needed, and a layer of SiO_2 must be thermally grown or deposited on the first polySi-layer to electrically-isolate it from subsequent-layers.

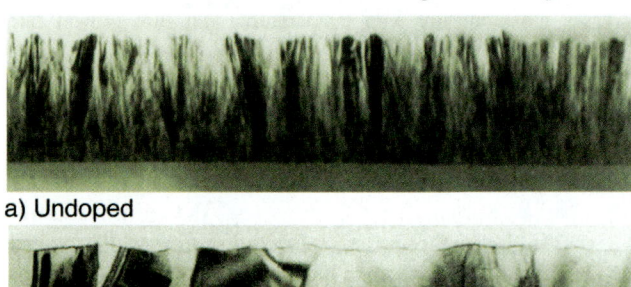

a) Undoped

b) Doped

c) Undoped and heat treated

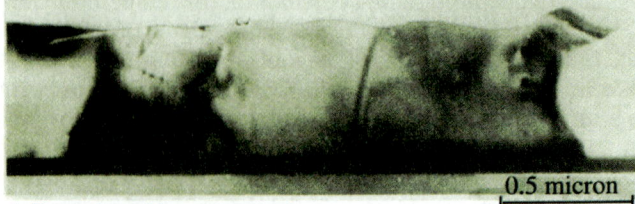

d) Doped and heat treated

Fig. 16-21 TEM cross-sections of polysilicon deposited at 625°C. (a) As-deposited, undoped-film, showing the thin grains in a columnar-structure. (b) As-deposited phosphorus-doped film, showing much larger grain-size. (c) Annealed (1000°C), undoped-film, showing little grain-growth as compared to (a). (d) Annealed (1000°C), phosphorus-doped film, showing evidence of grain-growth as compared to (b). Reprinted with permission of the Electrochemical Society.[8]

deposited by CVD is also used as the cell-plate material in DRAM-capacitors. In this section the properties of polySi-films and CVD methods for preparing such films are presented. A monograph on polySi by Kamins is a useful reference for more details.[9]

16.3.1 Properties of Polysilicon Thin-Films

Physical-Structure and Mechanical-Properties of Poly-Si: Thin-films of polycrystalline-silicon are made up of small (~100-nm) single-crystal regions (*grains*) separated by grain-boundaries (Fig. 16-21). Polycrystalline-silicon films exhibit many mechanical material-properties close to bulk single-crystal-silicon.

The grain-boundaries are composed of disordered atoms, and have large-numbers of defects due to incomplete-bonding. However, inside each grain, the Si-atoms are arranged in a periodic structure, and thus silicon material in the interior of the grains behaves much like that of bulk single-crystal-silicon. But, the defects at the grain-boundaries substantially alter the behavior of two important material-properties of polySi - namely its diffusion-characteristics and the dopant-distribution in polySi-films.

Diffusion in polySi-films is impacted because the diffusion-constants of dopants in the grain-boundary regions are significantly higher. This allows them to diffuse much more rapidly along grain-boundaries than through the grains. Thus, while the grain-boundaries occupy only a small-fraction of the film-volume, dopant-migration along these paths can dramatically increase dopant diffusion-rates in polySi. Likewise, *dopant-redistribution* in polySi-films is also affected. Most dopants will segregate from the single-crystal regions and collect along grain-boundaries, depleting the grains of dopants.

Electrical-Properties of Polysilicon: The electrical-properties of polySi-films are functions of both their semiconductor nature and of their polycrystalline structure and doping. That is, the single-crystal regions of polySi are assumed to behave electrically in a manner similar to bulk single-crystal silicon. As in single-crystal silicon, low-resistivity is obtained through heavy-doping with impurity atoms.

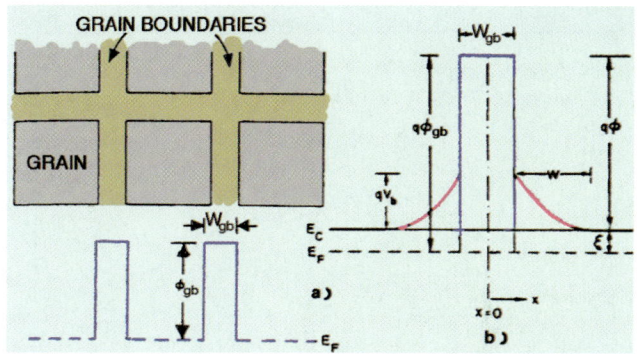

Fig. 16-22 (a) A schematic drawing of the potential-energy barriers generated by grain-boundaries. (b) Energy-band diagram near a grain-boundary under zero applied-voltage.[10] (© 1981 IEEE)

However, except at very-high dopant-concentrations (where the resistivity is only slightly greater), polySi at a given dopant-concentration, exhibits significantly-higher resistivity than does single-crystal silicon This is because some of the dopant-atoms (e.g., As and P, but not B) segregate during heat treatment to the grain-boundaries, where they do not effectively produce free-carriers. The concentration of dopant-atoms remaining in the single-crystal regions (which are still able to contribute charge-carriers), is decreased. Furthermore, the grain-boundaries are rich in incomplete-bonds, and these trap some free-carriers at the grain-boundaries. Such trapping decreases the overall free-carrier concentration (Fig. 16-22). Finally, the defects in the grain-boundaries decrease carrier-mobility. The sum of these effects leads to an increase in the film's resistivity. Nevertheless, in practice several techniques are available that can produce polySi thin-films with sheet-resistances as low as 10–30-Ω/sq.

16.3.2 Chemical Vapor Deposition of Polysilicon

Polysilicon is deposited by the thermal-decomposition (pyrolysis) of silane (SiH_4) in the temperature-range 580–650°C. LP-CVD is used because it produces films with good uniformity, purity, and economy. PolySi deposition is generally carried-out in a batch process, employing a low-pressure, hot-wall horizontal (or vertical) furnace. Single-wafer reactors for

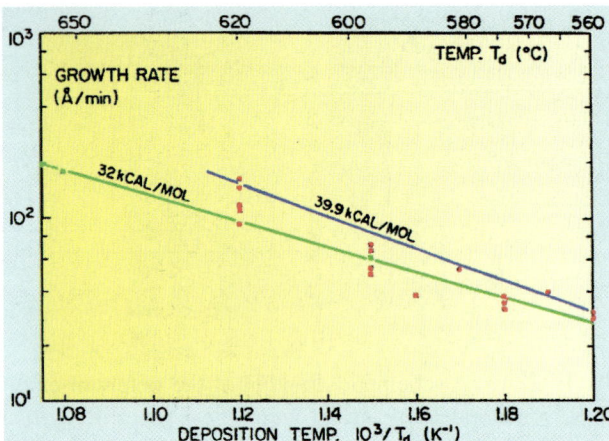

Fig. 16-23 Growth-rate as a function of deposition-temperature, T_d, for two different deposition-conditions:[11] • 350-mtorr, SiH_4 = 200-cm³/min; and ♦ 120-mtorr, SiH_4 = 50-cm³/min. Reprinted by permission of the publisher the Electrochemical Society, Inc.

depositing polySi were also introduced in the early 1990s. The overall deposition reaction is:

$$SiH_4 \; (vapor) \rightarrow Si \; (solid) + 2H_2 \; (gas) \quad (16.6)$$

Note that silane can also decompose in the gas phase. This is undesirable because such *gas-phase decomposition* together with *clustering* of silicon in the gas-phase produces silicon-particles. These will rain-down onto the growing polySi-film, causing a rough and porous layer that is unsuitable for IC- applications. Gas-phase decomposition occurs at high-silane concentrations. Hence, a carrier-gas (hydrogen) is used to dilute the silane-concentration. Gas-phase reactions occur more readily in inert carrier-gases (such as nitrogen and argon), than in hydrogen (which suppresses the decomposition-reaction because it is one of the reaction-products).

Deposition Parameters: For temperatures below 750°C, the CVD deposition-rate of Si shows an exponential-dependence on temperature (Figure 16-23). Depositions in LPCVD batch-processes are run in the 580–650°C range, since at higher-temperatures gas-phase reactions start to occur (leading to rough and poorly-adhering films), and below 580°C the rate is too-slow for practical use (< 5-nm/min).

A typical set of deposition-parameters for an LPCVD polySi batch-process is as follows: temperature = 620°C; pressure = 0.2–1.0-torr; SiH_4-flow-rate = 250-sccm; deposition-rate 8–10-nm/min. It takes about 2.5-hours (150-min) to deposit a 300-nm-thick polySi-film in a batch LPCVD-process (150-wafers). For single-wafer processing the 10–20-nm/min deposition-rate is too slow to give adequate-throughput to compete with the batch process. To increase the deposition-rate in single-wafer RTCVD-systems to about 200-nm/min, it is necessary to raise the temperature to 700°C, and to operate at pressures from 0.2-1.0-torr. This allows a cycle-time of 2 to 3 min per wafer, which makes the single-wafer throughput more comparable to that of a batch-process.

Polysilicon can be doped during deposition (*in-situ*), or after deposition (by diffusion or ion-implantation). The addition of diborane to the silane during deposition (boron-doping) causes rapid increases in the deposition-rate, while the addition of phosphine or arsine to the silane results in significant growth-rate reductions. Figure 16-24 shows a plot of the deposition-rate of polysilicon at 610°C as a function of the volume-fraction of dopant added to the silane.

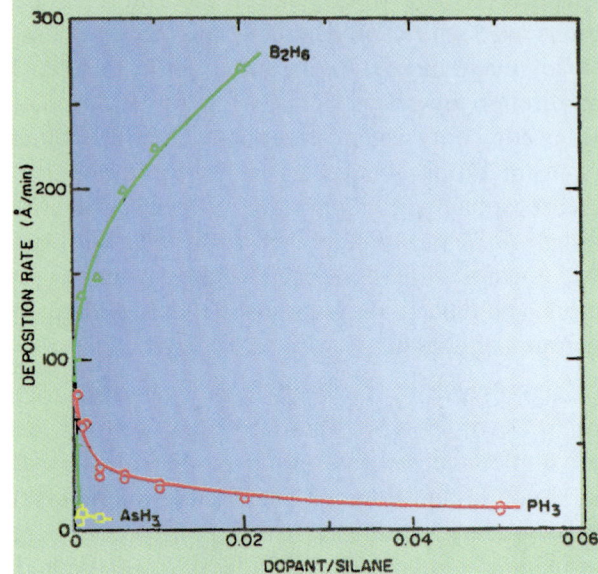

Fig. 16-24 The effect of adding dopants during deposition on the polysilicon deposition-rate at 610°C.[12]

In-situ doping of polysilicon has not been popular, even though it is considerably less complex than post-deposition doping (because of the difficulty of maintaining thickness and doping uniformity from wafer-to-wafer and across a wafer).

16.3.3 Doping Techniques for Polysilicon

Three techniques are used to dope polysilicon: **1)** diffusion; **2)** ion implantation; and **3)** *in-situ* doping. In most applications, poly is deposited undoped, and then doped by diffusion or implantation (Fig. 16-25).

Diffusion-Doping of Polysilicon: Diffusion-doping is carried out after deposition using a relatively high-temperature process (900–1000°C). For *n*-type doping, the undoped-polysilicon is exposed to a phosphorus-containing gas such as $POCl_3$ or PH_3. The advantage of this method is its ability to introduce very-high concentrations of dopants into the polySi-film, resulting in low-resistivity. The disadvantages of diffusion-doping are its high-temperature and the possibility of increasing film-surface roughness.

Ion-Implantation Doping of Polysilicon: The second doping-technique is ion implantation and subsequent anneal (Chap. 12). This method has the advantage of precise control of dopant-dose. Implantation energy is generally selected so that the impurity-peak is produced at the center of the film. A subsequent-anneal step (e.g., ~900°C, 30 min) redistributes and activates the implanted dopant. Rapid thermal processing (RTP) is used to anneal the film and electrically-activate the dopants. With the increased diffusion-rates of dopants along grain-boundaries, dopant-redistribution and activation of implanted polysilicon can be achieved by RTP in less than 30-seconds at 1150°C. The advantage of using RTP is its short-duration, which avoids redistributing dopants in the single-crystal-silicon substrate.

In-Situ Doping of Polysilicon: *In-situ* doping involves adding dopant-containing gases (such as diborane and phosphine) to the CVD-reactant-gases. Although combining doping and deposition in one step may appear simple, the control of film-thickness, dopant-

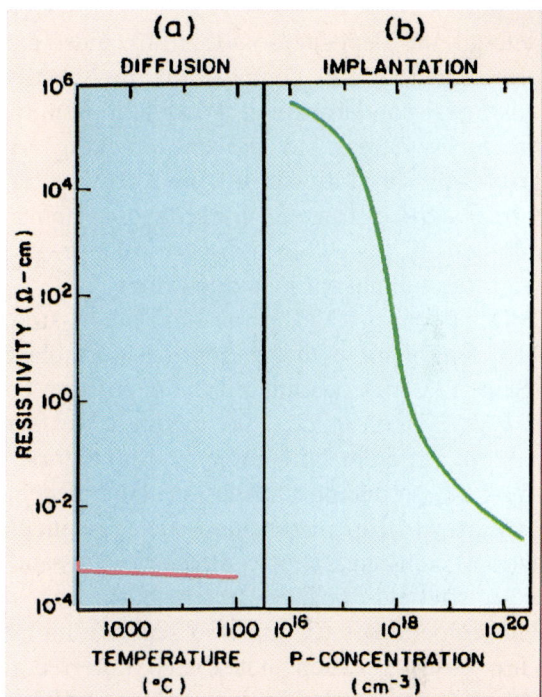

Fig. 16-25 Resistivity of phosphorus-doped polysilicon. (a) Diffusion, 1-h at the indicated temperature. (b) Ion-implantation, 1-h anneal at 1100°C.

uniformity, and deposition-rate is greatly complicated by the addition of the dopant-gases. Moreover, the physical properties of the film are affected. Adding phosphine can change the temperature-dependence of the polycrystalline film-structure, grain-size, and grain-orientation. *In-situ* doping with phosphorus, arsenic, and boron is feasible in single-wafer poly-deposition systems. Nevertheless, the proliferation of dual-doped-poly CMOS implies that both *p*-type and *n*-type poly are needed on the same circuit. Thus, ion-implantation doping of the polysilicon will likely predominate over *in-situ* doping.

16.4 PROPERTIES AND DEPOSITION OF CVD SiO_2

Chemically vapor deposited (CVD) SiO_2 films find wide-use in ULSI processing. These materials are used as: *insulating-layers (*between polysilicon and metal-layers); *insulating-layers (*between metal-layers in multilevel-metal systems); *capping-layers* (to prevent outdiffusion); and *final passivation-layers*.

In general, the deposited-oxide films must exhibit uniform-thickness and composition, low-particulate and chemical-contamination, good adhesion to the substrate, low-stress (to prevent cracking), good integrity for high-dielectric breakdown-strength, conformal step-coverage for multi-layer systems, low pinhole-density, low-k for high-performance devices, and high-throughput for manufacturing.

CVD silicon-dioxide has an amorphous-structure of SiO_4 tetrahedra with an empirical-formula SiO_2 (see Sect. 13.2). Depending on the deposition conditions, CVD silicon-oxides may have a lower-density and slightly different stoichiometry from thermal silicon-dioxide, producing changes in their mechanical and electrical film-properties (such as index-of-refraction, etch-rate, stress, dielectric-constant and high-electric-field breakdown strength). Deposition at high-temperatures (or use of a separate high-temperature post-deposition anneal-step, referred to as *densification*) can make the properties of CVD-oxide films approach those of thermal silicon-dioxide.

16.4.1 Chemical-Reactions for CVD SiO₂-Films

Various reactions can be used to prepare CVD SiO_2-films. The choice is dependent on the temperature requirements of the system, as well as on the equipment available for the process. The deposition-variables important for CVD-SiO_2 include: temperature, pressure, reactant-concentrations and their ratios, presence of dopant-gases, system-configuration, total gas-flow, and wafer-spacing. There are a number of CVD-SiO_2 processes that have been developed, each with its own chemical-reactions and reactor-configurations. They are divided into two categories: **1)** low-temperature deposition (300–450°C); and **2)** medium-temperature deposition (650–750°C).

16.4.2 Low-Temperature Silane-Based CVD SiO₂

The *low-temperature-deposition* of SiO_2 using silane as the silicon-source was the first CVD-SiO_2 process to be developed. It utilizes a reaction of silane and oxygen to form undoped-SiO_2 films. The deposition can be carried out in APCVD-reactors (primarily of the continuous-belt type), in distributed-feed LPCVD-reactors, or in PECVD-reactors. The addition of PH_3 to the gas-flow forms P_2O_5, which is incorporated into the SiO_2 film to produce a *phosphosilicate-glass* (PSG). The reactions when carried out in APCVD-reactors are given by:

$$SiH_4\,(gas) + O_2\,(gas) \rightarrow SiO_2\,(solid) + 2H_2 \quad \textbf{(16.7)}$$

$$4PH_3(gas) + 5O_2(gas) \rightarrow 2P_2O_5\,(solid) + 6H_2 \quad \textbf{(16.8)}$$

Highly-N_2-diluted SiH_4 is mixed with O_2 and is passed over heated wafers at 250–450°C. The reaction between silane and excess-oxygen forms SiO_2 by a heterogeneous surface-reaction. Homogeneous gas-phase-nucleation also occurs, leading to small SiO_2 particles that form a white-powder on the reaction-chamber walls (and on the injectors of the belt-type reactors). This can potentially give rise to particulate contamination in the deposited-films. The rate of SiO_2-deposition in APCVD-reactors can range up to 1400-nm/min, although rates of 200–500-nm/min are more-commonly used.

Silicon-dioxide films deposited at such low-temperatures exhibit lower-densities than thermal-SiO_2, and have an index-of-refraction of ~1.44. They also exhibit substantially-higher etch-rates in buffered hydrofluoric-acid (HF) solutions than do thermal SiO_2-films. Subsequent-heating of such films to temperatures between 700–1000°C causes *densification*. That is, this step causes the density of the material to increase from 2.1-g/cm³ to 2.2-g/cm³, the film-thickness to decrease, and the etch-rate in HF to decrease. It is believed that the densification proceeds by evolving H_2O (H-O-H) from the glass and increasing the number of bridging-oxygen atoms.

SiO_2 can also be deposited by a plasma-enhanced reaction between argon-diluted-SiH_4 and N_2O (*nitrous-oxide*) or NO (*nitric-oxide*) at temperatures between 200–400°C, according to:

$$\underset{(200\text{-}400°C,\,rf)}{SiH_4\,(gas) + 2N_2O\,(gas)} \rightarrow SiO_2\,(solid) + 2N_2\,(gas) + 2H_2\,(gas) \quad \textbf{(16.9)}$$

Nitrogen and/or hydrogen is often incorporated in PECVD-SiO_2. Silane-based PECVD oxides deposited

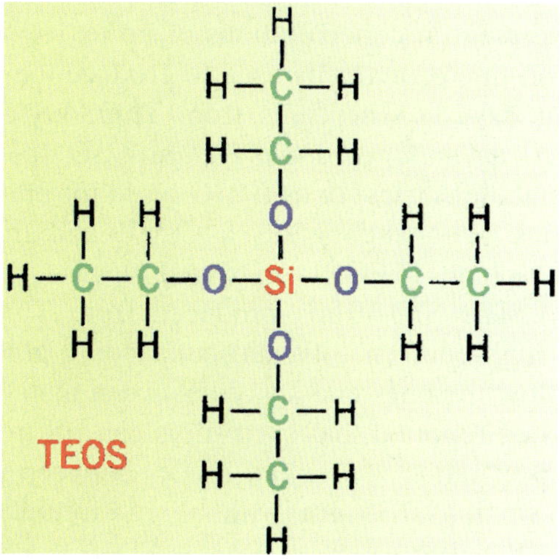

Fig. 16-26 The chemical structure of TEOS.

with conventional capacitively-coupled rf-diode plasma-sources have relatively poor step-coverage and hence are seldom used. With the advent of high-density-plasma sources, better step-coverage and film-properties are possible with silane-based chemistry.

16.4.3 Medium-Temperature LPCVD TEOS SiO$_2$

Use of silane as a process-gas involves an inherent safety-hazard. That is, silane is pyrophoric (i.e., it ignites spontaneously on contact with air). Thus, there has been an industry-wide shift away from silane. An alternative, safer, silicon-source for CVD-SiO$_2$ is tetraethoxysilane, Si(OC$_2$H$_5$)$_4$, also known as *TEtraethyl OrthoSilicate*, or *TEOS*. Besides added safety, using TEOS has the benefit of providing more-conformal films than those obtained using silane.

TEOS is a relatively-inert material, and is a liquid at room-temperature. TEOS-vapor can be supplied to the reaction-chamber using either a bubbler and nitrogen carrier-gas, or from a *direct-liquid-injection* system. The SiO$_2$ is formed by the decomposition of TEOS at elevated temperatures. The chemical-structure of the TEOS molecule is shown in Fig. 16-26. Films of undoped-SiO$_2$ from TEOS can be formed with adequate deposition-rates for IC-production if temperatures between 680 and 730°C are used (i.e., up to 25-nm/min). Such depositions are carried out in LPCVD tubular-hot-wall batch-reactors. LPCVD TEOS-oxide films are used for premetal-dielectrics (i.e., between polysilicon and metal-layers).

This is referred to as a *medium-temperature deposition-process* because the temperatures are low enough that redistribution of dopants in the substrate is not a concern, but the temperature is still too-high for use over aluminum-layers. As shown in Fig. 16-27b, the deposition-rate drops to unacceptably-small values for temperatures less than 600°C. The 675–695°C temperature range needed for practical LPCVD TEOS-processes precludes their use on wafers on where Al metallization already exists. The medium-temperature chemical-reaction is:

$$Si(OC_2H_5)_4 \; (liquid) \rightarrow SiO_2 \; (solid) + 4C_2H_4 \; (gas) + 2H_2O \; (gas) \quad (16.10)$$

The thermal TEOS-CVD process is carried out with highly-diluted TEOS/O$_2$ feedstock mixtures. The four oxygen atoms in TEOS allow deposition of SiO$_2$ by pyrolysis in the absence of O$_2$. But highly oxygen-rich mixtures are needed for good-quality films because TEOS contains carbon and hydrogen, which the O$_2$ burns to form CO (*gas*) and H$_2$O (*gas*).

Low-Temperature PECVD TEOS: Plasma-enhanced deposition of SiO$_2$ using TEOS as a source of silicon has been found to produce films at low-temperatures (i.e., <450°C) with better step-coverage and gap-filling characteristics than low-temperature APCVD silane-based oxides. The addition of plasma-energy reduces the temperature at which adequate SiO$_2$ film-deposition-rates can be achieved. As such, this process (which made its commercial-debut in the late-1980's) has found most application for forming intermetal dielectric-layers in multilevel-metal technologies. PECVD-TEOS deposition is carried out at temperatures ranging from 250–425°C and at pressures of 2-10 torr, with deposition-rates between 250 and 800-nm/min. An oxygen-nitrogen ambient is used with the TEOS. As is the case with thermal CVD of TEOS, it was found necessary to use O$_2$ (with O$_2$:TEOS ratios ranging from 10:1 to 20:1) to minimize the inclusion of traces of C and N in the films. In the presence of a

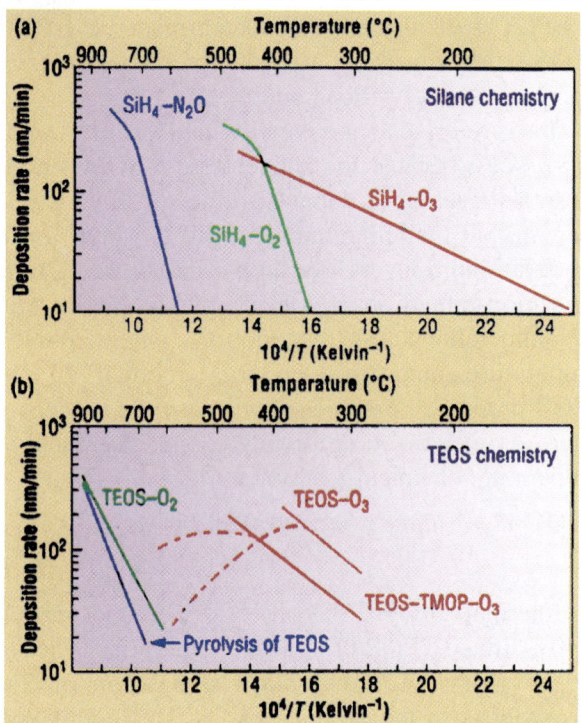

Fig. 16-27 CVD-oxide deposition-rate versus temperature for: (a) *silane*; and (b) *TEOS*.[13]

plasma, the TEOS reacts with oxygen according to:

$$Si(OC_2H_5)_4 \ (liquid) + O_2 \ (gas) \rightarrow SiO_2 \ (solid) + \text{by-products} \ (gas) \quad \textbf{(16.11)}$$

Doping of these TEOS-oxide films is done by adding trimethylborate (TMB) for boron-doping, and trimethylphosphite (TMP) for phosphorus-doping to the feed-gases.

Ozone TEOS: Under APCVD, the addition of O_2 to TEOS does not sufficiently raise the deposition-rate at temperatures below 500°C to make the process economically feasible for IC production. However, by adding ozone (O_3) to the TEOS vapor it is possible to obtain deposition-rates using APCVD which are comparable to those of SiH_4/O_2 at 400°C (and without the need for employing a plasma). For example, at 300°C, deposition-rates of 100–200-nm/min have been obtained with 4%-O_3, while at 400°C this rate occurs with 1–2%-O_3. Ozone produced from oxygen can be present at concentrations of up to 10%-O_3 in O_2. An in-line *ozonator* is used to produce the ozone (see Fig. 16-28). The films from TEOS/O_3 can be deposited as *undoped silicate-glass USG* (used for intermetal dielectrics) or as *BPSG* (used for pre-metal dielectrics, with such gases as TMB and triethyl-orthophosphate [TEOP] employed as the dopant-sources). The SiO_2-films produced from TEOS/O_3 exhibit high-conformality and outstanding ability to fill shallow trenches and spaces between metal-lines. It has been reported that such films can fill trenches with aspect ratios > 6:1, and spaces between metal-lines as small as 0.35-μm, without void formation.

16.4.4 Step-Coverage of CVD-SiO_2 Films

The manner in which a thin-film covers (or conforms to) the underlying features on a substrate is one of its important characteristics. The degree of coverage is a function of the film-species, reactor-type, and deposition-conditions. *Conformal-coverage* is defined as a condition where equal film-thickness exists over all substrate topography regardless of its slope (i.e., vertical and horizontal surfaces are coated with equal film-thickness, Fig. 16-29a). The thickness of a film at any given point in the mass-transport-limited regime is dependent on the reactant-flux arriving at that point. This is set by process-pressure and adatom-migration. The following model explains CVD-SiO_2 step-coverage in terms of these parameters.

If adsorbed-reactants are able to migrate rapidly across a substrate-surface, they will be found with equal probability on any part of the substrate, regardless of topography. This situation results in conformal step-coverage, as shown in Fig. 16-29a. Adatom mobility is a function of adatom-species and energy. Both higher substrate-temperature and ion-bombardment of adatoms enhance their surface-migration. Note that highly-conformal coverage is usually achieved under the relatively high-temperature conditions used to deposit LPCVD-poly-Si and Si_3N_4 films. In lower-temperature APCVD-deposition, non-conformal coverage is typically observed. The reasons for such non-conformality in APCVD-films

CHAPTER 16 CVD OF AMORPHOUS AND POLYCRYSTALLINE THIN-FILMS

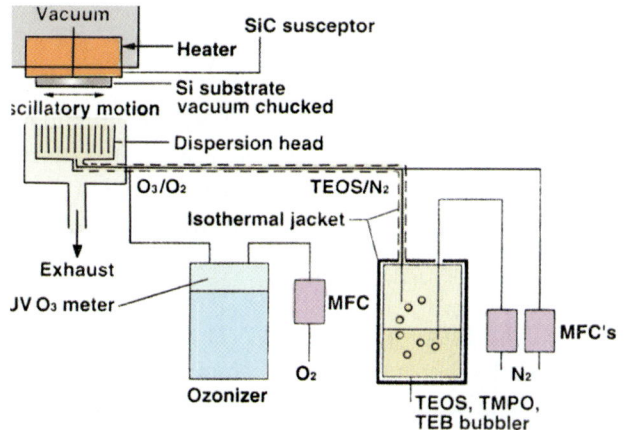

Fig. 16-28 Functional schematic of the Quester Technology APT-4800 TEOS/O₃ deposition system with an ozonator. Courtesy of Quester Technology, Inc.

will be discussed. Step-coverage in LPCVD-films and in TEOS/O$_3$-SiO$_2$ films will follow.

In the case of APCVD SiO$_2$-deposition using SiH$_4$ and O$_2$ chemistry, it has been found that the sticking coefficient (s_{SiH_4}) of silane is high (i.e., $s_{SiH_4} \sim 0.35$). In addition, once these species adsorb on the surface they are is essentially immobile. Since there is little, if any, reactant surface-migration in this case, the arrival-flux of gaseous-reactants at any given point becomes the most-important factor in determining the local film-thickness. In such CVD-processes, it is useful to introduce the concept of the *arrival-angle* to help model step-coverage. That is, the flux of reactant molecules arriving from an angle θ and θ + dθ can be expressed as P(θ). At atmospheric-pressure the mean-free-path, λ, for gas-molecules is very small (e.g., $\lambda \sim 1 \times 10^{-5}$ cm). Thus, the frequent-collisions of gas-molecules with each other completely randomize their velocity-directions. Under such conditions P(θ) is a constant, independent of θ. However, P(θ) is zero for values of θ which are blocked by the substrate. For example, in Fig. 16-30a, P(θ) is zero for any angle from 180–360°, and is constant for 0°< θ <180°.

In short, the value of the arriving-flux (and thus the eventual film-thickness), is directly proportional to the range of angles (i.e., the *arrival-angle*) for which P(θ) is not zero. As seen in Fig. 16-30b, at the top-corner of a step in an APCVD-process, P(θ) is non-zero

over a range of 270°. The resultant film-thickness is 270/180 (Fig. 16-30b), or 1.5-times greater than for the planar case of Fig. 16-30a, creating an *overhang* at the top-corner (see Fig. 16-29c). Similarly in Fig. 16-30c (i.e., at the bottom-corner of a step, or trench) the arrival angle is only 90°, and the film-thickness is 90/180, or one-half that of the planar-case. This explains the thickness cross-section observed when APCVD-SiO$_2$ is deposited over a sharp step (Fig. 16-31b). The film is thickest at the upper-corner and thinnest at the bottom of the step. This causes the slope in such SiO$_2$ films at the bottom corner to be >90° (i.e., so-called *re-entrant angle*s, are formed, see Fig. 16-31b) making deposition and anisotropic-etching of subsequently-deposited films extremely difficult.

On the other hand, if the mean-free-path of the reactant-gases is long (e.g., due to the low operating-pressure of such processes as LPCVD and the PVD-processes of evaporation or sputtering), and the surface-migration remains slow, a *shadowing* effect can occur. No longer do frequent-collisions over very-small distances randomize the gas-molecule velocity-vectors. Instead, reactant-molecules experience few collisions, and follow straight-line trajectories over distances comparable to substrate-topography dimen-

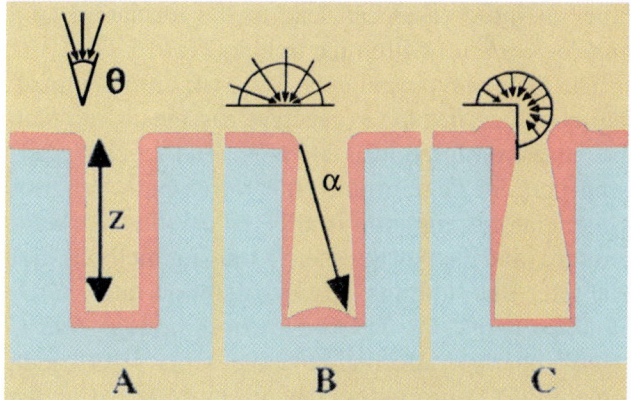

Fig. 16-29 Schematic diagrams showing types of step-coverage: (a) Conformal coverage resulting from rapid surface-migration; (b) Non-conformal step-coverage for long mean-free-path and no surface-migration. (c) Non-conformal coverage for short mean-free-path and no surface-migration (*overhang* forms at top corners).

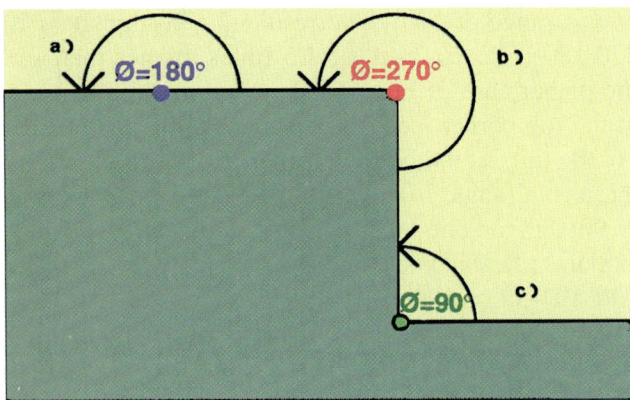

Fig. 16-30 The concept of *arrival-angle* and its effect on step-coverage: (a) 180°; (b) 270°; (c) 90°.

16.4.5 Applications of Undoped and Doped CVD-SiO$_2$ Films

Both undoped and doped films of SiO$_2$ from silane-chemistry are easily deposited at low-temperatures. Doping of TEOS-based SiO$_2$ films is somewhat more difficult. As shall be seen, doping of the SiO$_2$ can produce a variety of desirable film-properties for some applications.

Undoped CVD-SiO$_2$: Undoped CVD-SiO$_2$ (USG) is similar to its thermally-grown counterpart in many respects, with some exceptions being its as-deposited density and etch-rate, and a lower-quality SiO$_2$/Si interface. Due to this latter problem, USG is only used as a temporary-structure if it is used in contact with single-crystal silicon (e.g., as a capping-layer over doped regions to prevent outdiffusion during thermal-processes, or as an ion-implantation mask). However, it is mainly used to form permanent-structures. For example, it is deposited to increase the thickness of field-oxides, to provide isolation between conductors, and to form ILD-layers in Damascene-structures (see

sions. Substrate surface-features near to the points being struck can block the straight-paths of reactant-molecules. Thus, such points can be *shadowed* from the reactant-flux, and will experience less-deposition, (and less resulting film-thickness), than those points that are not shadowed. The arrival-angle for the given point now depends on the range of unobstructed *lines-of-sight* from that point to the reaction-chamber (see Fig. 16-29b). In such long mean-free-path cases the arrival-angle (and thus film-thickness), decreases with depth into a trench. As deposition proceeds, either of these cases can lead to the formation of a *void* (or *keyhole*) within the trench (Fig. 16-32).

The reactor-type and deposition-conditions impact step-coverage insofar as reactant-gas mean-free-path and adatom-migration is affected. The reactor operating-pressure determines mean-free-path lengths, while adatom-migration is affected by substrate-temperature (and also by the energy-transfer method). For example, due to high TEOS-migration-rates, SiO$_2$ films prepared by TEOS-decomposition at 645°C exhibit nearly conformal-coverage (Fig. 16-31a). In some PECVD-processes, the reactants arrive at the substrate with more energy (obtained from the glow-discharge) than in APCVD-processes. This results in increased adatom-migration. Thus, improved step-coverage by PECVD-SiO$_2$ films can sometimes be obtained compared to APCVD-depositions made at the same temperature.

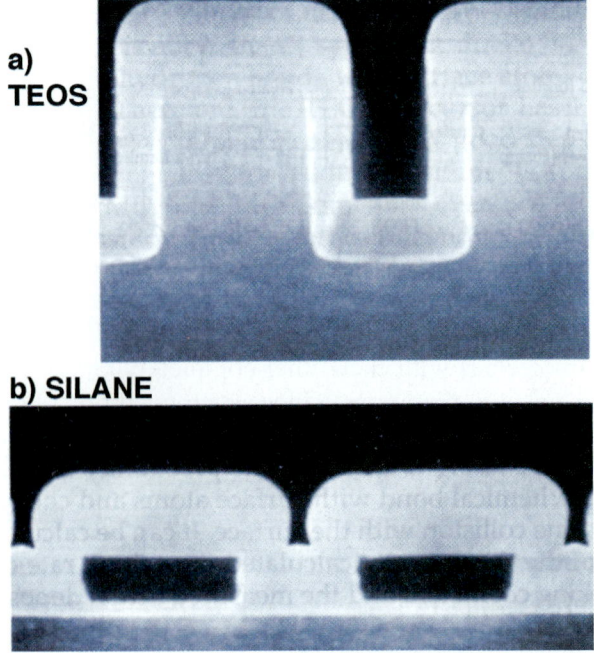

Fig. 16-31 Step-coverage of: (a) TEOS-based oxide-film; (b) silane-based oxide-film. Courtesy Applied Materials.

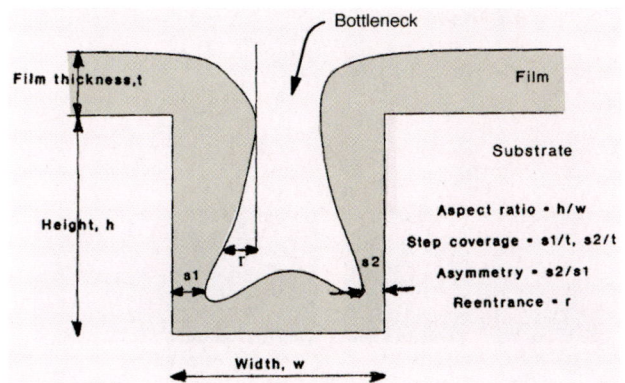

Fig. 16-32 Deposition of a film into a hole having a high aspect-ratio. A *bottleneck* may develop due to the buildup of an overhang on the top-corners.

Chap. 24). If the underlying-conductor is able to withstand high-temperatures (e.g., poly-silicon or a refractory-metal silicide) one of the LP-CVD methods may be utilized, because they produce films with excellent uniformity, good step-coverage, low particulate-contamination, and high-purity.

Often the film must be deposited over aluminum, and therefore the process must take place at less than 450°C. At these temperatures, LPCVD-TEOS reactions exhibit extremely-low deposition-rates (see Fig. 16-27). On the other hand the reaction of silane and oxygen mixtures at atmospheric pressure and temperatures from 300–500°C provides adequately-high deposition-rates. Therefore such APCVD-processes have been widely used, despite their particulate and step-coverage problems. USG can also be deposited by PECVD and APCVD TEOS/O_3 techniques, with generally improved step-coverage as compared to APCVD SiH_4/O_2-oxides.

Phosphosilicate Glass: Adding phosphorus-dopant during the deposition (typically in the form of phosphine, PH_3 or TMP), forms *phosphosilicate glass*, or PSG. Since PSG consists of two compounds, P_2O_5 and SiO_2, it is a *binary-glass* (or binary silicate), and some of its properties are considerably different from those of USG. That is, APCVD-PSG shows reduced stress and somewhat improved (though still relatively poor), step-coverage as compared to undoped CVD SiO_2. Although PSG is not a good diffusion-barrier to moisture, it getters alkali-ions. Finally PSG can be *flowed* at high-temperatures to create a smoother surface topography, which then facilitates the step coverage of subsequently-deposited films. The flow-step is performed at 1000–1100°C. The glass becomes viscous and responds to surface-tension forces, rounding sharp-corners. The extent of flow is often measured as the reduction in slope-angle of the flowed film-surface over an underlying step (Fig. 16-33), and is a function of flow-temperature and phosphorus-concentration. Increasing the temperature, duration of the flow-step, or phosphorus-concentration in the oxide will increase film-flow.

Borophosphosilicate Glass: ULSI processes require lower-temperatures than those needed for PSG- reflow (1000–1100°C) because such high-temperatures result in excessive-diffusion of shallow-junctions. But flowable-glass is still very desirable for easing film-coverage over abrupt steps in the substrate-topography. Glass-flow temperatures as low as 850°C can be obtained by adding boron dopant (e.g., B_2H_6) to the PSG gas-flow to form the *ternary* (three component) oxide-system B_2O_3-P_2O_5-SiO_2, *borophosphosilicate glass*, or BPSG (Fig. 16-34). Such films find wide use as the *interlevel dielectric-layer* (ILD) between poly-silicon and metal, and as dielectrics between stacked-capacitors and metal in DRAMs. Even lower glass-flow temperatures (i.e., 750°C) have been reported for BPSG films formed with TEOS/O_3.

BPSG-flow depends upon film-composition, flow-temperature, flow-time, and flow-ambient. It has been reported that an increase in boron-concentration of 1-wt% in BPSG decreases the required flow-temperature by ~40°C. A plot of required flow-temperatures versus BPSG dopant-concentrations in LPCVD-films is shown in Fig. 16-35. However, increasing the B-concentration beyond ~5-wt% does not further decrease BPSG flow-temperatures. An upper-limit on boron-concentration is also imposed by film-stability. That is, BPSG-films containing over 5-wt%-boron tend to be very hygroscopic and unstable. (Nucleation and precipitation of boric-acid [B_2O_3] and phosphoric-acid [P_2O_5] crystallites occurs, which can then lead to the formation of insoluble BPO_4-crystallites during the reflow-process). Such precipitation degrades the film properties since the acid-particles (though soluble) leave pits in the glass that are decorated later, and they leave local areas of low dopant-concentration that affect the flow-performance of the glass. The BPO_4-particles also remain behind as defects.

16.5 PROPERTIES AND CHEMICAL-VAPOR-DEPOSITION OF SILICON-NITRIDE

Silicon-nitride (SiN) films are amorphous, insulating-materials that find many applications in ULSI-fabrication, including: **1)** as a final-passivation and mechanical-protective-layer for ICs (especially for parts encapsulated in plastic-packages); **2)** as a mask for the selective-oxidation of silicon; **3)** as one of the dielectric-materials in the stacked oxide-nitride-oxide (ONO) layers in DRAM-capacitors; **4)** as sidewall-spacers in MOSFETs (where they are used to form lightly-doped-drain structures [LDD], and also serve as sidewall-passivation-structures during salicide-processes); **5)** as a CMP-stop layer in shallow-trench-isolation processes (see Chap. 4); and **6)** as an etch-stop layer in Damascene-structures (see Chap. 24).

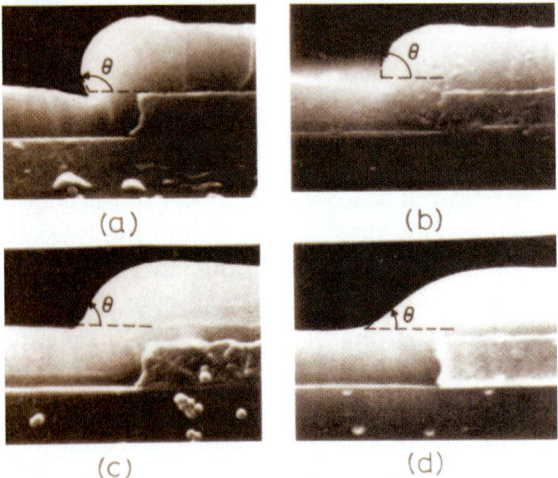

Fig. 16-33 SEM cross-sections (10,000X) of samples annealed in steam at 1100°C for 20-min for the following weight of phosphorus: (a) 0.0 wt.% P; (b) 2.2 wt.% P; (c) 4.6 wt.% P; and (d) 7.2 wt.% P.[14] Reprinted by permission of the publisher, the Electrochemical Society, Inc.

a) As deposited   b) After reflow

Fig. 16-34 Reflow of a BPSG-film containing 5%-B and 5%-P for 30-min in N_2: (a) Before reflow; (b) After reflow. Courtesy of Applied Materials

However, since silicon-nitride has a higher-dielectric-constant than SiO_2 (6–9 versus ~4.2), it is not used for interlevel-dielectric applications. (It would increase the value of the interlevel-capacitance, which would result in a reduction of the circuit-speed.)

As shown in Table 16-2, two CVD-techniques are used for depositing SiN. When used as a mask for selective-oxidation (or as a dielectric-layer in DRAM-capacitor-structures), SiN is generally-deposited by a medium-temperature (700-800°C) *LPCVD-technique* (for reasons of film-uniformity and lower-processing-cost). When used as a passivation-layer, the deposition-process must be compatible with such low-melting-point metals as aluminum. Thus, a lower-temperature process is needed (200–400°C). For such applications, *PECVD* is the deposition method-of-choice, as it can deposit SiN-films at 200–400°C. However, PECVD-SiN is non-stoichiometric and contains substantial-quantities of atomic-H (10–30-atomic %). Thus, it is sometimes chemically-represented as $Si_xN_yH_z$. Table 16-2 also compares the properties of LPCVD and PECVD silicon-nitride films.

LPCVD-silicon-nitride is most commonly deposited by reacting dichlorosilane ($SiCl_2H_2$) and ammonia (NH_3) at temperatures between 700-800°C in a hot-wall reactor according to the overall-reaction:

$$3SiCl_2H_2\ (gas) + 4NH_3\ (gas) \rightarrow Si_3N_4\ (solid) + 6HCl\ (gas) + 6H_2\ (gas) \quad (16.12)$$

Silicon-nitride depositions by LPCVD are controlled

by a large-number of deposition-parameters including: temperature, total-pressure, reactant-ratios, and temperature-gradients in the reactor. A temperature-ramp along the tube-length is required for obtaining uniform-depositions (as is discussed in the section on hot-wall tube-LPCVD-reactors, Sect 16.2.6).

In general, LPCVD-SiN-films have high-density (2.9–3.1-g/cm^3), a dielectric-constant of 6, and a hydrogen-content (up to 8-at%) that is lower than in PECVD-SiN-films. In addition, they exhibit excellent step-coverage and relatively-low particulate-contamination. Such films, however, have high-tensile-stresses (~1×10^{10} dynes/cm^2, about an order-of-magnitude higher than TEOS-deposited-SiO$_2$). This can cause cracking in LPCVD-SiN-films thicker than 200-nm.

Silicon-nitride-deposition by PECVD was first described in 1965, and SiN-films were the first-materials to be deposited by the PECVD-technique on a large-production-scale. The overall PECVD-deposition-reaction is written as:

$$SiH_4 \;(gas) + NH_3 \;(or\; N_2)\;(gas) \xrightarrow{(200\text{-}400°C,\; rf)} Si_xN_yH_z \;(solid) + H_2 \;(gas) \quad (16.13)$$

where silane and ammonia (or nitrogen) are reacted in a plasma at 200-400°C.

Hydrogen in PECVD-SiN can reach 18–22-at% in films deposited near 300°C. The presence of large quantities of hydrogen is harmful to IC-devices. It leads to significant threshold-voltage shifts, and the film-etching-characteristics (both wet and dry) are impacted. Processing-techniques that reduce H-concentration-levels in PECVD-SiN have been pursued.

Table 16-3 is a summary of the CVD-reactions for polysilicon, SiO$_2$, PSG, BPSG, and silicon-nitride.

16.6 SILICON-OXYNITRIDES DEPOSITED BY CVD

Materials can be prepared with characteristics between those of nitrides and oxides, and these are called *silicon-oxynitrides* [SiO$_x$N$_y$(H$_z$)]. Such films are less-permeable to moisture and other contaminants than deposited-oxides. They are formed by reacting SiH$_4$ with N$_2$O and NH$_3$, usually by PECVD. Properties can be tailored for improved-thermal-stability, low-stress, and crack-resistance. For example, the composition and properties of the SiO$_x$N$_y$-films can be varied smoothly by changing the flow-rates of the three-reactants.

16.7 CVD OF METALS, SILICIDES, AND NITRIDES FOR ULSI-APPLICATIONS

CVD has also been pursued as a deposition-technology for a number of metals used as ULSI-interconnects, including tungsten, aluminum, titanium, and copper. Of this group, only CVD of tungsten has found wide-acceptance as a production-process. CVD of the other metals has not been able to displace PVD (sputtering) as the main deposition-technology. Here we describe the deposition by CVD of tungsten and titanium-nitride. CVD of CU is covered in Chap. 24.

16.7.1 CVD of Tungsten (W)

Refractory-metals (i.e., W, Ti, Mo, and Ta) have been investigated for various applications in the intercon-

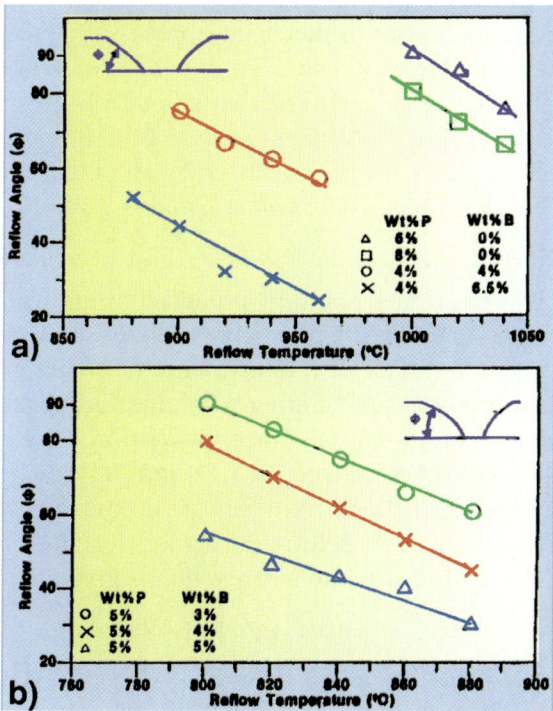

Fig. 16-35 (a) Reflow-angle vs. reflow-temperatures in a nitrogen-ambient (30 min). (b) Reflow-angle versus reflow-temperatures in a steam-ambient (30 min). Reprinted by permission of Solid State Technology.[15]

Property	HT-CVD—NP 900°C	PE-CVD—LP 300°C
Composition	Si_3N_4	$Si_xN_yH_z$
Si/N ratio	0.75	0.8–1.0
Density	2.8–3.1 g/cm^3	2.5–2.8 g/cm^3
Refractive index	2.0–2.1	2.0–2.1
Dielectric constant	6–7	6–9
Dielectric strength	1×10^7 V/cm	6×10^6 V/cm
Bulk resistivity	10^{15}–10^{17} ohms/cm	10^{15} ohms/cm
Surface resistivity	$>10^{13}$ ohms/square	1×10^{13} ohms/square
Stress at 23°C on Si	1.2–1.8×10^{10} dyn/cm^2 (tensile)	1–8×10^9 dyn/cm^2 (compressive)
Thermal expansion	4×10^{-6}/°C	$>4 < 7 \times 10^{-6}$/°C
Color, transmitted	None	Yellow
Step coverage	Fair	Conformal
H$_2$O permeability	Zero	Low–none
Thermal stability	Excellent	Variable > 400°C
Solution etch rate		
HFB 20–25°C	10–15 Å/min	200–300 Å/min
49% HF 23°C	80 Å/min	1500–3000 Å/min
85% H$_3$PO$_4$ 155°C	15 Å/min	100–200 Å/min
85% H$_3$PO$_4$ 180°C	120 Å/min	600–1000 Å/min
Plasma etch rate		
70% CF$_4$/30% O$_2$, 150 W, 100°C	200 Å/min	500 Å/min
Na$^+$ penetration	<100 Å	<100 Å
Na$^+$ retained in top 100 Å	>99%	>99%
IR absorption		
Si–N max	~870 cm^{-1}	~830 cm^{-1}
Si–H minor	—	2180 cm^{-1}

TABLE 16-2 PROPERTIES OF HIGH-TEMPERATURE-CVD-SILICON-NITRIDE AND PECVD-SILICON-NITRIDE

nect-systems of silicon-ICs. Their resistivities are higher than those of Al and its alloys, but lower than those of the refractory-metal-silicides and nitrides. Of these metals, tungsten (W) has adopted for several interconnect-applications,[16] the most important being that of a *plug* (i.e., a material that can completely-fill vias between aluminum-films, as well as contact-holes). It is used as a plug-material because CVD-W provides superior via-filling-capabilities than does PVD-aluminum. PVD-films cannot completely-fill contact-holes and vias. A method to *completely-fill* the contact-holes was therefore sought. CVD-W was able to do this. Blanket-W-CVD deposition-and-etch-back-processes are used for W-plug-applications.

CVD-tungsten exhibits high-thermal-stability (having the highest-melting-point of all metals - 3410°C), excellent conformal-step-coverage, and its thermal-expansion-coefficient closely matches that of silicon. In addition, it has excellent electromigration and corrosion-resistance. Some of its disadvantages include: **a**) its resistivity, although 200-times lower than that of heavily-doped-polysilicon, is still about twice as high as that of Al-alloy-films; **b**) W-films adhere poorly to oxides and nitrides; and **c**) oxides form on W-films when temperatures exceed 400°C (and thus care must be exercised to prevent oxidation, especially during subsequent dielectric-deposition).

16.7.2 The Chemistry of CVD-Tungsten

The chemical-vapor-deposition of tungsten is generally performed in cold-wall, low-pressure systems (an example is shown in Fig. 16-36). Although tungsten can be deposited either from WF$_6$ or WCl$_6$, tungsten-hexafluoride (WF$_6$) is better suited as the W-source gas, since it is a liquid that boils below room-temperature (17°C). On the other hand, WCl$_6$ is a solid that melts at 275°C. The low boiling-point makes WF$_6$ much easier to meter into process-chambers in a reproducible way. The main-drawback of WF$_6$ is its high-cost. Three reactions are used in CVD-W, namely reduction of WF$_6$ by: **1**) silicon; **2**) hydrogen; and **3**) silane (since it can be reduced by all of these materials). The *silicon-reduction reaction* is given by:

$$WF_6 \ (gas) + 3Si \rightarrow 2W \ (solid) + 3SiF_4 \ (gas) \quad \textbf{(16.14)}$$

This reaction is normally produced by allowing the WF$_6$-gas to react with regions of exposed solid-silicon on a wafer-surface at a temperature of about 300°C. The reaction is self-limiting when the film reaches a thickness of 10–15-nm, since the W-film serves as a diffusion-barrier between the Si and WF$_6$ once this thickness is reached. No deposit occurs on wafer-regions covered with SiO$_2$ during this reaction.

The *hydrogen-reduction-reaction* is given by:

$$WF_6 \ (gas) + 3H_2 \ (gas) \rightarrow W \ (solid) + 6HF \ (gas) \quad \textbf{(16.15)}$$

The process is carried out at reduced-pressures, usually at temperatures below 450°C. The resistivity of W-films deposited by hydrogen-reduction is in the 7–12-$\mu\Omega$-cm range. Because W does not adhere-well

Table 16-3 CVD DEPOSITION REACTIONS

PRODUCT	REACTANTS	METHOD	TEMP (°C)	COMMENTS
Polysilicon	SiH_4	LPCVD	580–650	may be *in situ* doped
Silicon Nitride	$SiH_4 + NH_3$	LPCVD	700–900	
	$SiCl_2H_2 + NH_3$	LPCVD	650–750	
	$SiH_4 + NH_3$	PECVD	200–350	
	$SiH_4 + N_2$	PECVD	200–350	
Silicon Dioxide (SiO_2)	$SiH_4 + O_2$	APCVD	300–500	poor step coverage
	$SiH_4 + O_2$	PECVD	200–350	good step coverage
	$SiH_4 + N_2O$	PECVD	200–350	
	$Si(OC_2H_5)_4$ [TEOS]	LPCVD	650–750	liquid source, conformal
	$SiCl_2H_2 + N_2O$	LPCVD	850–900	conformal
Doped SiO_2	$SiH_4 + O_2 + PH_3$	APCVD	300–500	PSG
	$SiH_4 + O_2 + PH_3$	PECVD	300–500	PSG
	$SiH_4 + O_2 + PH_3 + B_2H_6$	APCVD	300–500	BPSG, low temperature flow
	$SiH_4 + O_2 + PH_3 + B_2H_6$	PECVD	300–500	BPSG, low temperature flow

to SiO_2, an adhesion-layer is first deposited onto the SiO_2, and the W is then deposited onto it.

The overall *silane-reduction-reaction* is given by:

$$2WF_6 \,(gas) + 3SiH_4 \,(gas) \rightarrow 2W \,(solid) + 3SiF_4 \,(gas) + 6H_2 \,(gas) \quad (16.16)$$

This reaction (LPCVD at ~300°C) is used to produce a W-nucleation-layer for the hydrogen-reaction. Better-nucleation is consistently obtained with the silane-reduction on most surfaces, including TiN.

As noted earlier, CVD of tungsten is performed in cold-wall, low-pressure-CVD (LPCVD) reactors. The wafers are held on a heated-chuck opposite a showerhead through which a premixed-flow of WF_6 and one of the reducing-agent-gases (H_2, or SiH_4) is injected. Hot-wall systems are not used because W would deposit on the quartz furnace-tube-walls. Since W doesn't adhere to SiO_2, such films would soon-delaminate from the walls and create particles. Frequent-cleaning would be necessary to keep this problem under control. Furthermore, once the fused-silica furnace-walls are coated with W they become opaque. IR-radiation from the heating-coils would no longer be transmitted as efficiently through the walls as when the fused-silica was transparent.

16.7.3 Blanket CVD-W and Etchback

Tungsten can be deposited by CVD using either a selective or blanket-process. Only blanket W-deposition has emerged as a production-proven process (Fig. 16-37). Blanket-CVD-W-and-etchback (or CMP) has found widespread-use for contact-hole and via-filling applications in IC-technologies below about 1-μm. Both applications require adherent, low-cost-films. The plug-applications, however, call for high step-coverage and thickness-uniformity, but can tolerate higher-resistivity than is needed for W-films used as interconnects. For filling contact-holes, this W-plug-formation process has six steps (Fig. 16-38):

1. An *in-situ* surface-pre-clean is performed;

2. A contact-forming layer is deposited (usually a Ti-film formed by sputtering or CVD);

3. An adhesion/barrier-layer is deposited (typically a TiN-film formed by sputtering or CVD);

4. A blanket-CVD W-film is formed (typically with two-step CVD-process);

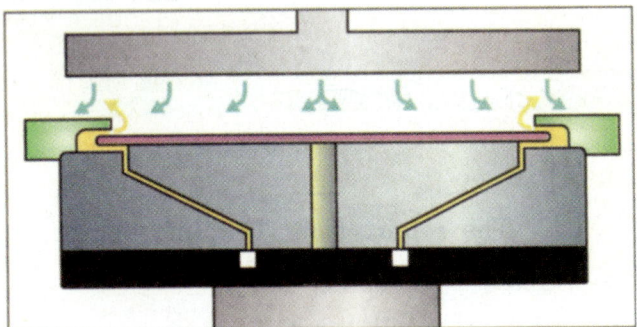

Fig. 16-36 Schematic drawing of a Novellus Concept Two® CVD tungsten-deposition system. Courtesy of Novellus Systems, Inc.

5. The W-film is etched back (or CMP is used);

6. The underlying adhesion-layer is removed from the top-SiO_2 surface by etch-back or CMP.

16.7.4 CVD of Titanium-Nitride (TiN)

In ULSI, global-interconnects (which use Al-alloy and Cu materials for the interconnect-lines, and possibly also for contact and via filling), as well as in blanket-W processes that form W-plugs and W-local-interconnects, supporting-role films are inevitably fabricated beneath these metals. Such films (consisting of refractory-metal silicides and nitrides) serve two major-purposes, depending on the metal they assist. For Al-alloy films they act as diffusion-barriers to prevent the formation of intermetallic-compounds that would destroy the contact-behavior (i.e., by shorting the shallow-junctions below the contact or increasing the series-resistance between the metal and silicon).

For blanket-layers of W and Cu, they serve not only as diffusion-barriers but also as adhesion-layers (i.e., films to which Cu and W adhere-well, and which in turn adhere-well to the oxide-below). Note that their role as diffusion-barriers in blanket-W processes has two purposes: **1)** preventing reaction of the contact-resistance-enhancing Ti-layer and WF_6; and **2)** protecting the Si-contact from damage by reaction with WF_6. In Cu-interconnects, such barriers must prevent Cu-diffusion into the underlying Si-substrate.

In any case, these barriers must retain their function over the full-range of temperatures encountered after deposition, and they must also satisfactorily perform this role when deposited into deep-submicron, high-aspect-ratio contact-holes and vias. Films that are required to perform both adhesion- and diffusion-barrier-functions are also termed *liners*. The deposition of liner-films is done by either PVD or CVD. PVD-technology has been the traditional deposition-method, and work to improve this method continues. Here we discuss the CVD of the most widely-used liner-film, *titanium-nitride* (TiN).

TiN is an attractive-material as a diffusion-barrier in silicon-ICs because it behaves not only as an impermeable-barrier to silicon, but also as a barrier to other substances attempting to diffuse through it. In the latter-cases, the activation-energy for the diffusion of other impurities in TiN is high (e.g., the activation energy for Cu-diffusion into TiN-thin-films is 4.3-eV, whereas the normal-value for diffusion of Cu into metals is only 1 to 2-eV). TiN is also chemically and thermodynamically very-stable (its melting-point is 2950°C), and when in thin-film-form it exhibits one of the lowest electrical-resistivities (25-75 $\mu\Omega$-cm) of the transition-metal carbides, borides, and nitrides.

TiN can be deposited by PVD (see Chap. 15), or by CVD. Deposition of CVD-TiN is done in two ways. The first involves the reaction of $TiCl_4$ with NH_3. But this can only be used for contact-hole liner-applications because it must be performed at temperatures above those that can be tolerated after Al has been

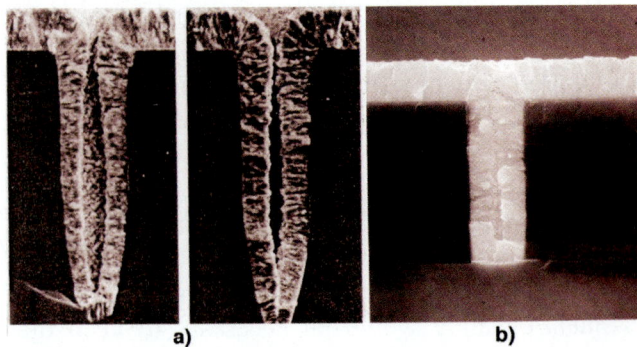

Fig. 16-37 (a) Blanket-CVD-W films deposited into a trench with a process that caused voids. (b) Blanket CVD-W film deposited into a trench without void formation.

deposited (Fig. 16-39). The other-method uses organo-metallic-precursors, which can be deposited at temperatures compatible with Al-interconnects. Thus, they can be used for vias as well as for contact-hole-liners. The two organo-metallic-sources of Ti are TDMAT and TDEAT.

REFERENCES

1. S.Wolf and R.N. Tauber, *Silicon Processing for the VLSI Era: Vol. 1 - Process Technology*, 2nd Ed., Lattice Press, Sunset Bch CA, 2000, Ch. 6 (Chemical Vapor Deposition)

2. S. Sivaram *Chemical Vapor Deposition,* McGraw-Hill, New York, 1995.

3. L.-Q. Xia, *et al.,* "Chemical Vapor Deposition," in *Handbook of Semiconductor Manufacturing Technology,* Y. Nishi and R. Doering, Eds., Marcel Dekker, New York, 2000, Ch. 11, p. 309.

4. D.L. Smith, *Thin-Film Deposition: Principles and Practice*, McGraw-Hill, New York, 1995.

5. A.S. Grove, "Mass Transfer in Semiconductor Technology," *Ind. & Eng. Chem.,* **58**, 48 (1966).

6. L. Sullivan and B. Han, "Vapor Delivery Methods for CVD: An Equipment Selection Guide," *Solid State Technology,* May 1996, p. 91.

7. A.R. Reinberg, "Dry Processing for Fabrication of

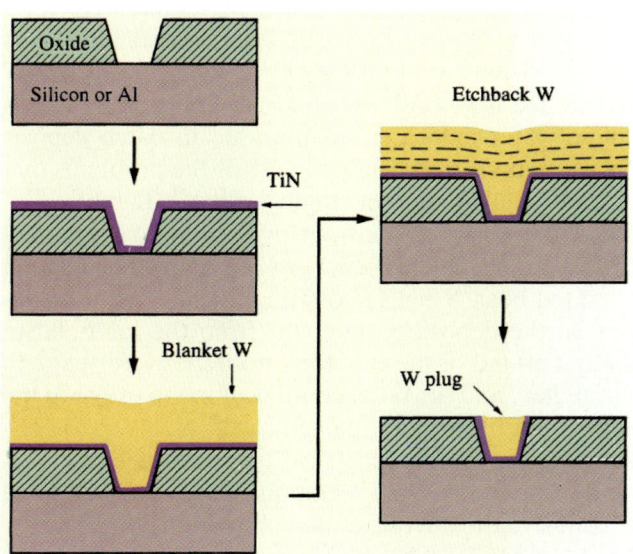

Fig. 16-38 Process sequence for forming tungsten-plugs.

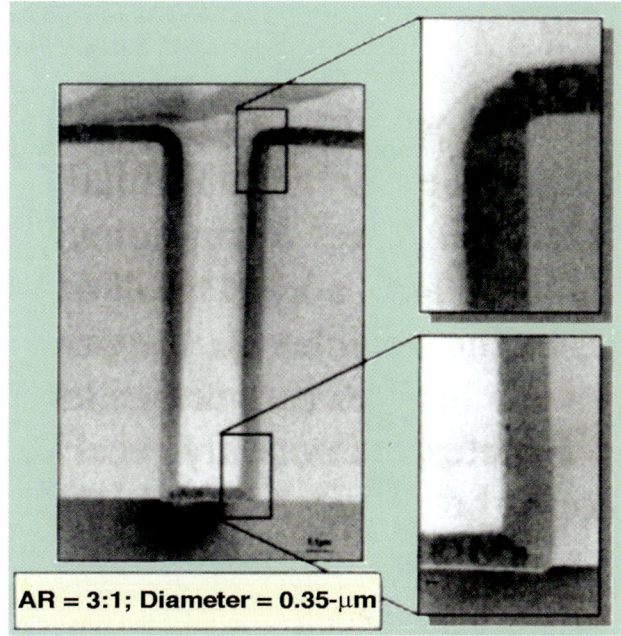

Fig. 16-39 SEM of TiN deposited by CVD into a high aspect-ratio contact-hole showing conformal-coverage.

VLSI Devices" in [N.G. Einspruch, Ed.] *VLSI Electronics-Microstructure Science,* Vol. 2, Academic Press, New York, 1981, Chap. 1.

8. R. Falkenberg, E. Doering, and H. Oppolzer, "Surface Roughness and Grain Growth of Thin P-Doped Polycrystalline Si-Films," *Proc. of Fall 1979 Electrochem. Soc. Meeting* (Los Angeles, October 1979), Abstract 570.

9. T. I. Kamins, *Polycrystalline Silicon for IC Applications,* Klewer Academic Publications, Boston, 1988.

10. M. M. Mandurah, K. C. Saraswat, and T. I. Kamins, "A Model for Conduction in Polycrystalline Silicon, Part I - Theory," *IEEE Trans. Electron Devices* **ED-28**, 1163 (Oct. 1981); "Comparison of Theory and Experiment," *ibid.* **ED-28**, 1171 (Oct. 1981).

11. G. Harbeke *et al.,* "Growth and Physical Properties of LPCVD Polycrystalline Silicon Films," *J. Electrochem. Soc.,* **131**, 675 (March 1984).

12. A.C. Adams, "Dielectric and Polysilicon Film Deposition," in *VLSI Technology*, S.M. Sze, Ed., Ch. 3, p. 93, McGraw-Hill Co., New York, 1983.

13. K. Maeda and S.M. Fisher, "CVD TEOS/O_3: Development History and Applications," *Solid State Technology,* June 1993, p. 83.

14. A.C. Adams and C.D. Capio, "The Deposition of Silicon Dioxide Films at a Reduced Pressure," *J. Electrochem. Soc.*, **126**, 1042 (1979).

15. J.E. Tong et al., "Process and Characterization of PECVD BPSG Films for VLSI Applications," *Solid State Technology*, January 1984, p. 161.

16. J.E. Schmitz, *Chemical Vapor Deposition of Tungsten and Tungsten Silicides*, Noyes Publications, 1992, Park Ridge, New Jersey.

PROBLEMS

1. Describe the steps of the CVD-process. What is the major difference between PVD and CVD?

2. List at least 5 CVD-films used in IC manufacture.

3. Sketch and name the major subsystems of a basic CVD-system.

4. Describe the differences between APCVD, LPCVD, and PECVD.

5. Why is the deposition temperature of the final passivation-layer limited to 450°C?

6. Why, and with what impurities, are CVD-oxides doped?

7. Describe the *reflow-process* of doped SiO_2-films (e.g., BPSG). Could the reflow-process be done using undoped-CVD-oxide (USG) films? What would be the limitation encountered if this was attempted?

8. A semiconductor company has a pilot-line production fab near the coast (at sea-level), and one of its high-volume manufacturing fabs in Colorado (at a 5000-foot elevation). It is found that some of the APCVD-processes developed at the pilot-line fab could not be directly-applied at the Colorado-fab. Why not?

9. Assume silicon is deposited at 1150°C at a rate of 500-nm/min with an activation energy, E_A, of 0.6-eV. By how much must the temperature be increased to double this rate? (Consult Appendices D and E.)

10. List 3 process-gases used for depositing Si_3N_4.

11. If a plasma-deposited silicon-nitride film contains 2×10^{21} hydrogen-atoms per cm^3, find the atomic-percent of hydrogen.

12. Why are *both* silicon-dioxide and silicon-nitride films sometimes used to make the final (multilayer) passivation film for an IC?

13. A trench in silicon is 4-μm. 0.5-μm wide, and 20-μm long. It is filled by successively and conformally depositing 50-nm-thick CVD polysilicon-films. What is the minimum total-film-thickness required to fill the trench? What problems could arise if the films are not deposited conformally?

14. For a process that is surface-reaction-rate limited, plot the deposition-rate versus temperature if the deposition-rate is 1-μm/minute at 1000°C and the activation energy is 1.6-eV (consult Appendix E).

15. Since in an LPCVD system, the gaseous diffusion constant is much larger than it is in an APCVD system, would deposition in the mass-transport-limited regime be expected to also be much higher? Explain your answer.

16. (a) Calculate the growth-rate of a silicon-layer grown using the $SiCl_4$ precursor at 1200°C. The vapor-phase mass-transfer coefficient of the reactor is h_g = 5-cm/sec, the surface reaction-rate coefficient k_s is given by $k_s = 10^7 \exp(-1.9\text{-eV}/kT)$-cm/sec, and the concentration of the $SiCl_4$ gas-molecules is $C_g = 3 \times 10^{16}/cm^3$. (b) What will be the change in growth rate if the reaction-temperature is increased by 1%. (Consult Appendix D for details on how to solve.)

17. In a silane-oxygen reaction to deposit undoped SiO_2 films, the deposition-rate is 150-Å/min at 425°C. What temperature is required to double the deposition rate? (The E_A of this reaction is 0.6-eV). Repeat the calculation for TEOS, in which the temperature for a 150-Å/min deposition-rate is 725°C (and the E_A of the TEOS-reaction is 1.9-eV).

18. Since new CVD reactors are being designed as single-wafer machines (as opposed to batch deposition systems), comment on some of the problems that have to be overcome in single-wafer CVD tools.

19. A process is carried out in a standard horizontal LPCVD tube, where the wafers are stacked on edge in a slotted boat. What factors might cause the a reduction in the deposition-rate: (a) from the inlet-end to the outlet-end of the tube; (b) from the wafer-edge to the center? and (c) What could be done to improve the deposition-uniformity in each case?

20. (a) What factors affect the step-coverage of CVD-oxide films? (b) What techniques have been adopted to improve the coverage of such oxide-films?

21. If LPCVD polysilicon-deposition has an E_A of 1.65-eV, and a deposition rate of 8-nm/min at 600°C, what is the deposition-rate at 620°C?

CHAPTER 16   CVD OF AMORPHOUS AND POLYCRYSTALLINE THIN-FILMS

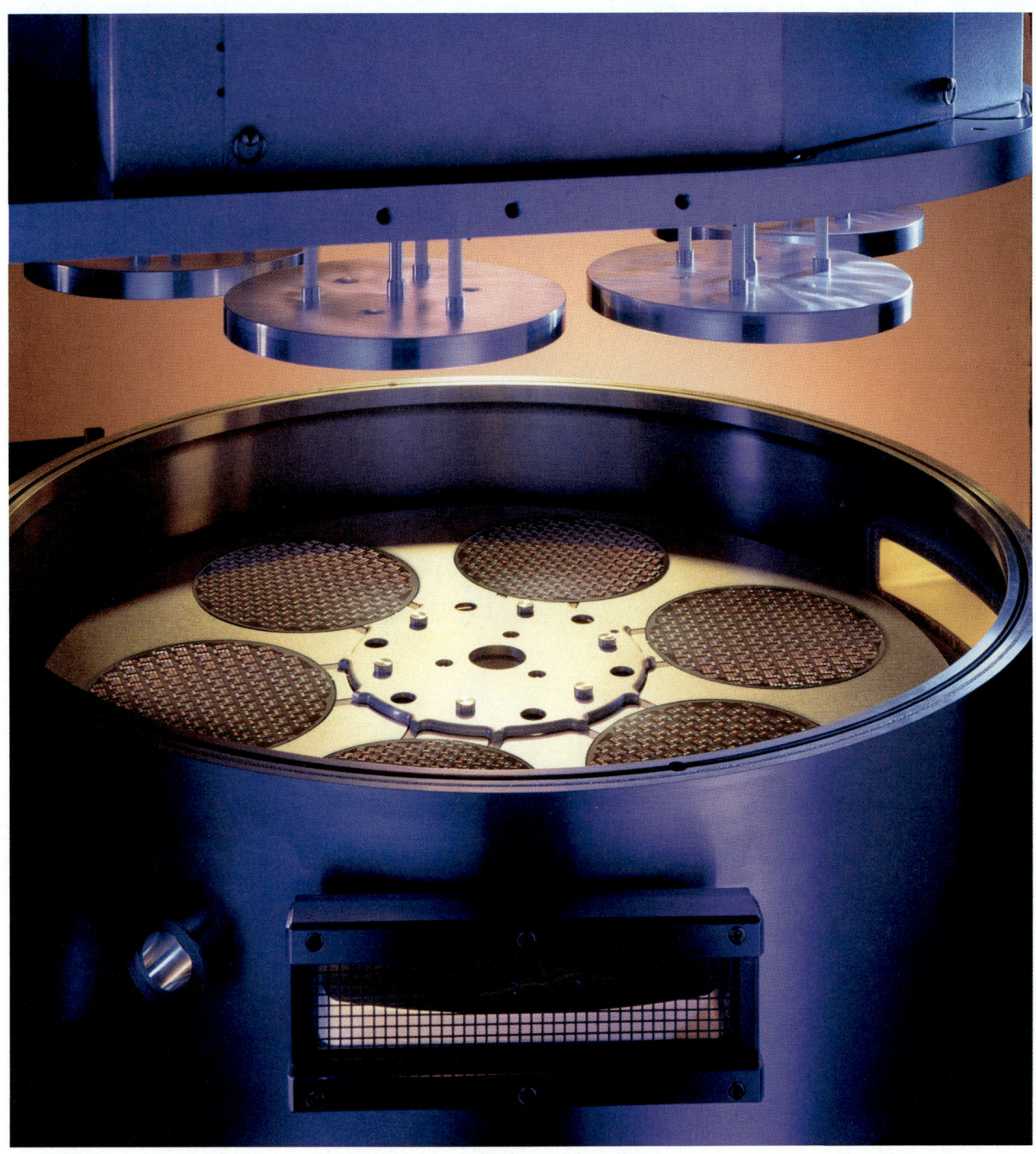

Photograph of a PECVD tool. Courtesy of Novellus Systems.

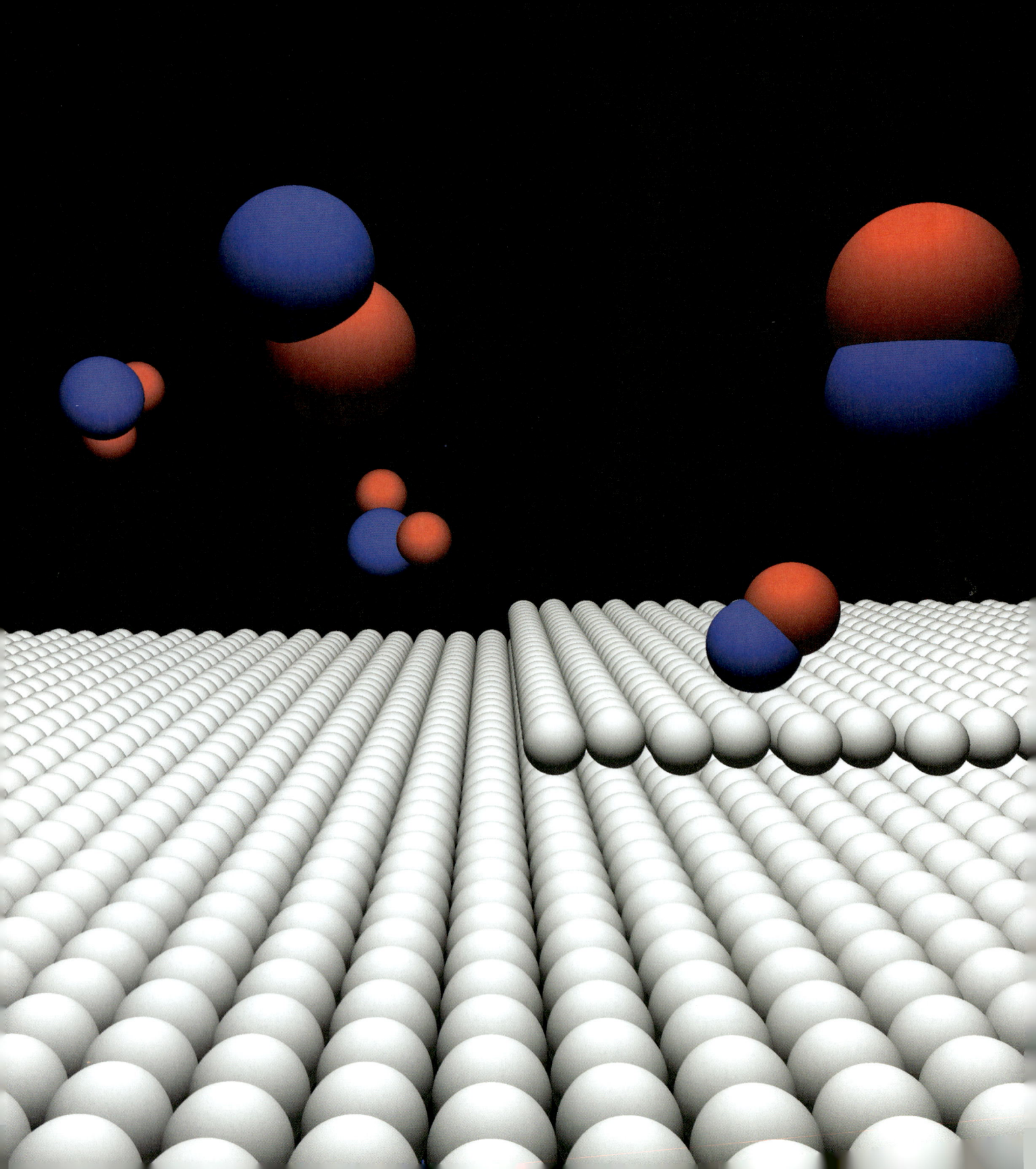

CHAPTER 17

SILICON EPITAXY AND SILICON-ON-INSULATOR TECHNOLOGY

CHAPTER CONTENTS

17.1 EPITAXY DEVICE APPLICATIONS

17.2 GROWTH OF EPITAXIAL LAYERS

17.3 CHEMICAL-REACTIONS OF SILICON

17.4 PROCESS CONSIDERATIONS FOR EPITAXIAL DEPOSITION

17.5 EPITAXIAL PROCESS-EQUIPMENT

17.6 CHARACTERIZING EPI-LAYERS

17.7 SILICON-ON-INSULATOR (SOI)

"I didn't fail 1000 times. The lightbulb was an invention with 1001 steps."

                              Thomas Edison

The *Molecular Dance Floor*.
Thanks to the graphic artists, Jerry Healey and Alan Gasperini.

Epitaxy is a process in which a single-crystal film is deposited upon the surface of another crystalline-substrate. The term *epitaxy* comes from two Greek words, *epi* ("upon") and *taxis* ("arranged," or "ordered"). Together they mean *arranged-upon,* and such films are also referred to as *epi-layers* (Fig. 17-1). If the grown-film consists of the same material as the substrate, the process is referred to as *homoepitaxy*. This is the most important use of epitaxy in silicon-technology. On the other hand, if the epitaxial-film is grown on a substrate of different material, the technique is called *heteroepitaxy*. An example of heteroepitaxy is the deposition of silicon on aluminum-oxide (*sapphire*), a method by which *silicon-on-sapphire* (*SOS*) wafers are fabricated. More details on Si epitaxial-growth can be found in Refs. 1 (and 3 to 6), listed at the end of the chapter.

Epitaxial-growth can be achieved by forming the film from: **1**) a vapor-phase (*vapor-phase epitaxy* [VPE]); **2**) a liquid-phase (*liquid-phase epi-*

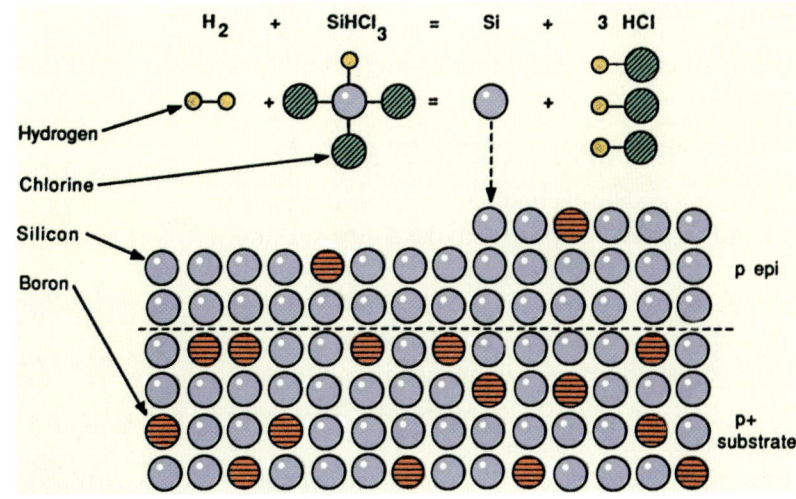

Fig. 17-1 Schematic drawing of a lightly-*p*-doped silicon-epitaxial-film on a heavily-*p*-doped silicon-substrate.

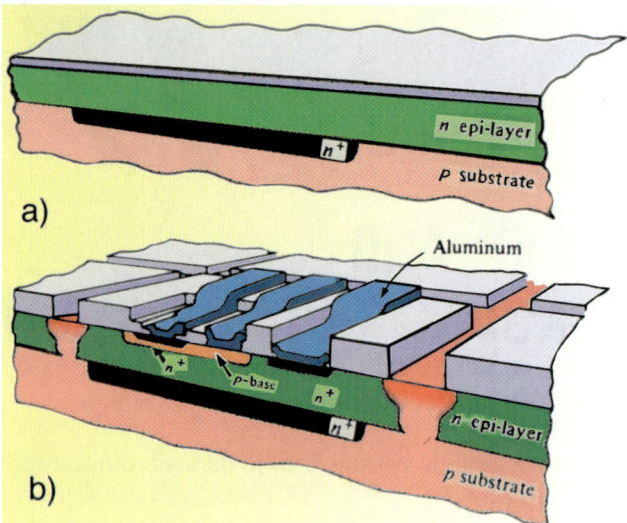

Fig. 17-2 (a) Lightly n-doped epitaxial-layer deposited over heavily n-doped buried-layers in a p-type Si-substrate. (b) Completed SBC-bipolar-transistor-structure built using an epitaxial-layer.

taxy [LPE]); or **3**) a solid-phase (*solid-phase epitaxy* [SPE]). High-temperature VPE (~1000°C) is used exclusively to produce epi-layers on silicon wafers because this technique offers tight-control of epi film-thickness and impurity-concentration, together with a high-degree of crystalline perfection.

This chapter describes: **1**) the uses of epitaxy in silicon-technology; **2**) the basics of epitaxial film-formation; **3**) epitaxial process-technology; and **4**) epitaxial film-characterization. It also briefly introduces silicon-on-insulator technology (SOI).

17.1 THE DEVICE APPLICATIONS OF EPITAXY

Epi-layers in silicon-technology are mostly used to form lightly-doped layers on heavily-doped substrates for device structures that require such doping-profiles at the wafer surface (see Fig. 17-1). Epitaxial-deposition is chosen because there is no other method in planar-processing for obtaining such lightly-doped layers on more heavily-doped sublayers. Two device structures need this kind of doping-profile: **1**) *standard–buried-collector* (SBC) *bipolar-transistor* structures; and **2**) certain types of *CMOS* ICs.

The BJT-SBC-structure increases the breakdown-voltage of the collector-base junction of bipolar-transistors. By growing a lightly-doped n-type epi-layer over a heavily-doped n^+-substrate, the bipolar-device is optimized for high breakdown-voltage of the collector-base junction, while preserving a low-value of collector-resistance. Low collector-resistance is needed to improve device operating-speeds at moderate-currents. Figure 17-2 depicts a cross-section of a bipolar-transistor showing the formation and role of the epi-layer. A key feature of bipolar-epitaxy is that the film is deposited over patterned, n^+-doped diffused-layers called *buried-layers* or *sub-collectors*. BiCMOS-devices are also be fabricated in epitaxial-films deposited over such patterned buried-layers.

Epi-layers are also widely-used in building CMOS integrated-circuits. In these CMOS-circuits the MOSFET-devices are fabricated in a thin (2-4-μm-thick), lightly-doped p-type epitaxial-layer deposited over a uniformly heavily-p^+-doped substrate. The role of such lightly-doped epi-layers on a heavily-doped substrate is to suppress the tendency of CMOS to undergo the failure-mode called *latchup*. Figure 17-3 shows a cross-section of a twin-well CMOS device depicting this type of epi-layer.

Epitaxial-layers also offer the important advantage of improving the reliability of thin gate-oxide films grown on a silicon-surface. This occurs because the epi-layer is free from oxygen, an impurity that exists in CZ-silicon as a result of the CZ crystal-growth process (see Chap. 9). With no oxygen in the epi-layer, oxygen precipitates cannot form (whereas they can occur near the surface of so-called *bulk-CZ-Si*

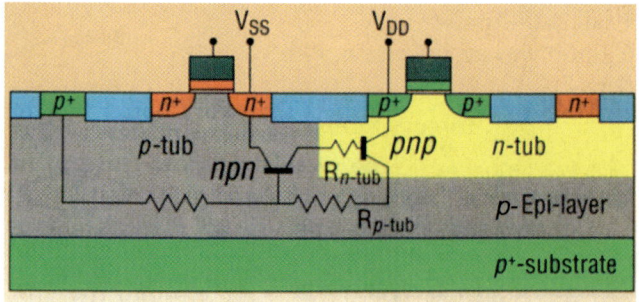

Fig. 17-3 Cross-section of a *twin-well CMOS structure* built in a p-epitaxial layer and a p^+ substrate.

wafers). The presence of such oxygen-precipitates degrades the electrical integrity of gate-oxides.

17.2 GROWTH OF EPITAXIAL-LAYERS

Epitaxial film-growth follows many of the same steps as the basic thin-film growth-sequence presented in Chap. 7. But other factors also play a role in the successful deposition of an epitaxial-film. These are described with the aid of Figs. 17-4 and 17-5.

Atoms from the vapor-phase are adsorbed on the crystalline silicon-surface and thus become *adatoms*. Such adatoms may migrate along the surface (Fig. 17-4). If this migration is rapid enough, the adatoms can orient themselves to the crystal-structure of the substrate surface. For example, if a migrating adatom encounters a growing *terrace* (such as position B in Fig. 17-5), or a *kink* (such as position C in Fig. 17-5), these are locations where the adatom will bond. In this way, they become an extension of the single-crystal lattice. Epi-growth continues, atom-by-atom, layer-by-layer, as additional adatoms migrate to terrace or kink positions and get incorporated in the crystalline-film that grows on the surface.

However, if adatom migration is constrained during growth, a polycrystalline-film is more likely to be deposited than an epitaxial-layer. This happens if the deposition-rate is too-high or the deposition-temperature is too-low. At any particular deposition-temperature there exists a maximum deposition-rate, above-

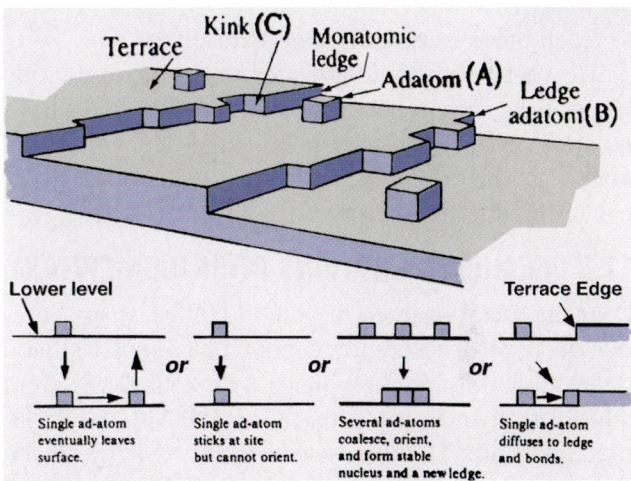

Fig. 17-5 Epitaxial growth-model showing step-wise growth with adatoms in terrace-positions A, adatoms at steps (or terrace-edges) B, and adatoms at stable kink-positions C.

which polycrystalline-films are formed, but below-which single-crystal epitaxial-films are deposited. Figure 17-6 shows a plot of the effect of growth-rate and temperature on the formation of single-crystal or polycrystalline-films. At high growth-rates there is not sufficient time for adatoms to migrate to the terrace (or kink) sites. Thus, only polycrystalline-growth can occur. But, as temperature is increased, the surface-migration-rate also rises, allowing time for adatoms

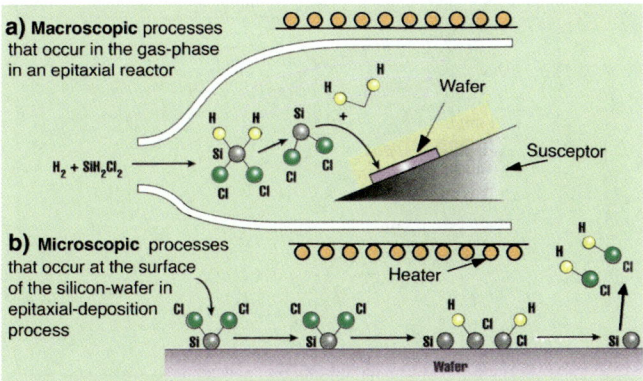

Fig. 17-4 Phenomena that occur during an epitaxial-deposition process at: (a) the *macroscopic* level; and (b) the *microscopic* level.

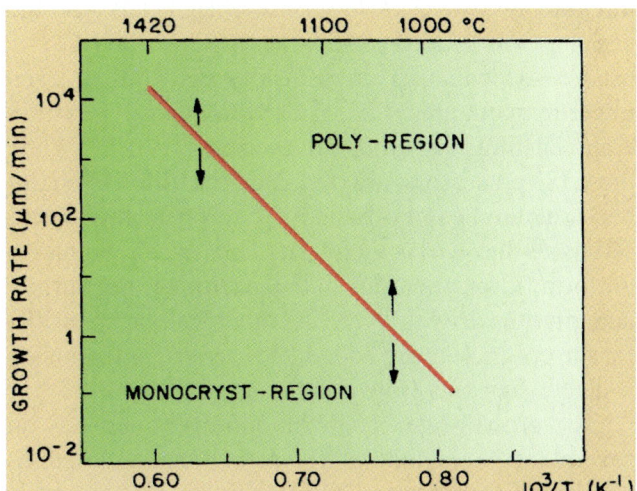

Fig. 17-6 Maximum growth-rate for the formation of *single-crystal silicon* as a function of temperature.[7]

to reach these sites and orient themselves.

However, if the silicon-wafer has a silicon-dioxide layer on its surface, the depositing adatoms will have no structure to align-to, and epitaxial-deposition will not occur. Instead, a polysilicon (or amorphous-silicon) film will be deposited.

17.3 CHEMICAL REACTIONS OF SILICON EPITAXY

Four chemical-sources of silicon (called *silicon-precursors*) are used commercially for epitaxial silicon deposition. All are compounds with a single Si-atom and four other atoms, either Cl or H. These silicon-precursors are: **1)** *silicon-tetrachloride* ($SiCl_4$); **2)** *trichlorosilane* ($SiHCl_3$ - called *TCS*); **3)** *dichlorosilane* (SiH_2Cl_2 - called *DCS*); and **4)** *silane* (SiH_4). Each silicon-precursor has advantages for specific deposition-conditions and epi-film applications.

$SiCl_4$ was the most widely-used precursor in the early days of IC-fabrication. But, because it needs the highest deposition-temperatures for practical epi-processes, $SiCl_4$ is not used much any more. Requirements for thinner epi-layers and lower deposition-temperatures have caused it to be largely replaced by SiH_2Cl_2, $SiHCl_3$, and SiH_4. For example, $SiHCl_3$ produces epi-films that are very similar in quality to those deposited by $SiCl_4$, but they are produced at temperatures about 100°C lower. Thus, $SiHCl_3$ has become popular for routine Si-epi-depositions. SiH_2Cl_2 is used to deposit thin, high-quality epi-films at even lower temperatures. SiH_4 is used to deposit very-thin epitaxial-films at temperatures below 900°C. When these Si-precursors are fed into the process-chamber, they are heavily-diluted in hydrogen, which also serves as the *carrier-gas*. The carrier-gas maintains uniform flow-conditions in the reactor before, during, and after the growth-cycle. As other gases responsible for the epi-growth are added, the carrier-gas maintains a steady-state gas-flow-condition.

The growth-rate of silicon epi-layers depends on several factors including: **a)** type of precursor used; **b)** deposition-temperature; **c)** mole-fraction of reactants present in the vapor-phase; and **d)** reactor-pressure.

Figure 17-7 illustrates the silicon growth-rate as a function of temperature for the four precursors listed above. Several key ideas can be learned from this figure. First, one of the factors that controls the deposition-rate is the type of silicon-precursor used. For example, use of SiH_4 provides the highest growth-rate at any temperature, followed by SiH_2Cl_2, and $SiHCl_3$ (with $SiCl_4$ the slowest). Second, two growth-regimes are evident. At temperatures where the deposition-rate increases relatively-slowly with temperature (Regime-B of Fig. 17-7) the mechanism that controls the growth-rate is *gas-phase mass-transport*. Such growth is said to *be mass-transport-limited* (see Ch. 16). At lower-temperatures, the growth-rate strongly depends on temperature, implying that it is controlled by a *thermally-activated process* (i.e., it is *surface-reaction-rate-limited* (Regime-A of Fig. 17-7).

17.4 PROCESS CONSIDERATIONS FOR EPITAXIAL-DEPOSITION

A variety of important process-considerations must be taken into account when depositing epi-layers for IC-device applications. They are covered in this section.

To begin, the epi-growth-rate is set by choosing the appropriate deposition-temperature and gas-flow conditions for the silicon-precursor being used. As shown in Fig. 17-8, the precursor-gas is diluted in a hydrogen carrier-gas, and its mole-fraction is selected to provide the desired epi-growth-rate.

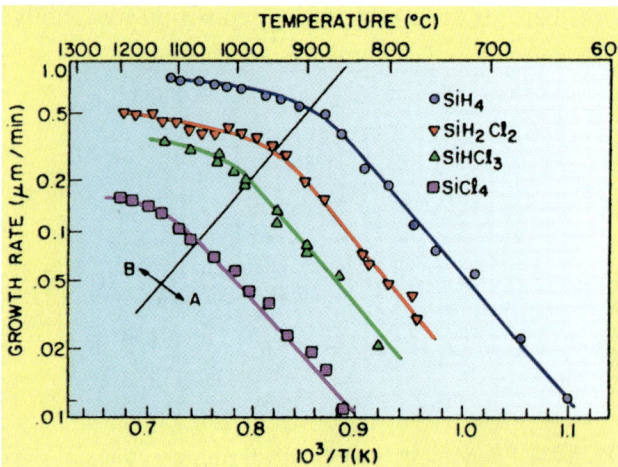

Fig. 17-7 Growth-rates of silicon films as a function of temperature for various silicon-sources.[8]

CHAPTER 17 SILICON EPITAXY AND SILICON-ON-INSULATOR TECHNOLOGY

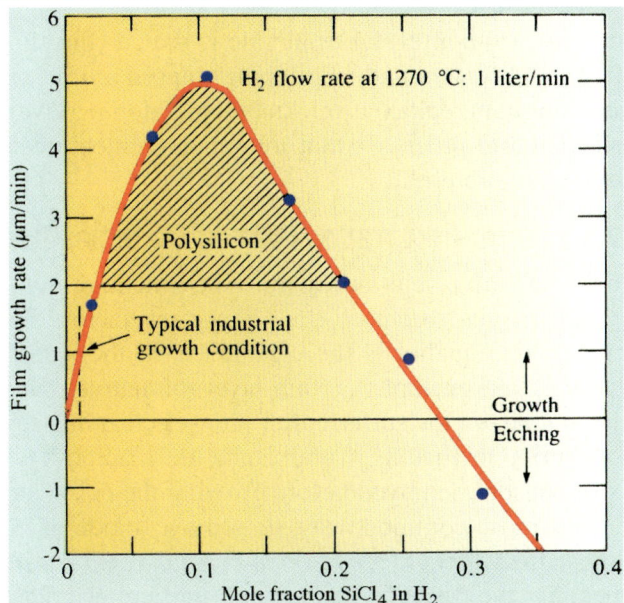

Fig. 17-8 Silicon growth-rate as a function of $SiCl_4$-concentration (mole-fraction, Y, of $SiCl_4$ in H_2).[9]

17.4.1 Intentional Doping of Epitaxial-Films

The doping-level in epitaxial-films is one of their most-important characteristics, and it must be tightly controlled. The correct doping-concentrations for a particular application are produced in epi-layers by adding gases containing the appropriate dopants to the reaction-gases during the epitaxial growth-process. For *p*-type-layers (boron-doping), diborane $[B_2H_6]$ is added to the reactant-gas. For *n*-type-layers either phosphine $[PH_3]$, or arsine $[AsH_3]$, is added for phosphorus or arsenic doping, respectively. At high-temperatures, these dopant-hydrides decompose and release boron, phosphorus, or arsenic into the growing epitaxial-film (Fig. 17-9). These gases are introduced into the reaction-chamber from gas-cylinders, where they are already diluted in hydrogen to concentrations of 20–100 parts-per-million (ppm). Such diluted-concentrations allow better control of the amount of dopant in the gas-stream (because smaller flow-variations are required). When dopant-gases are mixed with the main hydrogen-flow in the epi-reactor, their concentration is further diluted to the ppb-range necessary to produce the desired doping-levels in the epi-films.

Note that extraordinary caution must be exercised when handling the gas-cylinders containing these dopant-gases, since the gases are extremely-toxic.

17.4.2 Unintentional-Doping of Epi-Layers (Autodoping and Solid-State Diffusion)

Most epitaxial-applications in silicon technology call for a *lightly-doped layer* upon a *heavily-doped substrate*. Section 17.4.1 describes how the desired-concentration of dopants in the lightly-doped layers is controlled. However, during epitaxial-deposition, *unintentional-dopants* can also enter the growing-film in two other ways: **1)** dopants from the heavily-doped substrate can diffuse into the growing-film (an effect called *outdiffusion*); and **2)** dopants from the back of the heavily-doped substrate can diffuse out from the wafer, mix in the gas-stream, and become incorporated in the growing-film. This is called *vapor-phase-autodoping*. Outdiffusion and autodoping produce a *transition-layer* in the epi-layer in which the doping-concentration value lies between the substrate and desired epi-layer values. Without these effects, the carrier-concentration values would abruptly-change from those in the highly-doped substrate to those specified in the lightly-doped epi-layer (Fig. 17-10a).

The *transition-layer* between the heavily-doped substrate and the lightly-doped epi-regions is shown in Fig. 17-10b. The region of the epi-film close to the substrate shows higher doping-levels than intended,

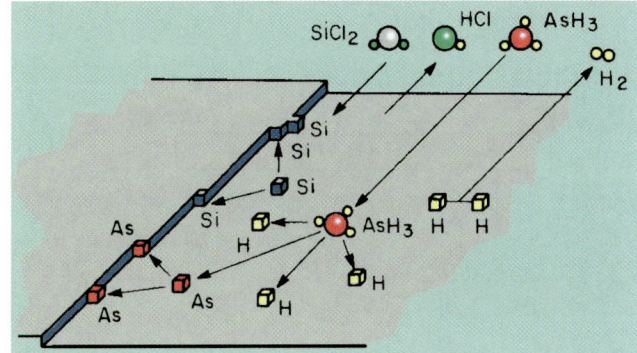

Fig. 17-9 Schematic of the epitaxial growth-process at the microscopic level showing the incorporation of Si and a dopant (in this case, *arsenic* [*As*]), into an epitaxial-film.

since dopant has diffused out of the substrate into the growing film during deposition (Region-A in Fig. 17-8b). However, the rapid growth of the film relative to the outdiffusion-front quickly limits this effect.

Thereafter, doping in the transition-layer is controlled by the incorporation of dopants from the vapor-phase. When dopant-atoms in the vapor-phase (evaporated from the substrate) exceed those intentionally-introduced, an *autodoping-tail* develops (Region-B in Fig. 17-10b). Since the high-concentration substrates quickly become covered with more lightly-doped material, the autodoping effect eventually ceases, and the desired concentration is reached. However, if the entire-substrate is heavily-doped, evaporation from the wafer-edges and backside can continue, even after the front-side is sealed (Fig. 17-10c). Outdiffusion and autodoping impose a limit on the minimum epitaxial-thickness and doping-level, and therefore processes that minimize unintentional-doping are preferred.

17.4.3 Wafer-Preparation Prior to Epi Deposition

To be effective, epi-layers must have a high-quality crystalline-structure. One important factor that impacts this quality is the cleanliness of the wafer-surface. For example, a thin-layer of native-oxide may remain on the surface after the wet-cleaning step done prior to loading wafers into the reactor. This oxide must be removed before growing the epi-layer.

The most common way to remove oxide is to expose the wafers to a *hydrogen pre-bake step* before injecting the reaction-gases into the epi-chamber. That is, wafers are baked in an ultrapure H_2-gas at a moderate-temperature (850–900°C) for a short period of time (2-3-min). In some cases, this step is performed at the same temperature as the epi deposition-step (e.g., to improve the throughput of single-wafer reactors). For this pre-bake step to be effective, the partial-pressure of oxygen within the reactor (from O_2 or H_2O) must be low enough so that SiO_2 is no longer thermodynamically-stable at the process-temperature. The pre-bake step works by converting SiO_2 to volatile sub-oxides (e.g., SiO) which sublime from the wafer-surface.

17.4.4 Batch Epitaxy Process-Sequence

The sequence of steps for depositing epi-layers in a batch-mode (Fig. 17-11), is identical for all types of batch-reactors. The wafers are carefully cleaned with a chemical pre-cleaning process just before loading them into the system. After the wafers are loaded into the reactor, it is closed and hydrogen is flowed in, to purge the air within. Next, the temperature is raised to 850-900°C. This pre-bake step in H_2 (for about 3-minutes) removes any native-oxide layer still present. Then the temperature is raised to 1150-1200°C, and HCl is flowed into the chamber for another 3-minutes. This slightly-etches the Si-surface to remove

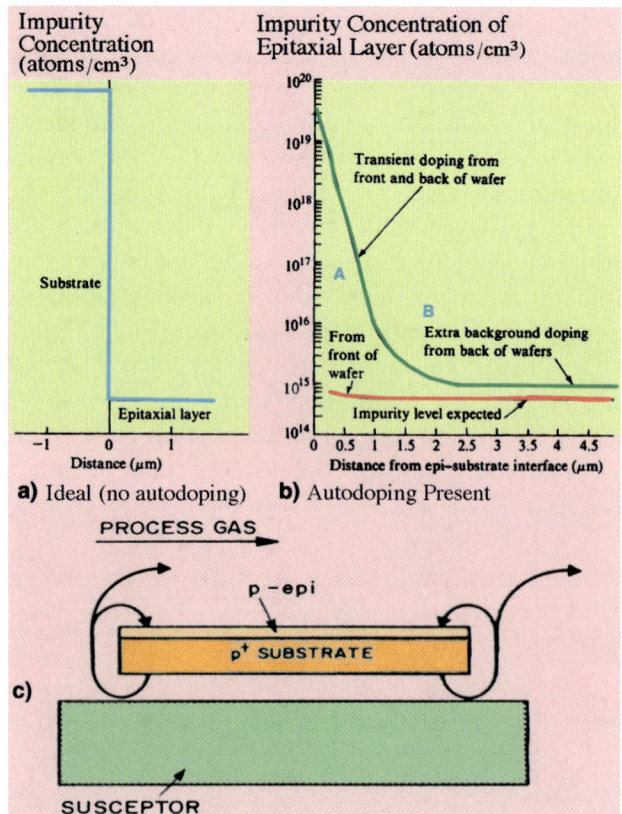

Fig. 17-10 Effect of autodoping on the concentration-profile near the interface between an epitaxial-layer and a heavily-doped-substrate: (a) Ideal-case (no autodoping); (b) Case when autodoping occurs; (c) Back-surface autodoping.

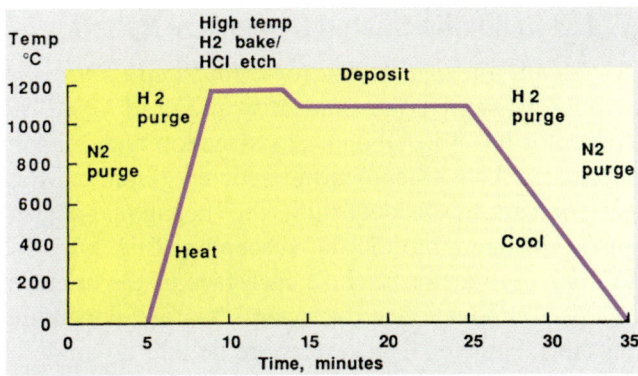

Fig. 17-11 Typical batch epitaxial-deposition cycle.

any other contamination (or lattice damage). Now the Si-surface is clean and ready for epitaxial-deposition. The chamber-temperature is adjusted to the deposition-temperature, and after stabilizing, the silicon-precursor, dopant, and carrier-gases are flowed into the chamber to grow the silicon epi-layer. After the film has been grown, the process-gases and heater-power are turned-off. H_2-gas is flowed to purge the chamber of the process-gases. When cooled sufficiently, N_2 purges the chamber to ambient-temperature. It can then be opened, unloaded, and reloaded. The whole sequence takes under an hour, and 10-28 wafers are processed at a time.

17.5 EPITAXIAL DEPOSITION EQUIPMENT

Silicon epitaxial-films are deposited in CVD-systems especially designed for epi-layer formation. A number of different reactor-designs have been developed. The two types in widest commercial-use today are: **1)** the multi-wafer IR-radiantly-heated barrel-reactors; and **2)** the single-wafer IR-lamp-heated reactors. Single-wafer reactors are most popular for large wafer-sizes ($\geq$ 200-mm).

Generally speaking, epitaxial deposition-systems all have the following subsystems: **a)** a reaction-chamber (which may be a quartz bell-jar or reactor-tube); **b)** a gas-distribution subsystem; **c)** a subsystem for heating the wafers; **d)** a susceptor that holds the wafers; **e)** an exhaust-system which removes the gases from the process-chamber; and **f)** a temperature-monitoring and control subsystem.

17.5.1 Batch Horizontal-Tube Epi-Reactors

The first epitaxial-reactors used for VPE in silicon technology were the batch, horizontal, fused-silica-tube reactors (Fig. 17-12a). In them, a number of wafers are placed on their backs on a flat susceptor (usually made of graphite, coated with silicon-carbide). The susceptor is heated by an rf-induced current. That is, rf-coils surround the tube, and rf-energy is passed through the walls of the fused-silica-tube without heating them. Instead, it is "coupled" to the molecules of the graphite, causing the susceptor to heat-up by rf-induction. The susceptor, in turn, heats the wafers by radiation and convection. The tube walls remain cool, making this a *cold-wall reactor*.

The process gases (at atmospheric-pressure) enter the inlet at the front of the tube. They flow over the wafers, and are exhausted from the outlet at the back. The reactor design is simple and less expensive than that of other epi-reactors. However, it is hard to maintain wafer-to-wafer uniformity. (Note that as shown in the figure, the susceptor is tilted a few degrees in an attempt to improve thickness-uniformity from wafer-to-wafer.) The throughput is also small, because the susceptor cannot hold many large-diameter wafers. For these reasons, batch, horizontal-tube epi-reactors no longer find much use.

17.5.2 Vertical Pancake Epi-Reactors

The so-called *induction-heated, vertical, pancake epi-reactor* is shown schematically in Figs. 17-12b, c, and d. The wafers are placed on a rotating silicon-carbide-coated graphite-susceptor, again heated inductively by underlying rf-coils. The reactant-gases enter the reaction-chamber from the center of the susceptor and are subsequently exhausted from the periphery. The gases are distributed symmetrically across the wafers, resulting in a constant growth-rate. The vertical gas-flow minimizes autodoping-effects. The pancake configuration, however, is susceptible to particulates falling on the wafer (and then becoming embedded in the growing film). The system is capable of processing wafers up to 200-mm in diameter, but the load-size decreases as the wafer-size is increased.

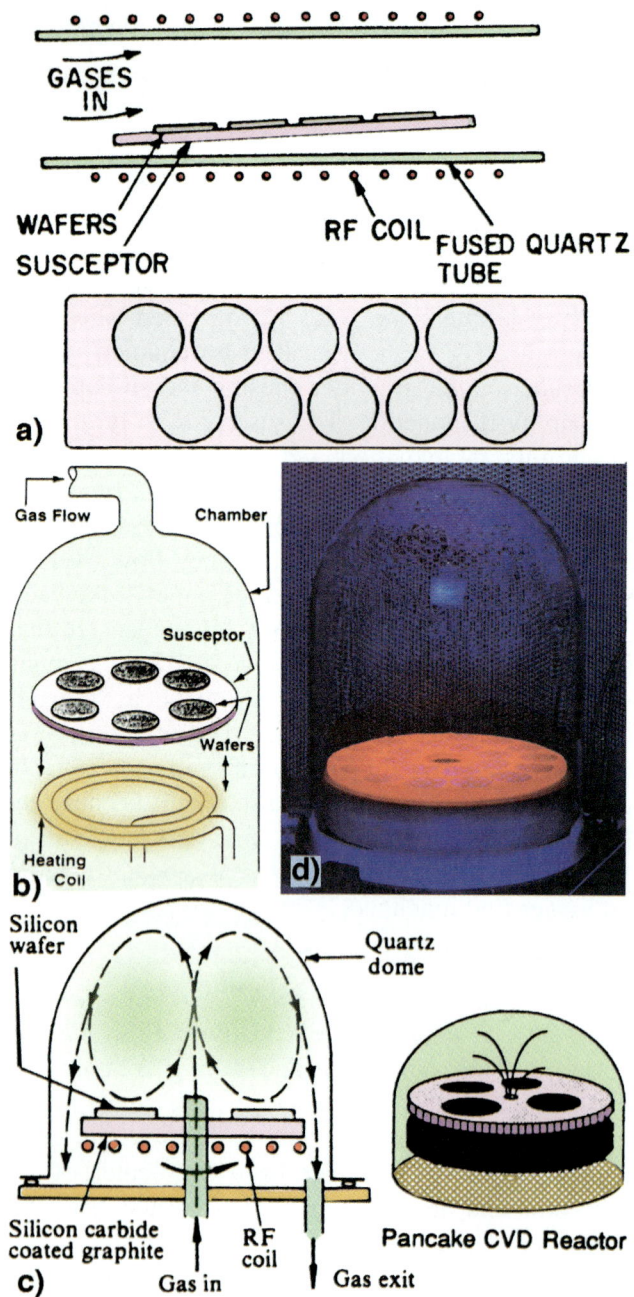

Fig. 17-12 (a) Horizontal-type epitaxial-reactor. (b) Pancake-type epitaxial-reactor. (c) Another schematic drawing of a pancake silicon-epitaxial-reactor. (d) Photograph of a pancake-type silicon-epitaxial-reactor.

17.5.3 Radiantly-Heated Cylindrical Reactors

A *radiantly-heated, barrel* (or *cylindrical*), *epitaxial-reactor* is shown schematically in Fig. 17-13. The susceptor has a hexagonal cross-section and is again fabricated from silicon-carbide-coated graphite. More wafers can be loaded onto the hexagonal-shaped susceptor than onto the flat-susceptor of the horizontal-tube epi-reactor because each face of the hexagon can accommodate one (or several) wafers, depending on their diameter. The wafers are held at an angle of 2.5° to the vertical, and the gases flow parallel to the wafer-surfaces. The reaction-chamber is a stainless-steel barrel, with high-intensity quartz-heaters placed about its inside surface. The susceptor rotates in the center of the barrel.

Reactant-gases are injected at the top of the chamber and exhausted from the bottom. Wafers are heated directly by water-cooled quartz-lamps lining the inside of the reactor. Heating the wafer from the front-side decreases the occurrence of thermally-induced slip. Rotation of the wafers in the reaction-gases produces a more uniform film-thickness, compared to batch horizontal-tube systems.

Cylindrical epi-reactors are capable of operating at atmospheric or reduced pressure (80–100-torr). Another advantage they have is that the wafers are held in near-vertical positions, so particles are less likely to land on them. Barrel-reactors are capable of resistivity- and thickness-uniformity better than ±5% across the wafer, wafer-to-wafer, and run-to-run. However, their throughput is low when processing large-diameter wafers (≥ 200-mm in diameter).

17.5.4 Single-Wafer Epitaxial Systems

Single-wafer epitaxial-reactors were introduced in the 1990's when 150- and 200-mm wafers first came into use. The single-wafer system becomes more economical than a batch-system for large wafers, and is now the most widely used type of epi-tool. Figures 17-14a & b show a cross-section of a single-wafer epi-reactor. The chamber is constructed from fused-silica. Because the process requires extremely-high temperatures (1050–1150°C) the chamber-walls could

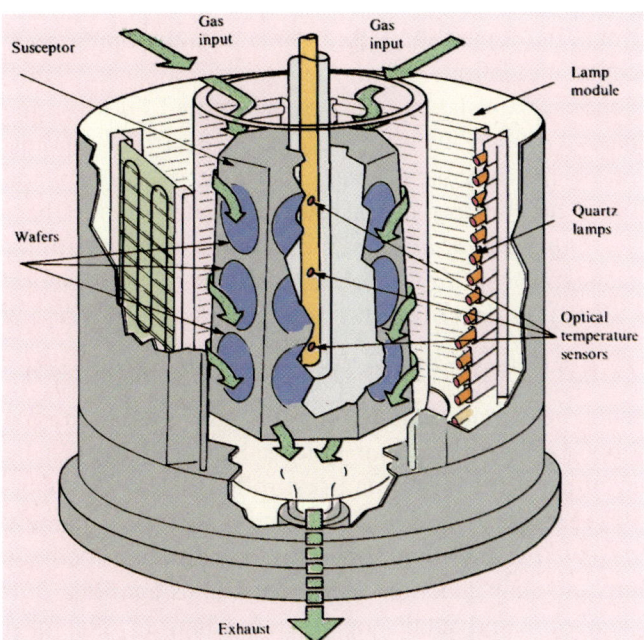

Fig. 17-13 Schematic of a radiantly-heated, cylindrical-type epitaxial reactor. Courtesy of Applied Materials, Inc.

then fed into the reactor from a side-port. (These gases are pre-mixed before they enter the chamber, along with the H_2 carrier-gas.) They flow over the wafer in a laminar manner and then exit the system from another side-port. By attaching a vacuum pump to the outlet of the reactor, it can also be operated at reduced-pressure.

When the deposition-process is complete, the flow of reactant and dopant gases is halted (but the H_2 flow continues). When the wafer has cooled to about 750°C it is taken from the reactor and moved into a cool-down chamber. This chamber rapidly cools the

become quite hot. To prevent them from becoming excessively-heated, water and air-cooling must be used. Wall-temperature is kept between 400–600°C.

Such systems have a high epi-layer growth-rates and better reliability and repeatability than batch epi-reactors. They can deposit high-quality films at atmospheric or reduced pressures. The susceptor (and wafer) is rotated to even-out temperature and gas-flow fluctuations. Wafer-rotation has been found to substantially improve process-uniformity.

The system rapidly heats the wafer (18°C/sec) to the process-temperature using quartz IR-lamps. The top-surface of the wafer is heated by the upper lamp-assembly and the bottom-surface through the suscep-tor (which is heated by a lower lamp-assembly). Thus, a uniform-temperature is maintained throughout the wafer, preventing slip-defects that could occur during such rapid-temperature ramp-rates. Once the wafer reaches temperature, it is stabilized and "pre-baked" in H_2 (for about 1-minute) to remove any native-oxide on the wafer-surface.

The process-gases (precursors and dopant) are

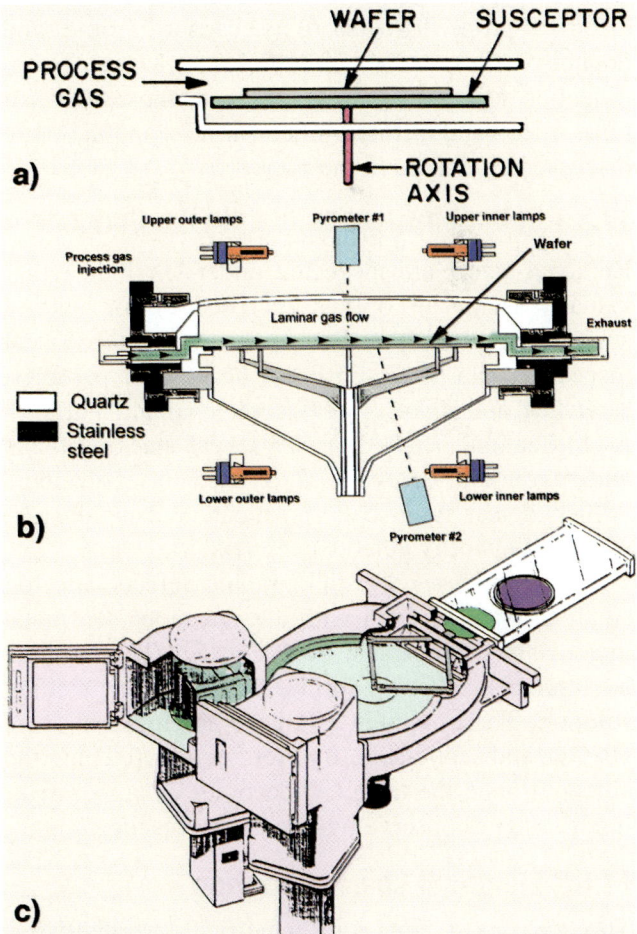

Fig. 17-14 (a) and (b) Cross-section of a single-wafer epitaxial reactor. Courtesy of Applied Materials, Inc. (c) Perspective drawing of a commercial single-wafer epitaxial reactor. Courtesy of ASM International.

wafer to <200°C, after which it is put back into the wafer cassette. Single-wafer epitaxial-reactors are manufactured by Applied Materials Inc. (Fig. 17-14b) and by ASM International (Fig. 17-14c).

17.6 CHARACTERIZATION OF EPITAXIAL-FILMS

Characterization of epitaxial-films is performed to monitor the epi-dependent properties of product-devices early in the fabrication sequence. If the epi-quality is poor, it is unlikely that high-quality devices can be made in the film. Epi-layers are inspected for: **a)** surface-quality; **b)** crystallographic-defects; **c)** electrical-properties (resistivity, carrier-lifetime, and dopant-profile); and **d)** film-thickness. Many characterization techniques are common to those used for bulk-silicon and diffused-silicon-layers, and these are discussed elsewhere in the book.

17.6.1 Optical Inspection of Epi-Film Surfaces

Epitaxial-layers are inspected under UV-light either visually (with an optical-microscope), or with a surface-scanning, laser-based, defect-counter. Visual inspection can spot slip-lines, haze, pits, particles, scratches, and spikes. Microscopic evaluation at low-magnification (75–200X) can reveal stacking-faults and pyramidal-defects. Automated laser-scanners (which detect light-scattering centers) can identify, map, and quantify defects in epitaxial-layers.

In some cases crystallographic-defects are too small to be seen even with a microscope. To make them visible, they must be *decorated*. This is done by immersing the epi-surface in a selective-etching solution. These contain HF, an oxidizing-acid (e.g., HNO_3), and acetic-acid. By preferentially-etching the silicon around the defect, it becomes enlarged, making it easier to observe by optical-microscopy.

17.6.2 Electrical Characterization

The two most important electrical-characteristics of epi-layers are their *film-resistivity* and *doping-concentration-profile*. Producing films with the specified resistivity is done by controlling dopant-gas concentrations during deposition. However, variations in dopant-incorporation and autodoping may cause deviations from the specified-values. Thus, deposited epi-layers must be monitored on a routine basis. *Film-resistivity* is determined from 4-point-probe measurements that yield sheet-resistivity, and these are performed on monitor-wafers (see Chap. 7). *Doping-profiles* are measured using spreading-resistance (SR), or secondary-ion mass-spectroscopy (SIMS), as described in Chaps. 11 and 24.

17.6.3 Epitaxial Film-Thickness Measurements

Thickness and thickness-uniformity are additional important properties of epitaxial-films that must be accurately measured and controlled. Epitaxial thickness-measurements fall into two categories, namely *destructive* and *non-destructive*. Destructive-methods include such junction-depth techniques as *angle-lap-and-stain*, and details are given in Chap. 11. The most common non-destructive method is *infrared reflectance*, as discussed in the next section.

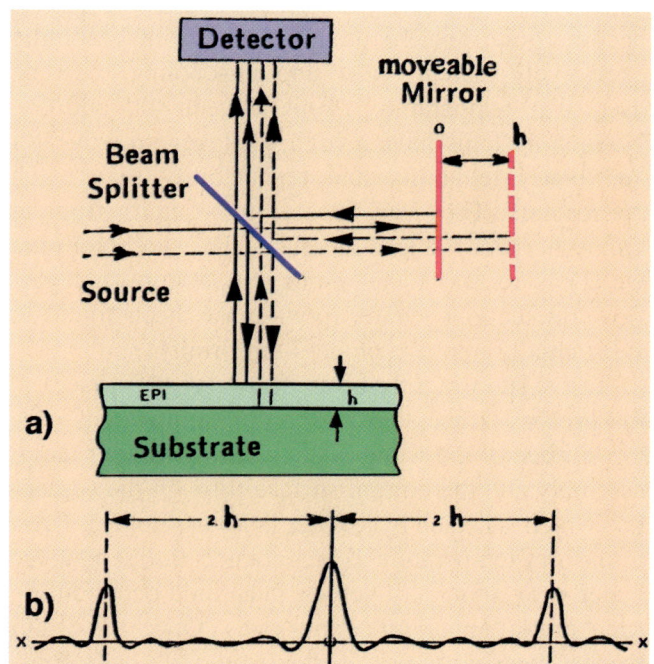

Fig. 17-15 Schematic representation of interferometric measurement of epitaxial-layer-thickness. (a) Interferometer showing path-lengths. (b) Interferogram showing a coherence peak and side-bursts.

17.6.4 Infrared-Reflectance Techniques

When electromagnetic-radiation (particularly in the infrared [IR] range) is incident upon a lightly-doped epitaxial-film on a heavily-doped Si-substrate, the following phenomenon takes place (Fig. 17-15). The incoming IR-light is partially-reflected from the air/film surface and the rest is transmitted through the film (where it strikes the surface of the heavily-doped substrate). The buried-surface reflects this light, and the reflected-beam combines with the radiation reflected from the exposed air/film-surface. The combined beams produce an interference-condition. As the wavelength of the IR-beam is varied, intensity-variations occur in the reflected-signal (depending on the wavelength of incident radiation). These can be correlated to the thickness of the film.

The IR-reflectance-technique is especially effective for lightly-doped epi-layers on heavily-doped substrates - which is exactly the film-structure encountered in silicon ICs. Lightly-doped epi-layers are relatively transparent to infrared-light having wavelengths from 2.5 to 50-μm. Accordingly, this is the range of the spectrum typically used in the IR-reflectance-measurement-technique. The heavily-doped substrate behaves as a reflective-surface for radiation in this wavelength-range.

Fourier-transform infrared-spectroscopy (FTIR), is another more sophisticated (and more accurate) non-destructive technique for measuring the thickness of epitaxial-films. More information about it can be found in Ref. 1.

17.7 SILICON-ON-INSULATOR TECHNOLOGY (SOI)

Silicon-on-insulator (*SOI*) circuits consist of ICs made from single-device islands, isolated from the substrate (and each other) by a dielectric region. Each island contains a single transistor. Figure 17-16a is a schematic depiction of MOSFETs fabricated on an SOI wafer. Note that a *buried-oxide* (*BOX*) beneath the device provides *vertical-isolation*, and the trench-oxide provides *lateral-isolation*. Since the trench-oxide extends all the way through the silicon device layer atop the BOX (and a thermal-oxide exists on the wafer top-surface), dielectric-material continuously surrounds each MOSFET on all its sides.

SOI-technology offers a number of significant advantages over bulk-CMOS technology, including: **1**) SOI-devices exhibit less parasitic-capacitance compared to bulk-Si devices. This reduces power-consumption or gives faster-performance; **2**) SOI-ICs have better immunity against radiation-induced failures, making them useful for space-applications; **3**) CMOS-SOI circuits do not undergo latchup; and **4**) SOI circuits can be manufactured with a fewer number of steps (up to 30% less). On the other hand, SOI-wafers are more costly and may have more defects than either bulk-silicon, or epi-silicon wafers.

SOI technology was first developed in the 1960's, but its early use was restricted to military-applications. However, as the need for higher-speed, lower-power-consumption, and cheaper fabrication-costs continued to drive the industry, chip manufacturers have recently pursued SOI as a technology that may

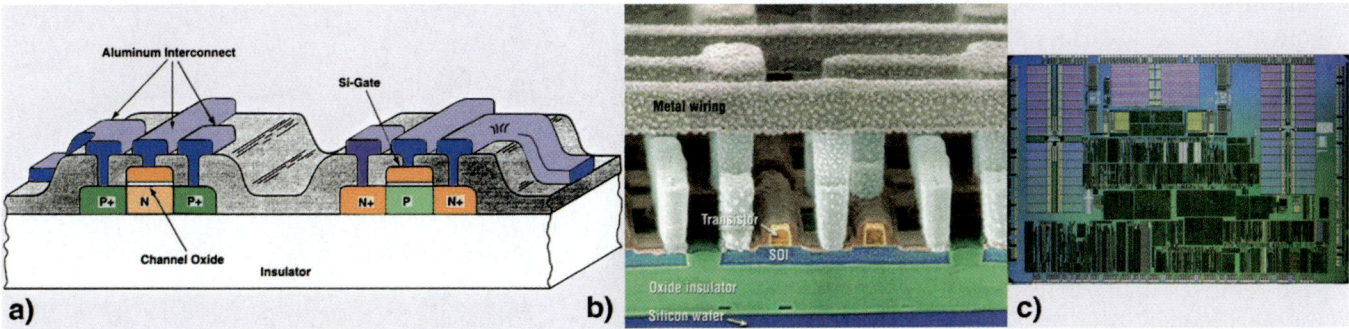

Fig. 17-16 (a) Schematic drawing of a CMOS structure built on an SOI-wafer. (b) SEM photograph of an SOI-IC. (c) Photograph of an SOI-IC. Courtesy of IBM.

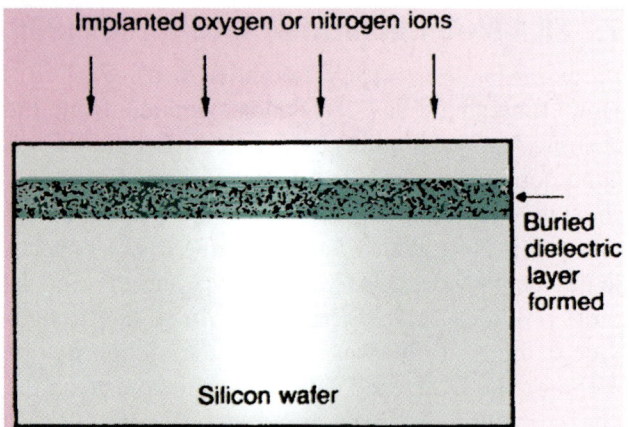

Fig. 17-17 Implanted oxygen-ion process for the formation of a SIMOX wafer.

help meet these goals (Figs. 17-16b & c).

Currently there are two main techniques for fabricating SOI-wafers: **1)** S*eparation by IMplanting OXygen* (*SIMOX*); and **2)** *wafer-bonding* (*WB*).[2,10]

17.7.1 Separation by IMplanting OXygen (SIMOX)

Separation by IMplanting OXygen (*SIMOX*) is the name of a technique that produces SOI-wafers by implanting a very-high dose of oxygen atoms into silicon to produce subsurface-layers of silicon-dioxide. Figure 17-17 shows a schematic of how such SIMOX buried-oxide-layers are fabricated:

1. A bare silicon-wafer is subjected to a high dose (~2×10^{18}/cm^2) oxygen (O$^+$) implant-step. The implant-energy (150–300-keV) locates the peak of the implant 0.3-0.5-microns beneath the silicon-surface. During the implant the wafers are held at >600°C to ensure that the silicon maintains its crystallinity.

2. The wafers are then given a post-implant anneal in N$_2$ for 3–5-hours at a very-high temperature (1300-1350°C). This causes a continuous *buried-oxide* (BOX) layer of silicon-dioxide to be formed near the peak of the implantation. It also removes many defects (dislocations) created by the implant. This BOX-layer is about 0.5-μm thick, and its depth is set by the location of the implant-peak.

3. If devices in the IC require a thicker Si-surface-layer than can be achieved with the depth of the ion-implanted oxygen, an epitaxial-Si layer can be deposited to provide this additional thickness.

SIMOX technology also has several shortcomings. First, fabricating SIMOX-wafers requires very-high doses of oxygen. Thus, long implantation-times are needed. Second, the implant-step may contaminate the wafer with transition metals. Such metals reduce the minority-carrier-lifetime of the devices. Third, despite the high-temperature anneal, the surface-silicon may still have a high-density of dislocation defects. Their presence can degrade gate-oxide integrity and increase junction-leakage. Finally, SIMOX-wafers are more expensive than bulk-wafers, although the difference in cost is expected to decrease with the growth in popularity of SIMOX.

The recent-introduction of SIMOX-dedicated oxygen-ion-implanters from IBIS Technology and

Fig. 17-18 Photograph of a SIMOX implanter wheel. Courtesy of Hitachi.

Hitachi have reduced the severity of these problems (Fig. 17-18). Such machines offer adequate throughput and acceptably low-levels of metallic-contamination. SIMOX materials have been used to build excellent-quality CMOS and bipolar ULSI-circuits.

17.7.2 Wafer-Bonding (Smart Cut)

Another popular technology for producing SOI wafers is *wafer-bonding*, in which two oxidized Si-wafers (or one-oxidized and one-bare wafer) are fused together, face-to-face, through a furnace process (Fig. 17-19). Wafer bonding forms BOX-layers much thicker than are possible with SIMOX (up to 50-μm thick).

A method of wafer-bonding that has been commercialized by SOITEC is called Smart Cut.® It allows the sacrificial wafer to be reused and only one wafer is required to make a bonded SOI-substrate. This one-wafer technology is termed Unibond.® Figure 17-20 shows how the process works. Two device-quality wafers (labeled A and B) start the process. Wafer-A is oxidized (to about 0.5-μm) to form what will become the buried-oxide. Then wafer-A is ion-implanted with H$^+$ to about 0.2-μm below the oxide-surface. The depth of the H ion-implantation determines the thickness of silicon above the BOX. Wafers-A and B are now cleaned and low-temperature bonded face-to-face. The bonded-wafers are next annealed (at ≤600°C). During this anneal the implanted-hydrogen forms a gas, and wafer-A delaminates at the peak of the implanted-hydrogen. Wafer-B, which now has a BOX-layer, becomes the SOI-wafer, while

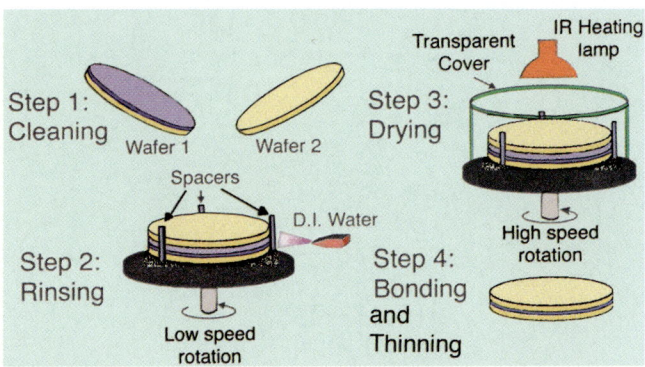

Fig. 17-19 Schematic of the wafer-bonding process.

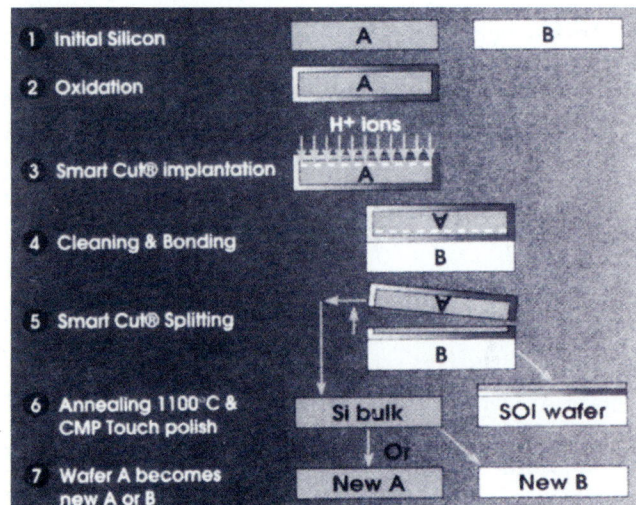

Fig. 17-20 Process flow for fabricating Smart Cut® wafers. Courtesy of SOITEC SA.

wafer-A becomes the sacrificial-wafer. Both wafers are annealed at 1100°C to remove defects. Wafer-A, which is now a fraction of a micron thinner than it was before the start of the process, is used as wafer-B when the process starts over. The method has been used to prepare commercially-available 200-mm, SOI-wafers with a silicon-thickness above the BOX of 0.2-μm, and a uniformity of better than 5%.

REFERENCES

1. S.Wolf and R.N. Tauber, *Silicon Processing for the VLSI Era: Vol. 1 - Process Technology*, 2nd Ed., Lattice Press, Sunset Beach CA, 2000, Ch. 7 (Si Epitaxial Growth and SOI).

2. S.Wolf, *Silicon Processing for the VLSI Era: Vol. 4 - Deep-Submicron Process Technology*, Lattice Press, Sunset Beach CA, 2002, Ch. 11 (Silicon-on-Insulator [SOI] Technology).

3. McD. Robinson, "Silicon Epitaxy," in *Microelectronic Materials and Processes*, R.A. Levy, Ed., Kluwer Academic Publishers, Dordrecht, 1989, Ch. 2, p. 25.

4. R.B. Herring, "Silicon Epitaxy," in *Handbook of Semiconductor Silicon Technology*, Noyes Publications, Park Ridge NJ, 1990. Ch. 5., p. 258.

5. B.J. Baliga, Ed. *Epitaxial Silicon Technology,* Academic Press, Orlando FL., 1986.

6. A. Thompson, R. Stall, and B. Kroll, "Advances in Epitaxial Deposition Technology," *Semiconductor International,* July 1994, p. 173.

7. J. Bloem, "Nucleation and Growth of Silicon by CVD," *J. Crst. Growth,* **50,** 581 (1980).

8. F.C. Eversteyn, "Chemical Reaction Engineering in the Semiconductor Industry," *Philips Res. Repts.,* **19** 45 (1974).

9. H.C. Theurer, "Epitaxial Silicon Films by the Reduction of $SiCl_4$," *j. Electrochem. Soc.,* **108,** 649 (1961).

10. J.P. Colinge, *Silicon-on-Insulator Technology,* 2nd Ed. Kluwer Academic Publishers, Dordrecht, 1997.

PROBLEMS

1. Describe what *epitaxy* is. What is the main advantage of epitaxy over other processes that makes it a process useful for some IC applications?

2. What is the role of epitaxy in: (a) bipolar transistor IC technology; and (b) CMOS IC technology? What is main benefit obtained by using a lightly-doped epi-layer on a heavily-doped substrate for a CMOS circuit?

3. List three silicon precursors used for depositing VPE Si-epitaxy layers.

4. List three dopant-gases used for intentionally adding dopants to an epitaxial layer. What are the safety-hazards associated with the use of these gases?

5. Why is epitaxy done in cold-wall reactors? (See p. 275 for a definition of this term.)

6. What are the advantages of *single-wafer* epitaxy reactors over *batch* epitaxy reactors?

7. Define the terms *outdiffusion* and *autodoping*. What measures are taken to suppress autodoping? Why is a highly-abrupt substrate/epitaxy interface difficult to obtain.

8. Referring to Fig. 17-8, determine the mole-fraction of $SiCl_4$ that will result in the maximum film growth-rate of a silicon-film. Why is epitaxial deposition not carried out under this condition?

9. Explain why the growth-curve shown in Fig. 17-8 becomes negative.

10. The intrinsic-diffusivity of boron in Si (see Appendix B) can be expressed as $D = 0.76 \exp(-3.46\text{-eV}/kT)$. Using this expression, calculate the minimum growth-rate that is required when a Si-epi film is grown on a heavily B-doped Si substrate for 20-min at 1200°C. Discuss why lower growth-temperatures must be used in order to obtain Si epi-layers that are less than 1-μm thick.

11. List some circuit applications that would benefit from using an SOI technology to fabricate them.

12. A high-energy (1-MeV) implanter is used to implant oxygen below the silicon-surface to form a buried-oxide layer (SIMOX process). Assume the desired buried-oxide layer is to be 0.2-μm thick: (a) What is the oxygen-dose that must be implanted into the silicon to form this layer? List the assumptions you made in determining this value. (b) What beam-current is required if a 200-mm wafer is to be implanted with this dose in 15-min? (c) How much power is being supplied to the ion beam? Discuss what effects this implantation may have on the wafer.

CHAPTER 17   SILICON EPITAXY AND SILICON-ON-INSULATOR TECHNOLOGY

Servicing a 300-mm etch tool. Photograph courtesy of Applied Materials, Inc.

CHAPTER 18

LITHOGRAPHY I: PHOTORESIST MATERIALS AND PROCESS TECHNOLOGY

CHAPTER CONTENTS

18.1 PHOTORESISTS

18.2 RESIST MATERIAL PARAMETERS

18.3 OPTICAL RESIST TYPES

18.4 PHOTORESIST PROCESSING

    Dehydration Baking and Priming
    Resist Processing: Spin Coating
    Soft-Bake
    Exposure
    Development
    After-Develop Inspection
    Post-Development-Bake

18.5 RESIST PROCESSING SYSTEMS

In the Beginning, God said,
"
$$\nabla \cdot E = \frac{\rho}{\epsilon_0} \quad \nabla \times E = -\frac{\partial B}{\partial t}$$
$$\nabla \cdot B = 0 \quad c^2 \nabla \times B = \frac{j}{\epsilon_0} + \frac{\partial E}{\partial t}$$
"

Exposing a photoresist film on a wafer. Courtesy of ASML.

The patterns that define structures in ICs are created by *lithographic-processes*. A layer of photoresist-material is first spin-coated onto the wafer-substrate. Next, the resist is selectively-exposed to some form of radiation, such as ultraviolet-light or electrons. An *exposure-tool* and *mask* are used to produce the desired selective-exposure. The patterns in the resist are created when the wafer undergoes a subsequent "development" step. The areas of resist that remain after development serve to protect the substrate-regions which they cover. Locations from which resist has been removed can be subjected to a variety subtractive or additive processes that transfer the pattern onto the substrate-surface. (The most common of these are *etching* and *ion-implantation*.) An advanced-IC can have 25 or more of such masking-layers. Approximately one-third of the total cost of semiconductor-manufacturing involves lithographic-processing.

    Three chapters of this text are devoted to the details of this critical technology. The first concerns the properties of *photoresist-materials* and the *resist-processing-technology* utilized in ULSI-fabrication. The discussion is restricted to resists exposed by optical (e.g., UV and DUV) radiation. The second deals with the *tools used to expose the resist (optical-aligning-equipment)*. In the third, *photomasks* as well as alternatives to optical-lithography (including *EUV* and *e-beam patterning-technology*) are described. Note that more details about resist-processing can be found in Refs. 1 - 4, listed at the end of the chapter.

18.1 PHOTORESISTS

Photoresists have been used in the printing-industry to make pre-coated lithographic printing-plates for more than a century. In the 1920s, photoresists were introduced to the printed-circuit-board industry. The semiconductor-industry adopted this technique for wafer-fabrication in the 1950s. By 1991 the semiconductor-industry was consuming about 2500- tons of photoresist per year, which represented sales of around $220 million. The selling price of photoresist in the late 1990's was about $900/gallon.

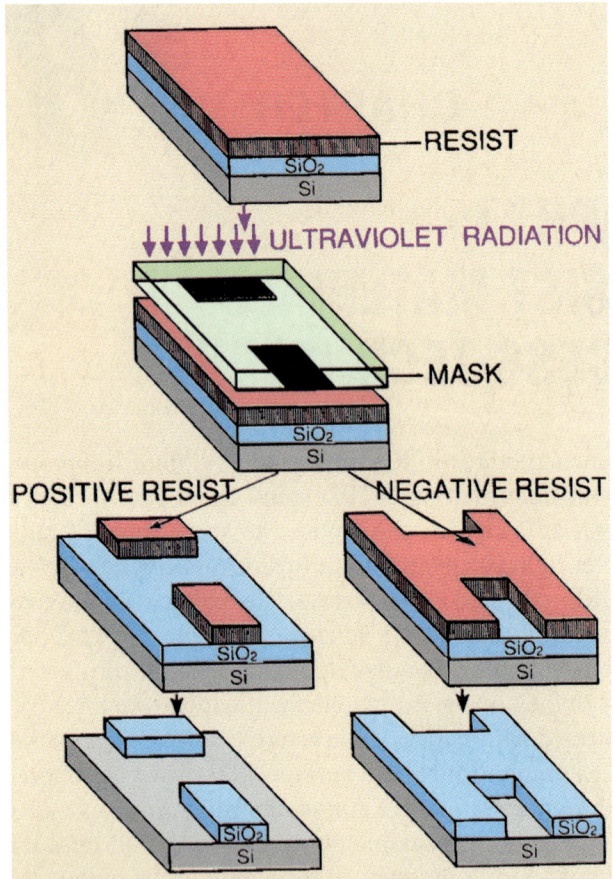

Fig. 18-1 Schematic of the lithographic-process used to transfer patterns from a mask to a thin-film on a wafer. Both *positive* and *negative* resist-behavior is illustrated.

In the early days of the IC-industry there were a large number of resist-suppliers. As the lithography-process matured, the number of resist-vendors consolidated, and by the end of the century a much-smaller number remained. In the U.S., there are two resist-vendors, Olin Microelectronic Materials (who purchased McDermitt, Ciba-Geigy and KTI {Kodak}), and the Shipley Company. Foreign suppliers include Clariant Corp. (formerly AZ Electronic Materials), Tokyo Ohka Kogyo, and JSR Microelectronics.

18.1.1 Basic Photoresist-Terminology

The basic steps of the lithographic-process are shown in Fig.18-1. The *photoresist* (*PR*), or *resist*, is applied as a thin-film onto a substrate (e.g., SiO_2 on Si). It is subsequently exposed through a *mask* (or, in step-and-repeat projection-systems, through a *reticle*). The mask contains *clear* and *opaque* features that define the pattern to be created in the PR-layer. The areas in the PR exposed to the light are made either soluble or insoluble in a specific solvent known as a *developer*. In the case when the *irradiated* (*exposed*) *regions* are soluble, a positive-image of the mask is produced in the resist. Such material is therefore termed a *positive-resist*. If the *non-irradiated regions* are dissolved by the developer, a negative-image results. The resist in this case is termed a *negative-resist*. Following development, regions of SiO_2 no longer covered by resist are removed by *etching*, thereby replicating the mask-pattern in that oxide-layer (Fig. 18-1).

The resist performs two roles in the lithography-process. First, it must respond to the exposing-radiation in such a way that the *mask-image* can be reproduced in the resist. Second, the areas of resist that remain after development, must protect the underlying substrate during subsequent processing. In fact, the second-half of its name (*resist*) evolved from the ability of these materials to "resist" etchants.

Although both negative and positive optical-resists are used to manufacture semiconductor-components, the higher-resolution capabilities of positive-resists made them the exclusive choice for IC-technologies from 2-μm down to about 0.5-μm. Such positive-resists are exposed with *g*-line (436-nm) and *i*-line (365-nm) UV-light. For smaller features than 0.5-μm, both positive and negative deep-UV (DUV) *chemically-amplified* (*CA*) resists have been developed. Conventional positive optical-lithographic processes and DUV-resists are capable of producing images on ULSI-substrates with dimensions smaller than 0.18-μm. For features smaller than about 0.1-μm, however, higher-resolution non-optical techniques may eventually replace optical-lithography. But, the jury is still out on this question.

Conventional optical-photoresists are three-component materials. They consist of: **1)** a *matrix material* (also called the *resin*), which serves as a binder, and establishes the mechanical-properties of the film;

2) an active ingredient, which is a *photoactive-compound* (*PAC*); and 3) a *solvent*. The solvent keeps the resist in a liquid-state until it is applied to the wafer being processed. The *matrix-resin* is usually inert to the incident-imaging-radiation and does not undergo chemical-change upon irradiation. It provides the resist with its adhesion and etch-resistance properties. It also determines other properties of the resist (such as thickness, flexibility, and thermal-flow stability).

The *PAC* is the component of the resist-material-that undergoes a chemical-reaction in response to the actinic-radiation. (The term *actinic* refers to light that drives chemical reactions in the photoresist.) In positive-resists, the PAC also acts as an *inhibitor* in unexposed-regions. That is, it slows down the rate at which the resist will dissolve when placed in a developing-solution. The PAC gives the resist its radiation-absorption properties and developer-resistance. The positive-resists used for g-line and i-line optical-lithography consist of a diazo-naphthoquinone (DNQ) PAC and a novolac-resin, and thus these are referred to as *DNQ/novolac-resists*.[5]

18.2 PHOTORESIST MATERIAL PARAMETERS

As previously mentioned, photoresist performs two primary functions: **1**) precise pattern-formation; and **2**) protection of the substrate during etch or ion-implantation. The material-properties possessed by a resist play a role in how effectively these functions are performed. The material-parameters can be grouped into three categories: **1**) *optical-properties*, including, resolution, photosensitivity, and index-of-refraction; **2**) *mechanical/chemical-properties*, including solids-content, viscosity, adhesion, etch-resistance, thermal-stability, flow-characteristics, and sensitivity to ambient (e.g., oxygen) gases; and **3**) *processing and safety-related-properties,* including cleanliness (particle count), metals-content, process-latitude, shelf-life, flashpoint, and threshold-limit-value (TLV, which is a measure of a chemical's toxicity). More details about these parameters can be found in Refs. 1 and 3.

18.3 OPTICAL PHOTORESIST-MATERIAL-TYPES

18.3.1 Positive Optical-Photoresists

As described earlier, g-line and i-line positive-photoresists for microelectronic-applications are three-component-materials (i.e., matrix, PAC, and solvent).[5] Their material-properties are altered by the photochemical-transformation of the PAC, from that of a *dissolution-inhibitor* to that of a *dissolution-enhancer*. The matrix-component of positive-resists is a low-molecular-weight *novolac-resin* that forms the resist film-properties. The generic term *novolac*

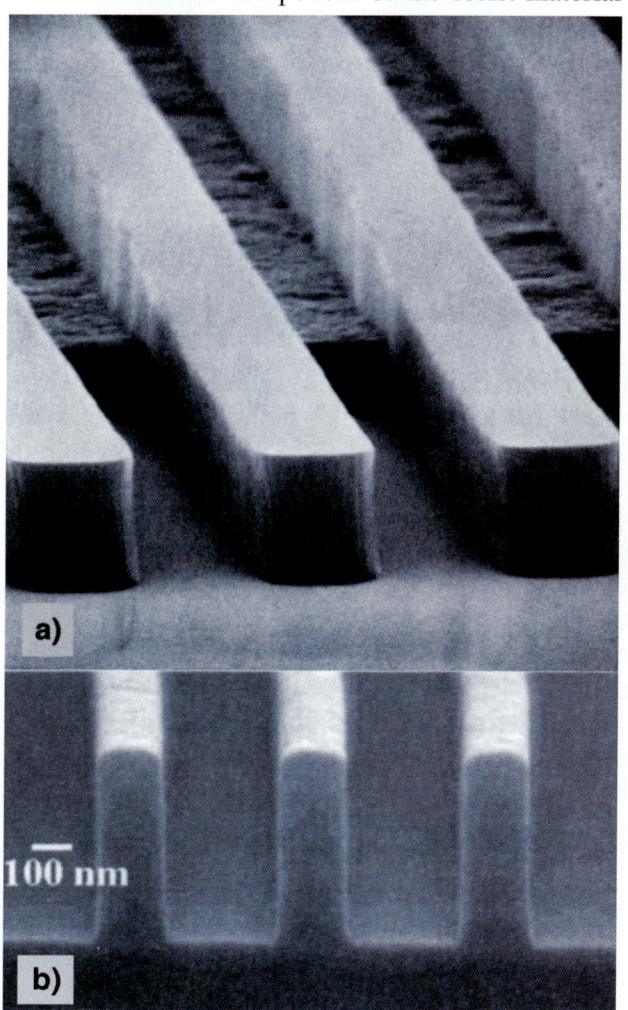

Fig. 18-2 SEM photographs showing examples of lines of photoresist on a wafer-surface after spin-on, exposure, and development. (a) Resist-lines are 1.0-μm wide. (b) Resist-lines are about 0.15-μm wide.

(from *new lacquer*) describes the purpose for which these resins were first developed. The vast-majority of novolac-resins produced are used for such applications as the adhesive in plywood. Only a very-small fraction is used for photoresist-materials. Novolac-resin will dissolve in an aqueous-base. For example, if a film consists only of novolac-resin (i.e., without the PAC that acts to inhibit-dissolution), such an adequately soft-baked film will dissolve in a basic-developer at a rate of ~15-nm/sec.

The *photoactive-compounds* (PACs) in positive-resists are *diazonaphthoquinones*. These substances are photosensitive, but are not soluble in an aqueous-developer-solution. They also, therefore, prevent the novolac-resin from being dissolved by the developer (i.e., their role is that of the *dissolution-inhibitor*). Unexposed positive-resists exhibit developer-attack-rates of only ~0.1-nm/sec. Typical ratios of resin to PAC are ~2-3:1. The *solvent*-systems for such resists are typically *propylene-glycol-monomethyl-ether* PGME, PGMEA (PGME-acetate), or ethyl-lactate.

Figure 18-3 provides a summary of the chemical-compositions and reactions that take place in DNQ/novolac-materials. Upon exposure to light, diazoquinones (which prior to exposure act as dissolution-inhibitors) photochemically-decompose. This leads to molecular-rearrangement and hydrolysis, with the end-product being *indene-carboxylic-acid* (ICA). The ICA-photoproduct is readily-soluble in basic-solutions, and acts as a *dissolution-enhancer*. Hence, in the exposed-regions of the film where complete-photodecomposition of the inhibitor has occurred, the material dissolves in aqueous-developer-solutions at rates equal to (or greater than) materials which consist of novolac-resin alone (e.g., ~100–200-nm/s vs. 15-nm/s). Meanwhile, the unexposed-regions remain much less-soluble in developer (~100 times less-soluble). This *differential-solubility* is the primary-mechanism for image-formation in positive-resists.

Figure 18-4 shows the ultraviolet absorption-spectrum of a typical diazonaphthoquinone (DNQ) and a common novolac-resin. The DNQ exhibits strong-absorbance at the 365-nm, 405-nm (and to a lesser extent the 436-nm) mercury-emission-lines.

Positive-photoresists have become the dominant-resists for *g*-line and *i*-line applications because they offer better-resolution than their negative-resist-counterparts. This capability arises from the fact that the unexposed-film-regions do not become permeated by the developer. As a result they closely retain the size possessed after exposure, even after being immersed in developer-solution. In conventional negative-resists, the developer permeates both the exposed and unexposed regions of the film. In the unexposed-regions such penetration leads to film-dissolution, but even in the exposed-areas (where little-dissolution occurs), the solvent-penetration causes the regions to swell and distort the resist-size. This degrades the resolution-capability of bis-arylazide-based negative-resists. In addition, positive-resists exhibit improved dry-etch resistance and better thermal-stability than do such negative-resists. Although positive-resists are slower than negative-resists (that is, they are less-photosensitive), in stepper-based PR-processes (in which throughput is dictated by alignment, focus, and wafer-handling), the lower-sensitivity does not significantly-reduce throughput in practical-applications.

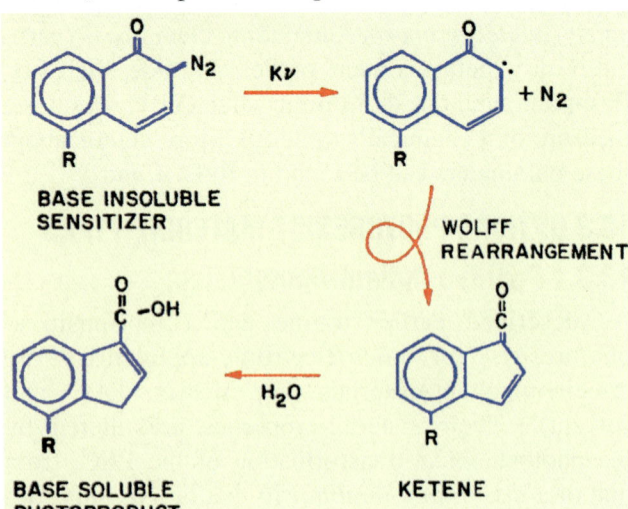

Fig. 18-3 Sequence of photochemical-transformations of the quinonediazide sensitizer (dissolution inhibitor) in DNQ positive-photoresists.

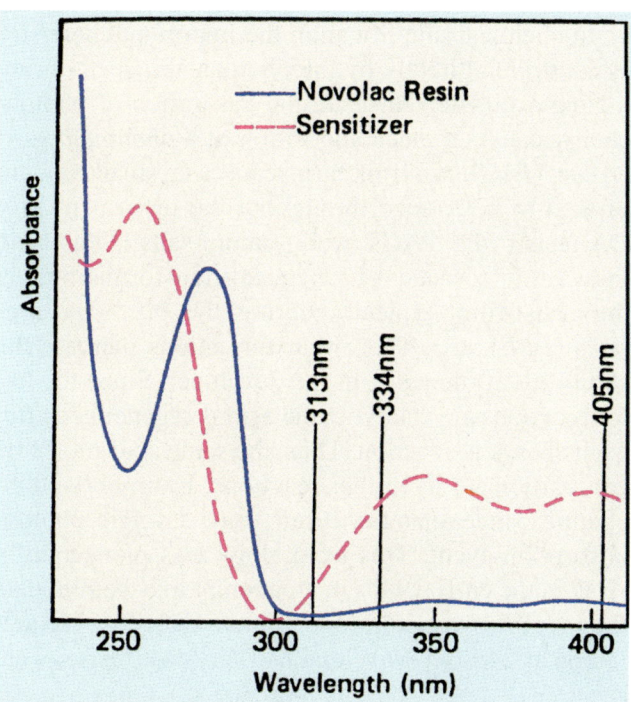

Fig. 18-4 Absorbance spectrum of a typical diazonaphthoquinone (DNQ) sensitizer and novolac film. Wavelengths of the principle mercury-emission-lines are also labeled.

18.3.2 Negative Optical-Photoresists

Negative-photoresist was the workhorse-resist in the early-days of the microelectronics industry. It was only replaced when resolution-requirements began to approach 2-μm. As noted above, the swelling of negative-resists during development makes them unsuitable for critical-dimensions smaller-than about 3-μm. Their early predominance, however, stemmed from several attractive characteristics that they possess for lithographic-applications. These advantages continue to keep negative-resist in use for some lower-resolution lithographic-processes. The advantages of negative- over positive-resist include the following: **1)** they adhere better to some substrate-surfaces; **2)** they exhibit faster-photospeed (which allows greater exposure-throughput - and thus lower-fabrication-costs); **3)** they have somewhat-greater process-latitude in terms of developer-dilutions and temperatures; and **4)** they cost-less (negative-resists are about one-third as expensive as positive-resists). However, another disadvantage is that they require solvent-based developers, and this creates potentially-severe environmental, health, and safety (EHS) problems.

Optical negative-resists, like their positive-counterparts, are three-component-materials (although their chemistry and photoactive-behavior is quite different). The film-forming component is a *cyclized-synthetic-rubber-resin*, which is radiation-insensitive but is also extremely-soluble in non-polar organic-solvents such as toluene and xylene. The photoactive-compound (PAC) is a *bis-arylazide*. A typical chemical-structure of the cyclized-rubber-matrix and a commonly-employed PAC are shown in Fig. 18-5. It should also be noted that prior to being applied to a wafer-surface, the negative-resist is in liquid-form, dissolved in an aromatic-solvent.

Most negative optical-resists function by becoming less-soluble in regions exposed to light. A photochemical-reaction generates a cross-linked, three-dimensional molecular-network that is insoluble in the developer. The transformations associated with the generation of these cross-linked-species is shown in Fig. 18-5. The most-important-reaction is the formation of nitrogen from the excited-state of the arylazide to form an extremely-reactive intermediate-compound called *nitrene*. Nitrene can undergo various-reactions which result in the formation of more stable-molecules (and in several cases, generation of polymer-polymer linkage, or *cross-linkage*).

18.3.3 Chemically-Amplified Deep-UV-Resists

As device-dimensions kept scaling to smaller feature-sizes, advances were made in both the DNQ/novolac-resists and in the lithography-process. These improvements allowed such resists to continue to be used to fabricate ever-smaller device-patterns. Specifically, by using *i*-line-illumination to replace *g*-line, by using larger-NA *i*-line lenses in the steppers, and by improving the novolac-resin and DNQ materials, the resolution of *i*-line-lithography kept increasing. Eventually, it became capable of printing the features needed for 0.35-μm CMOS-technology.

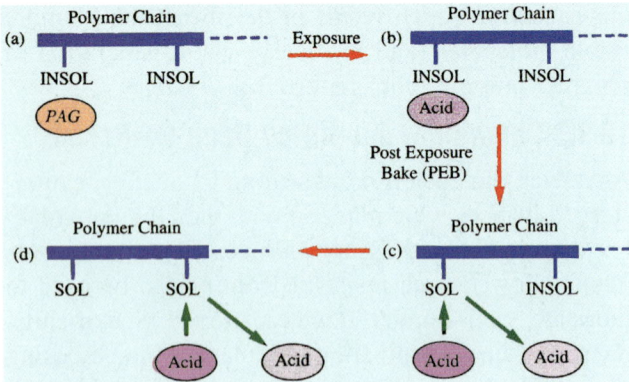

Fig. 18-5 Photochemical transformations of a *bisazide sensitizer* in negative-photoresists based on cyclized polyisoprene.

However, when the 0.25-μm-generation was reached (used for making 256-Mbit-DRAMs), it became clear that *i*–line-lithography could no-longer print 0.25-μm features (even when using such resolution-enhancement-techniques as phase-shift-masks and off-axis-illumination - see Chaps. 19 and 20). A shift from *i*–line to deep-UV-illumination had to be made. Since both novolac-resin and DNQ strongly-absorb UV-light at around 250-nm, DNQ/novolac-resists cannot operate properly in the DUV-region. Thus, a new resist-chemistry was needed, based on *chemical-amplification*, *CA* (see Fig. 18-6). This offers increased photospeed and contrast. Since CA-resists were compatible with DUV-illumination, they became the resist-materials of choice for advancing optical-lithography toward higher-resolution.[6]

In chemical-amplification the important chemistry is not driven directly by the photons absorbed during the exposure-step. Instead, the absorption of photon-energy causes the decomposition of a *photoacid-generator* (*PAG*). This in turn, causes a small-amount of acid to be formed throughout the resist. (In early CA-resists the PAGs were onium-salts.) The acid induces a cascade of chemical-transformations in the resist-film, typically during the *post-exposure-bake* (*PEB*) step. These transformations increase the solubility of the resist in the developer. Since the latter-reactions are catalytic, the acid is regenerated after each chemical-reaction. Thus, the same acid-molecule can participate in further reactions. From 500–1000-chemical-reactions can result from a single photon-absorption-event. This technology was pioneered by IBM in the early 1980s in an attempt to accommodate the low-intensity of illumination produced by Hg-arc-lamps at 248-nm-wavelengths. Very-fast resists were

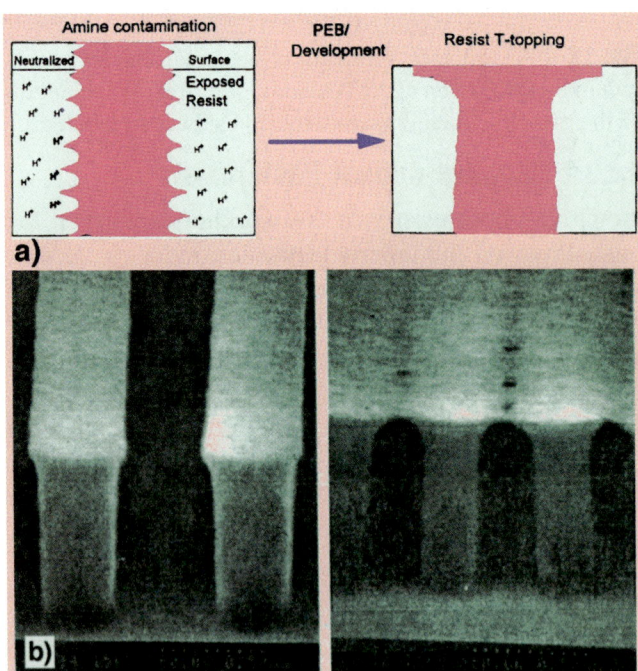

Fig. 18-6 Basic operation of a *chemically-amplified* (CA) resist. *PAG* is the photo-acid generator: *INSOL* and *SOL* are the *insoluble* and *soluble* portions of the polymer-base. Steps (c) and (d) may repeat tens or hundreds of times during the post-exposure bake (PEB).

Fig. 18-7 (a) Environmental-amine contamination of a chemically-amplified resist, resulting in "T-top" formation after development. (b) Photographs of "T-top" formation, courtesy of Moishe Preil.

needed, leading to development of the CA-reactions.

The biggest-advantage of CA-resists is their high-sensitivity (fast-photospeed, 10–50 mJ/cm^2), and they also exhibit a much-higher contrast than do the DNQ/novolac-resists. However, the catalytic-nature of the CA-process makes reproducibility a concern. If the catalytic-chain of reactions is somehow interrupted, many of the lithographically-important reactions will not occur, causing catastrophic resist-failure.

The most-common mechanism that quenches the catalytic-reactions occurs as a result of "environmental-contamination." Specifically, the culprits are trace-amounts (on the order of 10 ppb) of vapors of volatile-base-compounds (most commonly amines, ammonia, or N-methylpyrrolidone [NMP]). If the ambient contains any of these contaminant-vapors, the top-surface of the resist will absorb them during the time that elapses between exposure and post-exposure-bake (PEB). These base-molecules will then react with (and neutralize) the photo-generated-acid. The surface-region of the resist now lacks acid, and thus becomes less-soluble than the acid-containing-regions below. This less-soluble skin develops more-slowly, resulting in the formation of a T-top-structure (Fig. 18-7). Since the extent of such skin-formation is not controlled, this causes linewidth-variations from wafer-to-wafer. In addition to airborne-contamination, the acids also decay with time. Thus, any delay between exposure and PEB worsens the linewidth-control-problem.

The environmental-contamination problem is quite serious, as trace-quantities of basic (i.e., as opposed to acidic) vapors always exist in the semiconductor-fabrication ambient.[7] Several methods are used to combat this problem. The first involves filtration of the cleanroom-air with activated-charcoal. This has become standard-practice in the industry. Enclosing the photo-islands (where the resist-coating, exposure, and development-steps are performed), and purifying the air that enters with charcoal-filtering is even more effective (see Fig. 18-8).

CA-DUV-resists are expensive. Their cost in 1998 was $1500–$2000 per gallon, about 2–3-times-higher than DNQ/novolac-resists. Future CA-DUV-resist-formulations are predicted to be even more costly.

18.4 PHOTORESIST PROCESSING

The basic sequence of steps that comprise a complete photoresist-process is shown in Fig. 18-9. The remaining sections of this chapter describe these steps in detail. A few steps are covered in other chapters, including *Wafer Cleaning* (Chap. 8) and *Etching* (Chaps. 21 and 22). The steps of a complete resist-process are not independent of one another. That is, in specifying one step, several other steps will be impacted. As a result, establishing a resist-process requires substantial effort.

18 4.1 Resist Processing: Dehydration Baking and Priming

Good photoresist-adhesion to the wafer surface is critical for an effective lithographic-process. Poor

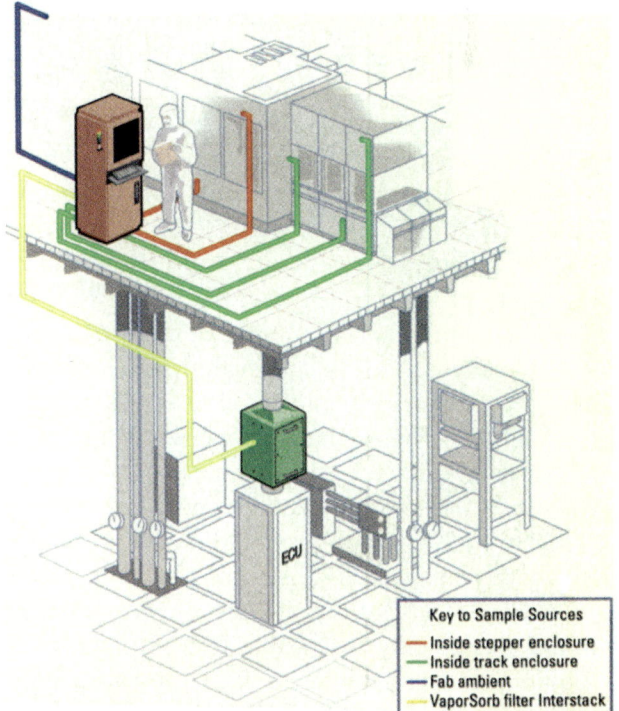

Fig. 18-8 Schematic of filtration-units connected to photo-island clusters to avoid amine-contamination of CA-resists. Courtesy of Extraction Systems, Inc.

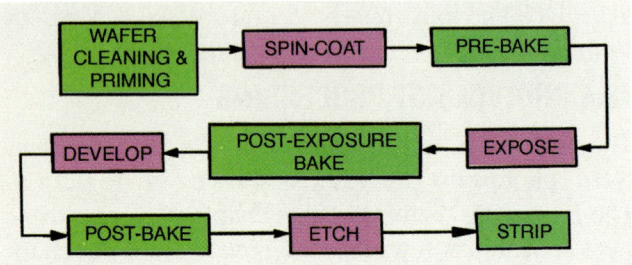

Fig. 18-9 Flow-chart of a typical resist-process.

adhesion is often caused by de-wetting of the resist-film, a phenomenon that arises when resist is applied to silicon-dioxide surfaces. Oxide is a *hydrophyllic* (water-attracting) surface, while resist is a *hydrophobic* (water-repelling) material. Since moisture from the atmosphere can be rapidly absorbed by oxides on a wafer surface, and such "water-containing" oxides have been shown to reduce adhesion (since no wetting will occurs with a hydrophobic-resist on a hydrophyllic-surface). Thus, a *dehydration-bake step* to evaporate this moisture is performed before priming and spin-coating a wafer with resist.

Following such a moisture-removal bake, a wafer is *primed* with a pre-resist-coating of a material designed to further-improve adhesion. The most widely used kind is *hexamethyldisilazane* (HMDS). The HMDS first reacts with any water remaining on the oxide to produce a *dehydrated-surface* (Fig. 18-10a). Next, additional HMDS (in the presence of heat) reacts with oxygen to form a trimethylsilyl ($Si[CH_3]_3$) oxide-species which chemically-binds to the surface (Fig. 18-10b). These reactions continue until the entire surface is covered with this species. In this way, HMDS serves as a surface-linking *adhesion-promoter* (e.g., SiO_2-surface to resist-surface linkage). Another such recently-introduced substance is *tri-methyl-silyl-di-ethyl-amine* (TMSDEA). Wafers should be coated with resist as quickly as possible after priming.

Equipment that combines vacuum-dehydration-baking of the wafers in the same chamber, prior to the introduction of HMDS-vapors, is also commercially-available. Such systems offer the opportunity to prime the wafers after vacuum dehydration-baking and without subsequently having to expose the wafers to atmospheric-moisture. The wafer comes out of the vapor-priming chamber at an elevated temperature. To cool it at a controlled-rate back down to a pre-determined ambient-temperature, the wafer is immediately loaded onto a *chill-plate*.

18.4.2 Resist Processing: Spin-Coating

Following cleaning, dehydration-baking, and priming, the wafers are ready to be coated with photoresist. The goal of the coating-step is to produce a uniform, adherent, defect-free polymeric-film of desired thickness over the entire wafer (0.5-1.5-μm-thick). *Spin-coating* is the standard method used to apply such films. This procedure is carried out by dispensing the resist-solution onto the wafer-surface, and then rapidly spinning the wafer until the resist is almost dry. In order to maintain reproducible linewidth, *resist-film uniformity* across the wafer (and from wafer-to-wafer) is required to be less than ±1.0-nm in 0.25-μm lithography.

The spin-coating procedure involves three stages: **1)** dispensing the resist-solution onto the wafer (which is held on a vacuum-chuck, Fig. 18-11); **2)** accelerating the wafer to the final rotational-speed; and **3)** spinning at constant-speed to establish the desired thickness and to dry the film (Fig. 18-12). Speeds of 3000-7000-rpm for 30-60-seconds are used. The final thickness depends on viscosity, and is inversely-pro-

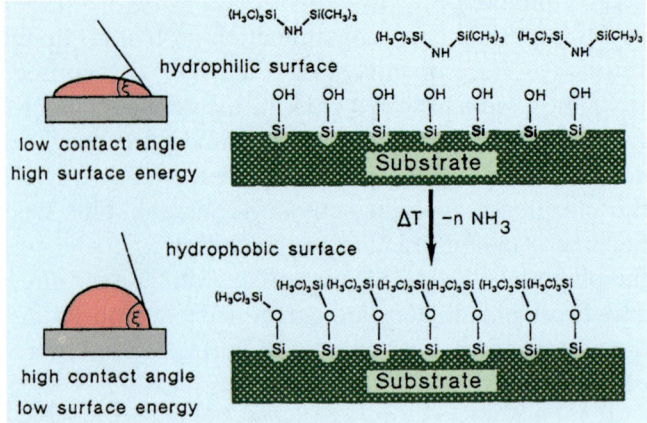

Fig. 18-10 HMDS bonding-mechanism with SiO_2.

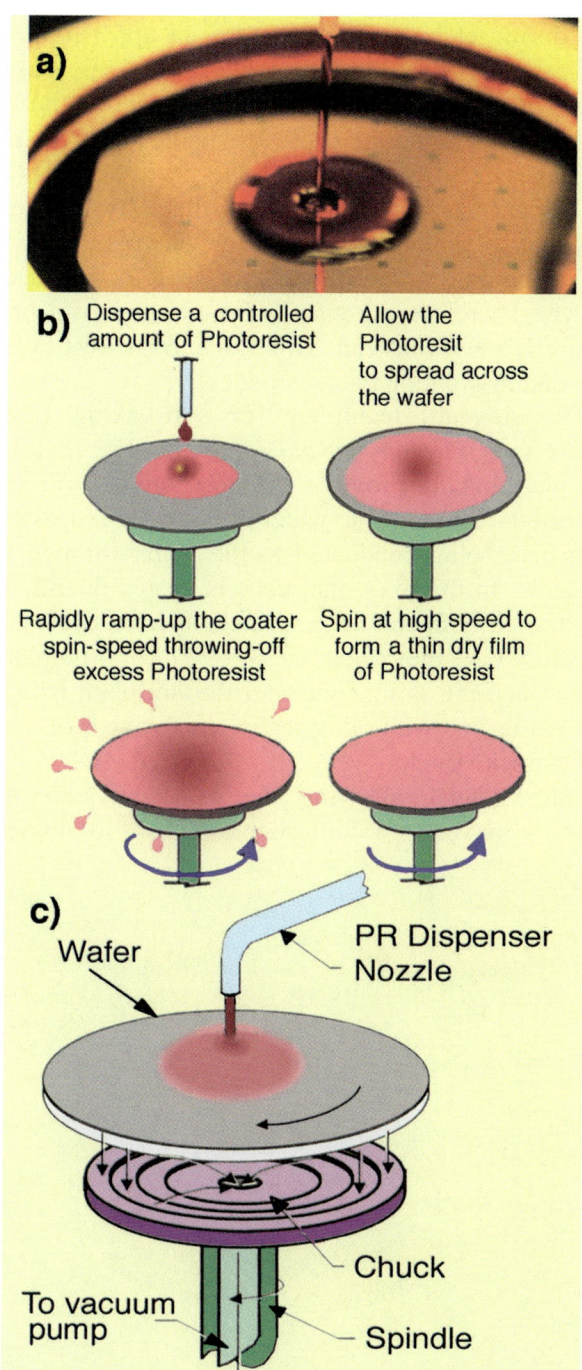

Fig. 18-11 (a) Photograph of resist being dispensed onto a wafer-surface. (b) Schematic of the resist-spinning process. (c) Wafer on a vacuum-chuck with resist being dispensed.

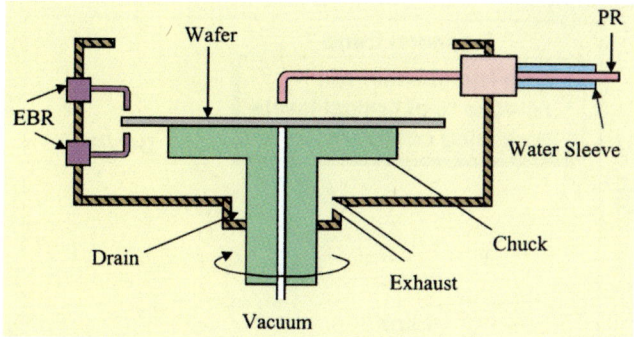

Fig. 18-12 Schematic of a photoresist spin-bowl.[11] Reprinted with permission.

portional to the square of spin-speed (Fig. 18-13).

The volume of resist dispensed for *g*-line resists is in the range 4-5-cm^3 per wafer. For *i*-line resists and 200-mm wafers, the dispensed volume was reduced to about 3.5-cm^3. With the advent of DUV, the cost of resists jumped dramatically (to $1500–2000/gallon, see Fig. 18-14). This spurred efforts to develop techniques that use a dispensed-resist-volume of only 2-cm^3 for 200-mm wafers. This cuts the resist-cost per wafer, and the burden of waste-disposal is reduced.

At completion of the resist-spinning step, a bead of resist is built up on the edge of the wafer, and it must be removed. If not removed, mechanical-handlers (such as robot-fingers) will crack the hardened, brittle resist and cause particulate-contamination. A chemical spray on the edge of the wafer removes this *edge bead* (*edge-bead removal* or *EBR,* see Fig. 18-15a).

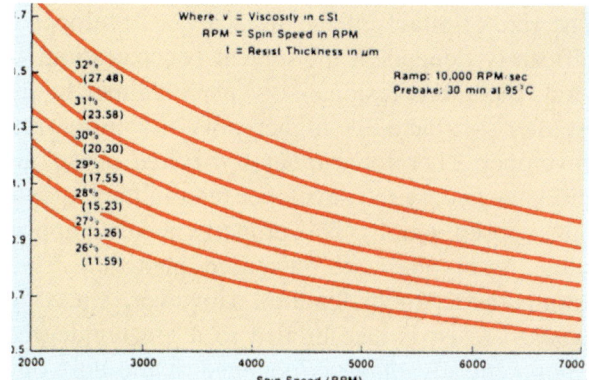

Fig. 18-13 Example curves of resist-thickness after dispense and spinning, as a function of spin-speed.

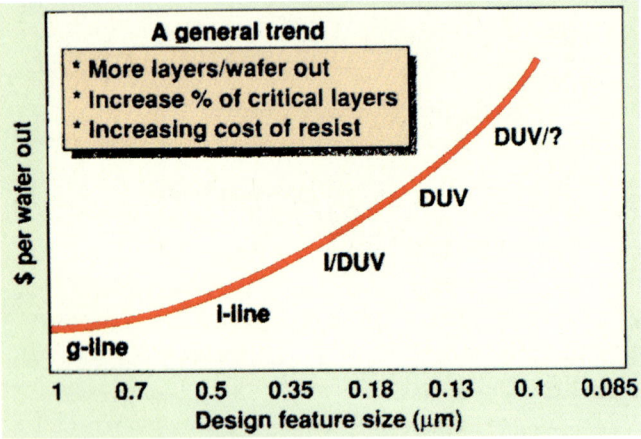

Fig. 18-14 Costs associated with photoresist are expected to rise dramatically as the industry pushes to smaller geometries. Courtesy of Semiconductor International.[12]

18.4.3 Resist Processing: Soft-Bake

After the wafers are coated with resist, they are subjected to a baking step called *soft-bake* (or *pre-bake*, or *post-apply bake*). Solvent is driven out of the spun-on resist, reducing its concentration in the film to about 5%, and the liquid-cast film is converted into solid form. (Note that in its liquid, pre-cast form, resist contains 65–85% solvent. After it is spin-cast, the solvent-content drops to 10–30%. But in this pre-soft-bake, spun-on state the resist is still "tacky," and is highly-susceptible to particulate contamination.) As a result of solvent-loss during soft-bake, the thickness of the resist is also reduced by about 10–20%.

The rate-of-attack of the resist by a developer is significantly dependent on solvent-concentration. In general, the more residual-solvent contained in the resist after soft-bake, the higher is its dissolution-rate in developer. Therefore, *under-soft-baked* resists are readily attacked by the developer in both the exposed and unexposed regions. This property makes it appear that the resist possesses increased photosensitivity (normally a desirable feature). However, since the unexposed-resist is also eroded to a greater degree, this apparent advantage is obtained at the cost of a thinner patterned-resist-layer. Thinner resist-films can lead to higher pinhole levels, or decreased protection during etching. There is also an upper-limit to soft-bake temperatures. Excessive soft-baking causes some (or all) of the PAC to undergo reaction during the bake, rendering the resist less-photosensitive during exposure. It is not desirable to try to drive all of the solvent from the resist during the soft-bake step. Residual-solvent is needed in DNQ/novolac films to aid in the conversion of the PAC to ICA (see Sect. 18.3.1). Therefore, a *controlled* solvent-removal-procedure is sought instead. This leads to the best exposure and dissolution characteristics.

The standard technique for soft-baking ULSI resists uses *vacuum-hot-plate baking* (Fig. 8-15b). Hot-plate baking provides the best temperature and uniformity control. One wafer at a time is processed, with heat being conducted to the wafer through its backside. In this way, the resist is heated quickly to the desired temperature, and the soft-bake cycle can be short (1-2-min/wafer at 100-120°C). Automatic wafer-handling is utilized, and the hot-plate is usually put in-line with other resist-processing tools in a wafer-track system.

Immediately following the soft-bake step, the wafer is moved to a chill-plate to cool it to ambient

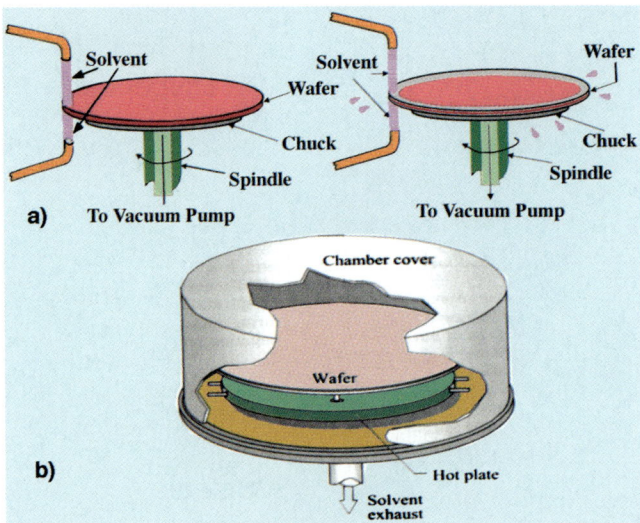

Fig. 18-15 (a) Chemical edge-bead removal.[11] Reprinted with permission. (b) Schematic of a hot-plate for soft-baking resist films.[13] Reprinted with permission.

temperature at a controlled-rate prior to exposure. The issues of temperature-control and uniformity become even more important in CA-DUV resists, mandating the use of hot-plate soft-baking for these materials. Temperature-control specifications for hot-plate baking in 0.25-μm lithography are in the ±0.1°C-accuracy-range (especially for post-exposure baking).

18.4.4 Resist Processing: Exposure

After a wafer has been coated with resist and suitably soft-baked, it is ready to be exposed to some form of radiation in order to create a *latent-image* in the resist. The degree of exposure is adjusted by controlling the energy impinging on the resist (a product of the *intensity* of the source and the *time-of-exposure*). An *energy-integrator* is used to detect the total-energy striking a unit-area of resist, and it automatically adjusts the time-of-exposure to compensate for any aging-variations in the light-source.

Photochemical-transformations occur within the resist during exposure. The goal of an optimized exposure-process is to produce the desired photochemical-effects in the shortest period of time (and in a highly reproducible manner). Successfully optimizing such exposure-steps is a complex task because it depends on other steps of the resist-processing sequence. In this section we consider the standing-waves in a resist that arise during exposure, and on anti-reflection coatings. The hardware used to perform the exposure-step is the subject of Chap. 19.

Standing-Waves: Standing-waves are caused when light-waves propagate through a resist-film down to the substrate, and are then reflected back up through the resist. The reflected-wave *constructively* or *destructively* interferes with the incident-wave and this creates zones of *high* and *low* exposure. This phenomenon causes two unwanted effects in resist-layers. First, the periodic-variation of light-intensity in the resist (the *standing-waves*) results in the resist receiving non-uniform doses of energy throughout the film-thickness (Fig. 18-16a). The resulting latent-image produced in the resist due to this phenomenon is shown in Fig. 18-16b. (Note that the distance between the depths of peak-exposure in the resist is on the order of ten's of nm, about one-tenth the value of the wavelength of the incident-light.) Second, it creates *linewidth-variations* as the resist crosses steps, which is due to the variation of energy coupled to the resist by interference-effects at different resist-thicknesses. Each of these effects contributes to resolution-loss, and they are significant for submicron-resolution.

Another problem is caused by off-normal light-rays reflecting from the substrate and exposing regions of the resist intended to be covered by a mask (Fig. 18-17). Highly-reflective substrates accentuate such off-normal reflection-effects (and standing-wave-effects). Because these effects give rise to significant CD-variations, a number of techniques have been developed to suppress them, including the use of *dyes, anti-reflective-coatings* (*ARCs*) below or above the resist layer, and *post-exposure-bakes* (*PEB*). The latter approach involves using a bake-step between exposure and development. The use of the above techniques has become essential to allow optical-lithography to be used for features smaller than 0.5-μm.

Anti-Reflective-Coatings (ARCs): Methods that reduce reflectivity at resist-interfaces can provide linewidth control with minimal-loss of resist-performance. For example, *anti-reflective-coatings* (*ARCs*) can be used to dampen reflections. The reflections are suppressed either by attenuating light that passes through the ARC, or by matching the index-of-refraction of the ARC to the resist-system at the exposure-wavelength

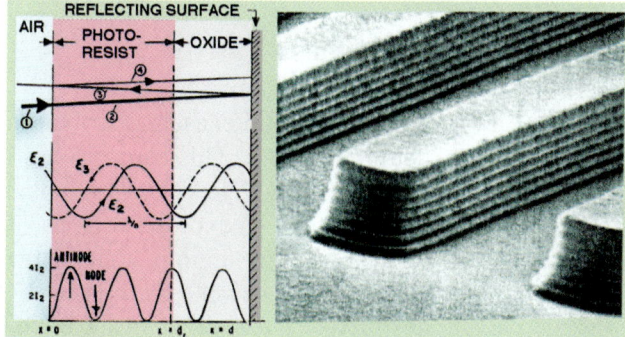

Fig. 18-16 (a) Standing light-waves in a resist-film caused by interference between incident and reflected light. (b) Photograph of *standing-wave-effects* in exposed resist.

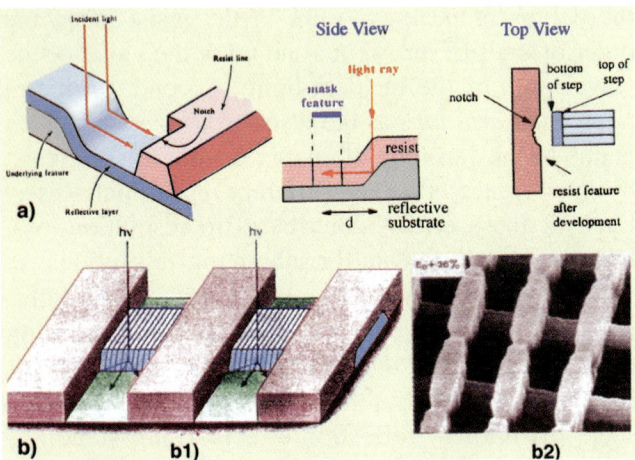

Fig. 18-17 Two cases of reflective-notching: (a) A step is perpendicular to the resist-line in which it causes a *notch*. (b) Lines run over steps (b1) and scattered-light may cause notching, as shown in (b2).

being employed (such that destructive-interference of the light passing through the ARC occurs). These layers can be deposited either on-top of resist-layers (forming a *Top-AR* coating or *TAR*), or on the surface of the wafer beneath the resist (to form a *Bottom-AR Coating* or *BARC* - see Fig. 18-18). For 0.18-μm lithography and below, BARCs must be able to suppress more than 99% of the substrate-reflected light to provide sufficient CD-control. Although it significantly increases process-complexity, use of *both* a BARC and a TAR have been reported for layers that require the tightest CD-control (i.e., poly-gate patterning). More details on ARCs are found in Ref. 1.

18.4.5 Resist Processing: Post-Exposure-Bake (PEB)

Standing-waves in the resist-film cause a distribution of light-intensity within the film. This causes a non-uniform distribution of the PAC in a DNQ/novolac positive-resist after exposure. Unexposed-regions of the resist contain PAC, exposed-regions contain the photo-product. The boundary between them is set by constructive and destructive interference of the exposure standing-wave.

A post-exposure (and pre-develop) bake-step (*PEB*) subjects the resist to a temperature 5-15°C higher than the soft-bake step. This causes the PAC from the unexposed-regions to diffuse through the resist and produce an averaging-effect across the exposed/unexposed boundary. The PEB time-and-temperature are selected to cause diffusion of the PAC to eliminate the ridges in the resist sidewall-profile caused by the standing-waves (Fig. 18-19). The diffusion-length of the PAC during the PEB must be greater than one-half the standing-wave period, but much less than the feature-size.

In DUV-CA, resists the PEB-step plays a different role. Here the PEB is used to accomplish the image-formation by producing the catalytic-acid-reactions that increase the solubility of the resist in the developer. In essence, the PEB-step thus produces the

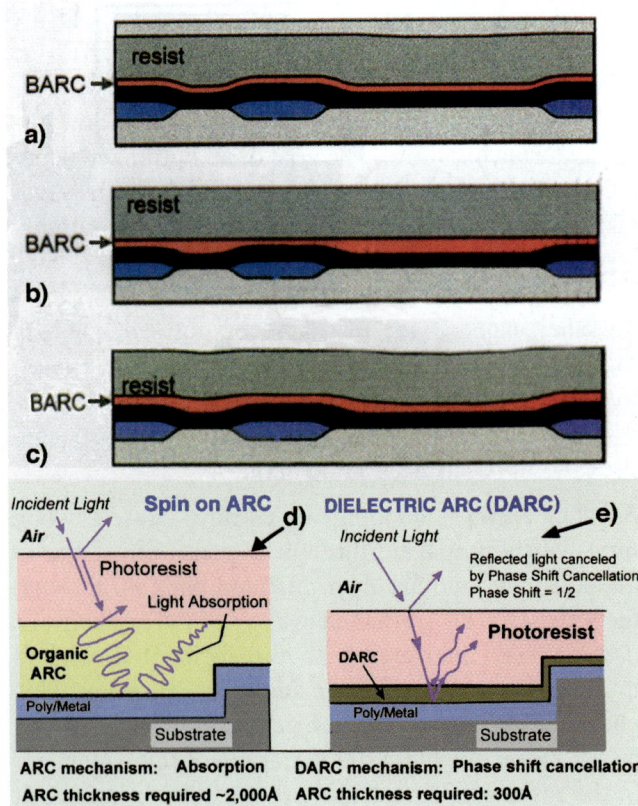

Fig. 18-18 (a) Conformal-BARC. (b) Planarizing-BARC. (c) Partially-planarizing-BARC. The principle of reflectivity-control that occurs in (d) *Spin-on* [planarizing] BARCs, and (e) *CVD-inorganic* (conformal) BARCs.

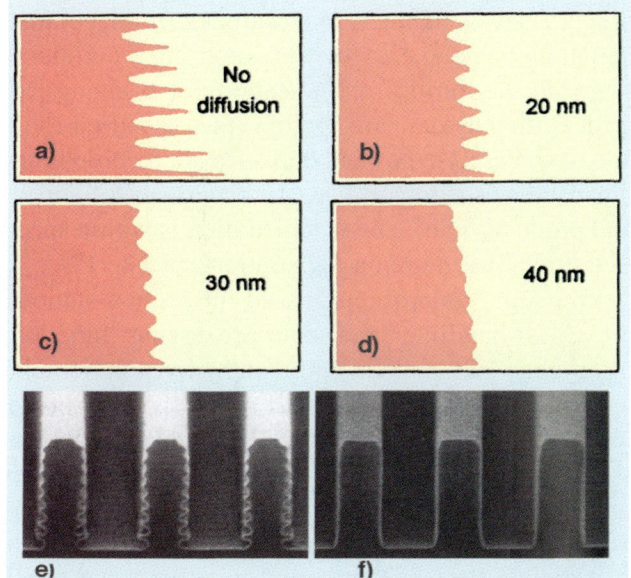

Fig. 18-19 (a)-(d) Simulated standing-wave-pattern in developed resist as a function of increasing PAC diffusion-length. SEM of resist images of 0.35-μm, i-line resist: (e) without; and (f) with, PEB.

latent-image in the resist. Therefore, it is as crucial as the exposure-step in terms of controlling the critical-dimension of the resist (CD). The PEB-step for DUV-CA resists must be tightly controlled, both in terms of the temperature and the duration of the baking.

18.4.6 Resist Processing: Development

Following exposure and post-exposure-baking, the resist-film must undergo *development* to turn the *latent-image* in the resist into the *final resist-image*. The resist-image which results after development serves as the mask in subsequent etching or ion-implantation steps.

The goals of an effective development-process include: **a)** the original film-thickness of unexposed (positive) resist should not be measurably reduced during development; **b)** the development-time should be short; **c)** development should cause minimum pattern-distortion or swelling; and **d)** specified pattern-dimensions should be precisely produced. After development, it is necessary to rinse and dry the substrate because the developing-action continues until the developer is completely removed.

It should be noted that *positive-resist-developers* are alkaline-solutions diluted with *water*. As a result, they have the advantage of requiring only a water rinse, while *negative-developers* for bis-arylazide-based resists are *organic-solvents*, and must be rinsed in other *organic-solvents* (e.g., n-butyl acetate).

There are three methods by which development is carried out: immersion, spray, and puddle developing. In *immersion-developing*, cassette-loaded wafers are batch-immersed and agitated in a bath of developer at a specific temperature and time, often using a mechanical agitating-arm. The advantages of immersion-development are high-throughput, good uniformity of development, low capital-equipment costs, and small clean-room footprint by the development-equipment. However, since batch, immersion-development processes are not compatible with the in-line wafers-tracks that now dominate production, this process has been phased out.

Spray-development processes can be carried out in *batch* or *single-wafer* modes, but the latter are now the standard. In *single-wafer spray-development systems*, the developer is sprayed onto a spinning-wafer, and each wafer is treated with a fresh dose of developing-solution (Fig. 18-20). The wafer is then rinsed, spun-dry, and transported to the hard-bake module.

In the *puddle-single-wafer technique*, a fixed amount of developer is first dispensed onto a static wafer. After the required develop-dwell-time, the developing-action is stopped by directing a stream of deionized-water onto the developed-wafer. A spin-dry step again follows the rinse-step. By the late 1990's, combination-development-processes also became popular, including such development-sequences as *spray-puddle*, and *puddle-spray-puddle*.

Most conventional-developers for positive-resists originally contained metal-ions, but concern about the possible contamination of the wafers from the developer residues led to the development of *metal-ion-free* developers (with less than 15-ppma of total metals). The metal-ion-free developer that is currently in widest use is tetramethyl-ammonium hydroxide (TMAH).

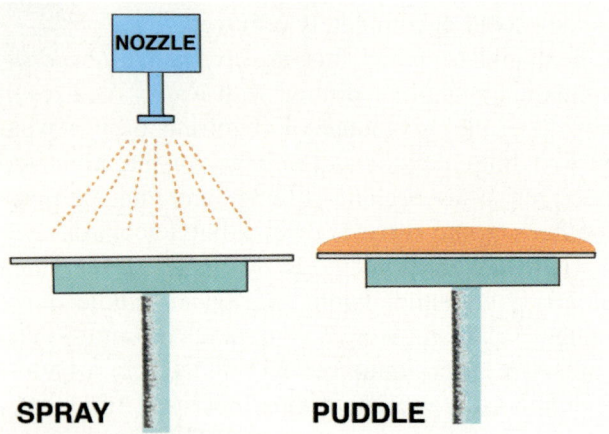

Fig. 18-20 Single-wafer resist-development methods: (a) spray; and (b) puddle.

TMAH developer-concentrations in the range of 0.2 to 0.3-N allow sufficient sensitivity with high-contrast and minimum erosion. A 0.26-N solution is becoming an industry-standard. *Surfactants* (which reduce surface tension), are also usually added to the TMAH to decrease development-time and scumming. Techniques to automatically control the development-process on an in-line basis are available.

18.4.7 Resist Processing: After-Develop Inspection

Following development, an inspection (sometimes referred to as an *after-develop-inspection*, or ADI) is performed. The purpose is to insure that the steps of the PR-process up to this point have been performed correctly, and within specified tolerances. Mistakes or unacceptable process-variations can still be corrected, since the resist-process has not yet produced any changes (e.g., through an etch-step) to the wafer itself. Thus, it is still possible to strip and rework any inadequately-processed wafers (known as *rejects*) detected by this inspection.

The wafers at ADI are typically inspected with an optical-microscope, SEM, or laser-based system. Some of the aspects of the resist-process that are monitored by the inspection-procedure are: **a**) the correct mask has been used; **b**) resist film-qualities are acceptable (i.e., the resist is free from contamination, scratches, bubbles, or striations); **c**) image-quality is adequate (i.e., look for good edge-definition, linewidth-uniformity or indications of bridging); **d**) critical-dimensions are within the specified-tolerances; **e**) defect-types and densities are recorded (and this data is used to correlate the occurrence of defects and product-yield): and **f**) registration is within specified limits (for more on registration, see Ch. 19).

On some microscopy-based inspection-stations, all wafer-handling and data-processing functions have been automated. Only human-vision remains as a non-automated aspect of the inspection-procedure. Wafers are transported by belts or vacuum-shuttle from an input-cassette to a pre-aligner, then onto an inspection-stage under the microscope (Fig. 18-21). Automatic handling allows the operator to concentrate on inspection, and minimizes the likeli-

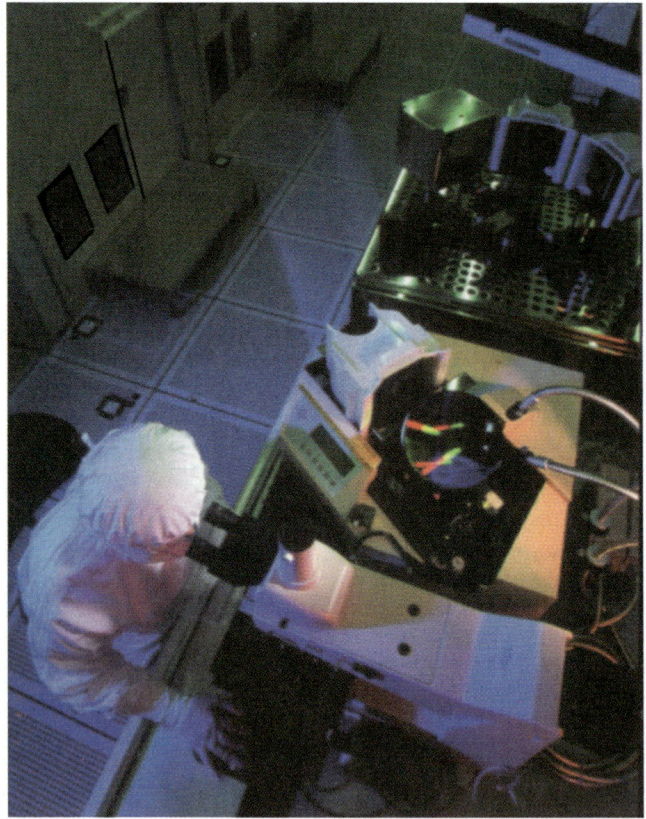

Fig. 18-21 Photograph of a wafer-inspection-station which utilizes an optical microscope.

hood of airborne or human-handling contamination. Inspection-data is entered with a keypad, and many stations include host-computer-interfacing capabilities for processing and storing this data.

In more-automated systems, the human-operator is completely removed from the defect-inspection task. *In-process, wafer-inspection-systems* (based on automatic image-processing), have been introduced. Defect-detection is accomplished either by die-to-die, or die-to-database comparison. Manufacturers of these systems claim defect-detection sensitivities well into the sub-micron range. Such instruments, however, often have difficulty detecting particles on substrates with surface granularity, or on wafers containing surface-topography. In addition, for particles near the minimum-size detection-limit, such machines can miss the presence of some particles, and signal the detection of others that may be non-existent.

Linewidth-Variation and Control: There are two aspects of feature-sizes that must be controlled in the lithographic/etching process: **1)** the absolute-size of a minimum-feature, including linewidth, spacing, or contact-dimensions (also referred to as a *critical dimension*, or *CD*); and **2)** the variations of the minimum-feature-sizes as they cross steps on wafer-surfaces. Linewidth (and spacing) measurements are regularly performed to determine the actual sizes of CDs at each masking-level of a process. The variation of linewidths over steps is also monitored.

Linewidth-control is impacted by a variety of factors, depending on hardware, processes, and materials. The degree of control is determined by measuring a series of test-structures with known feature-sizes across a wafer, and then plotting the feature-dimension as a function of location on the wafer. The standard-deviation at the *one-sigma* and *two-sigma* level then becomes a measure of the linewidth-control capability of a particular exposure/resist process. For example, in a process that has minimum linewidths and spaces of 0.35-μm, the controllability-specification might be a maximum 10% linewidth-variation, with 95% (2σ) confidence. Such data are then plotted as a function of time (Fig. 18-22) and are utilized to monitor the performance of a lithographic-line.

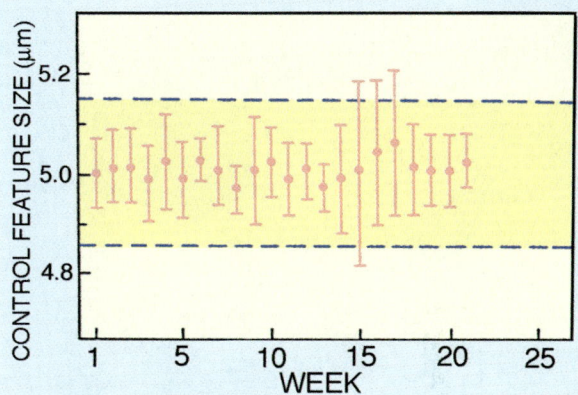

Fig. 18-22 Linewidth-control-data for a typical process-line. Weekly averages are shown for a 5.0-μm control-feature. The dashed-lines represent limits. Note weeks 15-17 represent a processing problem that was corrected in week 18.

Linewidth-Measurements: The width of features produced on wafers are measured in many phases of the fabrication-process, including: **a)** after development; **b)** after etching; and **c)** during photomask-production. A number of techniques are currently utilized to perform such feature-size measurements. For ULSI applications, the measurement-technique must be repeatable to less than 0.1-μm in order to verify that the ±10% size-tolerance specification (cited as an example in the previous section) is satisfied during fabrication. In fact, it is argued that the measurement uncertainty of the metrology-system must even be smaller. That is, in a 1.0-μm-wide polysilicon-line, if the uncertainty in the etching-process is ± 0.03-μm, then it must be necessary to determine that its photoresist-linewidth is 1.0-μm ± 0.07-μm.

Current optical-techniques based on ordinary microscopes are satisfactory for use with 1.0-μm (or larger) feature-sizes. These systems will continue to find wide applicability because of their low cost, ease of use, and high-throughput. Future development will be directed at improving these attributes, while maintaining the required accuracy and precision. For smaller geometries, SEM and laser-scanning techniques are more frequently employed.

Measurement of linewidth by optical-techniques is

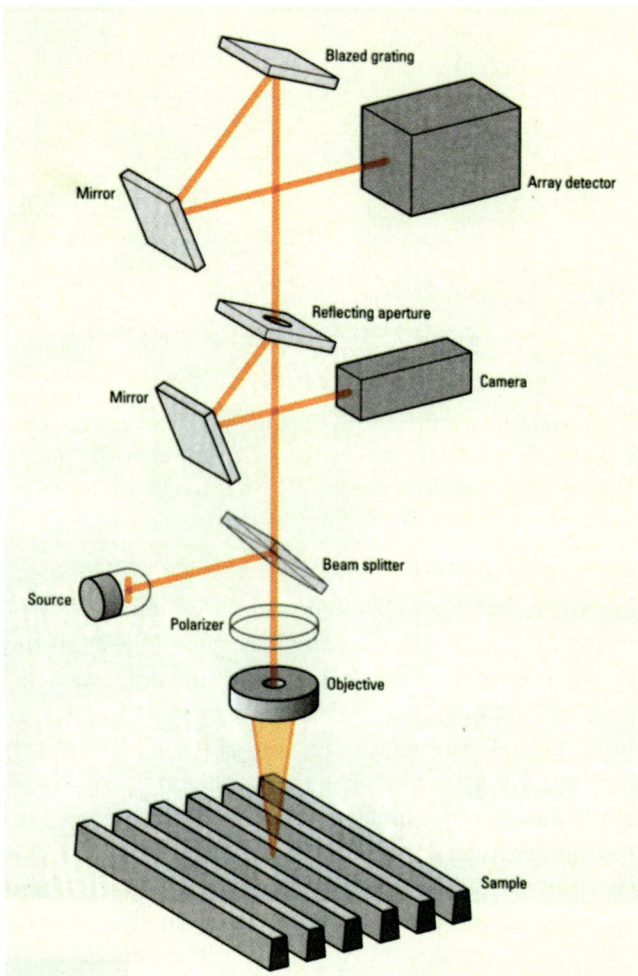

Fig. 18-23 The principle of the *laser-scanning* measurement technique is shown. Courtesy of Nanometrics.

accomplished with the following types of systems: **a)** mechanically scanning an optical-slit across the magnified-feature-image; **b)** video-scanning across the feature of interest; **c)** image-shearing; and **d)** scanning a laser-spot across the feature, and detecting the reflected-image (Fig. 18-23). In the scanning-slit technique, the light passing through the slit is measured by a photomultiplier-tube (PMT) to form a micro-densitometer profile, which is then used to perform the linewidth-measurements. The operator views the image through a binocular-microscope or color cathode-ray tube. The narrow slit (200-nm wide) is moved across the image (e.g., in 25-nm increments), and the intensity profile is acquired via the PMT. The profile is analyzed by an edge-sensing algorithm to determine the dimension. An auto-focus algorithm can be used to find the "best" focus, by moving the motorized-stage in the z-axis until the maximum-slope of the selected edge is located (or, in some systems, by use of an independent laser-focus technique). In video-based systems, a video-camera is used to capture and store the profile of the feature of interest (Fig. 18-24).

An operator moves a computer-controlled cursor to identify the specific-feature to be measured. The optical-profiles are then acquired and processed in a manner similar to the scanning-slit. In general, video systems are faster, but slit-scans offer greater sensitivity. The third optical technique is *image-shearing*, in which the edges of a pattern that has been sheared into two images is positioned. At the start of the measurement, the image-edges are butted against each other. The shearing control is then adjusted until the images are rejoined into one. The difference between the initial and final values of the shearing control-vernier is used to determine the linewidth. Slit-scan and video-scan techniques have largely replaced image-shearing methods because of their increased precision.

Linewidth-measurement systems based on scan-

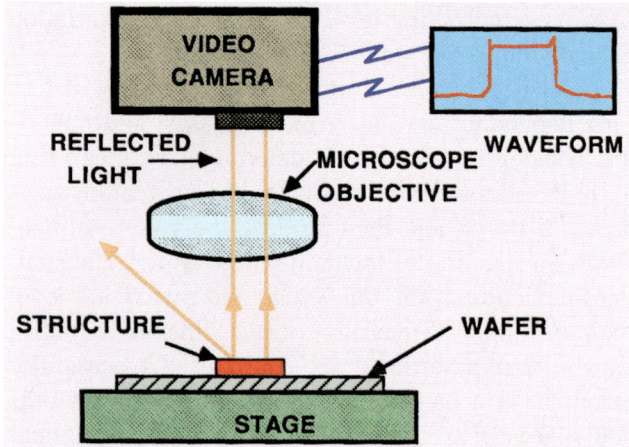

Fig. 18-24 The *video-scan technique* uses an image profile obtained by analyzing digital data obtained from a video-camera.

ning-electron-microscopes (SEMs) can overcome the accuracy-limitations of optical-techniques for submicron geometries. As the number of applications that demand such accuracy and precision continues to grow, use of SEMs for linewidth-measurement will also increase. SEMs provide data about the resist-profile (information that optical-techniques cannot provide). When low-selectivity dry-etch processes (see Chap. 22), or high-energy implants (see Chap. 12) are employed, knowledge about the shape of the resist-profile is quite valuable. To optimize throughput for production-line applications, totally dedicated single-purpose IC-metrology-SEMs are used.

18.4.8 Resist Processing: Post-Development-Bake

Post-development-baking (or hard-baking) is a process that subjects the resist to an elevated-temperature after completion of development, and is performed prior to etching or implant. Its functions are to remove residual-solvents, to improve adhesion, and to increase the etch-resistance of the resist. In addition, since hard-baking often causes the resist to flow, this effect is sometimes used to reduce the incidence of pinholes or thin-spots in the resist prior to etching. The onset of flow occurs at a temperature close to the temperature of the resist softening-point, known as the *glass-transition temperature*.

Hard-baking, like soft-baking is normally done using vacuum hot-plates. After soft-baking, the resist contains about 5% solvent, and hard-baking reduces this value even further. Additional solvent-removal is important if the patterned wafers must undergo ion-implantation or dry-etching. If solvents outgas from the film under vacuum, this can cause solvent-burst effects. Any remaining DNQ can lead to problems in subsequent process-steps. Such residual-DNQ is quickly decomposed at temperatures above 110°C.

Post-baking is generally required prior to all dry-etch processes. For fluorine-based plasmas used to etch SiO_2, a post-bake of 125–130°C may be adequate. For some Cl-based plasma-etch processes used to etch Al, however, post-bakes of up to 160°C may be needed. Since a purely thermal treatment causes resist to flow, an additional DUV-curing-step is used prior to such higher-temperature post-bake-steps.

18.5 PHOTORESIST-PROCESSING SYSTEMS

In the early days of IC-fabrication the pieces of equipment that performed the individual resist-processing-steps were stand-alone tools. As each step was completed, the wafers would be manually carried from one tool to the next. By about 1990, these tools were integrated into resist-processing systems that automatically moved the wafers from one tool to another without human-intervention. Such systems are now configured as either *wafer-tracks* or *wafer-clusters*. In either case, such resist-processing systems are then linked to a stepper to form an integrated *stepper-track photo-island*. A cassette of wafers is loaded onto the input-station of a photo-island, and 30–60-min later they emerge as patterned-wafers, ready to be sent to the next process (e.g., ion-implantation or etching).

Wafer-tracks originally consisted of two side-by-side lines of resist-processing tools (i.e., the prime, spin, and soft-bake modules on the line that fed the stepper, and the PEB, develop, and hard-bake modules on the other, Fig. 18-25). The wafers were transferred sequentially from one module to the next using a belt-transport mechanism (Fig.18-26). In today's track-systems, the individual-tools are still arranged in two lines, but a central robotic transfer-shuttle moves the wafers from one tool to another, offering the capability to randomly-access the tools. In *wafer-clusters* individual resist-processing tools are arranged in a circle around a central-robot that moves wafers among the tools, again allowing flexible access to each of the process-stations.

Stepper-track photo-islands are among the most complex systems in a wafer-fab (Fig. 18-27). They have up to 20 individual process-tools (besides the stepper) that must all work together. At any one moment, 15–20 wafers may be transferring through multi-process levels, putting resist on some, developing some, baking and chilling others. In addition to handling more wafers, newer lithography-processes require

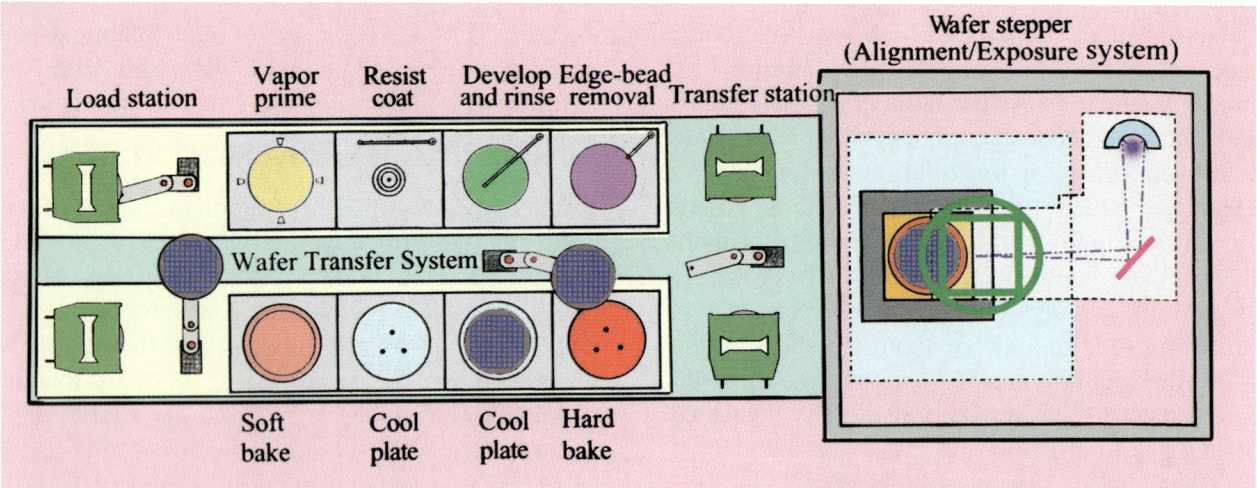

Fig. 18-25 Schematic drawing of a resist-track/stepper island.[13] Reprinted with permission.

the track to perform more steps per wafer. Newer processes include: edge-bead removal; applying anti-reflecting coatings; and processing multi-layer resists. Track-systems must ensure that each wafer (and each lot) has a consistent resist-coating. The largest manufacturers of such track and cluster systems are Tokyo Electron Ltd., Dainippon Screen, FSI International, and Silicon Valley Group (now ASML). More details resist track-systems are found in Ref. 1.

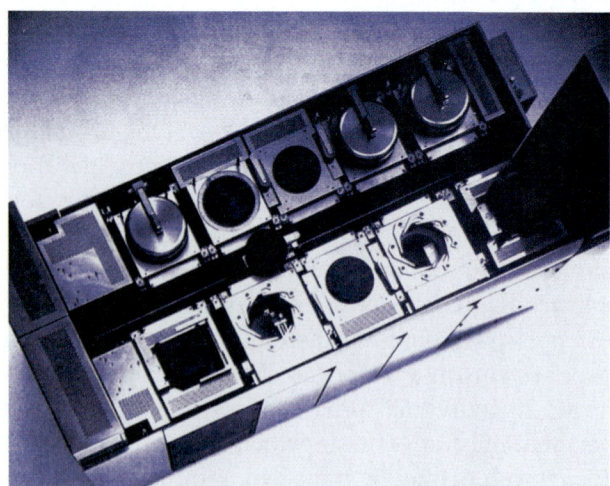

Fig. 18-26 Photograph of an in-line photoresist track-system, with resist spin, bake, and develop stations, and a central robot. Courtesy of ASM Lithography.

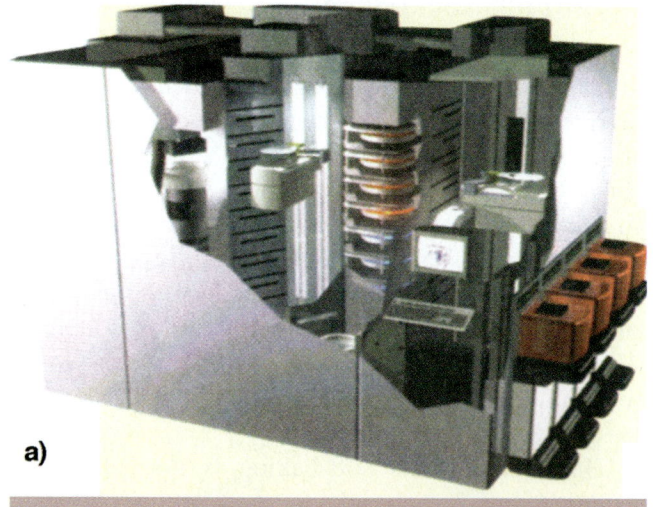

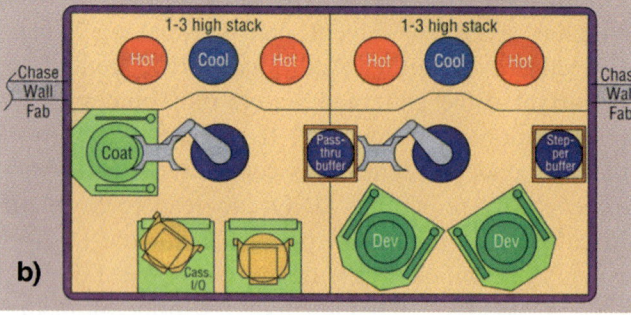

Fig. 18-27 Examples of modern resist-processing systems: (a) ProCell® cluster platform. Courtesy of ASM Lithography; (b) Schematic of the FSI Polaris resist-cluster tool. Courtesy of FSI.

REFERENCES

1. S.Wolf and R.N. Tauber, *Silicon Processing for the VLSI Era: Vol. 1 - Process Technology*, 2nd Ed., Lattice Press, Sunset Beach CA, 2000, Ch. 18 (Photoresist Processing).
2. S.Wolf, *Silicon Processing for the VLSI Era: Vol. 4 - Deep-Submicron Process Technology*, Lattice Press, Sunset Beach CA, 2002, Ch. 6 (Deep-Submicron Lithography-I).
3. H.J. Levinson, *Principles of Lithography*, SPIE Optical Engineering Press, Bellingham WA, 2001.
4. C. Garza et al., "Photoresist Materials and Processing," in *Handbook of Semiconductor Manufacturing Technology*, Y. Nishi and R. Doering, Eds., Marcel Dekker, Inc., New York, 2000. Chapter 17, p. 515-528.
5. T. Ueno, "Chemistry of Photoresist Materials," in *Microlithography: Science and Technology*, J.R. Sheats & B. Smith, Eds., Marcel Dekker, New York, 1998, p. 451.
6. A. Dammel, *Diazonaphthoquinone-Based Resists*, SPIE Optical Engineering Press, Bellingham, WA, 1993.
7. H. Ito, "Deep-UV Resists: Evolution and Status," *Solid State Technology*, June 1997, p. 115.
8. D. Ruede et al., "The Impact of Airborne Molecular Bases on DUV Photoresists," *Solid State Technology*, August 2001, p. 63.
9. B. Smith, "Resist Processing," in *Microlithography: Science and Technology*, J.R. Sheats & B. Smith, Eds., Marcel Dekker, New York, 1998, p. 529.
10. A. Braun, "Track Systems Meet Throughput & Productivity Challenges," *Semiconductor Intl.*, Febr. 1998, p. 63.
11. H. Xiao, *Introduction to Semiconductor Manufacturing Technology*, Prentice-Hall, 2001, p. 194 and p. 361.
12. B. Lorefice et al., "How to Minimize Resist Usage During Spin Coating," *Semicond. Intl.*, June 1998, p. 179.
13. M. Quirk and J. Serda, *Semiconductor Manufacturing Technology*, Prentice-Hall, 2001, p. 361 and p. 420.

PROBLEMS

1. Describe the role of photolithography in IC fabrication.
2. Explain the difference between *positive* and *negative* lithography. Describe the changes in positive and negative resist during exposure to light,
3. Define a *photoresist*. What are the two principle functions that a resist must perform? List the three components of a photoresist material.
4. What are the main reasons *negative-resist* was replaced with *positive-resist* during the evolution of IC fabrication technology.
5. Why did DNQ/novolac resists eventually become unable to used to process advanced IC-technologies?
6. How do *chemically-amplified* (CA) *resists* form a latent image? What is the role of the photoacid-generator (PAG) in CA-resists?
7. In what way are CA-resists sensitive to ambient-gas contamination?
8. Why are the following two processes carried-out prior to spinning resist onto a wafer: (a) *dehydration bake*; and (b) *applying a coating of HMDS*? What is the typical method used to apply HMDS?
9. List the factors that can impact PR spin-coating thickness and thickness uniformity.
10. Assume 2 cc (cubic-centimeters) of resist are dispensed onto a 200-mm wafer, and this results in a resist coating that is 1.0-μm thick. What percentage of the dispensed-resist remains on the wafer, assuming the resist is 20% solids?
11. List four reasons that soft-bake is performed. What are the results of overbaking or underbaking?
12. Describe the purpose of *edge-bead removal*, and the technique used to accomplish it.
13. What is the purpose of performing a *post-exposure bake* (PEB) step in: (a) DNQ/novolac; and (b) CA resist-processes?
14. Name two problems caused by subsurface reflection.
15. Explain how the use of anti-reflection coatings can help improve the resolution of lithographic-processes.
16. Describe three techniques used to measure the linewidth of patterned-features on a wafer. Why is accurate-measurement of linewidth more difficult on wafer-surfaces than on mask or reticle substrates?
17. List the three main methods used for the development of photoresists.
18. What is the purpose of the hard-bake step?

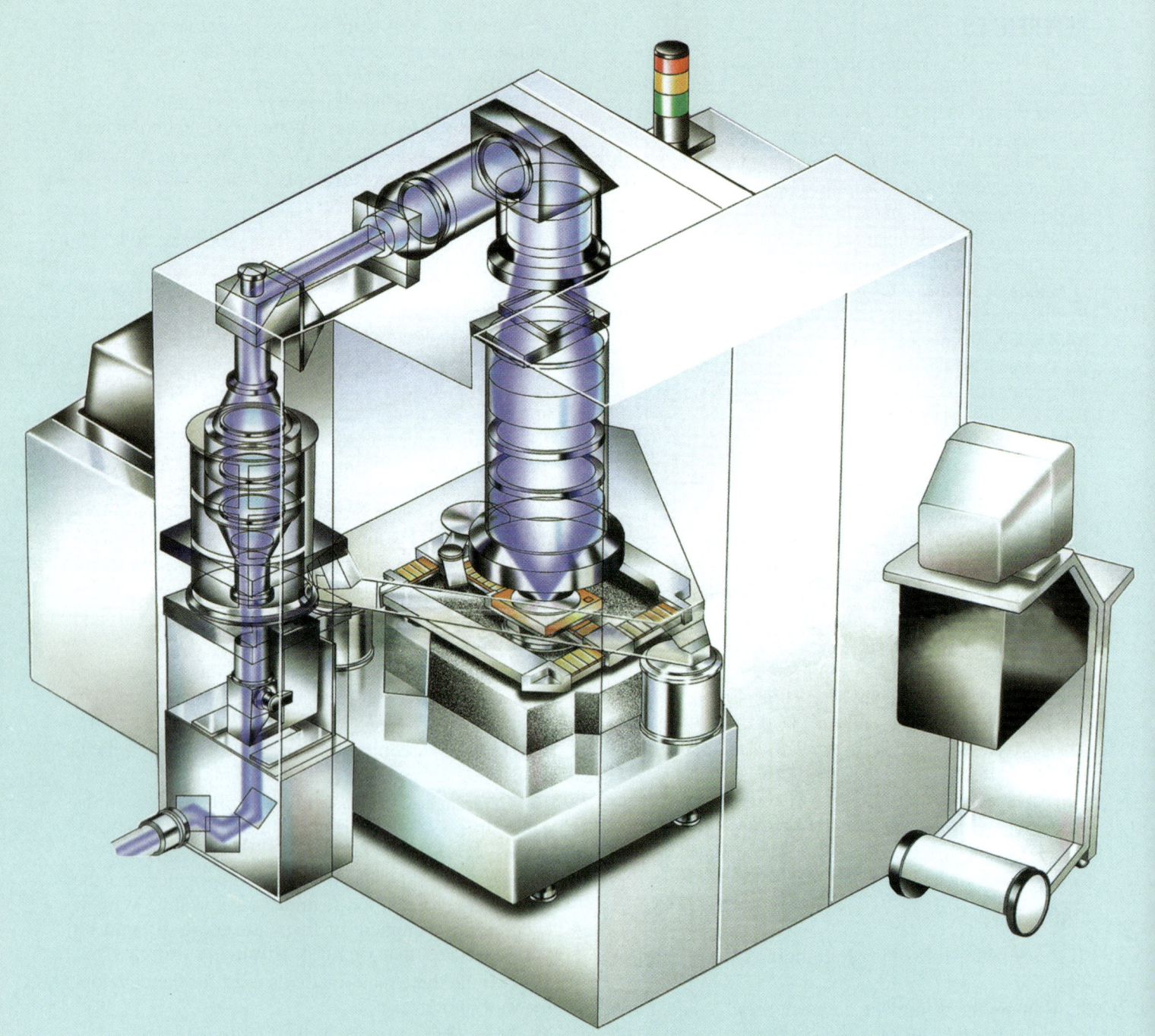

CHAPTER **19**

LITHOGRAPHY II: IMAGE-FORMATION AND OPTICAL-HARDWARE

CHAPTER CONTENTS

19.1 PRELIMINARIES: WAVE-MOTION AND THE BEHAVIOR OF LIGHT

19.2 RESOLUTION & DEPTH-OF-FOCUS IN MICROLITHOGRAPHY APPLICATIONS

19.3 LITHOGRAPHIC LIGHT-SOURCES

19.4 LITHOGRAPHIC EXPOSURE TOOLS

19.5 PROJECTION-PRINTERS
- Scanning Projection-Printing
- Step-and-Repeat Projection-Printing
- Step-and-Scan Projection-Printing

19.6 OVERLAY AND WAFER-STAGES

19.7 OFF-AXIS ILLUMINATION

"A little inaccuracy sometimes saves a ton of explanation."

H.H. Munro (Saki)

Illustration of a *step-and-repeat* exposure tool. Courtesy of ASML.

The microlithography process can be summarized as taking place in three stages (as shown in Fig. 19-1). Light is passed through a mask that contains the patterns of one layer of the IC-structure, and this causes specific areas of the photoresist film covering the wafer to be illuminated (Fig. 19-1a). The image of that pattern is captured in the thin resist-layer as a result of its being exposed to such actinic light (i.e, a *latent-image* is formed in the resist, Fig. 19-1b). This latent image is transformed into the desired resist-features by the chemical development-step (Fig. 19-1c). The image is then converted into a permanent part of the device by a series of chemical-etch or deposition processes.

This way of viewing the lithography process, however, does not give any insight as to *how the latent-image is created*, nor into the *role of the resist*. Thus, another way to perceive (projection) lithography processing is to look at the four major components that contribute to forming the final resist-feature. These are (as shown in Fig. 19-2): **1)** the *actinic illuminating light*; **2)** the *mask* (or *reticle*); **3)** the *lens* of the exposure-system; and **4)** the *photoresist film*.

In the previous chapter the material properties of photoresists and their processing technology were covered. This chapter describes a second group of topics involved in the microlithographic process of transferring patterns to silicon wafers, namely: **a)** an introduction to the formation of aerial images of the circuit patterns on the resist-surface; and **b)** the equipment used to project these images onto the resist (the *illumination sources*, and the *exposure-systems*). The latter discussion will include a description of the positioning-subsystems used in such exposure-tools.

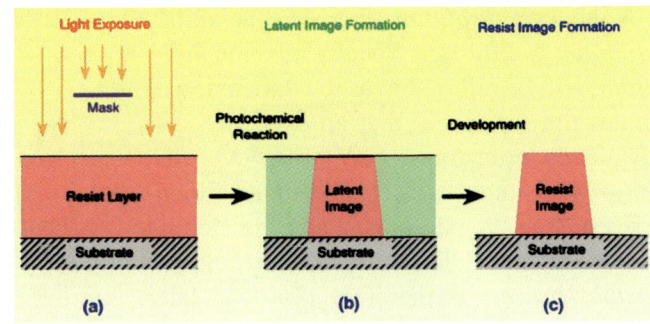

Fig. 19-1 The three basic steps of forming a positive-tone photoresist image.

341

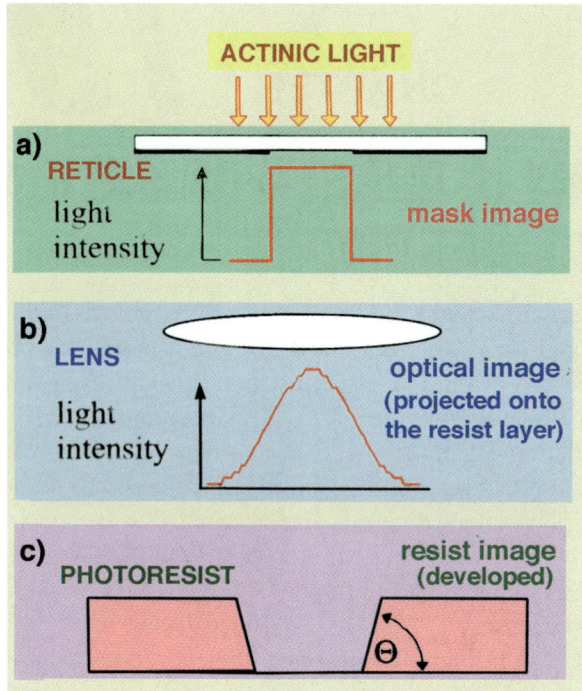

Fig. 19-2 The four components of a microlithography step that contribute to the overall *resolution-capability* of the process: (a) the *illuminating actinic light*; (b) the *reticle*; (c) the *lens;* and (d) the *photoresist*. Also shown are the forms taken by the image as it passes through the reticle and the lens, and how exposing the photoresist (and developing it) contributes to the shaping of the final resist-image-form.

Before beginning the discussion on lithography, it is useful to identify the key hardware-issues of this technology. The two most important characteristics of the machines and masks used to project patterns onto wafers are: **1)** their *resolution*; and **2)** their *pattern-registration capability* (alignment and overlay).

In general, the term *resolution* describes the ability to print a *minimum feature-size*. Specifically, the *minimum-resolution* will be referred to as the dimension of minimum-linewidth (or space) that an exposure-system can adequately print (or resolve). However, the ability to form IC-features of such minimum dimensions also depends on the photoresist and the etching technology. A more formal definition of the resolution of a *lithography-process* is therefore provided later (in Sect. 19.2.2). But, it is nevertheless important to emphasize at the outset that *high-resolution* is usually the most sought-after property of an aligner.

The *pattern-registration* capability is a measure of the degree to which the pattern being printed can be "fit" relative to a previously printed pattern. As the feature sizes are so small in microlithographic applications (and the number of layers that must be correctly placed on top of one another is large; e.g., 20–25 in ULSI), a very-tight fit indeed, is required. This subject is covered in more detail in Sect. 19.6.

19.1 PRELIMINARIES: WAVE-MOTION AND THE BEHAVIOR OF LIGHT

Light is a phenomenon of nature that exhibits *wave behavior* (i.e., oscillatory variation of some property with time at a given fixed-location in space). For example, a water wave has peaks and troughs present at some instant of time (Fig. 19-3a). The *amplitude* of

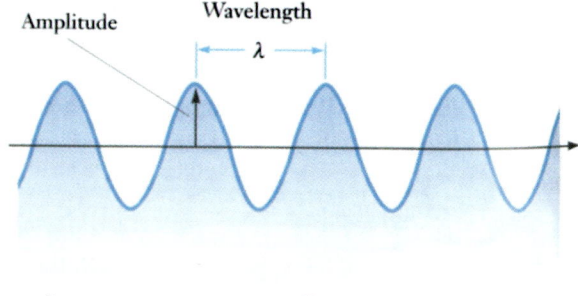

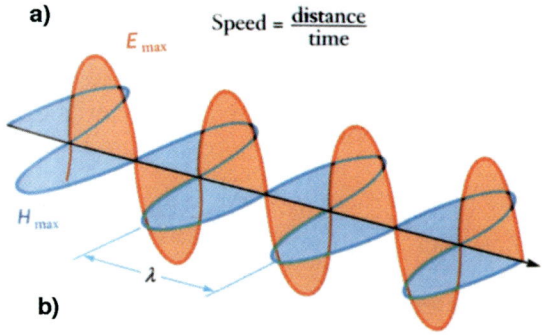

Fig. 19-3 (a) As a water-wave moves across an otherwise calm tank, its *amplitude* and *wavelength* can be determined. Its *speed* is found by dividing the travel distance of a particular wave-crest by the time elapsed. (b) *Light* consists of waves of oscillating *electric* (E) and *magnetic* (H) fields that are perpendicular to each other (and to the direction of propagation of the light).

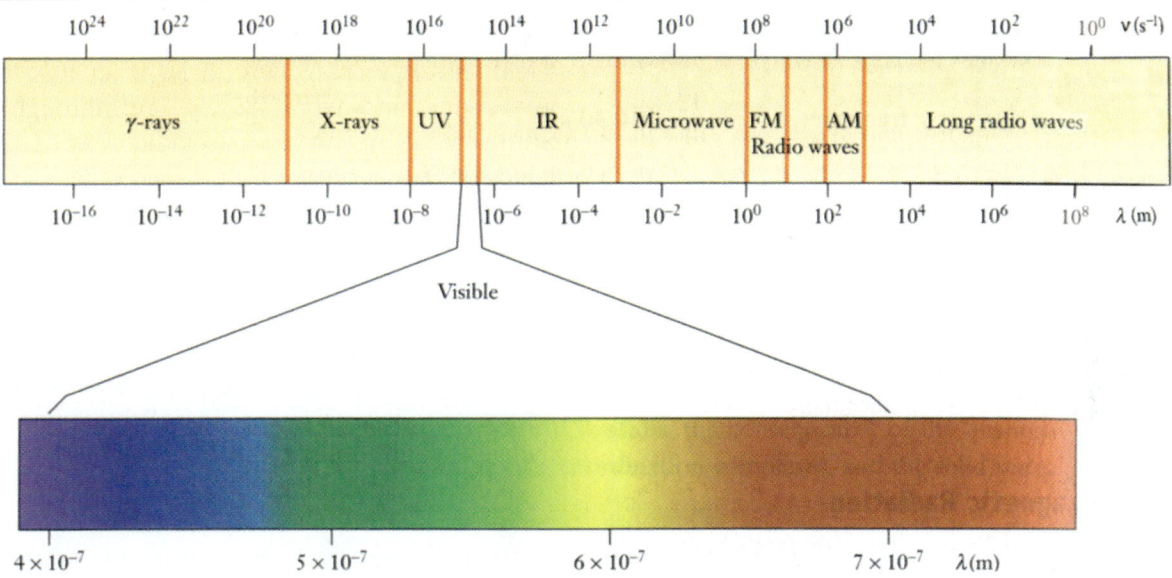

19-4 Regions of the *electromanetic spectrum*. This spectrum extends from the very-short wavelengths (very-high frequencies) of *gamma rays*, through the very-long wavelengths (very-low frequencies) of *radio waves*. The small fraction that is visible to the human eye is expanded to show the component colors.

the wave is the maximum-displacement of the water-surface over the undisturbed level of the water. The distance between two amplitude-peaks is called the *wavelength* λ (Greek *lambda*). Note that the wavelength is commonly expressed in *meters, m*. However, since very-short wavelengths are routinely encountered in lithography, the *micron* (μm, 10^{-6} m), the *nanometer* (nm, 10^{-9} m), and the *angstrom* (Å, 10^{-10} m) are also all widely used. The frequency, ν (Greek *nu*), of a water-wave can be measured by counting the number of peaks observed at a fixed point in space per second, and has units of *cycles per second* (in units of sec^{-1}, or *hertz* [*Hz*]). The wavelength and frequency of a wave are related through its speed - the rate at which it moves through the medium. The speed of a wave is the product of its wavelength and frequency, $\lambda * \nu$.

It was demonstrated by Maxwell in 1865 that light is *electromagnetic* (*EM*) radiation. A beam of light consists of oscillating *electric* and *magnetic* fields perpendicular to the direction in which the light is propagating (Fig. 19-3b). The speed *c* of light passing through a vacuum is a constant (called the *speed of light*), and is equal to the product λν:

$$c = \lambda\nu = 2.9979 \times 10^8 \text{ m/sec} \qquad (19.1)$$

The continuum of radiant-energy of electromagnetic waves (including both *visible-light* and the *non-visible* forms of electromagnetic radiation) is called the *electromagnetic spectrum* (Fig. 19-4). The waves in the spectrum all travel at the same speed, but differ in frequency and wavelength. This spectrum contains all the forms of radiant-energy, from those with very-short, to those with very-long wavelengths. The region visible to the eye is a very-small fraction of the spectrum (in the wavelength-range between 780-nm [*red*] and 450-nm [*violet*]), and comprises bands of colored light with particular wavelengths and frequencies. For example, *green-light* has a range of frequencies near wavelengths of 5.3×10^{-7} m (530-nm). Red-light has a longer wavelength than green, and violet has a shorter one. *White-light* contains the full range of visible-wavelengths. It can be resolved into its component wavelengths by passage through a prism.

Light used in photolithography has wavelengths shorter than visible-light, in the ranges of the EM-spectrum from *ultraviolet* (*UV*) light to *x-rays*. Since some of such light (i.e., UV-light within the range of 450 to 120 nm) can still be focused by the lenses used to focus visible-light, lithography processes employ-

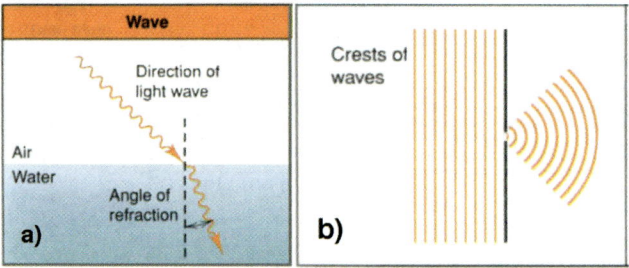

Fig. 19-5 (a) A *light-wave* passing from air to water is *refracted* (bent at an angle). (b) A (plane) light-wave is *diffracted through a small opening,* which gives rise to circular-waves on the other side. (The lines represent the peaks of the waves, as would be seen from above.)

ing UV (and deep-UV) light are still referred to as *optical lithography processes.* (Note that yellow-light is used in the lithography-areas of wafer fabs because it is in the portion of the spectrum with very-little UV, and therefore does not affect UV-photoresists.)

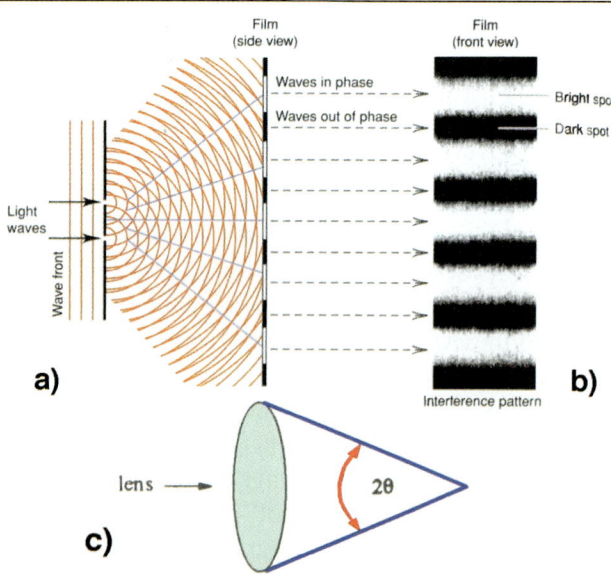

Fig. 19-6 The *diffraction-pattern* caused by light passing through two slits: (a) After passing through two closely-spaced slits, emerging circular waves interfere with each other to create a *diffraction (interference) pattern* on the film, of *bright-regions* (where peaks coincide and enhance each other - *in-phase*), and *dark-regions* (where peaks meet troughs and cancel each other - *out-of-phase*); (b) The interference pattern obtained on the film; and (c) Definition of *numerical-aperture* (NA).

19.1.2 Refraction and Diffraction of Light

Light travels more slowly through air than through a vacuum, and more slowly still through water. Similarly, it moves at particular speeds through quartz, different types of glass, and other transparent substances. Therefore, when a light-wave crosses a *boundary* from one medium to another, the speed of the light changes. If it strikes the boundary at an angle other than 90°, the wave bends and continues at a different angle, a phenomenon known as *refraction* (Fig. 19-5a). The new angle (*angle of refraction*) depends on the materials on either side of the boundary, and the wavelength of the light.

When a wave strikes the *edge* of an object, it bends around it, a phenomenon called *diffraction.* If the wave passes through a slit about as wide as its wavelength, it bends around both slit-edges and forms a semicircular wave on the other side of the opening (Fig. 19-5b). If waves of light pass through two adjacent slits, the emerging circular waves interact by a process called *interference,* creating a *diffraction-pattern* of brighter and darker regions (Figs. 19-6a & b).

As the feature-sizes of ICs shrink, the edge-acuity of the light-intensity distribution worsens because of such diffraction-effects. Consider the case of an isolated transparent-line on an opaque-mask. The intensity of the light projected onto the wafer-surface (as a function of the normalized-distance from the center of the line) is shown in Fig. 19-7 for: **1)** the ideal-case (i.e., *no diffraction occurs*); and **2)** the actual-case for

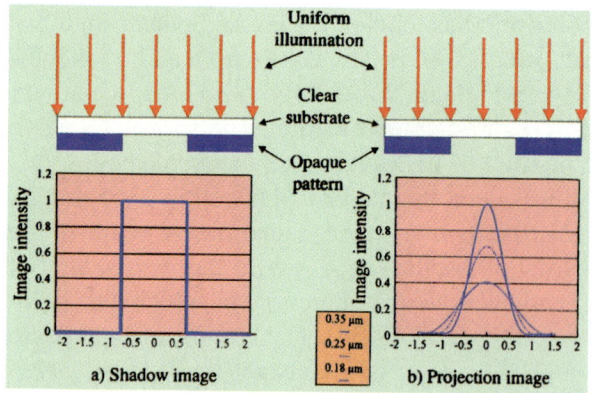

Fig. 19-7 Basic imaging characteristics. (a) Ideal shadow imaging. (b) Diffraction-broadened projection printing.

three different linewidths (0.35-μm, 0.25-μm, and 0.18-μm), when imaged by a specific optical-system containing a perfect (i.e., distortion-free) lens. In this example, the NA of the lens is 0.5, and the illuminating wavelength is 0.248-μm. (Even though all of these terms have not yet been defined, what is important to note is that for any specific optical-system, as the dimension of the line gets smaller, there is a degradation of the edge-acuity [or "sharpness"] of the *optical-intensity profile*, as compared to the ideal-case. At some point the edge-acuity becomes so degraded that one would have to declare this image is no longer "resolved.") However, it should be apparent from examining these intensity-profiles that in optical-lithography applications there is a gradual transition from a feature being "resolved" to "unresolved." Thus, the term resolution will have to be defined in a more rigorous manner later in upcoming sections (where it is seen that the *depth-of-focus* of the optical system also plays a role in this definition).

19.1.3 Resolution of Two Point-Sources of Light

The term resolution will first be considered for an optical-system that is being used to distinguish two point-sources of light (i.e., as is done when trying to resolve whether the image of a star consists of just one star, or two stars close to one another). In any such optical-system, one of the key-factors that limits its ability to resolve this type of image (i.e., its resolution capability), is the physical design of the *objective-lens* and its *numerical-aperture* (*NA*). The NA is a measure of a lens' capability to collect diffracted light from an object (photomask) and project it onto a wafer. The NA is defined as:

$$NA = n \sin \alpha \qquad (19.2)$$

where n is the *refractive-index* (typically n = 1.0 for air) and 2α equals the angle of acceptance of the lens (shown in Fig. 19-6c). The NA of lenses in projection-aligners ranges between about 0.16 and 0.76.

One way of defining the resolution of an optical-system is based on the *Rayleigh-Criterion*. This definition rests on the ability of a *telescope* to resolve two point-sources of light (typically two stars), that are separated by a small-angle. Since they are independent light-sources, there is no interference in the light from two stars. However, diffraction causes a lens to image an ideal, infinitesimally-small point-source into a blurry disc (Fig. 19-8a), called the *Airy-disc*. The light-intensity-distribution from such a point-source projected through a circular aperture in an opaque-screen is plotted in Fig. 19-8b. The *radius* of the Airy-disc is the *distance from the intensity-peak to the first zero* of this intensity-distribution profile.

When two points are so close that the two Airy-discs look like a single blurred image, the two light points (stars) cannot be resolved. The Rayleigh-Criterion defines two images as being *just-resolved* when the intensity-peak of the Airy-disc from one star falls on the first zero of the Airy-disc of the second star (Figs. 19-c2 and d2). The *Rayleigh-Criterion* deems that distance, δ, to be *the case when two point-sources are just-resolved*. The value of δ for a telescope can be found using the following expression:

$$\delta = 0.61 \, \lambda / NA \qquad (19.3)$$

where λ is the illuminating wavelength and NA is the numerical-aperture of the telescope lens. Figs. 19-8c and 19-8d demonstrate the resolving-power of a telescope on the images of two closely-spaced stars.

Note at this point, however, that the problem of resolution encountered in microlithography is quite different from that of resolving two closely-spaced stars. Thus, the Rayleigh-Criterion cannot be expected to provide a resolution-limit that is directly applicable to microlithography. Nevertheless, for a number of reasons, it is a useful guide to understanding the limits of resolution in microlithographic optical-systems. First, the Rayleigh Criterion provides information about how the resolution of an optical-system changes as certain optical parameters are varied. It indicates that the resolution is increased if either the *wavelength of the illuminating light* is reduced, or the *NA of the lens is made larger* (or if both of these situations occur). But, since properties of the photoresist also impact the minimum resolution, and the Rayleigh-Criterion is derived only on the basis of the characteristics of optical-systems, Eq. 19-3 does not provide a general-expression for minimum-resolution in lithography.

19.2 RESOLUTION IN MICROLITHOGRAPHY APPLICATIONS

19.2.1 Depth of Focus

The formation by a projection-system of an image on a mask has assumed up to this point, that the wafer-surface was in the *plane of best-focus*. However, defocus-effects also have an impact on the quality of the aerial image. Even from everyday experience it is known that the clarity of an object degrades if it is *out-of-focus*. Therefore, focus must also be taken into account. This factor has become especially important since feature-sizes have decreased below about 1-μm. Since that time, the *depth-of-focus* (*DOF*) has become a limiting-factor in the ability of optical-lithography to print smaller-geometry features. (The *depth-of-focus* is defined as *the distance along the optical-axis over which an image has resist profiles and line-widths that remain within specifications*.) Figure 19-9 qualitatively depicts the behavior of an optical-image through best-focus. The *Rayleigh DOF Criterion* defines DOF as *the range of focus over which the peak-intensity of a point-source remains within 20% of the peak-value*. For an optical-system with a given NA and λ, the *Raleigh-DOF* is calculated from:

$$\text{DOF} = \pm 0.5 \lambda/(\text{NA})^2 \quad (19.4)$$

One serious impact pointed out by the depth-of-focus analysis is that DOF *diminishes* as one attempts to print smaller-features. That is, from the Rayleigh-Criterion for *resolution*, there are two options for improving it: *decreasing the wavelength*, or *increas-*

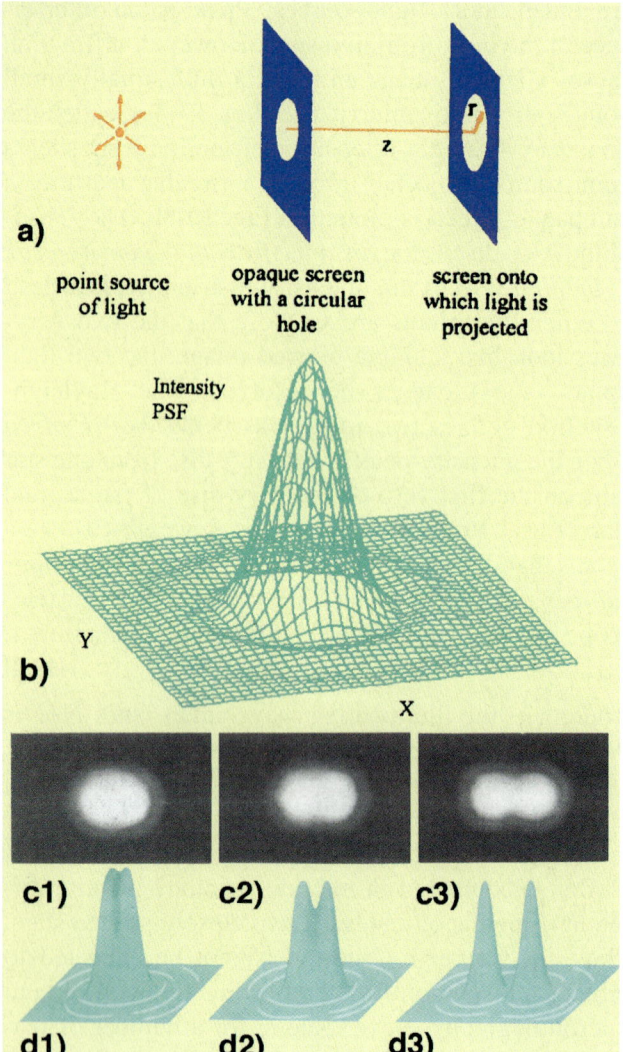

Fig. 19-8 (a) Light-intensity distribution from a point-source projected through a circular-aperture in an opaque-screen. (b) Three-dimensional representation of the point-source light-distribution projected on the screen shown in (a). (c) Three photos show the images of two point-sources (stars) formed by a converging-lens. Beneath each photograph (d) is shown a representation of the image-intensities. In (c-1) and (d-1), the angular-separation is too-small for them to be distinguished. In (c-2) and (d-2) they can be marginally distinguished. In (c-3) and (d-3) they can be clearly distinguished. The *Rayleigh-Criterion* is satisfied in (c-2) and (d-2), with the central intensity peak of each diffraction-pattern coinciding with the first-zero of the other.

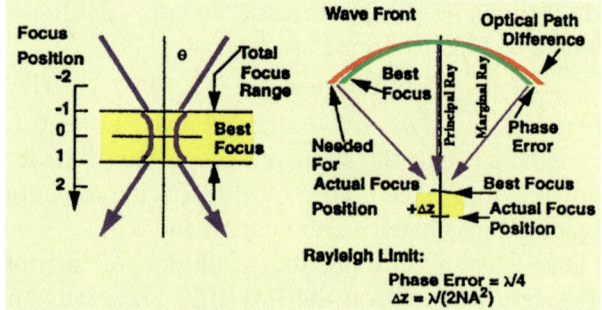

Fig. 19-9 Behavior of the optical-image through best-focus and the *quarter-wavelength phase-error* introduced by Rayleigh to define the focal-effect.

CHAPTER 19   LITHOGRAPHY II: IMAGE FORMATION AND OPTICAL HARDWARE

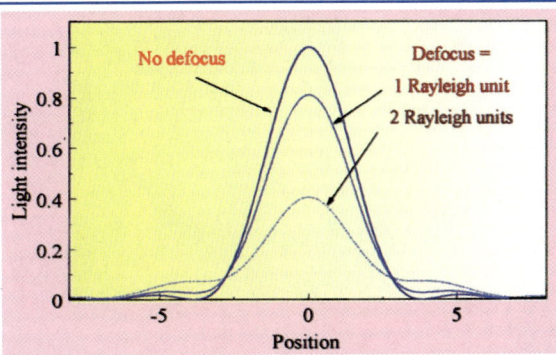

Fig. 19-10 Light-intensity profile of a point-source of light imaged by a circular diffraction-limited lens, at different *planes of focus*. The horizontal-axis is given in units of $2\pi NA/\lambda$. One Rayleigh-unit of defocus is $0.5\lambda/NA^2$.

ing NA. But, according to Eq. 19-4, doing either of these also *decreases DOF*.

For example, the light-intensity of profiles of an image from a point-source of light diffracted by a circular aperture (as discussed earlier) can also be calculated for different focal-planes. The results of this type of analysis are shown in Fig. 19-10. What can be seen is that with defocus, the *peak-intensity diminishes* and *more light is diffracted away from the center spot*. Figure 19-11 shows SEM photographs of cross-sections of resist lines and spaces as focus and exposure is varied. It indicates that the *linewidth varies* and the *sidewall-slope increases* for large defocus. Thus, there is limit to how *far out-of-focus* the wafer-surface can be, and still provide resist-features with adequate resolution. When a lithography process is being characterized, a plot of acceptable limits of focus and exposure is produced to allow lithography engineers to determine the outer performance limits of their exposure-tools that will still give resist-features with acceptable characteristics (Fig. 19-13). In summary, defocus-effects also play a role when a definition of the term "resolution" in lithography processes is formulated.

19.2.2 Definition of Lithographic Resolution

While the Rayleigh-Criterion provides insight about how the resolution of an exposure-system can be increased, the problem in lithography is broader. That is, the *resolution* goal of the lithography process is to produce resist features of the correct-size repeatedly, run-after-run. As was stated in the introduction, this not only involves the actinic-illumination (through λ) and the imaging-system (through NA), but also the reticle and the photoresist. Another aspect of the problem is that the resist features being produced are *three-dimensional forms* (see Fig. 18-2, in Chap. 18). Thus, not only do they have to have the correct linewidths and spacings (Fig. 19-12), but also the correct three-dimensional form. Specifically, the sidewall angle of the resist feature, and the final film-thickness after development also have to be within specification. With these factors in mind, we can now give a definition of the *resolution of a lithography process*.

> The *resolution* of a lithography process is defined as the ability to produce a *line* (or *line and space*, or *opening* - each type of feature will have its own set of specifications), that meets an acceptable set of criteria, including: **a)** *linewidth*; **b)** *sidewall-angle*; and **c)** *resist-thickness after develop*, in a repeatable manner. This ability is determined by the four components of the process:
>
> **1)** *Actinic Illumination Source* - through the wavelength of the light, intensity, spectral width, etc.
> **2)** *Mask* (or *Reticle*) - through correct pattern-dimensions on the mask (for *binary-masks*), and such RETs (see Chap. 20) as optical-proximity correction and phase-shift masking.
> **3)** *Lens of the Exposure System* - through its NA (to provide adequate minimum aerial-image size within the focus-exposure limits).
> **4)** *Photoresist Film* - through resist properties such as photospeed & contrast. Resist processing parameters must also be within spec (spin-speed, soft-bake, develop, hard-bake etc.).
>
> Note that all of the above factors play a part in having a resist process that can successfully meet the called-for resolution-specifications. All of these factors must also be controlled to within acceptable limits for the overall process to succeed.

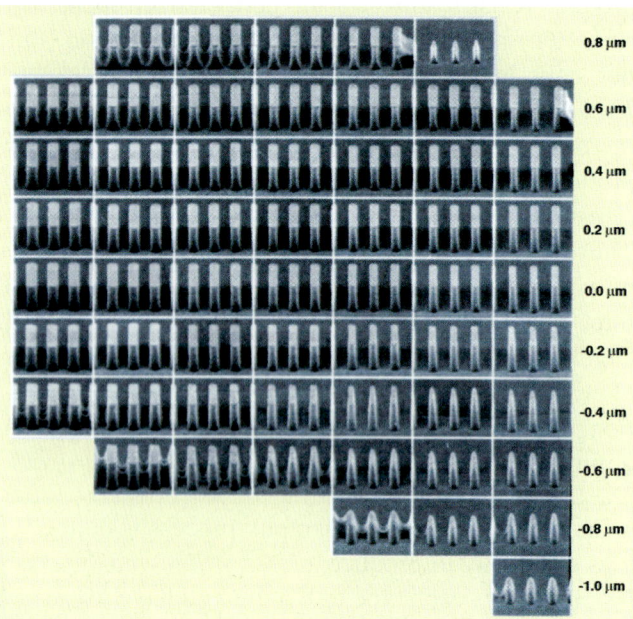

Fig. 19-11 Resist-profiles for imaging at various focal-planes and exposure-doses. Courtesy of Clariant Corp.

19.3 LIGHT-SOURCES FOR LITHOGRAPHY

Modern, high-throughput optical lithographic exposure-systems require light-sources of high-power and radiance. The optical power-density needed at the wafer plane can be calculated from a knowledge of the *resist-sensitivity* and the time allotted for the exposure-step. Resists used in *g*-line and *i*-line lithographic processes typically require about 100 mJ/cm^2 for their exposure. If the exposure-time is specified as 0.5-sec or less, a power-density of at least 200-mW/cm^2 is needed. Light-sources of steppers usually

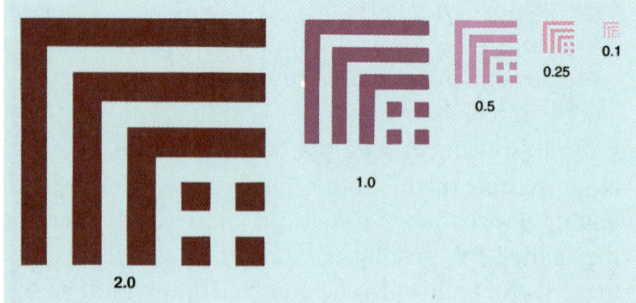

Fig. 19-12 Schematic drawing showing resolution-scaling of patterns printed on a wafer.

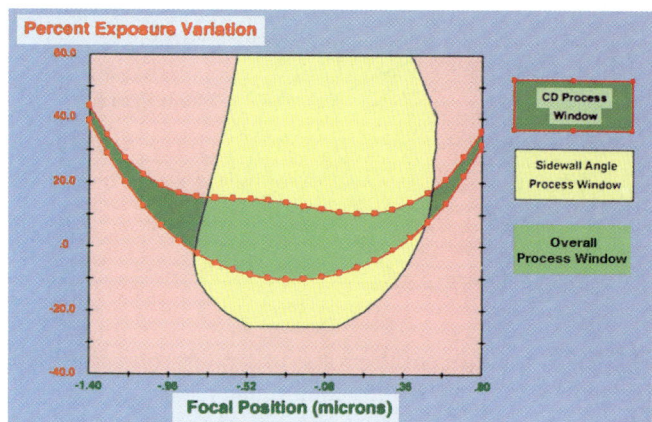

Fig. 19-13 The *focus-exposure process window* is constructed from contours of the specifications for linewidth (*yellow area*) and sidewall-angle (*dark-green areas*). The *light-green area* shows the overall process window.

provide power-densities at the wafer level in the range of 500-1000-mW/cm^2 (to allow for shorter exposure-times, or to accommodate less-sensitive resists). *Mercury-arc lamps* and *excimer-lasers* are the light-sources used in optical-microlithography. Their characteristics are discussed in this sections.

19.3.1 Mercury-Arc Lamps

Until 248-nm KrF-excimer-laser light was introduced for DUV optical-lithography, optical-aligners used high-pressure *mercury-vapor* (*Hg*) lamps as the illumination-energy source. These lamps consist of two conducting electrodes (separated by a gap of about 5-mm) sealed in a fused-silica bulb (which makes up the lamp-housing, Fig. 19-14). The lamp is filled with Hg-gas whose pressure (in cold lamps) is ~1-atm. The lamp is lit by applying a high-voltage spike

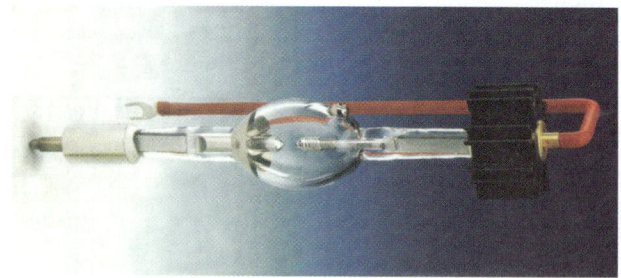

Fig. 19-14 Photograph of a typical high-pressure mercury-arc lamp. Courtesy of Osram Sylvania.

across the gap. The spike-voltage ionizes the gas, which establishes an arc-discharge (similar to the glow-discharge described in Chap. 14, but operated in a regime with a much higher current-density). The ionized Hg-gas gets very hot, and the pressure in the bulb during operation may reach 40-atm. In such lamps, a discharge-arc of the high-pressure Hg-vapor emits a characteristic spectrum. The *mercury-vapor emission spectrum* is shown in Fig. 19-15. It is not of uniform intensity at all wavelengths, and in fact, contains several intense, sharp-lines. In the UV-wavelength-range from 350 to 450-nm there are three strong lines, the *i*-line (365-nm), the *h*-line (405-nm), and the *g*-line (436-nm). Because of their optical-dispersion, refractive lithographic-lenses can use only a single emission-line. Each of these lines contain less than 2% of the total-power of the arc-lamp. Reduction-steppers were first designed to operate using the *g*-line (and these were used for feature-sizes down to ~0.8-μm), but later, *i*-line reduction-steppers were also developed (and were used for printing feature-sizes in the range of 0.4-0.8-μm). The 1X-steppers (which use a reflective optical-system) can utilize a wider-portion

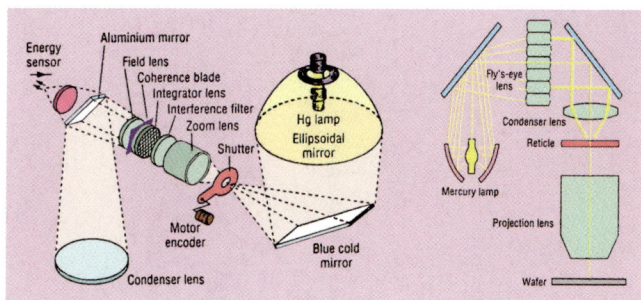

Fig. 19-16 (a) Schematic diagram of a mercury-arc lamp based illumination system. (b) Schematic diagram of an illumination system that uses a fly's eye lens.

of the spectrum than just a single-line (i.e., the entire wavelength range between 390–450 nm).

To obtain maximum light-intensity, high-power mercury-arc lamps are used (e.g., 200–2000-W). These lamps emit their light from a compact region a few millimeters in diameter. A large fraction of the total-power emerges as infrared and visible light-energy, which must be removed from the optical-path. This is done with multilayer dielectric-filters, which direct the energy to a liquid-cooled optical-path, designed to remove this heat-load from the system.

The Arc-Lamp Illumination-System: The task of an illumination-system is to collect the light emitted from an arc-lamp and project it through the mask into the entrance-pupil of the lithographic projection-optics. Illumination-systems must also create the same set of plane-waves for all points across the mask. Figure 19-16 is a schematic of such an illumination system.

19.3.2 Excimer-Laser DUV Light-Sources

Excimer-lasers are used to produce the light needed for deep-UV optical lithographic exposure-systems. The term *laser* is an acronym for *light amplification by stimulated emission of radiation*. The *krypton-fluoride* (*KrF*) excimer-laser is used to produce DUV-light with a wavelength of 0.248-μm (or 248-nm). *Argon-fluoride* (*ArF*) excimer-lasers are employed to produce DUV-light with a wavelength of 193-nm. No fundamental changes are necessary to allow a KrF-laser to operate as an ArF-laser, except to change the gas composition in the laser's plasma-tube and the

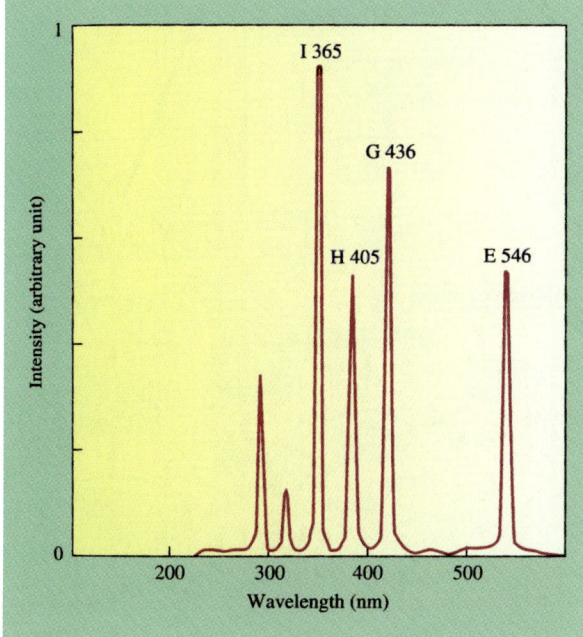

Fig. 19-15 Typical high-pressure mercury-arc spectrum.

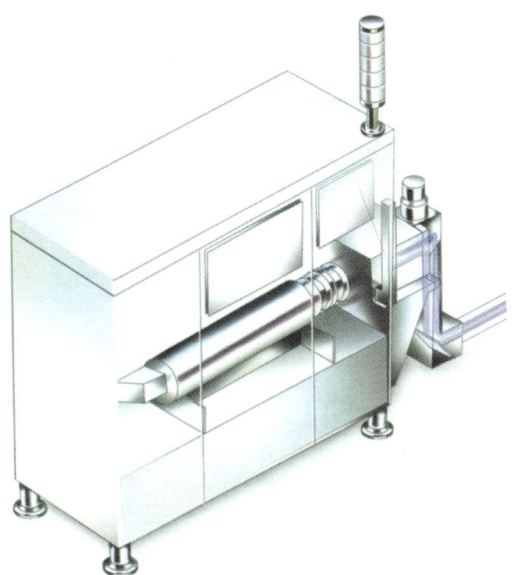

Fig. 19-17 Schematic drawing of an excimer-laser system used in stepper/scanner applications.

optically-based microlithography has proved to be a remarkably useful technology throughout the life of the IC-industry.

The heart of the photolithographic process is the *exposure-system*. This complex piece of machinery projects the image of a desired photomask-pattern onto the surface of the semiconductor-device being fabricated on a silicon-wafer. Such an exposure system consists of three parts: **1**) a *lithographic-lens*, **2**) an *illumination-system*, and **3**) a *wafer-positioning system*. Illumination-systems were described in Sect. 19.3, and wafer-stages will be discussed in Sect. 19.6. Here we examine the lens-subsystem, and the different types of exposure-systems.

The earliest masks for microlithographic applications were made by cutting and pasting a pattern of an opaque mylar-film (called *rubylith*). Optical-reduction techniques were then used to reduce this pattern in size. The result was a photomask that could produce images on a resist-covered wafer that had the same size as the patterns on the mask (*1X-printing*). By the

mirrors that make up the cavity. By 1998 KrF excimer-lasers were well-established in mainstream, high-volume, advanced IC production (i.e., for 0.35, 0.25, and 0.18 μm CMOS technologies, Fig. 19-17). At the same time, ArF-excimer lasers were being installed as the light-source for 193-nm stepper/scanners that were being used for advanced process-development applications, as well as for some pilot-line production (i.e., for sub-0.2-μm CMOS process-development lines).[6]

19.4 MICROLITHOGRAPHY ALIGNERS

Microlithography was introduced in the 1958 timeframe, as a way to be able to print the features of the *planar-transistors* needed in ICs (see Chap. 3, Sect 3.1). The evolution of microlithographic-technology is summarized in Fig. 19-18. The earliest IC-features patterned with optical-lithographic techniques had minimum-sizes in the range of *hundreds of microns*. However, since that time, the minimum feature-size has shrunk dramatically. By the year 2003, optical-lithography was being used to fabricate ICs with minimum feature-sizes as small as 0.09-μm (and development of optical-processes to reach even smaller dimensions was being pursued).[7,8] Thus,

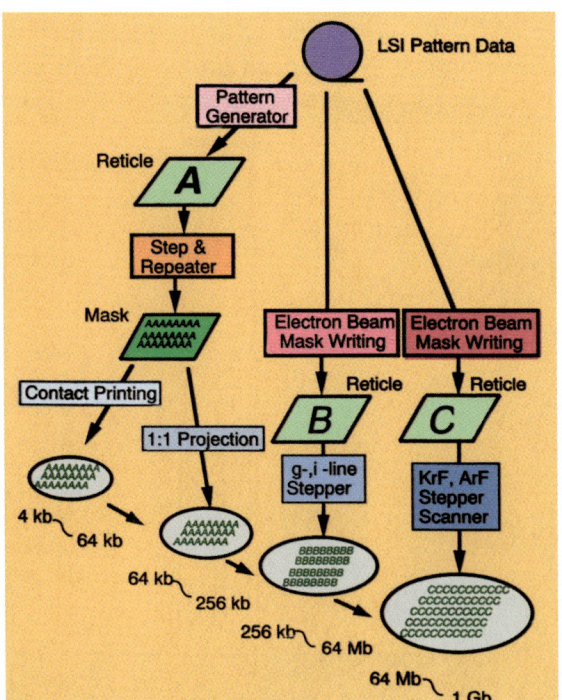

Fig. 19-18 The evolution of *microlithography* (shown as correlated to the evolution of DRAM technology).

early 1970's, the minimum feature-size on ICs (also called the *critical-dimension*, or *CD*) was in the range of 2–5-μm. At that time, wafer-imaging was carried out by replicating a mask using contact printing (as described next).

19.4.1 Contact-Printing

A resist-covered wafer was put into hard-contact with a photomask, and the resist was then exposed by near-UV light (*contact-printing*, as shown in Fig. 19-19a). This yields the most-faithful image-transfer (and also the best resolution) of any exposure method. Unfortunately, the repeated mask-to-wafer contacting steps generate defects on the mask (which print on the next-wafer exposed through the mask). To minimize this problem, the original emulsion-masks were replaced with chromium (Cr) masks, and these became the workhorse mask (even when projection eventually replaced contact-printing). Such Cr-masks could be inspected and cleaned regularly. But, when the defects could no longer be adequately removed by cleaning, even these masks had to be replaced. This problem led to the eventual abandonment of contact-printing, and to the development of lithographic methods where the wafer and mask do not make contact.

19.4.2 Proximity-Printing

The first non-contact printing technology was *proximity-printing* (Fig. 19-19b). The mask is placed in close-proximity to the wafer. This avoids the inevitable mask-damage that occurs in contact-printing. However, for wavelengths of light used in lithography (200-400 nm) and for practical gap-dimensions (10-μm), the minimum feature-size of optical-proximity lithography was about 2-μm. Thus, it could not be used for VLSI or ULSI (2-μm or less).

19.5 PROJECTION-PRINTING

To obtain feature-sizes smaller than 2-μm, proximity-printing was replaced with projection-printing (Fig. 19-19c). In *projection-printing*, a mask is illuminated, and the transmitted-light is collected by a lens, and imaged at the wafer-plane. The mask is the same type as that used by a proximity-printer (namely, a 1X

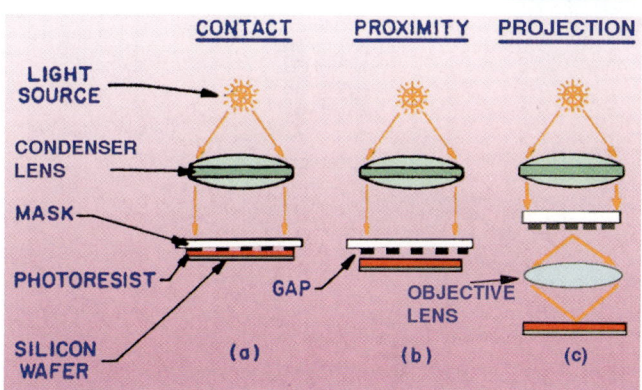

Fig. 19-19 Three methods of wafer-exposure: (a) *Contact-printing*; (b) *Proximity-printing*; (c) *Projection-printing*.

chrome-on-glass pattern that is large enough to cover the entire wafer). But, the use of a projection-system means that masks are no longer damaged by accidental (or deliberate) contact with the wafer-surface.

19.5.1 Scanning Projection-Printing

The first projection-printer (1974) used an all-reflective system (which allowed a broadband light-source to be utilized). It was a 1X-printer and could produce features down to about 1.5-μm. But, it is difficult to design a lens capable of projecting micron-scale images onto an entire 4-to-6 inch wafer in a single field-of-view. Thus, a clever design by the Perkin-Elmer Corporation allowed wafers of this size to be printed by simultaneously scanning the mask and wafer through a lens-field shaped like a narrow arc.

This lens-design takes advantage of the fact that most lens aberrations are functions of the radial position within the field-of-view. A lens with an extremely large circular-field can be designed with aberrations corrected only at a single-radius within this field. An aperture limits the exposure-field to a narrow-arc centered on this radius. Because the projector operates at 1X-magnification, a rather simple mechanical-system can scan the wafer and mask simultaneously through the *object* and *image* fields of this lens.

Figure 19-20 shows schematic drawings of the operating principle of such a projection-aligner system, which also relies on a reflective-lens system. These types of 1X-scanners were still in common use

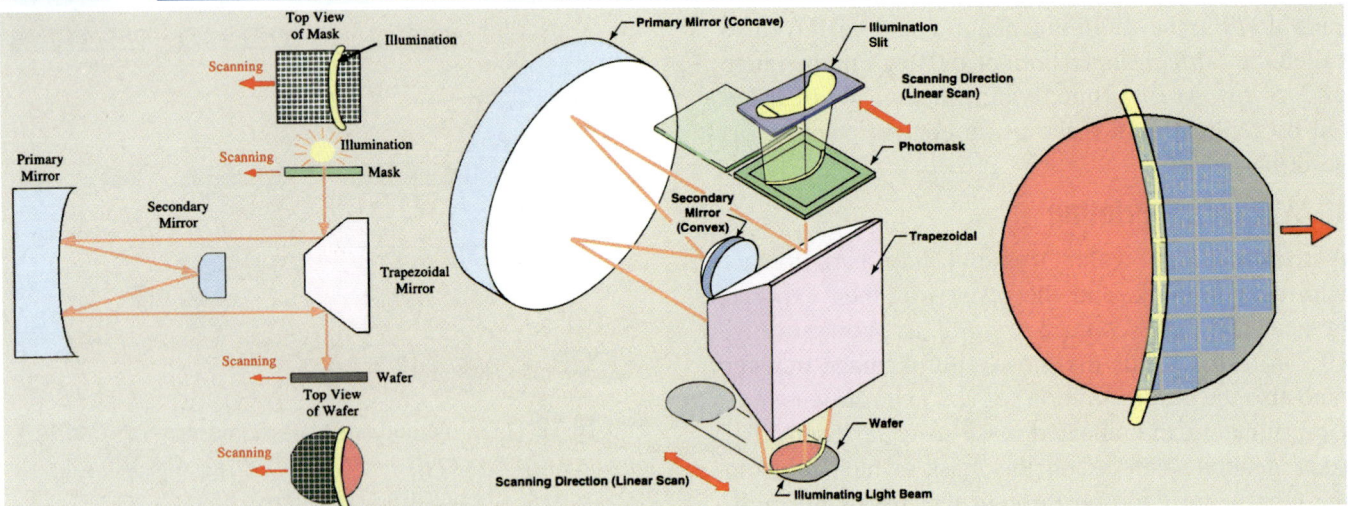

Fig. 19-20 *Scanning-projection printing*: (a) top-down view of a scanning-projection system; (b) Perspective view of the operation of a scanning-mirror projection-lithography system; (c) An image of a 1X-mask is projected into an arc-shaped slit. The wafer and mask are simultaneously scanned across the field-aperture (*yellow area*) until the entire wafer is exposed.

at the turn of the century throughout the world. One advantage still retained by such 1X-scanners is the immense size of the scanned-field. Some semiconductor devices (such as two-dimensional video-detector arrays), require this large field-size. However, in most cases the need for smaller images has driven lithography toward steppers (Sect. 19.5.2) or the newer step-and scan technology (Sect. 19.5.3).

19.5.2 Step-and-Repeat Projection-Printing

To image even smaller minimum feature-sizes than 1.0-μm it became necessary to employ *step-and-repeat, reduction projection-printing* (i.e., initially, reduction of 5X, and eventually 4X). The first step-and-repeat system (*stepper*) was introduced in 1978, and this technique remained viable for 6 generations of technology (i.e., from 1.2-μm down to 0.25-μm). Figure 19-21 illustrates the step-and-repeat principle.

The wafer-stage first moves the wafer into correct alignment with the reticle. That is, the wafer is positioned so that a projected-image of the layer to be exposed will properly overlay the patterns already on the wafer. (The systems that allow the stepper to bring the wafer into proper-position are discussed in Sect. 19.6.) Once the wafer is properly-positioned and brought into focus, a shutter in the illumination system is opened (see Fig. 19-16), allowing light to pass through the photomask. The entire pattern on the reticle is imaged by the lens onto the wafer (Fig. 19-21). However, the projected-image is reduced laterally (with respect to the pattern-size on the reticle) by the amount N:1 where N is the lens reduction-factor, most commonly equal to 4 or 5. On the most-

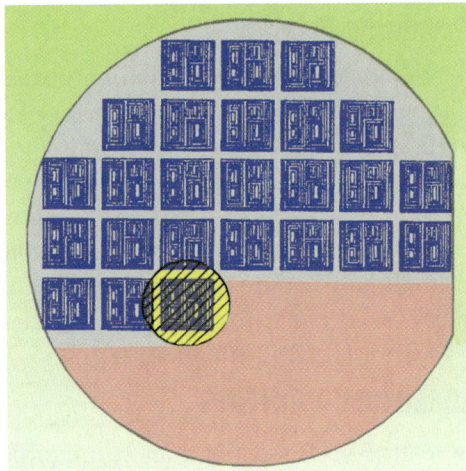

Fig. 19-21 *Step-and-repeat projection-systems (steppers)* employ reduction optics and expose only one-field at a time. The 4X or 5X reticle remains stationary with respect to the lens, whose maximum exposure-area is shown as the *yellow-area*.

CHAPTER 19 LITHOGRAPHY II: IMAGE FORMATION AND OPTICAL HARDWARE

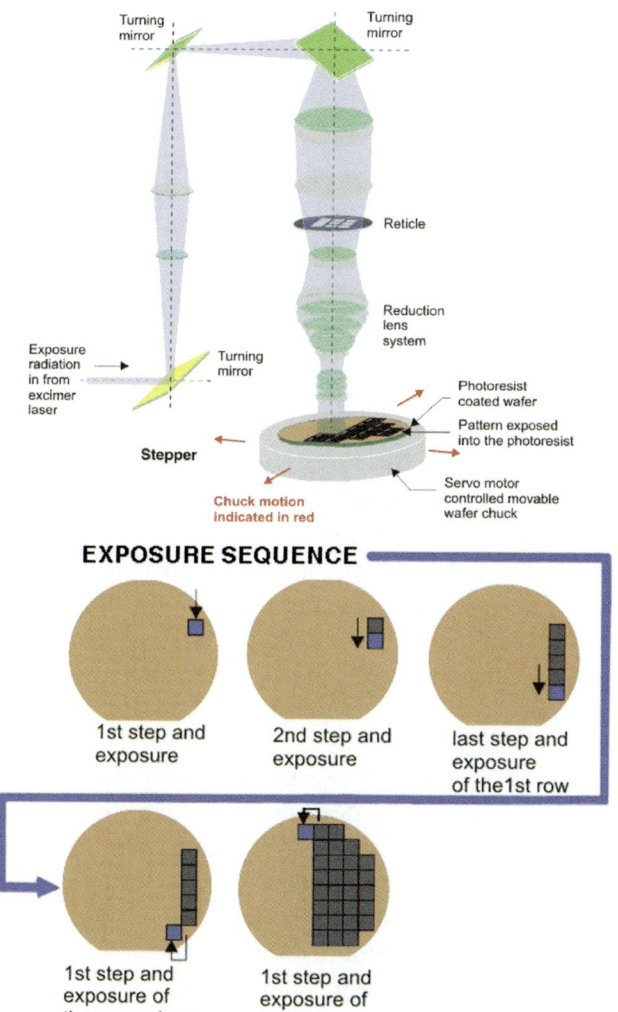

Fig. 19-22 After each field in is exposed (see Fig. 19-21), a high-precision-stage within the stepper moves the wafer to the position where the next exposure will occur. If the chip is small enough, two or more chips may be printed in each exposure. Drawings courtesy of *www.icknowledge.com*.

(or *steps*) the wafer to the next alignment-position, and exposure is performed again. This *stepping-and-exposing* is continued until all the fields on the wafer are exposed (Fig. 19-22b). In most steppers, the stage itself travels horizontally beneath a fixed, vertically mounted lens, as shown in Fig. 19-22a. However, in the Micrascan® stepper (and step-and-scan) systems of SVG Lithography (acquired by ASML in 2001), the wafer moves vertically (see Fig. 19-25b).

A cutaway-view of such a stepper is shown in Fig. 19-23 (and this same tool is shown from the rear in the figure on p. 340). It shows all of the major-subsystems: *reduction-lens* and *illuminator*, *excimer-laser light source*, *wafer-stage*, *wafer-cassettes*, and *operator workstation*.

Most stepper-lenses are *refractive* (see Sect. 19.1.2), meaning that the optical-elements are all made of transmitting-glass. The reduction-lenses are actually *compound-lenses*, that is, assemblies of multiple glass-elements (typically 25-35 in number) mounted properly (and held firmly), in a massive, steel cylindrical-jacket. This housing maintains their exact relative positions (see Fig. 19-24b).

A non-reduction (i.e., 1X) step-and-repeat system was also introduced in 1980 based on reflective optics. While it did not have as good a resolution-capability as the reduction steppers, economics made it attractive

advanced steppers used early in the decade starting in the year 2000, $N = 4$, whereas values of N of 1, 2, 2.5 were also found on steppers that were designed primarily for high-productivity.

Because of reduction by the projection-optics, only part of the wafer (an *exposure-field*) is exposed at any one-time by the wafer-stepper. After that exposure is completed, an extremely-precise x-y stage moves

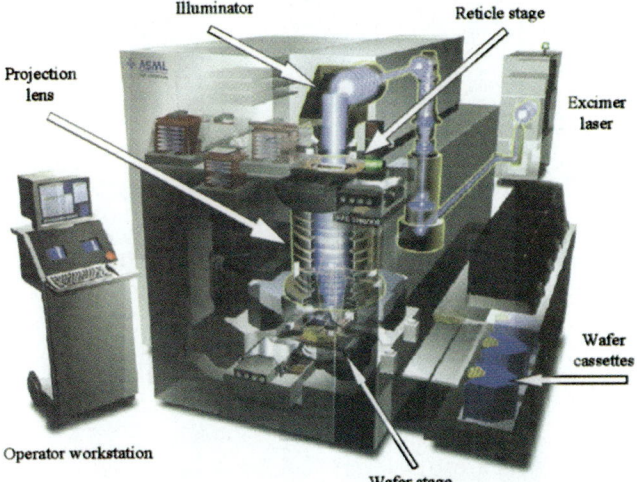

Fig. 19-23 Cutaway illustration of a *step-and-repeat* system. Shown is the ASML PAS5500® system. Courtesy ASML.

for printing non-critical levels of ICs that contained larger minimum-feature-sizes. Reduction-steppers (4X or 5X), however, are all-refractive optical systems that need a narrow-bandwidth light-source. But narrow-bandwidth illumination gives rise to multiple reflections in the photoresist (see Chap. 18), and thus the use of *anti-reflection coatings* became necessary.

To get the best resolution with such steppers, several approaches were pursued. First, *smaller wavelengths of light* were used. (After the *g*-line light of the mercury-arc lamp, came the use of steppers that employed *i*-line illumination. This was replaced by 248-nm light from a KrF excimer-laser light source.) Second, *larger numerical-aperture* (*NA*) lenses were used. The stepper-lenses could be made with a considerably-higher NA than is practical for the full-wafer scanner-lenses. The earliest steppers had an NA of 0.28, yielding a resolution of about 1.25-μm at an exposure-wavelength of 436-nm (the mercury *g*-line). By 2002 lenses with an NA greater than 0.7 became available. However, building large-NA lenses requires much-more-massive lens-assemblies. Early stepper-lenses weighed about 10-lbs (and resembled camera lenses). But these lenses grew in size (Fig. 19-24a), such that the largest of them approached one-ton in weight and were longer than 1-meter in length.

19.5.3 Step-and-Scan Projection Printing

When feature-size were reduced to less than 0.25-μm, even the step-and-repeat approach encountered difficulties. That is, larger-NA lenses were needed, and the chip-sizes themselves also increased (calling for larger exposure-fields). These combined needs were a serious challenge for lens-design and fabrication. One way to ease these demands was to return to scanning technology. Thus, in 1990 a step-and-scan approach was introduced, and by the 0.13-μm CMOS generation, this type of optical printing-tool was replacing step-and-repeat systems for critical masking-layers.

In the *step-and-scan* approach, a reduction-lens is used to scan the image of a large exposure-field onto a portion of the wafer. The wafer is then moved to a new position where the scanning process is repeated (Fig. 19-25a). The lens-field is required only to be a narrow slit (as in the older full-wafer scanners). This allows a scanned-exposure whose height is the diameter of the static lens-field, and whose length is limited only by the size of the mask and the travel of the wafer-stage. The first step-and-scan lithographic equipment was developed by Perkin-Elmer, using an arc-shaped slit. Step-and-scan systems were further pursued by Silicon Valley Group (SVGL), an industrial successor to the Perkin-Elmer lithographic-division. SVGL was later acquired by ASML (in 2001). The remaining manufacturers of lithographic tools also pursued this approach, and now offer step-and-scan systems.

19.6 OVERLAY AND WAFER-STAGES

To this point the quality of the lithographic imagery has been emphasized. Of comparable importance is the *accuracy with which an image can be positioned on the surface of a wafer*. Integrated-circuits are fabricated by patterning a sequence of masking-layers, and the features on successive-layers bear a spatial relationship to one another. As a part of the fabrication process, each level must be aligned to the previous

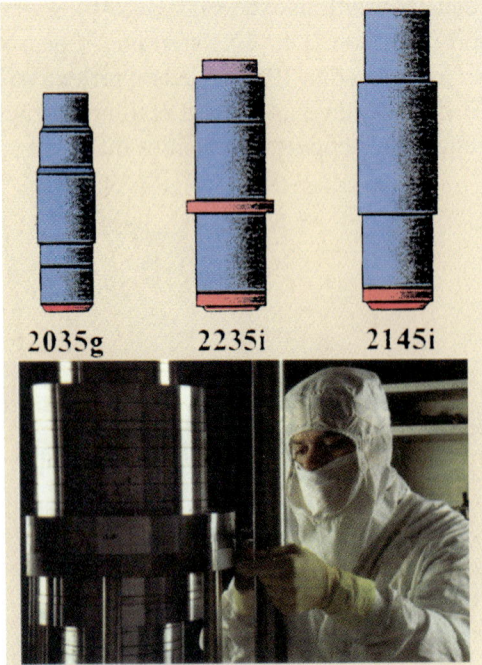

Fig. 19-24 (a) Evolution of the size of stepper-lenses from 1985-90. (b) Photo of technician adjusting a 325-kg lens.

levels. To do this *alignment-marks* must exist in the mask patterns (and therefore also on the wafer), and the wafer stage must be able to align these marks to within acceptable overlay specifications. These topics are discussed in this section.

19.6.1 Alignment-Marks

Alignment of the patterns on a mask or reticle to the pre-existing layers already patterned on the wafer is done with the use of *alignment-marks* (also called *targets*). These are special patterns (Fig. 19-26) which an operator (or an automatic alignment-system) uses to position a mark on the mask to a corresponding mark on the wafer pattern.

19.6.2 Wafer-Stages

In addition to the high-degree of optical-performance that a stepper must provide, it must also be able to *position* (and during exposure, *hold*) the wafer to a phenomenal degree of accuracy. The entire wafer must be physically in the proper-position at the time of each exposure, and this must be done rapidly and accurately. The task of bringing a 200-mm wafer to a position that is within 100-nm of the correct overall location (which is the allowed overlay error in lithography for 0.35-μm CMOS) is akin to being asked to move a 32-mile iceberg from one position to another, and then stop it within 1 inch of its desired location on an x and y grid (and with the correct θ as well).

The ability of the wafer-stage to position a wafer with such accuracy is based on the technology of the *laser heterodyne interferometer* (Fig. 19-26a). The light-output of this laser is extremely stable, with respect to the wavelength of the light it emits (632.8-nm). By counting the number of He-Ne laser-light wavelengths along an optical-path, the length of the path can be accurately determined. In the newest steppers and scanners, knowledge of the stage position is possible to a precision of $\lambda_{He-Ne}/256 = 2.5$-nm).

The design of stepper wafer-stages involves a number of tradeoffs, balancing the needs for high stage-speeds and positional-accuracy. High stage-acceleration and velocity are best met with a stage having a small-mass. But, wafer-size, thermal-stability, and dimensional-rigidity dictate a lower-limit on the mass. Stages are driven by *linear electric-motors*. To minimize the time between successive exposures on the same wafer, the settling-time must also be minimized (i.e., the time required to stop and settle the stage to a tolerable degree of vibration).

19.7 OFF-AXIS ILLUMINATION

Off-axis illumination (*OAI*) is a method for improving the resolution of a stepper without having to resort to a higher-NA lens or a shorter-wavelength of light. The method was introduced in 1989 to microlithography by Mack and Fehrs. Researchers at Canon, Inc., and

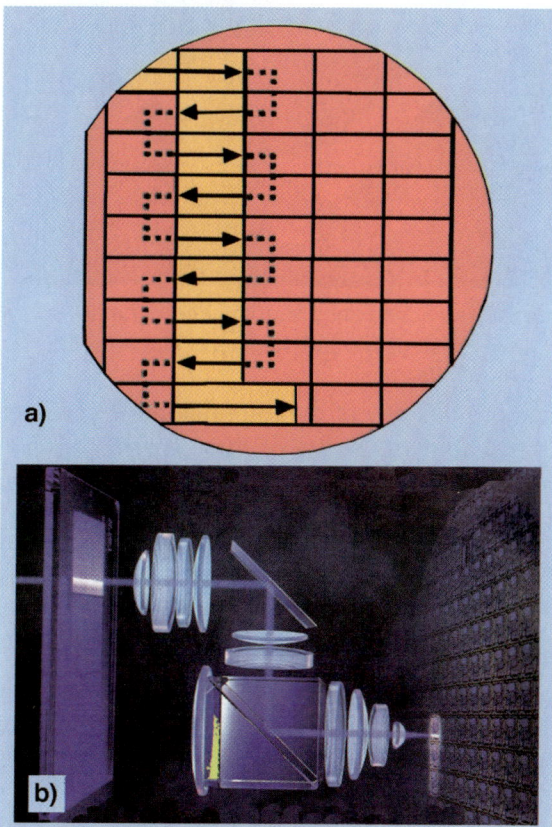

Fig. 19-25 (a) The *step-and-scan principle* combines the operations of a stepper and a scanner. The wafer and reticle are simultaneously scanned across the aperture. Stepping accomplishes the major moves from one exposure-field to the next. Within each exposure-field, the reticle-pattern is scanned across the field. (b) The *catadioptric-optics* of the ASML Micrascan® system. Courtesy of ASML, Inc.

Nikon Corporation incorporated it into the optical systems of steppers in 1992. The resolution-capability is improved by modifying the direction of incidence of the illumination on the reticle. It directs the beam of light through the reticle in a direction that makes it strike the projection-lens at the edge of the entrance pupil (rather than at the center). It produces the most resolution-enhancement for the case of dense-patterns of equal lines and spaces (such as occur in DRAM and SRAM interconnect-lines). It also improves the DOF of the stepper. One limitation is that it does not provide the same resolution enhancement for isolated lines as for dense lines and spaces. It is also necessary to compensate for loss of illumination.

Typically, symmetrical OAI sources are used. Two sources, placed on opposite sides of the pupil (Fig. 19-28a), or *quadrupole-illumination* (Fig. 19-28b) can be employed. An *annular-aperture* (Fig. 19-28c) improves features of all orientations, but at the price of milder improvements in the resolution and DOF.

OAI is easily implemented. Stepper manufactur-

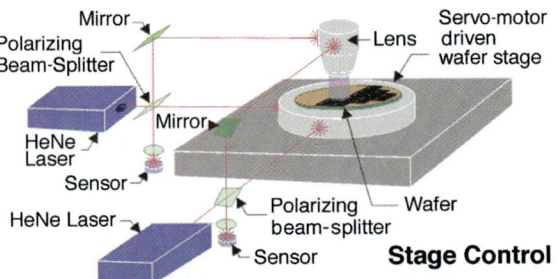
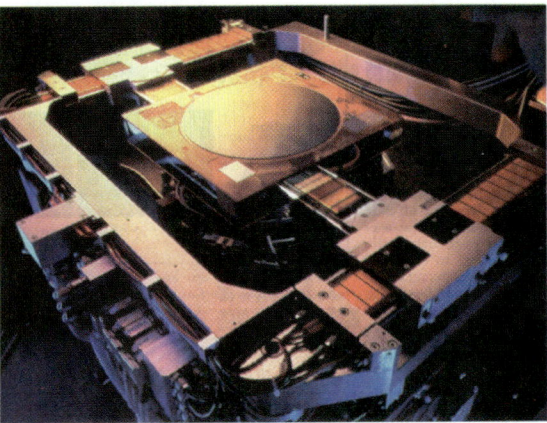

Fig. 19-27 (a) Schematic; and (b) Photograph of a *wafer-stage* used in steppers and scanners. Drawing courtesy of *www.icknowledge.com*.

ers can provide any of the illumination-aperture types described above. A series of these apertures can even be mounted on a turret, and the particular aperture

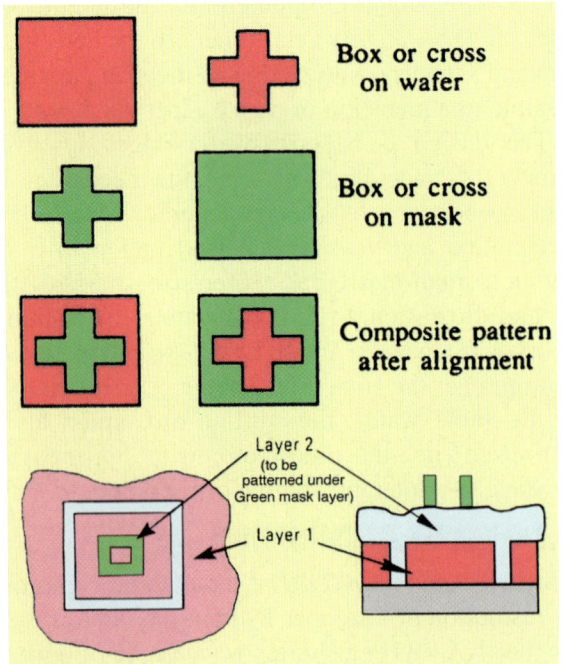

Fig. 19-26 (a) Typical cross-within-a-box overlay target designs. (b) Frame-within a frame overlay target design.

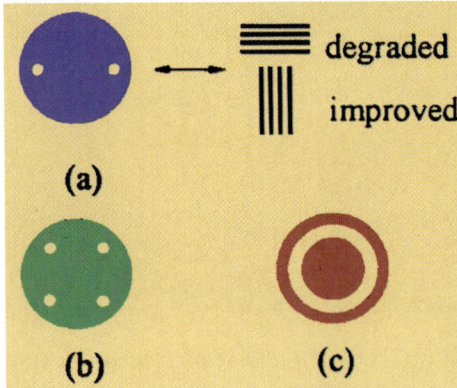

Fig. 19-28 Different types of *off-axis illumination apertures*. The apertures with holes at the edges or corners - (a) and (b) - will improve lines oriented in one direction, but will degrade imaging of lines in the orthogonal direction. (c) Annular off-axis illumination aperture.

desired for any mask can be automatically supplied by the stepper control-program. More information on off-axis illumination is given in Refs. 1, 2, 3, and 9.

REFERENCES

1. S.Wolf and R.N. Tauber, *Silicon Processing for the VLSI Era: Vol. 1 - Process Technology*, 2nd Ed., Lattice Press, Sunset Beach CA, 2000, Ch. 19 (Optical Hardware).

2. S.Wolf, *Silicon Processing for the VLSI Era: Vol. 4 - Deep-Submicron Process Technology*, Lattice Press, Sunset Beach CA, 2002, Ch. 6 (Deep-Submicron Lithography-II).

3. H.J. Levinson, *Principles of Lithography*, SPIE Optical Engineering Press, Bellingham WA, 2001.

4. M.S. Hibbs, "Exposure Systems," in *Microlithography: Science and Technology*, J.R. Sheats & B. Smith, Eds., Marcel Dekker, New York, 1998, Chap. 1, pp. 1-107.

5. *Sub-Half-Micron Lithography for ULSI*, K. Suzuki, S. Matsui, and Y. Ochiai, Eds., Cambridge Press, 2000.

6. O. Semprez, "Excimer Lasers for Future Lithography," *Solid State Technology*, July, 2000, p. 255.

7. P.M. Ware, "300-mm Lithography Enters Reality," *Semiconductor International*, July 2000, p. 128.

8. "Lithography Tools for 300-mm," *Solid-State Technology*, September 2000, p. 40.

9. A.K.-K. Wong, *Resolution Enhancement Techniques in Optical Lithography,*" SPIE Press, Bellingham WA, 2001.

PROBLEMS

1. List the types of *exposure-and-alignment* systems that require the use of *masks*. Then list the types of such systems that require use of *reticles*.

2. Name ten factors that have an impact on the resolution of a photoresist process.

3. Why do lithography processes for technologies with feature-sizes below 0.25-μm need CMP?

4. What is *light intensity* and *exposure dose*, and how are they related? Explain why exposure-and-alignment systems need high-intensity, short-wavelength light sources.

5. Calculate the minimum distance that can be just-resolved (*Rayleigh-Criterion*) by a 0.16-NA lens when: (a) *g*-line and; (b) *i*-line radiation from a mercury-arc spectrum is used. Repeat this exercise for a 0.28-NA and a 0.35 NA-lens.

6. As the NA of a lens is increased, the image-field diameter decreases. For a 150-mm diameter wafer, how many exposures would be necessary to completely expose a resist layer, if a step-and-repeat refractive projection-aligner with a lens of (a) 0.28-NA (with an image field-size of 1.96-cm^2), and (b) 0.35-NA (with an image field-size of 0.5-cm^2) was used?

7. Discuss the differences in the principles of operation of (a) scanning projection-aligners, (b) reduction refractive step-and-repeat aligners, and (c) 1X step-and-repeat aligners.

8. Define *depth-of-focus* (DOF), and *plane of best focus*. Write the equation for Rayleighs DOF Criterion. What happens to DOF as resolution increases.

9. Show that the optical resolution of a 0.54-ArF lens is the same as that of a 0.7-NA KrF lens. Which has the greater depth-of-focus?

10. An optical lithographic system has an *exposure-power* of 0.3-mW/cm^2. The resist *exposure-energy* for positive resist is 140-mJ/cm^2, and for negative resist is 9-mJ/cm^2. Assuming negligible times for loading and-unloading the wafers, compare the *throughput* for positive and negative photoresist.

11. For a reticle that is 152-mm x 152-mm, and assuming that a 10-mm border at the edges of the plate is required to hold the reticle, what is the largest field on the wafer that can be printed if the reduction factor is 10X, 5X, and 4X?

12. Give two reasons why alignment systems should be designed to function at wavelengths different than the ones used for patterning the resist.

13. Silicon wafers have a *coefficient of thermal expansion* (CTE, see Chap. 5) of 20-ppm/°C. Thus, what change in temperature will cause a 10-nm change in the distance between two points on opposite sides of a 300-mm wafer? What level of wafer-temperature control will be needed for 25-nm overlay budgets?

14. What happens to the resist sidewall-angle as the exposure-dose exceeds its optimum-value? Explain why this happens.

CHAPTER 20

LITHOGRAPHY III: PHOTOMASKS AND ADVANCED LITHOGRAPHY

CHAPTER CONTENTS

20.1 MASK & RETICLE FABRICATION
- Glass Quality and Preparation
- Glass-Coating (Chrome)
- Mask Imaging (Resist Application and Processing)
- Pattern-Generation
- Pellicles

20.2 RESOLUTION-ENHANCEMENT TECHNIQUES (RET)
- Optical Proximity Correction (OPC)
- Phase-Shift Masks (PSM)

20.3 LIMITS OF OPTICAL-LITHOGRAPHY

20.4 NEXT-GENERATION LITHOGRAPHIC TECHNOLOGIES (*NGL*)
- Extreme Ultra-Violet Lithography (EUV)
- E-Beam Projection-Lithography (EPL)

"The only way to discover the limits of the possible is to go beyond them, into the impossible."

Arthur C. Clarke

A worker inspects the *Engineering Test Stand* (ETS) - a full-scale prototype EUV lithography system - as it is assembled at Sandia National Laboratories. *The ETS was developed by the EUV Ltd. Liability Company (EUV LLC) under a Cooperative Research and Development Agreement (CRADA) with the Virtual National Laboratory (VNL).*

In this chapter we cover the remaining topics of *lithography* that were not treated in Chaps. 18 and 19, namely: **1**) photomasks and reticles; **2**) resolution-enhancement techniques that involve reticles (*optical-proximity-correction* [*OPC*] and *phase-shift masking* [*PSM*]); and **3**) *next-generation lithography* (*NGL*) techniques.

20.1 MASK AND RETICLE FABRICATION

The "patterned tooling" vehicles that carry IC circuit-designs are known as masks and reticles (Fig. 20-1). A *mask* is defined as a substrate that contains patterns that can be transferred to an *entire wafer (or to another mask)* in one exposure. A *reticle* is defined as a substrate containing a pattern that must be exposed *at each stepped-location* in order to create the image of that pattern a number of times over an entire wafer. Exposing an entire wafer with a reticle thus requires *multiple exposures*. Usually

Fig. 20-1 Semiconductor worker examining a *reticle*.

the pattern-size on a reticle is enlarged from four to five times the size of the image on the wafer (4X to 5X), but in some instances it is of equal size (1X). Reticles are used in two applications: **1)** for printing images directly onto wafers in *step-and-repeat* (and *step-and-scan*) aligners; and **2)** for printing images of patterns onto masks. In *1X wafer-steppers*, the pattern on the reticle is the same size as the image projected on the wafer (1X), while the patterns on reticles of *reduction-steppers* consist of enlarged versions of the actual-device patterns (e.g., by 4X or 5X).

The pattern on conventional optical-lithography reticles is formed in a thin chromium-layer that is deposited on a fused-silica (so-called "quartz") substrate. The pattern represents one-level of an integrated-circuit design. Typically, 20 to 25 different mask-levels are required for a complete ULSI-chip. Masks must be generated from an electronically-stored original pattern. For ULSI, the standard way of creating the pattern on a reticle or mask is by direct-writing this stored-pattern onto a *mask-blank* (i.e., a chrome-covered glass-plate coated with photoresist). The standard direct-writing techniques for ULSI employ either an *electron-beam* or *laser-beam* to form the pattern onto a 4X or 5X reticle. In this section, details of the steps necessary to generate high-quality masks and reticles are described. (More details about mask-making can be found in Ref. 1.) In Sect. 20.2, two *wavefront-engineering techniques* are covered. These two techniques, when incorporated into reticles, improve the optical-image so that optical-lithography can extend beyond the range of conventional patterning. These so-called *resolution-enhancing techniques (RETs)* are: **1)** *optical-proximity correction*; and **2)** *phase-shift masking*.

20.1.1 Terminology and History of Photomasks

The mask-writing process must provide: **1)** high-resolution; **2)** tight pattern-position accuracy; and 3) low-defect levels. One general mask and reticle-related term worth defining at this point is its *polarity*. That is, a mask or reticle is called a *clear-field* (or *positive*) *tool* if the field (or background) areas are transparent (Fig. 20-2a). It is called a *dark-field* (or *negative*) *tool* if the field is opaque (Fig. 20-2b). Another set of photomask-related terms has been established since the introduction of phase-shift masks. In conventional masks and reticles the patterned-area is either clear or opaque, and hence these masks are now termed *binary* masks (or reticles). In some phase-shift masks (PSM) however, there are regions covered with a film that only partially-transmits light, and these masks are called *halftone* or *embedded* masks, or reticles.

Up to about 1985, mask-makers had to match the demands of wafer-production stride-for-stride. Each time IC feature-sizes shrank, mask-makers had to develop ways to fabricate masks with correspondingly-reduced linewidths. This situation persisted until the minimum feature-size reached about 1-micron. At that point (which occurred in the mid-1980's), a period termed the *"maskmakers-vacation"* began, and it lasted for nearly a decade. During this time the further pursuit of mask-feature-size reduction was suspended. Several factors were the cause. The most important of these was that wafer-fabs switched from 1X-proximity-aligners and 1X-scanning-projection aligners to 5X-reduction-steppers. This meant that the features on the reticle could suddenly be five-times-larger than the features being produced on the wafer. The adoption of *pellicles* also reduced the demand for new reticles. (Pellicles are discussed in Sect. 20.1.4.) Finally, automation of electronic-designs reduced the number of test-iterations of a design prior to produc-

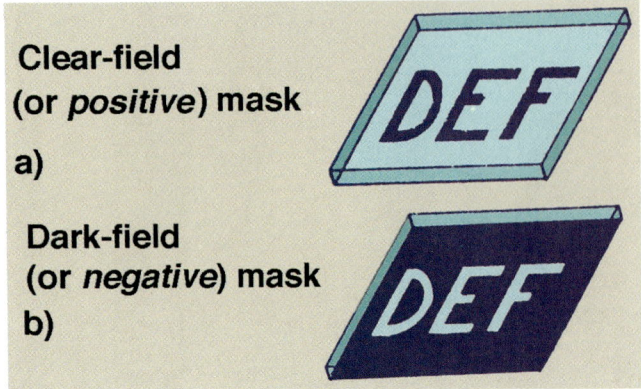

Fig. 20-2 Photomask polarity.

tion, and thus fewer reticles needed to be written. One result of these circumstances was that photomask costs declined. By 1995, this price-decline caused photomask-costs to drop to ~1% of the wafer selling-price (down from ~3% in 1985).

The lives of the pattern-generation machines developed in 1985 were stretched for three device-generations during the "maskmakers-vacation," but by 1995 they were no longer able to produce the leading-edge masks needed for new devices.[2] Although the mask-features were still typically 4 or 5 times larger than the images created on the wafer (due to the reduction lithographic-lens), the tolerances on the mask-dimensions have always been a much smaller-percentage of the feature-size (e.g., ±10%). Thus, for a 0.5-μm feature on a wafer, a 5X reticle-pattern would have a dimension of 2.5-μm, but this dimension would have to be controlled to within ±0.25-μm. The need to use resolution-enhancement-technology on the mask (i.e., *phase-shift masks* [*PSM*] and *optical-proximity correction* [*OPC*]), also required the mask to have features added to the primary-pattern smaller than the *primary feature-dimensions*. In fact, the introduction of these *secondary-features* has been causing the minimum feature-size on reticles to shrink faster than the feature-size on wafers. These secondary-patterns also make the time to write the mask longer (e.g., adding OPC-features to a reticle can increase the pattern-database by ~6X). As a result, by the late-1990s, photomask-costs rose dramatically. By 1998, reticle-costs averaged nearly 2% of the wafer selling-price. By 1998, mid-range-complexity reticles sold for about $1,500–$2,000 and high-end 0.35-μm reticles sold for $3,500–$9,000. But, prices of masks with 0.18-μm design-rules and OPC and PSM were being quoted at more than $35,000.[3] The worldwide merchant-market for masks in 1998 was $1.4 billion, and was expected to more than $2.4 billion by 2002.

By the late 1990's, *merchant-maskmakers* supplied about two-thirds of the reticles used by *semiconductor-chipmakers*, while *captive-mask-makers* supplied the rest. From 1985 to 1998 the photomask-industry matured and consolidated. By 1998 DuPont Photomasks and Photronics were the largest worldwide merchant-maskmakers, with Dai Nippon Printing and Toppan also having a large share in Asia. Captive mask-masking shops were being phased-out, as only the very-largest IC-manufacturers were able to justify the substantial investment of maskmaking. That is, the barriers to sustaining such a capability are formidable (as a result of the large capital-investment required to keep up with the rapid-pace of technology). Besides the mask-exposure tools, two mask-inspection tools must be purchased for every mask-lithography system. Mask-repair systems and a cleanroom are also needed. Electron-beam-lithography tools cost $7–15 million each, inspection-systems cost $3–4.5 million, and repair-tools cost about $3 million. Redundancy of key-equipment is also needed, making the overall costs even higher.

20.1.2 Fabrication of Photomasks and Reticles

A number of sequential steps are needed to fabricate the masks and reticles used chip-manufacturing. They include: **a)** preparation of high-quality glass-plates; **b)** glass-coating (with chrome); **c)** mask-imaging (resist-application and processing); and **d)** pattern-generation. These are described next in more detail.

Glass-Quality and Preparation: The substrates of ULSI-reticles are highly-polished, extremely-flat plates of *fused-silica* (SiO_2). The glass material used for this application must be free of defects on both surfaces (and internally), and should have high optical-transmission at the resist exposure-wavelength. Several types of glasses were used in earlier IC-fabrication for making photomasks, including soda-lime glass and borosilicate glass. However, their high thermal-coefficients-of-expansion (TCE) made them largely unsuitable for VLSI and ULSI applications. Fused-silica replaced them, due to its smaller TCE and the fact that the other glasses are not sufficiently transparent at 0.248-μm and 0.193-μm wavelengths. (The transmission of fused-silica is about 90%/cm and 85%/cm for these two wavelengths, respectively.) The low-TCE of fused-silica minimizes any temperature-variations during the mask-writing process and from thermal-effects caused by heat-transfer from the stepper. Nevertheless, even when fused-silica is used, a temperature-change of only 0.08°C will change the pattern-placement accuracy by 10% of the allowed-tolerance for 0.25-μm technology. The allowed temperature-variation within a pattern-generation environmental-chamber is typically specified at ±0.03°C, to prevent this error-value from being reached.

The fused-silica plates are prepared by cutting them from large glass-sheets. The inclusion-free plates are polished to remove any chips. Since fused-silica is a relatively hard material, it requires special effort to achieve the required flatness and surface-finish. The polishing-procedure goes through several iterations, using progressively finer-grade abrasives on both sides of the plate. The flatness of a typical reticle over the pattern-area is < 2-μm. Plates are cleaned, rinsed, and dried prior to inspection and coating.

Plates can also sag significantly under their own weight. Up to the mid-1990s, the standard reticle-substrate-size for 5X steppers was a 5"-square glass-plate. Thereafter, this size was increased to a 6"-square (in order to accommodate larger die-sizes). The minimum plate-thickness at that point also had to be increased to 0.250" (from the previous thicknesses of 0.090" and 0.120") to minimize pattern-placement errors due to the gravitational-sag of the glass-plate. By the late 1990's, however, die-designs with even larger dimensions appeared. The desire to stay with 6"-square glass-substrates caused the industry to move from 5X to 4X demagnification.[4] Increasing the plate-size to 9" has been proposed, but as of 1999 this change was still in the discussion stage.

Glass-Coating (Chrome): The glass-plates are next coated with a material in which the pattern is eventually formed. Today, the standard material used for this purpose is chromium, with an underlying-film that acts as a "glue" layer (consisting of a mixture of chromium, nitrogen, and oxygen), and a top-layer that acts as an anti-reflective layer (consisting, perhaps, of a thin [~20-nm] layer of Cr_2O_3). These films are deposited onto the plate by sputtering. Chrome was selected for this application because it is easy to deposit and etch, and is completely-opaque. Furthermore, techniques are available for repairing any pinholes detected in the Cr-layer. Generally, the thickness of the overall opaque-layer is about 100-nm. Sputtering has the advantage of high-throughput, good-adhesion, and excellent-thickness uniformity.

Mask Imaging (Resist Application and Processing): The application of resist and its subsequent processing to produce images (either with an e-beam or laser-beam) is similar to that used for wafers (see Chap. 18), except that the resist-coating is much thinner, and different types of exposure-equipment are used. *Laser-based mask patterning-systems* are used on less-complex designs as they are faster and less expensive to purchase and operate. *Electron-beam* (or *e-beam*) *systems* are more often employed for complex-reticle exposures, as they produce finer line-resolution (because the electron-beam has a much smaller spot-size than the laser-beam, as well as tighter overlay).[6] E-beam tools can also write larger die-sizes than the laser-tools.

Prior to coating, the blanks are cleaned and dried, and then spin-coated with filtered photoresist. The resist is rapidly soft-baked because of the thin-coatings used. Exposure-control is critical since the high-reflectance chrome-surface results in standing waves. The resist is then hard-baked prior to etching (to ~100°C). The process of transferring the pattern into the chrome has largely been accomplished using wet-etching. This approach is used because PBS-resists exhibit poor dry-etch resistance, and effective wet-etch processes have been developed for thin chrome-films. However, since wet-etching is an isotropic process, some undercutting of the chrome-features inevitably occurs.

Pattern-Generation: Pattern-generation creates an image of one-level of an IC-circuit on the reticle or photomask by exposing a set of accurately-positioned rectangles onto the resist-covered blank. Since the patterns generated by IC-designers for each level are, in general, polygons, these patterns must be decomposed into rectangles. The challenge of pattern-generation is to reduce any image to a set of rectangles with a defined height, width, and pattern-angle relative to a fixed-coordinate.

Most ULSI reticles are exposed using a direct-write electron-beam system. The *electron-beam exposure system* (EBES) was originally developed at Bell Laboratories during the 1970s. It utilizes raster-scan techniques to write the pattern. Commercial versions for mask-fabrication are available and find wide use at both commercial and in-house mask shops. The largest supplier of EBES systems is ETEC Systems (which was acquired by Applied Materials in 2000). Their MEBES 4500 and 5000 systems are very-well suited for 5X, 4X, and 1X reticle-fabrication and for 1X masters.[5] Figure 20-3 shows a schematic of such a system.

An alternative to e-beam pattern-generation of reticles (at least for 5X reticles), is a laser-based system.[7] By 1999 these could produce reticles with a minimum feature-size of 1.6-μm, a CD-uniformity of ±0.12-μm, an overlay-accuracy of ±0.12-μm, and a positioning-accuracy of ±0.25-μm over a field size of 128-mm^2. The advantage is that their price is about one-third that of e-beam systems. Furthermore, they are 5-10 times as fast, and use optical-resists.

20.1.3 Mask and Reticle Defects and Their Detection and Repair

Defects in masks and reticles have always been a source of yield-reduction in integrated-circuit manufacture. As the minimum pattern-sizes shrink below 1-μm, and circuits are designed with higher device-densities, defects that were once tolerable can no longer be accepted. For example, a single-defect that in the past killed only an individual-die, now becomes a repeating-defect in stepper-systems, and will kill every die in single-die reduction-reticles. Such defects can be due to incorrect-design of the mask patterns, or flaws introduced into the patterns during the pattern-generation process. Even if the design is correct, and the pattern-generation process is performed correctly (so that the desired mask-patterns are produced),

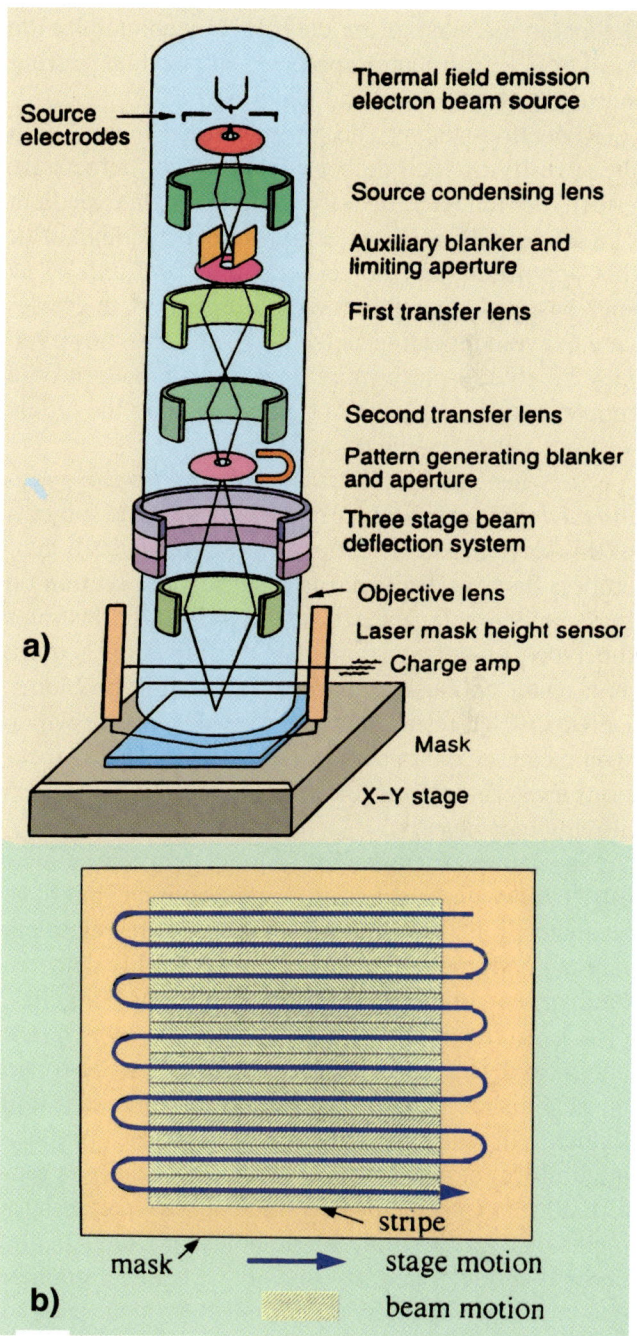

Fig. 20-3 (a) Electron-column of an EBES system. (b) The work-stage of an e-beam system. The electron-beam scans in the y-axis as the stage traverses under the column in the x-axis. Courtesy of Applied Materials.

defects in the mask or reticle can be generated by the mask/reticle fabrication-process, as well as during subsequent processing and handling.

Reduction-steppers do provide some reduction in the severity of such defect-effects. Many defects are effectively rendered invisible by the simple expedient of a 5X reduction. 1X-systems, of course, do not offer this opportunity for defect-reduction, and reticles in such systems must therefore be prepared with greater care to avoid repeating-defects. 1X-systems, however, may be able to place more than one die-pattern on a reticle, and in such cases a single-defect on the reticle will not kill every-die on a wafer.

It is the *mask-fabrication* and *wafer-processing-induced defects* (and their repair), that are the subjects of this section. The detection of design or pattern-generation flaws is discussed in a subsequent section on mask and reticle inspection. Figure 20-4 illustrates the types of defects that may be found on a mask as a result of mask and wafer fabrication problems. Defects which result in inoperative-devices (or which would cause a die to be rejected at final visual-inspection) are termed *fatal* (or *killer*) *defects*, while others are *nonfatal-defects*.

The causes of mask-fabrication defects are numerous.[8] *Raw glass-substrates* can contain bubbles, scratches, pits, and fractures. Bubbles are relatively rare in good-quality plates and can be readily detected (and plates containing bubbles, rejected). Scratches, pits, and fractures are surface-defects, which are harder to detect, and thus are more-likely to be found on a finished-mask. Pits inhibit chrome-adhesion (which will cause pinholes), while scratches produce nonuniform-etching in local areas - especially at pattern-edges. *Chrome-defects* include: **a)** particulate-inclusions in the film; **b)** pinholes or voids in the chrome-surface; and **c)** invisible chemical-anomalies (such as nitrides or carbides), which lead to erratic local-etching that produces undesired-patterns. *Resist-defects* include *voids* (which produce pinholes, that lead to chrome-spots), and *resist gels* that may locally-affect resist solubility. Masks and reticles made with e-beam lithography require the use of e-beam resists. Initially, e-beam processing had been more prone to resist-defects than conventional optical-resist processing, but advances in e-beam resist and material technology eventually made them about equal. Of course *dirt-particles* can also be introduced onto the mask-plate during any of the pattern-transfer steps, again leading to mask-defects.

Upon completion of the mask (or reticle) fabrication-sequence, a series of inspections to qualify the product are performed. The inspection-procedure examines several characteristics of the mask or reticle, including: **a)** linewidth-measurement; **b)** measurement of the registration among the arrayed die-patterns (if more than a single-die per mask or reticle is present); **c)** determining that all the features present in the design database have been transferred to the mask; and **d)** determining if any mask-fabrication defects have been produced. Different inspection-tools are utilized for each of the above inspections. Inspection-procedures and tools for detecting fabrication defects are discussed in this section, while the other inspections are discussed later.

In the past such inspections were carried out by human operators working with a microscope. As masks became more-complex, the task has been relegated to automatic defect-detection systems that perform it much more rapidly, and with fewer errors.[9,10]

The ITRS Roadmap specifies the maximum-size that defects on a reticle can have. For 0.25-μm CMOS-reticles, this size is 50-nm for 1X reticles, and 200-nm for 4X reticles. For 0.18-μm CMOS reticles, these sizes are 36-nm and 144-nm, respectively. Note that the defect-size corresponds to a printable-image on the wafer of 20% of the minimum feature-size.

There are several vendors of automatic mask/reticle defect-detection systems, including KLA-Tencor, Applied Materials, and Lasertec Corp, Japan. The most advanced-models have the capability of performing both die-to-die and die-to-database inspection (Fig. 20-5). The *die-to-die* inspection allows defects unique to an individual-die to be identified. *Die-to-database* inspection allows repeating-defects to be found (which would not be caught by a die-to-die comparison), as well as errors made in converting data from the computer-memory-format to the mask/reticle-pattern. With systems that can perform both inspections, only one-die of a mask-array must be inspected against the database, while die-to-die inspection checks each die for random-defects.

Repairing Defects in Masks and Reticles: Fatal defects in a mask or reticle are unacceptable. However, the cost and time involved in making a reticle are too great to allow it to be discarded for small defects. If such defects are found, attempts are made to repair the reticle, ultimately rendering it free of fatal-defects. Mask-repair methods for accomplishing this purpose have been developed. The repair of *opaque-defects* (i.e., removal of chromium-spots from areas in which they do not belong) has been performed with lasers for many years. A focused laser-beam merely evaporates unwanted material. One concern with laser evaporation is potential-damage of the glass substrate. Large chrome-spots may require several laser-pulses to remove them, and if damage occurs (*laser-burn*), it can become another printable-defect. Another opaque-defect removal-technique is *ion-milling*, in which a focused-beam of gallium-ions erodes the particle.

The repair of *clear-defects* (i.e., the deposition of chromium in areas from which it should be missing) is more difficult. One process is a chromium lift-off procedure (Fig. 20-6a). Several alternative-methods have also been developed, including one which uses a local *pyrolytic-decomposition* of a chromium-bearing gas at the spot where the clear-defect exists (Fig. 20-6b). *Focused ion-beam, mask-repair systems* are sold by FEI and Seiko Epson.

20.1.4 Pellicles

Even though masks and reticles used in projection-printing can be fabricated without defects (i.e., by utilizing repair-techniques if necessary), and no damage-creating contact between mask and wafer occurs, mask-defects due to handling or airborne-contamination (i.e., a speck of dust) can still be generated. In the regime of ≤1-μm resolution, particulates that are large

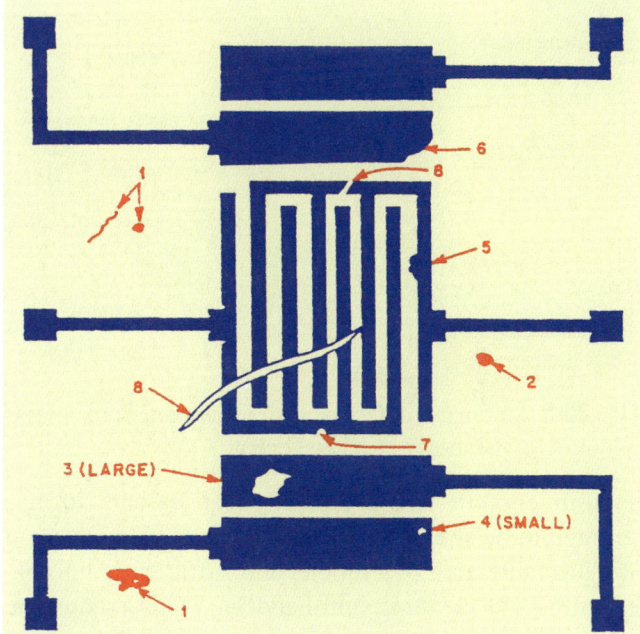

Fig. 20-4 Types of defects commonly encountered on a mask (or reticle): (1) contamination; (2) opaque-spot; (3) large-hole; (4) pin-hole; (5) excess-material; (6) lack of adhesion; (7) intrusion (*mouse-nip*); and (8) scratch.

enough to cause defects are also harder to detect and remove. Thus, even in projection-printing processes, a method for protecting masks and reticles against such defects is highly desirable. Such a method is available, and it involves the use of pellicles.[11,12]

A *pellicle* is a membrane that seals off the mask or reticle-surface from airborne-particulates and other forms of contamination. The membrane is mounted on a metal-frame which is securely attached to the chrome-side of the mask, and the membrane is thus suspended 5-to-10-mm above the mask-surface (as shown in Fig. 20-7c). Thin (0.090-inch-thick) reticles are sometimes given a pellicle on the back, as well as on the front-surface. Backside-pellicles are often not used on thick (0.250-inch-thick) reticles, because the back-surface of the reticle is already sufficiently-far from the focal-plane of the projection-optics. Particles (typically <100-μm) on the pellicle surface are kept so far out of the focal-plane that they are not imaged onto the wafer-surface (Fig. 20-7b). The membrane

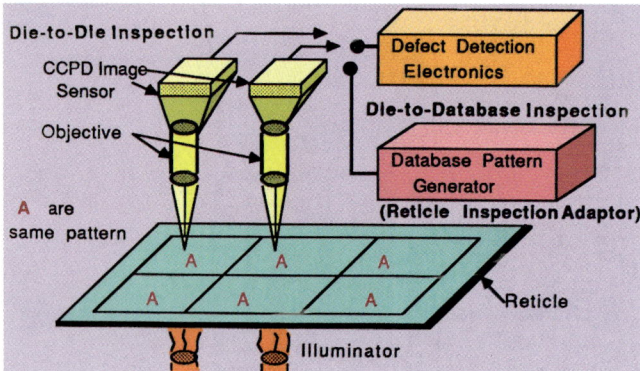

Fig. 20-5 Automatic defect-detection instrument for masks and reticles. Courtesy of KLA-Tencor.

is thin enough to be optically-transparent, strong-enough to be stretched across a support-ring covering the printable-area of a reticle, and sufficiently durable to withstand cleaning and handling. This membrane must also be stable enough to retain its shape over long periods of time and exposure to 100,000-flashes of UV-radiation, and be inexpensive enough to be cost-effective. Several materials have been used for pellicles, including *nitrocellulose-acetate* and *Teflon-like fluorocarbons*, with a thickness of 1-2-μm.

While pellicles adequately protect the wafer's-surface from being imaged by particles on the reticle, if a dust-particle is large enough, it can still cause a dark-spot on the aerial-image. As a result, some steppers offer a pellicle-inspection system that can detect such large dust-particles by scattered light. The pellicle can be inspected each time the mask is loaded to ensure that no such large dust-particles are present.

The optical-transmission losses of a pellicle material (due to reflection, absorption, and scattering), affects the exposure-time, and thus throughput. (In the 350-450-nm range a 2.0-μm-thick nitrocellulose pellicle, with an anti-reflection coating, has an average transmission of ~99%.) Nitrocellulose and mylar (another pellicle-material), however, both exhibit strong absorption near 300-nm, which limits their use for deep-UV applications. Optical interference-effects can also cause significant-variations in transmission through a pellicle, even if the pellicle-thickness exhibits submicron-nonuniformity. Fortunately, pellicles can be manufactured with thickness-uniformities of better than 1%, minimizing such problems.

Masks or reticles with pellicles attached, are cleaned with a DI-water rinse to remove most particulates. Fingerprints can be cleaned by using a mild surfactant and manual scrub. The glass (back) side of the mask is usually cleaned at the same time. Damaged pellicles can be successfully replaced if careful mask-cleaning and inspection is done prior to attaching the new unit. Use of pellicles has become an industry standard (Fig. 20-7d).

20.1.5 Critical-Dimension and Registration Inspection of Masks and Reticles

For ULSI-applications, reticles must be perfect. Any defects or mistakes in the pattern will destroy the functionality of a circuit printed with an imperfect-mask. Before a reticle is delivered to a semiconductor manufacturing-line, it must be inspected to verify that its fabrication was correctly carried out. The detection of fabrication process-induced defects was described earlier. In this section, inspection-procedures of the following mask and reticle characteristics are discussed: **a**) physical-characteristics (e.g., title correctness, polarity of the field of the primary and test patterns; **b**) critical-dimensions (CD); and **c**) registration. The ability to effectively perform the full gamut of inspections on ULSI-reticles involves the use of a set of complex and expensive automated mask-inspec-

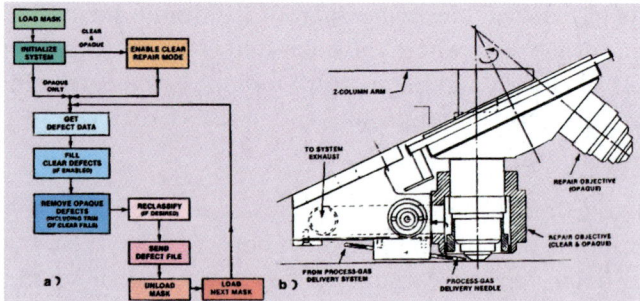

Fig. 20-6 (a) Typical photomask lift-off repair-process-cycle. (b) Opaque mask-defect repair-process by pyrolytic-decomposition of a chromium-bearing gas. Courtesy of Quantronix Corp.

CHAPTER 20 LITHOGRAPHY III: PHOTOMASKS AND ADVANCED LITHOGRAPHY

tion systems that search for any defects.

Checking *physical-characteristics* requires the least-complex inspection-equipment (e.g., an optical-microscope and video-measurement device), but any detected-flaws can still force the rejection of a finished-plate. Since this inspection can be easily performed, some users request a *check-plate* prior to initiating the full tooling-fabrication sequence. A check-plate can be made on a low-cost blank, allowing for early-detection and correction of physical-flaws.

The subject of linewidth-measurement on an IC-wafer was covered in Chap. 18, and the same types of linewidth-measurement techniques used to measure CDs in developed resist-patterns can be used to measure mask or reticle feature-dimensions. As pointed out earlier, the task is considerably easier on a mask or reticle, because the chrome pattern-layer is relatively-thin (making accurate edge-detection easier), and the topography under the chrome is flat (unlike the topography on a wafer near the end of the fabrication cycle). Linewidth-standards are available from the NBS for mask and reticle feature-dimension measurements. The ability to perform both inspection of *critical dimensions* and *registration* on ULSI masks or reticles with a single machine, however, requires sophisticated metrology instruments.

20.1.6 Storage, Transport, and Loading of Reticles into the Stepper

Wafer-fabs that have diverse or customized product-lines (e.g., ASIC fabs or wafer foundries) must manage thousands of active reticles. Custom fabrication-environments often need to make a reticle-change for nearly every lot. Management of the storage, transportation, qualification, and loading of reticles into steppers may require the efforts of several employees. But by the late-1990's these tasks had become increasingly automated.

The reticles in modern wafer-fabs are stored in ultraclean automated-enclosures called *reticle-stockers* or *reticle-management systems* (*RMS*). The ultraclean minienvironments in the RMS (i.e., Class-1 cleanroom-compatible) reduce defects on the reticles, and their on-board database allows rapid automatic-

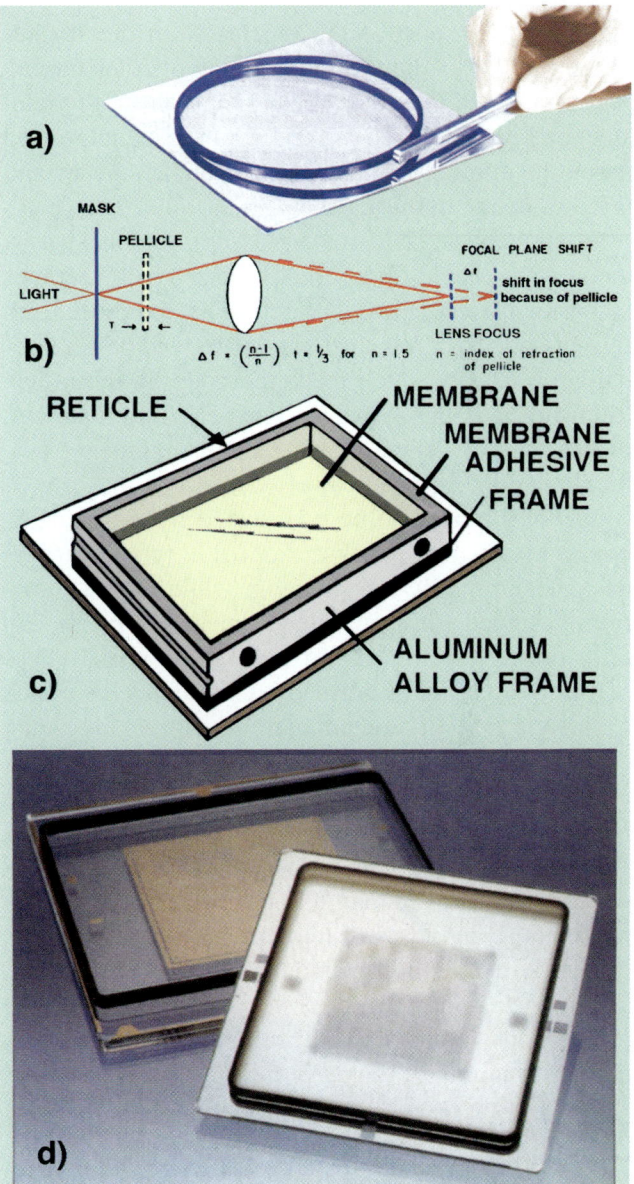

Fig. 20-7 (a) Attaching a pellicle. (b) Focal plane shift of a pellicle. (c) Construction of a pellicle. (d) Photograph of pellicles. Courtesy of Dai Nippon Printing Company.

retrieval of the desired reticle. Both the reticle-cassettes and the reticles themselves have barcodes with the name and serial-number of the reticle. Whenever a reticle is placed into a stepper or stocker, its barcode is read for positive identification. Fabs increasingly use automated-transport of the reticles from the stocker to the stepper-loading robot, using an overhead transport system. Figure 20-8 is a photograph of a reticle-stocker. Figure 20-9 shows the layout of an automated reticle-handling system, with one reticle-stocker and three steppers, in this case.[13]

Each stepper has its own loading-system. That is, the stepper reticle-library holds from 6 to 12 reticles in protective-cassettes. The reticle is removed from this cassette by a robot-arm and is moved past a laser bar-code reader, which reads the code on the reticle to ensure it is the requested one. The reticle is then placed onto the reticle-platen and the alignment of the reticle is done automatically.

Reticles must also be transported from the mask-house to the fab. Mask-makers use reticle-carriers designed primarily for shipping, while fabs use carriers designed for specific-equipment. Problems can arise during the transfer from shipping-box to stepper-type box. This is normally done manually, which creates a risk of contamination (or reticles being dropped). Consequently, a limited-number of personnel at a fab are certified to do such transfers. For these reasons, an effort has been undertaken to arrive at a standard reticle-carrier configuration. If such a universal carrier was available, it could be received from a mask-house and inserted into any machine in the fab, including a reticle-inspection tool, a reticle-stocker, a reticle cleaning-tool, or a stepper. Once inside the tool, robots would handle the reticles, eliminating any manual handling of them.

20.2 RESOLUTION-ENHANCEMENT TECHNIQUES

Techniques have been developed to extend optical-lithography beyond the range of conventional imaging. They are based on the idea that *the 2-D aerial-image formed by the projection-lens is a structure to be optimized*. Two of the ways that this can be done involve modifications to the conventional photomask-format, and they are: **1)** optical proximity-correction of the mask-patterns themselves; and **2)** phase-shift masking. These two techniques are described in the following sections.

20.2.1 Optical-Proximity Correction (OPC)

One way to optimize the quality of the aerial-image formed by a projection-lens is to *pre-compensate the reticle-pattern to account for expected pattern-distortion due to diffraction-effects*. In this method, a computer program is first used to simulate the 2-D aerial-image that is formed for a particular mask-feature (or a group of features).[14] Based on this predicted aerial-image, the reticle-pattern can be altered and simulated again to determine if the change has improved the 2-D aerial-image.[15] The features added to a reticle-pattern based on this procedure are called *optical-proximity correction* features, and the *optical-proximity-correction technique* is discussed next.

One effect observed in microlithography is that the linewidth of isolated and densely-grouped lines eventually varies by a significant amount from that of the predicted-linewidth (based on the pattern-size on the reticle) when the feature-sizes approach the wavelength of the illuminating-light. However, very-little difference between two such aerial-images is observed if the feature-sizes are large (i.e., > 1-μm). For example, assuming the use of positive-photoresist, 0.35-μm-wide *isolated* lines are 0.09-μm wider than *closely-spaced* 0.35-μm-wide lines (i.e., which might be part of a diffraction-grating pattern). Since the linewidth depends on the proximity of other features, this effect is called an *optical-proximity effect*. Such proximity-dependent linewidth-variation (specifically termed *iso-dense print bias*) may become significant in the interconnect-lines of a memory chip. That is, lines within the memory-array region of the chip would be regularly and closely spaced, but in the peripheral sense-amplifier and logic-circuitry, they could be spaced farther apart. If this dimensional-difference can be predicted, the width of the chrome-lines on the reticle can be adjusted to make isolated and densely-spaced printed lines on the wafer all the same width.[16] Such adjustments to the sizes of patterns on a reticle are termed *optical-proximity corrections* (OPC), and in this case such an OPC is performed as a *one-dimensional, pattern-dependent width-correction* to all the lines on a reticle.

Originally, OPC was considered to be just a method to correct the line-size variations as a function of the proximity of other features (as described above).[17] However, other pattern-modification techniques have been developed to improve the images of such two-dimensional IC-features as contact-holes (or vias) and short-lines of poly and metal. That is, corner-rounding and general loss of shape-fidelity in small-features is caused by the inability of the projection-lens to resolve details smaller than the diffraction-limit of the lens. For example, a *square* contact-hole pattern with a size close to the minimum resolution-limit of the optics, will print more nearly like a *circle*. Likewise, a short line (which should print with a rectangular-shape), instead prints as a feature with an *elliptical-shape*. These effects have therefore also been classified as another form of an *optical-proximity effect*.

The OPC-techniques used for modifying the 2-D reticle-patterns to alleviate such deviations in the aerial-images, involve appending *secondary, sub-resolution patterns* to the primary contact-hole (or short-poly and metal-line) patterns. For example,

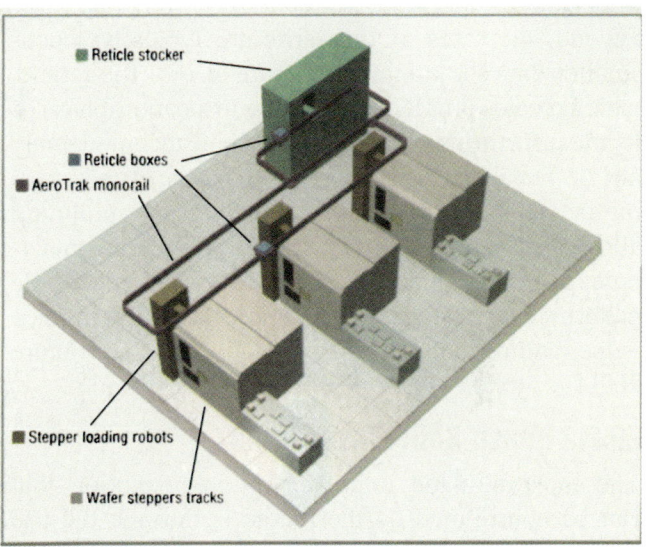

Fig. 20-9 Layout of an automated reticle-handling system. The system shown has one reticle-stocker and three steppers to which the reticles are delivered via vehicles on a monorail-track. Each stepper station has an adjacent stepper-loading robot.[13] Reprinted with permission of Solid State Technology.

by adding small secondary-patterns (called *serifs*) to the corners of a square contact-hole pattern (as shown in Fig, 20-10c) the aerial-image (and hence also the printed-image) becomes more square-like. This occurs because the serifs enhance the amount of light transmitted though the corners of transparent mask-features (e.g., a pattern of a contact-hole being imaged in a positive-resist). For short-lines, serifs (together with linewidth-changes) may be added as shown in Fig. 20-10e to improve the fidelity of the printed-feature. Finally, lines may be isolated during part of their run along the wafer-surface, but may be near other lines at other locations. In such cases, incremental linewidth-changes may be added to the linewidth (and these are also called *line-jogs*).

While OPC-techniques provide significant benefits, they are not gained without cost.[18] The adoption of OPC includes the following drawbacks. First, serifs add a large number of additional features to the database of the reticle-pattern. This increases the time needed to write a reticle (and therefore also its cost).

Fig. 20-8 Photograph of a *reticle stocker*.

Second, the sizes of the serifs are typically much-smaller than the minimum feature-size of the reticle, which creates problems during the inspection-phase of its manufacturing process. Finally, this technique may not be feasible in all situations. That is, if the spacing between two-contacts is already at the minimum allowed dimension, adding serifs to a square contact-hole pattern would bring parts of the contact-hole patterns to a distance below the minimally-allowed value, and in such cases, they cannot be used. Figure 20-11 is a photograph of serifs on mask patterns.

20.2.2 Phase-Shift Masks (PSM)

The most-dramatic improvement in resolution that can be contributed by the reticle is through the use of the phenomenon of phase-shifting. For example, if a lithographic-process can attain a resolution of 0.3-μm using a 0.248-nm KrF light-source with a stepper having an NA = 0.6, then by incorporating *phase-shifting techniques* on the reticle, the resolution can be improved to as much as 0.18-μm. The concept of increasing the resolution of a lithographic-image by modifying the optical-phase of the mask-transmission was first suggested in 1982.[19] While a number of different types of phase-shifting masks have subsequently been developed, they all employ the same basic concept, which is well-illustrated by the original version introduced by Levenson. His concept will be used here to demonstrate the principle of *phase-shift masking* (PSM). Several references review the topic of PSMs in more detail than are provided here.[20,21]

The phase-shifting of light as it passes through the clear-part of a mask is accomplished by adding an extra layer of transmissive-material to the optical-path (Fig. 20-12). For PSM-applications a phase-shift of 180° is usually desired, and this corresponds to an optical path-length-difference of $\lambda/2$ at the stepper-wavelength. Assume now that a grating (representing a *binary imaging-mask* (or BIM, Fig. 20-13a1) is being imaged by optics with a particular numerical-aperture and partially-coherent light. The (normalized) amplitude of the electric-field (*E*-field) at the mask is either +1 or 0, as shown in Fig. 20-13a2. However, the imaging is degraded at the wafer because light from the clear-areas on the mask is diffracted so that it strikes regions on the wafer that would ideally be completely dark. The nominally-dark regions have light diffracted onto them from the clear-spaces on both the right and left of them. The *E*-field amplitude on a wafer as a result of such diffraction is given in Fig. 20-13a3, and the light-intensity (which is the square of the *E*-field amplitude) is plotted on Fig. 20-13a4.

If a phase-shifting material is placed on alternating clear-spaces of a mask (as shown in Fig. 20-13b1), this forms a type of PSM called an *alternating-PSM*, or *ALT-PSM* (or sometimes even a *Levenson-PSM*). It should be noted that producing a *phase-shifter* on a clear-region of a mask may be done in one of two ways. In the first, an appropriate film of material can be deposited over the entire-mask, and then etched-away in the desired areas. In the second, the fused-silica-glass itself can be etched in the appropriate clear-areas to produce recesses in the glass. The latter-technique requires well-controlled etching, so that recesses of the appropriate depth can be produced with only a *timed-etch*. However, since deposition of a film different than fused-silica is not necessary, the

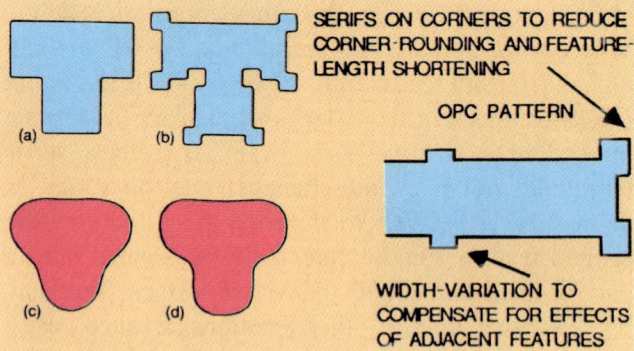

Fig. 20-10 A T-shaped feature (a) with dimensions near the resolution-limit of the projection-lens, is printed as a rather featureless-blob (c) on the wafer. Addition of serifs (b) brings the printed image (d) closer to the shape originally designed.[1] Reprinted with permission of Marcel Dekker. (e) Adjustments to the linewidth can correct for the proximity of nearby features, and serifs on the end of a line can correct line-shortening effects.

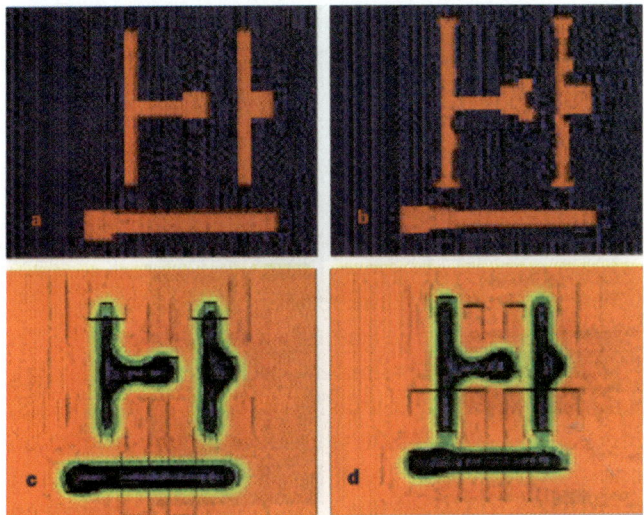

Fig. 20-11 Optical proximity correction - This example illustrates a layout of a polySi pattern of an 0.15-μm SRAM-cell. OPC tools examine the *original-layout* (a) and make corrections for subwavelength effects on the mask to produce a *final-geometry* (b). (Note the corrective-geometries added to outside and inside corners.) *Wafer-images - uncorrected* (c) and *corrected* (d) - show OPC effects.

use of this second-approach to produce ALT-PSMs avoids the potential-problem of the phase-shifter affecting either the light-amplitude or phase.

In ALT-PSMs, the light is phase-shifted by 180° when it passes through the spaces containing the PSM-material, and the *E*-field amplitude on the mask is shown in Fig. 20-13b2. Note that the *E*-field amplitude at the mask-spaces containing a phase-shifters is -1, but at those without a phase-shifter has a value of +1. The *E*-field-intensity imaged on the wafer using such a PSM is shown in Fig. 20-13b3, and the corresponding light-intensity on Fig. 20-13b4. It can be seen that the contrast in the light-intensity-distribution of the image on the wafer is greater when the PSM is used. *Contrast* is increased because the light diffracted into the nominally-dark area from the left clear-area will interfere *destructively* with the light diffracted from the right-clear area. All types of PSMs use this principle, in which *destructive-interference of light of opposite-phases* is used to improve image-contrast.

While the ALT-PSM mask gives the most improvement in resolution for the case when closely-spaced clear lines are separated on an opaque-background, this type of mask is not effective for some other types of patterns. One example is the case when tightly-packed features are layed-out in a pattern that resembles the stacking of bricks in a wall (Fig. 20-14). Since such a pattern has alternating-rows offset from one another, it is not possible to assign a phase value to each pattern that will make it 180° out-of-phase with all its neighboring-features. The resolution of isolated-features (such as contact-holes) is also not improved by ALT-PSMs. In general, non-repetitive patterns cannot be given phase-assignments that meet the ALT-PSM approach. However, other types of PSMs have been developed in an attempt to overcome the drawbacks of ALT-PSMs for some applications.

Before describing these other types of PSMs, it should be mentioned that the various types of PSMs are grouped into so-called *strong* and *weak* PSMs. Therefore, these terms must be defined. This distinction is based on the maximum-degree of improved *resolution* and *depth-of-focus* that each type can provide. Strong-PSMs (such as ALT- PSMs) can provide up to a 40% improvement in resolution, but a lesser-improvement in depth-of-focus. Weak-PSMs (such as the *rim-shift PSMs* and the *attenuated* [or *halftone*] *PSMs* which are described next) offer less-improvement in resolution, but more-improvement in depth-of-focus. The benefits each group offers are different, and this implies that the "best" type of PSM depends on the particular-application.

One of the alternative types of PSMs is the *rim-shift PSM*, which falls into the group of weak-PSMs. (see Fig. 20-15). In rim-shift PSMs, the phase-shifting takes place only at the rim of each mask-feature. As a result, such PSMs do not provide as much resolution-improvement as do the ALT-PSMs. However, they do not suffer from the pattern-restrictions that hinder ALT-PSMs. Thus, they can be used for arbitrary reticle-patterns, including isolated features. Among the issues of concern are rim-size optimization and the lack of inspection and repair of the phase-shifting materials of the mask.

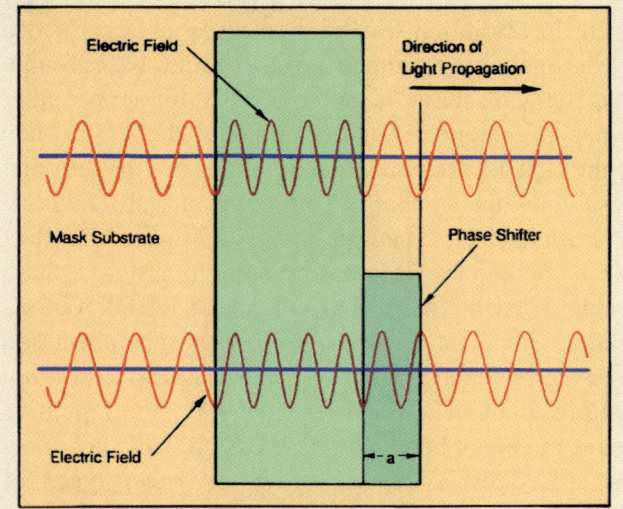

Fig. 20-12 A phase-shifting layer introduces a difference in optical path-length. Here the difference is $\lambda/2$.[20] (© IEEE (1993).

Another type of weak-PSM that can also be applied to arbitrary layout-patterns is the *attenuated (or halftone) PSM*.[22,23] Such *att-PSMs* are fabricated by replacing the opaque-part of a conventional reticle with a halftone-film (i.e., one that is partially transmissive, about 10%). For example, for *i*-line light such a film can be made from a very-thin layer of chromium-oxynitride or Mo-Si-ON (molybdenum silicon-oxynitride) having the proper-thickness to shift the phase of the light it transmits by 180° (Fig. 20-16). While some light will pass through the halftone-film, this light will be too weak to expose the photoresist to the degree necessary for it to be washed-away during development. But the negative-amplitude of this weak, phase-shifted light will destructively-interfere with the non-phase-shifted light so that a useful (if weak) improvement in the resolution is achieved. The att-PSMs are generally considered better than rim-shift PSMs. This is because att-PSMs are much easier to fabricate (since the halftone absorber-layer can be patterned just as if it was the opaque-layer of a conventional-mask). Isolated-defects, either clear or opaque can also be repaired with conventional repair-techniques. Because of their relative simplicity, att-PSMs have been the first type of PSM to be adopted in commercial IC-fabrication. One limitation of att-PSMs is that they are themselves subject to optical-proximity effects. One approach to reducing this problem is to combine OPC with att-PSMs. The reticle in this case must be made with three tones (clear, attenuated-chrome, and opaque-chrome). This technique is called *ternary-attenuated-PSM*.

Although PSMs were introduced in 1982, there was scant interest in the technique until about 1988. At that point, a large surge in research about PSMs began. While this resulted in dozens of papers dealing with this topic being published over the next ten years, PSMs were not widely-introduced into production until the late-1990s. Even by 1997 only about 1% of the world's mask-output contained PSMs. However, as 0.18-μm-CMOS was being moved into production in 1999, it became apparent that resolution-enhancement-techniques (including PSMs) would have to be used to enable optical-lithography to print such small-features. By 1998, weak-attenuating (or halftone) PSMs were being used on 0.18-μm contact-hole levels, and strong ALT-PSMs were being used on levels containing gates and interconnect-lines. ALT-PSMs are also well-suited for memory-circuits, in which the interconnect-lines are closely and regularly spaced. Reticle-inspection techniques for ALT-PSMs were being introduced by 1999, but repair techniques still lagged at that time. The combination of attenu-

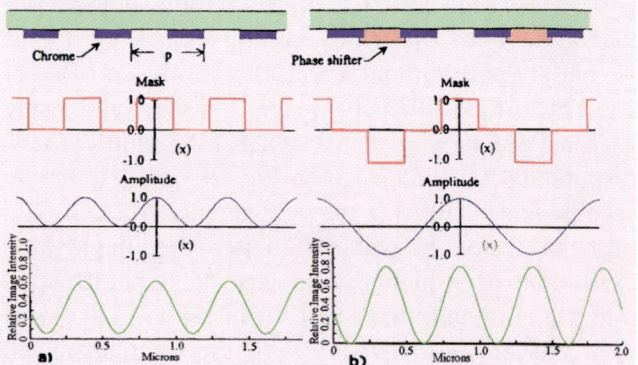

Fig. 20-13 Schematic of: (a) a *conventional binary-mask*; and (b) an *alternating phase-shift mask*. The *mask electric-field*, *amplitude*, and *image-intensity* is shown for each type of mask.[1] Reprinted with permission of Marcel Dekker.

CHAPTER 20 LITHOGRAPHY III: PHOTOMASKS AND ADVANCED LITHOGRAPHY

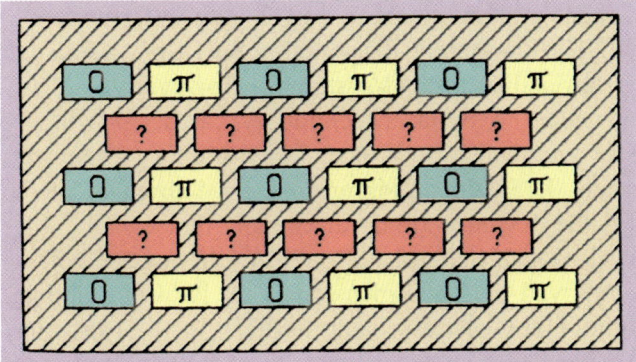

Fig. 20-14 Limitation of ALT-PSMs. There is no way to assign phases so that a 180° phase-difference occurs between adjacent transparent-features In this example, the phases have been assigned to the odd-numbered rows, but there is no way to assign phases consistently to the even-rows.[1] Reprinted with permission of Marcel Dekker.

ated PSMs and off-axis illumination provides an especially-effective resolution-enhancement strategy, as each method alone is only able to partially-reduce the zero-order light component.

20.3 MICROLITHOGRAPHY TRENDS

The advancement of IC-fabrication has depended heavily on the evolution of microlithographic-technology. However, the minimum feature-size of the most-advanced ULSIs reached the resolution-limits of *conventional* optical-lithography by the mid-1990's. Thus, the IC industry by 1999 faced a major crisis.[24] To overcome this barrier, many approaches (including *optical* and *non-optical* lithographic technologies) have come under intense investigation. In this section, the most important of these are described, as well as some predictions about the future directions in microlithography.

20.3.1 The Limits of Optical-Lithography

The demise of optical-lithography was expected to occur when IC feature-sizes reached about 1-μm. However, since optical-lithography is still being used for 0.1-μm IC-technologies, this prediction has proven to be incorrect. Nevertheless, some limits to optical-microlithography do exist, and perhaps another lithographic-technology will replace optical-lithography at some future date.

The resolution-limit of optical-lithography has generally been considered to be the wavelength of the exposure-light. However, by employing *resolution-enhancement technology* (RET) the minimum feature-size that can be practically imaged is about 30% smaller than the exposure-wavelength (e.g., 0.18-μm features can be produced with 0.248-μm KrF light).[26] The rate-of-shrinkage of the minimum feature-size has been faster than the reduction of the exposure-wavelength. For 0.25-μm technology, KrF excimer-laser-based lithography is employed (with the KrF-wavelength being 0.248-μm). But, to image 0.18-μm features without having to resort to such resolution-enhancement techniques as PSM, off-axis illumination, OPC, and multi-layer resists, a shorter-wavelength of light will be needed (e.g., ArF excimer-laser light, that has a wavelength of 0.193-μm). Since it is not clear whether ArF-technology (including appropriate light-sources, lens-materials, and resists), will be available in time for full-scale production of 0.13-μm ICs, adoption of such RETs onto KrF may be the way that 0.13-μm technology will initially be manufactured. By 1999, the first generation 0.193-μm ArF 4:1 stepper/scanners from ASML, Canon, and Nikon were being installed on 0.13-μm development lines. The anticipated average-price of ArF steppers is $15 million. ArF-resists from Olin, and others (based on cyclic-olefin polymer materials), were available for pilot-line use. But it appears that ArF excimer-light (or some other shorter-wavelength energy source), will certainly be needed to obtain adequate-resolution with sufficient depth-of-focus for the next generation-CMOS (0.09-μm). Note that at 0.13-μm, the dimension of the feature being printed is smaller than even the wavelength of ArF-excimer

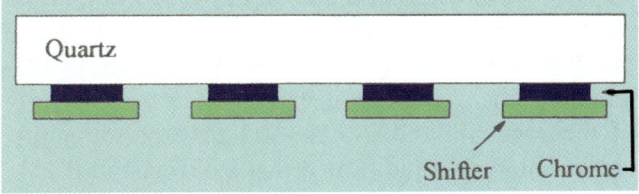

Fig. 20-15 Example of a *rim-shift phase-shifting mask*.

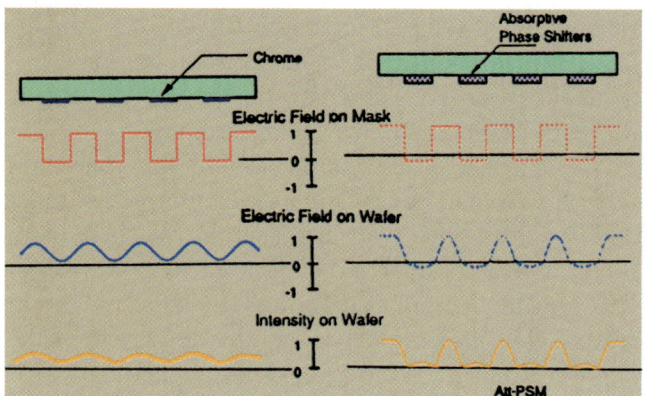

Fig. 20-16 Principle of the *attenuated-PSM (att-PSM)*.[21] © IEEE (1993).

light. Thus, RETs will also be needed. Since it appears that the limit of optical-lithography is now predicted to be about 0.09 μm, non-optical techniques will eventually be needed.[26] Such non-optical alternatives are covered in the next sections.

We noted it was predicted that optical-lithography would not be used for IC-technologies below about 1-μm. One of the chief reasons this estimate was incorrect had to do with unexpected advances in lens-making technology (in the decade starting in 1985). That is, lenses manufactured in 1985 with an NA = 0.3, had resolution-limits of 1-μm and a 1.5-μm depth-of-focus, respectively. But, by 1996 Nikon introduced a lens with an NA = 0.3 and it had resolution-limits of 0.7-μm and a depth of focus of 5-μm. This meant that lenses manufactured in 1996 were actually *diffraction-limited* (while those in the mid-1985's were not). Nevertheless, real diffraction-limits are now being encountered. Thus, the gains provided to lithography from ingenuity and innovation in lens-making will not be able to deliver the same benefits again.

Instead, to get to 0.1-μm with optical-lithography, shorter-wavelengths of light will be needed. UV-lasers operating at 0.157-μm (F_2) and 0.126-μm (Ar_2), are possible illumination-sources for such shorter-wavelengths. Since virtually all materials absorb-strongly in the range of 0.020-0.193-μm (except for some, like calcium-fluoride [CaF_2] that still transmit 0.157-μm light), it may be necessary to use all-reflective optical systems at these wavelengths. (Extreme-UV [EUV] lithography, discussed in Sect. 20.4, uses such all-reflective optical-systems). However, by 1999 advances in F_2 laser-technology, in the growing of CaF_2 crystals, and in 0.157-μm resist-chemistry, made 0.157-μm lithography a possible candidate to fill the 100-nm "lithography gap" that exists.[27] It is also probable that a mix-and-match strategy will be employed at this dimension. That is, a non-optical-lithography process (such as EUV, or e-beam) may be used for the layers with the smallest minimum-dimension, and optical-lithography techniques for the non-critical layers. Figure 20-17 shows the relationship between resolution and exposure-wavelength for various-types of microlithographic technologies.[28]

Overlay-limits are another issue to be considered with optical-systems. Stage-precision is the biggest practical-concern, but work is reportedly being carried out to solve this problem.[29] Another overlay-problem involves alignment-target acquisition. Metal and grainy-polysilicon have long presented difficulties for stepper-alignment systems. More recently, additional problems have been encountered on wafers that have undergone chemical-mechanical polishing. One solution has been to align all layers to a "zero-layer" having consistent alignment-targets.

The best level-to-level overlay capability of optical-steppers is estimated to be about 30-nm, which is adequate for 0.1-μm technology, but not for 0.07-μm. For 0.07-μm technology, it may be necessary to dedicate individual-steppers to a particular wafer-lot. This will reduce overall stepper-productivity and will introduce another source of additional costs.[30]

In summary, there are many challenges to be met to enable optical-lithography to be used at even smaller feature-size technologies. However, there is optimism that optical-lithography will continue to overcome these technologies, and still remain the dominant lithographic-approach for the foreseeable future.

20.4 NON-OPTICAL (OR NEXT-GENERATION) LITHOGRAPHIC TECHNOLOGIES (NGL)

In order to image features 0.1-μm and smaller it will probably be necessary to employ some form of non-

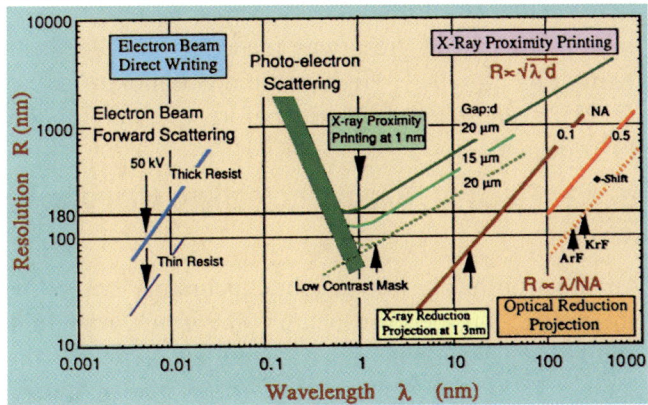

Fig. 20-17 Relation between resolution & exposure wavelength for various lithographic technologies.[28] (© 1996 IEEE).

optical technology (or *next generation lithography, NGL*).[36] The two NGLs described here are the leading candidates in 2003 to be an alternative to optical-lithography: **a)** *extreme ultra-violet* (*EUV*) lithography; and **b)** *electron-beam projection-lithography* (*EPL*). These approaches will be briefly introduced, and references provided. It should be noted that by 2001 the consensus was still that non-optical techniques were not-yet up to the overall production-capability of optical micro-lithography.[26] Two difficulties shared by the non-optical lithographic techniques are: **1)** the cost of a single-mask is expected to be in the $50,00-$100,000 range; and **2)** no pellicles can be mounted on the masks to protect them from defects. The latter problem makes masks for NGLs much-more vulnerable to particles and other defects.

20.4.1 Extreme Ultra-Violet Reflective Projection Lithography (EUV)

An attempt is being made to extend the principles of projection-lithography into the soft x-ray spectrum. That is, the region of the spectrum with photon-energies of about 100-eV and wavelengths between 10-15-nm (or 0.01-0.15-μm), can be considered to be either at the low-end of the x-ray spectrum (*soft x-rays*), or at the short-wavelength limit of the *extreme ultra-violet* (*EUV*). A number of interference-coatings that exhibit good-reflectivity of such EUV have been developed (i.e., silicon/molybdenum composites,

deposited by sputtering), and they can be used to produce EUV-mirrors. These coatings enable projection-systems that use all-reflective lithographic-components to image EUV to become feasible. The microlithographic-technology based on this approach has been dubbed *extreme-ultraviolet lithography* (*EUVL*).[31,32] Figure 20-18 shows a schematic of an EUV lithography-system. A laser generates EUV-radiation by bombarding a target material. This produces broadband-radiation with a significant EUV-component. This radiation is collected by a system of mirrors coated with EUV interference-films. The radiation is then used to illuminate an EUV reflection-reticle. The pattern on the reticle is imaged and de-magnified onto a resist-coated wafer. The entire reticle-pattern is exposed onto the wafer by synchronously scanning the mask and the wafer (i.e., a step-and-scan exposure is performed). Resolution of less than 0.1-μm is possible with EUVL if a wavelength of 13-nm is used. EUVL also offers the benefit of a large depth-of-focus for a system of such resolution (i.e., a DOF of greater than 1-micron).

There are a number of formidable challenges that must be overcome to make this technology practical. First, a modestly-priced, highly-reliable laser-driver is needed. Second, the fabrication of the aspheric mirrors is difficult (because of the extremely-tight

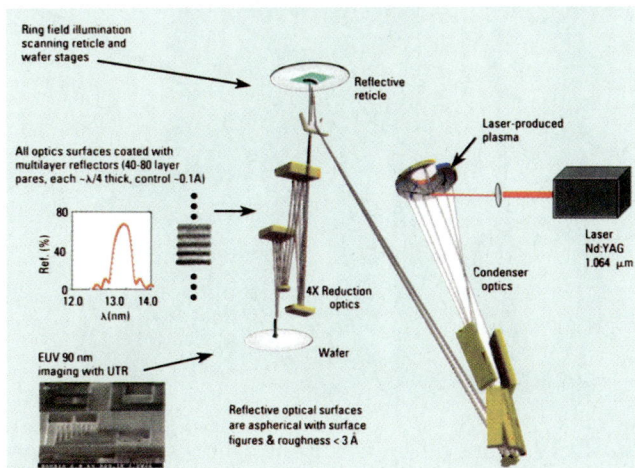

Fig. 20-18 Schematic drawing of the optics of an EUV system.

tolerance [i.e., better than 2-nm] on the *surface-figure* of these elements). Optics producers will need new metrology-techniques to measure the *figure* and *surface-roughness* of these mirrors to the required-accuracies. The stepper and the EUV beam-delivery system must also be operated in a vacuum to prevent absorption and scattering of the EUV-rays. The interference-films on the mirrors themselves must be fabricated in a multilayer-structure consisting of tens (or even hundreds) of films. Each layer must be deposited accurately to thicknesses of only a few-nanometers. But the reflectivity of these interference-films is relatively poor (e.g., 50-60%), so that if the system consists of more than two or three mirrors, the total transmission of the EUV is very low. The reflectivity will need to be improved to >70% to enable high-throughput, cost-effective lithography. These interference-films must also be able to withstand exposure to strong EUV-radiation without damage over long periods of time. While these challenges appear daunting, significant progress was made by 2001 to meet them (Fig. 20-19).[33] High-volume production using EUVL is predicted to begin before the start of the 2010-decade.

20.4.2 Electron Beam Projection Lithography (EPL)

An electron-beam lithographic technique based on projecting an electron-beam (*electron projection lithography, EPL*) has been developed at two companies: **1)** Lucent Technologies (now Agere), called *SCattering with Angular Limitation in Projection Electron beam Lithography* (or *SCALPEL*);[34,35] and **2)** IBM, called PREVAIL. In the EPL approach, a special scattering-mask is used to produce a high-contrast image with a technique commonly used in transmission-electron microscopy. That is, an electron-beam (coming down from the top of Fig. 20-20) is scanned across a special mask that consists of a thin silicon-nitride (SiN) blank (100–150-nm thick) having mask-patterns formed of very-thin (i.e., 25–50-nm-thick) tungsten or gold features. The electrons from the scanned-beam can either: **1)** strike regions of the mask covered with one of the metal-patterns; or **2)** strike a region of the mask *not* covered with metal. Electrons striking the SiN-film pass straight-through it (i.e., it only weakly-scatters the beam-electrons). On the other hand, electrons striking the metal-patterns are strongly-scattered as they pass through the heavy-metal of the mask-features. Those that pass unscattered through silicon-nitride are then focused by a lens so that they pass through the aperture of the *back focal-plane filter*. But most electrons that are scattered as they travel through the metal-patterns get blocked by the back focal-plane filter (and thus do not travel though its aperture). Electrons that make it through the filter-aperture are demagnified and strike the wafer-surface (whereupon they form a high-contrast, 4:1-reduced-image of the scattering-mask).

The resolution-limits of EPL can be extended below 0.1-μm. In addition, because the scattering-patterns are so thin, the demagnification of the mask can be made small-enough that chips can be fabri-

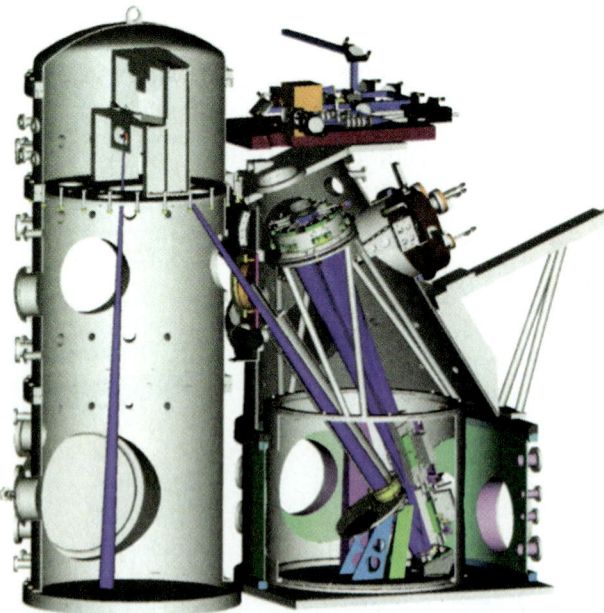

Fig. 20-19 In this illustration, EUV light is shown as a purple beam within EUVL equipment developed by the three members of the Virtual National Laboratory for EUV LLC. Courtesy of EUV LLC.

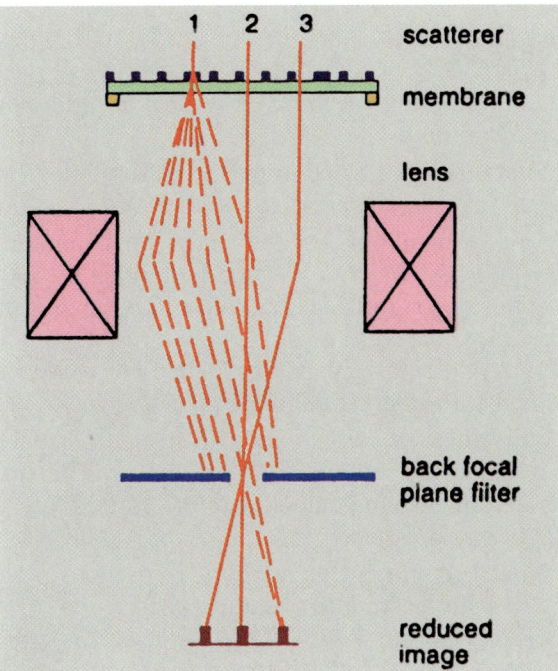

Fig. 20-20 Schematic of the SCALPEL-technique. Electrons (1) that hit the scatterer (the metal-patterns on the mask) are strongly scattered, and most are filtered out by the aperture. Electrons traveling through the silicon-nitride membrane (2, 3) are only weakly scattered and pass through the aperture. These are demagnified to form a high-contrast image on the substrate. Courtesy of Lucent Technologies, Inc.

cated with as many as 2×10^{10} pixels (defined here as a small-square whose side is equal to the minimum printable-linewidth). Since 1-Gigabit DRAM-patterns, for example, need roughly 10^{10} lithographic-pixels per chip, the EPL-technique is suitable for such chip-designs. The thin silicon-nitride mask-blank-substrate is supported on a silicon-wafer with periodic silicon support-struts. These struts are not imaged onto the wafer since the patterns are shifted into place as they are illuminated. While the mask-structure is similar to that used in proximity-x-ray masks, the silicon support-struts provide greater dimensional-stability of the patterns on EPL-masks, and the use of 4:1 reduction-optics makes mask-fabrication of the 4:1 reduction masks simpler.

REFERENCES

1. S. Rizvi and D. Van Den Broeke, "Mask Making," in *Handbook of Semiconductor Manufacturing Technology*, Y. Nishi and R. Doering, Eds., Marcel Dekker, New York, 2000, Ch. 20, p. 589.

2. D. Miller, "Maskmaking Gains New Prominence as Device Rules Shrink," *Solid State Technology*, November 1997, p. 69.

3. P. Heinz, "Lithography Arrives at the Crossroads," *Solid State Technology,* April 1997 p. 136.

4. M. Murakami and S. Matsushita, "Will 256 Mb Lithography Go with 4X Masks?" *Semiconductor International,* July 1996, p. 305.

5. P. Singer, "Merchant Mask-Making in State of Change," *Semiconductor International,* March 1986, p. 45.

6. R.C. Bracken and S.A. Rizvi, "Microlithography in Semiconductor Device Processing," in N.G. Einspruch Ed., *VLSI Electronics-Microstructure Science*, Vol. **6**, Academic Press, Orlando FL, 1983, Chap. 5, p. 256.

7. ETEC Systems, Hayward, CA.

8. A.C.Titus, "Photomask Defects: Causes and Solutions," *Semiconductor Intl.* October 1984, p. 94.

9. K.L. Harris, *et al.*, "Automated Wafer Inspector Characterization," *SPIE Proc*. Vol. **538,** *Optical Microlithography IV*, 1985, p. 138.

10. L.H. Lin, *et al.,* "A Holographic Photomask Defect Inspection System," *SPIE Proc.* Vol. **538,** *Optical Microlithography*, 1985, p. 110.

11. R. Herschel, "Pellicle Protection of Integrated Circuit Masks," *SPIE Proc.* Vol. **275**, *Semiconductor Microlithography VI*, 1981, p. 23.

12. J. Lent and S. Swayne, "Implementation of Pellicles Into An Established Production Area," *Proc. Kodak Interface*, 1982, p. 93.

13. C. Lambson *et al,* "Automated Reticle Transport & Stepper Loading," *Sol. State Technology,* Oct 1996, p 97.

14. R. DeJule, "OPC Simulation," *Semiconductor International,* September 1997, p. 56.

15. C. Mack, "Evaluating Proximity-Effects Using 3-D Optical Lithography Simulation," *Semiconductor International*, July 1996, p. 237.

16. C.A. Mack and P.M. Kaufman, "Mask Bias in Submicron Optical Lithography," *J. Vac. Sci. Technol.,* Vol. **B6** (Nov./Dec. 1998) p. 2213.

17. N. Shamma et al., "A Method of Correction of Proximity Effect in Optical Lithography," *KTI Microlithography Seminar Interface 1991*, p. 145.

18. A. Kornblit et al., "The Role of Etch Pattern Fidelity in the Printing of Optical Proximity Corrected Photomask," *J. Vac. Sci. Technol. B*, **13**(6), p. 2944 (1995).

19. M. Levenson et al., "Improving Resolution in Photolithography with a Phase-Shifting Mask," *IEEE Trans. Electron Dev.*, **ED-29**, 1828 (1982).

20. B.J. Lin, "Phase-Shifting Masks Gain an Edge," *IEEE Circuits & Devices Mag.*, March 1993, p. 28.

21. H.J. Levinson and W.H. Arnold, "Optical Lithography," in *Handbook of Microlithography, Micromachining, and Microfabrication*, Vol. 1, P.R. Choudhury, Ed., SPIE Press, Bellingham, WA 1997, Ch. 1, p. 74-82.

22. T. Terasawa et al., "Imaging Characteristics of Multi-Phase-Shifting Mask," *SPIE*, Vol. 108, 25 (1989).

23. B.J. Lin, "The Attenuated Phase-Shift Mask," *Solid State Technology*, January 1992, p. 43.

24. M. Sasago, "Lithography Solutions for Sub-0.1-μm," *Tech. Dig. 1998 Symp. on VLSI Technol.*, p. 6.

25. S. Okazaki, *Proc. of SPIE*, Vol. 2440, p. 18 (1995).

26. T. Brunner, "Pushing the Limits of Lithography for IC Production," *Tech. Dig. IEDM*, 1997, p. 9.

27. J.A. McClay and A.S.L. McIntyre, "157-nm Optical Lithography: The Accomplishments and the Challenges," *Solid State Technology*, June 1999, p. 57.

28. S. Okasaki, "Lithography Prospects for 0.18 μm Technology and Beyond," *Tech. Dig. IEDM*, 1996, p. 57.

29. S.A. Lis, "An Air-Turbulence Compensated Interferometer," *Proc. SPIE*, Vol. 2440, 9891 (1995).

30. *ibid.*, Ref. 21, p. 124.

31. A.M. Hawryluk et al., "EUV Lithography," *Solid State Technology*, July 1997, p. 151.

32. F. Zernike & D. Attwood, *Extreme Ultraviolet Lithography*, Optical Soc. America, Washington D.C., 1995.

33. A. Hand, "EUV Lithography Makes Serious Progress," *Semiconductor International*, June 2001, p. 54.

34. S. Berger et al., *SPIE Procs*, **Vol. 2322**, p. 434 (1994).

35. L.R. Harriott, *J. Vac. Sci. Technol.*, **B14**, 3825 (1996).

36. R. DeJule, "Next-Generation Lithography Tools," *Semiconductor International*, March 1999, p. 48.

PROBLEMS

1. Define the term *reticle*. How does a *reticle* differ from a *photomask*?

2. What material is used to make ULSI reticles? What opaque material is patterned on a reticle?

3. An L-shaped line is delineated on a photoresist layer. Indicate whether the line is defined by leaving resist inside or outside the shape for the following cases: (a) Clear-field mask, negative resist; (b) Clear-field mask, positive resist; (c) Dark-field mask, negative resist; (d) Dark-field mask, positive resist.

4. Why must the reticle used in a wafer-stepper be completely free of defects? Why can some defects be tolerated in systems that expose the entire-wafer in a single exposure-step?

5. What lithography technology is used to produce reticles with the smallest feature-sizes?

6. Explain the difference between the *die-to-die* and *die-to-database* methods of inspecting photomasks and reticles.

7. Explain why *clear defects* in a mask or reticle are more difficult to repair than *opaque defects*.

8. Draw a cross-section of a reticle with a *pellicle* attached. Identify all parts.

9. How does use of pellicles improve the manufacturing yield of ICs?

10. Describe the basic principles of *optical proximity projection* (OPC).

11. Explain the basis of *phase-shift masking* for enhancing resolution in projection-lithography.

12. List and describe the principles of two alternative lithography-technologies that are candidates to replace optical lithography.

13. What is the amount of misregistration caused across a 6-inch mask because of a 0.1°C temperature-change for a fused-silica reticle?

14. Reticle nonflatness consumes part of the depth-of-focus budget. It is desire that reticle non-flatness reduce the depth-of-focus budget by no more than 10%. For a 0.4-μm depth-of-focus, how flat must the reticle be, consistent with this 10% criterion and a 4X lens? 6X lens?

CHAPTER 20 LITHOGRAPHY III: PHOTOMASKS AND ADVANCED LITHOGRAPHY

Multilevel copper metallization of CMOS chip. This scanning electron micrograph of a CMOS-IC shows six-levels of copper metallization that are used to carry electrical-signals on the chip. The interlevel-dielectric insulators have been chemically etched away to reveal the copper interconnects. Photograph courtesy of IBM.

CHAPTER 21

TERMINOLOGY OF ETCHING AND WET-ETCHING TECHNOLOGY

CHAPTER CONTENTS

21.1 THE TERMINOLOGY OF ETCHING

21.2 ETCH PARAMETERS

    Etch-Rate
    Etch-Rate Uniformity
    Etch Profile
    Selectivity
    Etch Bias

21.3 WET-ETCHING TECHNOLOGY

    Wet-Etching Silicon
    Wet-Etching Silicon-Dioxide
    Wet-Etching Silicon-Nitride
    Wet-Etching Aluminum

"Change is not necessary. Survival is not mandatory."

W. Edwards Demming

Photograph of wafers being wet-etched.

After a photoresist-pattern has been formed on the surface of a wafer, the next process often involves permanently transferring that image into a layer under the resist by etching (see Fig 21-1). *The goal of an etching-process is to exactly transfer the image present in the resist-layer.* The degree of exactness is dependent on several factors that will be explored as a preparation for the discussion of the different available etch-methods. In this chapter, both the general-terminology of etching and a variety of wet-chemical-etching processes are described. Wet-chemical etching

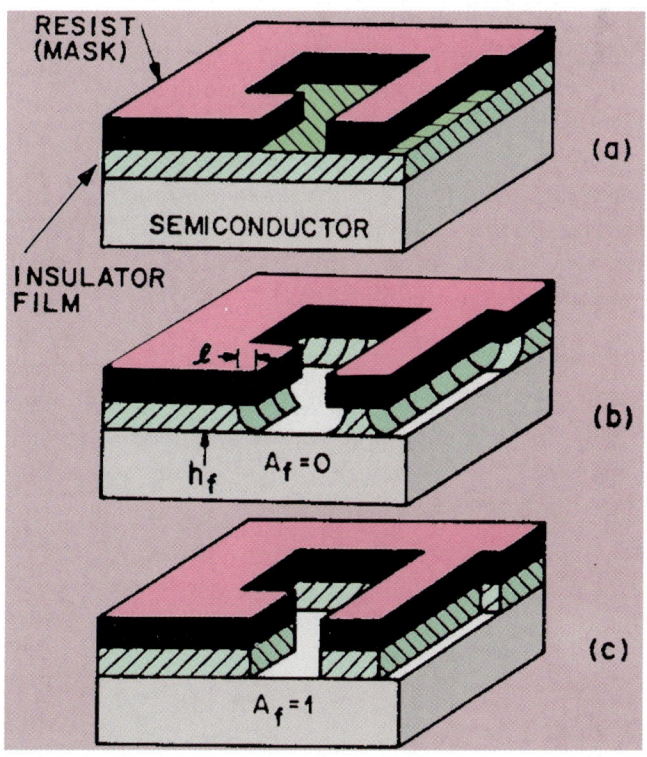

Fig. 21-1 (a) Resist-mask defining the region that is to be etched. Comparison of (b) *Isotropic-etching.* (c) *Completely-anisotropic etching.*

381

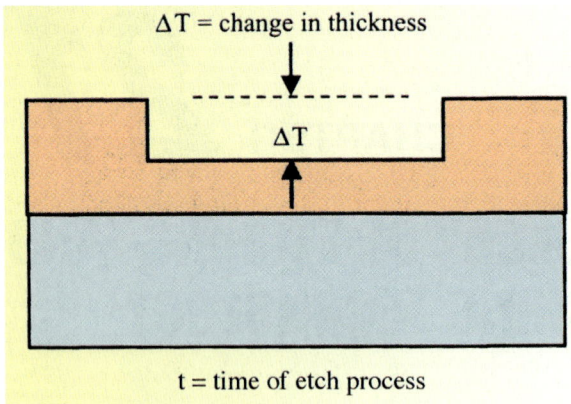

Fig. 21-2 Definition of *etch-rate*.

involves immersing wafers in liquid-reagents that attack the materials not protected by the etch-mask.

21.1 THE TERMINOLOGY OF ETCHING

Etching in microelectronic-fabrication is a process by which material is *selectively-removed* from the silicon-substrate or from thin-films on the substrate surface. When a *mask-layer* is used to protect specific regions of the wafer-surface, the goal of etching is to precisely remove the material not covered by the mask The two basic types of etch-processes used in semiconductor fabrication are *wet-chemical etching* and *dry-etching*. In this section the terms used to describe the basic aspects of etch-processes are discussed. Dry-etching is covered in Chap. 22.

21.2 ETCH PARAMETERS

There are a number of parameters used to characterize the effectiveness of an etching-process. They include the following:

1. Etch-rate
2. Etch-rate uniformity
3. Etch-profile (isotropic-etching and undercut)
4. Selectivity
5. Etch-bias

21.2.1 Etch-Rate

The rate at which material is removed from the film by an etch-process is known as the *etch-rate* (Fig. 21-2). The etch-rate is defined by:

Etch-rate = thickness-etched/etch-time = $\Delta T/t$ **(21.1)**

Generally, high etch-rates are desirable as they allow higher production-throughputs. The etch-rate is typically expressed in units such as $\text{Å}/sec$ or $\mu m/min$. Typically, desired etch-rates are hundreds (or thousands) of angstroms per minute ($\text{Å}/min$).

21.2.2 Etch-Rate Uniformity

Etch-rate uniformity is a measure of the ability of an etch-process to etch evenly. It is important to have highly-uniform etch-rates in an etch-process. The uniformity of the etch-rate is expressed for three conditions: **1)** *within a wafer*; **2)** from *wafer-to-wafer*; and **3)** from *run-to-run*).

The etch-rate uniformity within a wafer is determined by measuring the etch-rate at 5 to 9 locations on the wafer (see Fig. 21-3a), and the uniformity within the wafer can be calculated from this data. The *average etch-rate* for a wafer obtained from these measurements can also be compared to that from another wafer to get wafer-to-wafer-uniformity information (Fig. 21-3b).

21.2.3 Etch-Profile

In the case of an ideal etch-process, the mask-pattern would be precisely transferred to the underlying layer. This would also create a vertical edge-profile in the etched-layer coincident with original edge of the mask (Fig. 21-4a). This means that there would be no lateral-etching. On the other hand, if there is lateral-etching, film-material is etched to some degree under the mask, an effect called *undercut* (Fig. 21-4b).

When etching proceeds in all directions at the same rate, it is said to be *isotropic* (Fig. 21-4c). In this case, the etch-distance laterally under each-side of a mask is the same as vertically under the mask-opening. Most wet-etching processes (and some dry-etching processes) exhibit uniform etch-rates in all directions, and hence are isotropic. On the other hand, any etching that is not isotropic is (by-definition) *anisotropic*. If etching proceeds exclusively in one direction (e.g., only vertically), the etching-pro-

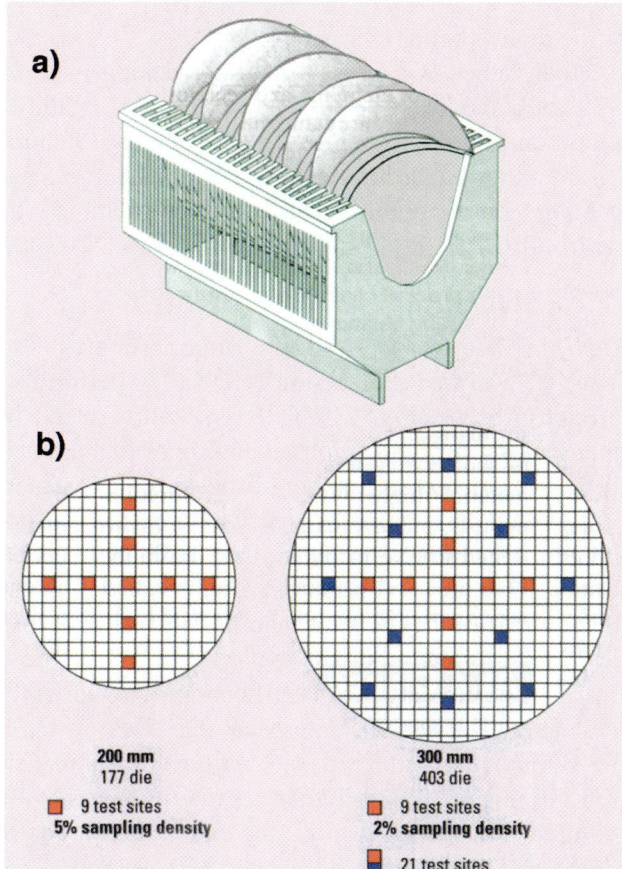

Fig. 21-3 Determining etch-rate uniformity: (a) Randomly select 3-5 wafers from a lot. (b) For 200-mm wafers, 9 test-sites must be measured to get a 5% sampling-density of the etch-rate. (c) For 300-mm wafers 21 test-sites must be selected and measured to get the same 5% sampling-density of etch-rate.

cess is said to be *completely-anisotropic*. Completely-anisotropic etching is desirable in many applications because it produces vertical-sidewalls and no undercutting. This permits features to be placed closer to one another, thus promoting higher IC-packing-densities. Completely-anisotropic etching is only possible with some dry-etch processes. Many etch-processes fall between the extremes of being isotropic and completely-anisotropic. These types of processes can also be advantageous for some applications. For example, an anisotropic tapered-profile is preferred for contacts and vias, because a tapered contact-hole (or via) has a larger arrival-angle, making it easier to fill such a hole with a tungsten-CVD-process without forming voids.

Since the thickness of the films used to fabricate ICs typically ranges from 0.5-1.0-μm, the isotropic nature of wet-etching becomes a fundamental limit. As shown in Fig. 21-5, if a film is 0.5-μm-thick, it will undercut the mask at both edges by 0.5-μm (assuming the etch is halted just as the film is completely cleared in the vertical-direction). Thus, a 3-μm-wide feature would only be 2-μm-wide at the top, as shown in Fig. 21-5a. If the feature is only 1-μm-wide, it will have the shape shown in Fig. 21-5c, which is not acceptable. For even smaller features, the etching will remove the material entirely (Fig. 21-5d). Thus, if films 0.5-1.0-μm-thick must be etched, an isotropic-process cannot be used if features decrease below about 2-μm. This is the main reason that wet-chemical-etching had to be abandoned, and replaced with processes that were capable of producing anisotropic-etching.

21.2.4 Selectivity

Three kinds of materials are involved in an etch-process: **1)** the *photoresist-mask*; **2)** the *film being etched*; and **3)** the *material under the film being etched* (e.g., the *substrate*). During an etch-process, all three could be attacked. The difference in their etch-rates is characterized by the *selectivity*. That is, *selectivity* is a

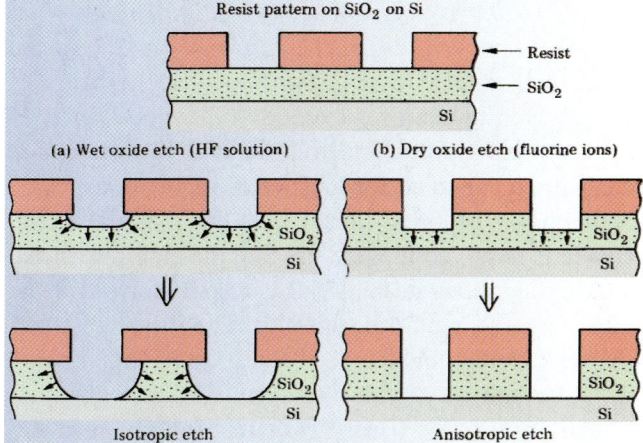

Fig. 21-4 (a) Wet (isotropic) etching of silicon-dioxide film, vs. (b) Dry (anisotropic) etching of a silicon-dioxide film.

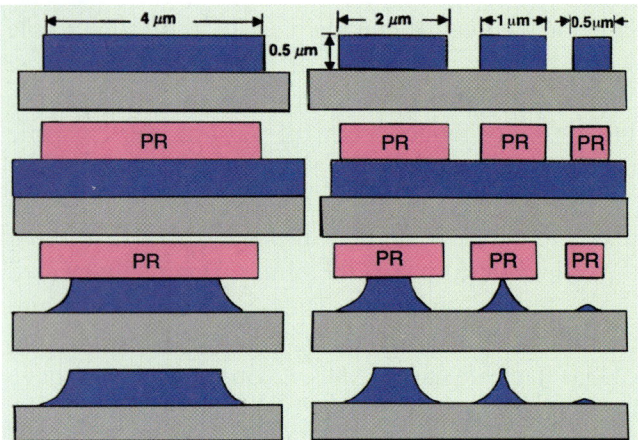

Fig. 21-5 Isotropic-etching of a 0.5-μm-thick film for features whose width is 4-μm, 2-μm, 1-μm, and 0.5-μm wide. Below about 1.0-μm-widths, the etched feature-shapes become unacceptable.

parameter that indicates how much faster one material etches than another under the same etch-conditions.

Two selectivity-parameters are of importance in integrated-circuit etching applications: **1)** *selectivity with respect to the substrate* S_S (Fig. 21-6a); and **2)** *selectivity with respect to the masking-material* S_m (Fig. 21-6b). High-selectivity of both kinds is necessary in most advanced-processes to ensure critical-dimension-control and profile-control. Selectivity is usually expressed as the *ratio of two etch-rates*: **1)** the *film being etched*; to **2)** the *material that we do not want removed* (either the *substrate* [in S_S], or the *masking material* [in S_m]). For large selectivities, it is said an etch-process for material-**1** *has good selectivity over* material-**2**. Selectivities of 25:1-to-50:1 are considered adequate for most etch-applications.

High-values of S_S mean that the material being etched is removed rapidly, but that the etch-rate of the underlying substrate-material is much slower. In the ideal case, S_S would be infinite. Then the substrate-material would not be etched at all. On the other hand, if S_S is small (say, with a ratio of 1:1), the substrate-selectivity is poor. This means the underlying-material is etched as fast as the material being etched. If the underlying-material is significantly-removed it may weaken the device-structure, leading to failure of the device being fabricated.

High values of S_m mean that the masking-material will not be attacked by the etch-process. This allows good-control of critical-dimensions during etching. On the other hand, low-values of S_m cause the mask to be eroded-away during the etch-step, making CD-control difficult, if not impossible (see Fig. 21-6b).

21.2.5 Etch-Bias

Etch-bias is a measure of the change in linewidth or space of a critical-dimension (CD) after performing an etch-process (Fig. 21-7). It is usually caused by undercutting. In fabrication-technologies that are performed using isotropic etching-processes, the problem of etch-bias is handled by specifying an appropriate amount of compensation in the mask-dimensions. For example, if the bias of an etch-process is 1-μm, a 6-μm feature on the *mask* can be used to produce a desired 5-μm feature on the *wafer* (for a 0.5-μm-thick film). Unfortunately, for VLSI-technologies in which the pattern-dimensions approach the thicknesses of the films being patterned, the margin for compensation diminishes, and a higher-degree of anisotropy is required. For practical purposes this situation arises-

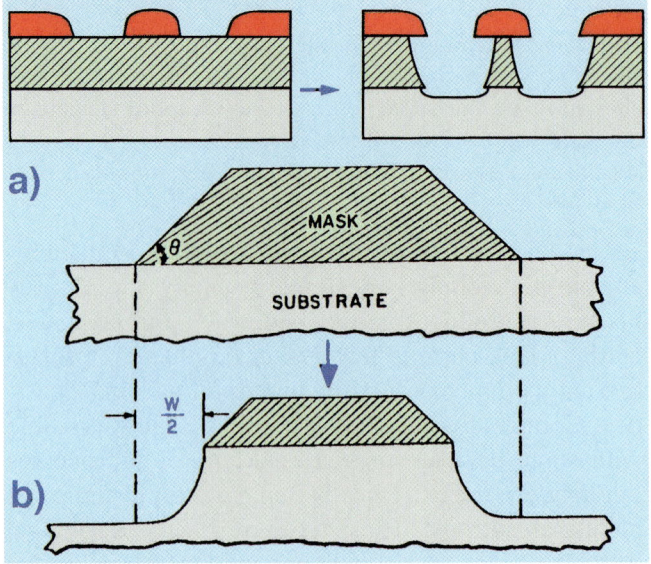

Fig. 21-6 (a) Selectivity with respect to the *substrate*; (b) Selectivity with respect to the *mask*.

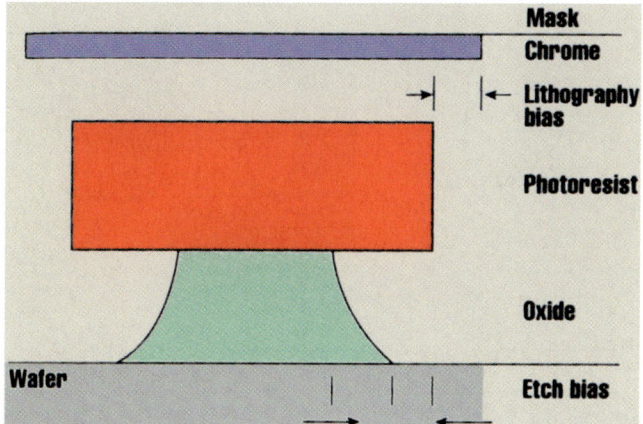

Fig.21-7 Typical isotropic etch-process showing *etch-bias*.

when pattern-features become smaller than ~3-μm. Under these circumstances, isotropic etching-processes become inadequate, and processes that provide higher-degrees of anisotropy need to be employed.

21.3 WET-ETCHING TECHNOLOGY

Wet-etching was the exclusive method used to etch patterns for the first 30-years of the semiconductor-industry.[1,2] It was adopted in the 1950's for transistor and IC-manufacture from the printing-industry, where wet-etching had been used for a long time. The emergence of feature-sizes smaller than 2-μm in the mid-1980's however, has led to a shift away from wet-etching to dry-etching techniques. As noted, this occurred because wet-etching-processes are usually isotropic. This undercuts the masking-photoresist and causes critical-dimension-loss. As depicted in Fig. 21-5, it makes wet-etching-processes inadequate for defining features that are less than about 2-μm-wide.

Nevertheless, for those processes that involve patterning-features whose minimum-dimensions are greater than 2-μm, wet-etching continues to be a viable technology (i.e., because it is cheaper and exhibits superior-selectivity compared to dry-etching). Since it turns out that a variety of semiconductor-products are still being fabricated with such larger geometries, wet-etching should not be ignored. Figure 21-8 shows a diagram of a modern wet-bench. In this section some of the more important aspects of wet-etching technology for current processing-needs are covered.

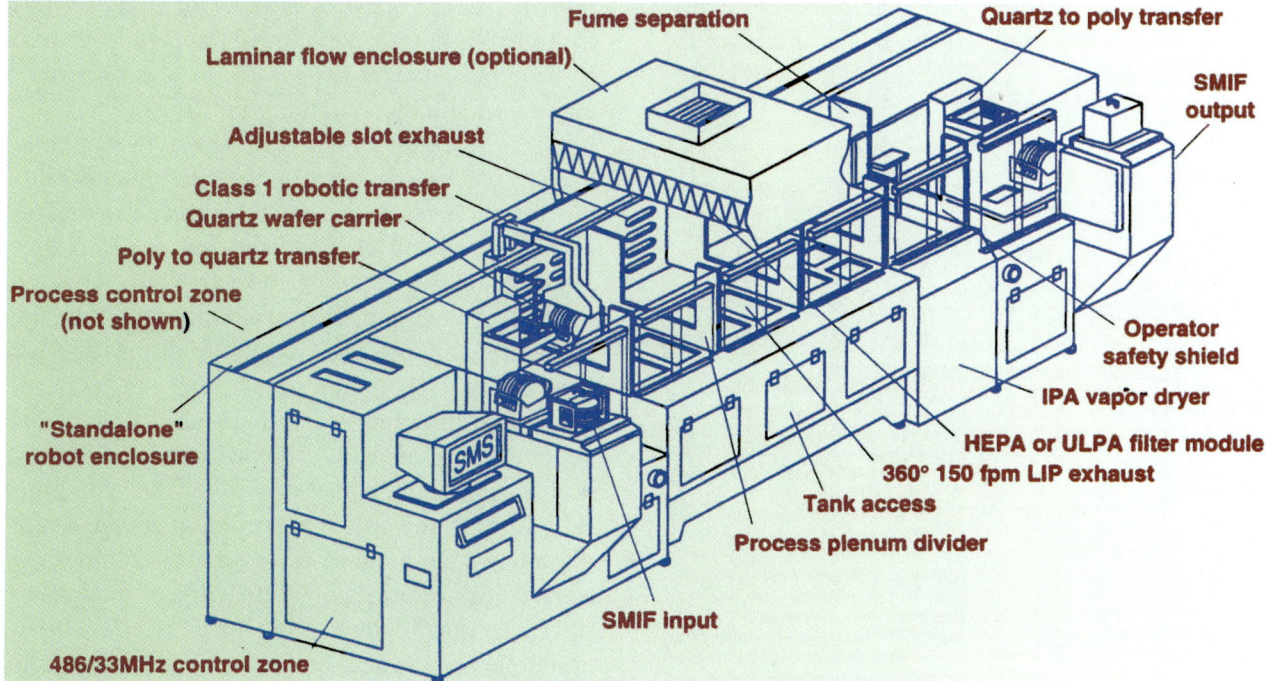

Fig. 21-8 Perspective drawing of an immersion wet-bench. Source: Semiconductor International.

Some recent refinements to wet-etching equipment have increased these advantages, including: **1)** the automation of wet-stations; **2)** using microprocessor-control to improve reproducibility of etching-conditions from run-to-run; **3)** point-of-use filtration of etchants prolongs their use by reducing etch-process-generated defects; and **4)** the development of spray etching. These advances will allow wet-etching to continue to find wide use in IC-fabrication.

On the other hand, besides the 3-μm limitation, wet-etching also has the following disadvantages: **1)** higher-cost of etchants and UP-water compared to dry-etch-gas costs; **2)** increased personnel safety-hazards from chemical-handling; **3)** exhaust-fumes and the potential of explosions; **4)** resist-adhesion problems; and **5)** bubble-formation and incomplete-wetting of wafer-surfaces by chemical-etchants, leading to incomplete-etching and etching non-uniformities.

In general, a wet-etch process can be broken down into three steps (Fig. 21-9): **1)** *diffusion of the reactant to the reacting-surface*; **2)** *reaction*; and **3)** *diffusion of the reaction-products away from the surface*. The slowest one of these steps will determine the etch-rate. The rate of that step will be that of the overall reaction. Following a wet-etch-process, the wafers are rinsed in DI-water and dried (Fig. 21-10).

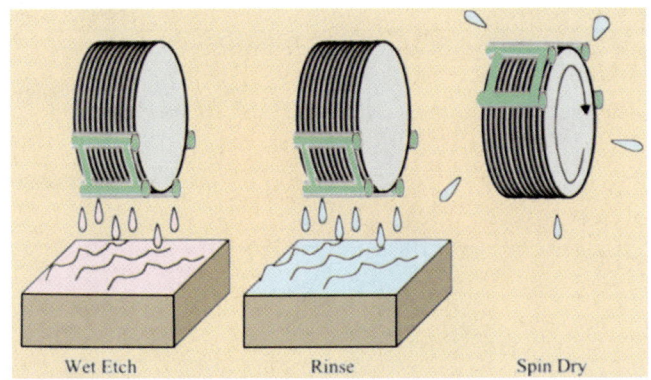

Fig. 21-10 Wet-etch process steps: (a) etch; (b) rinse; and (c) dry.[6] Used with permission.

Chemical-etching can occur by several processes. The simplest involves dissolution of the material in a liquid-solvent without any change in the chemical-nature of the dissolved species. Most etching processes, however, involve one or more chemical-reactions (Step-**2** above). Various types of reactions may take place, although one that is commonly encountered in semiconductor fabrication involves *oxidation-reduction* (or *redox*). That is, a layer of oxide is formed, then the oxide is dissolved and the next layer of oxide is formed. Such redox-reactions occur in the wet-etching of silicon and aluminum films.

In semiconductor-applications, wet-etching is used to produce patterns on the silicon-substrate or in thin-films. A mask is used to protect desired-surface regions from the etchant, and this mask is stripped after etching has been performed. Thus, when choosing a wet-etch process, in addition to selecting an etchant, a suitable masking-material must be picked. It must have good-adhesion to the underlying-films, good coating-integrity, and an ability to withstand attack by the etchant. *Photoresist* is the most-commonly-encountered masking-layer, but sometimes it falls short in this role. One problem encountered with photoresist as a mask-layer in wet-etching-applications is loss-of-adhesion at the edge of the mask-film-interface due to etchant-attack (Fig. 21-11). Edge-attack is controlled by using adhesion-promoters, such as hexamethyldisilazane (HMDS).

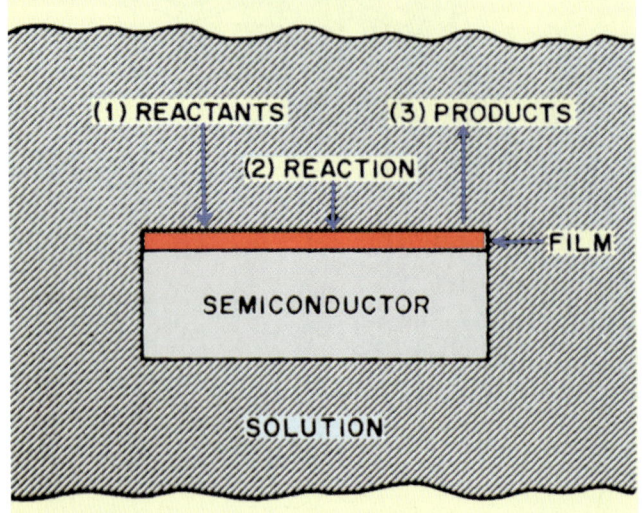

Fig. 21-9 The 3-basic steps of wet-chemical-etching.

Etching-processes which produce bubbles can lead to poor pattern-definition due to the clinging of bubbles to the substrate, Therefore, use of *wetting-agents* in the etchant (together with agitation) are measures taken to assist in dislodging such bubbles.

21.3.1 Wet-Etching Silicon

Both single-crystal-Si and poly-Si can be wet-etched in mixtures of nitric-acid (HNO_3) and hydrofluoric-acid (HF).[3] The reaction is initiated by the HNO_3 that forms a layer of silicon-dioxide on the Si, and the HF dissolves the oxide away. This process repeats itself again and again. The overall reaction is:

$$Si + HNO_3 + 6HF \rightarrow H_2SiF_6 + HNO_2 + H_2 + H_2O \quad (21.2)$$

In some applications, it is useful to etch Si more rapidly along some crystal-planes than others. This allows the etch to significantly slow-down or to etch specific-shapes or structures in the silicon. In the diamond-lattice the (111)-plane is more densely-packed than the (100)-plane, and thus etch-rates of (111)-oriented-surfaces are expected to be lower than those with (100)-orientations. An etchant that exhibits such orientation-dependent etching properties in silicon is a mixture of KOH and isopropyl-alcohol (23.4wt% KOH, 13.3wt% isopropyl-alcohol, and 63wt% H_2O). The etch-rate of this etchant is about 100-times faster along (100)-planes than along (111)-planes (at 80°C, the difference is 0.6-μm/min vs. 0.006-μm/min).[4]

Since this etchant contains no HF, a thermal-oxide can be used as a masking-layer. Thus, if features in a (100)-plane of Si are patterned with SiO_2, using this orientation-etchant will create precise V-shaped grooves in the silicon. The edges of the grooves will be (111)-planes at an angle of 54.7° from the (100)-surface (Fig. 21-12a). If (110)-surfaces are used, straight-walled-grooves with sides having (111)-planes are formed (Fig. 21-12b).

21.3.2 Wet-Etching Silicon-Dioxide

The most common wet-etched-layer is thermally grown silicon-dioxide, typically to open windows in SiO_2 surface-films down to a Si-substrate. Wet-etching of such SiO_2-films is carried out using various *hydrofluoric-acid* (*HF*) solutions.[5] This is because silicon-dioxide is readily attacked by room-temperature HF, while the underlying Si-substrate is not. Etching takes place according to the overall-equation:

$$SiO_2 + 6HF \rightarrow H_2SiF_6 + 2H_2O \quad (21.3)$$

H_2SiF_6 is water-soluble, therefore the HF-solution can etch-away the SiO_2.

The concentration of HF supplied by chemical-manufacturers is 49% in water. Such concentrated HF, however, etches SiO_2 too-quickly for good process-control (for instance, thermally grown SiO_2 is etched at approximately 300-Å/s at 25°C, so a 3000-Å oxide film would etch in only 10-sec). Thus, diluted-HF solutions are generally used instead. (Note that HF is corrosive, and poses a uniquely dangerous hazard if it contacts skin or the eyes. More information about HF safety-issues is found in Chap. 29.)

Since the reaction consumes HF, the reaction-rate will decrease with time. To avoid this, it is common to use HF with a *buffering-agent*, such as ammonium-fluoride (NH_4F). This keeps the concentration of HF in the solution constant (which maintains stable etching-characteristics). These solutions are known as *buffered-oxide etches* (*BOE*) or *buffered-HF* (*BHF*). A common BHF-mixture contains 6:1 (by volume) NH_4F:HF (40%:49%) which etches thermally-

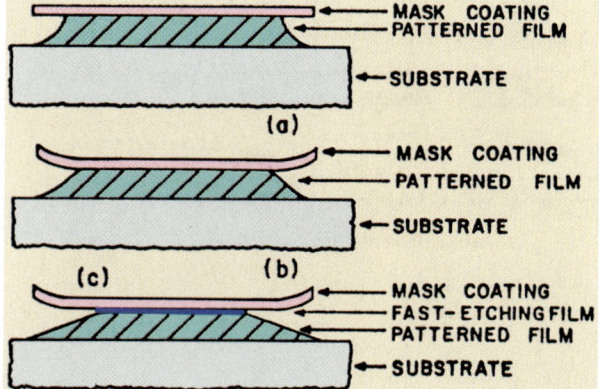

Fig. 21-11 Etch-profiles produced from various degrees of undercutting during wet etch. (a) Good mask-to-film adhesion. (b) Undercut occurring at mask-film interface. (c) Use of a fast-etching film to achieve controlled undercut.

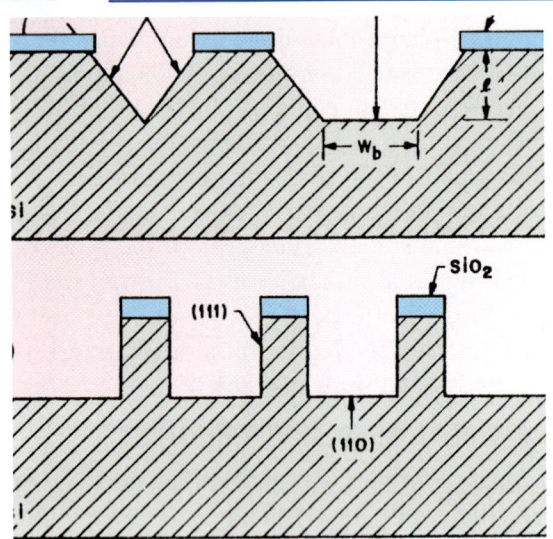

Fig. 21-12 Orientation-dependent Si-etching: (a) etch pattern profiles on <100>-Si; (b) etch pattern profiles on <110>-Si.

Fig. 21-13 Wet-etching silicon nitride, showing a hot phosphoric-acid module with a condensing/refluxing collar. Courtesy SCP-Global technologies.

grown SiO_2 at ~20-Å/sec (or ~1000-Å/min) at 25°C. Unbuffered HF-solutions diluted with water (e.g., 10:1 HF or 100:1 HF in H_2O) are commonly-used for oxide wet-etching CVD SiO_2 films (which typically have higher etch-rates than thermally-grown SiO_2).

21.3.3 Wet-Etching Silicon-Nitride

Silicon-nitride (Si_3N_4) is a difficult material to wet-etch. It etches very slowly by HF-solutions at room-temperature (less than 10-Å/min). However, it can be etched more rapidly using 85% phosphoric-acid heated to 180°C (Fig. 21-13). Since the acid evaporates rapidly at 180°C, a closed reflux-container with a cooled-lid to condense the vapors must be used.

The main problem of such wet-etching of Si_3N_4 is that photoresist is lifted during the process, and can-

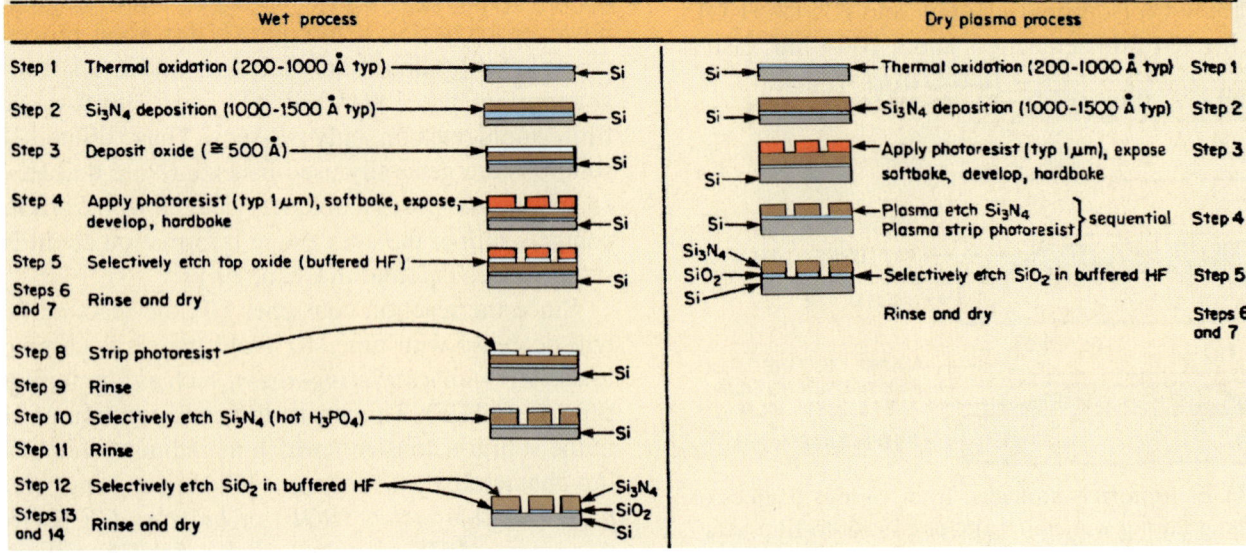

Fig. 21-14 Wet-etch process steps versus dry-etch process steps for Si_3N_4.

not be used as an etch-mask. Most wet nitride-etching processes thus use a thin SiO_2-layer (either thermally-grown, or deposited by CVD) to mask the nitride. The SiO_2-layer is first etched using a resist-mask, then the resist is stripped, and the patterned-oxide serves as the etch-mask for the nitride in the phosphoric-acid bath. The Si_3N_4 etch-rate in hot phosphoric-acid is about 100-Å/min, but only 0.25-Å/min for CVD-SiO_2.

The added complexity of using an SiO_2 etch-mask makes dry-etching of Si_3N_4 an attractive alternative. In fact, the first widely-used dry-etch process in microelectronics was developed for etching Si_3N_4 for this reason (Fig. 21-14).

21.3.4 Wet-Etching Aluminum (Al)

Wet-etching Al and Al-alloy films is done in heated-solutions (35-45°C) of 80% phosphoric-acid (H_3PO_4), 5% nitric-acid (HNO_3), 5% acetic-acid (CH_3COOH), and 10%-water. The etch-rate depends on several factors including etchant-composition and temperature, type of resist used, agitation of wafers during etch, and impurities or alloys present in the Al-film.

The chemical-mechanism of wet-etching Al are as follows: Nitric-acid oxidizes the Al to form aluminum-oxide (Al_2O_3), with H_2 as one of the by-products. The H_3PO_4 and water then dissolve the Al_2O_3.

One problem with wet-etching Al is that tiny H_2 gas-bubbles are formed. They cling tightly, particularly along pattern-edges and deny the etchant local-access to the film being etched. At such locations, etching temporarily slows-down or ceases (until the bubble is dislodged). The result is that the Al does not etch cleanly. Instead, ragged-edges (or even bridging) result. This may cause electrical-shorts between adjacent-leads. Mechanical-agitation during etching, and addition of wetting-agents (which lower the interfacial-tension) are used to control this problem.

REFERENCES

1. S.Wolf and R.N. Tauber, *Silicon Processing for the VLSI Era, Vol 1: Process Technology*, 1st Ed.. Lattice Press, Sunset Beach CA, 1986, Ch. 15 (Wet-Etching), p. 514.
2. W.A. Kern & C.A. Deckert, "Chemical Etching," in *Thin Film Processing*, J. Vossen, Ed., Academic, NY, 1978.
3. B. Schwartz and H. Robbins, "Chemical Etching of Silicon: Etching Technology," *J. Electrochem Soc.*, **123**, 1903 (1976).
4. K.E. Bean, "Anisotropic Etching of Silicon," *IEEE Trans. Electron Devices*, **ED-25**, 1185 (1978).
5. K. Christiansen *et al.*, "Better HF Etch-Uniformity (of SiO_2) With Single-Tank Approach, *Semiconductor International*, August 2002, p. 47.
6. H. Xiao, *Introduction to Semiconductor Manufacturing Technology*, Prentice-Hall, 2001, p. 194 and p. 361.

PROBLEMS

1. *Define* etching. What are the goals of etching?
2. Define *etch-rate* and write its formula. Why is it desirable to have high etch-rates?
3. An oxide film is 5000-Å-thick. After 30 sec of etching, the thickness is 2400-Å. Calculate the *etch-rate*.
4. Define *etch selectivity*. What is *high-selectivity*?
5. What are the differences between wet & dry etch?
6. Explain the difference between *isotropic* and *anisotropic etching*. What are desirable and undesirable aspects of isotropic and anisotropic etch profiles?
7. What are typical etchants used to wet etch: *silicon dioxide*; *silicon nitride*; and *aluminum* thin-films?
8. Describe some of the problems associated with the wet-etching of aluminum thin-films.
9. Assuming the mask and substrate materials are not attacked by a particular etchant., sketch the edge profile of an isotropically-etched feature of thickness h_f for: (a) etching just to completion; (b) 100% overetch; and 200% overetch. What shape does the profile tend toward as overetching proceeds?
10. A 6000-Å film of SiO_2 is to be etched with a buffered-HF solution with an etch-rate of 750-Å/min. If the oxide thickness varies by 10%, and the etch-rate varies by 15%, specify a time for the etch process; (b) By how much will the top of the film be undercut?
11. Given a 2500-Å-thick polysilicon film, with a thickness non-uniformity of 1.5%. The polySi etch-rate is 5000-Å/min, and the etch-rate non-uniformity is 5%. If only 5-Å can be lost of the 40-Å oxide film beneath the polySi, what is the minimum polySi-to-oxide selectivity during the overetch-step?
12. Why is dry-etching often used for Si_3N_4 films?

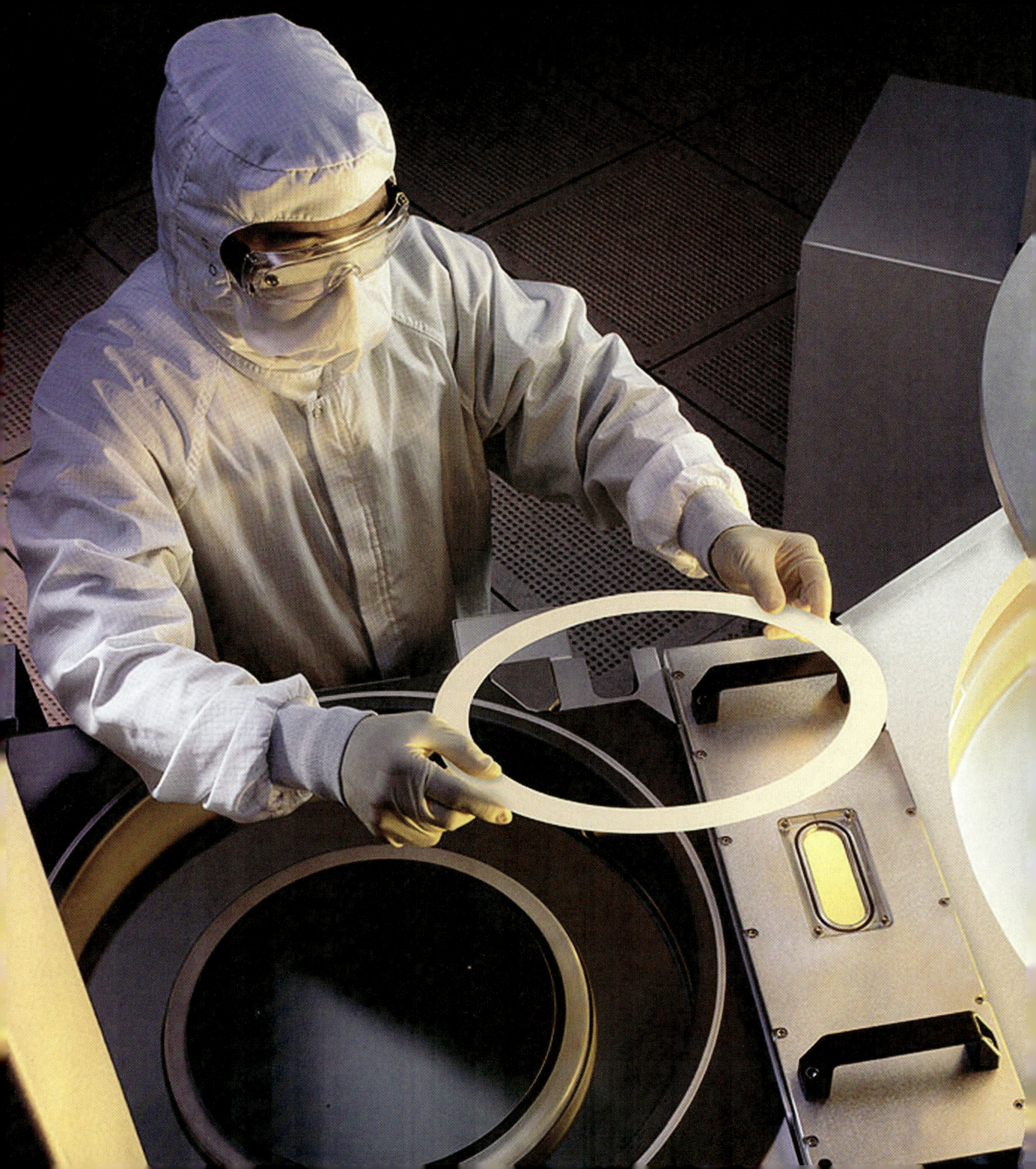

CHAPTER 22
DRY-ETCHING FOR ULSI

CHAPTER CONTENTS

22.1 TYPES OF DRY-ETCHING PROCESSES

22.2 THE PHYSICS & CHEMISTRY OF PLASMA-ETCHING

22.3 ETCHING SILICON AND SILICON DIOXIDE IN FLUOROCARBON PLASMAS

22.4 ANISOTROPIC ETCH-MECHANISMS

22.5 DRY-ETCHING OF VARIOUS TYPES OF MATERIALS IN ULSI PROCESSING
  Dry-Etching of Silicon Dioxide (SiO_2)
  Dry-Etching of Silicon Nitride
  Dry-Etching of Polysilicon
  Dry-Etching Aluminum
     and Aluminum-Alloys
  Dry-Etching of Organic Films

22.6 PROCESS MONITORING: ENDPOINT DETECTION

22.7 BATCH DRY-ETCH EQUIPMENT CONFIGURATIONS

22.8 SINGLE-WAFER ETCHERS

22.9 HIGH-DENSITY PLASMA SOURCES

22.10 DAMAGE FROM DRY-ETCHING

"Chance favors the trained mind."
            Louis Pasteur

9300 Exelan® dielectric-etch system. Photograph courtesy of Lam Research Corporation.

Wet-chemical etching (Chap. 21) was the standard pattern-transfer technique in early generations of ICs for several reasons. First, it was well known from its long-use in the printing-industry. Second, liquid-etchants can selectively remove materials being etched without affecting those beneath. Third, most do not attack photoresist (the most common etch-mask). However, such etching typically occurs in all directions at the same rate (*isotropically*). Thus, if the thickness of the film being etched is comparable to the minimum lateral pattern-dimension, the undercut-

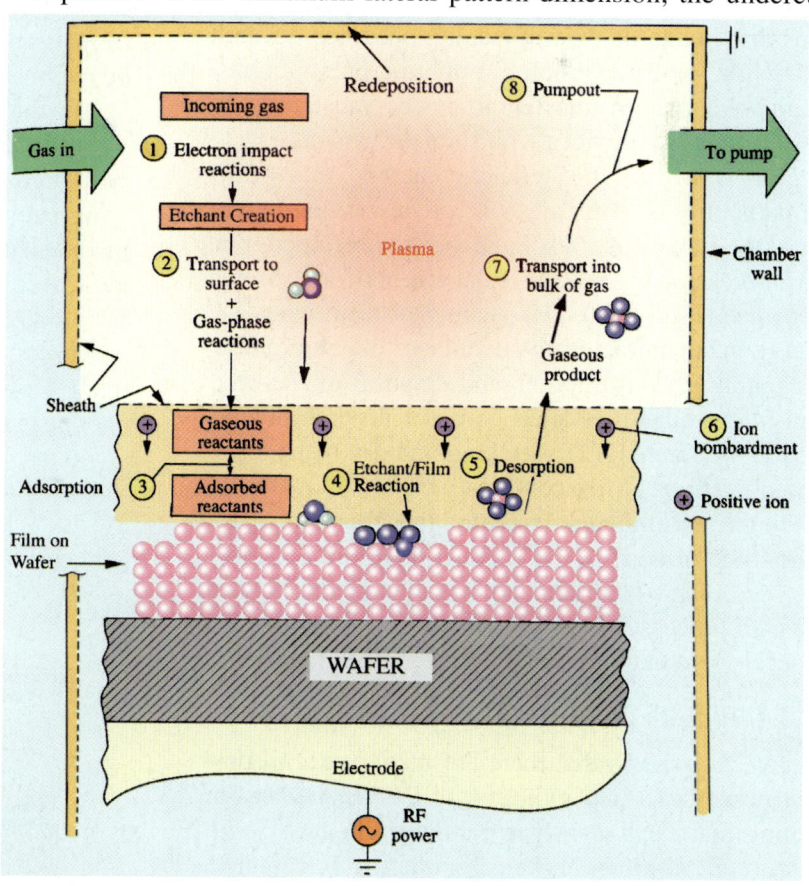

Fig. 22-1 Schematic view of the microscopic processes that occur during the *dry-etching* of a silicon wafer.

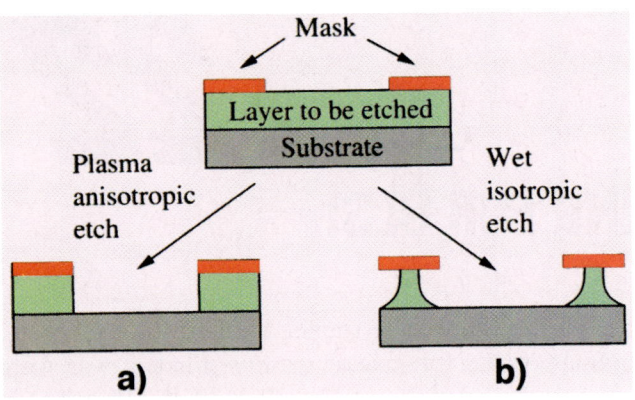

Fig. 22-2 Comparison between: (a) *completely anisotropic-etching*; and (b) *isotropic-etching*.

ting due to such isotropic-etching is unacceptable (see Fig. 22-2). Since many films used in ULSI-devices are 0.5–1.0-μm-thick, reproducible and controllable transfer of patterns by wet-etching becomes difficult at 1–2-μm, and impossible for submicron structures. Alternative pattern-transfer processes must thus be used to fabricate devices with such dimensions.

Dry-etching is an alternative method that offers the capability of non-isotropic (or *anisotropic*) etching. Dry-etch processes have been developed that can successfully serve as replacements for wet-etch processes (Fig. 22-1). Dry-etching also provides the important manufacturing-advantage of eliminating the handling, consumption, and disposal of the relatively-large quantities of dangerous acids and solvents used in wet-etching. Dry-etching and dry resist-stripping operations utilize comparatively small quantities of chemicals (although, some of those used are also quite toxic or corrosive, and must therefore be safely disposed). This chapter deals with the technology of dry-etch processes for ULSI fabrication. More details about dry-etching are found in References 1 thru 8.

22.1 TYPES OF DRY-ETCHING PROCESSES

Figure 22-3 shows that there are many types of dry-etch processes. Also indicated is the mechanism of etching that each such-type can have: **1**) a *physical-basis* (e.g., in glow-discharge sputtering [Chap. 15], or ion milling); **2**) *a chemical-basis* (e.g., in plasma-etching); or **3**) a combination of the two (termed *reactive-ion-etching* [*RIE*], or *ion-enhanced-etching*).

In processes that rely predominantly on the physical mechanism of sputtering, the strongly directional nature of the incident energetic-ions allows substrate material to be removed in a highly-anisotropic manner (essentially vertical-etch-profiles can be produced). Unfortunately, such material-removal mechanisms are also quite non-selective against both the masking-material and the materials underlying the layers being etched. The selectivity depends largely on sputter-yield differences between materials. Since sputter-yields for most materials are within a factor of three of each other, selectivities are typically not adequate. Furthermore, since the ejected-species are frequently non-volatile, redeposition and trenching can occur. Because of these issues, dry-etch processes for pattern-transfer based purely on *physical-removal-mechanisms* are not widely used in ULSI fabrication.

On the other hand, dry-processes relying *strictly on chemical-mechanisms for etching* can exhibit very-high-selectivities against both mask and underlying substrate-layers. Such purely chemical-etching mechanisms, however, typically etch in an isotropic-fashion. Some applications in ULSI fabrication which do not require anisotropic-etching, utilize such processes (e.g., photoresist stripping in oxygen-plasmas). However, etch-processes that rely primarily on chemical-reactions, are unable to solve the problem

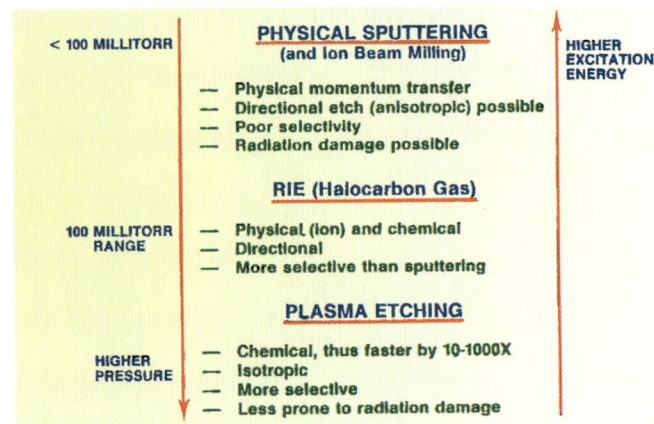

Fig. 22-3 The dry-etching spectrum.

of undercutting associated with isotropic-etching, and thus are not used to etch features smaller than 1-μm.

However, by adding a physical-component to a purely chemical-etching mechanism, the shortcomings of both sputter-based and purely-chemical dry-etching processes can be overcome. Dry-etch processes based on a *combination* of physical and chemical mechanisms offer the potential of controlled anisotropic-etching, together with adequate selectivity.

22.2 THE PHYSICS AND CHEMISTRY OF PLASMA-ETCHING

The basic concept of plasma-etching is rather simple (Fig. 22-1). An rf glow-discharge produces chemically-reactive species (atoms, radicals, and ions) from a relatively-inert *molecular-gas*. The etching-gas is selected to generate species which react chemically with the material to be etched, and whose reaction-product is volatile. As shown in Fig. 22-1, an ideal dry-etch process based solely on *chemical-mechanisms for material-removal* can thus be broken down into six steps: **1)** reactive-species are generated from the incoming-gases by the plasma; **2)** these species are transported by diffusion to the surface of the material being etched; **3)** they are then adsorbed on the surface; **4)** a chemical-reaction occurs, with the formation of a volatile by-product; **5)** the by-product is desorbed from the surface; and **6)** the desorbed species diffuse into the bulk of the gas, and are pumped from the chamber (#7 and #8 in Fig. 22-1). If any of these steps fails to occur, the etch-process ceases.

Step 5 (*product-desorption*) is a noteworthy step. Many reactive-species can react rapidly with a solid-surface, but unless the product has a reasonable vapor-pressure (so that desorption occurs), no etching takes place. Also note that steps **1**, **2**, and **6** involve events occurring in the *gas-phase* (and plasma), while steps **3**, **4**, and **5** take place *at the surface of the solid-layer being etched*. Hence, it is useful to briefly consider the physics and chemistry of events that involve the etching process that occur: **a)** in the *gas-phase* (and *plasma*); and **b)** on the *surface being etched*.

Finally, other significant effects can also occur, including: **1)** ion-bombardment of the wafer-surface and chamber-walls, which can assist in etch-product desorption from the wafers and erosion of materials from the walls; and **2)** deposition (or re-deposition) of gas-phase molecular-fragments and etching-products on the wafer surfaces and chamber walls.

22.2.1 The Reactive-Gas Glow-Discharge

In Chap. 14 methods for producing a glow-discharge using dc-diode and rf-diode configurations are described. In conventional plasma-etching processes (i.e., not a high-density-plasma process), an rf-diode configuration is used to establish the glow-discharge (for reasons listed in Chap. 14). The glow-discharge used in sputtering produces energetic-ions, which are then used to bombard target-surfaces and cause sputtering. The plasma-gases used in sputtering are thus *atomic, non-chemically-reactive gases* (such as Ar).

However, in *plasma-etching applications* the glow-discharge is not only used to produce energetic ionic-bombardment of the etched-surface, but it has another (and even more-important) role, namely that of *producing reactive-species for chemically-etching the surfaces of interest*. Thus, it is necessary to examine the properties of glow-discharges related to this function. Note that *molecular-gases* are used in dry-etching. The gases selected are either inherently reactive (such as Cl_2), or they can be dissociated into molecular-fragments (that are then reactive - for example, CF_4, a molecule that can be dissociated into such reactive-species as F, CF_3, CF_3^+, etc.).

Since plasmas consisting of fluorine-containing gases are extensively used for etching Si, SiO_2, Si_3N_4 (and other materials used in ULSI-fabrication), it is appropriate to study the glow-discharge created in a CF_4-gas as an example.

Before a glow-discharge is established in such a gas, the only species present are CF_4-molecules. Over the pressure-range at which an rf glow-discharge can be maintained: 1-Pa to 750-Pa (or 7.5-mtorr to 5.6-torr), the gas-density ranges from 2.7×10^{14}–2×10^{17} molecules/cm^3. When a glow-discharge has been formed, some fraction of the CF_4-molecules are dis-

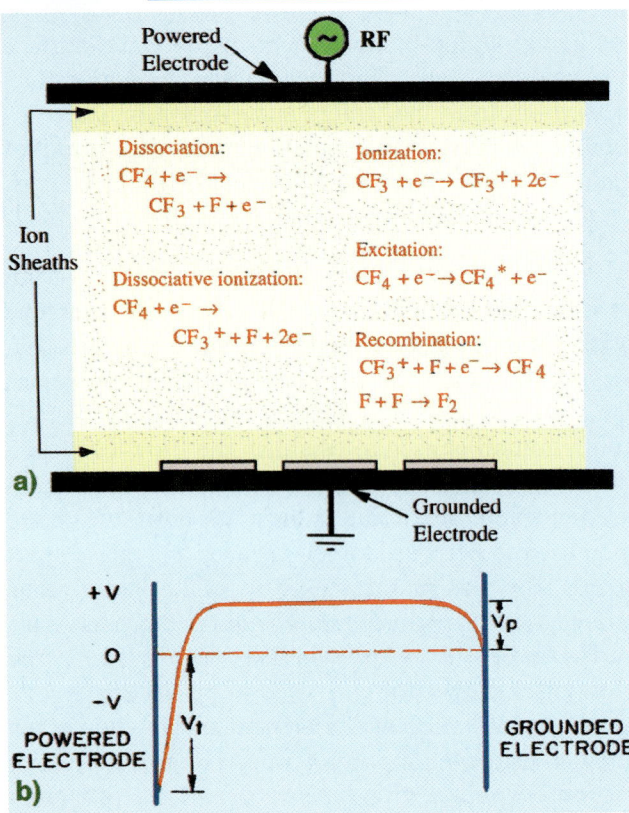

Fig. 22-4 (a) Schematic view of an rf glow-discharge, including the reactions and species present in a plasma used for dry-etching. b) *Potential-distribution* in a parallel-plate plasma-etcher having a *grounded-electrode* whose surface-area is larger than that of the *powered-electrode*.

sociated into other species.

A *plasma* is defined to be a partially-ionized gas composed of ions, electrons, and a variety of neutral-species (see Chap. 14). A *glow-discharge* is a plasma that exists in the pressure-range given above and contains approximately equal concentrations of *positive* (positive-ions) and *negative* particles (electrons and negative-ions). The density of these charged-particles in conventional (i.e., *low-density plasma*) glow-discharges ranges from 10^9–10^{11}/cm³. Thus, only one atom in 10,000 (to 1,000,000) of the gas is ionized in a low-density glow-discharge. The average-energy of electrons in glow-discharges is between 1-10-eV. The reactions that occur in the gas-phase (plasma) are called *homogeneous-reactions*, while those that occur at the surface are termed *heterogeneous-reactions*. Table 14-1 (in Chap. 14) lists the general types of homogeneous electron-impact reactions and heterogeneous surface-plasma reactions that can take place (see also Fig. 22-4a).

Ions are one of the reactive-species formed in such plasmas. For example, the most abundant *ionic-species* found in CF_4 plasmas is CF_3^+, and such ions are formed by the following electron-impact reaction:

$$e^* + CF_4 \rightarrow CF_3^+ + F + 2e \quad (22.1)$$

In addition to CF_4-molecules, ionic-species, and electrons, there are also a large number of radicals formed. A *radical* is an atom, or collection of atoms, which is electrically-neutral, but which also exists in a state of incomplete chemical-bonding - making it very chemically-reactive. Some examples of radicals include F, Cl, O, H, and CF_x, where $x = 1, 2,$ or 3. In CF_4-plasmas, the most-abundant radicals are CF_3 and F, formed by the reaction:

$$e^* + CF_4 \rightarrow CF_3 + F + e \quad (22.2)$$

In general, radicals are thought to exist in plasmas in much-higher concentrations than ions, because they are generated at a faster-rate and they survive longer than ions. Roughly speaking, the gas in an etch-chamber when conventional plasma-etching is underway, generally consists of the following-species (in order of decreasing-concentration, and estimated concentration-ranges): **a)** etch-gas molecules (70–98% of the total-species in the chamber); **b)** etch-product molecules (2–20%); **c)** radicals (0.1–20%); and **d)** charged-species (including positive-ions, electrons, and negative-ions [0.001-0.01%]).

The radicals, in fact, are responsible for most of the actual chemical-etching phenomena that occur at the surface of the material being etched. As will be described later, the ionic-species are believed primarily to enhance the etching that occurs by causing events that are not in-themselves chemical-reactions. Thus, the term *reactive-ion-etching* (that is commonly used to denote processes in which plasma-etching is accompanied by ionic-bombardment), is actually somewhat of a misnomer. Since the etching by the reactive-radicals is principally *enhanced* by ionic-bombardment, these processes should more aptly be described as *ion-assisted etching-processes*.

22.2.2 Electrical Aspects of Glow-Discharges

It is important to have information about the electrical potential-distribution in systems containing glow-discharges. In Chap. 14, details about this subject were given for both dc-diode and rf-diode glow-discharges. The information was used to explain how glow-discharge sputtering occurs. In plasma-etching systems, high-frequency (e.g., 13.56-MHz) rf-diode configurations are primarily used. (Readers are directed to Chap. 14 for general-information on rf glow-discharges.) In plasma-etching systems, knowledge about the potential-distribution is useful because the energy of particles impinging on the etched-surface depends on the potential-distribution. In addition, the plasma-potential determines the energy with which ions strike other surfaces in the chamber. High-energy bombardment of these surfaces can cause sputtering and consequent re-deposition of sputtered-material (in the form of contamination on the wafers).

As shown in Fig. 22-4b, the potential of the plasma is *positive* relative to that of the grounded-electrode (which is usually connected electrically to the chamber-walls, grounding them as well), and the powered-electrode develops a *negative*, dc self-bias-voltage relative to ground. The magnitude of the self-bias voltage depends on the amplitude of the rf-signal applied to the electrodes. If the electrodes of the rf-plasma system are of comparable area, the potential-difference across the dark-space of both electrodes will be equal. Since the powered-electrode develops a *negative dc-self-bias voltage*, in order for a potential-difference of equal magnitude to exist across the dark-space of the grounded-electrode, the plasma must assume a positive-potential of comparable magnitude. This means that even if wafers are placed on the grounded-electrode of such systems, they will be subjected to substantial energetic ion-bombardment. In systems where the area of the powered-electrode is *much-smaller than* that of the grounded-electrode, smaller potential-differences exist between the plasma and grounded-electrode, and thus grounded-surfaces receive less-energetic bombardment.

22.2.3 Heterogeneous (Surface) Reaction Considerations

The reactive-etchant species that undergo chemical-reactions at the surface to *produce* etching (as well as the ionic-species that bombard the surface to *enhance* such etching), are generated in the plasma. The events that take place at the surface are interactions between the gas-phase species and the solid-material to produce etching. The issues related to these events include: **a)** the sticking-probabilities of radicals and ions; **b)** the chemical-processes that: i) form films; ii) cause species to be adsorbed; or iii) lead to other gas-phase species; **c)** the desorption of species from the surface; and **d)** the effect of ion and electron fluxes on the surface. Some of the parameters that impact heterogeneous-reactions are: *surface-temperature; surface electrical-potential; the nature of the surface; & geometrical-aspects of the surface* (e.g., the angle of incidence of impinging-ions depends on whether they strike the bottom or sidewall of etched-features).

The gases adopted for plasma-etching processes are selected on the basis of their ability to form reactive-species in a plasma, which then react with the surface-materials being etched and lead to volatile-products. Table 22-1 lists the solid-gas systems for various solids to be etched in ULSI-fabrication, together with their resultant etch-products.

22.2.4 Parameter Control in Plasma-Processes

One of the more-challenging aspects of implementing a useful and reproducible etch-process involves

Table 22-1 EXAMPLES OF SOLID-GAS SYSTEMS USED IN PLASMA ETCHING

SOLID	ETCH GAS	ETCH PRODUCT
Si, SiO_2, Si_3N_4	CF_4, SF_6, NF_3	SiF_4
Si	Cl_2, CCl_2F_2	$SiCl_2$, $SiCl_4$
Al	BCl_3, CCl_4, $SiCl_4$, Cl_2	$AlCl_3$, Al_2Cl_6
Organic Solids	O_2	CO, CO_2, H_2O
	O_2 + CF_4	CO, CO_2, HF
Refractory Metals (W, Ta, Mo...)	CF_4	WF_6, ...

the *control* of the large number of parameters that affect the process. Figure 22-5 illustrates some of the parameters that impact the gas-phase interactions, as well as the surface-plasma interactions. Many macroscopic-parameters can be controlled, such as the type of feed-gas, power, and pressure. However, the precise effect of making any changes in these parameters is usually not well-understood. In fact, a change in a single macroscopic-parameter typically alters two or more basic *plasma-parameters*, and perhaps one or more of the *surface-parameters* (such as temperature or electrical-potential). This makes process-development in plasma-systems a challenge, and use of *factorial experimental-design techniques* very useful.

22.3 ETCHING SILICON AND SILICON-DIOXIDE IN FLUOROCARBON-PLASMAS

The dry-etching of silicon and SiO_2 is described here in some detail because these are very important processes in silicon ULSI-fabrication. When the mechanisms of plasma-etching were first being studied, the etching of silicon and SiO_2 in plasmas containing CF_4, mixtures of $CF_4 + O_2$, and mixtures of $CF_4 + H_2$ yielded important information about plasma-etching processes (as well as insights into the specific materials-system under investigation).

We start by describing several basic plasma-etching phenomena. It is known that in the absence of a glow-discharge, most gases commonly used in plasma-etching do not react with the surfaces to be etched. For example, CF_4 does not etch Si or SiO_2 without a discharge (because CF_4 does not chemisorb on Si and SiO_2 surfaces), and thus Step **3** of the dry-etching process does not occur.

The Etching of Silicon by Molecular-Fluorine: Fluorine, on the other hand, spontaneously etches Si (Fig. 22-6), even without the presence of a discharge. (However, etching in F_2 leaves the silicon-surface rough and pitted, and thus it is not used as a feedstock-gas in IC etching.) This implies that, when a discharge of CF_4 is created, it is not the CF_4-molecules that participate in the etching-reaction. Instead, the etching is done by the radical-species created from the dissociation of

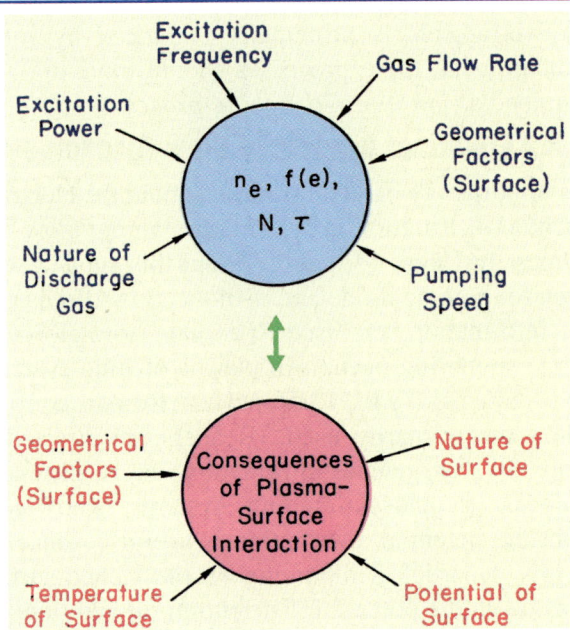

Fig. 22-5 Representation of the parameter problem in plasma etching systems.[6] (Note that n_e is the *electron-density*, f[e] is the *electron energy-distribution-function*, N is the gas-density, and τ is the gas-molecule *residence-time*.)

CF_4-molecules – namely, the fluorine-atoms.

The chemical-reaction of F with a Si-film that produces volatile-products is summarized in Fig. 22-7. Fluorine is first adsorbed on the silicon to form a stable "SiF_2-like" skin (2–5-atomic-layers thick). The SiF_2 does not readily desorb, since it is chemically-bonded to the wafer. But, additional impinging F-atoms penetrate this layer and attack the subsurface Si-Si bonds. Such penetration and bond-breaking by F-atoms continues until an SiF_4-entity is formed. This SiF_4-molecule can desorb from the surface with minimum-energy, since no more bonds connect it to the silicon-lattice. The etch-rate of silicon (and SiO_2) in pure CF_4-plasmas is relatively small (with the etch-rate of SiO_2 being much smaller, as shown Fig. 22-8, i.e., for the condition when the value of $H_2 = 0$).

Dry-Etching Silicon with O_2 Added to CF_4-Plasmas: If small-concentrations of O_2 are added to the CF_4-feed-gas (Fig. 22-8), the etch-rates of both Si and SiO_2 dramatically increase. (Note that the etch-rate of SiO_2 in pure-CF_4 is much smaller than that of Si - compare

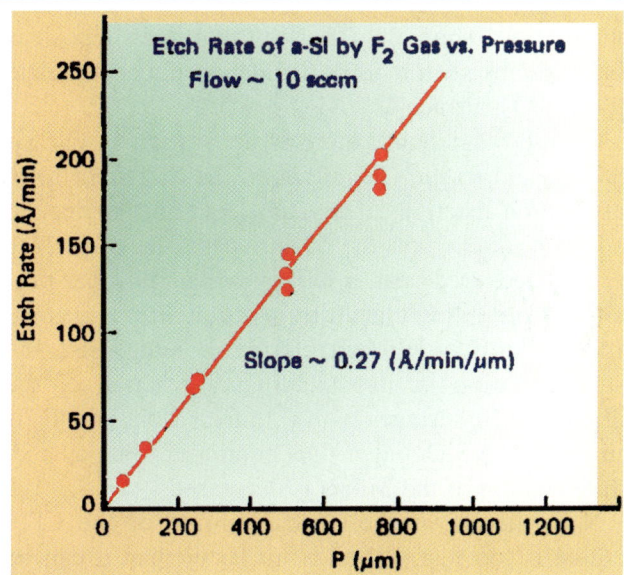

Fig. 22-6 The fluorine pressure-dependence of the etch rate of amorphous silicon at room-temperature.[9]

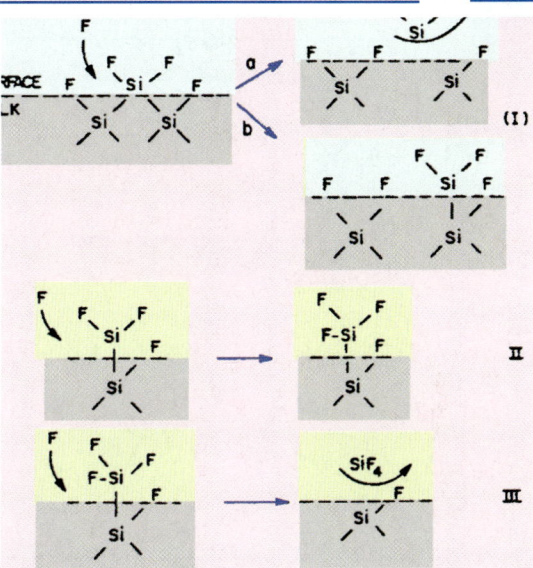

Fig. 22-7 Proposed mechanisms for F-atom reaction with a silicon-film leading to the products SiF_2 and SiF_4.[3]

their etch-rate values on the left-axis in Fig. 22-8). The addition of O_2 is accompanied by an increase in the density of F-atoms in the discharge. This causes higher Si and SiO_2 etch-rates, and such plasmas are therefore called *fluorine-rich*. The etch-rate of Si continues to increase until ~12% O_2 (by volume) is added. The etch-rate of Si reaches its maximum-value when ~20% O_2 is added. At greater-concentrations, the additional O_2 dilutes the F-concentration and this causes the etch-rates to decline.

Dry-Etching Silicon and SiO_2 with H_2 Added to CF_4-Plasmas: If H_2 is added to the CF_4-feed-gas, the etch-rate of Si decreases monotonically to almost zero for H_2 additions ≥40%. The etch-rate of SiO_2, however, remains relatively unchanged for H_2 additions of up to 40% (Fig. 22-9). The Si etch-rate decreases because the H_2 reacts with F to form HF, and this drastically reduces the F-atom concentration in the plasma. (It is said that the hydrogen *scavenges* F-atoms, and such plasmas are termed *fluorine-deficient*.) The effect of the Si etch-rate-decrease by itself may not be useful, but the fact that the SiO_2 etch-rate does not substantially decrease at the same time is valuable. This is because the SiO_2-to-Si *etch-rate-ratio* increases. A *higher-selectivity* (with respect to the substrate) can thus be achieved when etching SiO_2 over Si. This selectivity is necessary when SiO_2 films must be etched down to an underlying Si-layer, without significantly etching any Si exposed during the overetch-time.

Dry-Etching SiO_2 with CF_4/H_2 Plasmas: The mechanism responsible for the high SiO_2-to-Si etch-rate selectivity involves the combination of two phenomena:

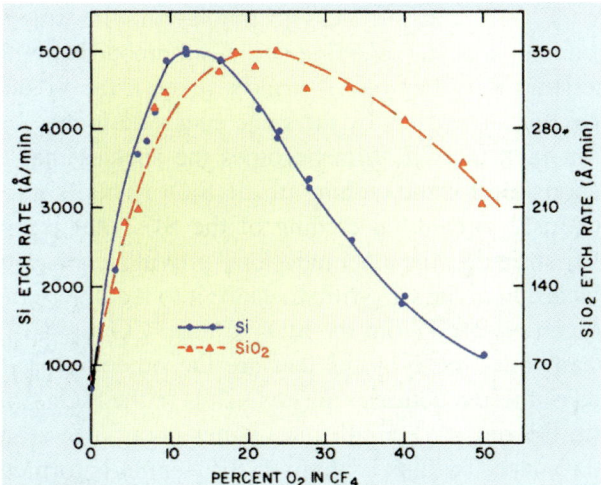

Fig. 22-8 The etch-rate of Si and SiO_2 versus the O_2 concentration in a CF_4-O_2 plasma.[10]

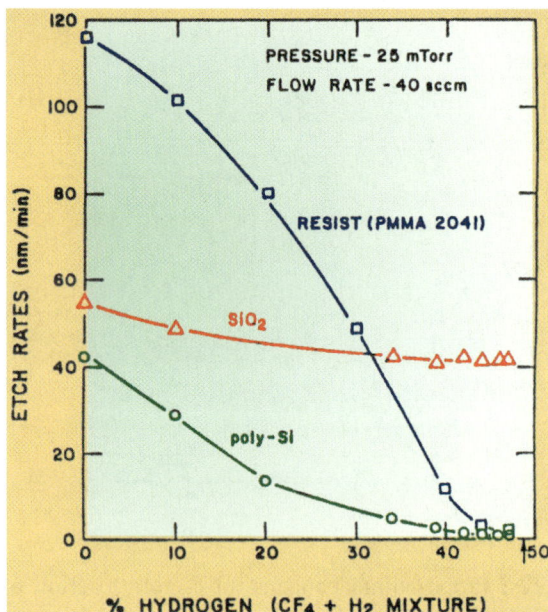

Fig. 22-9 The etch-rate of Si, resist, and SiO_2 (measured in a reactive-ion-etching configuration) as a function of the concentrations of H_2 in the CF_4-H_2 etch gas.[11]

1) the deposition of a nonvolatile-residue; and **2)** the role of oxygen in the etching of SiO_2. If a nonvolatile-layer (e.g., consisting of CF_x-fragments) deposits on a surface during etching, and it is not removed, etching will cease. While such CF_x-residues are found to deposit on all surfaces inside an etch-chamber containing $CF_4 + H_2$ plasmas, less accumulation with time is observed to occur on oxide-surfaces bombarded with energetic-ions than on Si-surfaces (even if the Si-surfaces are also struck by the same energetic-ions).

In fact, the CF_x-film becomes the species that is responsible for the etching of the SiO_2 (while simultaneously preventing etching of the Si). That is, on SiO_2-surfaces, the energetic-ions provide energy to the carbon in the CF_x-film to cause it to react with the oxygen released from the SiO_2 to form CO and CO_2. Meanwhile, the F-atoms that are the other by-products of these reactions replace O-sites in the SiO_2 film with F-atoms. When all four O-sites about a Si-atom are replaced with an F-atom, an SiF_4-entity is formed, and it then easily desorbs from the SiO_2-surface. The CO and CO_2 are also volatile at room-temperature

(and above). Thus, the CF_x-film that does deposit is volatilized by such reactions and does not accumulate on the SiO_2-surfaces.

On the other hand, there is no oxygen in the Si-regions, and so the CF_x-films on the Si cannot form gasification-reactions. Thus, CF_x continues to deposit and grow thicker on the Si-surfaces. The Si-surface becomes covered with a CF_x-polymer-film that protects it from being etched by F-atoms in the plasma. Since a fluorine-deficient plasma is being used, the F-concentration is also small, and reaction of the CF_x on the Si-surfaces with F to form volatile-CF_4 is minimal. The SiO_2-layer thus continues to be etched, while etching of the Si has is decreased.

Nevertheless, if the deposition-rate of the CF_x-residue is too fast, it builds up faster than it can be volatilized (even on SiO_2-surfaces), and etching will eventually stop on SiO_2-surfaces as well. Note that energetic-bombardment by ions is needed to drive the carbon-oxygen reactions, and thus only SiO_2-surfaces struck by a high ion-flux are etched at a significant-rate. This, in practice, implies that etching SiO_2 selectively over Si can only be done anisotropically. Furthermore, selective SiO_2-etching can only be done in fluorine-deficient plasmas, as the presence of F-atoms would spontaneously etch the Si. Thus, other gases which also consume F-atoms have been found to produce high SiO_2-to-Si selectivities (even without the use of H_2), including CHF_3, C_3F_8, and C_2F_6.

In practice, the exact process-conditions that produce selective-etching of SiO_2 over Si are experimentally found for each reactor (because high-selectivity requires the process be operated very-close to the demarcation between etching and polymerization - where even etching of the SiO_2 abruptly ceases). Although the adjustment of plasma-conditions to achieve high SiO_2/Si-selectivity was initially an art, such high-selectivity is now routinely achievable. For example, selectivities of >20:1 at oxide-etch-rates of 60–100-nm/min have been reported.

22.4 ANISOTROPIC DRY-ETCHING MECHANISMS

Up to this point, the etching of Si and SiO_2 in fluorocarbon-plasmas has been treated as a mechanism that proceeds by chemical-action alone (i.e., the reaction of Si by F-atoms generated by a plasma to form SiF_4). However, if the etching-action is purely chemical, the removal of material is isotropic, and no advantage in dimensional-control is gained over wet-etching. In such processes, the plasma plays no role other than to produce the etchant. The attraction of dry-etching for ULSI-patterning, however, is based on the possibility that it *can* etch in an anisotropic manner (Fig. 22-10). Thus, mechanisms that could produce anisotropic-etching need to be considered.

The ability to achieve anisotropic-etching is thought to depend in some way on the bombardment of the etched-surface with energetic-ions. Other parameters (such as the chemical-nature of the plasma), may influence the degree of anisotropy. But unless energetic-particles strike a surface, only isotropic-etching can be expected. The directional-etching effects in ion-assisted-etching processes, however, cannot be due to sputtering alone - as product-yields of over several-hundred substrate-atoms per incident-ion have been reported in ion-assisted etch-processes. Such product-yields are much higher than those of typical sputtering-yields (which are only about 2 sput-

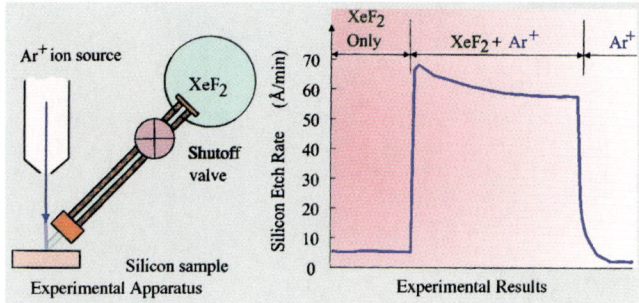

Fig. 22-11 An example of ion-assisted gas-surface chemistry in the etching of silicon with XeF_2. The XeF_2 flow is 2×10^{15} molecules/sec, and the Ar energy and current are 45-eV and 2.5-μA, respectively.[12]

tered-atoms-per-ion for 400-eV Ar^+ ions, see Ch. 15). This is lucky, because purely sputter-etch mechanisms result in processes with inadequate selectivities.

The fact that sputtering alone is not operative in such processes was elegantly demonstrated in an experimental manner by Coburn and Winters. They first exposed a Si-surface to a *gas* of XeF_2 (not a plasma of XeF_2), and observed a low etch-rate (Fig. 22-11). Next, while continuing to expose the surface to XeF_2, an Ar^+ ion-beam with an energy of 450-eV was directed at the Si. The observed etch-rate was ~10 times as great as with the XeF_2 alone. Finally, when the Ar^+-beam alone was directed at the surface, the smallest etch-rate of the three conditions was produced. The results of this experiment demonstrate that a strong cooperative-effect can result if the etching-surface is simultaneously exposed to a reactive-gas and bombardment by energetic-particles. The microscopic details of exactly how the ion-bombardment enhances the reaction between a reactive-gas and a surface, is the subject of substantial research efforts. Evidence indicates that different mechanisms exist for specific chemical-systems.

There are two principal mechanisms by which energetic-ions enhance the etch-rate of reactive-gases, and contribute to directional-etching (Fig. 22-12):

1. Relatively high-energy impinging-ions (>50-eV) produce lattice-damage at the surface being etched, which extends several-monolayers beneath the surface. Chemical-reactions at these

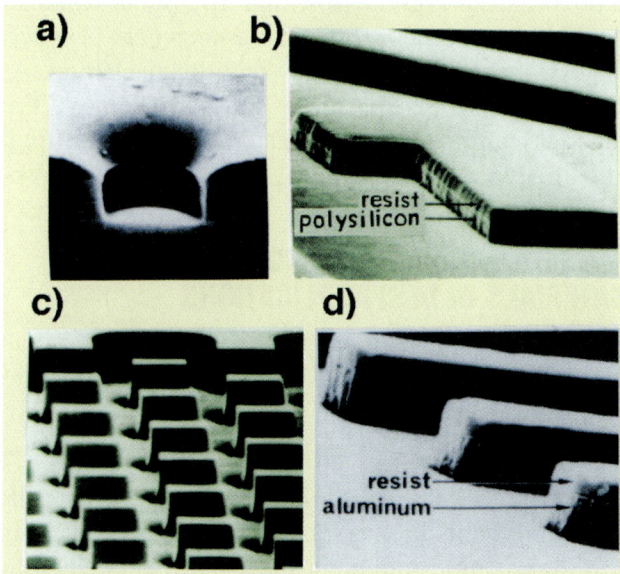

Fig. 22-10 Examples of anisotropic etch profiles that can be obtained when dry-etching various types of materials.

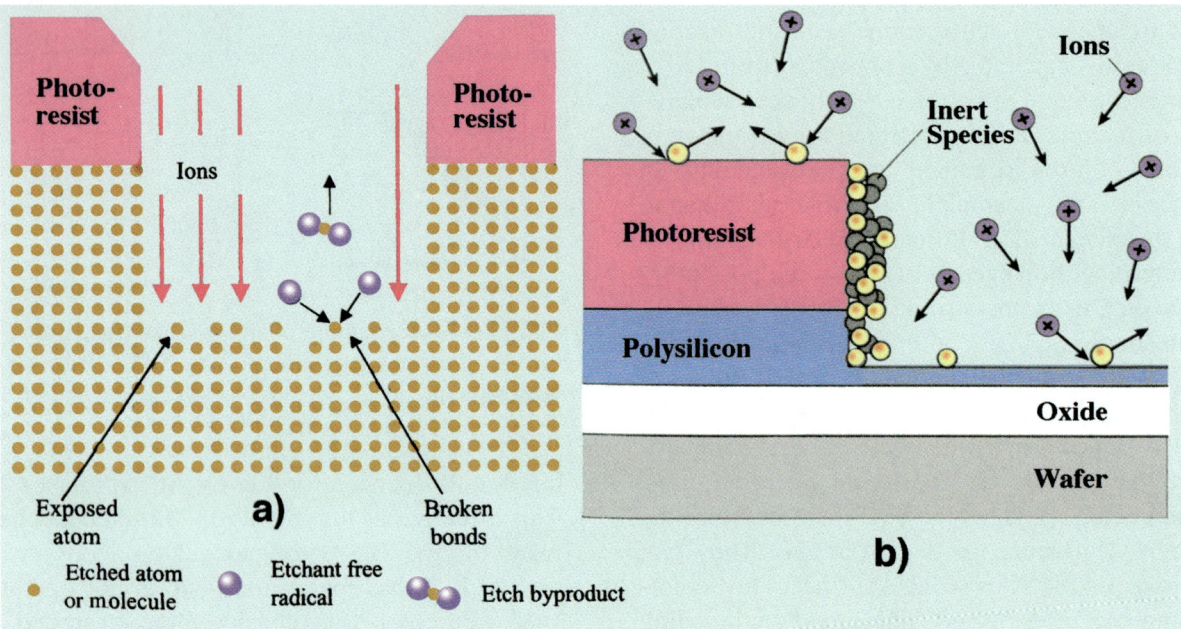

Fig. 22-12 (a) *Surface-damage mechanism*, and (b) *Surface-inhibitor mechanism*, for ion-assisted anisotropic etching.

damaged-sites are enhanced, compared to surfaces at which no damage has occurred (for instance, the feature-sidewalls receive a much smaller flux of bombarding energetic-ions, Fig. 22-12a).

2. Lower-energy ions (≤50-eV) provide enough energy to desorb *nonvolatile polymer-layers* (also referred to as *surface-inhibiting-*, or *blocking-layers*) that deposit on surfaces being etched. These deposited polymer-layers are only physisorbed on the surface, and require very-little energy to desorb them. In processes where polymer-depositions occur, surfaces not struck by ions retain their blocking-layer, and hence are protected against etching by the reactive-gas (Fig. 22-12b).

In features being etched on a wafer, the incident energetic-particles generally arrive in a direction perpendicular to the wafer-surface. Hence, they strike the bottom-surfaces of the etched-features. The sidewalls of the etched-features are subjected to little or no such bombardment. As a result, the bottoms of the features exhibit enhanced-etching. Ion-bombardment effects can be enhanced by decreasing the pressure in a high-frequency (> 5-MHz) plasma, or by decreasing the frequency of the discharge.

It should be noted that in an anisotropic-etch-process, if material is removed from a planar-region on a wafer, residual-material at steps has still not been removed (see Fig. 22-13a and b). Such residual-material (sometimes referred to as *stringers* or *picket fences*), must be removed by additional etching. As shown in Figs. 22-13c and d, failure to remove this material can lead to unwanted electrical-shorting-paths between adjacent-lines, for example, if a conductive-film is being etched.

22.5 DRY-ETCHING OF VARIOUS TYPES OF MATERIALS IN ULSI APPLICATIONS

In the following sections, details are provided about dry-etching processes used for various types of films used in IC-fabrication.

22.5.1 Dry-Etching of Silicon-Dioxide (SiO$_2$)

Contact-hole and via-hole-etching represent the most important applications involving SiO$_2$-etching in IC-fabrication (although there are others, such as pad-oxide and sacrificial-oxide-stripping, sidewall-spacer etching in MOSFETs, etching of trenches for

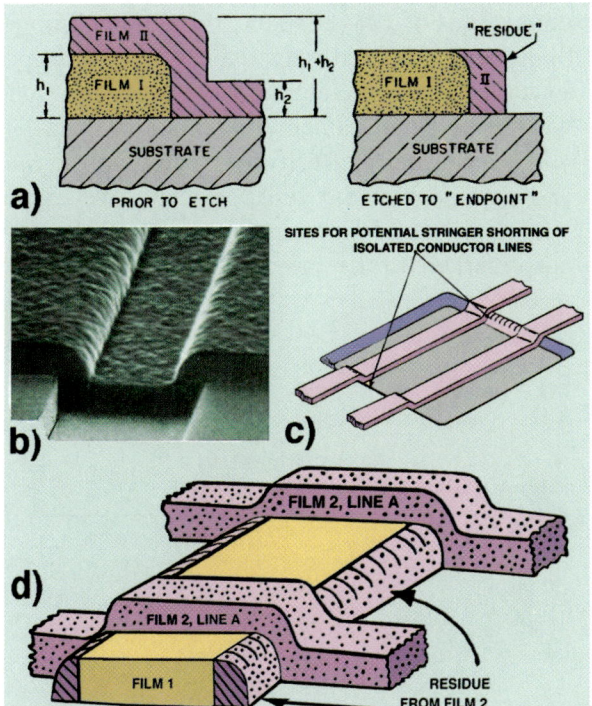

Fig. 22-13 (a) Residue at the base of steps after an anisotropic etch-process has removed the film material in the planar regions. (b) SEM of a polysilicon-film deposited over a step, showing the origin of the residue of part (a). (c) Stringer connecting two anisotropically dry-etched lines running down into a recessed wafer region. (d) Stringer connecting two anisotropically-etched lines running over another line.

Damascene-interconnects, and bond-pad etching).

The key issues of contact and via-etching are the following: **1)** sidewall-profile control; **2)** selectivity of the oxide-etch-process to the underlying silicon, polysilicon, or silicide; **3)** selectivity of the oxide-etch-process to the photoresist; **4)** etch-rate uniformity of densely-packed holes versus isolated-holes (*microloading*); **5)** etch-rate uniformity of holes with different aspect-ratios (*ARDE*); and **6)** etch-residue removal from contact-holes and vias. While these issues are briefly discussed here, more details can be found in Ref. 1 (listed at the end of the chapter).

It should be noted that *selectivity with respect to the silicon* is an especially crucial issue for oxide-etching. First, overetch is needed to clear all the contacts to the same level of silicon (e.g., the silicon-substrate), and the Si in contacts that have been cleared are exposed to the oxide-etching-plasma. Second, when etching contacts to both polySi and the Si-substrate in the same etch step, the oxide on top of the polysilicon-layer is several-hundred nanometers thinner than that over the silicon (Fig. 22-14). In such bi-level etch-processes, the exposed-polySi must not be significantly etched as the remainder of the oxide thickness is etched down to the Si-substrate. Oxide-to-silicon selectivities of greater than 25:1 are thus needed for 0.25-μm CMOS.

Shaping the Sidewalls of Contact-Holes and Vias by Dry-Etching: When contact-holes and vias are etched in SiO_2, the desired profile of their sidewalls will depend on the IC-technology (i.e., the minimum feature-size) and on the type of interconnect-structures being employed. Contact-holes and vias with non-vertical sidewall-profiles were used in IC-generations down to about 0.35-μm CMOS. After that, vertical-sidewall profiles were necessary (since even slightly-tapered contact-holes and vias would need too much chip-area, and this would represent an unacceptable functional-density penalty).

For ICs from 2.0–0.8-μm, that used 2 or 3 levels of metal (and in which Al is deposited into the contact-holes and vias), the vias were made easier to fill by creating contact-holes and vias with *wine-glass-shaped sidewalls* (see Fig. 22-15a and b). Such sidewall-profiles are created using a two-step etch-process. The top-half of the intermetal-dielectric film-thickness is etched with an isotropic etch-process, and the lower-half with an anisotropic one. The isotropic-process could be either a wet-etch or dry-etch step. Such a wine-glass shape makes it easier to fill vias with Al, but it uses considerably more chip-area than a vertical-via (with the same dimension as the bottom, anisotropically-etched portion of the via).

For ICs fabricated using 0.8–0.35-μm technologies, W-plugs are used to fill the contact-holes and vias. These do not require a wine-shaped-glass via-profile. That is, blanket CVD-W films can completely fill vias without voids or seams if the via-profile is

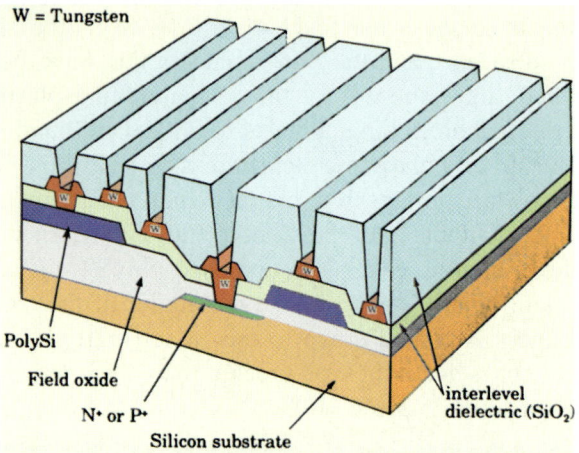

Fig. 22-14 Example which shows how much via-depths can vary in a double-level metal process if full-planarization is achieved.

slightly sloped (i.e., ~85° slope). Figure 22-15c shows a via with tapered-sidewalls (and a via with vertical sidewalls). Figure 22-15d shows a SEM of vertically-etched vias. One method for producing such a slope is by controlled-polymer-deposition on the contact-hole sidewalls (Fig. 22-16).

Via-Veil Removal After Via-Etching: Residues (called *via-veils*) are often left inside vias after etch. Such veils interfere with subsequent metal deposited into the vias, and result in poor-filling or voiding. They are also a source of contamination. Thus, these residues must be removed before the metal is deposited. The removal-process can be wet or dry (or some combination of both), and depends on many factors.

Such via-veils consist of a complex-mixture of materials that are sputtered onto via-sidewalls during the etch (and overetch time) from metals under the dielectric-layer (together with plasma-deposited-polymers from the etch-gases and eroded-photoresist). Via-veils generally contain oxides of Si and Al, as well as fluorocarbon-polymers. In addition, since Al rapidly becomes fluorinated upon exposure to fluorine, aluminum-fluorides may also be present. These materials are not deposited simultaneously. Instead, they are put down in layers, one after another, as the material from which they are derived are sequentially exposed. All of these factors make the organometallic-compounds of via-veils quite difficult to remove.

Many commercially-available resist-strippers (often based on organic-amines), and the positive photoresist-developer tetramethyl-ammonium hydroxide (TMAH) are usually effective in removing the *inorganic-component* of the veils. Specialty wet-chemicals (such as the amine-based solvent EKC 265®) are also marketed for removing the metal-containing

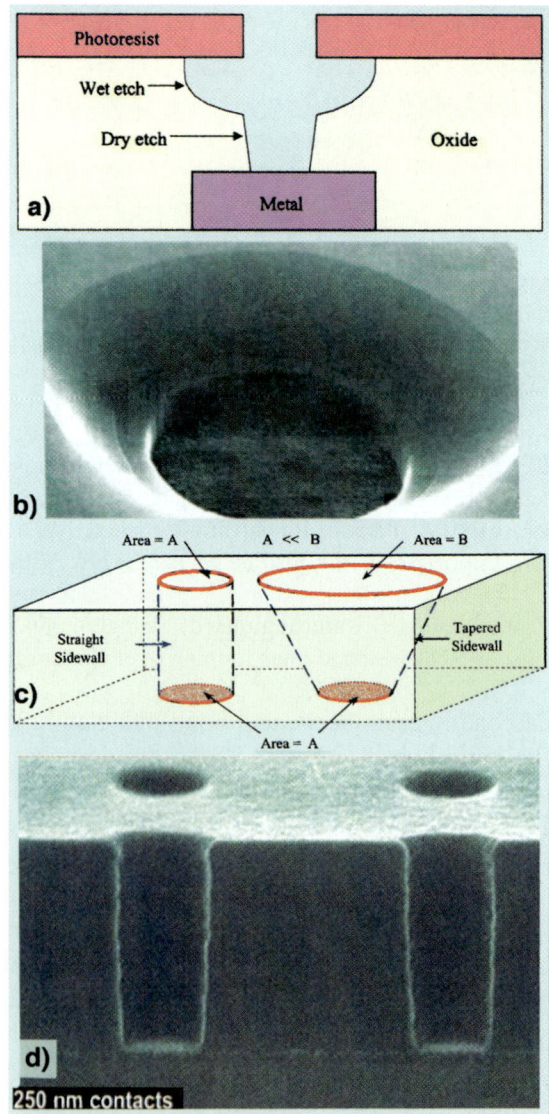

Fig. 22-15 (a) Drawing and (b) SEM of "champagne-glass" via-etch profile. (c) Schematic drawing of vertical and tapered-etched vias. (d) SEM of vertical-etched vias.

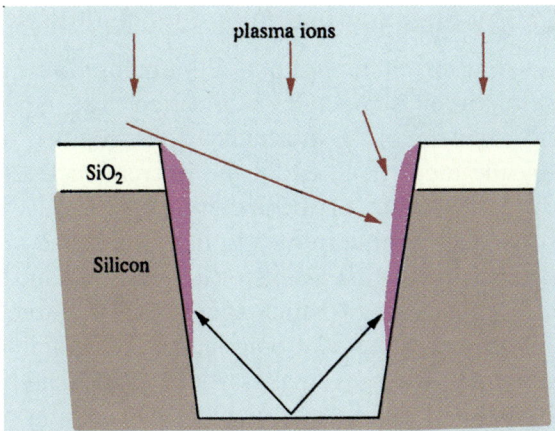

Fig. 22-16 The continuous buildup of sidewall-polymer-thickness in some dry-etch processes results in a slope of the sidewalls of the contact-hole being etched (if the etch process is anisotropic).

materials of the veils. Sometimes these wet-strippers will also remove the *organic-components* of the veils, but often they will not. In such cases dry-processes must be used for this task. The O_2-based plasmas are usually able to strip the organic-portion of the veils.

22.5.2 Dry-Etching of Silicon-Nitride

As described in Chap. 16, two types of silicon-nitride are used in ULSI fabrication. The first type is deposited by low-pressure CVD (LPCVD) at 700–800°C, and results in a stoichiometric Si_3N_4 film. The second is deposited by plasma-enhanced CVD (PECVD) techniques at ≤350°C, and such plasma-deposited nitrides are really polymer-like Si-N-H materials. Since fluorine-atoms isotropically-etch silicon-nitride, both nitrides are usually etched in CF_4-O_2 plasmas. PECVD nitride-films normally etch faster than LPCVD nitride-films in such plasmas.

22.5.3 Dry-Etching of Polysilicon

In CMOS-ICs the gate-length of the MOSFET is a small and critical-dimension (CD) that determines the channel-length of the devices. For CMOS technologies of 0.5-μm and smaller, the polySi is patterned before a silicide is formed on top of it, and in this section the process-issues associated with the etching of thin-films of polySi to form the gate of MOSFETs for such technologies are described.

As far as etching submicron-polySi-gates is concerned, it is crucial that the etched polySi linewidth-dimensions faithfully reproduce the dimension on the mask (e.g., to within ±5%). The polySi etch-process must therefore exhibit excellent linewidth-control and high-uniformity of etching. In addition, a high degree of anisotropy is also required to yield a gate-structure with vertical-sidewalls (Fig. 22-17). The doping of the source-drain and the polySi itself is typically done by ion-implantation. If the etch-process were to produce polySi-lines with sloped-sidewalls, then portions of the gate would not be thick enough to effectively mask the substrate against the implantation. This would result in devices whose channel-length would depend on the degree of sidewall-taper, and unless this could be accurately-controlled, a manufacturing control-problem would result.

Another key characteristic of the polysilicon etch-process is its selectivity to oxide. A thin gate-oxide (only 3.0–10.0-nm thick) lies beneath the polysilicon-film. A significant overetching-time is required to remove polysilicon-stringers at the base of steep steps in the underlying-topography (as shown in Fig. 22-13). If the thin oxide-layer is removed during this overetching-time, the underlying Si-substrate will be exposed and thereafter be rapidly etched by the same reactants that cause polysilicon-etching. Thus, a polysilicon etch-process with high-selectivity to oxide is needed. In addition, the resist etch-mask must remain substantially intact during etching to maintain high CD-control. Thus high-selectivity with respect to the mask-material is also required.

Fluorine-atoms etch Si isotropically and also attack photoresists, so F-based chemistries are not used to etch polySi gate-structures. Instead, plasmas containing Cl and Br gases are used. Chlorine (Cl or Cl_2) spontaneously etches undoped Si very slowly at temperatures less than about 100°C (i.e., <10-nm/min). However, in the presence of energetic ion-bombardment, Cl etches undoped-Si more rapidly. Thus, etching-polySi anisotropically is more feasible with Cl-based plasmas than with F-based plasmas.

Nevertheless, Cl-plasmas will etch heavily n-doped polySi even without ion-bombardment, and this leads to unacceptable undercutting of polysilicon-lines.

As a result, when etching heavily n-doped-poly in Cl-atom plasmas, an inhibitor-film must also form on the polySi-sidewalls. In some processes the resist-erosion will produce products that form such an inhibitor-film, but in others, additives such as C_2F_6 are included in the feedstock-gases to cause deposition of such protective-layers on the polySi-sidewalls.

More recently, plasmas containing Br-based gases, or mixtures of HBr and Cl-containing gases have become favored over plasmas containing only Cl-containing gases for etching polySi-films. The primary etch gas is Cl_2, while the HBr participates by creating $SiBr_x$-species which passivate the emerging polySi-sidewalls. This improves the etch-characteristics of selectivity to the resist and thin gate-oxide, and produces 100% anisotropic-etching of the poly features.

An actual polySi etch-process usually consists of several separate steps. The first is a "breakthrough" step designed to remove any native-oxide on the top-surface of the polysilicon-film, and it usually employs a fluorine-containing plasma. The second is the main etch-step, and it employs a plasma that contains Br and Cl gases. The final step is the overetch-step, and it is designed to remove the residual polysilicon-stringers, while providing a high-degree of selectivity against etching the underlying thin gate-oxide layer.

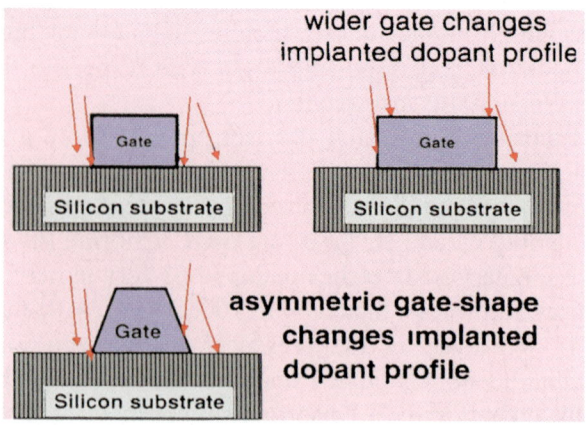

Fig. 22-17 Polysilicon etching.

22.5.4 Dry-Etching Aluminum (and Al-Alloys)

The etching of aluminum and aluminum-alloy films is an important step in IC fabrication. The device-density on many of the most advanced-circuits is limited by the area occupied by the interconnect paths. Anisotropic-etching of the metal-layers permits the use of smaller minimum metal-pitches (i.e., the *pitch* is the sum of the dimensions of a metal-line and the space between lines), which increases the interconnect-capability. The isotropic-nature of aluminum wet-etching processes makes them inadequate for ULSI-applications. A directional Al-dry-etching process is necessary. The metal-etch process must be able to anisotropically-etch through several different materials, with no residues or corrosion, minimal-undercut, good CD- control, and minimal loss of material underlying the interconnect-materials.

Fluorine-containing gases are not suitable for etching aluminum since the etch-product, AlF_3, has a very-low vapor-pressure at temperatures below 100°C (Fig. 22-18). Other halides of Al (such as $AlCl_3$, Fig. 22-18) have sufficiently high vapor-pressures to allow plasma-etching of Al, and thus Cl-containing gases are used in dry-etch processes for aluminum-films.

It has been determined that an aluminum-surface not covered by aluminum-oxide (Al_2O_3) will react spontaneously with Cl or molecular-Cl_2 to form $AlCl_3$, even in the absence of a plasma. If, however, the surface of the aluminum is covered with a thin-layer of Al_2O_3 (i.e., a native-oxide of ~3-nm), it will not react with Cl or Cl_2. Thus, etching of Al-films is two-step process, involving: **1)** removal of the native-oxide layer; and **2)** etching of the Al-film.

The successful removal of the native-oxide is an important step in achieving an effective aluminum-etching process, because removal of Al_2O_3 is far more difficult than the etching of pure-aluminum. The thickness of this oxide can also vary from run-to-run, depending on several factors.

An oxide-reducing species must thus be used to chemically-remove the Al_2O_3. Such oxidizing-species as BCl_3 or CCl_4, are capable of reducing Al_2O_3. However, if there is water-vapor present in the etch-chamber, it will scavenge the oxide-reducing species and react with the exposed-aluminum to form new Al_2O_3. Water-vapor must therefore be excluded from the etch-chamber to achieve reproducible Al-etch-processes. As shown in Fig. 22-19b the etch-rate of Al_2O_3 decreases rapidly with increasing partial-pressure of water-vapor. If an etch-chamber is exposed to ambient after an etch-run, moisture can be adsorbed on the chamber-walls. Water-vapor in the etch-chamber is most effectively reduced by load-locked chambers, and this is now the standard system-configuration on all single-wafer Al-etchers.

After the native-oxide is removed, the Al-etching proceeds at a rate that is affected by the gases used (e.g., Fig. 22-20). In general, the demands of high-throughput require the highest etch-rate consistent with good results. Processes with high-concentrations of Cl_2 in the feed-gas exhibit isotropic-etching. The anisotropic-etching of aluminum occurs as a result of the formation of an inhibiting-layer on the aluminum-surface, which is removed on surfaces struck by energetic-ions, allowing etching to occur there. Such inhibiting-layers are believed to come from the formation of chlorocarbon-polymers (e.g., CCl_x) originating from the resist etch-products.

Dry-Etching of Al:Cu-Alloys: Small quantities of other materials are added to aluminum to improve some of its properties. Silicon is often added in concentrations of 1–2 at% to prevent the aluminum from spiking through shallow-junctions (see Chap. 15). Copper (in concentrations of 1–2 at%) is added to enhance the electromigration-resistance and to suppress hillock-formation. Since $SiCl_4$ is volatile at room-temperature, Al-Si films are readily-etchable in chlorine-containing gases. Copper, on the other hand, forms an etch-product with chlorine (CuCl) that is relatively non-volatile below temperatures of 175°C (see Fig. 22-22). Thus, copper-containing residues remain after Al:Cu-alloy-films have been dry-etched. This makes Al-Cu more difficult to etch in chlorine-plasmas. The degree of difficulty increases with increasing Cu-concentration, and Al-films containing 4%-Cu pose a formidable etching-challenge.

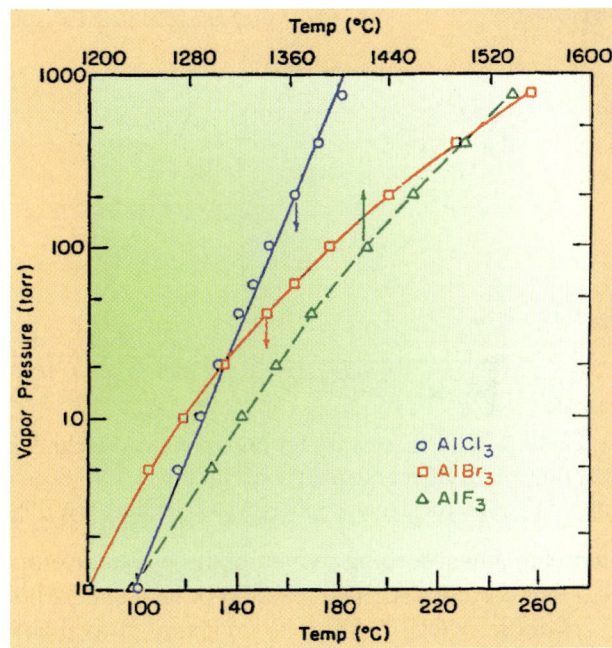

Fig. 22-18 Vapor-pressure of AlF_3, $AlCl_3$, and $AlBr_3$ as a function of temperature.[14]

Two methods are used to promote CuCl-desorption: **1)** increasing the substrate-temperature (up to the maximum temperature allowed with the resist-material being used); and **2)** enhancing the ionic-bombardment of the surface (so that significant sputtering of the Cu occurs). In batch-etchers, low-pressure-operation together with a slower etch-rate promotes high-energy bombardment while allowing more time for the sputtering to remove CuCl. In single-wafer etchers, higher-power is needed to produce sufficient ion-bombardment under the conditions of greater operating-pressure. This causes the unwanted effect of eroding the resist during etch. In addition, the etch-product, $AlCl_3$, is highly reactive, and it can also attack and degrade the resist. Thus, it is necessary to employ special UV-thermal-stabilization steps to enable the resist to withstand such harsh etching environments. See Chap. 18 for more information on resist-stabilization procedures by UV.

Post-Dry-Etch Aluminum Corrosion: Another Al-etch problem is *post-etch corrosion*, caused by $AlCl_3$ remaining on the film sidewalls, substrate, or resist

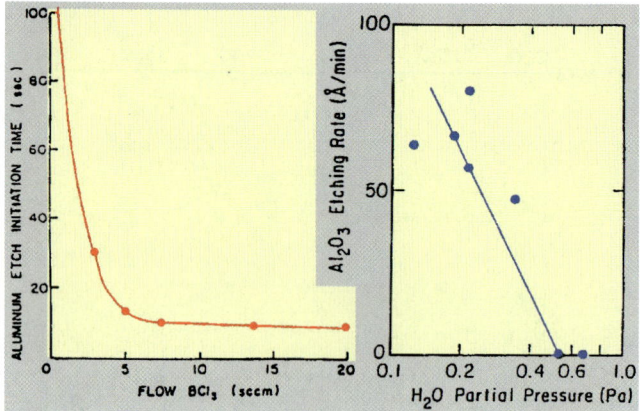

Fig. 22-19 (a) Time necessary for initiation of Al etching vs. BCl$_3$ flow.[15] (b) Effect of small concentrations of H$_2$O on the initiation period (designated as the Al$_2$O$_3$ etching rate).[16]

after etch. On absorbing moisture, these residues form HCl that corrodes Al. The reaction of HCl and Al produces more AlCl$_3$. As long as moisture is available, corrosion continues (Fig. 22-22). The problem is even more severe in Al-Cu alloys. The residual-Cl and ambient-moisture-levels necessary to induce corrosion are much lower when Cu is present in Al-films.

Various techniques are used to deal with this corrosion problem. As much of the chlorine as possible is removed from the wafers before they are exposed to air, which involves stripping some (if not all) of the photoresist before the wafers leave the etch chamber. Thus, most Al-etch tools have some kind of resist-stripping capability. Stripping the resist in an O$_2$-plasma (often in a separate-chamber of a cluster etch-tool) is one such approach. This is followed by a rinse in DI-water immediately after the wafer-cassette leaves the etch-tool. Additional practices that have proven effective involve exposing the etched and stripped wafers to a fluorine-containing plasma. The highly-reactive fluorine-radicals readily displace any chlorine that is bound to the aluminum. These Al-F bonds are very stable, and do not react with water.

22.5.5 Dry-Etching of Organic-Films

Organic-films are exposed to plasma-etching environments in many applications during ULSI fabrication. Photoresist is most commonly used as an etch-mask, and in such applications it is usually desired that the resist not be etched by the plasma. However, at the conclusion of the pattern-etching-step the resist must be removed, and this can be done by plasma-etching.

Plasmas containing pure-oxygen at moderate pressures produce species that attack organic-materials to form CO, CO$_2$, and H$_2$O as end-products. Such oxygen-plasmas provide a highly-selective method for removing organic-materials, since the O$_2$-plasmas do not etch Si, SiO$_2$, or Al.

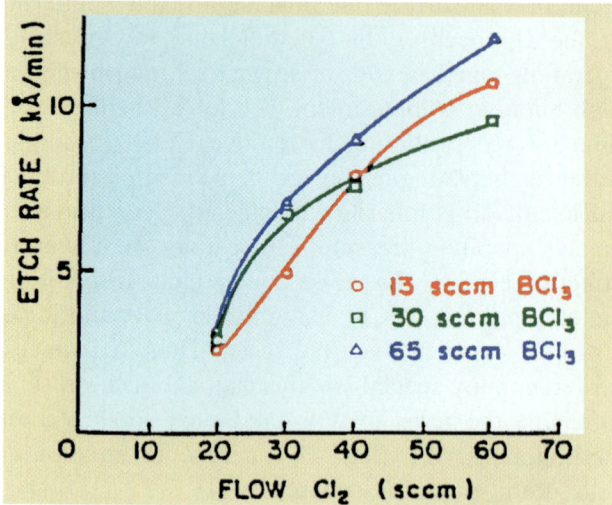

Fig. 22-20 Etch-rate of Al vs. Cl$_2$-flow for different BCl$_3$ flows. Other parameters include: 250-sccm He; 9.4-sccm CHCl$_3$; 160-Pa total pressure.[17]

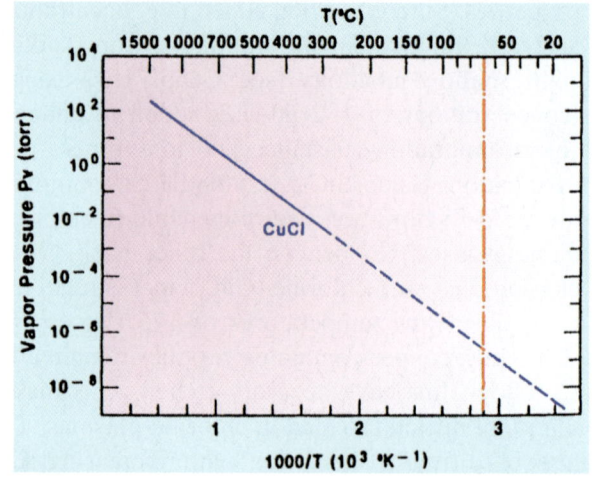

Fig. 22-21 Vapor pressure of CuCl as function of temperature.[18]

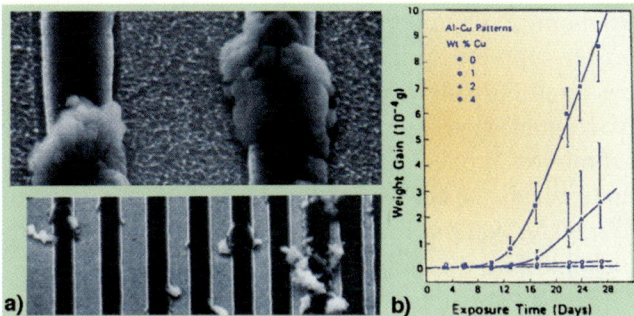

Fig. 22-22 Corrosion of etched Al lines.

22.6 PROCESS MONITORING AND ENDPOINT-DETECTION

Dry-etch equipment used in ULSI production requires effective diagnostic and etch-endpoint-detection tools. Extremely-tight-control of all process parameters must be maintained to ensure wafer-to-wafer reproducibility. In typical production facilities, some of these parameters can be controlled, while others cannot. For example, reactor wall-conditions can impact the etch-rate if the walls are exposed to atmosphere after every run. (This is one reason why single-wafer reaction-chambers are not exposed to the ambient between wafers.) Similarly, outgassing, virtual-leaks, and backstreaming from pumps can sufficiently change the etch-chemistry, so that relying on a calibrated etch-time generally does not provide enough control. Instead, techniques for directly determining the endpoint of a cycle are needed to reduce overetching, and to increase throughput and reproducibility. In this section two common methods for determining the endpoint of dry etch processes are described: **1)** laser-interferometry and reflectivity; and **2)** optical-emission-spectroscopy.

22.6.1 Laser-Interferometry & Laser-Reflectance

Laser-interferometry monitors the thickness of *optically-transparent-films* on reflective-substrates by making use of interference-effects. The principles of these measurement methods are the same as those used to measure the thickness of oxide-films (see Chap. 13). The *laser-reflectance method* exploits the difference in the reflectivity between a *non-transparent-material* being etched and the underlying-layer.

The same apparatus can be utilized to carry out both techniques, and is shown in Fig. 22-23. The system is designed to measure the intensity of light reflected from films being monitored. A major disadvantage of these techniques is that the information is obtained from only a small-area of that wafer, which may not be fully representative of other regions.

22.6.2 Optical-Emission-Spectroscopy

Optical-emission-spectroscopy is the most widely used method for endpoint-detection because it is easy to implement. It can offer high-sensitivity and provides useful information about both etching-species and etch-products. The technique relies on the *change in the emission-intensity of characteristic optical-radiation*, from either a reactant or product in a plasma. Light is emitted by excited-atoms or molecules when electrons relax from a higher energy-state to a lower one. Atoms and molecules emit a series of spectral-lines that is unique to each species. The emission-intensity is a function of the relative-concentration of a species in the plasma (see Fig. 14-3, Ch. 14). A typical apparatus utilized for endpoint-detection is shown in Fig. 22-24. It operates by recording the emission-spectrum during the etch-process in the presence and absence of the material that is to be etched. A detector is equipped with a filter that lets light of specific wavelengths pass through to be detected. To detect the endpoint, the emission-intensity of the process-sensi-

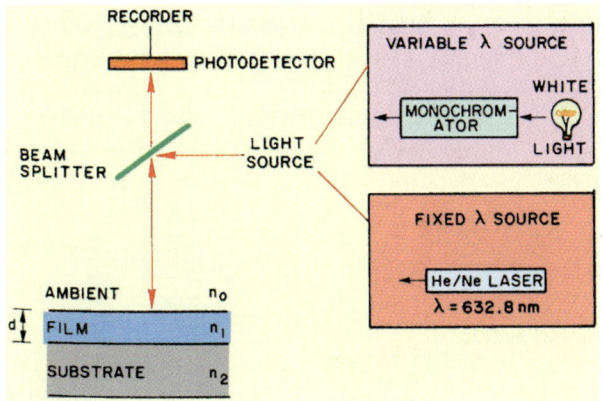

Fig. 22-23 Typical apparatus for the optical reflection method of end-point detection.

MICROCHIP MANUFACTURING

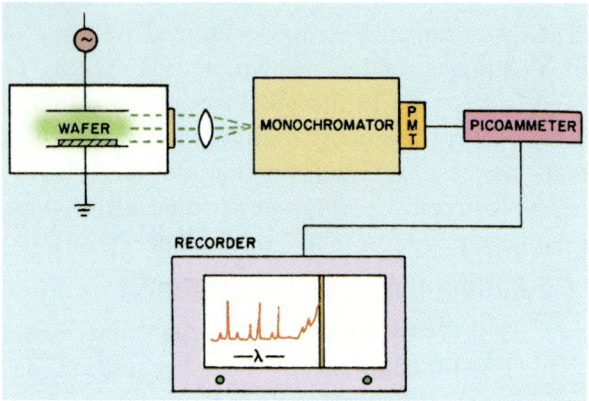

Fig. 22-24 Apparatus for using emission-spectroscopy as an end-point detector.

tive line (or band) is monitored at a fixed-wavelength. When the endpoint is reached, the emission-intensity changes. The change in emission-intensity at the endpoint depends on the species being monitored. The intensity due to reactive-species increases, while the intensity due to etch-products decreases. It is useful to monitor emission from both reactive-species and end-product species simultaneously (Table 22-2), because in some etching applications one or the other of these measurements may yield a stronger signal.

22.7 DRY-ETCH EQUIPMENT CONFIGURATIONS

Plasma-etching systems consist of the following components: **a)** an etch-chamber (evacuated to reduced-pressures); **b)** a pumping-system for establishing and maintaining reduced-pressure; **c)** pressure-gauges to monitor chamber-pressure; **d)** one (or two) rf-power-supplies to create the glow-discharge; **e)** a gas-handling-subsystem to meter and control flow of reactant-gases; and **f)** electrodes. Etch-tools for submicron applications also have a vacuum-load-lock that isolates the chamber from the ambient and a robot to transfer wafers from the cassettes through the load-lock and into the etch-chamber.

The most important commercially-available plasma-etch/RIE etch-system configurations are described here. Some of the information is historical, covering the batch-etching tools used for wafers up to 150-mm in size. These batch-tools include: **1)** barrel- etchers; **2)** parallel-electrode (planar) etchers; and **3)** hexode batch-etchers. Single-wafer etchers are discussed in Sect. 22.8. Etch-tools based on high-density plasma-sources (the newest types of dry-etching tools) are covered in Sect. 22.9.

22.7.1 Barrel-Etchers

The first (and simplest) plasma-etchers to be developed were *barrel-etchers* (Fig. 22-25). This configuration consists of a cylindrical reaction-vessel, usually made of quartz, with rf-power supplied by metal-electrodes placed on either side (and outside) of the cylinder. A perforated-metal cylindrical *etch-tunnel* is placed within the etch-chamber. This serves to confine the glow-discharge to the annular-region between the etch-tunnel and the chamber-wall. Wafers are placed

Table 22-2 SPECIES AND EMISSION WAVELENGTH FOR OPTICAL EMISSION ENDPOINT DETECTION

FILM	SPECIES MONITORED	WAVELENGTH (nm)
Resist	CO	297.7, 483.5, 519.8
	OH	308.9
	H	656.3
Silicon, Polysilicon	F	704
	SiF	777
Silicon Nitride	F	704
	CN	387
	N	674
Aluminum	AlCl	261.4
	Al	396

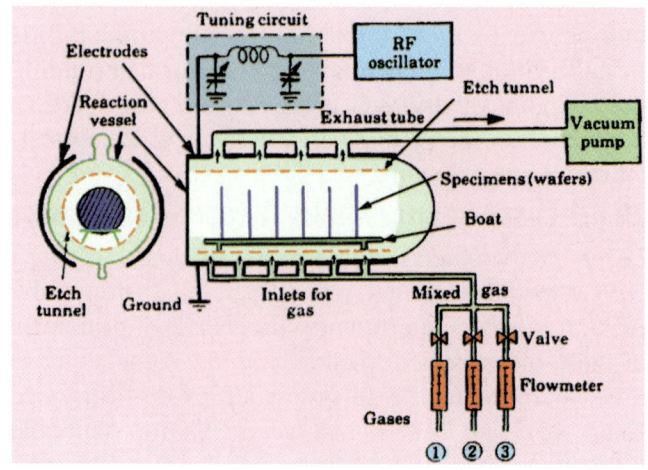

Fig. 22-25 Schematic of a barrel plasma-etching system.

in a holder at the center of the cylinder. Usually no electrical connection is made to them.

Reactive-species created by the discharge diffuse to the region within the etch-tunnel and surfaces to be etched, but the energetic-ions and electrons of the plasma do not. Since there is no ionic-bombardment, etching is almost purely-chemical, and thus isotropic. It is possible to obtain good selectivity with little or no radiation-damage. Most barrel-etchers operate in the high-pressure range of dry-etching (0.5–2.0-torr). However, the isotropic-nature of the etch limits barrel-etchers to such applications as resist-stripping.

22.7.2 Parallel-Electrode (Planar) Reactors

As described earlier, wafers exposed to energetic-ions of a plasma can be subjected to ion-assisted etching-processes. Etcher-configurations that utilize parallel-electrodes direct energetic-ions at the surfaces being etched, by causing them to be accelerated across the potential-difference that exists between the plasma and the electrode-surfaces (Fig. 22-26). As a result, both a physical-component and a chemical-component can impart directionality to the etch-process.

The parallel-electrode reactor was invented by Reinberg. In such parallel-electrode-systems (rf-diode), the electrodes have a planar, circular shape, and are of approximately the same size. One of the two electrodes of the planar-reactor configuration is connected to the rf-power-supply, and the other to ground. Wafers can be placed on either of the electrodes (Fig. 22-27). When wafers are etched in such systems by placing them on the grounded-electrode, the system is said to be operated in the *plasma-etch* (or *PE*) mode. When wafers are placed directly on the rf-powered

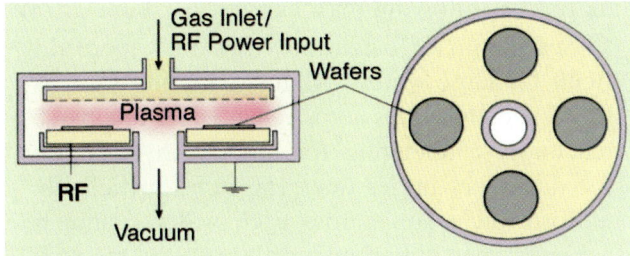

Fig. 22-26 Parallel-electrode (planar) type dry-etcher.

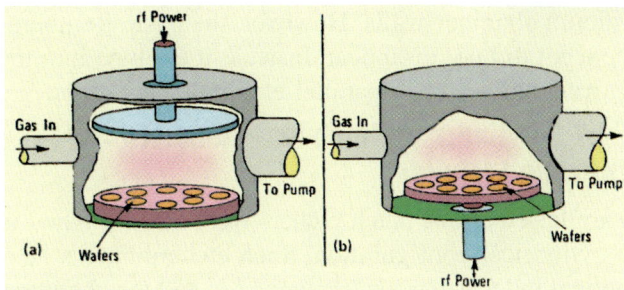

Fig. 22-27 (a) When wafers are placed on the grounded electrode, the system is configured in the *plasma-etch mode*. (b) When wafers are placed on the powered electrode, the system is operated in the *reactive-ion-etch (RIE) mode*.

electrode, these systems are said to be configured in a *reactive-ion-etch* (or *RIE*) mode. A potential-difference between the plasma and the grounded-electrode still exists, since the plasma-potential is always above ground-potential. Thus, even in the PE mode, wafers are subject to energetic-ion-bombardment (although usually to a lesser degree than in the RIE-mode). Energies of bombarding-ions are 10–100-eV in the PE-mode, and 100–1000-eV in the RIE-mode. In the RIE-mode, only the powered-electrode is subject to energetic, positive-ion bombardment. Nevertheless, etching in both modes in such systems is affected by the fact that both physical and chemical mechanisms are operative. Typically, etching in the RIE-mode is conducted at lower-pressures (<100-mtorr) than in the PE-mode (>100-mtorr).

Commercial-systems built in the parallel-electrode-configuration were initially batch-systems, with the wafers loaded manually onto electrodes of up to 90-cm in diameter. They were used primarily for processes involving wafers up to about 100-mm in size. The following drawbacks stopped their use for larger wafers: **1)** the chamber can only hold a small batch-size for larger wafers; **2)** the chamber has to be exposed to ambient during the loading and unloading after each run; **3)** automated loading and unloading of wafers to the chamber is difficult to implement; **4)** adequate etch-uniformity across larger-wafers is difficult to achieve, and **5)** a high-density of particles is created as a result of film-accumulation on the aluminum chamber-walls. However, the basic Reinberg-concept did not disappear. Instead, it evolved into the current single-wafer parallel-electrode etch-tools.

22.7.3 Cylindrical Batch Etch Reactors (Hexode-Etchers)

Parallel-electrode batch-etchers were also designed in a cylindrical-configuration. Such etchers have a hexagonal inner-electrode (the *hexode*), and the chamber-walls serve as the opposite electrode (Fig. 22-28). Up to 24, 100-mm (or 18, 150-mm) wafers-per-run can be mounted on the hexode (which is the powered-electrode, that also turns slowly during processing, to increase uniformity), while the other electrode (the chamber-wall) is grounded. The result is a system in which the area of the hexode is about one-half that of the grounded-electrode. This is the kind of *highly-asymmetrical electrode configuration* needed to cause ionic-bombardment of the powered-electrode, while minimizing ion-bombardment of the grounded-electrode and other chamber surfaces.

Thus, hexode-etchers are designed to operate effectively in the RIE-mode (but not in the PE-mode), and to provide good directional-etching capability. Since many wafers can be etched at once, the requirements for high etch-rates are somewhat relaxed. This results in lower bombarding-voltages and a reduced erosion-rate for resist. Hexode-reactors are typically operated at pressures between 10–50-mtorr.

The first-generation hexode models were not load-locked, and instead were manually-loaded. The second-generation models had load-locks and robotic-autoloading. In non-loadlocked models, more elaborate pumping-systems had to be provided to remove water-vapor from the chamber after exposure to atmosphere. Etching was conducted at low-pressures (20–100-mtorr) and low power-densities. For the reasons described below, hexode-etchers were not offered commercially for wafer sizes above 150-mm.

22.8 SINGLE-WAFER-ETCHERS

There are a number of different types of single-wafer etchers, including: **1)** conventional parallel-plate etchers; **2)** downstream-etchers; and **3)** magnetically-enhanced reactive-ion etchers. Single-wafer etchers replaced batch-etchers when wafer sizes reached 200-mm, for the following reasons (see also Table 22-3):

1. Scaling-up of batch-tools becomes more difficult for large wafers. Maintaining uniform gas-flows and etch-rates across each wafer in a batch-reactor is challenging for more than one-wafer for diameters larger than 150-mm. Sufficiently-uniform etch-rates across each wafer cannot be achieved in high-density-plasma or magnetically-enhanced-etchers if more than one wafer

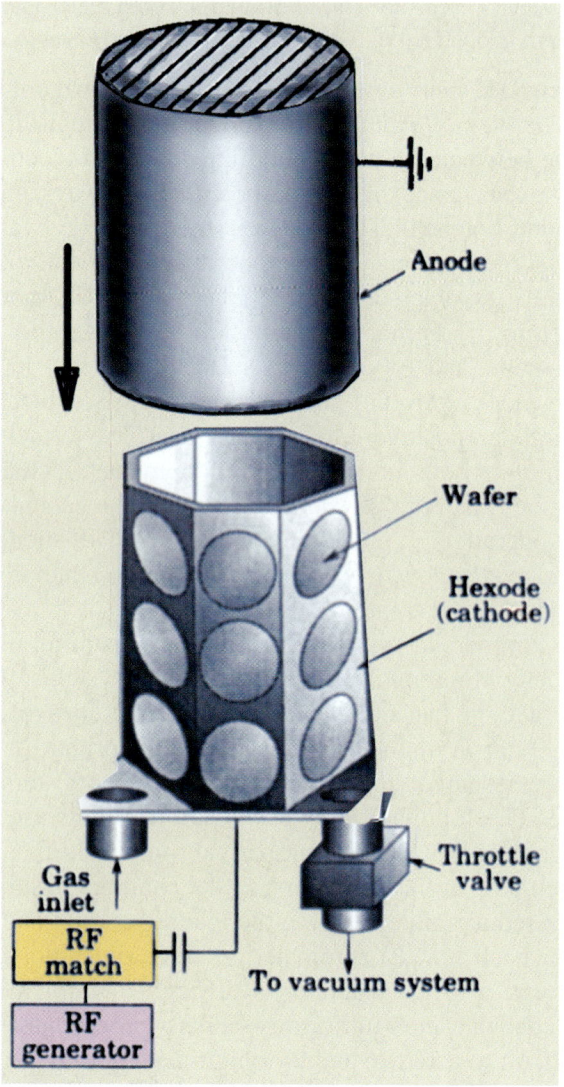

Fig. 22-28 Schematic drawing of the *hexode batch-etcher.*

Table 22-3 SINGLE-WAFER ETCHING VERSUS BATCH ETCHING

PROCESS CHARACTERISTIC	SINGLE	BATCH
Pressure	Low to High (often High)	Low
Throughput	Depends on Process	High
Etch Uniformity	Good	Adequate?
End Point Detection	Good Control	Averaged Over Batch
Automation	Relatively Easy	More Difficult
Laboratory Space	Smaller	Larger

at a time is etched in the chamber. Furthermore, the maximum-number of wafers in a batch-reactor decreases with increasing wafer-size, reducing the throughput-advantage of batch-reactors. At about the 200-mm wafer-size, the cost of etching per wafer becomes smaller in single-wafer etchers than in batch-etchers.

2. At design-rules smaller than 0.35-μm, feature-size-control cannot be adequately obtained across large wafers for more than one-wafer-at-a-time in the chamber.

3. Since only one wafer is at risk at any time, if a flaw in the process arises, it is possible to correct it before allowing the next wafer to be etched. Likewise, endpoint-detection, load-locking, and automation are more easily achieved.

4. Process-development for large wafers becomes increasingly expensive, and it becomes much less costly (and it can be completed more quickly) if performed on one-wafer-at-a-time.

However, single-wafer etchers must perform etching at higher-rates in order to achieve adequate throughput. There may be unwanted side-effects of such high-rate etching, including: **a**) excessive resist-erosion; **b**) reduced-selectivity; and **c**) multi-chamber systems are needed to carry out multi-step etch-processes.

22.8.1 Single-Wafer Parallel-Plate-Reactors

The Reinberg parallel-electrode-concept was adapted to single-wafer-reactors. In order to maximize throughput in such single-wafer etchers, a confined reactor-design was employed. In these chambers, the electrodes are very closely spaced (3–6-mm) to obtain a higher plasma-density. The total volume is also small, so that only short pumpdown-times are needed.

The gases enter through a perforated metal-plate (or *showerhead*), that also serves as the driven electrode. Obtaining good etch-uniformity requires an optimum configuration of the holes in the showerhead, and rings around the periphery of the wafer to reduce the plasma density.

Systems containing more than one etch-chamber in a modular integrated-platform have also become widely used. Systems with load-locks, and up to four individually-controlled and isolated chambers are available. The four chambers can be operated independently, allowing etching to continue in one or more chambers while wafer-transfer from other chambers to the load-lock can take place.

22.8.2 Magnetic-Enhanced RIE (MERIE)

Magnetic-enhanced reactive-ion-etchers (*MERIE*) are single-wafer machines with a magnetron-source similar to those employed in sputtering tools. Their use allows such plasma-etchers to operate at lower pressures. This produces higher-concentrations of reactant-species and more ion-bombardment of the surfaces being etched. Thus, some of the problems of single-wafer etchers arising from operation at high power-densities and pressure (such as radiation-damage and resist-erosion) can be reduced.

As shown in Fig. 22-29, a magnetic-field parallel to the wafer-surface is produced either by a group of permanent-magnets (located around the periphery of the wafer), or by two pairs of direct-current coils. These permit an ion-density in the plasmas of MERIE-reactors to be achieved that is higher than those in parallel-plate reactors (but lower than in high-density plasma-sources). Hence, they are termed *medium-density-plasmas*. To achieve sufficient plasma-uniformity, the magnetic-field is slowly rotated (either by mechanically rotating the permanent-magnet, or by modulating the current supplied to the coils). This produces a period of rotation that is short with respect to the total etch-time (~0.5 Hz).

MERIE-reactors have found widest use in dielectric (primarily oxide) etching. By the late 1990's, they were still being widely employed for dielectric-etching applications on 200-mm wafers and 0.18-μm

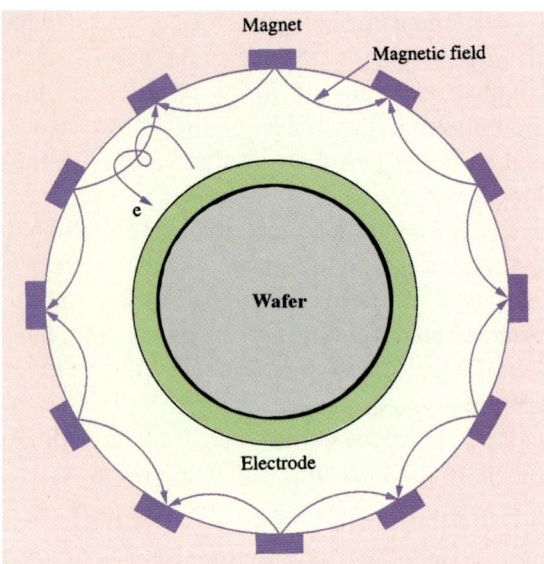

Fig. 22-29 MERIE-reactor with permanent-magnets located around the periphery of the wafer.

CMOS technologies. They remained popular because they could meet the dielectric-etching requirements of deep-submicron fabrication-processes at lower cost (since their plasma-source design is less complex than those in HDP-etchers).

22.9 HIGH-DENSITY PLASMA-SOURCES

To meet the demands of linewidth-control, selectivity, and low-damage in upcoming ULSI plasma-processes it is necessary to control the average ion-bombarding energy independently of the ion-flux. Since this is not possible in rf capacitively-coupled plasma-sources a variety of new electrodeless, lower-pressure (1.0-10-mtorr), *high-density* (n_e [cm^{-3}] = 10^{11}-10^{12}) *plasma-sources* (*HDP*) have been developed to achieve this goal (with typical HDP-processes run at pressures of 2–5-mtorr). In glow-discharge plasmas, created with electrodes in contact with the plasma, a high-voltage sheath is required to accelerate electrons to ionizing-energy. Some problems accompany the presence of such high-voltage sheaths, including: **1**) ions are accelerated by the sheath-voltage to the degree that damage is caused to the surfaces they bombard; **2**) a significant fraction of the power delivered by the power-supply is consumed in accelerating the ions, and is thus not available for forming the plasma (which limits the achievable power and plasma-density); and **3**) separate control of the plasma-density and ion-bombardment-energies is not possible.

By employing electrodeless excitation of the plasma, a much higher plasma-density can be achieved (with the *ionization-fraction* approaching unity when the pressure is low enough). The ion-bombarding energy can also be separately controlled by placing the wafers on an electrode that is biased with a separate rf capacitively-coupled power-supply.

There are four types of HDP-sources. All couple rf or microwave energy to the plasma through a dielectric-window (which is transparent to rf and electromagnetic radiation) from an external coil or antenna, instead of by direct-connection to an electrode in the plasma (hence, the name, *electrodeless* or [*remote*] *excitation*). The four types (shown in Fig. 22-30) are: **1**) *electron-cyclotron-resonance (ECR) sources*; **2**) *helicon-sources*; **3**) *inductively-coupled-plasma (ICP)* or *transformer-coupled-plasma (TCP) sources*, and **4**) *helical-resonator sources*. It appears the inductively-coupled sources will likely be the HDP-type most widely used in silicon IC-fabrication.

22.9.1 HDP-Sources in ULSI Fabrication

HDP-sources exhibit characteristics that offer significant advantages for ULSI PECVD and dry-etch applications. But, some issues still need to be resolved concerning their use in production. One involves process-uniformity over 200–300-mm wafer-diameters. While rf-diodes have a nearly one-dimensional geometry, the cylindrical HDP-sources have length-to-diameter ratios of 1:1 (or larger). Formation of the plasma and transport of the charged-species is thus radially non-uniform. Another problem is efficient power-transfer across dielectric-windows over a wide-range of plasma-parameters. The surface of the dielectric-window on the inside of the chamber may become degraded by the plasma, or deposits may buildup on it. This causes variations in the power coupled to the plasma over time, requiring frequent and costly cleaning-procedures. Operation of the plasma at low-pressures and high gas-throughputs requires use of large, expensive vacuum-pumps with high pumping-speeds.

Another difficulty is that the plasma and free-radical concentrations in the plasma depend strongly on the conditions of the surfaces in the reactor. This may cause etch-rates to vary with time. Finally, dc magnetic-fields are inherent to the operation of some types of these sources. These can lead to magnetic-field-induced process non-uniformities and damage.

22.9.2 Electrostatic-Chucks

Electrostatic-Chucks (*ESCs*) became the standard chuck for holding wafers in many process-tools in the 1990's, including plasma-etch systems, HDP-CVD systems, and ion-implanters (Fig. 22-31). Their use eliminates the need for front-side ring-clamps or clips that were previously employed to hold the wafer to the chuck. ESCs allow the wafer to be pulled completely flat (which boosts cooling-efficiency and uniformity, and also reduces particle-generation). The chucking-force on the wafer is very uniform, and it is applied without bowing the wafer.

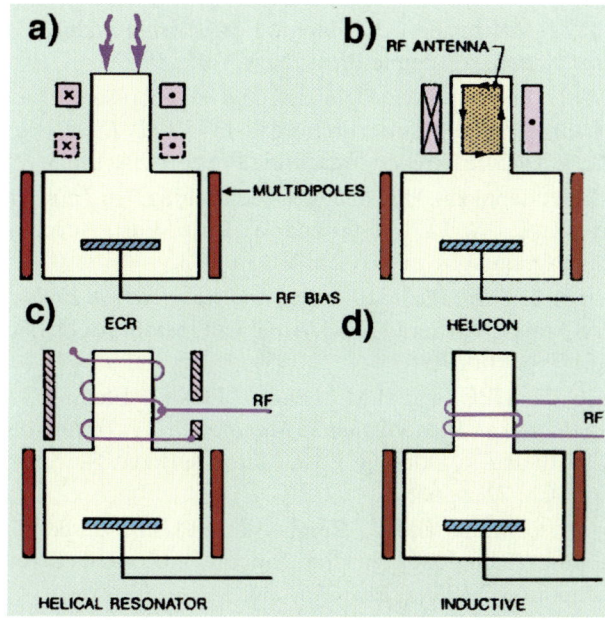

Fig. 22-30 Types of high-density plasma (HDP) sources: (a) *Electron cyclotron resonance* (ECR). (b) *Helicon*. (c) *Helical resonator*. (d) *Inductive*.[20] Reprinted with permission of Academic Press.

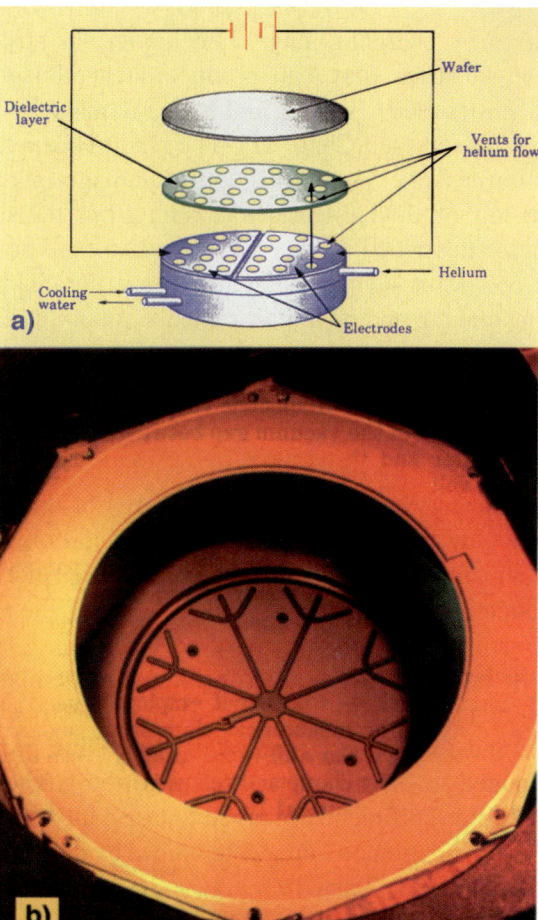

Fig. 22-31 Electrostatic chucks.

In ESCs a static-charge is generated by applying a voltage to one or more large surface-electrodes that are insulated from the rear of the wafer by a dielectric (e.g., a polyimde or ceramic material). The clamp-voltage induces an equal and opposite charge on the rear-surface of the wafer. The differences among the various ESCs are the number and configuration of their electrodes, the dielectric-material, and the mode of electrical-operation. The ideal-ESC should be rugged and cleanable, and should also quickly and consistently clamp (and then release) the wafer.

22.10 DAMAGE ARISING FROM DRY-ETCHING

Dry-etch processes can cause various types of damage to ICs. The first is caused by bombardment of wafer-surfaces by energetic-ions. If energetic-ions strike a silicon-substrate they can produce such damage as dislocations (see Chap. 9), and lattice-damage by ion-bombardment (see Chap. 12). The result is excessive junction-leakage. A second type of damage involves corrosion of device-features arising from the etch-process (described further in Sect. 22.6.5). The last type of plasma-etch damage involves wafer-charging during etching and its impact on oxide reliability. This issue is discussed next.

22.10.1 Oxide Damage During Polysilicon or Metal Etch Processes

Gate-oxides can be destroyed by dielectric-breakdown during reactive-ion-etching (RIE) of the poly-silicon gate-layer under some etching conditions. The mechanism of breakdown is not obvious, since the conductor is protected by a resist-mask after the gates become isolated. But, the culprit that causes this damage is thought to be non-uniformities in the plasma. This creates local, unbalanced ion and electron currents, and during a brief period of the etching-process (estimated to be only a few-seconds) the gate-oxide can become sufficiently charged to cause it to rupture. More information on damage to oxides by dry-etching is given in Refs. 2 and 19.

SUMMARY

Dry-etching processes (synonymous with *plasma-assisted etching*) are used to obtain highly-exact pattern transfers not obtainable with wet-etching. This chapter described the principles of dry-etching and the various types of dry-etching tools. These evolved from relatively-simple, parallel-plate configurations to complex systems with multiple-frequency generators and process-control sensors.

The future challenges of dry-etching are many, and they include: higher etch-selectivity; improved dimensional-control; and less plasma-induced damage and contamination. High-density plasma reactors (operated at low-pressures) will be required to meet these goals. Furthermore, as microchip manufacturing moves further toward 300-mm processing, better etch uniformity across the wafer will also be needed. In addition, new materials for gates to replace traditional polysilicon material and new capacitor electrodes will require new etch reactors and chemistries. Use of other novel materials, such as porous low-k dielectrics will also require innovative dry-etch developments.

The advancing technology roadmap demanded for progress and profitability will continue to drive advances in the sciences of plasma-etching. For example it is predicted that by 2007, it will be necessary to etch 64 Gb DRAMs and 0.07 μm logic devices in production. While details of how to do this have not yet been worked out, plasma etch technologists are confident that it can and will be done.

REFERENCES

1. S.Wolf and R.N. Tauber, *Silicon Processing for the VLSI Era: Vol. 1 - Process Technology*, 2nd Ed., Lattice Press, Sunset Beach CA, 2000, Ch. 14 (Dry Etching for ULSI).

2. S.Wolf, *Silicon Processing for the VLSI Era: Vol. 3 - The Submicron MOSFET*, Lattice Press, Sunset Beach CA, 1995, Ch. 7 (Thin Gate-Oxides: Growth and Reliability).

3. D.M. Manos and D.L. Flamm, Eds. *Plasma Etching: An Introduction,* Academic Press, New York, 1989.

4. D.L. Flamm, V.M. Donnelly, and D. Ibbotson, "Basic Principles of Plasma Etching," in *VLSI Electronics, Microstructure Science,* Academic Press, 1984, Ch. 8.

5. H.W. Lehmann, "Plasma Assisted Etching," in *Thin Film Processes - II,* J.L Vossen and W. Kern, Eds., Academic Press, Orlando FL, 1991, Ch. V-1, p. 673.

6. J.W. Coburn, *Plasma Etching and Reactive Ion Etching,* AVS Monogram Series, M-4, American Institute of Physics, Inc., New York, 1982,

7. G.W. Hills and J.M. Cook, "Plasma Etching," in *Handbook of Semiconductor Manufacturing Technology*, Y. Nishi and R. Doering, Eds., Marcel Dekker, New York, 2000, Ch. 21, p. 655.

8. G. Oehrlein and J. Rembetski, "Plasma-Based Dry Etching Techniques in the Silicon Integrated Circuit Technology," *IBM J. Res. Dev.*, **36**, 140 (1992).

9. M. Chen, V.J. Minkiewicz, and K. Lee, *J. Electrochem. Soc.*, **126**, 1946 (1979).

10. C.J. Mogab, A.C. Adams, and D.L. Flamm, "Plasma Etching of Si and SiO_2 - The Effect of Oxygen Additions to CF_4 Plasmas," *J. Appl. Phys.*, **49**, 3796 (1978).

11. L.M. Ephrath and E.J. Petrillo, *J. Electrochem. Soc.*, **129**, 2282 (1982).

12. J. Coburn, H. Winters, *J. Appl. Phys.*, **50**, 3189 (1979).

13. R. Gottshco *J. Vac. Sci. Technol.*, **B10**, 2133 (1992).

14. D.R. Stull, *Ind. Engr. Chem.*, **39**, 517 (1947).

15. R.H. Bruce and G.P. Malafsky, *J. Electrochem. Soc.*, **130**, 1369 (1983).

16. T. Tsukuda, et al., in *Plasma Processing*, Eds. J. Dieleman, R.G. Frieser, and G.S. Mathad, 1983, The Electrochemical Society Inc., Pennington, N.J.

17. R.H. Bruce, G. Malafsky, *Ext. Abs. of the Electrochem Soc. Meeting,* Fall, 1981, Denver, CO., Oct. 1981, Abs. No 288, Electrochemical Society, Pennington, N.J.

18. S. Broydo, "Important Considerations in Selecting Anisotropic Plasma-Etching Equipment," *Solid State Technology,* April 1983, p. 159.

19. G.S. Oehrlein, "Dry Etching Damage of Silicon: A Review, *Mat. Sci Eng. B* **4**:441 (1989).

20. M.A. Lieberman and R.A. Gottscho, "Design of High-Density Plasma Sources for Materials Processing," in *Plasma Sources for Thin-Film Deposition and Etching: Physics of Thin Films,* Vol. 18, Eds. M.H. Francombe, J.L. Vossen, Academic Press, San Diego, 1994, Ch. 1, p. 1.

PROBLEMS

1. Why is it necessary to use an anisotropic-etch process to etch the thin-films used in IC technologies having minimum feature-sizes smaller than 1-μm?

2. List four advantages of dry-etching compared to wet-etching. What are three disadvantages of dry-etching versus dry-etching?

3. Although anisotropic dry-etching is needed for many advanced IC etch applications, in others a purely chemical-etch process (i.e., isotropic) may be adequate. Explain why such a purely-chemical etch step is often suitable for etching silicon-nitride layers that define the field-oxide layer of the CMOS process-flow described in Chap. 4 (see Fig. 4-12).

4. Explain why *ion milling* and *reactive-ion beam etching* have found little application in the fabrication of VLSI and ULSI circuits.

5. Describe the two models that have been proposed to explain why some dry-etch processes can etch thin-film layers in an anisotropic fashion. Cite an example of an etch-process in which the mechanism causing anisotropic etching of each model is operative.

6. If a dry-etch process relies on ion-bombardment of the wafer surface, should the wafers be put *on* the electrode connected to the chamber-walls, or on the electrode that electrically *isolated from* the chamber-walls. Explain why one would make this choice.

7. Explain why the term *ion-assisted etching* is a more appropriate description of dry-etching processes that rely on both physical and chemical effects, than is the term *reactive-ion etching*.

8. Given a 1-μm-thick Al-film deposited on a field-oxide, and patterned with a photoresist layer. The exposed Al-film regions are then etched in a plasma consisting of a mixture of BCl_3/Cl_2 gases at a temperature of 70°C. In this process the selectivity of Al to the mask (resist) is 3:1. If a 30% overetch time is employed, what is the minimum resist-thickness that must be used to ensure that the top-surface of the metal is not etched?

9. Why does the polysilicon gate-etch process require high poly-to-oxide selectivity?

10. Why is fluorine not used as the main etch-gas in Al dry-etch processes?

11. Describe the approaches employed to prevent corrosion of Al-lines after they have been dry-etched.

12. Describe the subsystems of a basic dry-etch process tool. What gas chemistries are used to dry-etch: *silicon-dioxide, polysilicon, aluminum,* and *resist*?

13. What is *endpoint-detection* and why is it needed in dry-etch processes? What is the most common technique used to determine the endpoint of an etch process?

14. List the differences between conventional dry-etch tools, and those equipped with high-density plasmas.

15. The *etch-rate* of Si in a fluorine-plasma can be approximated as:

Etch-rate (nm/sec) = $4.67 \times 10^{-22} n_F T^{0.5} \exp(-E_A/kT)$

where E_A = 0.1075 eV, and n_F is the concentration of fluorine-atoms in the gas. Plot the etch-rate of Si (in nm/sec), as a function of 1/T in the range of 250 K to 450 K when $n_F = 10^{15}/cm^3$.

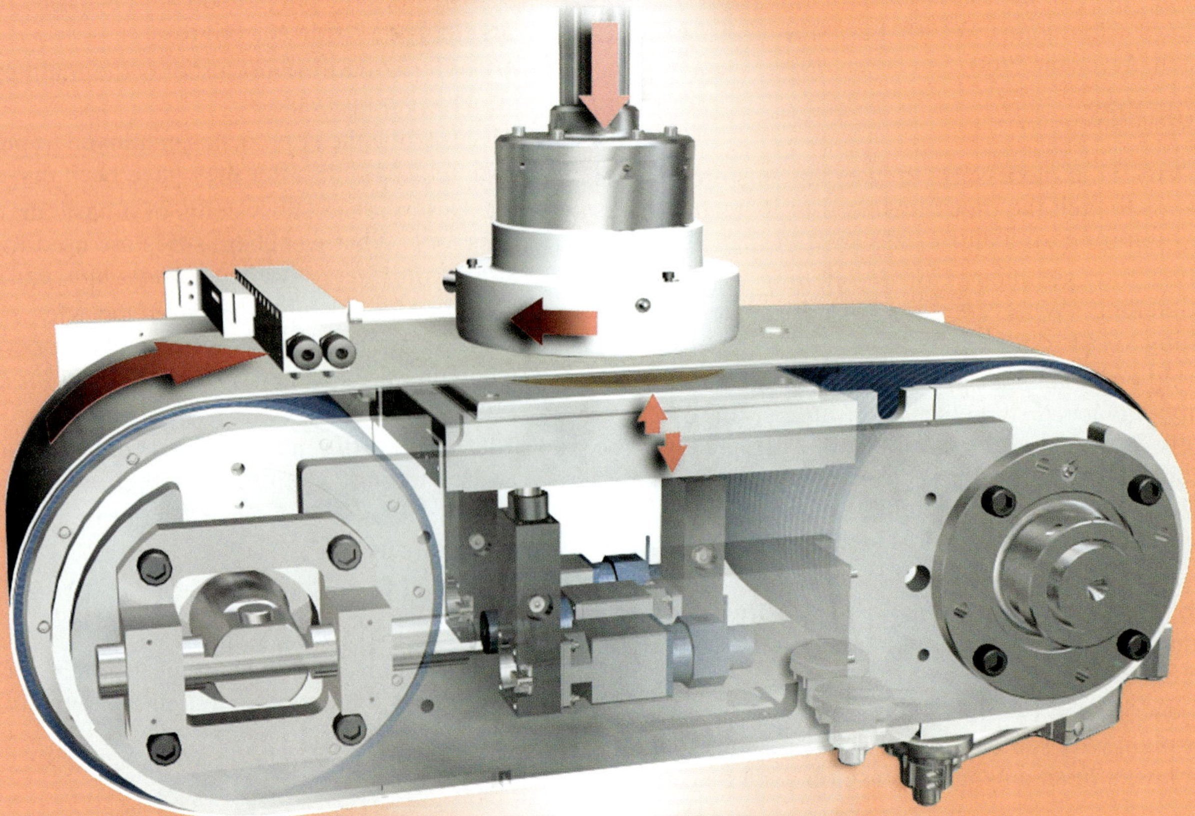

CHAPTER 23

CHEMICAL-MECHANICAL POLISHING (CMP)

CHAPTER CONTENTS

23.1 THE HISTORY OF CMP

23.2 THE MECHANISMS OF CMP
Metal CMP Mechanisms
Silicon Dioxide CMP Mechanisms
CMP of Low-*k* Dielectrics

23.3 CMP EQUIPMENT

23.4 CMP POLISHING TOOLS

23.5 CMP POLISHING PADS

23.6 CMP SLURRIES

23.7 CMP ENDPOINT DETECTION

23.8 CLEANING ISSUES IN CMP

23.9 CMP METROLOGY

23.10 DISHING PROBLEMS IN CMP

23.11 THICKNESS NON-UNIFORMITY WITHIN A WAFER AFTER CMP

23.12 ECONOMIC CONSIDERATIONS: THROUGHPUT AND COST OF OWNERSHIP

"Change before you have to."
   Jack Welch, CEO General Electric

ured (upper) is a top-down view of a TERES®
P system layout, showing the location of two
ar belts and one rotary process module, with
ntegrated clean system and in-line metrology.
w is a Linear Technology Planarization™
ule. Courtesy of Lam Research Corportation.

As the number of interconnect-layers in ICs has increased, the planarization of dielectric and metal layers has become more critical. In general, such *planarizing-techniques* as *thermal-flow*, *sacrificial resist-etchback*, and *spin-on-glass* are inadequate for interconnect-systems with more than three-layers of metal. All of these processes ultimately provide only a limited-degree of *smoothing* and *local-planarization* (i.e., over distances smaller than 10-μm). They are not able to provide *global-planarization* (Fig. 23-1e). Global-planarization of both dielectric and metal layers, however, *is* possible with *chemical-mechanical polishing* (*CMP*). In this technique, material from high-elevation features is selectively-removed

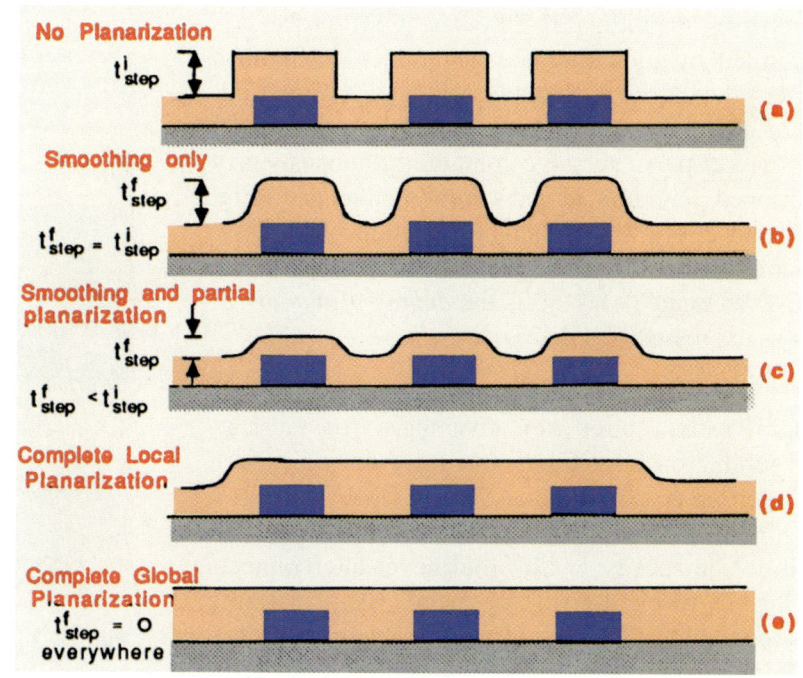

Fig. 23-1 The different degrees of planarization.

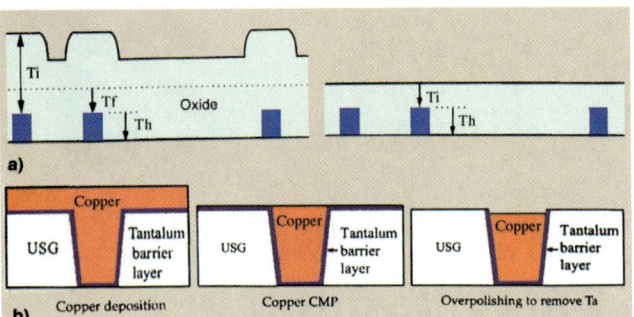

Fig. 23-2 The process of removing material by CMP falls into two categories: (a) Removal of homogeneous material but stopping before the entire surface-film is removed (e.g., planarization of an ILD-oxide); and (b) Removal of material, but not stopping until all material is removed on some regions of the wafer surface, but other material is allowed to remain in recessed regions (e.g., formation of recessed Cu-lines by CMP, in *Damascene* interconnect structures).

(that is, it is removed more rapidly than material at lower-elevations), resulting in a wafer-surface with improved planarity (see Fig. 23-2a [in which $h_{final} \ll h_{initial}$]). CMP is also used to remove blanket-deposited films, leaving undisturbed any material deposited in recesses (see Fig. 23-2b). The CMP-process is performed by mounting the wafer face-down on a carrier. The carrier is then pressed against a rotating-platen containing a polishing-pad. The carrier itself is also rotated. An abrasive-containing aqueous-slurry is dripped onto the table, saturating the pad (Fig. 23-3 and 23-4). A step-height-reduction of 90–95% is achievable using CMP, yielding values of β in the 0.90–0.95 range (where β is the *degree-of-planarization* as expressed by Eq. 23-1):

$$\beta = 1 - (h_{final}/h_{initial}) \qquad (23.1)$$

CMP offers three key-advantages for fabricating submicron-ICs. First, for logic-devices (such as microprocessors and ASICs), CMP makes it possible to achieve high-manufacturing-yields and high-device-speeds with multilevel-interconnects having up to eight-levels of metal. By planarizing IC-topography, CMP avoids the problem of metal-thinning over steep-topographies in multi-level interconnect-structures. Second, memory-devices (such as DRAMs) benefit from CMP, even though they use interconnect-systems with only two-levels of metal. For memories, the flat wafer-surface after CMP increases the *depth-of-focus budget* available for lithography. This allows memory-circuit-designers to employ smaller critical-dimensions, thus reducing chip-sizes without decreasing yield. Smaller chip-sizes provide a significant cost-advantage in the competitive realm of DRAM-manufacture. Third, CMP can reduce (rather than increase) defect-density.

In this chapter the following aspects of CMP are discussed: **a)** the history of CMP; **b)** mechanisms involved in CMP of metals and oxides; **c)** CMP equipment and consumables; **d)** CMP cleaning and metrology. For more details, consult Refs. 1-4.

23.1 THE HISTORY OF CMP

Chemical-mechanical-polishing was developed at IBM from 1983 to 1988.[5] During that time IBM

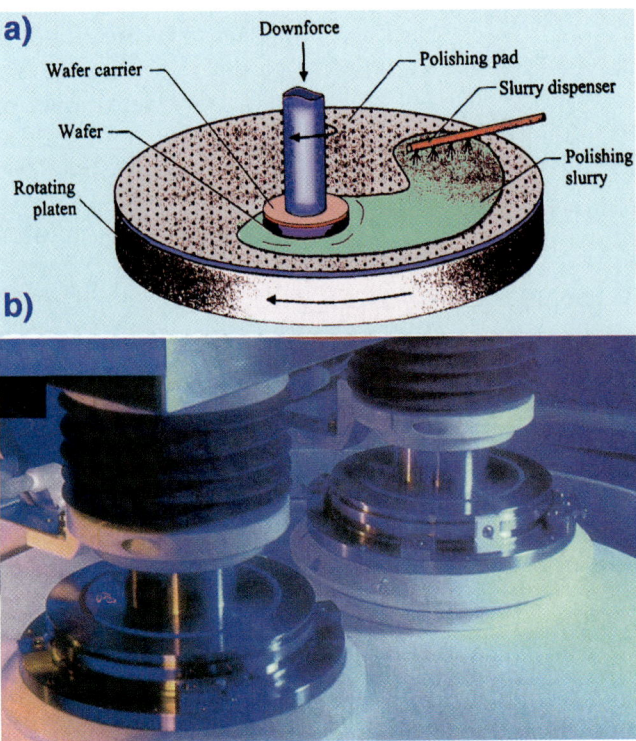

Fig. 23-3 (a) Schematic drawing of a rotary-table CMP system. (b) Photograph of a rotary-table CMP-tool. Courtesy of Strassbaugh.

removing it by CMP represents some difficulties. Cu is relatively soft and oxidizes easily with standard chemistry (such as hydrogen-peroxide or hydrogen-persulfate). On the other hand, Ta is very hard and is also much more chemically-inert. This typically leads to large differences in selectivity when simultaneously polishing Cu, Ta, and the underlying-oxide (i.e., the Ta removal-rate is only about 30-nm/min). This leads to significant *dishing* of wide Cu and oxide features (*dishing* is defined in Sect. 23-10). Damascene-CMP applications require low-selectivity between the primary conductor-material and barrier-films (1:1) and a high-selectivity between the primary-conductor and insulator (> 100:1) to essentially stop the process.

Multi-step polishing-processes are also used for CMP of Cu interconnects. As in W-plug CMP, a three-step damascene Cu-CMP process uses three platens, each with its own slurry-mixture. Copper is removed with the first platen, Ta with the second, and an oxide buffing-step is carried out on the third. It must be ensured that all the Cu in the field-regions is removed during the first-step, since the removal-rate of Cu in the second is deliberately kept very-small.

For multi-step Cu-CMP processes, one of 3 or 4 types of slurry might be best for the second polishing-step. The selection of a particular slurry might depend on the liner-layer used and the type of intermetal-dielectric beneath the metal-film. A large number of companies are developing slurries for Cu-CMP.

23.2.2 Silicon-Dioxide CMP-Mechanisms

The mechanisms that underlie CMP of silicon-dioxide are more complex than the mechanisms of metal-CMP.[8] Since SiO_2 is itself an oxide, the mechanism of oxidation used in metal-CMP is absent from the CMP of SiO_2. Like optical-glass polishing, much of the oxide-CMP process is still an art rather than a technology. However, since oxide-CMP is similar, some things about it can be learned from the long history of glass-polishing. Optical-glass and SiO_2-films on Si-wafers are polished with a slurry consisting of abrasive silica-particles (whose hardness equals that of the oxide) suspended in an aqueous alkaline-solution (typically KOH or NaOH, diluted in DI-water).

A relationship that models the *optical-glass polishing-rate*, R (defined as the thickness of glass removed per unit time), was developed by Preston, and is known as *Preston's law*.[9] This law states that R is proportional to the applied-pressure, P, and linear-velocity, v, at which a polishing-pad moves relative to the optical-glass specimen, or:

$$R = K_p P v \qquad (23.2)$$

where K_p (Preston's-coefficient) is a constant under a set of given polishing-conditions. K_p is a function of glass-hardness, Young's-modulus, polishing-slurry, and the composition and hardness of the polishing-pads. The same law applies to the rate of removal of oxide by CMP. (Typical values of P and v in CMP processes are 7-psi, and 30-40-ft/min, respectively.)

Various models have been proposed to explain why oxide-removal proceeds according to Preston's law, as well as how this mechanism allows CMP of oxide to selectively remove the "up-regions" of an oxide-film on wafers faster than the "down-regions" (thereby allowing planarization of the film to occur). One model proposes that the polishing of optical-glass occurs because water (under high-pressure) enters the SiO_2-surface and creates a softer surface-layer. This layer can then be removed by the abrasive-particles (silica) of the slurry.

While this model gives some insight into as to how oxide is polished with a silica-based slurry, it does not address the issue of how planarization of the oxide by such polishing is achieved. To obtain planarization of a film with topography, the surfaces of elevated-features ("up-regions") must be eroded faster than those at lower-elevations ("down-regions"). The selective-removal of material in "up-regions" is observed to depend on many factors, including the feature-dimensions, the structure (and surface-condition) of the polishing-pad, and the height of the step being removed.

Selective-removal of material that protrudes above the surrounding-topography is thought to arise from the compressive-nature of the polishing-pads. Figure 23-7a shows the localized-compression occurring within a pad as features on the wafer-surface are driven across it. The pad applies pressure to the wafer-

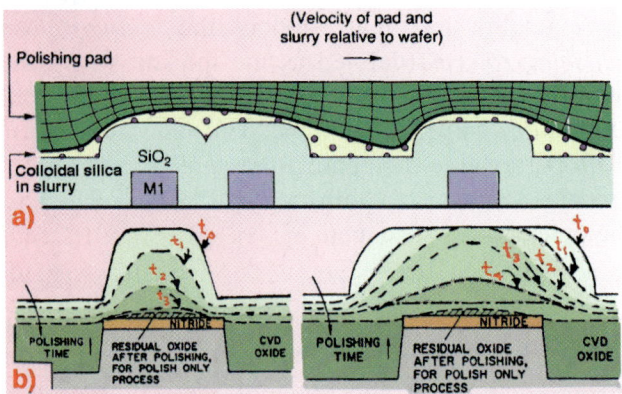

Fig. 23-7 (a) Depiction of how the leading-edge of a feature of an oxide-step is eroded by CMP (for example, the edge pressed against the pad by the combined movements of the carrier and platen). (b) Schematic cross-sections of *narrow* and *wide* oxide-steps being polished by CMP, as a function of time. Leading-edges are polished most rapidly (and features continually rotate - so all edges become the leading-edge for about the same length of time). Thus, narrow-feature planarize more rapidly than wider-ones.

surface through the slurry, with the largest-pressures being exerted where the pad is most compressed. Higher-pressure is applied to *elevated-regions* and *leading-edges* of the features, than to *lower-lying regions* and *trailing-edges*. The difference in local applied-pressure causes differences in material-removal rates. Smaller-features are rounded-off and polished faster than wider-features (see Fig. 23-7b).

In summary, planarization will occur most slowly for step-heights in an oxide over the largest or densest interconnect-structures adjacent to wide, unpatterned field-areas, scribe-lines, and wafer-edges. It will occur most rapidly for step-heights in the oxide over smaller-arrays, smaller-gaps, and narrow, isolated-features. It is also necessary to specify an acceptable *degree-of-planarization*, because in the latter stages of the CMP-process, material is removed nearly as fast from *down* as from *up* areas. Thereafter, step-heights no longer decrease at appreciable rates.

23.2.3 CMP of Low-*k*-Dielectrics

Other materials besides CVD-SiO_2 are being investigated for use as intermetal-dielectric films (especially those with a lower dielectric-constant, i.e., low-*k* dielectrics). In order to allow low-*k* dielectrics to be used in Damascene-interconnect structures, such materials must be compatible with CMP. However, low-*k* dielectric-films are eroded at much higher-rates by CMP compared to undoped TEOS-oxides. One way to avoid problems caused by this rapid-removal is to cap these layers with a CVD SiO_2-layer. This *cap-layer* must be thick enough so that the dielectric-stack is planarized by CMP, and the low-*k* layer is not reached. However, such cap-layers increase the overall dielectric-constant of the IMD-stack. Furthermore, depositing a stacked-film is a multi-step deposition-process. Another approach is to pursue the development of more chemically-stable low-*k* materials, and such efforts are continuing.

23.3 CMP EQUIPMENT

There are number of different tools and consumables used in a complete CMP process (Fig. 23-8), including: **1)** polishing-tools; **2)** pads; **3)** slurries; **4)** slurry-distribution-systems; **5)** endpoint-detection systems;

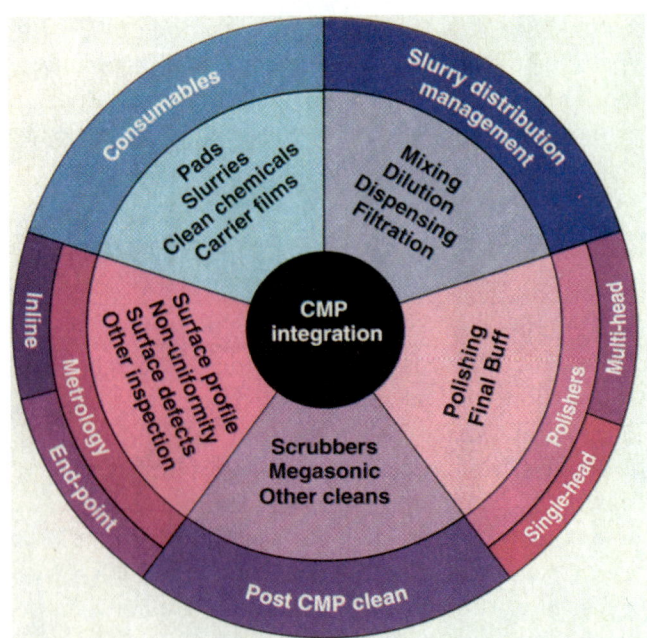

Fig. 23-8 Many components of CMP must be successfully integrated to develop a production-worthy CMP process.

CHAPTER 23 CHEMICAL-MECHANICAL POLISHING (CMP)

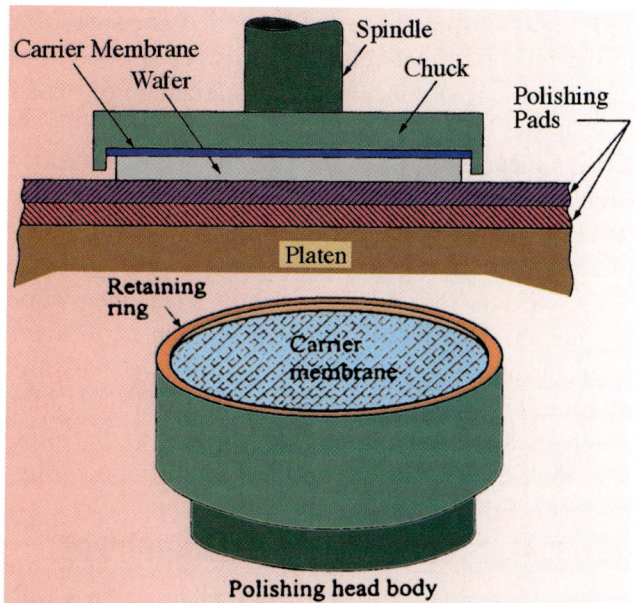

Fig. 23-9 Schematic a CMP wafer-carrier.

6) cleaning-tools; and 7) metrology-tools. These items are described in the following sections.

23.4 CMP POLISHING TOOLS

The basic tool used for CMP evolved from machines that had been used to polish bare-silicon substrate surfaces. When the idea for CMP was first conceived at IBM, they purchased silicon-polishing tools (used by wafer-manufacturers) from Strasbaugh and Westech (now SpeedFam). The basic configuration of these polishers was retained, but they were modified to make them suitable for developing CMP.

CMP-polishers differ from Si-wafer polishing tools in a number of ways. In wafer-polishing, several tens-of-microns of material are abraded, while in CMP processes, only about 0.5–1.0-μm of material is removed. The uniformity of material removal across wafers in CMP is also much-more stringent than in wafer polishing (±0.5-1.0-nm for CMP). Furthermore, CMP-tools require a higher-level of automation, throughput, reliability, and critical process-parameter control. Thus, despite more than 30-years of experience in Si-wafer polishing, new processes and equipment had to be developed for carrying out CMP.

In CMP-polishers, wafers are mounted upside-down on a wafer-carrier, and then pressed downward against a polishing-pad (see Figs. 23-3 and 23-9). The wafer is held in a pocket whose depth is chosen to permit the top-surface of the wafer to extend beyond the pocket, allowing its surface to be polished. Both the wafer-carrier and polishing-pad are rotated. Slurry is dripped onto the polishing-pad, which becomes saturated with slurry. The pad drags slurry beneath the rotating wafer-carrier, where it can polish the wafer-surface (as described in the section on mechanisms of CMP). The first polishers adapted for CMP (such as the IPEC 472® CMP-tool, Fig. 23-10), achieved wide-adoption during the mid-1990's. These tools were equipped with a single wafer-carrier.

While early CMP-polishers were single-wafer-carrier tools, more recent designs contain multiple-carriers (and multiple-platens). This increases the throughput of the polisher, and allows multi-step CMP-processes to be carried out on a single tool. The following tools are examples of rotary-motion CMP tools with mul-

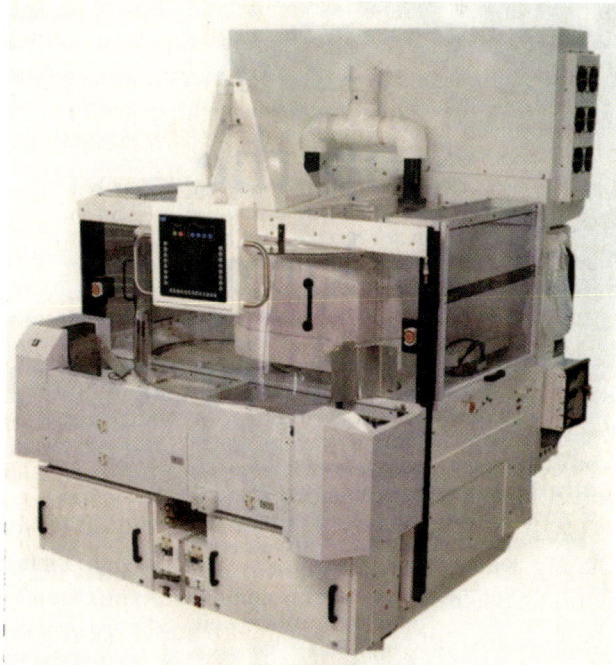

Fig. 23-10 Photo of the Speedfam 472® CMP tool. Courtesy of Novellus Systems.

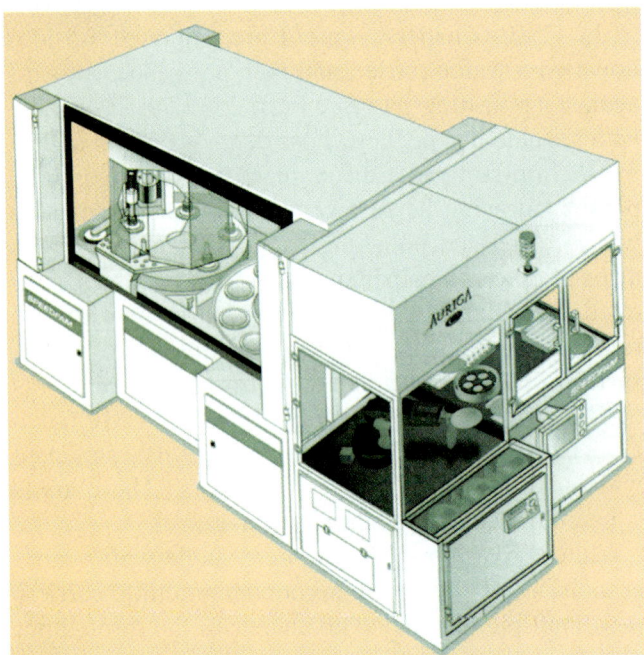

Fig. 23-11 (a) Detailed view of the SpeedFam Auriga® five-carrier polishing-tool. Courtesy of Novellus Systems.

kept high - despite the reduced applied-pressure - due to the increased pad-velocity). This set of conditions may be able to produce better-planarization of oxides in a shorter time and reduce the problem of dishing in CMP. Linear CMP-tools are offered by Lam Research (Teres,® Fig. 23-13 and p. 45x). The Teres® has two linear polishing-belts, and one station with a rotating-platen (for carrying out buffing/cleaning-steps). The wafer-carriers move independently from one station to another. The Teres® also employs an *air-bearing* beneath the polishing-belt. This allows the pressure-distribution across the wafers to be profiled. At a given pressure and speed, adjustments to the pressure-distribution profile can be made to reduce film-thickness non-uniformities.

23.5 CMP-CONSUMABLES (POLISHING-PADS)

Pads and slurries are the consumables of a CMP-process.[10] Historically, the polishing-tools were considered to be the technology's major contributors. But consumables are now being recognized as being equally important. Some CMP-customers select the slurries first, then review their compatibility with existing tools, and based on the results, select a polishing-tool. The importance of CMP-consumables is expected to increase, as more processes begin to incorporate Cu-CMP and low-*k*-CMP. The general characteristics of polishing-pads are described here, and those of slurries in the next section.

Polishing-pads are made of cast and sliced polyurethane (with filler), or polyurethane-impregnated felt (Fig. 23-14). It is generally agreed that the hardness and porosity of the pad are important parameters, and that the pad plays a crucial (but still poorly-understood) role in determining the final polish-results. Polishing-pads are supplied by a number of vendors, with Rodel Corp. (now a division of Rohm-Haas), having the largest market-share. Other suppliers include Cabot, Fujimi, and Teijin. The Rodel Suba IV (Fig. 23-15) and Rodel IC-1400 are families of polyurethane-pads with different hardness and porosity characteristics. The Rodel Politex pad is made from a flexible felt-like material.

tiple-wafer-carriers: Strasbaugh Symphony® which has four carriers and three platens; SpeedFam-IPEC Auriga® which has five wafer-carriers and two platens (for primary and secondary polishing, see Fig. 23-11); the SpeedFam-IPEC AvantGaard 776® which has four wafer-carriers and four platens; the Cybeq Nano Tech Isoplanar 8000® which has six wafer-carriers; and the Applied Materials Mirra,® which has four wafer-carriers and three platens (each with an *in-situ* pad-conditioning arm [Fig. 23-12]). In the Mirra,® a central-spindle rotates the wafer-carriers from one platen to the next. A three-platen configuration (such as is available on the Strasbaugh Symphony® [Fig. 23-3b] - and on the AMAT Mirra®), permits a three-step CMP-process to be performed on a single-tool.

CMP-tools that use a pad that moves past the rotating wafer-carrier in a linear (rather than rotary) motion have also been developed (*linear CMP-tools*). These tools offer the potential-benefits of higher linear-speeds (400–500-ft/min versus 34-ft/min in rotary CMP-tools), harder pad-materials, and reduced polishing-pressure (i.e., because the polish-rate can be

In many CMP-applications two pads are simulta-

polished before the pad is replaced.

One crucial-aspect of CMP is that unless a pad is continuously conditioned, removal-rates decrease rapidly with pad-usage. For example, starting with a new pad, the removal-rate can drop from 210-nm/min to 75-nm/min after only 50-wafers have been polished. This occurs because the surface of the pad rapidly glazes during the polishing-process. The pores of the pad become closed, reducing slurry-delivery to the wafer-surface. A consistent (and rapid) removal-rate is necessary for a production-worthy CMP-process. This can be achieved by pad-conditioning, which opens the pores of the pad by forming microscratches on the pad-surface. (Figure 23-17 shows SEM micrographs of the pad-surface with [a] and without [b] conditioning. The pores of the non-conditioned-pad are completely clogged.)

Pad-conditioning is carried out by sweeping the pad-surface with a wheel containing a diamond-abrasive-surface (Fig. 23-18). Alternative conditioning-abrasives have been pursued, since diamond-particles may break-off the conditioning-wheel and become embedded in the pad. This can lead to severe scratching of the surfaces being polished. Recently, however, superior-types of bonding-mechanisms have been developed, which chemically-bond the diamonds to the base of the conditioning-pad (Fig. 23-18b). Diamonds adhere better to such pads than to those which employ a plating-technique for bonding. Thus, diamond-loss (Fig. 23-18a) and resultant micro-scratching, is now much less of a problem. Many polishers incorporate a continuous *in-situ* pad-conditioning-capability (i.e., a unit that moves across the pad during polishing, see Fig. 23-19). This allows each wafer on the platen to be exposed to a pad with the same amount of conditioning, resulting in a more uniform removal-rate from wafer-to-wafer.

New pads also require a break-in conditioning-cycle to establish a constant removal-rate. In this procedure a set-number of *dummy-wafers* are polished to stabilize the pad-condition and performance. Break-in-cycles can consume a significant-fraction of the pad's useful-life, and the polishing-tool is not

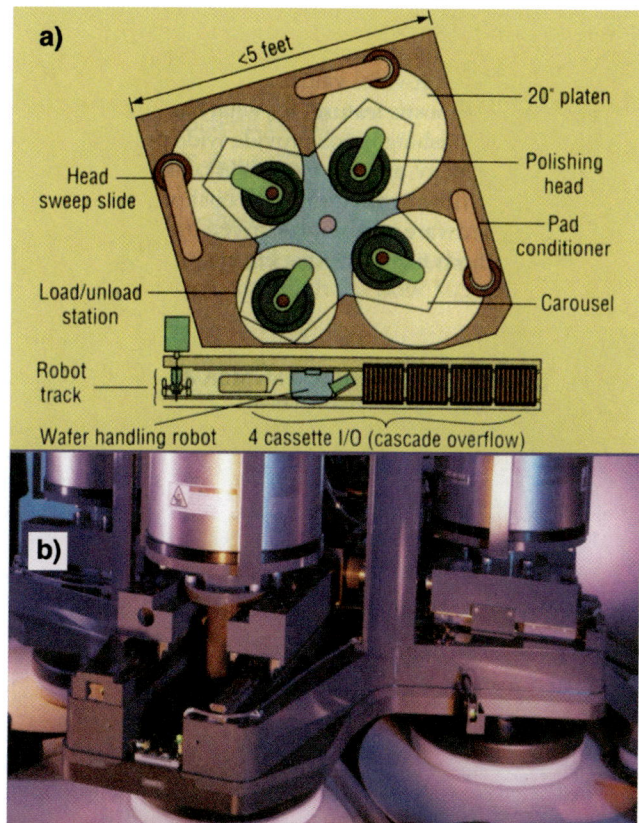

Fig. 23-12 (a) Architecture of the Mirra® three-platen CMP tool. (b) Photograph of the Mirra® tool. Courtesy of Applied Materials, Inc.

neously used, because a hard-pad gives better local (within-die) planarity, but a softer-pad gives better uniformity of material-removal across the entire wafer. Using a combination of two pads provides a compromise between these extremes. For example, a hard IC-1400 pad is mounted onto a softer SUBA IV pad to form a *stacked-pad*. As shown in Figs. 23-14 and 23-16, the surface of the pad is either cut with concentric-grooves, or contains perforations punched into the pad (1-mm-wide, and 250-μm-deep). These perforations form channels to help transport the slurry between the pad and wafer. Pores present on the pad-surface also aid in slurry-transport. Pad-life has been extended as CMP-process-understanding has improved. Early pads had to be changed about every 100-wafers, but now from 400 to 800-wafers can be

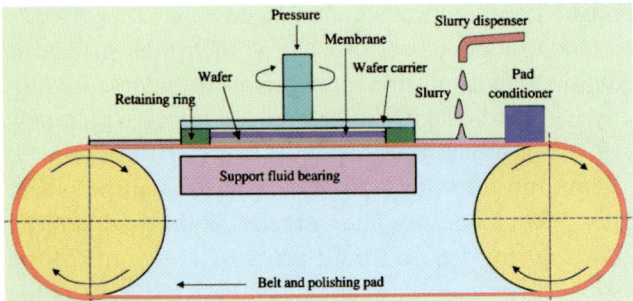

Fig. 23-13 Schematic drawing of the Lam Teres® linear CMP-tool. Courtesy of Lam Research, Inc.

available for productive use while this is in progress.

23.6 CMP-CONSUMABLES (SLURRIES)

The goal for each CMP-process is to find a slurry that produces high-removal-rates, good-planarity, local-film-uniformity, and high-selectivity. Also important are the ease with which the slurry can be cleaned from the wafer, the aggressiveness with which it attacks the polishing-equipment, and the cost of disposing of slurry-waste.

Slurries consist of small, abrasive-particles of specific-size (typically 10-100-nm) and shape, suspended in an aqueous-solution. The abrasive-particles have roughly the same hardness as the film being polished. Acids or bases are added to the solution, depending on the material to be polished. The slurry-parameters that impact the polishing-rate include the chemical-

Fig. 23-14 Photo of polishing pads. Courtesy of Rohm-Haas.

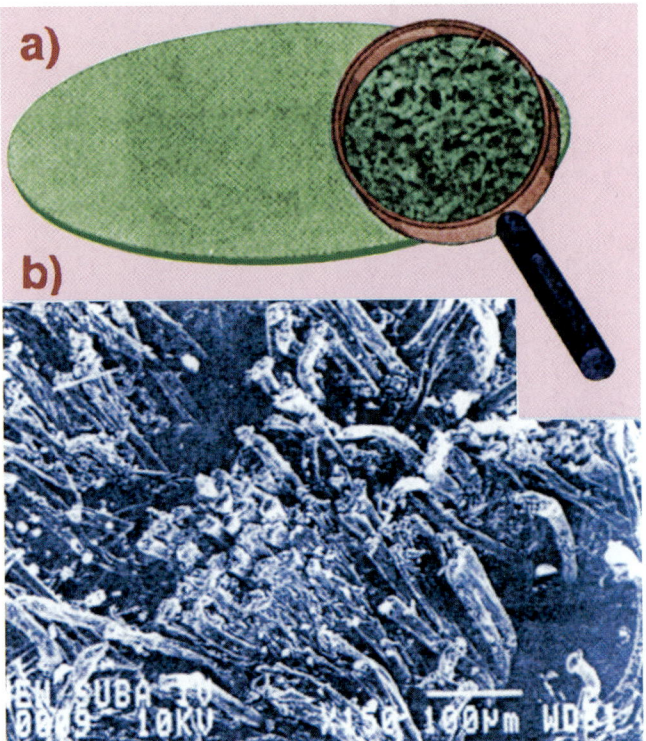

Fig. 23-15 (a) Schematic of a polishing pad; (b) SEM of a new Rodel Suba IV® pad. Courtesy of Rohm-Haas.

composition of the slurry-solution (and its pH) and the concentration (in wt %) of the solid-particles in the slurry. Also important are the shape, size, and size-distribution of the slurry-particles. An automatic slurry-feeding system is used to ensure uniform-wetting of the polishing-pad and proper delivery of the slurry. Slurries are marketed by a number of suppliers, including Cabot Microelectronics, Rodel, Fujimi, and Olin Microelectronic Materials.

Slurries for oxide-CMP consist of particles of col-

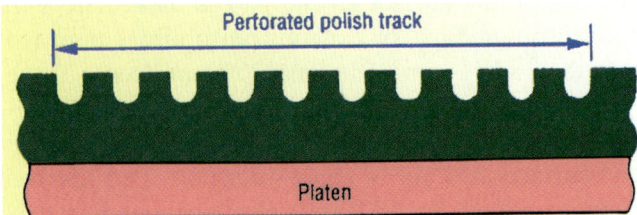

Fig. 23-16 Polishing-pad cross-section, showing grooves in pad surface.

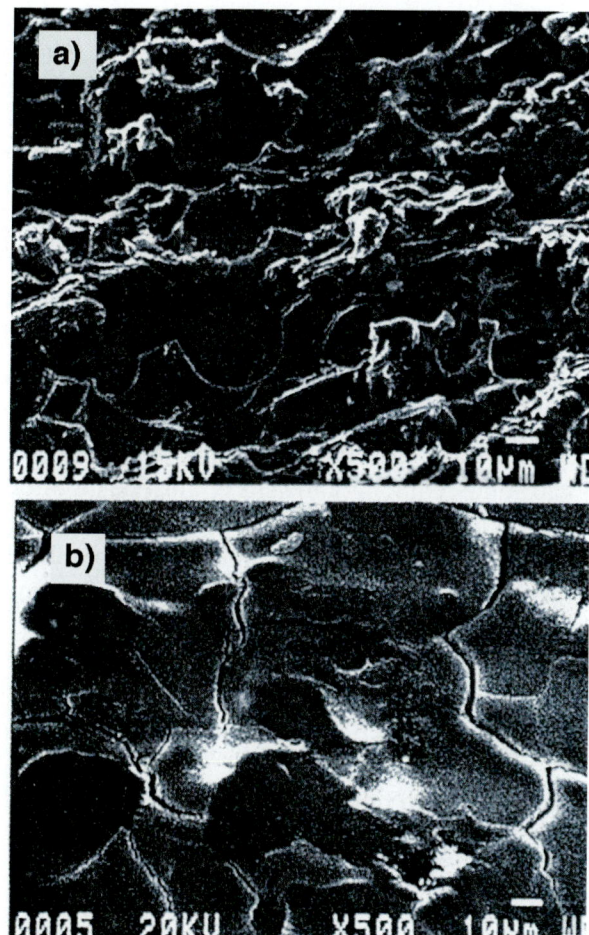

Fig. 23-17 SEM of: (a) new polishing pad; (b) surface of a glazed polishing-pad. Courtesy Rohm-Haas.

um-oxide particles) for CMP has also been pursued.

Slurries dry rapidly on a wafer once it leaves the wet-environment of the polishing-table. Once dried, it is very difficult (if not impossible) to remove all of the particles from the wafer-surface. To avoid this problem, wafers are quickly placed into wet-cleaning tanks to remove the slurry before it dries.

Slurry waste-disposal also must be addressed, because slurries contain aggressive-chemicals (and in some cases these are also highly-toxic). Toxic-material removed by a CMP-process may also end up in the expended-slurry and rinse-water (e.g., Cu). Therefore, fabs require waste-water treatment-systems to recycle the drain-water of CMP-processes. Similarly, as slurry-volumes go up, large amounts of solid-waste are generated (estimated to be 100,00–400,000 pounds per year, per typical fab), and the task of slurry-disposal must also be performed.

Slurryless CMP-processes have also been developed. A pad supplied by 3M embeds the abrasive-material in the pad itself (Fig. 23-21), and it is used for the CMP of oxides in interlevel-dielectrics and shallow-trench isolation applications. Such pads for metal-CMP processes are also in development. Slurryless-pads consist of four layers: a microreplicated-abrasive; a rigid-layer; a resilient-layer; and a self-adhesive-backing. A roll of abrasive pad-material

loidal (Fig. 23-20a) or fumed silica (Fig. 23-20b) suspended in an alkaline-solution (potassium-hydroxide [KOH] or ammonium-hydroxide [NH_4OH], diluted in DI-water). The pH is kept at 10–11, since slurry-particle agglomeration is minimized in this pH-range.

Slurries for metal-CMP applications typically use alumina-particles suspended in various liquids (depending on the specific-metal being polished). Alumina-based-slurries exhibit the drawback of poor colloidal-stability. This causes slurry-particle agglomerates to form, leading to unacceptable microscratching. Thus, development of slurries based on other particle-types (e.g., cerium-oxide and zirconi-

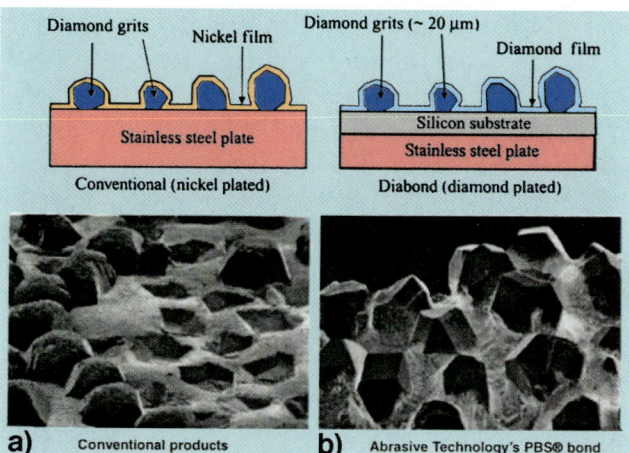

Fig. 23-18 Diamonds attached to the surface of a pad-conditioner: (a) Conventional plated-diamonds; (b) Chemically-bonded-diamonds. Courtesy of Abrasive Technology Inc.

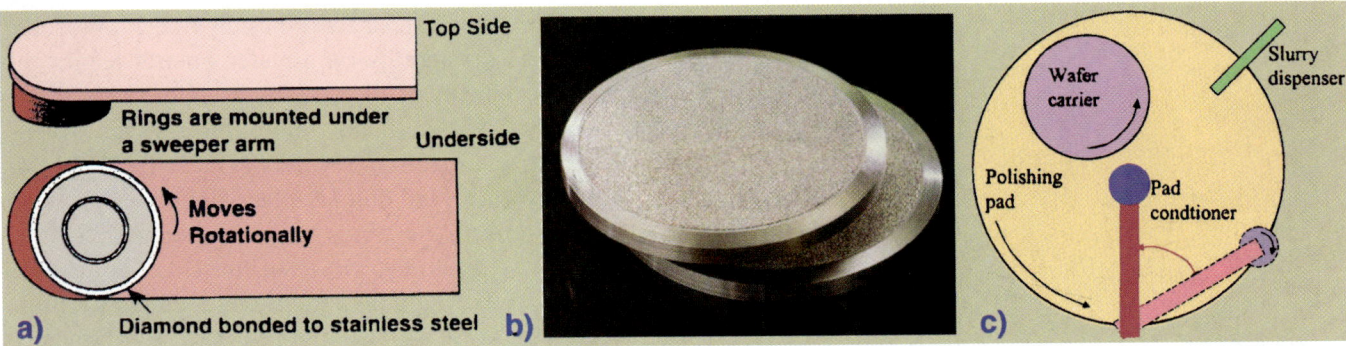

Fig. 23-19 (a) Schematic drawing of a pad-conditioning arm. (b) Photograph of pad-conditioner abrasive. (c) The operation of a pad-conditioning arm on a CMP-tool.

is spooled across a platen and incremented after each wafer (Fig. 23-22). No slurry (or pad-conditioning) is needed.

23.6.1 Slurry-Distribution Systems

Slurry-management (i.e., mixing, storage, distribution, and dispensing) is another key-issue of a CMP-process. Reliable slurry-distribution systems are important-components of the overall CMP equipment-set because precise-mixing and consistent batch-blends are necessary to achieve the necessary-repeatability. The slurry consists of particles in a liquid, and they must be kept suspended in it in a homogeneous-state. However, these particles tend to agglomerate into clumps, unless steps are taken to suppress this tendency. The slurry-pH plays a role in the dispersion of silica-particles. At a pH above 7.5, silica-particles attain sufficient surface-charge to generate electrostatic-repulsion which effectively-disperses the slurry. Some slurries (such as those used for Cu and W-CMP) require constant-agitation to maintain particle-suspension. Slurries also have a finite shelf-life (after which the agglomerate-level becomes excessive). This shelf-life ranges from less than one-week to about one-month. In any case, filtering is necessary to prevent any agglomerates (which might have formed during that time), from reaching the wafer being polished.

The distribution-system must continuously provide slurry to a number of polishers in a sustained manufacturing production-environment (i.e., 24-hours a day, seven-days a week, and in high-volumes). The pH and solids-content of the slurry must be monitored in-line to ensure consistency. Furthermore, since slurries are abrasive-particles contained in chemically-aggressive-liquids, they present a harsh-environment to the components of the distribution-system. That is, pumps, piping, valves, and measurement-apparatus in contact with slurry can be stressed to the point-of-failure in a short-period of time. Nevertheless, bulk slurry-distribution systems have been designed to handle these conditions (Fig. 23-23).

23.7 CMP ENDPOINT-DETECTION

Since CMP is a process for reducing-thickness at selected-locations on a wafer, it is necessary to be able to determine when the process has reached its *endpoint*. This is done by making measurements on the film to be polished before the CMP-process, and again after the wafer has been polished for some

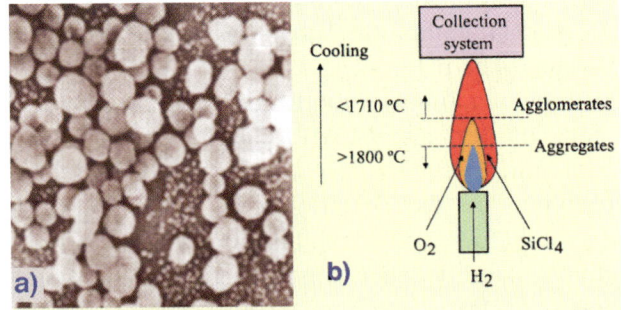

Fig. 23-20 (a) SEM of colloidal-silica slurry-particles. Courtesy of Fujimi Corp. (b) Fumed-silica slurry-particle formation.

Fig. 23-21 Example of a slurryfree-pad surface-structure. Courtesy of 3M Co.

pre-determined-time (based on the pre-polish film-measurements). This is termed an *in-line endpoint-detection* procedure. Another approach measures film-thickness as the polishing-process occurs. This is called a *real-time, in-situ endpoint-detection* method.

When *in-line endpoint-detection* is used for dielectric-CMP, a pre-CMP measurement of the dielectric-film to be polished is performed (Fig. 23-24). Based on the measured-value, the polishing-time needed to reduce the film to its final-thickness is calculated. The polishing-time is chosen from CMP calibration-data based on applied-pressure, rotation-speed, and average-pad-life-degradation. However, if the material-removed by this polishing-step exceeds the allowed-limit, the wafer must either be scrapped or more dielectric-material must be deposited. But, if the film remaining after polishing is too-thick, the wafer can always be returned to the polisher for more removal. In fact, a common manufacturing-strategy is to deliberately *underpolish* the SiO_2-layer. The film is then measured and polished a second-time, using an adjusted polishing-time. The cycle is repeated until the measured-thickness meets specification. The main drawback is that this is a time-consuming technique, which reduces the overall-time the polisher can be kept in use. In addition, there is risk of wafer-breakage during removal for measurement.

The use of a *real-time, in-situ endpoint-detection* method avoids these problems. *In-situ* endpoint-detection permits a process to be stopped when the film being polished has the correct-thickness. This improves throughput and widens the process-window. Several methods can perform *in-situ* endpoint-detection in CMP, including: **1)** front-side laser-interferometry or spectrometry (Fig. 23-25); and **2)** optical-sensing of multiple points on the wafer-backside. More details about these techniques are found in Refs. 1 and 11.

23.8 CLEANING-ISSUES IN CMP

CMP is an inherently-dirty process that can introduce yield-reducing defects to the wafer. Residual-slurry introduces foreign-particles, metal-contaminants, and chemicals, all undesirable for the remaining steps of chip fabrication. Early resistance to the adoption of CMP stemmed from such contamination-concerns. As a result, significant effort has been expended to develop post-polish cleaning techniques, so that wafers returned to the cleanroom after CMP are not contaminated. The main objective is to provide a wafer-surface that is free of slurry-particles, but the cleaning process must also remove organic-residues and trace metal-ions, without introducing further defects.

The current cleaning-technology for CMP uses *double-sided brush scrubbing* (see Fig. 8-26 in Chap. 8), although *megasonics-immersion* has also been used.[12] Such scrubbing must function reliably with bases (e.g., KOH and NH_4OH)

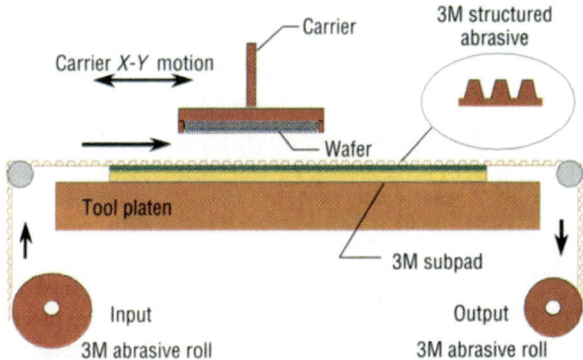

Fig. 23-22 Fixed-table with *slurry-free* roll-to-roll pad. Carrier rotates and also executes linear-motion.

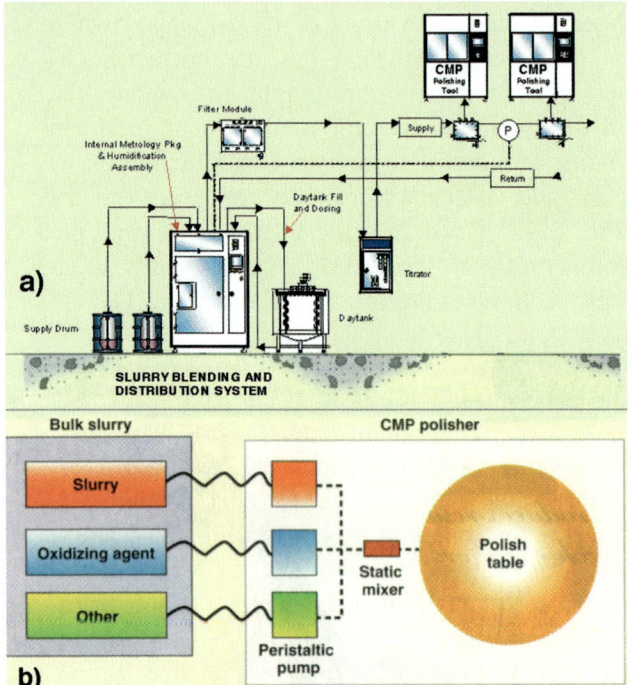

Fig. 23-23 (a) Schematic of a *slurry-management system* with metrology, auto-dosing, and filtration. Courtesy of BOC Edwards. (b) Point-of-use CMP chemical-mixing. Bulk-distributed slurry & oxidizing agents are mixed at polishing-tool for process-stability. Courtesy of Strasbaugh.

and acids (e.g., citric and HF) in order to optimally remove both metals and mobile-ions. For oxide-polishing, silica-based slurries such as NH_4OH and KOH decrease silica-particle adhesion (as long as the wafer is kept wet). The alumina-bearing slurries used for tungsten-polishing adhere more-tightly to a wafer-surface, and are generally more difficult to remove. The cleanliness of the DI-water used to rinse the wafers is also an issue. Tubs or baths in which wafers are kept wet, must be continuously-refreshed with new DI-water. The CMP-tool should automatically rinse all surfaces that come in contact with slurry, including the polishing-table, drain-tubs, carriers, and pad-conditioner. These areas should also be cleaned periodically to prevent dried-slurry buildup.

Wafer-rinsing should be done while the wafers are still on the polishing-pad. Spray-rinsing should be done as soon as the wafers are lifted from the pad. If wafers are exposed to air for even one or two minutes before such spraying (e.g., while waiting to be unloaded), subsequent cleaning becomes difficult (and perhaps impossible). Immediate removal of metal-polish slurries is also critical, because they can continue to react and remove material even after polishing has ceased. Double-sided brush-scrubbing follows this rinsing-step.

In most ULSI polishing-tools, the cleaners and scrubbers are integrated into the tool-design (Figs. 23-26 and 23-27). They usually have a megasonics-bath, two-double-sided brush-scrub modules, and a spin-rinse-dry module. Each module cleans both sides of the wafer, and all handling is done by edge-grip.

Scrubbing-brushes are made of a soft polyvinyl alcohol (PVA) material, brought into direct-contact with the wafer-surface. PVA-material has an open structure with interconnecting-cells that allow the brush to be constantly flushed with DI-water. The soft-brushes clean efficiently when compressed against a wafer. PVA-brushes are compatible with the base-liquids used in oxide-slurries (NH_4OH and KOH - pH of ~11). Adding NH_4OH to DI-water while scrubbing reduces residual-particles on the wafer.

In post-W-CMP cleaning, brush-scrubbing is also effective. In this application, the wafers come from an acidic-environment (2–4 pH) to a water-rinse and a PVA brush-clean. Rinsing the wafers with dilute-NH_4OH during post-W-CMP scrubbing is most effective in removing alumina-slurry-particles. An additional benefit of using NH_4OH is that it prevents loading of the PVA-brushes by alumina-particles or tungsten-oxides.

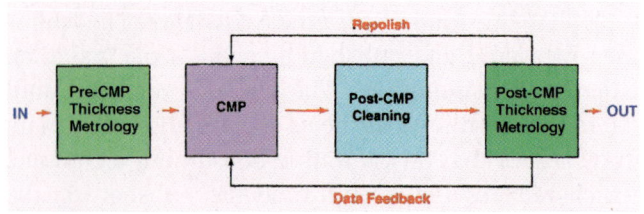

Fig. 23-24 *In-line* endpoint-detection method used in CMP (i.e., polishing, followed by measurement to confirm the thickness, and repolishing if necessary) is the cycle used in *in-line* endpoint-detection.

CHAPTER 23 CHEMICAL-MECHANICAL POLISHING (CMP)

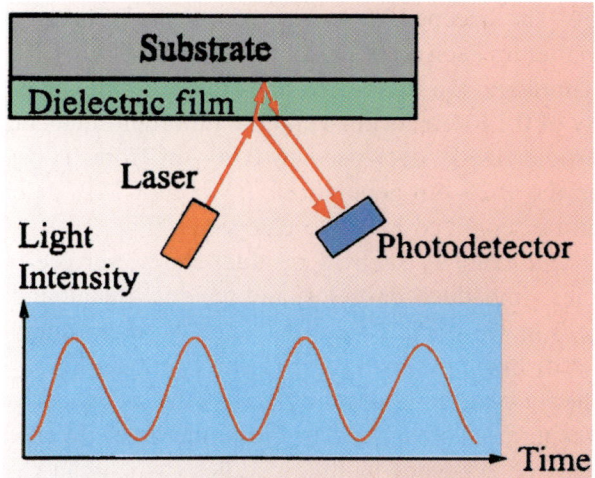

Fig. 23-25 Wafer-frontside laser-interferometry endpoint-detection for dielectric-CMP.

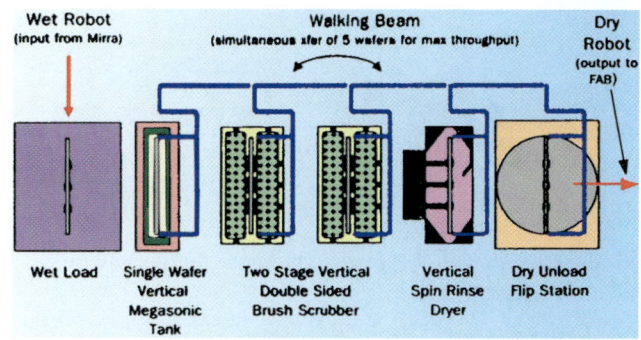

Fig. 23-26 Mirra Mesa® cleaning module layout. Courtesy of Applied Materials, Inc.

Cleaning wafers after copper-CMP presents a unique set of challenges. Not only must particles and chemicals from the slurry be removed, but also the Cu being polished must not be allowed to enter the fab-cleanroom. It must be removed from the front, back, and beveled-edge of the wafer. The copper-oxides formed and removed as part of the CMP-process do not dissolve in DI-water and may be easily transferred to the PVA brushes. If brushes become "loaded" with these copper-oxides, they may transfer them to subsequently-processed wafers, thus producing severe Cu-contamination in the fab. Proprietary chemicals have been developed to dissolve the copper-oxides and prevent such brush-loading (Fig. 23-28).

CMP-processes are generally located adjacent to the cleanroom, connected through a bulkhead-load, similar to etchers or CVD-systems. *Dry-in/dry-out CMP-systems* (in which the polisher and cleaning-system are completely integrated) are now the most widely-used CMP-systems. Wafers are polished, cleaned, dried, measured, and returned-back dry to the cassette located in the cleanroom.

23.9 CMP-METROLOGY

There are a number of metrology requirements in CMP.[11] First, it must be determined when the CMP process is completed (*endpoint-detection*, Sect. 23-7). Second, the thickness of the films after CMP must be measured. Third, the degree of film-planarity must be checked after CMP. Finally, films subjected to CMP must be inspected for defects and damage.

Film-thickness after CMP (and before CMP in in-line endpoint-detection), is performed with a number of different techniques, including: spectro-photometry, ellipsometry, and beam-profile-reflectometry. One problem associated with these techniques is that of *cycle-skipping* (or *order-skipping*). Different strategies by each measurement-technique are employed to avoid such problems. Ellipsometers use additional wavelengths, spectrophotometers rely on better signal-to-noise data-collection, and beam-profile reflectometry is combined with spectrophotometry to avoid cycle-skipping errors.

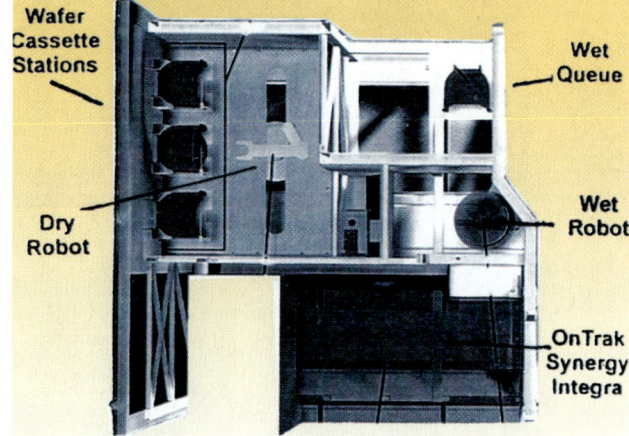

Fig. 23-27 Schematic drawing of the Lam Research Synergy® post-CMP cleaning-tool. Courtesy of Lam Research, Inc.

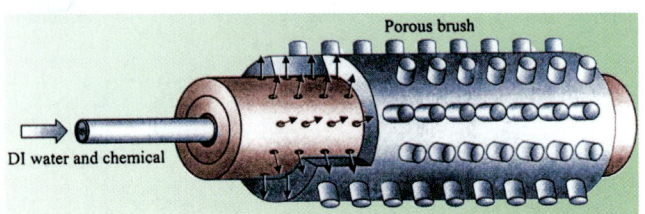

Fig. 23-28 *Through-the-brush* chemical delivery. Courtesy of Lam Research, Inc.

The *degree-of-planarity* achieved with a CMP-process is typically measured on a number of die on a wafer using either profilometry or atomic-force-microscopy. Such planarity is *location-* and *pattern-dependent*. Within-die planarity is influenced by the tendency of CMP to polish smaller, individual-features faster, and larger, densely-packed-features more slowly. Thus, CMP thickness-measurements made with optical-thickness tools commonly include several features at each selected-site (from the smallest that the tool can conveniently measure, to a large [field] area where polishing-rates are the lowest).

Detection of damage, contamination, and particles on post-CMP wafers is another aspect of CMP metrology. The acceptable density of defects is becoming smaller for each device-generation. In 0.18-μm CMOS technology, the SIA-Roadmap calls for a defect-density of 0.14-defects/cm^2, and the minimum-detectable particle-size is decreased to 0.06-μm from 0.08-μm for the 0.25-μm-generation. In oxide-CMP, defects can be separated into two categories (Fig. 23-30): **1)** *foreign-material* such as residual-slurry, surface-particles, embedded-particles, and residual-metal; and **2)** *voids in the surface-material* (such as microscratches and dishing). *Microscratches* can occur during CMP when a particle is dragged across the wafer-surface, causing a shallow-void in the oxide-layer. In metal-CMP, defect-types include puddles (which can cause shorts if they remain between metal-lines), surface-voids, residual-slurry, surface-particles, metal-filled micro-scratches, and plug-holes from which metal has been removed.

23.10 DISHING-PROBLEMS IN CMP

Dishing is a CMP-problem that involves the thinning of regions that remain exposed when a film is being polished down to a CMP-stop-layer (Fig. 23-30). One

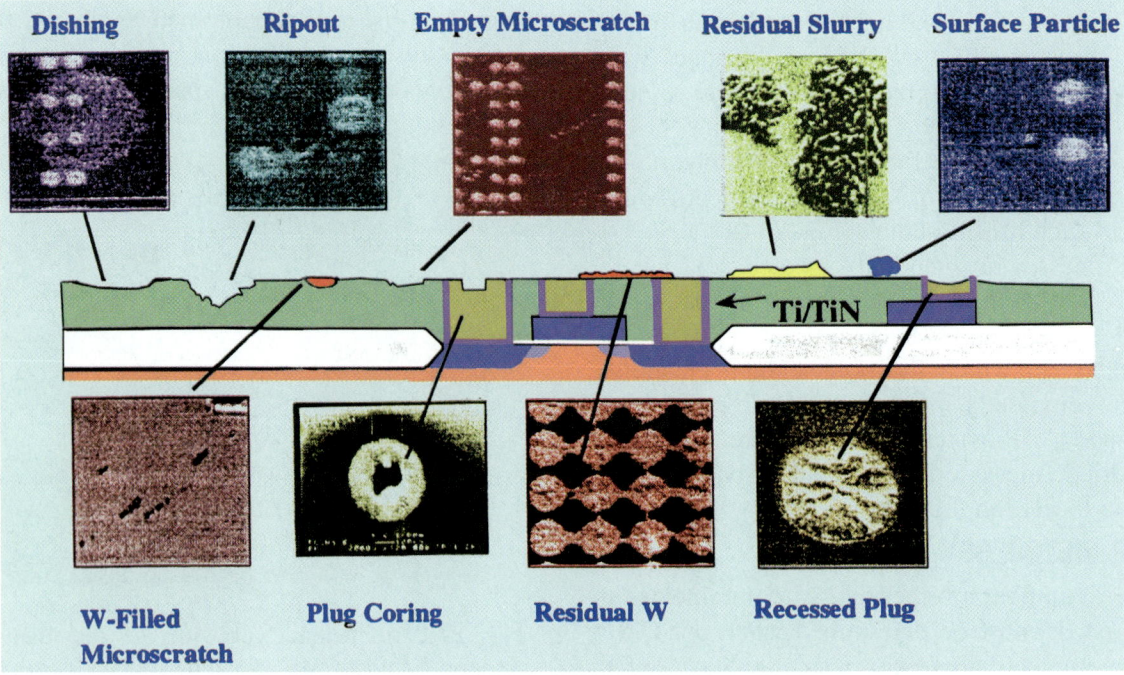

Fig. 23-29 Examples of defects that arise during W-CMP processes. Courtesy of KLA-Tencor.

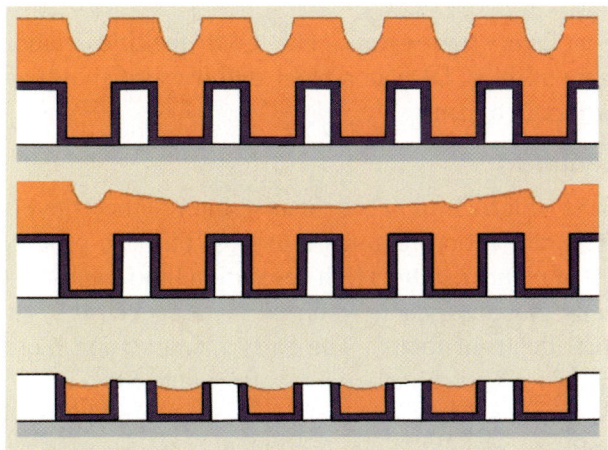

Fig. 23-30 Origin of the dishing-effect in CMP: (a) polishing-pad flexes into trench opening; (b) formation of concave-surface; (c) trench area remains "dished" after CMP.

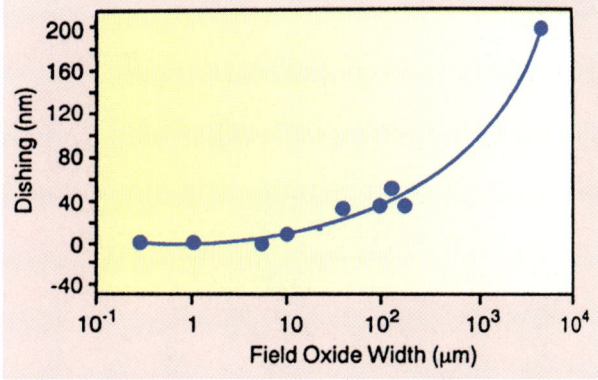

Fig. 23-31 Dishing versus field-oxide-region width for a specific oxide-CMP-process[17] (©1993 IEEE).

case where this occurs is in the CMP of a CVD-oxide-layer, deposited to fill trenches in the course of fabricating shallow-trench-isolation (STI) structures. A silicon-nitride layer is typically used in the trench-filling-and-polishing sequence to act as a CMP-stop-layer. The thinning of these exposed-areas (dishing) is known to be highly-dependent on feature-size, with the *dishing-effect* become larger as the width of the field-oxide-region is increased (Fig. 23-31).

Dishing occurs because the polishing-pad is not perfectly rigid. If a pad exhibited perfect-rigidity, CMP of the exposed oxide-regions would produce a planar-surface (because the exposed recessed-regions would not be subjected to any polishing-pressure). Since pads exhibit some flexibility, above-zero polishing-pressures are produced in recessed-areas, especially in the middle of wide exposed-regions. Thus, oxide is removed from such regions, and a "dished" area exists after planarization.

Dishing must be minimized to maintain planarity for subsequent wafer-processing. It has remained a serious impediment to the development of Damascene-processes (see Chap. 24). A number of approaches have been proposed to decrease dishing, including: **1)** combination CMP/RIE-processes; **2)** dummy active-regions; **3)** using linear polishing-tools; **4)** using rotating polishing-tools with a harder polishing-pad and an efficient endpoint-detection system. These techniques are discussed in more detail in Ref. 1.

23.11 THICKNESS NON-UNIFORMITY WITHIN A WAFER AFTER CMP

Rigid polishing-pads provide good-planarity within a die (chip), but poor thickness-uniformity across a wafer.[15] If wafers were perfectly flat, this problem would not occur. That is, a film on the surface of a flat-wafer would be polished to a uniform-thickness across the wafer (Fig. 23-32a). But not all wafers are equally flat. For example, if a wafer is tapered, and the pad is rigid, the film on the thicker-end of the wafer will be polished more than on the thinner-end (Fig. 23-32b). If the wafer-thickness fluctuates (with

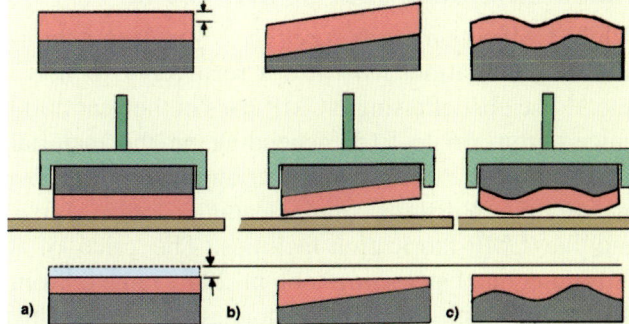

Fig. 23-32 If a rigid-pad is used for CMP: (a) Films on flat-wafers are uniformly polished; (b) Films on a tapered wafer are polished thinner over the thicker wafer-end; (c) For wafers with variations in wafer-thickness across the wafer, the film is polished thinner over the thick wafer-regions.

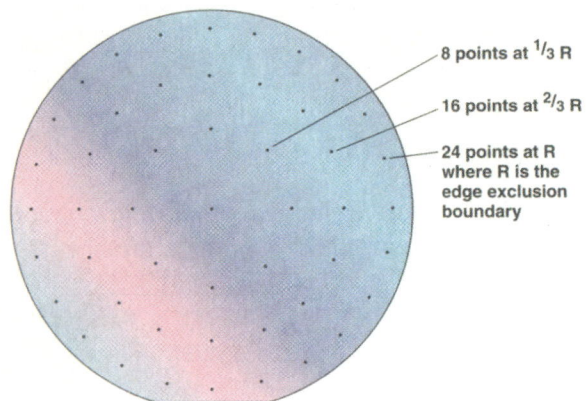

Fig. 23-33 Thickness-map of a wafer polished with CMP having a nonuniform film-thickness across the wafer.

some regions thicker than others), a film covering the thicker-wafer-regions will be polished more (and will end up thinner) than the film on the thin parts of the wafer (Fig. 23-32c). Softer (i.e., more-compliant) pads produce improved across-the-wafer thickness-uniformity, but at the price of degraded-planarity. One solution is to use a composite-pad (a rigid-pad backed by a softer one). Another is to use a linear polishing-tool with a flexible (e.g., air) bearing beneath the polishing-belt. This allows the belt to conform better to the wafer-surface. Figure 23-33 shows a map of a wafer with uneven film-removal across the wafer.

23.12 ECONOMIC CONSIDERATIONS: THROUGHPUT AND COST-OF-OWNERSHIP

While CMP is an *enabling-process* for deep-submicron IC-fabrication, it is also a relatively-expensive one.[13] The cost of a single CMP-pass in the year 2000 ranged from $6 to $15, depending on the material being polished. The main-contributors to this cost include the polishing (and cleaning) tools, slurry, and the CMP-fabrication-facilities. The price of a fully-integrated polishing-tool in 2000 ranged from $2–2.5 million, with the cleaning-module costing about $500,000 more. These tools are capable of polishing 50-60 200-mm wafers-per-hour. Slurries are also expensive. Oxide-slurries cost about $10/gallon, tungsten-slurries about $30/gallon, and Cu-slurries from $40–$50/gallon. One-gallon is consumed for about every ten-wafers. The CMP-facilities, labor-costs, utilities, DI-water, pads, and waste-disposal are additional expenses.

SUMMARY

ULSI microchips need many interconnect-levels. Their fabrication requires a process that can produce *global planarization*. High-resolution lithography also needs globally-planar wafer surfaces (because of the small depth-of-focus). The early planarization methods of resist-etchback, glass-reflow, and SOG cannot provide this. The *CMP technique*, pioneered by IBM, is able to achieve global (and local) planarization.

In this chapter, we discussed the aspects of CMP. The models for metal and oxide CMP were first described, as well as applications of these processes in IC fabrication. Next, process integration issues of CMP were covered. That is, the components of CMP systems were considered. These include: *polishing tools; CMP-consumables* (pads, pad-conditioners, and slurries); *end-point detection systems*; and *cleaning systems*. Finally, we examined production issues associated with CMP processes, including: *contamination; dishing* and *erosion*; and the *costs* of CMP.

CMP technology matured rapidly in the 1990s. It became a key enabling process in the drive toward deep-submicron ICs. CMP is now used for STI, W-plugs, Dual-Damascene copper interconnects, and ILD planarization. Some future challenges that lie ahead for CMP include: CMP of *low-k dielectric films*; CMP of *multilayer-metal films in Damascene interconnects*; and novel CMP methods for *high-k dielectrics* (such as Ta_2O_5 and BST) in Gb-DRAMs.

REFERENCES

1. S. Wolf, *Silicon Processing for the VLSI Era - Vol. 4: Deep-Submicron Process Technology*, Lattice Press, Sunset Beach CA 2002, Ch. 8 (CMP), p. 313.

2. J.M. Steigerwald, S.P. Murarka, and R.J. Gutmann, *Chemical Mechanical Planarization of Microelectronic Materials*, Wiley, New York, 1997.

3. G.B. Shinn, *et al.*, "Chemical Mechanical Polish," in *Handbook of Semiconductor Manufacturing Technology*, Y. Nishi and R. Doering, Eds., Marcel Dekker, New York,

2000, Ch. 15, p. 415.

4. *Chemical Mechanical Polishing in Silicon Processing,* S. Li & R. Miller, Eds., Academic Press, San Diego, 2000.

5. M.A. Fury, "The Early Days of CMP," *Solid State Technology,* May 1997, p. 81.

6. B. Davari *et al.*, "A New Planarization Technique, Using a Combination of RIE & Chemical-Mechanical Polishing," *Tech. Dig. IEDM,* 1989, p. 61.

7. F. Kaufman, "Chemical Mechanical Polishing for Fabricating W-Metal Features as Chip Interconnects," *J. Electrochem. Soc.,* **138**, November 1993, p. 3460.

8. M. Tomozawa, "Oxide CMP Mechanisms," *Solid State Technology,* July 1997, p. 169.

9. F. Preston, "Theory and Design of Plate Glass Polishing Machines," *J. Soc. Glass Tech.,* **11**, 214 (1927).

10. A. E. Braun, "Slurries and Pads Face 2001 Challenges," *Semiconductor International,* November 1998, p. 65.

11. W.L. Smith, *et al.*, "Film Thickness Measurements for Chemical Mechanical Polishing," *Solid State Technology,* January 1996, p. 77.

12. V.B. Menon, R.P. Donovan, and W. Kern, "Measurement and Control of Particulate Contaminants," in *Handbook of Wafer Cleaning Technology,* W. Kern Ed., Noyes Publications, Park Ridge NJ, 1993, Ch. 9, p. 379.

13. R. DeJule, "CMP Challenges Below a Quarter Micron," *Semiconductor International,* November 1997, p. 55.

14. K. Wijekoon, "Tungsten-CMP Process Developed," *Solid State Technology,* April 1998, p. 56.

15. T. Marbeiter, T. Cleary, and K. Sutter, "An Update: Transition to 300-mm CMP," *Semiconductor International,* November 1998, p. 78.

16. C. Hymes, "The Challenges of the Copper CMP Clean," *Semiconductor International,* June 1998, p. 118.

17. F.C. Fazan & V.K. Mathews, "A Highly Manufacturable Trench Isolation Process for Deep Submicron DRAMs," *Tech. Dig. IEDM,* 1993, p. 47.

PROBLEMS

1. Describe chemical mechanical planarization (CMP). How does CMP achieve planarity?

2. What are the two main applications of the CMP process in IC fabrication?

3. What other planarization processes were used before the CMP process became widely adopted in IC fabrication?

4. Define the term, *degree of planarization* (DP). Referring to Eq. 23-1, if $h_{initial}$ is 1.0-μm, and h_{final} is 0.1-μm, what is the DP?

5. Why does sub-quarter-micron IC manufacturing need CMP?

6. What are the advantages of CMP compared with other planarization methods?

7. Define the term *polishing-rate*.

8. Why does a *polishing-pad* need conditioning?

9. Why do oxide slurries need a solution with an alkaline nature (i.e., with a pH > 7.0)?

10. Why is *endpoint-detection* needed in CMP? List and describe the two-types of endpoint-detection techniques used in CMP.

11. Describe the basis of the CMP problems of: (a) *dishing*; and (b) *erosion*.

12. Explain the importance of the post-CMP cleaning process.

13. Explain the difference between planarity and film-thickness uniformity. How can a film on a wafer surface be *planar*, but not *uniform* in thickness?

14. How would the polishing-rate of an oxide film be expected to change if the CMP process-parameters were altered as follows: (a) the pressure applied by the wafer carrier, P, was doubled; (b) the linear speed of the pad relative to the oxide-film surface, v, was doubled; (c) v was doubled and P was reduced by half?

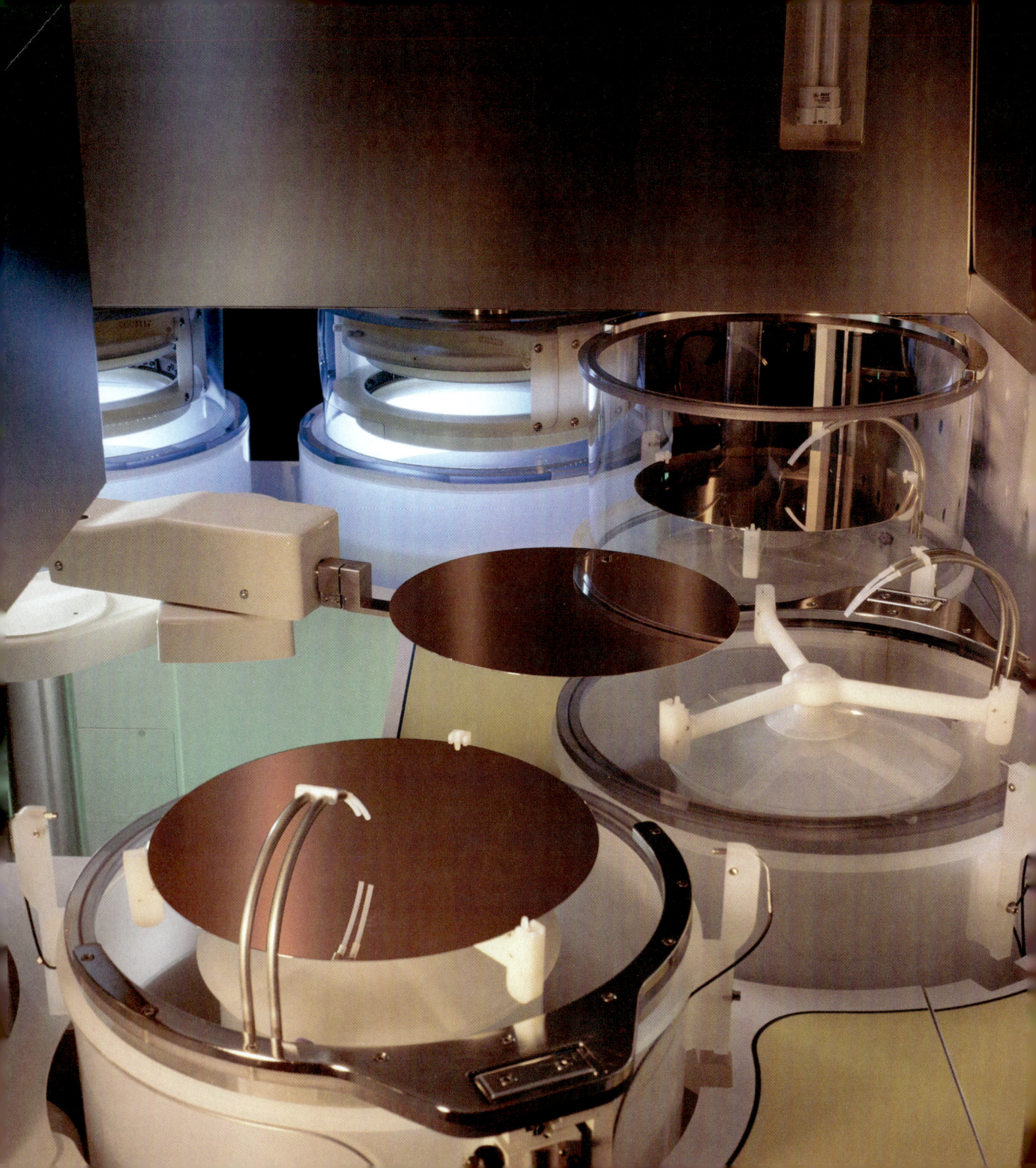

CHAPTER 24

MULTILEVEL INTERCONNECTS: COPPER, LOW-*k* DIELECTRICS, & DUAL-DAMASCENE STRUCTURES

The semiconductor-industry continues to pursue the development of ever more powerful ICs. This effort basically involves a two-pronged approach. The first approach seeks to increase the speed of *individual transistors* through a continual-reduction in the minimum-size of device features. Such scaling also provides an increase in the device-density on the chip. The second approach entails *developing increasingly-complex interconnect systems*, which now use multilayer-structures of metal wiring separated by interlayer-dielectrics (ILDs). This enables the higher-speed devices to be interconnected (resulting in ICs with enhanced-performance and system-functionality). In this chapter we cover the second of these approaches, focusing on the latest technologies for fabricating such interconnects. Thus, after an introduction to the general-topic of IC interconnects, we describe: **1)** *copper-technology*; **2)** *low-k dielectrics*; and **3)** *Dual-Damascene interconnect-structures*.

CHAPTER CONTENTS

24.1 THE NEED FOR MULTILEVEL-INTERCONNECT TECHNOLOGY

24.2 COPPER FOR ULSI INTERCONNECTS
   Process Integration Issues of Copper
   Electroplating of Copper

24.3 LOW-*k* DIELECTRICS
   First Gen Low-*k* Dielectrics: (2.8 < *k* < 3.5)
   Second-Gen Low-*k* Dielectrics: (2.5 < *k* < 2.8)
   2nd-Gen Spin-On Dielectrics with (2.5 < *k* < 2.8)
   2nd Gen-CVD Dielectrics with (2.5 < *k* < 2.8)
   Ultra-Low-*k* Dielectrics (*k* < 2.0)

24.4 DAMASCENE AND DUAL-DAMASCENE INTERCONNECT STRUCTURES

"But still try - for who knows what is possible?"
Michael Faraday

Copper electroplating tool. Shown is the Novellus SABRE® system. Courtesy of Novellus Systems.

24.1 THE NEED FOR MULTILEVEL-INTERCONNECT TECHNOLOGY

In this section we explain how the *speed* and *functional-density* of ICs are limited by chip-interconnects. Then we show how *multilevel-interconnects* can counteract these problems. That is, there are four limitations that interconnects can impose: **1)** the *minimum chip-area can become interconnect-limited*; **2)** the *speed-performance of the chip can be constrained by interconnects*; **3)** the *side-by-side (intralevel) capacitance-component dominates the interconnect capacitance when the line-spacing decreases below about 0.6-μm* (and it continues to increase as this spacing shrinks); and **4)** as *chip-sizes increase, they can become more costly to build* (because there become fewer-die per wafer). We explain how implementing multilevel-interconnects allows these limits to be overcome.

24.1.1 Interconnect Limitations of ULSI

ICs contain active-devices (transistors) which must be interconnected to implement circuits. When there were relatively smaller numbers of active-devices per chip, this could be done fairly easily with just one level

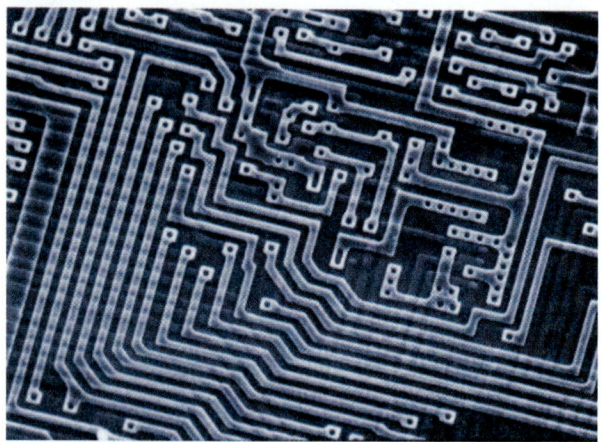

Fig. 24-1 SEM showing the patterned metal-layer on a circuit with only *one-level of metal*.

of metal (Fig. 24-1). However, as ICs evolved, the number of devices per chip increased steadily. But the area needed for interconnect lines grew more quickly than the area available for active-devices. Eventually, a condition was reached where the *minimum chip-area became interconnect-limited*. At that point, continued shrinking of the active-devices produced less

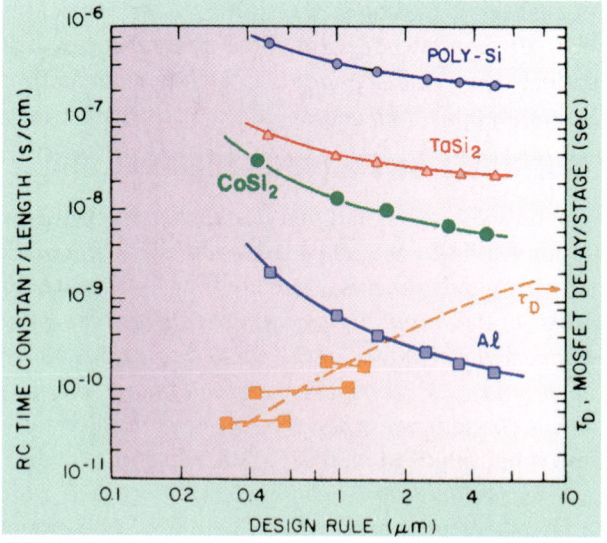

Fig. 24-2 *RC-time-constant* per unit length for several conductive materials as a function of feature-size. Also shown is the *delay-per-stage* of NMOS ring oscillators as a function of feature-size. The RC-time-constant is calculated assuming a field-oxide-thickness of 1-μm.[1] Copyright 1981 AIOP. Reprinted with permission.

improvement in circuit-performance.

One way to overcome this limitation is to employ additional levels of metal (i.e., multilevel-interconnect systems). In this way, the area needed by interconnects is shared among two or more levels of metal. This allows the fractional-area of the chip occupied by active-devices to be increased, making it possible to obtain higher functional-density.

Another problem occurs when device gate-lengths shrink below 0.35-μm. Then, the MOS transistor switching-speed itself no longer limits the logic-delay or access-time of an IC. Instead, the time required for the transistor to *charge capacitive-loads* limits the speed-performance of the IC. Furthermore, as devices shrink, the device-contribution to the *propagation-delay* of a digital-signal also decreases. But, scaling interconnect line-widths does not bring a corresponding decrease in propagation delay-time. Finally, as chip-sizes increase, interconnect-path lengths also increase. In fact, most large ULSI-circuits are also *interconnect propagation-delay-time limited*.

The latter-limit is reached when minimum feature-sizes decrease below about 1-μm (Fig. 24-2).[1] The delay resulting from various interconnect-materials per mm of length is graphed for some of these in Fig. 24-2 (assuming 1-cm-long interconnect lines).

For circuit-performance to be increased as device-dimensions shrink, two goals must therefore be met. First, the materials used for transmitting-signals over long-distances on a chip must have the lowest possible resistivity-values. This has been the key-motivation in the drive to replace polysilicon-interconnects (20-Ω/sq) with polycide-interconnects (1.7-Ω/sq for $TiSi_2$). It also explains one of the reasons why Cu is replacing Al wiring-material as devices shrink further.

Second, the length of interconnect-lines on a chip must be made as short as possible. The RC-delay is shown to be proportional to the square of the length of the interconnect line. That is:

$$R = (\rho \, l)/(w \, t_m) \quad (24.1)$$

and:

$$C = (\varepsilon \, w \, l)/t_{ox}, \quad (24.2)$$

and therefore:

$$RC = (\rho \varepsilon l^2)/ t_m t_{ox} \quad (24.3)$$

where ρ is the resistivity, l is the interconnect-line length, w is the line-width, ε is the permittivity, t_m is the thickness of the metal, and t_{ox} is the oxide thickness. Multi-level interconnect structures are an effective way to allow long-lines to be reduced in length.

Another issue arises with interconnect-delay, as shown in Fig. 24-3. If the space between interconnect-lines becomes smaller than ~0.6-μm, the total parasitic-capacitance of the lines will increase because the *side-by-side* (or *intralevel*) *capacitance* increases with decreasing line-pitch. The end-result is an overall-increase in the system cycle-time. Again, the use of a multilevel-metal interconnect system (in which the wires run orthogonally in adjacent-layers - as shown in Fig. 24-4 - can reduce this problem).

The final benefit of using multilevel-interconnects is economic. If multilevel-interconnect processes are used to fabricate integrated-circuits, the die-size should decrease. Thus, more die-per-wafer can be manufactured. If the manufacturing-cost per wafer remains the same (and the yield is not impacted by the implementation of a multilevel-interconnect process), the cost-per-chip will decrease. In fact, smaller

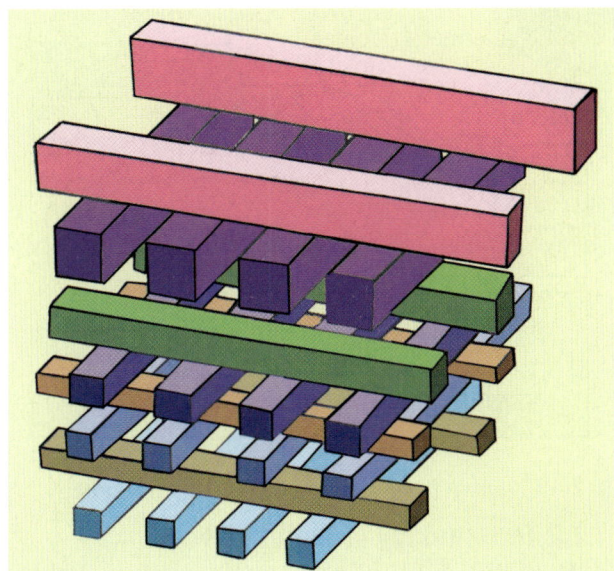

Fig. 24-4 Schematic showing the orthogonal-architecture of multilevel-interconnect lines.

die-sizes should imply higher-yields, and enhance the benefit of chip-size reduction. In addition, improved device-performance may allow the circuit to command a higher market-price.

However, the implementation of a multilevel-interconnect system requires that at least two additional masking-steps be used for each additional-level of interconnect. The extra process-steps add to the manufacturing-cost of each wafer. The number of defects/cm^2 is also generally proportional to the number of masking-steps. In addition, the manufacturing-yield and long-term-reliability for a multilevel-metal process are typically lower, since the process becomes more technically demanding. Hence, it must be determined whether the *chip-size-reduction* and *enhanced chip-value* will produce a margin-of-profit that is greater than the amount lost due to *additional incurred process-costs*, *yield-loss* and *lower-reliability*.

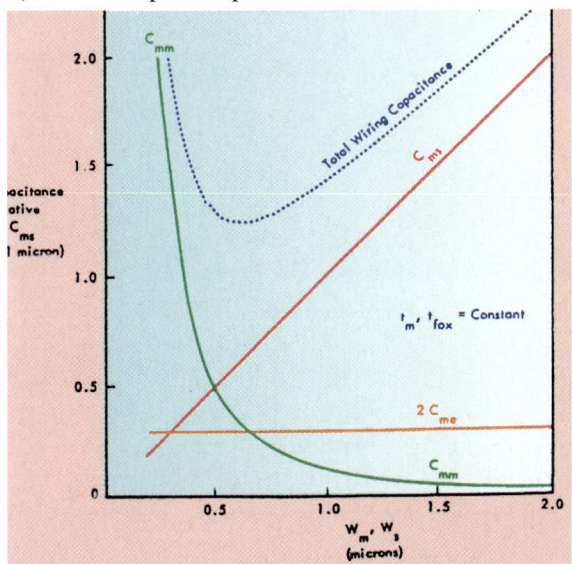

Fig. 24-3 Variation of the components of interconnect capacitance with design rules.[2] (© 1982 IEEE).

The problems related to multilevel-interconnects listed above show that benefits can be gained only by successfully pursuing a considerable technical-development effort. More specific details on these problems (and how they can be overcome) will be provided throughout the chapter.

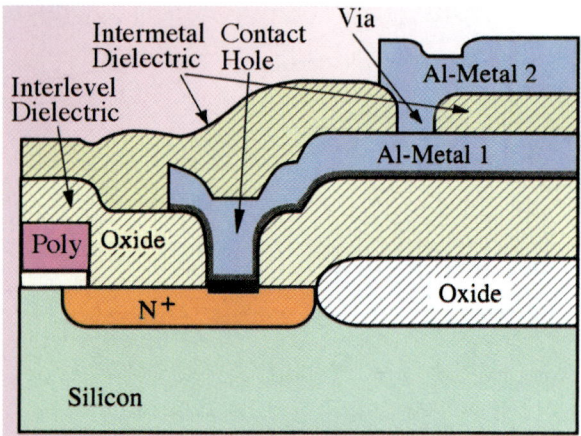

Fig. 24-5 *Double-level-metal* interconnect terminology.

24.1.2 Terminology of Multilevel-Interconnect Structures

Figure 24-5 shows the terminology associated with a double-level-metal structure for MOS-technologies. The MOS-structure has a dielectric-layer between the polysilicon gate/interconnect level and Metal-1, which is referred to as the *interlevel-dielectric* (or *ILD*). The dielectric-layers between metal-levels are called *intermetal-dielectrics* (*IMD*). The openings in the ILD are referred to as *contact-holes* (or *contacts*). They allow electrical-connections to be established between Metal-1 and the Si-substrate (and to polysilicon). Openings in the IMD are known as *vias*. These allow contact to be made between Metal-1 and Metal-2, Metal-2 and Metal-3, etc. Today, ICs are being fabricated with more than 6-levels of metal (Fig. 24-6).

Fig. 24-6 SEM photograph of an advanced IC-multilevel-interconnect structure.

24.2 COPPER FOR ULSI INTERCONNECTS

As of 2002, copper was poised to take over as the main on-chip conductor for all types of ICs. Copper is an attractive substitute for standard Al-interconnects for two main reasons: **1)** improved circuit-reliability; and **2)** increased speed-performance.

The improvement in reliability occurs because Cu-films exhibit superior electromigration (EM) resistance. The upper-limit of current-density to prevent electromigration for Cu is 5×10^6-A/cm^2 versus 2×10^5-A/cm^2 for Al. Thus, about an order of magnitude longer EM-lifetimes are obtained for Cu-lines of the same cross-section as Al lines (Fig. 24-7), primarily because Cu-atoms are more massive than Al-atoms. This characteristic can also be exploited in one of two ways to reduce the RC-time constant of a Cu interconnect-system. Either the total-resistance, R, of an interconnect-line can be reduced (by retaining the same metal-thickness), or the side-by-side capacitance, C_{intra}, of interconnect-lines can be reduced by decreasing the metal thickness (i.e., by keeping R constant, but reducing C, while carrying the same current in a smaller feature, without fear of EM-failure).

The primary clock-speed benefit derived from using Cu is gained by employing Cu at the upper

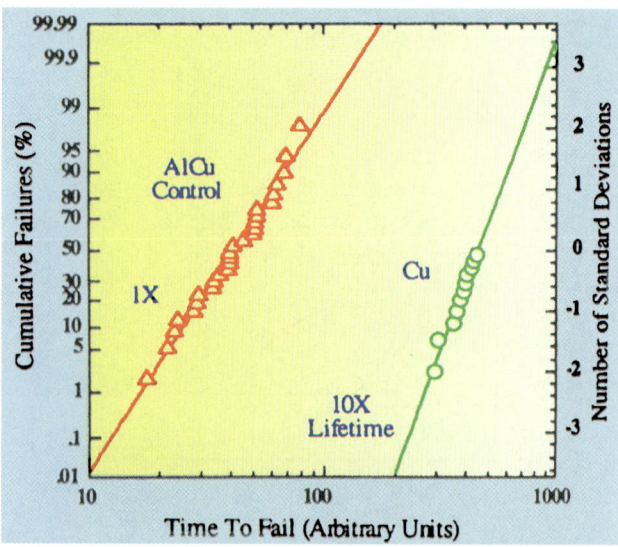

Fig. 24-7 Electromigration performance improvement using copper metallization.[3]

CHAPTER 24 MULTILEVEL INTERCONNECTS: COPPER, LOW-k, DUAL DAMASCENE

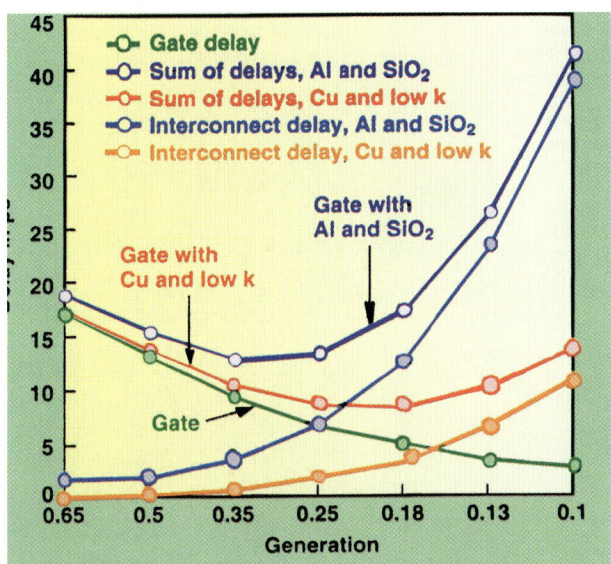

Fig. 24-8 Delays caused by interconnects are plotted as a function of feature size-for both Al/SiO$_2$ and Cu/low-k interconnects.[4] (© 1997 IEEE).

interconnect-levels (where conductor-lengths can be of the same order as the chip-size). But, for IC-generations at 0.18-μm (and smaller), current-densities in the lower interconnect-levels can also lead to electromigration-failure of Al:Cu conductor-structures. The increased electromigration-resistance of Cu helps overcome this limitation.

As shown in Fig. 24-8, by using Cu together with a low-k dielectric to replace Al and SiO$_2$, the overall delay (and power-consumption as well) in technologies below 0.25-μm is dramatically reduced. Finally, by using Cu, fewer-levels of metal are required in an IC, potentially by as much as one half (Fig. 24-9).

If Cu is substituted for Al, however, a new set of technical-challenges arise, including:

1. Cu does not form volatile copper fluorides or chlorides, making plasma-etching of Cu impractical at temperatures below 200°C. Thus, instead of patterning and etching Cu-films, it is more advantageous to *blanket-deposit Cu-films* into recesses in the dielectric-layers, and then *polish the Cu away by CMP* in the flat-areas above. This is the basis of *Damascene*-interconnect-structure formation (as discussed in Sect. 24.4).

2. Unlike Al, Cu does not form a dense, self-passivating oxide on its surface. Thus, chemical-reactions can continue to attack exposed Cu-lines, making them more vulnerable to corrosion than Al-lines.

3. Copper readily diffuses through SiO$_2$.

4. If Cu manages to diffuse through SiO$_2$ and enter the silicon lattice, it will create deep impurity-levels that degrade transistor-performance and cause junction-leakage. As a result, special Cu-processing areas are set up in wafer-fabs, to keep trace amounts of Cu from contaminating other process.

5. Unlike Al, Cu does not adhere well to SiO$_2$.

To counteract the last three problems, a film that is both an adhesion-layer and diffusion-barrier must be deposited between the Cu and the SiO$_2$. Such adhesion/diffusion-layers for inlaid-Cu processes must meet several criteria:

1. They must be *good adhesion-layers* for Cu onto SiO$_2$.

2. They must be *effective-barriers against Cu diffusing through them* at all post-copper-deposition temperatures.

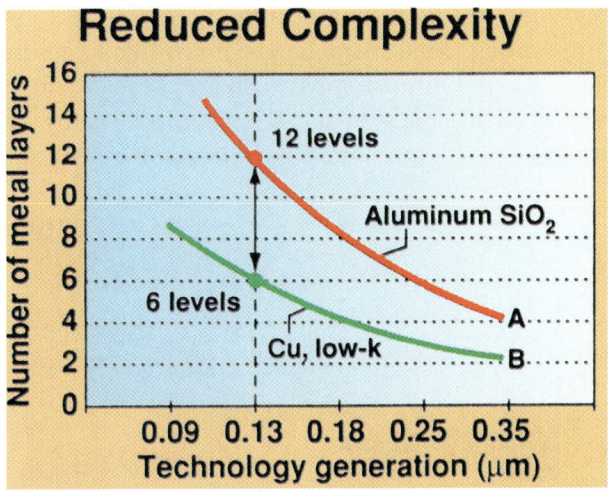

Fig. 24-9 One of the main benefits of copper, beyond the ability to increase chip-speed and power-consumption, is that the number of metal-levels can potentially be reduced by as much as half.[5] Reprinted with permission of Semiconductor International.

3. They should exhibit *low-resistivity*. That is, diffusion-barrier materials will be expected to exhibit resistivities of less than 1000-$\mu\Omega$-cm to meet the needs of advanced interconnect-processes. This will keep the total-resistivity of the composite Cu/barrier interconnect-structure below the resistivity of a comparably-thick, Al-alloy/barrier interconnect line.

4. They must retain their diffusion-barrier properties when they are very-thin, because they do not have the low-resistivity of Cu. That is, thinner diffusion-barriers for Cu will be needed than for Al, because inlaid Cu-interconnects require a barrier-layer to be formed on at least three sides of the Cu-conductor (Fig. 24-10). To keep the lower-conductivity layers from impacting the resistance of interconnect-lines, they must thus be thinner than such Cu-lines.

24.2.1 Process-Integration Issues of Copper

Thin, conformal-liners are needed to prevent Cu from diffusing (Fig. 24-11a). Materials that have been used for liners under Cu include: TiN, Ta, TaN, and TaSiN, deposited by sputtering or CVD. (Ta is an attractive barrier-material because it has a high-melting point, and does not mix significantly with Cu. It also forms strong metal-metal bonds, much like titanium.) Thus, Ta provides a low-resistance ohmic-contact, and excellent-adhesion to Cu above [and below] it. Doping Ta with a few-percent nitrogen blocks grain-

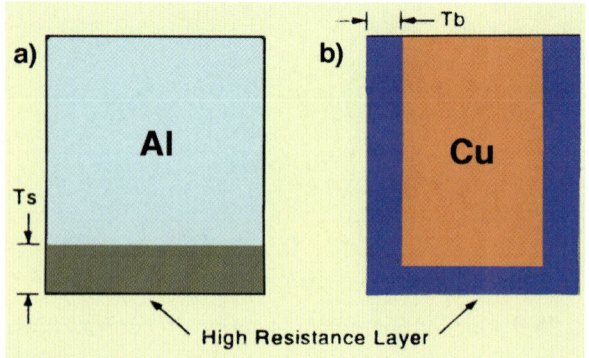

Fig. 24-10 Interconnect cross-sections showing the impact of higher-resistance shunt or barrier layers: (a) Al interconnect; (b) Cu interconnect.[6] (© 1995 IEEE).

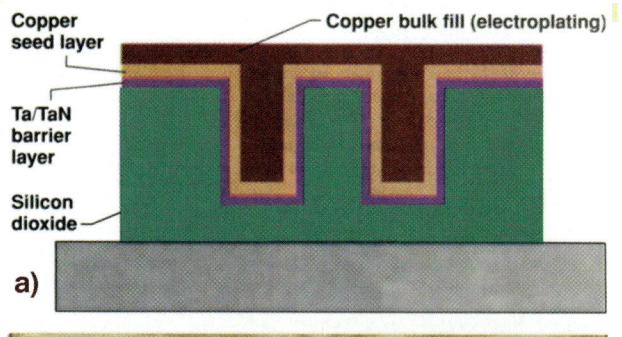

Fig. 24-11 (a) Copper interconnect-structure showing a PVD-deposited, Ta-based diffusion-barrier and PVD-copper-seed layer, and electroplated-copper fill. (b) Photo of a Cu-film filling a high-aspect-ratio trench without voiding.

boundary diffusion paths. More heavily-nitrided Ta (produced by reactive-sputtering of Ta in the presence of nitrogen), can also be used as a barrier-layer.

A *thin Cu seed-layer* (typically deposited by sputtering), must also be deposited on top of the liner. This layer is needed for the subsequent electroplating-operation (in which it provides a low-resistance conductor for the plating-current that drives the process, and also facilitates film-nucleation). The seed-layer carries current from the edge of the wafer to the center, allowing the plating-current source to contact the wafer only near the edge (see Fig. 24-12a). In addition, the seed-layer carries current from the top-surface into the bottoms of vias and trenches. If there is insufficient seed-layer thickness at the bottom of a trench or via, it will prematurely-close during deposition, leaving a center-void. To get adequate seed-thickness into high aspect-ratio vias, either an ionized-PVD-Cu (or CVD-Cu) process may be needed.

In the late 1990's, several chip-manufacturers announced the development of Cu-multilevel-interconnects that will be used on commercially available ICs. The resistivity of composite copper-interconnects in these reports (with a diffusion-barrier in place) did not exceed 2.2-$\mu\Omega$-cm, a 40% improvement over that of comparable Ti/Al:Cu-lines.

In the announced Cu-processes, tungsten-plugs are used in the *contact-holes* as a secondary-barrier that keeps the first-layer of Cu physically separated from the Si-substrate. Damascene-processes are used to pattern the Cu. In this approach, the Cu-lines residing in the SiO$_2$ recesses are capped with a thin-layer of low-stress silicon-nitride, so that the Cu is totally encapsulated (with a Ta or TaN diffusion-barrier on the sides and bottom). Manufacturing protocols must also be redesigned to tightly control Cu-cross-contamination. For example, sharing of process and metrology tools between front-end and back-end is prohibited (as opposed to Al-processing, where some tolerance of sharing such tools exists).

Copper-films can be deposited in a number of ways, including electroplating, PVD, and CVD. Electroplating of Cu-films is the main-deposition method, as described in Sect 24.2.2.

24.2.2 Electroplating of Copper

Electroplating of metals is a process used widely in the printed-circuit-board industry. It has recently been introduced into IC-manufacturing for depositing Cu-interconnect-films. It has become the main-method for creating such films because it can do this with high-throughput (> 50-wafers/hr) and low-cost. In addition, such processes can be carried out at low-temperatures ($\leq 40°C$), that are compatible with low-k. dielectrics. Electroplating-processes can also completely-fill high-aspect-ratio structures (for example, trenches with aspect-ratios of 5:1, see Fig. 24-11b).

Electroplating-processes utilize solutions containing ions of the metal to be deposited. In the case of copper, the solution contains copper-sulfate (CuSO$_4$), sulfuric-acid (H$_2$SO$_4$), and water. The CuSO$_4$ in this acidified-solution breaks up into Cu^{2+} (cupric) and SO$_4^{2-}$-ions. The wafers (with a conductive top-surface, which in the case of Cu plating, is the PVD-Cu *seed-layer* mentioned earlier) are immersed in the solution (Figs. 24-12b and 24-12c). The wafer-surface is also electrically-connected to the negative-side of an external dc-power-supply (Fig. 24-12a). Thus, it becomes the *cathode*. As the positive cupric-ions arrive at the negative-biased cathode they acquire two electrons and are reduced to Cu metal, which plates-out on the wafer-surface (with the reaction for Cu being Cu^{2-} + 2e$^-$ → Cu[O]).

All cupric-ions removed from solution by such plating are replaced by a reaction at a solid copper-anode, also immersed in the solution. This *anode* is attached to the positive-side of the power-supply, and the copper-atoms of the anode are oxidized to form the Cu^{2+}-ions. These ions are released from the anode and dissolve into the solution. The metal-ion formation-reaction is thus sustained by passing an electrical-current through the bath. The electroplating-rate is a direct function of the current-density. To obtain uniform-deposition across a wafer, the plat-

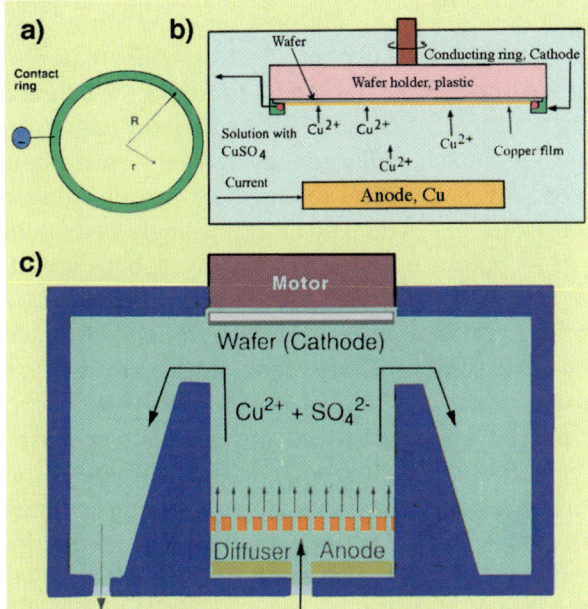

Fig. 24-12 (a) Contact-ring of a copper plating-cell. (b) Copper electroplating process. (c) Schematic-drawing of a cross-section of a Cu-electroplating cell.

444 MICROCHIP MANUFACTURING

Fig. 24-13 Photograph of a Novellus Sabre® Cu plating-tool. Courtesy of Novellus Systems.

ing-cell must be designed to give uniform potentials across the wafer-surface. Because of concerns about Cu-contamination, it is also critical that no copper be deposited on the wafer-backside.

Generally, electroplating-processes have high deposition-rates (e.g., 400-nm/min) and form desirable metallurgy. However, electroplating requires a thin seed-layer of Cu (about 50-nm-thick) which must first be deposited by some other method (e.g., CVD or PVD) onto the adhesion-layer. The correct choice of seed-layer is critical to complete recess-filling by electroplating. It would be ideal if a single-layer could serve as both a barrier and seed-layer, but currently two materials are used. An adhesion/barrier-layer like TiN, Ta/TaN or $TaSi_xN_y$ is first deposited, with a Cu seed-layer applied second. To ensure that low-resistance vias are fabricated with high-yield, the via pre-clean (carried out by Ar sputter-etching), barrier-layer deposition, and seed-layer deposition-steps are likely to be performed in a single vacuum-integrated cluster tool. A Cu-electroplating tool is shown in Fig. 24-13 (and in the chapter faceplate).[7] Figure 24-14 shows the layout of two commercial Cu-plating-tool systems.

An electroplating-process for depositing Cu in a Damascene-process was reported. It utilizes a Ta diffusion barrier-layer, a Cu-seed layer, deposited by collimated-PVD Cu, and a Cu-plated layer in which a sulfuric-acid/copper-sulfate solution was used as the plating-bath. The plated-Cu-films were formed to a thickness of about 1.5–2.5-μm. Then, CMP was used to remove the Cu in the field-regions, leaving Cu behind only in trenches and vias. Good trench-filling was observed for aspect-ratios of up to 3:1 and trench-widths of 0.4-μm. Figure 24-15 shows an SEM of a copper wiring-structure of one level of a Dual-Damascene interconnect, deposited by electroplating.

24.3 LOW-k DIELECTRICS

Interconnect-delay can be reduced not only by decreasing R of the conductor-structures in ICs, but also by decreasing C of the dielectric-layers. The value of C, in turn, can be reduced by using dielectric-

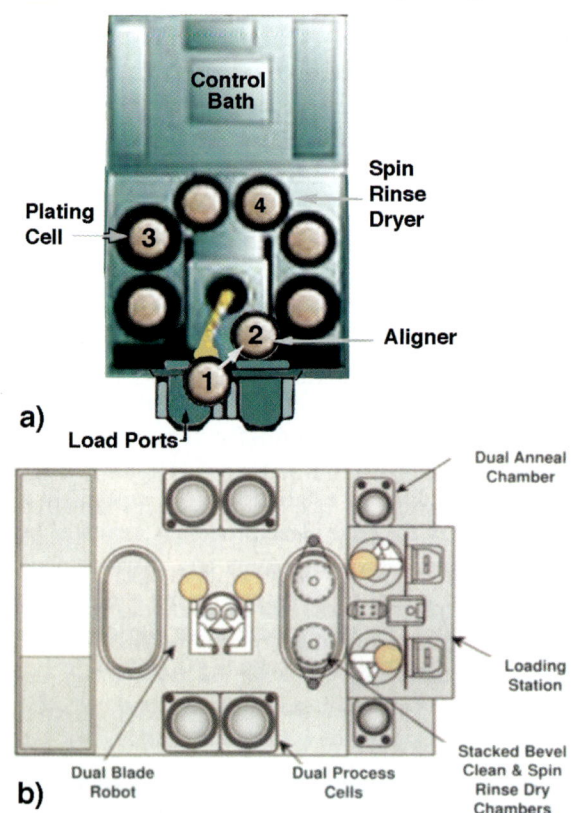

Fig. 24-14 Layout of two Cu-plating tool systems: (a) Novellus Sabre® tool. Courtesy Novellus Systems: (b) AMAT Electra® ECP System. Courtesy Applied Materials.

CHAPTER 24 MULTILEVEL INTERCONNECTS: COPPER, LOW-k, DUAL DAMASCENE

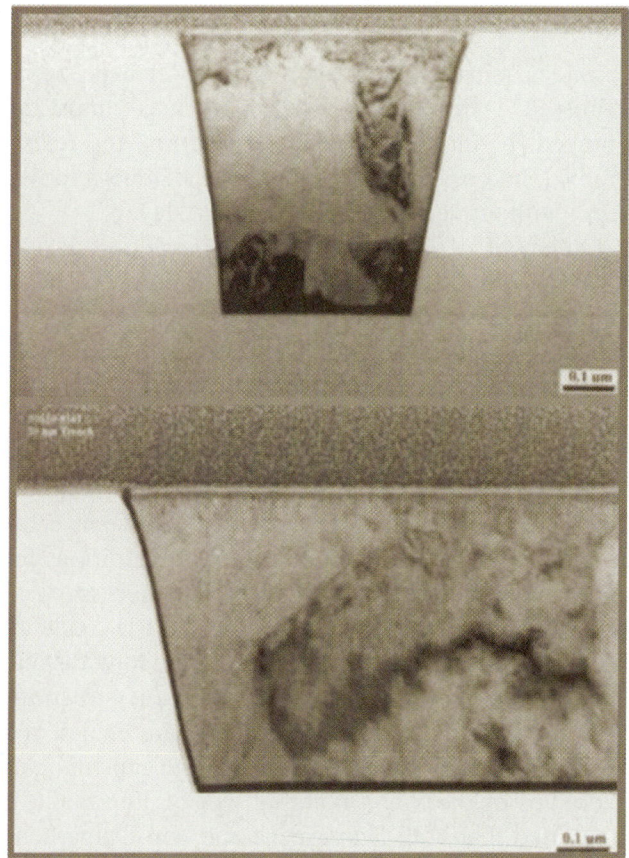

Fig. 24-15 SEMs of an in-laid copper-line in a Dual-Damascene interconnect-structure. Reprinted with permission of Semiconductor International.

materials with smaller permittivity values (i.e., *low-k dielectric materials*). The generic dielectric-film material for interconnects has been SiO_2, which has a k-value of ~4.0. (Note that the k-values of LPCVD and PECVD nitrides are even higher, ranging from 6.0–7.0 and 6.0–9.0, respectively.) Air has the lowest k-value, with a value of $k = 1.0$.

There has been a push to integrate materials with k-values smaller than that of SiO_2 into IC-interconnect structures. The ITRS has suggested that inter-level dielectrics with values of $k = 2.5$-3.0 be integrated into 0.18-μm device-technologies, and those with values of $k = 2.0$-2.5 into 0.15-μm ICs (Table 24-1). However, the industry has been more conservative. These technology generations will probably be implemented with dielectrics having ranges of $k = 3.0$-4.0, and $k = 2.5$-3.0, respectively. There are many process-integration issues that must be satisfied before new-materials are successfully incorporated into an IC-fabrication sequence. The new low-k materials must meet these requirements to be useful in IC-fabrication. Here such integration issues and the candidate low-k materials being analyzed for the suitability for this task are described. The classes of low-k dielectrics are divided into three k-ranges: **1)** $k = 2.8$–3.5; **2)** $k = 2.5$–2.8; and **3)** $k = <2.0$.

24.3.1 Process-Integration of Low-k Dielectrics

If a dielectric-material is to be successfully used in an IC, it must not only exhibit the desired k-value, but it must also: **1)** exhibit adequate material-characteristics (i.e., thermal, electrical, and mechanical); **2)** be able to work with the other materials of the interconnect structure; **3)** be compatible with the IC-processes of cleaning, etching, CMP, and thermal-treatments; **4)** be available in high-purity form and low-cost; **5)** be able to operate reliably over the product-life under the specified device-operating conditions (Table 24-2).

There are several important *thermal characteristics* of a dielectric-material that must be considered. First, it should be stable up to temperatures of 425°C for at least short-periods of time (to make it compatible with the thermal-excursions of metal-deposition, assembly, and packaging). Second, it is desirable that these dielectrics have a thermal-conductivity close to that of SiO_2 (12.0-mW/cm°C) for purposes of heat-dissipation during device-operation. Third, they should exhibit low thermal-expansion-coefficients (to prevent cracking during thermal-excursions, e.g., <

TABLE 24-1 ITRS Roadmap of the Effective Dielectric Constant Needed at Each Device Generation - Out to 2012

1997							
Year of first product	1997	1999	2001	2003	2006	2009	2012
Shipment technology node	250 nm	180 nm	150 nm	130 nm	100 nm	70 nm	50 nm
Interlevel metal insulator effective k	3.0-4.1	2.5-3.0	2.0-2.5	1.5-2.0	1.5-2.0	<1.5	<1.5
1999							
Year of first product		1999	2001	2002	2005	2008	2011
Shipment technology node		180 nm	150 nm	130 nm	100 nm	70 nm	50 nm
Interlevel metal insulator effective k		3.5-4.0	2.7-3.5	2.7-3.5	1.6-2.2	<1.5	<1.5
2000?							
Year of first product			2001	2003	2005	2007	2009	2011
Shipment technology node			180 nm	150 nm	130 nm	100 nm	70 nm	50 nm
Interlevel metal insulator effective k			3.5-4.0	2.7-3.5	2.7-3.5	2.2-2.7	2.2-2.7	1.6-2.2

TABLE 24-2 List of Desirable Properties of Low-k Materials

#	Property	Value
1	Dielectric constant (out-of-plane, 1MHz @ room temperature)	<3.2
2	Thermal stability (C)	>425
3	Moisture absorption	<2.0
4	Adhesion (ILD/,metal, metal/ILD, ILD/ILD)	No peel after 450C cycles
5	Thermal expansion coefficient (ppm/C)	-10 to +50
6	Solvent resistance (acid, bases, solvents, PR strippers)	No weight change after dip
7	Etch rate (wet or dry)	
8	Etch selectivity (ILD/metal, metal/ILD, PR/ILD)	
9	Stress (Pa, tensile or compressive)	<1E9
10	Gap fill	No void @0.35μm and AR=2
11	Tensile strength (MPa)	<200
	Tensile modulus (GPa)	<1
	Elegation at break (%)	>5%
12	Thickness uniformity over topography (before and after curing)	
13	Cost	IBO
14	Corrosion to Al-Cu	
15	Environmental, health and safety	Pass

10 ppm/°C). As far as *electrical-characteristics*, such films should have leakage-current and breakdown behavior (at least 2–3-MeV/cm) similar to SiO_2, as well as low charge-trapping characteristics. As for their *mechanical-properties*, the films should exhibit low film-stress, and have the ability to be deposited to a thickness greater than 2-μm without cracking.

The *material-compatibility* characteristics of such low-k dielectrics involve adhesion and chemical-reactivity issues. The films must adhere well to the materials that are deposited on them. They must also adhere-tightly to the materials on which they themselves are deposited. They should also not absorb moisture, nor exhibit excessive-permeability to moisture. Moisture can adversely-impact the other materials of an interconnect-structure, and it can also be outgassed during other processes, which degrades the process-ambient.

There are several-processes that need to be performed on dielectric-films after their deposition. The films must be able to be properly integrated into such processing. First, the films will undergo CMP, and such materials will thus need to be polishing-process-compatible. Next, vias (and perhaps grooves in a Damascene-technology) will have to be etched in the films. Issues of etch-chemistry, etch-selectivity to oxides, nitrides, and oxynitrides will need to be addressed. After etching, the resist-mask must be removed. It must be possible to perform any resist-removal processing (and any resist-residue-removal steps) without damaging the dielectric-layers.

Other *miscellaneous process-integration* issues of low-k dielectrics include the following. First, although the gap-fill characteristics of the low-k films are not a consideration for Damascene-structures (as the dielectric films are deposited on surfaces that have been planarized by CMP), they will need to be deposited in a single-pass at a thickness of about twice that used in dielectric-films in non-Damascene structures. Thus, they will need to exhibit crack-resistant characteristics for such film-thicknesses. In addition, for some dielectric-materials it might be necessary to use a capping-layer of conventional CVD-oxide or oxynitride to make the overall-layer able to withstand the Cu CMP-process. For non-Damascene structures (i.e., in which the metal [Al] patterns are etched and then covered with a dielectric layer), the gap-fill characteristics of the low-k material *will* be important. It is asserted that both Damascene and non-Damascene interconnect structures will continue to find use as devices are scaled, depending on the circuit-application. Finally, use of spin-on dielectric-films that evolve toxic-vapors will be avoided. Figure 24-16 summarizes the process-integration issues associated with using low-k dielectrics and Cu-films together.

24.3.2 First Generation Low-k Dielectrics: (2.8 < k < 3.5)

Dielectric-materials with k-values that are moderately lower than SiO_2 were the first materials to be adopted for this function. Three types of moderately low-k films found their way into IC-production by 1999: **1)** *HSQ films*; **2)** *fluorinated-oxide films* (also referred to as *fluorinated silica glass - FSG*); and **3)** *low-k SOGs*. Their main use is in 0.5-μm to 0.25-μm generation CMOS-technologies.

Spin-on HSQ films were used as replacement films for siloxane SOG-films in non-etchback SOG-pla-

CHAPTER 24 MULTILEVEL INTERCONNECTS: COPPER, LOW-k, DUAL DAMASCENE

Film properties	Manufacturing Integration
Dielectric constant: — Bulk: k=2.5-3.0 — Effective: k<3.0 Thermal stability: — High thermal conductivity — Tg>400°C, stable above 425° for short periods — Low expansion Electrical properties: — High reliability — Leakage current: similar to SiO_2 — Breakdown field: similar to SiO_2 — Dissipation factor: <0.01 — Low charge trapping Film composition: — Low film stress — >2 μm thick cracking threshold	—Good adhesion to metals (Ta, TaN, TiN, Cu), oxides/ nitrides —CMP compatible —Minimize need for liner/capping films — Etch selectivity to nitrides, oxides, oxynitrides — O_2 ash/solvent compatible — Avoid C_2H_6, C_3H_8 (CVD) — Avoid toxic solvents (spin-on dielectrics)

Fig. 24-16 Some of the many challenges of integrating low-k dielectrics in a copper Dual-Damascene process flow. Courtesy of Semiconductor International.

narization-processes. But they also exhibit a lower k-value than SiO_2 films ($k < 2.9$). When HSQ is cured, it has a built-in microporosity that results in a reduced dielectric-constant (Fig. 24-17). HSQ material is available from Dow-Corning under the trade name of FOx (Flowable Oxide). HSQ was integrated into products in the 0.5-μm generation, with one article describing its use in a five-layer Al-interconnect structure.[118] An oxide-cap was used on the HSQ-film to improve the stability of its low-k-value during W-CVD and Al-fill.

FSG-films are silicon-oxyfluorides (F_xSiO_y) deposited by CVD (Fig. 24-18). They exhibit k-values of ~3.5. FSG-films are formed in CVD-tools by adding SiF_4 to the gases used to deposit SiO_2 by CVD (silane and oxygen). By increasing the SiF_4-flowrate, more F is incorporated into the film. Higher-concentrations of F cause the value of k to decrease. A maximum of 6%-F can be added, because higher-concentrations cause F to be evolved during RIE. This corrodes the metal of interconnect-structures. FSG-films have been successfully integrated into 0.35 and 0.25-μm ICs.

Low-k SOD-films are available from Honeywell Electronics, with k-values of 2.9. They are inorganic siloxane polymers named Accuspin. When cured at 380-400°C, they can tolerate temperatures as high as 450°C before decomposing.

24.3.3 Second-Generation Low-k Dielectrics: (2.5 < k < 2.8)

The largest variety of low-k dielectrics is found in the second-generation of such substances, which exhibit k values between 2.5 and 2.8. These materials are incorporated into ICs with 0.18-μm (and smaller) feature-sizes. Both *spin-on* and *CVD* materials have been developed within this range of permittivities.[8]

2nd-Gen Spin-On Dielectrics with 2.5 < k < 2.8: Most spin-on, organic, low-k polymers are significantly different from their conventional SOG-counterparts, in that they do not evolve-moisture during curing. Because they exhibit better crack-resistance, they can also be applied in a single-pass to a greater thickness. The two candidate organic spin-on low-k dielectrics are: **1)** *poly(alylene) ethers (PAE)*; and **2)** *hydrido-organo siloxane polymers (HOSP)*. Spin-on dielectric films are deposited without the need for an expensive-CVD tool (instead they use a track cluster-tool with spinners and hot plates, see Fig. 24-19). Spin-on films also reduce the topology of any previous steps by more than 50%. However, the cost of the spin-on liquid is typically much higher per-wafer than the cost of the CVD reactant-gases. In addition, a 100-nm-thick capping SiO_2 layer (deposited by CVD) will probably still be needed with spin-on low-k layers.

PAE materials have k-values of about 2.5, and also

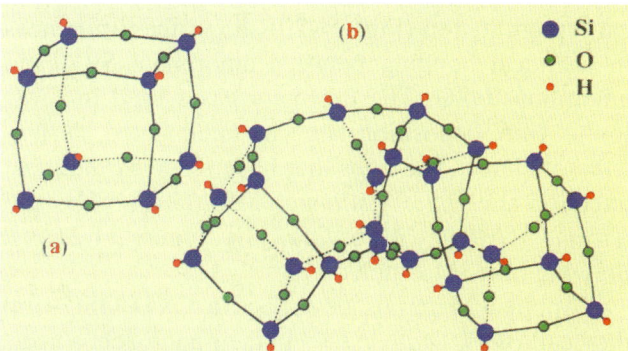

Fig. 24-17 Schematic representation of the molecular-structure of HSQ: (a) eight-corner oligomer; (b) resinous material.

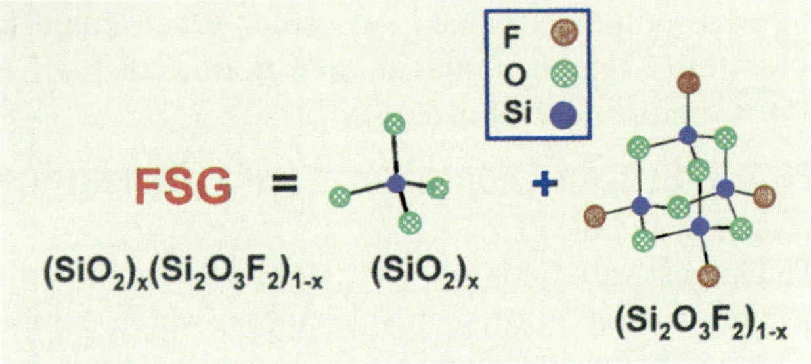

Fig. 24-18 Molecular structure of FSG-materials.

have the ability to withstand high-temperatures. A fluorinated-PAE product from Honeywell is named FLARE, and has a k-value of ~2.8. It was designed to have low outgassing-characteristics and high thermal and mechanical stability. Dow Chemical has a similar material named SiLK. In 2000, IBM announced plans to integrate SiLK into its 0.13-μm products.

The *SOG-like material* in this class is Accuspin T-24 from Honeywell. It is a *hydrido-organo siloxane polymer* (HOSP) which has a k-value of 2.5 after curing. This HOSP has a 12% carbon-content, which allows it to be formed into layers up to 1.2-μm thick without cracking. Its k-value remains stable at temperatures up to 380°C. HOSP also demonstrates good gap-fill and planarization capabilities.

2nd Gen-CVD Dielectrics with 2.5 < k < 2.8 (Black Diamond™ and Coral™): Several companies have developed low-k CVD films using a variety of carbon-containing precursors. The resulting *organosilicate-glass films* (OSG), also called carbon-doped oxides, have a composition of $Si_wC_xO_yH_z$. A graphical representation (Fig. 24-20) shows the differences in chemical bonding between amorphous-SiO_2 and amorphous-OSG. The precursor-gases used in CVD low-k film deposition are the methylated derivatives of silane. The ones that have received the most attention are *trimethylsilane* {3MS, $(CH_3)_3SiH$}, and *tetramethylsilane* {4MS, $(CH_3)_4Si$}. Figure 24-21 is a drawing of a 3MS-molecule.

Applied Materials developed a CVD OSG-film they call *Black Diamond*.™ In their approach, a near-room-temperature PECVD-process is used to deposit a silsesquioxane-film using organo-silane precursors, together with gaseous-oxidizers like O_2 or N_2O. This produces an as-deposited film that has a silsesquioxane-type porous-structure, and which exhibits a k-value of 2.7. However, since these films exhibit poor-step coverage and gap-fill capability, they can only be used as intermetal-dielectrics in dual-damascene interconnect structures.

Novellus calls their CVD-OSG films *Coral*.™ Their k-value is also 2.7. Novellus says that through process-optimization, the hardness of Coral films have been improved sufficiently so that they can be used in standard plastic-packages. To protect the OSG-layer from attack during resist-removal, a capping-layer of SiO_2 is CVD-deposited onto it (although SiC capping-layers are also being introduced). Figure 24-22 lists the many possible types of low-k materials that can be used in advanced ICs.

24.3.4 Ultra-Low-k Dielectrics: (k < 2.0)

By 2002, ultra-low-k dielectric-materials were still being investigated for their suitability as IC interconnect-layers. According to IRTS-2000, they will not be included in commercial products until at least 2005. Several different types were being considered, and are briefly mentioned here.[9,10]

The first kinds are *nanoporous silica* (SiO_2) *xerogel materials* whose k-value can be tuned from k = 1.3 to k = 2.5 (by varying the film-porosity). Since

Fig. 24-19 The spin-coating process for *spin-on dielectrics* (SODs) is optimized for uniform coverage across 300-mm substrates, as well as low-k materials. Courtesy Tokyo Electron.

CHAPTER 24 MULTILEVEL INTERCONNECTS: COPPER, LOW-k, DUAL DAMASCENE

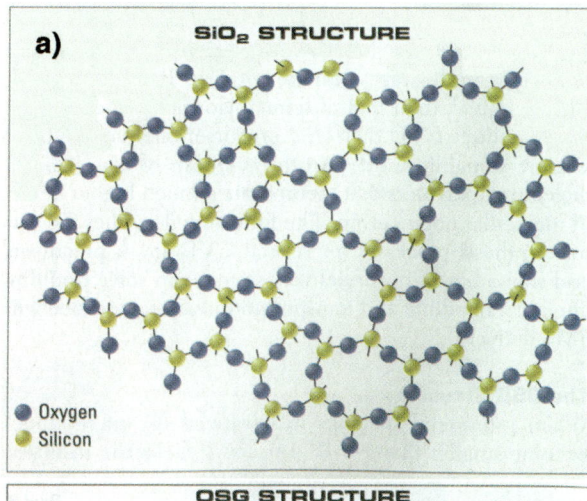

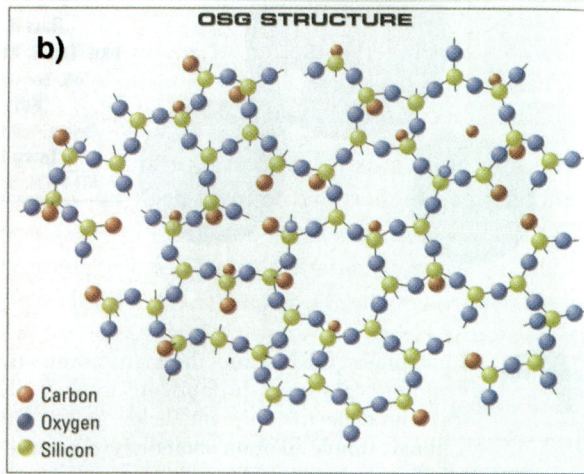

Fig. 24-20 Graphical representation of the difference in the chemical-bonding structure between: (a) SiO_2; and (b) an *organosilicate-glass* (*OSG*) with a Si:C ratio of 4:1.

the k of air is 1.0, by making a dielectric-layer porous (Fig. 24-23), more of it consists of air, thus decreasing its overall k-value. That is, in xerogels, air is trapped as bubbles in a solidified-gel. Xerogels are formed by evaporation-drying, with some shrinkage. One commercial version is offered by Honeywell, named Nanoglass.® In temperature-testing of nanoporous-silica, feasibility-studies indicated that such films are stable up to 500°C. (Note that all types of porous-materials - oxide or organic based - must withstand such process-steps as CMP, etching, and heat-treatments, without degradation of their pore-structure.)

A second type of ultra-low-k dielectric is formed with polyfluoro-tetraethylene (PTFE), a chemical related to Teflon (a DuPont trademame). PTFE is a chemical-compound consisting of carbon and fluorine, made up of uncrosslinked-CF_2-polymers (and also contains 67-at%-F). A prototype spin-on version of a PTFE-dielectric with k = 1.9 is being evaluated.

One big challenge of integrating ultra-low-k films into interconnect structures is that their mechanical strength and hardness decreases as the k-value is reduced (Fig. 24-24). This may require use of harder capping layers on top of the ultra-low-k films.

24.4 DAMASCENE AND DUAL-DAMASCENE INTERCONNECT STRUCTURES

As described in Chap. 4, the conventional way of fabricating an interconnect-structure for ICs is to first blanket-deposit a metal-film onto the wafer-surface. Next, this film is patterned by etching, and a dielectric-film is deposited over the etched metal-lines. Vias are then etched in the dielectric to allow connection passages for the next-layer of metal deposited on the dielectric-film. This approach requires etching metal-films and having deposition-processes with good gap-fill capability for both dielectric and metal-layers.

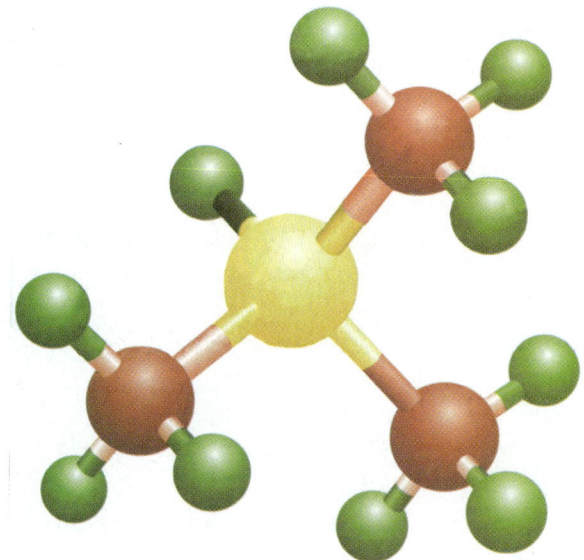

Fig. 24-21 Molecular structure of *trimethyl-silane*, $(CH_3)_3SiH$. Courtesy of Dow Corning.

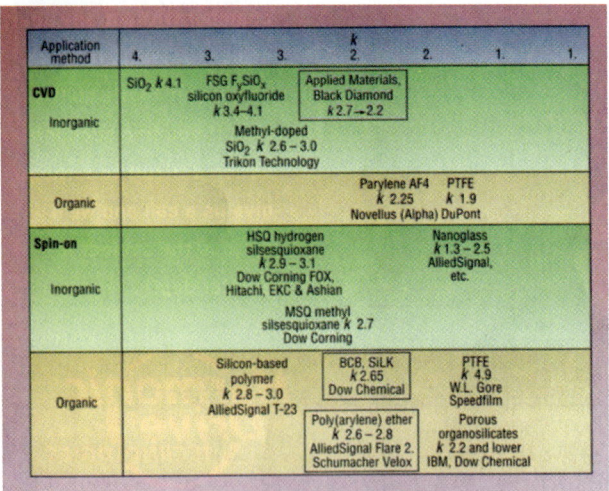

Fig. 24-22 This figure illustrates the many types of low-k materials that are being used in advanced ICs, together with the generations of ICs where each finds use.

An alternative interconnect technology has been developed in which no metal-etching-step is required. Also, gap-filling capability is only required of the metal (but not for the intermetal dielectric). Since metal-etch (especially etching of Cu) and dielectric gap-fill are viewed as two of the industry's greatest-challenges in the drive to smaller dimensions, the elimination of these steps represents a major benefit. The approach that offers this benefit is called the *Damascene structure,* and it was pioneered by IBM. Note that a third-advantage of Damascene-interconnects is that they circumvent some problems associat-

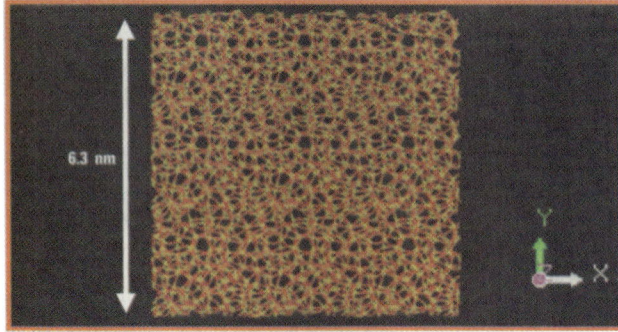

Fig. 24-23 Simulation of an amorphous-SiO_2 matrix shows microscopic voids, which can be considered pores. Courtesy of Honeywell Electronic Materials.

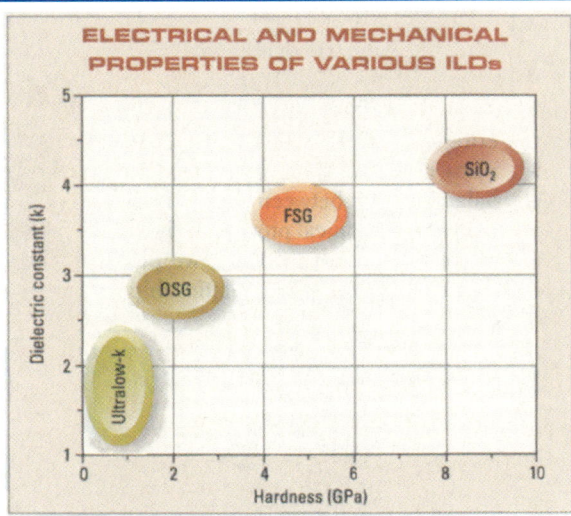

Fig. 24-24 Reducing the dielectric-constant is typically accompanied by reduced mechanical properties, including hardness. Courtesy of Air Products & Chemicals.

ed with lithographic-overlay tolerance, making it possible to achieve higher interconnect packing-density.

The Damascene interconnect-structure is named after the practice of creating metal-inlays, as developed by the artisans of Damascus in the Middle Ages. The process of creating a *single-damascene structure* consists of first-etching a trench or groove in a planarized dielectric-layer, and then filling that trench with a metal, such as aluminum or copper (Fig. 24-25). In a *Dual-Damascene (DD) structure*, a second

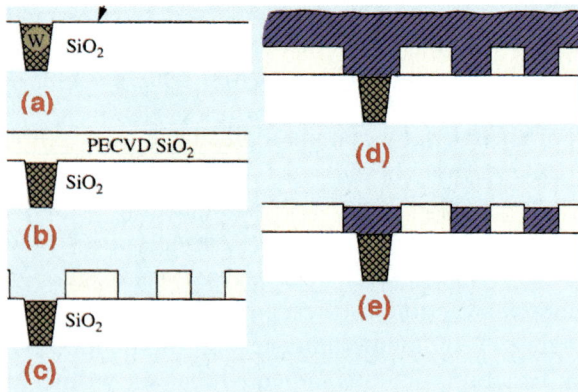

Fig. 24-25 *Single-Damascene process*: (a) W-plugs formed after SiO_2 is planarized; (b) IMD is deposited by CVD; (c) Trench is etched in IMD; (d) Metal is deposited to fill trenches; (e) Excess interconnect-metal removed by CMP.

level is involved where a series of holes (i.e., contacts or vias) are etched and filled, in addition to the trench. The vias and trenches are defined by using two lithography steps. The via is filled with metal in the same step as the metal-line. After filling, the excess-metal deposited outside the trench is removed by a CMP-process, and a planar-structure with metal inlays is achieved. Once a planarized surface is achieved, it is no longer necessary to perform CMP on the dielectric layers. Thus, one CMP-step is also eliminated. Another simplification is that the plug-formation process is also eliminated.

Although the filling of narrow spaces by the dielectric layer is eliminated, the filling of narrow, high-aspect-ratio holes with metal is still a formidable challenge. In conventional-interconnects, such holes are adequately filled with CVD-W. This can also be done in the Single-Damascene structure. But in Dual-Damascene interconnects, the metal that fills the holes is Al or Cu. (Although it should be noted that the *contact-holes* will likely be filled with CVD-W plugs in many Dual-Damascene structures, to keep Cu away from the Si-substrate.)

The Dual-Damascene structure can be fabricated in a number of different ways. In the first of these, called the *trench-first approach* (Fig. 24-26a-d), the trench-patterns are defined in the ILD after resist is spun-on, and the *trench-pattern mask* is used to expose the resist. After the trench is etched, the first resist-layer is stripped (see Figs. 24-26b and 24-26e). A second resist-layer is then spun on, and the *via-pattern-mask* is used to create openings in this resist-layer so that the vias can be etched (without further etching the dielectric in the trenches). After the resist is removed, the metal that fills both the vias and the trenches can be deposited, and polished back to create the Dual-Damascene interconnect-structure.

The major drawback of the trench-first sequence is that after the trench has been etched, the photoresist is applied for the via-patterning step. This resist will fill these trenches, creating local-regions of extra-thick resist, right in the regions where the small vias are to be patterned. Forming very-fine via-structures in such thick-resist is very difficult, and the process-margin for via-formation becomes unacceptably-small at deep-submicron geometries. As a result, a second approach to fabricating Dual-Damascene structures (called the *via-first* approach, as shown in Figs. 24-27 and 24-28) has therefore become the preferred-method for deep-

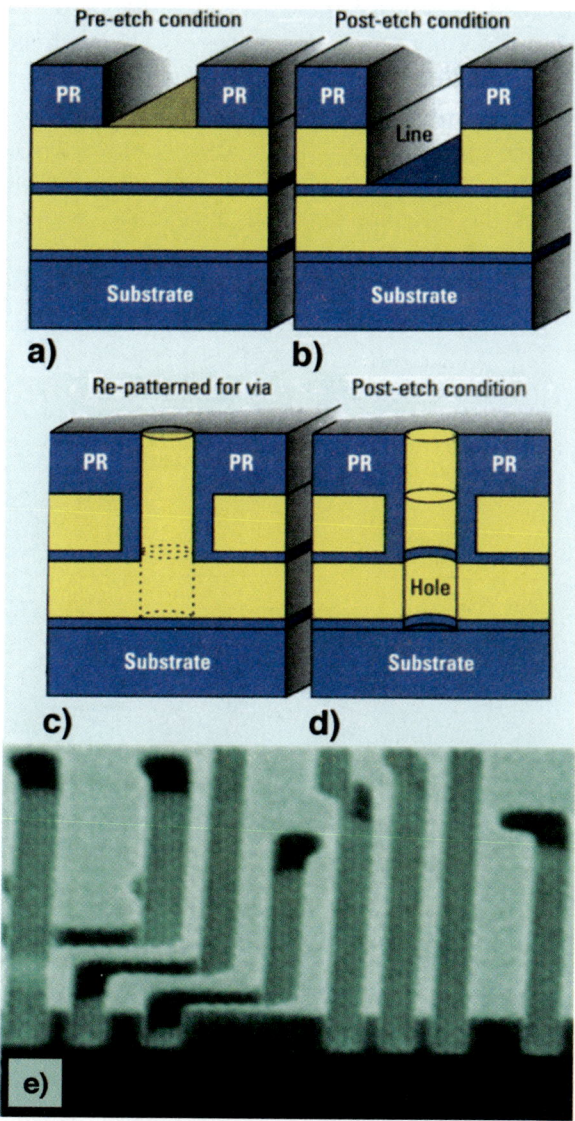

Fig. 24-26 *Trench-First* Dual-Damascene process: (a) IMD is deposited by CVD and planarized by CMP. Trenches are defined by PR #1; (b) Trenches are etched; (c) Vias are defined by PR #2; (d) Vias are etched, using PR #2 to protect other regions of the IMD from etching; (e) SEM of the etched-trenches after PR #1 has been stripped.

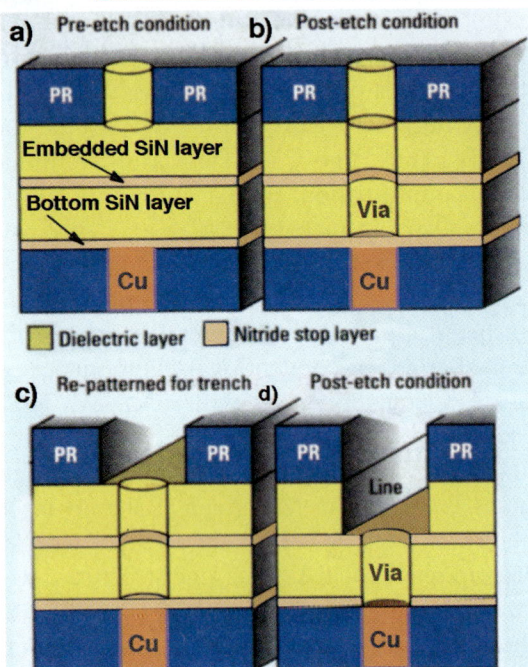

Fig. 24-27 *Via-First Dual Damascene process*: (a) IMD is deposited by CVD and planarized by CMP. Vias are defined by PR #1. (b) Vias are etched. (c) Trenches are defined by PR #2. (d) Trenches are etched, using PR #2 to protect other regions from being etched. Bottom-SiN layer is etched.

submicron technologies. It is described next. Figure 24-29 shows a multi-level Cu-based Dual-Damascene interconnect structure fabricated by IBM.[147]

The process-sequence for forming a *via-first Dual-Damascene* structure is now considered (Fig. 24-27). In this approach an IMD is first deposited (including an embedded *etch-stop layer*) over the silicon-nitride layer that encapsulates the underlying Cu-Damascene levels. Then the wafers are coated with resist and patterned with the *Via-Mask* (Fig. 24-27a). Next, an anisotropic etch cuts through the ILD (including the embedded SiN-layer. However, this etch stops on the bottom SiN-layer (Fig. 24-27b). It is important that this does not break through the bottom-layer, because, if it did, the following trench-etch step would sputter the now-exposed Cu onto the sidewalls of the via. This Cu would then quickly diffuse through the ILD into the Si wafer below, and cause device failure.

After the via resist-layer is stripped, a second resist-layer is applied and patterned with the *Trench-Mask* (Fig. 24-27c). Some of the second resist material is allowed to remain in the vias after development (to help protect the bottom SiN layer during the next etch step). That is, an anisotropic *trench-etch step* is performed to remove the material in the upper part of ILD-stack not covered with resist. This etch stops on the embedded SiN hard-mask layer, forming the trenches. The resist is stripped, and a third etch-step is used to open the bottom SiN layer (Fig. 24-27d), so that contact between metal-levels can occur there.

After the trenches and vias are formed using the

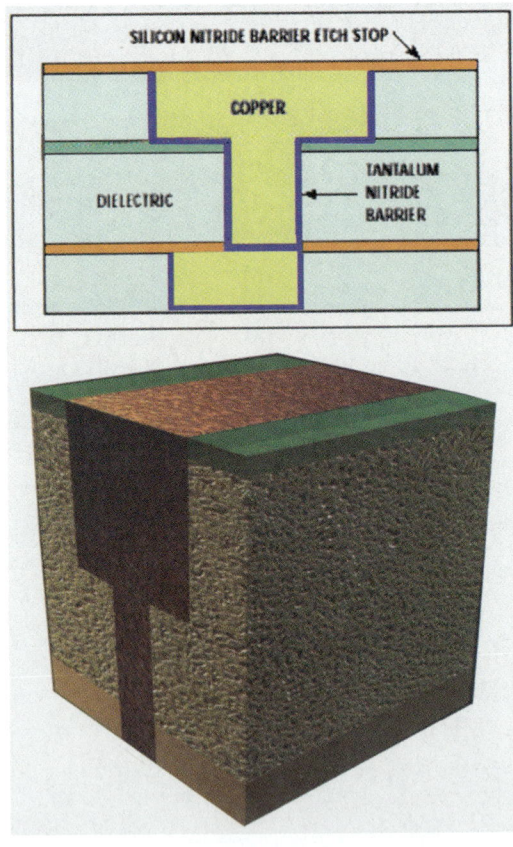

Fig. 24-28 (a) Schematic of a single-level, completed Cu-Dual-Damascene interconnect-structure. It shows the dielectric-stack (with via and trench etched in it), barrier-layer deposited under the Cu, the Cu-material that fills the via and trench, and the top diffusion-barrier-layer film (SiN). (b) Perspective drawing of a Cu-Dual-Damascene structure and porous low-*k*. Courtesy of Tokyo Electron.

CHAPTER 24   MULTILEVEL INTERCONNECTS: COPPER, LOW-*k*, DUAL DAMASCENE

Fig. 24-29 SEM of an IBM Cu-based Dual-Damascene interconnect structure.[5] Courtesy of IBM.

above process sequence, the recessed metal interconnect structures are created. That is, a barrier/liner layer is deposited by PVD, followed by the deposition of the seed-Cu layer (also by CVD). The bulk of the copper material that fills the recesses is next deposited by electroplating. CMP is then used to remove the metal from the top-surface above the trenches, leaving metal only in the recessed regions (see Fig. 24-15). Finally a SiN layer is deposited over the entire wafer-surface to seal the Cu in the recesses. Figure 24-28a shows a drawing of the completed recessed Cu-structure (i.e., after the top layer of SiN has been deposited), connected to a lower-level Damascene Cu-line.

REFERENCES

1. A.K. Sinha, *J. Vac. Sci Technol.*, **19**, 778 (1981).
2. D.J. McGreivy, in *VLSI Technologies*, D.J. McGreivy and K.A. Pickar, Eds., IEEE Computer Society Press, Los Angeles (1982), p. 185.
3. C.L. Hu and J. Harper, "Copper Interconnection and Reliability," *Mater. Chem. Phys.*, **52**, 5, (1998).
4. S.C. Sun, "Process Technologies for Advanced Metallization and Interconnect Systems," *Tech. Dig. IEDM,* 1997, p. 765.
5. P. Singer, "Tantalum, Copper, and Damascene: The Future Interconnects," *Semicond. Intnl.,* June 1998, p. 90.
6. M. Bohr, "Interconnect Scaling: The Real Limiter to High Performance USLI," *Tech. Dig. IEDM,* 1995, p. 241.
7. J.D. Reid *et al.,* "Factors Influencing Feature Filling Using Copper PVD and Electroplating," *Solid State Technology,* July 2000, p. 89.
8. L. Peters, "Low-*k* Dielectrics: Will Spin-On or CVD Prevail?" *Semiconductor International,* June 2000, p. 110.
9. J. H. Golden, C.J. Hawker, and P.S. Ho, "Designing Porous Low-*k* Dielectrics," *Semiconductor International,* May 2001, p. 79.
10. M. O'Neil *et al.,* "Low-*k* Materials by Design," *Semiconductor International,* June 2002, p. 93.

PROBLEMS

1. Why is the parasitic-capacitance per unit area for Metal-1 on field-oxide smaller than that of polysilicon on field oxide?

2. Compare the Al/W-plug metallization with that of a Dual-Damascene copper-interconnect approach. List three advantages of each over the other.

3. Metal lines of varying widths are to be fabricated. Before being patterned, the sheet-resistance of the metal-film is found to be 2.5-Ω/sq. Find the total resistance of 0.5, 1.0, 2.0, and 5.0-μm wide and 1-cm long lines. What other information is still needed before it can be determined what type of metal makes up these lines?

4. Why is the Dual-Damascene method used to form copper interconnects (instead of the subtractive method that is used to create Al-interconnect lines)?

5. What hampered the introduction of copper interconnects into IC processing?

6. List the locations in a copper/dielectric Dual-Damascene structure where a silicon-nitride layer is likely to exist. Explain the function of each Si_3N_4 layer.

7. Using the data in Fig. 24-7, it can be claimed that copper has a factor-of-10 greater electromigration resistance than aluminum. Interpret rgw data in this figure to explain how this conclusion is reached.

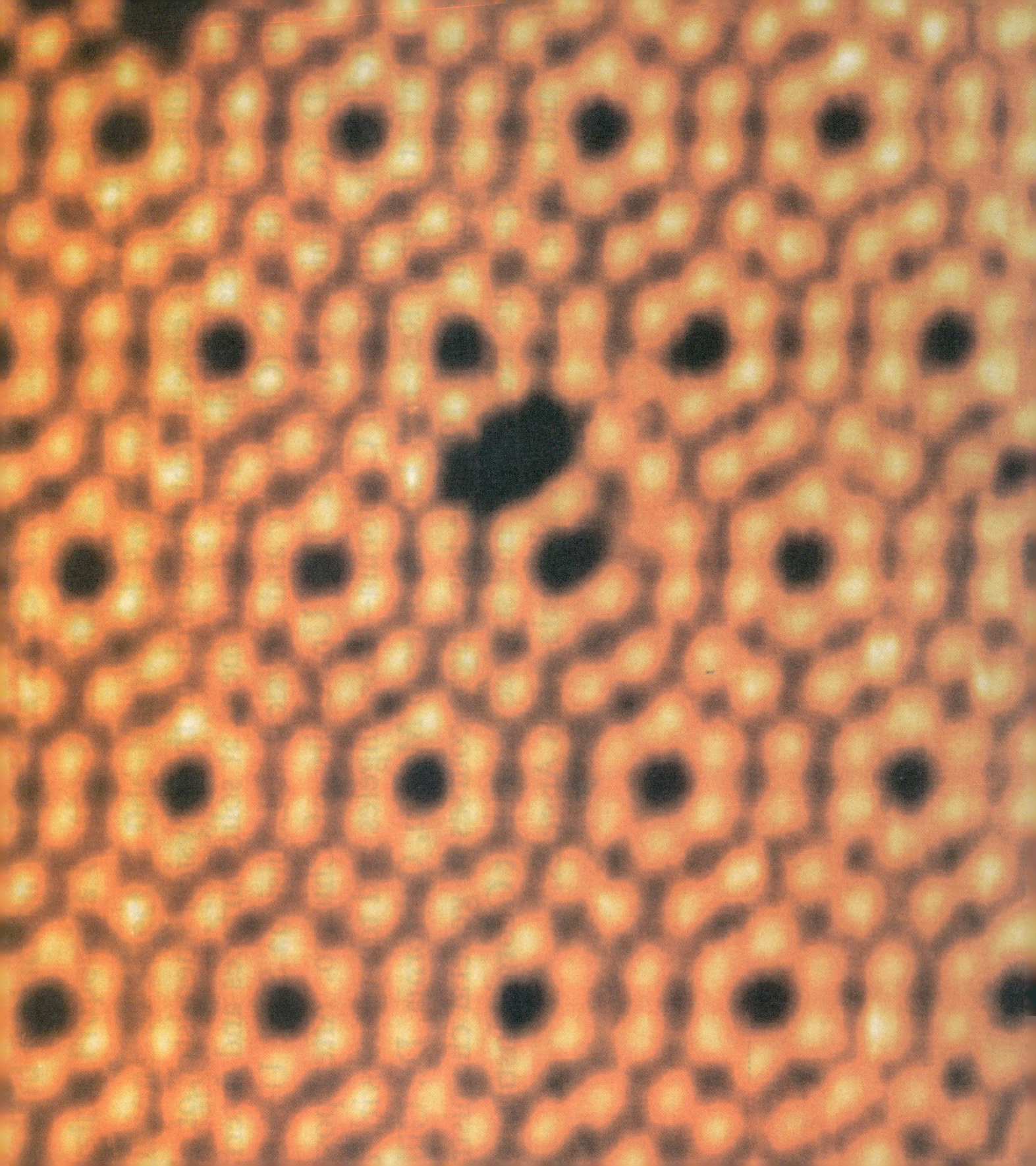

CHAPTER 25

MATERIALS CHARACTERIZATION TECHNIQUES FOR ULSI

CHAPTER CONTENTS

25.1 WHAT ARE WE TRYING TO DETECT AND HOW IS IT DONE

25.2 MICROSCOPY FOR ANALYZING IC-FEATURES

25.3 OPTICAL MICROSCOPES

25.4 SCANNING ELECTRON MICROSCOPY (SEM)

25.5 TRANSMISSION ELECTRON MICROSCOPY

25.6 ATOMIC-FORCE MICROSCOPY (AFM)

25.7 ELECTRON/X-RAY COMPOSITIONAL ANALYSIS TECHNIQUES

25.8 AUGER EMISSION SPECTROSCOPY (AES)

25.9 X-RAY EMISSION SPECTROSCOPY (XES)

25.10 X-RAY PHOTOELECTRON SPECTROSCOPY (XPS, ESCA)

25.11 X-RAY FLUORESCENCE (XRF)

25.12 ION-BEAM-EXCITED COMPOSITIONAL ANALYSIS

25.13 SECONDARY-ION MASS-SPECTROSCOPY (SIMS)

25.14 FOCUSED-ION-BEAM ANALYSIS

"…rors using inadequate data are much less than using no data at all."

George Babbage

…surface of a silicon-crystal imaged with …canning tunneling-microscope. Individual …ns are shown in red. The open-space between …ns is shown as the black background. …rtesy of Drs. Gilberto Medeiros-Ribeiro and …tanley Williams/Hewlett-Packard Research …oratories, Palo Alto, CA.

In order to fabricate complex integrated-circuits, many steps must be correctly performed using a variety of materials. Monitoring and characterizing each step during process-development and production requires the use of a host of measurement techniques. Most are discussed in the chapters that describe the processes that they commonly characterize (e.g., measuring the thickness and resistivity of films formed by sputtering, CVD, or ion-implantation).

The remaining techniques in wide use for identifying or characterizing the form, chemical-composition, and crystallographic-structure of materials are discussed in this chapter. A brief introduction to each is given, stat-

Fig. 25-1 SEM photograph of an IC-structure delineated with the aid of a focused ion-beam to reveal its structural cross-section. Courtesy of FEI Company.

ing the principles on which they are based. Then their capabilities and limitations for specific ULSI applications are listed. The purpose is to acquaint readers with them, so the appropriate ones can be chosen for particular analytical-applications. (For more information, readers can consult Refs. 1- 6, listed at the end of the chapter.) Figure 25-1 is an example of how several materials-characterization tools are used together. It shows a SEM-photograph of both the top-surface and the cross-section of an IC-structure delineated with help of a focused-ion-beam instrument.

25.1 WHAT ARE WE TRYING TO DETECT, AND HOW IS IT DONE?

Two types of material-properties are analyzed by the diagnostic-techniques discussed in this chapter:

1. The *form* and *structural-properties* of an IC device-feature, such as its thickness, step-coverage, or surface-roughness.

2. The *elemental* or *chemical composition* of a material.

The form and structure of an IC-feature is primarily observed with the various microscopy techniques - optical, scanning-electron (SEM), and transmission-electron (TEM). Crystal-structure and defects are analyzed mainly by x-ray-diffraction methods and TEM.

A variety of methods are used to determine the elemental and/or chemical-composition of materials. Most often such analysis is concerned with the presence of elements (and/or molecules) in the material being evaluated. Sometimes only *qualitative* information is required (*What* is present?), but other applications may require *quantitative* data (or *How much* of what elements is present?). Information about the composition of substances in differing kinds of locations on a substrate may also be needed. For instance, data might be needed on the composition of localized patterns (residue in contact-holes) or about regions of large-area films (perhaps the composition of deposited-films). Furthermore, data concerning the surface, or the bulk, might be of interest, or one might want information about the variation of the composition as a function of depth. Finally, there might be a need to know how a substance is distributed laterally in or on a film. In some cases this would require an instrument with fine lateral-resolution, but in others, a coarser lateral-resolution would be adequate.

How can we tell which materials are present on the sample of interest? In order to identify if a substance is present, it must be possible to identify some quality of an element that distinguishes it from others. There are two aspects unique to each element: **1**) its *atomic-mass*; and **2**) its *electronic-structure*. When an atom or molecule is ionized, the mass-to-charge-ratio of the ion can be used to selectively-filter only the desired mass-constituents, and thereby determine the presence of this species. The uniqueness of the electronic-structure of each element or molecule is used to identify the source of emitted-electrons and photons from materials under analysis, and in that way, to identify the elements present in the sample being analyzed.

25.1.1 Energy-Regimes and Energy-Levels in Materials Characterization

The energies used to describe the source-beams and emitted-species in materials-characterization techniques are usually expressed in electron-volts (eV) - which are *not* units of potential (or potential-difference), but units of *energy*, as defined by:

1 *electron-volt* (eV) = the amount of energy required to move a unit of electronic charge through a potential-difference of 1-volt.

Since 1 electronic-charge = 1.6×10^{-19} C, then

$$1\text{-eV} = 1.6 \times 10^{-19}\text{-C} \times 1\text{-V} = 1.6 \times 10^{-19}\text{-J} \quad (25.1)$$

The energy-ranges encountered in the materials-characterization are given in Table 25-1.

As mentioned earlier, one of the methods of determining the presence of elements in the materials being evaluated is through the detection of electrons or photons, emitted by the material under bombardment by a *primary* electron-beam or photon-beam (defined in the next section). The notation for describing the

CHAPTER 25   MATERIALS CHARACTERIZATION TECHNIQUES FOR ULSI

Table 25-1 ENERGY-RANGES ENCOUNTERED IN MATERIALS CHARACTERIZATION

1-eV	- the energy possessed by evaporated atoms arriving at a substrate
5-eV	- the energy possessed by sputtered-atoms arriving at a substrate
10-20-eV	- the energy required to ionize neutral atoms (Ar ≈ 15-eV)
20-eV-1-keV	- the energy possessed by emitted Auger electrons
1-20-keV	- the energy possessed by primary-beam species in SEM, AES, SIMS, XES
100-keV	- the energy possessed by the primary-beam of electrons in TEM (and the ions in ion-implantation.
1-3-MeV	- the energy possessed by primary ion-beam in RBS.

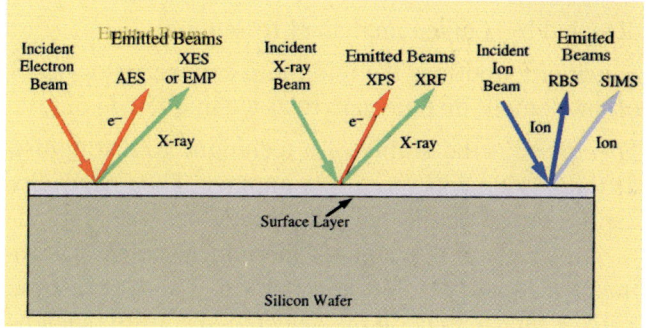

Fig. 25-3 Some of the *surface-analysis techniques* available to identify and quantify substances in IC-manufacturing. This figure also shows various incident (*primary*) and emitted (*secondary*) beams.

lower electronic energy-levels in an atomic-structure are shown in Fig. 25-2.

There is a unique-energy associated with each electronic-level in an atom, relative to free-space. For example, if an electron is present in the L_1 level, there is a specific minimum-energy that must be absorbed by that electron in order to be ejected into free-space. The detection of emitted-electrons or photons in response to specific incident-energy is one technique used to identify the presence of an element in the material being analyzed.

25.1.2 Some Definitions of Materials-Characterization Terminology

Primary-beam (electrons, ions, photons) - constituents of the beam which emanates from the instrument energy-source, and is then directed at the *test-sample* (see also Figs. 25-2 and 25-3).

Secondary-beam - objects (electrons, x-rays, ions) that are emitted by the material of the *test-sample* as a result of being struck by the *primary-beam*.

Auger-electrons - secondary-electrons emitted by indirect energy-transfer mechanisms (discussed in the section on Auger-emission-spectroscopy, AES, see also Fig. 25-3).

Matrix – the substance in which an impurity exists. For example, phosphorus, boron, and antimony can be impurities (or additives) in a silicon-matrix. Likewise, hydrogen is an impurity in the matrix of Si_3N_4.

Morphology - the *form* and *structure* of an object (e.g., the structural-aspects of a feature on an IC wafer, such as its thickness, width, shape, surface-structure, grain-structure, and interface-form).

Detectability - specifies the elements which can be

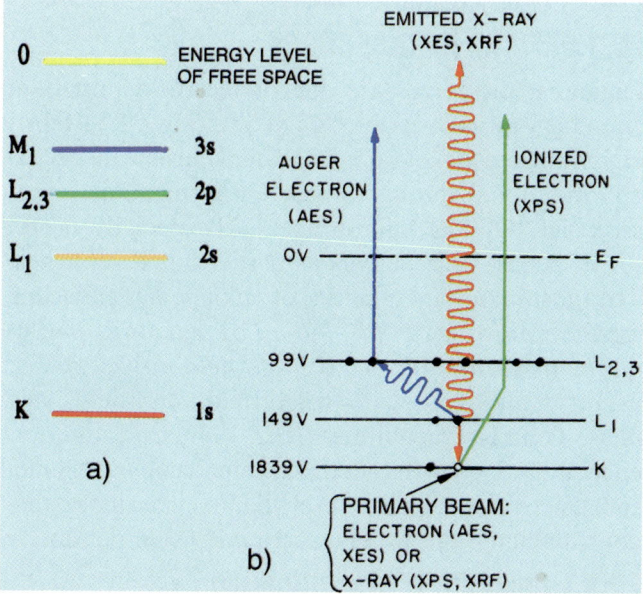

Fig. 25-2 (a) Notation for describing energy-levels in atomic structure. (b) A *primary-beam* (electron, ion, or x-ray) can excite several types of *secondary-species* (e.g., electrons from ionization-events, Auger-electrons, or x-rays).

detected by a given analytical-technique.

Sensitivity - defines the minimum concentration-level of an element that an analytical-technique can detect.

Specificity - the ability of a technique to distinguish different elements from one another. The specificity can be limited either by: **a)** aspects of the analytical-technique itself (such as an instrument-detector having inadequate resolution-capability); or **b)** by some kind of interference in the detected-spectrum (such as two elements that emit electrons with nearly identical energies, thus making it impossible for the measuring-instrument to tell them apart).

25.1.3 Vacuum-Requirements of Compositional-Analysis

Compositional-analysis measurements are performed in vacuum because the primary-beam objects must arrive at the sample without interacting with the gas-molecules in the space between the source of the primary-beam and the target. Such collisions would disturb the focus, energy, or ionic-charge of the primary-beam. There are two vacuum-ranges encountered in the use of compositional-analysis instruments: **1)** Those techniques that monitor (but do not dislodge elemental-material from) the surface of the samples (e.g., AES and XPS), place the most stringent require-ments on the cleanliness of the residual-vacuum in the sample-region. Surface-contaminants will weaken the analytical-signal emitted by the true sample, and will add their own characteristic-spectrum to the output-signal. To avoid this, the vacuum in AES and XPS instruments must be in the $1-3 \times 10^{-10}$-torr range; **2)** In SIMS, the surface is continually being sputtered, and consequently contaminants adsorbed from the residual-gases are continually removed. Thus, less-stringent requirements are placed on the SIMS vacuum system ($1 \times 10^{-8} - 1 \times 10^{-9}$-torr).

25.2 MICROSCOPY FOR ANALYZING IC FEATURES

The form and structure of VLSI features is typically obtained with one or another of three microscopic-techniques: **a)** *optical-microscopy*; **b)** *scanning-electron-microscopy* (*SEM*); and **c)** *transmission-electron-microscopy* (*TEM*). Since the maximum magnification-values of these three methods are 1000x, 100,000x, and 500,000x, respectively, and since the magnification-ranges overlap, it is possible to obtain images of any feature of interest.

25.3 OPTICAL MICROSCOPES

Optical microscopes are one of the most important analytical-tools available for monitoring ULSI fabrication-processes (Fig. 25-4). Wafers must be inspected and monitored throughout the entire manufacturing-sequence. Optical microscopes are especially useful for detecting such defects as particles and scratches. The most important qualities of an optical microscope are the following: **a)** *resolution*; **b)** *magnification*; **c)** *mechanical-stability*; and **d)** *wide-field-of-view*. Other useful features include brightfield, darkfield, and phase-contrast capabilities (i.e., Nomarski interference, Fig. 25-5), fluorescence-microscopy (organic-substances fluoresce more brilliantly than inorganic-constituents of the wafer), and a television option.

25.3.1 Resolution, Magnification, and Numerical-Aperture

In general, the term *resolution*, d, is defined as a measure of the ability to distinguish closely-spaced features. That is, it corresponds to the distance between

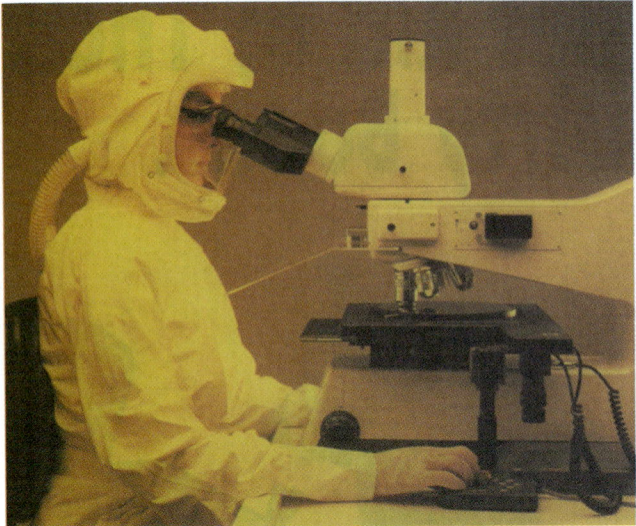

Fig. 25-4 Photograph of a fab-worker using a microscope.

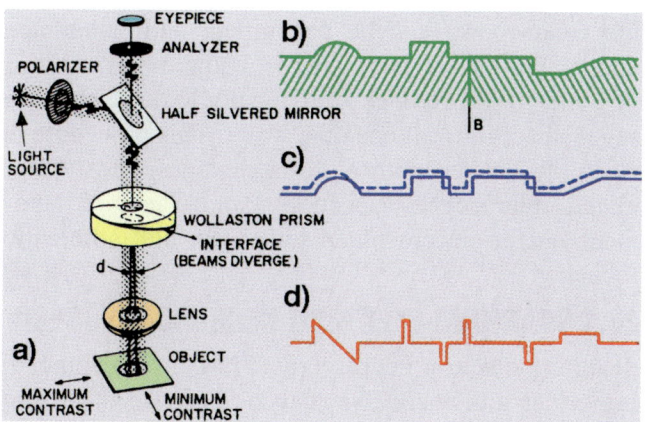

Fig. 25-5 (a) Features of a *Nomarski interference-contrast microscope* operating in the reflectance-mode. (b) Representation of a cross-section of a sample at a surface. (c) The wave-fronts of the reflected beams after emerging from the prism. (d) An intensity-distribution in the image-plane in a Nomarski interference-microscope.

two points in an image when those two points are recognized as being separated. The resolution of a microscope is calculated from:

$$d = \frac{\lambda}{2(NA)} \quad (25.2)$$

where λ is the wavelength of illumination, and (NA) is the numerical-aperture of the objective-lens of the microscope.

The resolution-limit of an optical microscope with an NA = 0.95 is approximately 0.25-μm if the illuminating-wavelength is 0.5-μm (this is the wavelength of green-light at the center of the visible-light spectrum). For routine inspections in a development-environment, optical microscopes are typically expected to resolve ~0.5-μm. The numerical-aperture (NA) is determined by the angle of the cone-of-light accepted by the objective of a microscope (angle-θ in Fig. 25-6a). It is defined by:

$$NA = n \sin\left(-\frac{\theta}{2}\right) \quad (25.3)$$

where n is the *index-of-refraction* of the medium between the objective and the specimen. Since ($\theta/2$) cannot exceed 90°, the theoretical-limit of $\sin(\theta/2)$ is 1.0. The highest NA of manufactured-lenses for microscopes in which air is the medium (and thus n = 1) is 0.95.

The *magnification*, M, describes the ability to enlarge a pattern (e.g., the ratio of the size of an image to the size of the corresponding-object), and the approximate maximum-magnification required for the eye to see all the possible detail that a microscope can reveal is:

$$M_{max} = 1000 \, (NA) \quad (25.4)$$

For an NA = 0.95, the maximum-magnification needed for optical microscopes is ~1000x. This upper magnification-limit arises from the fact that the resolution of the unaided human-eye is ~0.25-mm. At about 1000x, a 0.25-μm separation appears as 0.25-mm. Since the microscope cannot resolve any finer-features, at 1000x the eye sees all the microscope-resolvable detail. That is, a microscope with a higher-magnification would make images larger, but if the features were any closer than 0.25-μm apart, the image would only be a blur.

The magnification-range for the inspection of wafers and masks is 20x to about 1000x. Most microscopes sold for semiconductor-use can be provided with combinations of 10x to 20x *eyepieces*, and *objectives* in the range of 2x to 100x, to cover this entire magnification-range. On high-magnification microscopes, the *mechanical-stability* is as important as

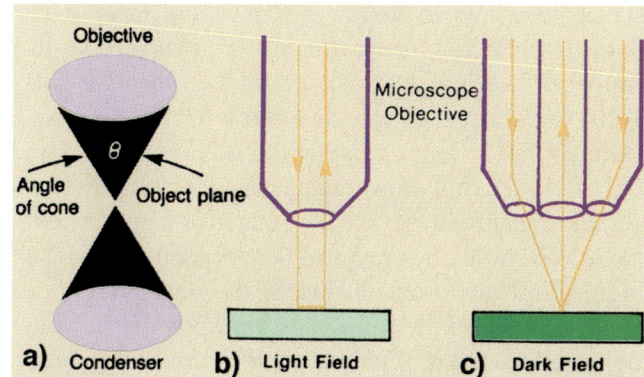

Fig. 25-6 (a) *Numerical-aperture* is determined by cone-of-light accepted by the microscope-objective. (b) Principle of *bright-field microscopy*. (c) Principle of *dark-field microscopy*.

the quality of the objective-lenses. Accordingly, such instruments are rather massive, and must be mounted on vibration-isolated tables.

In some applications (e.g., examination of large-areas for the presence of particulates or other surface-defects), optical microscopes may be more useful if they have a relatively low-magnification, but are equipped with lenses that have the same resolution as the high-magnification-lenses (but with a *field-of-view* that is 10-times as great).

25.3.2 Brightfield and Darkfield Illumination-Modes

In the *brightfield-mode*, the wafer under observation is seen by the light that it reflects. Light travels along the optical-axis of the microscope, through the objective-lens, to the sample being examined. The image is formed by the *reflection* of the light received by the sample, which then travels back through the same optical-elements. This is the most commonly-used mode, and for the majority of applications gives the best overall-image and information (Fig. 25-6b).

In the *darkfield-mode*, light is directed at the wafer from angles *outside* of the cone that the objective encompasses, so that it strikes the wafer surface obliquely (Fig. 25-6c). The light from a flat, featureless-surface is reflected at the same angle and misses the objective-lens (that is trying to collect the light). Hence, the field looks black (*darkfield*). Only when light is reflected or scattered by *features* on the wafer-surface is light collected by the objective-lens. Thus, the sample appears as a black background, and features that reflect or refract the light appear bright. Darkfield-illumination enhances the visibility of details that might be washed out in the brightfield-mode. Even small structural-details below the resolution-limit are often visible in the darkfield-mode, making it useful for quickly scanning wide-viewing-fields for particles, scratches, and chemical-residues.

25.3.3 Television System Interface-Capability

The ability to interface a microscope with a television is a useful option. For example, in the wafer-production-environment, remote consultation (e.g., out of the clean-room) can be performed, and video-tape documentation of the patterns being observed can be generated. The operator typically looks at three to seven specific-locations on a wafer. This procedure is easily automated with motorized stages. Most automated microscope-inspection-stations feature automatic wafer-placement on the stage, and automatic focusing.

25.4 SCANNING-ELECTRON-MICROSCOPY (SEM)

Scanning-electron microscopy (SEM) has become an important tool for ULSI analysis because it has the capability of providing much higher-magnification, resolution, and depth-of-field than optical microscopy (Fig. 25-7).[7] First, the *resolution* of an SEM can be up to 10Å (100Å is routinely obtained). The electron-beam has a much-smaller wavelength than the light used in an optical-microscope, so according to Eq. 25-2, the resolution can be better. Second, the *magnification* of an SEM ranges from 10x-100,000x. Third, the SEM has *depth-of-field* of 2-4-μm at 10,000x and 0.2-0.4-mm at 100x. (The depth-of-field relates to the ability of an optical-system to keep two planes in focus simultaneously.) In a photograph, if the subject is *in focus*, and the background is *out of focus*, the limit of the depth-of-field of the camera is exceeded. In a microscope, the depth-of-field decreases as the magnification increases. Thus, as the power is increased, it may not be possible to see the top and

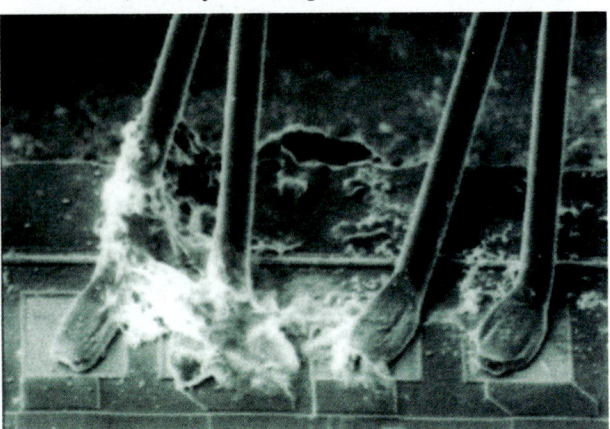

Fig. 25-7 Example of a SEM photograph, showing contamination on wire-bonds.

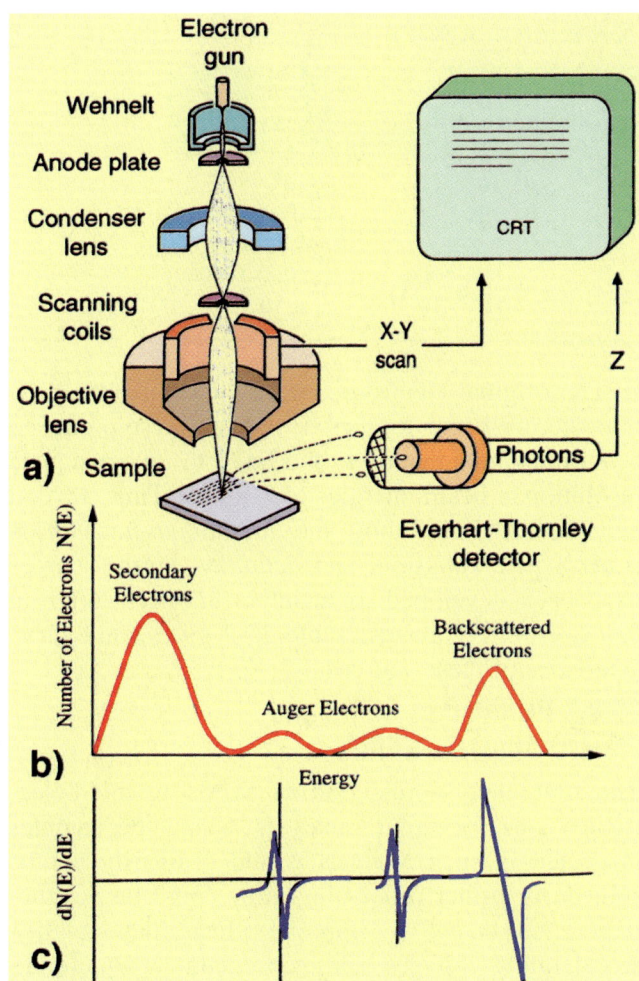

Fig. 25-8 (a) Block diagram of an electron optical-column of an SEM. (b) Energy-distribution of electrons emitted from a solid under electron-bombardment. (c) Derivative of the number of emitted-electrons with respect to energy.

A schematic-drawing of a SEM is shown in 25-8a. A source is used to create a beam of electrons that is accelerated to energies of 500-eV-to-40-keV, focused to a small diameter, and directed at the surface of a sample in a raster-scan pattern. The electrons striking the surface cause a number of phenomena to occur, the most important for SEM-applications being the emission of electrons and x-rays. (The emitted x-ray-signal is useful for chemical-analysis, and this is discussed in the section on XES.)

Figure 25-8b shows the energy-spectrum of electrons emitted from a surface bombarded by an electron-beam. We see that there are three predominant types of emitted-electrons: *lower-energy electrons* of 0-50-eV (peaking at about 5-eV); *higher-energy electrons* (with energies close to those of the primary-beam electrons); and *Auger-electrons*. The lower-energy electrons are called *secondary-electrons*. Because they possess such low energies, only secondary-electrons created close to the surface actually escape and are detected. *Note that it is these low-energy secondary-electrons which are generally the most useful for studies of ULSI features*. The higher-energy electrons are those that have undergone collisions with target-atoms and thus still possess most of their incident-energy. These are referred to as *backscattered-electrons*.

The detected-signal can be due to either secondary or backscattered electrons. This signal is used to modulate the intensity of the beam striking a CRT-screen. An image of the sample-surface is produced on the CRT-screen by synchronously raster-scanning the CRT-screen and the SEM electron-beam.

The *contrast* (the ability to see adjacent-points on an image that have different levels of dark and light) of the image depends on variations in the number of electrons arriving at the detector, and is thus related to the *number of emitted-electrons per incident-electron* at each spot on the sample. For secondary-electrons this number depends on the material, and is significantly higher for oxides than for silicon. This is another important factor that makes the use of secondary-electron SEM-imaging so valuable for ULSI studies. That

bottom surfaces in focus at the same time.) For optical-microscopes, the maximum-resolution and magnification-limits are about 0.5-μm (5,000-Å) and 1000x, and the depths-of-field are much shallower. Note that the high depth-of-field makes the SEM especially useful for high-magnification (i.e. >2000x) examination of ULSI device-surfaces, where film-thicknesses rarely exceed 1-μm. SEM-analysis yields information on linewidth, film-thickness, step-coverage, edge-profiles after etch, and other structural data.

TABLE 25-2 CHARACTERISTICS OF ELECTRON-BEAM SOURCES				
Type of Emission	Tungsten Hair Pin Thermionic (Heated)	Lanthanum Hexaboride Thermionic (Heated)	Field Emission (Room Temperature)	Schottky /Extended Field (Heated)
Brightness (A /cm^2/ster)	10^4	10^5	10^8	10^8
Effective Source Size (Å)	1,000,000	200,000	100	100
Energy Spread (eV)	3	3	0.2-0.3	0.28-0.38
Operating Life (hrs)	30-10	100-500	300-1000	2000-10000
Vacuum Required (torr)	10^{-3}-10^{-5}	10^{-5}-10^{-6}	10^{-9}-10^{-11}	$< 10^{-8}$

is, the effect makes metals, oxide, and silicon patterns readily distinguishable from one another when the SEM is used to produce images in this viewing-mode. The second-source of contrast in secondary-electron images is the dependence of *secondary-electron yield* on surface-curvature. Therefore, surfaces that differ significantly in slope can also be easily distinguished. Finally, surface-regions that face the detector appear brighter than other surface-regions.

The *resolution* of the SEM depends on several factors, including the type of sample under inspection and the incident-beam diameter (which is dependent on the electron-source, the focusing-optics, and the accelerating-voltage of the primary-beam). In early generation SEMs, high-voltages were needed to achieve the small spot-sizes for obtaining sufficient resolution for operation at high-magnification. Unfortunately, this causes insulating-surfaces to acquire excess negative-charge (which distorts the image). By applying a thin metallic-coating to the surface of the sample (e.g., 100-Å of gold), and attaching a ground-wire to the coating, an electrical-path to ground is provided. While this helps reduce such charging-effects, the Au-coating makes the sample unsuitable for further processing. Thus, this approach is not acceptable for applications where wafers from the fab-line must be inspected and returned to production (e.g., for linewidth-measurement with a SEM). Another disadvantage of high accelerating-voltages is that the electrons may damage the circuit under inspection. Therefore SEM manufacturers have developed techniques which use lower accelerating-voltages (i.e., 800-2000-eV), yet maintain high-resolution.

The original electron-sources were tungsten-hairpin electron-beam sources. However, these produced beam-diameters that were too large to give adequate resolution at beam-energies of 1-2-keV. Thus, several new sources, including the *lanthanum-hexaboride* (LaB$_6$), *field-emission*, and *Schottky-emitter* sources have been developed to maintain high-resolution at low operating-voltages. Table 25-2 gives some characteristics of these sources.

25.4.1 Production-SEMs and Failure-Analysis-SEMs

The first major-application of SEMs in microelectronics was for failure-analysis. Since the samples inspected in such analysis would ordinarily not be subject to further processing, they could be destructively gold-coated to improve SEM image-quality. Accelerating-voltages and subsequent sample-damage were therefore also less important. In addition, failure-analysis SEMs can be equipped with auxiliary analytical-capabilities such as x-ray-emission-spectroscopy (XES). This increases their flexibility to perform a wider variety of failure-analysis tasks. Finally, the time required to get samples into and out of such instruments is relatively long, making the rate of sample-inspection (throughput) quite low.

As features decrease to the micron and submicron regimes, optical-microscopes become less suitable for providing the resolution needed to give accurate and repeatable measurements of critical-dimensions. Therefore the use of SEMs as production-tools (especially for measuring the critical-dimensions of features – CDs) becomes more attractive. In order to

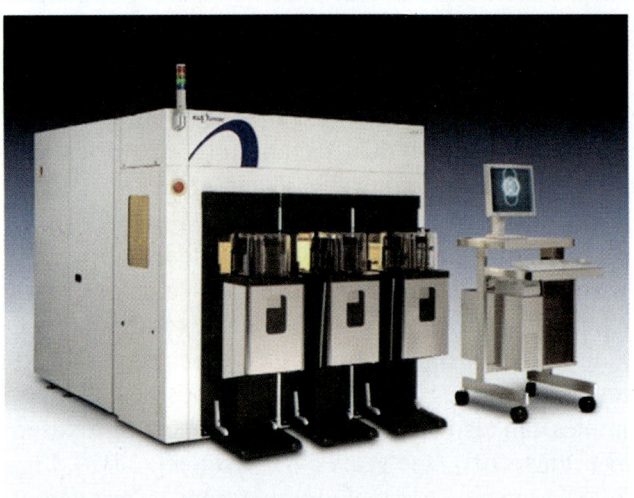

Fig. 25-9 Photograph of a production CD-SEM. Courtesy of KLA-Tencor.

make SEMs compatible with the production-environment, they have to be configured differently than the traditional failure-analysis instruments.[8,15]

Production-compatible SEMs (*CD-SEMs*) utilize electron-sources that operate at low enough voltages to give high-resolution on insulating-surfaces (resist, oxides, etc.) without charging or damaging the wafer-structures. They can rapidly evacuate the sample-chamber and quickly and easily change samples (cassette-to-cassette operation). They automatically and accurately position wafers to pre-selected measurement-locations. Throughput is as high as 70 wafers-per-hour at five measured-sites per wafer. CD-SEMs can also be used to perform defect-review and analysis. Finally, since highly-trained specialists of the failure-analysis staff are not available to operate production-tools, they need to be simple enough to be run by operators. Several manufacturers produce SEMs meeting these requirements (Fig. 25-9).

25.5 TRANSMISSION-ELECTRON MICROSCOPY

Just as shrinking-linewidths and vertical feature-sizes led to the replacement of optical -microscopes by the SEM, other ULSI applications for which the SEM had been adequate, now require an even higher-resolution technique - namely TEM (Fig. 25-10).[9] While maximum SEM-resolutions are in the 20-30Å range, TEM offers 2Å-resolution. The image in TEM is produced by electrons from an incident-beam (60-350-keV, electron-wavelength ~0.04Å) passing through very-thin-film samples. The sample must be thin enough to allow the electron-beam to pass through it, so that information caused by differences in sample-thickness, crystal-structure, and orientation is preserved. The limiting-thickness for TEM-imaging of a Si sample as a function of *accelerating-voltage* is shown in Fig. 25-11. For practical ULSI-analysis, the thickness is about 0.8-μm at 200-keV.

In a TEM, the electron-beam is focused by a condenser-lens, then passes through the sample and is imaged onto a photographic-plate or fluorescent-

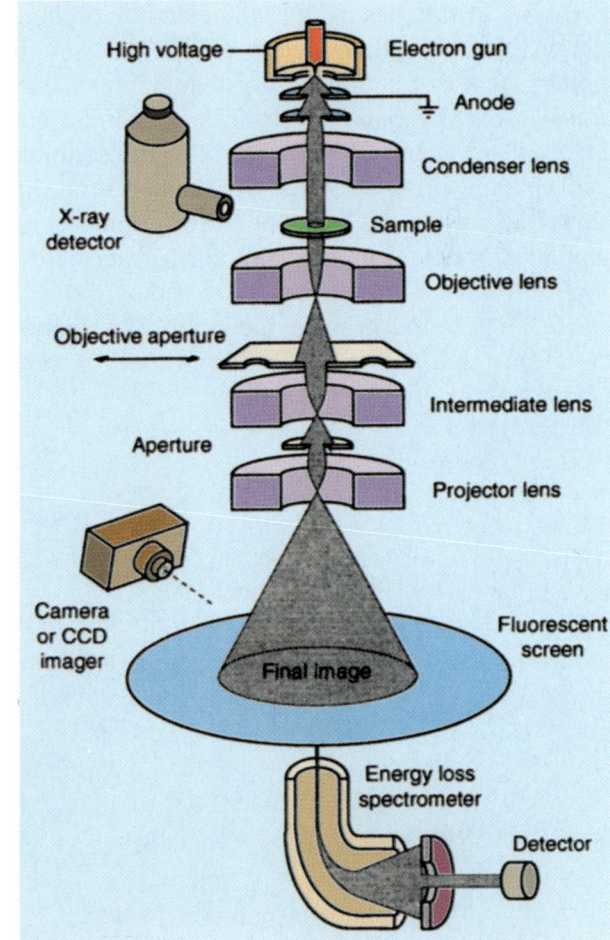

Fig. 25-10 Block diagram of TEM electron-column.

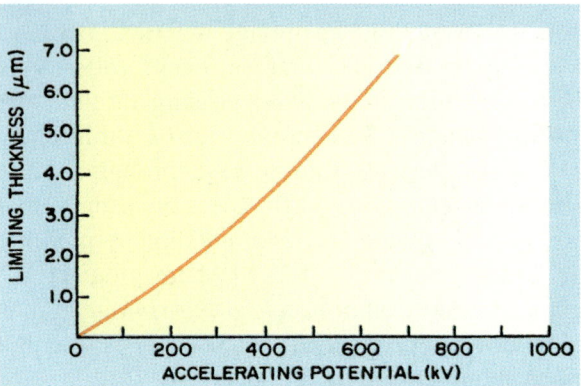

Fig. 25-11 Limiting-thickness of a silicon TEM-sample as a function of accelerating-voltage (potential).

screen. The *contrast* in a TEM-image arises for different reasons in samples of crystalline and amorphous materials. In crystalline-layers, abrupt changes in thickness, phase-structure, or crystallographic-orientation cause corresponding changes in contrast, and these features can be easily imaged at high-resolution. In amorphous-regions, contrast is obtained from differences in sample-thickness or from differences in chemical or phase composition. TEM-images from amorphous-materials (e.g., oxides, nitrides) are well imaged. TEM can also image the grains of polycrystalline-films, and is thus a very-useful tool for measuring grain-sizes in such thin-films.

25.5.1 TEM Sample-Preparation

There are two major factors that have prevented TEM from being more widely used, despite its excellent resolution and analytical capabilities: **1**) difficulties involved with preparing the required very-thin samples; and **2**) correctly interpreting TEM-images.

TEM-sample sections of most interest for VLSI studies are vertical cross-sections. Such samples are prepared as shown in Fig. 25-12.[11] Several hours are required to ion-mill samples to the necessary thickness, making their preparation a tedious task. TEM-samples can also be prepared with the help of focused ion-beams - FIB (see Fig. 25-13 and Sect. 25-14, Fig. 25-25). Several TEM-photographs appear throughout the text illustrating various TEM images and analysis applications (see also Fig. 25-14).

25.6 ATOMIC-FORCE MICROSCOPY

Another, non-optical microscopic-technique has recently been developed for microelectronic-fabrication applications, called *atomic-force microscopy* (AFM). The AFM is useful for measuring local-non-uniformities in film-thickness, such as surface-roughness. Here, an extremely-sharp tip is placed fractions of a nanometer from the sample-surface (Fig. 25-15a). The spacing is close enough for interactions between single-atoms and the tip to occur without

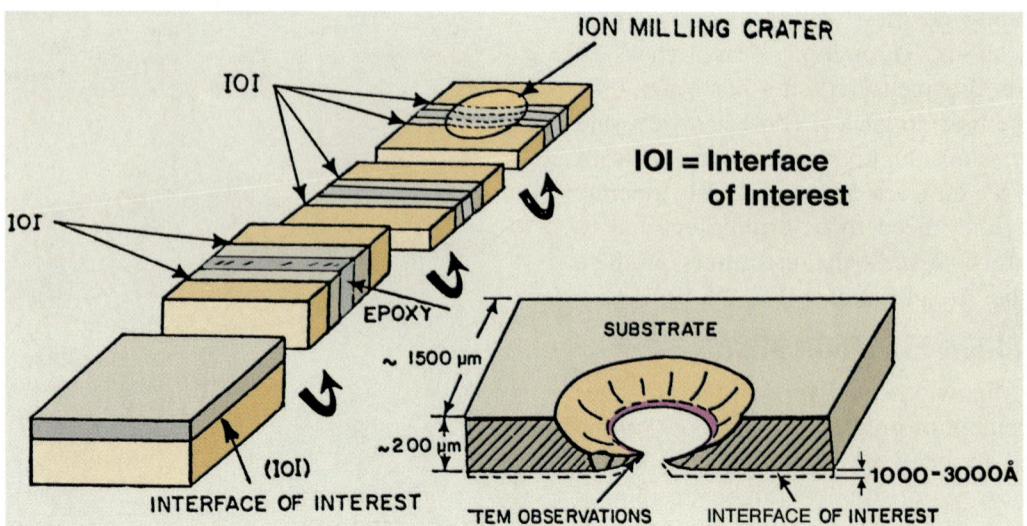

Fig. 25-12 Method for preparing a cross-sectional-sample for TEM study.

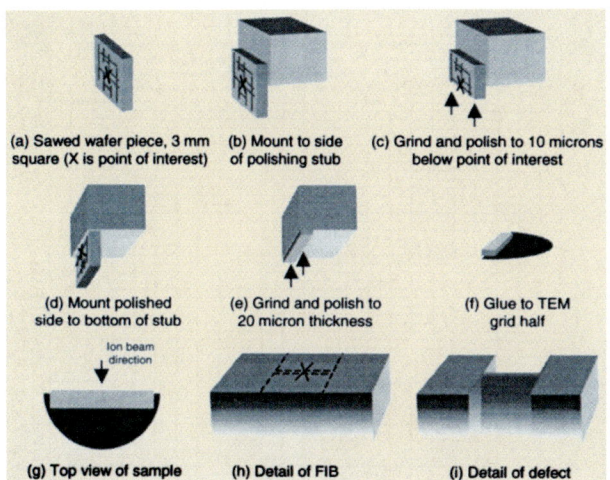

Fig. 25-13 Method for preparing a cross-sectional sample for TEM study with the aid of focused ion-beam-etching.

direct-contact taking place. The result may be a tiny current that flows between the tip and the surface, or a variety of minute-forces which relate to the properties of the surface. The tip is raster-scanned over the sample-surface. Its vertical-deflection is measured very accurately using laser-techniques (Fig. 25-15b), or piezoelectric-sensors. A highly-detailed 3-D-mapping of the surface can be obtained (Fig. 25-15c).

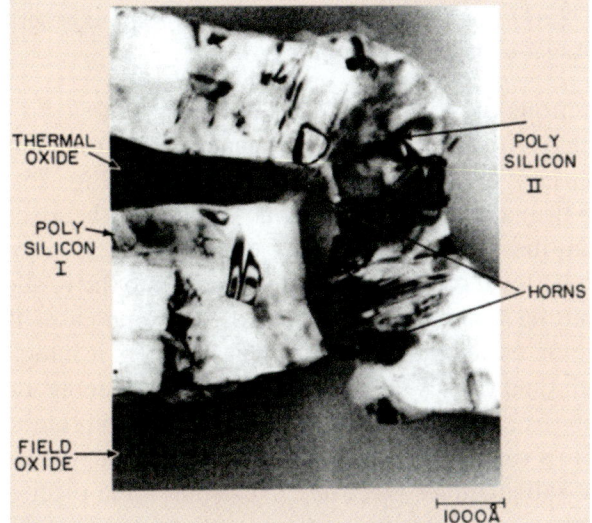

Fig. 25-14 TEM image of a cross-section through a double-polysilicon test-structure showing "horns" which constitute a failure-mode. This defect would be extremely difficult to detect with any other technique.[10]

However, AFM-measurements are very slow, which is a limitation for in-line measurements in a production environment. The chapter Faceplate is another example of an AFM-image.

25.7 ELECTRON/X-RAY COMPOSITIONAL-ANALYSIS TECHNIQUES

When the surface of a solid is struck by electrons

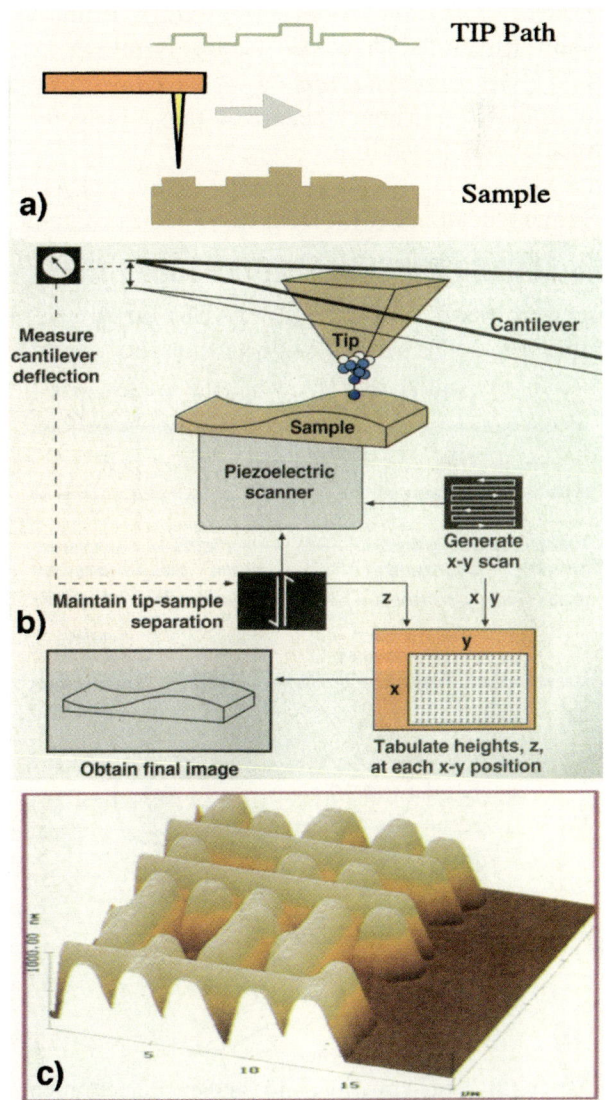

Fig. 25-15 (a) Principle of atomic-force microscopy (AFM). (b) Schematic of how signal detection is performed in AFM. (c) Example of an image of a surface obtained using AFM.

or x-rays, both x-rays and electrons are emitted in response. Some of these emitted-species contain energy-information about elements present at the surface and in the bulk of the bombarded-sample. The four analytical-techniques based on the detection of such emitted x-rays and electrons are (see Fig. 25-1):[12]

1. *Auger-Electron Spectroscopy*: Primary-beam: Electrons; Detected-Species: Auger Electrons.

2. *X-Ray-Emission Spectroscopy* (*XES*) - Primary-beam: Electrons; Detected-Species: X-rays.

3. *X-ray-Photoelectron Spectroscopy* (*XPS* or *ESCA*) - Primary-beam: X-rays; Detected-Species: Photoelectrons.

4. *X-Ray-Fluorescence Spectroscopy* (*XRF*) - Primary-beam: X-rays; Detected-Species: X-rays.

25.8 AUGER-EMISSION SPECTROSCOPY (AES)

Auger-emission spectroscopy (AES) involves the bombarding of a sample with an energetic-beam of

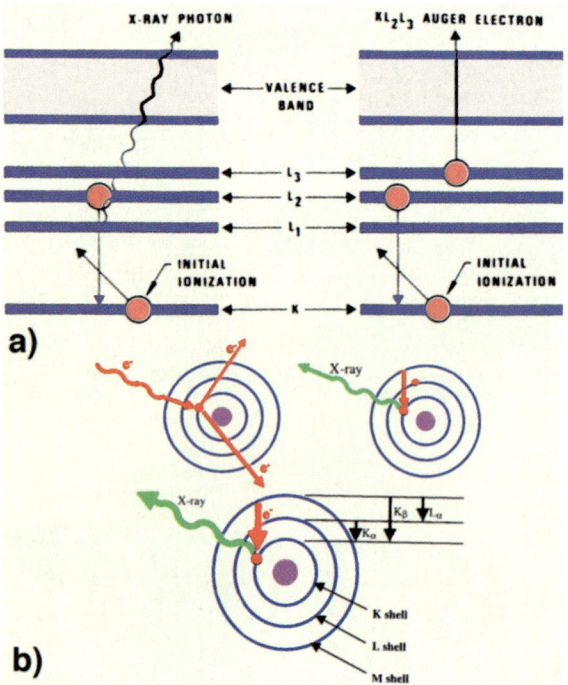

Fig. 25-16 (a) X-ray emission, and (b) Auger electron-emission during de-excitation of an atom after initial ionization caused by electron beam or x-ray bombardment. (b) Schematic diagram employing a picture of the atomic-structure to help illustrate the x-ray-emission process.

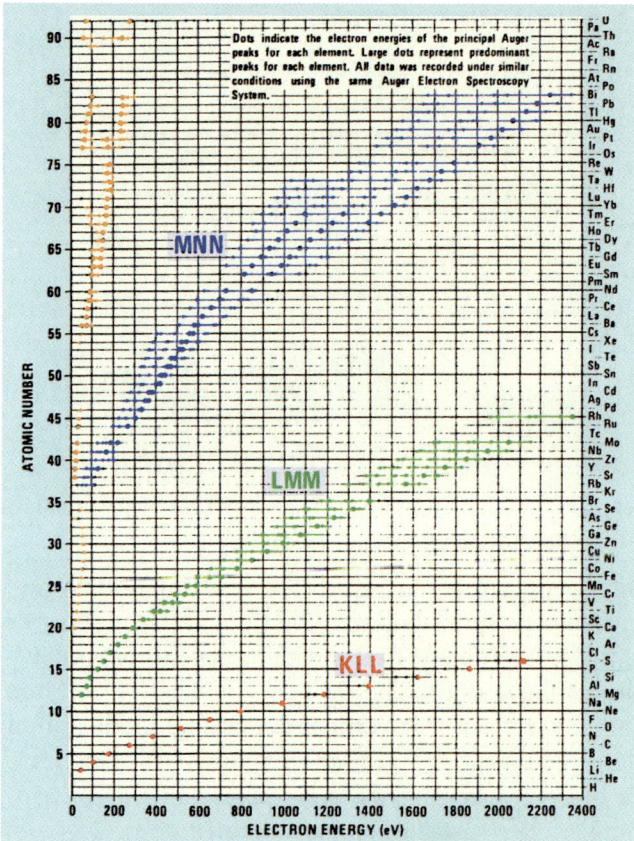

Fig. 25-17 The most prominent *Auger-transitions* observed in Auger-emission spectroscopy (AES).

electrons (up to 10-keV). A certain class of electrons, called *Auger-electrons* are generated (pronounced "ohzhay"). As shown in Fig. 25-16 (and Fig. 25-2), the primary-electron causes an electron to be ejected from the K-shell of a silicon-atom. An electron from the L_2-level of the same atom relaxes into the K-shell, emitting a photon in the process. In some cases this photon escapes the material, but in others it interacts with (and causes the ejection of) a lower-energy electron (from the L_3-level in this case). Electrons emitted from a sample according to such a sequence of events are called *Auger-electrons*. Note that three electrons of the sample-atom must be involved to create one Auger-electron. In this example, one was from the K-level, one from the L_2-level and another from the L_3-level. Accordingly, this Auger-electron is identified as a *KLL Auger-electron*. The energy of the emitted Auger-electrons is thus characteristic of the type of atom from which they came. There are several such characteristic Auger-transitions for most elements (Fig. 25-17). Note that since three electrons must be involved, elements with fewer than three electrons (i.e., H and He) cannot emit Auger-electrons, and hence cannot be detected with AES.

Most Auger-electron energies are between 20 and 2000-eV. The depth from which they can escape from the solid without losing a significant percentage of this energy (i.e., the specific energy-value that serves to identify their origin) is quite shallow - less than 50Å. Thus, AES is a technique that provides compositional-data only about the *surface-layers* of samples.

Many diagnostic-problems require information about the composition of the material at depths greater than the escape-depth of Auger-electrons. To obtain this information, AES data is taken from the bottom of a crater, ion-sputtered into the surface. This milling-process is stopped at regular intervals, during which an Auger-spectrum is taken (Fig. 25-18a). The Auger-peak-heights can be plotted as a function of milling-time and a depth-profile can thus be obtained.

The chief advantage of AES is its ability to provide excellent lateral-resolution (and consequent analysis of very-small areas of 1-μm^2 or less), together with acceptable sensitivity (~1% atomic). This can be combined with raster-scanning and ion-sputtering to yield three-dimensional maps of the distribution of the element of interest.

In the discussion of SEMs, it was noted that electrons emitted from solids struck by high-energy primary electron-beams have a wide energy-spectrum, and include secondary-electrons and backscattered-electrons (as well as Auger-electrons). The presence of Auger-electrons is manifested as small peaks in the total energy-distribution-function (see Fig. 25-8). All of these electrons can be collected, but in AES-analysis only the Auger-electrons are of interest. The remaining emitted-electrons constitute a strong, unwanted noise-source. To extract valid signal-information about emitted Auger-electrons from raw data, signal-processing methods must be performed to enhance the Auger-electron peaks. This signal-processing is performed with *lock-in amplifiers*. However, even with such enhancement, the presence of the background electron-signal places a lower bound on the sensitivity-limits of AES. Typically AES can only detect elements if their concentrations exceed 0.1-1% at the sample-surface.

As an example of how an Auger-spectrum is interpreted, consider Fig. 25-17b, which shows an Auger-spectrum of a P-doped SiO_2-film. The large peak in the spectrum at ~1600-eV is attributed to the *KLL*-Auger-transition of Si (see Fig. 25-17). Of the two peaks near 100-eV, the larger corresponds to the *LMM*-Auger-transition of Si, while the smaller peak (of slightly-higher energy) is due to the *LMM*-Auger-transition of phosphorus. The 500-eV peak is associated with the *KLL*-Auger-transition of oxygen.

The emission of Auger-electrons and x-rays from a solid are competitive-processes, with Auger-emission being dominant for low-Z (Z = atomic-number) elements, and x-ray-emission dominating in high-Z elements (Fig. 25-19). At Z equal to 33 (As) the probability is equal for both processes. As a result, AES is somewhat better at detecting low-Z elements.

AES suffers from the following limitations: **a)** some sample-surfaces (especially insulators) are prone to charging when struck by an electron-beam; **b)** some surfaces can be damaged from the high-energy primary electron-beam, especially organic-

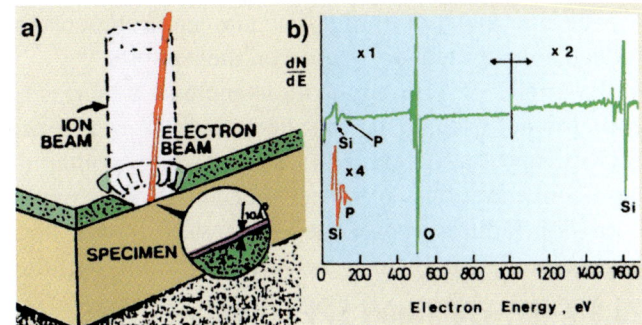

Fig. 25-18 (a) Ion milling a crater and performing AES on the exposed surface, thus allowing depth-profiling with AES. (b) *Auger-spectrum* of a P-doped SiO_2 film.

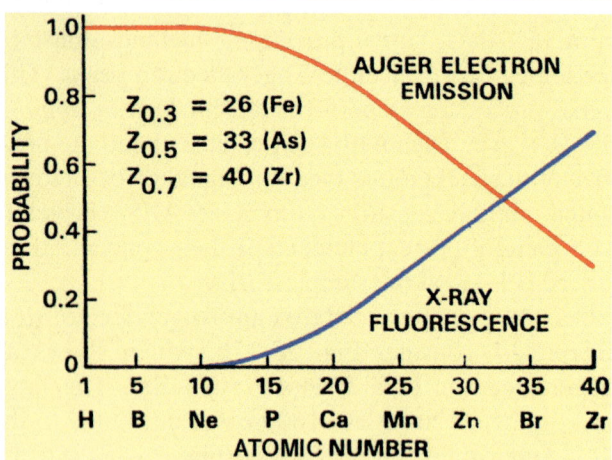

Fig. 25-19 Probability of X-ray-emission and Auger-electron-emission versus atomic-number.

materials; **c)** if depth-profiling is necessary, the ion-sputtering will destroy the sample in the local-area of the sample being examined; **d)** *matrix-effects* can occur. (The latter are signal-alterations which are observed when some elements are present in particular matrices. Some of these matrix-signal-shifts are well understood and can supply information about the chemical-state of the atoms present); **e)** the detection-limits depend on the area being examined, the time-duration of the acquiring-scan, and energy-range being scanned. For example, the larger the area the better the detectability. That is, 30-minute scans are more effective than 5-minute scans, and narrow-energy-range scans can detect lower-concentrations than wide-energy-range scans; and **f)** AES must be carried-out in an ultrahigh-vacuum environment to reduce contaminant-formation on the sample.

Examples of AES-applications include: **a)** surface-contamination-analysis; **b)** detection of very-thin SiO_2-layers (native-oxides); **c)** detecting contaminant-concentrations in barrier-metals; **d)** analysis of corrosion-failures in packaged integrated circuits; and **e)** determining P, B, As concentrations in SiO_2-films.

25.9 X-RAY-EMISSION SPECTROSCOPY (XES)

XES is a technique in which the sample is struck by an electron-beam, and the x-rays emitted in response are analyzed. The emitted x-ray-spectrum contains peaks characteristic of the sample, as well as a background-continuum. As x-rays are not readily absorbed by most materials, they are emitted from the material up to the depth to which the primary-electrons penetrate, see Fig. 25-20. (Note that some materials, such as lead, strongly absorb x-rays, making it possible to produce effective x-ray-shielding.) Hence, XES is not strictly a surface-analysis technique. In addition, the depth of the material that emits x-rays cannot be determined within the same resolution as AES or SIMs. As discussed in the AES section, the electron-beam spreads as it penetrates the sample, hence lateral-resolution is also much lower than in AES, even if the primary electron-beam has the same diameter.

Even though XES is generally not as useful as AES or SIMS, it is widely used because SEMs can be easily and inexpensively equipped with suitable detectors to allow XES-analysis. Since SEMs are commonly available at ULSI-fab-facilities, XES is also likely to serve as an "in-house" characterization-capability.

Two types of detectors are used for XES-studies (Fig. 25-21): **1)** *energy-dispersive x-ray-detectors* (*EDX*), and **2)** *wavelength-dispersive x-ray-detectors* (*WDX*). EDX-detectors convert an x-ray-photon to a voltage-pulse, and a thin-film of beryllium is usually used to provide vacuum-isolation from the SEM. Since the beryllium-window also absorbs low-energy

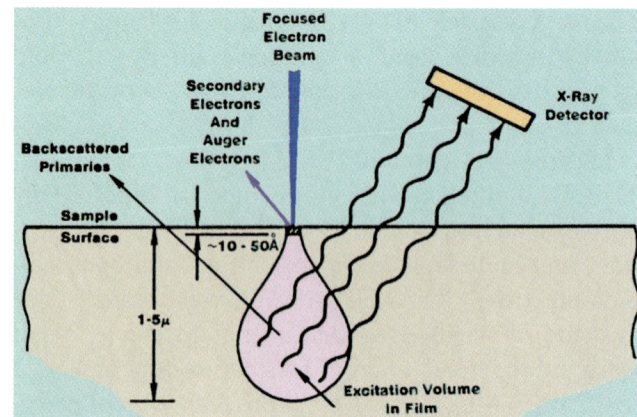

Fig. 25-20 Principle of EDX. This drawing also shows the spreading of the primary electron-beam below the specimen surface, showing the large volume from which backscattered-electrons can emanate.

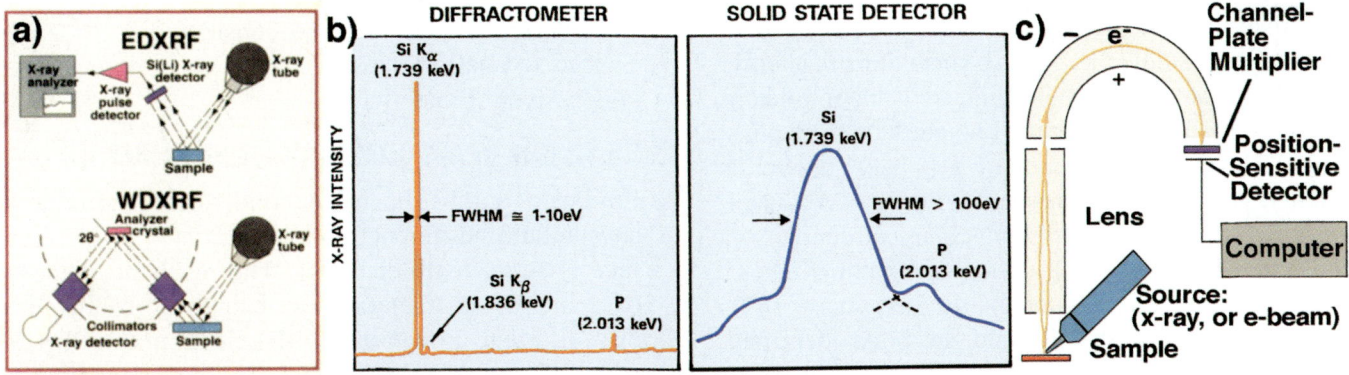

Fig. 25-21 (a) Principles of EDX and WDX. (b) Comparing energy-resolution achievable with energy-dispersive versus wavelength-dispersive analysis (EDX vs. WDX). (c) Detection-system used in EDX and WDX.

x-rays, it effectively prevents x-ray peaks from low-Z-elements (fluorine and below) from reaching the EDX-detector. Thus, EDX cannot detect such low-Z elements. A second drawback of EDX is its limited ability to distinguish between two adjacent x-ray-peaks. For example, since the resolution of EDX is about 150-eV, the K_α x-ray-peak from Si (at 1.74-keV) and the M_α x-ray-peak from Ta (at 1.71-keV) can not be distinguished from one another, since the energy-separation is only 30-eV. Thus, it would be very difficult to analyze the chemical-composition of a tantalum-silicide ($TaSi_x$) film with EDX.

On the other hand, WDX-detectors are used to detect one x-ray-peak at a time. They are crystal-analyzers that are sensitive to the Bragg-angle of the emitted x-rays. Thus, while WDX is more tedious to operate, its resolution is 5-10-eV. It also does not need to be isolated from vacuum, allowing it to be used to identify low-Z elements (from boron upwards).

SEMs can be equipped with both EDX/WDX-detectors, and having both of them available increases the characterization-capability of the instrument (Fig. 25-20c). The detection-limits for EDX/WDX are 0.3-1% (about the same as for AES and ESCA). XES is routinely used in such applications as finding the P-concentration in PSG, or the Cu-content of Al-alloys.

25.10 X-RAY-PHOTOELECTRON SPECTROSCOPY (XPS, ESCA)

Just as electron-bombardment of materials can produce emitted-electrons and x-rays, striking the material with x-rays can cause the same effects.[13] The technique of *XPS* (also known as *electron-spectroscopy for chemical-analysis*, or *ESCA*), uses low-energy x-rays (such as the K_α-line of aluminum, which has an energy of 1.487-keV), to cause photo-electron-emission. The emission of photoelectrons in XPS differs from Auger-emission in that it is the electron "knocked-out" of the atom that is analyzed in XPS. Therefore, the incident-x-ray-energy must be both monochromatic and have an accurately-known intensity. For these reasons the K_α-x-rays of Al and Mg are chosen for commercial XPS-systems.

As in AES, only those electrons from the top 1-10 atomic-layers of the film are emitted without significant energy-loss from collisions. Thus, even though the primary x-ray-beam penetrates deep into the sample-material, XPS is a *surface-analysis technique*. Electrons are detected in the same manner as in AES, and the collected-data (in the form of ESCA-scans), provides much the same information as AES. In fact, neither technique can detect H, and their detection-limits are comparable. Lateral spatial-resolution is relatively poor since the smallest x-ray beam-diameters are about 150-μm. However, such x-ray-beams allow a rapid XPS-depth-profile to be obtained, since ion-beams can easily sputter areas of that magnitude.

XPS is often used as a complement to AES-analysis, over which it has the following advantages: **a)** Some materials that dissociate and desorb from a surface when bombarded by electrons, are often undisturbed by x-rays. Thus, such materials can be non-destructively studied by XPS; **b)** Insulators that suffer charging-problems when irradiated by electron-beams are more easily characterized by neutral x-ray-beams; and **c)** The energy-resolution of XPS peaks due to the emitted-photoelectrons, is sharper than for AES (typically 0.5-eV). Since different chemical-bonds in a molecular-structure cause shifts in the binding-energy of atomic-electrons greater than 0.5-eV, these shifts can be detected with XPS and the bond identified. Thus, XPS can be used to obtain information about chemical-bonding (giving rise to the term, *ESCA*).

25.11 X-RAY FLUORESCENCE (XRF)

In *x-ray fluorescence* (*XRF*), both the primary-beam and the detected-signal are x-rays (Fig. 25-22). Methods for detecting and analyzing the emitted-x-rays are similar to those used in XES.

XRF, however, has some advantages and limitations relative to XES. One advantage of XRF is that it can be used to analyze layers that would either be electrically-charged or would decompose if bombarded by the electron-beam of XES (e.g., oxide or polymer-films). On the other hand, the large XRF-beam (150-μm to 1-mm) prevents it from being used to analyze the small-features of ULSI-circuits. In addition, the primary x-ray-beam penetrates deeper into the sample than the electron-beam of XES. This causes emission from material deeper into the substrate than XES. These effects must be considered when XRF is used to analyze substrates that have more than a single-layer of material.

25.12 ION-BEAM COMPOSITIONAL ANALYSIS

Ion-beam bombardment of solid-surfaces produces substantially different interactions than if the surface is struck with electrons. The resultant methods of materials-characterization using ion-beams are also different. In contrast to the electron/x-ray-based techniques (which rely only on the *unique electronic-structures of each element* for identification), ion-beam techniques identify the elements present through their *atomic-mass values*.

When-ion beams in the 1-30-keV energy-range are used to strike a sample-surface, and the sputtered positive or negative secondary-ions are analyzed using a mass-spectrometer, the technique is known as *secondary-ion-mass-spectroscopy* (*SIMS*). *Laser-ion-mass-spectroscopy* (*LIMS*), although not ion-beam-excited, is a related technique and is also covered.

25.13 SECONDARY-ION MASS-SPECTROSCOPY (SIMS)

In the SIMS technique, the bombardment of a material with energetic-ions (1-20-keV) causes *billiard-ball-like* collisions with atoms of the surface, leading to their ejection from the material (a process called *sputtering*, see Chap. 15).[14] The ions of the primary-beam are produced using a glow-discharge to create either O⁻ or Cs⁺ ions. The O⁻ or Cs⁺ ions (depending on the sample being analyzed) are then extracted and focused into a small spot with a magnet. This ion-beam is directed at the sample, causing sputtering.

A small-fraction of the atoms sputtered from the sample leave as secondary-ions. Over 90% of them come from the outer-two atomic-layers of the sample-surface. These sputtered-ions are collected by a mass-spectrometer for mass-to-charge separation and detection (Fig. 25-23a). The number of ions collected can also be digitally-counted to produce quantitative-data on the sample-composition. Thus, SIMS only analyzes the material removed by sputtering from a

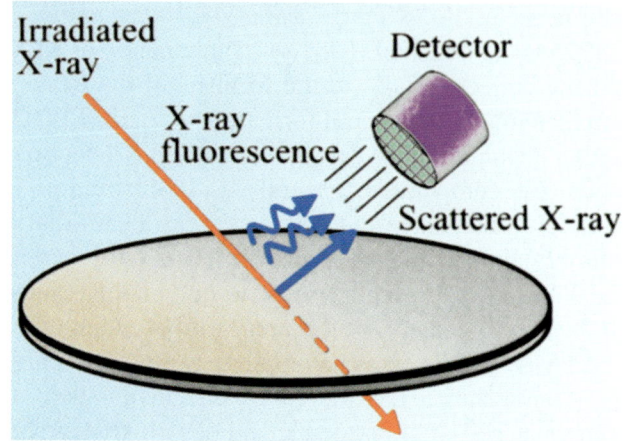

Fig. 25-22 Principle of *x-ray fluorescence spectroscopy* (*XRF*).

sample-surface - in contrast to AES, which analyzes the atomic-layers closest to the surface without substantial layer-removal.

The sputtering-process continuously removes surface-atoms. Therefore, the shallow analytical-zone advances into the sample as a function of sputtering-time. By monitoring the secondary-ion signals with time, a depth-profile can be produced. Sputter-rates of 2-5-Å/sec, at data-acquisition times of 10-sec, produce typical depth-increments (vertical-resolution) in the 20-50-Å range. Layers of up to 10,000-Å-thick can be depth-profiled with SIMS (Fig 25-22b). Usually the incident-beam is raster-scanned over a small area of the surface to create a crater with a nearly-flat bottom. Mass-analysis is only performed on the ionic-fraction of sputtered-material from the center of the crater.

SIMS has several unique capabilities that make it highly-useful for characterization-applications. First, it is capable of detecting *all* the elements (whereas AES and ESCA cannot detect H and He). Second, it can identify elements present in very low-concentrations. In fact, it is the only *surface-analysis-technique* with the ability to measure doping-level-concentrations in electronic-materials. It is especially sensitive to elements with low-ionization-potentials (e.g., Na, K), as well as elements with favorable electron-affinity (Group V & VI elements). For applications involving the detection of dopants and contaminants in Si, Table 25-3 lists the SIMS sensitivity-limits. These show that SIMS is an excellent-tool for generating concentration-profiles of dopants in Si at levels down to the 1×10^{15} cm^{-3} range (i.e., concentration and depth of dopants can be measured simultaneously).

Like all other analytical-methods, however, SIMS also has its drawbacks. First, the range of beam-diameters of SIMS is 1.0-200-μm, but maximum-sensitivity is achieved when wider-beams are used (e.g., 100-μm). As the beam is focused to a smaller spot, the sensitivity is correspondingly reduced (because fewer total-atoms are sputtered from the surface by the smaller-beam). In addition, SIMS is locally-destructive, and is subject to charging, especially when analyzing dielectric-layers.

TABLE 25-3 PARAMETERS OF SIMS RELEVANT TO VLSI MATERIALS ANALYSIS

Element (in Si matrix)	Primary Beam	Detected Element	Minimum Detectable Conc. (atoms/cm^3)
Arsenic	Cs$^+$	^{75}As$^-$	5×10^{14}
Phosphorus	Cs$^+$	^{31}P$^\pm$	5×10^{15}
Boron	O$_2^+$, O$^-$	^{11}B$^+$	1×10^{13}
Oxygen	Cs$^+$	^{16}O$^-$	1×10^{17}
Hydrogen	Cs$^+$	^{1}H$^-$	5×10^{18}

25.13.1 Laser-Ionization-Mass-Spectroscopy (LIMS) and Time-of-Flight SIMS (TOF)

LIMS is an analytical-technique in which a narrow laser-beam is used to remove material from a sample in one of two modes: **1)** laser-desorption (LD); and **2)** laser-ionization (LI). In both modes, ions produced are analyzed by a mass-spectrometer of the *time-of-flight* design, in which the transit-time of vaporized-ions accelerated across a high-vacuum region by a known electric-field determines their mass.

Laser-desorption produces ions only from substances adsorbed on the surface, and hence uses a lower-power laser. In *laser-ionization*, ions are produced during vaporization of the sample-surface. The ability to distinguish between adsorbed and bulk materials, and materials incorporated in the matrix, is a valuable characteristic of LIMS. Furthermore, another benefit of LIMS is that it can provide a mass-spectrum from

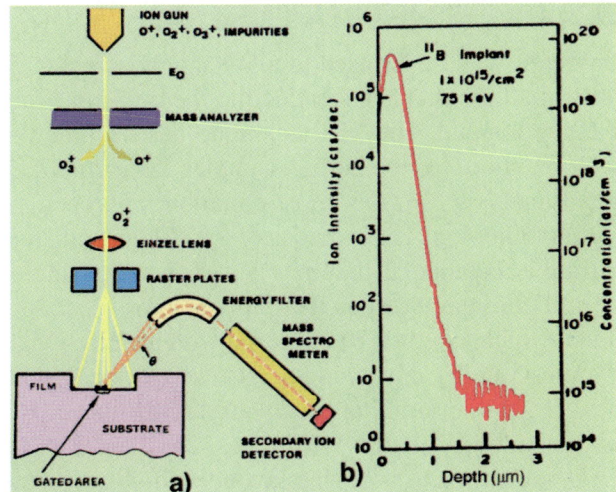

Fig. 25-23 (a) Schematic diagram of a secondary-ion mass-spectrometer. (b) Typical SIMS depth-profile for boron into boron-doped silicon.

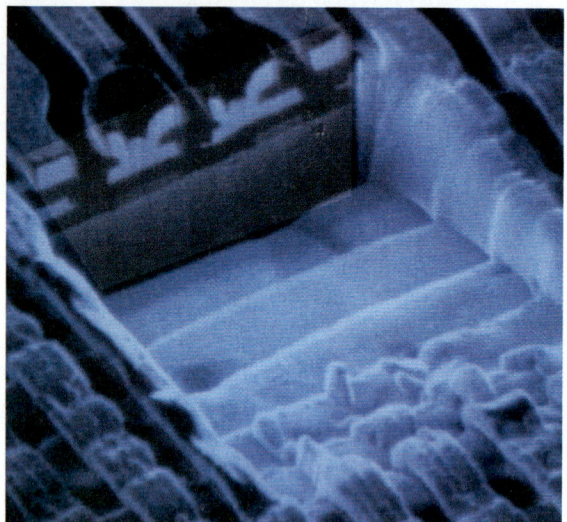

Fig. 25-24 SEM photograph of sample eroded by FIB to reveal the device cross-section. Courtesy of FEI Company.

a volume of material as small as 1-2-μm in diameter, and several monolayers thick (using LD).

Time-of-Flight SIMS (*TOF-SIMS*) uses an FIB-source, so it is capable of spatial-resolution below 0.1-μm. It also has a high-sensitivity and the ability to detect very-high-amu elements. TOF-SIMS uses the same approach to determining the mass of the elements in the sample as LIMS.

25.14 FOCUSED-ION-BEAM ANALYSIS

Ion-beams can be focused to allow sputter-etching of very-small areas. This is called the *focused-ion-beam* (*FIB*) technique. The FIB-technique has become the main method for preparing samples for cross-sectional-analysis. In the most-common approach, a beam of Ga$^+$-ions is rastered across a designated area, milling a rectangular-hole into a sample. One side of the milled-area becomes the exposed cross-sectional plane, which can then be observed with a SEM. Dual-column-systems are available which can mill with the FIB, while simultaneously observing the cross-section sidewall with a SEM (Figs. 25-1 and 25-24). CAD navigation uses device-layouts to quickly-locate an area to be sectioned. Improvements in ion-beam resolution enable the sectioning of submicron-features with precision and repeatability not achievable with mechanical-scanning sections. The FIB progressively-sections through a feature, efficiently providing a succession of views. This FIB-technique is replacing the traditional method of cleaving and polishing a wafer, and then analyzing its cross-section using an SEM.

However, FIB-sectioning can also be combined with cleaving or polishing. This is used when the sectioning-precision of the FIB is required, but the final-imaging needs to be a true-perpendicular-view of the cross-section. In this approach, the sample is cleaved or polished to within a few microns of the point of interest. FIB then mills out the material to the desired location. Using a similar approach, FIB can aid in TEM-sample-preparation, as shown in Fig. 25-25.

REFERENCES

1. *Handbook of Semiconductor Silicon Metrology*, A.C. Diebold, Ed., Marcel Dekker, New York, 2001.

2. *Encyclopedia of Materials Characterization*, Butterworth-Heinemann, Boston, 1992.

3. D.K. Schroder, *Semiconductor Material and Device*

Fig. 25-25 SEM photograph of a sample prepared for TEM with the aid of FIB. Courtesy of FEI Company.

Characterization, 2nd. Ed., McGraw-Hill, NY, 1998.

4. W.R. Runyan & T.J. Schaffner, *Semiconductor Measurements and Instrumentation*, McGraw-Hill, NY, 1998.

5. T.J. Schaffner and D.K. Schroder, "Electrical, Physical, and Chemical Characterization," in *Handbook of Semiconductor Maunufacturing Technology*, Y. Nishi & R. Doering, Eds., Marcel Dekker, NY, 2000, Ch. 28, p. 889.

6. *ibid.*, L. Wagner, "Failure Analysis," Ch. 29, p. 937.

7. *Electron Microscopy Society of America* (EMSA) 1998.

8. M. Gill and E. Woster, "The In-Fab SEM/EDX Integration Challenge," *Semiconductor International*, 1993, p. 78.

9. D.B. Williams and C.B. Carter, *Transmission Electron Microscopy*, Plenum Press New York, 1996.

10. R.B. Marcus and T.T. Sheng, *Transmission Electron Microscopy of Silicon VLSI Circuits and Structures*, John Wiley & Sons, New York, 1983.

11. *Specimen Preparation for Transmission Electron Microscopy of Materials - IV*, Materials Research Society, Warrendale PA, 1997.

12. D. Briggs and M.P. Sheah, "Practical Surface Analysis by Auger and X-Ray Photoelectron Spectroscopy," John Wiley & Sons, New York, 1984.

13. R. Jenkins, R.W. Gould, D. Gedke, *Quantitiative X-Ray Spectrometry*, Marcel Dekker, New York, 1997.

14. *Secondary Ion Mass Spectroscopy*, SIMS X, John Wiley & Sons, New York, 1994.

15. M. Davidson, A. Vladar, "The Physics of Metrology Instruments," *Solid State Technology*, June 1998, p. 135.

PROBLEMS

1. Explain the difference between operating a microscope in the *bright-field* and the *dark-field* mode.

2. Explain the basic operational principles of a SEM. What are the differences between a *failure-analysis SEM*, and a *CD-SEM*?

3. Explain the operational principles of TEM.

4. An AES characterization-tool must be operated under ultra-high vacuum conditions to prevent the formation of undesired atomic-layers on the surface of the sample being examined. What pressure is required if formation of a monolayer of contamination can be permitted after the sample has been in the sample-chamber for no less than 4-hours? (see Prob. 7, Ch. 6).

5. Why must surface-analysis compositional-tools be equipped with the capability of pumping the sample-analysis chamber to *very-high* or even *ultra-high* vacuum conditions? If every oxygen-atom that struck a surface were to stick to it, how long would it take to form a monolayer of oxygen, if it is present in the sample-analysis chamber at a pressure of 3×10^{-8} torr?

6. Why are optical-microscopes designed with a maximum-magnification of 1000X?

7. Explain the difference between the principles of how contrast arises in the images obtained in a SEM and in a TEM. Why is TEM able to achieve higher resolution than SEM?

8. Describe the difference between *cross-sectional* and *aerial* TEM, and list two applications in which each might find use.

9. In a SIMS-profile of an ion-implantation process, the peak-intensity is 10^5 counts/sec, and it occurs at 100-sec. The ion-implantation energy was 100-keV and the dose was 1×10^{15} cm^{-2}. If an ion-implantation with a dose of 4×10^{16}-ions/cm^2 is performed at an energy of 180-keV, estimate the peak-intensity and the time at which this intensity is reached.

10. How are depth-profiles calibrated in AES compositional analysis?

11. Describe the difference between the species that are detected in AES and XPS (ESCA). Can one of these techniques yield information about the materials being analyzed that the other cannot? If so, which one, and what is this information?

12. Explain the events that occur that gives rise to the emission of an *Auger-electron* by an atom. Give some explanation as to why AES is better suited to detecting lighter elements. Why can't AES or XES detect the presence of hydrogen or helium?

13. Explain the basic principles of *atomic-force microscopy* (AFM).

CHAPTER 26

WAFER AND CHIP TESTING

CHAPTER CONTENTS

26.1 THE PURPOSE AND PHILOSOPHY OF TESTING INTEGRATED CIRCUITS

26.2 IN-LINE PARAMETRIC TESTING

26.3 WAFER-SORT

26.4 BURN-IN

26.5 FINAL FUNCTIONAL TESTING

26.6 BINNING

26.7 MARK, PACK, AND SHIP

"Pray to God,
   But keep rowing for shore."
           Old Russian Proverb

Illustration of a probe test-card.

Semiconductor-devices are tested to verify that they meet design-specifications. In addition to checking that each chip performs properly, defect-data obtained from testing can be used by production-teams to correct fabrication-problems that caused the defects. Thus, IC-testing is an integral and important part of the overall manufacturing-process. The sequence of testing employed in wafer-fabrication is illustrated in Fig. 26-1.

26.1 PURPOSE & PHILOSOPHY OF TESTING INTEGRATED-CIRCUITS

ICs require testing because semiconductor-manufacturing processes (especially at the leading-edge of technology) produce a significant number of defective-parts. If all steps in wafer-fabrication were done correctly and all the materials used in chip-production were perfect, there would be no need for chip-testing. But such circumstances are highly unlikely, even in today's advanced wafer-fabs. Thus, unless appropriate tests are performed, defective-parts will find their way into final-products, and these will exhibit poor-quality. Customers have a quality-expectation for products they purchase, and they also want acceptable performance at a competitive-price. Testing helps ICs meet these criteria in two ways:

1. It monitors the effectiveness of each process-step, and checks that correct process-integration was accomplished. Test-data is used by production-teams to perform defect-reduction activities by analyzing

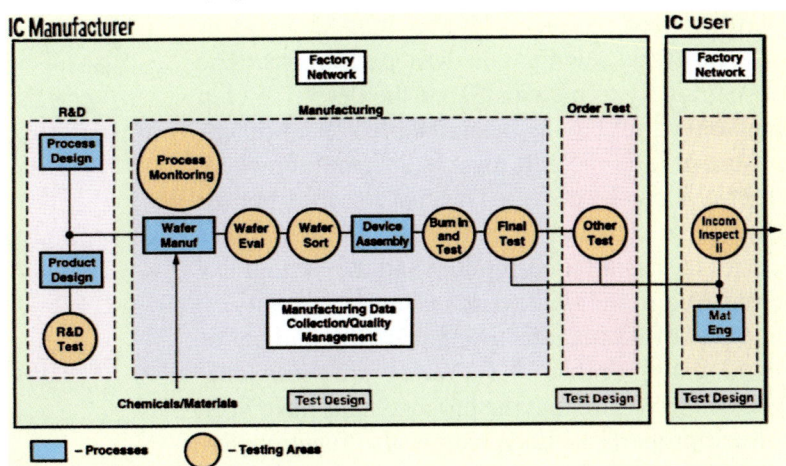

Fig. 26-1 Tests performed on ICs as they go through a manufacturing sequence.

476 MICROCHIP MANUFACTURING

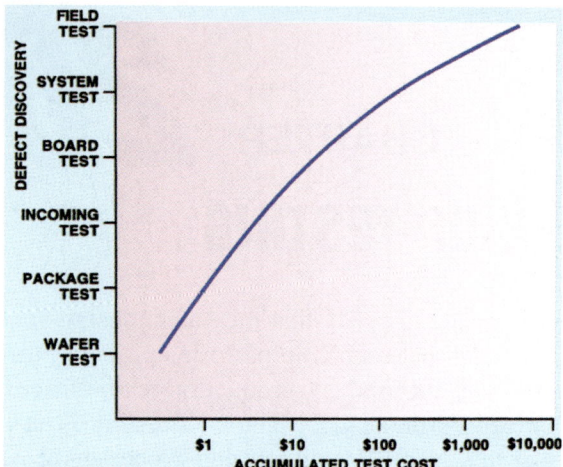

Fig. 26-2 *Accumulated test-costs* versus *defect-discovery*.

the data to determine the source of fabrication-problems. Actions can then be taken to correct them.

2. It identifies "good" (functional and reliable) ICs, as well as those that are "bad."

It is important to identify defective-ICs as early in the manufacturing-cycle of an electronic-product as possible. This is because the cost of detecting (and perhaps replacing) a defective-part increases dramatically at each stage of manufacturing an electronic-system (see Fig. 26-2). For example, detecting a bad-IC at the wafer-level may cost the chip-manufacturer only $0.10 per chip (i.e., just the cost of testing). The cost of detecting a failed-chip may rise to $1.00 if the part has been packaged (due to the added cost of assembly and packaging that was performed). The cost rises rapidly to perhaps $10 if the defective-chip is found only after having been mounted on a printed-circuit-board (i.e., the part must be located, removed, and replaced on the board). The cost rises further (to $100) at the electronics-system-level, or even higher if a customer has a failed-product and the system has to be repaired in the field (perhaps under warranty).

As a part of IC-manufacturing there are number of stages of testing (Table 26-1). Wafers are *parametrically-tested* to ensure that the process-steps have been performed properly.[1-3] Each chip is also functionally-tested twice. The first such test, called *wafer-sort* is done prior to sawing the wafer, and each chip is probed (using a custom *probe-card*). Electrical-signals are applied to the chip through the probes which contact the wire-bonding-area of each pad. The second such test, called *final-test* is done after the chips have been packaged, and the test-signals are then applied using the package-leads.

26.2 IN-LINE PARAMETRIC TESTING

After each process-step is completed, measurements are done to verify that it was done properly. These are referred to as *in-line parametric-tests*. At the completion of the wafer-processing-sequence, a *final-parametric-test* is carried out on each wafer to ensure that the whole process-flow meets the completed-wafer manufacturing-specifications. (In fact, in advanced ICs that have 4 or more levels-of-metallization, this *wafer-level parametric-test* is now carried out after the 1st-level of metal has been patterned. This allows only those wafers that pass this test to undergo the remainder of the metallization-process-steps.) Note that this is the first time the entire-wafer undergoes a comprehensive test.

The final parametric-test provides several kinds of useful data. First, wafers that fail the parametric-test

Table 26-1 Electrical-Tests for Integrated-Circuits

Test	Test Description
IC-Design-Verification	Characterize, debug, and verify new chip-designs.
In-Line Parametric-Test	Production-process verification-tests, used to monitor a process after each run.
Wafer-Sort (Probe)	Product functional-test to determine which die on a wafer function properly.
Burn-In (Reliability)	Die that pass Wafer-Sort are packaged and powered-up & tested at elevated temperatures to find early-failing chips.
Final-Test	Packaged-chips that pass *Burn-In* are tested for functionality, using product specifications

criteria are not processed further (since the likelihood of getting working parts from that wafer may low, or nil). This saves expending production-resources for naught. Second, parametric-test data gives warnings about fabrication-problems. It supplies information about how the wafers were processed and where there might be flaws in the process-flow. Yield-management teams can use such data to find the relationship (called a *correlation*) between the different parameters and the finished-product specification. This information can be used to optimize process-conditions.

Both the *in-line-parametric tests* and *final-parametric tests* are performed on special *test-structures*, designed and fabricated on the wafer for this purpose. That is, such *process-control-monitors* (*PCMs*) are test-patterns used to provide areas on a wafer for probing the electrical-characteristics after various process-steps. Such special test-structures are used because they avoid damaging the actual production-die in the course of parametric-testing.

Originally, PCMs were located within the *product-die* area. (That is, a product-die would be replaced with a PCM at a number of sites on the wafer, Fig. 26-3, and they were thus referred to as *test drop-ins*.) Today, most PCMs are placed in the scribe-line regions between the individual-die, and these are often called *scribe-line monitors*. This reserves the rest of the wafer-area just for product-die (Fig. 26-4).

To perform the tests, a probe-station lowers a ring of very fine needle-sharp probes (Fig. 26-6) into contact with the metal-pads on the test-die (or test-circuits in the scribe-lines). Test-equipment is connected

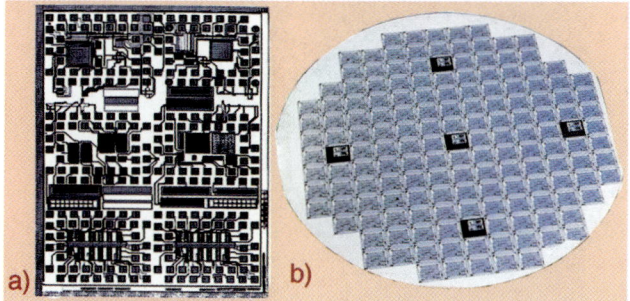

Fig. 26-3 Test *drop-ins* (a) Example of test-circuits used on *drop-ins*. (b) Wafer with *product-die* and *test-drop-ins*.

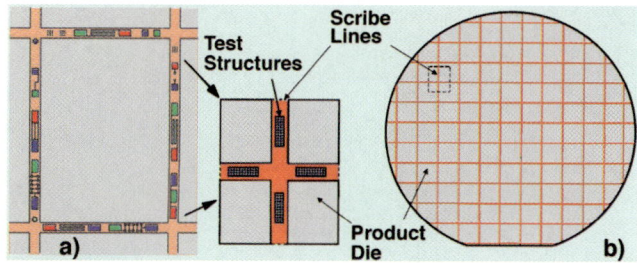

Fig. 26-4 (a) Example of parametric-test circuits fabricated in the scribe-lines of a wafer. (b) Wafer with parametric-test-structures within the scribe-lines.

to the test-structures and is controlled by a computer system. If the wafer-screening operation shows that the basic process and device parameters are within specification, the wafer is returned to the fab-line to fabricate the multilevel-metallization structures.

Parametric-test instrumentation are systems that automatically interface test-structures on a wafer to sophisticated hardware and software that performs the electrical-tests (Fig. 26-5). These parametric test-systems have the following subsystems: **1**) a probe-card-interface; **2**) a wafer-positioning-subsystem; **3**) tester-instrumentation (including power-supplies, meters that measure voltage, current, and capacitance, and a matrix of solid-state-switches that allow the tester to be configured flexibly to perform the desired tests); and **4**) a computer which directs the operations of the test-system.[4] A *test-algorithm* controls the test-instrumentation to perform measurements. This algorithm is written by the test-engineers for specific test-structures.[6] The tester also stores the test-results. Modern testers are generally connected to a network, allowing data-transfer to, and control-from, a host-computer.

Examples of the type of measurements made during the final parametric-test include: **a**) gate-oxide breakdown-voltage; **b**) drain-to-substrate breakdown and leakage; **c**) drain-current for 0-V and a specified "on-voltage" applied to the gate; **d**) MOSFET-transconductance at a specified operating-voltage; **e**) continuity and bridging in serpentine-structures over oxide-steps; **f**) contact-resistance of contact-hole and via-strings; and **g**) the capacitance of array-structures containing insulator-materials and gate-oxide-films.

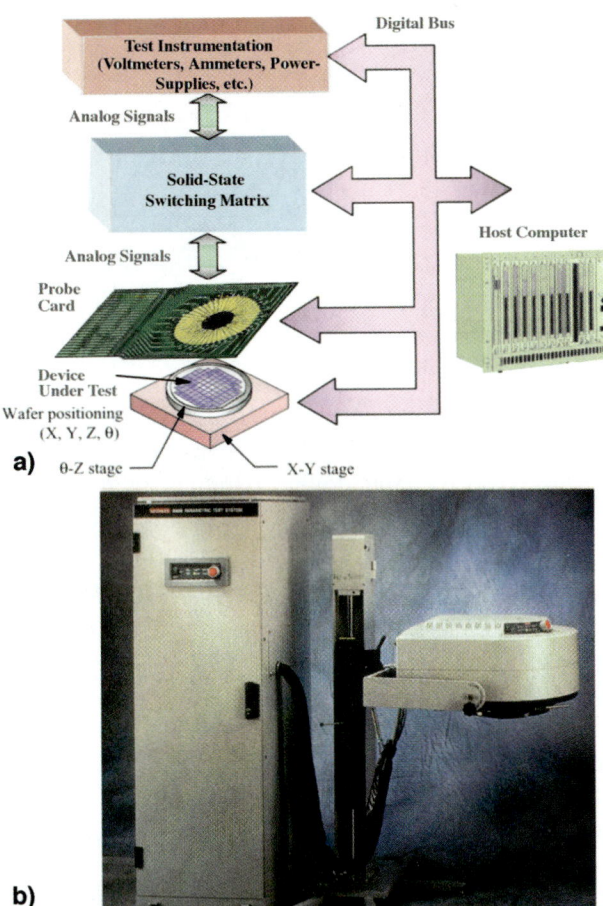

Fig. 26-5 (a) Schematic of an automatic parametric-test system. (b) Photograph of an *automatic parametric-test system*. Courtesy of Keithley Instruments.[5]

26.3 WAFER-SORT

At the end of wafer-fabrication, every die is also subjected to a functional-test while the wafer is still intact. This test is called *wafer-sort* or *wafer-probe*.[7] Each die on the wafer is probed, using a custom *probe-card*. Electrical-signals are applied to the chip using probes that contact the wire-bonding area of each pad. Die which pass the wafer-sort test are packaged and then are subjected to a *final-functional-test*. Note that at the wafer-sort step, functional-testing is primarily static (dc) in nature. High-speed dynamic-testing is difficult to do through probes, so speed-tests are not performed until after die-packaging is done.

When wafers arrive at the wafer-sort tester, the wafer-fabrication-sequence has been completed and the devices have been covered with a protective passivation-layer. Thus, the wafer-sort test can be performed in a facility that has a less-stringent cleanroom Class (see Chap. 8) than the wafer-fab. In fact, the test is typically conducted in a test-and-assembly plant. Until recently, test, assembly, and packaging processes were labor-intensive tasks. Therefore, IC-manufacturers historically built their test-and-assembly facilities in countries with lower labor-costs. However, with the advances made in test-and-assembly automation-technology, this trend may change.

26.3.1 Test Procedures for Wafer-Sort

The wafer-sort test begins by loading a *correlation-wafer* (which is a *known-good-wafer* with the same parts to be tested on it). Running the wafer-sort test first on this wafer ensures that the tester is working properly. Next, a finished-wafer is loaded onto a vacuum-chuck and the probes of the probe card (Fig. 26-7) are automatically aligned to the first-die on the

Fig. 26-6 *Probe-card* with probes in a test environment.

wafer to be tested. These probes make contact with the bond-pads of the chip. After the test on that die is done, the probe-station (under computer-control), steps across the wafer, performing functional-testing on each die. The tests performed are also under control of the *test-program* contained in the host-computer.

The ratio of *working-die*-to-*total-die* on the wafer gives the *sort-yield* for each wafer. (*Yield* is directly related to the ultimate-cost of the completed integrated-circuit, and is discussed more fully in Chap. 28.)

Die that fail the wafer-sort test are not packaged. These failed chips are identified in one of two ways:

1. By placing an *ink-dot* on the failed chip.
2. By storing the location of the failed part on the wafer in a computer.

The *ink-dot method* has been the traditional approach (Fig. 26-8). When the are being picked for placement onto the package, the die-attach tool "sees" the dot and does not pick up the inked-chip. However, inking wafers can cause contamination on good-die, and ink-reservoirs must be periodically refilled. As a result, in more recent wafer-sort testing-tools, information about failed-die is stored in a computer, and it is no longer necessary to ink bad-chips. The tester can also produce a wafer-map of the rejected-die.

26.4 BURN-IN

After packaging, many chip-types are subjected to a procedure called *burn-in* before performing final

Fig. 26-7 Photograph of a probe-card performing a wafer-sort test on a wafer.

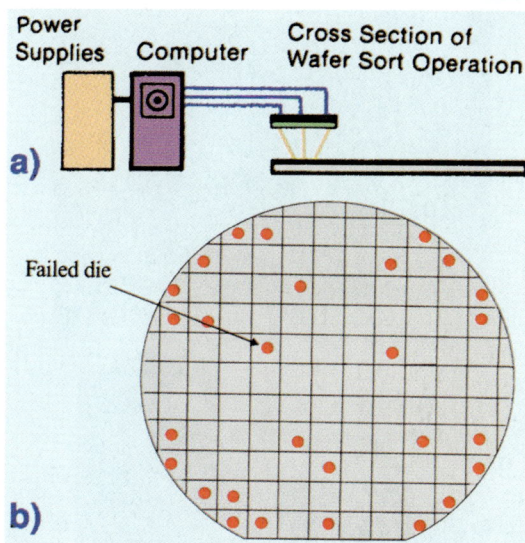

Fig. 26-8 (a) *Wafer-sort.* (b) Inked die that have failed wafer-sort testing.

functional-testing.[7] Note that not all chip-types are subjected to the burn-in step. That is, parts being made with a mature, high-yielding process may not need this stressing, as historical-data may have shown they exhibit low failure-rates, even without having been burned-in. Likewise, low-cost chips (such as DRAMs or SRAMs) may not be subjected to burn-in, as the cost is too high to warrant it.

In any case, during burn-in the ICs are loaded into sockets (Fig. 26-9a) in special burn-in-ovens (Fig. 26-9b), where they are simultaneously subjected to electrical-stressing and elevated-temperatures for several days. Stress-conditions are selected to force weak-chips to fail during burn-in, but not to damage properly-manufactured devices. That is, parts which might fail soon after being put into operation are weeded out by such stressing. Devices passing burn-in tests have been found to be statistically more reliable.

26.5 FINAL-FUNCTIONAL-TESTING

After assembly, packaging, and burn-in, the packaged-ICs are subjected to *final-functional-testing*, where it is verified that the parts can meet the product-specifications. This testing also serves to identify devices that have been damaged during the assembly operation. Furthermore, certain performance-charac-

480 MICROCHIP MANUFACTURING

Fig. 26-9 (a) Burn-in boards with test sockets. (b) Burn-in oven. Courtesy of Pycon, Inc.

teristics of the IC (such as speed), can only be measured with the chip in its package.

Note that in place of the probe-card used during wafer-sort, a *performance-board*, or *device-under-test board (DUT)* is used (with the package inserted into a special socket to obtain signals supplied by the tester). The performance-board in final-test provides an interface between the specific pinout, package and the general capabilities of the tester.

Packaged integrated-circuits are tested using specialized testers, referred to as *automated-test-equipment (ATE,* Fig. 26-10).[8] There are numerous kinds of ATE, but here we describe only *logic ATE*, as their operating-principles are shared by most other types. Logic-testers apply *test-programs* to one IC at a time. An ATE can be thought of as several instruments, including high-speed-waveform-generators and logic-analyzers, power-supplies, and precision measuring-units. The waveform-generation and logic-analyzer parts of ATE are organized in the form of *tester-channels*, which apply the signals to a particular device-pin. Signals applied to the pins are *functional-test-vectors* established either during the IC-design-phase from simulations of the circuit-operation during circuit-verification, or from *scan-vectors*, produced by *automatic-test-pattern-generation* software.

The total test-time is important, as ATE is expensive, and a long test-time can significantly increase the cost of manufacturing an IC. To increase tester-throughput, *design-for-test* strategies are used early in chip-design by taking into account its testability. One test used to rapidly evaluate CMOS ICs is the I_{DDQ}-test.[9] I_{DDQ} is the quiescent-current when CMOS-gates are not switching. If no logic-operations are being performed, the I_{DDQ}-current should be very small (due only to reverse-bias *pn*-junction leakage-current). However, in the presence of defective-circuitry, current will exceed the predicted I_{DDQ}-value. A drawback of I_{DDQ}-testing is that when chips fail this test, the root-cause of the failure is not revealed.

Parts which pass all tests are shipped to the electronic-system-manufacturers, where they are incorporated into products. Additional tests may be applied, either as incoming-inspection, or as board and system-tests. The quality-level of the parts is directly reflected in the number which fail. Often, defective-parts are sent back to the IC-maker for analysis, to determine the cause of failure. The results can be used by yield-engineers to help identify process-problems.

26.6 BINNING

Functional-testing of completed, packaged ICs can use dynamic-test-signals. The final-test can be used not only to ensure that they function properly, but also to gauge their performance-characteristics. For example, the speed of a microprocessor, or the access-time of a DRAM can be measured. Properly-functioning ICs can be further sorted according to predetermined performance-characteristics, a procedure called *binning*. ICs within each performance-range have their own *bin*. Binning is useful if faster parts can be sold for higher-prices than slower ones. The fastest microprocessor chips may be put into the most advanced computer products, which are sold at the highest prices. Slower-operating microprocessor-chips may still work well enough for some lower-performance applications, and hence they can still be sold as working parts (even if at a lower selling-price).

26.7 MARK, PACK, AND SHIP

Once the packaging of a chip has been completed, the package must be marked with key information (Fig. 27-11). Typical information put on the package is the company-logo, product-type, date, and lot-number. The main methods of marking are inking and laser-inscription. Regardless of the marking-method, all marks must meet the requirements of legibility (especially on smaller packages), and permanence when exposed to harsh-environments. After marking, the ICs can be packed for shipment to the customer.

SUMMARY

Integrated-circuits require more than just fabrication. They must first be designed. Then, after they are fabricated they must then be tested to ensure proper functionality. Such electrical-tests must be rapidly and accurate done. This chapter discusses the issues of parametric, wafer-sort and final-functional testing.

REFERENCES

1. W. Merkel, "Parametric Testing Improves Semiconductor Yields," *Semiconductor Online,* Mar, 1998, p. 3.

2. G. Pinkerton, "New Parametric-Test Technologies Meet Future Production Challenges," *Solid State Technology,* December 1996, p. 53.

3. R. DeJule, "Expanding Applications and Demands on Parametric Test," *Semiconductor Intl.,* June 1996, p. 110.

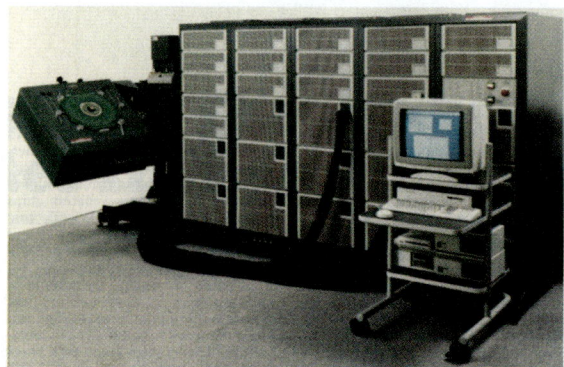

Fig. 26-10 Photograph showing example of *automatic test equipment* (ATE). Courtesy of Teradyne.

Fig. 26-11 Examples of markings on IC packages.

4. Keithley Vendor Literature, *S600 DC Parametric Test System,* Keithley Instruments.

5. V. Agarwal and S. Seth, *Test Generation for VLSI Chips,* Computer Society Press, Washington, DC, 1998.

6. R. Iscoff, "What's in the Cards for Wafer Probing?" *Semiconductor International,* June 1994, p. 77.

7. A. Righter *et al., CMOS IC Reliability Indicators and Burn-In Economics,* Proceedings of International Test Conference, Piscataway, NJ, IEEE, 1998 p. 194.

8. A.J. van de Goor, *Testing Semiconductor Memories: Theory and Practice,* John Wiley & Sons, 1991.

9. R. Rajsuman, *IDDQ Testing for CMOS VLSI,* Artech House, Boston, 1995.

PROBLEMS

1. Make a list of at least ten process or device parameters that could be monitored using a test-drop-in or scribe-line monitor on the wafer.

2. What aspects of the finished wafer are measured by the *wafer-sort test* (wafer thickness, defect-density, parametric data, or circuit-functionality)? Why are die that fail the wafer sort tests marked with an ink dot?

3. What is the purpose of the presence of *test drop-ins* (or *scribe-line monitors*) on a wafer?

4. What is the role of *burn-in* in IC manufacturing?

5. A simple microprocessor contains 115-flip-flops, and hence 2^{115} possible states. If a tester performs a new static-test every 100 nsec, how many years will it take to test every state in the microprocessor chip? If every wafer has 100 die, how long would it take to test the entire wafer?

6. A circuit will be built with a die size of 5 x 8 mm, on a 150-mm wafer. The yield at wafer-sort is 75%, and the wafer processing cost is $250/wafer. What will be the cost of the final product if testing and packaging adds $1.60 to the completed product?

CHAPTER 27
ASSEMBLY AND PACKAGING FOR ULSI

CHAPTER CONTENTS

27.1 DIE SEPARATION (DICING)

27.2 DIE-ATTACH

27.3 BONDPAD-PACKAGE CONNECTIONS
  Wire Bonding
  Tape Automated Bonding (TAB)
  Flip-Chip Bonding

27.4 INTRODUCTION TO CHIP PACKAGES

27.5 PACKAGING TECHNOLOGY I: HERMETIC PACKAGES (CERAMIC)

27.6 PACKAGING TECHNOLOGY II: PLASTIC PACKAGES

27.7 ULSI PACKAGE TYPES
  Through-Hole (TH) IC Packages
  Surface Mount (SM) IC Packages

27.8 TRENDS IN ULSI IC PACKAGES

"I wonder often what the Vintners buy, one half so precious as the goods they sell."

Omar Khayyam (*died 1123*)

Photograph of a packaged microchip which comprises more than six million transistors. Courtesy of Philips.

Since chip-packaging is the final procedure carried out in the manufacture of integrated circuits, it is one of the last topics covered in this text. It is a multi-disciplinary technology, involving many assembly and packaging steps. An awareness of packaging-technology is important for several reasons: **1**) packages can have a significant impact on the performance and reliability of the chips they house (and on the electronic end-products they inhabit); **2**) the cost of the package can be a large fraction of the total expense of manufacturing an IC; and **3**) each year, billions of packages are used to encapsulate the chips that are fabricated. Thus, the assembly and packaging business is a large and important part of the semiconductor-industry. For more details, consult References 1 - 3.

Our discussion begins with the processes involved in the assembly of ICs, and concludes with IC-package technology. Figure 27-1 shows the various steps performed on a chip after wafer-fabrication is completed.

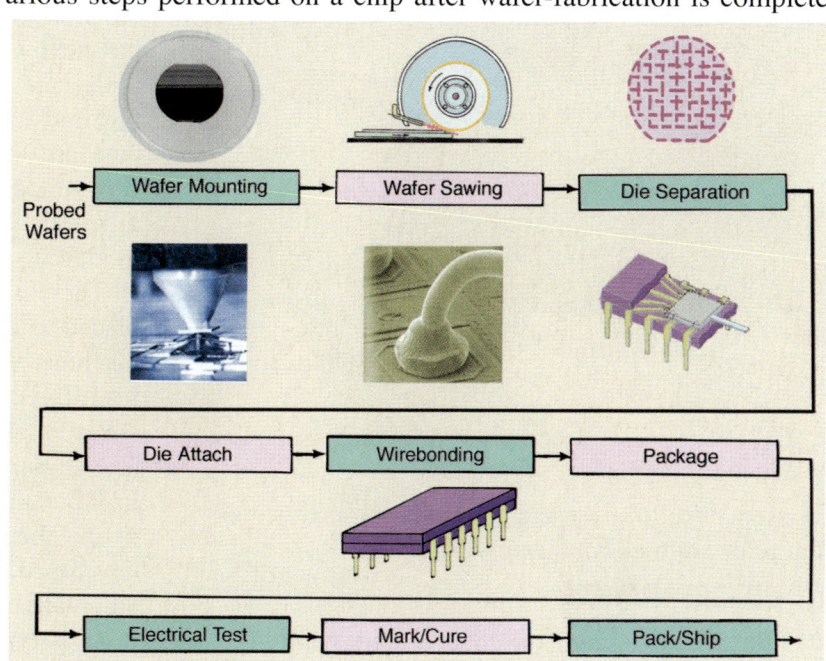

Fig. 27-1 Sequence of conventional assembly and packaging steps.

(*wafer-sort*), the wafer is cut apart (a process also known as *dicing*), so that each functional-die can be mounted into its own package. Die-separation is done with a circular diamond-impregnated dicing-saw, rotating at a speed of 20,000-rpm to cut through 100% of the wafer (Fig. 27-3). The difference between the hardness of the diamond-crystals embedded in the saw-blade and that of the Si-wafer, enables the blade to easily penetrate the silicon and make a clean cut.

The sawing operation itself consists of many sub-steps. Typically, one dicing-saw is able to support three to five die-bonders that, in turn, can keep 20 wire-bonders supplied. Dicing-saws generally range in price from $150,000–$180,000.

The wafer surface is prepared for the die-separation step by leaving enough room between adjacent-die to allow for the width of the saw-blade during cutting (*kerf-width*). These regions are referred to as *scribe-lines* or *scribe-streets*. They range in width from 60-μm to 150-μm, depending on the die-design.

Prior to dicing, the wafer is mounted onto a sticky, flexible tape (e.g., Mylar) that is itself attached to a rigid frame (Fig. 27-2). The tape continues to hold the die after the sawing-operation and during transport to the next assembly-step. The tape is about 0.1-mm thick, and the saw-cut penetrates about one third of this thickness. Adhesives that hold the wafer onto the tape are available with different adhesion-levels to accommodate various die-sizes. The frames are easier to handle than individual wafers, and can be marked to identify the wafer being carried.

The frames containing mounted-wafers are then placed into a cassette to be loaded into the dicing-machine. After such loading, each frame is moved from the cassette to an alignment-station, where the wafer is aligned with the cutting-blade (Fig. 27-3). The alignment must be accurate so the saw-cut occurs in the center of the scribe-street, and not in an area of the die. Automatic alignment-systems have been developed as part of the move toward implementing automation of the entire sawing-process. Possible process-errors that should be guarded against include:
a) saw-cuts that are too wide or cause excessive chip-

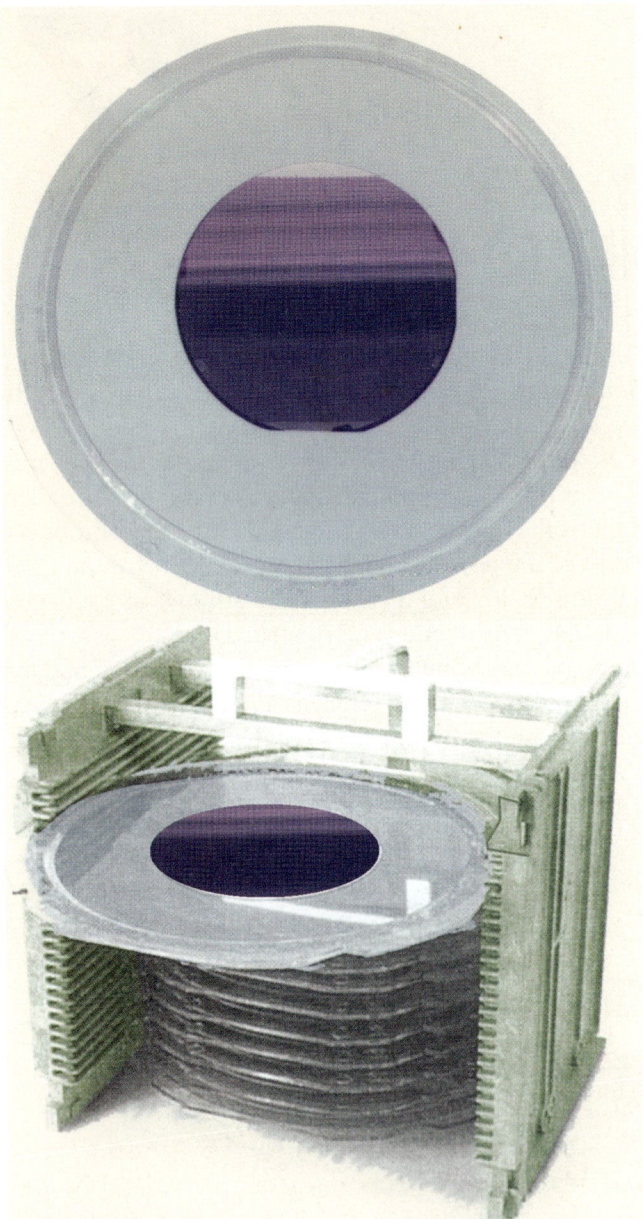

Fig. 27-2 Wafers are mounted on mylar-tape frames in preparation for sawing.

This sequence of test-and-assembly procedures is often referred to as *back-end processing*.

27.1 DIE-SEPARATION

After the *chips* (or *die*) on uncut-wafers have been subjected to preliminary testing for functionality

CHAPTER 27 ASSEMBLY AND PACKAGING FOR ULSI

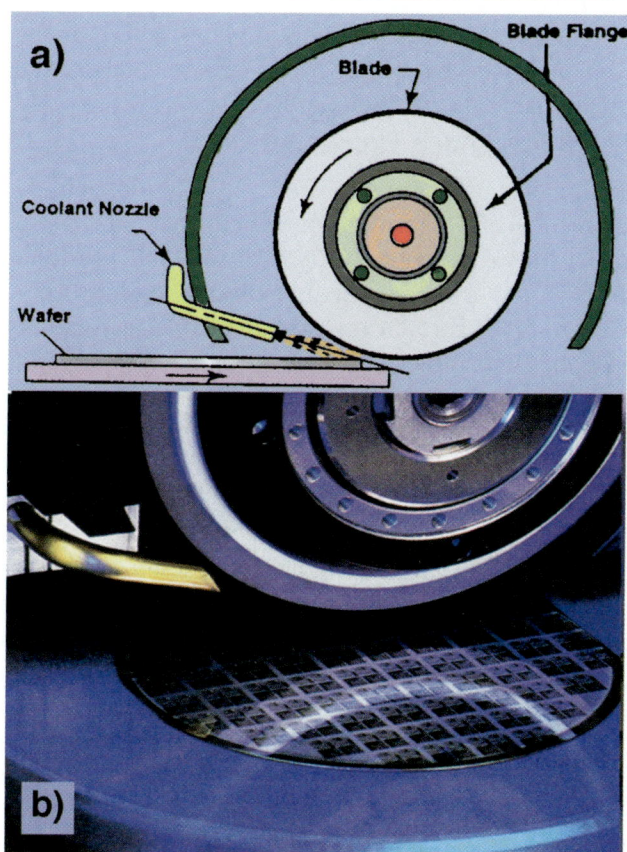

Fig. 27-3 The wafer-sawing operation: (a) Schematic drawing; (b) Photograph of a wafer-saw.

ping; **b)** saw-cuts that are too deep (or too shallow); and **c)** broken saw-blades.

After sawing, the wafers are moved to a cleaning-station. Here, the colloidal-silicon produced in the sawing-operation is removed. If this powdered-silicon was allowed to dry on the wafer-surface, bonding-pads could become contaminated with silicon-dust, creating wire-bonding problems. Also such residue could collect in the kerfs, producing die-attach-pickup-problems. The cleaning-process consists of water- spraying *during* sawing, followed by wafer-scrubbing *after* sawing. Advanced sawing-systems contain integrated cleaning-stations.

After the sawed-wafers are cleaned, and the die are still attached to the tape, the tape is usually stretched so the distance between adjacent-die is increased. This step is necessary when using some types of automatic die-bonders to allow them to pick up die from the tape.

27.2 DIE-ATTACH

After the wafer has been sawed, the *backside* of the now-separated, functional-die must be attached to a package or substrate, and this procedure is known as *die-attach* or *die-bonding*. The quality of this adhesive-bond is critical to the long-term stability of the assembled parts. The choice of die-bonding technology depends on tradeoffs between various packaging-issues such as: **1)** the thermal-stresses caused by mismatches between the *temperature-coefficient-of-expansion* (TCE) of the chip and packaging-materials; **2)** power-dissipation; **3)** electrical requirements; and **4)** cost. The connection between the die and substrate must be in the form of a total intimate-contact, without any voids that would lead to crack-causing stresses in the die. The intimate-contact is also necessary to assure adequate heat-dissipation during later circuit-operation, and to provide a firm foundation that will give sufficient support during subsequent wire-bonding operations. The increased use of larger die-sizes has introduced new challenges for die-attach technology. This results from even greater stresses and thermal-management problems that are associated with larger die. It is also important that the die-attach operation be carried out in a cost-effective manner, which today implies the use of high-speed, automatic die-bonding equipment.[4]

27.2.1 Die-Attach-Related Reliability Issues

The following reliability issues arise in connection with the die-attach process: **a)** corrosion of the die resulting from the die-attach step; **b)** inadequate heat-dissipation resulting from a poor die-bond; **c)** die-bonding failures; and **d)** die-cracking.[5]

27.2.2 Die-Attach Materials

Normally the materials used to connect chips to packages are electrically-conductive, so the low-resistance electrical-connection required in some applications can be made to the back of the die. Since electrical

(conductive-epoxies and polyimides) are commonly used in plastic-packages. More details on these die-attach materials are found in Ref. 1.

27.2.3 Die-Attach Equipment

Equipment used in the process of die-attach has evolved from operator-controlled manual-systems to the fully-automatic systems in standard use today (Fig. 27-4). Systems capable of die-bonding up to 4000 to 6000 parts/hr are now available, particularly for epoxy-attachment. In such die-bonders, the wafers are presented in sawed-wafer form, attached to an adhesive Mylar-tape. The die-bonder has a pick-and-place unit, and the pickup-arm then moves into position over the die to be attached. The arm is lowered so that a vacuum-collet can pick up the die (Fig. 27-5a). At the same time the collet lifts the die, an ejector-needle pushes the die from underneath, causing the die and the adhesive-Mylar-tape to disengage from one another. Some collet-designs make contact with the upper-edges of the die, while others have a rubber-tip that makes contact only with the top-surface of the die as it is lifted. Pattern-recognition systems enable the machines to recognize chipped or rejected-die. The die to be bonded is then placed on the package-substrate, onto which has previously been placed the

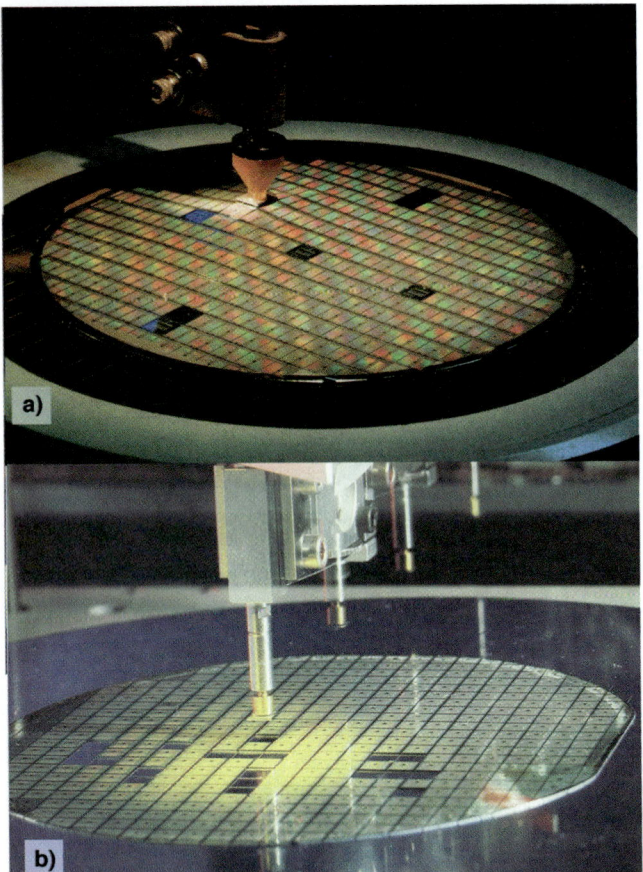

Fig. 27-4 The die-attach operation. (a) Photograph of a die being lifted from a mylar tape-carrier. (b) Die disengaged from the tape being lifted-away by the pickup-arm.

conductors are also good conductors of heat, they provide a good thermal-path between the die and the rest of the package. There are a variety of materials used to perform the die-attach function, including: **a)** *hard-solders* that form a eutectic die-bond; **b)** *silver-filled specialty-glasses*; **c)** *conductive-epoxies*; and **d)** *conductive-polyimides*. The selection of the optimum die-attach material for a particular application is determined by the hermeticity requirements of the package, thermal-dissipation, the heat being transferred through the die-bond, and the thermal-coefficient-of-expansion (TCE) of the chip-carrier media. Eutectic-bonds and silver-filled glasses are typically used in ceramic-packages, and organic-polymers

Fig. 27-5 Placing a die on the package for die-attach.

die-attach material. Die-placement pressure is applied by the bond-arm, programmed to accommodate different die-sizes. The die-bonder also contains a chip-attach material-supply station and a lead-frame loader and unloader. Die-bonders in 1999 ranged in price from $80,000–$250,000 for fully-automated systems.

27.3 BOND-PAD-TO-PACKAGE CONNECTIONS

After the chip has been attached to the package-substrate (or leadframe), electrical-connections must be made between the bonding-pads of the chip and the inner-leads of the package (which are themselves connected to the package-pins). Wire-bonding has been, and will likely remain, the most common technique for making the connections between the chip input/outputs (I/Os) and the package (especially for chips with up to about 224-I/Os). It is estimated that in 1996 over 4×10^{12} wire-bonds were formed in the fabrication of ICs. In this section wire-bonding is discussed first. The other two methods for making such connections are tape-automated-bonding (TAB), and flip-chip-bonding.[6-8]

27.3.1 Wire-Bonding

The wire-bonding process is carried out after the die-attach step. Flexible wires are attached, one at a time, from bonding-pads on top of the chip to the package. One of three methods is used to join the wire to the chip and substrate: *thermocompression (T/C) bonding*; *ultrasonic (U/S) bonding*; and *thermosonic (T/S) bonding*. T/C and T/S bonding can produce either a ball-bond (Fig. 27-6a) or a wedge-bond (Fig. 27-6b), while U/S bonding is only used to make wedge-bonds. The materials used for wire-bonding are either Au or Al because these metals are highly-conductive and ductile enough to withstand the deformation during the bonding-steps. They also bond well to the metallizations used on both the chip and package. In addition, the technology of manufacturing fine Au and Al wires (typically 25–30-μm in diameter) is mature, and metallurgically-stable bonds are formed by each process. Wire-bonds with Au use a ball-bond at one end and a wedge-bond at the other (the approach most

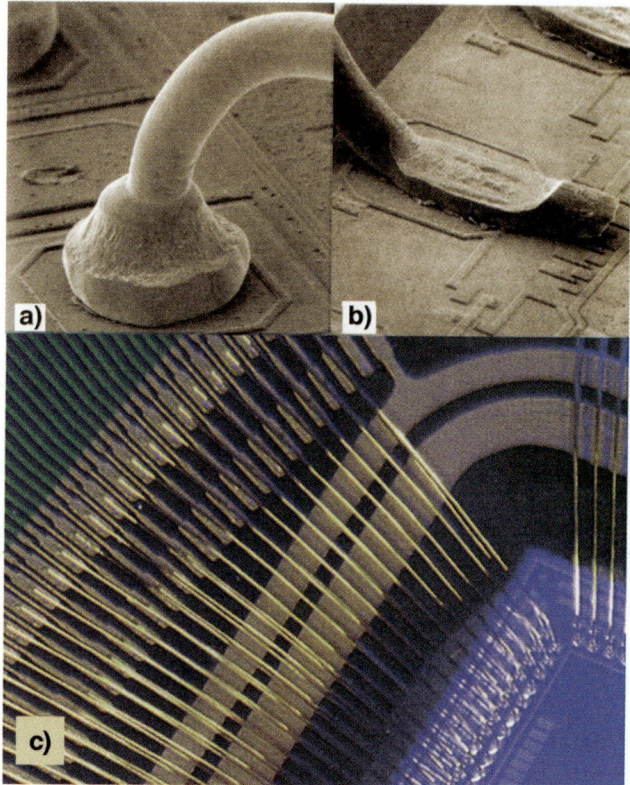

Fig. 27-6 Photograph of: (a) *ball wire-bond*; (b) *wedge wire-bond*; (c) Ball-bonds and wedge-bonds.

widely used, see Fig. 27-7). However, wire-bonding with Al-wire typically uses two-wedge bonds.

The wire-bonders used in the early days of the IC-industry were manually-operated and quite labor-intensive. These were replaced by fully-automated ball-and-wedge wire-bonders with pattern-recognition and computer-control. Such bonders now run at speeds of 6 wire-bonds per second (or 18,000 bonds per hour). Automated wire-bonders in 1999 ranged in price from $75,000 to more than $250,000.

One drawback of wire-bonding is that even after a wire-bond is formed on the Al-pads of a chip, a fairly-large area of the pad remains exposed. Aluminum is a highly chemically-reactive metal, and without its protective passivation-layer (SiO_2 or Si_3N_4), it reacts readily even with pure-water. In the other interconnection methods (i.e., TAB or flip chip) a thin interfacial-metal sandwich-layer (e.g., Cr/Cu/

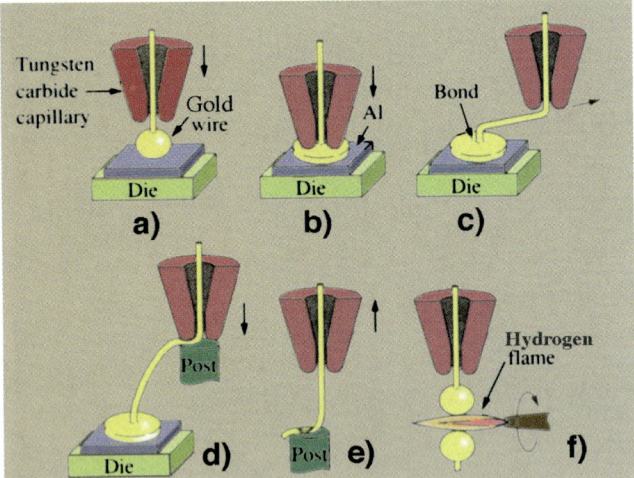

Fig. 27-7 Thermocompression (T/C) ball-wedge bonding of a gold wire: (a) A heated tungsten-carbide capillary with formed wire-ball is positioned over bond-pad; (b) A ball-bond is formed by simultaneously applying a vertical-load on the ball, while ultrasonically exciting the wire (the chip and substrate are heated to about 150°C); (c) Wire is looped to a position over the package bonding-pad; (d) A wedge-bond is formed under load and ultrasonic excitation; (e) Wire is withdrawn from package bonding-pad; (f) Ball-formation accomplished by heating the wire with an H_2 torch; Tungsten-capillary with ball is now ready to go to Step (a).

Au) is first deposited over the exposed Al bond-pad. Subsequently, a gold (Au) bump (in TAB), or a solder-ball (in flip-chip) is placed over the pad. This protects the corrosion-sensitive Al. ICs assembled with the latter methods exhibit reduced device failure-rates from corrosion compared to devices assembled with wire-bonding.

27.3.2 Wire-Bond Spacing

As the number of external connections to an IC increases, a larger number of bond-pads on the chip are needed. Since the number of such pads is increasing faster than the chip-size, the bond-pad *pitch* (bond-pad center-to-center distance) must decrease, especially if the bond-pads are along the periphery of the chip (as is the case with wire-bonds). In memory-chips, the pads are located along only two facing-sides of the chip.

In logic and microprocessor chips the I/O count is increasing, and the active-device size is decreasing. Thus, the interconnect-area represents a large fraction of the total Si-area. In some cases, it may even determine the chip-size (as extra-space is needed around the active-area for additional pad-placement). The minimum wire-bonding-pitch has been steadily declining, and by 1998 had decreased to 50–80-μm (with 23-μm wires). The wire-span from chip-to-package is typically 1 to 4-mm. Longer wires could lead to shorts to adjacent wires, as they would be more-likely to droop or deform. In plastic-packages, longer wires can also deform due to resin-flow during the molding-operation. To allow more wire-bonds to be made on a chip (while meeting the minimum-pad-pitch and wire-length requirements), staggered bond-pads and package-terracing are somctimes employed as a solution (Fig. 27-8).

27.3.3 Tape-Automated-Bonding (TAB)

For chips that require high pin-counts (e.g., more than about 208 pins), wire-bonding becomes undesirable (as the chip-size must grow to accommodate the space for the wire-bond pads). An alternative chip-pad-to-package-pad connection-method called *tape-automated-bonding* (TAB) has been developed to handle such applications.[9] In the TAB-method, the wires of wire-bonding are replaced with finely-patterned, thin metal-leads (usually made of Cu foil, plated with Au or Sn). The TAB-method involves bonding silicon-chips to patterned metal-on-polymer-tape by thermocompression-bonding (prior to the die-attach procedure). Usually, these leads are formed

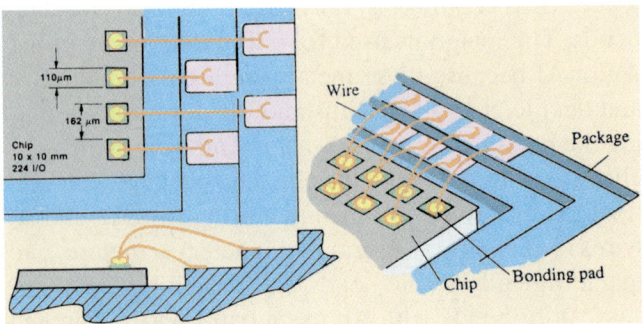

Fig. 27-8 Bond-pads staggered on different levels.

by depositing the metal-film onto a flexible strip of polymer-tape. Then this film is patterned by lithographic-techniques, similar to those used in the IC-fabrication-process. This forms the desired lead-pattern, and the result is a strip of tape containing many individual lead-systems (Fig. 27-9). The most popular way of connecting these leads to the chip involves the creation of bumps on the chip bonding-pads (formed by electroplating Au). The tape is moved by sprockets until one of the lead-systems is positioned exactly over the chip and the leads are then thermally-bonded to these bumps.

The fabrication of the bumps on the bond-pads is a multi-step-process (Fig. 27-10). First a multi-layer metal-film (consisting of Cr, Cu, and Au layers) is sputter-deposited in blanket-form onto the wafer-surface (which is covered with a passivation oxide-layer, except for the bond-pads). Lithography is then used to create patterns in a photoresist-film that will act as a 25–30-μm-thick mask for the Au electroplating-step. That is, openings in the resist are created over the bonding-pads, where electroplating is then used to deposit Au-bumps, 10-30-μm thick. Next, the resist is stripped. To complete the bump-formation, the diffusion-barrier film is etched. Au-bumps protect regions of the barrier-layer film underneath them from being etched during this step.

The bonding of the metal-leads to the bumps on the chip is done simultaneously by thermo-compression (Fig. 27-11). That is, a *heated-thermode* (300–360°C) is pressed against the leads, which are each pressing against a bump on the chip. The heated-thermode presses down for 1–2-seconds to form the bond between the lead and bump. This step is known as *inner-lead bonding*. At this point, the chip has been attached to the tape that carries the lead-pattern. Next, the chip and lead-pattern are encapsulated (while they are still attached to the strip of tape), usually by plastic-molding (Fig. 27-12). The package-chips are separated from the continuous-tape that carries the leads, to form an individual-packaged-part. These packaged-parts are finally attached to a *printed-wiring board* (*PWB*) by a process called *outer-lead bonding* (Fig. 27-13). That is, the packaged-chip with the leads in place is transferred to the PWB, where the leads are attached to the land-patterns on the board.

In addition to allowing bonding-pads with smaller pitches than the ones needed for wire-bonding, TAB offers the advantages of speed (since all the bonds are formed at once), and ease of automation offered by the tape-and-sprocket system. Since the leads have no loop (as do wire-bonds) TAB also offers a lower profile, which makes TAB applicable for smart-cards, watches, read/write-head circuitry, flash-memory modules, and the like. The disadvantages of TAB are: **1)** it requires a special tape-design for each different chip-design; **2)** there is less infrastructure to provide testability of packaged TAB-parts and for mounting the packaged TAB-parts to the next levels of packaging; and **3)** higher-cost, compared to wire-bonding.

27.3.4 Flip-Chip Bonding

In *flip-chip technologies*, solder-bumps are fabricated directly on the Al bonding-pads of the chip. The chip is then flipped face-down and aligned to the package or substrate. The bumps are bonded directly (and all at once) to the package (or substrate) pads by reflowing the solder-bumps. The process was introduced by IBM in 1964, and they called it C4 for *controlled-collapse-chip-connection*.[10,11] The advantages of flip-chip bonding are: **1)** the entire chip-surface can be covered with solder-bumps (making it an *area-array interconnect*). In other words, bonding-locations are not limited to the chip-perimeter, thus more I/O-capability is provided than by a perimeter-interconnect on

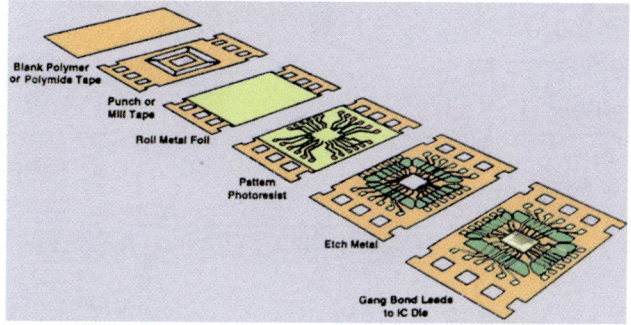

Fig. 27-9 Manufacturing sequence for tape-automated bonding (TAB).

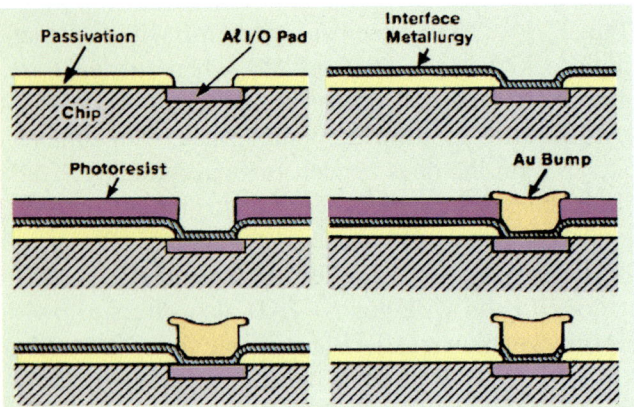

Fig. 27-9 Manufacturing sequence for TAB bond-pad bumps.

a die with the same size; and 2) the very-short lengths of the chip-to-package interconnect-paths minimizes their inductance.

As in TAB, the bumps are formed while the chip is still in wafer-form. The bump-formation process represents an added expense to overall wafer-fabrication cost. However, since this process eliminates the cost of wire-bonding, other production costs are reduced. Furthermore, the bump-metallurgy reseals the area over the bonding-pads that otherwise remains open after wire-bonding. Thus, flip-chip technology hermetically-seals the chip without the need for a package. This attribute is being exploited for so-called "packageless" attachment of die to substrates (as discussed in Section 27.10).

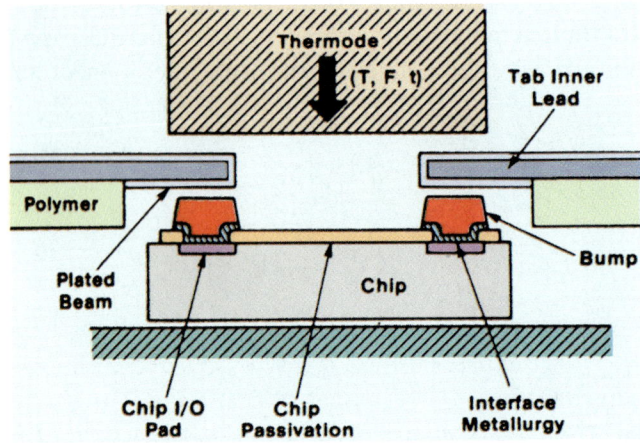

Fig. 27-11 TAB inner-lead bonding.

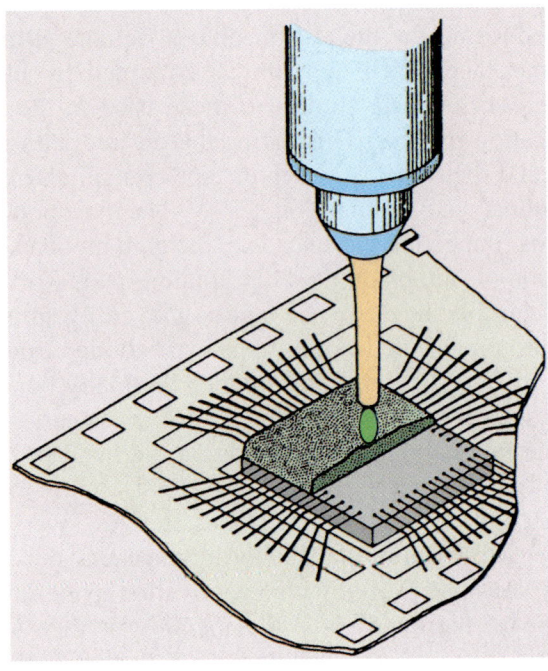

Fig. 27-12 TAB encapsulation.

The solder-bumps most widely used to attach chips to alumina ceramic-substrates have been high-lead (Pb) solders, especially 95 Pb:5 Sn. This solder melts at the relatively high-temperature of 315°C, which permits other lower-melting-point solders to be used in subsequent module-to-card, or card-to-board packaging-level processes without remelting the flip-chip bonds. A multi-layer film of metal (e.g., Cr-Cu-Au) is sandwiched between the Al chip-pad and solder bump to prevent the solder from interdiffusing into

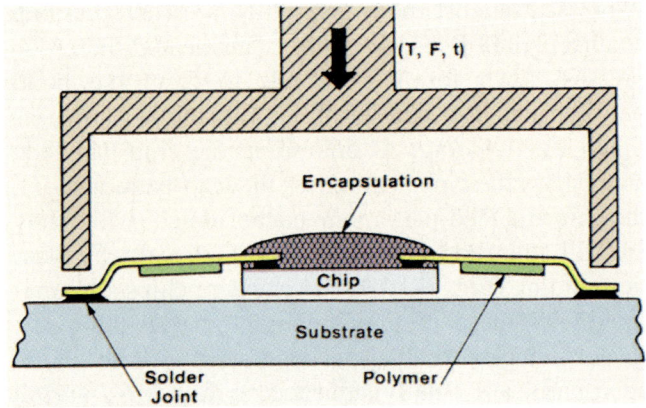

Fig. 27-13 TAB outer-lead bonding.

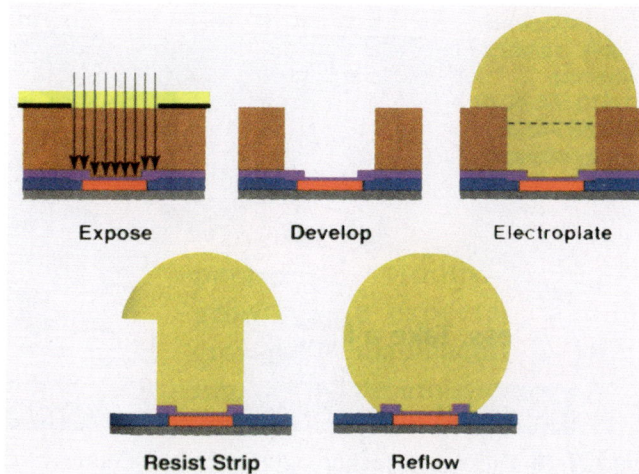

Fig. 27-14 Fabrication of solder-bumps for C4-technology.

die. Once flowed into place, the epoxy is cured to its solid-form by an additional heat-curing step.

The term "flip-chip" refers only to the method of attachment. If the chip is attached to a package, it is called *flip-chip-in-package* (FCIP) and this method applies to single-chip packages and multi-chip modules (MCMs). FCIP is used when the electrical performance of wire-bonds is inadequate or when the number of I/Os is too large for wire-bonding. Use of FCIP together with a ball-grid-array package (see Sect. 27-10) would represent a packaged-chip with a very-large I/O count. For example, the Intel Pentium II microprocessor is put into plastic and ceramic

the Al. The Cr-Cu-Au film forms a cap over each of the Al bonding-pads, and the size of each cap is also restricted by sequentially evaporating the Cr-Cu-Au film through a mask (Fig. 27-14a). The Pb:Sn is next evaporated onto the chip, again through a mask. The local-area of each Pb:Sn-layer is slightly larger than the area of the cap. Heating the wafer in an H_2-ambient at 350°C melts these localized Pb:Sn-layers. Surface-tension of the liquid-solder causes the film to recede from the oxide-surface, forming a solder-ball atop the Cr-Cu-Au-cap. The diameter of the base of the spherical bumps is determined by the dimension of the Cr-Cu-Au cap. This process is called *ball-limiting-metallization* or BLM (see Fig. 27-15).

Following reflow of the solder-bumps that connect the chip to the substrate, an underfill-adhesive is formed between the chip and substrate. Such underfilling is necessary to reduce the stress on the solder-bumps during temperature-cycling. When the underfill-adhesive is present, it shoulders a large fraction of the stress-burden, and stress on the bumps is significantly reduced. This decreases their tendency to fail through solder-bump stress-fatigue.

The underfill-adhesive (an epoxy or polyimide) is dispensed as a bead along the perimeter of the bump-attached chip. The chip is heated to 70–100°C and this lowers the viscosity of the underfill-epoxy so that capillary-force wicks it into the space underneath the

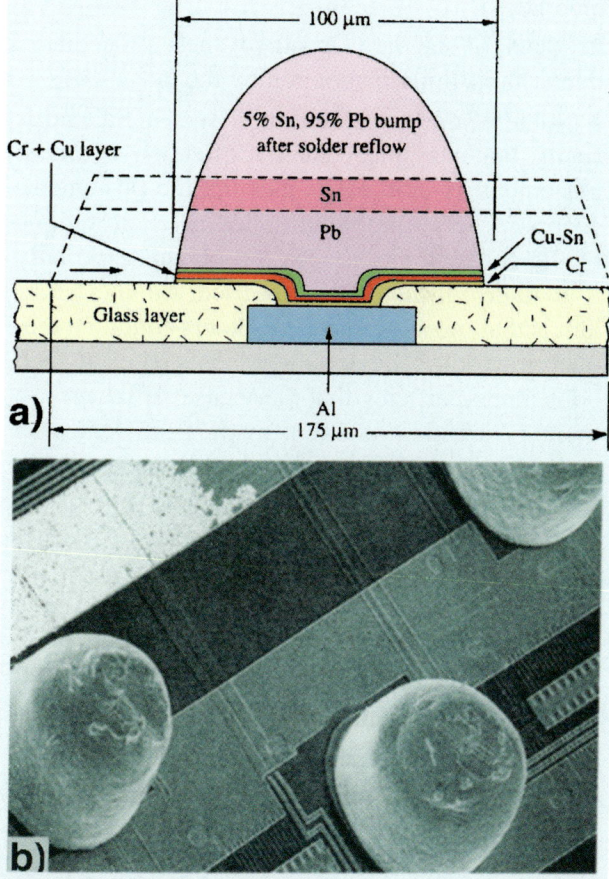

Fig. 27-15 (a) Cross section of a solder-bump used in flip-chip technologies and the deformation caused by reflow-soldering operations; (b) SEM view of the C4 solder-bumps on a chip. Reprinted with permission of IBM Corporation.

BGAs and MCMs with flip-chip-attach. *Direct-chip-attach* (DCA) refers to the direct-attachment of a chip to a PC-board (also called *flip-chip-on-board*). *Flip-chip DCA* (FC-DCA) bypasses the attachment of the chip to the package.

Flip-chip attachment, however, has some limitations, including the following:

1. The infrastructure for manufacturing, testing, and handling flip-chips from semiconductor vendors by 1998 was not yet mature. Yet, forecasts were that by 2001 the number of chips requiring FC bumping will have increased to more than 2.5 billion. By 1998, flip-chip-foundries had begun to emerge.

2. A lack of standards exists for these procedures, due to their limited adoption by the late 1990's.

3. Some of the manufacturing issues are also difficult. For example, visual-verification of bump placement is not easy, making die-placement-accuracy very important. Thermal-cycle-fatigue of the solder-joints is also a problem, especially if an underfill-material is not used. Another problem is that most ordinary Pb-solders contain trace-quantities of radioactive-elements (i.e., which emit alpha-particles that can cause soft-errors in memory-circuits). Solders made from lead that emits low-levels of alpha-particles are available for applications that cannot tolerate high soft-error-rates.

27.4 INTRODUCTION TO CHIP-PACKAGES

IC-chips (or die) are usually encapsulated in an IC-package prior to their being installed into electronic-systems. IC-packages perform four key functions:

1. They provide a sturdy set of leads that allow an IC to be connected to the system in which it will operate. These leads, however, should not significantly degrade the performance of the chip housed by the package.

2. They provide physical-protection for the chip against breakage or scratching.

3. They provide environmental protection against

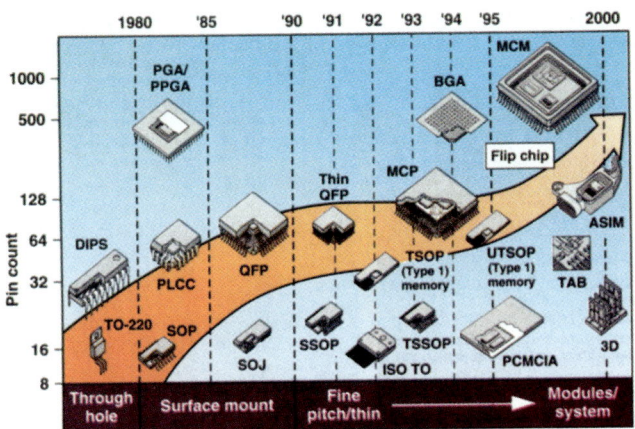

Fig. 27-16 Packaging-Technology Roadmap. Courtesy of National Semiconductor.[12]

damage from chemicals, moisture, or gases.

4. They dissipate the heat generated by the chip during operation. Chips generating large quantities of heat require additional-measures to be incorporated into the package-design.

In order to achieve the above objectives, packages are designed to have the following characteristics: **1**) low lead-capacitance (and inductance); **2**) material-compatibility; **3**) good thermal-conductivity; **4**) good hermetic-integrity; **5**) ease-of-manufacture; **6**) low-cost; and **7**) stress-levels that will not harm the chip or package. ICs in packages are also easier to handle and test than bare-die. They make it easier to compensate for the mismatches in coefficients-of-thermal-expansion among the materials used in the construction of electronic-systems. Chip-packages, however, also impact the cost, reliability, and sometimes even the performance of the IC (and maybe even the system in which the IC is being used).

Reduction in the prices of electronic-systems have primarily been driven by the shrinkage of the IC-devices themselves, but improvements in packaging have also helped decrease these costs. Packaging-cost is important because it can account for a significant fraction of the total-cost of the packaged-IC. This cost (as measured by *cost-per-pin*), has increased from about 1-cent-per-pin ($US) in small-scale- and medium-scale-ICs (*SSI* and *MSI*), to about 10-cents-per-pin for ULSI. For ULSI-devices in ceramic-packages, the package-cost may exceed the chip-cost by more than 2 times.

IC-package technology is generally divided into two categories, namely: *hermetic* and *non-hermetic* (i.e., plastic) packages. In hermetic-packages the chip is housed in an environment isolated from the external world by a vacuum-tight-enclosure. The package-material is usually ceramic-based, and the most widely used hermetic-package types are *ceramic-packages*. They are used in high-reliability applications and/or when the heat generated by the chip is too great for a plastic (i.e., non-hermetic) package. Since hermetic-packages are more costly than plastic, the applications that employ them must be able to justify such cost-penalties. For example, almost all microprocessors (which dissipate large amounts of power and have high *average-selling-prices* [ASPs]) go into ceramic packages.

In non-hermetic-packages, the chip is not completely isolated from the outside ambient (as moisture or other substances can penetrate the polymer-material - an epoxy resin - that encapsulates the chip). Such plastic-packages, however, are much cheaper than hermetic-packages, and their reliability and performance-characteristics have significantly improved since the early days of the IC-industry. Hence, they have become by far the most widely-used IC package-type (in terms of the number of chip-packages produced each year). Note that because ceramic chip-packages are much more expensive than plastic-packages, the sales-volume of ceramic-packages constituted roughly two-thirds of the total packaging-market in the early 1990's. The package-industry is a multi-billion-dollar industry worldwide, and in 1998, 62-billion IC-packages were produced. Figure 27-16 shows a roadmap of packaging-technology for ICs.[12]

27.5 PACKAGING-TECHNOLOGY I: HERMETIC-PACKAGES (CERAMIC)

A typical hermetic-package with a chip mounted in its cavity is shown in Fig. 27-17a. The steps used to fabricate such hermetic-packages are illustrated

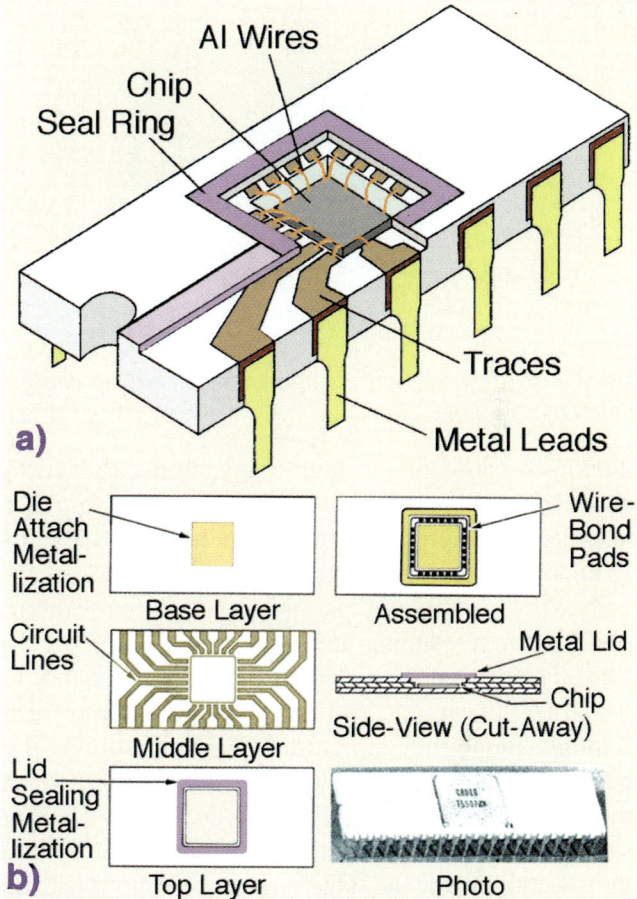

Fig. 27-17 (a) A typical ceramic package. (b) The parts of a ceramic dual-in-line (DIP) package.

in Fig. 27-17b. The base of the package is made of a multi-layer ceramic-material, and the chip sits in a cavity of this base. The layers of ceramic-material are formed by casting thin-sheets of alumina from a slurry consisting of alumina-powder, solvent, and resin-binder. After these sheets have dried, they are cut to size. Cavities and via-holes (i.e., holes extending through the ceramic-sheet so that interconnections can be made to the inside of the package), are then mechanically-punched into the sheets. Next, custom wiring-paths are screen-printed with a slurry of a refractory-metal powder (tungsten or molybdenum) onto some of the sheets, and the via-holes are filled with metal. Refractory-metals are used for these con-

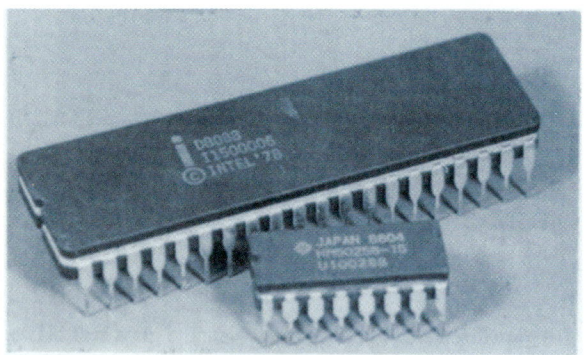

Fig. 27-18 A photograph of an assembled *ceramic dual in-line* (DIP) *package*.

ductors because of the high-temperatures needed to sinter the alumina. Several sheets are stacked together and placed in a press where they are laminated together under pressure and moderate-temperature.

Next, this structure is fired at high-temperature (1600°C) in a reducing-atmosphere (to prevent oxidation of the metal circuit-lines). This firing-operation densifies the ceramic and metal components to form a single monolithic-structure (which constitutes the package-base). Note that the wiring-paths formed on the surface of some of the ceramic-sheets are now sandwiched between layers of ceramic-material, and they "daylight" at the edges (sides or bottom) of the package. These metal-paths are now an integral part of the package-material.

However, after the firing-operation, the areas where such metal is exposed on the exterior of the package must be plated (e.g., with nickel) for protection against corrosion, and to allow attachment of external-leads to the package. For *leaded-packages*, external-pins are bonded to the metal-regions that penetrate the package-surface. In the *dual-in-line* ceramic-packages (DIPs, Fig. 27-18), leads made of a Fe-Ni-Co alloy (called *Kovar*) are brazed to the metal along the sides of the package, while in *pin-grid-array* ceramic-packages, leads are attached to the bottom of the package-base (Fig. 27-19). These leads are then plated with Au over Ni for bondability and environmental-protection. In *leadless-packages*, external metal-patterns are formed on the outside of the package. At that point, the package is ready for die-attach, wire-bonding and package-sealing.

After the chip-to-package connections are made, a package-sealing process completes the hermeticity. A metal or ceramic lid is placed over the cavity, and it is hermetically-sealed to the base. The most widely used method for this sealing-process is welding (because it is a rapid, reliable, and highly-reproducible technique). In welding, high-current-pulses produce local-heating of 1000-1500°C, fusing the lid (or the plating thereon) to the package. Since this heating is localized, damage to internal-components is avoided.

27.6 PACKAGING-TECHNOLOGY II: PLASTIC-PACKAGES

As noted earlier, the most widely-used packaging technology involves encapsulating the IC in molded-plastic, with more than 90% of ICs in the late 1990's being assembled this way.[13] The dominance of plastic-packages is primarily due to their much lower cost. However, this advantage is traded-off against the limitation of reduced-reliability. Plastic does not completely isolate the chip from the environment, and moisture or other substances can penetrate the plastic and cause corrosion of the metals on the chip. Nevertheless, the reliability of plastic-packages has increased to the degree that they can satisfy many ULSI-applications (especially consumer-products). For example, in the late 1970's it would take less than 1000-hours to induce a failure in a plastic-packaged

Fig. 27-19 Ceramic pin-grid array (PGA) package. (a) Cavity-up PGA configuration. (b) Cavity-down PGA configuration.

IC by an accelerated stress-test. By the late 1980's the parts could survive 5000-hours under the same test conditions. Such parts operate reliably for much longer periods of time under normal-operating-conditions (e.g., >10 years). Thus, it may now be more appropriate to characterize plastic-packages as being "less-hermetic."[14]

Figure 27-20 shows a typical plastic-DIP (see Sect. 27.9.1 for a description of DIPs). This type of package consists of a *lead-frame* (LF) and *plastic molding-material*. The lead-frame is made from a thin (0.25-mm thick) chemically-milled or stamped sheet of a Fe-Ni or Cu-based alloy. The most commonly used lead-frame material is Alloy 42 (42%Ni-58%Fe). The lead-frame is, in effect, a skeleton around which the packaging-material is formed. But it also provides the external-leads in the assembled package. The chip is die-attached to the lead-frame paddle, and the bond-pads on the chip are connected to the radiating lead-frame-terminals by gold wire-bonds. The lead-frame is then encapsulated with a thermoset epoxy-resin, which produces a rigid plastic-body that completely surrounds the chip. The most common method of such encapsulation is transfer-molding. After the plastic-encapsulation is completed, the leads are bent down to the configuration shown in Fig. 27-20.

The most widely-used materials for forming the rigid-body of the plastic-packages are epoxy novolac-based molding-compounds. These compounds typically consist primarily of novolac-epoxy-resin (25-30%); inert filler-material (68–70%); and flame-retarder (2%). The remaining small-percentage of material consists of an accelerator, curing-agent, mold-release agent, and colorant. The epoxy-resins originally used for IC packages had high-concentrations of chlorine (up to 3%), but by the late 1970's this was reduced to less than 30-ppm. Fillers are used to reduce the thermal-coefficient-of-expansion (TCE) of the compound and to increase the thermal-conductivity of the plastic. Powders of fused-silica are common filler-materials. Mold-release-agents prevent the cured-compounds from sticking to the mold cavity-walls, facilitating easy removal of the plastic-package. The following

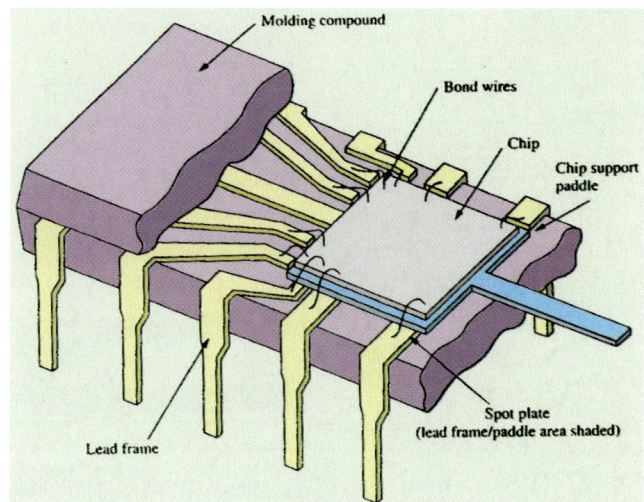

Fig. 27-20 A typical plastic package.

characteristics are desired of the plastic-molding-compound: low resin-viscosity; good-wetting of chip-components; good-adhesion to chip-components; short cure-time; low TCE; good dimensional-stability; and low moisture-permeability.

Transfer-molding is the standard technique for encapsulating ICs in plastic-packages. The process begins by loading the lead-frames (carrying the die-attached and wire-bonded chips) into the *mold-cavities* of a *transfer-molding-press* (Fig. 27-21). These cavities define the size, shape, and surface-finish of the plastic IC-package.

The mold has two halves (top and bottom), that open to receive the lead-frames. These are closed during the molding-cycle. A mold has multiple cavities (up to 400 for DIPs). A solid resin-pellet is put into a *pot* at the center of the mold, where it is heated until it melts. When the heated-resin has been liquefied to the correct viscosity, a hydraulically-driven piston forces it from the pot, through *runners*, to the mold-cavities. The resin flows into each cavity at a controlled-rate, determined by small openings called *gates*. (Since the bond-wires are long and fragile, care must be taken to prevent them from being damaged or deformed by the flowing-resin. This is accomplished by controlling the flow-velocity and the viscosity of the resin to within certain ranges.) Once resin has filled the cavity

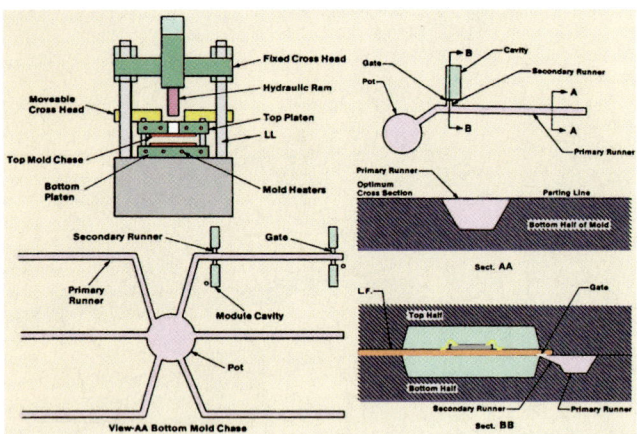

Fig. 27-21 The *transfer-molding process* for plastic IC-packages. (a) *Transfer-molding press* schematic. (b) *Bottom-mold chase*. (c) *Runner design*, in detail. (d) *Cavity* and *runner* cross-sections.

and surrounds the lead-frame and chip, curing within the mold takes place (typically for 1 to 5 minutes at 175°C and at high-pressure [e.g., 6 MPa]). During curing, some of the mold-release agent is extruded to the plastic/mold cavity-interface. After the molding-process is completed, the mold is opened and the packages are removed. The mold is cleaned of any remaining resin from the pot and runners.

Post-cure storage at 175°C after the packages are removed from the mold is needed to complete the curing of the epoxy-resin. The packages are then subjected to a *trim-and-form* procedure. That is, the outer-leads of DIPs and flat-packs are made with a *tie-bar* (which keeps the leads from being bent during the packaging-process). The *trimming-process* simultaneously cuts away the tie-bar and trims the leads to their proper length. In the *forming-process*, the leads are bent into their final position. A *deflashing*-step is performed to remove excess molding-material from the package-body. (*Flash* on molded-plastic parts is that small amount of excess-material which flows out through the *parting-line* where the mold-halves meet.) Deflashing is typically done by a physical-abrasion process on a machine similar to a sand-blaster, with plastic-beads as the abrasive.

Once completed, the package is identified with key information, in a process called *marking*. Typical information coded on the package is the product-type, device-specifications, date-of-manufacture, lot-number, and where it was made. The main methods of marking are *ink-printing* and *laser-inscription*. Laser-printing is well-suited for plastic-packages.

27.6.1 Plastic-Package Reliability-Problems

The failure-mechanisms of plastic-packaging are those due to mechanical-stresses, corrosion, and moisture-induced-failures. The stresses that cause IC- failure arise from epoxy-shrinkage during cure, and TCE-mismatches between the molding-compound, lead-frame, and chip-materials. When the resins cure, they shrink in volume by about 3 to 5%. Silicon has a TCE of $2.6 \times 10^{-6}/°C$. The TCE of a typical filled-epoxy-resin is $24 \times 10^{-6}/°C$. As packages cool from 175°C to room-temperature, a temperature-differential of over 100°C occurs, creating a large strain-mismatch. The resulting stresses can shear and completely lift-off wire-bonds. Measures to reduce these stresses include the use of compliant-materials between the plastic and the chip (such as polyimides), or the use of specially formulated low-stress molding- compounds.

The epoxy packing-material also contains filler-particles, which are non-spherical (e.g., they are tri-angular, or pointed instead). As a result they can exert a tremendous downward-force on the die, causing the passivation-layer to crack. A 5-μm-thick polyimide-layer is therefore applied to the die to act as a stress-relief-layer that can deform to accommodate forces exerted by the filler-particles. This radially dissipates such forces so they are not transmitted to the wafer-surface. If no polyimide-coating is used, severe cracking of the passivation-layer is likely.

When the resin cools, it shrinks and exerts compressive-forces around the chip, wires, and lead-frame. This forms a compressive-seal around the chip and lead-frame. However, this seal can be penetrated by moisture or other substances to cause corrosion of the chip-metallurgy. Water is transported to the chip from outside the plastic-package along two avenues: **1)** the interface between the lead-frame and plastic;

and **2**) by diffusion through the plastic-molding-material itself. The former problem can arise when the plastic and lead-frame materials separate to some degree. In a humid-environment, moisture and chemical-components from the PWB can migrate along the leads, climb across the wire-bonds, and finally arrive at the chip-metallization on the die-surface. Any chlorine present at the chip-surface can also react with the moisture to form HCl, which will corrode the Al metal on the pads.

27.7 ULSI PACKAGE-TYPES

ICs are also used in electronic-systems. The level of packaging in such systems above that of the chip-package, is the *printed-wiring-board* (PWB). The two ways IC-packages are attached to the PWB are: **1**) they are soldered into holes created in the PWB (Fig. 27-22); and **2**) they are soldered to the surface of the PWB (Fig. 27-24). Thus, two types of IC-packages are needed, depending on whether *through-hole* (TH) or *surface-mounting* (SM) of the IC packages is required by the PWB. Hermetic and plastic packages are available for both TH and SM PWB-applications.

27.7.1 Through-Hole (TH) IC-Packages

There are a number of popular TH package-configurations. The earliest TH IC-package was the *dual-in-line-package* (DIP). It became widely adopted due to its low-cost and ease of mounting. In fact, the DIP has become one of the icons of the IC-industry. DIPs are still being used, and hermetically-sealed ceramic-DIPs (Fig. 27-18) and plastic-DIPs (Fig. 27-20) are available. DIPs are thick, sturdy packages with two rows of outer-leads coming out of the sides of the package and bending down. While they are available with 4 to 80 leads, DIPs become less-satisfactory for ICs with larger pin-counts.

Fig. 27-22 Dual-in-line (DIP) *through-hole assembly*.

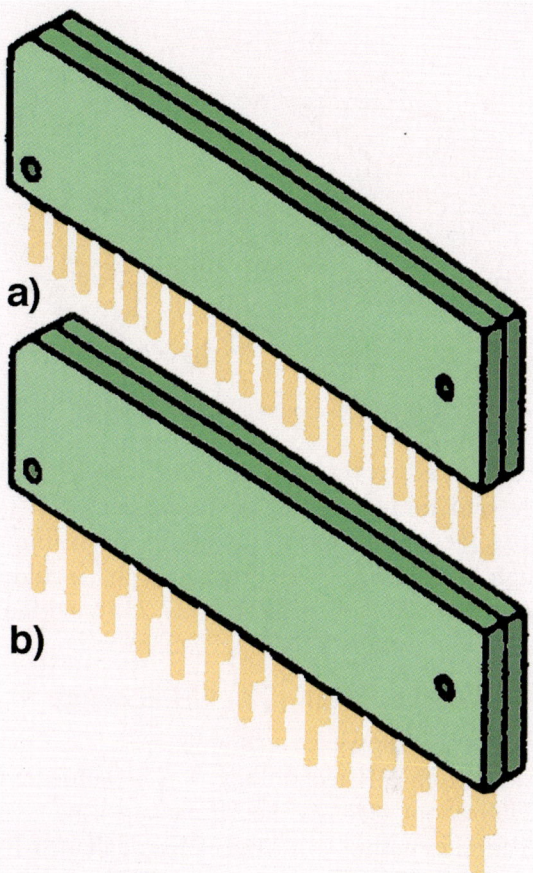

Fig. 27-23 (a) *Single-in-line* package (SIP). (b) Z*ig-zag-in-line* package (ZIP).

The *pin-grid-array* (PGA) has been developed to house such large I/O-circuits on TH-PWBs, and PGAs are available in ceramic or plastic package form. The leads come out of the bottom of the package in the form of pins (Fig. 27-19), and these pins are arranged in an *x-y*-array. PGAs can have hundreds of pins. The chip is attached in a cavity formed in either the top or bottom of the package-body.

TH-packages offer the advantages of easier placement for soldering and stronger solder-joints at the

Fig. 27-24 DIP surface-mount (SM) assembly.

PWB-level. However, they restrict PWB-design-flexibility, and limit the chip mounting-density. This is because there is a limit to the ability of drilling fine through-holes. Other TH package-types include the *single-in-line package* (SIP, Fig. 27-23a) and the *zig-zag-in-line package* (ZIP, Fig. 27-23b).

27.7.2 Surface-Mount (SM) IC-Packages

Figure 27-24 shows an example of surface-mounting. Although TH-technology was developed first, by 1998 about 80% of the IC-packages being used were surface-mounted. SM-packages enhance PWB-design-flexibility because it is not necessary to drill through-holes in the PWB. Finer-pitches of the PWB circuit-traces are possible because the pitch is limited by the etching-technology of Cu-foil, and not by the drilling of through-holes. Many of the first SM-packages were leadless-ceramic-packages mounted onto organic-PWBs. Unfortunately, they had a tendency to cause cracked solder-joints and broken-packages. As a result, SM-packages now use either *gull-wing-leads* (Fig. 27-25a) or *J-leads* (Fig. 27-25b). Where a smaller number of package-leads are needed, J-lead-packages are often preferred. The solder-joints of the J-leads are formed beneath the package, and hence they need less PWB-space than do gull-wing packages. The *small-outline-J-lead* (SOJ) and the *quad-flat-J-lead* (QFJ) are examples of such packages.

On the other hand, J-lead-bonds are more difficult to fabricate with high-yields at fine pitches. Thus, for circuits that have large pin-counts, gull-wing-lead-packages are more advantageous. Gull-wing-lead-joints are easier to form and inspect. Thus, finer-spacings between the joints allow circuits with larger pin-counts to be accommodated with smaller-packages. The *small-outline-package* (SOP) and the *quad-flat-package* (QFP) are examples of gull-wing

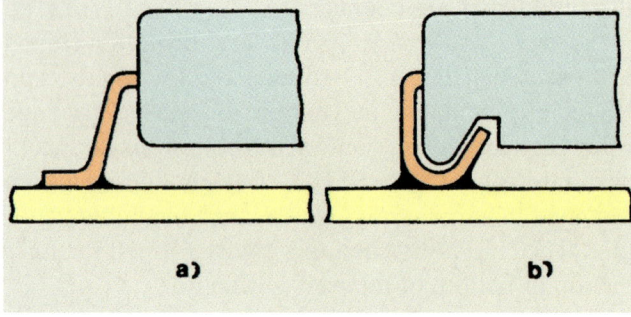

Fig. 27-25 (a) Surface-mount (SM) package with *gull-wing leads*. (b) Surface-mount (SM) package with *J-leads*.

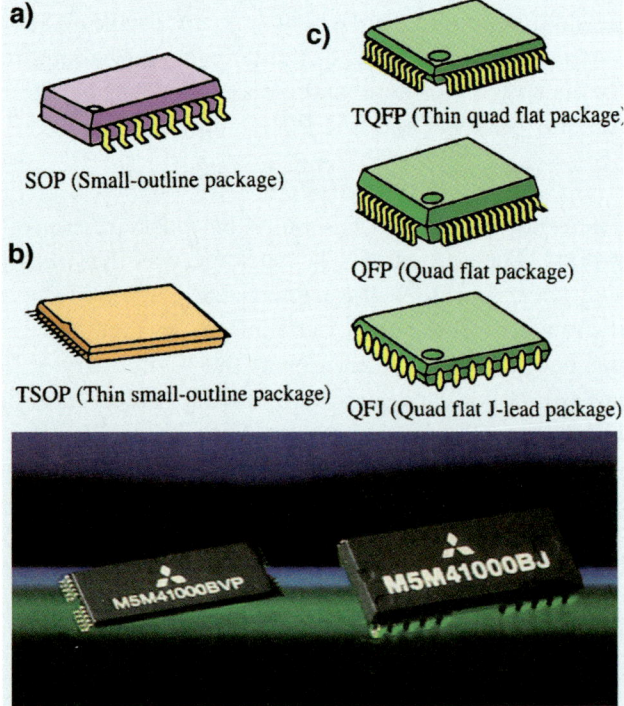

Fig. 27-26 Some of the many surface-mount packages that have been developed: (a) *Small-outline package* (SOP); (b) *Thin small-outline package* (TSOP); (c) *Quad flat-packs* (QPF); (d) Photograph of TSOP packages.

SM-packages. The *plastic-quad flat-pack* (PQFP) is an inexpensive version of the QFP. *Thin-small-outline-packages* (TSOP) have extended flat leads on two sides of the package (Figs. 27-26b & d).

27.8 TRENDS IN ULSI IC-PACKAGES

QFP and PQFP configurations become cumbersome for circuits with I/O-counts above about 208-pins. To accommodate chips needing larger lead-counts than this, QFPs must either increase in size (to where they take up too much board-space, or degrade circuit-performance), or the package-lead pitch must be decreased to the point where assembly-yields fall dramatically. There has thus been a move toward developing area-array-packaging, with the *plastic-ball-grid-array package* (PBGA) as an example (Fig. 27-27). The ball-grid-array package is a re-engineered pin-grid-array package, with the pins replaced by balls. The balls are usually 0.8-mm Pb:Sn-solder-bumps placed on the back-surface of the package at a spacing of 1-2-mm. The bumps are attached to the package after the plastic-molding process.

Another new family of IC-packages followed on the heels of the PBGA. They are collectively referred to as *chip-scale-packages* (CSP, Fig. 27-28). Their name stems from the fact that they have such a small form-factor (i.e., the footprint of a CSP is no larger than 1.2x that of the chip itself). CSPs appear to offer a path toward packages in which the cost will approach the 1-cent/lead cost of the SSI and MSI package-era. Their small size is also an aid to improving the package-performance. CSPs can be mounted to low-cost PWBs with standard SM-techniques. In 1998, CSPs were tagged as the fastest growing segment of the packaging-market.

A number of different approaches are being developed to produce chip-scale packages. Some use wire-bonding, others TAB, and still others flip-chip-techniques to interconnect the chip and package. Perimeter-leaded CSPs are suited for small-lead-count packages (e.g., 50–100-leads). High I/O CSPs use an area-array (e.g., a BGA) for the external-lead arrangement. It is predicted that such packages may eventually handle up to 1000-I/Os. Some packaging suppliers offer processing-services that prepare a finished-wafer for their package-type (e.g., they fabricate the TAB or flip-chip bumps on the wafer.) Another innovation is the use of a flexible *interposer* (the electrical interconnection structure between an IC and its package).

27.8.1 Packageless-Technologies

Despite the fact that IC-packages perform an important set of tasks in electronic-systems, they have also been viewed by some as merely a necessary evil. It has been argued that the IC-package adds no value to the product, but instead enlarges the physical presence of the IC and robs it of some of its performance-potential. As a result, some systems-designers have sought to get rid of the IC package by interconnecting ICs directly to PWBs. In fact, such *direct-chip attachment* (DCA) methods as flip-chip (or C4), and beam-leaded chip-technologies were introduced in the mid-1960's by IBM and AT&T, respectively. Yet, until recently the overall advantages offered by IC-packages prevented such "packageless" approaches

Fig. 27-27 *Ball-grid-array* (BGA) package.

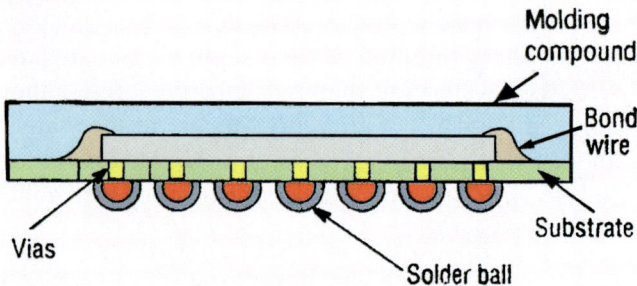

Fig. 27-28 Example of chip-scale-package (CSP).

from being widely implemented.[15,16]

By the late 1990's, however, the costs of IC-packages for ULSI had risen precipitously. Since packageless-technology appears to be a method in which costs can be reduced, interest in them has revived. In fact, this approach has been adopted for such applications as watches and smart-cards. Since rework of failed-circuits is not yet feasible in packageless-applications, this attachment-technology is usually reserved for mature, high-yielding circuits.

It has been noted by those in favor of CSP-technology that packaging-procedures are not actually eliminated by such techniques as FC-DCA. Instead, the packaging-steps are performed *in-situ* on the substrate where the die is flip-mounted and is then either encapsulated or underfilled. Thus, the costs of packaging are not completely eliminated. CSPs were described by their proponents as offering even lower-costs than packageless-technologies. Other experts, however, contend that FC-DCA is the ultimate IC-package-solution, and that CSPs are just stepping stones to FC-DCA. It seems, in any case, that CSP and FC-DCA will both be rapidly adopted for electronic-systems beyond the year 2000.

REFERENCES

1. S. Wolf and R.N. Tauber, *Silicon Processing for the VLSI Era, Vol. 1*, 2nd Ed., Lattice Press, 2000. Ch. 17, p. 841.

2. *Principles of Electronic Packaging*, Eds. D.P. Seraphim, R. Lasky, and C.-Y. Li, McGraw-Hill Book Company, New York, 1989.

3. *Microelectronic Packaging Handbook*, Eds. R.R. Tummala and E.J. Rymaszewski, Van Nostrand Reinhold, New York, 1989.

4. J. Baliga, "Package Styles Drive Advancements in Die Bonding," *Semiconductor International,* June 1997, p. 101 & Dec. 96. p. 89.

5. R. Shukla and N. Mencinger, "A Critical Review of VLSI Die-Attachment in High-Reliability Applications," *Solid State Technology,* July 1985, p. 67.

6. J. D'Ignazio, "Wirebonding's Reign Continues," *Semiconductor International,* June 1996, p. 117.

7. C.A. Steidel "Assembly Techniques and Packaging," in *VLSI Technology,* S.M. Sze Ed., p. 551, McGraw-Hill, New York, 1983.

8. L. Oboler, "Wire Bonding Still at the Head of the CLass," *Chip Scale Review,* July/August, 1999, p. 40.

9. G. Oswald and W.R. deMiranda, "Application of Tape Automated Bonding Technology to Hybrid Circuits," *Solid State Technology,* pp. 33-38, 1977.

10. D.L. Brownell and G.C. Waite, "Solder Bump Flip-Chip Fabrication Using Standard Chip & Wire Integrated Circuit Layout," *1974 ISHM Conf. Proc.,* p. 77, Sept. 1974.

11. P. Burggraaf, "Chip Scale and Flip Chip: Attractive Solutions," *Solid State Technology,* July 1998, p. 239.

12. G. Murakami, "Packaging Industry: Road Map to the Future," *Semicon/West Proc.,* p. 99-123, 1994.

13. P.V. Robock and L.T. Nguyen, "Plastic Packaging," in *Microelectronics Packaging Handbook,* R.R. Tummala and E.J. Rymaszewski, Eds., Ch. 8 pp. 523, Van Nostrand Reinhold, New York, 1989.

14. D.A. Jeannotte, L.S. Goldmann, and R.T. Howard, "Package Reliability," in *Microelectronics Packaging Handbook,* R.R. Tummala and E.J. Rymaszewski, Eds., Ch. 5, p. 225-359, Van Nostrand Reinhold, New York, 1989.

15. T. DiStefano and J. Fjelstad, "Chip Scale Packaging Meets Future Design Needs," *Solid State Technology*, April 1996, p. 82.

16. J. Fjelstad, "Trends in Low-Cost IC Packaging," *Semiconductor International,* December 1996, p. 73.

PROBLEMS

1. Name the four functions of a chip-package.

2. Describe the chip-to-package (*die-attach*), and bonding-pad-to-package (*wirebonding*) assembly processes.

3. What are the advantages and disadvantages of *ceramic-packages* versus *plastic-packages*. List the types of products that would call for ceramic-packages. Repeat for plastic-packages.

4. Describe *tape-automated bonding*.

5. Describe the *flip-chip* assembly process.

6. Describe the difference between *through-hole* and *surface mounting* of packaged chips to a PC-board.

7. Define a hermetic-package, and give an example of a hermetic and a non-hermetic package

8. What chip-size is required to place 352 pads (I/Os) on a pad-limited chip? Assume the chip is square, the bonding-pad pitch is 120 μm, and the chip-size is defined by the loci of bonding-pad centerlines.

9. If a package-cavity has a volume of 0.1 cm^3, and the water-vapor trapped in the cavity after the package is sealed is present at a concentration of 5000 ppm by volume. How much silicon (in grams) is required to react with all of this water? Assume atmospheric-pressure and room-temperature within the cavity.

CHAPTER 28

WAFER-FAB OPERATIONS AND YIELD

CHAPTER CONTENTS

28.1 WAFER-FAB
OPERATING COSTS
  Overhead-Costs
  Materials Costs
  Equipment Costs
  Labor Costs

28.2 WAFER-FAB LAYOUT
AND WAFER-TRANSPORT

28.3 STRATEGIES FOR
IMPROVING THE EFFICIENCY
OF IC-MANUFACTURING

28.4 YIELD-MANAGEMENT
& THE "LEARNING-CURVE"
  Yield Ramps
  Statistical Process-Control

28.5 ORGANIZATIONAL
STAFFING OF WAFER FABS

"Real men have fabs"
  Walter J. (Jerry) Sanders III,
  CEO AMD

21st Century wafer fab. Shown is the AMD-Dresden, Germany facility. Courtesy of Advanced Micro Devices.

Today's semiconductor industry is the foundation of the trillion-dollar electronics industry that globally provides tens of millions of jobs. It is the driver that has propelled high-technology to become the leading source of economic-growth in the industrialized world. These benefits have largely come from the ability to consistently double the number of transistors on a chip (Moore's Law, Chap. 1, Fig. 1-24).

The manufacturing facilities (wafer-fabs), in which chip-production occurs, are cleanrooms full of highly-specialized equipment, attended to by skilled production-workers. The cost to build the most advanced wafer-fabs exceeded $1 billion by the turn-of-the-century (see Fig. 1-28, Chap. 1). Along with these changes, have come rising materials and labor costs, and increases in process-complexity. The newer ICs also require more fabrication-steps. All of these cost-increases have been occurring as individual chip-prices continue to drop (through improved productivity and competition). To stay profitable, chip-manufacturers have had to relentlessly improve efficiency, yield, and cost-control at all levels.

Despite these challenges, the overall financial measure of a wafer-fab has remained the same: the *cost-per-functioning-die shipped out of fabrication*. The topics related to this *bottom-line issue* (of being able to make a profit by successfully operating a wafer-fab), are the subjects of this chapter. That is, it is mainly concerned with how manufacturing principles are applied to the technologies described earlier, so that they can be fabricated with a high-degree of success.

In this chapter we will discuss: **1)** *wafer-fabrication costs*; **2)** *layout of advanced wafer-fabs and the transport of wafers within a fab as they move through a process flow*; **3)** *yield* and *yield-modeling*; and **4)** the *organizational-staffing of wafer-fabs*.

As a final introductory note, the expenses of building and operating semiconductor-fabs have grown so large that many firms find the *cost of producing chips* in their own fabs to be less than the chip *selling-price*. Thus, they are increasingly turning to *wafer-foundries* to do some (or all) of their manufacturing. By sharing the cost of fab-operations, the *production-costs* are reduced to the degree that chip-firms can remain profitable. Those without any fabs are called *fabless semiconductor companies*.

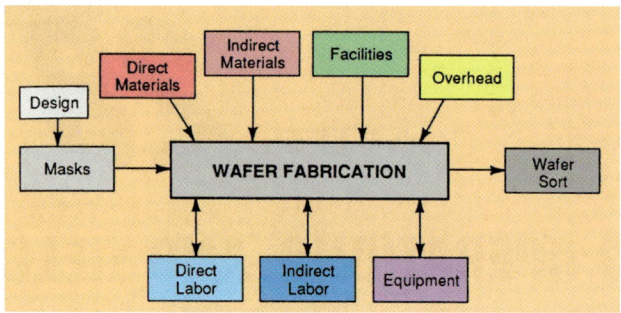

Fig. 28-1 Costs associated with operating a wafer fab.

28.1 WAFER-FAB OPERATING COSTS

As in any business venture, there are costs involved in operating a semiconductor-manufacturing company. The purpose of a business is to sell the products they make (in this case, *integrated circuits*), for more money than it costs to produce them - and thus make a profit. Figure 28-1 illustrates the main factors that contribute to production-costs of devices manufactured in wafer-fabs. A more detailed breakdown of these costs is given in Sections 28.1.1 through 28.1.4.

The *facility*, *overhead* (*administration*), *equipment*, and *labor costs* shown in Fig. 28-1 are *fixed-costs*. That is, they exist as a cost-burden whether

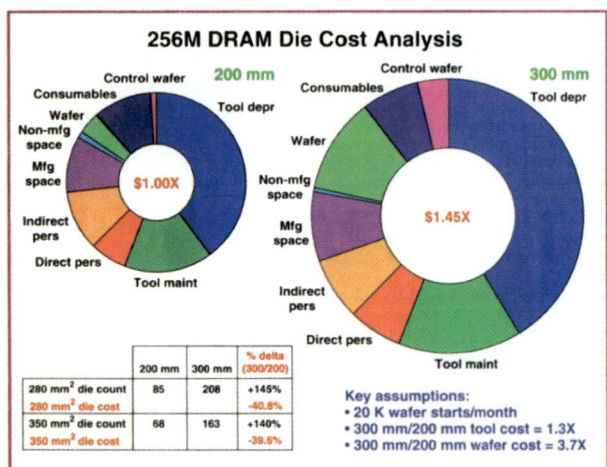

Fig. 28-2 Comparison of the cost-breakdown to manufacture 256-Mb DRAM-chips on 200-mm wafers versus 300-mm wafers. Although the cost-per-wafer to manufacture 300-mm wafers is 45% higher than for similarly-manufactured 200-mm wafers, the increased die-count on 300-mm wafers results in a lower cost-per-die.

or not any chips are produced. Over a longer time period, the *labor costs* also become *variable-costs*, as staff is added and let-go, to fit changing production-levels. The other *variable-costs* are the *materials consumed* in IC-fabrication, as these also fluctuate with the volume of wafers produced. Roughly speaking, *materials* represent the largest of the above costs, with *equipment* and *labor* being second and third. The least-major expenses are *overhead* and *utility* costs.

It should be noted that semiconductor-manufacturing costs also can be expressed in two other ways: **1)** the *production-cost-per-wafer* for a given fab; and **2)** the corresponding *cost-per-die*. For example, the cost-breakdowns of 256-Mb DRAM-chips on 200-mm and 300-mm wafers in 2002 are compared in Fig. 28-2. It can be seen that while the *cost-per-wafer* is higher for 300-mm wafers, the *cost-per-die* is lower.

28.1.1 Overhead-Costs

Overhead-costs are associated with paying for the administrative and executive staff of a wafer-fab, plus the costs of providing and maintaining a facility. One aspect of administrative-costs is that they rise faster than manufacturing-costs as a fab grows. This is because more information is generated about internal operations, and more data also needs to be handled from customers and suppliers. To be effective, this information must be available to an ever-growing staff. These two needs require more personnel to process information than to make products. (In excess of 50-percent of the work force in industrialized economies is involved with information-processing.) Another overhead-cost involves *chip-design* efforts. With expensive CAD-systems and large, professional design-groups, circuit-design costs are significant.

Providing and maintaining a *facility* is also a major expenditure. The cleanroom occupies only about 20-percent of the square-footage of a wafer-fab, yet it requires the majority of the expense to build and operate. Air-conditioning, utilities, chemical-storage and delivery, and the many other costs of building and operating a cleanroom are considerable. The cost just to build a modern IC-cleanroom are several-thousand dollars per square-foot.

28.1.2 Materials Costs

From an accounting perspective, the materials used in manufacturing chips are divided into two groups: *direct* and *indirect*. The materials that are considered to be *direct costs* are those that go directly into or on a chip. This includes the starting-wafers, added layer-materials, and packaging materials. The materials treated as *indirect costs* are those that support the process, but do not enter into the product (i.e., masks and reticles, cleaning chemicals, and photoresists).

28.1.3 Equipment Costs

The cost of equipment involves the tools used directly in the fabrication of wafers and ICs. It is included in cost-calculations either as a fixed overhead-cost, or as depreciation. The cost-per-tool has increased significantly with time. In 1975, the cost for a single-piece of wafer-processing equipment was about $10,000. By 1985, it had risen to about $100,000. In 1995, the price was between $500,000-$1 million. By 2003, the cost of 300-mm processing-tools ranged between $5-10 million per-tool.

28.1.4 Labor Costs

Labor costs have two components: *direct-labor costs* and *indirect-labor costs*. Direct-labor costs are associated with those workers that actually handle the wafers and equipment. Indirect-labor costs involve those personnel who support the production workers. The latter include supervisors, engineers, facility technicians, and office workers. More details on the staffing of a wafer-fab are given in Sect. 28-5.

28.2 WAFER-FAB LAYOUT AND WAFER-TRANSPORT

Wafer-fabs are constructed to allow wafers to be manufactured within them as efficiently as possible. Wafer-production itself is carried out in cleanrooms, as described in detail in Chap. 8. The cleanroom *layout* has evolved as the semiconductor industry has matured, and floorplans of early cleanrooms (and other aspects of their design) are described in Chap. 8, Sect. 8.4.1. Here we continue this discussion, focusing on *modern cleanroom-layouts* (i.e., those used to fabricate VLSI and ULSI ICs).[3]

The material movement in a VLSI wafer-fab is complex, because during each process-run some steps are skipped while others are repeated several times. (For example, there are more film-deposition steps than there are implant-steps.) Often, the wafers may not simply bypass one section; rather, they may take shortcuts and crisscross back and forth between various steps.

In any case, equipment with similar functions is generally clustered in one area (or in one *process bay*, or *bays*, as shown in Fig. 28-3). There are a number of reasons for organizing a cleanroom along functional-equipment lines rather than as sequential, in-line operations. First, the facility-support costs are lowered. Second, it allows for operator-specialization, with less training. The layout shown in Fig. 28-3 is typical of cleanrooms that were built up-until about the mid-1990s. The wafers in such fabs are moved

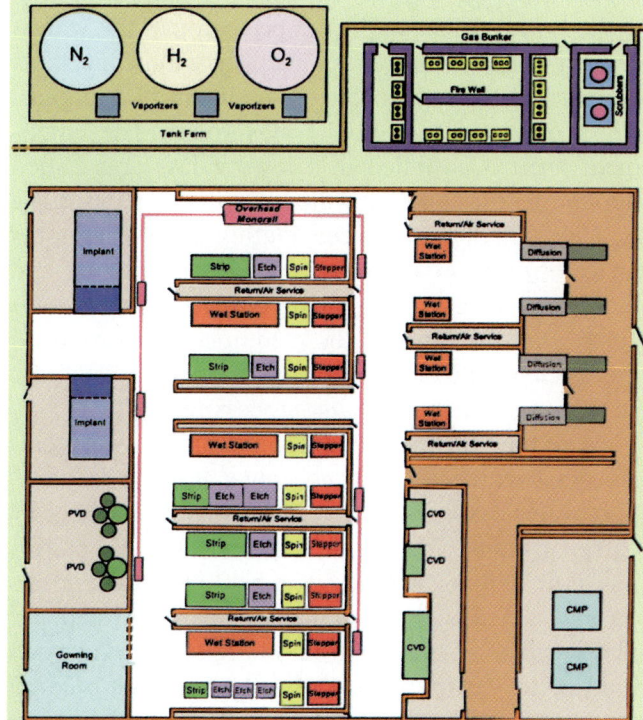

Fig. 28-3 Wafer fab floor-plan, circa 1990. Source: VLSI Research Inc.

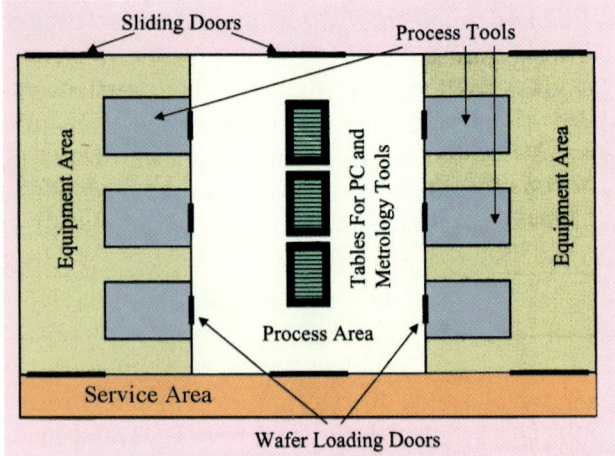

Fig. 28-4 Schematic showing the process (i.e., work) areas and the equipment (chase) areas of a wafer-fab of the kind shown in Fig. 28-3.

between process-tools by operators, who carry them in cassettes from one workstation to another.

A more-detailed drawing of the work-area in a process bay of such cleanrooms is shown in Fig. 28-4. In this schematic it can be seen that the wafers remain either in a process area, or within the tools themselves. These locations maintain the highest Class of cleanliness in the cleanroom. The equipment itself is installed in a lower-Class area called the *chase* (or sometimes the *gray area*). Chases are typically Class-1000 spaces. Putting the process tools in lower-class spaces reduces both the building and maintenance costs. Equipment engineers and technicians work primarily in these equipment areas, as do the tool manufacturers' support-staff while installing, starting up, and maintaining process-tools.

With the transition to 300-mm wafers, it was realized that workers would no longer be able to transport wafer-cassettes long distances between tools by carrying them. The weight of a cassette that carries 300-mm wafers is too heavy to be moved by workers on a sustained basis. (Note that there are two standard wafer-carrier sizes for 300-mm wafers, one which can hold up to 13-wafers, and the other that can hold up to 25-wafers. The weight of 25, 300-mm wafers alone is 3200-gm [or about 8-lb]). As a result, automated-transport of such wafer-carriers among the various process tools is necessary. Thus, cleanrooms in which 300-mm wafers are processed must be layed-out with this consideration in mind.

Figure 28-5 shows a popular layout (called a *spine layout*) for such 300-mm fabs. The wafers are transported among the bays by either an *overhead, monorail-loop automated transport system*, or *an automated guided vehicle (AGV) system*. Such inter-bay-systems are called *automated material handling systems* (AMHS), and they are located in the main central-aisle of the cleanroom. The overhead-loop systems are usually preferred, since they have a much higher throughput and require less floor-space. Usually there is only a single transport-loop running in one direction. But in fabs where very-high transport-volumes exist, there are multiple interbay-transport loops, running in opposite directions. Interbay-transport systems are judged on their ability to deliver wafer-lots to their intended destinations reliably and within the expected time-of-delivery, and without impacting fab-personnel safety or product-yield.

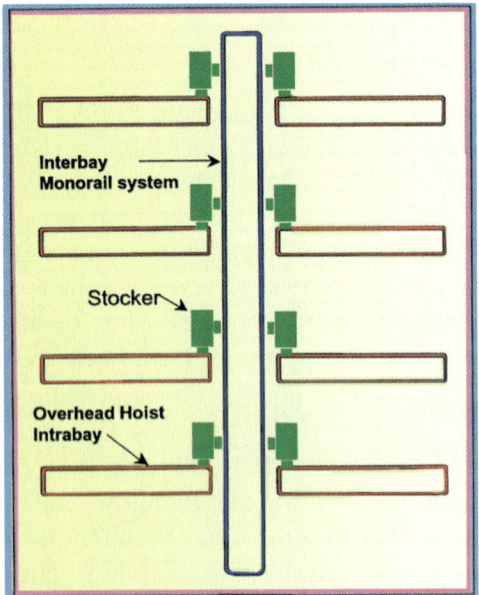

Fig. 28-5 Wafer-fab layout, circa 2000. An *overhead interbay-transport loop* in the central-aisle of the cleanroom moves wafers-carriers between each bay. *Stockers* at each process-bay (located at the entrance of the bay) are connected to this interbay-transport loop system.

At the entrance to each bay is a *stocker* which contains a multiplicity of vertically-stacked storage-locations (or bins). A computer-controlled system within the stocker can *pick-and-place* wafer-carriers from/into these bins, and then put them into input-output ports (which hand-off the carriers to AMHS transport elements, or to fab operating-personnel). That is, production workers access wafer-carriers at the stocker input/output-ports and then manually transport them the short-distance within the bay. However, some fabs also use automated *intrabay-transport* systems.

The role of such *automated intrabay-transport systems* is to move wafers to and between production-equipment within process-bays, as shown in Fig. 28-6. As can be seen, such systems can be either ceiling-based *overhead-hoist transport* (*OHT*) systems, or *automated-guided-vehicle* (*AGV*) and *rail-guided vehicle* (*RGV*) systems. The intrabay-vehicles deliver carriers of unprocessed-wafers to the *production-equipment loadports* for processing. Standardization of loadports for 300-mm tools will make intrabay-AMHS more effective. (See Fig. 8-13 in Chap. 8 for how intrabay transport-systems interface with the loadports of 300-mm tools.) Note that *manual-carts* serve as start-up intrabay-transport systems while a process and equipment are under development. They later become back-up transport systems whenever the intrabay system is down, either for scheduled maintenance or repair.

28.3 STRATEGIES FOR IMPROVING THE EFFICIENCY OF IC-MANUFACTURING

As described in Sect. 28.1, various costs are incurred in the process of producing the output of a wafer-fab (i.e., *microchips*). In order for a chip-manufacturing operation to be profitable, the income generated from the sale of these chips must exceed the cost to build them. To be profitable in today's competitive markets, a chip-manufacturing company must be able produce functional components at high-volume and sufficiently-low cost. Since many costs of running a fab are fixed, it becomes more profitable if the number of chips being built by a fab can be increased.

Wafer-fabs generally have a fixed-capacity (i.e., a

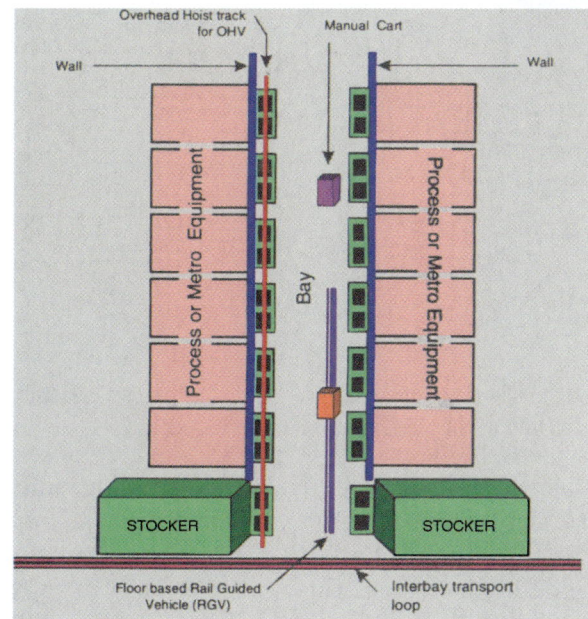

Fig. 28-6 Two different interbay transport technologies for wafer-fab layouts such as those shown in Fig. 28-5.

specific maximum-number of wafers per month can be run through a fab). Thus, an obvious way to improve the efficiency of its operation is to *increase the output of functional-die per wafer*. There are three ways this strategy can be pursued: **1)** By increasing the number of chips on a wafer *by making the product die as small as possible*. (A great deal of effort is expended to design ICs that can meet this goal, without violating the design-rules); **2)** By increasing the number of chips on a wafer by *increasing the wafer-size*; and **3)** By improving the *yield* of the chips produced per wafer. Of them, Approach-**2** is covered in Sect. 28.3.1, and Approach-**3** in Sect. 28.3.2. (Approach-**1** is covered in texts that deal with IC-design.)

28.3.1 The Economic Impact of Wafer-Size

With the passing of time, IC manufacturers have been fabricating ICs on wafers with ever-larger diameters. This has occurred for three reasons:

1. Since its area is increased, each wafer can carry more die (see, for example Fig. 28-2).

2. The percentage of chips on the periphery of a wafer that are *partial-die* declines as the wafer-

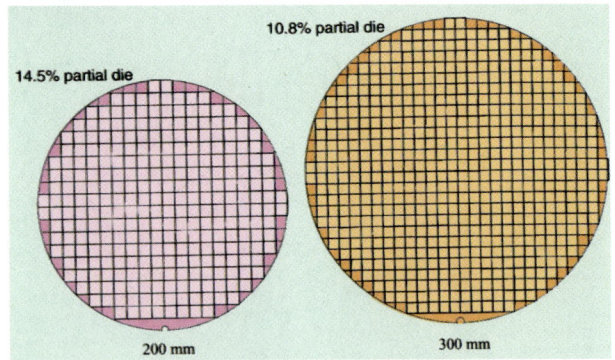

Fig. 28-7 As wafer sizes increase, the percentage of the partial-die at the wafer-periphery is reduced.

diameter is increased (Fig. 28-7). Partial-die-chips are non-functional. Thus, if their percentage on a wafer can be reduced, the number of potentially-good-die increases. This secondary-factor causes the usable die-count to increase even more rapidly than does the wafer-area.

3. It becomes economically more effective to manufacture ICs on bigger wafers. Let us examine why this is so in more detail.

As IC-technology has evolved, it has become more cost-effective to produce chips on larger-diameter wafers. For example, in 1985 it was advantageous to run 125-mm wafers. By 1990, fabs working on 150-mm wafers were the most efficient. In 1995, the most-profitable factories were running 200-mm wafers.

These increases in wafer-size about every five years suggests a trend. If it had continued, by the year 2000 the wafer-size should have been 250-mm. However, since the next wafer-size was chosen to be 300-mm, it would be reasonable to guess that in 2000 many facilities would still have been running 200-mm wafers (as was indeed the case). But, they would also be using many tools with a built-in capability to switch to 300-mm wafers, as the need arose. However, this analysis also indicates that 300-mm production may not reach its prime until about 2005.

28.3.2 IC Fabrication-Yield

Perhaps the most fruitful way of increasing the functional-die count is to use the third approach listed in Sect. 2.8 (*yield improvement*). If all the chips produced on a wafer met product-specifications, they could be sold (thus, maximizing income and profit). However, not all the chips produced work properly, and those that do not, must be scrapped. But, if the percentage of chips that are successfully fabricated, packaged, and tested can be increased (i.e., *yield* is improved), the profitability of the fab will also be increased. Furthermore, unlike moving to larger wafers, this increase in the number of saleable-units is achieved without making large capital-investments. In this section we discuss this issue in more detail.

The *cumulative* (or *overall*) die-yield Y is the product of several sub-yields, namely: *fab yield*, Y_f; *in-line/parametric-test yield*, Y_p; *wafer-sort yield*, Y_s, *assembly/package yield*, Y_a, and *final-test yield*, Y_t:

$$Y = Y_f Y_p Y_s Y_a Y_t \quad (28.1)$$

where Y_f is the fraction of wafers that *completes the entire process-flow* (i.e., there may wafer-breakage, or other mishaps that cause wafers to be scrapped during a process-run); Y_p is the fraction of wafers that pass the *parametric-test hurdle* at the end of the process-flow; Y_s is the fraction of die that *pass the wafer-sort test* (see Sect. 26.3 in Chap. 26); Y_a is the die-fraction that have *passed wafer-sort and are then successfully packaged*; and Y_t is the fraction of packaged-chips that passes *final-test*.

In most modern IC manufacturing-scenarios, the value of Y_a and Y_t are close to unity. In addition, the increased use of automated wafer-transport *among* process-tools and robotic-movement of wafers *within* tools has also reduced the incidence of wafer-breakage to the degree that Y_f can also be considered to be unity. Thus, to a first approximation, only the terms Y_p and Y_s contribute to the final-yield (which we will term the *random-defect process-yield*, Y_R). The rest of the discussion in this section thus concerns only Y_R.

Definition of Yield: For purposes of this discussion, *yield* is thus formally defined as the ratio of *good-chips* per-wafer to the *total-number of chips* per-wafer. A *good-chip* is defined as one that has passed all the parametric and functional tests that are specified for a product.

Definition of a Yield-Model: The historical meaning of the term *yield-model* refers to the mathematical relationship between the *random-defects* on a wafer and the number of otherwise good-die on the wafer that will fail as a result of the presence of defects. In practice, *yield-models* are *formulas* that predict what percentage of the die on a wafer will survive if there is some known-density of *randomly-distributed (killer) defects* on a wafer. Yield-model formulas are typically derived in the following way: First, it is presumed that any killer-defects on a wafer are distributed in a manner described by some statistical-distribution function (such as a *Poisson distribution*). Next, certain assumptions are made about the variations in the spatial-distribution of the *killer-defects*. The mathematical-formulas are then derived, based on these postulates.

Yield-modeling is an important tool for managing semiconductor-manufacturing operations mainly because it helps improve current-generations of processes and designs. Yield-models can be exploited to obtain such benefits in the following ways:

1. Yield-data can be used to partition yield-losses into groups, each caused by a different type of defect. This allows resources to be allocated to the yield-enhancement projects with the highest-payback. (For example, when it was learned that *contact-printing* was a yield-limiting source of killer-defects, work was done to develop *projection-printing* as a replacement.) Furthermore, these insights can help achieve a faster learning-rate for new products (as shown from the data in Fig. 28-10).
2. It makes possible accurate-forecasting of yields, which aids in production-run planning.
3. Yield-modeling can help to establish product-specifications that match process-capability.
4. It can be used to compare process-performance of different fabs, thus indicating where improvements in processing-facilities are needed.
5. With accurate yield-models, the cost and availability of future-circuits can be established (provided they are related in technology and design to the circuits used to develop the yield-modeling parameters). For example, yields of chips of different-sizes other than the ones used to develop the model, can be predicted.

28.3.3 Basic Mathematical Yield-Models

Using the above approach, a first-order approximation used to model the random-defect process-yield Y_R for a given process step is:

$$Y_R \cong 1/\exp(D_o A) \qquad (28.2)$$

where D_o is the average-number of killer-defects-per-unit-area introduced by that step, and A is area of the IC-chip (as defined in Fig. 28-8). Note that all yield-models predict that the yield decreases monotonically as the area of a chip increases.

As an example of using this formula, let us assume that a 10-masking-step process is to be performed (i.e., ten process-steps will occur, so $N = 10$). If each masking-level introduces a density of defects D_o onto the wafer, then the final-value of Y_R is

$$Y_R \cong 1/\exp(N D_o A) \qquad (28.3)$$

Figure 28-9 plots the mask-limited Y_R values for a 10-level lithographic-process as a function of chip-size for various values of defect-densities, using Eq. 28-2 for Y_R. For example, if $D_o = 0.5$ defects/cm^2, the yield is 22% for chips of 30-mm^2, but it drops to about 1% for larger-chips of 90-mm^2. This shows the need for cleanrooms (and inspection and cleaning of wafers),

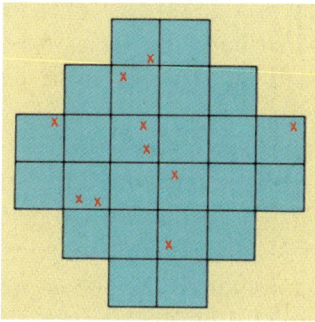

Fig. 28-8 Example of the definition of *defect-density*, D_o. Here there are 24 sites, each assumed to have an area of 1-cm^2. Since there are 10 *killer-defects* randomly distributed over the entire area, the defect-density is $D_o = 10/24 = 0.42$ defects/cm^2. Note that the area A is not defined as the entire wafer area, but just the portion susceptible to defects.

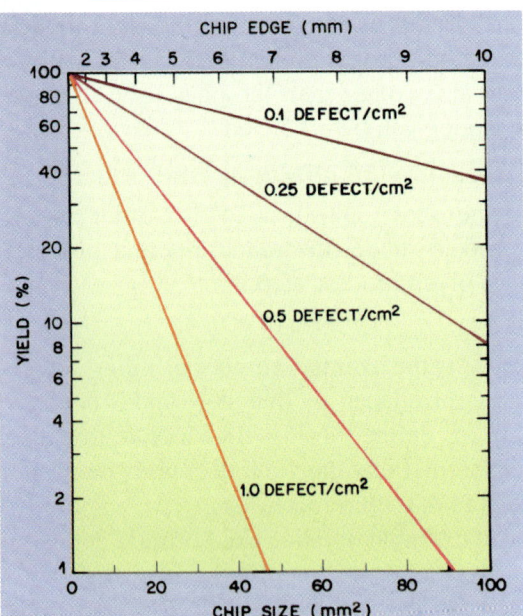

Fig. 28-9 Yield for a 10-mask lithographic-process with various defect-densities per level.[22] By permission of ATT Bell Telephone Laboratories.

if high-yields on large-chips are to be obtained.

28.3.4 Yield-Models for Large-Die-Size ICs

Although the above example is instructive, it should be understood that Eq. 28-2 makes an assumption that may not be valid, especially for chips with large die-sizes. (In fact, it has been found that using Eq. 28-2 typically underestimates the yield of ICs made on larger die.) That is, Eq. 28-2 assumes that the defects are randomly-distributed on the wafer-surface. Thus, the defect-density is assumed to be uniform across a wafer, and from wafer-to-wafer. However, D_o is generally not constant across an entire wafer (defects often cluster together), nor from wafer-to-wafer. Thus, if the value of D_o is assumed not to be constant at all locations, but instead has some statistical distribution, this can be incorporated into the expression for the yield-model as well. Two such distributions have been assumed: **1)** that the value of D_o across a wafer exhibits a *Gaussian distribution*; or **2)** that the value of D_o across a wafer has an *exponential distribution*.

If it is assumed that the D_o-variation has a Gaussian distribution, *Murphy's yield-model* is obtained:[1]

$$Y_R(\text{Murphy}) = ([1 - \exp\{D_o A\}]/D_o A)^2 \quad \textbf{(28.4)}$$

If it is assumed that the D_o-variation has an exponential distribution, *Seed's yield-model* is derived:

$$Y_R(\text{Seed}) = 1/\exp(\sqrt{D_o A}) \quad \textbf{(28.5)}$$

Thus, when calculating the yield of ICs with larger die-sizes, either the Murphy or Seeds model is generally used instead of Eq. 28-2.

It should also be pointed out, however, that while one or the other of the Y_R yield-models may provide accurate predictions for products, their predictions will be valid *only if the yields are being limited just by random-defects*. However, for a more complete predictive-capability, *systematic yield-losses* must also be taken into account. A yield-model that accounts for both kinds of defects provides the *total process-yield* Y_T for a given product, expressed as the product of the systematic-yield Y_{sys} and the random-yield Y_R

$$Y_T = Y_{sys} Y_R \quad \textbf{(28.6)}$$

Thus, it is useful to distinguish these two defect classes in more detail (*random* versus *systematic*).

The *random-yield* Y_R is only a function of *random-defects*. Such defects occur as a result of particle-deposition taking place during processing. The presence of such particles produces features on a die not intended in the design-layout. If random-defects occur in a critical-region of a die they can cause opens or shorts (or some other kind of killer-defect, see Fig. 8-6, in Chap. 8). On the other hand, *systematic yield*, Y_{sys}, is a function *systematic-defects*, which arise from process-variables not meeting specifications. This can be the result of photomask-misalignment, under-etching, or design-flaws (such as a design not meeting minimum-spacing rules).

Often, only the second-term in Eq. 28-6 is "modeled" (as was done in the example in Sect. 28.3.3, that used Eq. 28-3 to calculate the yield for the 10-masking-level process). A more complete yield model will employ both terms, and thus methods for estimating Y_{sys} are also needed. This topic is beyond the scope of this book, but it is discussed in Ref. 1.

28.3.5 Defect-Detection and Analysis

Some additional comments need to be made about the

defect-density term D_O introduced in Sect. 28.3.3. Its' value is determined from *in-line detection equipment* deployed at critical-points in the process-flow. Such inspection-tools provide size, number-density, and coordinate-location of detected defects. A yield-management software/hardware system collects and analyzes data from these *inspection-and-review* tools, and correlates them to electrical-test, parametric-test, and factory *computer-integrated-manufacturing* (CIM) data, such as the equipment and process recipe used.

The objectives of such inspection-programs are to provide a baseline defect-density profile of the process, and to detect defect-excursions immediately after they occur. Typically, 2-3 product-wafers from 50-100% of all lots are inspected on patterned-wafer inspection-tools (see Sect. 8.8.3, in Ch. 8). The same wafers are inspected after different steps, allowing for a calculation of the cumulative-number of defects in the line, and partitioning of defects from different process-segments. Partitioned defect-results can be displayed on an *in-line-defect pareto-chart* (Fig. 28-10). One point that becomes obvious is that the defect-density varies with the process-segment (unlike the assumption made in the example of Sect. 28.3.3). Pareto charts also indicate where allocation of resources to reduce defect-densities might best be applied. Figure 28-11 shows defect types that can lead to chip-failure at different layers in IC-structures.

28.4 YIELD-MANAGEMENT AND THE "LEARNING-CURVE"

The function of a yield-management program is to

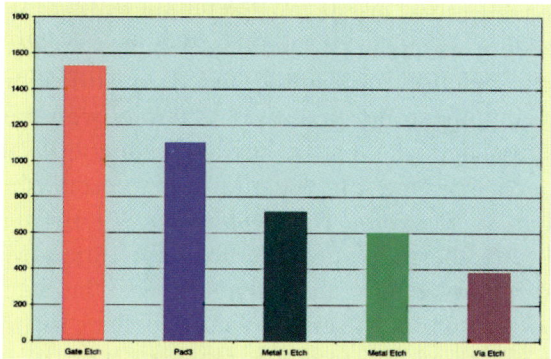

Fig. 28-10 Example of an in-line defect pareto-chart.

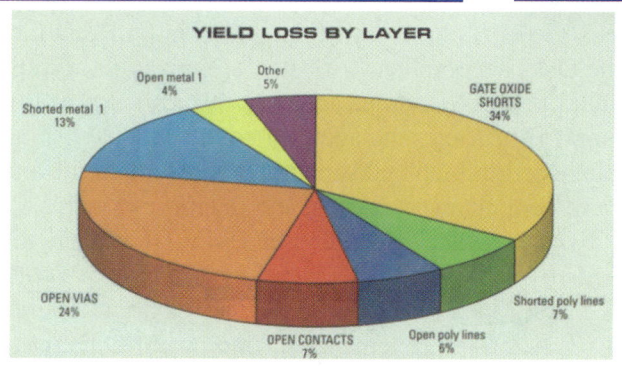

Fig. 28-11 Defects that cause yield-loss, as attributed to each layer. Courtesy of Kiethley Instruments, Inc.

improve yields by predicting, detecting, reducing, and preventing yield-losses in the shortest-possible time. For example, when a new-generation of technology is first introduced, IC-yields may be quite low (e.g., only 20%). Since in this case, 80% of the potential products being manufactured must be scrapped, the ability to quickly understand the origin of yield-losses (and to correct them), is critical to the economic-viability of a production-line. Not only does this understanding allow a chip-manufacturer to focus resources on the causes of the problems, but it also allows an accurate-projection of the supply of working parts (so that the proper number of wafers can be started in fabrication). The degree to which each of these goals is pursued depends on the maturity of the process.

28.4.1 Yield Ramps

As shown in Fig. 28-12, there are five stages of a *yield-ramp* of a new technology: **1)** *early development* (or *R&D*); **2)** *final development* (or *pilot-line production*); **3)** *technology-transfer and manufacturing start-up*; **4)** *volume-ramp*; and **5)** *high-volume production*. As the production of a new technology is ramped-up, the yield should also increase. This increase in yield with time-and-volume of production is referred to as the *learning-curve*. That is, as a new-technology matures, the major causes of yield-loss are identified and eliminated. When the process and design are both finally optimized, the remaining yield-loss mechanism is a low-level of random defects.

Yield Ramps - Stage-1: During Stage-1 (*early-process-*

development), the work is performed in an R&D fab. The tasks accomplished involve optimization of individual process-modules and their successful process-integration. Yield-management involves establishing an inspection-plan that supports process, material, and equipment characterization, and diagnoses *systematic* and *process-integration defects*. During this stage, the predictive yield-models (such as discussed in Sects. 28.3.3 & 4) are also defined. Successes are measured in terms of improvements in equipment-defectivity and reduction of integration-related yield-losses.

Yield Ramps - Stage-2: By Stage-**2** (*pilot-line production*) the process-recipes are more stabilized, and the causes of *systematic yield-losses* are identified. Thus, production of complete chips can be attempted. However, *yield* is likely to be low. The focus thus shifts to *product-manufacturability*. That is, yield-management now seeks to identify the cause of *random-defects*. To accomplish this, a large-number of inspection-steps are introduced into the process-flow to characterize random-defects arising from each process-module. A data-base of observed defects, their root-causes and problem solutions is created (Fig. 28-13). Yield-models developed in Stage-**1** are verified. During Stage-**2,** engineers from the production-fab arrive at the development-site to be trained on the new tools and methods.

Yield Ramps - Stage-3: In Stage-3, production is moved

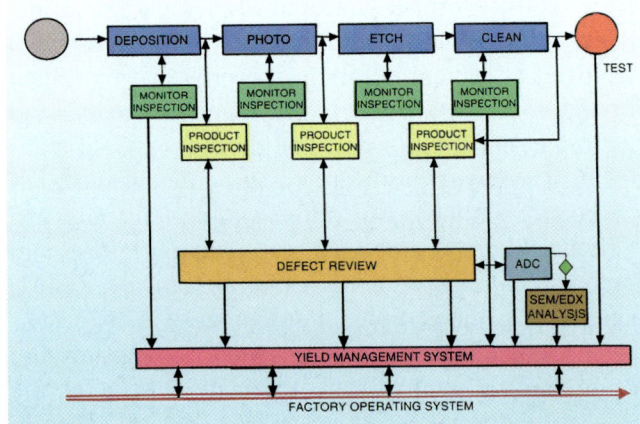

Fig. 28-13 Typical tool set and inspection locations for yield management.[2] By permission of Marcel Dekker.

from the pilot-line to a production-fab (i.e., *technology transfer* is carried-out, and *manufacturing start-up* occurs). In some companies, a *copy-exactly* strategy is employed for this stage. That is, the same equipment and processes used in the pilot-line are used in the production-fab. The same inspection-equipment set is also used, permitting inspection-recipes, defect-baselines, and yield-models to be used with minimum modification. This is designed to enable the same (or better) yields to be obtained in production immediately after technology-transfer. Key metrics for this stage are *yield-parity* (or better) with the pilot-line fab, and achievement of baseline defect-reduction goals.

Yield Ramps - Stage-4: In Stage-**4** (*volume-ramp*), the number of process-tools being used, and the number of lots in production increases dramatically. Three major objectives are pursued: **1**) rapid increase in the number of wafers in production; **2**) improvement over yields achieved in Stage-3; and **3**) qualification of new equipment and processes. A strong-well-trained yield-management team is key to meeting these goals (Fig. 24-14). One job of the team is to establish and lead *cross-functional yield-improvement programs* to deal with the problems that invariably arise. Major yield-improvements are often achieved in Stage-**4,** by solving the most serious yield-problems that have been identified by *defect pareto-charts* (Fig. 28-10).

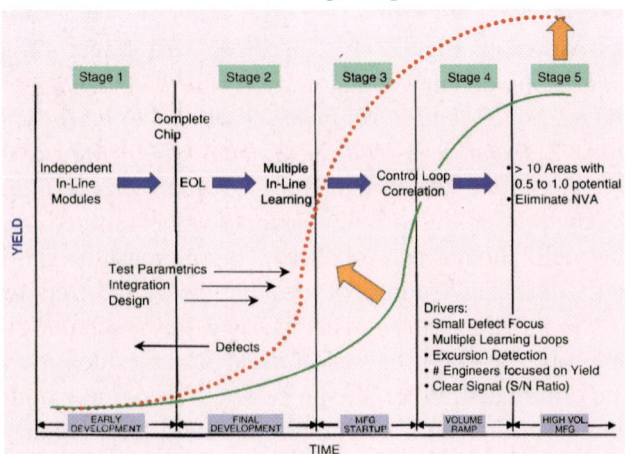

Fig. 28-12 Stages of yield ramp for new technologies.[2] Courtesy of Marcel Dekker.

Yield Ramps - Stage-5: In Stage-**5** (*high-volume manufacturing*), a fab has completed volume-ramp and the yield-curve starts to plateau. Additional yield-improvements require defect-reduction at many points in the process-flow. Management commitment to yield-improvement is probably the most-important factor at this point. Defect-excursions must be detected early to minimize the number of impacted lots. Corrective action may include shutting down, servicing, and correcting tools (see Fig. 28-15).

In Stage-**5,** additional yield-gains may require new equipment or circuit-redesigns. Nevertheless, a mature process should enter a period where good-die can be repeatedly produced with high-yield for the life-cycle of the product.

The Time-Compression of Yield-Ramps: The time to complete the entire *yield-ramp sequence* has been decreasing with time. This is due in part to the competitive-pressures that have reduced the life-cycle of IC products. Figure 28-16 demonstrates the dramatic reduction in the time required to complete a yield-ramp for DRAM chips. In 1985, during the era of 64-kb DRAMs, it took about 5-years to reach product maturity. But, by 2000, 256-Mb DRAMs were able to reach full production in just 1-year. Companies such as PDF Solutions and KLA-Tencor offer services and software to enable shorter yield-ramp times.

28.4.2 Statistical Process-Control

Statistical process-control techniques (SPC) are used to ensure that individual process-steps are being performed within specification. That is, SPC is a method of analyzing data to determine immediately if a process is under control or is in need of corrective action. For example, SPC is typically used to try and make each process step as consistent as possible. To initiate an SPC-program, trend-data that show the behavior of a process as a function of time must therefore first be gathered. Once enough data has been accumulated, they can be used to control the process. That is, the gathered-data can first be used to calculate the historic-average of the process-parameter being monitored. This value can then be plotted on an SPC-chart (see Fig. 28-16). Also plotted on each side of this average-value on the SPC-chart are the control-limits (e.g., the *upper control-limit*, ULL, and *lower-control-limit*, LCL). These *control-limit*s represent the limits that individual data-points will range between when the process is in control.

For instance, in a typical CVD-process, film thicknesses are measured on several samples of a batch of wafers immediately after deposition. The results are stored in a computer along with all previous data from the same process. The thickness-variation from wafer-to-wafer (and within a wafer), together with the refractive-index and other important thin-film parameters (vs. batch-number) are recorded. The results are then updated, and are available for display on the computer screen. By plotting the data from the

Fig. 28-15 Equipment technician servicing a process tool. Photo courtesy of Applied Materials.

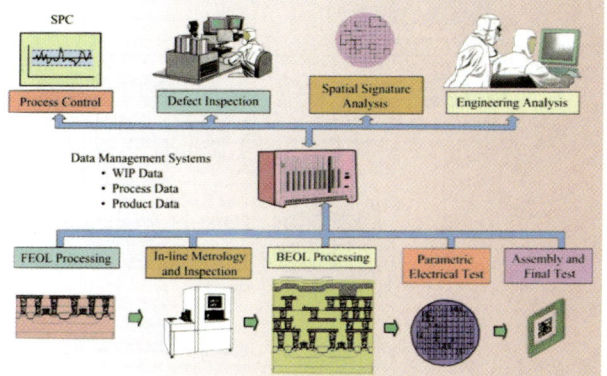

Fig. 28-14 The role of yield-management teams in yield-ramp campaigns.

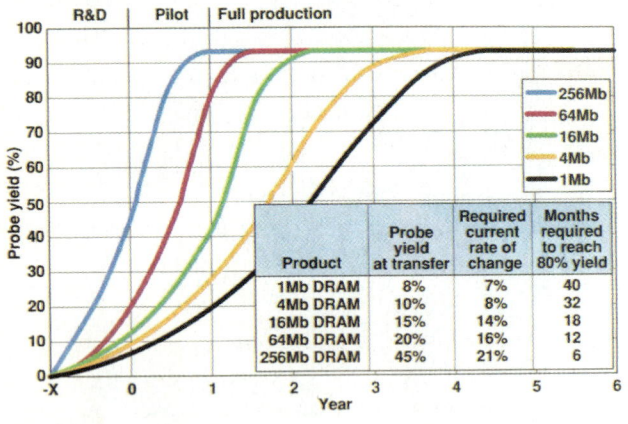

Fig. 28-16 Historical data of *yield-ramps* for DRAMs. Competitive pressure required that ramps of each new device-generation were steeper.

most-recent run along with data from previous runs, processes out of the control-limits are immediately identified. In addition, trends such as *long-term drift* in a process, become readily apparent.

An example of an SPC-chart is shown in Fig. 28-17 (see also Fig. 18-22 in Chap. 18). The data (e.g., the sheet-resistance in Fig. 28-17) should vary as a function of time in a regular manner within the control-limits about the average-value. This indicates that the process is in control. Data-points exceeding the control-limits indicate a problem. Data-points having a trend away from the average also point to a potentially out-of-control process.

28.5 ORGANIZATIONAL STAFFING OF WAFER FABS

An example of the manner in which the personnel that operate a wafer fab are organized is shown in Fig. 28-18. The primary responsibility rests with the wafer-fab manager. Reporting to this manager are engineering-supervisors, production-managers, and the equipment-maintenance group. The production-manager is responsible for processing the finished-wafers to the required specifications, at the planned cost, and on schedule. The engineering-group is charged with developing high-yield processes, with documenting these processes, and with helping sustain the daily-operations of the fab. Both the production and engineering staffs are partitioned into groups that focus on specific segments of the process flow. This organizational-form has the advantage of being able to tightly focus on the wafer fab's primary responsibility: manufacturing chips at a profit.

As can be seen in the organizational-chart, the routine tasks of operating a fab are performed by *operators*. Their jobs require the lowest-levels of edu-

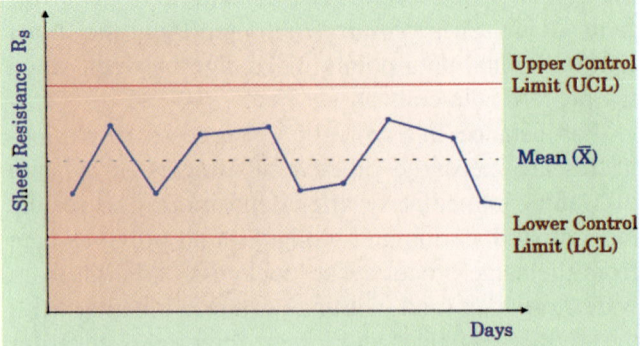

Fig. 28-17 Example of a *statistical process-control chart* (SPC). In this case the process is well-controlled, as the parameter being monitored stays within control-limits.

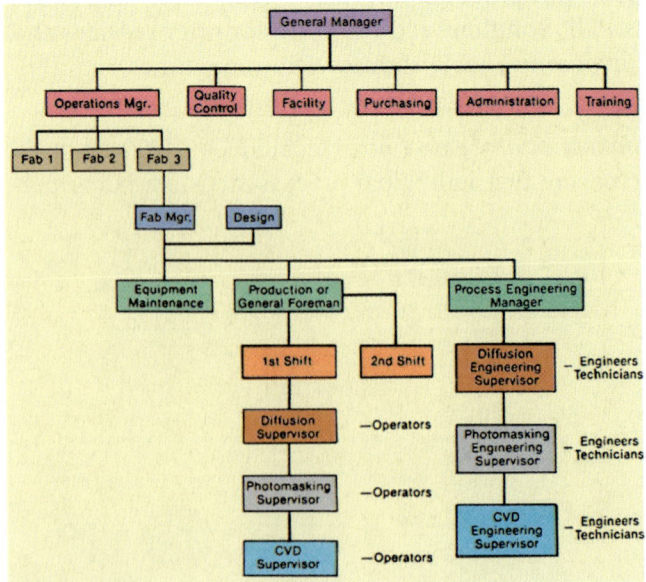

Fig. 28-18 Organization chart of personal that staff a semiconductor manufacturing-operation.

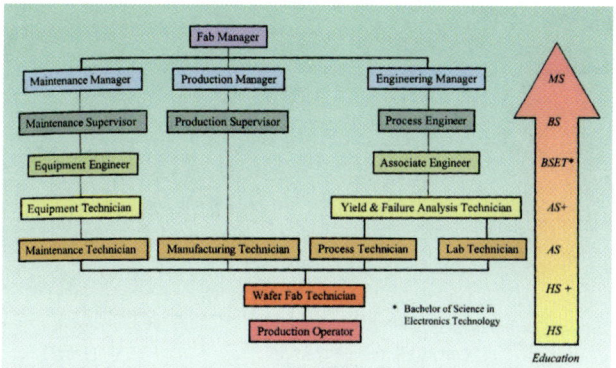

Fig. 28-19 Career paths in the semiconductor industry and their correlation with level of education.

cation and training to perform them. The next level of workers are *technicians*. As shown in Fig. 28-19, they normally require at least a 2-year degree from a college that has a technical-training program (Associates Degree). Above this group are the *engineering technologists*. They typically have a Bachelor of Science in Engineering Technology (B.S.E.T.), which is normally completed in four years of post-high-school education. Often these workers staff the maintenance-engineering groups of the fab. *Engineers* generally have completed at least a Bachelors of Science in Engineering (B.S.E.) program (e.g., in Chemical, Electrical, or Mechanical Engineering). A B.S.E. degree requires a higher level of mathematics, physics, and chemistry to be completed before studying engineering courses, than does the B.S.E.T. degree. Managers of the wafer-fab often have higher degrees (Masters of Science in Engineering [M.S.E.], Masters in Business Administration [M.B.A.], and/or a Doctorate in Engineering [Ph.D.]).

In general, one can say that increasing the level of one's education is the surest way to rise among the ranks of fab workers. But, it must also be noted that no matter what the level of work being performed in a wafer fab, learning should never stop. That is, process-technologies continually change (and usually become more advanced and complex). Thus, workers can only remain effective by keeping up with such changes. Such learning can be self-paced, or be a part of the training opportunities offered by the chip-man-

ufacturing company. That is, the training group may offer programs on-site, or workers can take classes at local educational institutions. In summary, continuous-learning strengthens the skills of fab-workers, which keeps the fab itself more competitive.

REFERENCES

1. R. Ross and N. Atchison, "Yield Modeling," in *Handbook of Semiconductor Manufacturing Technology,* R. Doering and Y. Nishi, Eds., Marcel Dekker, New York, 2000, Chap. 26, p. 851.

2. V.B. Menon, "Yield Management," *ibid*. 1, p. 869.

3. L. Foster and D. Pillai "Wafer Logistics & Automated Material Handling Systems," *ibid*. 1, p. 1067.

PROBLEMS

1. Indicate if yield will increase or decrease with the following changes: (a) Use larger-diameter wafers; (b) Design smaller die; (c) Increase process-steps; (d) decrease defect-density; and (e) Shrink feature-size.

2. Calculate the overall-yield for a process that has a 97% *fabrication-yield,* a 77% *sort-yield,* and a 92% *assembly and final-test yield.*

3. How many die of area 2-cm^2 can be placed on a 200-mm diameter wafer? Explain your assumptions on die-shape and unused wafer-perimeter.

4. What is the yield of the wafer shown in Fig. 28-8? (Note that partial [edge] die are not counted).

5. If a process has 300-steps and the yield from each step is 99%, what is the overall-yield?

6. Assume that Eq. 28-4 is used to predict the defect-limited yield, Y, of an IC process. If a 256-Mbit DRAM chip has a total-area of 3.25 cm^2 and it contains 3.5×10^8 thin-oxide gates, each with a size 0.25 μm x 0.5 μm, find the maximum "killer-defect" density that will still allow a 95% gate-oxide yield.

7. A given wafer-fab starts 1000 wafers/week. Every wafer has 100 chips, and they can be sold for $50 each (if functional). The yield is currently 50%. If the yield can be increased, the incremental-income is almost all profit (because all 100 chips are manufactured, regardless of whether or not they work). How much would the yield have to be increased to produce an *annual profit-increase* of $5 million?

8. Find the overall-yield for a 9-mask-level process (using Eq. 28-2), in which the average defect-density

CHAPTER 29

ENVIRONMENTAL, HEALTH, AND SAFETY ISSUES (EHS)

CHAPTER CONTENTS

29.1 SAFETY RECORD OF THE SEMICONDUCTOR INDUSTRY

29.2 SAFETY-HAZARDS IN WAFER FABS

29.3 GENERAL SAFETY PROCEDURES
    Warning Labels
    Hazard Information Labels
    Material Safety Data Sheets

29.4 HAZARDOUS PROCESS GASES

29.5 HAZARDOUS PROCESS CHEMICALS

29.6 ELECTRIC SHOCK HAZARDS AND ION-IMPLANTER SAFETY

29.7 WAFER-FAB RADIATION HAZARDS

29.8 FIRES IN WAFER FABS

29.9 ENVIRONMENTAL SAFETY ISSUES

"An ounce of prevention, is worth a pound of cure."
                             Old Proverb
or
"Be safe, and if you can't, be careful."
                        Ezra Hendrickson

Many potential safety hazards exist in wafer fabs. Drawing courtesy of Amtech Systems, Inc., a supplier of automation for diffusion furnaces.

The semiconductor wafer-fab is a complex manufacturing facility, that contains a variety of safety hazards. Many of the substances used to manufacture ICs are hazardous (i.e., toxic, flammable, or corrosive). Operating and maintaining fabrication-equipment can expose personnel to other risks, including electric-shock, burns, and exposure to radiation. To prevent accidents, workers need to be aware of these dangers, and must follow safe work-procedures. Wafer-fabs also produce waste that can pollute the environment. Measures must be used to prevent such pollution. This chapter covers both safety and pollution-abatement issues.

Note that the information given here provides an *awareness* of the perils associated with the equipment, processes, and materials used in IC-fabrication. It is not to be considered training for specific equipment or hazards. For more details on such EHS issues, consult Refs. 1 and 2.

29.1 THE SAFETY-RECORD OF THE SEMICONDUCTOR INDUSTRY

A semiconductor-fab, like any other manufacturing facility, has many kinds of dangers. But the semiconductor industry ranks high in terms of protect-

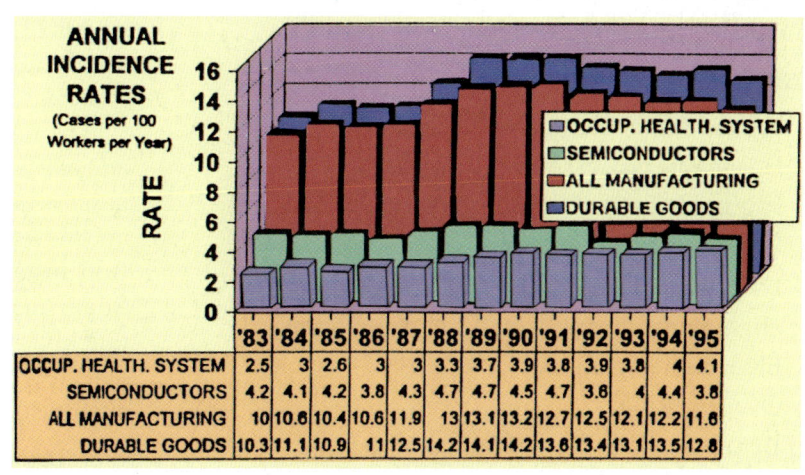

Fig. 29-1 Comparative annual incidence-rates for Occupational Health Systems, U.S. semiconductor-industry, durable-goods industries, and all manufacturing (1983-95). Source: Annual Surveys, Bureau of Labor Statistics.[3]

ing the health and safety of its workers. Among the manufacturing industries in the U.S., it ranks fourth-highest in terms of its safety-record. Workers are protected from harmful exposure to toxic-chemicals and physical-hazards in ways no large manufacturing work-force has previously ever been. Only aerospace and nuclear industries provide similar worker-exposure controls. While many of these protections were put in place with worker-safety in mind, concerns about product-purity also resulted in the installation of protection-systems that make wafer-fabs less dangerous places to work. Information on the incidence of work-injuries and illnesses in U.S. manufacturing industries comes from the U.S. Dept. of Labor, which publishes such statistics annually (see Fig. 29-1).

There is also an ongoing, industry-sponsored system of worker-injury surveillance, organized by the Semiconductor Safety Association, the International SEMATECH, and the Environmental Safety and Health Committee of the Semiconductor Industry Association (SIA). Several epidemiological studies (also sponsored by the industry) have been performed to find out how the reproductive-health of semiconductor-industry employees is impacted.

29.2 SAFETY-HAZARDS OF THE WAFER FAB

A list of some of the most common safety-risks in semiconductor manufacturing is given below.

Hazardous Materials and Physical Conditions

- Hazardous process-gases
- Hazardous process-chemicals (liquids)
- High-voltages (which can cause electric-shock)
- Radiation-hazards (Lasers, X-rays, Deep-UV)
- Fires
- High-temperatures (which can cause burns)
- Freezing-temperatures
- Over-exertion
- Being struck by flying, falling or moving objects

Hazardous Work-Operations

- Maintenance operations on hazardous equipment.
- Normal operations that give rise to fires, chemical-spills, or worker falls (see Faceplate).
- Wafer-fab operations that can harm the environment outside of the wafer fab.

29.3 GENERAL SAFETY PROCEDURES

Workers should be trained in safe-operating practices for specific processes.[3] In addition, they should adhere to such general-practices as not working alone when performing hazardous tasks. Wearing additional safety-apparel (including safety-glasses and goggles, protective-coats and aprons, and appropriate gloves for the job), also helps prevent accidents. A full-face shield should always be worn when pouring chemicals. In addition, it is good practice to wash hands with soap and water when leaving the work place.

Workstations should be monitored to ensure that safe-practices are being followed, and that unsafe conditions do not arise. Trained emergency-response-teams should be in place in case of accidents. Biomonitoring (such as radiation-badges) should be used for workers who might be exposed to chemicals (or physical-hazards) that could produce long-term ailments (provided such monitoring can help detect inadvertent-exposures to the hazardous-condition). Good lines of communication between workers and supervisors should exist to allow unsafe-conditions to be quickly identified and corrected. Thorough records should also be kept to help detect unusual conditions that might point to potential safety-problems.

The fab should be equipped with toxic-gas sensors and alarms that are set-off when the presence of toxic-gases is detected at ppm or ppb levels. Fail-safe mechanisms should be employed on hazardous-equipment (for instance, gas-valves and electrical-interlocks). Sophisticated ventilation-systems should exist to quickly remove toxic-gases. If unusual-odors are detected, workers should leave the area, notify their supervisors, and (if appropriate) sound an alarm.

Warning-Labels: Hazardous-areas and containers of hazardous-materials must have conspicuous warning-labels that draw attention to the hazard. Figure 29-2 gives examples of some standard warning-labels.

CHAPTER 29   ENVIRONMENTAL, HEALTH, AND SAFETY ISSUES (EHS)

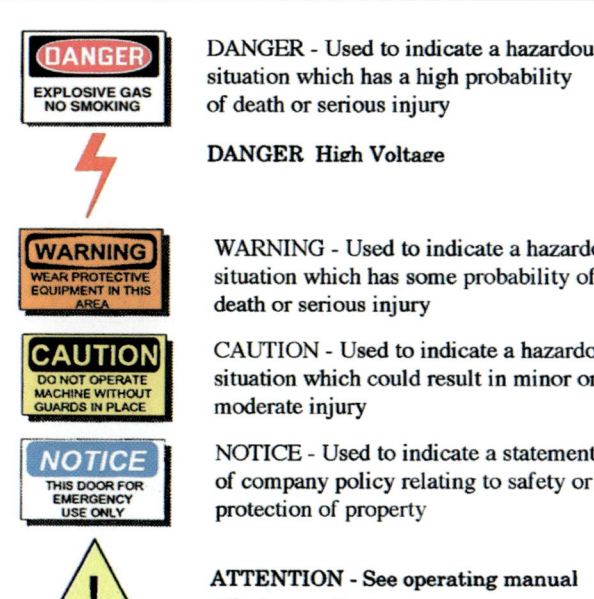

Fig. 29-2 Examples of Safety Warning Signs.

Hazard-Information Labels: *Hazard-Information Labels* should be posted on all equipment and containers holding gases and chemicals. Workers should understand how to interpret the information given on such labels (Fig. 29-3). For example, hazard-levels are posted with respect to their severity regarding the following hazard categories: **1)** fire; **2)** chemical-reactivity; **3)** health; and **4)** specific type of hazardous-condition. The classification numbers used on these labels put the hazard into one of five-levels of danger using the following categories:

4 Materials that could cause *death* or *major-injury* with very-short-exposure, even though prompt medical-treatment is given. This includes materials too dangerous to be approached without specialized-protective-equipment.

3 Materials that could cause *serious temporary or permanent-injury* with very-short-exposure, even though prompt medical-treatment is given. This includes those materials that require protection from all body-contact.

2 Materials that could cause *temporary incapacitation or possible injury* after intense or continued exposure, unless prompt medical-treatment is given. This includes those materials that require use of respiratory-protective-equipment with an independent air-supply

1 Materials that cause *irritation* or only *minor-injury* on exposure, even if no treatment is given. This includes those materials that require use of an approved canister-type gas-mask.

0 Materials that, on exposure, offer *no hazard beyond that of ordinary combustible-material*.

Material Safety Data Sheets: Safety information about chemicals is contained on *Materials Safety Data Sheets (MSDS)*. By law, an MSDS must be available in the fab for every hazardous-chemical being used in the facility. The suppliers of these materials provide this document. Some of the information provided by the MSDS is given here:

Chemical-Identity:	Common chemical-name, or trade-name (and synonyms)
Ingredients: (Principal Components of Mixture)	List of hazardous materials present over 1%, and all carcinogens over 0.1%
TLV and PEL:	Given in *parts-per-million* (ppm) and in *milligrams-per-cubic-meter* (mg/m^3)
Health Effects of Overexposure:	Specifies the human organs impacted by overexposure and their harmful effects

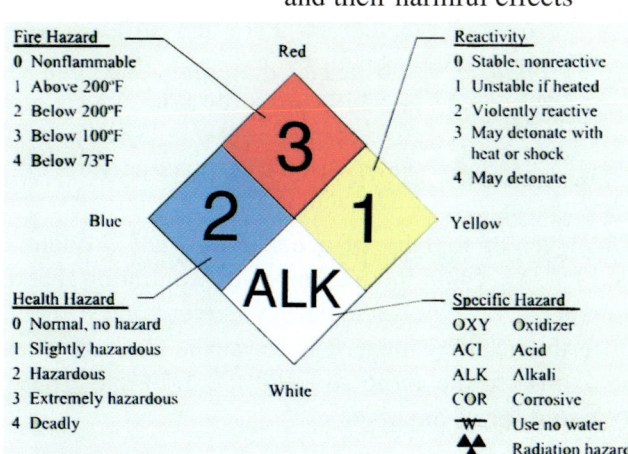

Fig. 29-3 Hazard Information Labels.

Physical/Chemical Characteristics: Gives such data as melting and boiling-points, vapor-pressure, and specific-gravity

Fire/Explosion Data: Lists *flash-point* (lowest temperature a substance can be ignited with a flame), and *autoignition-point* (lowest temperature a substance will ignite spontaneously in air)

Reactivity Hazard-Data: Describes whether a material is unstable, and under what conditions

29.4 HAZARDOUS PROCESS GASES

The process-gases used in IC manufacture have a number of hazardous characteristics, grouped into four categories:

Toxic: The quality of being dangerous to human-life. Examples include arsine (AsH_3), phosphine (PH_3) and diborane (B_2H_6).

Corrosive: Substances that destroy living-tissue and equipment that comes in contact with them. Examples include HCl and HF.

Flammable: Substances that give off vapors that can readily ignite if exposed to sparks, flames, and other sources of ignition. Examples, include H_2, PH_3, and $SiHCl_3$.

Pyrophoric: Substances that ignite spontaneously in air at (or below) 54°C. An example is *silane*, SiH_4, a gas widely used to deposit polysilicon, SiO_2, and Si_3N_4 by CVD.

The level of toxicity of a hazardous-gas is quantified by specifying the maximum-concentration that a human can be exposed to during a given time without suffering health damage. The lower the allowed-level of exposure, the more toxic is the material. These exposure levels are given as:

TLV-TWA: The *threshold limit value* (*TLV-TWA*) is specified by the American Conference of Governmental Industrial Hygienists. It specifies the level under which one can work for eight hours a day for an indefinite period without harmful effects.

TLV-STEL: The *threshold limit value-short term exposure limit* (*TLV-STEL*) is a 15-minute, time-weighted average-exposure. Exposures to the STEL should not be longer than 15 minutes, and not be repeated more than four times per day.

IDLH: *Immediately Dangerous to Life & Health (IDLH)*. This concentration represents a maximum-level for which one could be exposed to for 30-minutes without impeding escape, or causing any permanent health-effects.

PEL: The *permissible exposure limit* (*PEL*) is a standard for exposure set by the Occupational Safety and Health Act (OSHA). The PEL-value is a time-weighted average exposure-limit (typically for eight hours) or a ceiling exposure-limit.

Table 29-1 is a partial list of the process-gases used in IC fabrication and their maximum-exposure levels.

29.4.1 Reducing Toxic-Gas Hazards

A number of approaches have been pursued to reduce the hazards of using toxic-gases. First, adequate-training is imperative for personnel performing operations involving hazardous-gases. Examples of such operations include: **1)** change-out of toxic-gas cylinders; **2)** changing the oil of roughing-pumps in which toxic-gases have been dissolved; and **3)** working on the beamlines of ion-implanters (where the residue of toxic implant-gases can build-up on the inside walls). When performing operations involving such toxic-gases, workers should wear full-body chemical-protective coveralls, durable-gloves and use respiration precautions (e.g., a *Self-Contained Breathing Apparatus, SCBA*). Such work should be performed

small-leaks in the gas-system.

Other, more fundamental-approaches are also being developed. For example, less-hazardous replacement-gases are being sought. One such change involves using TEOS-gas instead of silane in some applications (partly because TEOS is not a pyrophoric-gas, as is silane). Another example involves using the so-called *safe-delivery system* (SDS^{TM}) for ion-implant applications (Fig. 29-6) instead of conventional gas-bottles.[6,7] That is, in the SDS-system, toxic-gases are physically-adsorbed onto the surface of an adsorber-material present inside the gas-cylinder. The pressure in a filled SDS-cylinder is just below atmospheric-pressure (650-torr) and gas is extracted from the SDS-system by the differential-pressure that exists between the ion-source-chamber (10^{-3}-torr) and the gas-cylinder (650-torr). Since pressure in the cylinder

Fig. 29-4 (a) Collars and caps (and placing cylinders in a box or rack) are common methods used to protect cylinder-valves from damage. (b) Ruptured gas cylinder.

by a team of at least two individuals. Gas-cylinders should be equipped to protect against valve-breakage in case a cylinder falls-over (Fig. 29-4a). While catastrophic cylinder failures are rare (Fig. 29-4b), their devastating effects can impact the industry. Figure 29-5 shows a gas-cylinder changeout operations.

To protect workers in the fab, toxic-monitoring systems should be installed, with both visible and audible alarms triggered if the level of toxic-gas (detected by gas-sensors) exceeds health-threatening-limits. Carefully interlocked exhaust-systems should remove any toxic-gases that may be present due to

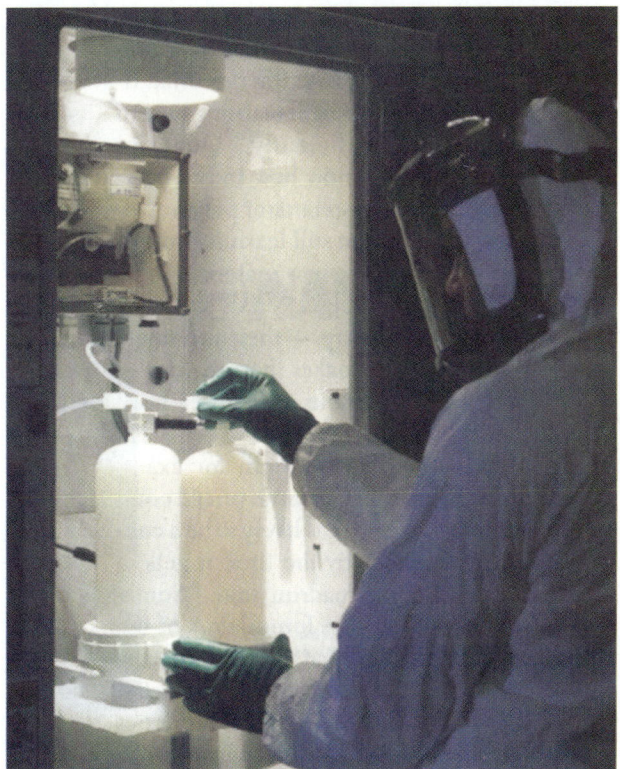

Fig. 29-5 Changing out a gas-cylinder in the chemical delivery system. Accidents can occur during such operations. Courtesy of BOC Edwards.

Table 29-1 Commonly Used Wafer Fab Chemicals and Their Safety Hazards

Chemical Name	Symbol	Combustible or Explosive	Health Hazard Class	TLV-TWA (ppm)	TLV-STEL (ppm)	IDHL (ppm)	Chemical Name	Symbol	Combustible or Explosive	Health Hazard Class	TLV-TWA (ppm)	TLV-STEL (ppm)	IDHL (ppm)
Ammonia	NH_3	X	2	25	35	500	Hydrogen Chloride	HCl		3	5	-	100
Argon	Ar		0	-	-	-	Nitrogen	N_2		0	-	-	-
Arsine	AsH_3	X	4	0.05	-	6	Nitrogen Trifluoride	NF_3		3	10	15	2000
Boron Trichloride	BCl_3		3	1	-	100	Nitrous Oxide	N_2O	X	2	50	-	-
Boron-Trifluoride	BF_3		3	1	-	100	Oxygen	O_2	X	0	-	-	-
Chlorine	Cl_2		3	0.5	1	30	Phosphine	PH_3	X	4	0.3	1	200
Carbon Dioxide	CO_2		1	5000	30000	30000	Silane	SiH_4	X	4	5	-	-
Diborane	B_2H_6		3	0.1	0.3	40	Silicon Tetrachloride	$SiCl_4$		3	5	-	-
Dichloro-Silane	SiH_2Cl_2		3	5	-	100	Sulfur Hexafluoride	SF_6		3	100	1250	-
Helium	He		0	-	-	-	Tetrafluoro-methane	CF_4	X	3			
Hydrogen	H_2	X	0	-	-	-	Tungsten Hexafluoride	WF_6		3	3	6	-
Hydrogen Bromide	HBr		3	3	-	50	TetraEthyl OrthoSilicate (TEOS)	$(C_2H_5)_4 SiO_4$	X	2	10	-	1000

is below atmosphere, this greatly reduces the risk of a catastrophic-rupture in a feed-line or manifold.

29.5 HAZARDOUS PROCESS CHEMICALS

Many of the liquid-chemicals used in a wafer-fab are hazardous and corrosive. Proper safety-procedures must be followed when handling them to avoid accidents and injuries.

Handling Corrosive Chemicals: *Corrosive-chemicals* can be either acids (pH < 7) or bases (pH > 7). When working with corrosive-chemicals, they should be positively identified before use. (For example, HF looks just like water.)

Incompatible chemicals should not be mixed. Appropriate eye-protection should be worn when handling hazardous-chemicals (i.e., safety-glasses, goggles, and/or full-face shields). Body-protecting apparel should also be worn, including acid-resistant aprons, gloves, and sleeve-guards. Boots that protect against chemical-spills should be worn. The chemicals should only be used under a fume-hood, as a measure to prevent breathing of their vapors. HF should be stored only in plastic containers (because HF attacks glass). Workers should be aware of the nearest eye-wash and chemical-shower.

Handling Solvents: Most solvents have harmful vapors, and many are flammable. As when working with corrosives, eye and body protection should be worn, and the solvent-fumes should not be inhaled. Solvents must be kept away from open-flames, sparks, and heat. Waste-solvents must be poured into waste-solvent-containers, not into acid-drains. Solvents should be stored in a flammable-materials storage-cabinet. Do not mix acid-waste with solvent-waste, because violent chemical-reactions can occur.

Hydrofluoric Acid (HF): Hydrofluoric-acid (HF) is widely used in fabs to wet-etch silicon-dioxide films,

CHAPTER 29 ENVIRONMENTAL, HEALTH, AND SAFETY ISSUES (EHS) 523

29.6 ELECTRIC-SHOCK HAZARDS AND ION-IMPLANTER SAFETY

When electric-current passes through the human-body, the effect it causes is called *electric-shock*. The damage caused by electric-shock can cause burns, muscle and nerve-damage, heart-failure, respiratory-paralysis, and death. The *threshold-of-perception* of current is about 1-mA. Current passing through the body at levels above 10-mA begins to cause involuntary muscular-contractions. Even though the pain is severe, the victim may be unable to release the grip on the electrical-conductor being held. If the current flowing in the body exceeds 100-mA, it begins to interfere with the coordinated-movement of the heart. If this causes the heart to enter a state of *fibrillation*, the heart is prevented from pumping blood - and death occurs within minutes. If proper first-aid is given, the shock may not be lethal, although severe-burns may have occurred. In fact, administering large current-pulses to fibrillating-hearts is used to restore them to their normal rhythm (*defibrillation*, Fig. 29-7).

The lethal-aspect of electric-shock is a function of the amount of current passed through the human conducting-path and exposure-time. It is not neces-

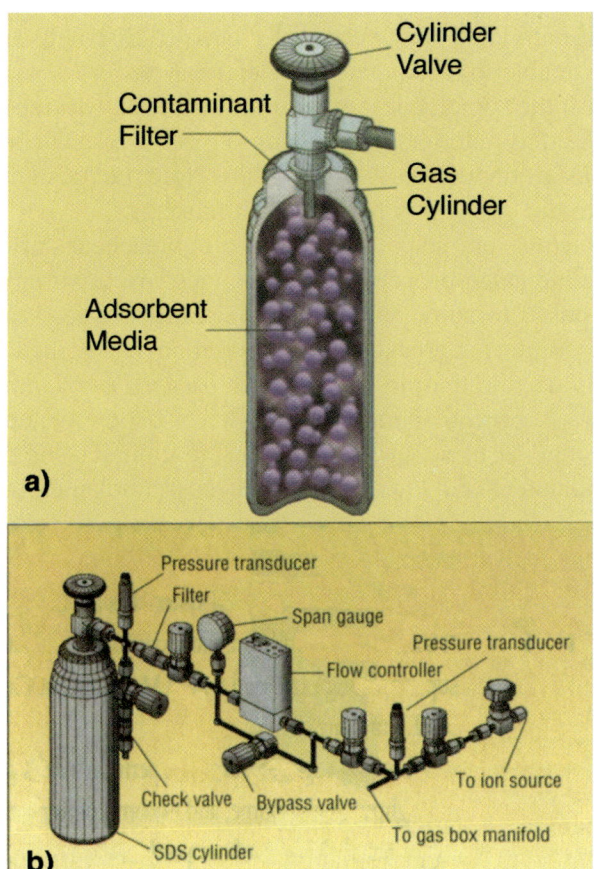

Fig. 29-6 (a) Schematic of a Safe-Delivery Gas-Source® (SDS). (b) SDS delivery scheme. Courtesy of ATMI.

and to clean diffusion-tubes and other glassware. Yet, this acid has its own unique and especially-dangerous safety-hazards. That is, no pain is felt when HF first makes contact with the skin. However, severe and painful burns occur later as the HF penetrates the skin and flesh, and then reacts with the calcium in the bone. Thus, serious-damage will occur if skin-contact with HF is not treated quickly in an appropriate manner. If it is suspected that contact has been made with HF, the skin should be thoroughly rinsed with water, and medical-help should be immediately sought.

As a result, when handling HF, proper gloves should be worn (in which it has been ensured that there are no pinholes or tears). Always treat any unknown clear-liquid as if it is HF, because HF and water look identical to one another.

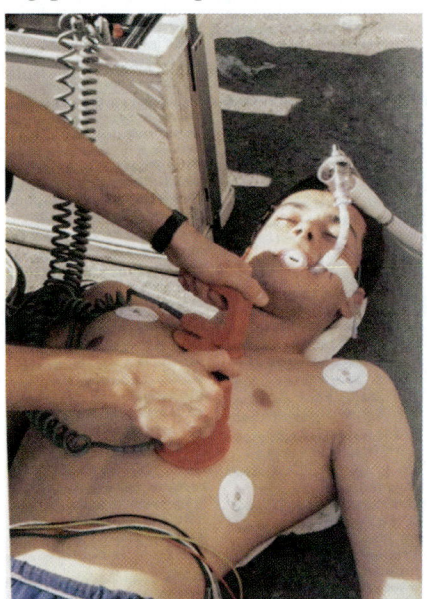

Fig. 29-7 To save a victim of *ventricular-fibrillation*, the heart must be shocked to re-establish its normal rhythm.

Fig. 29-8 Example of "Tag-Out" sign.

sarily dependent on the value of the applied-voltage. A shock from 100-V may turn out to be as deadly as a shock from 1000-V. However, statistics show that the death-rate for contact with a 250-V ac power-line is about 3%. As the voltage rises to 10-kV (or higher), the likelihood of death occurring rises dramatically.

Electric-shock can accidentally occur due to poor equipment-design, electrical-faults, human-error, or a combination of unfortunate circumstances. Some other factors that can cause electrical accidents involve *poor-housekeeping* (failure to repair equipment, broken-ground-wires, overloaded electrical-circuits, and flammable-materials near electricity) or *sloppy-wiring-practices*. Additional dangerous conditions include *inadequate safety-signs*, lack of *ground-fault indicators* (where electricity and water can mix), and insufficient lockout/tagout measures to keep personnel from unguarded energy-sources.

Some equipment in the process-fab has highly dangerous electrical-hazards, particularly ion implanters.[5] Maintenance-personnel must especially be protected against an accidental encounter with the high-voltages generated by the implanter. There are several high-voltage power-supplies used in implanters besides the one used to produce the acceleration-voltage. Auxiliary-supplies used for beam-extraction and scanning also produce maximum-voltages of several-kV. Therefore, all of these can also produce lethal electric-shocks. If contact with even low-voltage-conductors occurs, nerve-conductance may be blocked.

An individual may be thereby become disabled, and be unable to disconnect the electrical-power (or even push the "emergency-off" button). At least two-levels of safety-interlocks (utilizing keys, door-switches, and grounding-bars) should therefore be designed into the systems to prevent such mishaps.

However, since interlocks are breached during maintenance-procedures, personnel are invariably exposed to some risk. In such situations there is no substitute for precaution and common-sense, including the rule that *no work should be performed alone on hazardous-equipment!* The "buddy-system" should be used when changing the cylinders of the implanter feed-source gases. Because implanters are so large, people inside can easily be overlooked. Thus, before entering an implanter, it is important to work with a partner, and to "tag-out" the tool to be sure others know that someone is working inside the machine (Fig. 29-8). When entering the system, it is also important to take the door-key, so other workers can't lock the door and start it up.

29.7 RADIATION HAZARDS IN WAFER FABS

There are a variety of safety-hazards associated with light-sources and radiation effects in a wafer-fab. Some of the most common ones are listed here.

DUV Laser Light-Sources: The excimer-laser light-sources of DUV-steppers (see Chap. 19) represent a number of hazardous conditions. First, the laser light-source itself can cause damage to skin or eyes, if they are exposed to the DUV laser-beam. When DUV-lasers were located outside the cleanroom, workers could inadvertently walk into the beam-path. Elaborate laser-walls and curtains were needed during installation and maintenance to prevent such occurrences. Since these lasers are now installed and maintained inside the cleanroom (and are much closer to the stepper), this hazard has been decreased. The gases used in excimer-lasers are toxic (as they contain fluorine), and the laser-cavity must be periodically refilled with them. Safety-procedures must therefore be carefully followed during such refilling. Excimer-lasers also use high-voltage power-supplies. Over the years, their regulation, reliability, and safety have improved considerably, so they require little (if any) maintenance. Nonetheless, they must be treated with caution when work is being done on the laser.

X-Ray-Radiation: Small amounts of X-rays are generated in an implanter, and lead-shielding is used to absorb such x-rays (and to keep radiation-levels external to implanters below the recommended limit of 0.25-mrem/hr). Lead-shielding (>6.5-mm thick) is needed to reduce radiation to such levels for implanters. However, operators should wear *x-ray monitor-badges* and be tested on a routine-basis for *blood-hemoglobin-efficiency* (which is an early indicator of radiation-exposure). Periodic radiation-surveys should also be included in the implanter maintenance-schedule. By following the radiation-safety-guidelines furnished by the manufacturer and user, radiation-hazards of ion-implanters can be minimized.

29.8 FIRES IN WAFER FABS

Many things can cause fires in a wafer-fab. Flammable-chemicals and gases can be ignited by sparks, flames, or electric-heaters. Pyrophoric (and detonable) gases (such as silane) are in common use (Fig. 29-9). Surrounding materials can be set afire by electric-heaters, hot-plates and immersion-heaters.

Fig. 29-9 Photograph of a silane-flame in a gas cabinet. Courtesy of J. Dietz, ATMI.

Various safety-measures are employed to prevent fires, and to control or extinguish them if they ignite. Automatic-sprinklers are installed, backed by an adequate water-reservoir. The fab is built with non-combustible (or fire-resistant) materials, such as fire-retardant polypropylene (FRPP). Use of FRPP offers protection against fire. The air-handling system is designed to facilitate smoke-removal. Human-surveillance is used to watch over the fab, and early intervention plans are practiced to quickly douse any fire. Pyrophoric and flammable chemicals and gases are stored outside the fab.

29.9 ENVIRONMENTAL SAFETY ISSUES

Liquid-chemicals and gases not consumed by process-tools, vapors from process-chemicals, and process-byproducts all enter exhaust and drain systems of the fab. The wafer-fab exhaust-system consists of various subsystems: **a)** *general-exhaust*; **b)** *solvent-exhaust*; and **c)** *scrubbed-exhaust*.

The *general-exhaust-system* removes heat dissipated by the process-equipment, but this heated-air should not contain vapors from acids, caustics, or solvents. The *solvent-exhaust system* removes air containing the vapors of solvents from the process equipment (e.g., resist and developer solvents).

29.9.1 Scrubbed-Exhaust Systems

The *scrubbed-exhaust system* removes any air that contains vapors from acids and/or caustics, as well as the process-gases used in process equipment. The gases leaving a reaction-chamber are often toxic, corrosive, or flammable, and must therefore be treated to prevent them from leaving the wafer-fab in their hazardous state. (Since most process-gases are not completely consumed by a process step, they typically make up a substantial-fraction of the gases exhausted from the process chamber. In addition, the reaction by-products may also be hazardous.) These gases are treated in post-process, detoxifying-chambers. The two most common systems used in this role are *combustion-reactors* and *scrubbers*.

In most cases, these two systems are used together, with the exhaust of the combustion-chamber being fed

526 MICROCHIP MANUFACTURING

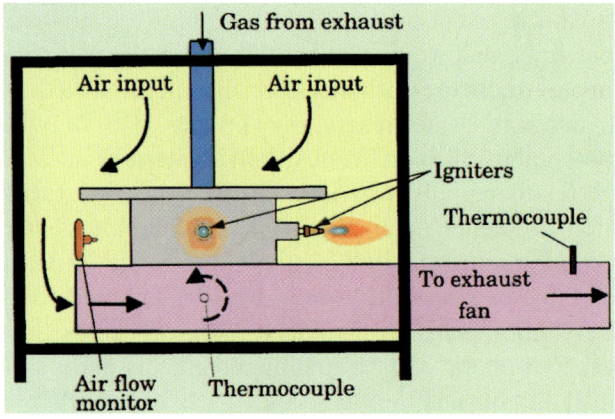

Fig. 29-10 Schematic of a burn-box.

to a scrubber. This is necessary because the exhaust-gases from a reactor may contain flammable as well as non-flammable gases. For example, the exhaust-gases of an epi-reactor include such flammable-gases as hydrogen and dopant-gases (diborane, arsine, or phosphine], and such non-flammable gases as HCl.

Combustion Reactors: *Combustion-reactors* are either *burn-boxes* or *flow-reactors,* both of which combust flammable-gases (including hydrogen, silane, and phosphine) by reacting them with oxygen (i.e., by *burning* them). The burning takes place far downstream from the process chamber. Volatile organic-compounds (for example, vapors from such solvents as such as isopropyl-alcohol and methanol) are also fed to a combustion-chamber.

The *burn-box* (Fig. 29-10) is typically used to burn pyrophoric-gases (such as silane), and highly-flammable-gases (such as hydrogen). *Flow-reactors* use direct-combustion by exposing less-flammable gases to oxygen and a direct-flame (Fig 29-11).

The by-products of these burning-reactions are usually less harmful than the gases entering the combustion-reactor. But, in some cases, the by-products may still be toxic. (For example, any compound containing arsenic is still hazardous.) However, since the by-products of the burning-process are often solid-particles, these may be collected by a special filter after leaving the combustion-chamber (and before this exhaust is fed to the wet-scrubber, Fig. 29-12). Such collected toxic-solid-particles can then be disposed of in special toxic-waste-sites.

Scrubbers: *Scrubbers* are used to treat the remaining by-products of the combustion-reactors, as well as other non-flammable process-gases. That is, such gases are diluted, cooled, and made to react in order to eliminate any harmful-effects on the environment. This threefold-task is accomplished by a *scrubber*, a large process-chamber outside of the fab-building (Fig. 29-13). Monitors and alarms warn of system-malfunctions. The figure shows their placement. One monitor is placed after the collector to warn if toxic-material concentrations exceed the TLV. Another after the scrubber is used to warn if the IDLH-concentration is exceeded. Still another monitors the pH-level of the scrubber-liquid (to warn if the pH drops below 6.0), thereby guarding against excessive-buildup of

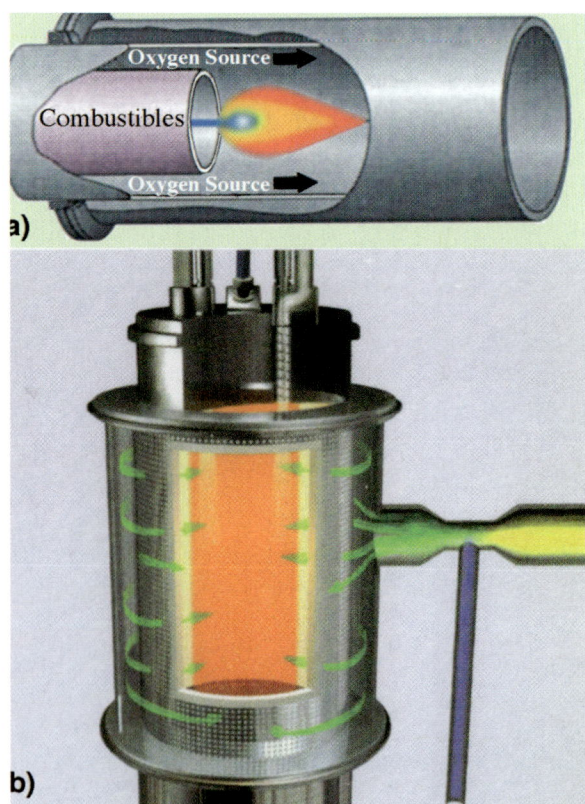

Fig. 29-11 Drawings of a flow-reactor. [Part-(b) Courtesy of Alveta Corp.]

CHAPTER 29 ENVIRONMENTAL, HEALTH, AND SAFETY ISSUES (EHS)

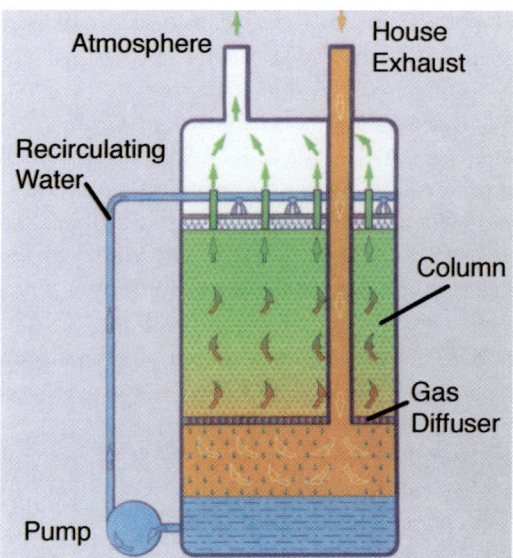

Fig. 29-12 Schematic drawings of a wet-scrubber. Courtesy of BOC Edwards.

arsenic-, boric-, or phosphoric-acid in the scrubber.

Gases are first diluted and cooled by adding large quantities of nitrogen. Next, they follow a long route through a continuous-spray or shower of water, in which the soluble-gases dissolve. The water is then treated to react the hazardous-gases dissolved in it (e.g., gases that have formed acids, such as HCl, are neutralized). The levels of any toxic-materials present in the waste-stream that cannot be treated are diluted by the water to levels that are not hazardous. The treated-water is then re-circulated or discharged. Water-scrubbing works well for materials that are highly-soluble in water, but is not of much use if the waste-stream is insoluble in water. (For example,

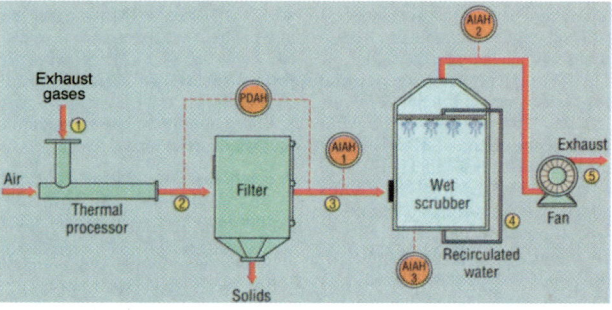

Fig. 29-12 Drawing of a complete exhaust-gas effluent-abatement system.

silane is not soluble in water.) That is why gases are usually subjected to a combustion-step prior to feeding them to the wet-scrubber.

29.9.2 Chemical Recycling

Large quantities of ultra-pure water and chemicals are consumed in the course of manufacturing ULSI-ICs, and this produces large volumes of chemical-waste. Strict environmental-regulations make it expensive to dispose of them. One way to reduce this cost (and the impact on the environment) is to recycle them.[9]

In addition to recycling ultra-pure water, efforts to commercialize HF-recycling-systems have been pursued. After HF has been used, it contains particles and metallic-contaminants. Such spent-HF represents about 40% of the total hazardous-waste produced by the semiconductor industry. Recent regulations have begun prohibiting landfill-disposal of HF, and so efforts to recycle this acid have been developed.[10]

The HF-recycling-process must be able to remove the contaminants and particles. The chemical-contaminants can be removed by distillation (which is process of boiling the liquid and then condensing the vapors on a cool-surface). Most of the contaminants present in HF do not vaporize and therefore are not present in the distilled and condensed HF. Ion-exchange can be used to remove the ions present in the spent-HF.

29.9.3 Perfluorocarbon-Compounds (Global Warming Gases)

PerFluoroCarbon-gases (PFCs) are used in dry-etch processes as etch-gases. However, a large-fraction of these gases leaves the etch-chambers unreacted (in excess of 50%). Previously, these were released into the atmosphere (since they are generally non-toxic gases). Unfortunately, their long-lifetimes in the atmosphere (and their strong infrared-absorption) make them potential *global-warming agents*. Thus, there is a global concerted-effort to substantially reduce the quantity of PFCs released into the atmosphere. The semiconductor-industry has pursued this goal in a number of ways: **1)** process-optimization to reduce the amount of PFCs exiting the process-chamber during each process-run; **2)** substitution of more-benign

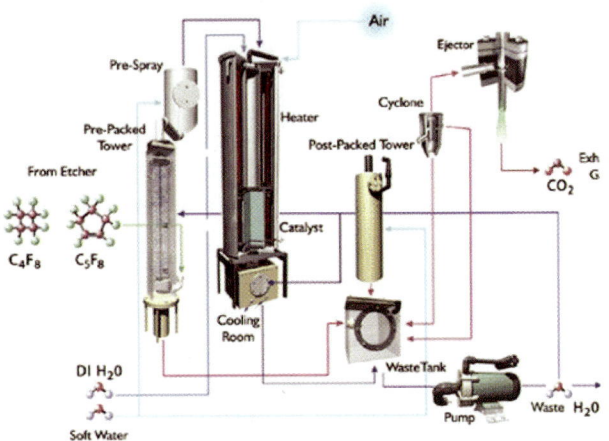

Fig. 29-14 PFC-abatement system for a wafer fab. Courtesy of Hitachi.

compounds for ones presently in use (e.g., using C_3F_8 instead of C_2F_6, because it breaks down more completely in the process-chamber); **3**) recycling the PFCs; and **4**) using special PFC-scrubbers to prevent the PFCs leaving the process-chambers from being released into the atmosphere (shown in Fig. 29-14).[11]

SUMMARY

Wafer-fabs contains a number of hazardous materials and process tools. Yet the semiconductor industry has built a reputation as an industry that provides a safe workplace for its employees (and is an environmentally-responsible industry). In this chapter, such wafer-fab hazards were described - as well as methods developed to keep workers safe while working amongst them. Techniques to abate environmental-pollution by wafer-fabs were also discussed.

REFERENCES

1. *Semiconductor Safety Handbook: Industrial Safety and Health*, R.A. Bolmen, Ed., Noyes, Park Ridge, NJ, 1998.

2. D.G. Baldwin, M.E. Williams, & P.L. Murphy, *Chemical Safety Handbook for the Semiconductor/Electronics Industry*, 2nd Ed. OME Press, Beverly MA 1996.

3. Annual Surveys, Bureau of Labor Statistics - U.S. Dept of Labor (DOL) and Office of Health and Labor (OHL). Injuries and Illnesses/Incident Rates by Industry/Semiconductors and Related Devices.

4. P. Singer, "Handling Hazardous Materials: What You Should Know," *Semiconductor Interntl.*, Dec. 1996, p. 63.

5. S. Roberge, H. Ryssel, and L. Frey, "Safety Considerations for Ion Implanters," in *Ion Implantation - 2000*, Ion Implantation Technology Co., 2000, pp. 642-680.

6. T. Romig, J. McManus, K. Olander, and R. Kirk, "Advances in Ion Implanter Productivity and Safety," *Solid State Technology,* December 1996, p. 69.

7. J. Mayer, K. Olander, and P.V. DeMers, "Analysis of the Advantages of an All SDS Gas Box," in *Proc. Ion Implantation Technology-98*, IEEE (1999).

8. B. Zorich and M. Majors, "Safety and Environmental Concerns of CVD Copper Precursors," *Solid State Technology,* September 1998, p. 101.

9. A. Kulkarni, D. Jukherjee, and W. Gill, "Membrane Reprocessing of Hydrofluoric Acid Solutions," *Semiconductor International,* July 1995.

10. R. Chiarello, "ESH Issues Make Progress," *Semiconductor International,* March 2001, p. 82.

11. W. Worth, "Reducing PFC Emissions: A Technology Update," *Future Fab International,* Issue 9, 2000, p. 57.

12. A. Hand, "Safety Engineers Exchange Old Fab Concerns for New," *Semiconductor Interntl.*, Dec. 2000, p. 67.

PROBLEMS

1. Define the terms: (a) toxic; (b) corrosive; (c) pyrolitic: (d) pyrophoric; and (e) carcinogenic.

2. Explain the purpose of "tagout." signs. Why should the "buddy system" always be used when a hazardous task is being performed?

3. What are the most toxic gases used in IC fabrication. What criteria did you use to decide your answer.

4. Discuss why hydrofluoric-acid (HF) is such an insidiously hazardous chemical to work with. What unique safety hazards does HF pose? State the special safety precautions appropriate when using HF acid.

5. Define the term *fibrillation*. How can this traumatic (and potentially lethal) accident arise from an electrically hazardous condition?

6. What is the job of a scrubber in the wafer fab?

7. Define the terms: (a) TLV; (b) IDLH; and (c) PEL.

8. List five safety measures that are employed to ensure that ion implanters are operated and maintained in a manner that is not hazardous to workers.

9. List appropriate safety precautions for the transport, pouring, and use of liquid etchants.

APPENDIX A
MODELING THERMAL OXIDATION OF SILICON

The most important model for predicting the growth of thermal-oxides on silicon is the *Deal-Grove model* (see also Sect. 13.3.1, in Chap. 13). This model accurately predicts the growth of thermal-SiO_2 on Si over a wide-range of thicknesses (30–500-nm), temperatures (700–1300°C), and oxidant partial-pressures (0.2–25-atm). We derive the model in this Appendix, and discuss its aspects in more detail than in Ch. 13.

A.1 DEAL-GROVE THERMAL OXIDATION MODEL

The *Deal-Grove model*[1] begins by assuming that silicon-oxidation depends on three-fluxes involving the oxidizing-species. The first two involve the movement of the *oxidizing-species* from the gas to the oxide-Si interface, where a third-flux exists (representing the consumption of these oxidizing species by reaction with the silicon substrate). Figure 13-4 (Chap. 13) depicts these fluxes. (*Flux* is defined as the number of objects crossing a unit-area in a unit-time.) In our case, we are interested in the flux of *atoms* (or *molecules*), so the units are, for example, *atoms/cm^2sec*.

These three-fluxes are explicitly defined as: **1)** F_1 is the flux of *oxidizing-species traveling from the main gas-flow region to the oxide-surface*; **2)** F_2 is the flux of *oxidizing-species diffusing through the existing-oxide film to the oxide-Si interface*; and **3)** F_3 is the flux arising from the *consumption of oxidizing-species by reaction with silicon at the oxide-Si interface*. This model also assumes that the oxidation-process is in a *steady-state growth condition*, in which case these three-fluxes must be equal (i.e., $F_1 = F_2 = F_3$). The Grove-Deal model mathematically approximates each flux with respect to the physical-phenomenon it represents, as described next.

Flux F_1 is postulated to be due to the *concentration-difference* that exists between the oxidizing-species in the main gas-flow region, C_g, and that at the top-surface of the oxide, C_s. It is modeled by assuming that this flux is *linearly-proportional to the concentration-difference* between C_g and C_s, and that

h_g (the *mass-transfer coefficient*) is the constant of proportionality. As a result, F_1 is expressed as:

$$F_1 = h_g (C_g - C_s) \quad \text{(A.1)}$$

The constant, h_g, is related to the diffusivity of the oxidizing species, D, and the thickness of the boundary-layer between the main gas-flow region and the solid oxide-surface. (Boundary layers are discussed in Ref. 1 of Chap. 12.) Note that since the three-fluxes are equal, it is useful to express each of them with same set of variables. This will allow us to reduce the oxide growth-model to a single-equation.

With this in mind, Eq. A-1 is rewritten in terms of the *equilibrium-concentration of the gas-species in the oxide*, instead of in terms of the *concentrations in the gas*. This is done with the aid *Henry's Law*, which states that the concentration of a species dissolved in a solid is proportional to the partial-pressure of the species in the surrounding gas (and the constant of proportionality is Henry's Law constant, H). Using Henry's Law, we can thus write:

$$C_o = HP_s \quad \text{(A.2)}$$

and

$$C^* = HP_g \quad \text{(A.3)}$$

where: P_s is the partial-pressure of the oxidizing-species in the gas-phase right-next to the oxide surface; P_g is the partial-pressure of the oxidizing-species in the main gas-flow region; C_o is the actual concentration of the oxidizing species at the oxide-surface (but within the oxide); and C^* represents what the concentration of the oxidizing-species *would be* in the oxide, *if the partial-pressure at the oxide-surface was P_g*. The driving-force for F_1 can therefore be treated as being proportional to the deviation from equilibrium of the concentration in the oxide (i.e., $C^* - C_o$).

Using the ideal gas law ($PV = kT$), C_g and C_s can be rewritten in terms of the partial-pressures P_g and P_s

$$C_g = \frac{P_g}{kT} \quad \text{(A.4)}$$

$$C_s = \frac{P_s}{kT} \quad (A.5)$$

By combining Eqs. A-1, A-2, A-3, A-4, and A-5, F_1 can be rewritten as:

$$F_1 = h(C^* - C_o) \quad (A.6)$$

where h is a newly-defined mass-transfer coefficient, related to h_g by the constant $(1/HkT)$, or:

$$h = \frac{h_g}{HkT} \quad (A.7)$$

Once inside the oxide, the oxidizing-species move towards the Si/SiO$_2$ interface by diffusion, and this movement is the second flux, F_2. This diffusional flux is represented by an equation based on Fick's first law, or:

$$F_2 = -D\frac{dC}{dx_{ox}} = D\frac{(C_o - C_i)}{x_{ox}} \quad (A.8)$$

where D is the diffusivity of the oxidizing-species in the oxide, C_i is the concentration of the oxidizing species at the oxide-Si interface, and x_{ox} is the thickness of the growing oxide.

The third flux, F_3 represents the consumption of oxidizing species as they react with the Si at the oxide-Si interface. It is expressed mathematically as:

$$F_3 = k_s C_i \quad (A.9)$$

where k_s is the chemical-reaction rate-constant of the reaction for oxidation at the oxide-Si interface.

As noted above, when a steady-state growth-condition is established, all three-fluxes must be equal. Thus, by setting Eqs. A-6, A-8, and A-9 equal, we can solve them for C_i and C_o to get:

$$C_i = \frac{C^*}{1 + \frac{k_s}{h} + \frac{k_s x_{ox}}{D}} \quad (A.10)$$

$$C_o = \frac{C^*(1 + \frac{k_s x_{ox}}{D})}{1 + \frac{k_s}{h} + \frac{k_s x_{ox}}{D}} \quad (A.11)$$

It is important to recognize at this point that the three-fluxes represent three-steps of the oxidation process that are in series. Thus, the slowest of them will control the growth-rate of the oxide (i.e., it will be the *rate-limiting step*). With this in mind, in thermal-oxidation of Si, the flux due to F_1 is rarely the rate-limiting step. Thus, it can be assumed that h is much larger than k_s. Using this assumption, Eqs. A-10 and A-11 can be further simplified to:

$$C_i = \frac{C^*}{1 + \frac{k_s x_{ox}(t)}{D}} \quad (A.12)$$

and

$$C^* = 0 \quad (A.13)$$

Combining Eqs. A-8, A-12, and A-13, and eliminating C_i from them, the flux F can be written as:

$$C_i = \frac{C^*}{1 + \frac{k_s x_{ox}(t)}{D}} \quad (A.14)$$

Thus, the rate at which the oxide-layer grows (dx_{ox}/dt) is given as:

$$\frac{dx_{ox}}{dt} = \frac{F}{N_1} = \frac{DC_o k_s}{N_1}\frac{1}{D + k_s x_{ox}} \quad (A.15)$$

where N_1 is the number of oxidant-molecules incorporated into a unit-volume of oxide. Recall from Chap. 2, Sect. 2.6, that there are 5×10^{22}-atoms of Si per cm^3 in a Si-lattice. The volume of silicon that is consumed to grow SiO$_2$ is 44% of the oxide-volume that is formed. Thus, 44% of the Si-atoms present in one-cm^3 of *Si* get incorporated into one-cm^3 of *oxide*. This means that the number of SiO$_2$-molecules per unit volume (e.g., 1-cm^3) is 2.2×10^{22}. Since two oxygen-atoms are incorporated per SiO$_2$-molecule, the value of N_1 for dry-oxidation (in which the oxidizing-species is O$_2$) is 2.2×10^{22}/cm^3, and for wet-oxidation it is 4.4×10^{22}/cm^3 (since two H$_2$O-molecules are needed to form one-molecule of SiO$_2$).

Equation A-15 is solved by applying the initial-condition that: $x_{ox}(t = 0) = x_i$. The value of x_i is the thickness of any oxide present on the Si-surface at the start of the oxidation process (e.g., a native-oxide, or an oxide-layer from a previous oxidation-step). This general-solution for Eq. A-15 is the Deal-Grove model, and is given by:

$$x_{ox}^2 + Ax_{ox} = B(t + \tau) \quad \text{(A.16)}$$

where:

$$A = 2D\left(\frac{1}{k_s} + \frac{1}{h}\right) \quad \text{(cm)} \quad \text{(A.17)}$$

$$B = \frac{2DC^*}{N_1} \quad \text{(cm}^2\text{/sec)} \quad \text{(A.18)}$$

and

$$\tau = \frac{x_i^2 + Ax_i}{B} \quad \text{(sec)} \quad \text{(A.19)}$$

In Eq. A-15, τ represents the time-factor needed in the solution to account for any oxide layer x_i present on the silicon-surface at the start of the oxidation process. Solving for the oxide-thickness, x_{ox}, as a function of oxidation time, we obtain:

$$x_{ox} = \frac{A}{2}\left\{\left(1 + \frac{t+\tau}{A^2/4B}\right)^{1/2} - 1\right\} \quad \text{(A.20)}$$

Equations A-15, A-16 and A-20 match the experimentally-obtained values of *oxide growth-rate* and *thickness* quite closely over a wide-range of growth-conditions. However, they fail to predict the initial-stage of oxidation ($x_{ox} < 30$-nm) when carried out by dry-oxidation. In such cases, the growth-rate is found to greatly exceed that found from Eq. A-15. Thus, modifications must be made to Eqs. A-15, A-16, and A-20 to get a model that works for this growth-regime. This topic (i.e., the *modeling of the growth of thin-oxides*) is discussed in the following section.

As noted earlier, Step-**1** of the three steps used in this model (i.e., due to F_1) is rarely the rate-limiting-step. Instead, one of the other two is typically the slowest. If Step-**2** is the one, the diffusivity of the oxidants, D, is *very-small* compared to $k_s x_{ox}(t)$. This will make $C_i \to 0$. Under such circumstances, oxide-growth is said to be *reaction-rate-limited*. Conversely, if D is *very-large* compared to $k_s x_{ox}(t)$, then $C_i \to C^* \cong C_o$. In such cases, the oxidation-process is said to be *mass-transport-limited*.

To help understand the fundamental-mechanisms underlying the Deal-Grove model, we examine two limiting-cases of Eq. A-16. The first case involves thick-oxide films (i.e., *long oxidation-times*). In such processes, t is much larger than τ, and thus the latter term can be neglected in Eq. A-16. Furthermore, for long oxidation-times, $Ax_{ox} << x_{ox}^2$. This is because $A = 0.1$-0.2-μm, and if given enough time, x_{ox} gets thicker than 0.2-μm. Thus, for processes in which the oxidation-time is very-long, the term Ax_{ox} can also be dropped from Eq. A-16. Consequently, for this limiting-case, Eq. A-16 is reduced to:

$$x_{ox}^2 = Bt \quad \text{(A.21)}$$

Equation A-21 is termed the *parabolic growth-law* (because the thickness of the oxide grown under this law is proportional to the *square-root* of the oxidation-time), and B is the *parabolic rate-constant*. From Eq. A-18, it is seen that B is proportional to the diffusion-constant, D. This is another indication that *oxide-growth in the parabolic growth-regime* is *mass-transport-limited*. That is, the oxide growth-rate is limited by the diffusion of oxidizing-species across the growing oxide-film. The thicker the oxide gets, the longer it takes for these species to cross it. Thus, in this limiting-case, as the oxide gets thicker, its growth-rate diminishes.

The second limiting-case involves very-short-time oxidation-processes. In such processes, $Ax_{ox} >> x_{ox}^2$ (since $A = 0.1$-0.2-μm, and this value is larger than the thickness of oxides grown for very-short periods of time). Thus, the x_{ox}^2 term in Eq. A-16 can be neglected in short-time oxidation-processes. The expression for this limiting-case then becomes:

$$x_{ox} = \frac{B}{A}(t + \tau) \quad \text{(A.22)}$$

Equation A-22 is termed the *linear growth-law* and (B/A) is the linear rate-constant. If Eqs. A-17 and A-18 are combined, it can be seen that (B/A) does not depend on the diffusion coefficient D. That is, B/A is expressed as:

$$\frac{B}{A} = \frac{k_s h}{k_s + h}\left(\frac{C^*}{N_1}\right) \quad \text{(A.23)}$$

Since in most typical oxidation-processes the mass-transfer coefficient h is about 1000 times as large the reaction-rate constant, k_s, h can be eliminated from Eq. A-23. Then the linear-rate constant, B/A, can be written as:

$$B/A = k_s (C^*/N_1) \qquad (A.24)$$

The key-point made by Eq. A-24 is that for short oxidation-times, the growth-rate depends only on k_s (i.e., diffusion of oxidizing-species across the oxide-layer has no impact on the oxide growth-rate). The physical-significance of the rate-constants B and B/A will now be discussed, starting with the rate-constant B.

As noted previously, the parabolic growth-constant B is linked to the diffusion of oxidizing-species across the oxide, while the linear-growth constant B/A depends on the chemical-reaction-rate at the oxide-Si interface. Deal and Grove compared the model with experimental data taken on (111)-lightly boron-doped silicon-wafers. They extracted the values of B and B/A for both dry and wet oxidations from many measurements taken over a broad range of temperatures. These values of B and B/A were graphed using an Arrhenius-plot (see Appendix E for a discussion of Arrhenius-plots) to obtain the activation-energies for these constants. Figure A-1 shows the parabolic rate-constant B plotted on an Arrhenius-plot for both wet and dry oxidation. The results can be written as:

$$B = B_0 e^{-E_A/kT} \qquad (A.25)$$

where B_0 is the pre-exponential term, E_A is the activation-energy for the process, k is Boltzmann's constant, and T is the temperature (in K). The pre-exponential term is related to C^* and N_1 through Eq. A-18. The activation-energy values calculated from the data in Fig. A-1 are 1.24-eV for dry-oxidation, and 0.74-eV for wet-oxidation. Since B and D (the diffusivity of the oxidizing-species through the oxide) are proportional (see Eq. D-18), it is expected that the activation-energy for both D and B should be similar. But, there is no simple-method for measuring the diffusivity of oxidizing-species in thin-oxides. Consequently, the results must be compared to those from bulk fused-silica diffusion-data. The bulk-diffusivity values are 1.17-eV for O_2 and 0.80-eV for H_2O. These values show good agreement with those found for B, confirming the validity of the Deal-Grove model.

The *linear rate-constant B/A* is considered next. Figure A-2 shows an Arrhenius-plot for the B/A constant. From the slope of the plots, the activation-energy for dry and wet oxidation, respectively, is determined to be 2.0-eV and 1.96-eV. Unlike in the case of the parabolic growth-regime, the activation-energies for both wet and dry oxidation in the linear growth-regime are comparable. If the activation-energies of different oxidizing-species have a similar value, it is likely that both have the same limiting mechanism.

From Eq. A-24 it is seen that B/A is proportional to the reaction-rate constant k_s. Assuming that oxidant scan only react with *free* (i.e., *unbonded*) Si-atoms, it is plausible that the limiting mechanism for k_s (and thus also for B/A) is the energy required to break the Si–Si bonds at the silicon-surface. That is, only after such a bond is broken, does a silicon-atom become available to react with an incoming oxidant (to form an SiO_2-molecule). If the above scenario is indeed correct, the activation-energy for the process should

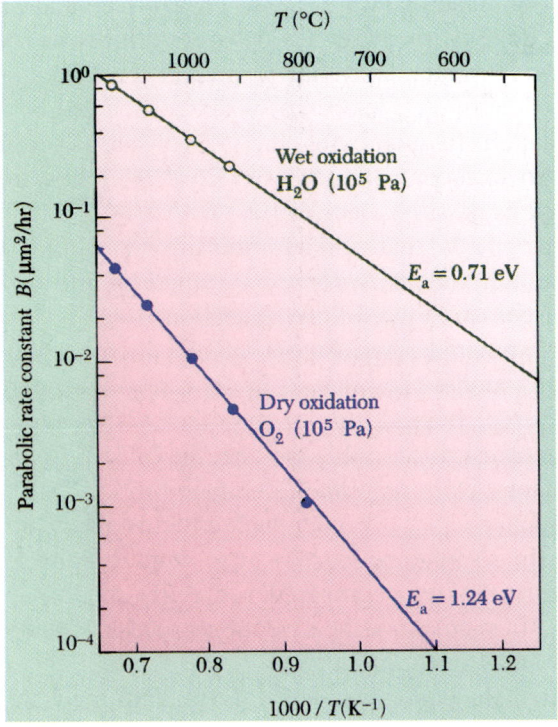

Fig. A-1 Parabolic rate constant versus temperature.[1]

APPENDIX A

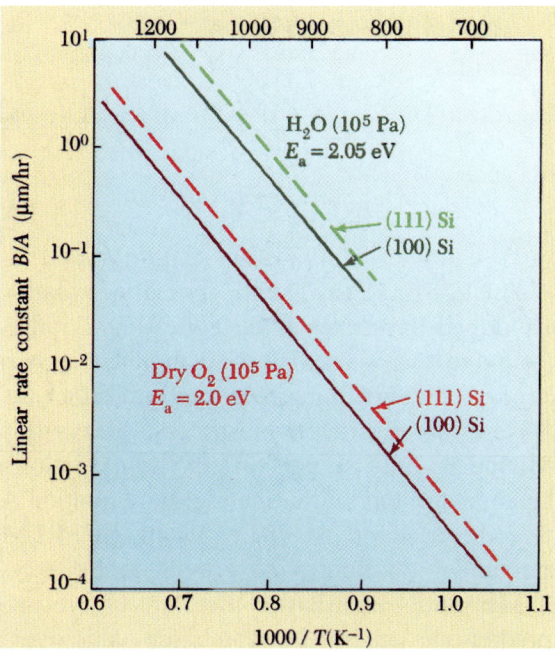

Fig. A-2 Linear rate constant versus temperature.[1]

not depend on the oxidizing-species, but should only be a function of the energy needed to break Si–Si bonds. To validate this premise, the energy for Si-Si bond-breaking was calculated. It was found to be 1.83-eV/molecule. This value is considered a good-fit to the experimentally-determined value for B/A.

Tables 13-2 and 13-3 (in Chap. 13) list the measured values for B, B/A, and τ for wet-oxidation (Table 13-2) and dry-oxidation (Table 13-3) for several representative-temperatures. Note that at any given temperature the B/A and B constants are significantly larger for wet- than for dry-oxidation. As a result, growth-rates of oxides are higher for wet-oxidation. There are two reasons for such enhanced wet-oxidation growth-rates. First, the concentration of the oxidizing-species dissolved in the oxide, C^*, is much higher for H_2O than for O_2 (i.e., $C^*[H_2O] = 3 \times 10^{19}/cm^3$, and $C^*[O_2] = 5.2 \times 10^{16}/cm^3$). Second, in the parabolic growth-regime, the activation-energy for diffusion of H_2O is much smaller than for O_2. (The H_2O-molecules are smaller than O_2 molecules, and can move more easily through the SiO_2 network.)

It is also important to note that crystal-orientation dependence of the parabolic and linear rate-constants indicates that the B/A for (111)-silicon is an average of 1.68 times that of B/A for (100)-silicon. Grove and Deals data given in Figs. A-1 and A-2 was obtained from oxidation of (111)-silicon wafers. The values of B/A given Tables 13-2 and 13-3 in Chap. 13 are given in terms of oxide-growth on (100)-silicon.

A.2 MODELS OF THIN-OXIDE GROWTH

The Deal-Grove model described above for the growth of SiO_2 provides excellent agreement with experimental-observations for thick-oxides. That is, the rate-constants derived in the Deal-Grove model allow the thickness of oxide-films >350-Å to be well predicted as a function of temperature, furnace-ambient, Si-doping-concentration, and Si-crystal-orientation. However, the model does not give a detailed understanding of the mechanisms that produce these dependencies, nor is it valid when oxides are thinner than 200-Å under all growth conditions. Specifically, the experimentally determined oxide-growth-rates in dry-O_2 are not accurately predicted by the model for oxides less than about 200-Å (20-nm). This is the so-called *anomalous-regime* of the Deal-Grove model, and this anomaly is particularly troublesome because gate-oxides of thicknesses of ≤ 100-Å (grown in dry-O_2) are used for sub-micron MOSFETs.

Several physical-mechanisms have been proposed as models for the enhanced-oxidation observed when oxides thinner than 200-Å are grown in dry-O_2. But none of them has yet been shown to more correct than the others. We are left with several empirical models that yield reasonable-agreement with measured values of oxide thickness in the thin-oxide regime, but with no completely satisfactory physical explanation. It is hoped that additional work in the future will help clarify some of these issues.

One of the empirical models for thin-oxide growth was proposed by Reisman and Nicollian in 1987. By analyzing a vast amount of data, they derived an empirical expression that calculates the oxide-thickness versus time using a general power-law of the form:

$$t_{ox} = a(t_g + t)^b \quad \text{(A.26)}$$

where a and b are constants, t_g is the growth-time measured in a given experiment, and t is the time to grow an oxide of thickness t_{ox_i} already present on the surface. Their model can fit all published dry-O_2 data. In fitting this equation to experimental-data, they extracted b values between 0.25 and 1.0, depending on temperature and oxidant partial-pressure. This equation has two fitting parameters (a and b), just as the Deal-Grove model has two (A and B/A). However, there are significant differences between the models. First, the Reisman-Nicollian model is claimed to be able to fit data down to oxide-thicknesses of essentially zero-thickness. There is no anomalous-regime, as exists with the Deal-Grove model. Second, each model has its own physical-basis for the oxide-growth process. The Deal-Grove model is based on the idea of oxidant-diffusion and an interface-reaction - with each process dominating the growth under different conditions. Reisman and Nicollian's model suggests that the interface-reaction actually controls the oxidation-process at all times, and that the volume-expansion necessary at that interface to accommodate the growing oxide was provided by viscous-flow (relaxation) of the oxide layer. The time-dependent viscous-flow of the oxide in the Reisman model is used to explain the extracted pressure and temperature dependence of the parameters in that model (a and b). A third model by Han and Helms (which is not considered in detail here), has also been proposed to predict thin-oxide growth.

Despite the fact that the Deal-Grove model in its original form does a poor job at predicting dry-oxidation-growth in the thin-oxide regime, their model can be "fixed" to do a much better job. Such a "fix" was developed by Massoud who demonstrated that a much-better fit can be accomplished than with the Deal-Grove model alone (by adding an additional-term that decays exponentially with thickness, to the growth-rate (see Ref. 8 in Chap. 13). The expression for the oxide growth-rate in the classical Deal-Grove model dx_{ox}/dt (see Vol. 1, Chap. 7) given by:

$$(dx_{ox}/dt) = B/(2x_{ox} + A) \quad \text{(A.27)}$$

then becomes:

$$(dx_{ox}/dt) = \{B/(2x_{ox} + A)\} + C \exp(-x_{ox}/L_2) \quad \text{(A.28)}$$

where

$$C = C^0 \exp(-E_A/k_B T) \quad \text{(A.29)}$$

and $C^0 \sim 3.6 \times 10^8$-μm/hr, $E_A \sim 2.35$-eV, and L_2 is the characteristic decay length. For dry-oxidation of lightly-doped substrates in the 800-1000°C range, L_2 was found to be $\cong 7$-nm, independent of surface-orientation [i.e., the above numbers apply to either (100)- or (111)-oriented silicon-substrates]. Heavy-doping was found to only slightly affect the oxidation-rate enhancement in the thin-oxide regime. Equation A-28 has been inserted into SUPREM III and SUPREM-IV as the model for thin-oxide growth in dry-oxygen. With Massoud's addition to the Deal-Grove model, its predictions accurately match the data over the entire thickness-range. Note that it was decided to implement the Massoud model (but not the Reisman model) in SUPREM III and IV. This is because many other oxidation-effects have also been tied to the basic Deal-Grove model, and these effects are also modeled in SUPREM III and IV.

While the debate continues about the physical mechanisms responsible for thin-oxide kinetics, the problem has taken on a new importance. This is because deep-submicron MOSFET gate-oxides are routinely grown with thicknesses in the anomalous regime. This is an example of an industrial application outpacing basic scientific-understanding. From the industrial view, it is more important to be able to grow thin-oxide layers reproducibly and uniformly (and with good electrical properties), than it is to understand the governing physical principles.

REFERENCES

1. B.E. Deal and A.S. Grove, "General Relationship for the Thermal Oxidation of Silicon," *J. Appl. Phys.*, **36**, 3770, (1965).

APPENDIX B

MATHEMATICAL MODELS OF DIFFUSION IN SILICON

In this Appendix, the basic mathematical-relationships that model diffusion in silicon are presented. These models allow doping-profiles to be analytically-calculated for some diffusion-processes used in Si-device-fabrication. Today, however, simulation-software packages are more commonly employed to perform these calculations (as they provide more-accurate results, when compared to experimental-data). However, by working with the analytical relationships given here, a better insight into Si-diffusion processes (and the calculations used in the process-simulators to obtain their solutions) is gained.

B.1 FICK'S LAWS OF DIFFUSION

The basic mathematical-tools for treating diffusion were formulated by Fick in 1855. He suggested that equations analogous to Fourier's heat-flow equations might also apply to diffusion-phenomena in matter. He proposed two relationships to model diffusion, which came to be known as *Fick's Laws of Diffusion*.

Fick postulated that if a *concentration-gradient* of an impurity, $\partial C/\partial x$, exists in a substance, there is a tendency for the impurity to move so as to decrease the magnitude of the gradient. That is, if the impurity is free to move, it will tend to move such that the gradient is reduced. This type of movement is termed *diffusion*. If diffusion continues long enough, the impurity-concentration becomes uniform throughout the substance, and at that point the net-flow of matter at all points (i.e., diffusion) will cease.

The impurity-concentration, C, at any point in a substance can be expressed as a function of position and time, $C(x,y,z,t)$. For a one-dimensional case, concentration-profiles of the impurity can then be plotted terms of $C(x,t)$, as is shown in Fig. 11-11 (Chap. 11).

B.2 FICK'S FIRST LAW OF DIFFUSION (IN ONE DIMENSION)

Fick's First Law of Diffusion states that the flow of an impurity (represented by the *particle-flux, F*) is directly-proportional to $\partial C/\partial x$, and the constant of proportionality is termed the *diffusion-coefficient D* of the impurity in that particular substance. In the case of one-dimensional particle-flow, this statement is expressed mathematically as:

$$F = -D\frac{\partial C(x,t)}{\partial x} \qquad \text{(B.1)}$$

where F is defined as number of particles crossing a unit-area in a unit-time. The terms in Eq. B-1 are generally expressed in units of *cm* and *sec* when describing diffusion in semiconductors (although other units could also be used). In our case, the quantities in Eq. B-1 are thus expressed as: $C(x,t)$ - number of particles/cm^3; D - cm^2/sec; F - number of particles/cm^{2}sec; and $\partial C/\partial x$ - number of particles/cm^4. The diffusion-coefficient, D, depends on the temperature, the diffusing-species, and its concentration.

Note that there is a negative-sign in Eq. B-1. This is because, for a decreasing-concentration with depth (i.e., from left to right), $\partial C/\partial x$ will be negative. However, by including a negative-sign in Eq. B-1, a particle-flux moving from left to right will yield a positive value. Thus, the presence of the negative-sign is needed to make Eq. B-1 reflect the fact that net particle-movement occurs in the direction of decreasing-concentration.

An example of *one-dimensional diffusion* in a solid is depicted in Fig. 11-5 (Chap. 11). In that example, a higher-concentration of impurity initially exists at the surface of a solid (Fig. 11-5a). But as time goes on, diffusion reduces the concentration-gradient.

Eventually, the impurity-concentration is uniform throughout the solid. At such time, a gradient no longer exists (i.e., $\partial C/\partial x = 0$, Fig. 11-5d), and diffusion ceases to occur.

B.3 FICK'S SECOND LAW OF DIFFUSION (IN ONE-DIMENSION)

While *Fick's First Law* is important, it is not sufficient. It does not give any insight as to how the impurity-concentration in a substance changes with time. In order to obtain such information, an additional relationship is needed. *Fick's Second Law* addresses this need, and the one-dimensional form of this expression is derived from the First Law.

In doing this, we make reference to Fig. B-1. There, a *volume-element* V (of a substance in which an impurity exists) is shown. This volume-element has a thickness Δx, and a *unit* cross-sectional area ($A = 1 \cdot 1$). Thus, the volume of $V = 1 \cdot 1 \cdot \Delta x = \Delta x$.

We make the assumption that the impurity can enter or leave this volume only in one-dimension (i.e., only in the $\pm x$ direction). We define F_1 as the flux of material *entering* the volume-element, $V = \Delta x$, and F_2 as the flux *leaving* this same volume-element. Fick's Second Law (for the case of one-dimensional flux) says that the change in the concentration of the impurity in the volume-element is simply the difference between the flux *into* and *out-of* it. In other words, "what goes into the volume (Δx) - and does not come out - must have stayed there" (and if more *went-in* than *came-out*, then the concentration in the volume-element (Δx) will have increased in that *time*).

This concept can be written mathematically as:

$$\frac{\Delta C}{\Delta t} = \frac{\Delta F}{\Delta x} = \frac{F_1 - F_2}{\Delta x} \quad \text{(B.2)}$$

Now, if the volume-element is very thin (i.e., $\Delta x \to 0$), then the term on the right of Eq. B-2 can be written as:

$$\frac{F_2 - F_1}{\Delta x} = -\left(\frac{\partial F}{\partial x}\right) \quad \text{(B.3)}$$

and in the time-limit of $\Delta t \to 0$, the term on the left hand side of Eq. B-2 becomes $\Delta C/\Delta t \to \partial C/\partial t$. Thus,

if both limits are invoked, Eq. B-2 can be rewritten as:

$$\frac{\partial C(x,t)}{\partial t} = -\frac{\partial F}{\partial x} \quad \text{(B.4)}$$

By substituting F from Eq. B-1 into Eq. B-4, *Fick's Second Law* (in one-dimension) is obtained:

$$\frac{\partial C(x,t)}{\partial t} = \frac{\partial}{\partial x}\left(D\frac{\partial C}{\partial x}\right) \quad \text{(B.5)}$$

Equation B-5 is the most general-representation of Fick's Second Law. However, if the diffusion-coefficient D is independent of position, then D in Eq. B-5 can be taken outside the differential, and the equation assumes a simpler form:

$$\frac{\partial C(x,t)}{\partial t} = D\frac{\partial^2 C(x,t)}{\partial x^2} \quad \text{(B.6)}$$

For low impurity-concentrations the assumption that D is independent of position is correct, and Eq. B-6 can be used. But, for high impurity-concentrations, this is not valid, and Eq. B-5 must be solved instead.

B.4 SOLUTIONS TO FICK'S SECOND LAW

To obtain predictive-information about diffusion of impurities in Si-devices under various process conditions, Eq. B-6 (or Eq. B-5) must be solved. These are both second-order differential equations, and if analytical (or exact) solutions are available, they still require the knowledge of one initial-condition and two boundary-conditions (i.e., one for time, and two for position). Note, however, that when: **1)** D is not a constant of position; and/or **2)** the simple boundary-conditions required for an exact-solution are not satisfied, then Eq. B-5 must be solved (using numerical-solutions), instead of Eq. B-6. That is, to calculate doping-profiles for such cases, numerical analysis must be employed. This is usually performed by computer-aided process-simulation software (such as SUPREM III or SUPREM IV, see Sect. B.6).

For the cases of analytical-solutions to Fick's Second Law, there are two sets of boundary-conditions which are of practical use in IC-processing (and which lead to exact-solutions of Eq. B-6). An examination of these solutions, in fact, serves two useful purposes. First, it leads to a basic understanding of diffusion-phenomena in Si-processing. Second, the solutions provide rough-approximations of actual impurity-profiles after specific process-steps.

The first such exact-solution of Eq. B-6 is used to treat the case in which impurities are introduced into the top surface of a Si-wafer. It is assumed that the supply of dopant at this surface is above the solid-solubility limit of the dopant for the duration of the step. Thus, there is a *constant-supply of dopant* being introduced into the Si during the entire-process, and the quantity of dopant in the Si continues to increase as long as the process goes on. This type of step is carried out as the first-step of a two-step diffusion process, and is called a *pre-deposition* (or *pre-dep*) step. It is used to bring a known quantity of a dopant (called the *dose*), into a Si-wafer. (As noted in Ch. 11, this step was once carried out with chemical-vapor sources, but is now largely done using ion-implant.)

The second exact-solution of Eq. B-6 is used to predict diffusion-phenomena that occur when an initial amount of impurity (perhaps introduced earlier into the Si by the *pre-dep step*) is diffused subject to specific boundary-conditions. In this case, the amount of dopant in the Si remains fixed for the duration of the diffusion-step, and this type of diffusion-process is called a *drive-in diffusion-step*.

As noted earlier, most processes used to introduce dopants into a Si-wafer use both of the above steps (i.e., a *two-step diffusion process* is performed):

1) The first one (*pre-dep*) is a short, *constant-dopant-supply* step (carried out either with a *chemical-diffusion from a gas-ambient*, or with a *shallow ion-implantation step*). Its function is to bring a specific-dose of dopant into the silicon-wafer, localized very-near to the top-surface.

2) This is followed by a longer, *fixed-dopant-dose* diffusion-step, called *drive-in*. Its role is to move the *diffusion-front* of the dopant brought into the Si by the *pre-dep* to the desired depth.

Use of such *two-step diffusion-processes* allows more flexibility to set the doping-concentrations and final junction-depths, than if just a *one-step diffusion-process* was used.

B.4.1 The Mathematics of Pre-Dep Diffusion

An exact solution to Eq. B-6 is available by invoking one initial-condition and two boundary-conditions encountered in *pre-dep* (together with the assumption that the impurity is being diffused into the surface of a semi-infinite solid.) The *initial-condition* invoked states that there is no dopant anywhere in the Si at t = 0. This is written as:

$$C(x, 0) = 0 \quad \text{(B.7)}$$

The two *boundary-conditions* for this process-step are established next. The first one (Eq. B-8) says that the concentration of dopant at $x = 0$ (i.e., at the surface of the Si), for any time during the process remains constant at a value C_s for the entire *pre-dep* diffusion-time. This written is as:

$$C(0, t) = C_s \quad \text{(B.8)}$$

The second one states that the concentration of dopant at all times at $x = \infty$ is equal to zero, or:

$$C(\infty, t) = 0 \quad \text{(B.9)}$$

The solution to Eq. B-6 for these conditions is given by Eq. B-10:

$$C(x, t) = C_s \text{erfc}\left(\frac{x}{2\sqrt{Dt}}\right) \quad \text{(B.10)}$$

where *erfc* is a mathematical-function known as the *complementary-error-function*. The complementary-error-function is tabulated in many handbooks for various values (and in Appendix F). The data in such Tables is used to evaluate the *erfc* in Example B-2. (Note that the highest surface-concentration encountered in diffusion problems is in the range of 10^{20} atoms/cm^3, and the lowest typical wafer background-level is about 5×10^{14} atoms/cm^3. Thus, to be able to plot a diffusion profile over this range will require a range of values over about six orders of magnitude. Thus, from Eq. B.10, the value of the error function must range from 1 to 10^{-6}. The w-range (see Table in Appendix F) corresponding to this is 0.0 to 4.0, is covered in the table in Appendix F.

The term in the denominator of Eq. B-10, $\sqrt{Dt}$, is referred to as the *diffusion-length*. It is a characteristic of how far atoms move during a given thermal-step. Figure 11-11 shows an *erfc-doping-profile* (in one dimension) versus time. One can see that: **1)** the concentration at the surface remains constant throughout the entire *pre-dep* step; **2)** the total-quantity of dopants introduced into the silicon continues to increase as time progresses; and **3)** the junction-depth gets deeper as time increases.

The *total-quantity of dopant* that accumulates in the silicon during *pre-deposition*, Q_o, is calculated by integrating Eq. B-10 with respect to x over all space (in one-dimension) to obtain Eq. B-11.

$$Q_o = \int_{-\infty}^{+\infty} C_s \mathrm{erfc}\left(\frac{x}{2\sqrt{Dt}}\right) dx = \frac{2}{\sqrt{\pi}} C_s \sqrt{Dt} \quad \text{(B.11)}$$

The dose, Q_o, is expressed in units of *impurities per unit area* (typically per cm^2). Since the profile depth is typically < 1-μm, a dose of 10^{15}-cm^{-2} will produce a large *volume-concentration* (10^{19}-cm^{-3}). According to Eq. B-11, since the surface-concentration remains constant in *pre-deposition-diffusions*, the total-dose increases as the *square-root* of the time.

B.4.2 The Mathematics of Drive-In Diffusion

To allow an exact-solution to the diffusion-equation to exist for the second-step of a two-step diffusion-process (*drive-in*), a preliminary premise is made. That is, an assumption is made that the doping-profile of the impurity-dose introduced during *pre-dep* is a delta-function (i.e., it is located within an infinitely-thin rectangular-region at the wafer-surface). The magnitude of this delta-function is equal to the total dose, Q_o. Next this initial-condition, and the two boundary-conditions for this process are mathematically expressed. The delta-function, initial-condition stated above is given by Eq. B-12 as:

$$C(x, 0) = 0 \quad \text{for } x > \delta \quad \text{(B.12)}$$

The two boundary-conditions are embodied in the form of Eqs. B-13 and B-14:

$$\int_o^{+\infty} C(x,t) dx = Q_o \quad \text{(B.13)}$$

and

$$C(\infty, t) = 0 \quad \text{(B.14)}$$

The boundary-condition of Eq. B-13, states that the total-quantity of dopant stays fixed at Q_o. Under these conditions, the solution of Eq. B-6 is given by:

$$C(x,t) = \frac{Q_o}{\sqrt{\pi Dt}} \exp\left(\frac{-x^2}{4Dt}\right) \quad \text{(B.15)}$$

This solution has the form of a *Gaussian-distribution*. The surface-concentration, C_S (which in this case decreases as a function of time) is determined from Eq. B-16 (i.e., at $x = 0$):

$$C_S = C(0, t) = \frac{Q_o}{\sqrt{\pi Dt}} \quad \text{(B.16)}$$

Since the *dose* remains constant throughout the duration of the *drive-in*, as the diffusion-front moves deeper into the wafer, the surface-concentration must decrease (so the area under the curve can remain constant with time). Equation B-16 expresses this decrease quantitatively. The junction-depth x_j for the drive-in diffusion case is given by:

$$x_j = 2\sqrt{Dt}\left(\ln\frac{Q_o}{C_{sub}\sqrt{\pi Dt}}\right)^{1/2} \quad \text{(B.17)}$$

where C_{sub} is the substrate background-doping-concentration. Figure 11-14 (in Chap. 11) shows a plot of the concentration-profiles $C(x)$ for several diffusion-times at a constant-temperature (*Gaussian curves*). Note that in practical situations, if the *Dt-product* for a *drive-in step* is *much-greater* than the *Dt-product* for the *pre-dep* step, then the initial-condition used in the *drive-in* model is valid. The resulting impurity-profile is then closely approximated by a *Gaussian*

TABLE B-1 Typical Diffusion Coefficient Values for a Number of Impurities.

Element	$D_0(cm^2/sec)$	$E_A(eV)$
B	10.5	3.69
Al	8.00	3.47
Ga	3.60	3.51
In	16.5	3.90
P	10.5	3.69
As	0.32	3.56
Sb	5.60	3.95

distribution (Eq. B-15). However, if the *Dt-product* for the *drive-in-step* is *less-than* (or *comparable-to*) the *Dt-product* of the *pre-dep step*, then the resulting impurity-profile will be more-closely approximated by the *erfc-expression* (Eq. B-10).

B.4.3 Drive-In Diffusion From An Ion-Implantation Pre-Dep Step

If the *pre-dep* step is carried out by ion-implantation, the initial doping-profile for the *drive-in* step is a near-Gaussian distribution (located close to the wafer-surface). This Gaussian-profile will spread-out when exposed to the high-temperature drive-in step (see Fig. 12-18, Chap. 12). The solution to Eq. B-6 with an *initial Gaussian-profile* (instead of a *delta-function profile*) has been reported for the cases of *drive-ins* performed in a neutral-ambient and an oxidizing-ambient. For drive-in diffusions in an oxidizing ambient, the solution to Eq. B-6 is difficult to obtain in closed-form, since it involves a moving-boundary problem. Consequently, one must resort to numerical-methods to obtain solutions for this type of process.

B.5 INTRINSIC DIFFUSION-COEFFICIENTS OF DOPANTS IN SILICON

The doping-concentrations in silicon-devices that exist after various thermal-processing steps are found by solving Fick's Second Law. One quantity needed for such solutions is the correct value of D for impurities present as the process is carried out. The diffusion-coefficients of the common-impurities in silicon generally follow the Arrhenius behavior described in Appendix E. Thus, their value depends exponentially on temperature, which is therefore expressed mathematically as:

$$D = D_o \exp(-E_A/kT) \quad \text{(B.18)}$$

where E_A is the activation-energy (in eV), D_o is the prefactor, k is the Boltzmann constant, and T is the temperature (in Kelvins).

The impurities of most interest are the substitutional-dopants (e.g., B, P, As, and Sb), as these are the ones introduced into Si to form *pn*-junctions and other device-regions. The activation-energy, E_A, for diffusion of these dopants in Si is in the range of 3.5-4.5-eV. Table B-1 lists the values of E_A and the prefactor, D_o (which is a nearly temperature-independent term) for these dopants under intrinsic diffusion-conditions. (Note that when the doping-concentrations in Si are high, these values of D_o and E_A are not necessarily valid. That is, the diffusion-process may be operating under what are termed *extrinsic-diffusion-conditions*. The values of these factors for the latter conditions are described in Sect. B.7.)

Here we continue with the discussion of intrinsic-diffusion conditions. *Intrinsic-diffusion conditions* exist when the *intrinsic carrier-concentration* in a volume of Si is higher than the doping-concentration at that position. The intrinsic carrier-concentration is exponentially-dependent on temperature, and its value can be quite high at typical diffusion-temperatures (see any basic semiconductor device-physics text for the relationship used to calculate n_i, but for example, $n_i = 7 \times 10^{19}/cm^3$ at 1000°C). Thus, for diffusions occurring at a temperature of 1000°C (or higher), as long as N_D or N_A is less than this value, the diffusion-behavior is accurately predicted if intrinsic-diffusion-coefficient values are used when solving the diffusion-equation. Thus, there are many processes in which the dopant diffusion can be accurately-calculated using intrinsic diffusion-coefficient values.

EXAMPLE B-1: Calculate the intrinsic diffusion-coefficient D for boron (B) in Si at 1000°C.

SOLUTION: The values of D_o and E_A for B in Si are found from Table B-1 ($D_o = 10.5$-cm²/sec, and $E_A = 3.69$-eV), and T = 1000°C = 1373 K. Using these numbers in Eq. B-19 we get;

$D = 10.5 \exp\{-3.69/[(8.614 \times 10^{-5})(1373)]\}$

$= 2.96 \times 10^{-13}$ cm²/sec.

EXAMPLE B-2: A two-step boron-diffusion is made into a 0.18-ohm-cm *n*-type silicon-wafer. The first-step is a solid-solubility-limited B-*pre-dep* carried-out at 900°C for 15-minutes. The second-step is a *drive-in* performed at 1100°C for 5-hours. Find the surface-concentration and junction-depth at the conclusion of: **a)** *pre-dep*; and **b)** *drive-in*.

SOLUTION: Since the *pre-dep* step is carried out at the solid-solubility limit, we can use Fig. 9-12 (in Chap. 9), to find this limit of B at 900°C. This gives the surface-concentration after pre-dep, or 1.1×10^{20} cm^{-3}. The value of D_1 is found next, using Table B-1 and Eq. B-18 (and noting that 900°C = 1173 K). Thus, D_1 is (following Example B-1):

$$D_1 = 1.45 \times 10^{-15} \text{ cm}^2/\text{sec}.$$

Since this *pre-dep* is a constant-source diffusion, an *erfc-profile* results. This concentration-profile can thus be expressed using Eq. B-10, as:

$$C(x) = 1.1 \times 10^{20} \, erfc(x/2\sqrt{D_1 t_1}) \text{ boron atoms/cm}^3$$

To find the junction-depth, x_j we must find the depth at which the concentration $C(x)$ equals the background-concentration C_{sub}. Using Fig. 2-22 (in Chap. 2), we find that a 0.18-Ω-cm, *n*-type wafer has a background doping-concentration of $C_{sub} = 3 \times 10^{16}/\text{cm}^3$. Thus, at x_{j1}, the doping-profile has this doping-concentration value, or:

$$1.1 \times 10^{20} \, erfc\,(x_{j1}/2\sqrt{D_1 t_1}) = 3 \times 10^{16}$$

Solving this expression for x_j yields:

$$x_{j1} = 2\sqrt{D_1 t_1} \, erfc^{-1}(0.000273) = (2\sqrt{D_1 t_1})(w)$$

EVALUATING THE TERM "$erfc^{-1}(x_{j1}/2\sqrt{Dt})$"

Evaluating the term "$erfc^{-1}$" in the above equations deserves special mention, because it is not a straightforward operation. To begin with, it should be noted that "$erfc^{-1}$" refers to the *inverse complementary error function*. That is, the -1 exponent *does not imply* the reciprocal, but rather an *inverse function* (just as tan^{-1}, also written *arc tan*) is the *inverse* tangent function.

Next, we should define what we mean by the term. That is, the symbol w in the right-hand term of the equation just above this box, is a number whose value, if inserted into the upper limit of the integral (Eq. F.2) of Appendix F, and the mathematical operations on the right-side of Eq. F-2 are carried out, the result is v

---

(which, in our problem is $v = x_{j1}/2\sqrt{Dt}$). [This rather lengthy verbal-description is written mathematically as: $w = erfc^{-1}(v)$].

In any case, one could perform the above operations to find (v), but this would entail some serious computation. Instead, if one knows the value of w, they can look up the value of v in a Table, or a graph - if such information is handy. This is just what is provided in the Table given in Appendix F.

In principle, if one knew v, they could also use the Table to find w. This what is done when evaluating "$erfc^{-1}$", using the data in Appendix F, as follows:

As indicated in Appendix F, the complementary error function $erfc\,(w)$ is defined as

$$erfc\,(w) = 1 - erf\,(w).$$

This, in turn implies that,

$$erfc^{-1}(v) = erf^{-1}(1 - v) = erf^{-1}(y).$$

Therefore, the term $erfc^{-1}$ (0.000273) can be evaluated with the aid of the Table in Appendix F in the following way: First, for v = 0.000273, (1 - v) = (y) = 0.999727. Then, from the Table it is found that

$$w = erfc^{-1}(0.000273) = erf^{-1}(0.999727) = 2.57$$

This result can now be used to calculate x_{j1}.

$$x_{j1} = (2\sqrt{1.31 \times 10^{-12}})(2.57) \text{ cm} = 0.0587\text{-}\mu m$$

Next, we must calculate the dose of impurities, Q_o, introduced during pre-dep (as this is needed for the analysis of the drive-in step). Q_o equals:

$$Q_o = 2C_s\sqrt{D_1 t_1}/\sqrt{\pi}$$
$$= 2\,(1.1 \times 10^{20})\,\sqrt{(1.45 \times 10^{-15})\,(900)}/\sqrt{\pi}$$
$$= 1.42 \times 10^{14} \text{ boron-atoms/cm}^2$$

At the drive-in temperature of 1100°C (1373 K), $D_2 = 2.96 \times 10^{-13}$ cm^2/sec, and the drive-in time is $t_2 = 5$-hr = 18,000-sec. Since this is a fixed-dopant-dose diffusion, the impurity-profile after this drive-in step is a Gaussian-distribution, which can be expressed using Eq. B-15 as:

$$C(x) = 1.1 \times 10^{18} \exp\,(-x^2/4D_2 t_2) \text{ boron-atoms/cm}^2$$

Setting this expression equal to the background doping concentration, C_{sub}, yields the final junction depth x_{j2} of:

$$x_{j2} = 2.77\text{-}\mu m.$$

B.6 DIFFUSION MODELING WITH PROCESS SIMULATORS

The movement of dopant-atoms in Si during thermal-processing can also be found with the aid of process-simulators. Using process-simulation software-packages allows engineers doing process-development and process-integration to substantially reduce the number of experiments they must carry out during technology-development. There are several such simulation programs available, but the most widely-used is SUPREM (Stanford University PRocess Engineering Model). Modeling diffusion in *one-dimension* is performed by SUPREM III, while extension to *two-dimensions* is done with SUPREM IV.

B.7 SUPREM-III MODELS FOR: BORON, ARSENIC, PHOSPHORUS, AND ANTIMONY DIFFUSION

The models used in SUPREM III are based on the *vacancy-model under non-oxidizing conditions*, proposed by Fair and Tsai. This model, however, does not accurately reflect what is occurring on an atomic-scale (since diffusion is related to *both* dopant/interstitialcy and dopant/vacancy interactions). However, the Fair-Tsai model is nonetheless very useful, since it provides an accurate-representation of the diffusion-profile for the common-dopants in silicon.

The *Fair-Tsai Model* assumes that the diffusivity of an ionized-dopant atom is based on the sum of the diffusivities of neutral-vacancies and ionized vacancies, weighted by the probability of their existence. Accordingly, there are four possible states of a vacancy: **1)** *neutral-vacancy* V^o; **2)** *single-negatively-charged vacancy* V^{-1}; **3)** *double-negatively-charged vacancy* V^{-2}; and **4)** *single-positively-charged vacancy* V^{+1}. During *extrinsic-diffusion* each contribution must be modified by the ratio of the doping-level to the intrinsic-carrier-concentration raised to the power (including the correct-sign) of the charge-state. For example, the contribution of the double-negatively-charged vacancy is modified by $(n/n_i)^2$, while that of the single-positively-charged vacancy by (n_i/n). Thus, the effective-diffusion-coefficient under can be calculated from the sum of all the individual vacancy components. The effective D is given by:

$$D = D^o + D^-\left(\frac{n}{n_i}\right) + D^{2-}\left(\frac{n}{n_i}\right)^2 + D^+\left(\frac{n_i}{n}\right) \quad \text{(B.19)}$$

where the individual diffusivities on the right side of the equation correspond to the interaction between dopant atoms and neutral or charged vacancies.

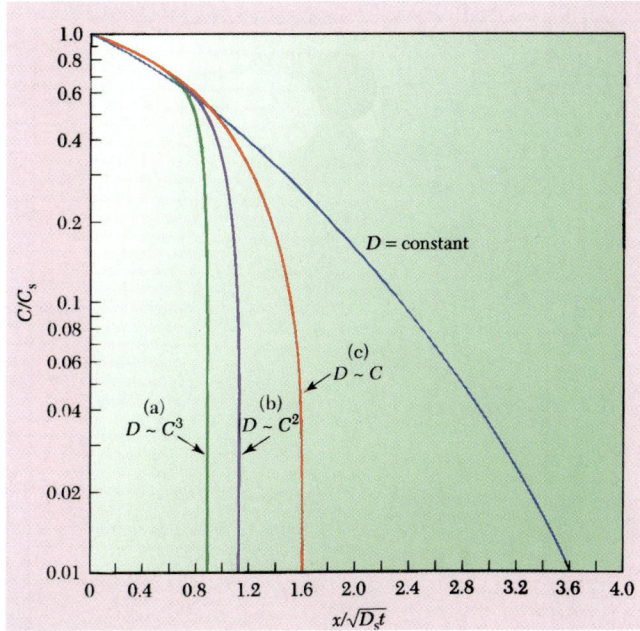

Fig. B-1 Normalized diffusion profiles for extrinsic diffusion where the diffusion coefficient becomes concentration dependent.[1]

B.7.1 Extrinsic Diffusion Coefficients for the Common Dopants (As, P, and B)

Extensive work has been done with respect to modeling extrinsic-diffusion of the common dopants in Si (As, P, and B). However, for our purposes, the dependence of their diffusion coefficients on increased doping concentration in Si can be treated simply, in terms of a parameter γ. That is the intrinsic diffusion coefficient is multiplied by a term $(C/C_s)^\gamma$, where C is the doping concentration at the location of interest, and C_s is the surface concentration, and γ is the parameter that describes the concentration for the

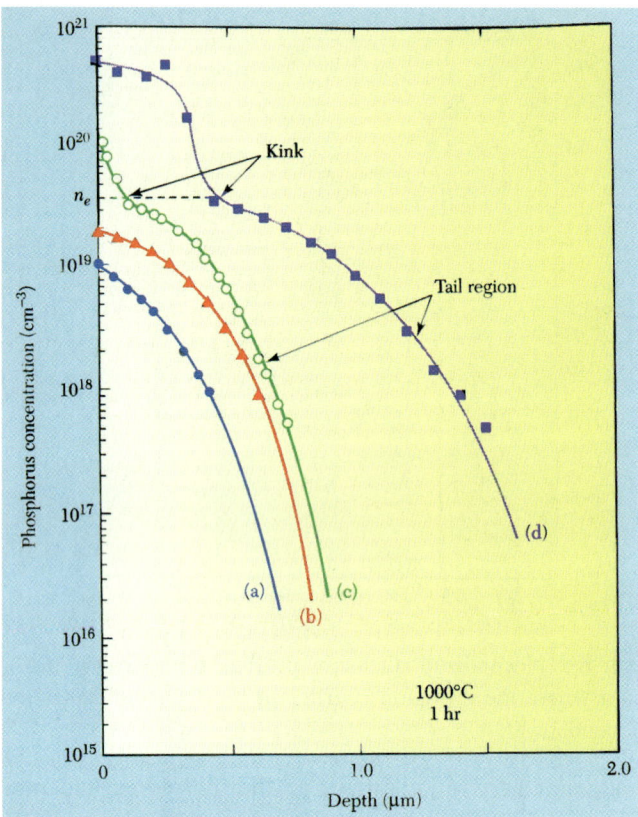

Fig. B-2 Phosphorus diffusion profiles for various concnetrations after diffusion into silicon for 1 hour at 1000°C.

negatively-charged vacancy of Eq. B-19. The diffusion coefficient of P at high concentrations thus varies as C^2. In addition, because of a dissociation effect, the phosphorus diffusion profile exhibits anomalous behavior. Figure B-2 shows phosphorus diffusion profiles for various surface concentrations after diffusion into silicon for 1 hour at 1000°C. When the surface concentration is low, corresponding to the intrinsic diffusion region, the diffusion profile is given by an erfc (curve a). At very-high surface concentrations a kink occurs and is followed by a rapid diffusion in the tail region. Because of its high diffusivity, phosphorus is commonly used to form deep junctions, such as the n-wells in CMOS.

REFERENCES

1. T.Y. Tan and U. Gosele, *Appl. Phys.*, **A37**, 1 (1985).

2. R.B. Fair, "Concentration Profiles of Diffused Dopants," in F.Y. Wang, Ed. *Impurity Doping Processes in Silicon*, North Holland, Amsterdam, 1981.

particular dopant type. For $\gamma = 0$, diffusion follows the ordinary diffusion theory as outlined earlier in the chapter (*intrinsic diffusion*). For $\gamma > 0$, the diffusivity decreases as the concentration decreases, and increasingly steep and boxlike concentration profiles result for increasing γ. Therefore, highly abrupt junctions are formed when diffusions are made into a background of an opposite impurity type. The abruptness of the doping profile results in a junction depth virtually independent of the background concentration.

The measured diffusion coefficients of boron and arsenic in silicon have a concentration dependence with $\gamma \sim 1$. Their concentration profiles are abrupt, as depicted by curve c in Fig. B-1. The diffusion of phosphorus in silicon is associated with the doubly-

APPENDIX C

MATHEMATICAL MODELING OF IMPURITY-PROFILES OF IMPLANTED IONS

In order to benefit from the ability to control the number of impurities implanted into a substrate, it is necessary to know where the implanted-atoms are located after implantation (i.e., it must be possible to predict the *depth-distribution*, or *implanted-profile* of the as-implanted atoms). For example, this information is necessary for selecting appropriate *doses* and *energies* when designing a fabrication process-sequence for new or modified integrated-circuit devices. What is needed to make accurate predictions of implantation profiles is a *theoretical-model* (or *models*) based on the energy-interaction mechanisms between the impinging-ions and the substrate. In this section the topic of how such theoretical-models have been developed, and under what conditions they provide accurate predictions of implantation-profiles, will be addressed.

Despite the fact that the derivation of the models is quite mathematical and complex, the scope of this Appendix is limited to a somewhat more-qualitative discussion. Even on a largely qualitative-level such a presentation is valuable. That is, it provides readers with an appreciation of the intellectual-underpinning of *ion-implantation profile-prediction*, and also serves as an introduction to other physical-mechanisms associated with ion-implantation. These include *channeling-effects during implantation*, *substrate-damage from implantation*, and *recoil-effects* (that occur when implantations are done through thin-layers present on the substrate-surface). Readers interested in gaining a deeper and more-quantitative understanding of implantation-profile models can refer to references given at the end of Chap. 12.

The definitions associated with ion-implantation profiles are given in Sect. 12.1.1 in Chap. 12 of this text. Readers should review that section before continuing with the discussion of this Appendix.

C.1 THE THEORY OF ION-STOPPING

As energetic, implanted-ions move through a solid-target, they transfer energy by collisions with the target nuclei (*nuclear-collisions*) and by coulombic-interaction with the electrons in the target-material. In the latter mechanism, the energy transferred to electrons can lead to exciting the electrons to higher energy-levels (*excitation*), or to the ejection of electrons from their atomic-orbits (*ionization*). The energy-loss due to such target-interactions gradually slows-down these ions, eventually bringing them to a stop. If the energy of an ion at any point along its trajectory in the target is given by E, the process of energy-loss through nuclear-collisions can be characterized by an *energy-loss per unit length due to nuclear-stopping*, $S_n(E)$, and an energy-loss from interactions with target electrons by an *energy-loss per unit length due to electronic stopping*, $S_e(E)$. The total-rate of energy-loss $(dE/dx)_{total}$ is given by the sum of these stopping mechanisms:

$$\left[\frac{dE}{dx}\right]_{total} = S_n(E) + S_e(E) \quad \text{(C.1)}$$

If the *total distance* that the ion travels before coming to a complete-stop is given by R, then:

$$R = \int_0^{E_o} \frac{dR}{(dE/dR)} \quad \text{(C.2)}$$

where E_o is the initial incident-ion-energy.

The *nuclear-stopping process* can be visualized with the aid of the simplification that treats the event as a *collision between two hard-spheres* (see Fig. 12-4, in Ch. 12). However, a more-correct view assumes that such scattering is described by a *Coulombic force-at-a-distance interaction*. In the latter description, an appropriate *atomic-scattering-potential* $V(r)$ must be used. The most successful model for

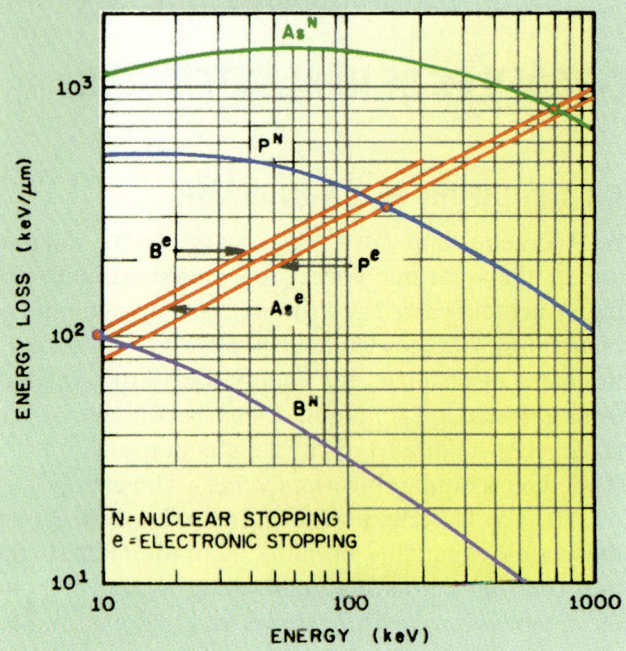

Fig. C-1 Calculated values of DE/dx for As, P, and B at various energies. The nuclear (N) and electronic (e) components are shown. Note the points (o) at which nuclear and elecctronic stopping are equal.

predicting implantation-profiles based on the ion-stopping approach is the so called the *LSS model*, discussed in the following section. The LSS model utilizes a *modified Thomas-Fermi screened-potential* for V(r). Calculations using this model show that nuclear-stopping increases linearly at low-energies, reaches a maximum at some intermediate-energy, and decreases at higher-energies (because ions move past target-nuclei too quickly to efficiently-transfer energy to them). Values of $S_n(E)$ for B, P, and As are shown in Fig. C-1. Also note that $S_n(E)$ increases with the mass of the implanted-ion, and thus heavy ions (such as arsenic) will transfer much-more of their energy through nuclear-collisions than will B-atoms.

The *electronic-stopping-process* can be considered as similar to the stopping of a projectile in a viscous medium, and the stopping-magnitude can be approximated to be proportional to the square-root of the ion-energy:

$$S_e(E) = k_e (E)^{1/2} \quad \text{(C.3)}$$

where k_e is a constant that depends weakly on the ion and target atomic-masses and numbers. There is a crossover-energy at which electronic-stopping becomes more effective than nuclear stopping. As can be seen from Fig. C-1, this cross-over energy is higher for heavier ions. For example, for boron $S_e(E)$ is the predominant energy-loss mechanism down to ~10-keV, while for P and As the energy-loss due to nuclear-stopping predominates for energies up to 130-keV and 700-keV, respectively.

C.2 MODELS FOR IMPLANTATION PROFILES IN AMORPHOUS-SOLIDS

The range-energy relation given by Eq. C-1 was reformulated by Lindhard, Scharff, and Schiott (LSS) for implantation into amorphous material in terms of the reduced parameters ε and ρ, as:

$$\frac{d\varepsilon}{d\rho} = \left(\frac{d\varepsilon}{d\rho}\right)_n + k_\varepsilon (\varepsilon)^{1/2} \quad \text{(C.4)}$$

where ε and ρ are dimensionless variables related to the range R, and the incident-energy E_o by:

$$\rho = \frac{(RNM_1 M_2 4\pi a^2)}{(M_1 + M_2)} \quad \text{(C.5a)}$$

and

$$\varepsilon = \frac{E_o a M_2}{[Z_1 Z_2 q (M_1 + M_2)]} \quad \text{(C.5b)}$$

where: M_1 and M_2 are the mass of the incident-ions and target-atoms, respectively; N is the number of atoms per unit volume; and **a** is the screening-length.

LSS used a *modified Thomas-Fermi screened-potential* to calculate energy-loss due to nuclear-stopping, together with the assumption that energy-loss due to electronic-stopping is given by Eq. C-3 (Fig. C-2). Using this approach, they calculated values of ρ for different-values of ε. (Note that these calculations are quite complex, and are found in the classic paper by LSS.) The value of ρ was then converted to R (using Eq. C-5a), and finally a value for R_p was obtained from the approximate expression:

$$R_p \cong \frac{R}{1+[\frac{M_2}{3M_1}]} \quad \text{(C.6)}$$

LSS assumed that the distribution of the implanted-ions in amorphous-materials could be described by a symmetrical Gaussian-curve. If this assumption is valid then the implanted ion-concentration, n, as a function of depth, x, can be described by:

$$n(x) = \frac{\phi}{\sqrt{2\pi}\Delta R_p} \exp[\frac{-(x-R_p)^2}{2\Delta R_p^2}] \quad \text{(C.7)}$$

where: ϕ is the dose (in number of implanted ions/cm^2), and ΔR_p is the *standard-deviation* of the Gaussian-distribution (or *projected-straggle* of the distribution in the direction of incidence of the beam). The value of ΔR_p is calculated in terms of R_p and the mass of the implanted-ions M_1 and the target atoms M_2, by the approximate expression:

$$\Delta R_p \cong \frac{2R_p}{3}[\frac{\sqrt{M_1 M_2}}{M_1 + M_2}] \quad \text{(C.8)}$$

The concentration is maximum at R_p, and Eq. C-7 at $x = R_p$ reduces to:

$$n(x = R_p) = \frac{\phi}{\sqrt{2\pi}\Delta R_p} \cong \frac{0.4\phi}{\Delta R_p} \quad \text{(C.9)}$$

The assumption by LSS that the distribution of implanted-atoms in amorphous materials is well-fit by a Gaussian-curve is not completely correct, but is nevertheless very-useful as a first-order description. Indeed, the fit to experimental-data is almost always good for implantations near the peak. The peak-value predicted by Eq. C-9 is generally within 1% of the measured-value (except for shallow [*low-energy*] implants). On the other hand, significant-asymmetries appear in experimental implant-profiles in amorphous targets once the concentration-levels drop by a factor of 10 below the peak value (and which are not taken into account by a symmetrical Gaussian approximation). Therefore, various distribution-curves with a higher-number of moments than a Gaussian (which can be described with only two moments, R_p and ΔR_p), have been examined for their ability to fit the experimental implant-profiles. In addition, LSS-theory and the Gaussian-curve fail to account for several effects that occur when implants are made into single-crystal material. Thus, modifications (or even other analytical-models) must be used to obtain a good fit to data obtained in such situations.

Nevertheless, in practice the Gaussian-distribution is still commonly-used to provide quick estimates of doping-distributions into amorphous and single-crystal targets. The higher-moment distributions (or alternate models) are subsequently utilized to fine-tune the dose or energy to obtain better results. Several workers have calculated R_p and ΔR_p values for many elements commonly implanted into Si and SiO$_2$ (using LSS theory to perform the calculations). Fig. 12-3a (in Ch. 12) is an example of such R_p-data for B, P, and As into silicon. Values for projected-lateral-straggle, $\Delta R_\perp$, have also been compiled, and are given in Fig. 12-3b (together with ΔR_p).

Example: A 150-mm wafer is implanted with 10-keV boron ions to a dose of 5x10^{14} ions/cm^2:
1) Determine the *projected-range, projected straggle & peak-concentration* using Figs. 12-3.
2) If the implantation-time is 1-min, calculate the required *ion beam-current*, I.

Solution: 1) From Fig. 12-3 it is found that the *projected-range* and *projected-straggle* are 0.32-μm and 0.07-μm, respectively. The *peak-concen-*

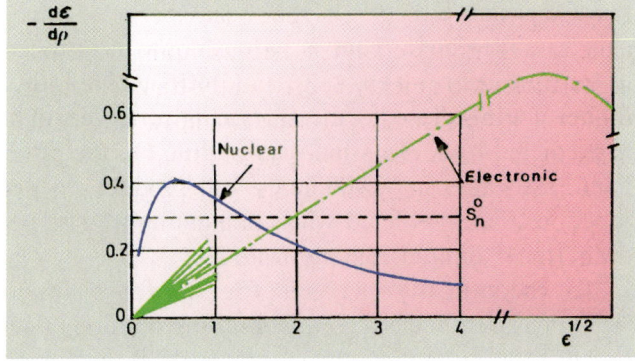

Fig. C-2 Nuclear Stopping power for Thomas-Fermi potential (solid line) and electronic stopping (dash-dot line) based on the LSS theory.

tration, $N(x = R_p)$ is calculated from Eq. C-9:

$$N(x = R_p) = \frac{0.4 \phi}{\Delta R_p} = \frac{0.4 (5 \times 10^{14} \text{ cm}^{-2})}{(0.07 \times 10^{-4} \text{ cm})}$$
$$= 2.8 \times 10^{19} \text{ ions/cm}^3$$

2) To find the *ion beam-current*, I, first calculate Q (the total-number of implanted ions, where Q = dose ∗ wafer area):

$Q = 5 \times 10^{14} \text{ ions/cm}^2 * [\pi (7.5 \text{ cm})^2] = 8.8 \times 10^{16}$ ions

Then, the required ion-beam-current is determined by dividing the total charge, qQ, by the time of implantation:

$I = qQ/t = (1.6 \times 10^{-19}) * (8.8 \times 10^{16})/60\text{-s} = 0.23\text{-mA}$

Higher-Moment Distributions for Implant-Profiles in Amorphous Material: As noted earlier, even when implanting into amorphous-material, the experimental profiles exhibit some asymmetry, or *skewness*. This is not surprising if one considers the forward momentum of the ions. That is, when relatively-light atoms make collisions with target-atoms (e.g., B in Si), they experience a significant-degree of *backscattering*. Hence, more will come to rest at a distance closer to the surface than R_p (causing the concentration near the surface to be higher). On the other hand, heavier-atoms will undergo little backscattering, and the concentration on the deeper-side of the peak will be higher. Thus, even if a Gaussian-distribution is used to approximate the implantation-profile, such non-Gaussian effects on the behavior of devices can be anticipated. That is, when boron is utilized to implant deep *p*-wells (e.g., in CMOS-technology), higher-doping close to the surface is observed than is predicted by a Gaussian-distribution. On the other hand, skewness in As-implants will produce deeper junctions than predicted when implanting n^+-source/drain (or n^+ emitter-regions) with As.

To theoretically account for skewness found in measured-data, a probability-distribution with higher-order moments must be used. It has been found that use of a distribution with a third central-moment closely approximates depth-profiles if the asymmetry of the profile is not excessive (i.e., the value of the third central-moment is less than ΔR_p). When this approach is taken, the distribution is actually represented by two half-Gaussian profiles, each with their own projected straggle ΔR_{p1} and ΔR_{p2}, joined together at the depth of their modal-range, $R_m = R_p - 0.8(\Delta R_{p1} - \Delta R_{p2})$. The joined half-Gaussian approximation produces a good-fit to the profiles of phosphorus and arsenic atoms implanted into Si. The concentration-values of such distributions as a function of depth can be calculated from:

$$n(x) = \frac{2\phi}{\sqrt{2\pi}(\Delta R_{p1} + \Delta R_{p2})} \exp\left[\frac{-(x - R_M)^2}{2\Delta R_{p1}^2}\right] \quad \begin{array}{l} x \geq R_M \\ \textbf{(C.10a)} \end{array}$$

and

$$n(x) = \frac{2\phi}{\sqrt{2\pi}(\Delta R_{p1} + \Delta R_{p2})} \exp\left[\frac{-(x - R_M)^2}{2\Delta R_{p2}^2}\right] \quad \begin{array}{l} x \leq R_M \\ \textbf{(C.10b)} \end{array}$$

An approach representing the implanted-profile with a distribution described by *four-moments* is more exact than the three-moment approach (and is applicable when the third central-moment is large). Hofker demonstrated that excellent agreement with measured B-profiles in amorphous-Si is obtained by assuming an implant-profile can be described by a Pearson-IV distribution-function (see Ref. 7 in Ch. 12), a type of distribution-function which can be specified by four-moments. These moments describe various characteristics of the implant-profile curve: **1)** μ_1 (*mean-range*); **2)** μ_2 (straggle); **3)** γ_1 (skewness); and **4)** β (kurtosis - which characterizes the *tail*-aspect of the distribution). It was later shown that equally-good agreement is found using Pearson-IV curves for other implanted-species into amorphous-targets.

APPENDIX D
MATHEMATICAL MODELING OF CVD PROCESSES

As noted in Sect. 16-1 (Chap. 16), a CVD-process can be assumed to consist of five sequential-steps. Thus, in steady-state-CVD film-growth, the slowest-step will determine the overall-rate of film-deposition, and is referred to as the *growth-rate-limiting step*. Since high growth-rates are essential for economically-viable processes, the determination of the slowest-step can be valuable. This knowledge may allow process-modifications to be developed that will increase the overall deposition-rate.

A model that allows the growth-rate of CVD films to be predicted would also be useful for developing CVD-processes. However, deriving a such mathematical-relationship based on all five-steps has proven difficult. Instead, less-complex growth-rate models have been created, based on the observation that these steps can be grouped into two categories: **1)** those occurring in the gas-phase (*gas-phase processes*); and **2)** those occurring on the substrate-surface or chamber-wall surface (*surface processes*). Based on this observation, Grove developed a simple CVD growth-rate model in 1966 that is still widely used (Ref. 5, Chap. 16). His model assumes that only *one* of the gas-phase steps (i.e., transport of the reactants across the boundary-layer [Step-**2**]), or only *one* of the surface-processes (i.e., the surface chemical-reaction [Step-**4**]) is *the growth-rate-limiting step*. Despite these simplifying assumptions, the model explains many CVD-phenomena, and predicts the growth-rates of many CVD-films quite well..

Figure 16-3 (in Chap. 16) is a schematic depicting the essentials of Grove's model. The flux F_1 is approximated by assuming that it is linearly-proportional to the concentration-difference between the reactant in the bulk of the gas C_g and that at the surface of the substrate, C_s. (Note that flux is the number of atoms or molecules crossing a unit-area in a unit time; e.g., atoms/cm²sec.) The constant of proportionality is termed the *gas-phase mass-transfer coefficient*, h_g.

The relationship for F_1 is written as:

$$F_1 = h_g (C_g - C_s) \quad (D.1)$$

The flux F_2 is assumed to be linearly-proportional to the surface-concentration of the reactant (implying the reaction obeys first-order kinetics), and the constant-of-proportionality is the *chemical-surface-reaction rate-constant*, k_s. The expression for F_2 is:

$$F_2 = k_s C_s \quad (D.2)$$

Note that the flux of reaction by-products away from the substrate is neglected in this model.

Under steady-state growth-conditions (no build-up or depletion of material at any point) these two fluxes must be equal: $F_1 = F_2 = F$. By setting Eqs. D-1 and D-2 equal, we obtain an expression for the surface concentration of reactants, C_s:

$$C_s = \frac{C_g}{1 + \frac{k_s}{h_g}} \quad (D.3)$$

There are two limiting-cases of Eq. D-3. If $h_g \gg k_s$, the value of C_s approaches C_g. This is termed the *surface-reaction-rate limited* case. However, if $h_g \ll k_s$, the value of C_s approaches zero. This condition is known as the *mass-transfer-rate limited* case.

If N_1 is given in (cm^{-3}) and F in (number of atoms/cm²/sec), the growth-rate of a thin-film is given by G in units of (cm/sec):

$$G = \frac{F}{N_1} \quad (D.4)$$

where N_1 is the number of atoms incorporated into a unit-volume of the film. (Recall from Chap. 2, that the value of N_1 for silicon is 5×10^{22} atoms/cm³.) Substituting for F_2 ($F_1 = F_2 = F$) from Eq. D-2 and for C_s from Eq. D-3 into Eq. D-4, we obtain:

$$G = \frac{F}{N_1} = \frac{k_s h_g}{k_s + h_g} \left(\frac{C_g}{N_1}\right) \quad (D.5)$$

As noted earlier, in many CVD-processes the reactant-gas is diluted in an inert *carrier* (or *diluent*) gas. In such cases, the concentration of the reactant in the gas-phase can be defined as:

$$C_g = C_T Y \quad \text{(D.6)}$$

where Y is the mole-fraction of the reactant-species and C_T is the total-number of molecules (including the carrier-gas) per cm³ in the gas. Substituting Eq. D-6 into Eq. D-5, the *general expression for the film growth-rate* for Grove's CVD-model is obtained:

$$G = \frac{k_s h_g}{k_s + h_g} \frac{C_T}{N_1} Y \quad \text{(D.7)}$$

Two important effects are predicted by Eqs. D-5 and D-7. First, they indicate that the *growth-rate should be proportional to either*: **a**) *the reactant-gas concentration C_g* (i.e., as is stated in Eq. D-5, which applies when no diluent-gas is used, such as in the case of LPCVD); or **b**) to the *mole-fraction of the reacting-species in the gas-phase Y* (as is stated by Eq. D-7, which applies when a diluent-gas is used, such as in APCVD). This is in agreement with experimental observations.

The second important effect predicted by Eqs. D-5 and Eq. D-7 is that the growth-rate at constant C_g (or Y) is controlled (in the limits) by the smaller-value of k_s and h_g. In these limiting-cases the growth-rates are given either by:

$$G = (C_T k_s Y)/N_1 \quad k_s \ll h_g$$
(surface-reaction-rate-limited case) **(D.8)**

or by:

$$G = (C_T h_g Y)/N_1 \quad h_g \ll k_s$$
(mass-transfer-rate-limited case) **(D.9)**

The surface-reaction-rate constant k_s describes the kinetics of the chemical-reaction at the substrate surface. Chemical-reactions are often thermally-activated, and if this is the case, they can be represented by an Arrhenius-type equation (see Appendix E). Assuming that the reactions at the surface exhibit such behavior, k_s can be written as:

$$k_s = k_o e^{-E_A/kT} \quad \text{(D.10)}$$

where k_o is a temperature-independent *frequency-factor*, and E_A is the *activation-energy* of the reaction. We see from Eq. D-10, that if the growth-rate is operating in the reaction-rate-limited regime, then it is very-sensitive to variations in the temperature. This also implies that as the temperature is reduced, the surface-reaction-rate is reduced. Thus, at sufficiently-low temperatures, the arrival-rate of reactants eventually exceeds the rate at which they are consumed at the surface, making deposition-rate surface-reaction-rate-limited.

Also, according to Eq. D-10 the surface-reaction-rate increases exponentially with increasing temperature. For a given surface-reaction, if the temperature rises high-enough, the reaction-rate will exceed the rate at which reactant-species arrive at the surface. In such cases, the reaction cannot proceed any more rapidly than the rate at which reactant-gases are supplied to the substrate by mass-transport (no matter how high the temperature is increased). Then the growth-rate becomes mass-transport-limited.

TO SUMMARIZE: At high-temperatures, the deposition is usually mass-transport-limited, while at lower-temperatures it is usually surface-reaction-rate-limited (see Fig. 16-4, in Chap. 16). In actual processes, the temperature at which the deposition-condition moves from one of these growth-regimes to the other is dependent on the *activation-energy* of the reaction and the *gas-flow conditions* in the reactor.

The mass-transfer coefficient, h_g, depends on gas-phase phenomena. The gas-phase mechanism of most interest to the CVD-process is the rate at which molecules impinge on the substrate. This is modeled by the rate at which the reactant-species cross the boundary-layer separating the main gas-flow region from the substrate-surface. Such transport-processes occur by diffusion (which is proportional to the *diffusivity*, D, of the reactant-gas, and the concentration-gradient across the boundary layer, dC/dx). The rate of mass-transport is only weakly-influenced by temperature ($D \propto T^{1.5-2.0}$), but depends mainly on the gas-flow

APPENDIX D

(fluid-dynamics) in the reactor.

Figure 17-7 in Chap. 17 shows growth-rate data of Si-films for several Si-sources. It can be seen that at low-temperatures the growth-rate follows the exponential-law (Arrhenius-behavior) with $E_A = 1.9$-eV, and $k_o = 1 \times 10^7$-cm/sec. At higher-temperatures, the growth-rate tends to become more temperature-insensitive. This growth-rate data can be explained with the use of Eqs. D-7, D-8, D-9, and D-10. That is, at low-temperatures $h_g \gg k_s$, and the growth-rate is limited by k_s. Equation D-8 then describes the growth-rate, and Eq. D-10 provides the value of k_s. Since k_s increases rapidly with temperature, the reactant-gas supply reaching the surface (which is controlled by h_g), eventually cannot keep up with the demand of the reaction, and the reaction-rate tends to level-off. At high-temperatures, $h_g \ll k_s$ and the growth-rate is thus limited by the mass-transfer of gas-molecules across the boundary-layer (and the value of G is found by using Eq. D-9). The value of h_g in Eqs. D-1, D-7, and D-9 is relatively temperature-insensitive. At intermediate-temperatures both h_g and k_s contribute and the growth-rate does not increase as rapidly as it does at lower-temperatures.

The Grove model is thus a simplified-approach to predicting the growth-rate of films deposited by CVD since it neglects the flux of reaction-products, and assumes the reaction-rate flux depends linearly on the surface-concentration. As we have observed, the latter is true only for low-values of Y. The model also neglects the effect of any temperature-gradients on the gas-phase mass-transfer. Despite these limitations, the Grove model predicts the two-regions of the growth process (i.e., the mass-transfer-rate-limited and the surface-reaction-rate-limited growth-rate regions). It also provides reasonable-estimates of the values of k_s and h_g from growth-rate data.

The Grove model can also give insight into CVD-processes. That is, in CVD-processes run under surface-reaction-rate-limited conditions, the process-temperature is the key parameter. In such processes, uniform deposition-rates throughout a reactor require conditions that maintain a constant-surface-reaction rate. This, in turn, implies that a constant-temperature must also exist everywhere at all wafer-surfaces, and that controlling the temperature becomes the main issue in the reactor-design. Under such conditions, the rate at which reactant-species arrive at the surface is not as important (since their concentration does not limit the growth-rate). Thus, it is not as critical that a reactor be designed to supply an equal-flux of reactants to all locations of a wafer surface. Hence, in low-pressure-CVD (LPCVD) reactors, wafers can be stacked at very-close spacing because such systems operate in a surface-reaction-rate-limited mode (as is described further in Sect. 16.2.6, in Chap. 16).

In deposition-processes that are mass-transport-limited, the temperature-control is not nearly as critical because the mass-transport process (which limits the growth-rate) is only weakly-dependent on temperature. In these processes, it is very important that the same concentration of reactants be present in the main (or bulk) gas-flow regions adjacent to all locations of a wafer, since the arrival-rate is directly-proportional to the concentration in the bulk-gas. Thus, to insure uniform film-thickness across a wafer, reactors which are operated in the mass-transport-limited regime must be designed so that all locations of wafer surfaces are supplied with an equal-flux of reactant species. Atmospheric-pressure CVD (APCVD) reactors that deposit SiO_2 at ~400°C and epitaxial-reactors operating at ≥ 1000°C, operate in the mass-transport-limited regime. The most widely used APCVD-SiO_2 reactor-designs provide a uniform-supply of reactants by horizontally-positioning the wafers and moving them under a reactant-gas stream (see Sect. 16.2.5, in Chap. 16).

APPENDIX E

ARRHENIUS BEHAVIOR

Many of the chemical-reactions that take place during the course of carrying out VLSI-fabrication processes occur at reaction-rates that depend significantly on temperature. In many cases it is found that the reaction-rate dependence is well predicted by a relationship known as the Arrhenius equation. This equation was formulated by the Englishman J. Hood, and is named after the Swedish chemist Svante Arrhenius. The equation is written as:

$$R(T) = R_o e^{(E_A/kT)} \qquad \text{(E.1)}$$

where, $R(T)$ is the rate-constant at temperature T (in units of absolute-temperature [kelvin, or K]), R_o is the frequency factor (also referred-to in some cases as the pre-exponential factor), k is Boltzmann's constant (8.6×10^{-5} eV/K), and E_A is the activation-energy (expressed in units of eV). The activation-energy, E_A, and the frequency-factor, R_o, are also termed Arrhenius parameters. The values of E_A and R_o can often be used to gain insight into the physical or chemical mechanisms that impact the reaction-kinetics of the process being investigated. Several examples of Arrhenius behavior are cited in the text.

If a reaction-rate depends on temperature, it is necessary to determine if it also obeys the Arrhenius equation. If the reaction-rate does indeed exhibit Arrhenius behavior, it is valid to extract values of E_A and R_o from measured reaction-rate data. To verify that the reaction-rates follow the Arrhenius relationship, the rate-constant for a particular reaction is measured at several different-temperatures. The measured-values are plotted on special graph-paper on which one axis is expressed in terms of the log of the rate-constant, and the other axis in terms of the reciprocal of the temperature (usually 1000/T [K] - see Fig. E-1). If the measured rate-constant values fall along a straight-line on this plot, it can be concluded that the reaction obeys the Arrhenius equation. The plot of the rate-constants on this type of graph paper is termed an Arheniius plot. If the plot does not yield a straight line, this is evidence that a more complex reaction is taking place than that described by the Arrhenius equation.

The values of E_A and R_o can be determined in one of two ways: 1) graphically; or 2) analytically. In the graphical technique, the slope of the plotted data, S, represents $S = -E_A/k$. The value of R_o is then found by using this value of E_A, together with a value of $R(T)$ in the Arrhenius equation.

In the analytical technique, the rate constants at two temperatures are used to create two simultaneous equations:

$$R_1(T) = R_o e^{(E_A/kT_1)} \qquad \text{(E.2)}$$

and

$$R_2(T) = R_o e^{(E_A/kT_2)} \qquad \text{(E.3)}$$

From the above equations, E_A is found from:

$$E_a = \frac{k \ln\left(\frac{R_2(T)}{R_2(T)}\right)}{\left(\frac{1}{T_1}\right) - \left(\frac{1}{T_2}\right)} \qquad \text{(E.4)}$$

and R_o is computed in the same manner as above, once the value of E_A has been determined.

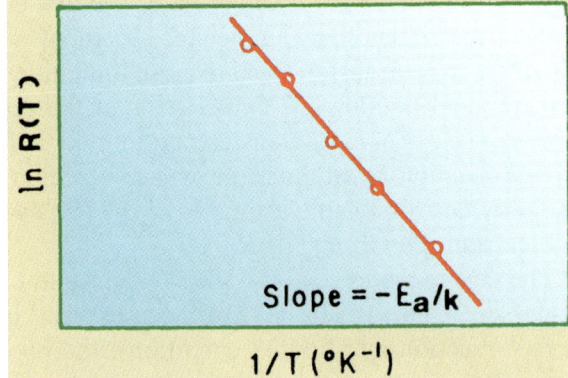

Fig. E-1 Arrhenius plot

Some Properties of the Error Function

APPENDIX F

ERROR FUNCTION TABLES (erf w)

The error function is defined by

$$\text{erf } w = \frac{2}{\sqrt{\pi}} \int_0^w e^{-z^2} \, dz = \frac{2}{\sqrt{\pi}} \left(w - \frac{w^3}{3 \times 1!} + \frac{w^5}{5 \times 2!} - \cdots \right) \quad \text{F.1}$$

so that

$$\text{erfc } w = 1 - \text{erf } w = \frac{2}{\sqrt{\pi}} \int_w^{\infty} e^{-z^2} \, dz \quad \text{F.2}$$

The error function is an odd function: $\text{erf}(-w) = -\text{erf } w$

which also means $\text{erf}(0) = 0$

The limit as $w \longrightarrow \infty$ is $\text{erf}(\infty) = 1,$

which also means $\text{erfc}(0) = 1, \quad \text{erfc}(\infty) = 0$

Assymptotic approximations

$$\text{erf } w \approx \frac{2w}{\sqrt{\pi}} \quad \text{for} \quad w \ll 1 \qquad \text{erfc } w \approx \frac{1}{\sqrt{\pi}} \frac{e^{-w^2}}{w} \quad \text{for} \quad u \gg 1$$

Error Function erf(w)

w	erf(w)	w	erf(w)	w	erf(w)	w	erf(w)	w	erf(w)	w	erf(w)
0.00	0.000 000	0.32	0.349 126	0.64	0.634 586	0.96	0.825 424	1.28	0.929 734	1.60	0.976 348
0.01	0.011 283	0.33	0.359 279	0.65	0.642 029	0.97	0.829 870	1.29	0.931 899	1.61	0.977 207
0.02	0.022 565	0.34	0.369 365	0.66	0.649 377	0.98	0.834 232	1.30	0.934 008	1.62	0.978 038
0.03	0.033 841	0.35	0.379 382	0.67	0.656 628	0.99	0.838 508	1.31	0.936 063	1.63	0.978 843
0.04	0.045 111	0.36	0.389 330	0.68	0.663 782	1.00	0.842 701	1.32	0.938 065	1.64	0.979 622
0.05	0.056 372	0.37	0.399 206	0.69	0.670 840	1.01	0.846 810	1.33	0.940 015	1.65	0.980 376
0.06	0.067 622	0.38	0.409 009	0.70	0.677 801	1.02	0.850 838	1.34	0.941 914	1.66	0.981 105
0.07	0.078 858	0.39	0.418 739	0.71	0.684 666	1.03	0.854 784	1.35	0.943 762	1.67	0.981 810
0.08	0.090 078	0.40	0.428 392	0.72	0.691 433	1.04	0.858 650	1.36	0.945 561	1.68	0.982 493
0.09	0.101 281	0.41	0.437 969	0.73	0.698 104	1.05	0.862 436	1.37	0.947 312	1.69	0.983 153
0.10	0.112 463	0.42	0.447 468	0.74	0.704 678	1.06	0.866 144	1.38	0.949 016	1.70	0.983 790
0.11	0.123 623	0.43	0.456 887	0.75	0.711 156	1.07	0.869 773	1.39	0.950 673	1.71	0.984 407
0.12	0.134 758	0.44	0.466 225	0.76	0.717 537	1.08	0.873 326	1.40	0.952 285	1.72	0.985 003
0.13	0.145 867	0.45	0.475 482	0.77	0.723 822	1.09	0.876 803	1.41	0.953 852	1.73	0.985 578
0.14	0.156 947	0.46	0.484 655	0.78	0.730 010	1.10	0.880 205	1.42	0.955 376	1.74	0.986 135
0.15	0.167 996	0.47	0.493 745	0.79	0.736 103	1.11	0.883 533	1.43	0.956 857	1.75	0.986 672
0.16	0.179 012	0.48	0.502 750	0.80	0.742 101	1.12	0.886 788	1.44	0.958 297	1.76	0.987 190
0.17	0.189 992	0.49	0.511 668	0.81	0.748 003	1.13	0.889 971	1.45	0.959 695	1.77	0.987 691
0.18	0.200 936	0.50	0.520 500	0.82	0.753 811	1.14	0.893 082	1.46	0.961 054	1.79	0.988 641
0.19	0.211 840	0.51	0.529 244	0.83	0.759 524	1.15	0.896 124	1.47	0.962 373	1.80	0.989 091
0.20	0.222 703	0.52	0.537 899	0.84	0.765 143	1.16	0.899 096	1.48	0.963 654	1.81	0.989 525
0.21	0.233 522	0.53	0.546 464	0.85	0.770 668	1.17	0.902 000	1.49	0.964 898	1.82	0.989 943
0.22	0.244 296	0.54	0.554 939	0.86	0.776 110	1.18	0.904 837	1.50	0.966 105	1.83	0.990 347
0.23	0.255 023	0.55	0.563 323	0.87	0.781 440	1.19	0.907 608	1.51	0.967 277	1.84	0.990 736
0.24	0.265 700	0.56	0.571 616	0.88	0.786 687	1.20	0.910 314	1.52	0.968 413	1.85	0.991 111
0.25	0.276 326	0.57	0.579 816	0.89	0.719 843	1.21	0.912 956	1.53	0.969 516	1.86	0.991 472
0.26	0.286 900	0.58	0.587 923	0.90	0.796 908	1.22	0.915 534	1.54	0.970 586	1.87	0.991 821
0.27	0.297 418	0.59	0.595 936	0.91	0.801 883	1.23	0.918 050	1.55	0.971 623	1.88	0.992 156
0.28	0.307 880	0.60	0.603 856	0.92	0.806 768	1.24	0.920 505	1.56	0.972 628	1.89	0.992 479
0.29	0.318 283	0.61	0.611 681	0.93	0.811 564	1.25	0.922 900	1.57	0.973 603	1.90	0.992 790
0.30	0.328 627	0.62	0.619 411	0.94	0.816 271	1.26	0.925 236	1.58	0.974 547	1.91	0.993 090
0.31	0.338 908	0.63	0.627 046	0.95	0.820 891	1.27	0.927 514	1.59	0.975 462	1.92	0.993 378

Error Function erf(w)

w	erf(w)	w	erf(w)	w	erf(w)	w	erf(w)	w	erf(w)
1.93	0.993 656	2.46	0.999 497	3.00	0.999 977 91	3.53	0.999 999 403	3.95	0.999 999 977
1.94	0.993 923	2.47	0.999 523	3.01	0.999 979 26	3.54	0.999 999 445	3.96	0.999 999 979
1.95	0.994 179	2.48	0.999 547	3.02	0.999 980 53	3.55	0.999 999 485	3.97	0.999 999 980
1.96	0.994 426	2.49	0.999 571	3.03	0.999 981 73	3.56	0.999 999 521	3.98	0.999 999 982
1.97	0.994 664	2.50	0.999 593	3.04	0.999 982 86	3.57	0.999 999 555	3.99	0.999 999 983
1.98	0.994 892	2.51	0.999 614	3.05	0.999 983 92	3.58	0.999 999 587		
1.99	0.995 111	2.52	0.999 634	3.06	0.999 984 92	3.59	0.999 999 617		
2.00	0.995 322	2.53	0.999 654	3.07	0.999 985 86	3.50	0.999 999 257		
2.01	0.995 525	2.54	0.999 672	3.08	0.999 986 74	3.51	0.999 999 309		
2.02	0.995 719	2.55	0.999 689	3.09	0.999 987 57	3.52	0.999 999 358		
2.03	0.995 906	2.56	0.999 706	3.10	0.999 988 35	3.53	0.999 999 403		
2.04	0.996 086	2.57	0.999 722	3.11	0.999 989 08	3.54	0.999 999 445		
2.05	0.996 258	2.58	0.999 736	3.12	0.999 989 77	3.55	0.999 999 485		
2.06	0.996 423	2.59	0.999 751	3.13	0.999 990 42	3.56	0.999 999 521		
2.07	0.996 582	2.60	0.999 764	3.14	0.999 991 03	3.57	0.999 999 555		
2.08	0.996 734	2.61	0.999 777	3.15	0.999 991 60	3.58	0.999 999 587		
2.09	0.996 880	2.62	0.999 789	3.16	0.999 992 14	3.59	0.999 999 617		
2.10	0.997 021	2.63	0.999 800	3.17	0.999 992 64	3.60	0.999 999 644		
2.11	0.997 155	2.64	0.999 811	3.18	0.999 993 11	3.61	0.999 999 670		
2.12	0.997 284	2.65	0.999 822	3.19	0.999 993 56	3.62	0.999 999 694		
2.13	0.997 407	2.66	0.999 831	3.20	0.999 993 97	3.63	0.999 999 716		
2.14	0.997 525	2.67	0.999 841	3.21	0.999 994 36	3.64	0.999 999 736		
2.15	0.997 639	2.68	0.999 849	3.22	0.999 994 73	3.65	0.999 999 756		
2.16	0.997 747	2.69	0.999 858	3.23	0.999 995 07	3.66	0.999 999 773		
2.17	0.997 851	2.70	0.999 866	3.24	0.999 995 40	3.67	0.999 999 790		
2.18	0.997 951	2.71	0.999 873	3.25	0.999 995 70	3.68	0.999 999 805		
2.19	0.998 046	2.72	0.999 880	3.26	0.999 995 98	3.69	0.999 999 820		
2.20	0.998 137	2.73	0.999 887	3.27	0.999 996 24	3.70	0.999 999 833		
2.21	0.998 224	2.74	0.999 893	3.28	0.999 996 49	3.71	0.999 999 845		
2.22	0.998 308	2.75	0.999 899	3.29	0.999 996 72	3.72	0.999 999 857		
2.23	0.998 388	2.76	0.999 905	3.30	0.999 996 94	3.73	0.999 999 867		
2.24	0.998 464	2.77	0.999 910	3.31	0.999 997 15	3.74	0.999 999 877		
2.25	0.998 537	2.78	0.999 916	3.32	0.999 997 34	3.75	0.999 999 886		
2.26	0.998 607	2.79	0.999 920	3.33	0.999 997 51	3.76	0.999 999 895		
2.27	0.998 674	2.80	0.999 925	3.34	0.999 997 68	3.77	0.999 999 903		
2.28	0.998 738	2.81	0.999 929	3.35	0.999 997 838	3.78	0.999 999 910		
2.29	0.998 799	2.82	0.999 933	3.36	0.999 997 983	3.79	0.999 999 917		
2.30	0.998 857	2.83	0.999 937	3.37	0.999 998 120	3.80	0.999 999 923		
2.31	0.998 912	2.85	0.999 944	3.38	0.999 998 247	3.81	0.999 999 929		
2.32	0.998 966	2.86	0.999 948	3.39	0.999 998 367	3.82	0.999 999 934		
2.33	0.999 016	2.87	0.999 951	3.40	0.999 998 478	3.83	0.999 999 939		
2.34	0.999 065	2.88	0.999 954	3.41	0.999 998 582	3.84	0.999 999 944		
2.35	0.999 111	2.89	0.999 956	3.42	0.999 998 679	3.85	0.999 999 948		
2.36	0.999 155	2.90	0.999 959	3.43	0.999 998 770	3.86	0.999 999 952		
2.37	0.999 197	2.91	0.999 961	3.44	0.999 998 855	3.87	0.999 999 956		
2.38	0.999 237	2.92	0.999 964	3.45	0.999 998 934	3.88	0.999 999 959		
2.39	0.999 275	2.93	0.999 966	3.46	0.999 999 008	3.89	0.999 999 962		
2.40	0.999 311	2.94	0.999 968	3.47	0.999 999 077	3.90	0.999 999 965		
2.41	0.999 346	2.95	0.999 970	3.48	0.999 999 141	3.91	0.999 999 968		
2.42	0.999 379	2.96	0.999 972	3.49	0.999 999 201	3.92	0.999 999 970		
2.43	0.999 411	2.97	0.999 973	3.50	0.999 999 257	3.93	0.999 999 973		
2.44	0.999 441	2.98	0.999 975	3.51	0.999 999 309	3.94	0.999 999 975		
2.45	0.999 469	2.99	0.999 976	3.52	0.999 999 358				

INDEX and GLOSSARY

This *INDEX* also serves a second purpose, that of a *GLOSSARY*. Whenever an entry is followed by a *def,* a definition of the term is given in the text on that page.

Absolute temperature scale, def, 73-
Absorption, def, 92
Acceleration column, 200
Acceptor, def, 28
Acid, def, 76-7
   inorganic, 77
   organic, 77
   used in chip making, list, 77
Actinic light, 323, 341, 347
Actinic absorbance, 323
Activation energy, def, 181, 550
Adatom, def, 113, 292-3, 294
Adsorption, def, 92, 113
Adhesion, see Thin Films
Adhesion layers, Cu, 441-2, 449
Adhesion promoter, 328
Aerial image, def, 346, 369
After-develop inspection, 334-7
Airy disc, def, 345
Alignment, 342
   marks, 354-6
Aligners, see Contact, Proximity and Projection Printing
Aluminum oxide (Al_2O_3), 406
Alkali, def, 77
Aluminum trichloride, $AlCl_3$, 405-8
Al:Cu alloys, 248, 406-8
   dry-etching of, 406-8
Al:Si alloys, 248
Alternating phase shift masks (AltPSM), 371-3
Aluminum, 248
Aluminum spiking, 248
Amine contamination, 326-7
Ammonium hydroxide, 134
Amorphous, def, 145
   layers, 195
   threshold dose, 197
Analog integrated circuits, 44-5
Analyzer, ion
   in ion implanters, 200
   in RGAs, 106-7
   in SIMS
Angle lap and stain, 185-6, 208
Angstrom (Å) def, 343
Anisotropic etching, 382, 392, 399-401
   aluminum, of, 406-8
   defined, 382-3

mechanisms of, in dryetching, 399-401
problems arising from, 400-401
techniques to control wall-profile, 23, 392, 399-401, 403
wet etching of Si, 387
Anti-reflective coating, 331-3, 354
American Society of Testing and Materials (ASTM), 168
Annealing, see Rapid Thermal Processing
Antimony, 151, 184, 224
Aqueous solution, def, 75,77
Arc-discharge, 238
Arc-lamps, mercury, 348-9
Argon (Ar), 241, 251, 257, 259
Argon-fluoride. ArF excimer laser, 349-50
Arrival angle, 262, 293-4
Arrhenius behavior, 532-3, 550
Arrhenius plot, 550
Arsenic (As), 183, 184, 199, 309
Arsine (AsH_3), 79, 183, 198, 309
Aspect ratio, def, 261
Assembly, ICs, 483-92, (Chap 27)
Atmospheric-pressure, def, 89
Atomic structure, 15-16
Atomic force microscopy AFM 266, 464-5
Atomic (mass) number (amu), def, 16, 17, 107
Atomistic epitaxy model, 307-8
Atomistic models of diffusion, 178-9
Attenuating phase-shift masks, 373-4
Auger electron emission, def, 252, 457, 461, 466-8
Auger electronspectroscopy (AES), 466-8
Autodoping, see Epitaxial Growth
Automatic guided vehicle (AGV), 507
Automated material handling systems (AMHS), 506-7
Automated test equipment (ATE), 480-81
Avogadro's number, def, 88

Backscattered electrons in SEM, 461
Backend of line (BEOL), def, 133
Backing-pressure, 95, 96
Backing pump, def, 96, 102
Backstreaming, def, 94, 95-6
Bacteria, in DI water, 84, 122, 123, 138
Ball-bonds, 487-8

Ball grid array (BGA), 499
Ball limiting metallization (BLM), 491
Ballroom (cleanroom) layout, 126
Band, energy, def, 24
Band-gap, def, 24
Bar (pressure unit) def, 89
BARC bottom anti-reflective coating, 332
Barrel epitaxial-reactor, 312-13
Barrel etching-reactor, 410
Barrier metals, 65, 262, 264, 300-301, 442, 444
Base (BJT), def, 39
Base (chemical), def, 77
   used in chip making, list, 77
Batch processing,
   epitaxy, 311-12
   LPCVD, 280-81
   oxidation, 227, 229-30
Bay and Chase Layout, 126
Bay, def, 505
Bayard-Alpert gauge, 105
Bernas source, 200
Beam-neutralization, 206
BEOL (Back end of line), 133
Bias-errors, etch, 384-5
BiCMOS, 42, 287
Binary-collisions, 193-4
Binary-masks, def, 360, 371
Binning, def, 481
Bipolar junction transistors, BJTs, def, 39-40
   operation of, 40-41
   process flow, to build, 41
Bird's beak, def, 57-8
BLM (ball-limiting metallization), 491
Blanket deposition, 60, 62, 65, 300
Blocking capacitor, 244
Boats, diffusion furnace, 225, 277, 280
Body-centered cubic lattice, 146
Bohr's model of the atom, 16
Bonding-pads, 68
Boron depletion during oxidation, 224
Boron-nitride (BN), 184
Boron-trichloride (BCl_3) 80, 183, 406, 407
Boron-trifluoride (BF_3), 80, 183
Borophosphosilicate glass (BPSG), 64, 292, 295-297
Bottom anti-reflective coating (BARC),

INDEX AND GLOSSARY

def, 332
Boundary layer, 270
BOX (buried-oxide), 315-17
Bridging-oxygen, def, 215
Brightfield-mode, microscope, 460
Brush, scrubbing, 136-7, 431-2
Bubble-formation in wet etching, 386, 389
Bubbler system, 183, 226, 274, 275, 291
Buffered hydrofluoric acid (BHF), 387-8
Bulk chemical-distribution (BCD), 78
Bulk gas-distribution (BGD), 79
Bulk gases, 79
   list of, 79
Bumped chip, 488, 490
Buried amorphous layer, 197
Buried collector, 41, 306
Buried oxide (BOX), 315-17
Burn-box, 226, 526
Burn-in, 479-80

Calcium-fluoride (CaF_2), 375
Cantilever loading, 225, 227
Capacitance, def, 35
Capacitive pressure-sensor, 105
Capacitor, def, 35
Capping layer, 422
Carbon tetrafluoride (CF_4), 393-99
Carrier-gas, def, 272, 273, 308, 311
Career ladder, wafer-fab, 515
Carrier-mobility, def, 30
Carrier, wafer, 131
Cascade-rinser, 138
Cassette, wafer, 131
Cathode, def, 238
Cavitation, 136
CD-SEM, 337, 462-3
Celcius, temperature-scale, def, 73-4
Ceramic packaging, 493-4
C4 (see controlled-collapse chip connection)
Channel, MOSFET, 42
Channels, in Si lattice, 194
Channeling, ion implantation, 194-5
Channel-stop implantation, 56-7
Charged-vacancy, 541
Charging of surfaces,
   during SEM and AES, 462
   during implantation, 207
Chase, in cleanroom, def, 126, 506
Chemical amplification (CA), 325-6
Chemical-bonding model, def, 22-3, 25
Chemical Mechanical Polishing (CMP),
   64-5, 65, 69, 417-34 (Chap 23)
   cleaning process, 431-2
   endpoint-detection, 430-31
   of low-k-dielectrics, 422
   of metals, 419-20
   of SiO_2, 421022
   tools, 423-25
Chemical Vapor Deposition (CVD), 112,
   (see Chap. 16)
   atmospheric-deposition (APCVD), 277-9, 293
   boundary-layers, 270
   epitaxial-deposition, 308-14
   growth-model, 271-2
   high-density plasmas (HDP), 244-5
   low pressure (LPCVD), 279-81
   mass-transfer-controlled process, def, 272-3
   surface-reaction-controlled process, def, 272-3
   tungsten, 298-300
   tungsten silicide, 299, 300
Chip-scale package (CSP), 499
Chromium masks, 351, 362
Chemical Vapor Depostion - see CVD
CVD of copper, 442-3
CVD growth rate model,
   mathematics of, 547-8
CVD polysilicon, 288-90
CVD reactors, 276-86
   atmospheric, 277-9
   cold-wall, def, 275-6
   distributed-feed, 279-80
   end feed, 279-80
   epitaxial, 311-14
   hot-wall, def, 275-6
   low-pressure (LPCVD), 277, 279-81, 293
   plasma-enhanced (PECVD), 277, 281-86
   planar, 283-6
   single-wafer, 280-1
   tube, hot-wall, 280-1
   tungsten deposition reactors, 299-300
   vertical-tube, 280
CVD SiO_2,
   atmospheric deposited, 290-91
   borophosphosilicate glass (BPSG), 295-97
   PECVD, 291
   phosphorus-doped (PSG), 290, 295
   properties of, 290
   reflow, 295-97
   tetraethylorthosilicate (TEOS), 291
Circuit design, 52
Circular magnetrons, 253-6
Class number (cleanroom), 127-29
Cleaning, 133-39
   BEOL, 133
   FEOL, 133
   HCl, in epitaxy, 310
   pre-furnace, 133-4
   RCA-clean, 133-4
   removing particulates, 135-7
   resist-stripping, 134-5
   wafer, after sawing, 485
Cleanrooms, 125-32
   classification, 127-8
   garments, 128-30
   gowning room, 130
   protocols, 131
Clear-field mask, def, 360
Closed-vacuum systems, 91, 93
Cluster-tools, def, 93, 258
CMOS, technology, 51 68
   IC process-flow, 55-68
   inverter, 54-5
Coating, resist (see Spin coating)
Coalescence (thin-film growth), 113
Cobalt-silicide ($CoSi_2$), 228
Coefficient of thermal expansion (CTE), def, 74-5, 115
   plastic package failures, due to, 496
Cold-wall reactor, def, 275, 299, 311
   PECVD, 283-4
Collimated-sputtering, def, 262-4
Color chart, SiO_2 thickness, 230-1
Color order, 231
Combustion reactor, 526
Complementary error function (erfc), 180, 537, 551
   def, 537
   evaluating, 540
   Tables of, 551
Compound, chemical, def, 19
Compound semiconductors, 301
Compression-ratio, pump, 98, 102
Compressive-stress in thin films, 115-6
   in thin films, 115-6
   measuring, 116
Computer integrated manufacturing (CIM), 511
Concentration-dependent diffusivities, 539-41
Concentration-gradient, def, 218

INDEX AND GLOSSARY

Condensation, def, 113
Conditioner, pad (CMP), 427-8
Conduction band, def, 24
Conductivity, def, 20
   of various materials, 21
Conductor, def, 20
Conformality, 292-3. 294
Contact holes, def, 440
   etching of, 401-4
Contact-printing, def, 351
Contamination, 122-5
   chemical removal of, 133-5
   effects on ULSI devices, 123
   in ion-implantation, 206, 207
   mobile ion, 123-4, 222
   particle-prevention,
      in a sputtering process, 252
   sources in IC fabrication, 122-3
   types, 121, 122
Contrast in resist, 347, 372
Contrast in SEM, 460-62
Contour-map, def, 208-9
Controlled collapse-chip connection (C4), 489
Control-limits, SPC, 514
Copper, 440-44
   CMP of, 420-21
   electrochemical plating, 443-4
   interconnections, 440-44
   seed-layers, 442-3
   slurry for CMP of, 421
Copper chloride (CuCl), 407-8
Copy exactly, technology-transfer model, 513
Corrosion,
   in Al etching, 407-8
   in plastic packaging, 497
Corrosive, def, 520
   chemicals, 522
Covalent-bonding, def, 23, 28
Cracking-pattern (RGA spectrum), 107
Critical-dimension, def, 335
   on masks, 368-9
Crossover-pressure, def, 96
Crucible, Si ingot-growth, 150, 153-4
Cryopumps, 99-101, 201
   regeneration of, 101
Crystals, def, 145
   directions, def, 146, 194
   orientation, 146
   planes, 146
Crystalline defects in Si
   (see Defects crystalline)
Crystalline material, def, 145
Cubic-lattices, def, 146
Cylinder change-out, 83
Cylinder, gas, safety, 521
Czochralski (CZ) silicon, 149-54
   growth, 149-50
   incorporating impurities in, 150-52
   ingot analysis, 163
   pullers, 153-4

Damage,
   dry etch, 414-15
   in CMP processes, 431-2
   ion implantation (see Chap. 12), 195-7
Damascene process, 449-52
Dangling bonds, 158, 222
Dark-field mask, def, 360
Dark-field mode, microscope, 460
Dark-space, 242
   shielding, 243, 260
Dark-space, plasma, 242
Dash-process, crystal growth, 150
DCA (Direct chip attach), 492, 499-500
Deal-Grove oxidation model, 216-19, 529-35
Deep-UV resists, 325-6
Defect-analysis, 512
Defect-density, def, 509
Defects, crystalline, 154-59
   area defects, 157-8
      stacking faults, 157
         extrinsic, 158
         intrinsic, 158
         oxidation induced, 158
   bulk defects, 159-60
      growth model of, 158
      heterogeneous nucleation of, 158
      homogeneous nucleation of, 158
      nuclei, 158
      oxygen precipitates in Si, 159
      precipitates, 158-9
   dislocations, 155-57, 198
      decoration of, 157, 314
      edge, 156
      loops, 156
      misfit, 156
      slip, 156
         during epitaxy, 314
         thermally induced, 157
         influence on device properties,157
      wafer resistance to warpage, 156
         due to implant damage, 196-7
   point, 155 Frenkel, def, 155
      intrinsic, def, 155
      Schottky, def, 155
      self-interstitial, def, 155
      substitutional, def,28
   removal by gettering, 159-61
Defects, mask, 364-6
Deflashing, plastic packages, 496
Dehydration baking, 328
Deionized water (DI), 22, def, 83-5
   resistivity of, 22
Delta-doping, def, 538
Density of materials, def, 75
Denuded zone, def, 159
Depletion of boron during oxidation, 224
Depletion region (pn-junction), def, 28
Depth of field, in microscopy, def, 460-1
Depth of focus, def, 346-7, 418
   improvement by off-axis illuminaton, 356
   improvement by phase-shift masking, 373
Design-for-test, def, 481
Designing CMOS ICs, 52-5
Design-rules, 53
Desorption, def, 92
Development, resist, 333-4
   immersion, 333
   puddle, 333-4
   spray,334
Device under test (DUT), def, 480
Diamond cubic lattice, 23, 146-7
Diatomic molecules, def, 19
Diazonaphthoquinone (DNQ), 323-5
Diborane, 79, 183, 309
Dicing operation, 484
Dichlorosilane, SiH_2Cl_2, 80, 308
Die-attach, 485-7
Die, def, 484
   size, 507
   testing, 475-81
   yield, 479, 507-11
Die-to-database inspection, 335, 365-6
Die-to-die inspection, 335, 365-6
Dielectric constant, def, 35
Dielectrics, def, 22
Dielectrics, low-k, 444-49
   CMP of, 422
Differential solubility, 324

INDEX AND GLOSSARY

Diffraction, of light, def, 344-45
Diffused resistors, 34
Diffusion (see Chap. 11),
    atomistic models of, 178-9
    basic concepts of, 176-8
    constant (coefficient), def, 181, 535
    extrinsic, def, 541
    heavy-doping effects on, 541
    intrinsic, 539-41
    polysilicon, in, 287
    in SiO_2, 182
    length, def, 538
    of implanted impurities, 198-9
    profiles, 179-80
    sources, 182-4
    transient-enhanced, 199
Diffusivity, 181 (see also Diffusion, constant)
    of dopants in oxide, 223
Digital circuits, 44, 45
Diluent, 272, 273
Diodes, def, 36
    light-emitting (LEDs), 31
    operation of *pn* diodes, 37-8
    photo, 38-9
    pn-junction, 36-8
Direct chip attach (DCA), def, 492, 499-500
Direct liquid-injection system, 275, 291
Direct vapor-pressure system, 275
Discrete-device, def, 9
Dishing (in CMP), 421, def, 433-4
Dislocations, 155-57, 198
    loops, 156
Dissociation, def, 241, 393-4
Distributed feed, LPCVD reactors, 281
DNQ/novolac resist, 323-5
Donor, def, 28
Dopants, def, 22
    in Si, 27
    n-typed, def, 28
    p-typed, def, 28
Doping,
    concentration, def, 29
    in epitaxial growth, 309-10
Doping profile, def, 179-80
    in ion implantation, 191-2
Dose,
    ion implantation, def, 190, 202, 203
    measurement of, 207-8
    pre-dep diffusion,
Drive-in diffusion, 58-9, 182, 538-9
    mathematics of, 538-9
    well-formation process, 58-9
Drop-ins, test, 477
Dry etching, 391-416 (see Chap 22)
    aluminum, 405-8
    contamination in, 415
    damage in, 414-15
    end-point detection, 408-9
    equipment configurations, 409-15
    goals of, 392-93
    organic films, of, 408
    polysilicon films, of, 404-5,
    Si_3N_4 films, 404
    SiO_2 films, 397-99
Drying of wafers, 138-9
Dry rough-pumps, 98-9
Dry-stripping processes, 408
Dry-in, dry-out (CMP), 431-2
Dry-oxidation, 216, 219, 221-2
Dual-damascene process, 449-52
Dual-in-line packages (DIPs) def, 493, 497
Dump-rinsers, 138
Dynamic RAMs (DRAMs), 46-7

Edge-bead, def, 329
    removal, 329-30
Edge-die, def, 508
Edge-rounding of wafers, 166-7
EDX (energy-dispersive analysis), 469
Elastic-collision, def, 240
Electricity,
    basic, 4-5
    defined, 4, 6
    history of, 5-7
Electric-shock hazard, 523
Electromagnetic spectrum, 343
Electromigration, 440-41
Electron cyclotron resonance (ECR) sources, 414-5
Electron-hole pair, def, 27
Electronic grade Si (EGS), def, 147-49
Electronics, history of, 4-12
Electronic stopping, of ions, 193-4, 543-4
Electron impact-ionization, def, 240-41
Electron projection lithography EPL 377-8
Electron sources, 461-2
Electron spectroscopy for chemical analysis (ESCA), 469-70
Electron-volt, unit (eV), def, 20, 456
Electroplating, 112, 443-4, 489
Electrostatic-chucks, def, 414-15
Ellipsometers, def, 232-3
Embedded etch-stop layer, 451-2

End-feed reactor-tubes, def. 279-80
End-point detection, CMP, 430-31
End-point detection, dry etch, 408-9
End-stations, ion implant, 201-2
Endura, sputtering tool, 258
Energy-band diagram, def, 24, 28
Energy-band model, def, 22, 23-25
Energy dispersive spectroscopy(EDX) 469
Energy-gap, def, 24
    in insulators, 25
    in semiconductors, 24, 25
Energy-levels, atomic, def, 24, 456-7
ENIAC, computer, 7
Epitaxial growth, (Chap. 17) and 114
    atomistic model of growth, 307-8
    autodoping, 309-10
    cleaning prior to, 310
    characterizing epi layers, 314-5
        electrical, 314
        optical, 314
        thickness, 314-15
    defined, 305
    equipment, 311-14
        barrel-reactors, 312
        pancake-reactors, 311-12
    growth model of, 307-8
        kink-position, 307
    process-sequence, 310-11
    solid-state diffusion in, 309-10
    solid-phase epitaxy, 306
Epoxy die-attach, 486
EPROMs, 47
EEPROMS, 47
erfc, 180, 537, 540, 551
Erosion effect (CMP), def, 433-4
ESCA, 469-70
Etching (see Chaps. 21 and 22)
Etch-rate uniformity, def, 382
Eutectic die-attach, 486
Evaporation, def, 92
Excimer-laser sources, 349-50
    safety, 525
Exponential yield-model, def, 509-10
Exposure-field, def, 353
Exposure, resist, 331-32
Extreme-UV (EUV), def, 376-78
Extrinsic-gettering, def, 159-61
Extrinsic-semiconductor, def, 26-27

Fabless semiconductor company, 13, 503
Fabrication costs, 12-13, 504-7
Face-centered-cubic lattice (FCC), 146
    interpenetrating, 147

INDEX AND GLOSSARY

Faceting, 250
Fahrenheit, temperature scale, def, 74
Fair's vacancy model, 541
Faraday cup, 106, 201, 206
Fast-ramp furnaces, 225, 228
Faults, stacking, def, 157
Federal Standard-209, def, 127
Fibrillation, def, 523
Fick's Laws, 180, 535-7
    Fick's First law, 180, 535
    Fick's Second law, 181, 536-8
Field-implant, def, 56
Field-oxide, 56
Fillers, plastic, packaging, 495, 496
Film-stress, def, 75
Fires, in wafer fabs, 525
Fixed oxide-charge, def, 222, 230
Flash-memory, 47
Flat-zone, def, 225, 281
Flats in wafers, 163-4, 169
Flip-chip technology, 489-90
Float-zone crystal-growth, 147
Flood-gun, electron, 207
Flow-reactors, 526-7
Fluorescence microscopy, 459
Fluorinated oxides (FSG), 446-7
Fluorine-etching of silicon, 396-99
Flux, def, 271, 529, 535
Focused-ion beam (FIB), 455, 472-3
Focus-exposure window, def, 348
Forbidden-gap, def, 24
Forepressure, def, 95, 96
Forepump, def, 96
Forward-bias (of diode), 38
Foundry, wafer, def, 13
FOUP (front opening unified pod) def, 127
Fourier-transform infrared,
    spectroscopy, 315
Four-point probe, def, 118, 184, 207, 266
Frequency factor, def, 550
Frenkel defect, def, 155
Frequency factor, 550
Front-opening unified pod (FOUP), 127
Full-face erosion, sputtering target, 254
Fumed-silica slurry, 428
Functional testing, 480-81
Furnace flat-zone, 225, 281
Fused-silica, def, 215
Fused silica, mask blanks, 361

g-line (Hg-arc lamp) 322, 348-9, 354
Gap-fill, 292
Gas, def, 19

    average speed in, 88
    ideal gas model, 87
Gas-cabinet, 81-2, 226
Gas-cylinders, 81-2, 199, 309
Gas-delivery systems,
Gas-depletion effects in LPCVD, 281
Gas jungle, 226
Gas manifold, 82, 274
Gas-phase-mass transfer coefficient, def,
Gas-phase nucleation, def, 271, 290
Gas purge, 82
Gas stick, 82
Gate-delay, def, 438-9, 441
Gate oxidation, 60, 221
Gate-oxide integrity (GOI) def, 221, 306-7
Gaussian distribution,
    diffusion, 182
    drive-in, 358-9
    ion implantation, 192, 198, 545
        joined-half Gaussians, 546
    in yield models, 510
Germanium, 25, 26
Gettering, 159-61
    basic principles, 159-60
    denuded zones, 159
    extrinsic, 160
    intrinsic, 160-61
Glass, photomask substrate, 361-2
Glass transition temperature, 337
Global planarization, def, 417-8
Glow discharge, 238-45, 282
    dc, 238-43
    self-sustained, 241, 242
Gowning, cleanroom, 127-131
Gradient, def, 176, 218
Grain boundary, def, 158, 287
Grain-growth in thin films, 114
Gram-molecule, def, 88
Groove and stain, 186
Gull-wing packages, 498

Halftone-photomasks, def, 360, 373
Halogen-lamps, 229, 276, 286
Hard-bake, def, 337
Hazard information labels, def, 519
Haze, 141
Heating-lamp array, 229, 276, 286
Helium-neon laser, 355
Henry's Law, def, 529
HEPA filter, def, 126
Hermetically-sealed packages, def, 493-4
Hertz (Hz), def, 343
Heteroepitaxy, def, 114, 305

Heterogeneous-reactions, def, 271, 290
Heterogeneous reactions in dry etching, 395-6
Hexamethyldisilazane (HMDS), 386
Hexode etchers, 411-12
High-angle implanters, 204-5, 206
High-current implanters, 201, 203-4
High-density plasma etchers, 413-14
High-density plasma sources, 244-5
High-efficiency particulate attenuation
    filters (HEPA), 126
High-energy implantation, 210
High-energy implanters, 202, 204, 205
High-volume manufacturing, 513
Hillocks, def, 116-7, 249, 285
Hexamethyldisilazane (HMDS), 80
Hollow-cathode sputtering source, 265
Homoepitaxy, def, 114, 305, 328
Homogeneous gas-phase reactions, def, 271, 290
Hoerni, Jean, 10
Hole-current in semiconductors, 28-9
Holes, def, 27, 28, 30, 37
Horizontal-furnaces, 225-6
Hot-plate, soft-baking, 330-31
Hydrofluoric acid (HF), 77, 290, 314, 387-8
Recycling, 521
    safety issues, 521
Hydrogen,
    annealing, 66, 223
    effect on etch rate of Si and SiO_2
        in CF_4 plasmas, 398-9
Hydrogen passivation, 66, 223
Hydrogen-peroxide, H_2O_2, 134
Hydrophobic, def, 328
Hydrophyilic, def, 328
Hydroxyl (OH) def, 77, 216

Ideal gas model, def, 87
i-line, def, 348, 354
i-line resists, 322
I_{DDQ} testing, 481
Illumination systems, 349
Immediately-Dangerous to Life & Health
    (IDLH), def, 519-20
Impact-ionization, def, 240
Impedance matching network, def, 244
Implantation damage, 195-7
Implantation masks, 208
Impurities in Si, def, 27
Impuritiy-activation, def, 197-8
Indium implants, 209

Index-of-refraction, def, 232, 345, 459
Induction-heating, def, 276
Inductively-coupled plasma, 414-15
Inelastic collision, def, 240
Infra-red reflectance, 315
Ingot, def, 150
Initial growth-regime, oxides, def, 221-2
Initiation period, Al etching, 405-6
Ink mark, 478-9
In-line defect pareto, def, 511
In-line parametric test, def, 476-8
Inner-lead bonding, def, 489
In-situ, def.
    endpoint detection, CMP, 430-31
Inspection, after develop, def, 334-7
Insulators, def, 22, 23, 24
    perfect, 23
Integrated-circuits, def, 9, 33
    invention of, 9-11
    monolithic, def, 10
    price per bit, 12, 13
Interbay transport, def, 506
Interconnect-limited, def, 437, 438
Interface trap-charge in SiO_2, def, 222-3
Interlevel dielectric (ILD), def, 64, 440
Intermetal dielectric (IMD), def, 67, 440
Interstitial impurities, def, 55
    self-interstitials, def, 155
Intra-level capacitance, def, 437, 439
Intrinsic carrier-concentration, def, 27
Intrinsic-gettering, def, 159
Intrinsic semiconductor, def, 26
Ion, def, 18
Ion-beam compositional analysis, 470
Ion-beam milling, def, 392-3
Ionic-bonds, def, 18, 22
Ion implantation, see Chap. 12,
    def, 189
Ionized metal plasma, def, 264-5
Ion sources, 199-200
Ionization, 18, 105-6, 107, 240, 242
Ionization pressure gauge, 105-6
IPA vapor-dry, def, 139
Island-growth, def, 113
Isochronal annealing, def, 197-8
Isolation structures,
    LOCOS, 56-7
    silicon-on-insulator (SOI), 315-17
    STI, 68-9
Isopropyl alcohol (IPA), 138
    vapor dryer, 139
Isotropic etching, def, 382-4, 391

ITRS Roadmap, 214
J-lead package, 498
Junction depth, def, 180
    measurements of, 185-6
Junction, *pn*, def, 36-38
Junction-spiking, def, 248
Kelvin temperature scale, def, 73-4
Kerf, def, 165, 484
Kilby Jack, 10-11
Kick-out mechanism of diffusion in Si,
    def, 179
Kink-sites, def, 307-8
Knudsen-flow, of gases, def, 91
Kovar, 494
Krypton-fluorine (KrF) laser, 349-50, 354
Kurtosis, 546
Laminar flow-see Viscous flow
Lamps,
    halogen, 229, 276, 286
    mercury arc, 348-9
Lanthanum hexaboride, LaB_6, 462
Lapping, def, 166
Large-angle tilt implantation (LATID), 205
Laser interferometry, 355, 408-9, 430-31
Laser-ionized mass spectroscopy (LIMS),
    def, 471
Laser marking, packages, 182, 481
Laser marking, wafers, 166
Latent image, in resist, def, 333, 341
Lateral etching, def, 382
Latchup, def, 210, 306
Lattice, def, 146
Lattice-constant, def, 146
Layout, IC, 53
LDD (see lightly-doped drain)
Leadframe, def, 495
Lead shielding, 525
Learning-curve, def, 511-12
Lens field-size, 352-4
Light,
    basic behavior of, 342-3
    diffraction of, def, 344-5
    emission by glow-discharge, 234
    intensity, 345
    interference behavior, 231, 331, 344-5, 372
    in phase-shift masking, 372
    infra-red reflectance, 315
    mercury-arc spectrum, 348-9
    refraction of, def, 344-5
    sources, 348-9
    spectrum, def, 343

    speed of, def, 343
    standing waves in, def, 331-3
    UV, def, 343, 349
    visible, def, 343
    wavelength, def, 343
Light-emitting diodes (LEDs), 31
Lightly-doped drain (LDD), def, 61-2
Linear-growth regime, def, 217-18
Linear polishing tool (CMP), 425
Linear rate constant (B/A), def, 531-4
Liner films, def, 300
Linewidth control, def, 335
Linewidth measurement, 335-9
    image shearing, 337
    laser, 337
    scanning slit, 33
    TV scan, 338
Liquid, def, 19
Liquid sources, diffusion,
    CVD, 183,
    diffusion, 274-5
Load locks, def, 92-3
Load-ports, def, 507
LOCOS, def, 56-7, 214
Logic-design, 52
Long-throw sputtering, def, 262, 264
Low-energy implanters, 202, 204
Low-k dielectrics, def, 444-49
    CMP of, 422
    process integration issues of, 445-6
Low pressure CVD (LPCVD), def, 279-81
Low-vacuum, def, 90
LSI (large-scale integration), def, 11
LSS Model, 542-4

Magnetically-enhanced reactive ion etching (MERIE), def, 244, 413-14
Magnetron sputtering, def, 244
Magnetrons, def, 253
    circular, 253
    planar, 253
    principle of, 253
Magnification, def, 458-9
Majority-carrier, def, 28, 38-9
Manifold, gas-supply, def, 274
Marking packages, 481, 496
Mask, def, 53, 359
    binary, def, 360
    clear-field, def, 360
    dark-field, def, 360
    fabrication of, 361-6
    inspection of, 365-6
    halftone, def, 360, 373

INDEX AND GLOSSARY

1X, 360
   pattern generation (PG) in,
      e-beam, 362-3
      laser, 362-3
   polarity, def, 360
   reduction, def, 360
   repair of defects, 363, 364-5
Mask-blanks, def, 361-2, 365
Mass-flow controllers (MFCs), 82, 107-8
Mass-spectroscopy, 106-7
Mass-spectrum (RGA), def, 106
Mass-to-charge ratio, def, 106
Mass-transfer coefficient, def, 529
Mass-transfer-rate limited step, def, 272-3, 308, 547
Mass-transport in LPCVD, 279,
Material Safety Data Sheets, def, 519
Matrix, def, 323-4, 457
Mean-free-path, def, 90, 91, 95, 102, 240, 252, 293-4
Medium-current implanters, 201, 202-3
Megasonic cleaning, def, 136, 431
Memory-cell capacitor, def, 36
   stacked, 36
   trenched, 36
Memory ICs, def, 45-6
Mercury-arc lamps, 348-9
Mercury-arc spectrum, def, 349
Mercury barometer, def, 89
Metal CVD, 298-301
Metallic-bonding, def, 23
Metallurgical-grade Si, def, 148
Metal-oxide semiconductor capacitor (MOS-C), 35-6
Metal-oxide semiconductor FET (MOSFET), def, 35, 233
Microchips, price of, 12
Micron (unit of length), def, 12, 343
Microprocessors, def, 45
Mini-batch radial PECVD reactors, 284-5
Minienvironments, def, 127-8
Minority-carrier, def, 28, 38-9
Mixture, chemical, def, 19, 75
Mobile-carriers, def, 23, 28
Mobile-ionic charge in SiO_2, 222
Mobility, carrier, def, 30
   electron, 30
   hole, 30
Mole, def, 108
Molecule, def, 19
Molecular-flow, in gases, def, 91, 95, 102,
Molecular formulas, def, 19

Monitor-wafers, dfe, 207
Monolayer, def, 252
Monorail wafer-transport system, 506-7
Moore, Gordon, 11
Moore's Law, 11-12Morphology, def, 451
MOS technology, 42-3
MOSFET, device operation, 42-44
   SOI, 315
   threshold voltage of, def, 44
MSI (medium-scale integration), def, 11
Murphy's Yield Model, def, 510
Multi-chip-module (MCM), def, 491
Multilevel-metalization, def, 67, 438-9
   terminology of, 440

NaCl, structure, 18
Nanometer, def, 343
Nanoporous silica, 448-9
Native oxide, 213, 243
Negative-resists, def, 32
   e-beam, 362-3
   optical, 325
Next generation lithography (NGL), 375-6
Electron-projection lithography (EPL), 377-8
Extreme UV (EUV), 376-8
   PREVAIL, 378
   SCALPEL, 378
Nitric acid, 77, 314
NMOS, process flow, 43
   structure, 44
Nomarski interference microscopy, 458-9
Non-volatile memory, 47
Notch on wafers, 163-4, 169,
Novolac, 324
Novolac resin for plastic packages, 495
Noyce, Robert, 10-11
npn BJT transistor, 40-41
n-type dopant, def, 28
n-type semiconductor, def, 28
Nuclear stopping of ions, 193-4, 543
Nucleation, def, 113
Numerical aperture, NA, def, 344, 345, 346, 347, 354
of microscope lens, 458-9

Off-axis illumination, def, 355-6
Ohmic contacts, 248
Ohm's Law, 34
Ohms, unit, def, 20
Oil-sealed roughing pumps, 97-8
Open-flow vacuum systems, 91, 93, 273
Optical dosimetry, 207-8

Optical emission spectroscopy, 409
Optical interference technique, 231-2
Optical-lithography, limits of, 374-5
Optical microscopy, 314, 458-60
Optical-proximity correction (OPC), 370-1
Organo-metallic precursors, 301
Out-diffusion, in epitaxy, 309-10
Outer-lead bonding, 489
Outgassing, def, 92, 93
   during implantation, 206-7
Overetch, 401
Overflow-rinser, 138
Overhang, 262, 265, 293, 295
Overhead-hoist system (OHT), 507
Overhead-loop systems, 506-7
Overlay accuracy, 354-5, 376
Oxidation furnaces, 225-29
Oxidation-induced stacking faults - see Defects, crystalline,
Oxidation of polysilicon, 224-5
Oxidation of silicon, thermal (see Silicon dioxide),
Oxidation-reduction (redox), def, 386
Oxide-integrity, gate (GOI), 221, 306-7
Oxide-spacers, 62
Oxygen in silicon, 159
   concentration in Si, 152, 159
   denuded-zones, 159
   measurement of in Si, 159
Oxide-trench-and-refill process (STI), 68-9
Overflow-rinser, 138
Ozonator, 292-3
Ozone-TEOS, 292

Packaging, IC, 492-500
   ceramic, 493-4
   plastic, 494-7
Packageless technologies, 499
Pad-conditioner, 427-8
Pad-oxide, def, 55, 69
Pads, polishing, CMP, def, 425-8
   conditioners, 427-8
   slurry-free, 428-9
   stacked, 426-7
Pancake reactors, epitaxy, 311-12
Parabolic constant, B, 217-18, 531-4
Parallel-electrode planar reactors, 283-6
Parallel plate etchers, 410-14
Parametric-testing, def, 476-8
   instrumentation, 477-8
Partial-die, 508
Pareto-charts (yield analysis), def, 511
Partial-pressure, def, 100

INDEX AND GLOSSARY

RGA-measurement of, 106
Particle-counters, automatic, 140-41
Particle-detection, 139-40
Particle-removal, 135-37
Particles-per-wafer-per-pass (PPWP)
    def, 140
Particulates, 124
    sizes of, 125
Parts per billion (ppb),
    def, 75, 132, 148, 257, 309
Parts per million atoms (ppma),
    def, 75, 147, 309
Pascal (pressure unit), def, 89
Paschen's curve, 240-41
Passivation layer, def, 68
Pattern registration, def, 342
Pearson-IV distribution, def, 546
PEB (post-exposure bake), def, 327
PEL (permissible exposure limit), 519-20
Pellicles, def, 367-8
Pentium-IV microprocessor, 45
Perfluorocarbon (PFC) gas abatement,
    527-8
Periodic Table of the Elements, def, 17, 18
Permittivity, def, 35
PE-TEOS, 291-2
Phase diagram, Al:Si, def, 248
Phase-shifting masks (PSM), def, 371-4
pH-scale, def, 78
    of CMP slurries, 428, 429
    of corrosive chemicals, 522
    of DI water, 84
Phenol-free strippers, 135
Phosphine, 79, 183, 198, 289, 309
Phosphoric-acid, 77, 388-9
Phosphosilicate-glass (PSG), def,
    64, 68, 290, 295
Photo-acid generator (PAG), def, 326-7
Photoacoustics, 265-6
Photo-active compound (PAC), def, 323-4
Photodiodes, 38-9
Photoelectrons, def, 469
Photoresist (see Chap. 18)
Photoresist-stripping, 134-5
    dry-etching, 135, 408
    inorganic strippers, 134, 135
    organic strippers, 135
Photostabilization, def, 337
Physical-vapor-deposition (PVD) 112,
    (see Chap. 15)
    def, 247-8
Physics of sputtering, 249-52

Physisorption, def, 135
Pileup, dopant during oxidation, def, 224
Pilot-line production, 512
Pin-grid array (PGA), def, 494, 407
Piranha, def, 134, 135
Pirani gauge, def, 105
Pitch, def, 405
Planarization, def, 64-5, 417-18
Planar magnetrons, 253
Planar process, 9 def, 33-4
Plasma, def, 237-8 (see Chap. 14)
Plasma-enhanced CVD, def, 281-86
Plasma etching,
    parameter control, 396
    potential of plasmas, def, 394, 410-11
    reactive-gas glow discharge, def, 393-4
    steps in, 393-4
Plastic packaging, def, 494-7
Platen, CMP, def, 418-19
Plating, electro (Cu), 443-4
    tools, 443-4
Plenum, def, 278
PMOSFETS, def, 42, 44
Plugs, tungsten, def, 65-7, 420, 443, 450
pn junctions, def, 36-8
$POCL_3$, 183, 289
Point-defects, see defects, point
Point-of-use chemical generation
    (POUCG), def, 132
Point-of-use (POU) delivery, def, 78
Point-of-use (POU) filtration, def, 386
Point-of-use filtration of UPW, 84
Poisson's Model (Yield), def, 509
Polarization, def, 232-3
Polishing pads, CMP, def, 425-8
Polish-rate, def, 421-22
Poly (alylene) ethers (PAE) 447, 448
Polycides, def, 63-4
Polycrystalline material, def, 145
Polycrystalline-silicon, see polysilicon,
Polysilicon, 286-90
    applications of, 286
    CVD deposition of, 60, 288-90
    diffusion in, 287
    doping of CVD deposited, 60-1, 289-90
    dry-etching of, 60-61
    in-situ doping of, def, 289-90
    properties of thin film, def, 286-8
    structure of, 286-7
    thermal oxidation of, 225-6
Polyvinyl alcohol (PVA), def, 136, 431-2
Positive resists,

optical, 322, 323-5
    e-beam, 362-3
Post-baking, resists, def, 337
Post-CMP clean, def, 431-2
Post-etch corrosion, 407
Post-exposure bake (PEB),
    def, 327, 331, 332-3
    in DUV resists, def, 327
Post-oxidation anneal (POA), def, 230
ppma, def, 75
Pre-amorphization, def, 195
Precipitates, def, 158-9
Precursors, def, 301, 308
Predeposition diffusion, def, 181
    by ion implantation, 189-90
    mathematics of, 537-8
Pressure, def, 88
    measurement of,
    partial, 106-7
        residual gas analyzers (RGA) 106-7
    total, 104-6
        capacitance-manometer, 104-5
        ionization gauge, 105
        Pirani gauge, 105
        ranges of total-pressure gauges,
            103
        thermocouple gauge, 104-5
Preston's Law, def, 421-22
PREVAIL, 378
Primary-beam, def, 457
Priming, prior to resist coating, def, 328
Printed circuit board (PCB), 2, 3
Probe card, def, 477-9
Process control monitor (PCM), def, 477
Process-flow, CMOS, 55-68
Process variation, def, 514
Product life cycles, def, 513-14
Profile, etch, def, 382
Profilometer, def, 264-5
Profit margin, def, 503-4
Projected lateral straggle (ion
    implantation), def, 192
Projected range, R_p (ion implantation)
    def, 191
Projected straggle ΔR_p (ion implantation)
    def, 192
Projection-printing, def, 351, 351-54
    non-reduction step & repeat (1X),
        350, 351, 353
    reduction step and repeat, (5X),
        351-2, 362, 364
    scanning projection printers, def, 351-2

step-and-scan projection, def, 353-4
Propagation delay, def, 438
Proximity effects (OPC), def, 370-1
Proximity-printing, def, 351
p-type dopant, def, 28
p-type semiconductor, def, 28
Puddle developing, def, 233-4
Pullers, ingot, def, 153-4
Pump-fluids, def, 99
Pumps, vacuum (see Vacuum pumps)
Punchthrough, def, 61
Punchthrough-prevention implant, 59
Purity,
    of chemicals, def, 75, 132
    of gases, def, 79
    of sputtering targets, def, 256-7
Purging gas lines, def, 82
PVD (see Physical vapor deposition)
Pyrogenic steam-oxidation, def, 226-7
Pyrolytic, def, 271
Pyrometer, optical, 229
Pyrophoric, def, 274, 291, 520

Quad flat pack (QFP), def, 499
Quartzware, def, 225, 277, 280

Race track, def, 255
Radical, def, 241, 283
Rapid thermal oxidation (RTO), def, 228
Rapid thermal processing, 62, 225, 228-30
    implant annealing, 198, 199
    oxidation, 225
    shallow-junction formation, 209-10
    systems for, 228-9
Radio-frequency (RF), def, 243-4
Random-defect process yield, def 509, 512
Range, R (ion implantation), def, 191
Raster scanning, def, 363-4, 461
Rate,
    equation, Arrhenius, def, 550
    etch, def, 382
    growth, CVD, def, 272, 548
    limiting step of,
        CVD, def, 271-2, 548
        oxide-growth, def, 216-19, 530
    temperature ramp, def, 228, 229
Ratio, aspect (See aspect ratio),
Rayleigh's (Depth of focus) Criterion,
    def, 346-7
Rayleigh's (Resolution) Criterion,
    def, 345-6, 347
RCA pre-furnace clean, def, 133-4, 188
RC-time delay, def, 438-9. 441

Reactive-gas glow discharge, def, 393-4
Reaction-rate limited, def, 271
Reactive-ion etching (RIE), 392, 394, 410
Reactive-sputtering, def, 260-1
Re-cycling, chemical, def, 527
Re-deposition in sputtering, def, 254, 264
Redistribution, during oxidation, def 223-4
Reduction-oxidation (redox), def, 386
Reduction, printing, def, 351-2, 364
Reduction reticles, def, 362
Re-entrant angle, def, 294
Reflective notching, def, 332
Reflow, 295-7 angle, def, 296-7
    process, 296
Refractive index, def, 232, 345, 459
Refractive, lenses, def, 353
Refractory metal silicides, def, 295
    silicide formation, 63-4
Regeneration, cryopump, def, 101
Registration, litho, def, 342
    automatic, 354
    during mask making, 364
    inspection of masks, 365-6
Reisman-Nicollian model of thin oxide
    growth, 533-4
Reject water, def, 85
Relaxation, electron, def, 240
Reliability problems,
    electromigration, 440, 441
    gate oxide integrity, 221. 306
    plastic packaging, 496-7
Reprocessing,
    chemicals, 132
    HF, 521
    water, 84
Residual gas analyzer (RGA), def, 106-7
Resin, plastic package material, def, 497
Resistance, def, 20, 34
    sheet, def, 117
Resistivity, def, 20, 21, 34
    measurement procedure, 118
    of deionized water, def, 22, 84
    of extrinsic Si,
        doping dependence of, 29
    of intrinsic Si, def, 27
    of metals, 29
    of silicon-dioxide, 22
Resistor, def, 21
    diffused, def, 34
    thin-film, def, 34-5
Resist processing (See Chap. 18)
Resolution, def, 342, 345

    in lithography processes, def, 347
    of SEM, 462
Resolution enhancement techniques
    (RETs), 355-6, 369-7
    off-axis illumination (OAI), def, 355-6
    optical proximity correction (OPC),
        def, 370-71
    phase-shift masking (PSM), def, 371-4
Reticle, def, (see also Mask),
    342, 347, 359-67, def, 359
Reticle-stocker, def, 368-9
Retrograde wells, def, 204, 210
Reverse-bias, def, 38
    leakage current, def, 36, 38
Reverse osmosis (RO), def, 84-5
Rework, photoresist, def, 334
Rf-heating coils, def, 276, 311
Rf-generator in PECVD systems, 284,
    glow-discharge, 243-4
RIE (reactive ion etching),
    392, 394, (def), 410
Rim phase-shift masks, def, 373-4
Rinse, def, 137-8
Rinser-driers, def, 138-9
Robotic wafer handling, 508-9
Roots-pumps, def, 98-9, 104
Rotary-piston pumps, def, 97-8
Rotary-vane pumps, def, 97-8
Roughing pumps, def, 97-8
Roughness, film surface, def, 113-4
Rubylith, 356

S-Gun, 257
Safe delivery system (SDS), def, 199, 521
Safety (see Chap. 29), 517-29
    HF, 523
    ion implantation, 523-4
    record of semiconductor industry,
        517-18
Salicides, def, 63, 64
Sawing wafers, 484-5
Scaling, MOSFET, def, 12
SCALPEL, 378
Scanning auger microscopy, def, 460-63
Scanning electron microscopy (SEM), 337
Scanning methods in ion implanters,
    def, 200-201, 204
sccm (standard cubic centimeter per
    minute), def, 108
Scribe-line, 477-8, (def), 484
Scribe-line monitors, def, 477-8
Scrubber, effluent gas, def, 183, 274, 526-7
Scrubbing brush, def, 136-7

INDEX AND GLOSSARY

Scrubbing, particle removal in CMP, 431-2
Secondary-electron flood gun, def, 207
Schottky field-emission electron source, def, 462
Screen-oxides, def, 195
Secondary-beam, def, 457
Secondary-electron emission, def, 243, 252, 461, 466-8
   Auger electron emission (AES), def, 466-8
      crossover point, def, 467
      during sputtering, 252
      during SEM, 461-2
Secondary-ion mass-spectroscopy (SIMS), def, 208, 314, 467, 470-71
Seed layer, Cu, def, 442-3
Seeds' Yield Model, def, 510
Segregation coefficient,
   during Si ingot growth, def, 150-2
   during oxide growth, def, 223-4
Selectivity, during etching, def, 383-4
   etching of polysilicon over SiO_2, 397-99
   etching SiO_2 over Si, 397-99
   with respect to mask (S_m), def, 383-4
   with respect to substrate (S_s), def, 383-4
Self-aligned silicides (*Salicides*) def, 63-64
Self-aligned gates, 43
Self-bias, in rf-glow discharges, 395
Self-bias, in rf plasmas, def, 242
Self-interstitials, def, 155
Semiconductor, def, 15, 74
Semiconductor industry, 1-4, 11-13
   safety record, 517-18
Sensitivity, in compositional analysis, def, 458
Sensitivity of resists, def, 348
Serifs (OPC features), def, 370-71
Shallow-junction formation, 61, 199, 204, 209-10
Shallow-trench isolation (STI), def, 68-9
Sheath voltages, def, 242
Sheet-resistance, def, 117
   measurement of, four-point probe, 117-8, 184
   measurement of, spreading resistance, 184-5, 208, 314
Shelf-life (photoresist), 323
Shockley, William, 8-9
Siemens process for producing EGS, 149
Sidewall slope, of photoresist, def, 347
Sidewall spacers, def, 62, 296
Silane, SiH_4, 79, 274, 288, 290, 291, 293, 294, 308

Safety, 525 (see Chap. 29)
Silica, fused, def, 215
Silicides, def, 63
Silicon, 25
   extrinsic, def, 26
   intrinsic, def, 26
   number of atoms/cm^3, 25, 26
Silicon dioxide, SiO_2 (See Chap. 13)
   CMP of, 421-22
   chemically vapor deposited - see CVD SiO_2 (Chap. 16)
   thermal,
     applications of, 214
     Deal-Grove model of, 216-19, 529-35
     dopant-redistribution during oxidation, 223-4
     dry-etching, 397-9, 401-2
     dry-oxidation, 216, 219, 221-2
     fixed oxide charge, 222
     growth-rate dependence, 217-19
       chlorine dependence, 221
       dopant effects, 220
       pressure effects, 220-21
     initial oxidation stage, 221-22
     interface-trap charge in, 222
     linear rate-constant, B/A, 217-8
     masking, properties of, 209
     mobile ionic-charge, 222
     nature of the Si/SiO_2 interface, 222-23
     of polysilicon, 224-5
     oxidation growth-model, 216-19, 529-35
     oxide trapped-charge, 222
     parabolic rate-constant, B, 217-18
     properties of silica glass, 214-16
     thickness measurement, 230-3
     thin-oxide growth, 221-2
     wet etching, 387
     wet oxidation, 216, 219, 226-7
Silicon-doped aluminum, 248
Silicon-gate MOS process flow, 43
Silicon Nitride, Si_3N_4, SiNH, 56-7, 62, 68, 69, 296-8
   CVD of, 297-8
   hydrogen in, 297
   LPCVD of, 297
   PECVD of, 297
   plasma etching of, 388, 404
   properties of, 296-7
Silicon-on-insulator (SOI), def, 315-17
Silicon oxynitrides, def, 297-8

Silicon tetrachloride, $SiCl_4$, 79, 308-9
Silicon tetrafluoride, SiF_4, 396-8
SIMOX, def, 316
SIMS (secondary-ion mass-spectroscopy) def, 467, 470-71
Single-crystal, def, 146
Single-in-line package (SIP), def, 498
Single-wafer epitaxy reactors, 312-13
Single-wafer etchers, 412-14
Single-wafer PECVD Systems, 285
Sinter, def, 66
Skewness, def, 546
Slim-rod, def, 149
Slip (see Defects, dislocations),
Slurry, CMP, def, 428-9
Slurry distribution system, def, 429-30
Slurryless pads, CMP, def, 428-9
Small batch, fast-ramp furnaces, 225, 228
Small outline IC (SOIC), def, 498
Smart-Cut, def, 317
SMIF (standard mechanical interface), 127
Sodium, Na, contamination,
   in SiO_2, 222
Soft-baking, def, 330-31, 337
Solder-bumps, def, 490
Solid, def, 19
Solid-phase epitaxy, def, 306
Solid-solubility, def, 152, 182
   of elements in silicon, 152
Solid-state-electronic devices, def, 8, 20
Solid-state electronics, Age of, 8-9
Solutes, def, 75
Solution, chemical, def, 75, 77
Solvents, def, 75, 77
   used in chip making, list, 78
   safety, 521
Sources, chemical,
   diffusion, 182-4
   ion implantation, 199-200
Source-drain implantation, 61-2
Sources, electron, see Electron Sources
Spacers, sidewall, 62
SPC (statistical process-control), 514
Specialty gases, def, 74
   distribution, 81-2
   list of, used in chip making, 79-80
Specificity, of compositional analysis techniques, def, 458
Spike-annealing, 199
Spin-coating, resist, def, 328-9
Spin dryer, def, 138-9
Spine-layout of a fab, def, 506

INDEX AND GLOSSARY

Spin-on, film deposition by, def, 112-3
Spin-on-glass (SOG), def, 447, 448
Spin-rinse dryers, def, 138-9
Spin-wheel (implant disk), 201, 203, 204
Spray,
    cleaning, def, 134
    develop, def, 233-4
    rinser, def, 138-9
Spray etching, def, 521
Spreading-resistance measurements, def, 184-5, 208, 314
Sputtering - (See also Chap. 15)
    def, 247-8
    sputtering equipment, 253-58
        components of, 256
Sputtering-yield, def, 250-51
SRAM (static RAMs), def, 47
SSI (small-scale integration), def, 11
Stage, wafer-stepper, def, 355-6
Standard-buried collector BJT, def, 41
Standard Clean-1 (RCA), def, 134, 135-6
Standard Clean-2 (RCA), def, 134, 136
Standard cubic centimeter per minute (sccm), def, 108
Standard mechanical interface (SMIF), 127
Standard temperature and pressure (STP), def, 88
Standing waves, in exposed resist, def, 331-3
Starting wafers, for CMOS ICs, 55
States-of-matter, def, 19-20
Statistical process-control (SPC), 335, 514
Steam oxidation, def, 216, 219, 226-7
Step-and-repeat, def, 352-3
Step-and-scan, def, 353-4
Step coverage, def, 261, 294, 295
Step height, def, 265
Steppers, def, 352-3
Sticking-coefficient, def, 293
Stockers, def, 368-9, 506-7
Stopping mechanisms, ion, def, 193-4, 543-5
Straggle, def, 192
Static RAMs (SRAMs), def, 47
Stress,
    in thin-films, def, 75
    in Si_3N_4, def, 296-7
Stringers, def, 400-401
Stylus profilometers, def, 265-6
Sublimation, def, 92
Submerged-electrode arc furnace, def, 148

Subresolution patterns (OPC), def, 371
Substitutional impurity, def, 180
Supersaturation, def, 152
SUPREM, 181, 534, 536, 541
Surface defect detection, def, 139-41
Surface-migration, def, 113, 292, 294
Surface-mounting (SM) of packaged ICs, def, 498-9
Surface reaction-rate constant, def, 272
Surface-reaction rate limited, processes, def, 271, 308-9, 547-8
Surface roughness, def, 113-4
Susceptors, def, 276, 311, 313
Systematic defects, in yield models, def, 511, 512

Tantalum barrier layers, def, 442, 444
Tantalum-nitride, def, 260
Tape-automated bonding TAB def, 488-89
Tagout signs, def, 526
Tape-test (adhesion of thin films), def, 115
Targets, sputtering, def. 250, 253-4, 256-7
Technicians, job tasks, 515
Technology-transfer, def, 512
Temperature, def, 73-4
    control, 229
    ramping, in LPCVD, 281
    ramp-rates, 225, 228
    scales, def, 73-4, 228
Tensile-stress, def, in thin-films, 116
TEOS, def, 80, 274, 291-2, 294, 295
    ozone, 292
    PECVD, 292
Test die, def, 476-7
Test drop-ins, def, 477
Testing ICs, 475-81 (see Chap. 26)
    philosophy of, 475-6
Test structures, 477
Test-vectors, def, 480
Tetraethylorthosilicate (TEOS), 80, 274, 291-2, 294, 295
Tetramethyl-ammonium hydroxide (TMAH), 334, 404
Tetramethyl-silane (TMS), 448
Thermal expansion coefficient (TCE) def, 74-5
Thermal-wave system, def, 208,
Thermocouple,
    furnace, 226
    pressure gauge, 104-5
Thermodynamic equilibrium, def, 92
Thermocompression bonding, def, 487-8

Thickness measirements,
    thin-film metals, 265-6
    thin-film oxides, 230-33
Thin films (see Chap. 7, 111-118)
    adhesion of, def, 115
    electrical properties of, def, 117-8
    measurement of resistivity in, 118
    measurement of sheet-resistance (see Sheet resistance),
    grain boundaries in, def, 113-4
        diffusion along, diff, 287
    growth, 113-14
        nucleation, def, 113
    stress in, def, 115
        measurement of, 115
Thin-film resistors, def, 34-5
Thin-oxide growth model, def, 533-4
    Massoud's model, 534
    Reisman-Nicollian model, 533-4
Thin-small outline package (TSOP), 498-9
Thomas-Fermi screened potential, def, 544
Threshold-limit value (TLV), def, 323
Threshold voltage, def, 44
    control by ion implantation, 59, 209
Through-hole mounted (TH) packages, 497
Throughput,
ion-implantation, 202, 203
photolithography,
RTP, 230
single wafer vs. batch etching,
Tilt, wafer, in ion implantation, 194-5
Time of flight (TOF) SIMS, 472
Titanium (Ti), 260, 263, 300
Titanium-nitride (TiN), 65, 249, 260-1, 263, 300-301
Properties of, 300-1
Titanium silicide ($TiSi_2$) 63, 228
Titanium-tungsten (Ti:W), 259
TLV (threshold-limit value) def, 519-20
Tolerance, mask, 360-61
Torr, def, 89
Tower, wafer, 227, 229
Toxic, def, 520
Track systems, resist, 337-9
Transfer-molding, def, 495
Transient-enhanced diffusion TED def 199
Transistor, invention of, 8
Transmission electron microscopy (TEM), def, 208, 233, 463-5
    sample-preparation, 464

Trench-etching, in Si, 69
Trench-first, Dual Damascene, def, 451-2
Trench-isolation, shallow (STI), 68-9
Trichlorosilane, SiHCl$_3$, 79, 149, 308-9
Trim-and-form packaging, def, 496
Trimethylborate (TMB), 184, 292
Trimethyl silane (3MS), def, 448
"T-top" formation, def, 326
Tungsten, 65, 298-300
    blanket-deposition of, def, 65, 298-300
    selective-deposition of, def, 300
Tungsten field-emission source, 462
Tungsten-halogen lamps, 229, 276, 286
Tungsten hexafluoride (WF$_6$), 80, 298-99, 300-301
Tungsten plugs, def, 65, 67, 298, 300, 301, 420, 443, 450
Turbomolecular pumps, def, 101-104
    Mag-lev, def, 104
    Molecular-drag stage, def, 104
Turbulent flow - see Viscous flow,
Twin-well CMOS, def, 55, 58-9

ULSI, def, 11
Ultra-low k dielectrics, def, 448-9
Ultra-low-particle air-handling (ULPA), def, 227
Ultra-pure chemicals, def, 132
Ultrapure water (UPW), def, 83-5
Ultrasonic scrubbing, def, 136
Undercutting, def, 387
Uniformity etch rate, def, 382
Unit-cell, def, 146
UV-light, def, 343-4, 349
UV resist-stabilization, def, 337
UV sources, 348-9

Vacancy-interstitial model, def, 541
Vacancy model, def, 541
Vacuum, def, 89
    ranges, 90
    Vacuum gauges, 102-107
        partial pressure (RGA), 106-7
            residual gas analyzers, 106-7
        total-pressure,
            capacitance manometer, def, 105
            ionization gauge, def, 105
            Pirani gauge, def, 104-5
            ranges of pressure gauges, 104
            thermocouple gauges, def, 104-5
    Vacuum pumps,
        compression pumps, def, 94
        cryopumps, def, 99-101, 201
        dry-mechanical rough pumps, def, 98-9
        entrapment pumps, def, 95, 99
        inlet, def, 94
        oil-sealed roughing pumps, def, 96
        outlet of, def, 94
        pressure ranges of, 95
        Roots-pumps, def, 98-99, 104
        rotary-piston pumps, def, 96-7
        rotary-vane pumps, def, 96-7
        roughing-pumps, def, 95
        specifications of vacuum pumps, 94-6
        turbomolecular pumps, def, 101-4
Vacuum-pump oils, def, 99
Vacuum,
    backing pressure, def 95, 96
    closed system, def, 91
    cryocondensation, def, 100
    cryosorption, def, 100
        regeneration, def, 101
    inlet pressure, def,
    open-flow, def, 91, 93
    outlet pressure, def, 96
Vaporization, def, 92
Vertical-furnace, def, 227
Via-first, Dual-Damascene, def, 451-2
Vias, def, 440
Via-veil removal, 403
Visible-light, def, 343
Viscous-flow, def, 91, 95, 103
    laminar-flow, 83, 91
    turbulent-flow, 91
VLSI (very-large-scale integration) def, 11
Voids, formation, 262, 294, 300

Wafers
    bonding, def, 316-17
    bow, def, 169
    costs, 171
    diameter increase with time, 149, 171
    edge rounding of, 166-7
    flats, def, 163-4, 169
    flatness, def, 170
    lapping, def, 166
    marking, def, 166
    orientation, def, 170
    sawing, def, 164-5
    specifications of, 168-70
    total thickness variation (TTV) def, 169
Wafering process, 163-4
Wafer fab, 3, 4, 12-13
    foundries, def, 13
Wafer fab layouts, 505-7
Wafer fab personnel staffing, 515
Wafer operation costs, 504-5
Wafer sawing, def, 484-5
Wafer-sort, def, 476, 478-9. 484
Wafer stages (litho), def, 350, 355-6
Wafer transport automated handling systems (AMHS), 506-7
    interbay, 506
    intrabay, 507
Warning labels, def, 518
Warp, def, 169
Water, deionized (DI), def, 20
Wavelength, def, 342
Wavelength dispersive spectroscopy, WDX, def, 469
Wedge bonds, def, 488
Well-formation, CMOS, def, 58-9
Wet bench, def, 140, 385
Wet etching, 381-90 (Chap 21)
    aluminum, def, 389
        bubbles in, 336, 389
    hydrofluoric acid, 387-8
    silicon, def, 387
    Si$_3$N$_4$, def, 388-9
    SiO$_2$, def, 387-8
    steps of, def, 386
White elephant, def, 225
Wineglass-contact, def, 401-2
Wirebonding, def, 487-8
Wire-saw, def, 165
Within-wafer-non-uniformity (WWNU), in CMP, def, 433-4

Xerogels, def, 449
X-ray emission spectroscopy, def, 468-9
X-ray fluoresence, def, 470
X-ray radiation, safety, def, 525
X-ray photoelectron spectroscopy (XPS), 469-70

Yield, def, 479, 507-11
    random defect, def, 510
    sort-yield, def, 479
    systematic defect, def, 510
Yield-management, def, 511-14
    team, def, 513
Yield models, def, 509-11
Yield-ramps, def, 512-14

Zig-Zag in line package (ZIP), def, 498
Zone, flat, furnace, def, 225, 281